STUDENT'S SOLUTIONS MANUAL

JUDITH A. PENNA
Indiana University Purdue University Indianapolis

PREALGEBRA AND INTRODUCTORY ALGEBRA
SECOND EDITION

Marvin L. Bittinger
Indiana University Purdue University Indianapolis

David J. Ellenbogen
Community College of Vermont

Barbara L. Johnson
Indiana University Purdue University Indianapolis

PEARSON

Addison
Wesley

Boston San Francisco New York
London Toronto Sydney Tokyo Singapore Madrid
Mexico City Munich Paris Cape Town Hong Kong Montreal

Reproduced by Pearson Addison-Wesley from electronic files supplied by the author.

Copyright © 2008 Pearson Education, Inc.
Publishing as Pearson Addison-Wesley, 75 Arlington Street, Boston, MA 02116.

ISBN-13: 978-0-321-34833-3
ISBN-10: 0-321-34833-8

4 5 6 BB 09 08

Contents

Chapter 1
Whole Numbers

Exercise Set 1.1

1. 2 3 $\boxed{5}$, 8 8 8

The digit 5 means 5 thousands.

3. 1 , 4 8 8 , $\boxed{5}$ 2 6

The digit 5 means 5 hundreds.

5. 1 , 5 8 $\boxed{2}$, 3 7 0

The digit 2 names the numbers of thousands.

7. $\boxed{1}$, 5 8 2 , 3 7 0

The digit 1 names the number of millions.

9. 5702 = 5 thousands + 7 hundreds + 0 tens + 2 ones, or 5 thousands + 7 hundreds + 2 ones

11. 93,986 = 9 ten thousands + 3 thousands + 9 hundreds + 8 tens + 6 ones

13. 2058 = 2 thousands + 0 hundreds + 5 tens + 8 ones, or 2 thousands + 5 tens + 8 ones

15. 1268 = 1 thousand + 2 hundreds + 6 tens + 8 ones

17. 519,955 = 5 hundred thousands + 1 ten thousand + 9 thousands + 9 hundreds + 5 tens + 5 ones

19. 308,845 = 3 hundred thousands + 0 ten thousands + 8 thousands + 8 hundreds + 4 tens + 5 ones, or 3 hundred thousands + 8 thousands + 8 hundreds + 4 tens + 5 ones

21. 4,302,737 = 4 millions + 3 hundred thousands + 0 ten thousands + 2 thousands + 7 hundreds + 3 tens + 7 ones, or 4 millions + 3 hundred thousands + 2 thousands + 7 hundreds + 3 tens + 7 ones

23. A word name for 85 is eighty-five.

25.

27.

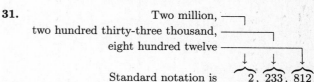

29.

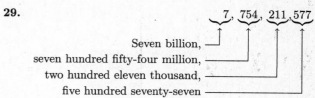

Seven billion,
seven hundred fifty-four million,
two hundred eleven thousand,
five hundred seventy-seven

31.

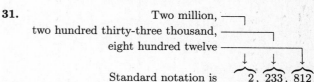

Two million,
two hundred thirty-three thousand,
eight hundred twelve

Standard notation is 2, 233, 812.

33.

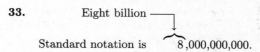

Eight billion

Standard notation is 8,000,000,000.

35.

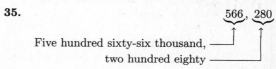

Five hundred sixty-six thousand,
two hundred eighty

37.

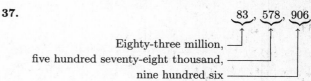

Eighty-three million,
five hundred seventy-eight thousand,
nine hundred six

39.

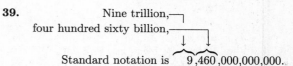

Nine trillion,
four hundred sixty billion,

Standard notation is 9,460,000,000,000.

41.

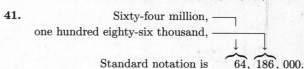

Sixty-four million,
one hundred eighty-six thousand,

Standard notation is 64, 186, 000.

43. Discussion and Writing Exercise

45. First consider the whole numbers from 100 through 199. The 10 numbers 102, 112, 122, . . . , 192 contain the digit 2. In addition, the 10 numbers 120, 121, 122, . . . , 129 contain the digit 2. However, we do not count the number 122 in this group because it was counted in the first group of ten numbers. Thus, 19 numbers from 100 through 199 contain the digit 2. Using the same type of reasoning for the whole numbers from 300 to 400, we see that there are also 19 numbers in this group that contain the digit 2.

Finally, consider the 100 whole numbers 200 through 299. Each contains the digit 2.

Thus, there are 19 + 19 + 100, or 138 whole numbers between 100 and 400 that contain the digit 2 in their standard notation.

Exercise Set 1.2

1. One truck   The other   Altogether,
        hauls   truck hauls   they haul
        6 cu yd.   8 cu yd.   14 cu yd.

   6 cu yd +   8 cu yd   = 14 cu yd

3. One parcel   The other   The total
        contains   parcel contains   purchase is
        500 acres.   300 acres.   800 acres.

   500 acres +   300 acres   = 800 acres

5.
```
    3 6 4
  +   2 3
  -------
    3 8 7
```
Add ones, add tens, then add hundreds.

7.
```
     1
    1 7 1 6
  + 3 4 8 2
  ---------
    5 1 9 8
```
Add ones: We get 8. Add tens: We get 9 tens. Add hundreds: We get 11 hundreds, or 1 thousand + 1 hundred. Write 1 in the hundreds column and 1 above the thousands. Add thousands: We get 5 thousands.

9.
```
     1
      8 6
  +   7 8
  -------
    1 6 4
```
Add ones: We get 14 ones, or 1 ten + 4 ones. Write 4 in the ones column and 1 above the tens. Add tens: We get 16 tens.

11.
```
     1
      9 9
  +     1
  -------
    1 0 0
```
Add ones: We get 10 ones, or 1 ten + 0 ones. Write 0 in the ones column and 1 above the tens. Add tens: We get 10 tens.

13.
```
    8 1 1 3
  +   3 9 0
  ---------
    8 5 0 3
```
Add ones: We get 3. Add tens: We get 10 tens, or 1 hundred + 0 tens. Write 0 in the tens column and 1 above the hundreds. Add hundreds: We get 5. Add thousands: We get 8.

15.
```
     1
      3 5 6
  + 4 9 1 0
  ---------
    5 2 6 6
```
Add ones: We get 6. Add tens: We get 6. Add hundreds: We get 12 hundreds, or 1 thousand + 2 hundreds. Write 2 in the hundreds column and 1 above the thousands. Add thousands: We get 5.

17.
```
    1 2 1
    3 8 7 0
      9 2
        7
  +   4 9 7
  ---------
    4 4 6 6
```
Add ones: We get 16 ones, or 1 ten + 6 ones. Write 6 in the ones column and 1 above the tens. Add tens: We get 26 tens, or 2 hundreds + 6 tens. Write 6 in the tens column and 2 above the hundreds. Add hundreds: We get 14 hundreds, or 1 thousand + 4 hundreds. Write 4 in the hundreds column and 1 above the thousands. Add thousands: We get 4.

19.
```
    1 1
    4 8 2 5
  + 1 7 8 3
  ---------
    6 6 0 8
```
Add ones: We get 8. Add tens: We get 10 tens. Write 0 in the tens column and 1 above the hundreds. Add hundreds: We get 16 hundreds. Write 6 in the hundreds column and 1 above the thousands. Add thousands: We get 6 thousands.

21.
```
     1 1 1
    2 3, 4 4 3
  + 1 0, 9 8 9
  -----------
    3 4, 4 3 2
```
Add ones: We get 12 ones, or 1 ten + 2 ones. Write 2 in the ones column and 1 above the tens. Add tens: We get 13 tens. Write 3 in the tens column and 1 above the hundreds. Add hundreds: We get 14 hundreds. Write 4 in the hundreds column and 1 above the thousands. Add thousands: We get 4 thousands. Add ten thousands: We get 3 ten thousands.

23.
```
    1 1 1 1
    7 7, 5 4 3
  + 2 3, 7 6 7
  ------------
  1 0 1, 3 1 0
```
Add ones: We get 10 ones, or 1 ten + 0 ones. Write 0 in the ones column and 1 above the tens. Add tens: We get 11 tens. Write 1 in the tens column and 1 above the hundreds. Add hundreds: We get 13 hundreds. Write 3 in the hundreds column and 1 above the thousands. Add thousands: We get 11 thousands. Write 1 in the thousands column and 1 above the ten thousands. Add ten thousands: We get 10 ten thousands.

25. We look for pairs of numbers whose sums are 10, 20, 30, and so on.

```
   45 -------→ 70
   25
   36 -------→ 80
   44
  +80 -------→ 80
  ---          ---
  230          230
```

27.
```
     1 1
    1 2, 0 7 0
       2 9 5 4
  +    3 4 0 0
  -----------
    1 8, 4 2 4
```
Add ones: We get 4. Add tens: We get 12 tens, or 1 hundred + 2 tens. Write 2 in the tens column and 1 above the hundreds. Add hundreds: We get 14 hundreds, or 1 thousand + 4 hundreds. Write 4 in the hundreds column and 1 above the hundreds. Add thousands: We get 8 thousands. Add ten thousands: We get 1 ten thousand.

29.
```
    3 1 2
    4 8 3 5
      7 2 9
    9 2 0 4
    8 9 8 6
  + 7 9 3 1
  ---------
  3 1, 6 8 5
```
Add ones: We get 25. Write 5 in the ones column and 2 above the tens. Add tens: We get 18 tens. Write 8 in the tens column and 1 above the hundreds. Add hundreds: We get 36 hundreds. Write 6 in the hundreds column and 3 above the thousands. Add thousands: We get 31 thousands.

31. We regroup.

$$(2 + 5) + 4 = 2 + (5 + 4)$$

33. We regroup:

$$6 + (3 + 2) = (6 + 3) + 2$$

35. We reverse the order of the addends.

$$2 + 7 = 7 + 2$$

37. We reverse the order of the addends.

$$6 + 1 = 1 + 6$$

39. We reverse the order of the addends.

$$2 + 9 = 9 + 2$$

41. Add from the top.

We first add 7 and 9, getting 16; then 16 and 4, getting 20; then 20 and 8, getting 28.

$$\boxed{\begin{array}{c}7\\9\end{array}} \to \boxed{\begin{array}{c}16\\4\end{array}} \to \boxed{\begin{array}{c}20\\8\end{array}} \to 28$$

$$\begin{array}{r}7\\9\\4\\+\ 8\\\hline 28\end{array}$$

Check by adding from the bottom.

We first add 8 and 4, getting 12; then 12 and 9, getting 21; then 21 and 7, getting 28.

$$\begin{array}{r}7\\9\\4\\+\ 8\\\hline 28\end{array}$$

$$\boxed{\begin{array}{c}7\\21\end{array}} \to 28 \qquad \boxed{\begin{array}{c}9\\12\end{array}} \qquad \boxed{\begin{array}{c}4\\8\end{array}}$$

43. Add from the top.

$$\boxed{\begin{array}{c}8\\6\end{array}} \to \boxed{\begin{array}{c}14\\2\end{array}} \to \boxed{\begin{array}{c}16\\3\end{array}} \to \boxed{\begin{array}{c}19\\7\end{array}} \to 26$$

$$\begin{array}{r}8\\6\\2\\3\\+\ 7\\\hline 26\end{array}$$

Check:

$$\boxed{\begin{array}{c}8\\18\end{array}} \to 26 \qquad \boxed{\begin{array}{c}6\\12\end{array}} \qquad \boxed{\begin{array}{c}2\\10\end{array}} \qquad \boxed{\begin{array}{c}3\\7\end{array}}$$

$$\begin{array}{r}8\\6\\2\\3\\+\ 7\\\hline 26\end{array}$$

45. Perimeter = 14 mi + 13 mi + 8 mi + 10 mi + 47 mi + 22 mi

We carry out the addition.

$$\begin{array}{r}{\scriptstyle 2}\\1\ 4\\1\ 3\\8\\1\ 0\\4\ 7\\+\ 2\ 2\\\hline 1\ 1\ 4\end{array}$$

The perimeter of the figure is 114 mi.

47. Perimeter = 200 ft + 85 ft + 200 ft + 85 ft

We carry out the addition.

$$\begin{array}{r}{\scriptstyle 1\ 1}\\2\ 0\ 0\\8\ 5\\2\ 0\ 0\\+\ \ \ 8\ 5\\\hline 5\ 7\ 0\end{array}$$

The perimeter of the hockey rink is 570 ft.

49. Discussion and Writing Exercise

51. 4 $\boxed{8}$ 6, 2 0 5

The digit 8 tells the number of ten thousands.

53. Discussion and Writing Exercise

55. $5,987,943 + 328,959 + 49,738,765$

Using a calculator to carry out the addition, we find that the sum is 56,055,667.

57. One method is described in the answer section in the text. Another method is: $1 + 100 = 101$, $2 + 99 = 101$, $\ldots$, $50 + 51 = 101$. Then $50 \cdot 101 = 5050$.

Exercise Set 1.3

1.

Number to begin with		Number spent		Number left
20	−	4	=	16

3.

Amount to begin with		Amount sold		Amount left
126 oz	−	13 oz	=	113 oz

5. $7 - 4 = 3$

This number gets added (after 3).

$$7 = 3 + 4$$

(By the commutative law of addition, $7 = 4 + 3$ is also correct.)

7. $13 - 8 = 5$

This number gets added (after 5).

$$13 = 5 + 8$$

(By the commutative law of addition, $13 = 8 + 5$ is also correct.)

9. $23 - 9 = 14$

This number gets added (after 14).

$$23 = 14 + 9$$

(By the commutative law of addition, $23 = 9 + 14$ is also correct.)

11. $43 - 16 = 27$

This number gets added (after 27).

$$43 = 27 + 16$$

(By the commutative law of addition, $43 = 16 + 27$ is also correct.)

13. $6 + 9 = 15$ $6 + 9 = 15$

This addend gets   This addend gets
subtracted from   subtracted from
the sum.   the sum.
$6 = 15 - 9$ $9 = 15 - 6$

15. $8 + 7 = 15$ $8 + 7 = 15$

This addend gets   This addend gets
subtracted from   subtracted from
the sum.   the sum.
$8 = 15 - 7$ $7 = 15 - 8$

17. $17 + 6 = 23$ $17 + 6 = 23$

This addend gets   This addend gets
subtracted from   subtracted from
the sum.   the sum.
$17 = 23 - 6$ $6 = 23 - 17$

19. $23 + 9 = 32$ $23 + 9 = 32$

This addend gets   This addend gets
subtracted from   subtracted from
the sum.   the sum.
$23 = 32 - 9$ $9 = 32 - 23$

21. We first write an addition sentence. Keep in mind that all numbers are in millions.

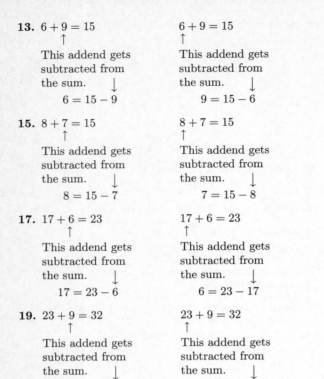

Now we write a related subtraction sentence.

$17 \ + \ \boxed{} \ = \ 32$

$\boxed{} \ = \ 32 - 17$ The addend 17 gets subtracted.

We have $32 - 17 = 15$.

23. We first write an addition sentence.

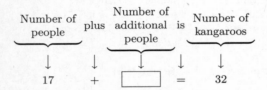

Now we write a related subtraction sentence.

$10 \ + \ \boxed{} \ = \ 23$

$\boxed{} \ = \ 23 - 10$ The addend 10 gets subtracted.

We have $23 - 10 = 13$.

25.
$$\begin{array}{r} 6\,5 \\ -\,2\,1 \\ \hline 4\,4 \end{array}$$
Subtract ones, then subtract tens.

27.
$$\begin{array}{r} 8\,6\,6 \\ -\,3\,3\,3 \\ \hline 5\,3\,3 \end{array}$$
Subtract ones, subtract tens, then subtract hundreds.

29.
$$\begin{array}{r} \overset{7\ \ 16}{\cancel{8}\ \cancel{0}} \\ -\,4\,7 \\ \hline 3\,9 \end{array}$$
We cannot subtract 7 ones from 6 ones. Borrow 1 ten to get 16 ones. Subtract ones, then subtract tens.

31.
$$\begin{array}{r} \overset{7\ \ 11}{9\ \cancel{8}\ \cancel{1}} \\ -\,7\,4\,7 \\ \hline 2\,3\,4 \end{array}$$
We cannot subtract 7 ones from 1 one. Borrow 1 ten to get 11 ones. Subtract ones, subtract tens, then subtract hundreds.

33.
$$\begin{array}{r} \overset{6\ \ 16}{7\ 7\ \cancel{6}\ 9} \\ -\,2\,3\,8\,7 \\ \hline 5\,3\,8\,2 \end{array}$$
Subtract ones. We cannot subtract 8 tens from 6 tens. Borrow 1 hundred to get 16 tens. Subtract tens, subtract hundreds, then subtract thousands.

35.
$$\begin{array}{r} \overset{6\ \ 16\ \ 3\ \ 10}{7\ \cancel{6}\ \cancel{4}\ \cancel{0}} \\ -\,3\,8\,0\,9 \\ \hline 3\,8\,3\,1 \end{array}$$
We cannot subtract 9 ones from 0 ones. Borrow 1 ten to get 10 ones. Subtract ones, then tens. We cannot subtract 8 hundreds from 6 hundreds. Borrow 1 thousand to get 16 hundreds. Subtract hundreds, then thousands.

37.
$$\begin{array}{r} \overset{11\ 15\ 13}{\overset{1\ \ 5\ \ 3\ \ 17}{1\,\cancel{2},\cancel{6}\,\cancel{4}\,7}} \\ -\ \ \ 4\,8\,9\,9 \\ \hline 7\,7\,4\,8 \end{array}$$

39.
$$\begin{array}{r} \overset{8\ 10\ \ \ \ 2\ \ 17}{\cancel{9}\,\cancel{0},2\,\cancel{3}\,\cancel{7}} \\ -\,4\,7,2\,0\,9 \\ \hline 4\,3,0\,2\,8 \end{array}$$

41.
$$\begin{array}{r} \overset{7\ \ 10}{\cancel{8}\ \cancel{0}} \\ -\,2\,4 \\ \hline 5\,6 \end{array}$$

43.
$$\begin{array}{r} \overset{8\ \ 10}{6\ \cancel{9}\ \cancel{0}} \\ -\,2\,3\,6 \\ \hline 4\,5\,4 \end{array}$$

45.
$$\begin{array}{r} \overset{7\ \ 9\ \ 18}{6\ \cancel{8}\,\cancel{0}\,\cancel{8}} \\ -\,3\,0\,5\,9 \\ \hline 3\,7\,4\,9 \end{array}$$
We have 8 hundreds or 80 tens. We borrow 1 ten to get 18 ones. We then have 79 tens. Subtract ones, then tens, then hundreds, then thousands.

47.
$$\begin{array}{r} \overset{2\;9\;10}{2\,3\,\cancel{0}\,\cancel{0}} \\ -\;\;\;1\,0\,9 \\ \hline 2\,1\,9\,1 \end{array}$$
We have 3 hundreds or 30 tens. We borrow 1 ten to get 10 ones. We then have 29 tens. Subtract ones, then tens, then hundreds, then thousands.

49.
$$\begin{array}{r} \overset{15}{} \\ \overset{\cancel{5}\;10}{\cancel{1}\,\cancel{6}\,\cancel{0}} \\ -\;\;\;7\,4 \\ \hline 8\,6 \end{array}$$

51.
$$\begin{array}{r} \overset{3\;10}{7\,8\,\cancel{4}\,\cancel{0}} \\ -\;3\,0\,2\,7 \\ \hline 4\,8\,1\,3 \end{array}$$

53.
$$\begin{array}{r} \overset{7\;\;\cancel{3}\;13}{5\,\cancel{8}\,\cancel{4}\,\cancel{3}} \\ -\;\;\;\;\;9\,8 \\ \hline 5\,7\,4\,5 \end{array}$$

55.
$$\begin{array}{r} \overset{10\;16}{} \\ \overset{9\;\cancel{0}\;\cancel{6}\;13}{\cancel{1}\,\cancel{0}\,1,7\,\cancel{3}\,4} \\ -\;\;\;\;\;\;5\,7\,6\,0 \\ \hline 9\,5,9\,7\,4 \end{array}$$

57.
$$\begin{array}{r} \overset{9\;9\;9\;14}{\cancel{1}\,\cancel{0},\cancel{0}\,\cancel{0}\,\cancel{4}} \\ -\;\;\;\;\;\;\;\;2\,9 \\ \hline 9\,9\,7\,5 \end{array}$$
We have 1 ten thousand, or 1000 tens. We borrow 1 ten to get 10 ones. We then have 999 tens. Subtract ones, then tens, then hundreds, then thousands.

59.
$$\begin{array}{r} \overset{8\;9\;17}{8\,3,\cancel{9}\,\cancel{0}\,7} \\ -\;\;\;\;\;\;8\,9 \\ \hline 8\,3,8\,1\,8 \end{array}$$

61.
$$\begin{array}{r} \overset{6\;9\;9\;10}{\cancel{7}\,\cancel{0}\,\cancel{0}\,\cancel{0}} \\ -\;2\,7\,9\,4 \\ \hline 4\,2\,0\,6 \end{array}$$
We have 7 thousands or 700 tens. We borrow 1 ten to get 10 ones. We then have 699 tens. Subtract ones, then tens, then hundreds, then thousands.

63.
$$\begin{array}{r} \overset{7\;9\;9\;10}{4\,\cancel{8},\cancel{0}\,\cancel{0}\,\cancel{0}} \\ -\;3\,7,6\,9\,5 \\ \hline 1\,0,3\,0\,5 \end{array}$$
We have 8 thousands or 800 tens. We borrow 1 ten to get 10 ones. We then have 799 tens. Subtract ones, then tens, then hundreds, then thousands, then ten thousands.

65. Discussion and Writing Exercise

67.
$$\begin{array}{r} \overset{1\;1}{9\,4\,6} \\ +\;\;\;7\,8 \\ \hline 1\,0\,2\,4 \end{array}$$
Add ones: We get 14. Write 4 in the ones column and 1 above the tens. Add tens: We get 12. Write 2 in the tens column and 1 above the hundreds. Add hundreds: We get 10 hundreds.

69.
$$\begin{array}{r} \overset{1\;\;1\;\;\;\;1}{5\,7,8\,7\,7} \\ +\;3\,2,4\,0\,6 \\ \hline 9\,0,2\,8\,3 \end{array}$$
Add ones: We get 13. Write 3 in the ones column and 1 above the tens. Add tens: We get 8. Add hundreds: We get 12. Write 2 in the hundreds column and 1 above the thousands. Add thousands: We get 10. Write 0 in the thousands column and 1 above the ten thousands. Add ten thousands: We get 9 ten thousands.

71.
$$\begin{array}{r} \overset{1\;\;1}{5\,6\,7} \\ +\;7\,7\,8 \\ \hline 1\,3\,4\,5 \end{array}$$
Add ones: We get 15. Write 5 in the ones column and 1 above the tens. Add tens: We get 14. Write 4 in the tens column and 1 above the hundreds. Add hundreds: We get 13 hundreds.

73.
$$\begin{array}{r} \overset{1\;\;1\;\;\;\;1}{1\,2,8\,8\,5} \\ +\;\;\;9\,8\,0\,7 \\ \hline 2\,2,6\,9\,2 \end{array}$$
Add ones: We get 12. Write 2 in the ones column and 1 above the tens. Add tens: We get 9. Add hundreds: We get 16. Write 6 in the hundreds column and 1 above the thousands. Add thousands: We get 12. Write 2 in the thousands column and 1 above the ten thousands. Add ten thousands. We get 2 ten thousands.

75.

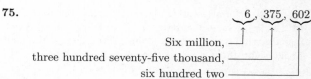

Six million, —
three hundred seventy-five thousand, —
six hundred two —

77. Discussion and Writing Exercise

79. $3,928,124 - 1,098,947$

Using a calculator to carry out the subtraction, we find that the difference is 2,829,177.

81.
$$\begin{array}{r} 9,\_4\,8,6\,2\,1 \\ -\;2,0\,9\,7,\_8\,1 \\ \hline 7,2\,5\,1,1\,4\,0 \end{array}$$

To subtract tens, we borrow 1 hundred to get 12 tens.

$$\begin{array}{r} \overset{\;\;\;\;\;\;\;\;\;\;5\;12}{9,\_4\,8,\cancel{6}\,\cancel{2}\,1} \\ -\;2,0\,9\,7,\_8\,1 \\ \hline 7,2\,5\,1,1\,4\,0 \end{array}$$

In order to have 1 hundred in the difference, the missing digit in the subtrahend must be 4 ($5 - 4 = 1$).

$$\begin{array}{r} \overset{\;\;\;\;\;\;\;\;\;\;5\;12}{9,\_4\,8,\cancel{6}\,\cancel{2}\,1} \\ -\;2,0\,9\,7,4\,8\,1 \\ \hline 7,2\,5\,1,1\,4\,0 \end{array}$$

In order to subtract ten thousands, we must borrow 1 hundred thousand to get 14 ten thousands. The number of hundred thousands left must be 2 since the hundred thousands place in the difference is 2 ($2 - 0 = 2$). Thus, the missing digit in the minuend must be $2 + 1$, or 3.

$$\begin{array}{r} \overset{2\;14\;\;\;\;\;\;5\;12}{9,\cancel{3}\,\cancel{4}\,8,\cancel{6}\,\cancel{2}\,1} \\ -\;2,0\,9\,7,4\,8\,1 \\ \hline 7,2\,5\,1,1\,4\,0 \end{array}$$

Exercise Set 1.4

1. Round 48 to the nearest ten.

4 $\boxed{8}$
↑

The digit 4 is in the tens place. Consider the next digit to the right. Since the digit, 8, is 5 or higher, round 4 tens up to 5 tens. Then change the digit to the right of the tens digit to zero.

The answer is 50.

3. Round 467 to the nearest ten.

4 6 $\boxed{7}$
↑

The digit 6 is in the tens place. Consider the next digit to the right. Since the digit, 7, is 5 or higher, round 6 tens up to 7 tens. Then change the digit to the right of the tens digit to zero.

The answer is 470.

5. Round 731 to the nearest ten.

7 3 $\boxed{1}$
↑

The digit 3 is in the tens place. Consider the next digit to the right. Since the digit, 1, is 4 or lower, round down, meaning that 3 tens stays as 3 tens. Then change the digit to the right of the tens digit to zero.

The answer is 730.

7. Round 895 to the nearest ten.

8 9 $\boxed{5}$
↑

The digit 9 is in the tens place. Consider the next digit to the right. Since the digit, 5, is 5 or higher, we round up. The 89 tens become 90 tens. Then change the digit to the right of the tens digit to zero.

The answer is 900.

9. Round 146 to the nearest hundred.

1 $\boxed{4}$ 6
↑

The digit 1 is in the hundreds place. Consider the next digit to the right. Since the digit, 4, is 4 or lower, round down, meaning that 1 hundred stays as 1 hundred. Then change all digits to the right of the hundreds digit to zeros.

The answer is 100.

11. Round 957 to the nearest hundred.

9 $\boxed{5}$ 7
↑

The digit 9 is in the hundreds place. Consider the next digit to the right. Since the digit, 5, is 5 or higher, round up. The 9 hundreds become 10 hundreds. Then change all digits to the right of the hundreds digit to zeros.

The answer is 1000.

13. Round 9079 to the nearest hundred.

9 0 $\boxed{7}$ 9
↑

The digit 0 is in the hundreds place. Consider the next digit to the right. Since the digit, 7, is 5 or higher, round 0 hundreds up to 1 hundred. Then change all digits to the right of the hundreds digit to zeros.

The answer is 9100.

15. Round 32,850 to the nearest hundred.

3 2, 8 $\boxed{5}$ 0
↑

The digit 8 is in the hundreds place. Consider the next digit to the right. Since the digit, 5, is 5 or higher, round 8 hundreds up to 9 hundreds. Then change all digits to the right of the hundreds digit to zero.

The answer is 32,900.

17. Round 5876 to the nearest thousand.

5 $\boxed{8}$ 7 6
↑

The digit 5 is in the thousands place. Consider the next digit to the right. Since the digit, 8, is 5 or higher, round 5 thousands up to 6 thousands. Then change all digits to the right of the thousands digit to zeros.

The answer is 6000.

19. Round 7500 to the nearest thousand.

7 $\boxed{5}$ 0 0
↑

The digit 7 is in the thousands place. Consider the next digit to the right. Since the digit, 5, is 5 or higher, round 7 thousands up to 8 thousands. Then change all the digits to the right of the thousands digit to zeros.

The answer is 8000.

21. Round 45,340 to the nearest thousand.

4 5, $\boxed{3}$ 4 0
↑

The digit 5 is in the thousands place. Consider the next digit to the right. Since the digit, 3, is 4 or lower, round down, meaning that 5 thousands stays as 5 thousands. Then change all the digits to the right of the thousands digit to zeros.

The answer is 45,000.

23. Round 373,405 to the nearest thousand.

3 7 3, $\boxed{4}$ 0 5
↑

The digit 3 is in the thousands place. Consider the next digit to the right. Since the digit, 4, is 4 or lower, round down, meaning that 3 thousands stays as 3 thousands. Then change all the digits to the right of the thousands digit to zeros.

The answer is 373,000.

25.

	Rounded to the nearest ten
7 8 + 9 7	8 0 + 1 0 0
	1 8 0 ← Estimated answer

27.

	Rounded to the nearest ten
8 0 7 4 − 2 3 4 7	8 0 7 0 − 2 3 5 0
	5 7 2 0 ← Estimated answer

29.

	Rounded to the nearest ten
4 5 7 7 2 5 + 5 6	5 0 8 0 3 0 + 6 0
3 4 3	2 2 0 ← Estimated answer

The sum 343 seems to be incorrect since 220 is not close to 343.

31.

	Rounded to the nearest ten
6 2 2 7 8 8 1 + 1 1 1	6 2 0 8 0 8 0 + 1 1 0
9 3 2	8 9 0 ← Estimated answer

The sum 932 seems to be incorrect since 890 is not close to 932.

33.

	Rounded to the nearest hundred
7 3 4 8 + 9 2 4 7	7 3 0 0 + 9 2 0 0
	1 6 , 5 0 0 ← Estimated answer

35.

	Rounded to the nearest hundred
6 8 5 2 − 1 7 4 8	6 9 0 0 − 1 7 0 0
	5 2 0 0 ← Estimated answer

37. We round the cost of each option to the nearest hundred and add.

7 4 5 0 1 5 9 5 1 5 4 0 + 6 2 5	7 5 0 0 1 6 0 0 1 5 0 0 + 6 0 0
	1 1 , 2 0 0

The estimated cost is $11,200.

39. We round the cost of each option to the nearest hundred and add.

8 8 2 0 2 8 7 0 6 2 4 5 + 9 8 5	8 8 0 0 2 9 0 0 6 2 0 0 + 1 0 0 0
	1 8 , 9 0 0

The estimated cost is $18,900. Since this is more than Sara and Ben's budget of $17,700, they cannot afford their choices.

41. Answers will vary depending on the options chosen.

43.

	Rounded to the nearest hundred
2 1 6 8 4 7 4 5 + 5 9 5	2 0 0 1 0 0 7 0 0 + 6 0 0
1 6 4 0	1 6 0 0 ← Estimated answer

The sum 1640 seems to be correct since 1600 is close to 1640.

45.

	Rounded to the nearest hundred
7 5 0 4 2 8 6 3 + 2 0 5	8 0 0 4 0 0 1 0 0 + 2 0 0
1 4 4 6	1 5 0 0 ← Estimated answer

The sum 1446 seems to be correct since 1500 is close to 1446.

47.

	Rounded to the nearest thousand
9 6 4 3 4 8 2 1 8 9 4 3 + 7 0 0 4	1 0 , 0 0 0 5 0 0 0 9 0 0 0 + 7 0 0 0
	3 1 , 0 0 0 ← Estimated answer

49.

	Rounded to the nearest thousand
9 2 , 1 4 9 − 2 2 , 5 5 5	9 2 , 0 0 0 − 2 3 , 0 0 0
	6 9 , 0 0 0 ← Estimated answer

51.

Since 0 is to the left of 17, $0 < 17$.

53.

Since 34 is to the right of 12, $34 > 12$.

55.

Since 1000 is to the left of 1001, $1000 < 1001$.

57.

Since 133 is to the right of 132, $133 > 132$.

59.

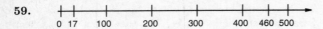

Since 460 is to the right of 17, $460 > 17$.

61.

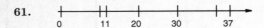

Since 37 is to the right of 11, $37 > 11$.

63. Since 2,083,660 lies to the left of 2,296,335 on the number line, we can write $2,083,660 < 2,296,335$.

Conversely, since 2,296,335 lies to the right of 2,083,660 on the number line, we could also write $2,296,335 > 2,083,660$.

65. Since 6482 lies to the right of 4641 on the number line, we can write $6482 > 4641$.

Conversely, since 4641 lies to the left of 6482 on the number line, we can also write $4641 < 6482$.

67. Discussion and Writing Exercise

69. $7992 = 7$ thousands $+ 9$ hundreds $+ 9$ tens $+ 2$ ones

71.

$$246,\overbrace{605},\overbrace{004},\overbrace{032}$$

Two hundred forty-six billion, — ⌐
six hundred five million, ——————
four thousand, ——————
thirty-two ——————

73.
$$\begin{array}{r} \overset{1\ 1\ 1\ 1}{6\,7,7\,8\,9} \\ +\,1\,8,9\,6\,5 \\ \hline 8\,6,7\,5\,4 \end{array}$$
Add ones. We get 14. Write 4 in the ones column and 1 above the tens. Add tens: We get 15 tens. Write 5 in the tens column and 1 above the hundreds. Add hundreds: We get 17 hundreds. Write 7 in the hundreds column and 1 above the thousands. Add thousands: We get 16 thousands. Write 6 in the thousands column and 1 above the ten thousands. Add ten thousands: We get 8 ten thousands.

75.
$$\begin{array}{r} \overset{16}{\overset{5\ \cancel{6}\ 17}{\cancel{6}\,\cancel{7},7\,8\,9}} \\ -\,1\,8,9\,6\,5 \\ \hline 4\,8,8\,2\,4 \end{array}$$
Subtract ones: We get 4. Subtract tens: We get 2. We cannot subtract 9 hundreds from 7 hundreds. We borrow 1 thousand to get 17 hundreds. Subtract hundreds. We cannot subtract 8 thousands from 6 thousands. We borrow 1 ten thousand to get 16 thousands. Subtract thousands, then ten thousands.

77. Discussion and Writing Exercise

79. Using a calculator, we find that the sum is 30,411. This is close to the estimated sum found in Exercise 47.

81. Using a calculator, we find that the difference is 69,594. This is close to the estimated difference found in Exercise 49.

Exercise Set 1.5

1. Think of a rectangular array consisting of 21 rows with 21 objects in each row.

$21 \cdot 21 = 441$

3. Repeated addition fits best in this case.

$$\boxed{12\ \text{oz}} + \boxed{12\ \text{oz}} + \boxed{12\ \text{oz}} + \cdots + \boxed{12\ \text{oz}}$$
8 addends
$8 \cdot 12\ \text{oz} = 96\ \text{oz}$

5. Think of a rectangular array consisting of 4800 rows with 1200 objects in each row.

$4800 \cdot 1200 = 5,760,000$

7.
$$\begin{array}{r} 8\,7 \\ \times\ 1\,0 \\ \hline 8\,7\,0 \end{array}$$
Multiplying by 1 ten (We write 0 and then multiply 87 by 1.)

9.
$$\begin{array}{r} 2\,3\,4\,0 \\ \times\ 1\,0\,0\,0 \\ \hline 2,3\,4\,0,0\,0\,0 \end{array}$$
Multiplying by 1 thousand (We write 000 and then multiply 2340 by 1.)

11.
$$\begin{array}{r} \overset{4}{6\,5} \\ \times\ \ \ 8 \\ \hline 5\,2\,0 \end{array}$$
Multiplying by 8

13.
$$\begin{array}{r} \overset{2}{9\,4} \\ \times\ \ \ 6 \\ \hline 5\,6\,4 \end{array}$$
Multiplying by 6

15.
$$\begin{array}{r} \overset{2}{5\,0\,9} \\ \times\ \ \ \ 3 \\ \hline 1\,5\,2\,7 \end{array}$$
Multiplying by 3

17.
$$\begin{array}{r} \overset{1\ 2\ 6}{9\,2\,2\,9} \\ \times\ \ \ \ \ 7 \\ \hline 6\,4,6\,0\,3 \end{array}$$
Multiplying by 7

19.
$$\begin{array}{r} \overset{2}{5\,3} \\ \times\ \ 9\,0 \\ \hline 4\,7\,7\,0 \end{array}$$
Multiplying by 9 tens (We write 0 and then multiply 53 by 9.)

21.
$$\begin{array}{r} \overset{2}{\overset{3}{8\,5}} \\ \times\ 4\,7 \\ \hline 5\,9\,5 \\ 3\,4\,0\,0 \\ \hline 3\,9\,9\,5 \end{array}$$
Multiplying by 7
Multiplying by 40
Adding

23.
```
      2
    6 4 0
  ×   7 2
  ───────
    1 2 8 0    Multiplying by 2
  4 4 8 0 0    Multiplying by 70
  ─────────
  4 6,0 8 0    Adding
```

25.
```
    1 1
    1 1
    4 4 4
  ×   3 3
  ───────
    1 3 3 2    Multiplying by 3
  1 3 3 2 0    Multiplying by 30
  ─────────
  1 4,6 5 2    Adding
```

27.
```
        3
        7
      5 0 9
  ×   4 0 8
  ─────────
      4 0 7 2    Multiplying by 8
  2 0 3 6 0 0    Multiplying by 4 hundreds (We write 00
  ───────────    and then multiply 509 by 4.)
  2 0 7,6 7 2
```

29.
```
      4 2
      1
      3 1
      8 5 3
  ×   9 3 6
  ─────────
      5 1 1 8    Multiplying by 6
    2 5 5 9 0    Multiplying by 30
  7 6 7 7 0 0    Multiplying by 900
  ───────────
  7 9 8,4 0 8    Adding
```

31.
```
        1   2
            1
            1
        1 1 3
        6 4 2 8
  ×     3 2 2 4
  ─────────────
      2 5 7 1 2      Multiplying by 4
    1 2 8 5 6 0      Multiplying by 20
  1 2 8 5 6 0 0      Multiplying by 200
  1 9 2 8 4 0 0 0    Multiplying by 3000
  ───────────────
  2 0,7 2 3,8 7 2    Adding
```

33.
```
      1 3
      3 4 8 2
  ×     1 0 4
  ───────────
    1 3 9 2 8      Multiplying by 4
  3 4 8 2 0 0      Multiplying by 1 hundred (We write 00
  ───────────      and then multiply 3482 by 1.)
  3 6 2,1 2 8
```

35.
```
        2
        4
      5 0 0 6
  ×   4 0 0 8
  ───────────
      4 0 0 4 8        Multiplying by 8
  2 0 0 2 4 0 0 0      Multiplying by 4 thousands  (We
  ───────────────      write 000 and then multiply 5006 by
  2 0,0 6 4,0 4 8      4.)
```

37.
```
          2   3
          3   4
        5 6 0 8
  ×     4 5 0 0
  ─────────────
    2 8 0 4 0 0 0       Multiplying by 5 hundreds (We write
  2 2 4 3 2 0 0 0       00 and then multiply 5608 by 5.)
  ───────────────       Multiplying by 4000
  2 5,2 3 6,0 0 0       Adding
```

39.
```
        2 1
        3 2
        3 3
        8 7 6
  ×     3 4 5
  ───────────
      4 3 8 0      Multiplying by 5
    3 5 0 4 0      Multiplying by 40
  2 6 2 8 0 0      Multiplying by 300
  ───────────
  3 0 2,2 2 0      Adding
```

41.
```
          5 5 5
          1 1 1
          1 1 1
          3 3 3
          7 8 8 9
  ×       6 2 2 4
  ───────────────
        3 1 5 5 6      Multiplying by 4
      1 5 7 7 8 0      Multiplying by 20
    1 5 7 7 8 0 0      Multiplying by 200
  4 7 3 3 4 0 0 0      Multiplying by 6000
  ───────────────
  4 9,1 0 1,1 3 6      Adding
```

43.
```
                Rounded to
                the nearest ten

    4 5              5 0
  × 6 7            × 7 0
  ─────           ───────
                  3 5 0 0  ← Estimated answer
```

45.
```
                Rounded to
                the nearest ten

    3 4              3 0
  × 2 9            × 3 0
  ─────           ───────
                    9 0 0  ← Estimated answer
```

47.
```
                Rounded to
                the nearest hundred

    8 7 6              9 0 0
  × 3 4 5            × 3 0 0
  ───────           ─────────
                    2 7 0,0 0 0  ← Estimated answer
```

49.
```
                Rounded to
                the nearest hundred

    4 3 2              4 0 0
  × 1 9 9            × 2 0 0
  ───────           ─────────
                      8 0,0 0 0  ← Estimated answer
```

51. a) First we round the cost of the car and the destination charges to the nearest hundred and add.
```
    2 7,8 9 6          2 7,9 0 0
  +      5 4 0       +      5 0 0
  ───────────       ───────────
                      2 8,4 0 0
```

The number of sales representatives, 112, rounded to the nearest hundred is 100. Now we multiply the rounded total cost of a car and the rounded number of representatives.

$$
\begin{array}{r}
2\,8,4\,0\,0 \\
\times\qquad 1\,0\,0 \\
\hline
2,8\,4\,0,0\,0\,0
\end{array}
$$

The cost of the purchase is approximately $2,840,000.

b) First we round the cost of the car to the nearest thousand and the destination charges to the nearest hundred and add.

$$
\begin{array}{rr}
2\,7,8\,9\,6 & 2\,8,0\,0\,0 \\
+\qquad 5\,4\,0 & +\qquad 5\,0\,0 \\
\cline{2-2}
 & 2\,8,5\,0\,0
\end{array}
$$

From part (a) we know that the number of sales representatives, rounded to the nearest hundred, is 100. We multiply the rounded total cost of a car and the rounded number of representatives.

$$
\begin{array}{r}
2\,8,5\,0\,0 \\
\times\qquad 1\,0\,0 \\
\hline
2,8\,5\,0,0\,0\,0
\end{array}
$$

The cost of the purchase is approximately $2,850,000.

53. $A = 728 \text{ mi} \times 728 \text{ mi} = 529,984$ square miles

55. $A = l \times w = 6 \times 3 = 18$ square feet

57. $A = l \times w = 11 \text{ yd} \times 11 \text{ yd} = 121$ square yards

59. $A = l \times w = 48 \text{ mm} \times 3 \text{ mm} = 144$ square millimeters

61. $A = l \times w = 90 \text{ ft} \times 90 \text{ ft} = 8100$ square feet

63. Discussion and Writing Exercise

65.
$$
\begin{array}{r}
\overset{1\quad\ 1}{} \\
4\,9\,0\,8 \\
5\,6\,6\,7 \\
+\ 2\,1\,1\,0 \\
\hline
1\,2,6\,8\,5
\end{array}
$$
Add ones: We get 15. Write 5 in the ones column and 1 above the tens. Add tens: We get 8. Add hundreds: We get 16. Write 6 in the hundreds column and 1 above the thousands. Add thousands: We get 12 thousands.

67.
$$
\begin{array}{r}
\overset{1\quad\ 1\ 1\ 1}{} \\
3\,4\,0,7\,9\,8 \\
+\ \ 8\,6,6\,7\,9 \\
\hline
4\,2\,7,4\,7\,7
\end{array}
$$
Add ones: We get 17. Write 7 in the ones column and 1 above the tens. Add tens: We get 17. Write 7 in the tens column and 1 above the hundreds. Add hundreds: We get 14. Write 4 in the hundreds column and 1 above the thousands. Add thousands: We get 7. Add ten thousands: We get 12. Write 2 in the ten thousands column and 1 above the hundred thousands. Add hundred thousands: We get 4 hundred thousands.

69.
$$
\begin{array}{r}
\overset{8\ 10}{} \\
4\,\cancel{9}\,\cancel{0}\,8 \\
-\ 3\,6\,6\,7 \\
\hline
1\,2\,4\,1
\end{array}
$$
Subtract ones. We cannot subtract 6 tens from 0 tens. We have 9 hundreds or 90 tens. We borrow 1 hundred to get 10 tens. We have 8 hundreds. Subtract tens, hundreds, and thousands.

71.
$$
\begin{array}{r}
\overset{\ \ 13}{} \\
\overset{2\ \cancel{3}\ 10\ \ 8\ 18}{} \\
\cancel{3}\,\cancel{4}\,\cancel{0},7\,\cancel{9}\,\cancel{8} \\
-\ \ \ 8\,6,6\,7\,9 \\
\hline
2\,5\,4,1\,1\,9
\end{array}
$$
We cannot subtract 9 ones from 8 ones. Borrow 1 ten to get 18 ones. Subtract ones. Then subtract tens and hundreds. We cannot subtract 6 thousands from 0 thousands. We have 4 ten thousands or 40 thousands. We borrow 1 ten thousand to get 10 thousands. Subtract thousands. We cannot subtract 8 ten thousands from 3 ten thousands. We borrow 1 hundred thousand to get 13 ten thousands. Subtract ten thousands and then hundred thousands.

73. Round $6,\,3\,7\,5,\ \boxed{6}\,0\,2$ to the nearest thousand.
↑

The digit 5 is in the thousands place. Consider the next digit to the right. Since the digit 6 is 5 or higher, round 5 thousands to 6 thousands. Then change all digits to the right of the thousands digit to zero.

The answer is 6,376,000.

75. Discussion and Writing Exercise

77. Use a calculator to perform the computations in this exercise.

First find the total area of each floor:
$$A = l \times w = 172 \times 84 = 14,448 \text{ square feet}$$

Find the area lost to the elevator and the stairwell:
$$A = l \times w = 35 \times 20 = 700 \text{ square feet}$$

Subtract to find the area available as office space on each floor:
$$14,448 - 700 = 13,748 \text{ square feet}$$

Finally, multiply by the number of floors, 18, to find the total area available as office space:
$$18 \times 13,748 = 247,464 \text{ square feet}$$

79. First, find the area of the photo.
$$A = l \times w = 8 \times 10 = 80 \text{ square inches}$$

From Exercise 5 we know that the printer prints 5,760,000 dots per square inch. We multiply to find the total number of dots.
$$80 \times 5,760,000 = 460,800,000$$

The printer will print 460,800,000 dots.

Exercise Set 1.6

1. Think of an array with 4 rows. The number of pounds in each row will go to a mule.

How many in each row?

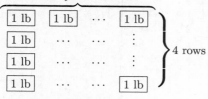

$$760 \div 4 = 190$$

3. Think of an array with 5 rows. The number of mL in each row will go in a beaker.

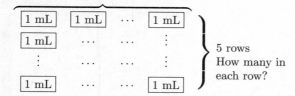

$$455 \div 5 = 91$$

5. $18 \div 3 = 6$ The 3 moves to the right. A related multiplication sentence is $18 = 6 \cdot 3$. (By the commutative law of multiplication, there is also another multiplication sentence: $18 = 3 \cdot 6$.)

7. $22 \div 22 = 1$ The 22 on the right of the $\div$ symbol moves to the right. A related multiplication sentence is $22 = 1 \cdot 22$. (By the commutative law of multiplication, there is also another multiplication sentence: $22 = 22 \cdot 1$.)

9. $54 \div 6 = 9$ The 6 moves to the right. A related multiplication sentence is $54 = 9 \cdot 6$. (By the commutative law of multiplication, there is also another multiplication sentence: $54 = 6 \cdot 9$.)

11. $37 \div 1 = 37$ The 1 moves to the right. A related multiplication sentence is $37 = 37 \cdot 1$. (By the commutative law of multiplication, there is also another multiplication sentence: $37 = 1 \cdot 37$.)

13. $9 \times 5 = 45$

Move a factor to the other side and then write a division.

$9 \times 5 = 45$ $9 \times 5 = 45$

$9 = 45 \div 5$ $5 = 45 \div 9$

15. Two related division sentences for $37 \cdot 1 = 37$ are:

$37 = 37 \div 1$ ($37 \cdot 1 = 37$)

and

$1 = 37 \div 37$ ($37 \cdot 1 = 37$)

17. $8 \times 8 = 64$

Since the factors are both 8, moving either one to the other side gives the related division sentence $8 = 64 \div 8$.

19. Two related division sentences for $11 \cdot 6 = 66$ are:

$11 = 66 \div 6$ ($11 \cdot 6 = 66$)

and

$6 = 66 \div 11$ ($11 \cdot 6 = 66$)

21.

$$
\begin{array}{r}
1\,2 \\
6\,\overline{)7\,2} \\
6\,0 \\
\hline
1\,2 \\
1\,2 \\
\hline
0
\end{array}
$$

Think: 7 tens $\div$ 6. Estimate 1 ten.
Think: 12 ones $\div$ 6. Estimate 2 ones.

The answer is 12.

23. $\dfrac{23}{23} = 1$ Any nonzero number divided by itself is 1.

25. $22 \div 1 = 22$ Any number divided by 1 is that same number.

27. $\dfrac{16}{0}$ is not defined, because division by 0 is not defined.

29.

$$
\begin{array}{r}
5\,5 \\
5\,\overline{)2\,7\,7} \\
2\,5\,0 \\
\hline
2\,7 \\
2\,5 \\
\hline
2
\end{array}
$$

Think: 2 hundreds $\div$ 5. There are no hundreds in the quotient.
Think: 27 tens $\div$ 5. Estimate 5 tens.
Think: 27 ones $\div$ 5. Estimate 5 ones.

The answer is 55 R 2.

31.

$$
\begin{array}{r}
1\,0\,8 \\
8\,\overline{)8\,6\,4} \\
8\,0\,0 \\
\hline
6\,4 \\
6\,4 \\
\hline
0
\end{array}
$$

Think: 8 hundreds $\div$ 8. Estimate 1 hundred.
Think: 6 tens $\div$ 8. There are no tens in the quotient (other than the tens in 100). Write a 0 to show this.
Think: 64 ones $\div$ 8. Estimate 8 ones.

The answer is 108.

33.

$$
\begin{array}{r}
3\,0\,7 \\
4\,\overline{)1\,2\,2\,8} \\
1\,2\,0\,0 \\
\hline
2\,8 \\
2\,8 \\
\hline
0
\end{array}
$$

Think: 12 hundreds $\div$ 4. Estimate 3 hundreds.
Think: 2 tens $\div$ 4. There are no tens in the quotient (other than the tens in 300). Write a 0 to show this.
Think: 28 ones $\div$ 4. Estimate 7 ones.

The answer is 307.

35.

$$
\begin{array}{r}
7\,5\,3 \\
6\,\overline{)4\,5\,2\,1} \\
4\,2\,0\,0 \\
\hline
3\,2\,1 \\
3\,0\,0 \\
\hline
2\,1 \\
1\,8 \\
\hline
3
\end{array}
$$

Think: 45 hundreds $\div$ 6. Estimate 7 hundreds.
Think: 32 tens $\div$ 6. Estimate 5 tens.
Think: 21 ones $\div$ 6. Estimate 3 ones.

The answer is 753 R 3.

37.
```
        1 7 0 3
    5 ) 8 5 1 5
        5 0 0 0
        3 5 1 5
        3 5 0 0
            1 5
            1 5
             0
```
Think: 8 thousands ÷ 5. Estimate 1 thousand.
Think: 35 hundreds ÷ 5. Estimate 7 hundreds.
Think: 1 ten ÷ 5. There are no tens in the quotient (other than the tens in 1700). Write a 0 to show this.
Think: 15 ones ÷ 5. Estimate 3 ones.

The answer is 1703.

39.
```
        9 8 7
    9 ) 8 8 8 8
        8 1 0 0
          7 8 8
          7 2 0
            6 8
            6 3
             5
```
Think: 88 hundreds ÷ 9. Estimate 9 hundreds.
Think: 78 tens ÷ 9. Estimate 8 tens.
Think: 68 ones ÷ 9. Estimate 7 ones.

The answer is 987 R 5.

41.
```
           1 2,7 0 0
    1 0 ) 1 2 7,0 0 0
          1 0 0,0 0 0
            2 7,0 0 0
            2 0,0 0 0
               7 0 0 0
               7 0 0 0
                     0
```
Think: 12 ten thousands ÷ 10. Estimate 1 ten thousand.
Think: 27 thousands ÷ 10. Estimate 2 thousands.
Think: 70 hundreds ÷ 10. Estimate 7 hundreds.
Since the difference is 0, there are no tens or ones in the quotient (other than the tens and ones in 12,700). We write zeros to show this.

The answer is 12,700.

43.
```
              1 2 7
    1 0 0 0 ) 1 2 7,0 0 0
              1 0 0,0 0 0
                2 7,0 0 0
                2 0,0 0 0
                   7 0 0 0
                   7 0 0 0
                         0
```
Think: 1270 hundreds ÷ 1000. Estimate 1 hundred.
Think: 2700 tens ÷ 1000. Estimate 2 tens.
Think: 7000 ones ÷ 1000. Estimate 7 ones.

The answer is 127.

45.
```
           5 2
    7 0 ) 3 6 9 2
          3 5 0 0
            1 9 2
            1 4 0
              5 2
```
Think: 369 tens ÷ 70. Estimate 5 tens.
Think: 192 ones ÷ 70. Estimate 2 ones.

The answer is 52 R 52.

47.
```
           2 9
    3 0 ) 8 7 5
          6 0 0
          2 7 5
          2 7 0
              5
```
Think: 87 tens ÷ 30. Estimate 2 tens.
Think: 275 ones ÷ 30. Estimate 9 ones.

The answer is 29 R 5.

49.
```
               3
    1 1 1 ) 3 2 1 9
            3 3 3 0
```
Round 111 to 100.
Think: 321 tens ÷ 100. Estimate 3 tens.

Since we cannot subtract 3330 from 3219, the estimate is too high.

```
              2 9
    1 1 1 ) 3 2 1 9
            2 2 2 0
              9 9 9
              9 9 9
                  0
```
Think: 321 tens ÷ 100. Estimate 2 tens.
Think: 999 ones ÷ 100. Estimate 9 ones.

The answer is 29.

51.
```
        1 0 5
    8 ) 8 4 3
        8 0 0
          4 3
          4 0
           3
```
Think: 8 hundreds ÷ 8. Estimate 1 hundred.
Think: 4 tens ÷ 8. There are no tens in the quotient (other than the tens in 100). Write a 0 to show this.
Think: 43 ones ÷ 8. Estimate 5 ones.

The answer is 105 R 3.

53.
```
        1 6 0 9
    5 ) 8 0 4 7
        5 0 0 0
        3 0 4 7
        3 0 0 0
            4 7
            4 5
             2
```
Think: 8 thousands ÷ 5. Estimate 1 thousand.
Think: 30 hundreds ÷ 5. Estimate 6 hundreds.
Think: 4 tens ÷ 5. There are no tens in the quotient (other than the tens in 1600). Write a 0 to show this.
Think: 47 ones ÷ 5. Estimate 9 ones.

The answer is 1609 R 2.

55.
```
        1 0 0 7
    5 ) 5 0 3 6
        5 0 0 0
            3 6
            3 5
             1
```
Think: 5 thousands ÷ 5. Estimate 1 thousand.
Think: 0 hundreds ÷ 5. There are no hundreds in the quotient (other than the hundreds in 1000). Write a 0 to show this.
Think: 3 tens ÷ 5. There are no tens in the quotient (other than the tens in 1000). Write a 0 to show this.
Think: 36 ones ÷ 5. Estimate 7 ones.

The answer is 1007 R 1.

57.
```
            2 2
    4 6 ) 1 0 5 8
          9 2 0
          1 3 8
            9 2
            4 6
```
Round 46 to 50.
Think: 105 tens ÷ 50. Estimate 2 tens.
Think: 138 ones ÷ 50. Estimate 2 ones.

Since 46 is not smaller than the divisor, 46, the estimate is too low.

```
        2 3
  4 6 ⟌1 0 5 8
        9 2 0
        1 3 8    Think: 138 ones ÷ 50. Estimate 3 ones.
        1 3 8
            0
```

The answer is 23.

59.
```
        1 0 7     Round 32 to 30.
  3 2 ⟌3 4 2 5    Think: 34 hundreds ÷ 30. Estimate
        3 2 0 0   1 hundred.
          2 2 5   Think: 22 tens ÷ 30. There are no
          2 2 4   tens in the quotient (other than the
              1   tens in 100). Write 0 to show this.
                  Think: 225 ones ÷ 30. Estimate 7
                  ones.
```

The answer is 107 R 1.

61.
```
          4       Round 24 to 20.
  2 4 ⟌8 8 8 0    Think: 88 hundreds ÷ 20. Estimate
        9 6 0 0   4 hundreds.
```

Since we cannot subtract 9600 from 8880, the estimate
is too high.

```
         3 8      Think: 88 hundreds ÷ 20. Estimate
  2 4 ⟌8 8 8 0    3 hundreds.
        7 2 0 0
        1 6 8 0   Think: 168 tens ÷ 20. Estimate 8
        1 9 2 0   tens.
```

Since we cannot subtract 1920 from 1680, the estimate
is too high.

```
        3 7 0     Think: 168 tens ÷ 20. Estimate 7
  2 4 ⟌8 8 8 0    tens.
        7 2 0 0
        1 6 8 0   Think: 0 ones ÷ 20. There are no
        1 6 8 0   ones in the quotient (other than the
              0   ones in 370). Write a 0 to show this.
```

The answer is 370.

63.
```
          5       Round 28 to 30.
  2 8 ⟌1 7, 0 6 7 Think: 170 hundreds ÷ 30. Esti-
        1 4 0 0 0 mate 5 hundreds.
          3 0  6 7
```

Since 30 is larger than the divisor, 28, the estimate is
too low.

```
         6 0 8    Think: 170 hundreds ÷ 30. Esti-
  2 8 ⟌1 7, 0 6 7 mate 6 hundreds.
        1 6 8 0 0
            2 6 7 Think: 26 tens ÷ 30. There are no
            2 2 4 tens in the quotient (other than the
              4 3 tens in 600.) Write a zero to show
                  this.

                  Think: 267 ones ÷ 30. Estimate 8
                  ones.
```

Since 43 is larger than the divisor, 28, the estimate is
too low.

```
         6 0 9
  2 8 ⟌1 7, 0 6 7
        1 6 8 0 0
            2 6 7 Think: 267 ones ÷ 30. Estimate 9
            2 5 2 ones.
              1 5
```

The answer is 609 R 15.

65.
```
          3 0 4   Think: 243 hundreds ÷ 80. Esti-
  8 0 ⟌2 4, 3 2 0 mate 3 hundreds.
        2 4 0 0 0 Think: 32 tens ÷ 80. There are no
            3 2 0 tens in the quotient (other than the
            3 2 0 tens in 300). Write a 0 to show this.
                0 Think: 320 ones ÷ 80. Estimate 4
                  ones.
```

The answer is 304.

67.
```
            3 5 0 8
  2 8 5 ⟌9 9 9, 9 9 9
          8 5 5 0 0 0
          1 4 4 9 9 9
          1 4 2 5 0 0
              2 4 9 9
              2 2 8 0
                2 1 9
```

The answer is 3508 R 219.

69.
```
            8 0 7 0
  4 5 6 ⟌3, 6 7 9, 9 2 0
          3 6 4 8 0 0 0
              3 1 9 2 0
              3 1 9 2 0
                    0
```

The answer is 8070.

71. Discussion and Writing Exercise

73. The distance around an object is its perimeter.

75. For large numbers, digits are separated by commas into
groups of three, called periods.

77. In the sentence 28 ÷ 7 = 4, the dividend is 28.

79. The minuend is the number from which another number
is being subtracted.

81. Discussion and Writing Exercise

83.

a	b	$a \cdot b$	$a + b$
	68	3672	
84			117
		32	12

To find a in the first row we divide $a \cdot b$ by b:

$$3672 \div 68 = 54$$

Then we add to find $a + b$:

$$54 + 68 = 122$$

To find b in the second row we subtract a from $a + b$:

$$117 - 84 = 33$$

Then we multiply to find $a \cdot b$:

$$84 \cdot 33 = 2772$$

To find a and b in the last row we find a pair of numbers whose product is 32 and whose sum is 12. Pairs of numbers whose product is 32 are 1 and 32, 2 and 16, 4 and 8. Since $4 + 8 = 12$, the numbers we want are 4 and 8. We will let $a = 4$ and $b = 8$. (We could also let $a = 8$ and $b = 4$.)

The completed table is shown below.

a	b	$a \cdot b$	$a + b$
54	68	3672	122
84	33	2772	117
4	8	32	12

85. We divide 1231 by 42:

```
        2 9
  4 2 ⟌ 1 2 3 1
        8 4 0
        3 9 1
        3 7 8
        1 3
```

The answer is 29 R 13. Since 13 students will be left after 29 buses are filled, then 30 buses are needed.

Exercise Set 1.7

1. $x + 0 = 14$

We replace x by different numbers until we get a true equation. If we replace x by 14, we get a true equation: $14 + 0 = 14$. No other replacement makes the equation true, so the solution is 14.

3. $y \cdot 17 = 0$

We replace y by different numbers until we get a true equation. If we replace y by 0, we get a true equation: $0 \cdot 17 = 0$. No other replacement makes the equation true, so the solution is 0.

5.
$$13 + x = 42$$
$$13 + x - 13 = 42 - 13 \qquad \text{Subtracting 13 on both sides}$$
$$0 + x = 29 \qquad \text{13 plus } x \text{ minus 13 is } 0 + x.$$
$$x = 29$$

Check: $\dfrac{13 + x = 42}{}$

$$13 + 29 \ ? \ 42$$
$$42 \ \mid \quad \text{TRUE}$$

The solution is 29.

7.
$$12 = 12 + m$$
$$12 - 12 = 12 + m - 12 \qquad \text{Subtracting 12 on both sides}$$
$$0 = 0 + m \qquad \text{12 plus } m \text{ minus 12 is } 0 + m.$$
$$0 = m$$

Check: $\dfrac{12 = 12 + m}{}$

$$12 \ ? \ 12 + 0$$
$$\mid \ 12 \qquad \text{TRUE}$$

The solution is 0.

9. $3 \cdot x = 24$

$$\frac{3 \cdot x}{3} = \frac{24}{3} \qquad \text{Dividing by 3 on both sides}$$
$$x = 8 \qquad \text{3 times } x \text{ divided by 3 is } x.$$

Check: $\dfrac{3 \cdot x = 24}{}$

$$3 \cdot 8 \ ? \ 24$$
$$24 \ \mid \quad \text{TRUE}$$

The solution is 8.

11. $112 = n \cdot 8$

$$\frac{112}{8} = \frac{n \cdot 8}{8} \qquad \text{Dividing by 8 on both sides}$$
$$14 = n$$

Check: $\dfrac{112 = n \cdot 8}{}$

$$112 \ ? \ 14 \cdot 8$$
$$\mid \ 112 \qquad \text{TRUE}$$

The solution is 14.

13. $45 \times 23 = x$

To solve the equation we carry out the calculation.

```
      4 5
    × 2 3
    1 3 5
    9 0 0
  1 0 3 5
```

We can check by repeating the calculation. The solution is 1035.

15. $t = 125 \div 5$

To solve the equation we carry out the calculation.

```
        2 5
    5 ⟌ 1 2 5
        1 0 0
        2 5
        2 5
        0
```

We can check by repeating the calculation. The solution is 25.

17. $p = 908 - 458$

To solve the equation we carry out the calculation.

$$\begin{array}{r} 9\,0\,8 \\ -\,4\,5\,8 \\ \hline 4\,5\,0 \end{array}$$

We can check by repeating the calculation. The solution is 450.

19. $x = 12,345 + 78,555$

To solve the equation we carry out the calculation.

$$\begin{array}{r} 1\,2,3\,4\,5 \\ +\,7\,8,5\,5\,5 \\ \hline 9\,0,9\,0\,0 \end{array}$$

We can check by repeating the calculation. The solution is 90,900.

21. $3 \cdot m = 96$

$\dfrac{3 \cdot m}{3} = \dfrac{96}{3}$ Dividing by 3 on both sides

$m = 32$

Check: $\overline{3 \cdot m = 96}$

$3 \cdot 32\ ?\ 96$

$96\ \big|\ $ TRUE

The solution is 32.

23. $715 = 5 \cdot z$

$\dfrac{715}{5} = \dfrac{5 \cdot z}{5}$ Dividing by 5 on both sides

$143 = z$

Check: $\overline{715 = 5 \cdot x}$

$715\ ?\ 5 \cdot 143$

$\big|\ 715\ \ $ TRUE

The solution is 143.

25. $10 + x = 89$

$10 + x - 10 = 89 - 10$

$x = 79$

Check: $\overline{10 + x = 89}$

$10 + 79\ ?\ 89$

$89\ \big|\ $ TRUE

The solution is 79.

27. $61 = 16 + y$

$61 - 16 = 16 + y - 16$

$45 = y$

Check: $\overline{61 = 16 + y}$

$61\ ?\ 16 + 45$

$\big|\ 61\ \ $ TRUE

The solution is 45.

29. $6 \cdot p = 1944$

$\dfrac{6 \cdot p}{6} = \dfrac{1944}{6}$

$p = 324$

Check: $\overline{6 \cdot p = 1944}$

$6 \cdot 324\ ?\ 1944$

$1944\ \big|\ $ TRUE

The solution is 324.

31. $5 \cdot x = 3715$

$\dfrac{5 \cdot x}{5} = \dfrac{3715}{5}$

$x = 743$

The number 743 checks. It is the solution.

33. $47 + n = 84$

$47 + n - 47 = 84 - 47$

$n = 37$

The number 37 checks. It is the solution.

35. $x + 78 = 144$

$x + 78 - 78 = 144 - 78$

$x = 66$

The number 66 checks. It is the solution.

37. $165 = 11 \cdot n$

$\dfrac{165}{11} = \dfrac{11 \cdot n}{11}$

$15 = n$

The number 15 checks. It is the solution.

39. $624 = t \cdot 13$

$\dfrac{624}{13} = \dfrac{t \cdot 13}{13}$

$48 = t$

The number 48 checks. It is the solution.

41. $x + 214 = 389$

$x + 214 - 214 = 389 - 214$

$x = 175$

The number 175 checks. It is the solution.

43. $567 + x = 902$

$567 + x - 567 = 902 - 567$

$x = 335$

The number 335 checks. It is the solution.

45. $18 \cdot x = 1872$

$\dfrac{18 \cdot x}{18} = \dfrac{1872}{18}$

$x = 104$

The number 104 checks. It is the solution.

47. $40 \cdot x = 1800$

$\dfrac{40 \cdot x}{40} = \dfrac{1800}{40}$

$x = 45$

The number 45 checks. It is the solution.

49. $2344 + y = 6400$

$2344 + y - 2344 = 6400 - 2344$

$y = 4056$

The number 4056 checks. It is the solution.

51. $8322 + 9281 = x$
         $17,603 = x$ Doing the addition
   The number 17,603 checks. It is the solution.

53. $234 \cdot 78 = y$
         $18,252 = y$ Doing the multiplication
   The number 18,252 checks. It is the solution.

55. $58 \cdot m = 11,890$
         $\dfrac{58 \cdot m}{58} = \dfrac{11,890}{58}$
         $m = 205$
   The number 205 checks. It is the solution.

57. Discussion and Writing Exercise

59. $7 + 8 = 15$ $7 + 8 = 15$
            ↑   ↑
   This number gets   This number gets
   subtracted from the   subtracted from the
   sum.   ↓   sum.   ↓
         $7 = 15 - 8$ $8 = 15 - 7$

61. Since 123 is to the left of 789 on the number line,
   $123 < 789$.

63. Since 688 is to the right of 0 on the number line, $688 > 0$.

65.
$$
\begin{array}{r}
1\,4\,2 \\
9\,\overline{\smash)1\,2\,8\,3} \\
9\,0\,0 \\
\hline
3\,8\,3 \\
3\,6\,0 \\
\hline
2\,3 \\
1\,8 \\
\hline
5
\end{array}
$$
   Think: 12 hundreds ÷ 9. Estimate 1 hundred.
   Think: 38 tens ÷ 9. Estimate 4 tens.
   Think: 23 ones ÷ 9. Estimate 2 ones.

   The answer is 142 R 5.

67.
$$
\begin{array}{r}
3\,3\,4 \\
1\,7\,\overline{\smash)5\,6\,7\,8} \\
5\,1\,0\,0 \\
\hline
5\,7\,8 \\
5\,1\,0 \\
\hline
6\,8 \\
6\,8 \\
\hline
0
\end{array}
$$
   Think 56 hundreds ÷ 17. Estimate 3 hundreds.
   Think 57 tens ÷ 17. Estimate 3 tens.
   Think 68 ones ÷ 17. Estimate 4 ones.

   The answer is 334.

69. Discussion and Writing Exercise

71. $23,465 \cdot x = 8,142,355$
         $\dfrac{23,465 \cdot x}{23,465} = \dfrac{8,142,355}{23,465}$
         $x = 347$ Using a calculator to divide
   The number 347 checks. It is the solution.

Exercise Set 1.8

1. *Familiarize.* We visualize the situation. We are combining quantities, so addition can be used.

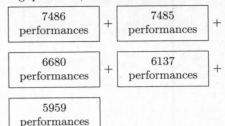

   Let $p =$ the total number of performances of all five shows.

   Translate. We translate to an equation.

   $7486 + 7485 + 6680 + 6137 + 5959 = p$

   Solve. We carry out the addition.

$$
\begin{array}{r}
{\scriptstyle 2\ \ 3\ 2} \\
7\,4\,8\,6 \\
7\,4\,8\,5 \\
6\,6\,8\,0 \\
6\,1\,3\,7 \\
+\ 5\,9\,5\,9 \\
\hline
3\,3,7\,4\,7
\end{array}
$$

   Thus, $33,747 = p$.

   Check. We can repeat the calculation. We can also estimate by rounding, say to the nearest thousand.

   $7486 + 7485 + 6680 + 6137 + 5959$

   $\approx 7000 + 7000 + 7000 + 6000 + 6000$

   $\approx 33,000 \approx 33,747$

   Since the estimated answer is close to the calculated answer, our result is probably correct.

   State. The five longest-running Broadway shows had a total of 33,747 performances.

3. *Familiarize.* We visualize the situation. Let $c =$ the number by which the performances of *The Phantom of the Opera* exceeded the performances of *A Chorus Line*.

Chorus Line performances	Excess *Phantom* performances
6137	c
Number of *Phantom* performances	
7486	

   Translate. We see this as a "how many more" situation.

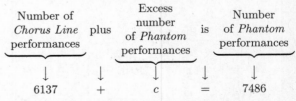

   Solve. We subtract 6137 on both sides of the equation.

   $6137 + c = 7486$

   $6137 + c - 6137 = 7486 - 6137$

   $c = 1349$

Check. We can add the difference, 1349, to the subtrahend, 6137: $6137 + 1349 = 7486$. We can also estimate:

$$7486 - 6137 \approx 7500 - 6100$$
$$\approx 1400 \approx 1349$$

The answer checks.

State. There were 1349 more performances of *The Phantom of the Opera* than *A Chorus Line*.

5. Familiarize. We visualize the situation. Let $m =$ the number of miles by which the Canadian border exceeds the Mexican border.

Mexican border 1933 mi	Excess miles in Canadian border m
Canadian border 3987 mi	

Translate. We see this as a "how many more" situation.

Length of Mexican border	plus	Excess length of Canadian border	is	Length of Canadian border
↓	↓	↓	↓	↓
1933	+	m	=	3987

Solve. We subtract 1933 on both sides of the equation.

$$1933 + m = 3987$$
$$1933 + m - 1933 = 3987 - 1933$$
$$m = 2054$$

Check. We can add the difference, 2054, to the subtrahend, 1933: $1933 + 2054 = 3987$. We can also estimate:

$$3987 - 1933 \approx 4000 - 2000$$
$$\approx 2000 \approx 2054$$

The answer checks.

State. The Canadian border is 2054 mi longer than the Mexican border.

7. Familiarize. We first make a drawing. Let $r =$ the number of rows.

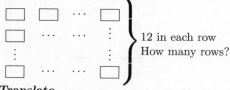

12 in each row
How many rows?

Translate.

Number of holes	divided by	Number per row	is	Number of rows
↓	↓	↓	↓	↓
216	÷	12	=	r

Solve. We carry out the division.

$$\begin{array}{r} 1\,8 \\ 12\overline{\smash{)}2\,1\,6} \\ \underline{1\,2\,0} \\ 9\,6 \\ \underline{9\,6} \\ 0 \end{array}$$

Thus, $18 = r$, or $r = 18$.

Check. We can check by multiplying: $12 \cdot 18 = 216$. Our answer checks.

State. There are 18 rows.

9. Familiarize. We visualize each situation. We are combining quantities, so addition can be used.

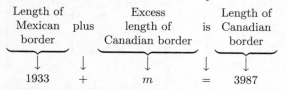

| 451,097 | + | 341,219 |

degrees awarded to men in 1970   degrees awarded to women in 1970

Let $x =$ the total number of bachelor's degrees awarded in 1970.

| 775,424 | + | 573,079 |

degrees awarded to women in 2003   degrees awarded to men in 2003

Let $y =$ the total number of bachelor's degrees awarded in 2003.

Translate. We translate each situation to an equation.

For 1970: $451,097 + 341,219 = x$

For 2003: $775,424 + 573,079 = y$

Solve. We carry out the additions.

$$\begin{array}{r} {}^{1\ 1} \\ 4\,5\,1,0\,9\,7 \\ +\ 3\,4\,1,2\,1\,9 \\ \hline 7\,9\,2,3\,1\,6 \end{array}$$

Thus, $792,316 = x$.

$$\begin{array}{r} {}^{1}\quad{}^{1\ 1} \\ 7\,7\,5,4\,2\,4 \\ +\ 5\,7\,3,0\,7\,9 \\ \hline 1,3\,4\,8,5\,0\,3 \end{array}$$

Thus, $1,348,503 = y$.

Check. We will estimate.

For 1970: $451,097 + 341,219$
$$\approx 450,000 + 340,000$$
$$\approx 790,000 \approx 792,316$$

For 2003: $775,424 + 573,079$
$$\approx 780,000 + 570,000$$
$$\approx 1,350,000 \approx 1,348,503$$

The answer checks.

State. In 1970 a total of 792,316 bachelor's degrees were awarded; the total in 2003 was 1,348,503.

11. Familiarize. We visualize the situation. Let $w =$ the number by which the degrees awarded to women in 2003 exceeded those awarded to men.

Men's degrees 573,079	Excess women's degrees w
Women's degrees 775,424	

Translate. We see this as a "how much more" situation.

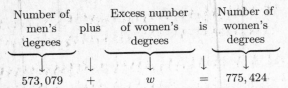

Number of men's degrees	plus	Excess number of women's degrees	is	Number of women's degrees
↓	↓	↓	↓	↓
573,079	+	w	=	775,424

Solve. We subtract 573,079 on both sides of the equation.

$$573{,}079 + w = 775{,}424$$
$$573{,}079 + w - 573{,}079 = 775{,}424 - 573{,}079$$
$$w = 202{,}345$$

Check. We will estimate.

$$775{,}424 - 573{,}079$$
$$\approx 780{,}000 - 570{,}000$$
$$\approx 210{,}000 \approx 202{,}345$$

The answer checks.

State. There were 202,345 more bachelor's degrees awarded to women than to men in 2003.

13. Familiarize. We visualize the situation. Let $m =$ the median mortgage debt in 2004.

Debt in 1989 $39,802	Excess debt in 2004 $48,388
Debt in 2004 m	

Translate. We translate to an equation.

Debt in 1989	+	Excess debt in 2004	is	Debt in 2004
↓	↓	↓	↓	↓
39,802	+	48,388	=	m

Solve. We carry out the addition.

$$\begin{array}{r} {\scriptstyle 1\ \ 1\ \ \ \ 1} \\ 3\,9{,}8\,0\,2 \\ +\ 4\,8{,}3\,8\,8 \\ \hline 8\,8{,}1\,9\,0 \end{array}$$

Thus, $88{,}190 = m$.

Check. We can estimate.

$$39{,}802 + 48{,}388$$
$$\approx 40{,}000 + 48{,}000$$
$$\approx 88{,}000 \approx 88{,}190$$

The answer checks.

State. In 2004 the median mortgage debt was $88,190.

15. Familiarize. We visualize the situation. Let $l =$ the excess length of the Nile River, in miles.

Length of Missouri-Mississippi 3710 miles	Excess length of Nile l
Length of Nile 4180 miles	

Translate. This is a "how much more" situation. We translate to an equation.

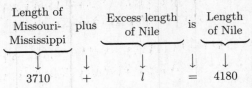

Length of Missouri-Mississippi	plus	Excess length of Nile	is	Length of Nile
↓	↓	↓	↓	↓
3710	+	l	=	4180

Solve. We subtract 3710 on both sides of the equation.

$$3710 + l = 4180$$
$$3710 + l - 3710 = 4180 - 3710$$
$$l = 470$$

Check. We can check by adding the difference, 470, to the subtrahend, 3710: $3710 + 470 = 4180$. Our answer checks.

State. The Nile River is 470 mi longer than the Missouri-Mississippi River.

17. Familiarize. We first draw a picture. Let $h =$ the number of hours in a week. Repeated addition works well here.

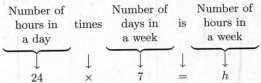

24 hours	+	24 hours	$+ \cdots +$	24 hours

7 addends

Translate. We translate to an equation.

Number of hours in a day	times	Number of days in a week	is	Number of hours in a week
↓	↓	↓	↓	↓
24	×	7	=	h

Solve. We carry out the multiplication.

$$\begin{array}{r} 2\,4 \\ \times\ \ \ 7 \\ \hline 1\,6\,8 \end{array}$$

Thus, $168 = h$, or $h = 168$.

Check. We can repeat the calculation. We an also estimate:

$$24 \times 7 \approx 20 \times 10 = 200 \approx 168$$

Our answer checks.

State. There are 168 hours in a week.

19. Familiarize. We first draw a picture. Let $s =$ the number of squares in the puzzle. Repeated addition works well here.

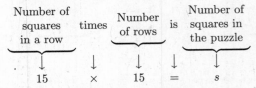

15 squares	+	15 squares	$+ \cdots +$	15 squares

15 addends

Translate. We translate to an equation.

Number of squares in a row	times	Number of rows	is	Number of squares in the puzzle
↓	↓	↓	↓	↓
15	×	15	=	s

Solve. We carry out the multiplication.

$$\begin{array}{r} 1\,5 \\ \times\ 1\,5 \\ \hline 7\,5 \\ 1\,5\,0 \\ \hline 2\,2\,5 \end{array}$$

Thus, $225 = s$.

Check. We can repeat the calculation. The answer checks.

State. There are 225 squares in the crossword puzzle.

21. **Familiarize**. We draw a picture of the situation. Let $c =$ the total cost of the purchase. Repeated addition works well here.

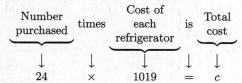

Translate. We translate to an equation.

Number purchased	times	Cost of each refrigerator	is	Total cost
↓	↓	↓	↓	↓
24	×	1019	=	c

Solve. We carry out the multiplication.

$$
\begin{array}{r}
\overset{1}{\overset{3}{}} \\
1\,0\,1\,9 \\
\times\quad 2\,4 \\
\hline
4\,0\,7\,6 \\
2\,0\,3\,8\,0 \\
\hline
2\,4,4\,5\,6
\end{array}
$$

Thus, $24,456 = c$.

Check. We can repeat the calculation. We can also estimate: $24 \times 1019 \approx 24 \times 1000 \approx 24,000 \approx 24,456$. The answer checks.

State. The total cost of the purchase is $24,456.

23. **Familiarize**. We first draw a picture. Let $w =$ the number of full weeks the episodes can run.

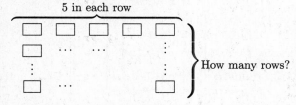

Translate. We translate to an equation.

Number of episodes	divided by	Number shown per week	is	Number of weeks
↓	↓	↓	↓	↓
177	÷	5	=	w

Solve. We carry out the division.

$$
\begin{array}{r}
3\,5 \\
5\overline{)1\,7\,7} \\
1\,5\,0 \\
\hline
2\,7 \\
2\,5 \\
\hline
2
\end{array}
$$

Check. We can check by multiplying the number of weeks by 5 and adding the remainder, 2:

$$5 \cdot 35 = 175, \qquad 175 + 2 = 177$$

State. 35 full weeks will pass before the station must start over. There will be 2 episodes left over.

25. **Familiarize**. We first draw a picture of the situation. Let $g =$ the number of gallons that will be used in 6136 mi of city driving.

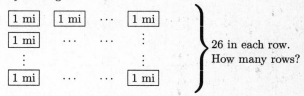

Translate. We translate to an equation.

Number of miles	divided by	Number of mpg	is	Number of gallons
↓	↓	↓	↓	↓
6136	÷	26	=	g

Solve. We carry out the division.

$$
\begin{array}{r}
2\,3\,6 \\
26\overline{)6\,1\,3\,6} \\
5\,2\,0\,0 \\
\hline
9\,3\,6 \\
7\,8\,0 \\
\hline
1\,5\,6 \\
1\,5\,6 \\
\hline
0
\end{array}
$$

Thus, $236 = g$.

Check. We can check by multiplying the number of gallons by the number of miles per gallon: $26 \cdot 236 = 6136$. The answer checks.

State. The Hyundai Tucson GLS will use 236 gal of gasoline in 6136 mi of city driving.

27. **Familiarize**. We visualize the situation. Let $d =$ the number of miles by which the nonstop flight distance of the Boeing 747 exceeds the nonstop flight distance of the Boeing 777.

777 distance 5210 mi	Excess 747 distance d
747 distance 8826 mi	

Translate. This is a "how much more" situation.

777 distance	plus	Excess 747 distance	is	747 distance
↓	↓	↓	↓	↓
5210	+	d	=	8826

Solve. We subtract 5210 on both sides of the equation.

$$5210 + d = 8826$$
$$5210 + d - 5210 = 8826 - 5210$$
$$d = 3616$$

Check. We can estimate.

$8826 - 5210 \approx 8800 - 5200 \approx 3600 \approx 3616$

The answer checks.

State. The Boeing 747's nonstop flight distance is 3616 mi greater than that of the Boeing 777.

29. Familiarize. We draw a picture. Let $g =$ the number of gallons of fuel needed for a 4-hr flight of the Boeing 747. Repeated addition works well here.

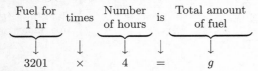

Translate. We translate to an equation.

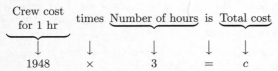

Solve. We carry out the multiplication.

$$\begin{array}{r} 3\,2\,0\,1 \\ \times \quad\quad 4 \\ \hline 1\,2,8\,0\,4 \end{array}$$

Thus, $12,804 = g$.

Check. We can repeat the calculation. The answer checks.

State. For a 4-hr flight of the Boeing 747, 12,804 gal of fuel are needed.

31. Familiarize. This is a multistep problem. First we will find the cost for the crew. Then we will find the cost for the fuel. Finally, we will find the total cost for the crew and the fuel. Let $c =$ the cost of the crew, $f =$ the cost of the fuel, and $t =$ the total cost.

Translate.

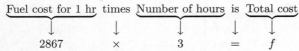

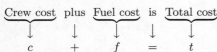

Solve. First we carry out the multiplications to solve the first two equations.

$$\begin{array}{r} {\scriptstyle 2\,1\,2} \\ 1\,9\,4\,8 \\ \times \quad\quad 3 \\ \hline 5\,8\,4\,4 \end{array}$$

Thus, $5844 = c$.

$$\begin{array}{r} {\scriptstyle 2\,2\,2} \\ 2\,8\,6\,7 \\ \times \quad\quad 3 \\ \hline 8\,6\,0\,1 \end{array}$$

Thus, $8601 = f$.

Now we substitute 5844 for c and 8601 for f in the third equation and carry out the addition.

$$c + f = t$$
$$5844 + 8601 = t$$
$$14,445 = t$$

Check. We repeat the calculations. The answer checks.

State. The total cost for the crew and the fuel for a 3-hr flight of the Boeing 747 is $14,445.

33. Familiarize. We first draw a picture. We let $x =$ the amount of each payment.

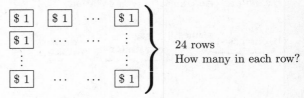

Translate. We translate to an equation.

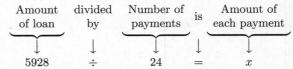

Solve. We carry out the division.

$$\begin{array}{r} 2\,4\,7 \\ 2\,4\,\overline{)5\,9\,2\,8} \\ 4\,8\,0\,0 \\ \hline 1\,1\,2\,8 \\ 9\,6\,0 \\ \hline 1\,6\,8 \\ 1\,6\,8 \\ \hline 0 \end{array}$$

Thus, $247 = x$, or $x = 247$.

Check. We can check by multiplying 247 by 24: $24 \cdot 247 = 5928$. The answer checks.

State. Each payment is $247.

35. Familiarize. We first draw a picture. Let $A =$ the area and $P =$ the perimeter of the court, in feet.

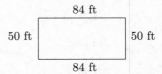

Translate. We write one equation to find the area and another to find the perimeter.

a) Using the formula for the area of a rectangle, we have
$$A = l \cdot w = 84 \cdot 50$$

b) Recall that the perimeter is the distance around the court.
$$P = 84 + 50 + 84 + 50$$

Solve. We carry out the calculations.

a)
$$\begin{array}{r} 5\,0 \\ \times\,8\,4 \\ \hline 2\,0\,0 \\ 4\,0\,0\,0 \\ \hline 4\,2\,0\,0 \end{array}$$

Thus, $A = 4200$.

b) $P = 84 + 50 + 84 + 50 = 268$

Check. We can repeat the calculation. The answers check.

State. a) The area of the court is 4200 square feet.

b) The perimeter of the court is 268 ft.

37. Familiarize. We visualize the situation. Let $a =$ the number of dollars by which the imports exceeded the exports.

Exports $2,596,000,000	Excess amount of imports a
Imports $31,701,000,000	

Translate. This as a "how much more" situation.

Exports plus Excess imports is Imports

$$2,596,000,000 \;+\; a \;=\; 31,701,000,000$$

Solve. We subtract 2,596,000,000 on both sides of the equation.

$$2,596,000,000 + a = 31,701,000,000$$
$$2,596,000,000 + a - 2,596,000,000 = 31,701,000,000 - 2,596,000,000$$
$$a = 29,105,000,000$$

Check. We can estimate.

$$31,701,000,000 - 2,596,000,000$$
$$\approx 32,000,000,000 - 3,000,000,000$$
$$\approx 29,000,000,000 \approx 29,105,000,000$$

The answer checks.

State. Imports exceeded exports by $29,105,000,000.

39. Familiarize. We visualize the situation. Let $p =$ the Colonial population in 1680.

Population in 1680 p	Increase in population 2,628,900
Population in 1780 2,780,400	

Translate. This is a "how much more" situation.

Population in 1680 plus Increase in population is Population in 1780

$$p \;+\; 2,628,900 \;=\; 2,780,400$$

Solve. We subtract 2,628,900 on both sides of the equation.

$$p + 2,628,900 = 2,780,400$$
$$p + 2,628,900 - 2,628,900 = 2,780,400 - 2,628,900$$
$$p = 151,500$$

Check. Since $2,628,900 + 151,500 = 2,780,400$, the answer checks.

State. In 1680 the Colonial population was 151,500.

41. Familiarize. We draw a picture of the situation. Let $n =$ the number of 20-bar packages that can be filled.

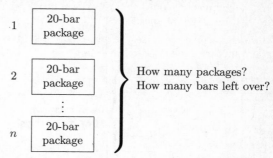

Translate. We translate to an equation.

Number of bars divided by Number per package is Number of packages

$$11,267 \;\div\; 20 \;=\; n$$

Solve. We carry out the division.

$$\begin{array}{r} 5\,6\,3 \\ 20\,\overline{)\,1\,1,2\,6\,7} \\ \underline{1\,0,0\,0\,0} \\ 1\,2\,6\,7 \\ \underline{1\,2\,0\,0} \\ 6\,7 \\ \underline{6\,0} \\ 7 \end{array}$$

Thus, $n = 563$ R 7.

Check. We can check by multiplying the number of packages by 20 and then adding the remainder, 7:

$$20 \cdot 563 = 11,260 \qquad 11,260 + 7 = 11,267$$

The answer checks.

State. 563 packages can be filled. There will be 7 bars left over.

43. Familiarize. First we find the distance in reality between two cities that are 6 in. apart on the map. We make a drawing. Let $d =$ the distance between the cities, in miles. Repeated addition works well here.

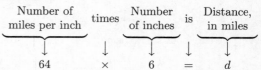

6 addends

Translate.

Number of miles per inch times Number of inches is Distance, in miles

$$64 \;\times\; 6 \;=\; d$$

Solve. We carry out the multiplication.

$$\begin{array}{r} 6\,4 \\ \times\,6 \\ \hline 3\,8\,4 \end{array}$$

Thus, $384 = d$.

Check. We can repeat the calculation or estimate the product. Our answer checks.

State. Two cities that are 6 in. apart on the map are 384 miles apart in reality.

Next we find distance on the map between two cities that, in reality, are 1728 mi apart.

Familiarize. We visualize the situation. Let $m =$ the distance between the cities on the map.

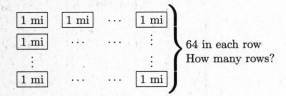

Translate.

Number of miles	divided by	Number of miles per inch	is	Distance, in inches.
↓	↓	↓	↓	↓
1728	÷	64	=	m

Solve. We carry out the division.

$$\begin{array}{r} 27 \\ 64\overline{\smash{)}1728} \\ \underline{1280} \\ 448 \\ \underline{448} \\ 0 \end{array}$$

Thus, $27 = m$, or $m = 27$.

Check. We can check by multiplying: $64 \cdot 27 = 1728$. Our answer checks.

State. The cities are 27 in. apart on the map.

45. Familiarize. First we draw a picture. Let $c =$ the number of columns. The number of columns is the same as the number of squares in each row.

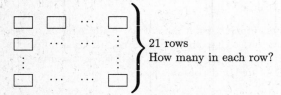

Translate. We translate to an equation.

Number of squares	divided by	Number of rows	is	Number of columns.
↓	↓	↓	↓	↓
441	÷	21	=	c

Solve. We carry out the division.

$$\begin{array}{r} 21 \\ 21\overline{\smash{)}441} \\ \underline{420} \\ 21 \\ \underline{21} \\ 0 \end{array}$$

Thus, $21 = c$.

Check. We can check by multiplying the number of rows by the number of columns: $24 \cdot 21 = 441$. The answer checks.

State. The puzzle has 21 columns.

47. Familiarize. We visualize the situation as we did in Exercise 39. Let $c =$ the number of cartons that can be filled.

Translate.

Number of books	divided by	Number per carton	is	Number of full cartons.
↓	↓	↓	↓	↓
1355	÷	24	=	c

Solve. We carry out the division.

$$\begin{array}{r} 56 \\ 24\overline{\smash{)}1355} \\ \underline{1200} \\ 155 \\ \underline{144} \\ 11 \end{array}$$

Check. We can check by multiplying the number of cartons by 24 and adding the remainder, 11:

$$24 \cdot 56 = 1344, \qquad 1344 + 11 = 1355$$

Our answer checks.

State. 56 cartons can be filled. There will be 11 books left over. If 1355 books are to be shipped, it will take 57 cartons.

49. Familiarize. This is a multistep problem.

We must find the total price of the 5 video games. Then we must find how many 10's there are in the total price. Let $p =$ the total price of the games.

To find the total price of the 5 video games we can use repeated addition.

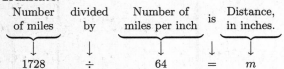

Translate.

Price per game	times	Number of games	is	Total price of games
↓	↓	↓	↓	↓
64	·	5	=	p

Solve. First we carry out the multiplication.

$$64 \cdot 5 = p$$
$$320 = p$$

The total price of the 5 video games is $320. Repeated addition can be used again to find how many 10's there are in $320. We let $x =$ the number of $10 bills required.

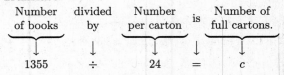

Translate to an equation and solve.

$$10 \cdot x = 320$$

$$\frac{10 \cdot x}{10} = \frac{320}{10}$$

$$x = 32$$

Check. We repeat the calculations. The answer checks.

State. It takes 32 ten dollar bills.

51. Familiarize. This is a multistep problem. We must find the total amount of the debits. Then we subtract this amount from the original balance and add the amount of the deposit. Let a = the total amount of the debits. To find this we can add.

Translate.

First debit	plus	Second debit	plus	Third debit	is	Total amount
↓	↓	↓	↓	↓	↓	↓
46	+	87	+	129	=	a

Solve. First we carry out the addition.

```
    1 2
      4 6
      8 7
  + 1 2 9
  ---------
    2 6 2
```

Thus, $262 = a$.

Now let b = the amount left in the account after the debits.

Amount left	is	Original amount	minus	Amount of debits
↓	↓	↓	↓	↓
b	=	568	−	262

We solve this equation by carrying out the subtraction.

```
    5 6 8
  - 2 6 2
  ---------
    3 0 6
```

Thus, $b = 306$.

Finally, let f = the final amount in the account after the deposit is made.

Final amount	is	Amount after debits	plus	Amount of deposit
↓	↓	↓	↓	↓
f	=	306	+	94

We solve this equation by carrying out the addition.

```
    1 1
    3 0 6
  +   9 4
  ---------
    4 0 0
```

Thus, $f = 400$.

Check. We repeat the calculations. The answer checks.

State. There is $400 left in the account.

53. Familiarize. This is a multistep problem. We begin by visualizing the situation.

One pound 3500 calories			
100 cal	100 cal	...	100 cal
8 min	8 min		8 min

Let x = the number of hundreds in 3500. Repeated addition applies here.

Translate. We translate to an equation.

100 calories	times	How many 100's	is 3500?
↓	↓	↓	↓ ↓
100	·	x	= 3500

Solve. We divide by 100 on both sides of the equation.

$$100 \cdot x = 3500$$

$$\frac{100 \cdot x}{100} = \frac{3500}{100}$$

$$x = 35$$

We know that running for 8 min will burn 100 calories. This must be done 35 times in order to lose one pound. Let t = the time it takes to lose one pound. We have:

$$t = 35 \times 8$$

$$t = 280$$

Check. $280 \div 8 = 35$, so there are 35 8's in 280 min, and $35 \cdot 100 = 3500$, the number of calories that must be burned in order to lose one pound. The answer checks.

State. You must run for 280 min, or 4 hr, 40 min, at a brisk pace in order to lose one pound.

55. Familiarize. This is a multistep problem. We begin by visualizing the situation.

One pound 3500 calories			
100 cal	100 cal	...	100 cal
15 min	15 min		15 min

From Exercise 53 we know that there are 35 100's in 3500. From the chart we know that doing aerobic exercise for 15 min burns 100 calories. Thus we must do 15 min of exercise 35 times in order to lose one pound. Let t = the number of minutes of aerobic exercise required to lose one pound.

Translate. We translate to an equation.

Number of times	times	Number of minutes	is	Total time
↓	↓	↓	↓	↓
35	×	15	=	t

```
    1 5
  × 3 5
  -------
    7 5
  4 5 0
  -------
  5 2 5
```

Thus, $525 = t$.

Check. $525 \div 15 = 35$, so there are 35 15's in 525 min, and $35 \cdot 100 = 3500$, the number of calories that must be burned in order to lose one pound. The answer checks.

State. You must do aerobic exercise for 525 min, or 8 hr, 45 min, in order to lose one pound.

57. Familiarize. This is a multistep problem. We will find the number of bones in both hands and the number in both feet and then the total of these two numbers. Let $h =$ the number of bones in two human hands, $f =$ the number of bones in two human feet, and $t =$ the total number of bones in two hands and two feet.

Translate. We translate to three equations.

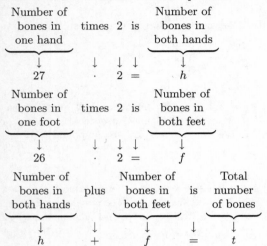

Solve. We solve each equation.

$$27 \cdot 2 = h \qquad 26 \cdot 2 = f$$
$$54 = h \qquad\quad 52 = f$$

$$h + f = t$$
$$54 + 52 = t$$
$$106 = t$$

Check. We repeat the calculations. The answer checks.

State. In all, a human has 106 bones in both hands and both feet.

59. Familiarize. This is a multistep problem. First we find the writing area on one card and then the total writing area on 100 cards. Let $a =$ the writing area on one card, in square inches, and $t =$ the total writing area on 100 cards. Keep in mind that we can write on both sides of each card. Recall that the formula for the area of a rectangle is length × width.

Translate. We translate to two equations.

Writing area on one card is Two times Length times Width

$$a = 2 \cdot 5 \cdot 3$$

Writing area on 100 cards is 100 times Writing area on one card

$$t = 100 \cdot a$$

Solve. First we carry out the multiplication in the first equation.

$$a = 2 \cdot 5 \cdot 3$$
$$a = 30$$

Now substitute 30 for a in the second equation and carry out the multiplication.

$$t = 100 \cdot a$$
$$t = 100 \cdot 30$$
$$t = 3000$$

Check. We can repeat the calculations. The answer checks.

State. The total writing area on 100 cards is 3000 square inches.

61. Discussion and Writing Exercise

63. Round 234,562 to the nearest hundred.

$$2\,3\,4,5\,\boxed{6}\,2$$
$$\uparrow$$

The digit 5 is in the hundreds place. Consider the next digit to the right. Since the digit, 6, is 5 or higher, round 5 hundreds up to 6 hundreds. Then change all digits to the right of the hundreds place to zeros.

The answer is 234,600.

65. Round 234,562 to the nearest thousand.

$$2\,3\,4,\,\boxed{5}\,6\,2$$
$$\uparrow$$

The digit 4 is in the thousands place. Consider the next digit to the right. Since the digit, 5, is 5 or higher, round 4 thousands up to 5 thousands. Then change all digits to the right of the thousands place to zeros.

The answer is 235,000.

67.
Rounded to the nearest thousand

$$
\begin{array}{r}
2\,8,4\,3\,0 \\
-\,1\,1,9\,7\,7 \\
\hline
\end{array}
\qquad
\begin{array}{r}
2\,8,0\,0\,0 \\
-\,1\,2,0\,0\,0 \\
\hline
1\,6,0\,0\,0 \leftarrow \text{Estimated answer}
\end{array}
$$

69.
Rounded to the nearest thousand

$$
\begin{array}{r}
5\,8\,0\,0 \\
-\,2\,1\,0\,0 \\
\hline
\end{array}
\qquad
\begin{array}{r}
6\,0\,0\,0 \\
-\,2\,0\,0\,0 \\
\hline
4\,0\,0\,0 \leftarrow \text{Estimated answer}
\end{array}
$$

71.
Rounded to the nearest hundred

$$
\begin{array}{r}
7\,9\,9 \\
\times\,8\,8\,7 \\
\hline
\end{array}
\qquad
\begin{array}{r}
8\,0\,0 \\
\times\quad 9\,0\,0 \\
\hline
7\,2\,0,0\,0\,0 \leftarrow \text{Estimated answer}
\end{array}
$$

73. Discussion and Writing Exercise

75. Familiarize. This is a multistep problem. First we will find the differences in the distances traveled in 1 second. Then we will find the differences for 18 seconds. Let $d =$ the difference in the number of miles light would travel per second in a vacuum and in ice. Let g = the difference in the number of miles light would travel per second in a vacuum and in glass.

Translate. Each is a "how much more" situation.

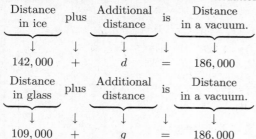

$$142,000 + d = 186,000$$

$$109,000 + g = 186,000$$

Solve. We begin by solving each equation.

$$142,000 + d = 186,000$$
$$142,000 + d - 142,000 = 186,000 - 142,000$$
$$d = 44,000$$

$$109,000 + g = 186,000$$
$$109,000 + g - 109,000 = 186,000 - 109,000$$
$$g = 77,000$$

Now to find the differences in the distances in 18 seconds, we multiply each solution by 18.

For ice: $18 \cdot 44,000 = 792,000$

For glass: $18 \cdot 77,000 = 1,386,000$

Check. We repeat the calculations. Our answers check.

State. In 18 seconds light travels 792,000 miles farther in ice and 1,386,000 miles farther in glass than in a vacuum.

Exercise Set 1.9

1. Exponential notation for $3 \cdot 3 \cdot 3 \cdot 3$ is 3^4.

3. Exponential notation for $5 \cdot 5$ is 5^2.

5. Exponential notation for $7 \cdot 7 \cdot 7 \cdot 7 \cdot 7$ is 7^5.

7. Exponential notation for $10 \cdot 10 \cdot 10$ is 10^3.

9. $7^2 = 7 \cdot 7 = 49$

11. $9^3 = 9 \cdot 9 \cdot 9 = 729$

13. $12^4 = 12 \cdot 12 \cdot 12 \cdot 12 = 20,736$

15. $11^2 = 11 \cdot 11 = 121$

17. $\begin{aligned} 12 + (6+4) &= 12 + 10 \quad \text{Doing the calculation inside the parentheses} \\ &= 22 \quad \text{Adding} \end{aligned}$

19. $\begin{aligned} 52 - (40-8) &= 52 - 32 \quad \text{Doing the calculation inside the parentheses} \\ &= 20 \quad \text{Subtracting} \end{aligned}$

21. $1000 \div (100 \div 10)$
$\quad = 1000 \div 10 \quad$ Doing the calculation inside the parentheses
$\quad = 100 \quad$ Dividing

23. $(256 \div 64) \div 4 = 4 \div 4 \quad$ Doing the calculation inside the parentheses
$\qquad = 1 \quad$ Dividing

25. $(2+5)^2 = 7^2 \quad$ Doing the calculation inside the parentheses
$\qquad = 49 \quad$ Evaluating the exponential expression

27. $(11-8)^2 - (18-16)^2$
$\quad = 3^2 - 2^2 \quad$ Doing the calculations inside the parentheses
$\quad = 9 - 4 \quad$ Evaluating the exponential expressions
$\quad = 5 \quad$ Subtracting

29. $16 \cdot 24 + 50 = 384 + 50 \quad$ Doing all multiplications and divisions in order from left to right
$\qquad = 434 \quad$ Doing all additions and subtractions in order from left to right

31. $83 - 7 \cdot 6 = 83 - 42 \quad$ Doing all multiplications and divisions in order from left to right
$\qquad = 41 \quad$ Doing all additions and subtractions in order from left to right

33. $10 \cdot 10 - 3 \times 4$
$\quad = 100 - 12 \quad$ Doing all multiplications and divisions in order from left to right
$\quad = 88 \quad$ Doing all additions and subtractions in order from left to right

35. $4^3 \div 8 - 4$
$\quad = 64 \div 8 - 4 \quad$ Evaluating the exponential expression
$\quad = 8 - 4 \quad$ Doing all multiplications and divisions in order from left to right
$\quad = 4 \quad$ Doing all additions and subtractions in order from left to right

37. $17 \cdot 20 - (17 + 20)$
$\quad = 17 \cdot 20 - 37 \quad$ Carrying out the operation inside parentheses
$\quad = 340 - 37 \quad$ Doing all multiplications and divisions in order from left to right
$\quad = 303 \quad$ Doing all additions and subtractions in order from left to right

39. $6 \cdot 10 - 4 \cdot 10$
$\quad = 60 - 40 \quad$ Doing all multiplications and divisions in order from left to right
$\quad = 20 \quad$ Doing all additions and subtractions in order from left to right

41. $300 \div 5 + 10$
$\quad = 60 + 10 \quad$ Doing all multiplications and divisions in order from left to right
$\quad = 70 \quad$ Doing all additions and subtractions in order from left to right

43. $3 \cdot (2 + 8)^2 - 5 \cdot (4 - 3)^2$

$\quad = 3 \cdot 10^2 - 5 \cdot 1^2$ Carrying out operations inside parentheses

$\quad = 3 \cdot 100 - 5 \cdot 1$ Evaluating the exponential expressions

$\quad = 300 - 5$ Doing all multiplications and divisions in order from left to right

$\quad = 295$ Doing all additions and subtractions in order from left to right

45. $4^2 + 8^2 \div 2^2 = 16 + 64 \div 4$

$\qquad\qquad\quad = 16 + 16$

$\qquad\qquad\quad = 32$

47. $10^3 - 10 \cdot 6 - (4 + 5 \cdot 6) = 10^3 - 10 \cdot 6 - (4 + 30)$

$\qquad\qquad\qquad\qquad\quad = 10^3 - 10 \cdot 6 - 34$

$\qquad\qquad\qquad\qquad\quad = 1000 - 10 \cdot 6 - 34$

$\qquad\qquad\qquad\qquad\quad = 1000 - 60 - 34$

$\qquad\qquad\qquad\qquad\quad = 940 - 34$

$\qquad\qquad\qquad\qquad\quad = 906$

49. $6 \times 11 - (7 + 3) \div 5 - (6 - 4) = 6 \times 11 - 10 \div 5 - 2$

$\qquad\qquad\qquad\qquad\qquad\qquad = 66 - 2 - 2$

$\qquad\qquad\qquad\qquad\qquad\qquad = 64 - 2$

$\qquad\qquad\qquad\qquad\qquad\qquad = 62$

51. $\quad 120 - 3^3 \cdot 4 \div (5 \cdot 6 - 6 \cdot 4)$

$= 120 - 3^3 \cdot 4 \div (30 - 24)$

$= 120 - 3^3 \cdot 4 \div 6$

$= 120 - 27 \cdot 4 \div 6$

$= 120 - 108 \div 6$

$= 120 - 18$

$= 102$

53. $2^3 \cdot 2^8 \div 2^6 = 8 \cdot 256 \div 64$

$\qquad\qquad\quad = 2048 \div 64$

$\qquad\qquad\quad = 32$

55. We add the numbers and then divide by the number of addends.

$$\frac{\$64 + \$97 + \$121}{3} = \frac{\$282}{3} = \$94$$

57. We add the numbers and then divide by the number of addends.

$$\frac{320 + 128 + 276 + 880}{4} = \frac{1604}{4} = 401$$

59. $8 \times 13 + \{42 \div [18 - (6 + 5)]\}$

$= 8 \times 13 + \{42 \div [18 - 11]\}$

$= 8 \times 13 + \{42 \div 7\}$

$= 8 \times 13 + 6$

$= 104 + 6$

$= 110$

61. $[14 - (3 + 5) \div 2] - [18 \div (8 - 2)]$

$= [14 - 8 \div 2] - [18 \div 6]$

$= [14 - 4] - 3$

$= 10 - 3$

$= 7$

63. $(82 - 14) \times [(10 + 45 \div 5) - (6 \cdot 6 - 5 \cdot 5)]$

$= (82 - 14) \times [(10 + 9) - (36 - 25)]$

$= (82 - 14) \times [19 - 11]$

$= 68 \times 8$

$= 544$

65. $4 \times \{(200 - 50 \div 5) - [(35 \div 7) \cdot (35 \div 7) - 4 \times 3]\}$

$= 4 \times \{(200 - 10) - [5 \cdot 5 - 4 \times 3]\}$

$= 4 \times \{190 - [25 - 12]\}$

$= 4 \times \{190 - 13\}$

$= 4 \times 177$

$= 708$

67. $\{[18 - 2 \cdot 6] - [40 \div (17 - 9)]\}+$

$\qquad\quad \{48 - 13 \times 3 + [(50 - 7 \cdot 5) + 2]\}$

$= \{[18 - 12] - [40 \div 8]\}+$

$\qquad\quad \{48 - 13 \times 3 + [(50 - 35) + 2]\}$

$= \{6 - 5\} + \{48 - 13 \times 3 + [15 + 2]\}$

$= 1 + \{48 - 13 \times 3 + 17\}$

$= 1 + \{48 - 39 + 17\}$

$= 1 + 26$

$= 27$

69. Discussion and Writing Exercise

71. $\qquad x + 341 = 793$

$\quad x + 341 - 341 = 793 - 341$

$\qquad\qquad\quad x = 452$

The solution is 452.

73. $\quad 7 \cdot x = 91$

$\quad \dfrac{7 \cdot x}{7} = \dfrac{91}{7}$

$\qquad x = 13$

The solution is 13.

75. $\qquad 3240 = y + 898$

$\quad 3240 - 898 = y + 898 - 898$

$\qquad\quad 2342 = y$

The solution is 2342.

77. $\quad 25 \cdot t = 625$

$\quad \dfrac{25 \cdot t}{25} = \dfrac{625}{25}$

$\qquad t = 25$

The solution is 25.

79. *Familiarize*. We first make a drawing.

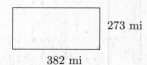

273 mi

382 mi

Translate. We use the formula for the area of a rectangle.

$$A = l \cdot w = 382 \cdot 273$$

Solve. We carry out the multiplication.

$$A = 382 \cdot 273 = 104,286$$

Check. We repeat the calculation. The answer checks.

State. The area is 104,286 square miles.

81. Discussion and Writing Exercise

83. $1 + 5 \cdot 4 + 3 = 1 + 20 + 3$

$\qquad\qquad = 24 \qquad$ Correct answer

To make the incorrect answer correct we add parentheses:

$1 + 5 \cdot (4 + 3) = 36$

85. $12 \div 4 + 2 \cdot 3 - 2 = 3 + 6 - 2$

$\qquad\qquad\qquad = 7 \qquad$ Correct answer

To make the incorrect answer correct we add parentheses:

$12 \div (4 + 2) \cdot 3 - 2 = 4$

Chapter 1 Review Exercises

1. $4, 6\ 7\ \boxed{8}\ , 9\ 5\ 2$

The digit 8 means 8 thousands.

2. $1\ \boxed{3}\ , 7\ 6\ 8\ , 9\ 4\ 0$

The digit 3 names the number of millions.

3. $2793 = 2$ thousands $+ 7$ hundreds $+ 9$ tens $+ 3$ ones

4. $56,078 = 5$ ten thousands $+ 6$ thousands $+ 0$ hundreds $+ 7$ tens $+ 8$ ones, or 5 ten thousands $+ 6$ thousands $+ 7$ tens $+ 8$ ones

5. $4,007,101 = 4$ millions $+ 0$ hundred thousands $+ 0$ ten thousands $+ 7$ thousands $+ 1$ hundred $+ 0$ tens $+ 1$ one, or 4 millions $+ 7$ thousands $+ 1$ hundred $+ 1$ one

6. $\underbrace{67}\ ,\underbrace{819}$

Sixty-seven thousand, —↑

eight hundred nineteen —↑

7. $\underbrace{2}\ ,\underbrace{781}\ ,\underbrace{427}$

Two million, —↑

seven hundred eighty-one thousand, —↑

four hundred twenty-seven —↑

8. $\underbrace{1}\ ,\underbrace{065}\ ,\underbrace{070}\ ,\underbrace{607}$

One billion, —↑

sixty-five million, —↑

seventy thousand, —↑

six hundred seven —↑

9. Four hundred seventy-six thousand, —

five hundred eighty-eight —

↓   ↓

Standard notation is $\underbrace{476}\ ,\underbrace{588}$.

10. Two billion, —

four hundred thousand, —

↓   ↓

Standard notation is $\underbrace{2}\ ,000,\underbrace{400},000$.

11.
$$
\begin{array}{r}
\overset{1\ \ \ 1}{7\ 3\ 0\ 4} \\
+\ 6\ 9\ 6\ 8 \\
\hline
1\ 4,2\ 7\ 2
\end{array}
$$

12.
$$
\begin{array}{r}
\overset{1\ \ 1\ \ \ 1}{2\ 7,6\ 0\ 9} \\
+\ 3\ 8,4\ 1\ 5 \\
\hline
6\ 6,0\ 2\ 4
\end{array}
$$

13.
$$
\begin{array}{r}
\overset{1\ \ \ \ \ 1}{2\ 7\ 0\ 3} \\
4\ 1\ 2\ 5 \\
6\ 0\ 0\ 4 \\
+\ 8\ 9\ 5\ 6 \\
\hline
2\ 1,7\ 8\ 8
\end{array}
$$

14.
$$
\begin{array}{r}
\overset{1\ 1}{9\ 1,4\ 2\ 6} \\
+\ \ \ 7,4\ 9\ 5 \\
\hline
9\ 8,9\ 2\ 1
\end{array}
$$

15. $10 - 6 = 4$

↑

This number gets added (after 4).

↓

$10 = 6 + 4$

(By the commutative law of addition, $10 = 4 + 6$ is also correct.)

16. $8 + 3 = 11$ $\qquad\qquad\qquad$ $8 + 3 = 11$

↑ $\qquad\qquad\qquad\qquad\qquad$ ↑

This addend gets $\qquad\qquad$ This addend gets

subtracted from $\qquad\qquad$ subtracted from

the sum. $\qquad$ ↓ $\qquad\qquad$ the sum. $\qquad$ ↓

$8 = 11 - 3$ $\qquad\qquad\qquad$ $3 = 11 - 8$

17.
$$
\begin{array}{r}
\overset{7\ \ 9\ \overset{13}{\cancel{3}}\ 15}{\cancel{8\ 0\ 4\ 5}} \\
-\ 2\ 8\ 9\ 7 \\
\hline
5\ 1\ 4\ 8
\end{array}
$$

18.
$$
\begin{array}{r}
\overset{8\ \ 9\ \ 9\ \ 11}{\cancel{9\ 0\ 0\ 1}} \\
-\ 7\ 3\ 1\ 2 \\
\hline
1\ 6\ 8\ 9
\end{array}
$$

19.
$$
\begin{array}{r}
\overset{5\ \ 9\ \ 9\ \ 13}{\cancel{6\ 0\ 0\ 3}} \\
-\ 3\ 7\ 2\ 9 \\
\hline
2\ 2\ 7\ 4
\end{array}
$$

20.
$$
\begin{array}{r}
\overset{\ \ \ \ 16\ 13}{\overset{2\ \ \cancel{6}\ \ \cancel{3}\ \ 9\ \ 15}{\cancel{3\ 7,4\ 0\ 5}}} \\
-\ 1\ 9,6\ 4\ 8 \\
\hline
1\ 7,7\ 5\ 7
\end{array}
$$

21. Round 345,759 to the nearest hundred.

$3\ 4\ 5,7\ \boxed{5}\ 9$

↑

The digit 7 is in the hundreds place. Consider the next digit to the right. Since the digit, 5, is 5 or higher, round 7 hundreds up to 8 hundreds. Then change the digits to the right of the hundreds digit to zero.

The answer is 345,800.

22. Round 345,759 to the nearest ten.

$$3\ 4\ 5,\ 7\ 5\ \boxed{9}$$
$$\uparrow$$

The digit 5 is in the tens place. Consider the next digit to the right. Since the digit, 9, is 5 or higher, round 5 tens up to 6 tens. Then change the digit to the right of the tens digit to zero.

The answer is 345,760.

23. Round 345,759 to the nearest thousand.

$$3\ 4\ 5,\ \boxed{7}\ 5\ 9$$
$$\uparrow$$

The digit 5 is in the thousands place. Consider the next digit to the right. Since the digit, 7, is 5 or higher, round 5 thousands up to 6 thousands. Then change the digits to the right of the thousands digit to zero.

The answer is 346,000.

24. Round 345,759 to the nearest hundred thousand.

$$3\ \boxed{4}\ 5,\ 7\ 5\ 9$$
$$\uparrow$$

The digit 3 is in the hundred thousands place. Consider the next digit to the right. Since the digit, 4, is 4 or lower, round down, meaning that 3 hundred thousands stays as 3 hundred thousands. Then change the digits to the right of the hundred thousands digit to zero.

The answer is 300,000.

25.
```
                    Rounded to
                 the nearest hundred
      4 1, 3 4 8        4 1, 3 0 0
    + 1 9, 7 4 9      + 1 9, 7 0 0
                       6 1, 0 0 0 ← Estimated answer
```

26.
```
                    Rounded to
                 the nearest hundred
      3 8, 6 5 2        3 8, 7 0 0
    - 2 4, 5 4 9      - 2 4, 5 0 0
                       1 4, 2 0 0 ← Estimated answer
```

27.
```
                    Rounded to
                 the nearest hundred
        3 9 6            4 0 0
      × 7 4 8          × 7 0 0
                    2 8 0, 0 0 0 ← Estimated answer
```

28. Since 67 is to the right of 56 on the number line, $67 > 56$.

29. Since 1 is to the left of 23 on the number line, $1 < 23$.

30.
```
        2
      1 7, 0 0 0
    ×      3 0 0     Multiplying by 300
  5, 1 0 0, 0 0 0    (Write 00 and then
                     multiply 17,000 by 3.)
```

31.
```
      6 3 4
      7 8 4 6
    ×     8 0 0      Multiplying by 800
  6, 2 7 6, 8 0 0    (Write 00 and then
                     multiply 7846 by 8.)
```

32.
```
        1 3
        2 5
        2 4
        7 2 6
      × 6 9 8
      5 8 0 8    Multiplying by 8
    6 5 3 4 0    Multiplying by 9
  4 3 5 6 0 0    Multiplying by 6
  5 0 6, 7 4 8
```

33.
```
        3 2
        6 4
        5 8 7
      ×   4 7
      4 1 0 9    Multiplying by 7
    2 3 4 8 0    Multiplying by 4
    2 7, 5 8 9
```

34.
```
        8 3 0 5
      ×   6 4 2
      1 6 6 1 0
    3 3 2 2 0 0
  4 9 8 3 0 0 0
  5, 3 3 1, 8 1 0
```

35. $56 \div 7 = 8$ The 7 moves to the right. A related multiplication sentence is $56 = 8 \cdot 7$. (By the commutative law of multiplication, there is also another multiplication sentence: $56 = 7 \cdot 8$.)

36. $13 \cdot 4 = 52$

Move a factor to the other side and then write a division.

$$13 \cdot 4 = 52 \qquad\qquad 13 \cdot 4 = 52$$

$$13 = 52 \div 4 \qquad\qquad 4 = 52 \div 13$$

37.
```
        1 2
    5 ) 6 3
        5 0
        1 3
        1 0
         3
```
The answer is 12 R 3.

38.
```
          5
    1 6 ) 8 0
          8 0
           0
```
The answer is 5.

39.
```
        9 1 3
    7 ) 6 3 9 4
        6 3 0 0
          9 4
          7 0
          2 4
          2 1
           3
```
The answer is 913 R 3.

40.
$$\begin{array}{r} 3\,8\,4 \\ 8\overline{\smash)3\,0\,7\,3} \\ 2\,4\,0\,0 \\ \hline 6\,7\,3 \\ 6\,4\,0 \\ \hline 3\,3 \\ 3\,2 \\ \hline 1 \end{array}$$

The answer is 384 R 1.

41.
$$\begin{array}{r} 4 \\ 6\,0\overline{\smash)2\,8\,6} \\ 2\,4\,0 \\ \hline 4\,6 \end{array}$$

The answer is 4 R 46.

42.
$$\begin{array}{r} 5\,4 \\ 7\,9\overline{\smash)4\,2\,6\,6} \\ 3\,9\,5\,0 \\ \hline 3\,1\,6 \\ 3\,1\,6 \\ \hline 0 \end{array}$$

The answer is 54.

43.
$$\begin{array}{r} 4\,5\,2 \\ 3\,8\overline{\smash)1\,7{,}1\,7\,6} \\ 1\,5\,2\,0\,0 \\ \hline 1\,9\,7\,6 \\ 1\,9\,0\,0 \\ \hline 7\,6 \\ 7\,6 \\ \hline 0 \end{array}$$

The answer is 452.

44.
$$\begin{array}{r} 5\,0\,0\,8 \\ 1\,4\overline{\smash)7\,0{,}1\,1\,2} \\ 7\,0\,0\,0\,0 \\ \hline 1\,1\,2 \\ 1\,1\,2 \\ \hline 0 \end{array}$$

The answer is 5008.

45.
$$\begin{array}{r} 4\,3\,8\,9 \\ 1\,2\overline{\smash)5\,2{,}6\,6\,8} \\ 4\,8\,0\,0\,0 \\ \hline 4\,6\,6\,8 \\ 3\,6\,0\,0 \\ \hline 1\,0\,6\,8 \\ 9\,6\,0 \\ \hline 1\,0\,8 \\ 1\,0\,8 \\ \hline 0 \end{array}$$

The answer is 4389.

46.
$$46 \cdot n = 368$$
$$\frac{46 \cdot n}{46} = \frac{368}{46}$$
$$n = 8$$

Check:
$$\begin{array}{c} 46 \cdot n = 368 \\ \hline 46 \cdot 8 \ ? \ 368 \\ \hline 368 \ \big| \quad \text{TRUE} \end{array}$$

The solution is 8.

47.
$$47 + x = 92$$
$$47 + x - 47 = 92 - 47$$
$$x = 45$$

Check:
$$\begin{array}{c} 47 + x = 92 \\ \hline 47 + 45 \ ? \ 92 \\ \hline 92 \ \big| \quad \text{TRUE} \end{array}$$

The solution is 45.

48.
$$1 \cdot y = 58$$
$$y = 58 \qquad (1 \cdot y = y)$$

The number 58 checks. It is the solution.

49.
$$24 = x + 24$$
$$24 - 24 = x + 24 - 24$$
$$0 = x$$

The number 0 checks. It is the solution.

50. Exponential notation for $4 \cdot 4 \cdot 4$ is 4^3.

51. $10^4 = 10 \cdot 10 \cdot 10 \cdot 10 = 10,000$

52. $6^2 = 6 \cdot 6 = 36$

53.
$$\begin{aligned} 8 \cdot 6 + 17 &= 48 + 17 \quad \text{Multiplying} \\ &= 65 \qquad\quad\ \text{Adding} \end{aligned}$$

54.
$$\begin{aligned} 10 \cdot 24 &- (18 + 2) \div 4 - (9 - 7) \\ &= 10 \cdot 24 - 20 \div 4 - 2 \quad \text{Doing the calculations} \\ &\qquad\qquad\qquad\qquad\qquad \text{inside the parentheses} \\ &= 240 - 5 - 2 \qquad\quad\ \text{Multiplying and dividing} \\ &= 235 - 2 \qquad\qquad\quad \text{Subtracting from} \\ &= 233 \qquad\qquad\qquad\ \text{left to right} \end{aligned}$$

55.
$$\begin{aligned} 7 + (4 + 3)^2 &= 7 + 7^2 \\ &= 7 + 49 \\ &= 56 \end{aligned}$$

56.
$$\begin{aligned} 7 + 4^2 + 3^2 &= 7 + 16 + 9 \\ &= 23 + 9 \\ &= 32 \end{aligned}$$

57.
$$\begin{aligned} (80 \div 16) &\times [(20 - 56 \div 8) + (8 \cdot 8 - 5 \cdot 5)] \\ &= 5 \times [(20 - 7) + (64 - 25)] \\ &= 5 \times [13 + 39] \\ &= 5 \times 52 \\ &= 260 \end{aligned}$$

58. We add the numbers and divide by the number of addends.
$$\frac{157 + 170 + 168}{3} = \frac{495}{3} = 165$$

59. *Familiarize*. Let $x =$ the additional amount of money, in dollars, Natasha needs to buy the desk.

***Translate*.** This is a "how much more" situation.

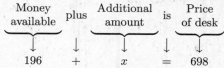

Money available	plus	Additional amount	is	Price of desk
↓	↓	↓	↓	↓
196	+	x	=	698

Solve. We subtract 196 on both sides of the equation.

$$196 + x = 698$$
$$196 + x - 196 = 698 - 196$$
$$x = 502$$

Check. We can estimate.

$$196 + 502 \approx 200 + 500 \approx 700 \approx 698$$

The answer checks.

State. Natasha needs $502 dollars.

60. *Familiarize*. Let b = the balance in Taneesha's account after the deposit.

Translate.

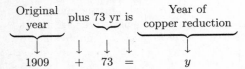

$$
\begin{array}{ccccc}
406 & + & 78 & = & b
\end{array}
$$

Solve. We add on the left side.

$$406 + 78 = b$$
$$484 = b$$

Check. We can repeat the calculation. The answer checks.

State. The new balance is $484.

61. *Familiarize*. Let y = the first year in which the copper content of pennies was reduced.

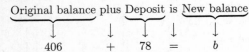

$$
\begin{array}{ccccc}
1909 & + & 73 & = & y
\end{array}
$$

Solve. We add on the left side.

$$1909 + 73 = y$$
$$1982 = y$$

Check. We can estimate.

$$1909 + 73 \approx 1910 + 70 \approx 1980 \approx 1982$$

The answer checks.

State. The copper content of pennies was first reduced in 1982.

62. *Familiarize*. We first make a drawing. Let c = the number of cartons filled.

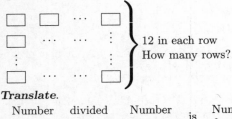

12 in each row
How many rows?

Translate.

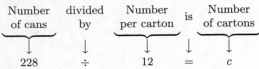

$$
\begin{array}{ccccc}
228 & \div & 12 & = & c
\end{array}
$$

Solve. We carry out the division.

$$
\begin{array}{r}
1\,9 \\
12\,\overline{)2\,2\,8} \\
1\,2\,0 \\
\hline
1\,0\,8 \\
1\,0\,8 \\
\hline
0
\end{array}
$$

Thus, $19 = c$, or $c = 19$.

Check. We can check by multiplying: $12 \cdot 19 = 228$. Our answer checks.

State. 19 cartons were filled.

63. *Familiarize*. Let b = the number of beehives the farmer needs.

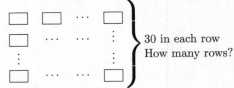

30 in each row
How many rows?

Translate.

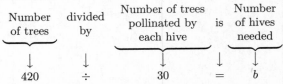

$$
\begin{array}{ccccc}
420 & \div & 30 & = & b
\end{array}
$$

Solve. We carry out the division.

$$
\begin{array}{r}
1\,4 \\
30\,\overline{)4\,2\,0} \\
3\,0\,0 \\
\hline
1\,2\,0 \\
1\,2\,0 \\
\hline
0
\end{array}
$$

Thus, $14 = b$, or $b = 14$.

Check. We can check by multiplying: $30 \cdot 14 = 420$. The answer checks.

State. The farmer needs 14 beehives.

64. *Familiarize*. This is a multistep problem. Let s = the cost of 13 stoves, r = the cost of 13 refrigerators, and t = the total cost of the stoves and refrigerators.

Translate.

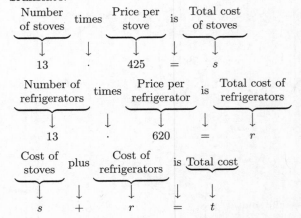

$$
\begin{array}{ccccc}
13 & \cdot & 425 & = & s
\end{array}
$$

$$
\begin{array}{ccccc}
13 & \cdot & 620 & = & r
\end{array}
$$

$$
\begin{array}{ccccc}
s & + & r & = & t
\end{array}
$$

Solve. We first carry out the multiplications in the first two equations.

$$13 \cdot 425 = s \qquad 13 \cdot 620 = r$$
$$5525 = s \qquad 8060 = r$$

Now we substitute 5525 for s and 8060 for r in the third equation and then add on the left side.

$$s + r = t$$
$$5525 + 8060 = t$$
$$13,585 = t$$

Check. We repeat the calculations. The answer checks.

State. The total cost was $13,585.

65. *Familiarize*. This is a multistep problem. Let $b =$ the total amount budgeted for food, clothing, and entertainment and let $r =$ the income remaining after these allotments.

Translate.

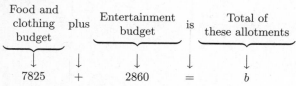

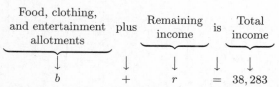

Solve. We add on the left side to solve the first equation.

$$7825 + 2860 = b$$
$$10,685 = b$$

Now we substitute 10,685 for b in the second equation and solve for r.

$$b + r = 38,283$$
$$10,685 + r = 38,283$$
$$10,685 + r - 10,685 = 38,283 - 10,685$$
$$r = 27,598$$

Check. We repeat the calculations. The answer checks.

State. After the allotments for food, clothing, and entertainment, $27,598 remains.

66. *Familiarize*. We make a drawing. Let $b =$ the number of beakers that will be filled.

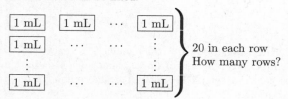

Translate.

Amount of alcohol	divided by	Amount per beaker	is	Number of beakers filled
↓	↓	↓	↓	↓
2753	÷	20	=	b

Solve. We carry out the division.

```
      1 3 7
2 0 ) 2 7 5 3
      2 0 0 0
      -------
        7 5 3
        6 0 0
      -------
        1 5 3
        1 4 0
      -------
          1 3
```

Thus, 137 R 13 = b.

Check. We can check by multiplying the number of beakers by 137 and then adding the remainder, 13.

$$137 \cdot 20 = 2740 \text{ and } 2740 + 13 = 2753$$

The answer checks.

State. 137 beakers can be filled; 13 mL will be left over.

67. $A = l \cdot w = 14 \text{ ft} \cdot 7 \text{ ft} = 98 \text{ square ft}$

Perimeter $= 14 \text{ ft} + 7 \text{ ft} + 14 \text{ ft} + 7 \text{ ft} = 42 \text{ ft}$

68. *Discussion and Writing Exercise*. A vat contains 1152 oz of hot sauce. If 144 bottles are to be filled equally, how much will each bottle contain? Answers may vary.

69. *Discussion and Writing Exercise*. No; if subtraction were associative, then $a - (b - c) = (a - b) - c$ for any a, b, and c. But, for example,

$$12 - (8 - 4) = 12 - 4 = 8,$$

whereas

$$(12 - 8) - 4 = 4 - 4 = 0.$$

Since $8 \neq 0$, this examples shows that subtraction is not associative.

70.
```
      9 d
  ×   d 2
  -------
  8 0 3 6
```

By using rough estimates, we see that the factor $d2 \approx 8100 \div 90 = 90$ or $d2 \approx 8000 \div 100 = 80$. Since $99 \times 92 = 9108$ and $98 \times 82 = 8036$, we have $d = 8$.

71.
```
            9 a 1
2 b 1 ) 2 3 6,4 2 1
```

Since $250 \times 1000 = 250,000 \approx 236,421$ we deduce that $2b1 \approx 250$ and $9a1 \approx 1000$. By trial we find that $a = 8$ and $b = 4$.

72. At the beginning of each day the tunnel reaches 500 ft − 200 ft, or 300 ft, farther into the mountain than it did the day before. We calculate how far the tunnel reaches into the mountain at the beginning of each day, starting with Day 2.

Day 2: 300 ft

Day 3: 300 ft + 300 ft = 600 ft

Day 4: 600 ft + 300 ft = 900 ft

Day 5: 900 ft + 300 ft = 1200 ft

Day 6: 1200 ft + 300 ft = 1500 ft

We see that the tunnel reaches 1500 ft into the mountain at the beginning of Day 6. On Day 6 the crew tunnels an additional 500 ft, so the tunnel reaches 1500 ft + 500 ft, or 2000 ft, into the mountain. Thus, it takes 6 days to reach the copper deposit.

Chapter 1 Test

1. $\boxed{5}$ 4 6, 7 8 9
The digit 5 tells the number of hundred thousands.

2. 8843 = 8 thousands + 8 hundreds + 4 tens + 3 ones

3.
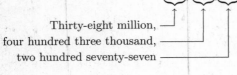

Thirty-eight million, ┐
four hundred three thousand, ┐
two hundred seventy-seven ┘

4.
```
      6 8 1 1
   +  3 1 7 8
   ─────────
      9 9 8 9
```
Add ones, add tens, add hundreds, and then add thousands.

5.
```
     1 1   1
    4 5, 8 8 9
 + 1 7, 9 0 2
 ───────────
   6 3, 7 9 1
```

6. We look for pairs of numbers whose sums are 10, 20, 30, and so on.

```
 12  ──→  20
  8
  3  ──→  10
  7
 +4  ──→   4
───      ───
 34       34
```

7.
```
       6 2 0 3
    +  4 3 1 2
   ───────────
    1 0, 5 1 5
```

8.
```
      7 9 8 3
   -  4 3 5 3
   ─────────
      3 6 3 0
```
Subtract ones, subtract tens, subtract hundreds, and then subtract thousands.

9.
```
        6 14
     2 9 7 4
   - 1 9 3 5
   ─────────
     1 0 3 9
```

10.
```
      8  9 17
     8 9 0 7
   - 2 0 5 9
   ─────────
     6 8 4 8
```

11.
```
        12
      1 2 9 16
     2 3, 0 0 7
   - 1 7, 8 9 2
   ───────────
       5 1 7 5
```

12.
```
       5 6 7
      4 5 6 8
   ×        9
   ─────────
   4 1, 1 1 2
```

13.
```
       5 4 3
      8 8 7 6
   ×    6 0 0
   ───────────
   5, 3 2 5, 6 0 0
```
Multiply by 6 hundreds (We write 00 and then multiply 8876 by 6.)

14.
```
        6 5
   ×    3 7
   ───────
       4 5 5      Multiplying by 7
     1 9 5 0      Multiplying by 30
   ─────────
     2 4 0 5      Adding
```

15.
```
          6 7 8
   ×      7 8 8
   ─────────────
        5 4 2 4
      5 4 2 4 0
    4 7 4 6 0 0
   ─────────────
    5 3 4, 2 6 4
```

16.
```
         3
     4 ⟌ 1 5
         1 2
       ─────
           3
```
The answer is 3 R 3.

17.
```
          7 0
     6 ⟌ 4 2 0
         4 2 0
       ───────
             0
             0
           ───
             0
```
The answer is 70.

18.
```
            9 7
   8 9 ⟌ 8 6 3 3
         8 0 1 0
       ─────────
           6 2 3
           6 2 3
         ───────
               0
```
The answer is 97.

19.
```
              8 0 5
   4 4 ⟌ 3 5, 4 2 8
         3 5 2 0 0
       ───────────
             2 2 8
             2 2 0
           ───────
                 8
```
The answer is 805 R 8.

20. *Familiarize.* Let n = the number of 12-packs that can be filled. We can think of this as repeated subtraction, taking successive sets of 12 snack cakes and putting them into n packages.

Translate.

Number of cakes	divided by	Number in each package	is	Number of 12-packs
↓	↓	↓	↓	↓
22, 231	÷	12	=	n

Solve. We carry out the division.

$$
\begin{array}{r}
1\,8\,5\,2 \\
1\,2\,\overline{\smash{)}2\,2,2\,3\,1} \\
1\,2\,0\,0\,0 \\
\hline
1\,0\,2\,3\,1 \\
9\,6\,0\,0 \\
\hline
6\,3\,1 \\
6\,0\,0 \\
\hline
3\,1 \\
2\,4 \\
\hline
7
\end{array}
$$

Then 1852 R 7 = n.

Check. We multiply the number of cartons by 12 and then add the remainder, 7.

$$12 \cdot 1852 = 22,224$$

$$22,224 + 7 = 22,231$$

The answer checks.

State. 1852 twelve-packs can be filled. There will be 7 cakes left over.

21. Familiarize. Let a = the total land area of the five largest states, in square meters. Since we are combining the areas of the states, we can add.

Translate.

$$571,951 + 261,797 + 155,959 + 145,552 + 121,356 = a$$

Solve. We carry out the addition.

$$
\begin{array}{r}
{\scriptstyle 2\ 1\ 3\ 3\ 2} \\
5\,7\,1,9\,5\,1 \\
2\,6\,1,7\,9\,7 \\
1\,5\,5,9\,5\,9 \\
1\,4\,5,5\,5\,2 \\
+\,1\,2\,1,3\,5\,6 \\
\hline
1,2\,5\,6,6\,1\,5
\end{array}
$$

Then $1,256,615 = a$.

Check. We can repeat the calculation. We can also estimate the result by rounding. We will round to the nearest ten thousand.

$$571,951 + 261,797 + 155,959 + 145,552 + 121,356$$

$$\approx 570,000 + 260,000 + 160,000 + 150,000 + 120,000$$

$$= 1,260,000$$

Since $1,260,000 \approx 1,256,615$, we have a partial check.

State. The total land area of Alaska, Texas, California, Montana, and New Mexico is $1,256,615$ m^2.

22. a) We will use the formula Perimeter = $2 \cdot$ length + $2 \cdot$ width to find the perimeter of each pool table in inches. We will use the formula Area = length $\cdot$ width to find the area of each pool table, in in^2.

For the 50 in. by 100 in. table:

$$\text{Perimeter} = 2 \cdot 100 \text{ in.} + 2 \cdot 50 \text{ in.}$$

$$= 200 \text{ in.} + 100 \text{ in.}$$

$$= 300 \text{ in.}$$

$$\text{Area} = 100 \text{ in.} \cdot 50 \text{ in.} = 5000 \text{ in}^2$$

For the 44 in. by 88 in. table:

$$\text{Perimeter} = 2 \cdot 88 \text{ in.} + 2 \cdot 44 \text{ in.}$$

$$= 176 \text{ in.} + 88 \text{ in.}$$

$$= 264 \text{ in.}$$

$$\text{Area} = 88 \text{ in.} \cdot 44 \text{ in.} = 3872 \text{ in}^2$$

For the 38 in. by 76 in. table:

$$\text{Perimeter} = 2 \cdot 76 \text{ in.} + 2 \cdot 38 \text{ in.}$$

$$= 152 \text{ in.} + 76 \text{ in.}$$

$$= 228 \text{ in.}$$

$$\text{Area} = 76 \text{ in.} \cdot 38 \text{ in.} = 2888 \text{ in}^2$$

b) Let a = the number of square inches by which the area of the largest table exceeds the area of the smallest table. We subtract to find a.

$$a = 5000 \text{ in}^2 - 2888 \text{ in}^2 = 2112 \text{ in}^2$$

23. Familiarize. Let v = the number of Nevada voters who voted early in the 2000 presidential election.

Translate.

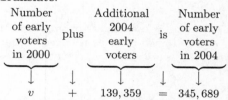

Solve. We subtract 139,359 on both sides of the equation.

$$v + 139,359 = 345,689$$

$$v + 139,359 - 139,359 = 345,689 - 139,359$$

$$v = 206,330$$

Check. We can add the difference, 206,330, to the subtrahend, 139,359: $139,359 + 206,330 = 345,689$. The answer checks.

State. 206,330 voters voted early in Nevada in 2000.

24. Familiarize. There are three parts to this problem. First we find the total weight of each type of fruit and then we add. Let x = the total weight of the oranges, y = the total weight of the apples, and t = the total weight of both fruits together.

Translate.

For the oranges:

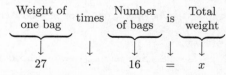

For the apples:

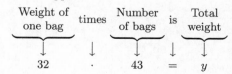

For the total weight of both fruits:

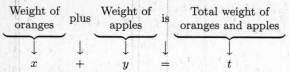

$$x \quad + \quad y \quad = \quad t$$

Solve. We solve the first two equations and then add the solutions.

$$27 \cdot 16 = x$$
$$432 = x$$

$$32 \cdot 43 = y$$
$$1376 = y$$

$$x + y = t$$
$$432 + 1376 = t$$
$$1808 = t$$

Check. We repeat the calculations. The answer checks.

State. The total weight of 16 bags of oranges and 43 bags of apples is 1808 lb.

25. *Familiarize*. Let $s =$ the number of staplers that can be filled. We can think of this as repeated subtraction, taking successive sets of 250 staples and putting them into s staplers.

Translate.

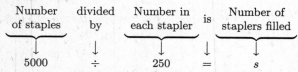

$$5000 \quad \div \quad 250 \quad = \quad s$$

Solve. We carry out the division.

$$
\begin{array}{r}
2\,0 \\
2\,5\,0\,\overline{)\,5\,0\,0\,0} \\
5\,0\,0\,0 \\
\hline
0 \\
0 \\
\hline
0
\end{array}
$$

Then $20 = s$.

Check. We can multiply the number of staplers filled by the number of staples in each one.

$$20 \cdot 250 = 5000$$

The answer checks.

State. 20 staplers can be filled from a box of 5000 staples.

26. $28 + x = 74$

$28 + x - 28 = 74 - 28$ Subtracting 28 on both sides

$x = 46$

Check: $\dfrac{28 + x = 74}{28 + 46 \;?\; 74}$

$74 \;\bigg|\;$ TRUE

The solution is 46.

27. $169 \div 13 = n$

We carry out the division.

$$
\begin{array}{r}
1\,3 \\
1\,3\,\overline{)\,1\,6\,9} \\
1\,3\,0 \\
\hline
3\,9 \\
3\,9 \\
\hline
0
\end{array}
$$

The solution is 13.

28. $38 \cdot y = 532$

$\dfrac{38 \cdot y}{38} = \dfrac{532}{38}$ Dividing by 38 on both sides

$y = 14$

Check: $\dfrac{38 \cdot y = 532}{38 \cdot 14 \;?\; 532}$

$532 \;\bigg|\;$ TRUE

The solution is 14.

29. $381 = 0 + a$

$381 = a$ Adding on the right side

The solution is 381.

30. Round 34,578 to the nearest thousand.

$$3\,4,\;\boxed{5}\;7\,8$$
$$\uparrow$$

The digit 4 is in the thousands place. Consider the next digit to the right, 5. Since 5 is 5 or higher, round 4 thousands up to 5 thousands. Then change all the digits to the right of thousands to zeros.

The answer is 35,000.

31. Round 34,578 to the nearest ten.

$$3\,4,\;5\,7\;\boxed{8}$$
$$\uparrow$$

The digit 7 is in the tens place. Consider the next digit to the right, 8. Since 8 is 5 or higher, round 7 tens up to 8 tens. Then change the digit to the right of tens to zero.

The answer is 34,580.

32. Round 34,578 to the nearest hundred.

$$3\,4,\;5\;\boxed{7}\;8$$
$$\uparrow$$

The digit 5 is in the hundreds place. Consider the next digit to the right, 7. Since 7 is 5 or higher, round 5 hundreds up to 6 hundreds. Then change all the digits to the right of hundreds to zeros.

The answer is 34,600.

33.

Rounded to
the nearest hundred

$$
\begin{array}{r}
2\,3,\,6\,4\,9 \\
+\;5\,4,\,7\,4\,6 \\
\hline
\end{array}
\qquad
\begin{array}{r}
2\,3,\,6\,0\,0 \\
+\;5\,4,\,7\,0\,0 \\
\hline
7\,8,\,3\,0\,0 \;\leftarrow \text{Estimated answer}
\end{array}
$$

34.

$$\begin{array}{r}\text{Rounded to}\\\text{the nearest hundred}\end{array}$$

5 4, 7 5 1	5 4, 8 0 0
− 2 3, 6 4 9	− 2 3, 6 0 0
	3 1, 2 0 0 ← Estimated answer

35.

$$\begin{array}{r}\text{Rounded to}\\\text{the nearest hundred}\end{array}$$

8 2 4	8 0 0
× 4 8 9	× 5 0 0
	4 0 0, 0 0 0 ← Estimated answer

36. Since 34 is to the right of 17 on the number line, $34 > 17$.

37. Since 117 is to the left of 157 on the number line, $117 < 157$.

38. Exponential notation for $12 \cdot 12 \cdot 12 \cdot 12$ is 12^4.

39. $7^3 = 7 \cdot 7 \cdot 7 = 343$

40. $10^5 = 10 \cdot 10 \cdot 10 \cdot 10 \cdot 10 = 100,000$

41. $25^2 = 25 \cdot 25 = 625$

42.
$$35 - 1 \cdot 28 \div 4 + 3$$
$= 35 - 28 \div 4 + 3$ Doing all multiplications and
$= 35 - 7 + 3$ divisions in order from left to right
$= 28 + 3$ Doing all additions and subtractions
$= 31$ in order from left to right

43.
$$10^2 - 2^2 \div 2$$
$= 100 - 4 \div 2$ Evaluating the exponential expressions
$= 100 - 2$ Dividing
$= 98$ Subtracting

44.
$$(25 - 15) \div 5$$
$= 10 \div 5$ Doing the calculation inside the parentheses
$= 2$ Dividing

45.
$$8 \times \{(20 - 11) \cdot [(12 + 48) \div 6 - (9 - 2)]\}$$
$= 8 \times \{9 \cdot [60 \div 6 - 7]\}$
$= 8 \times \{9 \cdot [10 - 7]\}$
$= 8 \times \{9 \cdot 3\}$
$= 8 \times 27$
$= 216$

46.
$$2^4 + 24 \div 12$$
$= 16 + 24 \div 12$ Evaluating the exponential expression
$= 16 + 2$ Dividing
$= 18$ Adding

47. We add the numbers and then divide by the number of addends.
$$\frac{97 + 98 + 87 + 86}{4} = \frac{368}{4} = 92$$

48. *Familiarize.* We make a drawing.

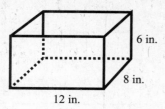

Observe that the dimensions of two sides of the container are 8 in. by 6 in. The area of each is 8 in. $\cdot$ 6 in. and their total area is $2 \cdot 8$ in. $\cdot$ 6 in. The dimensions of the other two sides are 12 in. by 6 in. The area of each is 12 in. $\cdot$ 6 in. and their total area is $2 \cdot 12$ in. $\cdot$ 6 in. The dimensions of the bottom of the box are 12 in. by 8 in. and its area is 12 in. $\cdot$ 8 in. Let $c =$ the number of square inches of cardboard that are used for the container.

Translate. We add the areas of the sides and the bottom of the container.
$$2 \cdot 8 \text{ in.} \cdot 6 \text{ in.} + 2 \cdot 12 \text{ in.} \cdot 6 \text{ in.} + 12 \text{ in.} \cdot 8 \text{ in.} = c$$

Solve. We carry out the calculation.
$$2 \cdot 8 \text{ in.} \cdot 6 \text{ in.} + 2 \cdot 12 \text{ in.} \cdot 6 \text{ in.} + 12 \text{ in.} \cdot 8 \text{ in.} = c$$
$$96 \text{ in}^2 + 144 \text{ in}^2 + 96 \text{ in}^2 = c$$
$$336 \text{ in}^2 = c$$

Check. We can repeat the calculations. The answer checks.

State. 336 in^2 of cardboard are used for the container.

49. We can reduce the number of trials required by simplifying the expression on the left side of the equation and then using the addition principle.

$$359 - 46 + a \div 3 \times 25 - 7^2 = 339$$
$$359 - 46 + a \div 3 \times 25 - 49 = 339$$
$$359 - 46 + \frac{a}{3} \times 25 - 49 = 339$$
$$359 - 46 + \frac{25 \cdot a}{3} - 49 = 339$$
$$313 + \frac{25 \cdot a}{3} - 49 = 339$$
$$264 + \frac{25 \cdot a}{3} = 339$$
$$264 + \frac{25 \cdot a}{3} - 264 = 339 - 264$$
$$\frac{25 \cdot a}{3} = 75$$

We see that when we multiply a by 25 and divide by 3, the result is 75. By trial, we find that $\frac{25 \cdot 9}{3} = \frac{225}{3} = 75$, so $a = 9$. We could also reason that since $75 = 25 \cdot 3$ and $9/3 = 3$, we have $a = 9$.

50. *Familiarize.* First observe that a 10-yr loan with monthly payments has a total of $10 \cdot 12$, or 120, payments. Let $m =$ the number of monthly payments represented by \$9160 and let $p =$ the number of payments remaining after \$9160 has been repaid.

Translate. First we will translate to an equation that can be used to find m. Then we will write an equation that can be used to find p.

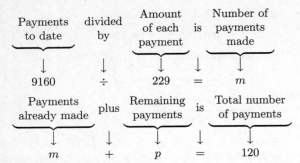

$$m + p = 120$$

Solve. To solve the first equation we carry out the division.

$$
\begin{array}{r}
40 \\
229\overline{\smash{\big)}\,9160} \\
\underline{9160} \\
0 \\
0 \\
\underline{} \\
0
\end{array}
$$

Thus, $m = 40$.

Now we solve the second equation.

$$m + p = 120$$
$$40 + p = 120 \qquad \text{Substituting 40 for } m$$
$$40 + p - 40 = 120 - 40$$
$$p = 80$$

Check. We can approach the problem in a different way to check the answer. In 10 years, Cara's loan payments will total $120 \cdot \$229$, or $\$27{,}480$. If $\$9160$ has already been paid, then $\$27,480 - \$9160$, or $\$18{,}320$, remains to be paid. Since $80 \cdot \$229 = \$18,320$, the answer checks.

State. 80 payments remain on the loan.

51. $\dfrac{90 + 90 + 90 + 80 + 80 + 80 + 80 + 74}{8} = \dfrac{664}{8} = 83$

Chapter 2

Introduction to Integers and Algebraic Expressions

Exercise Set 2.1

1. The integer $-34,000,000$ corresponds to a $34 million fine.

3. The integer 40 corresponds to receiving $40 for a ton of paper; the integer -15 corresponds to paying $15 to get rid of a ton of paper.

5. The integer 820 corresponds to receiving an $820 refund; the integer -541 corresponds to owing $541.

7. The integer -280 corresponds to 280 ft below sea level; the integer 14,491 corresponds to an elevation of 14,491 ft.

9. Since -8 is to the left of 0, we have $-8 < 0$.

11. Since 9 is to the right of 0, we have $9 > 0$.

13. Since 8 is to the right of -8, we have $8 > -8$.

15. Since -6 is to the left of -4, we have $-6 < -4$.

17. Since -8 is to the left of -5, we have $-8 < -5$.

19. Since -13 is to the left of -9, we have $-13 < -9$.

21. Since -3 is to the right of -4, we have $-3 > -4$.

23. The distance from 57 to 0 is 57, so $|57| = 57$.

25. The distance from 0 to 0 is 0, so $|0| = 0$.

27. The distance from -24 to 0 is 24, so $|-24| = 24$.

29. The distance from 53 to 0 is 53, so $|53| = 53$.

31. This distance from -8 to 0 is 8, so $|-8| = 8$.

33. To find the opposite of x when x is -7, we reflect -7 to the other side of 0. We have $-(-7) = 7$. The opposite of -7 is 7.

35. To find the opposite of x when x is 7, we reflect 7 to the other side of 0. We have $-(7) = -7$. The opposite of 7 is -7.

37. When we try to reflect 0 to the other side of 0, we go nowhere. The opposite of 0 is 0.

39. To find the opposite of x when x is -19, we reflect -19 to the other side of 0. We have $-(-19) = 19$.

41. To find the opposite of x when x is 42, we reflect 42 to the other side of 0. We have $-(42) = -42$.

43. The opposite of -8 is 8. $-(-8) = 8$

45. The opposite of 7 is -7. $-(7) = -7$

47. The opposite of -29 is 29. $-(-29) = 29$

49. The opposite of -22 is 22. $-(-22) = 22$

51. The opposite of 1 is -1. $-(1) = -1$

53. We replace x by 7. We wish to find $-(-7)$. Reflecting 7 to the other side of 0 gives us -7 and then reflecting back gives us 7. Thus, $-(-x) = 7$ when x is 7.

55. We replace x by -9. We wish to find $-(-(-9))$. Reflecting -9 to the other side of 0 gives us 9 and then reflecting back gives us -9. Thus, $-(-x) = -9$ when x is -9.

57. We replace x by -17. We wish to find $-(-(-17))$. Reflecting -17 to the other side of 0 gives us 17 and then reflecting back gives us -17. Thus $-(-x) = -17$ when x is -17.

59. We replace x by 23. We wish to find $-(-23)$. Reflecting 23 to the other side of 0 gives us -23 and then reflecting back gives us 23. Thus, $-(-x) = 23$ when x is 23.

61. We replace x by -1. We wish to find $-(-(-1))$. Reflecting -1 to the other side of 0 gives us 1 and then reflecting back gives us -1. Thus, $-(-x) = -1$ when x is -1.

63. We replace x by 85. We wish to find $-(-85)$. Reflecting 85 to the other side of 0 gives us -85 and then reflecting back gives us 85. Thus, $-(-x) = 85$ when x is 85.

65. We have $-|-x| = -|-47|$. Since $|-47| = 47$, it follows that $-|-47| = -47$.

67. We have $-|-x| = -|-345|$. Since $|-345| = 345$, it follows that $-|-345| = -345$.

69. We have $-|-x| = -|-0|$. Since $|-0| = |0| = 0$, it follows that $-|-0| = -|0| = -0 = 0$.

71. We have $-|-x| = -|-(-8)| = -|8|$. Since $|8| = 8$, it follows that $-|-(-8)| = -8$.

73. Discussion and Writing Exercise

75.
$$\begin{array}{r} \overset{1\ 1}{3\ 2\ 7} \\ +\ 4\ 9\ 8 \\ \hline 8\ 2\ 5 \end{array}$$

77.
$$\begin{array}{r} \overset{\ \ 2}{\underset{3}{2\ 0\ 9}} \\ \times\ \ \ 3\ 4 \\ \hline 8\ 3\ 6 \\ 6\ 2\ 7\ 0 \\ \hline 7\ 1\ 0\ 6 \end{array}$$

836   Multiplying 209 by 4
6270   Multiplying 209 by 30
7106   Adding

79. $9^2 = 9 \cdot 9 = 81$

81. $5(8-6) = 5(2) = 10$

83. Discussion and Writing Exercise

85. Answers may vary. On many scientific calculators we would add 549 and 387 and then take the opposite of the sum.

$$\boxed{5}\,\boxed{4}\,\boxed{9}\,\boxed{+}\,\boxed{3}\,\boxed{8}\,\boxed{7}\,\boxed{=}\,\boxed{+/-}$$

87. $|-5| = 5$ and $|-2| = 2$. Since 5 is to the right of 2, we have $|-5| > |-2|$.

89. $|-8| = 8$ and $|8| = 8$, so $|-8| = |8|$.

91. $|-8| = 8$, so $-|-8| = -(8) = -8$.

93. $|7| = 7$, so $-|7| = -(7) = -7$.

95. The integers whose distance from 0 is less than 2 are -1, 0, and 1. These are the solutions.

97. First note that $2^{10} = 1024$, $|-6| = 6$, $|3| = 3$, $2^7 = 128$, $7^2 = 49$, and $10^2 = 100$. Listing the entire set of integers in order from least to greatest, we have $-100, -5, 0, |3|$, $4, |-6|, 7^2, 10^2, 2^7, 2^{10}$.

Exercise Set 2.2

1. Add: $-7 + 2$

Start   Move 2 units
at -7. to the right.

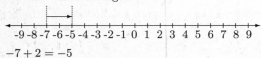

$-7 + 2 = -5$

3. Add: $-9 + 5$

Start   Move 5 units
at -9.   to the right.

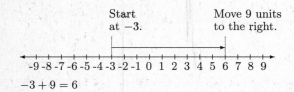

$-9 + 5 = -4$

5. Add: $-3 + 9$

Start   Move 9 units
at -3.   to the right.

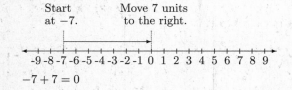

$-3 + 9 = 6$

7. $-7 + 7$

Start   Move 7 units
at -7.   to the right.

$-7 + 7 = 0$

9. $-3 + (-1)$

Move 1 unit   Start
to the left.   at -3.

$-3 + (-1) = -4$

11. $4 + (-9)$

Move 9 units   Start
to the left.   at 4.

$4 + (-9) = -5$

13. $-7 + (12)$

Start   Move 12 units
at -7.   to the right.

$-7 + (12) = 5$

15. $-3 + (-9)$ Two negative integers

Add the absolute values: $3 + 9 = 12$
Make the answer negative: $-3 + (-9) = -12$

17. $-6 + (-5)$ Two negative integers

Add the absolute values: $6 + 5 = 11$
Make the answer negative: $-6 + (-5) = -11$

19. $5 + (-5) = 0$

For any integer a, $a + (-a) = 0$.

21. $-2 + 2 = 0$

For any integer a, $-a + a = 0$.

23. $0 + 6 = 6$

For any integer a, $0 + a = a$.

25. $13 + (-13) = 0$

For any integer a, $a + (-a) = 0$.

27. $-25 + 0 = -25$

For any integer a, $a + 0 = a$.

29. $0 + (-27) = -27$

For any integer a, $0 + a = a$.

31. $-31 + 31 = 0$

For any integer a, $-a + a = 0$.

33. $-8 + 0 = -8$

When 0 is added to any number, that number remains unchanged.

35. $9 + (-4)$ The absolute values are 9 and 4. The difference is 5. The positive number has the larger absolute value, so the answer is positive. $9 + (-4) = 5$

37. $-4 + (-5)$ Two negative integers

Add the absolute values: $4 + 5 = 9$
Make the answer negative: $-4 + (-5) = -9$

39. $0 + (-5) = -5$

For any integer a, $0 + a = a$.

41. $14 + (-5)$ The absolute values are 14 and 5. The difference is 9. The positive number has the larger absolute value, so the answer is positive. $14 + (-5) = 9$

43. $-11 + 8$ The absolute values are 11 and 8. The difference is 3. Since the negative number has the larger absolute value, the answer is negative. $-11 + 8 = -3$

45. $-19 + 19 = 0$

For any integer a, $-a + a = 0$.

47. $-17 + 7$ The absolute values are 17 and 7. The difference is 10. Since the negative number has the larger absolute value, the answer is negative. $-17 + 7 = -10$

49. $-17 + (-7)$ Two negative integers

Add the absolute values: $17 + 7 = 24$
Make the answer negative: $-17 + (-7) = -24$

51. $11 + (-16)$ The absolute values are 11 and 16. The difference is 5. Since the negative number has the larger absolute value, the answer is negative. $11 + (-16) = -5$

53. $-15 + (-6)$ Two negative integers

Add the absolute values: $15 + 6 = 21$
Make the answer negative: $-15 + (-6) = -21$

55. $11 + (-9)$ The absolute values are 11 and 9. The difference is 2. The positive number has the larger absolute value, so the answer is positive. $11 + (-9) = 2$

57. $-11 + 17$ The absolute values are 11 and 17. The difference is 6. The positive number has the larger absolute value, so the answer is positive. $-11 + 17 = 6$

59. We will add from left to right.
$$-15 + (-7) + 1 = -22 + 1$$
$$= -21$$

61. We will add from left to right.
$$30 + (-10) + 5 = 20 + 5$$
$$= 25$$

63. We will add from left to right.
$$-23 + (-9) + 15 = -32 + 15$$
$$= -17$$

65. We will add from left to right.
$$40 + (-40) + 6 = 0 + 6$$
$$= 6$$

67. $12 + (-65) + (-12)$

Note that $12 + (-12) = 0$. Then we have $0 + (-65)$, so the sum is -65.

69. We will add from left to right.
$$-24 + (-37) + (-19) + (-45) + (-35)$$
$$= \quad -61 + (-19) + (-45) + (-35)$$
$$= \quad\quad\quad -80 + (-45) + (-35)$$
$$= \quad\quad\quad\quad\quad -125 + (-35)$$
$$= \quad\quad\quad\quad\quad\quad\quad -160$$

71. $28 + (-44) + 17 + 31 + (-94)$

a) $28 + 17 + 31 = 76$ Adding the positive numbers

b) $-44 + (-94) = -138$ Adding the negative numbers

c) $76 + (-138) = -62$ Adding the results

73. $-19 + 73 + (-23) + 19 + (-73)$

a) $-19 + 19 = 0$ Adding one pair of opposites

b) $73 + (-73) = 0$ Adding the other pair of opposites

c) We have $0 + (-23) = -23$

75. Discussion and Writing Exercise

77.
$$\begin{array}{r} \overset{3}{}\overset{13}{} \\ 5\,\cancel{4}\,\cancel{3} \\ -\ 2\,1\,9 \\ \hline 3\,2\,4 \end{array}$$

79.
$$\begin{array}{r} \overset{8}{}\overset{11}{} \\ 2\,8\,\cancel{9}\,\cancel{1} \\ -\ 1\,4\,0\,7 \\ \hline 1\,4\,8\,4 \end{array}$$

81. 3 ten thousands + 9 thousands + 4 hundreds + 1 ten + 7 ones

83. a) Locate the digit in the thousands place.

$$3\,2,\boxed{8}\,3\,1$$
$$\uparrow$$

b) Then consider the next digit to the right.

c) Since the digit is 5 or higher, round 2 thousands up to 3 thousands.

d) Change all digits to the right of thousands to zeros. The answer is 33,000.

85.
$$\begin{array}{r} 3\,2 \\ 9\,\overline{)2\,8\,8} \\ 2\,7\,0 \\ \hline 1\,8 \\ 1\,8 \\ \hline 0 \end{array}$$

The answer is 32.

87. Discussion and Writing Exercise

89. $-|27| + (-|-13|) = -27 + (-13) = -40$

91. We use a calculator.

$-3496 + (-2987) = -6483$

93. We use a calculator.

$-7846 + 5978 = -1868$

95. If $-x$ is positive, it is the reflection of a negative number x across 0 on the number line. Thus, $-x$ is positive for all negative numbers x.

97. If n is positive, $-n$ is negative. Then $-n + m$, the sum of two negative numbers, is negative.

99. If n is negative and m is less than n, then m is also negative. Then $n + m$, the sum of two negative numbers, is negative.

Exercise Set 2.3

1. $2 - 7 = 2 + (-7) = -5$

3. $0 - 8 = 0 + (-8) = -8$

5. $-7 - (-4) = -7 + 4 = -3$

7. $-11 - (-11) = -11 + 11 = 0$

9. $13 - 17 = 13 + (-17) = -4$

11. $20 - 27 = 20 + (-27) = -7$

13. $-9 - (-4) = -9 + 4 = -5$

15. $-40 - (-40) = -40 + 40 = 0$

17. $7 - 7 = 7 + (-7) = 0$

19. $7 - (-7) = 7 + 7 = 14$

21. $8 - (-3) = 8 + 3 = 11$

23. $-6 - 8 = -6 + (-8) = -14$

25. $-3 - (-9) = -3 + 9 = 6$

27. $1 - 9 = 1 + (-9) = -8$

29. $-6 - (-5) = -6 + 5 = -1$

31. $8 - (-10) = 8 + 10 = 18$

33. $0 - 10 = 0 + (-10) = -10$

35. $-5 - (-2) = -5 + 2 = -3$

37. $-7 - 14 = -7 + (-14) = -21$

39. $0 - (-5) = 0 + 5 = 5$

41. $-8 - 0 = -8 + 0 = -8$

43. $7 - (-5) = 7 + 5 = 12$

45. $6 - 25 = 6 + (-25) = -19$

47. $-42 - 26 = -42 + (-26) = -68$

49. $-72 - 9 = -72 + (-9) = -81$

51. $24 - (-92) = 24 + 92 = 116$

53. $-50 - (-50) = -50 + 50 = 0$

55. $-30 - (-85) = -30 + 85 = 55$

57. $7 - (-5) + 4 - (-3) = 7 + 5 + 4 + 3 = 19$

59.
$$\begin{aligned}
& -31 + (-28) - (-14) - 17 \\
&= -31 + (-28) + 14 + (-17) \\
&= -31 + (-28) + (-17) + 14 \quad \text{Using a commutative} \\
& \hspace{9.5cm} \text{law} \\
&= -76 + 14 \quad \text{Adding the negative numbers} \\
&= -62
\end{aligned}$$

61. $-34 - 28 + (-33) - 44 = (-34) + (-28) + (-33) + (-44) = -139$

63.
$$\begin{aligned}
& -93 - (-84) - 41 - (-56) \\
&= -93 + 84 + (-41) + 56 \\
&= -93 + (-41) + 84 + 56 \quad \text{Using a commutative} \\
& \hspace{9.5cm} \text{law} \\
&= -134 + 140 \quad \text{Adding negatives and adding} \\
& \hspace{7.3cm} \text{positives} \\
&= 6
\end{aligned}$$

65.
$$\begin{aligned}
& -5 - (-30) + 30 + 40 - (-12) \\
&= -5 + 30 + 30 + 40 + 12 \\
&= -5 + 112 \quad \text{Adding the positive numbers} \\
&= 107
\end{aligned}$$

67. $132 - (-21) + 45 - (-21) = 132 + 21 + 45 + 21 = 219$

69. We subtract the beginning page number from the final page number.
$$62 - 37 = 25$$
Alicia read 25 pages.

71. The integer 8 corresponds to 8 lb above the ideal weight, and -9 corresponds to 9 lb below it. We subtract the lower weight from the higher weight:
$$8 - (-9) = 8 + 9 = 17$$
Rod lost 17 lb.

73. We subtract the lower temperature from the higher temperature.
$$27 - (-128) = 27 + 128 = 155$$
The difference in temperatures is 155°C.

75. We subtract the initial reading from the final reading.
$$29 - (-21) = 29 + 21 = 50$$
50 minutes were recorded. Thus, the entire 60-minute show was not recorded.

77. We start with the original temperature, add the rise in temperature, and subtract the drop in temperature.
$$32 + 15 - 50 = 32 + 15 + (-50) = 47 + (-50) = -3$$
The final temperature was $-3°$.

79. The integer -5000 represents a loss of \$5000, and the integer 8000 represents a profit of \$8000. We subtract the amount of the loss from the profit.
$$8000 - (-5000) = 8000 + 5000 = 13,000$$
The store made \$13,000 more in 2005 than in 2004.

81. To find the elevation that is 2293 ft deeper than -7718 ft, we subtract the additional depth from the original depth.
$$-7718 - 2293 = -7718 + (-2293) = -10,011$$
In 2005 the elevation of the deepwater drilling record was $-10,011$ ft.

83. First we subtract the cost of the tolls from the original balance in the account.

$$13 - 20 = 13 + (-20) = -7$$

Then we subtract the cost of the fines and fees from the new balance in the account.

$$-7 - 80 = -7 + (-80) = -87$$

The Murrays were $87 in debt as a result of their travel on toll roads.

85. Discussion and Writing Exercise

87. $4^3 = 4 \cdot 4 \cdot 4 = 64$

89. $1^7 = 1 \cdot 1 \cdot 1 \cdot 1 \cdot 1 \cdot 1 \cdot 1 = 1$

91. *Familiarize*. Let $n =$ the number of 12-oz cans that can be filled. We think of an array consisting of 96 oz with 12 oz in each row.

The number n corresponds to the number of rows in the array.

Translate and Solve. We translate to an equation and solve it.

$$96 \div 12 = n \qquad \begin{array}{r} 8 \\ 12\overline{)96} \\ \underline{96} \\ 0 \end{array}$$

Check. We multiply the number of cans by 12: $8 \cdot 12 = 96$. The result checks.

State. Eight 12-oz cans can be filled.

93.
$$\begin{aligned} &5 + 4^2 + 2 \cdot 7 \\ &= 5 + 16 + 2 \cdot 7 \\ &= 5 + 16 + 14 \\ &= 21 + 14 \\ &= 35 \end{aligned}$$

95. $(9 + 7)(9 - 7) = (16)(2) = 32$

97. Discussion and Writing Exercise

99. Use a calculator to do this exercise.

$$123,907 - 433,789 = -309,882$$

101. False; $3 - 0 \neq 0 - 3$.

103. True

105. True

107. a is the number we add to -57 to get -34. If we think of starting at -57 on the number line and moving to -34, we move 17 units to the right, so $a = 17$.

109. The changes during weeks 1 to 5 are represented by the integers $-13, -16, 36, -11,$ and 19, respectively. We add to find the total rise or fall:

$$-13 + (-16) + 36 + (-11) + 19 = 15$$

The market rose 15 points during the 5 week period.

1. $-2 \cdot 8 = -16$

3. $-9 \cdot 2 = -18$

5. $8 \cdot (-6) = -48$

7. $-10 \cdot 3 = -30$

9. $-3 \cdot (-5) = 15$

11. $-9 \cdot (-2) = 18$

13. $(-6)(-7) = 42$

15. $-10(-2) = 20$

17. $12(-10) = -120$

19. $-6(-50) = 300$

21. $(-72)(-1) = 72$

23. $(-20)17 = -340$

25. $-47 \cdot 0 = 0$

27. $0(-14) = 0$

29.
$$\begin{aligned} &3 \cdot (-8) \cdot (-1) \\ &= -24 \cdot (-1) \qquad \text{Multiplying the first two numbers} \\ &= 24 \end{aligned}$$

31.
$$\begin{aligned} &7(-4)(-3)5 \\ &= 7 \cdot 12 \cdot 5 \qquad \text{Multiplying the negative numbers} \\ &= 84 \cdot 5 \\ &= 420 \end{aligned}$$

33.
$$\begin{aligned} &-2(-5)(-7) \\ &= 10 \cdot (-7) \qquad \text{Multiplying the first two numbers} \\ &= -70 \end{aligned}$$

35.
$$\begin{aligned} &(-5)(-2)(-3)(-1) \\ &= 10 \cdot 3 \qquad \text{Multiplying the first two numbers and the last two numbers} \\ &= 30 \end{aligned}$$

37.
$$\begin{aligned} &(-15)(-29)0 \cdot 8 \\ &= 435 \cdot 0 \qquad \text{Multiplying the first two numbers and the last two numbers} \\ &= 0 \end{aligned}$$

(We might have noted at the outset that the product would be 0 since one of the numbers in the product is 0.)

39.
$$\begin{aligned} &(-7)(-1)(7)(-6) \\ &= 7(-42) \qquad \text{Multiplying the first two numbers and the last two numbers} \\ &= -294 \end{aligned}$$

41. $(-6)^2 = (-6)(-6) = 36$

43.
$$\begin{aligned} (-5)^3 &= (-5)(-5)(-5) \\ &= 25(-5) \\ &= -125 \end{aligned}$$

45. $(-10)^4 = (-10)(-10)(-10)(-10)$
$= 100 \cdot 100$
$= 10,000$

47. $-2^4 = -1 \cdot 2^4$
$= -1 \cdot 2 \cdot 2 \cdot 2 \cdot 2$
$= -1 \cdot 4 \cdot 4$
$= -1 \cdot 16$
$= -16$

49. $(-3)^5 = (-3)(-3)(-3)(-3)(-3)$
$= 9 \cdot 9 \cdot (-3)$
$= 81(-3)$
$= -243$

51. $(-1)^{12}$
$= (-1) \cdot (-1) \cdot (-1) \cdot (-1) \cdot (-1) \cdot (-1) \cdot (-1) \cdot (-1) \cdot$
$(-1) \cdot (-1) \cdot (-1) \cdot (-1)$
$= 1 \cdot 1 \cdot 1 \cdot 1 \cdot 1 \cdot 1$
$= 1 \cdot 1 \cdot 1$
$= 1 \cdot 1$
$= 1$

53. -3^6
$= -1 \cdot 3^6$
$= -1 \cdot 3 \cdot 3 \cdot 3 \cdot 3 \cdot 3 \cdot 3$
$= -1 \cdot 9 \cdot 9 \cdot 9$
$= -9 \cdot 81$
$= -729$

55. $-4^3 = -1 \cdot 4^3$
$= -1 \cdot 4 \cdot 4 \cdot 4$
$= -4 \cdot 16$
$= -64$

57. -8^4 is read "the opposite of eight to the fourth power."

59. $(-9)^{10}$ is read "negative nine to the tenth power."

61. Discussion and Writing Exercise

63. a) Locate the digit in the hundreds place.

$5\ 3\ 2,\ 4\ \boxed{5}\ 1$
$\uparrow$

b) Then consider the next digit to the right.

c) Since that digit is 5 or higher, round 4 hundreds up to 5 hundreds.

d) Change all digits to the right of hundreds to zeros.

The answer is 532,500.

65.
$$\begin{array}{r} 80 \\ 36\overline{\smash)2880} \\ \underline{2880} \\ 0 \\ \underline{0} \\ 0 \end{array}$$

The answer is 80.

67. $10 - 2^3 + 6 \div 2$
$= 10 - 8 + 6 \div 2$ Evaluating the exponential expression
$= 10 - 8 + 3$ Dividing
$= 2 + 3$ Adding and subtracting in
$= 5$ order from left to right

69. Familiarize. We first make a drawing.

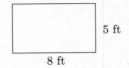

Let $A =$ the area.

Translate. Using the formula for area, we have
$A = l \cdot w = 8 \cdot 5$.

Solve. We carry out the multiplication.

$A = 8 \cdot 5 = 40$

Check. We repeat our calculation.

State. The area of the rug is 40 ft^2.

71. Familiarize. Let $n =$ the number of trips the ferry will make carrying 12 cars. We can think of this problem as repeated subtraction.

Translate.

Number of cars	divided by	Number per trip	is	Number of 12-car trips
↓	↓	↓	↓	↓
53	÷	12	=	n

Solve. We carry out the division.

$$\begin{array}{r} 4 \\ 12\overline{\smash)53} \\ \underline{48} \\ 5 \end{array}$$

$53 \div 12 = n$
$4 \text{ R } 5 = n$

After the ferry makes 4 trips, carrying 12 cars on each trip, 5 cars remained to be ferried. Thus, a fifth trip will be required to ferry the remaining cars.

Check. We can check by multiplying the number of full trips by 12 and then adding the remainder, 5.

$4 \cdot 12 = 48$
$48 + 5 = 53$

Since 53 cars were to be ferried, the answer checks.

State. A total of 5 trips will be required.

73. Discussion and Writing Exercise

75. $(-3)^5(-1)^{379} = -243(-1) = 243$

77. $-9^4 + (-9)^4 = -6561 + 6561 = 0$

79.
$$|(-2)^5 + 3^2| - (3 - 7)^2$$
$$= |-32 + 9| - (-4)^2$$
$$= |-23| - 16$$
$$= 23 - 16$$
$$= 7$$

81. Use a calculator. On many scientific calculators the keystrokes are $\boxed{4}\ \boxed{7}\ \boxed{x^2}\ \boxed{+/-}$. We get -2209.

83. Use a calculator. On many scientific calculators the keystrokes are $\boxed{1}\ \boxed{9}\ \boxed{+/-}\ \boxed{x^y}\ \boxed{4}\ \boxed{=}$. We get 130,321. (Some calculators have a $\boxed{y^x}$ key rather than an $\boxed{x^y}$ key.)

85. Use a calculator. On many scientific calculators the keystrokes are $\boxed{(}\ \boxed{7}\ \boxed{3}\ \boxed{-}\ \boxed{8}\ \boxed{6}\ \boxed{)}\ \boxed{x^y}\ \boxed{3}\ \boxed{=}$. We get -2197. (Some calculators have a $\boxed{y^x}$ key rather than an $\boxed{x^y}$ key.)

87. Use a calculator. On many scientific calculators the keystrokes are $\boxed{9}\ \boxed{3}\ \boxed{5}\ \boxed{+/-}\ \boxed{\times}\ \boxed{5}\ \boxed{+/-}\ \boxed{x^y}\ \boxed{3}\ \boxed{=}$. We get 116,875. (Some calculators have a $\boxed{y^x}$ key rather than an $\boxed{x^y}$ key.)

89. The new balance will be
$68 - 7(\$13) = \$68 - \$91 = -\$23$.

91. a) If $[(-5)^m]^n$ is to be negative, first m must be an odd number so that $(-5)^m$ is negative. Similarly, n must also be odd in order for $[(-5)^m]^n$ to be negative. Thus, both m and n must be odd numbers.

b) If $[(-5)^m]^n$ is to be positive, at least one of m and n must be an even number. For example, if m is even then $(-5)^m$ is positive and so is $[(-5)^m]^n$ regardless of whether n is even or odd. If m is odd, then $(-5)^m$ is negative and n must be even in order for $[(-5)^m]^n$ to be positive.

Exercise Set 2.5

1. $28 \div (-4) = -7$ Check: $-7(-4) = 28$

3. $\dfrac{28}{-2} = -14$ Check: $-14(-2) = 28$

5. $\dfrac{18}{-2} = -9$ Check: $-9(-2) = 18$

7. $\dfrac{-48}{-12} = 4$ Check: $4(-12) = -48$

9. $\dfrac{-72}{8} = -9$ Check: $-9 \cdot 8 = -72$

11. $-100 \div (-50) = 2$ Check: $2(-50) = -100$

13. $-344 \div 8 = -43$ Check: $-43 \cdot 8 = -344$

15. $\dfrac{200}{-25} = -8$ Check: $-8(-25) = 200$

17. $\dfrac{-56}{0}$ is undefined.

19. $\dfrac{88}{-11} = -8$ Check: $-8(-11) = 88$

21. $-\dfrac{276}{12} = \dfrac{-276}{12} = -23$ Check: $-23 \cdot 12 = -276$

23. $\dfrac{0}{-2} = 0$ Check: $0 \cdot (-2) = 0$

25. $\dfrac{19}{-1} = -19$ Check: $-19(-1) = 19$

27. $-41 \div 1 = -41$ Check: $-41 \cdot 1 = -41$

29. $5 - 2 \cdot 3 - 6 = 5 - 6 - 6$ Multiplying
$\qquad\qquad\quad = -1 - 6$ Doing all additions and subtractions in order
$\qquad\qquad\quad = -7$ from left to right

31. $9 - 2(3 - 8) = 9 - 2(-5)$ Subtracting inside parentheses
$\qquad\qquad\quad = 9 + 10$ Multiplying
$\qquad\qquad\quad = 19$ Adding

33. $16 \cdot (-24) + 50 = -384 + 50$ Multiplying
$\qquad\qquad\qquad\quad = -334$ Adding

35. $\quad 40 - 3^2 - 2^3$
$= 40 - 9 - 8$ Evaluating the exponential expressions
$= 31 - 8$ Doing all additions and subtractions
$= 23$ in order from left to right

37. $\quad 4 \cdot (6 + 8)/(4 + 3)$
$= 4 \cdot 14/7$ Adding inside parentheses
$= 56/7$ Doing all multiplications and divisions
$= 8$ in order from left to right

39. $4 \cdot 5 - 2 \cdot 6 + 4 = 20 - 12 + 4$ Multiplying
$\qquad\qquad\qquad\quad = 8 + 4$
$\qquad\qquad\qquad\quad = 12$

41. $\dfrac{9^2 - 1}{1 - 3^2}$
$= \dfrac{81 - 1}{1 - 9}$ Evaluating the exponential expressions
$= \dfrac{80}{-8}$ Subtracting in the numerator and in the denominator
$= -10$

43. $8(-7) + 6(-5) = -56 - 30$ Multiplying
$\qquad\qquad\qquad = -86$

45. $20 \div 5(-3) + 3 = 4(-3) + 3$ Dividing
$\qquad\qquad\qquad = -12 + 3$ Multiplying
$\qquad\qquad\qquad = -9$ Adding

47. $18 - 0(3^2 - 5^2 \cdot 7 - 4)$

Observe that $a \cdot 0 = 0$ for an integer a. Then $0(3^2 - 5^2 \cdot 7 - 4) = 0$, so the result is $18 - 0$, or 18.

49. $\quad 4 \cdot 5^2 \div 10$
$= 4 \cdot 25 \div 10$ Evaluating the exponential expression
$= 100 \div 10$ Multiplying
$= 10$ Dividing

51. $(3-8)^2 \div (-1)$
$= (-5)^2 \div (-1)$ Subtracting inside parentheses
$= 25 \div (-1)$ Evaluating the exponential
                expression
$= -25$

53. $17 - 10^3$
$= 17 - 1000$ Evaluating the exponential
                expression
$= -983$ Subtracting

55. $2 + 10^2 \div 5 \cdot 2^2$
$= 2 + 100 \div 5 \cdot 4$ Evaluating the exponential
                    expressions
$= 2 + 20 \cdot 4$ Doing all multiplications and di-
$= 2 + 80$ visions in order from left to right
$= 82$ Adding

57. $12 - 20^3 = 12 - 8000$
$= -7988$

59. $2 \times 10^3 - 5000 = 2 \times 1000 - 5000$
$= 2000 - 5000$
$= -3000$

61. $6[9 - (3-4)] = 6[9 - (-1)]$ Subtracting inside the
                    innermost parentheses
$= 6[9 + 1]$
$= 6[10]$
$= 60$

63. $-1000 \div (-100) \div 10 = 10 \div 10$ Doing the divi-
                        sions in order
$= 1$ from left to right

65. $8 - |7 - 9| \cdot 3 = 8 - |-2| \cdot 3$
$= 8 - 2 \cdot 3$
$= 8 - 6$
$= 2$

67. $9 - |7 - 3^2| = 9 - |7 - 9|$
$= 9 - |-2|$
$= 9 - 2$
$= 7$

69. $\dfrac{6^3 - 7 \cdot 3^4 - 2^5 \cdot 9}{(1 - 2^3)^3 + 7^3}$
$= \dfrac{216 - 7 \cdot 81 - 32 \cdot 9}{(1 - 8)^3 + 343}$
$= \dfrac{216 - 567 - 288}{(-7)^3 + 343}$
$= \dfrac{-351 - 288}{-343 + 343}$
$= -\dfrac{639}{0}$

Since division by 0 is not defined, this expression is not defined.

71. $\dfrac{2 \cdot 3^2 \div (3^2 - (2+1))}{5^2 - 6^2 - 2^2(-3)}$
$= \dfrac{2 \cdot 3^2 \div (3^2 - 3)}{25 - 36 - 4(-3)}$
$= \dfrac{2 \cdot 3^2 \div (9 - 3)}{25 - 36 + 12}$
$= \dfrac{2 \cdot 3^2 \div 6}{-11 + 12}$
$= \dfrac{2 \cdot 9 \div 6}{1}$
$= \dfrac{18 \div 6}{1}$
$= \dfrac{3}{1}$
$= 3$

73. $\dfrac{(-5)^3 + 17}{10(2-6) - 2(5+2)}$
$= \dfrac{-125 + 17}{10(2-6) - 2(5+2)}$ Evaluating the exponential
                        expression
$= \dfrac{-125 + 17}{10(-4) - 2 \cdot 7}$ Doing the calculations within
                        parentheses
$= \dfrac{-125 + 17}{-40 - 14}$ Multiplying
$= \dfrac{-108}{-54}$ Adding and subtracting
$= 2$

75. $\dfrac{2 \cdot 4^3 - 4 \cdot 32}{19^3 - 17^4}$
$= \dfrac{2 \cdot 64 - 4 \cdot 32}{6859 - 83,521}$ Evaluating the exponential ex-
                        pressions
$= \dfrac{128 - 128}{6859 - 83,521}$ Multiplying
$= \dfrac{0}{-76,662}$ Subtracting
$= 0$ Dividing

77. Discussion and Writing Exercise

79. *Familiarize*. We first make a drawing.

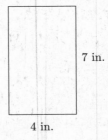

7 in.

4 in.

Let $A =$ the area.

Translate. Using the formula for area, we have $A = l \cdot w = 7 \cdot 4$.

Solve. We carry out the multiplication.

$A = 7 \cdot 4 = 28$

Check. We repeat the calculation.

State. The area of the ad was 28 in^2.

81. **Familiarize**. We let $g =$ the number of gallons needed to travel 384 miles. Think of a rectangular array consisting of 384 miles with 32 miles in each row. The number g is the number of rows.

Translate.

The number of miles	divided by	the number of miles per gallon	is	the number of gallons needed.
↓	↓	↓	↓	↓
384	÷	32	=	g

Solve. We carry out the division.

$$\begin{array}{r} 1\,2 \\ 32\overline{)3\,8\,4} \\ 3\,2\,0 \\ \hline 6\,4 \\ 6\,4 \\ \hline 0 \end{array}$$

Check. We multiply the number of gallons by the number of miles per gallon:

$$12 \cdot 32 = 384$$

We get the number of miles to be traveled, so the answer checks.

State. It will take 12 gallons of gasoline to travel 384 miles.

83. **Familiarize**. We let $c =$ the number of calories in a 1-oz serving. Think of a rectangular array consisting of 1050 calories arranged in 7 rows. The number c is the number of calories in each row.

Translate.

Total calories	divided by	number of ounces	is	number of calories in 1 oz.
↓	↓	↓	↓	↓
1050	÷	7	=	c

Solve. We carry out the division.

$$\begin{array}{r} 1\,5\,0 \\ 7\overline{)1\,0\,5\,0} \\ 7\,0\,0 \\ \hline 3\,5\,0 \\ 3\,5\,0 \\ \hline 0 \\ 0 \\ \hline 0 \end{array}$$

Check. We multiply the number of calories in 1 oz by 7:

$$150 \cdot 7 = 1050$$

The result checks

State. There are 150 calories in a 1-oz serving.

85. **Familiarize**. Let $p =$ the number of whole pieces of gum each person will receive. We can think of this problem as repeated subtraction.

Translate.

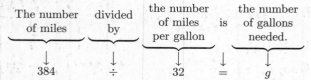

Total number of sticks	divided by	number of people	is	number of whole pieces per person.
↓	↓	↓	↓	↓
18	÷	4	=	p

Solve. We carry out the division.

$$\begin{array}{r} 4 \\ 4\overline{)1\,8} \\ 1\,6 \\ \hline 2 \end{array}$$

$$18 \div 4 = p$$

$$4 \text{ R } 2 = p$$

Check. We multiply the number of whole pieces of gum per person by the number of people and then add the number of remaining pieces.

$$4 \cdot 4 = 16$$

$$16 + 2 = 18$$

We got the number of sticks of gum in a package, 18, so the answer checks.

State. Each person will receive 4 pieces of gum, and there will be 2 extra pieces remaining.

87. Discussion and Writing Exercise

89.
$$\frac{9 - 3^2}{2 \cdot 4^2 - 5^2 \cdot 9 + 8^2 \cdot 7}$$
$$= \frac{9 - 9}{2 \cdot 16 - 25 \cdot 9 + 64 \cdot 7}$$
$$= \frac{0}{32 - 225 + 448}$$
$$= \frac{0}{-193 + 448}$$
$$= \frac{0}{255}$$
$$= 0$$

91.
$$\frac{(25 - 4^2)^3}{17^2 - 16^2} \cdot ((-6)^2 - 6^2) = \frac{(25 - 16)^3}{289 - 256} \cdot (36 - 36)$$
$$= \frac{9^3}{289 - 256} \cdot 0$$
$$= 0$$

93. Use a calculator.
$$\frac{19 - 17^2}{13^2 - 34} = -2$$

95. Use a calculator.
$$28^2 - 36^2/4^2 + 17^2 = 992$$

97. (1 5 x^2 – 5 x^y 3) ÷ (3 x^2 + 4 x^2) = (Some calculators have a y^x key rather than an x^y key.)

99. Entering the given keystrokes and then pressing = , we get 5.

101. $-n$ and m are both negative, so $\dfrac{-n}{m}$ is the quotient of two negative numbers and, thus, is positive.

103. $\dfrac{-n}{m}$ is positive (see Exercise 101), so $-\left(\dfrac{-n}{m}\right)$ is the opposite of a positive number and, thus, is negative.

105. $-n$ is negative and $-m$ is positive, so $\dfrac{-n}{-m}$ is the quotient of a negative and a positive number and, thus, is negative. Then $-\left(\dfrac{-n}{-m}\right)$ is the opposite of a negative number and, thus, is positive.

Exercise Set 2.6

1. $12n = 12 \cdot 2 = 24\cent$

3. $\dfrac{x}{y} = \dfrac{6}{-3} = -2$

5. $\dfrac{2q}{p} = \dfrac{2 \cdot 3}{6} = \dfrac{6}{6} = 1$

7. $\dfrac{72}{r} = \dfrac{72}{4} = 18$ yr

9. $3 + 5 \cdot x = 3 + 5 \cdot 2 = 3 + 10 = 13$

11. $2l + 2w = 2 \cdot 3 + 2 \cdot 4 = 6 + 8 = 14$ ft

13. $2(l + w) = 2(3 + 4) = 2 \cdot 7 = 14$ ft

15. $7a - 7b = 7 \cdot 5 - 7 \cdot 2 = 35 - 14 = 21$

17. $7(a - b) = 7(5 - 2) = 7 \cdot 3 = 21$

19. $16t^2 = 16 \cdot 5^2 = 16 \cdot 25 = 400$ ft

21. $\begin{aligned} a + (b - a)^2 &= 6 + (4 - 6)^2 \\ &= 6 + (-2)^2 \\ &= 6 + 4 \\ &= 10 \end{aligned}$

23. $\begin{aligned} 9a + 9b &= 9 \cdot 13 + 9(-13) \\ &= 117 - 117 \\ &= 0 \end{aligned}$

25. $\begin{aligned} \dfrac{n^2 - n}{2} &= \dfrac{9^2 - 9}{2} \\ &= \dfrac{81 - 9}{2} \\ &= \dfrac{72}{2} \\ &= 36 \end{aligned}$

27. $\begin{aligned} m^3 - m^2 &= 5^3 - 5^2 \\ &= 125 - 25 \\ &= 100 \end{aligned}$

29. $-\dfrac{a}{b}, \dfrac{-a}{b}$, and $\dfrac{a}{-b}$ all represent the same number. Thus we can also write $\dfrac{-5}{t}$ as $\dfrac{-5}{t}$ and $\dfrac{5}{-t}$.

31. $-\dfrac{a}{b}, \dfrac{-a}{b}$, and $\dfrac{a}{-b}$ all represent the same number. Thus we can also write $\dfrac{-n}{b}$ as $-\dfrac{n}{b}$ and $\dfrac{n}{-b}$.

33. $-\dfrac{a}{b}, \dfrac{-a}{b}$, and $\dfrac{a}{-b}$ all represent the same number. Thus we can also write $\dfrac{9}{-p}$ as $-\dfrac{9}{p}$ and $\dfrac{-9}{p}$.

35. $-\dfrac{a}{b}, \dfrac{-a}{b}$, and $\dfrac{a}{-b}$ all represent the same number. Thus we can also write $\dfrac{-14}{w}$ as $-\dfrac{14}{w}$ and $\dfrac{14}{-w}$.

37. $\dfrac{-a}{b} = \dfrac{-45}{9} = -5;$

$\dfrac{a}{-b} = \dfrac{45}{-9} = -5;$

$-\dfrac{a}{b} = -\dfrac{45}{9} = -5$

39. $\dfrac{-a}{b} = \dfrac{-81}{3} = -27;$

$\dfrac{a}{-b} = \dfrac{81}{-3} = -27;$

$-\dfrac{a}{b} = -\dfrac{81}{3} = -27$

41. $(-3x)^2 = (-3 \cdot 2)^2 = (-6)^2 = 36;$
$-3x^2 = -3(2)^2 = -3 \cdot 4 = -12$

43. $5x^2 = 5(3)^2 = 5 \cdot 9 = 45;$
$5x^2 = 5(-3)^2 = 5 \cdot 9 = 45$

45. $x^3 = 6^3 = 6 \cdot 6 \cdot 6 = 216;$
$x^3 = (-6)^3 = (-6) \cdot (-6) \cdot (-6) = -216$

47. $x^8 = 1^8 = 1 \cdot 1 \cdot 1 \cdot 1 \cdot 1 \cdot 1 \cdot 1 \cdot 1 = 1;$
$x^8 = (-1)^8 =$
$(-1) \cdot (-1) \cdot (-1) \cdot (-1) \cdot (-1) \cdot (-1) \cdot (-1) \cdot (-1) = 1$

49. $a^5 = 2^5 = 2 \cdot 2 \cdot 2 \cdot 2 \cdot 2 = 32;$
$a^5 = (-2)^5 = (-2)(-2)(-2)(-2)(-2) = -32$

51. $5(a + b) = 5 \cdot a + 5 \cdot b = 5a + 5b$

53. $4(x + 1) = 4 \cdot x + 4 \cdot 1 = 4x + 4$

55. $2(b + 5) = 2 \cdot b + 2 \cdot 5 = 2b + 10$

57. $7(1 - t) = 7 \cdot 1 - 7 \cdot t = 7 - 7t$

59. $6(5x - 2) = 6 \cdot 5x - 6 \cdot 2 = 30x - 12$

61. $8(x + 7 + 6y) = 8 \cdot x + 8 \cdot 7 + 8 \cdot 6y = 8x + 56 + 48y$

63. $-7(y - 2) = -7 \cdot y - (-7) \cdot 2 = -7y - (-14) = -7y + 14$

65. $(x + 2)3 = x \cdot 3 + 2 \cdot 3 = 3x + 6$

67. $-4(x - 3y - 2z) = -4 \cdot x - (-4)3y - (-4)2z =$
$-4x - (-12y) - (-8z) = -4x + 12y + 8z$

69. $8(a - 3b + c) = 8 \cdot a - 8 \cdot 3b + 8 \cdot c =$
$8a - 24b + 8c$

71. $4(x - 3y - 7z) = 4 \cdot x - 4 \cdot 3y - 4 \cdot 7z =$
$4x - 12y - 28z$

73. $(4a - 5b + c - 2d)5 = 4a \cdot 5 - 5b \cdot 5 + c \cdot 5 - 2d \cdot 5 =$
$20a - 25b + 5c - 10d$

75. Discussion and Writing Exercise

77. Twenty-three million, forty-three thousand, nine hundred twenty-one

79.
$$
\begin{array}{r}
5\,2\,8\,3 \\
-\,2\,4\,7\,5 \\
\hline
\end{array}
\qquad
\begin{array}{r}
5\,2\,8\,0 \\
-\,2\,4\,8\,0 \\
\hline
2\,8\,0\,0
\end{array}
$$

81. *Familiarize.* Since we are combining snowfall amounts, addition can be used. We let $s =$ the total snowfall.

Translate. We translate to an equation.

$$9 + 8 = s$$

Solve. We carry out the addition.

$$9 + 8 = 17$$

Thus, $17 = s$, or $s = 17$.

Check. We repeat the calculation.

State. It snowed 17 in. altogether.

83. *Familiarize.* This is a multistep problem. We let $x =$ the total price of the plain pizzas, $y =$ the total price of the pepperoni pizzas, and $t =$ the total price of the pizza.

Translate.

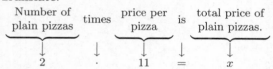

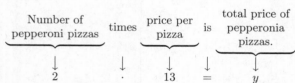

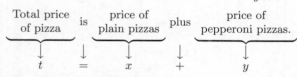

Solve. We solve each equation.

$$2 \cdot 11 = x \qquad 2 \cdot 13 = y \qquad t = x + y$$
$$22 = x \qquad\quad 26 = y \qquad\quad t = 22 + 26$$
$$t = 48$$

Check. We can repeat the calculations. We can also check by rounding, multiplying, and adding. The answer checks.

State. Brett's wife paid \$48 for the pizza.

85. Discussion and Writing Exercise

87. We substitute 370 for C in the formula in Example 5.

$$\frac{9C}{5} + 32 = \frac{9 \cdot 370}{5} + 32 = \frac{3330}{5} + 32 = 666 + 32 = 698$$

The temperature 370°Celsius corresponds to a Fahrenheit temperature of 698°.

89. Use a calculator.

$$a - b^3 + 17a = 19 - (-16)^3 + 17 \cdot 19 = 4438$$

91. Use a calculator.

$$r^3 + r^2 t - rt^2 = (-9)^3 + (-9)^2 \cdot 7 - (-9) \cdot 7^2 = 279$$

93.
$$
\begin{aligned}
a^{1996} - a^{1997} &= (-1)^{1996} - (-1)^{1997} \\
&= 1 - (-1) \\
&= 1 + 1 \\
&= 2
\end{aligned}
$$

95.
$$
\begin{aligned}
(m^3 - mn)^m &= (4^3 - 4 \cdot 6)^4 \\
&= (64 - 4 \cdot 6)^4 \\
&= (64 - 24)^4 \\
&= 40^4 \\
&= 2,560,000
\end{aligned}
$$

97. $-32 \boxed{\times} (88 \boxed{-} 29) = -1888$

99. True

101. True

103. Discussion and Writing Exercise

Exercise Set 2.7

1. $2a + 5b - 7c = 2a + 5b + (-7c)$
The terms are $2a$, $5b$, and $-7c$.

3. $9mn - 6n + 8 = 9mn + (-6n) + 8$
The terms are $9mn$, $-6n$, and 8.

5. $3x^2y - 4y^2 - 2z^3 = 3x^2y + (-4y^2) + (-2z^3)$
The terms are $3x^2y$, $-4y^2$, and $-2z^3$.

7. $5x + 9x = (5 + 9)x = 14x$

9. $10a - 13a = (10 - 13)a = -3a$

11. $2x + 6z + 9x = 2x + 9x + 6z$
$= (2 + 9)x + 6z$
$= 11x + 6z$

13. $27a + 70 - 40a - 8 = 27a - 40a + 70 - 8$
$= (27 - 40)a + 70 - 8$
$= -13a + 62$

15. $9 + 5t + 7y - t - y - 13$
$= 9 - 13 + 5t - 1 \cdot t + 7y - 1 \cdot y$
$= (9 - 13) + (5 - 1)t + (7 - 1)y$
$= -4 + 4t + 6y$

17. $\quad a + 3b + 5a - 2 + b$
$= a + 5a + 3b + b - 2$
$= (1 + 5)a + (3 + 1)b - 2$
$= 6a + 4b - 2$

19. $-8 + 11a - 5b + 6a - 7b + 7$
$= 11a + 6a - 5b - 7b - 8 + 7$
$= (11 + 6)a + (-5 - 7)b + (-8 + 7)$
$= 17a - 12b - 1$

21. $8x^2 + 3y - x^2 = 8x^2 - x^2 + 3y$
$\qquad\qquad\quad = (8 - 1)x^2 + 3y$
$\qquad\qquad\quad = 7x^2 + 3y$

23. $11x^4 + 2y^3 - 4x^4 - y^3 = 11x^4 - 4x^4 + 2y^3 - y^3$
$\qquad\qquad\qquad\qquad\quad = (11 - 4)x^4 + (2 - 1)y^3$
$\qquad\qquad\qquad\qquad\quad = 7x^4 + y^3$

25. $9a^2 - 4a + a - 3a^2 = 9a^2 - 3a^2 - 4a + a$
$\qquad\qquad\qquad\qquad\quad = (9 - 3)a^2 + (-4 + 1)a$
$\qquad\qquad\qquad\qquad\quad = 6a^2 - 3a$

27. $\quad x^3 - 5x^2 + 2x^3 - 3x^2 + 4$
$= x^3 + 2x^3 - 5x^2 - 3x^2 + 4$
$= (1 + 2)x^3 + (-5 - 3)x^2 + 4$
$= 3x^3 - 8x^2 + 4$

29. $\quad 7a^3 + 4ab - 5 - 7ab + 8$
$= 7a^3 + 4ab - 7ab - 5 + 8$
$= 7a^3 + (4 - 7)ab + (-5 + 8)$
$= 7a^3 - 3ab + 3$

31. $\quad 9x^3y + 4xy^3 - 6xy^3 + 3xy$
$= 9x^3y + (4 - 6)xy^3 + 3xy$
$= 9x^3y - 2xy^3 + 3xy$

33. $\quad 3a^6 - 9b^4 + 2a^6b^4 - 7a^6 - 2b^4$
$= 3a^6 - 7a^6 - 9b^4 - 2b^4 + 2a^6b^4$
$= (3 - 7)a^6 + (-9 - 2)b^4 + 2a^6b^4$
$= -4a^6 - 11b^4 + 2a^6b^4$

35. $P = 2 \cdot (l + w)$ Perimeter of a rectangle
$\quad P = 2 \cdot (3 \text{ ft} + 2 \text{ ft})$
$\quad P = 2 \cdot (3 + 2) \text{ ft}$
$\quad P = 2 \cdot 5 \text{ ft}$
$\quad P = 10 \text{ ft}$

37. $\qquad$ Perimeter
$= 7 \text{ km} + 7 \text{ km} + 7 \text{ km} + 7 \text{ km} + 7 \text{ km} + 7 \text{ km}$
$= (7 + 7 + 7 + 7 + 7 + 7) \text{ km}$
$= 42 \text{ km}$

39. Perimeter $= 3 \text{ m} + 1 \text{ m} + 3 \text{ m} + 1 \text{ m}$
$\qquad\qquad\quad = (3 + 1 + 3 + 1) \text{ m}$
$\qquad\qquad\quad = 8 \text{ m}$

41. A singles court is 78 ft by 27 ft.
$\quad P = 2l + 2w = 2 \cdot 78 \text{ ft} + 2 \cdot 27 \text{ ft}$
$\qquad\qquad = 156 \text{ ft} + 54 \text{ ft}$
$\qquad\qquad = 210 \text{ ft}$

43. The rectangle formed by the services lines and the singles sideline is 42 ft by 27 ft.
$\quad P = 2l + 2w = 2 \cdot 42 \text{ ft} + 2 \cdot 27 \text{ ft}$
$\qquad\qquad = 84 \text{ ft} + 54 \text{ ft}$
$\qquad\qquad = 138 \text{ ft}$

45. $\quad P = 2(l + w) = 2(10 \text{ ft} + 8 \text{ ft})$
$\qquad\quad = 2 \cdot 18 \text{ ft} = 36 \text{ ft}$

47. $\quad P = 4s$
$\qquad\quad = 4 \cdot 14 \text{ in.} = 56 \text{ in.}$

49. $\quad P = 4s$
$\qquad\quad = 4 \cdot 65 \text{ cm} = 260 \text{ cm}$

51. $\quad P = 2(l + w) = 2(20 \text{ ft} + 12 \text{ ft})$
$\qquad\quad = 2 \cdot 32 \text{ ft} = 64 \text{ ft}$

53. Discussion and Writing Exercise

55. *Familiarize.* Let s = the number of servings of Shaw's Corn Flakes in one box. Visualize a rectangular array consisting of 510 grams with 30 grams in each row. Then s is the number of rows.

Translate.

Total weight	divided by	weight of one serving	is	number of servings.
↓	↓	↓	↓	↓
510	÷	30	=	s

Solve. We carry out the division.

$$\begin{array}{r} 1\ 7 \\ 3\,0\,\overline{)5\,1\,0} \\ 3\,0\,0 \\ \hline 2\,1\,0 \\ 2\,1\,0 \\ \hline 0 \end{array}$$

We have $17 = s$, or $s = 17$.

Check. We multiply the number of servings by the weight of a serving.

$$17 \cdot 30 = 510$$

We get 510 oz, the total weight of the corn flakes, so the answer checks.

State. There are 17 servings in a box of Shaw's Corn Flakes.

57. $\quad 5 + 3 \cdot 2^3$
$= 5 + 3 \cdot 8$ Evaluating the exponential expression
$= 5 + 24$ Multiplying
$= 29$ Adding

59. $\quad 12 \div 3 \cdot 2$
$= 4 \cdot 2$ Dividing and multiplying in order
$= 8$ from left to right

61. $\quad 15 - 3 \cdot 2 + 7$
$= 15 - 6 + 7$ Multiplying
$= 9 + 7$ Subtracting and adding in order
$= 16$ from left to right

63. $\qquad 25 = t + 9$
$\quad 25 - 9 = t + 9 - 9$
$\qquad 16 = t$

The solution is 16.

65. $45 = 3x$

$\dfrac{45}{3} = \dfrac{3x}{3}$

$15 = x$

The solution is 15.

67. Discussion and Writing Exercise

69. $5(x+3) + 2(x-7) = 5x + 15 + 2x - 14 = 7x + 1$

71. $2(3-4a) + 5(a-7) = 6 - 8a + 5a - 35 = -3a - 29$

73. $-5(2+3x+4y) + 7(2x-y) =$
$-10 - 15x - 20y + 14x - 7y = -10 - x - 27y$

75. *Familiarize*. First we will find the amount of sealant needed to caulk each door and each window, keeping in mind that the bottom of each door requires no caulk. Thus, for each door we add the lengths of the other three sides and for each window we find the perimeter. Then we will find the amount of caulk required for all the doors and windows. Next we will determine how many sealant cartridges are needed and, finally, we will find the cost of the sealant.

Translate.

The amount of caulk required for each door is given by

7 ft + 3 ft + 7 ft.

The perimeter of each window is given by

$P = 2(l + w) = 2(3 \text{ ft} + 4 \text{ ft})$.

Solve. First we do the calculations in the Translate step.

For each door: 7 ft + 3 ft + 7 ft = 17 ft

For each window: $P = 2(3 \text{ ft} + 4 \text{ ft}) = 2 \cdot 7 \text{ ft} = 14 \text{ ft}$

We multiply to find the amount of caulk required for 3 doors and of 13 windows.

Doors: $3 \cdot 17 \text{ ft} = 51 \text{ ft}$

Windows: $13 \cdot 14 \text{ ft} = 182 \text{ ft}$

We add to find the total of the perimeters:

51 ft + 182 ft = 233 ft

Next we divide to determine how many sealant cartridges are needed:

$$\begin{array}{r} 4 \\ 5\,6\,\overline{\smash{\big)}\,2\,3\,3} \\ \underline{2\,2\,4} \\ 9 \end{array}$$

The answer is 4 R 9. Since 9 ft will be left unsealed after 4 cartridges are used, Andrea should buy 5 sealant cartridges.

Finally, we multiply to find the cost of 5 sealant cartridges.

$5 \cdot \$5.95 = \$29.75$

Check. We repeat the calculations. The result checks.

State. It will cost Andrea $29.75 to seal the windows and doors.

77. *Familiarize*. The inside of the rack is a square whose side has a length that is the total diameter of 4 balls, or $4 \cdot 57$ mm, or 228 mm. We find the perimeter of a square with side 228 mm.

Translate.

$P = 4s = 4 \cdot 228$ mm

Solve. We calculate the perimeter.

$P = 4 \cdot 228$ mm = 912 mm

Check. We repeat the calculation. The answer checks.

State. The inside perimeter of the storage rack is 912 mm.

Exercise Set 2.8

1. $2x = 10$ $\qquad$ $5x = 25$

$\dfrac{2x}{2} = \dfrac{10}{2}$ $\qquad$ $\dfrac{5x}{5} = \dfrac{25}{5}$

$x = 5$ $\qquad\quad$ $x = 5$

We see that $2x = 10$ and $5x = 25$ are equivalent equations.

3. Combining like terms in $4a - 3 + 3a$, we have

$4a - 3 + 3a = (4+3)a - 3$

$= 7a - 3$.

We see that $7a - 3$ and $4a - 3 + 3a$ are equivalent expressions.

5. Combining like terms in $8 + 4r - 5$, we have

$8 + 4r - 5 = 4r + (8 - 5)$

$= 4r + 3$.

We see that $4r + 3$ and $8 + 4r - 5$ are equivalent expressions.

7. $x - 9 = 8$ $\qquad$ $x + 3 = 20$

$x - 9 + 9 = 8 + 9$ $\qquad$ $x + 3 - 3 = 20 - 3$

$x = 17$ $\qquad\qquad$ $x = 17$

We see that $x - 9 = 8$ and $x + 3 = 20$ are equivalent equations.

9. $3(t+2) = 3t + 6$

$5 + 3t + 1 = 3t + (5+1) = 3t + 6$

We see that $3(t+2)$ and $5 + 3t + 1$ are equivalent expressions.

11. $x + 4 = -8$ $\qquad$ $2x = -24$

$x + 4 - 4 = -8 - 4$ $\qquad$ $\dfrac{2x}{2} = -\dfrac{24}{2}$

$x = -12$ $\qquad\qquad$ $x = -12$

We see that $x + 4 = -8$ and $2x = -24$ are equivalent equations.

13. $x - 6 = -9$

$x - 6 + 6 = -9 + 6$ $\quad$ Adding 6 to both sides

$x + 0 = -3$

$x = -3$

Check: $\dfrac{x - 6 = -9}{-3 - 6 \;?\; -9}$

$\qquad\qquad -9 \;\big|\; -9$ TRUE

The solution is -3.

15. $x - 4 = -12$

$x - 4 + 4 = -12 + 4$ Adding 4 to both sides

$x + 0 = -8$

$x = -8$

Check: $\dfrac{x - 4 = -12}{}$

$\begin{array}{c|c} -8 - 4 \ ? \ -12 \\ \hline -12 & -12 \quad \text{TRUE} \end{array}$

The solution is -8.

17. $a + 7 = 25$

$a + 7 - 7 = 25 - 7$ Subtracting 7 from both sides

$a + 0 = 18$

$a = 18$

The solution is 18.

19. $x + 8 = -6$

$x + 8 - 8 = -6 - 8$ Subtracting 8 from both sides

$x + 0 = -14$

$x = -14$

The solution is -14.

21. $24 = t - 8$

$24 + 8 = t - 8 + 8$ Adding 8 to both sides

$32 = t + 0$

$32 = t$

The solution is 32.

23. $-12 = x + 5$

$-12 - 5 = x + 5 - 5$ Subtracting 5 from both sides

$-17 + 0 = x$

$-17 = x$

The solution is -17.

25. $-5 + a = 12$

$5 - 5 + a = 5 + 12$ Adding 5 to both sides

$0 + a = 12$

$a = 17$

The solution is 17.

27. $-8 = -8 + t$

$8 - 8 = 8 - 8 + t$ Adding 8 to both sides

$0 = 0 + t$

$0 = t$

The solution is 0.

29. $6x = -24$

$\dfrac{6x}{6} = \dfrac{-24}{6}$ Dividing both sides by 6

$x = -4$

The solution is -4.

31. $-3t = 42$

$\dfrac{-3t}{-3} = \dfrac{42}{-3}$ Dividing both sides by -3

$t = -14$

The solution is -14.

33. $-7n = -35$

$\dfrac{-7n}{-7} = \dfrac{-35}{-7}$ Dividing both sides by -7

$n = 5$

The solution is 5.

35. $0 = 6x$

$\dfrac{0}{6} = \dfrac{6x}{6}$ Dividing both sides by 6

$0 = x$

The solution is 0.

37. $55 = -5t$

$\dfrac{55}{-5} = \dfrac{-5t}{-5}$ Dividing both sides by -5

$-11 = t$

The solution is -11.

39. $-x = 56$

$\dfrac{-x}{-1} = \dfrac{56}{-1}$ Dividing both sides by -1

$x = -56$

The solution is -56.

41. $n(-4) = -48$

$\dfrac{n(-4)}{-4} = \dfrac{-48}{-4}$ Dividing both sides by -4

$n = 12$

The solution is 12.

43. $-x = -390$

$\dfrac{-x}{-1} = \dfrac{-390}{-1}$ Dividing both sides by -1

$x = 390$

The solution is 390.

45. $t - 6 = -2$

To undo the addition of -6, or the subtraction of 6, we subtract -6, or simply add 6, to both sides.

$t - 6 = -2$

$t - 6 + 6 = -2 + 6$

$t + 0 = 4$

$t = 4$

The solution is 4.

47. $6x = -54$

To undo multiplication by 6, we divide both sides by 6.

$6x = -54$

$\dfrac{6x}{6} = \dfrac{-54}{6}$

$x = -9$

The solution is -9.

49.
$$15 = -x$$
$$-1 \cdot 15 = -1 \cdot (-x) \quad \text{Multiplying both sides by } -1$$
$$-15 = x$$
The solution is -15.

51.
$$-21 = x + 5$$
$$-21 - 5 = x + 5 - 5 \quad \text{Subtracting 5 from both sides}$$
$$-26 = x$$
The solution is -26.

53. $35 = -7t$

To undo multiplication by -7, we divide both sides by -7.
$$35 = -7t$$
$$\frac{35}{-7} = \frac{-7t}{-7}$$
$$-5 = t$$
The solution is -5.

55. $-17x = 68$

To undo multiplication by -17, we divide both sides by -17.
$$-17x = 68$$
$$\frac{-17x}{-17} = \frac{68}{-17}$$
$$x = -4$$
The solution is -4.

57. $18 + t = -160$

To undo the addition of 18, we subtract 18 from both sides.
$$18 + t = -160$$
$$18 + t - 18 = -160 - 18$$
$$t + 0 = -178$$
$$t = -178$$
The solution is -178.

59. $-27 = x + 23$

To undo the addition of 23, we subtract 23 from both sides.
$$-27 = x + 23$$
$$-27 - 23 = x + 23 - 23$$
$$-50 = x + 0$$
$$-50 = x$$
The solution is -50.

61.
$$5x - 1 = 34$$
$$5x - 1 + 1 = 34 + 1 \quad \text{Adding 1 to both sides}$$
$$5x + 0 = 35$$
$$5x = 35$$
$$\frac{5x}{5} = \frac{35}{5} \quad \text{Dividing both sides by 5}$$
$$x = 7$$
Check:
$$\begin{array}{c|c} \multicolumn{2}{c}{5x - 1 = 34} \\ \hline 5 \cdot 7 - 1 \; ? \; 34 & \\ 35 - 1 & \\ 34 & 34 \quad \text{TRUE} \end{array}$$
The solution is 7.

63.
$$4t + 2 = 14$$
$$4t + 2 - 2 = 14 - 2 \quad \text{Subtracting 2 from both sides}$$
$$4t + 0 = 12$$
$$4t = 12$$
$$\frac{4t}{4} = \frac{12}{4} \quad \text{Dividing both sides by 4}$$
$$t = 3$$
Check:
$$\begin{array}{c|c} \multicolumn{2}{c}{4t + 2 = 14} \\ \hline 4 \cdot 3 + 2 \; ? \; 14 & \\ 12 + 2 & \\ 14 & 14 \quad \text{TRUE} \end{array}$$
The solution is 3.

65.
$$6a + 1 = -17$$
$$6a + 1 - 1 = -17 - 1 \quad \text{Subtracting 1 from both sides}$$
$$6a + 0 = -18$$
$$6a = -18$$
$$\frac{6a}{6} = \frac{-18}{6} \quad \text{Dividing both sides by 6}$$
$$a = -3$$
The solution is -3.

67.
$$2x - 9 = -23$$
$$2x - 9 + 9 = -23 + 9 \quad \text{Adding 9 to both sides}$$
$$2x + 0 = -14$$
$$2x = -14$$
$$\frac{2x}{2} = \frac{-14}{2} \quad \text{Dividing both sides by 2}$$
$$x = -7$$
The solution is -7.

69.
$$-2x + 1 = 17$$
$$-2x + 1 - 1 = 17 - 1 \quad \text{Subtracting 1 from both sides}$$
$$-2x + 0 = 16$$
$$-2x = 16$$
$$\frac{-2x}{-2} = \frac{16}{-2} \quad \text{Dividing both sides by } -2$$
$$x = -8$$
The solution is -8.

71.
$$-8t - 3 = -67$$
$$-8t - 3 + 3 = -67 + 3 \quad \text{Adding 3 to both sides}$$
$$-8t + 0 = -64$$
$$-8t = -64$$
$$\frac{-8t}{-8} = \frac{-64}{-8} \quad \text{Dividing both sides by } -8$$
$$t = 8$$
The solution is 8.

73.
$$-x + 9 = -15$$
$$-x + 9 - 9 = -15 - 9 \quad \text{Subtracting 9 from both sides}$$
$$-x + 0 = -24$$
$$-x = -24$$
$$-1(-x) = -1(-24) \quad \text{Multiplying both sides by } -1$$
$$x = 24$$
The solution is 24.

75.
$$7 = 2x - 5$$
$$7 + 5 = 2x - 5 + 5 \quad \text{Adding 5 to both sides}$$
$$12 = 2x + 0$$
$$12 = 2x$$
$$\frac{12}{2} = \frac{2x}{2} \quad \text{Dividing both sides by 2}$$
$$6 = x$$
The solution is 6.

77.
$$13 = 3 + 2x$$
$$13 - 3 = 3 + 2x - 3 \quad \text{Subtracting 3 from both sides}$$
$$10 = 2x + 0$$
$$10 = 2x$$
$$\frac{10}{2} = \frac{2x}{2} \quad \text{Dividing both sides by 2}$$
$$5 = x$$
The solution is 5.

79.
$$13 = 5 - x$$
$$13 - 5 = 5 - x - 5 \quad \text{Subtracting 5 from both sides}$$
$$8 = 0 - x$$
$$8 = -x$$
$$-1 \cdot 8 = -1(-x) \quad \text{Multiplying both sides by } -1$$
$$-8 = x$$
The solution is -8.

81. Discussion and Writing Exercise

83. A <u>polygon</u> is a closed geometric figure.

85. Numbers we multiply together are called <u>factors</u>.

87. The result of an addition is a <u>sum</u>.

89. The <u>absolute value</u> of a number is its distance from zero on a number line.

91. Discussion and Writing Exercise

93.
$$2x - 7x = -40$$
$$-5x = -40 \quad \text{Collecting like terms}$$
$$\frac{-5x}{-5} = \frac{-40}{-5}$$
$$x = 8$$
The solution is 8.

95.
$$17 - 3^2 = 4 + t - 5^2$$
$$17 - 9 = 4 + t - 25$$
$$8 = t - 21 \quad \text{Collecting like terms}$$
$$8 + 21 = t - 21 + 21 \quad \text{Adding 21 to both sides}$$
$$29 = t$$
The solution is 29.

97.
$$(-7)^2 - 5 = t + 4^3$$
$$49 - 5 = t + 64$$
$$44 = t + 64$$
$$44 - 64 = t + 64 - 64 \quad \text{Subtracting 64 from both sides}$$
$$-20 = t$$
The solution is -20.

99.
$$x - (19)^3 = -18^3$$
$$x - 6859 = -5832$$
$$x - 6859 + 6859 = -5832 + 6859$$
$$x = 1027$$
The solution is 1027.

101.
$$35^3 = = -125t$$
$$42,875 = -125t$$
$$\frac{42,875}{-125} = \frac{-125t}{-125}$$
$$-343 = t$$
The solution is -343.

103.
$$529 - 143x = -1902$$
$$529 - 143x - 529 = -1902 - 529$$
$$-143x = -2431$$
$$\frac{-143x}{-143} = \frac{-2431}{-143}$$
$$x = 17$$
The solution is 17.

Chapter 2 Review Exercises

1. The integer 527 corresponds to having \$527 in an account; the integer -53 corresponds to a \$53 debt.

2. Since 0 is to the right of -5, we have $0 > -5$.

3. Since -7 is to the left of 6, we have $-7 < 6$.

4. Since -4 is to the right of -19, we have $-4 > -19$.

5. The distance from -39 to 0 is 39, so $|-39| = 39$.

6. The distance from 23 to 0 is 23, so $|23| = 23$.

7. The distance from 0 to 0 is 0, so $|0| = 0$.

8. When $x = -72$, $-x = -(-72) = 72$.

9. When $x = 59$, $-(-x) = -(-59) = 59$.

10. $-14+5$ The absolute values are 14 and 5. The difference is 9. The negative number has the larger absolute value, so the answer is negative. $-14 + 5 = -9$

11. $-5 + (-6)$

Add the absolute values: $5 + 6 = 11$

Make the answer negative: $-5 + (-6) = -11$

12. $14 + (-8)$ The absolute values are 14 and 8. The difference is 6. The positive number has the larger absolute value, so the answer is positive. $14 + (-8) = 6$

13. $0 + (-24) = -24$

When 0 is added to any number, that number remains unchanged.

14. $17 - 29 = 17 + (-29) = -12$

15. $9 - (-14) = 9 + 14 = 23$

16. $-8 - (-7) = -8 + 7 = -1$

17. $-3 - (-10) = -3 + 10 = 7$

18.
$$-3 + 7 + (-8)$$
$$= -3 + (-8) + 7 \quad \text{Using a commutative law}$$
$$= -11 + 7$$
$$= -4$$

19.
$$8 - (-9) - 7 + 2$$
$$= 8 + 9 + (-7) + 2$$
$$= 19 + (-7) \qquad \text{Adding the positive numbers}$$
$$= 12$$

20. $-23 \cdot (-4) = 92$

21. $7(-12) = -84$

22.
$$2(-4)(-5)(-1)$$
$$= -8 \cdot 5 \quad \text{Multiplying the first two numbers and the last two numbers}$$
$$= -40$$

23. $15 \div (-5) = -3$ Check: $-3(-5) = 15$

24. $\dfrac{-55}{11} = -5$ Check: $-5 \cdot 11 = -55$

25. $\dfrac{0}{7} = 0$ Check: $0 \cdot 7 = 0$

26.
$$7 \div 1^2 \cdot (-3) - 4$$
$$= 7 \div 1 \cdot (-3) - 4 \quad \text{Evaluating the exponential expression}$$
$$= 7 \cdot (-3) - 4 \qquad \text{Dividing}$$
$$= -21 - 4 \qquad \text{Multiplying}$$
$$= -25 \qquad \text{Subtracting}$$

27.
$$(-3)|4 - 3^2| - 5$$
$$= (-3)|4 - 9| - 5$$
$$= (-3)|-5| - 5$$
$$= -3 \cdot 5 - 5$$
$$= -15 - 5$$
$$= -20$$

28. $3a + b = 3 \cdot 4 + (-5) = 12 + (-5) = 7$

29.
$$\frac{-x}{y} = \frac{-30}{5} = -6$$
$$\frac{x}{-y} = \frac{30}{-5} = -6$$
$$-\frac{x}{y} = -\frac{30}{5} = -6$$

30. $4(5x + 9) = 4 \cdot 5x + 4 \cdot 9 = 20x + 36$

31. $3(2a - 4b + 5) = 3 \cdot 2a - 3 \cdot 4b + 3 \cdot 5 = 6a - 12b + 15$

32. $5a + 12a = (5 + 12)a = 17a$

33. $-7x + 13x = (-7 + 13)x = 6x$

34.
$$9m + 14 - 12m - 8$$
$$= 9m - 12m + 14 - 8$$
$$= (9 - 12)m + (14 - 8)$$
$$= -3m + 6$$

35. $P = 2l + 2w = 2 \cdot 10 \text{ in.} + 2 \cdot 8 \text{ in.}$
$$= 20 \text{ in.} + 16 \text{ in.} = 36 \text{ in.}$$

36. $P = 4s = 4 \cdot 25 \text{ cm} = 100 \text{ cm}$

37.
$$x - 9 = -17$$
$$x - 9 + 9 = -17 + 9$$
$$x = -8$$
The solution is -8.

38.
$$-4t = 36$$
$$\frac{-4t}{-4} = \frac{36}{-4}$$
$$t = -9$$
The solution is -9.

39.
$$13 = -x$$
$$-1 \cdot 13 = -1 \cdot (-x)$$
$$-13 = x$$
The solution is -13.

40.
$$56 = 6x - 10$$
$$56 + 10 = 6x - 10 + 10$$
$$66 = 6x$$
$$\frac{66}{6} = \frac{6x}{6}$$
$$11 = x$$
The solution is 11.

41.
$$-x + 3 = -12$$
$$-x + 3 - 3 = -12 - 3$$
$$-x = -15$$
$$\frac{-x}{-1} = \frac{-15}{-1}$$
$$x = 15$$
The solution is 15.

42.
$$18 = 4 - 2x$$
$$18 - 4 = 4 - 2x - 4$$
$$14 = -2x$$
$$\frac{14}{-2} = \frac{-2x}{-2}$$
$$-7 = x$$

The solution is -7.

43. *Discussion and Writing Exercise.* Equivalent expressions are expressions that have the same value when evaluated for various replacements of the variable(s). Equivalent equations are equations that have the same solution(s).

44. *Discussion and Writing Exercise.* A number's absolute value is the number itself if the number is nonnegative, and the opposite of the number if the number is negative. In neither case is the result less than the number itself, so "no," a number's absolute value is never less than the number itself.

45. *Discussion and Writing Exercise.* The notation "$-x$" means "the opposite of x." If x is a negative number, then $-x$ is a positive number. For example, if $x = -2$, then $-x = 2$.

46. *Discussion and Writing Exercise.* The expressions $(a-b)^2$ and $(b-a)^2$ are equivalent for all choices of a and b because $a - b$ and $b - a$ are opposites. When opposites are raised to an even power, the results are the same.

47.
$$87 \div 3 \cdot 29^3 - (-6)^6 + 1957$$
$$= 87 \div 3 \cdot 24,389 - 46,656 + 1957$$
$$= 29 \cdot 24,389 - 46,656 + 1957$$
$$= 707,281 - 46,656 + 1957$$
$$= 660,625 + 1957$$
$$= 662,582$$

48.
$$1969 + (-8)^5 - 17 \cdot 15^3$$
$$= 1969 + (-32,768) - 17 \cdot 3375$$
$$= 1969 + (-32,768) - 57,375$$
$$= -30,799 - 57,375$$
$$= -88,174$$

49.
$$\frac{113 - 17^3}{15 + 8^3 - 507} = \frac{113 - 4913}{15 + 512 - 507}$$
$$= \frac{-4800}{527 - 507}$$
$$= \frac{-4800}{20}$$
$$= -240$$

50. $8 + x^3$ will be negative for all values of x for which x^3 is less than -8. Thus, $8 + x^3$ will be negative for $x < -2$.

51. $|x| > x$ for all negative values of x, or for $x < 0$.

Chapter 2 Test

1. The integer -542 corresponds to selling 542 fewer shirts than expected; the integer 307 corresponds to selling 307 more shirts than expected.

2. Since -14 is to the right of -21, we have $-14 > -21$.

3. The distance from -739 to 0 is 739, so $|-739| = 739$.

4. When $x = -19$, $-(-x) = -(-(-19)) = -(19) = -19$.

5. $6 + (-17)$ The absolute values are 6 and 17. The difference is 11. The negative number has the larger absolute value, so the answer is negative. $6 + (-17) = -11$

6. $-9 + (-12)$

Add the absolute values: $9 + 12 = 21$

Make the answer negative: $-9 + (-12) = -21$

7. $-8 + 17$ The absolute values are 8 and 17. The difference is 9. The positive number has the larger absolute value, so the answer is positive. $-8 + 17 = 9$

8. $0 - 12 = 0 + (-12) = -12$

When 0 is added to any number, that number remains unchanged.

9. $7 - 22 = 7 + (-22) = -15$

10. $-5 - 19 = -5 + (-19) = -24$

11. $-8 - (-27) = -8 + 27 = 19$

12.
$$31 - (-3) - 5 + 9$$
$$= 31 + 3 + (-5) + 9$$
$$= 43 + (-5) \quad \text{Adding the positive numbers}$$
$$= 38$$

13. $(-4)^3 = -4(-4)(-4) = 16(-4) = -64$

14. $27(-10) = -270$

15. $-9 \cdot 0 = 0$

16. $-72 \div (-9) = 8$ Check: $8(-9) = -72$

17. $\frac{-56}{7} = -8$ Check: $-8 \cdot 7 = -56$

18.
$$8 \div 2 \cdot 2 - 3^2 = 8 \div 2 \cdot 2 - 9$$
$$= 4 \cdot 2 - 9$$
$$= 8 - 9$$
$$= -1$$

19.
$$29 - (3 - 5)^2 = 29 - (-2)^2$$
$$= 29 - 4$$
$$= 25$$

20. We subtract the lower temperature from the higher temperature.

$$-67 - (-81) = -67 + 81 = 14$$

The average high temperature is 14°F higher than the average low temperature.

21. We subtract the final mark from the first mark.

$$8 - (-15) = 8 + 15 = 23$$

Thus, 23 min of tape were rewound.

22. $\dfrac{a-b}{6} = \dfrac{-8-10}{6} = \dfrac{-18}{6} = -3$

23. $7(2x + 3y - 1) = 7 \cdot 2x + 7 \cdot 3y - 7 \cdot 1 = 14x + 21y - 7$

24. $9x - 14 - 5x - 3 = 9x - 5x - 14 - 3$
$$= (9-5)x + (-14-3)$$
$$= 4x - 17$$

25. $\quad -7x = -35$
$$\frac{-7x}{-7} = \frac{-35}{-7}$$
$$x = 5$$

The solution is 5.

26. $\quad a + 9 = -3$
$$a + 9 - 9 = -3 - 9$$
$$a = -12$$

The solution is -12.

27. The amount of trim needed is given by the perimeter of the room, less the 3 ft width of the door, plus the lengths of the three sides of the door that will get trim.

Perimeter of room: $P = 2(l + w)$
$$= 2(14 \text{ ft} + 12 \text{ ft})$$
$$= 2(26 \text{ ft})$$
$$= 52 \text{ ft}$$

Subtract the width of the door: $52 \text{ ft} - 3 \text{ ft} = 49 \text{ ft}$

Trim on door: $7 \text{ ft} + 3 \text{ ft} + 7 \text{ ft} = 17 \text{ ft}$

Total length of trim: $49 \text{ ft} + 17 \text{ ft} = 66 \text{ ft}$

28. $\quad 9 - 5[x + 2(3 - 4x)] + 14$
$$= 9 - 5[x + 6 - 8x] + 14$$
$$= 9 - 5(-7x + 6) + 14$$
$$= 9 + 35x - 30 + 14$$
$$= 35x - 7$$

29. $\quad 15x + 3(2x - 7) - 9(4 + 5x)$
$$= 15x + 6x - 21 - 36 - 45x$$
$$= -24x - 57$$

30. $\quad 49 \cdot 14^3 \div 7^4 + 1926^2 \div 6^2$
$$= 49 \cdot 2744 \div 2401 + 3{,}709{,}476 \div 36$$
$$= 134{,}456 \div 2401 + 3{,}709{,}476 \div 36$$
$$= 56 + 3{,}709{,}476 \div 36$$
$$= 56 + 103{,}041$$
$$= 103{,}097$$

31. $\quad 3487 - 16 \div 4 \cdot 4 \div 2^8 \cdot 14^4$
$$= 3487 - 16 \div 4 \cdot 4 \div 256 \cdot 38{,}416$$
$$= 3487 - 4 \cdot 4 \div 256 \cdot 38{,}416$$
$$= 3487 - 16 \div 256 \cdot 38{,}416$$
$$= 3487 - 2401 \quad \text{Dividing and then multiplying}$$
$$= 1086$$

Chapter 3

Fraction Notation: Multiplication and Division

Exercise Set 3.1

1.
$1 \cdot 7 = 7$	$6 \cdot 7 = 42$
$2 \cdot 7 = 14$	$7 \cdot 7 = 49$
$3 \cdot 7 = 21$	$8 \cdot 7 = 56$
$4 \cdot 7 = 28$	$9 \cdot 7 = 63$
$5 \cdot 7 = 35$	$10 \cdot 7 = 70$

3.
$1 \cdot 20 = 20$	$6 \cdot 20 = 120$
$2 \cdot 20 = 40$	$7 \cdot 20 = 140$
$3 \cdot 20 = 60$	$8 \cdot 20 = 160$
$4 \cdot 20 = 80$	$9 \cdot 20 = 180$
$5 \cdot 20 = 100$	$10 \cdot 20 = 200$

5.
$1 \cdot 3 = 3$	$6 \cdot 3 = 18$
$2 \cdot 3 = 6$	$7 \cdot 3 = 21$
$3 \cdot 3 = 9$	$8 \cdot 3 = 24$
$4 \cdot 3 = 12$	$9 \cdot 3 = 27$
$5 \cdot 3 = 15$	$10 \cdot 3 = 30$

7.
$1 \cdot 12 = 12$	$6 \cdot 12 = 72$
$2 \cdot 12 = 24$	$7 \cdot 12 = 84$
$3 \cdot 12 = 36$	$8 \cdot 12 = 96$
$4 \cdot 12 = 48$	$9 \cdot 12 = 108$
$5 \cdot 12 = 60$	$10 \cdot 12 = 120$

9.
$1 \cdot 10 = 10$	$6 \cdot 10 = 60$
$2 \cdot 10 = 20$	$7 \cdot 10 = 70$
$3 \cdot 10 = 30$	$8 \cdot 10 = 80$
$4 \cdot 10 = 40$	$9 \cdot 10 = 90$
$5 \cdot 10 = 50$	$10 \cdot 10 = 100$

11.
$1 \cdot 25 = 25$	$6 \cdot 25 = 150$
$2 \cdot 25 = 50$	$7 \cdot 25 = 175$
$3 \cdot 25 = 75$	$8 \cdot 25 = 200$
$4 \cdot 25 = 100$	$9 \cdot 25 = 225$
$5 \cdot 25 = 125$	$10 \cdot 25 = 250$

13. We divide 61 by 3.

$$
\begin{array}{r}
2\,0 \\
3\,\overline{)\,6\,1} \\
6\,0 \\
\hline
1 \\
0 \\
\hline
1
\end{array}
$$

Since the remainder is not 0 we know that 61 is not divisible by 3.

15. We divide 527 by 7.

$$
\begin{array}{r}
7\,5 \\
7\,\overline{)\,5\,2\,7} \\
4\,9\,0 \\
\hline
3\,7 \\
3\,5 \\
\hline
2
\end{array}
$$

Since the remainder is not 0 we know that 527 is not divisible by 7.

17. We divide 8127 by 9.

$$
\begin{array}{r}
9\,0\,3 \\
9\,\overline{)\,8\,1\,2\,7} \\
8\,1\,0\,0 \\
\hline
2\,7 \\
2\,7 \\
\hline
0
\end{array}
$$

The remainder of 0 indicates that 8127 is divisible by 9.

19. Because $8 + 4 = 12$ and 12 is divisible by 3, 84 is divisible by 3.

21. 5553 is not divisible by 5 because the ones digit is neither 0 nor 5.

23. 671,500 is divisible by 10 because the ones digit is 0.

25. Because $1 + 7 + 7 + 3 = 18$ and 18 is divisible by 9, 1773 is divisible by 9.

27. 21,687 is not divisible by 2 because the ones digit is not even.

29. 32,109 is not divisible by 6 because it is not even.

31. 6825 is not divisible by 2 because the ones digit is not even.

Because $6 + 8 + 2 + 5 = 21$ and 21 is divisible by 3, 6825 is divisible by 3.

6825 is divisible by 5 because the ones digit is 5.

6825 is not divisible by 6 because it is not even.

Because $6 + 8 + 2 + 5 = 21$ and 21 is not divisible by 9, 6825 is not divisible by 9.

6825 is not divisible by 10 because the ones digit is not 0.

33. 119,117 is not divisible by 2 because the ones digit is not even.

Because $1 + 1 + 9 + 1 + 1 + 7 = 20$ and 20 is not divisible by 3, then 119,117 is not divisible by 3.

119,117 is not divisible by 5 because the ones digit is neither 0 nor 5.

119,117 is not divisible by 6 because it is not even.

Because $1 + 1 + 9 + 1 + 1 + 7 = 20$ and 20 is not divisible by 9, then 119,117 is not divisible by 9.

119,117 is not divisible by 10 because the ones digit is not 0.

35. 127,575 is not divisible by 2 because the ones digit is not even.

Because $1 + 2 + 7 + 5 + 7 + 5 = 27$ and 27 is divisible by 3, then 127,575 is divisible by 3.

127,575 is divisible by 5 because the ones digit is 5.

127,575 is not divisible by 6 because it is not even.

Because $1 + 2 + 7 + 5 + 7 + 5 = 27$ and 27 is divisible by 9, then 127,575 is divisible by 9.

127,575 is not divisible by 10 because the ones digit is not 0.

37. 9360 is divisible by 2 because the ones digit is even.

Because $9 + 3 + 6 + 0 = 18$ and 18 is divisible by 3, 9360 is divisible by 3.

9360 is divisible by 5 because the ones digit is 0.

We saw above that 9360 is even and that it is divisible by 3, so it is divisible by 6.

Because $9 + 3 + 6 + 0 = 18$ and 18 is divisible by 9, 9360 is divisible by 9.

9360 is divisible by 10 because the ones digit is 0.

39. A number is divisible by 3 if the sum of the digits is divisible by 3.

46 is not divisible by 3 because $4 + 6 = 10$ and 10 is not divisible by 3.

224 is not divisible by 3 because $2 + 2 + 4 = 8$ and 8 is not divisible by 3.

19 is not divisible by 3 because $1 + 9 = 10$ and 10 is not divisible by 3.

555 is divisible by 3 because $5 + 5 + 5 = 15$ and 15 is divisible by 3.

300 is divisible by 3 because $3 + 0 + 0 = 3$ and 3 is divisible by 3.

36 is divisible by 3 because $3 + 6 = 9$ and 9 is divisible by 3.

45,270 is divisible by 3 because $4 + 5 + 2 + 7 + 0 = 18$ and 18 is divisible by 3.

4444 is not divisible by 3 because $4 + 4 + 4 + 4 = 16$ and 16 is not divisible by 3.

85 is not divisible by 3 because $8 + 5 = 13$ and 13 is not divisible by 3.

711 is divisible by 3 because $7 + 1 + 1 = 9$ and 9 is divisible by 3.

13,251 is divisible by 3 because $1 + 3 + 2 + 5 + 1 = 12$ and 12 is divisible by 3.

254,765 is not divisible by 3 because $2+5+4+7+6+5 = 29$ and 29 is not divisible by 3.

256 is not divisible by 3 because $2 + 5 + 6 = 13$ and 13 is not divisible by 3.

8064 is divisible by 3 because $8 + 0 + 6 + 4 = 18$ and 18 is divisible by 3.

1867 is not divisible by 3 because $1 + 8 + 6 + 7 = 22$ and 22 is not divisible by 3.

21,568 is not divisible by 3 because $2 + 1 + 5 + 6 + 8 = 22$ and 22 is not divisible by 3.

41. A number is divisible by 10 if its ones digit is 0.

Of the numbers under consideration, only 300 and 45,270 have one digits of 0. Therefore, only 300 and 45,270 are divisible by 10.

43. For a number to be divisible by 6, the sum of the digits must be divisible by 3 and the ones digit must be 0, 2, 4, 6 or 8 (even). It is most efficient to determine if the ones digit is even first and then, if so, to determine if the sum of the digits is divisible by 3.

46 is not divisible by 6 because 46 is not divisible by 3.

$$4 + 6 = 10$$
$$\uparrow$$
$$\text{Not divisible by 3}$$

224 is not divisible by 6 because 224 is not divisible by 3.

$$2 + 2 + 4 = 8$$
$$\uparrow$$
$$\text{Not divisible by 3}$$

19 is not divisible by 6 because 19 is not even.

$$19$$
$$\uparrow$$
$$\text{Not even}$$

555 is not divisible by 6 because 555 is not even.

$$555$$
$$\uparrow$$
$$\text{Not even}$$

300 is divisible by 6.

$$300 \qquad 3 + 0 + 0 = 3$$
$$\uparrow \qquad\qquad\quad \uparrow$$
$$\text{Even} \qquad\quad \text{Divisible by 3}$$

36 is divisible by 6.

$$36 \qquad 3 + 6 = 9$$
$$\uparrow \qquad\qquad \uparrow$$
$$\text{Even} \qquad \text{Divisible by 3}$$

45,270 is divisible by 6.

$$45,270 \qquad 4 + 5 + 2 + 7 + 0 = 18$$
$$\uparrow \qquad\qquad\qquad\qquad \uparrow$$
$$\text{Even} \qquad\qquad\quad \text{Divisible by 3}$$

4444 is not divisible by 6 because 4444 is not divisible by 3.

$$4 + 4 + 4 + 4 = 16$$
$$\uparrow$$
$$\text{Not divisible by 3}$$

85 is not divisible by 6 because 85 is not even.

$$85$$
$$\uparrow$$
$$\text{Not even}$$

711 is not divisible by 6 because 711 is not even.

$$711$$
$$\uparrow$$
$$\text{Not even}$$

13,251 is not divisible by 6 because 13,251 is not even.

$$13,251$$
$$\uparrow$$
$$\text{Not even}$$

254,765 is not divisible by 6 because 254,765 is not even.

$$254{,}765$$
$$\uparrow$$
Not even

256 is not divisible by 6 because 256 is not divisible by 3.

$$2 + 5 + 6 = 13$$
$$\uparrow$$
Not divisible by 3

8064 is divisible by 6.

$$8064 \qquad 8 + 0 + 6 + 4 = 18$$
$$\uparrow \qquad\qquad\qquad \uparrow$$
Even   Divisible by 3

1867 is not divisible by 6 because 1867 is not even.

$$1867$$
$$\uparrow$$
Not even

21,568 is not divisible by 6 because 21,568 is not divisible by 3.

$$2 + 1 + 5 + 6 + 8 = 22$$
$$\uparrow$$
Not divisible by 3

45. A number is divisible by 2 if its <u>ones digit</u> is even.

5<u>6</u> is divisible by 2 because <u>6</u> is even.
32<u>4</u> is divisible by 2 because <u>4</u> is even.
78<u>4</u> is divisible by 2 because <u>4</u> is even.
55,55<u>5</u> is not divisible by 2 because <u>5</u> is not even.
20<u>0</u> is divisible by 2 because <u>0</u> is even.
4<u>2</u> is divisible by 2 because <u>2</u> is even.
50<u>1</u> is not divisible by 2 because <u>1</u> is not even.
300<u>9</u> is not divisible by 2 because <u>9</u> is not even.

7<u>5</u> is not divisible by 2 because <u>5</u> is not even.
81<u>2</u> is divisible by 2 because <u>2</u> is even.
234<u>5</u> is not divisible by 2 because <u>5</u> is not even.
200<u>1</u> is not divisible by 2 because <u>1</u> is not even.
3<u>5</u> is not divisible by 2 because <u>5</u> is not even.
40<u>2</u> is divisible by 2 because <u>2</u> is even.
111,11<u>1</u> is not divisible by 2 because <u>1</u> is not even.
100<u>5</u> is not divisible by 2 because <u>5</u> is not even.

47. A number is divisible by 5 if the ones digit is 0 or 5.

5<u>6</u> is not divisible by 5 because the ones digit (6) is not 0 or 5.

32<u>4</u> is not divisible by 5 because the ones digit (4) is not 0 or 5.

78<u>4</u> is not divisible by 5 because the ones digit (4) is not 0 or 5.

55,55<u>5</u> is divisible by 5 because the ones digit is 5.

20<u>0</u> is divisible by 5 because the ones digit is 0.

4<u>2</u> is not divisible by 5 because the ones digit (2) is not 0 or 5.

50<u>1</u> is not divisible by 5 because the ones digit (1) is not 0 or 5.

300<u>9</u> is not divisible by 5 because the ones digit (9) is not 0 or 5.

7<u>5</u> is divisible by 5 because the ones digit is 5.

81<u>2</u> is not divisible by 5 because the ones digit (2) is not 0 or 5.

234<u>5</u> is divisible by 5 because the ones digit is 5.

200<u>1</u> is not divisible by 5 because the ones digit (1) is not 0 or 5.

3<u>5</u> is divisible by 5 because the ones digit is 5.

40<u>2</u> is not divisible by 5 because the ones digit (2) is not 0 or 5.

111,11<u>1</u> is not divisible by 5 because the ones digit (1) is not 0 or 5.

100<u>5</u> is divisible by 5 because the ones digit is 5.

49. A number is divisible by 9 if the sum of the digits is divisible by 9.

56 is not divisible by 9 because $5 + 6 = 11$ and 11 is not divisible by 9.

324 is divisible by 9 because $3 + 2 + 4 = 9$ and 9 is divisible by 9.

784 is not divisible by 9 because $7 + 8 + 4 = 19$ and 19 is not divisible by 9.

55,555 is not divisible by 9 because $5 + 5 + 5 + 5 + 5 = 25$ and 25 is not divisible by 9.

200 is not divisible by 9 because $2 + 0 + 0 = 2$ and 2 is not divisible by 9.

42 is not divisible by 9 because $4 + 2 = 6$ and 6 is not divisible by 9.

501 is not divisible by 9 because $5 + 0 + 1 = 6$ and 6 is not divisible by 9.

3009 is not divisible by 9 because $3 + 0 + 0 + 9 = 12$ and 12 is not divisible by 9.

75 is not divisible by 9 because $7 + 5 = 12$ and 12 is not divisible by 9.

812 is not divisible by 9 because $8 + 1 + 2 = 11$ and 11 is not divisible by 9.

2345 is not divisible by 9 because $2 + 3 + 4 + 5 = 14$ and 14 is not divisible by 9.

2001 is not divisible by 9 because $2 + 0 + 0 + 1 = 3$ and 3 is not divisible by 9.

35 is not divisible by 9 because $3 + 5 = 8$ and 8 is not divisible by 9.

402 is not divisible by 9 because $4 + 0 + 2 = 6$ and is not divisible by 9.

111,111 is not divisible by 9 because $1 + 1 + 1 + 1 + 1 + 1 = 6$ and 6 is not divisible by 9.

1005 is not divisible by 9 because $1 + 0 + 0 + 5 = 6$ and 6 is not divisible by 9.

51. Discussion and Writing Exercise

53.
$$16 \cdot t = 848$$
$$\frac{16 \cdot t}{16} = \frac{848}{16} \quad \text{Dividing by 16 on both sides}$$
$$t = 53$$

The solution is 53.

55.
$$23 + x = 15$$
$$23 + x - 23 = 15 - 23 \quad \text{Subtracting 23 on both sides}$$
$$x = -8$$

The solution is -8.

57. *Familiarize*. This is a multistep problem. Find the total cost of the sweaters and the total cost of the jackets and then find the sum of the two.

We let $s =$ the total cost of the sweaters and $t =$ the total cost of the jackets.

Translate. We write two equations.

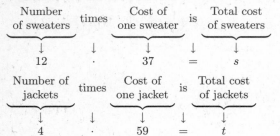

Solve. We carry out the multiplication.

$$12 \cdot 37 = s$$
$$444 = s \quad \text{Doing the multiplication}$$

The total cost of the 12 sweaters is \$444.

$$4 \cdot 59 = t$$
$$236 = t \quad \text{Doing the multiplication}$$

The total cost of the 4 jackets is \$236.

Now we find the total amount spent. We let $a =$ this amount.

$$
\underbrace{\text{Total cost}}_{\text{of sweaters}} \; \underbrace{\text{plus}}_{} \; \underbrace{\text{Total cost}}_{\text{of jackets}} \; \underbrace{\text{is}}_{} \; \underbrace{\text{Total amount}}_{\text{spent}}
$$
$$
444 \quad + \quad 236 \quad = \quad a
$$

To solve the equation, carry out the addition.

$$
\begin{array}{r}
4\,4\,4 \\
+\,2\,3\,6 \\
\hline
6\,8\,0
\end{array}
$$

Check. We can repeat the calculations. The answer checks.

State. The total cost is \$680.

59. $5^3 = 5 \cdot 5 \cdot 5 = 125$

61. $4^5 = 4 \cdot 4 \cdot 4 \cdot 4 \cdot 4 = 1024$

63. $\underbrace{9 \cdot 9 \cdot 9 \cdot 9 \cdot 9}_{\text{5 factors}} = 9^5$

65. Discussion and Writing Exercise

67. When we use a calculator to divide the largest five-digit number, 99,999, by 47 we get 2127.638298. This tells us that 99,999 is not divisible by 47 but that 2127×47, or 99,969, is divisible by 47 and that it is the largest such five-digit number.

69. We list multiples of 2, 3, and 5 and find the smallest number that is on all 3 lists.

Multiples of 2: $2, 4, 6, 8, 10, 12, 14, 16, 18, 20, 22, 24, 26,$ $28, \underline{30}, 32, \cdots$

Multiples of 3: $3, 6, 9, 12, 15, 18, 21, 24, 27, \underline{30}, 33, \cdots$

Multiples of 5: $5, 10, 15, 20, 25, \underline{30}, 35, \cdots$

The smallest number that is simultaneously a multiple of 2, 3, and 5 is 30.

71. We list multiples of 4, 6, and 10 and find the smallest number that is on all 3 lists.

Multiples of 4: $4, 8, 12, 16, 20, 24, 28, 32, 36, 40, 44, 48,$ $52, 56, \underline{60}, 64, \cdots$

Multiples of 6: $6, 12, 18, 24, 30, 36, 42, 48, 54, \underline{60}, 66, \cdots$

Multiples of 10: $10, 20, 30, 40, 50, \underline{60}, 70, \cdots$

The smallest number that is simultaneously a multiple of 4, 6, and 10 is 60.

73. First note that 85 is a multiple of 17 ($85 = 5 \cdot 17$). Thus, any multiple of 85 will also be a multiple of 17. Then, using a calculator, we list multiples of 43 and 85 and find the smallest number that is on both lists. We find that this number is 3655.

75. First note that 120 is a multiple of 30 ($120 = 4 \cdot 30$). Thus, any multiple of 120 is also a multiple of 30. Now we list multiples of 70 and 120 and find the smallest number that is on both lists.

Multiples of 70: $70, 140, 210, 280, 350, 420, 490, 560, 630,$ $700, 770, \underline{840}, 910, \ldots$

Multiples of 120: $120, 240, 360, 480, 600, 720, \underline{840}, 960, \ldots$

The smallest number that is simultaneously a multiple of 30, 70, and 120 is 840.

77. Discussion and Writing Exercise

79. The sum of the given digits is $9 + 5 + 8$, or 22. If the number is divisible by 99, it is also divisible by 9 since 99 is divisible by 9. The smallest number that is divisible by 9 and also greater than 22 is 27. Then the sum of the two missing digits must be at least $27 - 22$, or 5. We try various combinations of two digits whose sum is 5, using a calculator to divide the resulting number by 99:

95,058 is not divisible by 99.

95,148 is not divisible by 99.

95,238 is divisible by 99.

Thus, the missing digits are 2 and 3 and the number is 95,238.

Exercise Set 3.2

1. Since 18 is even, we know that 2 is a factor. Since the sum of the digits is 9 and 9 is divisible by both 3 and 9, we know that 3 and 9 are both factors. Also, since 2 and 3 are factors, 6 is a factor as well. We write a list of factorizations.

$18 = 1 \cdot 18 \qquad 18 = 3 \cdot 6$
$18 = 2 \cdot 9$

Factors: 1, 2, 3, 6, 9, 18

3. Since 54 is even, we know that 2 is a factor. Since the sum of the digits is 9 and 9 is divisible by both 3 and 9, we know that 3 and 9 are both factors. Also, since 2 and 3 are factors, 6 is a factor as well. We write a list of factorizations.

$54 = 1 \cdot 54$ $54 = 3 \cdot 18$
$54 = 2 \cdot 27$ $54 = 6 \cdot 9$

Factors: 1, 2, 3, 6, 9, 18, 27, 54

5. Since 9 is divisible by 3, we know that 3 is a factor. We write a list of factorizations.

$9 = 1 \cdot 9$
$9 = 3 \cdot 3$

Factors: 1, 3, 9

7. The number 13 is prime. It has only 1 and 13 as factors.

9. The number 17 is prime. It has only the factors 1 and 17.

11. The number 22 has factors 1, 2, 11, and 22. Since it has at least one factor other than itself and 1, it is composite.

13. The number 48 has factors 1, 2, 3, 4, 6, 8, 12, 16, 24, and 48. Since it has at least one factor other than itself and 1, it is composite.

15. The number 53 is prime. It has only the factors 1 and 53.

17. 1 is neither prime nor composite.

19. The number 81 has factors 1, 3, 9, 27, and 81.

Since it has at least one factor other than itself and 1, it is composite.

21. The number 47 is prime. It has only the factors 1 and 47.

23. The number 29 is prime. It has only the factors 1 and 29.

25.
$$\begin{array}{r} 3 \\ 3\,\overline{)\,9} \\ 3\,\overline{)\,2\,7} \end{array}$$ $\leftarrow$ 3 is prime.

$27 = 3 \cdot 3 \cdot 3$

27.
$$\begin{array}{r} 7 \\ 2\,\overline{)\,1\,4} \end{array}$$ $\leftarrow$ 7 is prime.

$14 = 2 \cdot 7$

29.
$$\begin{array}{r} 5 \\ 2\,\overline{)\,1\,0} \\ 2\,\overline{)\,2\,0} \\ 2\,\overline{)\,4\,0} \\ 2\,\overline{)\,8\,0} \end{array}$$ $\leftarrow$ 5 is prime.

$80 = 2 \cdot 2 \cdot 2 \cdot 2 \cdot 5$

We can also use a factor tree.

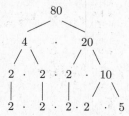

31.
$$\begin{array}{r} 5 \\ 5\,\overline{)\,2\,5} \end{array}$$ $\leftarrow$ 5 is prime. (25 is not divisible by 2 or 3. We move to 5.)

$25 = 5 \cdot 5$

33.
$$\begin{array}{r} 3\,1 \\ 2\,\overline{)\,6\,2} \end{array}$$ $\leftarrow$ 31 is prime.

$62 = 2 \cdot 31$

35.
$$\begin{array}{r} 5 \\ 5\,\overline{)\,2\,5} \\ 2\,\overline{)\,5\,0} \\ 2\,\overline{)\,1\,0\,0} \end{array}$$ $\leftarrow$ 5 is prime.

$100 = 2 \cdot 2 \cdot 5 \cdot 5$

We can also use a factor tree.

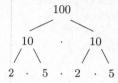

37.
$$\begin{array}{r} 1\,3 \\ 1\,1\,\overline{)\,1\,4\,3} \end{array}$$ $\leftarrow$ 13 is prime. (143 is not divisible by 2, 3, 5, or 7. We move to 11.)

$143 = 11 \cdot 13$

39.
$$\begin{array}{r} 1\,1 \\ 1\,1\,\overline{)\,1\,2\,1} \end{array}$$ $\leftarrow$ 11 is prime. (121 is not divisible by 2, 3, 5, or 7. We move to 11.)

$121 = 11 \cdot 11$

41.
$$\begin{array}{r} 1\,3 \\ 7\,\overline{)\,9\,1} \\ 3\,\overline{)\,2\,7\,3} \end{array}$$ $\leftarrow$ 13 is prime. (273 is not divisible by 2. We move to 3.)

$273 = 3 \cdot 7 \cdot 13$

43.
$$\begin{array}{r} 7 \\ 5\,\overline{)\,3\,5} \\ 5\,\overline{)\,1\,7\,5} \end{array}$$ $\leftarrow$ 7 is prime. (175 is not divisible by 2 or 3. We move to 5.)

$175 = 5 \cdot 5 \cdot 7$

45.
$$\begin{array}{r} 1\,9 \\ 1\,1\,\overline{)\,2\,0\,9} \end{array}$$ $\leftarrow$ 19 is prime. (209 is not divisible by 2, 3, 5, or 7. We move to 11.)

$209 = 11 \cdot 19$

47.
$$\begin{array}{r} 4\,3 \\ 2\,\overline{)\,8\,6} \end{array}$$ $\leftarrow$ 43 is prime.

$86 = 2 \cdot 43$

49.
$$\begin{array}{r} 3\,1 \\ 7\,\overline{)\,2\,1\,7} \end{array}$$ $\leftarrow$ 31 is prime. (217 is not divisible by 2, 3, or 5. We move to 7.)

$217 = 7 \cdot 31$

51.
$$
\begin{array}{r}
7 \quad\leftarrow\;7\text{ is prime.}\\
5\,\overline{)\,35}\\
5\,\overline{)\,175}\\
2\,\overline{)\,875}\\
2\,\overline{)\,1\,750}\\
2\,\overline{)\,3\,500}\\
2\,\overline{)\,7\,000}
\end{array}
$$
$7000 = 2 \cdot 2 \cdot 2 \cdot 5 \cdot 5 \cdot 5 \cdot 7$

53.
$$
\begin{array}{r}
1\,7 \quad\leftarrow\;17\text{ is prime.}\\
11\,\overline{)\,187}\\
3\,\overline{)\,561}\\
2\,\overline{)\,1\,122}
\end{array}
$$
$1122 = 2 \cdot 3 \cdot 11 \cdot 17$

55. Since 100 is even we know that 2 is a factor. Using other tests for divisibility, we determine that 5 and 10 are also factors. We write a list of factorizations.

$100 = 1 \cdot 100 \qquad 100 = 5 \cdot 20$
$100 = 2 \cdot 50 \qquad 100 = 10 \cdot 10$
$100 = 4 \cdot 25$

Factors: 1, 2, 4, 5, 10, 20, 25, 50, 100

57. Using tests for divisibility we determine that 5 is a factor. We write a list of factorizations.

$385 = 1 \cdot 385 \qquad 385 = 7 \cdot 55$
$385 = 5 \cdot 77 \qquad 385 = 11 \cdot 35$

Factors: 1, 5, 7, 11, 35, 55, 77, 385

59. Using tests for divisibility we determine that 3 and 9 are factors. We write a list of factorizations.

$81 = 1 \cdot 81 \qquad 81 = 9 \cdot 9$
$81 = 3 \cdot 27$

Factors: 1, 3, 9, 27, 81

61. Using tests for divisibility we determine that 3, 5, and 9 are factors. We write a list of factorizations.

$225 = 1 \cdot 225 \qquad 225 = 9 \cdot 25$
$225 = 3 \cdot 75 \qquad 225 = 15 \cdot 15$
$225 = 5 \cdot 45$

Factors: 1, 3, 5, 9, 15, 25, 45, 75, 225

63. Discussion and Writing Exercise

65. $-2 \cdot 13 = -26$ (The signs are different, so the answer is negative.)

67. $-17 + 25$ The absolute values are 17 and 25. The difference is 8. The positive number has the larger absolute value, so the answer is positive. $-17 + 25 = 8$

69. $53 \div 53 = 1$

71. $0 \div 22 = 0$ (0 divided by a nonzero number is 0.)

73. $-42 \div 1 = -42$ (The signs are different, so the answer is negative.)

75. Discussion and Writing Exercise

77. Discussion and Writing Exercise

79. Using a calculator to perform successive divisions by prime numbers, we find that $102,971 = 11 \cdot 11 \cdot 23 \cdot 37$.

81. Using a calculator to perform successive divisions by prime numbers, we find that $168,840 = 2 \cdot 2 \cdot 2 \cdot 3 \cdot 3 \cdot 5 \cdot 7 \cdot 67$

83. Answers may vary. One arrangement is a 3-dimensional rectangular array consisting of 2 tiers of 12 objects each where each tier consists of a rectangular array of 4 rows with 3 objects each.

85. The factors of 63 whose sum is 16 are 7 and 9.

The factors of 36 whose sum is 20 are 2 and 18.

The factors of 72 whose sum is 38 are 2 and 36.

The factors of 140 whose sum is 24 are 10 and 14.

The factors of 96 whose sum is 20 are 8 and 12.

The factors of 48 whose sum is 14 are 6 and 8.

The factors of 168 whose sum is 29 are 8 and 21.

The factors of 110 whose sum is 21 are 10 and 11.

The factors of 90 whose sum is 19 are 9 and 10.

The factors of 432 whose sum is 42 are 18 and 24.

The factors of 63 whose sum is 24 are 3 and 21.

Exercise Set 3.3

---

1. The top number is the numerator, and the bottom number is the denominator.

$$\frac{3}{4} \quad\begin{array}{l}\leftarrow\text{ Numerator}\\\leftarrow\text{ Denominator}\end{array}$$

3. $\dfrac{7}{-9} \quad\begin{array}{l}\leftarrow\text{ Numerator}\\\leftarrow\text{ Denominator}\end{array}$

5. $\dfrac{2x}{3z} \quad\begin{array}{l}\leftarrow\text{ Numerator}\\\leftarrow\text{ Denominator}\end{array}$

7. The dollar is divided into 4 parts of the same size, and 2 of them are shaded. This is $2 \cdot \dfrac{1}{4}$ or $\dfrac{2}{4}$. Thus, $\dfrac{2}{4}$ (two-fourths) of the dollar is shaded.

9. The yard is divided into 8 parts of the same size, and 1 of them is shaded. Thus, $\dfrac{1}{8}$ (one-eighth) of the yard is shaded.

11. The window is divided into 9 parts of the same size, and 4 of them are shaded. Thus, $\dfrac{4}{9}$ (four-ninths) of the window is shaded.

13. The acre is divided into 4 parts of the same size, and 3 of them are shaded. This is $3 \cdot \dfrac{1}{4}$ or $\dfrac{3}{4}$ of the acre.

15. The pie is divided into 8 equal parts. The unit is $\dfrac{1}{8}$. The denominator is 8. We have 4 parts shaded. This tells us that the numerator is 4. Thus, $\dfrac{4}{8}$ is shaded.

17. The square mile is divided into 12 equal parts. The unit is $\dfrac{1}{12}$. The denominator is 12. We have 6 parts shaded. This tells us that the numerator is 6. Thus, $\dfrac{6}{12}$ is shaded.

19. Each inch on the ruler is divided into 16 equal parts. The shading extends to the 12th mark, so $\frac{12}{16}$ is shaded.

21. Each inch on the ruler is divided into 16 equal parts. The shading extends to the 38th mark, so $\frac{7}{16}$ is shaded.

23. There are 8 circles, and 5 are shaded. Thus, $\frac{5}{8}$ of the circles are shaded.

25. There are 7 objects in the set, and 4 of the objects are shaded. Thus, $\frac{4}{7}$ of the set is shaded.

27. The gas gauge is divided into 8 equal parts.

    a) The needle is 2 marks from the E (empty) mark, so the amount of gas in the tank is $\frac{2}{8}$ of a full tank.

    b) The needle is 6 marks from the F (full) mark, so $\frac{6}{8}$ of a full tank of gas has been burned.

29. The gas gauge is divided into 8 equal parts.

    a) The needle is 3 marks from the E (empty) mark, so the amount of gas in the tank is $\frac{3}{8}$ of a full tank.

    b) The needle is 5 marks from the F (full) mark, so $\frac{5}{8}$ of a full tank of gas has been burned.

31. We have 2 gold bars, each divided into 8 parts. We take 9 of those parts. This is $9 \cdot \frac{1}{8}$, or $\frac{9}{8}$. Thus, $\frac{9}{8}$ of a gold bar is shaded.

33. We have 3 feet, each divided into 6 equal parts. We take 7 of those parts. This is $7 \cdot \frac{1}{6}$, or $\frac{7}{6}$. Thus, $\frac{7}{6}$ of a foot is shaded.

35. We have 2 spools, each divided into 5 parts. We take 7 of those parts. This is $7 \cdot \frac{1}{5}$, or $\frac{7}{5}$. Thus, $\frac{7}{5}$ of a spool is shaded.

37. a) The ratio is $\frac{390}{13}$.

    b) The ratio is $\frac{13}{390}$.

39. The ratio is $\frac{850}{1000}$.

41. a) There are 7 people in the set and 3 are women, so the desired ratio is $\frac{3}{7}$.

    b) There are 3 women and 4 men, so the ratio of women to men is $\frac{3}{4}$.

    c) There are 7 people in the set and 4 are men, so the desired ratio is $\frac{4}{7}$.

    d) There are 4 men and 3 women, so the ratio of men to women is $\frac{4}{3}$.

43. a) In Orlando there are 35 police officers per 10,000 residents, so the ratio is $\frac{35}{10,000}$.

    b) In New York there are 50 police officers per 10,000 residents, so the ratio is $\frac{50}{10,000}$.

    c) In Detroit there are 44 police officers per 10,000 residents, so the ratio is $\frac{44}{10,000}$.

    d) In Washington there are 63 police officers per 10,000 residents, so the ratio is $\frac{63}{10,000}$.

    e) In St. Louis there are 43 police officers per 10,000 residents, so the ratio is $\frac{43}{10,000}$.

    f) In Santa Fe there are 21 police officers per 10,000 residents, so the ratio is $\frac{21}{10,000}$.

45. Remember: $\frac{0}{n} = 0$, for any integer n that is not 0.

$$\frac{0}{17} = 0$$

Think of dividing an object into 17 parts and taking none of them. We get 0.

47. Remember: $\frac{n}{1} = n$.

$$\frac{15}{1} = 15$$

Think of taking 15 objects and dividing them into 1 part. (We do not divide them.) We have 15 objects.

49. Remember: $\frac{n}{n} = 1$, for any integer n that is not 0.

$$\frac{20}{20} = 1$$

If we divide an object into 20 parts and take 20 of them, we get all of the object (1 whole object).

51. Remember: $\frac{n}{n} = 1$, for any integer n that is not 0.

$$\frac{-14}{-14} = 1$$

53. Remember: $\frac{0}{n} = 0$, for any integer n that is not 0.

$$\frac{0}{-234} = 0$$

55. Remember: $\frac{n}{n} = 1$, for any integer n that is not 0.

$$\frac{3n}{3n} = 1$$

57. Remember: $\frac{n}{n} = 1$, for any integer n that is not 0.

$$\frac{9x}{9x} = 1$$

59. Remember: $\frac{n}{1} = n$

$$\frac{-63}{1} = -63$$

61. Remember: $\frac{0}{n} = 0$, for any integer n that is not 0.

$$\frac{0}{2a} = 0$$

63. Remember: $\frac{n}{0}$ is not defined.

$\frac{52}{0}$ is undefined.

65. Remember: $\frac{n}{1} = n$

$$\frac{7n}{1} = 7n$$

67. $\frac{6}{7-7} = \frac{6}{0}$

Remember: $\frac{n}{0}$ is not defined. Thus, $\frac{6}{7-7}$ is undefined.

69. Discussion and Writing Exercise

71. $-7(30) = -210$
(The signs are different, so the answer is negative.)

73. $(-71)(-12)0 = -71 \cdot 0 = 0$
(We might have observed at the outset that the answer is 0 since one of the factors is 0.)

75. *Familiarize*. Let c = the number of excess calories in the Burger King meal.

Translate. We can think of this as a "how much more" situation.

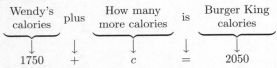

Solve. We subtract 1750 on both sides of the equation.
$$1750 + c = 2050$$
$$1750 + c - 1750 = 2050 - 1750$$
$$c = 300$$

Check. Since $1750 + 300 = 2050$, the answer checks.

State. The Burger King meal has 300 more calories than the Wendy's meal.

77. Discussion and Writing Exercise

79. $365 = 52 \cdot 7 + 1$, so in one year there are 52 full weeks plus one additional day. Since 2006 began on a Sunday, the additional day is not a Monday. (It is a Sunday.) Thus, of the 365 days in 2006, 52 were Mondays, so $\frac{52}{365}$ were Mondays.

81. The surface of the earth is divided into $3+1$, or 4 parts. Three of them are taken up by water, so $\frac{3}{4}$ is water. One of them is land, so $\frac{1}{4}$ is land.

83. Since the denominators are all the same, the numerators tell us the relative sizes of the fractions. The smallest fraction is the one with the smallest numerator and so on. Accordingly, we label the smallest sector "No working

TV." Then the next smallest sector is labeled "One TV," the next smallest "Two TVs," and the largest sector is labeled "Three or more TVs."

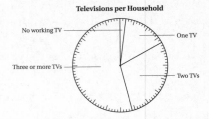

85. We can think of the object as being divided into 6 sections, each the size of one of the sections shaded. Since 2 sections are shaded, $\frac{2}{6}$ of the object is shaded. We could also express this as $\frac{1}{3}$.

87. We can think of the object as being divided into 16 sections, each the size of one of the sections shaded. Since a portion of the object that is the equivalent of 6 sections is shaded, $\frac{6}{16}$ of the object is shaded. We could also express this as $\frac{3}{8}$.

Exercise Set 3.4

1. $3 \cdot \frac{1}{8} = \frac{3 \cdot 1}{8} = \frac{3}{8}$

3. $(-5) \times \frac{1}{6} = \frac{-5 \times 1}{6} = \frac{-5}{6}$, or $-\frac{5}{6}$

5. $\frac{2}{3} \cdot 7 = \frac{2 \cdot 7}{3} = \frac{14}{3}$

7. $(-1)\frac{7}{9} = \frac{(-1)7}{9} = \frac{-7}{9}$, or $-\frac{7}{9}$

9. $\frac{5}{6} \cdot x = \frac{5 \cdot x}{6} = \frac{5x}{6}$

11. $\frac{2}{5}(-3) = \frac{2(-3)}{5} = \frac{-6}{5}$, or $-\frac{6}{5}$

13. $a \cdot \frac{2}{7} = \frac{a \cdot 2}{7} = \frac{2a}{7}$

15. $17 \times \frac{m}{6} = \frac{17 \times m}{6} = \frac{17m}{6}$

17. $-3 \cdot \frac{-2}{5} = \frac{-3}{1} \cdot \frac{-2}{5} = \frac{-3(-2)}{1 \cdot 5} = \frac{6}{5}$

19. $-\frac{2}{7}(-x) = \frac{-2}{7} \cdot \frac{-x}{1} = \frac{-2(-x)}{7 \cdot 1} = \frac{2x}{7}$

21. $\frac{1}{3} \cdot \frac{1}{5} = \frac{1 \cdot 1}{3 \cdot 5} = \frac{1}{15}$

23. $\left(-\frac{1}{4}\right) \times \frac{1}{10} = -\frac{1 \times 1}{4 \times 10} = -\frac{1}{40}$, or $\frac{-1}{40}$

25. $\dfrac{2}{3} \times \dfrac{1}{5} = \dfrac{2 \times 1}{3 \times 5} = \dfrac{2}{15}$

27. $\dfrac{2}{y} \cdot \dfrac{x}{9} = \dfrac{2 \cdot x}{y \cdot 9} = \dfrac{2x}{9y}$

29. $\left(-\dfrac{3}{4}\right)\left(-\dfrac{3}{4}\right) = \dfrac{(-3)(-3)}{4 \cdot 4} = \dfrac{9}{16}$

31. $\dfrac{2}{3} \cdot \dfrac{7}{13} = \dfrac{2 \cdot 7}{3 \cdot 13} = \dfrac{14}{39}$

33. $\dfrac{1}{10}\left(\dfrac{-3}{5}\right) = \dfrac{1(-3)}{10 \cdot 5} = \dfrac{-3}{50}$, or $-\dfrac{3}{50}$

35. $\dfrac{7}{8} \cdot \dfrac{a}{8} = \dfrac{7 \cdot a}{8 \cdot 8} = \dfrac{7a}{64}$

37. $\dfrac{1}{y} \cdot \dfrac{1}{100} = \dfrac{1 \cdot 1}{y \cdot 100} = \dfrac{1}{100y}$

39. $\dfrac{-14}{15} \cdot \dfrac{13}{19} = \dfrac{-14 \cdot 13}{15 \cdot 19} = \dfrac{-182}{285}$, or $-\dfrac{182}{285}$

41. *Familiarize*. We draw a picture. We let h = the amount of sliced almonds needed.

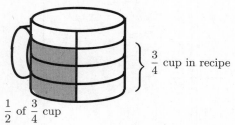

$\dfrac{3}{4}$ cup in recipe

$\dfrac{1}{2}$ of $\dfrac{3}{4}$ cup

Translate. We are finding $\dfrac{1}{2}$ of $\dfrac{3}{4}$, so the multiplication sentence $\dfrac{1}{2} \cdot \dfrac{3}{4} = h$ corresponds to the situation.

Solve. We multiply:

$$\dfrac{1}{2} \cdot \dfrac{3}{4} = \dfrac{1 \cdot 3}{2 \cdot 4} = \dfrac{3}{8}$$

Check. We repeat the calculation. The answer checks.

State. $\dfrac{3}{8}$ cup of sliced almonds is needed.

43. *Familiarize*. Recall that area is length times width. We draw a picture. We will let A = the area of the table top.

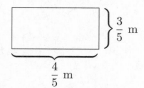

$\dfrac{3}{5}$ m

$\dfrac{4}{5}$ m

Translate. Then we translate.

$$\begin{array}{ccccc} \text{Area} & \text{is} & \text{length} & \text{times} & \text{width} \\ \downarrow & \downarrow & \downarrow & \downarrow & \downarrow \\ A & = & \dfrac{4}{5} & \times & \dfrac{3}{5} \end{array}$$

Solve. The sentence tells us what to do. We multiply.

$$\dfrac{4}{5} \times \dfrac{3}{5} = \dfrac{4 \times 3}{5 \times 5} = \dfrac{12}{25}$$

Check. We repeat the calculation. The answer checks.

State. The area is $\dfrac{12}{25}$ m².

45. *Familiarize*. We know that $\dfrac{4}{5}$ of all municipal waste is dumped in landfills. We also know that $\dfrac{1}{10}$ of the waste in landfills is landscape trimmings. We let y = the fractional part of municipal waste that is landscape trimmings that are landfilled.

Translate. The multiplication sentence $\dfrac{1}{10} \cdot \dfrac{4}{5} = y$ corresponds to this situation.

Solve. We carry out the multiplication.

$$\dfrac{1}{10} \cdot \dfrac{4}{5} = \dfrac{1 \cdot 4}{10 \cdot 5} = \dfrac{4}{50}$$

Check. We repeat the calculation. The answer checks.

State. $\dfrac{4}{50}$ of municipal waste is landscape trimmings that are landfilled.

47. *Familiarize*. A picture of the situation appears in the text. Let f = the fraction of the floor that has been tiled.

Translate. The multiplication sentence $\dfrac{3}{5} \cdot \dfrac{3}{4} = f$ corresponds to the situation.

Solve. We multiply.

$$\dfrac{3}{5} \cdot \dfrac{3}{4} = \dfrac{3 \cdot 3}{5 \cdot 4} = \dfrac{9}{20}$$

Check. We repeat the calculation. The answer checks.

State. $\dfrac{9}{20}$ of the floor has been tiled.

49. *Familiarize*. We know that 1 of 35 high school basketball players plays college basketball. That is, $\dfrac{1}{35}$ of high school basketball players play college basketball. In addition, we know that 1 of 75 college players plays professional basketball. That is, $\dfrac{1}{75}$ of college basketball players play professional basketball or $\dfrac{1}{75}$ of the $\dfrac{1}{35}$ of high school players who play college basketball also play professionally. We let f = the fractional part of high school basketball players that plays professional basketball.

Translate. The multiplication sentence $f = \dfrac{1}{75} \cdot \dfrac{1}{35}$ corresponds to this situation.

Solve. We carry out the multiplication.

$$\dfrac{1}{75} \cdot \dfrac{1}{35} = \dfrac{1 \cdot 1}{75 \cdot 35} = \dfrac{1}{2625}$$

Check. We repeat the calculation. The answer checks.

State. The fractional part of high school players that plays professional basketball is $\dfrac{1}{2625}$.

51. Discussion and Writing Exercise

53.

$$20\overline{\smash{\big)}\,180}$$
$$\underline{180}$$
$$0$$

The answer is 9.

55. $450 \div (-9) = -50$

(The signs are different, so the answer is negative.)

57. $\dfrac{-35}{5} = -7$

(The signs are different, so the answer is negative.)

59. $\dfrac{-65}{-5} = 13$

(The signs are the same, so the answer is positive.)

61. $4, 6\,7\,\boxed{8}\,,9\,5\,2$

The digit 8 means 8 thousands.

63. $7\,1\,4\,\boxed{8}$

The digit 8 means 8 ones.

65. Discussion and Writing Exercise

67. *Familiarize*. Let $t =$ the number of gallons of two-cycle oil in a freshly filled chainsaw.

Translate. The multiplication sentence $\dfrac{1}{16} \cdot \dfrac{1}{5} = t$ corresponds to this situation.

Solve. We carry out the multiplication.

$$\frac{1}{16} \cdot \frac{1}{5} = \frac{1 \cdot 1}{16 \cdot 5} = \frac{1}{80}$$

Check. We repeat the calculation. The answer checks.

State. There is $\dfrac{1}{80}$ gal of two-cycle oil in a freshly filled chainsaw.

69. Use a calculator.

$$\left(-\frac{57}{61}\right)^3 = -\frac{185,193}{226,981}, \text{ or } \frac{-185,193}{226,981}$$

71. $\left(-\dfrac{1}{2}\right)^5 \left(\dfrac{3}{5}\right) = -\dfrac{1}{32}\left(\dfrac{3}{5}\right) = -\dfrac{3}{160}, \text{ or } \dfrac{-3}{160}$

73. $-\dfrac{2}{3}xy = -\dfrac{2}{3} \cdot \dfrac{2}{5}\left(-\dfrac{1}{7}\right)$

$$= -\frac{4}{15}\left(-\frac{1}{7}\right)$$

$$= \frac{4}{105}$$

75. Use a calculator.

$$-\frac{4}{7}ab = -\frac{4}{7} \cdot \frac{93}{107} \cdot \frac{13}{41} = -\frac{4836}{30,709}$$

Exercise Set 3.5

1. Since $10 \div 2 = 5$, we multiply by $\dfrac{5}{5}$.

$$\frac{1}{2} = \frac{1}{2} \cdot \frac{5}{5} = \frac{1 \cdot 5}{2 \cdot 5} = \frac{5}{10}$$

3. Since $-48 \div 4 = -12$, we multiply by $\dfrac{-12}{-12}$.

$$\frac{3}{4} = \frac{3}{4}\left(\frac{-12}{-12}\right) = \frac{3(-12)}{4(-12)} == \frac{-36}{-48}$$

5. Since $50 \div 10 = 5$, we multiply by $\dfrac{5}{5}$.

$$\frac{7}{10} = \frac{7}{10} \cdot \frac{5}{5} = \frac{7 \cdot 5}{10 \cdot 5} = \frac{35}{50}$$

7. Since $5t \div 5 = t$, we multiply by $\dfrac{t}{t}$.

$$\frac{11}{5} = \frac{11}{5} \cdot \frac{t}{t} = \frac{11 \cdot t}{5 \cdot t} = \frac{11t}{5t}$$

9. Since $48 \div 12 = 4$, we multiply by $\dfrac{4}{4}$.

$$\frac{5}{12} = \frac{5}{12} \cdot \frac{4}{4} = \frac{5 \cdot 4}{12 \cdot 4} = \frac{20}{48}$$

11. Since $54 \div 18 = 3$, we multiply by $\dfrac{3}{3}$.

$$-\frac{17}{18} = -\frac{17}{18} \cdot \frac{3}{3} = -\frac{17 \cdot 3}{18 \cdot 3} = -\frac{51}{54}$$

13. Since $-40 \div -8 = 5$, we multiply by $\dfrac{5}{5}$.

$$\frac{3}{-8} = \frac{3}{-8} \cdot \frac{5}{5} = \frac{3 \cdot 5}{-8 \cdot 5} = \frac{15}{-40}$$

15. Since $132 \div 22 = 6$, we multiply by $\dfrac{6}{6}$.

$$\frac{-7}{22} = \frac{-7}{22} \cdot \frac{6}{6} = \frac{-7 \cdot 6}{22 \cdot 6} = \frac{-42}{132}$$

17. Since $8x \div 8 = x$, we multiply by $\dfrac{x}{x}$.

$$\frac{5}{8} = \frac{5}{8} \cdot \frac{x}{x} = \frac{5x}{8x}$$

19. Since $7a \div 7 = a$, we multiply by $\dfrac{a}{a}$.

$$\frac{10}{7} = \frac{10}{7} \cdot \frac{a}{a} = \frac{10a}{7a}$$

21. Since $9ab \div 9 = ab$, we multiply by $\dfrac{ab}{ab}$.

$$\frac{4}{9} \cdot \frac{ab}{ab} = \frac{4ab}{9ab}$$

23. Since $27b \div 9 = 3b$, we multiply by $\dfrac{3b}{3b}$.

$$\frac{4}{9} = \frac{4}{9} \cdot \frac{3b}{3b} = \frac{12b}{27b}$$

25. $\dfrac{2}{4} = \dfrac{1 \cdot 2}{2 \cdot 2}$ ⟵ Factor the numerator
⟵ Factor the denominator

$= \dfrac{1}{2} \cdot \dfrac{2}{2}$ ⟵ Factor the fraction

$= \dfrac{1}{2} \cdot 1$ ⟵ $\dfrac{2}{2} = 1$

$= \dfrac{1}{2}$ ⟵ Removing a factor of 1

27. $-\dfrac{6}{9} = -\dfrac{2 \cdot 3}{3 \cdot 3}$ ⟵ Factor the numerator
⟵ Factor the denominator

$= -\dfrac{2}{3} \cdot \dfrac{3}{3}$ ⟵ Factor the fraction

$= -\dfrac{2}{3} \cdot 1$ ⟵ $\dfrac{3}{3} = 1$

$= -\dfrac{2}{3}$ ⟵ Removing a factor of 1

29. $\dfrac{10}{25} = \dfrac{2 \cdot 5}{5 \cdot 5}$ ⟵ Factor the numerator
⟵ Factor the denominator

$= \dfrac{2}{5} \cdot \dfrac{5}{5}$ ⟵ Factor the fraction

$= \dfrac{2}{5} \cdot 1$ ⟵ $\dfrac{5}{5} = 1$

$= \dfrac{2}{5}$ ⟵ Removing a factor of 1

31. $\dfrac{27}{-3} = \dfrac{9 \cdot 3}{-1 \cdot 3} = \dfrac{9}{-1} \cdot \dfrac{3}{3} = \dfrac{9}{-1} \cdot 1 = -9$

33. $\dfrac{27}{36} = \dfrac{9 \cdot 3}{9 \cdot 4} = \dfrac{9}{9} \cdot \dfrac{3}{4} = 1 \cdot \dfrac{3}{4} = \dfrac{3}{4}$

35. $-\dfrac{24}{14} = -\dfrac{12 \cdot 2}{7 \cdot 2} = -\dfrac{12}{7} \cdot \dfrac{2}{2} = -\dfrac{12}{7}$

37. $\dfrac{16n}{48n} = \dfrac{1 \cdot 16n}{3 \cdot 16n} = \dfrac{1}{3} \cdot \dfrac{16n}{16n} = \dfrac{1}{3}$

39. $\dfrac{-17}{51} = \dfrac{-1 \cdot 17}{3 \cdot 17} = \dfrac{-1}{3} \cdot \dfrac{17}{17} = \dfrac{-1}{3}$

41. $\dfrac{420}{480} = \dfrac{2 \cdot 2 \cdot 3 \cdot 5 \cdot 7}{2 \cdot 2 \cdot 2 \cdot 2 \cdot 2 \cdot 3 \cdot 5}$

$= \dfrac{2}{2} \cdot \dfrac{2}{2} \cdot \dfrac{3}{3} \cdot \dfrac{5}{5} \cdot \dfrac{7}{2 \cdot 2 \cdot 2}$

$= \dfrac{7}{2 \cdot 2 \cdot 2}$

$= \dfrac{7}{8}$

43. $\dfrac{153}{136} = \dfrac{3 \cdot 3 \cdot 17}{2 \cdot 2 \cdot 2 \cdot 17}$

$= \dfrac{3 \cdot 3}{2 \cdot 2 \cdot 2} \cdot \dfrac{17}{17}$

$= \dfrac{3 \cdot 3}{2 \cdot 2 \cdot 2}$

$= \dfrac{9}{8}$

45. $\dfrac{132}{143} = \dfrac{11 \cdot 12}{11 \cdot 13} = \dfrac{11}{11} \cdot \dfrac{12}{13} = \dfrac{12}{13}$

47. $\dfrac{221}{247} = \dfrac{13 \cdot 17}{13 \cdot 19} = \dfrac{13}{13} \cdot \dfrac{17}{19} = \dfrac{17}{19}$

49. $\dfrac{3ab}{8ab} = \dfrac{3 \cdot a \cdot b}{8 \cdot a \cdot b} = \dfrac{3}{8} \cdot \dfrac{a}{a} \cdot \dfrac{b}{b} = \dfrac{3}{8}$

51. $\dfrac{9xy}{6x} = \dfrac{3 \cdot 3 \cdot x \cdot y}{2 \cdot 3 \cdot x} = \dfrac{3 \cdot y}{2} \cdot \dfrac{3}{3} \cdot \dfrac{x}{x} = \dfrac{3y}{2}$

53. $\dfrac{-18a}{20ab} = \dfrac{-9 \cdot 2 \cdot a}{10 \cdot 2 \cdot a \cdot b} = \dfrac{-9}{10 \cdot b} \cdot \dfrac{2}{2} \cdot \dfrac{a}{a} = \dfrac{-9}{10b}$

55. We multiply these   We multiply these
two numbers: two numbers:

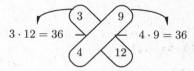

$3 \cdot 12 = 36$ $4 \cdot 9 = 36$

Since $36 = 36$, $\dfrac{3}{4} = \dfrac{9}{12}$.

57. We multiply these   We multiply these
two numbers: two numbers:

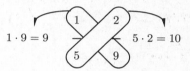

$1 \cdot 9 = 9$ $5 \cdot 2 = 10$

Since $9 \neq 10$, $\dfrac{1}{5}$ and $\dfrac{2}{9}$ do not name the same number.

Thus, $\dfrac{1}{5} \neq \dfrac{2}{9}$.

59. We multiply these   We multiply these
two numbers: two numbers:

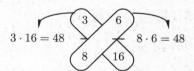

$3 \cdot 16 = 48$ $8 \cdot 6 = 48$

Since $48 = 48$, $\dfrac{3}{8} = \dfrac{6}{16}$.

61. We multiply these   We multiply these
two numbers: two numbers:

$2 \cdot 7 = 14$ $5 \cdot 3 = 15$

Since $14 \neq 15$, $\dfrac{2}{5}$ and $\dfrac{3}{7}$ do not name the same number.

Thus, $\dfrac{2}{5} \neq \dfrac{3}{7}$.

63.

We multiply these two numbers: We multiply these two numbers:

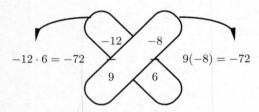

$-3 \cdot 12 = -36$ $10(-4) = -40$

Since $-36 \neq -40$, $\dfrac{-3}{10}$ and $\dfrac{-4}{12}$ do not name the same number. Thus, $\dfrac{-3}{10} \neq \dfrac{-4}{12}$.

65. We rewrite $-\dfrac{12}{9}$ as $\dfrac{-12}{9}$ and check cross products.

We multiply these two numbers: We multiply these two numbers:

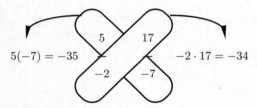

$-12 \cdot 6 = -72$ $9(-8) = -72$

Since $-72 = -72$, $-\dfrac{12}{9} = \dfrac{-8}{6}$.

67. We rewrite $-\dfrac{17}{7}$ as $\dfrac{17}{-7}$ and check cross products.

We multiply these two numbers: We multiply these two numbers:

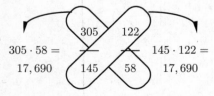

$5(-7) = -35$ $-2 \cdot 17 = -34$

Since $-35 \neq -34$, $\dfrac{5}{-2}$ and $\dfrac{17}{-7}$ do not name the same number. Thus, $\dfrac{5}{-2} \neq \dfrac{17}{-7}$, or $\dfrac{5}{-2} \neq -\dfrac{17}{7}$.

69.

We multiply these two numbers: We multiply these two numbers:

$305 \cdot 58 =$ $145 \cdot 122 =$
$17{,}690$ $17{,}690$

Since $17{,}690 = 17{,}690$, $\dfrac{305}{145} = \dfrac{122}{58}$.

71. Discussion and Writing Exercise

73. *Familiarize*. We make a drawing. We let $A =$ the area.

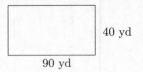

40 yd

90 yd

Translate. Using the formula for area, we have
$$A = l \cdot w = 90 \cdot 40.$$

Solve. We carry out the multiplication.

$$\begin{array}{r} 4\,0 \\ \times\ \ 9\,0 \\ \hline 3\,6\,0\,0 \end{array}$$

Thus, $A = 3600$.

Check. We repeat the calculation. The answer checks.

State. The area is 3600 yd^2.

75. $-12(-5) = 60$

(The signs are the same, so the product is positive.)

77. $-9 \cdot 7 = -63$

(The signs are different, so the product is negative.)

79. $\quad 30 \cdot x = 150$

$\dfrac{30 \cdot x}{30} = \dfrac{150}{30}$ Dividing both sides by 30

$\quad x = 5$

The solution is 5.

81. $\quad\quad 5280 = 1760 + t$

$5280 - 1760 = 1760 + t - 1760$ Subtracting 1760 from both sides

$\quad\quad 3520 = t$

The solution is 3520.

83. Discussion and Writing Exercise

85. $\dfrac{391}{667} = \dfrac{17 \cdot 23}{23 \cdot 29} = \dfrac{17}{29} \cdot \dfrac{23}{23} = \dfrac{17}{29}$

87. $-\dfrac{1073x}{555y} = -\dfrac{29 \cdot 37 \cdot x}{15 \cdot 37 \cdot y} = -\dfrac{29 \cdot x}{15 \cdot y} \cdot \dfrac{37}{37} = -\dfrac{29x}{15y}$

89. $\dfrac{4247}{4619} = \dfrac{31 \cdot 137}{31 \cdot 149} = \dfrac{31}{31} \cdot \dfrac{137}{149} = \dfrac{137}{149}$

91. The part of the population that is shy is $\dfrac{4}{10}$. We simplify:

$\dfrac{4}{10} = \dfrac{2 \cdot 2}{5 \cdot 2} = \dfrac{2}{5} \cdot \dfrac{2}{2} = \dfrac{2}{5}$

Since 4 out of 10 people are shy, then $10 - 4$, or 6, are not shy. The part of the population that is not shy is $\dfrac{6}{10}$. We simplify:

$\dfrac{6}{10} = \dfrac{3 \cdot 2}{5 \cdot 2} = \dfrac{3}{5} \cdot \dfrac{2}{2} = \dfrac{3}{5}$

93. Derrek Lee's batting average was $\dfrac{199}{594}$; Michael Young's batting average was $\dfrac{221}{668}$. We test these fractions for equality:

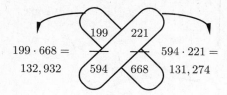

$$199 \cdot 668 = \frac{199}{594} \quad \frac{221}{668} \quad 594 \cdot 221 =$$
$$132,932 \qquad \qquad \qquad 131,274$$

Since $132,932 \neq 131,274$, $\frac{199}{594}$ and $\frac{221}{668}$ do not name the same number. Thus, $\frac{199}{594} \neq \frac{221}{668}$ and the batting averages are not the same.

Exercise Set 3.6

1. $\frac{3}{8} \cdot \frac{7}{3} = \frac{3 \cdot 7}{8 \cdot 3} = \frac{3}{3} \cdot \frac{7}{8} = 1 \cdot \frac{7}{8} = \frac{7}{8}$

3. $\frac{7}{8} \cdot \frac{-1}{7} = \frac{7(-1)}{8 \cdot 7} = \frac{7}{7} \cdot \frac{-1}{8} = \frac{-1}{8}$, or $-\frac{1}{8}$

5. $\frac{1}{8} \cdot \frac{6}{7} = \frac{1 \cdot 6}{8 \cdot 7} = \frac{1 \cdot 2 \cdot 3}{2 \cdot 4 \cdot 7} = \frac{2}{2} \cdot \frac{1 \cdot 3}{4 \cdot 7} = \frac{3}{28}$

7. $\frac{1}{6} \cdot \frac{4}{3} = \frac{1 \cdot 4}{6 \cdot 3} = \frac{1 \cdot 2 \cdot 2}{2 \cdot 3 \cdot 3} = \frac{2}{2} \cdot \frac{1 \cdot 2}{3 \cdot 3} = \frac{2}{9}$

9. $\frac{12}{-5} \cdot \frac{9}{8} = \frac{12 \cdot 9}{-5 \cdot 8} = \frac{4 \cdot 3 \cdot 9}{-5 \cdot 2 \cdot 4} = \frac{4}{4} \cdot \frac{3 \cdot 9}{-5 \cdot 2} = \frac{3 \cdot 9}{-5 \cdot 2} =$
$\frac{27}{-10}$, or $-\frac{27}{10}$

11. $\frac{5x}{9} \cdot \frac{4}{5} = \frac{5x \cdot 4}{9 \cdot 5} = \frac{5 \cdot x \cdot 4}{9 \cdot 5} = \frac{5}{5} \cdot \frac{x \cdot 4}{9} = \frac{4x}{9}$

13. $\frac{1}{4} \cdot 12 = \frac{1 \cdot 12}{4} = \frac{12}{4} = \frac{4 \cdot 3}{4 \cdot 1} = \frac{4}{4} \cdot \frac{3}{1} = \frac{3}{1} = 3$

15. $21 \cdot \frac{1}{3} = \frac{21 \cdot 1}{3} = \frac{21}{3} = \frac{3 \cdot 7}{3 \cdot 1} = \frac{3}{3} \cdot \frac{7}{1} = \frac{7}{1} = 7$

17. $-16\left(-\frac{3}{4}\right) = \frac{16 \cdot 3}{4} = \frac{4 \cdot 4 \cdot 3}{4 \cdot 1} = \frac{4}{4} \cdot \frac{4 \cdot 3}{1} =$
$\frac{4 \cdot 3}{1} = \frac{12}{1} = 12$

19. $\frac{3}{8} \cdot 8a = \frac{3 \cdot 8a}{8} = \frac{3 \cdot 8 \cdot a}{8 \cdot 1} = \frac{8}{8} \cdot \frac{3 \cdot a}{1} = \frac{3a}{1} = 3a$

21. $\left(-\frac{3}{8}\right)\left(-\frac{8}{3}\right) = \frac{3 \cdot 8}{8 \cdot 3} = \frac{3 \cdot 8}{3 \cdot 8} = 1$

23. $\frac{a}{b} \cdot \frac{b}{a} = \frac{a \cdot b}{b \cdot a} = \frac{a \cdot b}{a \cdot b} = 1$

25. $\frac{1}{26} \cdot 143a = \frac{1 \cdot 143a}{26} = \frac{1 \cdot 11 \cdot 13 \cdot a}{2 \cdot 13} =$
$\frac{13}{13} \cdot \frac{1 \cdot 11a}{2} = \frac{11a}{2}$

27. $176\left(\frac{1}{-6}\right) = \frac{176 \cdot 1}{-6} = \frac{2 \cdot 88 \cdot 1}{-3 \cdot 2} = \frac{2}{2} \cdot \frac{88 \cdot 1}{-3} =$
$\frac{88}{-3}$, or $-\frac{88}{3}$

29. $-8x \cdot \frac{1}{-8x} = \frac{8x \cdot 1}{8x \cdot 1} = 1$

31. $\frac{2x}{9} \cdot \frac{27}{2x} = \frac{2x \cdot 27}{9 \cdot 2x} = \frac{2x \cdot 9 \cdot 3}{9 \cdot 2x \cdot 1} =$
$\frac{2x \cdot 9}{2x \cdot 9} \cdot \frac{3}{1} = 3$

33. $\frac{7}{10} \cdot \frac{34}{150} = \frac{7 \cdot 34}{10 \cdot 150} = \frac{7 \cdot 2 \cdot 17}{2 \cdot 5 \cdot 150} = \frac{2}{2} \cdot \frac{7 \cdot 17}{5 \cdot 150} = \frac{119}{750}$

35. $\frac{36}{85} \cdot \frac{25}{-99} = \frac{9 \cdot 4 \cdot 5 \cdot 5}{5 \cdot 17 \cdot 9(-11)} = \frac{9 \cdot 5}{9 \cdot 5} \cdot \frac{4 \cdot 5}{17(-11)} =$
$\frac{20}{-187}$, or $-\frac{20}{187}$

37. $\frac{-98}{99} \cdot \frac{27a}{175a} = \frac{7(-14) \cdot 9 \cdot 3 \cdot a}{9 \cdot 11 \cdot 7 \cdot 25 \cdot a} =$
$\frac{7 \cdot 9 \cdot a}{7 \cdot 9 \cdot a} \cdot \frac{-14 \cdot 3}{11 \cdot 25} = \frac{-42}{275}$, or $-\frac{42}{275}$

39. $\frac{110}{33} \cdot \frac{-24}{25x} = \frac{2 \cdot 5 \cdot 11 \cdot 3 \cdot 8}{3 \cdot 11 \cdot 5 \cdot 5 \cdot x} = -\frac{2 \cdot 8}{5 \cdot x} \cdot \frac{3 \cdot 5 \cdot 11}{3 \cdot 5 \cdot 11} =$
$\frac{-16}{5x}$, or $-\frac{16}{5x}$

41. $\left(-\frac{11}{24}\right)\frac{3}{5} = -\frac{11 \cdot 3}{24 \cdot 5} = -\frac{11 \cdot 3}{3 \cdot 8 \cdot 5} = \frac{3}{3}\left(-\frac{11}{8 \cdot 5}\right) =$
$-\frac{11}{40}$

43. $\frac{10a}{21} \cdot \frac{3}{8b} = \frac{10a \cdot 3}{21 \cdot 8b} = \frac{2 \cdot 5 \cdot a \cdot 3}{3 \cdot 7 \cdot 2 \cdot 4 \cdot b} = \frac{2 \cdot 3}{2 \cdot 3} \cdot \frac{5 \cdot a}{7 \cdot 4 \cdot b} = \frac{5a}{28b}$

45. **Familiarize.** Let $n =$ the number of inches the screw will go into the piece of oak when it is turned 10 complete rotations.

Translate. We write an equation.

Total distance	is	Distance for one revolution	times	Number of revolutions
↓	↓	↓	↓	↓
n	$=$	$\frac{1}{16}$	$\cdot$	10

Solve. We carry out the multiplication.
$$n = \frac{1}{16} \cdot 10 = \frac{1 \cdot 10}{16}$$
$$= \frac{1 \cdot 2 \cdot 5}{2 \cdot 8} = \frac{2}{2} \cdot \frac{1 \cdot 5}{8}$$
$$= \frac{5}{8}$$

Check. We can repeat the calculation. We can also determine that the answer seems reasonable since we multiplied 10 by a number less than 10 and the result is less than 10. The answer checks.

State. The screw will go $\frac{5}{8}$ in. into the piece of oak when it is turned 10 complete rotations.

47. **Familiarize.** Let $s =$ the swimming speed of a dolphin, in mph.

Translate. We write an equation.

$$\underbrace{\text{Dolphin's speed}}_{\downarrow} \text{ is } \overset{\downarrow}{\tfrac{3}{5}} \overset{\downarrow}{\text{ of }} \underbrace{\text{Whale's speed}}_{\downarrow}$$
$$s \quad = \tfrac{3}{5} \cdot \quad 30$$

Solve. We carry out the multiplication.

$$s = \frac{3}{5} \cdot 30 = \frac{3 \cdot 30}{5}$$
$$= \frac{3 \cdot 5 \cdot 6}{5 \cdot 1} = \frac{5}{5} \cdot \frac{3 \cdot 6}{1}$$
$$= 18$$

Check. We can repeat the calculation. We can also determine that the answer seems reasonable since we multiplied 30 by a number less than 30 and the result is less than 30. The answer checks.

State. The swimming speed of a dolphin is 18 mph.

49. *Familiarize*. We visualize the situation. We let $n =$ the number of addresses that will be incorrect after one year.

Mailing list 2500 addresses		
1/4 of the addresses n		

Translate.

$$\underbrace{\text{Number incorrect}}_{\downarrow} \overset{\downarrow}{\text{ is }} \overset{\downarrow}{\tfrac{1}{4}} \overset{\downarrow}{\text{ of }} \underbrace{\text{Number of addresses}}_{\downarrow}$$
$$n \quad = \tfrac{1}{4} \cdot \quad 2500$$

Solve. We carry out the multiplication.

$$n = \frac{1}{4} \cdot 2500 = \frac{1 \cdot 2500}{4} = \frac{2500}{4}$$
$$= \frac{4 \cdot 625}{4 \cdot 1} = \frac{4}{4} \cdot \frac{625}{1}$$
$$= 625$$

Check. We can repeat the calculation. We can also determine that the answer seems reasonable since we multiplied 2500 by a number less than 1 and the result is less than 2500. The answer checks.

State. After one year 625 addresses will be incorrect.

51. *Familiarize*. Let $f =$ the fraction of Americans who eat breakfast at home. This fraction is $\frac{3}{4}$ of $\frac{2}{3}$.

Translate. We translate to a multiplication sentence.

$$\frac{3}{4} \cdot \frac{2}{3} = f$$

Solve. We multiply and simplify.

$$f = \frac{3}{4} \cdot \frac{2}{3} = \frac{3 \cdot 2}{4 \cdot 3} = \frac{3 \cdot 2 \cdot 1}{2 \cdot 2 \cdot 3} = \frac{2 \cdot 3}{2 \cdot 3} \cdot \frac{1}{2} = \frac{1}{2}$$

Check. We can repeat the calculation. The answer checks.

State. $\frac{1}{2}$ of Americans eat breakfast at home.

53. *Familiarize*. We draw a picture.

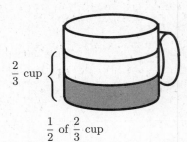

$\frac{2}{3}$ cup

$\frac{1}{2}$ of $\frac{2}{3}$ cup

We let $n =$ the amount of flour the chef should use.

Translate. The multiplication sentence

$$\frac{1}{2} \cdot \frac{2}{3} = n$$

corresponds to the situation.

Solve. We multiply and simplify:

$$n = \frac{1}{2} \cdot \frac{2}{3} = \frac{1 \cdot 2}{2 \cdot 3} = \frac{2}{2} \cdot \frac{1}{3} = \frac{1}{3}$$

Check. We can repeat the calculation. We can also determine that the answer seems reasonable since we multiplied $\frac{2}{3}$ by a number less than 1 and the result is less than $\frac{2}{3}$. The answer checks.

State. The chef should use $\frac{1}{3}$ cup of flour.

55. *Familiarize*. We visualize the situation. Let $a =$ the assessed value of the house.

Value of house $154,000	
3/4 of the value $a	

Translate. We write an equation.

$$\underbrace{\text{Assessed value}}_{\downarrow} \overset{\downarrow}{\text{ is }} \overset{\downarrow}{\tfrac{3}{4}} \overset{\downarrow}{\text{ of }} \underbrace{\text{the value of the house}}_{\downarrow}$$
$$a \quad = \tfrac{3}{4} \cdot \quad 154{,}000$$

Solve. We carry out the multiplication.

$$a = \frac{3}{4} \cdot 154{,}000 = \frac{3 \cdot 154{,}000}{4}$$
$$= \frac{3 \cdot 4 \cdot 38{,}500}{4 \cdot 1} = \frac{4}{4} \cdot \frac{3 \cdot 38{,}500}{1}$$
$$= 115{,}500$$

Check. We can repeat the calculation. We can also determine that the answer seems reasonable since we multiplied 154,000 by a number less than 1 and the result is less than 154,000. The answer checks.

State. The assessed value of the house is $115,500.

57. *Familiarize.* We draw a picture.

$\dfrac{2}{3}$ in.

1 in.
240 miles

We let n = the number of miles represented by $\dfrac{2}{3}$ in.

Translate. The multiplication sentence

$$n = \frac{2}{3} \cdot 240$$

corresponds to the situation.

Solve. We multiply and simplify:

$$n = \frac{2}{3} \cdot 240 = \frac{2 \cdot 240}{3} = \frac{2 \cdot 3 \cdot 80}{1 \cdot 3}$$

$$= \frac{3}{3} \cdot \frac{2 \cdot 80}{1} = \frac{2 \cdot 80}{1}$$

$$= 160$$

Check. We can repeat the calculation. We can also determine that the answer seems reasonable since we multiplied 240 by a number less than 1 and the result is less than 240.

State. $\dfrac{2}{3}$ in. on the map represents 160 miles.

59. *Familiarize.* This is a multistep problem. First we find the amount of each of the given expenses. Then we find the total of these expenses and take it away from the annual income to find how much is spent for other expenses.

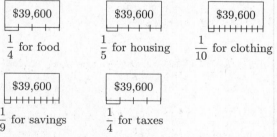

$39,600

$\dfrac{1}{4}$ for food

$39,600

$\dfrac{1}{5}$ for housing

$39,600

$\dfrac{1}{10}$ for clothing

$39,600

$\dfrac{1}{9}$ for savings

$39,600

$\dfrac{1}{4}$ for taxes

We let f, h, c, s, and t represent the amounts spent on food, housing, clothing, savings, and taxes, respectively.

Translate. The following multiplication sentences correspond to the situation.

$$\frac{1}{4} \cdot 39,600 = f \qquad \frac{1}{9} \cdot 39,600 = s$$

$$\frac{1}{5} \cdot 39,600 = h \qquad \frac{1}{4} \cdot 39,600 = t$$

$$\frac{1}{10} \cdot 39,600 = c$$

Solve. We multiply and simplify.

$$f = \frac{1}{4} \cdot 39,600 = \frac{39,600}{4} = \frac{4 \cdot 9900}{4 \cdot 1} = \frac{4}{4} \cdot \frac{9900}{1} =$$

9900

$$h = \frac{1}{5} \cdot 39,600 = \frac{39,600}{5} = \frac{5 \cdot 7920}{5 \cdot 1} = \frac{5}{5} \cdot \frac{7920}{1} =$$

7920

$$c = \frac{1}{10} \cdot 39,600 = \frac{39,600}{10} = \frac{10 \cdot 3960}{10 \cdot 1} = \frac{10}{10} \cdot \frac{3960}{1} =$$

3960

$$s = \frac{1}{9} \cdot 39,600 = \frac{39,600}{9} = \frac{9 \cdot 4400}{9 \cdot 1} = \frac{9}{9} \cdot \frac{4400}{1} =$$

4400

$$t = \frac{1}{4} \cdot 39,600 = \frac{39,600}{4} = \frac{4 \cdot 9900}{4 \cdot 1} = \frac{4}{4} \cdot \frac{9900}{1} =$$

9900

We add to find the total of these expenses.

```
$ 9 9 0 0
  7 9 2 0
  3 9 6 0
  4 4 0 0
  9 9 0 0
$ 3 6, 0 8 0
```

We let m = the amount spent on other expenses and subtract to find this amount.

Annual income	minus	Total of itemized expenses	is	Total spent on other expenses
↓	↓	↓	↓	↓
$39,600	−	$36,080	=	m
		$3520	=	m Subtracting

Check. We repeat the calculations. The results check.

State. $9900 is spent for food, $7920 for housing, $3960 for clothing, $4400 for savings, $9900 for taxes, and $3520 for other expenses.

61. $A = \dfrac{1}{2} \cdot b \cdot h$ \qquad Area of a triangle

$A = \dfrac{1}{2} \cdot 15$ in. $\cdot\, 8$ in. \qquad Substituting 15 in. for b and 8 in. for h

$A = \dfrac{15 \cdot 8}{2}$ in^2

$A = 60$ in^2

63. $A = \dfrac{1}{2} \cdot b \cdot h$ \qquad Area of a triangle

$A = \dfrac{1}{2} \cdot 5$ mm $\cdot\, \dfrac{7}{2}$ mm \qquad Substituting 5 mm for b and $\dfrac{7}{2}$ mm for h

$A = \dfrac{5 \cdot 7}{2 \cdot 2}$ mm^2

$A = \dfrac{35}{4}$ mm^2

65.
$A = \dfrac{1}{2} \cdot b \cdot h$ Area of a triangle

$A = \dfrac{1}{2} \cdot \dfrac{9}{2} \text{ m} \cdot \dfrac{7}{2} \text{ m}$ Substituting $\dfrac{9}{2}$ m for b and

                        $\dfrac{7}{2}$ m for h

$A = \dfrac{9 \cdot 7}{2 \cdot 2 \cdot 2} \text{ m}^2$

$A = \dfrac{63}{8} \text{ m}^2$

67. *Familiarize.* We look for figures whose areas we can calculate using area formulas we already know.

Translate. The figure consists of a rectangle with a length of 10 mi and a width of 8 mi and of a triangle with a base of $13 - 10$, or 3 mi, and a height of 8 mi. We use the formula $A = l \cdot w$ for the area of a rectangle and the formula $A = \dfrac{1}{2} \cdot b \cdot h$ for the area of a triangle and add the two areas.

Solve. For the rectangle: $A = l \cdot w = 10 \text{ mi} \cdot 8 \text{ mi} = 80 \text{ mi}^2$

For the triangle: $A = \dfrac{1}{2} \cdot b \cdot h = \dfrac{1}{2} \cdot 3 \text{ mi} \cdot 8 \text{ mi} = 12 \text{ mi}^2$

Then we add: $80 \text{ mi}^2 + 12 \text{ mi}^2 = 92 \text{ mi}^2$

Check. We repeat the calculations.

State. The area of the figure is 92 mi^2.

69. *Familiarize.* We look for figures with areas we can calculate using area formulas that we already know. We let $T =$ the area of the front of the tie.

Translate. The front of the tie can be thought of as two triangles, each with base 4 cm + 1 cm, or 5 cm, and height $\dfrac{3}{2}$ cm. We can use the formula $A = \dfrac{1}{2} \cdot b \cdot h$ for the area of a triangle and then multiply by 2.

$\underbrace{\text{Area of tie}}$ is   twice $\underbrace{\text{Area of triangle}}$.

$\downarrow$ $\downarrow$ $\downarrow$ $\downarrow$

T $=$ $2 \cdot$ $\dfrac{1}{2} \cdot (5 \text{ cm}) \cdot \left(\dfrac{3}{2} \text{ cm} \right)$

Solve. We carry out the calculation.

$T = 2 \cdot \dfrac{1}{2} \cdot (5 \text{ cm}) \cdot \left(\dfrac{3}{2} \text{ cm} \right)$

$= 1 \cdot 5 \text{ cm} \cdot \dfrac{3}{2} \text{ cm}$

$= \dfrac{5 \cdot 3}{2} \text{ cm}^2$

$= \dfrac{15}{2} \text{ cm}^2$

Check. We can repeat the calculations. The answer checks.

State. The area of the front of the tie is $\dfrac{15}{2}$ cm^2.

71. Discussion and Writing Exercise

73.
$48 \cdot t = 1680$

$\dfrac{48 \cdot t}{48} = \dfrac{1680}{48}$

$t = 35$

The solution is 35.

75.
$3125 = 25 \cdot t$

$\dfrac{3125}{25} = \dfrac{25 \cdot t}{25}$ Dividing by 25 on both sides

$125 = t$

The solution is 125.

77. $t + 28 = 5017$

      $t = 5017 - 28$

      $t = 4989$

The solution is 4989.

79.
          $8797 = y + 2299$

$8797 - 2299 = y + 2299 - 2299$ Subtracting 2299

                             on both sides

          $6498 = y$

The solution is 6498.

81. Discussion and Writing Exercise

83.
$\dfrac{201}{535} \cdot \dfrac{4601}{6499} = \dfrac{201 \cdot 4601}{535 \cdot 6499}$

$= \dfrac{3 \cdot 67 \cdot 43 \cdot 107}{5 \cdot 107 \cdot 67 \cdot 97}$

$= \dfrac{67 \cdot 107}{67 \cdot 107} \cdot \dfrac{3 \cdot 43}{5 \cdot 97}$

$= \dfrac{129}{485}$

85.
$\dfrac{667}{899} \cdot \dfrac{558}{621} = \dfrac{667 \cdot 558}{899 \cdot 621}$

$= \dfrac{23 \cdot 29 \cdot 2 \cdot 3 \cdot 3 \cdot 31}{29 \cdot 31 \cdot 3 \cdot 3 \cdot 3 \cdot 23}$

$= \dfrac{3 \cdot 3 \cdot 23 \cdot 29 \cdot 31}{3 \cdot 3 \cdot 23 \cdot 29 \cdot 31} \cdot \dfrac{2}{3}$

$= \dfrac{2}{3}$

87. *Familiarize.* We know that $\dfrac{7}{8}$ of the students are high school graduates and $\dfrac{1}{7}$ of all students are left-handed. Also, if we divide the group of students into 3 equal parts and take 2 of them, we have the fractional part of the students who are over the age of 20. Then the 1 part remaining, or $\dfrac{1}{3}$ of the students, are 20 yr old or younger. Thus, we want to find $\dfrac{7}{8}$ of $\dfrac{1}{7}$ of $\dfrac{1}{3}$ of 480 students. We let $s =$ this number.

Translate. The multiplication sentence

$s = \dfrac{7}{8} \cdot \dfrac{1}{7} \cdot \dfrac{1}{3} \cdot 480$

corresponds to this situation.

Solve. We carry out the multiplication.

$s = \dfrac{7}{8} \cdot \dfrac{1}{7} \cdot \dfrac{1}{3} \cdot 480 = \dfrac{7 \cdot 1 \cdot 1 \cdot 480}{8 \cdot 7 \cdot 3}$

$= \dfrac{7 \cdot 1 \cdot 1 \cdot 3 \cdot 8 \cdot 20}{8 \cdot 7 \cdot 3 \cdot 1}$

$= \dfrac{7 \cdot 3 \cdot 8}{7 \cdot 3 \cdot 8} \cdot \dfrac{1 \cdot 1 \cdot 20}{1}$

$= 20$

Check. We can repeat the calculations. The answer checks.

State. 20 students are left-handed high school graduates 20 yr old or younger.

89. Area of each triangular end:
$$A = \frac{1}{2} \cdot b \cdot h = \frac{1}{2} \cdot 30 \text{ mm} \cdot 26 \text{ mm} = 390 \text{ mm}^2$$
Area of each rectangular side:
$$A = l \cdot w = 140 \text{ mm} \cdot 30 \text{ mm} = 4200 \text{ mm}^2$$
Total area: $2 \cdot 390 \text{ mm}^2 + 3 \cdot 4200 \text{ mm}^2 =$
$780 \text{ mm}^2 + 12,600 \text{ mm}^2 = 13,380 \text{ mm}^2$

91. From Exercise 70 we know that the total area of the sides and entrances of the building, including the area of the windows and doors, is 6800 ft². The area of each window is
$$A = l \cdot w = 4 \text{ ft} \cdot 3 \text{ ft} = 12 \text{ ft}^2.$$
The area of each entrance is
$$A = l \cdot w = 8 \text{ ft} \cdot 6 \text{ ft} = 48 \text{ ft}^2.$$
From the drawing we see that there are 26 windows and 2 entrances. Then the total area of the windows and entrances is
$$26 \cdot 12 \text{ ft}^2 + 2 \cdot 48 \text{ ft}^2 = 312 \text{ ft}^2 + 96 \text{ ft}^2 = 408 \text{ ft}^2.$$
We subtract to find the area of the sides and ends of the building excluding the area of the windows and entrances.
$$6800 \text{ ft}^2 - 408 \text{ ft}^2 = 6392 \text{ ft}^2$$
The building requires 6392 ft² of siding.

Exercise Set 3.7

1. $\frac{7}{3}$ Interchange the numerator and denominator.

The reciprocal of $\frac{7}{3}$ is $\frac{3}{7}$. $\left(\frac{7}{3} \cdot \frac{3}{7} = \frac{21}{21} = 1\right)$

3. Think of 9 as $\frac{9}{1}$.

$\frac{9}{1}$ Interchange the numerator and denominator.

The reciprocal of 9 is $\frac{1}{9}$. $\left(\frac{9}{1} \cdot \frac{1}{9} = \frac{9}{9} = 1\right)$

5. $\frac{1}{7}$ Interchange the numerator and denominator.

The reciprocal of $\frac{1}{7}$ is 7. $\left(\frac{7}{1} = 7; \frac{1}{7} \cdot \frac{7}{1} = \frac{7}{7} = 1\right)$

7. $-\frac{10}{3}$ Interchange the numerator and denominator.

The reciprocal of $-\frac{10}{3}$ is $-\frac{3}{10}$. $\left(-\frac{10}{3}\left(-\frac{3}{10}\right) = \frac{30}{30} = 1\right)$

9. $\frac{3}{17}$ Interchange the numerator and denominator.

The reciprocal of $\frac{3}{17}$ is $\frac{17}{3}$. $\left(\frac{3}{17} \cdot \frac{17}{3} = \frac{51}{51} = 1\right)$

11. $\frac{-3n}{m}$ Interchange the numerator and denominator.

The reciprocal of $\frac{-3n}{m}$ is $\frac{m}{-3n}$.
$$\left(\frac{-3n}{m} \cdot \frac{m}{-3n} = \frac{-3mn}{-3mn} = 1\right)$$

13. $\frac{8}{-15}$ Interchange the numerator and denominator.

The reciprocal of $\frac{8}{-15}$ is $\frac{-15}{8}$. $\left(\frac{8}{-15}\left(\frac{-15}{8}\right) = \frac{-120}{-120} = 1\right)$

15. Think of $7m$ as $\frac{7m}{1}$.

$\frac{7m}{1}$ Interchange the numerator and denominator.

The reciprocal of $7m$ is $\frac{1}{7m}$. $\left(\frac{7m}{1} \cdot \frac{1}{7m} = \frac{7m}{7m} = 1\right)$

17. $\frac{1}{4a}$ Interchange the numerator and denominator.

The reciprocal of $\frac{1}{4a}$ is $\frac{4a}{1}$, or $4a$.
$$\left(\frac{1}{4a} \cdot \frac{4a}{1} = \frac{4a}{4a} = 1\right)$$

19. The reciprocal of $-\frac{1}{3z}$ is $-\frac{3z}{1}$, or $-3z$.
$$\left(-\frac{1}{3z} \cdot (-3z) = \frac{3z}{3z} = 1\right)$$

21. $\frac{3}{7} \div \frac{3}{4} = \frac{3}{7} \cdot \frac{4}{3}$ Multiplying by the reciprocal of the divisor

$= \frac{3 \cdot 4}{7 \cdot 3}$ Multiplying numerators and denominators

$= \frac{3}{3} \cdot \frac{4}{7} = \frac{4}{7}$ Removing a factor equal to 1

23. $\frac{7}{6} \div \frac{5}{-3} = \frac{7}{6} \cdot \frac{-3}{5}$ Multiplying by the reciprocal of the divisor

$= \frac{7(-1)(3)}{2 \cdot 3 \cdot 5}$ Factoring

$= \frac{3}{3} \cdot \frac{7(-1)}{2 \cdot 5}$

$= \frac{-7}{10}$, or $-\frac{7}{10}$

25. $\frac{4}{3} \div \frac{1}{3} = \frac{4}{3} \cdot 3 = \frac{4 \cdot 3}{3} = \frac{3}{3} \cdot 4 = 4$

27. $\left(-\frac{1}{3}\right) \div \frac{1}{6} = -\frac{1}{3} \cdot 6 = -\frac{1 \cdot 2 \cdot 3}{1 \cdot 3} = -\frac{1 \cdot 3}{1 \cdot 3} \cdot 2 = -2$

29. $\left(-\dfrac{10}{21}\right) \div \left(-\dfrac{2}{15}\right) = \left(-\dfrac{10}{21}\right) \cdot \left(-\dfrac{15}{2}\right) = \dfrac{10 \cdot 15}{21 \cdot 2} =$

$\dfrac{2 \cdot 5 \cdot 3 \cdot 5}{3 \cdot 7 \cdot 2} = \dfrac{2 \cdot 3}{2 \cdot 3} \cdot \dfrac{5 \cdot 5}{7} = \dfrac{25}{7}$

31. $\dfrac{3}{8} \div 24 = \dfrac{3}{8} \cdot \dfrac{1}{24} = \dfrac{3 \cdot 1}{8 \cdot 3 \cdot 8} = \dfrac{3}{3} \cdot \dfrac{1}{8 \cdot 8} = \dfrac{1}{64}$

33. $\dfrac{12}{7} \div (4x) = \dfrac{12}{7} \cdot \dfrac{1}{4x} = \dfrac{4 \cdot 3 \cdot 1}{7 \cdot 4 \cdot x} = \dfrac{4}{4} \cdot \dfrac{3 \cdot 1}{7 \cdot x} = \dfrac{3}{7x}$

35. $(-12) \div \dfrac{3}{2} = -12 \cdot \dfrac{2}{3} = -\dfrac{3 \cdot 4 \cdot 2}{3 \cdot 1}$

$\qquad = -\dfrac{3}{3} \cdot \dfrac{4 \cdot 2}{1} = -\dfrac{8}{1} = -8$

37. $28 \div \dfrac{4}{5a} = 28 \cdot \dfrac{5a}{4} = \dfrac{28 \cdot 5a}{4} = \dfrac{4 \cdot 7 \cdot 5 \cdot a}{4 \cdot 1} = \dfrac{4}{4} \cdot \dfrac{7 \cdot 5 \cdot a}{1}$

$\qquad = 35a$

39. $\left(-\dfrac{5}{8}\right) \div \left(-\dfrac{5}{8}\right) = -\dfrac{5}{8}\left(-\dfrac{8}{5}\right) = \dfrac{5 \cdot 8}{8 \cdot 5} = \dfrac{5 \cdot 8}{5 \cdot 8} = 1$

41. $\dfrac{-8}{15} \div \dfrac{4}{5} = \dfrac{-8}{15} \cdot \dfrac{5}{4} = \dfrac{-8 \cdot 5}{15 \cdot 4} = \dfrac{-2 \cdot 4 \cdot 5}{3 \cdot 5 \cdot 4} =$

$\dfrac{4 \cdot 5}{4 \cdot 5} \cdot \dfrac{-2}{3} = \dfrac{-2}{3},$ or $-\dfrac{2}{3}$

43. $\dfrac{77}{64} \div \dfrac{49}{18} = \dfrac{77}{64} \cdot \dfrac{18}{49} = \dfrac{7 \cdot 11 \cdot 2 \cdot 9}{2 \cdot 32 \cdot 7 \cdot 7} =$

$\dfrac{2 \cdot 7}{2 \cdot 7} \cdot \dfrac{11 \cdot 9}{32 \cdot 7} = \dfrac{99}{224}$

45. $120a \div \dfrac{45}{14} = 120a \cdot \dfrac{14}{45} = \dfrac{8 \cdot 15 \cdot a \cdot 14}{3 \cdot 15} =$

$\dfrac{15}{15} \cdot \dfrac{8 \cdot a \cdot 14}{3} = \dfrac{112a}{3}$

47. $\dfrac{\frac{2}{5}}{\frac{3}{7}} = \dfrac{2}{5} \div \dfrac{3}{7} = \dfrac{2}{5} \cdot \dfrac{7}{3} = \dfrac{2 \cdot 7}{5 \cdot 3} = \dfrac{14}{15}$

49. $\dfrac{\frac{7}{20}}{\frac{8}{5}} = \dfrac{7}{20} \div \dfrac{8}{5} = \dfrac{7}{20} \cdot \dfrac{5}{8} = \dfrac{7 \cdot 5}{20 \cdot 8} = \dfrac{7 \cdot 5}{4 \cdot 5 \cdot 8} =$

$\dfrac{5}{5} \cdot \dfrac{7}{4 \cdot 8} = \dfrac{7}{32}$

51. $\dfrac{-\frac{15}{8}}{\frac{9}{10}} = -\dfrac{15}{8} \div \dfrac{9}{10} = -\dfrac{15}{8} \cdot \dfrac{10}{9} = -\dfrac{15 \cdot 10}{8 \cdot 9} =$

$-\dfrac{3 \cdot 5 \cdot 2 \cdot 5}{2 \cdot 4 \cdot 3 \cdot 3} = -\dfrac{5 \cdot 5}{4 \cdot 3} \cdot \dfrac{3 \cdot 2}{3 \cdot 2} = -\dfrac{25}{12}$

53. $\dfrac{-\frac{9}{16}}{-\frac{6}{5}} = -\dfrac{9}{16} \div \left(-\dfrac{6}{5}\right) = -\dfrac{9}{16} \cdot \left(-\dfrac{5}{6}\right) =$

$\dfrac{9 \cdot 5}{16 \cdot 6} = \dfrac{3 \cdot 3 \cdot 5}{16 \cdot 2 \cdot 3} = \dfrac{3}{3} \cdot \dfrac{3 \cdot 5}{16 \cdot 2} = \dfrac{15}{32}$

55. Discussion and Writing Exercise

57. The equation $14 + (2 + 30) = (14 + 2) + 30$ illustrates the <u>associative</u> law of addition.

59. A natural number that has exactly two different factors, only itself and 1, is called a <u>prime</u> number.

61. Since $a + 0 = a$ for any number a, the number 0 is the <u>additive</u> identity.

63. The sum of 6 and -6 is 0; we say that 6 and -6 are <u>opposites</u> of each other.

65. Discussion and Writing Exercise

67. $\left(\dfrac{4}{15} \div \dfrac{2}{25}\right)^2 = \left(\dfrac{4}{15} \cdot \dfrac{25}{2}\right)^2$

$\qquad = \left(\dfrac{4 \cdot 25}{15 \cdot 2}\right)^2$

$\qquad = \left(\dfrac{2 \cdot 2 \cdot 5 \cdot 5}{3 \cdot 5 \cdot 2}\right)^2$

$\qquad = \left(\dfrac{2 \cdot 5}{2 \cdot 5} \cdot \dfrac{2 \cdot 5}{3}\right)$

$\qquad = \left(\dfrac{10}{3}\right)^2$

$\qquad = \dfrac{100}{9}$

69. $\left(\dfrac{9}{10} \div \dfrac{2}{5} \div \dfrac{3}{8}\right)^2 = \left(\dfrac{9}{10} \cdot \dfrac{5}{2} \div \dfrac{3}{8}\right)^2$

$\qquad = \left(\dfrac{9 \cdot 5}{10 \cdot 2} \div \dfrac{3}{8}\right)^2$

$\qquad = \left(\dfrac{9 \cdot 5}{2 \cdot 5 \cdot 2} \div \dfrac{3}{8}\right)^2$

$\qquad = \left(\dfrac{9}{2 \cdot 2} \div \dfrac{3}{8}\right)^2$

$\qquad = \left(\dfrac{9}{2 \cdot 2} \cdot \dfrac{8}{3}\right)^2$

$\qquad = \left(\dfrac{9 \cdot 8}{2 \cdot 2 \cdot 3}\right)^2$

$\qquad = \left(\dfrac{3 \cdot 3 \cdot 2 \cdot 2 \cdot 2}{2 \cdot 2 \cdot 3 \cdot 1}\right)^2$

$\qquad = \left(\dfrac{3 \cdot 2}{1}\right)^2$

$\qquad = 6^2$

$\qquad = 36$

71. $\left(\dfrac{14}{15} \div \dfrac{49}{65} \cdot \dfrac{77}{260}\right)^2 = \left(\dfrac{14}{15} \cdot \dfrac{65}{49} \cdot \dfrac{77}{260}\right)^2$

$\qquad = \left(\dfrac{2 \cdot 7 \cdot 5 \cdot 13 \cdot 7 \cdot 11}{3 \cdot 5 \cdot 7 \cdot 7 \cdot 2 \cdot 2 \cdot 5 \cdot 13}\right)^2$

$\qquad = \left(\dfrac{2 \cdot 5 \cdot 7 \cdot 7 \cdot 13}{2 \cdot 5 \cdot 7 \cdot 7 \cdot 13} \cdot \dfrac{11}{2 \cdot 3 \cdot 5}\right)^2$

$\qquad = \left(\dfrac{11}{30}\right)^2$

$\qquad = \dfrac{121}{900}$

73. Use a calculator.

$\dfrac{711}{1957} \div \dfrac{10{,}033}{13{,}081} = \dfrac{711}{1957} \cdot \dfrac{13{,}081}{10{,}033}$

$\qquad = \dfrac{711 \cdot 13{,}081}{1957 \cdot 10{,}033}$

$\qquad = \dfrac{3 \cdot 3 \cdot 79 \cdot 103 \cdot 127}{19 \cdot 103 \cdot 79 \cdot 127}$

$\qquad = \dfrac{79 \cdot 103 \cdot 127}{79 \cdot 103 \cdot 127} \cdot \dfrac{3 \cdot 3}{19}$

$\qquad = \dfrac{9}{19}$

75. Use a calculator.

$\dfrac{451}{289} \div \dfrac{123}{340} = \dfrac{451}{289} \cdot \dfrac{340}{123}$

$\qquad = \dfrac{451 \cdot 340}{289 \cdot 123}$

$\qquad = \dfrac{11 \cdot 41 \cdot 17 \cdot 20}{17 \cdot 17 \cdot 3 \cdot 41}$

$\qquad = \dfrac{41 \cdot 17}{41 \cdot 17} \cdot \dfrac{11 \cdot 20}{17 \cdot 3}$

$\qquad = \dfrac{220}{51}$

Exercise Set 3.8

1. $\qquad \dfrac{4}{5}x = 12$

$\dfrac{5}{4} \cdot \dfrac{4}{5}x = \dfrac{5}{4} \cdot 12 \qquad$ The reciprocal of $\dfrac{4}{5}$ is $\dfrac{5}{4}$.

$1x = \dfrac{5 \cdot 4 \cdot 3}{4}$

$x = 15 \qquad$ Removing the factor $\dfrac{4}{4}$

Check: $\qquad \dfrac{4}{5}x = 12$

$\qquad \dfrac{4}{5} \cdot 15 \ ? \ 12$

$\qquad \dfrac{4 \cdot 3 \cdot 5}{5 \cdot 1}$

$\qquad 12 \ \Big|\ 12 \qquad$ TRUE

The solution is 15.

3. $\qquad \dfrac{7}{3}a = 21$

$\dfrac{3}{7} \cdot \dfrac{7}{3}a = \dfrac{3}{7} \cdot 21 \qquad$ The reciprocal of $\dfrac{7}{3}$ is $\dfrac{3}{7}$.

$1a = \dfrac{3 \cdot 3 \cdot 7}{7}$

$a = 9 \qquad$ Removing the factor $\dfrac{7}{7}$

Check: $\qquad \dfrac{7}{3}a = 21$

$\qquad \dfrac{7}{3} \cdot 9 \ ? \ 21$

$\qquad \dfrac{7 \cdot 3 \cdot 3}{3 \cdot 1}$

$\qquad 21 \ \Big|\ 21 \qquad$ TRUE

The solution is 9.

5. $\qquad \dfrac{2}{9}x = -10$

$\dfrac{9}{2} \cdot \dfrac{2}{9}x = \dfrac{9}{2}(-10) \qquad$ The reciprocal of $\dfrac{2}{9}$ is $\dfrac{9}{2}$.

$1x = -\dfrac{9 \cdot 2 \cdot 5}{2 \cdot 1}$

$x = -45 \qquad$ Removing the factor $\dfrac{2}{2}$

Check: $\qquad \dfrac{2}{9}x = -10$

$\qquad \dfrac{2}{9}(-45) \ ? \ -10$

$\qquad -\dfrac{2 \cdot 5 \cdot 9}{9 \cdot 1}$

$\qquad -10 \ \Big|\ -10 \qquad$ TRUE

The solution is -45.

7. $\qquad 6a = \dfrac{12}{17}$

$\dfrac{1}{6} \cdot 6a = \dfrac{1}{6} \cdot \dfrac{12}{17} \qquad$ The reciprocal of 6 is $\dfrac{1}{6}$.

$1a = \dfrac{2 \cdot 6}{6 \cdot 17}$

$a = \dfrac{2}{17} \qquad$ Removing the factor $\dfrac{6}{6}$

Check: $\qquad 6a = \dfrac{12}{17}$

$\qquad 6 \cdot \dfrac{2}{17} \ ? \ \dfrac{12}{17}$

$\qquad \dfrac{12}{17} \ \Big|\ \dfrac{12}{17} \qquad$ TRUE

The solution is $\dfrac{2}{17}$.

9. $\frac{1}{4}x = \frac{3}{5}$

$\frac{4}{1} \cdot \frac{1}{4}x = \frac{4}{1} \cdot \frac{3}{5}$

$x = \frac{12}{5}$

$\frac{12}{5}$ checks and is the solution.

11. $\frac{3}{2}t = -\frac{8}{7}$

$\frac{2}{3} \cdot \frac{3}{2}t = \frac{2}{3}\left(-\frac{8}{7}\right)$

$t = -\frac{16}{21}$

$-\frac{16}{21}$ checks and is the solution.

13. $\frac{4}{5} = -10a$

$-\frac{1}{10} \cdot \frac{4}{5} = -\frac{1}{10}(-10a)$

$-\frac{2 \cdot 2}{2 \cdot 5 \cdot 5} = a$

$-\frac{2}{25} = a$

$-\frac{2}{25}$ checks and is the solution.

15. $\frac{9}{5}x = \frac{3}{10}$

$\frac{5}{9} \cdot \frac{9}{5}x = \frac{5}{9} \cdot \frac{3}{10}$

$x = \frac{5 \cdot 3 \cdot 1}{3 \cdot 3 \cdot 2 \cdot 5}$

$x = \frac{1}{6}$

$\frac{1}{6}$ checks and is the solution.

17. $-\frac{9}{10}x = 8$

$-\frac{10}{9}\left(-\frac{9}{10}x\right) = -\frac{10}{9} \cdot 8$

$x = -\frac{10 \cdot 8}{9}$

$x = -\frac{80}{9}$

$-\frac{80}{9}$ checks and is the solution.

19. $a \cdot \frac{9}{7} = -\frac{3}{14}$

$a \cdot \frac{9}{7} \cdot \frac{7}{9} = -\frac{3}{14} \cdot \frac{7}{9}$

$a \cdot 1 = -\frac{3 \cdot 7 \cdot 1}{2 \cdot 7 \cdot 3 \cdot 3}$

$a = -\frac{1}{6}$

$-\frac{1}{6}$ checks and is the solution.

21. $-x = \frac{7}{13}$

$-1(-x) = -1 \cdot \frac{7}{13}$

$x = -\frac{7}{13}$

$-\frac{7}{13}$ checks and is the solution.

23. $-x = -\frac{27}{31}$

$-1(-x) = -1\left(-\frac{27}{31}\right)$

$x = \frac{27}{31}$

$\frac{27}{31}$ checks and is the solution.

25. $7t = 6$

$\frac{1}{7} \cdot 7t = \frac{1}{7} \cdot 6$

$t = \frac{6}{7}$

$\frac{6}{7}$ checks and is the solution.

27. $-24 = -10a$

$-\frac{1}{10}(-24) = -\frac{1}{10}(-10a)$

$\frac{2 \cdot 12}{2 \cdot 5} = a$

$\frac{12}{5} = a$

$\frac{12}{5}$ checks and is the solution.

29. $-\frac{14}{9} = \frac{10}{3}t$

$\frac{3}{10}\left(-\frac{14}{9}\right) = \frac{3}{10} \cdot \frac{10}{3}t$

$-\frac{3 \cdot 2 \cdot 7}{2 \cdot 5 \cdot 3 \cdot 3} = t$

$-\frac{7}{15} = t$

$-\frac{7}{15}$ checks and is the solution.

31. $n \cdot \frac{4}{15} = \frac{12}{25}$

$n \cdot \frac{4}{15} \cdot \frac{15}{4} = \frac{12}{25} \cdot \frac{15}{4}$

$n = \frac{4 \cdot 3 \cdot 3 \cdot 5}{5 \cdot 5 \cdot 4}$

$n = \frac{9}{5}$

$\frac{9}{5}$ checks and is the solution.

33.
$$-\frac{7}{20}x = -\frac{21}{10}$$

$$-\frac{20}{7}\left(-\frac{7}{20}x\right) = -\frac{20}{7}\left(-\frac{21}{10}\right)$$

$$x = \frac{2\cdot 10\cdot 3\cdot 7}{7\cdot 10}$$

$$x = 6$$

6 checks and is the solution.

35.
$$-\frac{25}{17} = -\frac{35}{34}a$$

$$-\frac{34}{35}\left(-\frac{25}{17}\right) = -\frac{34}{35}\left(-\frac{35}{34}a\right)$$

$$\frac{2\cdot 17\cdot 5\cdot 5}{5\cdot 7\cdot 17} = a$$

$$\frac{10}{7} = a$$

$\frac{10}{7}$ checks and is the solution.

37. Familiarize. We draw a picture. Let $t =$ the number of times Benny will be able to brush his teeth.

t brushings

Translate. The multiplication that corresponds to the situation is
$$\frac{2}{5}\cdot t = 30.$$

Solve. We solve the equation by dividing on both sides by $\frac{2}{5}$ and carrying out the division:

$$t = 30 \div \frac{2}{5} = 30\cdot\frac{5}{2} = \frac{2\cdot 15\cdot 5}{2\cdot 1} = \frac{2}{2}\cdot\frac{15\cdot 5}{1} = 75$$

Check. We repeat the calculation. The answer checks.

State. Benny can brush his teeth 75 times with a 30-g tube of toothpaste.

39. Familiarize. Let $g =$ the number of gallons of gasoline the tanker holds when it is fully loaded.

Translate. We translate to an equation.

$$\underbrace{1400\text{ gal}}_{1400} \text{ is } \frac{7}{9} \text{ of } \underbrace{\text{a full load}}_{g}$$

$$1400 = \frac{7}{9}\cdot g$$

Solve. We solve the equation.

$$1400 = \frac{7}{9}\cdot g$$

$$1400 \div \frac{7}{9} = g$$

$$1400 \cdot \frac{9}{7} = g$$

$$\frac{1400\cdot 9}{7} = g$$

$$\frac{7\cdot 200\cdot 9}{7\cdot 1} = g$$

$$\frac{7}{7}\cdot\frac{200\cdot 9}{1} = g$$

$$1800 = g$$

Check. $\frac{7}{9}$ of 1800 gal is $\frac{7}{9}\cdot 1800 = \frac{7\cdot 1800}{9} = \frac{7\cdot 9\cdot 200}{9\cdot 1} = \frac{9}{9}\cdot\frac{7\cdot 200}{1} = 1400$ gal. The answer checks.

State. The tanker holds 1800 gal of gasoline when it is full.

41. Familiarize. Let $w =$ the number of worker bees it takes to produce $\frac{3}{4}$ tsp of honey.

Translate.

Amount produced by one bee	times	Number of bees	is	$\frac{3}{4}$ tsp
↓	↓	↓	↓	↓
$\frac{1}{12}$	$\cdot$	w	$=$	$\frac{3}{4}$

Solve. We solve the equation.

$$\frac{1}{12}\cdot w = \frac{3}{4}$$

$$w = \frac{3}{4}\div\frac{1}{12} = \frac{3}{4}\cdot\frac{12}{1} = \frac{3\cdot 12}{4\cdot 1}$$

$$= \frac{3\cdot 3\cdot 4}{4\cdot 1} = \frac{3\cdot 3}{1}\cdot\frac{4}{4} = \frac{3\cdot 3}{1}$$

$$= 9$$

Check. Since $\frac{1}{12}\cdot 9 = \frac{9}{12} = \frac{3}{4}$, the answer checks.

State. It takes 9 worker bees to produce $\frac{3}{4}$ tsp of honey.

43. Familiarize. We make a drawing. Let $p =$ the number of packages that can be made from 15 lb of cheese.

p packages

Translate. The problem translates to the following equation:

$$p = 15 \div \frac{3}{4}.$$

Solve. We carry out the division.

$$p = 15 \div \frac{3}{4}$$

$$= 15 \cdot \frac{4}{3}$$

$$= \frac{3 \cdot 5 \cdot 4}{3 \cdot 1} = \frac{3}{3} \cdot \frac{5 \cdot 4}{1}$$

$$= 20$$

Check. If 20 packages, each containing $\frac{3}{4}$ lb of cheese, are made, a total of

$$20 \cdot \frac{3}{4} = \frac{5 \cdot 4 \cdot 3}{4 \cdot 1} = \frac{4}{4} \cdot \frac{5 \cdot 3}{1} = 15,$$

or 15 lb of cheese is used. The answer checks.

State. 20 packages can be made.

45. Familiarize. Let c = the amount of clay each art department will receive, in tons.

Translate. The problem translates to the following equation:

$$c = \frac{3}{4} \div 6.$$

Solve. We carry out the division.

$$c = \frac{3}{4} \div 6$$

$$= \frac{3}{4} \cdot \frac{1}{6}$$

$$= \frac{3 \cdot 1}{4 \cdot 2 \cdot 3} = \frac{3}{3} \cdot \frac{1}{4 \cdot 2}$$

$$= \frac{1}{8}$$

Check. If each of 6 art departments get $\frac{1}{8}$ T of clay, the total amount of clay is

$$6 \cdot \frac{1}{8} = \frac{6 \cdot 1}{8} = \frac{2 \cdot 3 \cdot 1}{2 \cdot 4} = \frac{3}{4} \text{ T.}$$

The answer checks.

State. Each art department will receive $\frac{1}{8}$ T of clay.

47. Familiarize. We make a drawing. Let w = the number of walkways that can be covered with 6 yd of gravel.

$$\underbrace{\boxed{\frac{3}{4} \text{ yd}} \quad \boxed{\frac{3}{4} \text{ yd}} \quad \cdots \quad \boxed{\frac{3}{4} \text{ yd}}}_{w \text{ walkways}}$$

Translate. The problem translates to the following situation:

$$w = 6 \div \frac{3}{4}.$$

Solve. We carry out the division.

$$w = 6 \div \frac{3}{4}$$

$$= 6 \cdot \frac{4}{3}$$

$$= \frac{2 \cdot 3 \cdot 4}{3 \cdot 1} = \frac{3}{3} \cdot \frac{2 \cdot 4}{1}$$

$$= 8$$

Check. If each of 8 walkways is covered with $\frac{3}{4}$ yd of gravel, a total of

$$8 \cdot \frac{3}{4} = \frac{2 \cdot 4 \cdot 3}{4 \cdot 1} = \frac{4}{4} \cdot \frac{2 \cdot 3}{1} = 6,$$

or 6 yd of gravel is used. The answer checks.

State. 8 walkways can be covered with one dump truck load of gravel.

49. Familiarize. We make a drawing. Let c = the number of customers that can be accommodated with a 30 yd batch of mulch.

$$\underbrace{\boxed{\frac{2}{3} \text{ yd}} \quad \boxed{\frac{2}{3} \text{ yd}} \quad \cdots \quad \boxed{\frac{2}{3} \text{ yd}}}_{c \text{ customers}}$$

Translate. The problem translates to the following situation.

$$c = 30 \div \frac{2}{3}.$$

Solve. We carry out the division.

$$c = 30 \div \frac{2}{3}$$

$$= 30 \cdot \frac{3}{2}$$

$$= \frac{2 \cdot 15 \cdot 3}{2 \cdot 1} = \frac{2}{2} \cdot \frac{15 \cdot 3}{1}$$

$$= 45$$

Check. If each of 45 customers gets $\frac{2}{3}$ yd of mulch, a total of

$$45 \cdot \frac{2}{3} = \frac{3 \cdot 15 \cdot 2}{3 \cdot 1} = \frac{3}{3} \cdot \frac{15 \cdot 2}{1} = 30,$$

or 30 yd of mulch is used. The answer checks.

State. 45 customers can be accommodated with 30 yd of mulch.

51. Familiarize. We draw a picture.

$$\underbrace{\boxed{}\boxed{}\boxed{} \cdots \boxed{}}_{\frac{3}{4} \text{ yd per pair}} \Big\} \text{ 24 yd makes how many pairs?}$$

We let s = the number of pairs of basketball shorts that can be made.

Translate. The problem translates to the following equation:

$$s = 24 \div \frac{3}{4}.$$

Solve. We carry out the division.

$$s = 24 \div \frac{3}{4}$$

$$= 24 \cdot \frac{4}{3}$$

$$= \frac{3 \cdot 8 \cdot 4}{1 \cdot 3} = \frac{3}{3} \cdot \frac{8 \cdot 4}{1}$$

$$= 32$$

Check. If each of 32 pairs of shorts requires $\frac{3}{4}$ yd of nylon, a total of

$$32 \cdot \frac{3}{4} = \frac{32 \cdot 3}{4} = \frac{4 \cdot 8 \cdot 3}{4} = 8 \cdot 3,$$

or 24 yd of nylon is needed. Our answer checks.

State. 32 pairs of basketball shorts can be made from 24 yd of nylon.

53. Familiarize. Let p = the pitch of the screw, in inches. The distance the screw has traveled into the wallboard is found by multiplying the pitch by the number of complete rotations.

Translate. We translate to an equation.

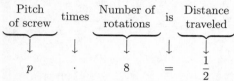

$$p \cdot 8 = \frac{1}{2}$$

Solve. We divide on both sides of the equation by 8 and carry out the division.

$$p = \frac{1}{2} \div 8 = \frac{1}{2} \cdot \frac{1}{8} = \frac{1 \cdot 1}{2 \cdot 8} = \frac{1}{16}$$

Check. We repeat the calculation. The answer checks.

State. The pitch of the screw is $\frac{1}{16}$ in.

55. Discussion and Writing Exercise

57. $-23 + 49 = 26$

(Find the difference of the absolute values. The positive integer has the larger absolute value, so the answer is positive.)

59. $-38 - 29 = -67$

(Add the absolute values. The answer is negative.)

61. $\begin{aligned} 36 \div (-3)^2 \times (7 - 2) &= 36 \div (-3)^2 \times 5 \\ &= 36 \div 9 \times 5 \\ &= 4 \times 5 \\ &= 20 \end{aligned}$

63. $13x + 4x = (13 + 4)x = 17x$

65. $\begin{aligned} 2a + 3 + 5a &= 2a + 5a + 3 \\ &= (2 + 5)a + 3 \\ &= 7a + 3 \end{aligned}$

67. Discussion and Writing Exercise

69. $\begin{aligned} 2x - 7x &= -\frac{10}{9} \\ -5x &= -\frac{10}{9} \\ -\frac{1}{5}(-5x) &= -\frac{1}{5}\left(-\frac{10}{9}\right) \\ x &= \frac{2 \cdot 5}{5 \cdot 9} \\ x &= \frac{2}{9} \end{aligned}$

71. Familiarize. Let w = the weight of the package when it is completely filled.

Translate.

$$\underbrace{\frac{3}{4}}_{\frac{3}{4}} \text{ of } \underbrace{\text{total weight}}_{w} \text{ is } \underbrace{\frac{21}{32}}_{\frac{21}{32}} \text{ lb}$$

$$\frac{3}{4} \cdot w = \frac{21}{32}$$

Solve. We solve the equation.

$$\frac{3}{4} \cdot w = \frac{21}{32}$$

$$\frac{4}{3} \cdot \frac{3}{4} \cdot w = \frac{4}{3} \cdot \frac{21}{32}$$

$$w = \frac{4 \cdot 3 \cdot 7}{3 \cdot 4 \cdot 8} = \frac{4 \cdot 3}{4 \cdot 3} \cdot \frac{7}{8}$$

$$w = \frac{7}{8}$$

Check. We find $\frac{3}{4}$ of $\frac{7}{8}$ lb.

$$\frac{3}{4} \cdot \frac{7}{8} = \frac{3 \cdot 7}{4 \cdot 8} = \frac{21}{32} \text{ lb}$$

The answer checks.

State. The package could hold $\frac{7}{8}$ lb of coffee beans when it is completely filled.

73. Familiarize. Let x = the number of slices yielded by the $\frac{3}{32}$-in. cuts and y = the number of slices yielded by the $\frac{5}{32}$-in. cuts. Half the block is $\frac{1}{2} \cdot 12$ in., or 6 in.

Translate. The problem translates to the following situations.

$$x = 6 \div \frac{3}{32} \quad \text{and} \quad y = 6 \div \frac{5}{32}$$

Solve. We carry out the division.

$$\begin{aligned} x &= 6 \div \frac{3}{32} \\ &= 6 \cdot \frac{32}{3} \\ &= \frac{2 \cdot 3 \cdot 32}{3 \cdot 1} = \frac{3}{3} \cdot \frac{2 \cdot 32}{1} \\ &= 64 \end{aligned}$$

$$\begin{aligned} y &= 6 \div \frac{5}{32} \\ &= 6 \cdot \frac{32}{5} \\ &= \frac{192}{5} \\ &= 38 \text{ R } 2 \end{aligned}$$

The $\frac{3}{32}$-in. cuts yield 64 slices. The $\frac{5}{32}$-in. cuts yield 38 slices that are $\frac{5}{32}$ in. thick and an additional slice (indicated by the remainder) that is less than $\frac{5}{32}$ in., so this cutting yields 39 slices. Then the total number of slices is $64 + 39$, or 103.

Check. We repeat the calculations. The answer checks.

State. The cutting will yield 103 slices of cheese.

75. *Familiarize*. Let w = the number of walkways that can be covered with 6 yd of gravel. Let c = the cost of covering each walkway. We will find w and c and then multiply to find how much Eric will receive.

Translate. The problem translates to the following situations.

$$w = 6 \div \frac{3}{5} \quad \text{and} \quad c = 85 \cdot \frac{3}{5}$$

Solve. We carry out the calculations.

$$w = 6 \div \frac{3}{5}$$
$$= 6 \cdot \frac{5}{3}$$
$$= \frac{2 \cdot 3 \cdot 5}{3 \cdot 1} = \frac{3}{3} \cdot \frac{2 \cdot 5}{1}$$
$$= 10$$

Eric will use his full load and will cover 10 walkways.

$$c = 85 \cdot \frac{3}{5}$$
$$= \frac{5 \cdot 17 \cdot 3}{5 \cdot 1} = \frac{5}{5} \cdot \frac{17 \cdot 3}{1}$$
$$= 51$$

If Eric is paid \$51 for each of 10 walkways, then he receives \$51 · 10 = \$510. (Note that, since we know Eric will use the full load of 6 yd of gravel, we could also have found this result by multiplying \$85 · 6 to get \$510.)

Check. We repeat the calculations. The answer checks.

State. Eric will receive \$510 for a full load.

77. *Familiarize*. First we find the number n of customers who can be provided with $\frac{3}{4}$ yd of mulch from a 25-yd batch. Then we will multiply to find the amount a that Green Season Gardening would receive for the mulch.

Translate. We have

$$n = 25 \div \frac{3}{4}.$$

Solve.

$$n = 25 \div \frac{3}{4}$$
$$= 25 \cdot \frac{4}{3}$$
$$= \frac{100}{3}$$
$$= 33 \text{ R } 1$$

Thus, 33 customers can receive $\frac{3}{4}$ yd of mulch. Now we multiply to find the amount Green Season Gardening would receive for the mulch.

$$a = 33 \cdot \frac{3}{4} \cdot 65$$
$$= \frac{6435}{4}$$
$$= 1608.75$$

Check. We repeat the calculations. The answer checks.

State. Green Season Gardening would receive \$1608.75 for the mulch. (Note that when each of 33 customers receives $\frac{3}{4}$ yd of mulch, then $33 \cdot \frac{3}{4}$, or $\frac{99}{4}$ yd is used. Since $25 = 25 \cdot \frac{4}{4} = \frac{100}{4}$, this means that $\frac{1}{4}$ yd is left unused after the $\frac{99}{4}$ yd are distributed.)

Chapter 3 Review Exercises

1.
$1 \cdot 8 = 8$	$6 \cdot 8 = 48$
$2 \cdot 8 = 16$	$7 \cdot 8 = 56$
$3 \cdot 8 = 24$	$8 \cdot 8 = 64$
$4 \cdot 8 = 32$	$9 \cdot 8 = 72$
$5 \cdot 8 = 40$	$10 \cdot 8 = 80$

2. 3920 is even because the ones digit is even; $3+9+2+0 = 14$ and 14 is not divisible by 3, so 3920 is not divisible by 3. Since 3920 is not divisible by 3, it is not divisible by 6.

3. Because $6 + 8 + 5 + 3 + 7 = 29$ and 29 is not divisible by 3, then 68,537 is not divisible by 3.

4. 673 is not divisible by 5 because the ones digit is neither 0 nor 5.

5. 4936 is divisible by 2 because the ones digit is even.

6. Because $5 + 2 + 3 + 8 = 18$ and 18 is divisible by 9, then 5238 is divisible by 9.

7. Since the ones digit of 60 is 0 we know that 2, 5, and 10 are factors. Since the sum of the digits is 6 and 6 is divisible by 3, then 3 is a factor. Since 2 and 3 are factors, 6 is also a factor. We write a list of factorizations.

$60 = 1 \cdot 60$	$60 = 4 \cdot 15$
$60 = 2 \cdot 30$	$60 = 5 \cdot 12$
$60 = 3 \cdot 20$	$60 = 6 \cdot 10$

Factors: 1, 2, 3, 4, 5, 6, 10, 12, 15, 20, 30, 60

8. 176 is even so 2 is a factor. None of the other tests for divisibility yields additional factors, so we find as many two-factor factorizations as we can:

$176 = 1 \cdot 176$	$176 = 8 \cdot 22$
$176 = 2 \cdot 88$	$176 = 11 \cdot 16$
$176 = 4 \cdot 44$	

Factors: 1, 2, 4, 8, 11, 16, 22, 44, 88, 176

9. The only factors of 37 are 1 and 37, so 37 is prime.

10. 1 is neither prime nor composite.

11. The number 91 has factors 1, 7, 13, and 91, so it is composite.

12.
$$\begin{array}{r} 7 \quad \leftarrow \text{ 7 is prime.} \\ 5 \overline{\smash{\big)}\, 3\,5} \\ 2 \overline{\smash{\big)}\, 7\,0} \end{array}$$

$70 = 2 \cdot 5 \cdot 7$

13.
$$
\begin{array}{r}
3 \leftarrow \text{3 is prime.} \\
3\,\overline{\big)\,9} \\
2\,\overline{\big)\,18} \\
2\,\overline{\big)\,36} \\
2\,\overline{\big)\,72}
\end{array}
$$
$$72 = 2 \cdot 2 \cdot 2 \cdot 3 \cdot 3$$

14.
$$
\begin{array}{r}
5 \leftarrow \text{5 is prime.} \\
3\,\overline{\big)\,15} \\
3\,\overline{\big)\,45}
\end{array}
$$
$$45 = 3 \cdot 3 \cdot 5$$

15.
$$
\begin{array}{r}
5 \leftarrow \text{5 is prime.} \\
5\,\overline{\big)\,25} \\
3\,\overline{\big)\,75} \\
2\,\overline{\big)\,150}
\end{array}
$$
$$150 = 2 \cdot 3 \cdot 5 \cdot 5$$

16.
$$
\begin{array}{r}
3 \leftarrow \text{3 is prime.} \\
3\,\overline{\big)\,9} \\
3\,\overline{\big)\,27} \\
3\,\overline{\big)\,81} \\
2\,\overline{\big)\,162} \\
2\,\overline{\big)\,324} \\
2\,\overline{\big)\,648}
\end{array}
$$
$$648 = 2 \cdot 2 \cdot 2 \cdot 3 \cdot 3 \cdot 3 \cdot 3$$

17.
$$
\begin{array}{r}
5 \leftarrow \text{5 is prime.} \\
5\,\overline{\big)\,25} \\
3\,\overline{\big)\,75} \\
2\,\overline{\big)\,150} \\
2\,\overline{\big)\,300} \\
2\,\overline{\big)\,600} \\
2\,\overline{\big)\,1200}
\end{array}
$$
$$1200 = 2 \cdot 2 \cdot 2 \cdot 2 \cdot 3 \cdot 5 \cdot 5$$

18. The top number is the numerator, and the bottom number is the denominator.
$$\begin{array}{l} 9 \leftarrow \text{Numerator} \\ \overline{7} \leftarrow \text{Denominator} \end{array}$$

19. The object is divided into 8 equal parts. The unit is $\frac{1}{8}$. The denominator is 8. We have 3 parts shaded. This tells us that the numerator is 3. Thus, $\frac{3}{8}$ is shaded.

20. We can regard this as 2 bars of 6 parts each and take 7 of those parts. The unit is $\frac{1}{6}$. The denominator is 6 and the numerator is 7. Thus, $\frac{7}{6}$ is shaded.

21. a) The ratio is $\frac{3}{5}$.

b) The ratio is $\frac{5}{3}$.

c) There are $3 + 5$, or 8, members of the committee. The desired ratio is $\frac{3}{8}$.

22. $\frac{0}{n} = 0$, for any number n that is not 0.
$$\frac{0}{6} = 0$$

23. $\frac{n}{n} = 1$, for any number n that is not 0.
$$\frac{74}{74} = 1$$

24. $\frac{n}{1} = n$, for any number n.
$$\frac{48}{1} = 48$$

25. Remember: $\frac{n}{n} = 1$ for any number n that is not 0.
$$\frac{7x}{7x} = 1$$

26. $-\frac{10}{15} = -\frac{2 \cdot 5}{3 \cdot 5} = -\frac{2}{3} \cdot \frac{5}{5} = -\frac{2}{3}$

27. $\frac{7}{28} = \frac{7 \cdot 1}{7 \cdot 4} = \frac{7}{7} \cdot \frac{1}{4} = \frac{1}{4}$

28. $\frac{-42}{42} = \frac{-1 \cdot 42}{1 \cdot 42} = \frac{-1}{1} \cdot \frac{42}{42} = \frac{-1}{1} = -1$

29. $\frac{9m}{12m} = \frac{3 \cdot 3 \cdot m}{3 \cdot 4 \cdot m} = \frac{3 \cdot m}{3 \cdot m} \cdot \frac{3}{4} = \frac{3}{4}$

30. $\frac{12}{30} = \frac{2 \cdot 6}{5 \cdot 6} = \frac{2}{5} \cdot \frac{6}{6} = \frac{2}{5}$

31. Remember: $\frac{n}{0}$ is not defined.
$$\frac{-27}{0} \text{ is undefined.}$$

32. Remember: $\frac{n}{1} = n$
$$\frac{6x}{1} = 6x$$

33. $\frac{-9}{-27} = \frac{-9 \cdot 1}{-9 \cdot 3} = \frac{-9}{-9} \cdot \frac{1}{3} = \frac{1}{3}$

34. Since $21 \div 7 = 3$, we multiply by $\frac{3}{3}$.
$$\frac{5}{7} = \frac{5}{7} \cdot \frac{3}{3} = \frac{5 \cdot 3}{7 \cdot 3} = \frac{15}{21}$$

35. Since $55 \div 11 = 5$, we multiply by $\frac{5}{5}$.
$$\frac{-6}{11} = \frac{-6}{11} \cdot \frac{5}{5} = \frac{-6 \cdot 5}{11 \cdot 5} = \frac{-30}{55}$$

36. 3 and 100 have no prime factors in common, so $\frac{3}{100}$ cannot be simplified.
$$\frac{8}{100} = \frac{2 \cdot 4}{25 \cdot 4} = \frac{2}{25} \cdot \frac{4}{4} = \frac{2}{25}$$
$$\frac{10}{100} = \frac{10 \cdot 1}{10 \cdot 10} = \frac{10}{10} \cdot \frac{1}{10} = \frac{1}{10}$$
$$\frac{15}{100} = \frac{3 \cdot 5}{20 \cdot 5} = \frac{3}{20} \cdot \frac{5}{5} = \frac{3}{20}$$

21 and 100 have no prime factors in common, so $\frac{21}{100}$ cannot be simplified.

43 and 100 have no prime factors in common, so $\frac{43}{100}$ cannot be simplified.

37. We multiply these   We multiply these
         two numbers: two numbers:

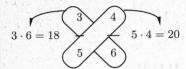

$$3 \cdot 6 = 18 \qquad 5 \cdot 4 = 20$$

Since $18 \neq 20$, $\frac{3}{5}$ and $\frac{4}{6}$ do not name the same number. Thus, $\frac{3}{5} \neq \frac{4}{6}$.

38. We multiply these   We multiply these
         two numbers: two numbers:

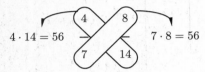

$$4 \cdot 14 = 56 \qquad 7 \cdot 8 = 56$$

Since $56 = 56$, $\frac{4}{7} = \frac{8}{14}$.

39. We multiply these   We multiply these
         two numbers: two numbers:

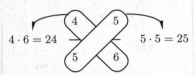

$$4 \cdot 6 = 24 \qquad 5 \cdot 5 = 25$$

Since $24 \neq 25$, $\frac{4}{5}$ and $\frac{5}{6}$ do not name the same number. Thus, $\frac{4}{5} \neq \frac{5}{6}$.

40. We multiply these   We multiply these
         two numbers: two numbers:

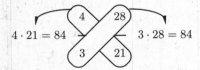

$$4 \cdot 21 = 84 \qquad 3 \cdot 28 = 84$$

Since $84 = 84$, $\frac{4}{3} = \frac{28}{21}$.

41. Interchange the numerator and denominator.

The reciprocal of $\frac{2}{13}$ is $\frac{13}{2}$. $\left(\frac{2}{13} \cdot \frac{13}{2} = \frac{26}{26} = 1 \right)$

42. Think of -7 as $\frac{-7}{1}$. Interchange the numerator and denominator.

The reciprocal of -7 is $\frac{1}{-7}$, or $\frac{-1}{7}$, or $-\frac{1}{7}$.

$\left(-7 \cdot \left(-\frac{1}{7} \right) = \frac{7}{7} = 1 \right)$

43. Interchange the numerator and denominator.

The reciprocal of $\frac{1}{8}$ is $\frac{8}{1}$, or 8. $\left(\frac{1}{8} \cdot 8 = \frac{8}{8} = 1 \right)$

44. Interchange the numerator and denominator.

The reciprocal of $\frac{3x}{5y}$ is $\frac{5y}{3x}$. $\left(\frac{3x}{5y} \cdot \frac{5y}{3x} = \frac{15xy}{15xy} = 1 \right)$

45. $\frac{2}{9} \cdot \frac{7}{5} = \frac{2 \cdot 7}{9 \cdot 5} = \frac{14}{45}$

46. $\frac{3}{x} \cdot \frac{y}{7} = \frac{3 \cdot y}{x \cdot 7} = \frac{3y}{7x}$

47. $\frac{3}{4} \cdot \frac{8}{9} = \frac{3 \cdot 8}{4 \cdot 9} = \frac{3 \cdot 2 \cdot 4}{4 \cdot 3 \cdot 3} = \frac{3 \cdot 4}{3 \cdot 4} \cdot \frac{2}{3} = \frac{2}{3}$

48. $-\frac{5}{7} \cdot \frac{1}{10} = -\frac{5 \cdot 1}{7 \cdot 10} = -\frac{5 \cdot 1}{7 \cdot 2 \cdot 5} = -\frac{1}{7 \cdot 2} \cdot \frac{5}{5} = -\frac{1}{14}$

49. $\frac{3a}{10} \cdot \frac{2}{15a} = \frac{3a \cdot 2}{10 \cdot 15a} = \frac{3 \cdot a \cdot 2 \cdot 1}{2 \cdot 5 \cdot 3 \cdot 5 \cdot a} = \frac{2 \cdot 3 \cdot a}{2 \cdot 3 \cdot a} \cdot \frac{1}{5 \cdot 5} = \frac{1}{25}$

50. $\frac{4a}{7} \cdot \frac{7}{4a} = \frac{4a \cdot 7}{7 \cdot 4a} = 1$

51. $9 \div \frac{5}{3} = \frac{9}{1} \cdot \frac{3}{5} = \frac{9 \cdot 3}{1 \cdot 5} = \frac{27}{5}$

52. $\frac{3}{14} \div \frac{6}{7} = \frac{3}{14} \cdot \frac{7}{6} = \frac{3 \cdot 7}{14 \cdot 6} = \frac{3 \cdot 7 \cdot 1}{2 \cdot 7 \cdot 2 \cdot 3} = \frac{3 \cdot 7}{3 \cdot 7} \cdot \frac{1}{2 \cdot 2} = \frac{1}{2 \cdot 2} = \frac{1}{4}$

53. $120 \div \frac{3}{5} = 120 \cdot \frac{5}{3} = \frac{120 \cdot 5}{3} = \frac{3 \cdot 40 \cdot 5}{3 \cdot 1} = \frac{3}{3} \cdot \frac{40 \cdot 5}{1} = \frac{40 \cdot 5}{1} = 200$

54. $-\frac{5}{36} \div \left(-\frac{25}{12} \right) = -\frac{5}{36} \cdot \left(-\frac{12}{25} \right) = \frac{5 \cdot 12}{36 \cdot 25} = \frac{5 \cdot 12 \cdot 1}{3 \cdot 12 \cdot 5 \cdot 5} = \frac{5 \cdot 12}{5 \cdot 12} \cdot \frac{1}{3 \cdot 5} = \frac{1}{15}$

55. $21 \div \frac{7}{2a} = \frac{21}{1} \cdot \frac{2a}{7} = \frac{3 \cdot 7 \cdot 2a}{1 \cdot 7} = \frac{7}{7} \cdot \frac{3 \cdot 2a}{1} = 6a$

56. $-\frac{23}{25} \div \frac{23}{25} = -\frac{23}{25} \cdot \frac{25}{23} = -\frac{23 \cdot 25}{25 \cdot 23} = -1$

57. $\dfrac{\frac{21}{30}}{\frac{14}{15}} = \frac{21}{30} \cdot \frac{15}{14} = \frac{21 \cdot 15}{30 \cdot 14} = \frac{3 \cdot 7 \cdot 3 \cdot 5}{2 \cdot 3 \cdot 5 \cdot 2 \cdot 7} = \frac{3 \cdot 5 \cdot 7}{3 \cdot 5 \cdot 7} \cdot \frac{3}{2 \cdot 2} = \frac{3}{4}$

58. $\dfrac{-\frac{3}{40}}{-\frac{35}{54}} = -\frac{3}{40} \cdot \left(-\frac{35}{54} \right) = \frac{3 \cdot 35}{40 \cdot 54} = \frac{3 \cdot 5 \cdot 7}{5 \cdot 8 \cdot 3 \cdot 18} = \frac{3 \cdot 5}{3 \cdot 5} \cdot \frac{7}{8 \cdot 18} = \frac{7}{144}$

59. $A = \frac{1}{2} \cdot b \cdot h$

$A = \frac{1}{2} \cdot 14 \text{ m} \cdot 6 \text{ m}$

$A = \frac{14 \cdot 6}{2} \text{ m}^2$

$A = 42 \text{ m}^2$

60. $A = \dfrac{1}{2} \cdot b \cdot h$

$A = \dfrac{1}{2} \cdot \dfrac{7}{2} \text{ ft} \cdot 10 \text{ ft}$

$A = \dfrac{7 \cdot 10}{2 \cdot 2} \text{ ft}^2$

$A = \dfrac{35}{2} \text{ ft}^2$

61. $\dfrac{2}{3}x = 160$

$\dfrac{3}{2} \cdot \dfrac{2}{3}x = \dfrac{3}{2} \cdot 160$

$1x = \dfrac{3 \cdot 2 \cdot 80}{2}$

$x = 240$

The solution is 240.

62. $\dfrac{3}{8} = -\dfrac{5}{4}t$

$-\dfrac{4}{5} \cdot \dfrac{3}{8} = -\dfrac{4}{5}\left(-\dfrac{5}{4}t\right)$

$-\dfrac{4 \cdot 3}{5 \cdot 2 \cdot 4} = 1t$

$-\dfrac{3}{10} = t$

The solution is $-\dfrac{3}{10}$.

63. $-\dfrac{1}{7}n = -4$

$-7\left(-\dfrac{1}{7}n\right) = -7(-4)$

$n = 28$

The solution is 28.

64. *Familiarize*. Let $d =$ the number of days it will take to repave the road.

Translate.

Number of miles repaved each day	times	Number of days	is	Total number of miles repaved
↓	↓	↓	↓	↓
$\dfrac{1}{12}$	$\cdot$	d	$=$	$\dfrac{3}{4}$

Solve. We divide by $\dfrac{1}{12}$ on both sides of the equation.

$d = \dfrac{3}{4} \div \dfrac{1}{12}$

$d = \dfrac{3}{4} \cdot \dfrac{12}{1} = \dfrac{3 \cdot 12}{4 \cdot 1} = \dfrac{3 \cdot 3 \cdot 4}{4 \cdot 1}$

$= \dfrac{4}{4} \cdot \dfrac{3 \cdot 3}{1} = \dfrac{3 \cdot 3}{1} = 9$

Check. We repeat the calculation. The answer checks.

State. It will take 9 days to repave the road.

65. *Familiarize*. Let $s =$ the amount of sugar required for $\dfrac{1}{2}$ of the recipe, in cups. We want to find $\dfrac{1}{2}$ of $\dfrac{3}{4}$ cup.

Translate. We write a multiplication sentence.

$\dfrac{1}{2} \cdot \dfrac{3}{4} = s$

Solve. We carry out the multiplication.

$s = \dfrac{1}{2} \cdot \dfrac{3}{4} = \dfrac{1 \cdot 3}{2 \cdot 4} = \dfrac{3}{8}$

Check. We repeat the calculation. The answer checks.

State. $\dfrac{3}{8}$ cup of sugar should be used for $\dfrac{1}{2}$ of the recipe.

66. *Familiarize*. This is a multistep problem. First we find the length of the total trip. Then we find how many kilometers were left to drive. We draw a picture. We let $n =$ the length of the total trip.

$$\dfrac{5}{8} \text{ of the trip}$$

$$\underbrace{\hspace{2cm}180 \text{ km}\hspace{2cm}}$$

$$\underbrace{\hspace{4cm}n \text{ km}\hspace{4cm}}$$

Translate. We translate to an equation.

Fraction of trip completed	times	Total length of trip	is	Amount already traveled
↓	↓	↓	↓	↓
$\dfrac{5}{8}$	$\cdot$	n	$=$	180

Solve. We solve the equation as follows:

$\dfrac{5}{8} \cdot n = 180$

$n = 180 \div \dfrac{5}{8} = 180 \cdot \dfrac{8}{5} = \dfrac{5 \cdot 36 \cdot 8}{5 \cdot 1}$

$= \dfrac{5}{5} \cdot \dfrac{36 \cdot 8}{1} = \dfrac{36 \cdot 8}{1} = 288$

The total trip was 288 km.

Now we find how many kilometers were left to travel. Let $t =$ this number.

Length of total trip	minus	Distance traveled	is	Distance left to travel
↓	↓	↓	↓	↓
288	$-$	180	$=$	t

We carry out the subtraction:

$288 - 180 = t$

$108 = t$

Check. We repeat the calculation. The results check.

State. The total trip was 288 km. There were 108 km left to travel.

67. *Familiarize*. Let $d =$ the distance each person will swim, in miles.

Translate.

Number of swimmers	times	Distance each swims	is	Total distance
↓	↓	↓	↓	↓
4	$\cdot$	d	$=$	$\dfrac{2}{3}$

Solve. We solve the equation.

$$4 \cdot d = \frac{2}{3}$$

$$\frac{1}{4} \cdot 4 \cdot d = \frac{1}{4} \cdot \frac{2}{3}$$

$$d = \frac{1 \cdot 2}{4 \cdot 3} = \frac{1 \cdot 2}{2 \cdot 2 \cdot 3}$$

$$= \frac{2}{2} \cdot \frac{1}{2 \cdot 3} = \frac{1}{6}$$

Check. Since $4 \cdot \frac{1}{6} = \frac{4}{6} = \frac{2}{3}$, the answer checks.

State. Each person will swim $\frac{1}{6}$ mi.

68. Familiarize. Let $c = $ the number of metric tons of corn produced in the U.S. in 2003.

Translate.

$$\underbrace{\text{U.S. corn production}}_{\downarrow} \quad \underset{\downarrow\ \downarrow\ \downarrow}{\text{is } \frac{2}{5} \text{ of}} \quad \underbrace{\text{Total world corn production}}_{\downarrow}$$

$$c \quad = \frac{2}{5} \cdot \quad 640{,}000{,}000$$

Solve. We carry out the multiplication.

$$c = \frac{2}{5} \cdot 640{,}00{,}000 = \frac{2 \cdot 640{,}000{,}000}{5}$$

$$= \frac{2 \cdot 5 \cdot 128{,}000{,}000}{5 \cdot 1} = \frac{5}{5} \cdot \frac{2 \cdot 128{,}000{,}000}{1}$$

$$= \frac{2 \cdot 128{,}000{,}000}{1} = 256{,}000{,}000$$

Check. We repeat the calculation. The answer checks.

State. The U.S. produced 256,000,000 metric tons of corn in 2003.

69. *Discussion and Writing Exercise*. Taking $\frac{1}{2}$ of a number is equivalent to multiplying the number by $\frac{1}{2}$. Dividing by $\frac{1}{2}$ is equivalent to multiplying by the reciprocal of $\frac{1}{2}$, or 2. Thus taking $\frac{1}{2}$ of a number is not the same as dividing by $\frac{1}{2}$.

70. *Discussion and Writing Exercise*. Because $\frac{2}{8}$ simplifies to $\frac{1}{4}$, it is incorrect to suggest that $\frac{2}{8}$ is the simplified form of $\frac{20}{80}$.

71.

$$\frac{15x}{14z} \cdot \frac{17yz}{35xy} \div \left(-\frac{3}{7}\right)^2$$

$$= \frac{15x}{14z} \cdot \frac{17yz}{35xy} \div \frac{9}{49}$$

$$= \frac{15x \cdot 17yz}{14z \cdot 35xy} \div \frac{9}{49}$$

$$= \frac{15x \cdot 17yz}{14z \cdot 35xy} \cdot \frac{49}{9}$$

$$= \frac{15x \cdot 17yz \cdot 49}{14z \cdot 35xy \cdot 9}$$

$$= \frac{3 \cdot 5 \cdot x \cdot 17 \cdot y \cdot z \cdot 7 \cdot 7}{2 \cdot 7 \cdot z \cdot 5 \cdot 7 \cdot x \cdot y \cdot 3 \cdot 3}$$

$$= \frac{3 \cdot 5 \cdot 7 \cdot 7 \cdot x \cdot y \cdot z}{3 \cdot 5 \cdot 7 \cdot 7 \cdot x \cdot y \cdot z} \cdot \frac{17}{2 \cdot 3}$$

$$= \frac{17}{6}$$

72. The digit must be even and the sum of the digits must be divisible by 3. Let $d = $ the digit to be inserted. Then $5 + 7 + 4 + d$, or $16 + d$, must be divisible by 3. The only even digits for which $16 + d$ is divisible by 3 are 2 and 8.

73. 13 and 31 are both prime numbers, so 13 is a palindrome prime.

19 is prime but 91 is not ($91 = 7 \cdot 13$), so 19 is not a palindrome prime.

16 is not prime ($16 = 2 \cdot 8 = 4 \cdot 4$), so it is not a palindrome prime.

11 is prime and when its digits are reversed we have 11 again, so 11 is a palindrome prime.

15 is not prime ($15 = 3 \cdot 5$), so it is not a palindrome prime.

24 is not prime ($24 = 2 \cdot 12 = 3 \cdot 8 = 4 \cdot 6$), so it is not a palindrome prime.

29 is prime but 92 is not ($92 = 2 \cdot 46 = 4 \cdot 23$), so 29 is not a palindrome prime.

101 is prime and when its digits are reversed we get 101 again, so 101 is a palindrome prime.

201 is not prime ($201 = 3 \cdot 67$), so it is not a palindrome prime.

37 and 73 are both prime numbers, so 37 is a palindrome prime.

74.

$$\frac{19}{24} \div \frac{a}{b} = \frac{19}{24} \cdot \frac{b}{a} = \frac{19 \cdot b}{24 \cdot a} = \frac{187{,}853}{268{,}224}$$

Then, assuming the quotient has not been simplified, we have

$$19 \cdot b = 187{,}853 \quad \text{and} \quad 24 \cdot a = 268{,}224$$

$$b = \frac{187{,}853}{19} \quad \text{and} \quad a = \frac{268{,}224}{24}$$

$$b = 9887 \quad \text{and} \quad a = 11{,}176.$$

75.

$$\frac{1751}{267}x = \frac{3193}{2759}$$

$$\frac{267}{1751} \cdot \frac{1751}{267}x = \frac{267}{1751} \cdot \frac{3193}{2759}$$

$$x = \frac{267 \cdot 3193}{1751 \cdot 2759}$$

$$x = \frac{3 \cdot 89 \cdot 31 \cdot 103}{17 \cdot 103 \cdot 31 \cdot 89}$$

$$x = \frac{89 \cdot 31 \cdot 103}{89 \cdot 31 \cdot 103} \cdot \frac{3}{17}$$

$$x = \frac{3}{17}$$

The solution is $\frac{3}{17}$.

Chapter 3 Test

1. Because $5 + 6 + 8 + 2 = 21$ and 21 is divisible by 3, then 5682 is divisible by 3.

2. 7018 is not divisible by 5 because the ones digit is neither 0 nor 5.

3. Since the ones digit of 90 is 0 we know that 2, 5, and 10 are factors. Since the sum of the digits is 9 and 9 is divisible by both 3 and 9, we know that 3 and 9 are factors. Since 2 and 3 are factors, 6 is also a factor. We write a list of factorizations.

$90 = 1 \cdot 90 \qquad 90 = 5 \cdot 18$
$90 = 2 \cdot 45 \qquad 90 = 6 \cdot 15$
$90 = 3 \cdot 30 \qquad 90 = 9 \cdot 10$

Factors: 1, 2, 3, 5, 6, 9, 10, 15, 18, 30, 45, 90

4. The number 93 has factors 1, 3, 31, and 93. Since it has at least one factor other than itself and 1, it is composite.

5.
```
          3   ← 3 is prime.
      3 ⌐ 9
    2 ⌐ 1 8
    2 ⌐ 3 6
```
$36 = 2 \cdot 2 \cdot 3 \cdot 3$

6. We use a factor tree.

```
        60
       /  \
      6    10
     / \   / \
    2   3 2   5
```

$60 = 2 \cdot 3 \cdot 2 \cdot 5$, or $2 \cdot 2 \cdot 3 \cdot 5$

7. $\dfrac{4}{9}$ ← Numerator
← Denominator

8. The figure is divided into 4 equal parts, so the unit is $\frac{1}{4}$ and the denominator is 4. Three of the units are shaded, so the numerator is 3. Thus, $\frac{3}{4}$ is shaded.

9. There are 7 objects in the set, so the denominator is 7. Three of the objects are shaded, so the numerator is 3. Thus, $\frac{3}{7}$ of the set is shaded.

10. a) The ratio is $\dfrac{1112}{1202}$.

b) The ratio is $\dfrac{90}{1202}$.

11. Remember: $\dfrac{n}{1} = n$.

$$\frac{32}{1} = 32$$

12. Remember: $\dfrac{n}{n} = 1$ for any integer n that is not 0.

$$\frac{-12}{-12} = 1$$

13. Remember: $\dfrac{0}{n} = 0$ for any integer n that is not 0.

$$\frac{0}{16} = 0$$

14. $\dfrac{-8}{24} = \dfrac{-1 \cdot 8}{3 \cdot 8} = \dfrac{-1}{3} \cdot \dfrac{8}{8} = \dfrac{-1}{3}$

15. $\dfrac{9x}{45x} = \dfrac{9 \cdot x \cdot 1}{5 \cdot 9 \cdot x} = \dfrac{9 \cdot x}{9 \cdot x} \cdot \dfrac{1}{5} = \dfrac{1}{5}$

16. $\dfrac{7}{63} = \dfrac{7 \cdot 1}{7 \cdot 9} = \dfrac{7}{7} \cdot \dfrac{1}{9} = \dfrac{1}{9}$

17. We multiply these two numbers: We multiply these two numbers:

$3 \cdot 8 = 24$ $\qquad 4 \cdot 6 = 24$

Since $24 = 24$, $\dfrac{3}{4} = \dfrac{6}{8}$.

18. We multiply these two numbers: We multiply these two numbers:

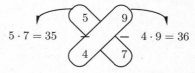

$5 \cdot 7 = 35$ $\qquad 4 \cdot 9 = 36$

Since $35 \neq 36$, $\dfrac{5}{4}$ and $\dfrac{9}{7}$ do not name the same number. Thus, $\dfrac{5}{4} \neq \dfrac{9}{7}$.

19. Since $40 \div 8 = 5$, we multiply by $\dfrac{5}{5}$.

$$\frac{3}{8} = \frac{3}{8} \cdot \frac{5}{5} = \frac{3 \cdot 5}{8 \cdot 5} = \frac{15}{40}$$

20. Interchange the numerator and denominator.

The reciprocal of $\dfrac{a}{42}$ is $\dfrac{42}{a}$. $\left(\dfrac{a}{42} \cdot \dfrac{42}{a} = \dfrac{42a}{42a} = 1 \right)$

21. Think of -9 as $\dfrac{-9}{1}$. Interchange the numerator and denominator.

The reciprocal of -9 is $\dfrac{1}{-9}$, or $-\dfrac{1}{9}$, or $\dfrac{-1}{9}$.

$$\left(-9\cdot\left(-\dfrac{1}{9}\right)=\dfrac{9\cdot 1}{9}=1\right)$$

22. $\dfrac{5}{7}\cdot\dfrac{7}{2}=\dfrac{5\cdot 7}{7\cdot 2}=\dfrac{7}{7}\cdot\dfrac{5}{2}=\dfrac{5}{2}$

23. $\dfrac{2}{11}\div\dfrac{3}{4}=\dfrac{2}{11}\cdot\dfrac{4}{3}=\dfrac{2\cdot 4}{11\cdot 3}=\dfrac{8}{33}$

24. $3\cdot\dfrac{x}{8}=\dfrac{3\cdot x}{8}=\dfrac{3x}{8}$

25. $\dfrac{\frac{4}{7}}{-\frac{8}{3}}=\dfrac{4}{7}\cdot\left(-\dfrac{3}{8}\right)=-\dfrac{4\cdot 3}{7\cdot 8}=-\dfrac{4\cdot 3}{7\cdot 2\cdot 4}=$

$-\dfrac{3}{7\cdot 2}\cdot\dfrac{4}{4}=-\dfrac{3}{14}$

26. $12\div\dfrac{2}{3}=12\cdot\dfrac{3}{2}=\dfrac{12\cdot 3}{2}=\dfrac{2\cdot 6\cdot 3}{2\cdot 1}=\dfrac{2}{2}\cdot\dfrac{6\cdot 3}{1}=\dfrac{6\cdot 3}{1}=18$

27. $\dfrac{4a}{13}\cdot\dfrac{9b}{30ab}=\dfrac{4a\cdot 9b}{13\cdot 30ab}=\dfrac{2\cdot 2\cdot a\cdot 3\cdot 3\cdot b}{13\cdot 2\cdot 3\cdot 5\cdot a\cdot b}=$

$\dfrac{2\cdot 3\cdot a\cdot b}{2\cdot 3\cdot a\cdot b}\cdot\dfrac{2\cdot 3}{13\cdot 5}=\dfrac{6}{65}$

28. *Familiarize.* Let $c=$ the number of pounds of cheese each person receives.

Translate. We write a division sentence.

$$c=\dfrac{3}{4}\div 5$$

Solve. We carry out the division.

$$c=\dfrac{3}{4}\div 5=\dfrac{3}{4}\cdot\dfrac{1}{5}=\dfrac{3\cdot 1}{4\cdot 5}=\dfrac{3}{20}$$

Check. Since $\dfrac{3}{20}\cdot 5=\dfrac{3\cdot 5}{20}=\dfrac{3\cdot 5}{4\cdot 5}=\dfrac{3}{4}$, the answer checks.

State. Each person receives $\dfrac{3}{20}$ lb of cheese.

29. *Familiarize.* Let $w=$ Monroe's weight, in pounds. We want to find $\dfrac{5}{7}$ of 175 lb.

Translate. We write a multiplication sentence.

$$w=\dfrac{5}{7}\cdot 175$$

Solve. We carry out the multiplication.

$$w=\dfrac{5}{7}\cdot 175=\dfrac{5\cdot 175}{7}=\dfrac{5\cdot 7\cdot 25}{7\cdot 1}$$
$$=\dfrac{7}{7}\cdot\dfrac{5\cdot 25}{1}$$
$$=125$$

Check. We can repeat the calculation. The answer checks.

State. Monroe weighs 125 lb.

30. $\dfrac{7}{8}\cdot x=56$

$\qquad x=56\div\dfrac{7}{8}\qquad$ Dividing by $\dfrac{7}{8}$ on both sides

$\qquad x=56\cdot\dfrac{8}{7}$

$\qquad=\dfrac{56\cdot 8}{7}=\dfrac{7\cdot 8\cdot 8}{7\cdot 1}=\dfrac{7}{7}\cdot\dfrac{8\cdot 8}{1}=\dfrac{8\cdot 8}{1}=64$

The solution is 64.

31.
$$\dfrac{7}{10}=\dfrac{-2}{5}\cdot t$$
$$\dfrac{5}{-2}\cdot\dfrac{7}{10}=\dfrac{5}{-2}\cdot\dfrac{-2}{5}\cdot t$$
$$\dfrac{5\cdot 7}{-2\cdot 10}=1t$$
$$\dfrac{5\cdot 7}{-2\cdot 2\cdot 5}=t$$
$$\dfrac{7}{-2\cdot 2}\cdot\dfrac{5}{5}=t$$
$$\dfrac{7}{-4}=t$$

The solution is $\dfrac{7}{-4}$, or $-\dfrac{7}{4}$, or $\dfrac{-7}{4}$.

32. $A=\dfrac{1}{2}\cdot b\cdot h$

$A=\dfrac{1}{2}\cdot 13\text{ m}\cdot 7\text{ m}$

$A=\dfrac{13\cdot 7}{2}\text{ m}^2=\dfrac{91}{2}\text{ m}^2$

33. *Familiarize.* This is a multistep problem. First we will find half the amount of salt for one batch of pancakes. Then we will find 5 times this amount. Let $s=$ half the amount of salt in a single batch, in teaspoons.

Translate. We translate to an equation.

$$s=\dfrac{1}{2}\cdot\dfrac{3}{4}$$

Solve. We carry out the multiplication.

$$s=\dfrac{1}{2}\cdot\dfrac{3}{4}=\dfrac{1\cdot 3}{2\cdot 4}=\dfrac{3}{8}$$

Half the amount of salt in a single batch of pancakes is $\dfrac{3}{8}$ tsp. Let $p=$ the number of teaspoons of salt in 5 batches. The equation that corresponds to this situation is

$$p=5\cdot\dfrac{3}{8}.$$

We solve the equation by carrying out the multiplication.

$$p=5\cdot\dfrac{3}{8}=\dfrac{5\cdot 3}{8}=\dfrac{15}{8}$$

Check. We repeat the calculations. The answer checks.

State. Jacqueline will need $\dfrac{15}{8}$ tsp of salt.

34. *Familiarize.* This is a multistep problem. First we will find the number of acres Karl received. Then we will find how much of that land Irene received. Let $k=$ the number of acres of land Karl received.

Translate. We translate to an equation.

$$k = \frac{7}{8} \cdot \frac{2}{3}$$

Solve. We carry out the multiplication.

$$k = \frac{7}{8} \cdot \frac{2}{3} = \frac{7 \cdot 2}{8 \cdot 3} = \frac{7 \cdot 2}{2 \cdot 4 \cdot 3} = \frac{2}{2} \cdot \frac{7}{4 \cdot 3} = \frac{7}{12}$$

Karl received $\frac{7}{12}$ acre of land. Let $a =$ the number of acres Irene received. An equation that corresponds to this situation is

$$a = \frac{1}{4} \cdot \frac{7}{12}.$$

We solve the equation by carrying out the multiplication.

$$a = \frac{1}{4} \cdot \frac{7}{12} = \frac{1 \cdot 7}{4 \cdot 12} = \frac{7}{48}$$

Check. We repeat the calculations. The answer checks.

State. Irene received $\frac{7}{48}$ acre of land.

35. First we will evaluate the exponential expression; then we will multiply and divide in order from left to right.

$$\left(-\frac{3}{8}\right)^2 \div \frac{6}{7} \cdot \frac{2}{9} \div (-5) = \frac{9}{64} \div \frac{6}{7} \cdot \frac{2}{9} \div (-5)$$

$$= \frac{9}{64} \cdot \frac{7}{6} \cdot \frac{2}{9} \div (-5)$$

$$= \frac{9 \cdot 7}{64 \cdot 6} \cdot \frac{2}{9} \div (-5)$$

$$= \frac{9 \cdot 7 \cdot 2}{64 \cdot 6 \cdot 9} \div (-5)$$

$$= \frac{9 \cdot 7 \cdot 2}{64 \cdot 6 \cdot 9} \cdot \left(-\frac{1}{5}\right)$$

$$= -\frac{9 \cdot 7 \cdot 2 \cdot 1}{64 \cdot 6 \cdot 9 \cdot 5}$$

$$= -\frac{9 \cdot 7 \cdot 2 \cdot 1}{64 \cdot 2 \cdot 3 \cdot 9 \cdot 5}$$

$$= -\frac{9 \cdot 2}{9 \cdot 2} \cdot \frac{7 \cdot 1}{64 \cdot 3 \cdot 5}$$

$$= -\frac{7}{960}, \text{ or } \frac{-7}{960}$$

36.

$$\frac{33}{38} \cdot \frac{34}{55} = \frac{17}{35} \cdot \frac{15}{19} x$$

$$\frac{33 \cdot 34}{38 \cdot 55} = \frac{17 \cdot 15}{35 \cdot 19} x$$

$$\frac{3 \cdot 11 \cdot 2 \cdot 17}{2 \cdot 19 \cdot 5 \cdot 11} = \frac{17 \cdot 3 \cdot 5}{5 \cdot 7 \cdot 19} x$$

$$\frac{2 \cdot 11}{2 \cdot 11} \cdot \frac{3 \cdot 17}{19 \cdot 5} = \frac{5}{5} \cdot \frac{17 \cdot 3}{7 \cdot 19} x$$

$$\frac{3 \cdot 17}{19 \cdot 5} = \frac{17 \cdot 3}{7 \cdot 19} x$$

$$\frac{3 \cdot 17}{19 \cdot 5} \div \frac{17 \cdot 3}{7 \cdot 19} = x \quad \text{Dividing by } \frac{17 \cdot 3}{7 \cdot 19} \text{ on both sides}$$

$$\frac{3 \cdot 17}{19 \cdot 5} \cdot \frac{7 \cdot 19}{17 \cdot 3} = x$$

$$\frac{3 \cdot 17 \cdot 7 \cdot 19}{19 \cdot 5 \cdot 17 \cdot 3} = x$$

$$\frac{3 \cdot 17 \cdot 19}{3 \cdot 17 \cdot 19} \cdot \frac{7}{5} = x$$

$$\frac{7}{5} = x$$

The solution is $\frac{7}{5}$.

Chapter 4

Fraction Notation: Addition and Subtraction

Exercise Set 4.1

In this section we will find the LCM using the multiples method in Exercises 1 - 19 and the prime factorization method in Exercises 21 - 43.

1. 1. 10 is the larger number and is a multiple of 5, so it is the LCM.

The LCM = 10.

3. 1. 25 is the larger number, but it is not a multiple of 10.

2. Check multiples of 25:

$2 \cdot 25 = 50$ A multiple of 10

The LCM = 50.

5. 1. 40 is the larger number and is a multiple of 20, so it is the LCM.

The LCM = 40.

7. 1. 27 is the larger number, but it is not a multiple of 18.

2. Check multiples of 27:

$2 \cdot 27 = 54$ A multiple of 18

The LCM = 54.

9. 1. 50 is the larger number, but it is not a multiple of 30.

2. Check multiples of 50:

$2 \cdot 50 = 100$ Not a multiple of 30
$3 \cdot 50 = 150$ A multiple of 30

The LCM = 150.

11. 1. 40 is the larger number, but it is not a multiple of 30.

2. Check multiples of 40:

$2 \cdot 40 = 80$ Not a multiple of 30
$3 \cdot 40 = 120$ A multiple of 30

The LCM = 120.

13. 1. 24 is the larger number, but it is not a multiple of 18.

2. Check multiples of 24:

$2 \cdot 24 = 48$ Not a multiple of 18
$3 \cdot 24 = 72$ A multiple of 18

The LCM = 72.

15. 1. 70 is the larger number, but it is not a multiple of 60.

2. Check multiples of 70:

$2 \cdot 70 = 140$ Not a multiple of 60
$3 \cdot 70 = 210$ Not a multiple of 60
$4 \cdot 70 = 280$ Not a multiple of 60
$5 \cdot 70 = 350$ Not a multiple of 60
$6 \cdot 70 = 420$ A multiple of 60

The LCM = 420.

17. 1. 36 is the larger number, but it is not a multiple of 16.

2. Check multiples of 36:

$2 \cdot 36 = 72$ Not a multiple of 16
$3 \cdot 36 = 108$ Not a multiple of 16
$4 \cdot 36 = 144$ A multiple of 16

The LCM = 144.

19. 1. 20 is the larger number, but it is not a multiple of 18.

2. Check multiples of 20:

$2 \cdot 20 = 40$ Not a multiple of 18
$3 \cdot 20 = 60$ Not a multiple of 18
$4 \cdot 20 = 80$ Not a multiple of 18
$5 \cdot 20 = 100$ Not a multiple of 18
$6 \cdot 20 = 120$ Not a multiple of 18
$7 \cdot 20 = 140$ Not a multiple of 18
$8 \cdot 20 = 160$ Not a multiple of 18
$9 \cdot 20 = 180$ A multiple of 18

The LCM = 180.

21. 1. Write the prime factorization of each number. Because 2, 3, and 7 are all prime we write $2 = 2$, $3 = 3$, and $7 = 7$.

2. a) None of the factorizations contains the other two.

b) We begin with 2. Since 3 contains a factor of 3, we multiply by 3:

$2 \cdot 3$

Next we multiply $2 \cdot 3$ by 7, the factor of 7 that is missing:

$2 \cdot 3 \cdot 7$

The LCM is $2 \cdot 3 \cdot 7$, or 42.

3. To check, note that 2, 3, and 7 appear in the LCM the greatest number of times that each appears as a factor of 2, 3, or 7. The LCM is $2 \cdot 3 \cdot 7$, or 42.

23. 1. Write the prime factorization of each number.

$3 = 3$
$6 = 2 \cdot 3$
$15 = 3 \cdot 5$

2. a) None of the factorizations contains the other two.

b) We first consider 3 and 6. Since the factorization of 6 contains 3, we next multiply $2 \cdot 3$ by the factor of 15 that is missing, 5. The LCM is $2 \cdot 3 \cdot 5$, or 30.

3. To check, note that 2, 3, and 5 appear in the LCM the greatest number of times that each appears as a factor of 3, 6, or 15. The LCM is $2 \cdot 3 \cdot 5$, or 30.

25. 1. Write the prime factorization of each number.

$24 = 2 \cdot 2 \cdot 2 \cdot 3$
$36 = 2 \cdot 2 \cdot 3 \cdot 3$
$12 = 2 \cdot 2 \cdot 3$

2. a) None of the factorizations contains the other two.

   b) We begin with the factorization of 24, $2 \cdot 2 \cdot 2 \cdot 3$. Since 36 contains a second factor of 3, we multiply by another factor of 3:

   $$2 \cdot 2 \cdot 2 \cdot 3 \cdot 3$$

   Next we look for factors of 12 that are still missing. There are none. The LCM is

   $2 \cdot 2 \cdot 2 \cdot 3 \cdot 3$, or 72.

3. To check, note that 2 and 3 appear in the LCM the greatest number of times that each appears as a factor of 24, 36, or 12. The LCM is $2 \cdot 2 \cdot 2 \cdot 3 \cdot 3$, or 72.

27. 1. Write the prime factorization of each number.

   $$5 = 5$$
   $$12 = 2 \cdot 2 \cdot 3$$
   $$15 = 3 \cdot 5$$

2. a) None of the factorizations contains the other two.

   b) We begin with the factorization of 12, $2 \cdot 2 \cdot 3$. Since 5 contains a factor of 5, we multiply by 5:

   $$2 \cdot 2 \cdot 3 \cdot 5$$

   Next we look for factors of 15 that are still missing. There are none. The LCM is $2 \cdot 2 \cdot 3 \cdot 5$, or 60.

3. The result checks.

29. 1. Write the prime factorization of each number.

   $$9 = 3 \cdot 3$$
   $$12 = 2 \cdot 2 \cdot 3$$
   $$6 = 2 \cdot 3$$

2. a) None of the factorizations contains the other two.

   b) We begin with the factorization of 12, $2 \cdot 2 \cdot 3$. Since 9 contains a second factor of 3, we multiply by another factor of 3:

   $$2 \cdot 2 \cdot 3 \cdot 3$$

   Next we look for factors of 6 that are still missing. There are none. The LCM is $2 \cdot 2 \cdot 3 \cdot 3$, or 36.

3. The result checks.

31. 1. Write the prime factorization of each number.

   $$180 = 2 \cdot 2 \cdot 3 \cdot 3 \cdot 5$$
   $$100 = 2 \cdot 2 \cdot 5 \cdot 5$$
   $$450 = 2 \cdot 3 \cdot 3 \cdot 5 \cdot 5$$

2. a) None of the factorizations contains the other two.

   b) We begin with the factorization of 450, $2 \cdot 3 \cdot 3 \cdot 5 \cdot 5$. Since 180 contains another factor of 2, we multiply by 2:

   $$2 \cdot 3 \cdot 3 \cdot 5 \cdot 5 \cdot 2$$

   Next we look for factors of 100 that are still missing. There are none. The LCM is $2 \cdot 3 \cdot 3 \cdot 5 \cdot 5 \cdot 2$, or 900.

3. The result checks.

33. 1. Write the prime factorization of each number.

   $$75 = 3 \cdot 5 \cdot 5$$
   $$100 = 2 \cdot 2 \cdot 5 \cdot 5$$

2. a) Neither factorization contains the other.

   b) We begin with the factorization of 100, $2 \cdot 2 \cdot 5 \cdot 5$. Since 75 contains a factor of 3, we multiply by 3:

   $$2 \cdot 2 \cdot 5 \cdot 5 \cdot 3$$

   The LCM is $2 \cdot 2 \cdot 5 \cdot 5 \cdot 3$, or 300.

3. The result checks.

35. 1. We have the following factorizations:

   $$ab = a \cdot b$$
   $$bc = b \cdot c$$

2. a) Neither factorization contains the other.

   b) Consider the factorization of ab, $a \cdot b$. Since bc contains a factor of c, we multiply by c.

   $$a \cdot b \cdot c$$

   The LCM is $a \cdot b \cdot c$, or abc.

3. The result checks.

37. 1. We have the following factorizations:

   $$3x = 3 \cdot x$$
   $$9x^2 = 3 \cdot 3 \cdot x \cdot x$$

2. a) One factorization, $3 \cdot 3 \cdot x \cdot x$, contains the other. Thus the LCM is $3 \cdot 3 \cdot x \cdot x$, or $9x^2$.

39. 1. We have the following factorizations:

   $$4x^3 = 2 \cdot 2 \cdot x \cdot x \cdot x$$
   $$x^2y = x \cdot x \cdot y$$

2. a) Neither factorization contains the other.

   b) Consider the factorization of $4x^3$, $2 \cdot 2 \cdot x \cdot x \cdot x$. Since x^2y contains a factor of y, we multiply by y.

   $$2 \cdot 2 \cdot x \cdot x \cdot x \cdot y$$

   The LCM is $2 \cdot 2 \cdot x \cdot x \cdot x \cdot y$, or $4x^3y$.

3. The result checks.

41. 1. We have the following factorizations:

   $$6r^3st^4 = 2 \cdot 3 \cdot r \cdot r \cdot r \cdot s \cdot t \cdot t \cdot t \cdot t$$
   $$8rs^2t = 2 \cdot 2 \cdot 2 \cdot r \cdot s \cdot s \cdot t$$

2. a) Neither factorization contains the other.

   b) Consider the factorization of $6r^3st^4$, $2 \cdot 3 \cdot r \cdot r \cdot r \cdot s \cdot t \cdot t \cdot t \cdot t$. Since $8rs^2t$ contains two more factors of 2 and one more factor of s, we multiply by $2 \cdot 2 \cdot s$.

   $$2 \cdot 3 \cdot r \cdot r \cdot r \cdot s \cdot t \cdot t \cdot t \cdot t \cdot 2 \cdot 2 \cdot s$$

   The LCM is $2 \cdot 3 \cdot 2 \cdot 2 \cdot r \cdot r \cdot r \cdot s \cdot s \cdot t \cdot t \cdot t \cdot t$, or $24r^3s^2t^4$.

3. The result checks.

43. 1. We have the following factorizations:

   $$a^3b = a \cdot a \cdot a \cdot b$$
   $$b^2c = b \cdot b \cdot c$$
   $$ac^2 = a \cdot c \cdot c$$

2. a) No one factorization contains the others.

   b) Consider the factorization of a^3b,
   $a \cdot a \cdot a \cdot b$. Since b^2c contains another factor of b and a factor of c, we multiply by $b \cdot c$.

   $$a \cdot a \cdot a \cdot b \cdot b \cdot c$$

   Now consider ac^2. Since ac^2 contains another factor of c, we multiply by c.

   $$a \cdot a \cdot a \cdot b \cdot b \cdot c \cdot c$$

   The LCM is $a \cdot a \cdot a \cdot b \cdot b \cdot c \cdot c$, or $a^3b^2c^2$.

3. The result checks.

45. We find the LCM of the number of years it takes Jupiter and Saturn to make a complete revolution around the sun.

   Jupiter: $12 = 2 \cdot 2 \cdot 3$

   Saturn: $30 = 2 \cdot 3 \cdot 5$

   The LCM $= 2 \cdot 2 \cdot 3 \cdot 5$, or 60. Thus, Jupiter and Saturn will appear in the exact same direction in the night sky as seen from Earth tonight once every 60 years.

47. Discussion and Writing Exercise

49. $-38 + 52$

   The absolute values are 38 and 52. The difference is 14. The positive number has the larger absolute value, so the answer is positive.

   $$-38 + 52 = 14$$

51.
$$\begin{array}{r} {\scriptstyle 1} \\ {\scriptstyle 1\,1} \\ 3\,4\,5 \\ \times\quad 2\,3 \\ \hline 1\,0\,3\,5 \\ 6\,9\,0\,0 \\ \hline 7\,9\,3\,5 \end{array}$$

53. $\dfrac{4}{5} \div \left(-\dfrac{7}{10}\right) = \dfrac{4}{5} \cdot \left(-\dfrac{10}{7}\right) = -\dfrac{4 \cdot 10}{5 \cdot 7} = -\dfrac{4 \cdot 2 \cdot 5}{5 \cdot 7} =$

$-\dfrac{4 \cdot 2}{7} \cdot \dfrac{5}{5} = -\dfrac{8}{7}$

55. Discussion and Writing Exercise

57. Discussion and Writing Exercise

59. a) 7800 is the larger number, but it is not a multiple of 2700.

   b) Check multiples using a calculator:

$2 \cdot 7800 = 15,600$	Not a multiple of 2700
$3 \cdot 7800 = 23,400$	Not a multiple of 2700
$4 \cdot 7800 = 31,200$	Not a multiple of 2700
$5 \cdot 7800 = 39,000$	Not a multiple of 2700
$6 \cdot 7800 = 46,800$	Not a multiple of 2700
$7 \cdot 7800 = 54,600$	Not a multiple of 2700
$8 \cdot 7800 = 62,400$	Not a multiple of 2700
$9 \cdot 7800 = 70,200$	A multiple of 2700

   c) The LCM is 70,200.

61. a) 24,339 is the larger number, but it is not a multiple of 17,385.

   b) Check multiples using a calculator:

$2 \cdot 24,339 = 48,678$	Not a multiple of 17,385
$3 \cdot 24,339 = 73,017$	Not a multiple of 17,385
$4 \cdot 24,339 = 97,356$	Not a multiple of 17,385
$5 \cdot 24,339 = 121,695$	A multiple of 17,385

   c) The LCM is 121,695.

63. The smallest number of strands that can be used is the LCM of 10 and 3. The prime factorizations are $10 = 2 \cdot 5$ and $3 = 3$. Thus, the LCM is $2 \cdot 5 \cdot 3$, or 30, so the smallest number of strands that can be used is 30.

65. We find the LCM of 30 and 14.
   $$30 = 2 \cdot 3 \cdot 5$$
   $$14 = 2 \cdot 7$$

   We form the LCM using the greatest power of each factor.

   The LCM is $2 \cdot 3 \cdot 5 \cdot 7$, or 210, so the prescriptions will both be refilled on the same day in 210 days.

67. 1. From Example 9 we know that the LCM of 27, 90, and 84 is $2 \cdot 3 \cdot 3 \cdot 5 \cdot 3 \cdot 2 \cdot 7$, so the LCM of 27, 90, 84, 210, 108, and 50 must contain at least these factors. We write the prime factorizations of 210, 108, and 50:

   $$210 = 2 \cdot 3 \cdot 5 \cdot 7$$
   $$108 = 2 \cdot 2 \cdot 3 \cdot 3 \cdot 3$$
   $$50 = 2 \cdot 5 \cdot 5$$

   2. a) Neither of the four factorizations above contains the other three.

   b) Begin with the LCM of 27, 90, and 84, $2 \cdot 3 \cdot 3 \cdot 5 \cdot 3 \cdot 2 \cdot 7$. Neither 210 nor 108 contains any factors that are missing in this factorization. Next we look for factors of 50 that are missing. Since 50 contains a second factor of 5, we multiply by 5:

   $$2 \cdot 3 \cdot 3 \cdot 5 \cdot 3 \cdot 2 \cdot 7 \cdot 5$$

   The LCM is $2 \cdot 3 \cdot 3 \cdot 5 \cdot 3 \cdot 2 \cdot 7 \cdot 5$, or 18,900.

   3. The result checks.

69. Answers may vary.

   $$56 = 2 \cdot 2 \cdot 2 \cdot 7$$

   Three pairs are $2 \cdot 2 \cdot 2$ and 7, or 8 and 7; $2 \cdot 2 \cdot 2$ and $2 \cdot 7$, or 8 and 14; and $2 \cdot 2 \cdot 2$ and $2 \cdot 2 \cdot 7$, or 8 and 28.

Exercise Set 4.2

1. $\dfrac{4}{9} + \dfrac{1}{9} = \dfrac{4+1}{9} = \dfrac{5}{9}$

3. $\dfrac{4}{7} + \dfrac{3}{7} = \dfrac{4+3}{7} = \dfrac{7}{7} = 1$

5. $\dfrac{7}{10} + \dfrac{3}{-10} = \dfrac{7}{10} + \dfrac{-3}{10} = \dfrac{7+(-3)}{10} = \dfrac{4}{10} = \dfrac{2 \cdot 2}{2 \cdot 5} =$

$\dfrac{2}{2} \cdot \dfrac{2}{5} = 1 \cdot \dfrac{2}{5} = \dfrac{2}{5}$

7. $\dfrac{9}{a} + \dfrac{4}{a} = \dfrac{9+4}{a} = \dfrac{13}{a}$

9. $\dfrac{-7}{11} + \dfrac{3}{11} = \dfrac{-7+3}{11} = \dfrac{-4}{11}$, or $-\dfrac{4}{11}$

11. $\dfrac{2}{9}x + \dfrac{5}{9}x = \left(\dfrac{2}{9} + \dfrac{5}{9}\right)x = \dfrac{7}{9}x$

13. $\dfrac{3}{32}t + \dfrac{13}{32}t$

$= \left(\dfrac{3}{32} + \dfrac{13}{32}\right)t$

$= \dfrac{16}{32}t$

$= \dfrac{16 \cdot 1}{16 \cdot 2}t$

$= \dfrac{16}{16} \cdot \dfrac{1}{2}t$

$= \dfrac{1}{2}t$

15. $-\dfrac{2}{x} + \left(-\dfrac{7}{x}\right) = \dfrac{-2}{x} + \dfrac{-7}{x} = \dfrac{-2+(-7)}{x} = \dfrac{-9}{x}$,

or $-\dfrac{9}{x}$

17. $\dfrac{1}{8} + \dfrac{1}{6}$ $8 = 2 \cdot 2 \cdot 2$ and $6 = 2 \cdot 3$, so the LCD is $2 \cdot 2 \cdot 2 \cdot 3$, or 24

$= \dfrac{1}{8} \cdot \dfrac{3}{3} + \dfrac{1}{6} \cdot \dfrac{4}{4}$

Think: $6 \times \square = 24$. The answer is 4, so we multiply by 1, using $\dfrac{4}{4}$.

Think: $8 \times \square = 24$. The answer is 3, so we multiply by 1, using $\dfrac{3}{3}$.

$= \dfrac{3}{24} + \dfrac{4}{24}$

$= \dfrac{7}{24}$

19. $\dfrac{-4}{5} + \dfrac{7}{10}$ 5 is a factor of 10, so the LCD is 10.

$= \dfrac{-4}{5} \cdot \dfrac{2}{2} + \dfrac{7}{10}$ ← This fraction already has the LCD as denominator.

Think: $5 \times \square = 10$. The answer is 2, so we multiply by 1, using $\dfrac{2}{2}$.

$= \dfrac{-8}{10} + \dfrac{7}{10}$

$= \dfrac{-1}{10}$, or $-\dfrac{1}{10}$

21. $\dfrac{7}{12} + \dfrac{3}{8}$ $12 = 2 \cdot 2 \cdot 3$ and $8 = 2 \cdot 2 \cdot 2$, so the LCD is $2 \cdot 2 \cdot 2 \cdot 3$, or 24.

$= \dfrac{7}{12} \cdot \dfrac{2}{2} + \dfrac{3}{8} \cdot \dfrac{3}{3}$

Think: $8 \times \square = 24$. The answer is 3, so we multiply by 1, using $\dfrac{3}{3}$.

Think: $12 \times \square = 24$. The answer is 2, so we multiply by 1, using $\dfrac{2}{2}$.

$= \dfrac{14}{24} + \dfrac{9}{24} = \dfrac{23}{24}$

23. $\dfrac{3}{20} + 4$

$= \dfrac{3}{20} + \dfrac{4}{1}$ Rewriting 4 in fractional notation

$= \dfrac{3}{20} + \dfrac{4}{1} \cdot \dfrac{20}{20}$ The LCD is 20.

$= \dfrac{3}{20} + \dfrac{80}{20}$

$= \dfrac{83}{20}$

25. $\dfrac{5}{-8} + \dfrac{5}{6}$

$= \dfrac{-5}{8} + \dfrac{5}{6}$ Recall that $\dfrac{m}{-n} = \dfrac{-m}{n}$. The LCD is 24. (See Exercise 17.)

$= \dfrac{-5}{8} \cdot \dfrac{3}{3} + \dfrac{5}{6} \cdot \dfrac{4}{4}$

$= \dfrac{-15}{24} + \dfrac{20}{24}$

$= \dfrac{5}{24}$

27. $\dfrac{3}{10}x + \dfrac{7}{100}x$

$= \dfrac{3}{10} \cdot \dfrac{10}{10} \cdot x + \dfrac{7}{100}x$ 10 is a factor of 100, so the LCD is 100.

$= \dfrac{30}{100}x + \dfrac{7}{100}x$

$= \dfrac{37}{100}x$

29. $\dfrac{5}{12} + \dfrac{8}{15}$ $12 = 2 \cdot 2 \cdot 3$ and $15 = 3 \cdot 5$, so the LCM is $2 \cdot 2 \cdot 3 \cdot 5$, or 60.

$= \dfrac{5}{12} \cdot \dfrac{5}{5} + \dfrac{8}{15} \cdot \dfrac{4}{4}$

$= \dfrac{25}{60} + \dfrac{32}{60} = \dfrac{57}{60}$

$= \dfrac{3 \cdot 19}{3 \cdot 20} = \dfrac{3}{3} \cdot \dfrac{19}{20}$

$= \dfrac{19}{20}$

31. $\dfrac{-7}{10} + \dfrac{-29}{100}$ 10 is a factor of 100, so the LCD is 100.

$= \dfrac{-7}{10} \cdot \dfrac{10}{10} + \dfrac{-29}{100}$

$= \dfrac{-70}{100} + \dfrac{-29}{100} = \dfrac{-99}{100},$ or $- \dfrac{99}{100}$

33. $-\dfrac{1}{10}x + \dfrac{1}{15}x$

$= -\dfrac{1}{2 \cdot 5}x + \dfrac{1}{3 \cdot 5}x$ The LCD is $2 \cdot 5 \cdot 3$.

$= -\dfrac{1}{2 \cdot 5} \cdot \dfrac{3}{3}x + \dfrac{1}{3 \cdot 5} \cdot \dfrac{2}{2}x$

$= -\dfrac{3}{30}x + \dfrac{2}{30}x$

$= -\dfrac{1}{30}x$

35. $-5t + \dfrac{2}{7}t$

$= \dfrac{-5}{1}t + \dfrac{2}{7}t$ The LCD is 7.

$= \dfrac{-5}{1} \cdot \dfrac{7}{7} \cdot t + \dfrac{2}{7}t$

$= \dfrac{-35}{7}t + \dfrac{2}{7}t$

$= \dfrac{-33}{7}t,$ or $- \dfrac{33}{7}t$

37. $-\dfrac{5}{12} + \dfrac{7}{-24}$

$\dfrac{-5}{12} + \dfrac{-7}{24}$ 12 is a factor of 24, so the LCD is 24.

$= \dfrac{-5}{12} \cdot \dfrac{2}{2} + \dfrac{-7}{24}$

$= \dfrac{-10}{24} + \dfrac{-7}{24}$

$= \dfrac{-17}{24},$ or $- \dfrac{17}{24}$

39. $\dfrac{4}{10} + \dfrac{3}{100} + \dfrac{7}{1000}$ 10 and 100 are factors of 1000, so the LCD is 1000.

$= \dfrac{4}{10} \cdot \dfrac{100}{100} + \dfrac{3}{100} \cdot \dfrac{10}{10} + \dfrac{7}{1000}$

$= \dfrac{400}{1000} + \dfrac{30}{1000} + \dfrac{7}{1000}$

$= \dfrac{437}{1000}$

41. $\dfrac{3}{10} + \dfrac{5}{12} + \dfrac{8}{15}$

$= \dfrac{3}{2 \cdot 5} + \dfrac{5}{2 \cdot 2 \cdot 3} + \dfrac{8}{3 \cdot 5}$ Factoring the denominators

The LCD is $2 \cdot 5 \cdot 2 \cdot 3$.

$= \dfrac{3}{2 \cdot 5} \cdot \dfrac{2 \cdot 3}{2 \cdot 3} + \dfrac{5}{2 \cdot 2 \cdot 3} \cdot \dfrac{5}{5} + \dfrac{8}{3 \cdot 5} \cdot \dfrac{2 \cdot 2}{2 \cdot 2}$

In each case we multiply by 1 to obtain the LCD.

$= \dfrac{3 \cdot 2 \cdot 3}{2 \cdot 5 \cdot 2 \cdot 3} + \dfrac{5 \cdot 5}{2 \cdot 2 \cdot 3 \cdot 5} + \dfrac{8 \cdot 2 \cdot 2}{3 \cdot 5 \cdot 2 \cdot 2}$

$= \dfrac{18}{2 \cdot 5 \cdot 2 \cdot 3} + \dfrac{25}{2 \cdot 5 \cdot 2 \cdot 3} + \dfrac{32}{2 \cdot 5 \cdot 2 \cdot 3}$

$= \dfrac{75}{2 \cdot 5 \cdot 2 \cdot 3}$

$= \dfrac{3 \cdot 5 \cdot 5}{2 \cdot 5 \cdot 2 \cdot 3} = \dfrac{3 \cdot 5}{3 \cdot 5} \cdot \dfrac{5}{2 \cdot 2}$

$= \dfrac{5}{4}$

43. $\dfrac{5}{6} + \dfrac{25}{52} + \dfrac{7}{4}$

$= \dfrac{5}{2 \cdot 3} + \dfrac{25}{2 \cdot 2 \cdot 13} + \dfrac{7}{2 \cdot 2}$ LCD is $2 \cdot 3 \cdot 2 \cdot 13$.

$= \dfrac{5}{2 \cdot 3} \cdot \dfrac{2 \cdot 13}{2 \cdot 13} + \dfrac{25}{2 \cdot 2 \cdot 13} \cdot \dfrac{3}{3} + \dfrac{7}{2 \cdot 2} \cdot \dfrac{3 \cdot 13}{3 \cdot 13}$

$= \dfrac{5 \cdot 2 \cdot 13}{2 \cdot 3 \cdot 2 \cdot 13} + \dfrac{25 \cdot 3}{2 \cdot 2 \cdot 13 \cdot 3} + \dfrac{7 \cdot 3 \cdot 13}{2 \cdot 2 \cdot 3 \cdot 13}$

$= \dfrac{130}{2 \cdot 3 \cdot 2 \cdot 13} + \dfrac{75}{2 \cdot 3 \cdot 2 \cdot 13} + \dfrac{273}{2 \cdot 3 \cdot 2 \cdot 13}$

$= \dfrac{478}{2 \cdot 3 \cdot 2 \cdot 13}$

$= \dfrac{2 \cdot 239}{2 \cdot 3 \cdot 2 \cdot 13} = \dfrac{2}{2} \cdot \dfrac{239}{3 \cdot 2 \cdot 13}$

$= \dfrac{239}{78}$

45. $\dfrac{2}{9} + \dfrac{7}{10} + \dfrac{-4}{15}$

$= \dfrac{2}{3 \cdot 3} + \dfrac{7}{2 \cdot 5} + \dfrac{-4}{3 \cdot 5}$ LCD is $3 \cdot 3 \cdot 2 \cdot 5$.

$= \dfrac{2}{3 \cdot 3} \cdot \dfrac{2 \cdot 5}{2 \cdot 5} + \dfrac{7}{2 \cdot 5} \cdot \dfrac{3 \cdot 3}{3 \cdot 3} + \dfrac{-4}{3 \cdot 5} \cdot \dfrac{3 \cdot 2}{3 \cdot 2}$

$= \dfrac{2 \cdot 2 \cdot 5}{3 \cdot 3 \cdot 2 \cdot 5} + \dfrac{7 \cdot 3 \cdot 3}{2 \cdot 5 \cdot 3 \cdot 3} + \dfrac{-4 \cdot 3 \cdot 2}{3 \cdot 5 \cdot 3 \cdot 2}$

$= \dfrac{20}{3 \cdot 3 \cdot 2 \cdot 5} + \dfrac{63}{3 \cdot 3 \cdot 2 \cdot 5} + \dfrac{-24}{3 \cdot 3 \cdot 2 \cdot 5}$

$= \dfrac{59}{3 \cdot 3 \cdot 2 \cdot 5}$

$= \dfrac{59}{90}$

47.

$$-\frac{3}{4} + \frac{1}{5} + \frac{-7}{10}$$

$$= \frac{-3}{4} + \frac{1}{5} + \frac{-7}{10}$$

$$= \frac{-3}{2 \cdot 2} + \frac{1}{5} + \frac{-7}{2 \cdot 5} \qquad \text{The LCD is } 2 \cdot 2 \cdot 5.$$

$$= \frac{-3}{2 \cdot 2} \cdot \frac{5}{5} + \frac{1}{5} \cdot \frac{2 \cdot 2}{2 \cdot 2} + \frac{-7}{2 \cdot 5} \cdot \frac{2}{2}$$

$$= \frac{-15}{2 \cdot 2 \cdot 5} + \frac{4}{5 \cdot 2 \cdot 2} + \frac{-14}{2 \cdot 5 \cdot 2}$$

$$= \frac{-25}{2 \cdot 2 \cdot 5} = \frac{-5 \cdot 5}{2 \cdot 2 \cdot 5} = \frac{-5}{2 \cdot 2} \cdot \frac{5}{5}$$

$$= \frac{-5}{4}, \text{ or } -\frac{5}{4}$$

49. Since there is a common denominator, compare the numerators.

$$3 > 2, \text{ so } \frac{3}{8} > \frac{2}{8}.$$

51. The LCD is 6. We multiply $\frac{2}{3}$ by 1 to make the denominators the same.

$$\frac{2}{3} \cdot \frac{2}{2} = \frac{4}{6}$$

The denominator of $\frac{5}{6}$ is the LCD.

Since $4 < 5$, it follows that $\frac{4}{6} < \frac{5}{6}$, so $\frac{2}{3} < \frac{5}{6}$.

53. The LCD is 21. We multiply by 1 to make the denominators the same.

$$\frac{-2}{3} \cdot \frac{7}{7} = \frac{-14}{21}$$

$$\frac{-5}{7} \cdot \frac{3}{3} = \frac{-15}{21}$$

Since $-14 > -15$, it follows that $\frac{-14}{21} > \frac{-15}{21}$, so $\frac{-2}{3} > \frac{-5}{7}$.

55. The LCD is 30. We multiply by 1 to make the denominators the same.

$$\frac{9}{15} \cdot \frac{2}{2} = \frac{18}{30}$$

$$\frac{7}{10} \cdot \frac{3}{3} = \frac{21}{30}$$

Since $18 < 21$, it follows that $\frac{18}{30} < \frac{21}{30}$, so $\frac{9}{15} < \frac{7}{10}$.

57. Express $-\frac{1}{5}$ as $\frac{-1}{5}$. The LCD is 20. Multiply by 1 to make the denominators the same.

$$\frac{3}{4} \cdot \frac{5}{5} = \frac{15}{20}$$

$$\frac{-1}{5} \cdot \frac{4}{4} = \frac{-4}{20}$$

Since $15 > -4$, it follows that $\frac{15}{20} > \frac{-4}{20}$, so $\frac{3}{4} > -\frac{1}{5}$. We might have observed at the outset that one number is positive and the other is negative and it follows that the positive number is greater than the negative number.

59. The LCD is 60. We multiply by 1 to make the denominators the same.

$$\frac{-7}{20} \cdot \frac{3}{3} = \frac{-21}{60}$$

$$\frac{-6}{15} \cdot \frac{4}{4} = \frac{-24}{60}$$

Since $-21 > -24$, it follows that $\frac{-21}{60} > \frac{-24}{60}$, so $\frac{-7}{20} > \frac{-6}{15}$.

61. The LCD is 60. We multiply by 1 to make the denominators the same.

$$\frac{3}{10} \cdot \frac{6}{6} = \frac{18}{60}$$

$$\frac{5}{12} \cdot \frac{5}{5} = \frac{25}{60}$$

$$\frac{4}{15} \cdot \frac{4}{4} = \frac{16}{60}$$

Since $16 < 18$ and $18 < 25$, when we arrange $\frac{18}{60}$, $\frac{25}{60}$, and $\frac{16}{60}$ from smallest to largest we have $\frac{16}{60}$, $\frac{18}{60}$, $\frac{25}{60}$. Then it follows that when we arrange the original fractions from smallest to largest we have $\frac{4}{15}$, $\frac{3}{10}$, $\frac{5}{12}$.

63. *Familiarize*. We draw a picture. We let $p =$ the number of pounds of candy Todd bought.

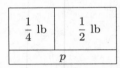

***Translate*.** The problem can be translated to an equation as follows:

Pounds of gumdrops	plus	Pounds of caramels	is	Total pounds of candy
↓	↓	↓	↓	↓
$\frac{1}{4}$	$+$	$\frac{1}{2}$	$=$	p

***Solve*.** We carry out the addition. Since 4 is a multiple of 2, the LCM of the denominators is 4.

$$\frac{1}{4} + \frac{1}{2} = p$$

$$\frac{1}{4} + \frac{1}{2} \cdot \frac{2}{2} = p$$

$$\frac{1}{4} + \frac{2}{4} = p$$

$$\frac{3}{4} = p$$

***Check*.** We check by repeating the calculation. We also note that the sum is larger than either of the individual weights, so the answer seems reasonable.

***State*.** Todd bought $\frac{3}{4}$ lb of candy.

65. *Familiarize*. We draw a picture. We let D = the total distance walked.

$$\underbrace{\underbrace{\frac{7}{8} \text{ mi}} \quad \underbrace{\frac{2}{5} \text{ mi}}}_{D}$$

***Translate*.** The problem can be translated to an equation as follows:

Distance to student union	plus	Distance to class	is	Total distance
↓	↓	↓	↓	↓
$\frac{7}{8}$	$+$	$\frac{2}{5}$	$=$	D

***Solve*.** To solve the equation, carry out the addition. Since $8 = 2 \cdot 2 \cdot 2$ and $5 = 5$, the LCM of the denominators is $2 \cdot 2 \cdot 2 \cdot 5$, or 40.

$$\frac{7}{8} + \frac{2}{5} = D$$
$$\frac{7}{8} \cdot \frac{5}{5} + \frac{2}{5} \cdot \frac{8}{8} = D$$
$$\frac{35}{40} + \frac{16}{40} = D$$
$$\frac{51}{40} = D$$

***Check*.** We repeat the calculation. We also note that the sum is larger than either of the original distances, so the answer seems reasonable.

***State*.** Kate walked $\frac{51}{40}$ mi.

67. *Familiarize*. We draw a picture and let f = the total amount of flour used.

$\frac{1}{2}$ lb	$\frac{1}{4}$ lb	$\frac{1}{3}$ lb
	f	

***Translate*.** The problem can be translated to an equation as follows:

Amount for rolls	plus	Amount for donuts	plus	Amount for cookies	is	Total amount
↓	↓	↓	↓	↓	↓	↓
$\frac{1}{2}$	$+$	$\frac{1}{4}$	$+$	$\frac{1}{3}$	$=$	f

***Solve*.** We carry out the addition. Since $2 = 2$, $4 = 2 \cdot 2$, and $3 = 3$, the LCM of the denominators is $2 \cdot 2 \cdot 3$, or 12.

$$\frac{1}{2} + \frac{1}{4} + \frac{1}{3} = f$$
$$\frac{1}{2} \cdot \frac{6}{6} + \frac{1}{4} \cdot \frac{3}{3} + \frac{1}{3} \cdot \frac{4}{4} = f$$
$$\frac{6}{12} + \frac{3}{12} + \frac{4}{12} = f$$
$$\frac{13}{12} = f$$

***Check*.** We repeat the calculation. We also note that the sum is larger than any of the individual amounts, as expected.

***State*.** $\frac{13}{12}$ lb of flour was used.

69. *Familiarize*. We draw a picture and let r = the total rainfall, in inches.

$\frac{1}{2}$ in.	$\frac{3}{8}$ in.
	r

***Translate*.** The problem can be translated to an equation as follows:

Morning rain	plus	Afternoon rain	is	Total rainfall
↓	↓	↓	↓	↓
$\frac{1}{2}$	$+$	$\frac{3}{8}$	$=$	r

***Solve*.** We carry out the addition. Since 8 is a multiple of 2, the LCM of the denominators is 8.

$$\frac{1}{2} + \frac{3}{8} = r$$
$$\frac{1}{2} \cdot \frac{4}{4} + \frac{3}{8} = r$$
$$\frac{4}{8} + \frac{3}{8} = r$$
$$\frac{7}{8} = r$$

***Check*.** We repeat the calculations. We also note that the sum is larger than either of the individual amounts, so the answer seems reasonable.

***State*.** Altogether it rained $\frac{7}{8}$ in.

71. *Familiarize*. We draw a picture and let d = the number of miles the naturalist hikes.

Lookout   Nest   Campsite

$$\underbrace{\frac{3}{5} \text{ mi} \quad \frac{3}{10} \text{ mi} \quad \frac{3}{4} \text{ mi}}_{d}$$

***Translate*.** We translate to an equation as follows:

$$\underbrace{\text{Miles to} \atop \text{lookout}} \quad \text{plus} \quad \underbrace{\text{Miles to} \atop \text{nest}} \quad \text{plus} \quad \underbrace{\text{Miles to} \atop \text{campsite}} \quad \text{is} \quad \underbrace{\text{Total} \atop \text{miles}}$$

$$\frac{3}{5} \quad + \quad \frac{3}{10} \quad + \quad \frac{3}{4} \quad = \quad d$$

Solve. We carry out the addition. Since $5 = 5$, $10 = 2 \cdot 5$, and $4 = 2 \cdot 2$, the LCM of the denominators is $2 \cdot 2 \cdot 5$, or 20.

$$\frac{3}{5} + \frac{3}{10} + \frac{3}{4} = d$$

$$\frac{3}{5} \cdot \frac{4}{4} + \frac{3}{10} \cdot \frac{2}{2} + \frac{3}{4} \cdot \frac{5}{5} = d$$

$$\frac{12}{20} + \frac{6}{20} + \frac{15}{20} = d$$

$$\frac{33}{20} = d$$

Check. We repeat the calculation. We also note that the sum is larger than any of the individual distances, as expected.

State. The naturalist hiked a total of $\frac{33}{20}$ mi.

73. Familiarize. First we will find the amount of liquid needed. Let $l =$ this amount, in quarts.

Translate.

$$\underbrace{\text{Amount of} \atop \text{ginger ale}} \quad \text{plus} \quad \underbrace{\text{Amount of} \atop \text{strawberry} \atop \text{soda}} \quad \text{is} \quad \underbrace{\text{Total} \atop \text{amount of} \atop \text{liquid}}$$

$$\frac{1}{5} \quad + \quad \frac{3}{5} \quad = \quad l$$

Solve. We carry out the addition.

$$\frac{1}{5} + \frac{3}{5} = l$$

$$\frac{4}{5} = l$$

Let $d =$ the number of quarts of liquid needed if the recipe is doubled. We write a multiplication sentence and carry out the multiplication.

$$d = 2 \cdot \frac{4}{5}$$

$$d = \frac{2 \cdot 4}{5} = \frac{8}{5}$$

Let $h =$ the number of quarts of liquid needed if the recipe is halved. We write another multiplication sentence and carry out the multiplication.

$$h = \frac{1}{2} \cdot \frac{4}{5}$$

$$h = \frac{1 \cdot 4}{2 \cdot 5} = \frac{1 \cdot 2 \cdot 2}{2 \cdot 5} = \frac{2}{2} \cdot \frac{1 \cdot 2}{5} = \frac{2}{5}$$

Check. We repeat the calculations. The answers check.

State. The recipe requires $\frac{4}{5}$ qt of liquid. If the recipe is doubled, $\frac{8}{5}$ qt is required; $\frac{2}{5}$ qt is required if the recipe is halved.

75. Familiarize. We draw a picture. We let $t =$ the total thickness.

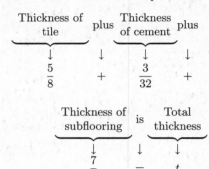

Translate. We translate to an equation.

$$\underbrace{\text{Thickness of} \atop \text{tile}} \quad \text{plus} \quad \underbrace{\text{Thickness} \atop \text{of cement}} \quad \text{plus}$$

$$\frac{5}{8} \quad + \quad \frac{3}{32} \quad +$$

$$\underbrace{\text{Thickness of} \atop \text{subflooring}} \quad \text{is} \quad \underbrace{\text{Total} \atop \text{thickness}}$$

$$\frac{7}{8} \quad = \quad t$$

Solve. We carry out the addition. The LCD is 32 since 8 is a factor of 32.

$$\frac{5}{8} + \frac{3}{32} + \frac{7}{8} = t$$

$$\frac{5}{8} \cdot \frac{4}{4} + \frac{3}{32} + \frac{7}{8} \cdot \frac{4}{4} = t$$

$$\frac{20}{32} + \frac{3}{32} + \frac{28}{32} = t$$

$$\frac{51}{32} = t$$

Check. We repeat the calculation. We also note that the sum is larger than any of the individual thicknesses, as expected.

State. The result is $\frac{51}{32}$ in. thick.

77. Discussion and Writing Exercise

79. $-7 - 6 = -7 + (-6) = -13$

81. $9 - 17 = 9 + (-17) = -8$

83. $\dfrac{x - y}{3} = \dfrac{7 - (-3)}{3} = \dfrac{7 + 3}{3} = \dfrac{10}{3}$

85. Familiarize. Let $x =$ the amount by which spending on books in 2005 exceeded spending on books in 2004.

Translate. We consider this to be a "missing addend" situation.

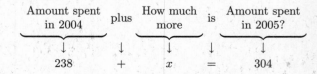

Solve. We subtract 238 on both sides of the equation.

$$238 + x = 304$$
$$238 + x - 238 = 304 - 238$$
$$x = 66$$

Check. We can repeat the calculation. We can also round and estimate: $238 + 66 \approx 240 + 70 \approx 310 \approx 304$. The answer checks.

State. Freshmen planned to spend \$66 more on textbooks in 2005 than in 2004.

87. *Familiarize*. Let s = the amount by which spending on shoes in 2004 exceeded spending on shoes in 2005.

Translate. We consider this to be a "missing addend" situation.

Amount spent in 2005	plus	How much more	is	Amount spent in 2004?
↓	↓	↓	↓	↓
36	+	s	=	93

Solve. We subtract 36 on both sides of the equation.

$$36 + s = 93$$
$$36 + s - 36 = 93 - 36$$
$$s = 57$$

Check. We can repeat the calculation. The answer checks.

State. Freshmen planned to spent \$57 more on shoes in 2004 than in 2005.

89. *Familiarize*. Let t = the total planned expenditure in 2004. We are combining amounts, so we add.

Translate. We translate to an equation.

$$238 + 760 + 58 + 83 + 32 + 93 = t$$

Solve. We carry out the addition.

$$
\begin{array}{r}
{\scriptstyle 3\ 2} \\
2\ 3\ 8 \\
7\ 6\ 0 \\
5\ 8 \\
8\ 3 \\
3\ 2 \\
+\ \ \ 9\ 3 \\
\hline
1\ 2\ 6\ 4
\end{array}
$$

Check. We can repeat the calculation. The answer checks.

State. The total planned expenditure in 2004 was \$1264.

91. Discussion and Writing Exercise

93.
$$\frac{3}{10}t + \frac{2}{7} + \frac{2}{15}t + \frac{3}{5}$$
$$= \left(\frac{3}{10} + \frac{2}{15}\right)t + \left(\frac{2}{7} + \frac{3}{5}\right)$$
$$= \left(\frac{3}{10} \cdot \frac{3}{3} + \frac{2}{15} \cdot \frac{2}{2}\right)t + \left(\frac{2}{7} \cdot \frac{5}{5} + \frac{3}{5} \cdot \frac{7}{7}\right)$$
$$= \left(\frac{9}{30} + \frac{4}{30}\right)t + \left(\frac{10}{35} + \frac{21}{35}\right)$$
$$= \frac{13}{30}t + \frac{31}{35}$$

95.
$$5t^2 + \frac{6}{a}t + 2t^2 + \frac{3}{a}t$$
$$= (5 + 2)t^2 + \left(\frac{6}{a} + \frac{3}{a}\right)t$$
$$= 7t^2 + \frac{9}{a}t$$

97. Use a calculator to do this exercise. First, add on the left.

$$\frac{12}{169} + \frac{53}{103} = \frac{10,193}{17,407}$$

Now compare $\dfrac{10,193}{17,407}$ and $\dfrac{10,192}{17,407}$. The denominators are the same. Since $10,193 > 10,192$, it follows that $\dfrac{10,193}{17,407} > \dfrac{10,192}{17,407}$, so $\dfrac{12}{169} + \dfrac{53}{103} > \dfrac{10,192}{17,407}$.

99. *Familiarize*. First we find the fractional part of the band's pay that the guitarist received. We let f = this fraction.

Translate. We translate to an equation.

One-third	of	one-half	plus	one-fifth	of	one-half	is	fractional part
↓	↓	↓	↓	↓	↓	↓	↓	↓
$\frac{1}{3}$	$\cdot$	$\frac{1}{2}$	$+$	$\frac{1}{5}$	$\cdot$	$\frac{1}{2}$	$=$	f

Solve. We carry out the calculation.

$$\frac{1}{3} \cdot \frac{1}{2} + \frac{1}{5} \cdot \frac{1}{2} = f$$
$$\frac{1}{6} + \frac{1}{10} = f \qquad \text{LCD is 30.}$$
$$\frac{1}{6} \cdot \frac{5}{5} + \frac{1}{10} \cdot \frac{3}{3} = f$$
$$\frac{5}{30} + \frac{3}{30} = f$$
$$\frac{8}{30} = f$$
$$\frac{4}{15} = f$$

Now we find how much of the \$1200 received by the band was paid to the guitarist. We let p = the amount.

Four-fifteenths	of	\$1200	=	guitarist's pay
↓	↓	↓	↓	↓
$\frac{4}{15}$	$\cdot$	1200	=	p

We solve the equation.

$$\frac{4}{15} \cdot 1200 = p$$
$$\frac{4 \cdot 1200}{15} = p$$
$$\frac{4 \cdot 3 \cdot 5 \cdot 80}{3 \cdot 5} = p$$
$$320 = p$$

Check. We repeat the calculations.

State. The guitarist received $\frac{4}{15}$ of the band's pay. This was \$320.

101. Trying various placements and using a calculator to evaluate them yields $4 + \dfrac{6}{3} \cdot 5$, or $4 + \dfrac{5}{3} \cdot 6$. Each is equal to 14.

103. First we find the LCM of the denominators.

$$4 = 2 \cdot 2$$
$$21 = 3 \cdot 7$$
$$15 = 3 \cdot 5$$
$$9 = 3 \cdot 3$$
$$17 = 17$$
$$12 = 2 \cdot 2 \cdot 3$$
$$22 = 2 \cdot 11$$

The LCM is $2 \cdot 2 \cdot 3 \cdot 3 \cdot 5 \cdot 7 \cdot 11 \cdot 17 = 235,620$.

Now write each fraction with this denominator.

$$\frac{3}{4} = \frac{176,715}{235,620}$$
$$\frac{17}{21} = \frac{190,740}{235,620}$$
$$\frac{13}{15} = \frac{204,204}{235,620}$$
$$\frac{7}{9} = \frac{183,260}{235,620}$$
$$\frac{15}{17} = \frac{207,900}{235,620}$$
$$\frac{13}{12} = \frac{255,255}{235,620}$$
$$\frac{19}{22} = \frac{203,490}{235,620}$$

Now arrange the numerators in order from smallest to largest. This yields the following arrangement of the original fractions.

$$\frac{3}{4}, \frac{7}{9}, \frac{17}{21}, \frac{19}{22}, \frac{13}{15}, \frac{15}{17}, \frac{13}{12}$$

Exercise Set 4.3

1. When denominators are the same, subtract the numerators and keep the denominator.

$$\frac{5}{6} - \frac{1}{6} = \frac{5-1}{6} = \frac{4}{6} = \frac{2 \cdot 2}{2 \cdot 3} = \frac{2}{2} \cdot \frac{2}{3} = \frac{2}{3}$$

3. When denominators are the same, subtract the numerators and keep the denominator.

$$\frac{9}{16} - \frac{13}{16} = \frac{9-13}{16} = \frac{-4}{16} = \frac{-1 \cdot 4}{4 \cdot 4} = \frac{-1}{4} \cdot \frac{4}{4} = \frac{-1}{4},$$
or $-\dfrac{1}{4}$

5. $\dfrac{8}{a} - \dfrac{6}{a} = \dfrac{8-6}{a} = \dfrac{2}{a}$

7. $-\dfrac{2}{9} - \dfrac{5}{9} = \dfrac{-2-5}{9} = \dfrac{-7}{9}$, or $-\dfrac{7}{9}$

9. $-\dfrac{3}{8} - \dfrac{1}{8} = \dfrac{-3-1}{8} = \dfrac{-4}{8} = \dfrac{-1 \cdot 4}{2 \cdot 4} = \dfrac{-1}{2} \cdot \dfrac{4}{4} = \dfrac{-1}{2}$, or $-\dfrac{1}{2}$

11. $\dfrac{10}{3t} - \dfrac{4}{3t} = \dfrac{10-4}{3t} = \dfrac{6}{3t} = \dfrac{3}{3} \cdot \dfrac{2}{t} = \dfrac{2}{t}$

13. $\dfrac{3}{5a} - \dfrac{7}{5a} = \dfrac{3-7}{5a} = \dfrac{-4}{5a}$, or $-\dfrac{4}{5a}$

15. The LCM of 8 and 16 is 16.

$$\frac{7}{8} - \frac{1}{16} = \frac{7}{8} \cdot \frac{2}{2} - \frac{1}{16} \longleftarrow \text{This fraction already has the LCM as the denominator.}$$

Think: $8 \times \square = 16$. The answer is 2, so we multiply by 1, using $\dfrac{2}{2}$.

$$= \frac{14}{16} - \frac{1}{16} = \frac{13}{16}$$

17. The LCM of 15 and 5 is 15.

$$\frac{7}{15} - \frac{4}{5} = \frac{7}{15} - \frac{4}{5} \cdot \frac{3}{2}$$

Think: $5 \times \square = 15$. The answer is 3, so we multiply by 1, using $\dfrac{3}{3}$.

This fraction already has the LCM as the denominator.

$$= \frac{7}{15} - \frac{12}{15}$$
$$= \frac{-5}{15} = \frac{5}{5} \cdot \frac{-1}{3}$$
$$= \frac{-1}{3}, \text{ or } -\frac{1}{3}$$

19. The LCM of 4 and 20 is 20.

$$\frac{3}{4} - \frac{1}{20} = \frac{3}{4} \cdot \frac{5}{5} - \frac{1}{20}$$
$$= \frac{15}{20} - \frac{1}{20} = \frac{14}{20}$$
$$= \frac{2 \cdot 7}{2 \cdot 10} = \frac{2}{2} \cdot \frac{7}{10}$$
$$= \frac{7}{10}$$

21. The LCM of 15 and 12 is 60.

$$\frac{2}{15} - \frac{5}{12} = \frac{2}{15} \cdot \frac{4}{4} - \frac{5}{12} \cdot \frac{5}{5}$$
$$= \frac{8}{60} - \frac{25}{60} = \frac{8-25}{60}$$
$$= \frac{-17}{60}, \text{ or } -\frac{17}{60}$$

23. The LCM of 10 and 100 is 100.

$$\frac{7}{10} - \frac{23}{100} = \frac{7}{10} \cdot \frac{10}{10} - \frac{23}{100}$$
$$= \frac{70}{100} - \frac{23}{100} = \frac{47}{100}$$

25. The LCM of 15 and 25 is 75.

$$\frac{7}{15} - \frac{3}{25} = \frac{7}{15} \cdot \frac{5}{5} - \frac{3}{25} \cdot \frac{3}{3}$$

$$= \frac{35}{75} - \frac{9}{75} = \frac{26}{75}$$

27. The LCM of 10 and 100 is 100.

$$\frac{69}{100} - \frac{9}{10} = \frac{69}{100} - \frac{9}{10} \cdot \frac{10}{10}$$

$$= \frac{69}{100} - \frac{90}{100} = \frac{69 - 90}{100}$$

$$= \frac{-21}{100}, \text{ or } -\frac{21}{100}$$

29. The LCM of 8 and 3 is 24.

$$\frac{1}{8} - \frac{2}{3} = \frac{1}{8} \cdot \frac{3}{3} - \frac{2}{3} \cdot \frac{8}{8}$$

$$= \frac{3}{24} - \frac{16}{24}$$

$$= \frac{-13}{24}, \text{ or } -\frac{13}{24}$$

31. The LCM of 10 and 25 is 50.

$$-\frac{3}{10} - \frac{7}{25} = -\frac{3}{10} \cdot \frac{5}{5} - \frac{7}{25} \cdot \frac{2}{2}$$

$$= -\frac{15}{50} - \frac{14}{50}$$

$$= -\frac{29}{50}, \text{ or } -\frac{29}{50}$$

33. The LCM of 3 and 5 is 15.

$$\frac{2}{3} - \frac{4}{5} = \frac{2}{3} \cdot \frac{5}{5} - \frac{4}{5} \cdot \frac{3}{3}$$

$$= \frac{10}{15} - \frac{12}{15}$$

$$= \frac{-2}{15}, \text{ or } -\frac{2}{15}$$

35. The LCM of 18 and 24 is 72.

$$\frac{-5}{18} - \frac{7}{24} = \frac{-5}{18} \cdot \frac{4}{4} - \frac{7}{24} \cdot \frac{3}{3}$$

$$= \frac{-20}{72} - \frac{21}{72}$$

$$= \frac{-41}{72}, \text{ or } -\frac{41}{72}$$

37. The LCM of 90 and 120 is 360.

$$\frac{13}{90} - \frac{17}{120} = \frac{13}{90} \cdot \frac{4}{4} - \frac{17}{120} \cdot \frac{3}{3}$$

$$= \frac{52}{360} - \frac{51}{360}$$

$$= \frac{1}{360}$$

39. The LCM of 3 and 9 is 9.

$$\frac{2}{3}x - \frac{4}{9}x = \frac{2}{3} \cdot \frac{3}{3} \cdot x - \frac{4}{9}x$$

$$= \frac{6}{9}x - \frac{4}{9}x$$

$$= \frac{2}{9}x$$

41. The LCM of 5 and 4 is 20.

$$\frac{2}{5}a - \frac{3}{4}a = \frac{2}{5} \cdot \frac{4}{4} \cdot a - \frac{3}{4} \cdot \frac{5}{5}a$$

$$= \frac{8}{20}a - \frac{15}{20}a$$

$$= \frac{-7}{20}a, \text{ or } -\frac{7}{20}a$$

43.
$$x - \frac{4}{9} = \frac{3}{9}$$

$$x - \frac{4}{9} + \frac{4}{9} = \frac{3}{9} + \frac{4}{9} \quad \text{Adding } \frac{4}{9} \text{ to both sides}$$

$$x + 0 = \frac{7}{9}$$

$$x = \frac{7}{9}$$

The solution is $\frac{7}{9}$.

45.
$$a + \frac{2}{11} = \frac{6}{11}$$

$$a + \frac{2}{11} - \frac{2}{11} = \frac{6}{11} - \frac{2}{11} \quad \text{Subtracting } \frac{2}{11} \text{ from both sides}$$

$$a + 0 = \frac{4}{11}$$

$$a = \frac{4}{11}$$

The solution is $\frac{4}{11}$.

47.
$$x + \frac{1}{3} = \frac{7}{9}$$

$$x + \frac{1}{3} - \frac{1}{3} = \frac{7}{9} - \frac{1}{3} \quad \text{Subtracting } \frac{1}{3} \text{ from both sides}$$

$$x + 0 = \frac{7}{9} - \frac{1}{3} \cdot \frac{3}{3} \quad \text{The LCD is 9. We multiply by 1 to get the LCD.}$$

$$x = \frac{7}{9} - \frac{3}{9} = \frac{4}{9}$$

The solution is $\frac{4}{9}$.

49. $a - \dfrac{3}{8} = \dfrac{3}{4}$

$a - \dfrac{3}{8} + \dfrac{3}{8} = \dfrac{3}{4} + \dfrac{3}{8}$ Adding $\dfrac{3}{8}$ on both sides

$a + 0 = \dfrac{3}{4} \cdot \dfrac{2}{2} + \dfrac{3}{8}$ The LCD is 8. We multiply by 1 to get the LCD.

$a = \dfrac{6}{8} + \dfrac{3}{8} = \dfrac{9}{8}$

The solution is $\dfrac{9}{8}$.

51. $\dfrac{2}{3} + x = \dfrac{4}{5}$

$\dfrac{2}{3} + x - \dfrac{2}{3} = \dfrac{4}{5} - \dfrac{2}{3}$ Subtracting $\dfrac{2}{3}$ on both sides

$x + 0 = \dfrac{4}{5} \cdot \dfrac{3}{3} - \dfrac{2}{3} \cdot \dfrac{5}{5}$ The LCD is 15. We multiply by 1 to get the LCD.

$x = \dfrac{12}{15} - \dfrac{10}{15} = \dfrac{2}{15}$

The solution is $\dfrac{2}{15}$.

53. $\dfrac{3}{8} + a = \dfrac{1}{12}$

$\dfrac{3}{8} + a - \dfrac{3}{8} = \dfrac{1}{12} - \dfrac{3}{8}$ Subtracting $\dfrac{3}{8}$ on both sides

$a + 0 = \dfrac{1}{12} \cdot \dfrac{2}{2} - \dfrac{3}{8} \cdot \dfrac{3}{3}$ The LCD is 24. We multiply by 1 to get the LCD.

$a = \dfrac{2}{24} - \dfrac{9}{24} = \dfrac{2 - 9}{24}$

$a = \dfrac{-7}{24}$, or $-\dfrac{7}{24}$

The solution is $-\dfrac{7}{24}$.

55. $n - \dfrac{3}{10} = -\dfrac{1}{6}$

$n - \dfrac{3}{10} + \dfrac{3}{10} = -\dfrac{1}{6} + \dfrac{3}{10}$ Adding $\dfrac{3}{10}$ to both sides

$n + 0 = -\dfrac{1}{6} \cdot \dfrac{5}{5} + \dfrac{3}{10} \cdot \dfrac{3}{3}$ The LCD is 30. We multiply by 1 to get the LCD.

$n = -\dfrac{5}{30} + \dfrac{9}{30}$

$n = \dfrac{4}{30}$

$n = \dfrac{2 \cdot 2}{2 \cdot 15} = \dfrac{2}{2} \cdot \dfrac{2}{15}$

$n = \dfrac{2}{15}$

The solution is $\dfrac{2}{15}$.

57. $x + \dfrac{3}{4} = -\dfrac{1}{2}$

$x + \dfrac{3}{4} - \dfrac{3}{4} = -\dfrac{1}{2} - \dfrac{3}{4}$ Subtracting $\dfrac{3}{4}$ on both sides

$x + 0 = -\dfrac{1}{2} \cdot \dfrac{2}{2} - \dfrac{3}{4}$ The LCD is 4. We multiply by 1 to get the LCD.

$x = -\dfrac{2}{4} - \dfrac{3}{4} = \dfrac{-2}{4} - \dfrac{3}{4}$

$x = \dfrac{-2 - 3}{4}$

$x = \dfrac{-5}{4}$, or $-\dfrac{5}{4}$

The solution is $-\dfrac{5}{4}$.

59. *Familiarize.* We visualize the situation. Let $d =$ the distance that remains to be swum.

$\dfrac{1}{5}$ mi d

$\dfrac{1}{2}$ mi

Translate. This is a "missing addend" situation that can be translate as follows:

Distance already swum	plus	Distance remaining to be swum	is	Total distance
↓	↓	↓	↓	↓
$\dfrac{1}{5}$	$+$	d	$=$	$\dfrac{1}{2}$

Solve. We subtract $\dfrac{1}{5}$ from both sides of the equation.

$\dfrac{1}{5} + d - \dfrac{1}{5} = \dfrac{1}{2} - \dfrac{1}{5}$

$d + 0 = \dfrac{1}{2} \cdot \dfrac{5}{5} - \dfrac{1}{5} \cdot \dfrac{2}{2}$ The LCD is 10. We multiply by 1 to get the LCD.

$d = \dfrac{5}{10} - \dfrac{2}{10} = \dfrac{3}{10}$

Check. We return to the original problem and add.

$\dfrac{1}{5} + \dfrac{3}{10} = \dfrac{1}{5} \cdot \dfrac{2}{2} + \dfrac{3}{10} = \dfrac{2}{10} + \dfrac{3}{10} = \dfrac{5}{10} = \dfrac{5}{5} \cdot \dfrac{1}{2} = \dfrac{1}{2}$

The answer checks.

State. Deb should swim $\dfrac{3}{10}$ mi farther.

61. *Familiarize.* We visualize the situation. Let $a =$ the amount of cheese left in the bowl, in cups.

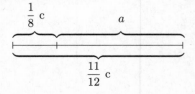

$\dfrac{1}{8}$ c a

$\dfrac{11}{12}$ c

Translate. This is a "how much more" situation.

Amount served	plus	Amount left	is	Original amount
↓	↓	↓	↓	↓
$\frac{1}{8}$	$+$	a	$=$	$\frac{11}{12}$

Solve. We subtract $\frac{1}{8}$ on both sides of the equation.

$$\frac{1}{8} + a - \frac{1}{8} = \frac{11}{12} - \frac{1}{8}$$

$$a + 0 = \frac{11}{12} \cdot \frac{2}{2} - \frac{1}{8} \cdot \frac{3}{3} \quad \text{The LCD is 24. We multiply by 1 to get the LCD.}$$

$$a = \frac{22}{24} - \frac{3}{24}$$

$$a = \frac{19}{24}$$

Check. We return to the original problem and add.

$$\frac{1}{8} + \frac{19}{24} = \frac{1}{8} \cdot \frac{3}{3} + \frac{19}{24} = \frac{3}{24} + \frac{19}{24} = \frac{22}{24} = \frac{2 \cdot 11}{2 \cdot 12} = \frac{2}{2} \cdot \frac{11}{12} = \frac{11}{12}$$

The answer checks.

State. There is $\frac{19}{24}$ cup of cheese left in the bowl.

63. *Familiarize.* Using the label on the drawing in the text, we let r = the amount by which the board should be planed down, in inches.

Translate. This is a "missing addend" situation.

Desired thickness	plus	Excess amount	is	Original thickness
↓	↓	↓	↓	↓
$\frac{3}{4}$	$+$	r	$=$	$\frac{15}{16}$

Solve. We subtract $\frac{3}{4}$ from both sides of the equation.

$$\frac{3}{4} + r - \frac{3}{4} = \frac{15}{16} - \frac{3}{4}$$

$$r + 0 = \frac{15}{16} - \frac{3}{4} \cdot \frac{4}{4} \quad \text{The LCD is 16. We multiply by 1 to get the LCD.}$$

$$r = \frac{15}{16} - \frac{12}{16}$$

$$r = \frac{3}{16}$$

Check. We return to the original problem and add.

$$\frac{3}{4} + \frac{3}{16} = \frac{3}{4} \cdot \frac{4}{4} + \frac{3}{16} = \frac{12}{16} + \frac{3}{16} = \frac{15}{16}$$

The answer checks.

State. The board should be planed down $\frac{3}{16}$ in.

65. *Familiarize.* We visualize the situation. Let c = the amount of cheese remaining, in pounds.

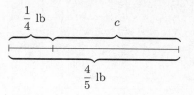

Translate. This is a "how much more" situation.

Amount served	plus	Amount remaining	is	Original amount
↓	↓	↓	↓	↓
$\frac{1}{4}$	$+$	c	$=$	$\frac{4}{5}$

Solve. We subtract $\frac{1}{4}$ on both sides of the equation.

$$\frac{1}{4} + c - \frac{1}{4} = \frac{4}{5} - \frac{1}{4}$$

$$c + 0 = \frac{4}{5} \cdot \frac{4}{4} - \frac{1}{4} \cdot \frac{5}{5} \quad \text{The LCD is 20.}$$

$$c = \frac{16}{20} - \frac{5}{20}$$

$$c = \frac{11}{20}$$

Check. Since $\frac{1}{4} + \frac{11}{20} = \frac{1}{4} \cdot \frac{5}{5} + \frac{11}{20} = \frac{5}{20} + \frac{11}{20} = \frac{16}{20} = \frac{4 \cdot 4}{4 \cdot 5} = \frac{4}{4} \cdot \frac{4}{5} = \frac{4}{5}$, the answer checks.

State. $\frac{11}{20}$ lb of cheese remains on the wheel.

67. *Familiarize.* We visualize the situation. Let t = the time spent on country driving, in hours.

Translate. This is a "how much more" situation.

City driving time	plus	Country driving time	is	Total time
↓	↓	↓	↓	↓
$\frac{2}{5}$	$+$	t	$=$	$\frac{3}{4}$

Solve. We subtract $\frac{2}{5}$ from both sides of the equation.

$$\frac{2}{5} + t - \frac{2}{5} = \frac{3}{4} - \frac{2}{5}$$

$$t + 0 = \frac{3}{4} \cdot \frac{5}{5} - \frac{2}{5} \cdot \frac{4}{4} \quad \text{The LCD is 20.}$$

$$t = \frac{15}{20} - \frac{8}{20} = \frac{7}{20}$$

Check. We return to the original problem and add.

$$\frac{2}{5} + \frac{7}{20} = \frac{2}{5} \cdot \frac{4}{4} + \frac{7}{20} = \frac{8}{20} + \frac{7}{20} = \frac{15}{20} = \frac{3 \cdot 5}{4 \cdot 5} = \frac{3}{4}$$

The answer checks.

State. Jorge spent $\frac{7}{20}$ hr on country driving.

69. Familiarize. We visualize the situation. Let s = the amount of syrup that should be added, in cups.

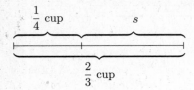

Translate. This is a "missing addend" situation.

$$\underbrace{\text{Original amount}}_{\downarrow} \quad \underbrace{\text{plus}}_{\downarrow} \quad \underbrace{\text{Additional amount}}_{\downarrow} \quad \underbrace{\text{is}}_{\downarrow} \quad \underbrace{\text{Total amount}}_{\downarrow}$$
$$\frac{1}{4} \qquad + \qquad s \qquad = \qquad \frac{2}{3}$$

Solve. We subtract $\frac{1}{4}$ from both sides of the equation.

$$\frac{1}{4} + s - \frac{1}{4} = \frac{2}{3} - \frac{1}{4}$$

$$s + 0 = \frac{2}{3} \cdot \frac{4}{4} - \frac{1}{4} \cdot \frac{3}{3} \quad \begin{array}{l}\text{The LCD is 12. We mul-}\\ \text{tiply by 1 to get the LCD.}\end{array}$$

$$s = \frac{8}{12} - \frac{3}{12} = \frac{5}{12}$$

Check. We return to the original problem and add.

$$\frac{1}{4} + \frac{5}{12} = \frac{1}{4} \cdot \frac{3}{3} + \frac{5}{12} = \frac{3}{12} + \frac{5}{12} = \frac{8}{12} = \frac{4}{4} \cdot \frac{2}{3} = \frac{2}{3}$$

The answer checks.

State. Blake should add $\frac{5}{12}$ cup of syrup to the batter.

71. Discussion and Writing Exercise

73. $\dfrac{3}{7} \div \dfrac{9}{4} = \dfrac{3}{7} \cdot \dfrac{4}{9} = \dfrac{3 \cdot 4}{7 \cdot 9}$

$\qquad = \dfrac{3 \cdot 4}{7 \cdot 3 \cdot 3} = \dfrac{3}{3} \cdot \dfrac{4}{7 \cdot 3}$

$\qquad = \dfrac{4}{21}$

75. $7 \div \dfrac{1}{3} = 7 \cdot \dfrac{3}{1} = \dfrac{7 \cdot 3}{1}$

$\qquad = \dfrac{21}{1} = 21$

77. Familiarize. Let d = the number of days it will take to fill an order for 66,045 tire gauges.

Translate.

$$\underbrace{\begin{array}{c}\text{Gauges produced}\\ \text{per day}\end{array}}_{\downarrow} \quad \underbrace{\text{times}}_{\downarrow} \quad \underbrace{\begin{array}{c}\text{Number}\\ \text{of days}\end{array}}_{\downarrow} \quad \underbrace{\text{is}}_{\downarrow} \quad \underbrace{\begin{array}{c}\text{Total}\\ \text{order}\end{array}}_{\downarrow}$$
$$3885 \qquad \cdot \qquad d \qquad = \qquad 66,045$$

Solve. We divide by 3885 on both sides of the equation.

$$\frac{3885 \cdot d}{3885} = \frac{66,045}{3885}$$
$$d = 17$$

Check. Since $3885 \cdot 17 = 66,045$, the answer checks.

State. It will take 17 days to produce 66,045 tire gauges.

79.
$$3x - 8 = 25$$
$$3x - 8 + 8 = 25 + 8 \quad \text{Adding 8 to both sides}$$
$$3x + 0 = 33$$
$$3x = 33$$
$$\frac{3x}{3} = \frac{33}{3} \quad \text{Dividing both sides by 3}$$
$$x = 11$$

The solution is 11.

81. Discussion and Writing Exercise

83. The LCM of 8, 4, and 16 is 16.

$$\frac{7}{8} - \frac{3}{4} - \frac{1}{16}$$
$$= \frac{7}{8} \cdot \frac{2}{2} - \frac{3}{4} \cdot \frac{4}{4} - \frac{1}{16}$$
$$= \frac{14}{16} - \frac{12}{16} - \frac{1}{16}$$
$$= \frac{14 - 12 - 1}{16}$$
$$= \frac{1}{16}$$

85. $\dfrac{2}{5} - \dfrac{1}{6}(-3)^2 = \dfrac{2}{5} - \dfrac{1}{6} \cdot 9$

$\qquad = \dfrac{2}{5} - \dfrac{9}{6}$

$\qquad = \dfrac{2}{5} \cdot \dfrac{6}{6} - \dfrac{9}{6} \cdot \dfrac{5}{5} \quad \text{The LCD is 30.}$

$\qquad = \dfrac{12}{30} - \dfrac{45}{30}$

$\qquad = \dfrac{-33}{30} = \dfrac{3}{3} \cdot \dfrac{-11}{10}$

$\qquad = \dfrac{-11}{10}, \text{ or } -\dfrac{11}{10}$

87. $-4 \cdot \dfrac{3}{7} - \dfrac{1}{7} \cdot \dfrac{4}{5} = \dfrac{-12}{7} - \dfrac{4}{35}$

$\qquad = \dfrac{-12}{7} \cdot \dfrac{5}{5} - \dfrac{4}{35} \quad \text{The LCD is 35.}$

$\qquad = \dfrac{-60}{35} - \dfrac{4}{35}$

$\qquad = \dfrac{-64}{35}, \text{ or } -\dfrac{64}{35}$

89. $\left(-\dfrac{2}{5}\right)^3 - \left(-\dfrac{3}{10}\right)^3$

$\qquad = -\dfrac{8}{125} - \left(-\dfrac{27}{1000}\right)$

$\qquad = -\dfrac{8}{125} + \dfrac{27}{1000}$

$\qquad = -\dfrac{8}{125} \cdot \dfrac{8}{8} + \dfrac{27}{1000} \quad \text{The LCD is 1000.}$

$\qquad = -\dfrac{64}{1000} + \dfrac{27}{1000}$

$\qquad = -\dfrac{37}{1000}$

91. *Familiarize*. Let $t =$ the fractional portion of the business Trey owns.

Translate. This is a "how much more" situation.

Becky's portion	plus	Clay's portion	plus	Trey's portion	is	One entire business
↓	↓	↓	↓	↓	↓	↓
$\frac{1}{3}$	$+$	$\frac{1}{2}$	$+$	t	$=$	1

Solve. First we collect like terms on the left side.

$$\frac{1}{3} + \frac{1}{2} + t = 1$$

$$\frac{1}{3} \cdot \frac{2}{2} + \frac{1}{2} \cdot \frac{3}{3} + t = 1 \quad \text{The LCD is 6.}$$

$$\frac{2}{6} + \frac{3}{6} + t = 1$$

$$\frac{5}{6} + t = 1$$

$$\frac{5}{6} + t - \frac{5}{6} = 1 - \frac{5}{6}$$

$$t + 0 = \frac{6}{6} - \frac{5}{6}$$

$$t = \frac{1}{6}$$

Check. We return to the original problem and add.

$$\frac{1}{3} + \frac{1}{2} + \frac{1}{6} = \frac{1}{3} \cdot \frac{2}{2} + \frac{1}{2} \cdot \frac{3}{3} + \frac{1}{6} = \frac{2}{6} + \frac{3}{6} + \frac{1}{6} = \frac{6}{6} = 1$$

The answer checks.

State. Trey owns $\frac{1}{6}$ of the business.

93. *Familiarize*. This is a multistep problem. First we find the portion of shoppers who stay for 1-2 hr. Let $s =$ this portion of the shoppers.

Translate. This is a "missing addend" situation.

Portions shown in graph	plus	Remaining portion	is	One entire group
↓	↓	↓	↓	↓
$\frac{25}{50} + \frac{5}{50} + \frac{2}{50}$	$+$	s	$=$	1

Solve. First we collect like terms on the left.

$$\frac{26}{50} + \frac{5}{50} + \frac{2}{50} + s = 1$$

$$\frac{33}{50} + s = 1$$

$$\frac{33}{50} + s - \frac{33}{50} = 1 - \frac{33}{50} \quad \text{Subtracting } \frac{33}{50} \text{ from both sides}$$

$$s + 0 = 1 \cdot \frac{50}{50} - \frac{33}{50} \quad \text{The LCD is 50.}$$

$$s = \frac{50}{50} - \frac{33}{50}$$

$$s = \frac{17}{50}$$

Now we add the portion of shoppers who stay less than one hour and the portion who stay 1-2 hr to find the portion of shoppers who stay 0-2 hr.

$$\frac{26}{50} + \frac{17}{50} = \frac{43}{50}$$

Check. We repeat the calculation.

State. $\frac{43}{50}$ of shoppers stay 0-2 hr when visiting a mall.

95. First, we solve an equation to find the portion of the tape used at the 4-hr speed.

Tape speed	times	Portion used	is	Time used
↓	↓	↓	↓	↓
4	$\cdot$	x	$=$	$\frac{1}{2}$

$$x = \frac{1}{2} \div 4$$

$$x = \frac{1}{2} \cdot \frac{1}{4}$$

$$x = \frac{1}{8}$$

Thus, $\frac{1}{8}$ of the tape was used at the 4-hr speed.

Now we solve an equation to find the portion of the tape used at the 2-hr speed.

Tape speed	times	Portion used	is	Time used
↓	↓	↓	↓	↓
2	$\cdot$	y	$=$	$\frac{3}{4}$

$$y = \frac{3}{4} \div 2$$

$$y = \frac{3}{4} \cdot \frac{1}{2}$$

$$y = \frac{3}{8}$$

Thus, $\frac{3}{8}$ of the tape is used at the 2-hr speed.

Next, we solve an equation to find the portion of the tape that is unused.

Portion used at 4-hr speed	+	Portion used at 2-hr speed	+	Unused portion	=	Entire tape
↓	↓	↓	↓	↓	↓	↓
$\frac{1}{8}$	$+$	$\frac{3}{8}$	$+$	p	$=$	1

$$p = 1 - \frac{1}{8} - \frac{3}{8}$$

$$p = \frac{4}{8} = \frac{1}{2}$$

One-half of the tape is unused.

To find how much time is left on the tape at the 6-hr speed, we solve an equation.

$$\underbrace{\text{Unused portion}} \quad \text{of} \quad \underbrace{\text{Tape speed}} \quad \text{is} \quad \underbrace{\text{Time left}}$$
$$\downarrow \qquad\qquad \downarrow \qquad\qquad \downarrow \qquad\qquad \downarrow \qquad \downarrow$$
$$\frac{1}{2} \qquad \cdot \qquad\quad 6 \qquad\quad = \qquad\quad t$$

$$\frac{6}{2} = t$$
$$3 = t$$

There are three hours left at the 6-hr speed.

97. Familiarize. Let $d =$ the fractional portion of the dealership that Paul owns. Then Ella owns the same portion and together the Romanos and the Chrenkas own $\frac{7}{12} + \frac{1}{6} + d + d$.

Translate.

$$\underbrace{\text{Total portions}} \quad \text{are} \quad \underbrace{\text{1 dealership}}$$
$$\downarrow \qquad\qquad\quad \downarrow \qquad\quad \downarrow$$
$$\frac{7}{12} + \frac{1}{6} + d + d \;\; = \qquad\quad 1$$

Solve. First we collect like terms on the left.

$$\frac{7}{12} + \frac{1}{6} + d + d = 1$$
$$\frac{7}{12} + \frac{1}{6} \cdot \frac{2}{2} + 2d = 1 \quad \text{The LCD is 12.}$$
$$\frac{7}{12} + \frac{2}{12} + 2d = 1$$
$$\frac{9}{12} + 2d = 1$$
$$\frac{9}{12} + 2d - \frac{9}{12} = 1 - \frac{9}{12}$$
$$2d + 0 = 1 \cdot \frac{12}{12} - \frac{9}{12}$$
$$2d = \frac{3}{12}$$
$$\frac{1}{2} \cdot 2d = \frac{1}{2} \cdot \frac{3}{12}$$
$$d = \frac{3}{24} = \frac{3}{3} \cdot \frac{1}{8}$$
$$d = \frac{1}{8}$$

Check. We return to the original problem and add.
$$\frac{7}{12} + \frac{1}{6} + \frac{1}{8} + \frac{1}{8} = \frac{7}{12} \cdot \frac{2}{2} + \frac{1}{6} \cdot \frac{4}{4} + \frac{1}{8} \cdot \frac{3}{3} + \frac{1}{8} \cdot \frac{3}{3} =$$
$$\frac{14}{24} + \frac{4}{24} + \frac{3}{24} + \frac{3}{24} = \frac{24}{24} = 1$$
The answer checks.

State. Paul owns $\frac{1}{8}$ of the dealership.

99. Use a calculator.
$$x + \frac{16}{323} = \frac{10}{187}$$
$$x + \frac{16}{323} - \frac{16}{323} = \frac{10}{187} - \frac{16}{323}$$
$$x + 0 = \frac{10}{11 \cdot 17} - \frac{16}{17 \cdot 19}$$
$$x = \frac{10}{11 \cdot 17} \cdot \frac{19}{19} - \frac{16}{17 \cdot 19} \cdot \frac{11}{11} \quad \begin{array}{l} \text{The LCD is} \\ 11 \cdot 17 \cdot 19. \end{array}$$
$$x = \frac{190}{11 \cdot 17 \cdot 19} - \frac{176}{17 \cdot 19 \cdot 11}$$
$$x = \frac{14}{11 \cdot 17 \cdot 19}$$
$$x = \frac{14}{3553}$$

The solution is $\frac{14}{3553}$.

101.
$$\frac{10 + a}{23} = \frac{330}{391} - \frac{a}{17}$$
$$\frac{10 + a}{23} \cdot \frac{17}{17} = \frac{330}{391} - \frac{a}{17} \cdot \frac{23}{23}$$
$$\frac{17(10 + a)}{391} = \frac{330}{391} - \frac{23a}{391}$$
$$\frac{170 + 17a}{391} = \frac{330 - 23a}{391}$$

Since the denominators are the same, the numerators are the same.

$$170 + 17a = 330 - 23a$$
$$170 + 17a + 23a = 330 - 23a + 23a$$
$$170 + 40a = 330$$
$$170 + 40a - 170 = 330 - 170$$
$$40a = 160$$
$$\frac{40a}{40} = \frac{160}{40}$$
$$a = 4$$

103. Use the two cuts to cut the bar into three pieces as follows: one piece is $\frac{1}{7}$ of the bar, one is $\frac{2}{7}$ of the bar, and then the remaining piece is $\frac{4}{7}$ of the bar. On Day 1, give the contractor $\frac{1}{7}$ of the bar. On Day 2, have him/her return the $\frac{1}{7}$ and give him/her $\frac{2}{7}$ of the bar. On Day 3, add $\frac{1}{7}$ to what the contractor already has, making $\frac{3}{7}$ of the bar. On Day 4, have the contractor return the $\frac{1}{7}$ and $\frac{2}{7}$ pieces and give him/her the $\frac{4}{7}$ piece. On Day 5, add the $\frac{1}{7}$ piece to what the contractor already has, making $\frac{5}{7}$ of the bar. On Day 6, have the contractor return the $\frac{1}{7}$ piece and give him/her the $\frac{2}{7}$ to go with the $\frac{4}{7}$ piece he/she also has, making $\frac{6}{7}$ of the bar. On Day 7, give him/her the $\frac{1}{7}$

piece again. Now the contractor has all three pieces, or the entire bar. This assumes that he/she does not spend any part of the gold during the week.

Exercise Set 4.4

1.
$$6x - 3 = 15$$
$$6x - 3 + 3 = 15 + 3 \quad \text{Using the addition principle}$$
$$6x + 0 = 18$$
$$6x = 18$$
$$\frac{1}{6} \cdot 6x = \frac{1}{6} \cdot 18 \quad \text{Using the multiplication principle}$$
$$1x = \frac{18}{6}$$
$$x = 3$$

Check:
$$\begin{array}{c|c} 6x - 3 = 15 \\ \hline 6 \cdot 3 - 3 \ ? \ 15 \\ 18 - 3 \\ 15 & 15 \quad \text{TRUE} \end{array}$$

The solution is 3.

3.
$$5x + 7 = -8$$
$$5x + 7 - 7 = -8 - 7 \quad \text{Using the addition principle}$$
$$5x = -15$$
$$\frac{1}{5} \cdot 5x = \frac{1}{5} \cdot (-15) \quad \text{Using the multiplication principle}$$
$$1x = \frac{-15}{5}$$
$$x = -3$$

Check:
$$\begin{array}{c|c} 5x + 7 = -8 \\ \hline 5(-3) + 7 \ ? \ -8 \\ -15 + 7 \\ -8 & -8 \quad \text{TRUE} \end{array}$$

The solution is −3.

5.
$$31 = 3x - 5$$
$$31 + 5 = 3x - 5 + 5 \quad \text{Using the addition principle}$$
$$36 = 3x + 0$$
$$36 = 3x$$
$$\frac{1}{3} \cdot 36 = \frac{1}{3} \cdot 3x \quad \text{Using the multiplication principle}$$
$$\frac{36}{3} = 1x$$
$$12 = x$$

Check:
$$\begin{array}{c|c} 31 = 3x - 5 \\ \hline 31 \ ? \ 3 \cdot 12 - 5 \\ & 36 - 5 \\ 31 & 31 \quad \text{TRUE} \end{array}$$

The solution is 12.

7.
$$4x - 5 = \frac{1}{3}$$
$$4x - 5 + 5 = \frac{1}{3} + 5 \quad \text{Using the addition principle}$$
$$4x + 0 = \frac{1}{3} + 5 \cdot \frac{3}{3}$$
$$4x = \frac{1}{3} + \frac{15}{3}$$
$$4x = \frac{16}{3}$$
$$\frac{1}{4} \cdot 4x = \frac{1}{4} \cdot \frac{16}{3} \quad \text{Using the multiplication principle}$$
$$1x = \frac{1 \cdot 16}{4 \cdot 3} = \frac{1 \cdot 4 \cdot 4}{4 \cdot 3}$$
$$x = \frac{4}{4} \cdot \frac{1 \cdot 4}{3} = \frac{4}{3}$$

Check:
$$\begin{array}{c|c} 4x - 5 = \frac{1}{3} \\ \hline 4 \cdot \frac{4}{3} - 5 \ ? \ \frac{1}{3} \\ \frac{16}{3} - 5 \\ \frac{16}{3} - \frac{15}{3} \\ \frac{1}{3} & \frac{1}{3} \quad \text{TRUE} \end{array}$$

The solution is $\frac{4}{3}$.

9.
$$\frac{3}{2}t - \frac{1}{4} = \frac{1}{2}$$
$$\frac{3}{2}t - \frac{1}{4} + \frac{1}{4} = \frac{1}{2} + \frac{1}{4} \quad \text{Adding } \frac{1}{4} \text{ to each side}$$
$$\frac{3}{2}t + 0 = \frac{2}{4} + \frac{1}{4}$$
$$\frac{3}{2}t = \frac{3}{4}$$
$$\frac{2}{3} \cdot \frac{3}{2}t = \frac{2}{3} \cdot \frac{3}{4} \quad \text{Multiplying both sides by } \frac{2}{3}$$
$$1t = \frac{\cancel{2} \cdot \cancel{3}}{3 \cdot \cancel{2} \cdot 2}$$
$$t = \frac{1}{2}$$

Check:
$$\begin{array}{c|c} \frac{3}{2}t - \frac{1}{4} = \frac{1}{2} \\ \hline \frac{3}{2} \cdot \frac{1}{2} - \frac{1}{4} \ ? \ \frac{1}{2} \\ \frac{3}{4} - \frac{1}{4} \\ \frac{2}{4} \\ \frac{1}{2} & \frac{1}{2} \quad \text{TRUE} \end{array}$$

The solution is $\frac{1}{2}$.

11.
$$\frac{2}{5}x + \frac{3}{10} = \frac{3}{5}$$

$$\frac{2}{5}x + \frac{3}{10} - \frac{3}{10} = \frac{3}{5} - \frac{3}{10} \qquad \text{Subtracting } \frac{3}{10} \text{ from both sides}$$

$$\frac{2}{5}x + 0 = \frac{6}{10} - \frac{3}{10}$$

$$\frac{2}{5}x = \frac{3}{10}$$

$$\frac{5}{2} \cdot \frac{2}{5}x = \frac{5}{2} \cdot \frac{3}{10} \qquad \text{Multiplying both sides by } \frac{5}{2}$$

$$1x = \frac{\not{5} \cdot 3}{2 \cdot \not{5} \cdot 2}$$

$$x = \frac{3}{4}$$

Check:
$$\frac{2}{5}x + \frac{3}{10} = \frac{3}{5}$$

$$\frac{2}{5} \cdot \frac{3}{4} + \frac{3}{10} \; ? \; \frac{3}{5}$$

$$\frac{\not{2} \cdot 3}{5 \cdot \not{2} \cdot 2} + \frac{3}{10}$$

$$\frac{3}{10} + \frac{3}{10}$$

$$\frac{6}{10}$$

$$\frac{3}{5} \quad \Big| \quad \frac{3}{5} \qquad \text{TRUE}$$

The solution is $\frac{3}{4}$.

13.
$$5 - \frac{3}{4}x = 3$$

$$5 - \frac{3}{4}x - 5 = 3 - 5 \qquad \text{Subtracting 5 from both sides}$$

$$-\frac{3}{4}x = -2$$

$$-\frac{4}{3}\left(-\frac{3}{4}x\right) = -\frac{4}{3}(-2) \qquad \text{Multiplying both sides by } -\frac{4}{3}$$

$$1x = \frac{8}{3}$$

$$x = \frac{8}{3}$$

The number $\frac{8}{3}$ checks and is the solution.

15.
$$-1 + \frac{2}{5}t + 1 = -\frac{4}{5}$$

$$-1 + \frac{2}{5}t + 1 = -\frac{4}{5} + 1 \qquad \text{Adding 1 to both sides}$$

$$\frac{2}{5}t = -\frac{4}{5} + \frac{5}{5}$$

$$\frac{2}{5}t = \frac{1}{5}$$

$$\frac{5}{2} \cdot \frac{2}{5}t = \frac{5}{2} \cdot \frac{1}{5}$$

$$1t = \frac{\not{5} \cdot 1}{2 \cdot \not{5}}$$

$$t = \frac{1}{2}$$

The number $\frac{1}{2}$ checks and is the solution.

17.
$$12 = 8 + \frac{7}{2}t$$

$$12 - 8 = 8 + \frac{7}{12}t - 8 \qquad \text{Subtracting 8 from both sides}$$

$$4 = \frac{7}{2}t$$

$$\frac{2}{7} \cdot 4 = \frac{2}{7} \cdot \frac{7}{2}t \qquad \text{Multiply both sides by } \frac{2}{7}$$

$$\frac{8}{7} = t$$

The number $\frac{8}{7}$ checks and is the solution.

19.
$$-4 = \frac{2}{3}x - 7$$

$$-4 + 7 = \frac{2}{3}x - 7 + 7 \qquad \text{Adding 7 to both sides}$$

$$3 = \frac{2}{3}x$$

$$\frac{3}{2} \cdot 3 = \frac{3}{2} \cdot \frac{2}{3}x \qquad \text{Multiplying both sides by } \frac{3}{2}$$

$$\frac{9}{2} = x$$

The number $\frac{9}{2}$ checks and is the solution.

21.
$$7 = a + \frac{14}{5}$$

$$7 - \frac{14}{5} = a + \frac{14}{5} - \frac{14}{5} \qquad \text{Using the addition principle}$$

$$7 \cdot \frac{5}{5} - \frac{14}{5} = a + 0$$

$$\frac{35}{5} - \frac{14}{5} = a$$

$$\frac{21}{5} = a$$

The number $\frac{21}{5}$ checks and is the solution.

23.
$$\frac{2}{5}t - 1 = \frac{7}{5}$$

$$\frac{2}{5}t - 1 + 1 = \frac{7}{5} + 1 \qquad \text{Using the addition principle}$$

$$\frac{2}{5}t + 0 = \frac{7}{5} + 1 \cdot \frac{5}{5}$$

$$\frac{2}{5}t = \frac{7}{5} + \frac{5}{5}$$

$$\frac{2}{5}t = \frac{12}{5}$$

$$\frac{5}{2} \cdot \frac{2}{5}t = \frac{5}{2} \cdot \frac{12}{5} \qquad \text{Using the multiplication principle}$$

$$1t = \frac{5 \cdot 12}{2 \cdot 5}$$

$$t = \frac{5 \cdot 2 \cdot 6}{2 \cdot 5 \cdot 1} = \frac{5 \cdot 2}{5 \cdot 2} \cdot \frac{6}{1}$$

$$t = 6$$

The number 6 checks and is the solution.

25.
$$\frac{39}{8} = \frac{11}{4} + \frac{1}{2}x$$

$$\frac{39}{8} - \frac{11}{4} = \frac{11}{4} + \frac{1}{2}x - \frac{11}{4} \qquad \text{Using the addition principle}$$

$$\frac{39}{8} - \frac{11}{4} \cdot \frac{2}{2} = \frac{1}{2}x + 0$$

$$\frac{39}{8} - \frac{22}{8} = \frac{1}{2}x$$

$$\frac{17}{8} = \frac{1}{2}x$$

$$2 \cdot \frac{17}{8} = 2 \cdot \frac{1}{2}x \qquad \text{Using the multiplication principle}$$

$$\frac{2 \cdot 17}{8} = 1x$$

$$\frac{2 \cdot 17}{2 \cdot 4} = x$$

$$\frac{17}{4} = x$$

The number $\frac{17}{4}$ checks and is the solution.

27.
$$\frac{13}{3}x + \frac{11}{2} = \frac{35}{4}$$

$$\frac{13}{3}x + \frac{11}{2} - \frac{11}{2} = \frac{35}{4} - \frac{11}{2} \qquad \text{Using the addition principle}$$

$$\frac{13}{3}x + 0 = \frac{35}{4} - \frac{11}{2} \cdot \frac{2}{2}$$

$$\frac{13}{3}x = \frac{35}{4} - \frac{22}{4}$$

$$\frac{13}{3}x = \frac{13}{4}$$

$$\frac{3}{13} \cdot \frac{13}{3}x = \frac{3}{13} \cdot \frac{13}{4} \qquad \text{Using the multiplication principle}$$

$$1x = \frac{3 \cdot 13}{13 \cdot 4}$$

$$x = \frac{3}{4}$$

The number $\frac{3}{4}$ checks and is the solution.

29.
$$\frac{1}{2}x - \frac{1}{4} = \frac{1}{2}$$

$$4\left(\frac{1}{2}x - \frac{1}{4}\right) = 4 \cdot \frac{1}{2} \qquad \text{Multiplying both sides by the LCD, 4}$$

$$\frac{4 \cdot 1}{2}x - 4 \cdot \frac{1}{4} = \frac{4 \cdot 1}{2}$$

$$\frac{2 \cdot 2}{2}x - 1 = \frac{2 \cdot 2}{2}$$

$$2x - 1 = 2$$

$$2x - 1 + 1 = 2 + 1 \qquad \text{Adding 1 to both sides}$$

$$2x = 3$$

$$\frac{2x}{2} = \frac{3}{2} \qquad \text{Dividing both sides by 2}$$

$$x = \frac{3}{2}$$

The number $\frac{3}{2}$ checks and is the solution.

31.
$$7 = \frac{4}{9}t + 5$$

$$9 \cdot 7 = 9\left(\frac{4}{9}t + 5\right) \qquad \text{Multiplying both sides by the LCD, 9}$$

$$63 = \frac{9 \cdot 4}{9}t + 9 \cdot 5$$

$$63 = 4t + 45$$

$$63 - 45 = 4t + 45 - 45 \qquad \text{Subtracting 45 from both sides}$$

$$18 = 4t$$

$$\frac{18}{4} = \frac{4t}{4} \qquad \text{Dividing both sides by 4}$$

$$\frac{9 \cdot 2}{2 \cdot 2} = 1t$$

$$\frac{9}{2} = t$$

The number $\frac{9}{2}$ checks and is the solution.

33.
$$-3 = \frac{3}{4}t - \frac{1}{2}$$

$$4(-3) = 4\left(\frac{3}{4}t - \frac{1}{2}\right) \qquad \text{Multiplying both sides by the LCD, 4}$$

$$-12 = \frac{4 \cdot 3}{4}t - \frac{4 \cdot 1}{2}$$

$$-12 = \frac{4 \cdot 3}{4}t - \frac{2 \cdot 2}{2}$$

$$-12 = 3t - 2$$

$$-12 + 2 = 3t - 2 + 2 \qquad \text{Adding 2 to both sides}$$

$$-10 = 3t$$

$$\frac{-10}{3} = \frac{3t}{3} \qquad \text{Dividing both sides by 3}$$

$$-\frac{10}{3} = t$$

The number $-\frac{10}{3}$ checks and is the solution.

35.
$$\frac{4}{3} - \frac{5}{6}x = \frac{3}{2}$$

$$6\left(\frac{4}{3} - \frac{5}{6}x\right) = 6 \cdot \frac{3}{2} \qquad \text{Multiplying both sides by the LCD, 6}$$

$$\frac{6 \cdot 4}{3} - \frac{6 \cdot 5}{6}x = \frac{6 \cdot 3}{2}$$

$$\frac{2 \cdot \cancel{3} \cdot 4}{\cancel{3}} - \frac{\cancel{6} \cdot 5}{\cancel{6}}x = \frac{\cancel{2} \cdot 3 \cdot 3}{\cancel{2}}$$

$$8 - 5x = 9$$

$$8 - 5x - 8 = 9 - 8 \qquad \text{Subtracting 8 from both sides}$$

$$-5x = 1$$

$$\frac{-5x}{5} = \frac{1}{-5} \qquad \text{Dividing both sides by } -5$$

$$x = -\frac{1}{5}$$

The number $-\frac{1}{5}$ checks and is the solution.

37.
$$-\frac{3}{4} = -\frac{5}{6} - \frac{1}{2}x$$

$$12\left(-\frac{3}{4}\right) = 12\left(-\frac{5}{6} - \frac{1}{2}x\right) \qquad \text{Multiplying both sides by the LCD, 12}$$

$$-\frac{12 \cdot 3}{4} = -\frac{12 \cdot 5}{6} - \frac{12 \cdot 1}{2}x$$

$$-\frac{\cancel{4} \cdot 3 \cdot 3}{\cancel{4}} = -\frac{2 \cdot \cancel{6} \cdot 5}{\cancel{6}} - \frac{\cancel{2} \cdot 6}{\cancel{2}}x$$

$$-9 = -10 - 6x$$

$$-9 + 10 = -10 - 6x + 10 \qquad \text{Adding 10 to both sides}$$

$$1 = -6x$$

$$\frac{1}{-6} = \frac{-6x}{-6} \qquad \text{Dividing both sides by } -6$$

$$-\frac{1}{6} = x$$

The number $-\frac{1}{6}$ checks and is the solution.

39.
$$\frac{4}{3} - \frac{1}{5}t = \frac{3}{4}$$

$$60\left(\frac{4}{3} - \frac{1}{5}t\right) = 60 \cdot \frac{3}{4} \qquad \text{Multiplying both sides by the LCD, 60}$$

$$\frac{60 \cdot 4}{3} - \frac{60 \cdot 1}{5}t = \frac{60 \cdot 3}{4}$$

$$\frac{\cancel{3} \cdot 20 \cdot 4}{\cancel{3}} - \frac{\cancel{5} \cdot 12}{\cancel{5}}t = \frac{\cancel{4} \cdot 15 \cdot 3}{\cancel{4}}$$

$$80 - 12t = 45$$

$$80 - 12t - 80 = 45 - 80 \qquad \text{Subtracting 80 from both sides}$$

$$-12t = -35$$

$$\frac{-12t}{-12} = \frac{-35}{-12}$$

$$t = \frac{35}{12}$$

The number $\frac{35}{12}$ checks and is the solution.

41. Discussion and Writing Exercise

43. $39 \div (-3) = -13$

Think: What number multiplied by -3 gives 39? The number is -13.

45. $(-72) \div (-4) = 18$

Think: What number multiplied by -4 gives -72? The number is 18.

47. -200 represents the \$200 withdrawal, 90 represents the \$90 deposit, and -40 represents the \$40 withdrawal. We add these numbers to find the change in the balance.

$$-200 + 90 + (-40) = -110 + (-40) = -150$$

The account balance decreased by \$150.

49.
$$\frac{10}{7} \div 2m = \frac{10}{7} \cdot \frac{1}{2m} = \frac{10 \cdot 1}{7 \cdot 2m} = \frac{2 \cdot 5 \cdot 1}{7 \cdot 2 \cdot m} =$$
$$\frac{2}{2} \cdot \frac{5 \cdot 1}{7 \cdot m} = \frac{5}{7m}$$

51. Discussion and Writing Exercise

53. Use a calculator.
$$\frac{553}{2451}a - \frac{13}{57} = \frac{29}{43}$$

$$\frac{553}{2451}a - \frac{13}{57} + \frac{13}{57} = \frac{29}{43} + \frac{13}{57}$$

$$\frac{553}{2451}a = \frac{29}{43} + \frac{13}{57}$$

$$\frac{553}{2451}a = \frac{29}{43} \cdot \frac{57}{57} + \frac{13}{57} \cdot \frac{43}{43}$$

$$\frac{553}{2451}a = \frac{1653}{43 \cdot 57} + \frac{559}{57 \cdot 43}$$

$$\frac{553}{2451}a = \frac{2212}{2451}$$

$$\frac{2451}{553} \cdot \frac{553}{2451}a = \frac{2451}{553} \cdot \frac{2212}{2451}$$

$$a = 4$$

The solution is 4.

55. Use a calculator.
$$\frac{71}{73} = \frac{19}{47} - \frac{53}{91}t$$

$$\frac{71}{73} - \frac{19}{47} = \frac{19}{47} - \frac{53}{91}t - \frac{19}{47}$$

$$\frac{71}{73} \cdot \frac{47}{47} - \frac{19}{47} \cdot \frac{73}{73} = -\frac{53}{91}t$$

$$\frac{3337}{3431} - \frac{1387}{3431} = -\frac{53}{91}t$$

$$\frac{1950}{3431} = -\frac{53}{91}t$$

$$-\frac{91}{53} \cdot \frac{1950}{3431} = -\frac{91}{53}\left(-\frac{53}{91}\right)t$$

$$-\frac{177,450}{181,843} = t$$

The solution is $-\frac{177,450}{181,843}$.

57. $-\dfrac{a}{5} + \dfrac{31}{4} = \dfrac{16}{3}$

$\quad -\dfrac{1}{5}a + \dfrac{31}{4} = \dfrac{16}{3} \qquad \left(-\dfrac{1}{5}\cdot a = -\dfrac{a}{5}\right)$

$\quad\quad -\dfrac{1}{5}a = \dfrac{16}{3} - \dfrac{31}{4}$

$\quad\quad -\dfrac{1}{5}a = \dfrac{16}{3}\cdot\dfrac{4}{4} - \dfrac{31}{4}\cdot\dfrac{3}{3}$

$\quad\quad -\dfrac{1}{5}a = \dfrac{64}{12} - \dfrac{93}{12}$

$\quad\quad -\dfrac{1}{5}a = -\dfrac{29}{12}$

$\quad -5\left(-\dfrac{1}{5}a\right) = -5\left(-\dfrac{29}{12}\right)$

$\quad\quad\quad a = \dfrac{145}{12}$

The solution is $\dfrac{145}{12}$.

59. $\dfrac{49}{8} + \dfrac{2x}{9} = 4$

$\quad\quad \dfrac{2x}{9} = 4 - \dfrac{49}{8}$

$\quad\quad \dfrac{2x}{9} = \dfrac{32}{8} - \dfrac{49}{8}$

$\quad\quad \dfrac{2x}{9} = -\dfrac{17}{8}$

$\quad\quad\quad x = \dfrac{9}{2}\left(-\dfrac{17}{8}\right)$

$\quad\quad\quad x = -\dfrac{153}{16}$

The solution is $-\dfrac{153}{16}$.

61. $5 + 2 + 4 + 5 + 2 + 2x = 21$

$\quad\quad\quad 18 + 2x = 21$

$\quad 18 + 2x - 18 = 21 - 18$

$\quad\quad\quad\quad 2x = 3$

$\quad\quad\quad\quad \dfrac{2x}{2} = \dfrac{3}{2}$

$\quad\quad\quad\quad x = \dfrac{3}{2}$

The value of x is $\dfrac{3}{2}$ cm.

63. $7n + 6 + 5n = 15$

$\quad\quad 12n + 6 = 15$

$\quad 12n + 6 - 6 = 15 - 6$

$\quad\quad\quad 12n = 9$

$\quad\quad\quad \dfrac{12n}{12} = \dfrac{9}{12}$

$\quad\quad\quad n = \dfrac{\cancel{3}\cdot 3}{\cancel{3}\cdot 4}$

$\quad\quad\quad n = \dfrac{3}{4}$

The value of n is $\dfrac{3}{4}$ cm.

1. $\boxed{b} \quad\quad \boxed{a}$ Multiply: $3 \cdot 7 = 21$.

$\quad 7\dfrac{2}{3} = \dfrac{23}{3} \quad \boxed{b}$ Add: $21 + 2 = 23$.

$\quad \boxed{a} \quad\quad \boxed{c}$ Keep the denominator.

3. $\boxed{b} \quad\quad \boxed{a}$ Multiply: $6 \cdot 4 = 24$.

$\quad 6\dfrac{1}{4} = \dfrac{25}{4} \quad \boxed{b}$ Add: $24 + 1 = 25$.

$\quad \boxed{a} \quad\quad \boxed{c}$ Keep the denominator.

5. $-20\dfrac{1}{8} = -\dfrac{161}{8} \quad (20 \cdot 8 = 160; 160 + 1 = 161;$ include the negative sign)

7. $5\dfrac{1}{10} = \dfrac{51}{10} \quad (5 \cdot 10 = 50; 50 + 1 = 51)$

9. $20\dfrac{3}{5} = \dfrac{103}{5} \quad (20 \cdot 5 = 100; 100 + 3 = 103)$

11. $-8\dfrac{2}{7} = -\dfrac{58}{7} \quad (8 \cdot 7 = 56; 56 + 2 = 58;$ include the negative sign)

13. $6\dfrac{9}{10} = \dfrac{69}{10} \quad (6 \cdot 10 = 60; 60 + 9 = 69)$

15. $-12\dfrac{3}{4} = -\dfrac{51}{4} \quad (12 \cdot 4 = 48; 48 + 3 = 51;$ include the negative sign)

17. $5\dfrac{7}{10} = \dfrac{57}{10} \quad (5 \cdot 10 = 50; 50 + 7 = 57)$

19. $-5\dfrac{7}{100} = -\dfrac{507}{100} \quad (5 \cdot 100 = 500; 500 + 7 = 507;$ include the negative sign)

21. To convert $\dfrac{16}{3}$ to a mixed numeral, we divide.

$$\begin{array}{r} 5 \\ 3\,\overline{)1\,6} \\ 1\,5 \\ \hline 1 \end{array} \qquad \dfrac{16}{3} = 5\dfrac{1}{3}$$

23. To convert $\dfrac{45}{6}$ to a mixed numeral, we divide.

$$\begin{array}{r} 7 \\ 6\,\overline{)4\,5} \\ 4\,2 \\ \hline 3 \end{array} \qquad \dfrac{45}{6} = 7\dfrac{3}{6} = 7\dfrac{1}{2}$$

25. $\begin{array}{r} 5 \\ 1\,0\,\overline{)5\,7} \\ 5\,0 \\ \hline 7 \end{array} \qquad \dfrac{57}{10} = 5\dfrac{7}{10}$

27.
$$9\overline{)65}$$
$$\underline{63}$$
$$2$$
with quotient 7, and $\dfrac{65}{9} = 7\dfrac{2}{9}$

29.
$$6\overline{)33}$$
$$\underline{30}$$
$$3$$
quotient 5, and $\dfrac{33}{6} = 5\dfrac{3}{6} = 5\dfrac{1}{2}$

Since $\dfrac{33}{6} = 5\dfrac{1}{2}$, we have $-\dfrac{33}{6} = -5\dfrac{1}{2}$.

31.
$$4\overline{)46}$$ quotient 11
$$\underline{40}$$
$$6$$
$$\underline{4}$$
$$2$$
$\dfrac{46}{4} = 11\dfrac{2}{4} = 11\dfrac{1}{2}$

33.
$$8\overline{)12}$$ quotient 1
$$\underline{8}$$
$$4$$
$\dfrac{12}{8} = 1\dfrac{4}{8} = 1\dfrac{1}{2}$

Since $\dfrac{12}{8} = 1\dfrac{1}{2}$, we have $-\dfrac{12}{8} = -1\dfrac{1}{2}$.

35.
$$5\overline{)307}$$ quotient 61
$$\underline{300}$$
$$7$$
$$\underline{5}$$
$$2$$
$\dfrac{307}{5} = 61\dfrac{2}{5}$

37.
$$50\overline{)413}$$ quotient 8
$$\underline{400}$$
$$13$$
$\dfrac{413}{50} = 8\dfrac{13}{50}$

Since $\dfrac{413}{50} = 8\dfrac{13}{50}$, we have $-\dfrac{413}{50} = -8\dfrac{13}{50}$.

39. We first divide as usual.
$$8\overline{)869}$$ quotient 108
$$\underline{800}$$
$$69$$
$$\underline{64}$$
$$5$$

The answer is 108 R 5. We write a mixed numeral for the quotient as follows: $108\dfrac{5}{8}$.

41. We first divide as usual.
$$7\overline{)6345}$$ quotient 906
$$\underline{6300}$$
$$45$$
$$\underline{42}$$
$$3$$

The answer is 906 R 3. We write a mixed numeral for the quotient as follows: $906\dfrac{3}{7}$.

43.
$$21\overline{)852}$$ quotient 40
$$\underline{840}$$
$$12$$

We get $40\dfrac{12}{21}$. This simplifies as $40\dfrac{4}{7}$.

45. First we find $302 \div 15$.
$$15\overline{)302}$$ quotient 20
$$\underline{300}$$
$$2$$
$$\underline{0}$$
$$2$$
$\dfrac{302}{15} = 20\dfrac{2}{15}$

Since $302 \div 15 = 20\dfrac{2}{15}$, we have $-302 \div 15 = -20\dfrac{2}{15}$.

47. First we find $471 \div 21$.
$$21\overline{)471}$$ quotient 22
$$\underline{420}$$
$$51$$
$$\underline{42}$$
$$9$$
$\dfrac{471}{21} = 22\dfrac{9}{21} = 22\dfrac{3}{7}$

Since $471 \div 21 = 22\dfrac{3}{7}$, we have $471 \div (-21) = -22\dfrac{3}{7}$.

49. There are 5 health organizations in the list. We add the expenses for these organizations and then divide by 5.
$$\frac{\$4 + \$6 + \$7 + \$10 + \$14}{5} = \frac{\$41}{5} = \$8\frac{1}{5}$$

51. We add the expenses for the first 6 organizations and then divide by 6.
$$\frac{\$1 + \$4 + \$5 + \$6 + \$7 + \$8}{6} = \frac{\$31}{6} = \$5\frac{1}{6}$$

53. Discussion and Writing Exercise

55.
$$\frac{7}{9} \cdot \frac{24}{21} = \frac{7 \cdot 24}{9 \cdot 21}$$
$$= \frac{7 \cdot 3 \cdot 8}{3 \cdot 3 \cdot 3 \cdot 7}$$
$$= \frac{3 \cdot 7}{3 \cdot 7} \cdot \frac{8}{3 \cdot 3}$$
$$= \frac{8}{9}$$

57.
$$\frac{7}{10} \cdot \frac{5}{14} = \frac{7 \cdot 5}{10 \cdot 14} = \frac{7 \cdot 5 \cdot 1}{2 \cdot 5 \cdot 2 \cdot 7} = \frac{7 \cdot 5}{7 \cdot 5} \cdot \frac{1}{2 \cdot 2} = \frac{1}{2 \cdot 2} = \frac{1}{4}$$

59.
$$-\frac{17}{25} \cdot \frac{15}{34} = -\frac{17 \cdot 15}{25 \cdot 34}$$
$$= -\frac{17 \cdot 3 \cdot 5}{5 \cdot 5 \cdot 2 \cdot 17}$$
$$= -\frac{17 \cdot 5}{17 \cdot 5} \cdot \frac{3}{5 \cdot 2}$$
$$= -\frac{3}{10}$$

61. Discussion and Writing Exercise

63. Use a calculator.

$$\frac{128,236}{541} = 237\frac{19}{541}$$

65. $\dfrac{56}{7} + \dfrac{2}{3} = 8 + \dfrac{2}{3}$ $\qquad (56 \div 7 = 8)$

$$= 8\frac{2}{3}$$

67. $\dfrac{12}{5} + \dfrac{19}{15} = \dfrac{36}{15} + \dfrac{19}{15} = \dfrac{55}{15}$

$$
\begin{array}{r}
3 \\
1\,5\,\overline{)\,5\,5} \\
4\,5 \\
\hline
1\,0
\end{array}
\qquad \frac{55}{15} = 3\frac{10}{15} = 3\frac{2}{3}
$$

Thus, $\dfrac{12}{5} + \dfrac{19}{15} = 3\dfrac{2}{3}$.

69. There are 365 days in a year and 7 days in a week, so we write a mixed numeral for $\dfrac{365}{7}$.

$$
\begin{array}{r}
5\,2 \\
7\,\overline{)\,3\,6\,5} \\
3\,5\,0 \\
\hline
1\,5 \\
1\,4 \\
\hline
1
\end{array}
\qquad \frac{365}{7} = 52\frac{1}{7}
$$

There are $52\dfrac{1}{7}$ weeks in a year.

Exercise Set 4.6

1.
$$
\begin{array}{r}
6 \\
+5\dfrac{2}{5} \\
\hline
11\dfrac{2}{5}
\end{array}
$$

3.
$$
\begin{array}{r}
2\dfrac{7}{8} \\
+6\dfrac{5}{8} \\
\hline
8\dfrac{12}{8} = 8 + \dfrac{12}{8} \\
= 8 + 1\dfrac{1}{2} \\
= 9\dfrac{1}{2}
\end{array}
$$

To find a mixed numeral for $\dfrac{12}{8}$ we divide:

$$
\begin{array}{r}
1 \\
8\,\overline{)\,1\,2} \\
8 \\
\hline
4
\end{array}
\qquad \frac{12}{8} = 1\frac{4}{8} = 1\frac{1}{2}
$$

5. The LCD is 12.

$$
\begin{array}{r}
4\,\boxed{\dfrac{1}{4} \cdot \dfrac{3}{3}} = \quad 4\dfrac{3}{12} \\
+1\,\boxed{\dfrac{2}{3} \cdot \dfrac{4}{4}} = +1\dfrac{8}{12} \\
\hline
5\dfrac{11}{12}
\end{array}
$$

7. The LCD is 12.

$$
\begin{array}{r}
7\,\boxed{\dfrac{3}{4} \cdot \dfrac{3}{3}} = \quad 7\dfrac{9}{12} \\
+5\,\boxed{\dfrac{5}{6} \cdot \dfrac{2}{2}} = +5\dfrac{10}{12} \\
\hline
12\dfrac{19}{12} = 12 + \dfrac{19}{12} \\
= 12 + 1\dfrac{7}{12} \\
= 13\dfrac{7}{12}
\end{array}
$$

9. The LCD is 10.

$$
\begin{array}{r}
3\,\boxed{\dfrac{2}{5} \cdot \dfrac{2}{2}} = \quad 3\dfrac{4}{10} \\
+8\,\dfrac{7}{10} = +8\dfrac{7}{10} \\
\hline
11\dfrac{11}{10} = 11 + \dfrac{11}{10} \\
= 11 + 1\dfrac{1}{10} \\
= 12\dfrac{1}{10}
\end{array}
$$

11. The LCD is 24.

$$
\begin{array}{r}
6\,\boxed{\dfrac{3}{8} \cdot \dfrac{3}{3}} = \quad 6\dfrac{9}{24} \\
+10\,\boxed{\dfrac{5}{6} \cdot \dfrac{4}{4}} = +10\dfrac{20}{24} \\
\hline
16\dfrac{29}{24} = 16 + \dfrac{29}{24} \\
= 16 + 1\dfrac{5}{24} \\
= 17\dfrac{5}{24}
\end{array}
$$

13. The LCD is 10.

$$
\begin{array}{r}
18\,\boxed{\dfrac{4}{5} \cdot \dfrac{2}{2}} = \quad 18\dfrac{8}{10} \\
+2\,\dfrac{7}{10} = +2\dfrac{7}{10} \\
\hline
20\dfrac{15}{10} = 20 + \dfrac{15}{10} \\
= 20 + 1\dfrac{5}{10} \\
= 21\dfrac{5}{10} \\
= 21\dfrac{1}{2}
\end{array}
$$

15. The LCD is 8.

$$14\frac{5}{8} \quad = \quad 14\frac{5}{8}$$

$$+13\boxed{\frac{1}{4}\cdot\frac{2}{2}} = +13\frac{2}{8}$$

$$\rule{2cm}{0.4pt}$$

$$27\frac{7}{8}$$

17.

$$4\frac{1}{5} = 3\frac{6}{5}$$

$$-2\frac{3}{5} = -2\frac{3}{5}$$

$$\rule{2cm}{0.4pt}$$

$$1\frac{3}{5}$$

Since $\frac{1}{5}$ is smaller than $\frac{3}{5}$, we cannot subtract until we borrow:

$$4\frac{1}{5} = 3+\frac{5}{5}+\frac{1}{5} = 3+\frac{6}{5} = 3\frac{6}{5}$$

19. The LCD is 10.

$$9\boxed{\frac{3}{5}\cdot\frac{2}{2}} = \quad 9\frac{6}{10}$$

$$-3\boxed{\frac{1}{2}\cdot\frac{5}{5}} = -3\frac{5}{10}$$

$$\rule{2cm}{0.4pt}$$

$$6\frac{1}{10}$$

21. The LCD is 24.

$$34\boxed{\frac{1}{3}\cdot\frac{8}{8}} = \quad 34\frac{8}{24} = \quad 33\frac{32}{24}$$

$$-12\boxed{\frac{5}{8}\cdot\frac{3}{3}} = -12\frac{15}{24} = -12\frac{15}{24}$$

$$\rule{3cm}{0.4pt}$$

$$21\frac{17}{24}$$

$\left(\text{Since } \frac{8}{24} \text{ is smaller than } \frac{15}{24}, \text{ we cannot subtract until we}\right.$

borrow: $34\frac{8}{24} = 33+\frac{24}{24}+\frac{8}{24} = 33+\frac{32}{24} = 33\frac{32}{24}.\Big)$

23.

$$19 \quad = \quad 18\frac{4}{4} \quad \left(19 = 18+1 = 18+\frac{4}{4} = 18\frac{4}{4}\right)$$

$$-\ 5\frac{3}{4} = -\ 5\frac{3}{4}$$

$$\rule{2cm}{0.4pt}$$

$$13\frac{1}{4}$$

25.

$$34 \quad = \quad 33\frac{8}{8} \quad \left(34 = 33+1 = 33+\frac{8}{8} = 33\frac{8}{8}\right)$$

$$-18\frac{5}{8} = -18\frac{5}{8}$$

$$\rule{2cm}{0.4pt}$$

$$15\frac{3}{8}$$

27. The LCD is 12.

$$21\boxed{\frac{1}{6}\cdot\frac{2}{2}} = \quad 21\frac{2}{12} = \quad 20\frac{14}{12}$$

$$-13\boxed{\frac{3}{4}\cdot\frac{3}{3}} = -13\frac{9}{12} = -13\frac{9}{12}$$

$$\rule{3cm}{0.4pt}$$

$$7\frac{5}{12}$$

$\left(\text{Since } \frac{2}{12} \text{ is smaller than } \frac{9}{12}, \text{ we cannot subtract until we}\right.$

borrow: $21\frac{2}{12} = 20+\frac{12}{12}+\frac{2}{12} = 20+\frac{14}{12} = 20\frac{14}{12}.\Big)$

29. The LCD is 18.

$$25\boxed{\frac{1}{9}\cdot\frac{2}{2}} = \quad 25\frac{2}{18} = \quad 24\frac{20}{18}$$

$$-13\boxed{\frac{5}{6}\cdot\frac{3}{3}} = -13\frac{15}{18} = -13\frac{15}{18}$$

$$\rule{3cm}{0.4pt}$$

$$11\frac{5}{18}$$

$\left(\text{Since } \frac{2}{18} \text{ is smaller than } \frac{15}{18}, \text{ we cannot subtract until we}\right.$

borrow: $25\frac{2}{18} = 24+\frac{18}{18}+\frac{2}{18} = 24+\frac{20}{18} = 24\frac{20}{18}.\Big)$

31.

$$1\frac{3}{14}t + 7\frac{2}{21}t$$

$$= \left(1\frac{3}{14} + 7\frac{2}{21}\right)t \quad \text{Using the distributive law}$$

$$= \left(1\frac{9}{42} + 7\frac{4}{42}\right)t \quad \text{The LCD is 42.}$$

$$= 8\frac{13}{42}t \qquad\qquad \text{Adding}$$

33.

$$9\frac{1}{2}x - 7\frac{3}{8}x$$

$$= \left(9\frac{1}{2} - 7\frac{3}{8}\right)x \quad \text{Using the distributive law}$$

$$= \left(9\frac{4}{8} - 7\frac{3}{8}\right)x \quad \text{The LCD is 8.}$$

$$= 2\frac{1}{8}x \qquad\qquad \text{Subtracting}$$

35.

$$5\frac{9}{10}t + 2\frac{7}{8}t$$

$$= \left(5\frac{9}{10} + 2\frac{7}{8}\right)t \quad \text{Using the distributive law}$$

$$= \left(5\frac{36}{40} + 2\frac{35}{40}\right)t \quad \text{The LCD is 40.}$$

$$= 7\frac{71}{40}t = 8\frac{31}{40}t$$

37. $\quad 37\frac{5}{9}t - 25\frac{4}{5}t$

$\quad = \left(37\frac{5}{9} - 25\frac{4}{5}\right)t \qquad$ Using the distributive law

$\quad = \left(37\frac{25}{45} - 25\frac{36}{45}\right)t \qquad$ The LCD is 45.

$\quad = \left(36\frac{70}{45} - 25\frac{36}{45}\right)t$

$\quad = 11\frac{34}{45}t$

39. $\quad 2\frac{5}{6}x + 3\frac{1}{3}x$

$\quad = \left(2\frac{5}{6} + 3\frac{1}{3}\right)x \qquad$ Using the distributive law

$\quad = \left(2\frac{5}{6} + 3\frac{2}{6}\right)x \qquad$ The LCD is 6.

$\quad = 5\frac{7}{6}x = 6\frac{1}{6}x$

41. $\quad 1\frac{3}{11}x + 8\frac{2}{3}x$

$\quad = \left(1\frac{3}{11} + 8\frac{2}{3}\right)x \qquad$ Using the distributive law

$\quad = \left(1\frac{9}{33} + 8\frac{22}{33}\right)x \qquad$ The LCD is 33.

$\quad = 9\frac{31}{33}x$

43. Familiarize. Let f = the number of yards of fabric needed to make the outfit.

Translate. We write an equation.

$$\underbrace{\text{Fabric for dress}}_{\downarrow} + \underbrace{\text{Fabric for band}}_{\downarrow} + \underbrace{\text{Fabric for jacket}}_{\downarrow} \underset{\downarrow}{\text{ is }} \underbrace{\text{Total fabric}}_{\downarrow}$$
$$1\frac{3}{8} \quad + \quad \frac{5}{8} \quad + \quad 3\frac{3}{8} \quad = \quad f$$

Solve. We add.

$$\begin{array}{r} 1\ \dfrac{3}{8} \\[4pt] \dfrac{5}{8} \\[4pt] +\ 3\ \dfrac{3}{8} \\[2pt] \hline 4\ \dfrac{11}{8} = 4 + \dfrac{11}{8} \\[6pt] = 4 + 1\dfrac{3}{8} \\[6pt] = 5\dfrac{3}{8} \end{array}$$

Check. We can repeat the calculation. Also note that the answer is reasonable since it is larger than any of the individual amounts of fabric.

State. The outfit requires $5\frac{3}{8}$ yd of fabric.

45. Familiarize. We let w = the total weight of the meat.

Translate. We write an equation.

$$\underbrace{\text{Weight of one package}}_{\downarrow} \underset{\downarrow}{\text{ plus }} \underbrace{\text{Weight of second package}}_{\downarrow} \underset{\downarrow}{\text{ is }} \underbrace{\text{Total weight}}_{\downarrow}$$
$$1\frac{2}{3} \quad + \quad 5\frac{3}{4} \quad = \quad w$$

Solve. We carry out the addition. The LCD is 12.

$$\begin{array}{r} 1\ \boxed{\dfrac{2}{3} \cdot \dfrac{4}{4}} = \ 1\ \dfrac{8}{12} \\[8pt] +5\ \boxed{\dfrac{3}{4} \cdot \dfrac{3}{3}} = +5\ \dfrac{9}{12} \\[2pt] \hline 6\ \dfrac{17}{12} = 6 + \dfrac{17}{12} \\[6pt] = 6 + 1\dfrac{5}{12} \\[6pt] = 7\dfrac{5}{12} \end{array}$$

Check. We repeat the calculation. We also note that the answer is larger than either of the individual weights, so the answer seems reasonable.

State. The total weight of the meat was $7\frac{5}{12}$ lb.

47. Familiarize. We let h = Juan's excess height.

Translate. We have a missing addend situation.

$$\underbrace{\text{Height of daughter}}_{\downarrow} \underset{\downarrow}{\text{ plus }} \underbrace{\text{How much more height}}_{\downarrow} \underset{\downarrow}{\text{ is }} \underbrace{\text{Juan's height}}_{\downarrow}$$
$$180\frac{3}{4} \quad + \quad h \quad = \quad 187\frac{1}{10}$$

Solve. We solve the equation as follows:

$$h = 187\frac{1}{10} - 180\frac{3}{4}$$

$$187\ \boxed{\dfrac{1}{10} \cdot \dfrac{2}{2}} = \ 187\ \dfrac{2}{20}$$

$$180\ \boxed{\dfrac{3}{4} \cdot \dfrac{5}{5}} = \ 180\ \dfrac{15}{20}$$

$$\begin{array}{r} 187\ \dfrac{1}{10} = \ 187\ \dfrac{2}{20} = \ 186\ \dfrac{22}{20} \\[8pt] -\ 180\ \dfrac{3}{4} = -\ 180\ \dfrac{15}{20} = -\ 180\ \dfrac{15}{20} \\[2pt] \hline 6\ \dfrac{7}{20} \end{array}$$

Thus, $h = 6\frac{7}{20}$.

Check. We add Juan's excess height to his daughter's height:

$$180\frac{3}{4} + 6\frac{7}{20} = 180\frac{15}{20} + 6\frac{7}{20} = 186\frac{22}{20} = 187\frac{2}{20} = 187\frac{1}{10}$$

The answer checks.

State. Juan is $6\frac{7}{20}$ cm taller.

49. *Familiarize*. We draw a picture, letting x = the amount of pipe that was used, in inches.

$$\vdash\!\!-\!\!-\!\!-10\tfrac{5}{16}\text{ in.}\!\!-\!\!-\!\!-\!\!\dashv\!\!-\!\!-8\tfrac{3}{4}\text{ in.}\!\!-\!\!-\!\!\dashv$$
$$\vdash\!\!-\!\!-\!\!-\!\!-\!\!-\!\!-\!\!-x\!\!-\!\!-\!\!-\!\!-\!\!-\!\!-\!\!-\!\!\dashv$$

Translate. We write an addition sentence.

First length	plus	Second length	is	Total length
↓	↓	↓	↓	↓
$10\tfrac{5}{16}$	$+$	$8\tfrac{3}{4}$	$=$	x

Solve. We carry out the addition. The LCD is 16.

$$10\,\tfrac{5}{16} \;=\; 10\,\tfrac{5}{16}$$
$$+\;8\,\boxed{\tfrac{3}{4}\cdot\tfrac{4}{4}} = +\;8\,\tfrac{12}{16}$$
$$\rule{3cm}{0.4pt}$$
$$18\,\tfrac{17}{16} = 18 + \tfrac{17}{16}$$
$$= 18 + 1\tfrac{1}{16}$$
$$= 19\tfrac{1}{16}$$

Check. We repeat the calculation. We also note that the total length is larger than either of the individual lengths, so the answer seems reasonable.

State. The plumber used $19\tfrac{1}{16}$ in. of pipe.

51. *Familiarize*. We let f = the total number of yards of fabric Art bought.

Translate.

Length of first piece	plus	Length of second piece	is	Total length
↓	↓	↓	↓	↓
$9\tfrac{1}{4}$	$+$	$10\tfrac{5}{6}$	$=$	f

Solve. We carry out the addition.

$$9\,\boxed{\tfrac{1}{4}\cdot\tfrac{3}{3}} \;=\; 9\,\tfrac{3}{12}$$
$$+\,10\,\boxed{\tfrac{5}{6}\cdot\tfrac{2}{2}} = +\,10\,\tfrac{10}{12}$$
$$\rule{3cm}{0.4pt}$$
$$19\,\tfrac{13}{12} = 19 + \tfrac{13}{12}$$
$$= 19 + 1\tfrac{1}{12}$$
$$= 20\tfrac{1}{12}$$

Check. We can subtract one of the shorter lengths from the total and check to determine if we get the other length.
$$20\tfrac{1}{12} - 9\tfrac{1}{4} = 20\tfrac{1}{12} - 9\tfrac{3}{12} = 19\tfrac{13}{12} - 9\tfrac{3}{12} = 10\tfrac{10}{12} = 10\tfrac{5}{6}$$
The answer checks.

State. Art bought a total of $20\tfrac{1}{12}$ yd of fabric.

53. *Familiarize*. This is a multistep problem. First we find the number of gallons of fertilizer left in the tank after the application. Let x = this amount.

Translate.

Amount applied	plus	How much more fertilizer	is	Amount originally in tank
↓	↓	↓	↓	↓
$178\tfrac{2}{3}$	$+$	x	$=$	$283\tfrac{5}{8}$

Solve. We solve the equation.

$$x = 283\tfrac{5}{8} - 178\tfrac{2}{3}$$

We have:

$$283\,\boxed{\tfrac{5}{8}\cdot\tfrac{3}{3}} \;=\; 283\,\tfrac{15}{24}$$
$$-178\,\boxed{\tfrac{2}{3}\cdot\tfrac{8}{8}} = -178\,\tfrac{16}{24}$$

Now we carry out the subtraction.

$$283\,\tfrac{15}{24} = 282\,\tfrac{39}{24}$$
$$-\,178\,\tfrac{16}{24} = -\,178\,\tfrac{16}{2}$$
$$\rule{3cm}{0.4pt}$$
$$104\,\tfrac{23}{24}$$

Next we let f = the number of gallons of fertilizer in the tank after the delivery.

Amount after application	plus	Amount delivered	is	Final amount in tank
↓	↓	↓	↓	↓
$104\tfrac{23}{24}$	$+$	250	$=$	f

We carry out the addition.

$$104\,\tfrac{23}{24}$$
$$+\,250$$
$$\rule{2cm}{0.4pt}$$
$$354\,\tfrac{23}{24}$$

Check. We repeat the calculations. The answer checks.

State. After the delivery the tank contains $354\tfrac{23}{24}$ gal of fertilizer.

55. *Familiarize*. Let m = how much farther Angela will run in the marathon, in miles.

Translate.

Fun Run distance	plus	Additional marathon distance	is	Marathon distance
↓	↓	↓	↓	↓
$6\tfrac{1}{5}$	$+$	m	$=$	$26\tfrac{7}{32}$

Solve. We solve the equation as follows:

$$m = 26\frac{7}{32} - 6\frac{1}{5}$$

$$26\;\boxed{\frac{7}{32} \cdot \frac{5}{5}} = \quad 26\frac{35}{160}$$

$$6\;\boxed{\frac{1}{5} \cdot \frac{32}{32}} = -\;6\frac{32}{160}$$

$$\rule{3cm}{0.4pt}\qquad\rule{2cm}{0.4pt}$$

$$20\frac{3}{160}$$

Check. We add the additional marathon distance to the Fun Run distance:

$$6\frac{1}{5} + 20\frac{3}{160} = 6\frac{32}{160} + 20\frac{3}{160} = 26\frac{35}{160} = 26\frac{7}{32}$$

The answer checks.

State. Angela will run $20\frac{3}{160}$ mi farther in the marathon.

57. *Familiarize*. Let $x =$ the number of cups of ingredients. This is the sum of the measures listed.

Translate.

$$\underbrace{\text{Number of cups}}\; \text{is}\; \underbrace{\text{Sum of measures listed}}$$
$$\downarrow \qquad\qquad \downarrow \qquad\qquad \downarrow$$
$$x \qquad = 1\frac{1}{2} + 1\frac{1}{2} + 2\frac{1}{2} + 1\frac{1}{2} + \frac{3}{4}$$

Solve.

$$x = 1\frac{1}{2} + 1\frac{1}{2} + 2\frac{1}{2} + 1\frac{1}{2} + \frac{3}{4}$$

$$x = 1\frac{2}{4} + 1\frac{2}{4} + 2\frac{2}{4} + 1\frac{2}{4} + \frac{3}{4}$$

$$x = 5\frac{11}{4} = 5 + \frac{11}{4} = 5 + 2\frac{3}{4}$$

$$x = 7\frac{3}{4}$$

Check. We repeat the calculations. The answer checks.

State. There are $7\frac{3}{4}$ cups of ingredients listed.

59. *Familiarize*. We make a drawing. We let $t =$ the number of hours Sue worked on the third day.

$$\vdash\!\!-2\tfrac{1}{2}\text{ hr}\,-\!\!\vdash\!\!-4\tfrac{1}{5}\text{ hr}\,-\!\!\vdash\!\!- t \,-\!\!\dashv$$
$$\vdash\!\!\!-\!\!\!-\!\!\!-\!\!\!- 10\tfrac{1}{2}\text{ hr}\,-\!\!\!-\!\!\!-\!\!\!-\!\!\!-\!\!\dashv$$

Translate. We write an addition sentence.

$$2\frac{1}{2} + 4\frac{1}{5} + t = 10\frac{1}{2}$$

Solve. This is a two-step problem.

First we add $2\frac{1}{2} + 4\frac{1}{5}$ to find the time worked on the first two days. The LCD is 10.

$$2\;\boxed{\frac{1}{2} \cdot \frac{5}{5}} = \quad 2\frac{5}{10}$$

$$+\;4\;\boxed{\frac{1}{5} \cdot \frac{2}{2}} = +\;4\frac{2}{10}$$

$$\rule{3cm}{0.4pt}\qquad\rule{2cm}{0.4pt}$$

$$6\frac{7}{10}$$

Then we subtract $6\frac{7}{10}$ from $10\frac{1}{2}$ to find the time worked on the third day. The LCD is 10.

$$6\frac{7}{10} + t = 10\frac{1}{2}$$

$$t = 10\frac{1}{2} - 6\frac{7}{10}$$

$$10\;\boxed{\frac{1}{2} \cdot \frac{5}{5}} = \quad 10\frac{5}{10} = \quad 9\frac{15}{10}$$

$$-\;6\frac{7}{10} \qquad = -\;6\frac{7}{10} = -\;6\frac{7}{10}$$

$$\rule{3cm}{0.4pt}\quad\rule{2.5cm}{0.4pt}\quad\rule{2.5cm}{0.4pt}$$

$$3\frac{8}{10} = 3\frac{4}{5}$$

Check. We repeat the calculations.

State. Sue worked $3\frac{4}{5}$ hr the third day.

61. The figure is equivalent to a rectangle with length $16\frac{1}{2}$ in. and width $9\frac{1}{4}$ in. We add to find the perimeter.

$$16\frac{1}{2} + 9\frac{1}{4} + 16\frac{1}{2} + 9\frac{1}{4}$$

$$= 16\frac{2}{4} + 9\frac{1}{4} + 16\frac{2}{4} + 9\frac{1}{4}$$

$$= 50\frac{6}{4} = 50 + \frac{6}{4}$$

$$= 50 + 1\frac{2}{4} = 50 + 1\frac{1}{2}$$

$$= 51\frac{1}{2}$$

The perimeter is $51\frac{1}{2}$ in.

63. We add to find the perimeter.

$$4 + 3\frac{3}{4} + 6\frac{3}{4} + 4 + 3\frac{3}{4} + 5\frac{1}{2}$$

$$= 4 + 3\frac{3}{4} + 6\frac{3}{4} + 4 + 3\frac{3}{4} + 5\frac{2}{4}$$

$$= 25\frac{11}{4} = 25 + \frac{11}{4}$$

$$= 25 + 2\frac{3}{4} = 27\frac{3}{4}$$

The perimeter is $27\frac{3}{4}$ ft.

65. We add to find the perimeter.

$$1\frac{5}{12} + \frac{17}{24} + 1 + 1 + \frac{17}{24}$$

$$= 1\frac{10}{24} + \frac{17}{24} + 1 + 1 + \frac{17}{24}$$

$$= 3\frac{44}{24} = 3 + \frac{44}{24}$$

$$= 3 + 1\frac{20}{24} = 3 + 1\frac{5}{6}$$

$$= 4\frac{5}{6}$$

The perimeter is $4\frac{5}{6}$ ft.

67. We see that d and the two smallest distances combined are the same as the largest distance. We translate and solve.

$$2\frac{3}{4} + d + 2\frac{3}{4} = 12\frac{7}{8}$$

$$d = 12\frac{7}{8} - 2\frac{3}{4} - 2\frac{3}{4}$$

$$= 10\frac{1}{8} - 2\frac{3}{4} \quad \text{Subtracting } 2\frac{3}{4} \text{ from } 12\frac{7}{8}$$

$$= 7\frac{3}{8} \quad \quad \text{Subtracting } 2\frac{3}{4} \text{ from } 10\frac{1}{8}$$

The length of d is $7\frac{3}{8}$ ft.

69. *Familiarize*. We let b = the length of the bolt.

Translate. From the drawing we see that the length of the small bolt is the sum of the diameters of the two tubes and the thicknesses of the two washers and the nut. Thus, we have

$$b = \frac{1}{2} + \frac{1}{16} + \frac{3}{4} + \frac{1}{16} + \frac{3}{16}.$$

Solve. We carry out the addition. The LCD is 16.

$$b = \frac{1}{2} + \frac{1}{16} + \frac{3}{4} + \frac{1}{16} + \frac{3}{16} =$$

$$\frac{1}{2} \cdot \frac{8}{8} + \frac{1}{16} + \frac{3}{4} \cdot \frac{4}{4} + \frac{1}{16} + \frac{3}{16} =$$

$$\frac{8}{16} + \frac{1}{16} + \frac{12}{16} + \frac{1}{16} + \frac{3}{16} = \frac{25}{16} = 1\frac{9}{16}$$

Check. We repeat the calculation.

State. The smallest bolt is $1\frac{9}{16}$ in. long.

71. $8\frac{3}{5} - 9\frac{2}{5} = 8\frac{3}{5} + \left(-9\frac{2}{5}\right)$

Since $9\frac{2}{5}$ is greater than $8\frac{3}{5}$, the answer will be negative.
The difference in absolute values is

$$9\frac{2}{5} = \quad 8\frac{7}{5}$$

$$\underline{-8\frac{3}{5} = -8\frac{3}{5}}$$

$$\frac{4}{5}$$

so $8\frac{3}{5} - 9\frac{2}{5} = -\frac{4}{5}$.

73. $3\frac{1}{2} - 6\frac{3}{4} = 3\frac{1}{2} + \left(-6\frac{3}{4}\right)$

Since $6\frac{3}{4}$ is greater than $3\frac{1}{2}$, the answer will be negative.
The difference in absolute values is

$$6\frac{3}{4} = \quad 6\frac{3}{4} \quad = \quad 6\frac{3}{4}$$

$$\underline{-3\frac{1}{2} = -3\boxed{\frac{1}{2} \cdot \frac{1}{2}} = -3\frac{2}{4}}$$

$$3\frac{1}{4}$$

so $3\frac{1}{2} - 6\frac{3}{4} = -3\frac{1}{4}$.

75. $3\frac{4}{5} - 7\frac{2}{3} = 3\frac{4}{5} + \left(-7\frac{2}{3}\right)$

Since $7\frac{2}{3}$ is greater than $3\frac{4}{5}$, the answer will be negative.
The difference in absolute values is

$$7\frac{2}{3} = \quad 7\boxed{\frac{2}{3} \cdot \frac{5}{5}} = \quad 7\frac{10}{15} = \quad 6\frac{25}{15}$$

$$\underline{-3\frac{4}{5} = -3\boxed{\frac{4}{5} \cdot \frac{3}{3}} = -3\frac{12}{15} = -3\frac{12}{15}}$$

$$3\frac{13}{15}$$

so $3\frac{4}{5} - 7\frac{2}{3} = -3\frac{13}{15}$.

77. $-3\frac{1}{5} - 4\frac{2}{5} = -3\frac{1}{5} + \left(-4\frac{2}{5}\right)$

We add the absolute values and make the answer negative.

$$3\frac{1}{5}$$

$$\underline{+4\frac{2}{5}}$$

$$7\frac{3}{5}$$

Thus, $-3\frac{1}{5} - 4\frac{2}{5} = -7\frac{3}{5}$.

79. $-4\frac{2}{5} - 6\frac{3}{7} = -4\frac{2}{5} + \left(-6\frac{3}{7}\right)$

We add the absolute values and make the answer negative.

$$4\frac{2}{5} = \quad 4\boxed{\frac{2}{5} \cdot \frac{7}{7}} = \quad 4\frac{14}{35}$$

$$\underline{+6\frac{3}{7} = +6\boxed{\frac{3}{7} \cdot \frac{5}{5}} = +6\frac{15}{35}}$$

$$10\frac{29}{35}$$

Thus, $-4\frac{2}{5} - 6\frac{3}{7} = -10\frac{29}{35}$.

81. $-6\frac{1}{9} - \left(-4\frac{2}{9}\right) = -6\frac{1}{9} + 4\frac{2}{9}$

Since $-6\frac{1}{9}$ has the greater absolute value, the answer will be negative. The difference in absolute values is

$$\begin{array}{rl} 6\frac{1}{9} = & 5\frac{10}{9} \\ -4\frac{2}{9} = & -4\frac{2}{9} \\ \hline & 1\frac{8}{9} \end{array}$$

so $-6\frac{1}{9} - \left(-4\frac{2}{9}\right) = -1\frac{8}{9}$.

83. Discussion and Writing Exercise

85. *Familiarize.* We visualize the situation. Repeated subtraction, or division, works well here.

$$\boxed{\frac{3}{4} \text{ lb}} \quad \boxed{\frac{3}{4} \text{ lb}} \cdots \boxed{\frac{3}{4} \text{ lb}}$$

12 lb fills how many packages?

Let $n =$ the number of packages that can be made.

Translate. We translate to an equation.

$$n = 12 \div \frac{3}{4}$$

Solve. We carry out the division.

$$n = 12 \div \frac{3}{4} = 12 \cdot \frac{4}{3} = \frac{12 \cdot 4}{3}$$
$$= \frac{3 \cdot 4 \cdot 4}{3 \cdot 1} = \frac{3}{3} \cdot \frac{4 \cdot 4}{1}$$
$$= 16$$

Check. If each of 16 packages contains $\frac{3}{4}$ lb of cheese, a total of

$$16 \cdot \frac{3}{4} = \frac{16 \cdot 3}{4} = \frac{4 \cdot 4 \cdot 3}{4} = 4 \cdot 3,$$

or 12 lb of cheese is used. The answer checks.

State. 16 packages of cheese can be made from a 12-lb slab.

87. The sum of the digits is $9 + 9 + 9 + 3 = 30$. Since 30 is divisible by 3, then 9993 is divisible by 3.

89. The sum of the digits is $2 + 3 + 4 + 5 = 14$. Since 14 is not divisible by 9, then 2345 is not divisible by 9.

91. The ones digit of 2335 is not 0, so 2335 is not divisible by 10.

93. 18,888 is even because the ones digit is even. Because $1 + 8 + 8 + 8 + 8 = 33$ and 33 is divisible by 3, then 18,888 is divisible by 3. Thus, 18,888 is divisible by 6.

95. $\frac{15}{9} \cdot \frac{18}{39} = \frac{15 \cdot 18}{9 \cdot 39} = \frac{3 \cdot 5 \cdot 2 \cdot 3 \cdot 3}{3 \cdot 3 \cdot 3 \cdot 13}$
$$= \frac{3 \cdot 3 \cdot 3}{3 \cdot 3 \cdot 3} \cdot \frac{5 \cdot 2}{13}$$
$$= \frac{10}{13}$$

97. Discussion and Writing Exercise

99. Use a calculator.

$$\begin{array}{ll} 3289\frac{1047}{1189} = & 3289\ \frac{1047}{1189} = 3289\frac{1047}{1189} \\ +5278\frac{32}{41} = +5278\ \boxed{\frac{32}{41} \cdot \frac{29}{29}} = +5278\frac{928}{1189} \\ \hline & 8567\frac{1975}{1189} = 8568\frac{786}{1189} \end{array}$$

101. Use a calculator.

$$5848\frac{17}{29} - 4230\frac{19}{73} = 5848\frac{1241}{2117} - 4230\frac{551}{2117} =$$
$$1618\frac{690}{2117}, \text{ so } 4230\frac{19}{73} - 5848\frac{17}{29} = -1618\frac{690}{2117}$$

103. $35\frac{2}{3} + n = 46\frac{1}{4}$

$$n = 46\frac{1}{4} - 35\frac{2}{3}$$
$$n = 46\frac{3}{12} - 35\frac{8}{12}$$
$$n = 45\frac{15}{12} - 35\frac{8}{12}$$
$$n = 10\frac{7}{12}$$

105. $-15\frac{7}{8} = 12\frac{1}{2} + t$

$$-15\frac{7}{8} - 12\frac{1}{2} = t$$
$$-28\frac{3}{8} = t$$

107. The resulting rectangle will have length $\left(8\frac{1}{2} + 1\frac{1}{8} + 8\frac{1}{2}\right)$ in. and width $9\frac{3}{4}$ in. We add to find the perimeter.

$$8\frac{1}{2} + 1\frac{1}{8} + 8\frac{1}{2} + 9\frac{3}{4} + 8\frac{1}{2} + 1\frac{1}{8} + 8\frac{1}{2} + 9\frac{3}{4}$$
$$= 8\frac{4}{8} + 1\frac{1}{8} + 8\frac{4}{8} + 9\frac{6}{8} + 8\frac{4}{8} + 1\frac{1}{8} + 8\frac{4}{8} + 9\frac{6}{8}$$
$$= 52\frac{30}{8} = 52 + \frac{30}{8}$$
$$= 52 + 3\frac{6}{8} = 52 + 3\frac{3}{4}$$
$$= 55\frac{3}{4}$$

The perimeter is $55\frac{3}{4}$ in.

Exercise Set 4.7

1. $16 \cdot 1\frac{2}{5}$

$$= \frac{16}{1} \cdot \frac{7}{5} \quad \text{Writing fraction notation}$$
$$= \frac{16 \cdot 7}{1 \cdot 5} = \frac{112}{5}$$
$$= 22\frac{2}{5}$$

3. $6\dfrac{2}{3} \cdot \dfrac{1}{4}$

$= \dfrac{20}{3} \cdot \dfrac{1}{4}$ Writing fraction notation

$= \dfrac{20 \cdot 1}{3 \cdot 4} = \dfrac{4 \cdot 5 \cdot 1}{3 \cdot 4} = \dfrac{4}{4} \cdot \dfrac{5 \cdot 1}{3} = \dfrac{5}{3} = 1\dfrac{2}{3}$

5. $20\left(-2\dfrac{5}{6}\right) = \dfrac{20}{1} \cdot \left(-\dfrac{17}{6}\right) = -\dfrac{20 \cdot 17}{1 \cdot 6} = -\dfrac{2 \cdot 10 \cdot 17}{2 \cdot 3} =$

$\dfrac{2}{2}\left(-\dfrac{10 \cdot 17}{3}\right) = -\dfrac{170}{3} = -56\dfrac{2}{3}$

7. $3\dfrac{1}{2} \cdot 4\dfrac{2}{3} = \dfrac{7}{2} \cdot \dfrac{14}{3} = \dfrac{7 \cdot 14}{2 \cdot 3} = \dfrac{7 \cdot 2 \cdot 7}{2 \cdot 3} = \dfrac{2}{2} \cdot \dfrac{7 \cdot 7}{3} =$

$\dfrac{49}{3} = 16\dfrac{1}{3}$

9. $-2\dfrac{3}{10} \cdot 4\dfrac{2}{5} = -\dfrac{23}{10} \cdot \dfrac{22}{5} = -\dfrac{23 \cdot 22}{10 \cdot 5} = -\dfrac{23 \cdot 2 \cdot 11}{2 \cdot 5 \cdot 5} =$

$\dfrac{2}{2}\left(-\dfrac{23 \cdot 11}{5 \cdot 5}\right) = -\dfrac{253}{25} = -10\dfrac{3}{25}$

11. $\left(-6\dfrac{3}{10}\right)\left(-5\dfrac{7}{10}\right) = \dfrac{63}{10} \cdot \dfrac{57}{100} = \dfrac{3591}{100} = 35\dfrac{91}{100}$

13. $30 \div 2\dfrac{3}{5}$

$= 30 \div \dfrac{13}{5}$ Writing fractional notation

$= 30 \cdot \dfrac{5}{13}$ Multiplying by the reciprocal

$= \dfrac{30 \cdot 5}{13} = \dfrac{150}{13} = 11\dfrac{7}{13}$

15. $8\dfrac{2}{5} \div 7$

$= \dfrac{42}{5} \div 7$ Writing fractional notation

$= \dfrac{42}{5} \cdot \dfrac{1}{7}$ Multiplying by the reciprocal

$= \dfrac{42 \cdot 1}{5 \cdot 7} = \dfrac{6 \cdot 7}{5 \cdot 7} = \dfrac{7}{7} \cdot \dfrac{6}{5} = \dfrac{6}{5} = 1\dfrac{1}{5}$

17. $5\dfrac{1}{4} \div 2\dfrac{3}{5} = \dfrac{21}{4} \div \dfrac{13}{5} = \dfrac{21}{4} \cdot \dfrac{5}{13} = \dfrac{21 \cdot 5}{4 \cdot 13} =$

$\dfrac{105}{52} = 2\dfrac{1}{52}$

19. $-5\dfrac{1}{4} \div 2\dfrac{3}{7} = -\dfrac{21}{4} \div \dfrac{17}{7} = -\dfrac{21}{4} \cdot \dfrac{7}{17} = -\dfrac{21 \cdot 7}{4 \cdot 17} =$

$-\dfrac{147}{68} = -2\dfrac{11}{68}$

21. $5\dfrac{1}{10} \div 4\dfrac{3}{10} = \dfrac{51}{10} \div \dfrac{43}{10} = \dfrac{51}{10} \cdot \dfrac{10}{43} = \dfrac{51 \cdot 10}{10 \cdot 43} =$

$= \dfrac{10}{10} \cdot \dfrac{51}{43} = \dfrac{51}{43} = 1\dfrac{8}{43}$

23. $20\dfrac{1}{4} \div (-90) = \dfrac{81}{4} \div (-90) = \dfrac{81}{4}\left(-\dfrac{1}{90}\right) = -\dfrac{81 \cdot 1}{4 \cdot 90} =$

$-\dfrac{9 \cdot 9 \cdot 1}{4 \cdot 9 \cdot 10} = \dfrac{9}{9} \cdot \left(-\dfrac{9 \cdot 1}{4 \cdot 10}\right) = -\dfrac{9}{40}$

25. $lw = 2\dfrac{3}{5} \cdot 9$

$= \dfrac{13}{5} \cdot 9$

$= \dfrac{117}{5} = 23\dfrac{2}{5}$

27. $rs = 5 \cdot 3\dfrac{1}{7}$

$= 5 \cdot \dfrac{22}{7}$

$= \dfrac{110}{7} = 15\dfrac{5}{7}$

29. $mt = 6\dfrac{2}{9}\left(-4\dfrac{3}{5}\right)$

$= \dfrac{56}{9}\left(-\dfrac{23}{5}\right)$

$= -\dfrac{1288}{45} = -28\dfrac{28}{45}$

31. $R \cdot S \div T = 4\dfrac{2}{3} \cdot 1\dfrac{3}{7} \div (-5)$

$= \dfrac{14}{3} \cdot \dfrac{10}{7} \div (-5)$

$= \dfrac{14 \cdot 10}{3 \cdot 7} \div (-5)$

$= \dfrac{2 \cdot 7 \cdot 10}{3 \cdot 7} \div (-5)$

$= \dfrac{7}{7} \cdot \dfrac{2 \cdot 10}{3} \div (-5)$

$= \dfrac{20}{3} \div (-5)$

$= \dfrac{20}{3} \cdot \left(-\dfrac{1}{5}\right)$

$= -\dfrac{20 \cdot 1}{3 \cdot 5}$

$= -\dfrac{4 \cdot 5}{3 \cdot 5}$

$= -\dfrac{4}{3}$

$= -1\dfrac{1}{3}$

33. $r + ps = 5\dfrac{1}{2} + 3 \cdot 2\dfrac{1}{4}$

$= 5\dfrac{1}{2} + \dfrac{3}{1} \cdot \dfrac{9}{4}$

$= 5\dfrac{1}{2} + \dfrac{27}{4}$

$= 5\dfrac{1}{2} + 6\dfrac{3}{4}$

$= 5\dfrac{2}{4} + 6\dfrac{3}{4}$

$= 11\dfrac{5}{4}$

$= 12\dfrac{1}{4}$

35.
$$m + n \div p = 7\frac{2}{5} + 4\frac{1}{2} \div 6$$
$$= 7\frac{2}{5} + \frac{9}{2} \div 6$$
$$= 7\frac{2}{5} + \frac{9}{2} \cdot \frac{1}{6}$$
$$= 7\frac{2}{5} + \frac{9 \cdot 1}{2 \cdot 6} = 7\frac{2}{5} + \frac{3 \cdot 3 \cdot 1}{2 \cdot 2 \cdot 3}$$
$$= 7\frac{2}{5} + \frac{3 \cdot 1}{2 \cdot 2} = 7\frac{2}{5} + \frac{3}{4}$$
$$= 7\frac{8}{20} + \frac{15}{20} = 7\frac{23}{20}$$
$$= 8\frac{3}{20}$$

37. *Familiarize.* Let b = the number of beagles registered with The American Kennel Club.

Translate.

$3\frac{4}{9}$ times	Number of beagles registered	is	Number of Labrador retrievers registered
↓ ↓	↓	↓	↓
$3\frac{4}{9}$ ·	b	=	155,000

Solve. We divide by $3\frac{4}{9}$ on both sides of the equation.

$$b = 155,000 \div 3\frac{4}{9}$$
$$b = 155,000 \div \frac{31}{9}$$
$$b = 155,000 \div \frac{9}{31} = \frac{155,000 \cdot 9}{31}$$
$$b = \frac{31 \cdot 5000 \cdot 9}{31 \cdot 1} = \frac{31}{31} \cdot \frac{5000 \cdot 9}{1} = \frac{5000 \cdot 9}{1}$$
$$b = 45,000$$

Check. Since $3\frac{4}{9} \cdot 45,000 = \frac{31}{9} \cdot 45,000 = 155,000$, the answer checks.

State. There are 45,000 beagles registered with The American Kennel Club.

39. *Familiarize.* Let s = the number of teaspoons 10 average American women consume in one day.

Translate. A multiplication corresponds to this situation.

$$s = 10 \cdot 1\frac{1}{3}$$

Solve. We carry out the multiplication.

$$s = 10 \cdot 1\frac{1}{3} = 10 \cdot \frac{4}{3} = \frac{10 \cdot 4}{3} = \frac{40}{3} = 13\frac{1}{3}$$

Check. We repeat the calculation. The answer checks.

State. In one day 10 average American women consume $13\frac{1}{3}$ tsp of sodium.

41. *Familiarize, Translate, and Solve.* To find the ingredients for $\frac{1}{2}$ recipe, we multiply each ingredient by $\frac{1}{2}$.

$$1\frac{2}{3} \cdot \frac{1}{2} = \frac{5}{3} \cdot \frac{1}{2} = \frac{5 \cdot 1}{3 \cdot 2} = \frac{5}{6}$$
$$3 \cdot \frac{1}{2} = \frac{3 \cdot 1}{2} = \frac{3}{2} = 1\frac{1}{2}$$
$$4\frac{1}{2} \cdot \frac{1}{2} = \frac{9}{2} \cdot \frac{1}{2} = \frac{9 \cdot 1}{2 \cdot 2} = \frac{9}{4} = 2\frac{1}{4}$$
$$1 \cdot \frac{1}{2} = \frac{1 \cdot 1}{2} = \frac{1}{2}$$
$$3\frac{3}{4} \cdot \frac{1}{2} = \frac{15}{4} \cdot \frac{1}{2} = \frac{15 \cdot 1}{4 \cdot 2} = \frac{15}{8} = 1\frac{7}{8}$$
$$\frac{3}{4} \cdot \frac{1}{2} = \frac{3 \cdot 1}{4 \cdot 2} = \frac{3}{8}$$
$$1\frac{1}{2} \cdot \frac{1}{2} = \frac{3}{2} \cdot \frac{1}{2} = \frac{3 \cdot 1}{2 \cdot 2} = \frac{3}{4}$$

Check. We repeat the calculations.

State. The ingredients for $\frac{1}{2}$ recipe are $\frac{5}{6}$ cup water, $1\frac{1}{2}$ tablespoons canola oil, $2\frac{1}{4}$ teaspoons sugar, $\frac{1}{2}$ teaspoon salt, $1\frac{7}{8}$ cups bread flour, $\frac{3}{8}$ cup Grape-Nuts cereal, and $\frac{3}{4}$ teaspoon active dry yeast.

Familiarize, Translate and Solve. To find the ingredients for 3 recipes, we multiply each ingredient by 3.

$$1\frac{2}{3} \cdot 3 = \frac{5}{3} \cdot 3 = \frac{5 \cdot \cancel{3}}{\cancel{3} \cdot 1} = 5$$
$$3 \cdot 3 = 9$$
$$4\frac{1}{2} \cdot 3 = \frac{9}{2} \cdot 3 = \frac{9 \cdot 3}{2} = \frac{27}{2} = 13\frac{1}{2}$$
$$1 \cdot 3 = 3$$
$$3\frac{3}{4} \cdot 3 = \frac{15}{4} \cdot 3 = \frac{15 \cdot 3}{4} = \frac{45}{4} = 11\frac{1}{4}$$
$$\frac{3}{4} \cdot 3 = \frac{3 \cdot 3}{4} = \frac{9}{4} = 2\frac{1}{4}$$
$$1\frac{1}{2} \cdot 3 = \frac{3}{2} \cdot 3 = \frac{3 \cdot 3}{2} = \frac{9}{2} = 4\frac{1}{2}$$

Check. We can repeat the calculations.

State. The ingredients for 3 recipes are 5 cups water, 9 tablespoons canola oil, $13\frac{1}{2}$ teaspoons sugar, 3 teaspoons salt, $11\frac{1}{4}$ cups bread flour, $2\frac{1}{4}$ cups Grape-Nuts cereal, and $4\frac{1}{2}$ teaspoons active dry yeast.

43. *Familiarize.* Let h = the number of hours spent shopping on the Internet in 2005.

Translate.

$9\frac{1}{8}$ times	Internet shopping time	is	Television time
↓ ↓	↓	↓	↓
$9\frac{1}{8}$ ·	h	=	1825

Solve. We solve the equation.

$$9\frac{1}{8} \cdot h = 1825$$

$$\frac{73}{8} \cdot h = 1825$$

$$\frac{8}{73} \cdot \frac{73}{8} \cdot h = \frac{8}{73} \cdot 1825$$

$$1 \cdot h = \frac{8 \cdot 1825}{73}$$

$$h = \frac{8 \cdot 25 \cdot \cancel{73}}{\cancel{73} \cdot 1}$$

$$h = 200$$

Check. Since $9\frac{1}{8} \cdot 200 = \frac{73}{8} \cdot 200 = \frac{73 \cdot 200}{8} = \frac{73 \cdot 8 \cdot 25}{8 \cdot 1} = $ 1825, the answer checks.

State. The average person spent about 200 hr shopping on the Internet in 2005.

45. *Familiarize*. We let t = the Fahrenheit temperature.

Translate.

Celsius temperature	times $1\frac{4}{5}$	plus	32°	is	Fahrenheit temperature
↓	↓	↓	↓	↓	↓
20	$\cdot \ 1\frac{4}{5}$	+	32	=	t

Solve. We multiply and then add, according to the rules for order of operations.

$$t = 20 \cdot 1\frac{4}{5} + 32 = \frac{20}{1} \cdot \frac{9}{5} + 32 = \frac{20 \cdot 9}{1 \cdot 5} + 32 =$$

$$\frac{4 \cdot 5 \cdot 9}{1 \cdot 5} + 32 = \frac{5}{5} \cdot \frac{4 \cdot 9}{1} + 32 = 36 + 32 = 68$$

Check. We repeat the calculation.

State. 68° Fahrenheit corresponds to 20° Celsius.

47. *Familiarize*. First we will find the total width of the columns and determine if this is less than or greater than $8\frac{1}{2}$ in. Then, if the total is less than $8\frac{1}{2}$ in., we will find the difference between $8\frac{1}{2}$ in. and the total width and then divide by 2 to find the width of each margin. Let w = the total width of the columns.

Translate. First we write an equation for finding the total width of the columns.

$$w = 2 \cdot 1\frac{1}{2} + 5 \cdot \frac{3}{4}$$

Solve. We solve the equation.

$$w = 2 \cdot 1\frac{1}{2} + 5 \cdot \frac{3}{4}$$

$$w = 2 \cdot \frac{3}{2} + 5 \cdot \frac{3}{4}$$

$$w = \frac{2 \cdot 3}{2} + \frac{5 \cdot 3}{4}$$

$$w = 3 + \frac{15}{4} = 3 + 3\frac{3}{4}$$

$$w = 6\frac{3}{4}$$

Since the total width of the columns is less than $8\frac{1}{2}$ in., the table will fit on a piece of standard paper. Let l = the number of inches by which the width of the paper exceeds the width of the table. Then we have:

$$l = 8\frac{1}{2} - 6\frac{3}{4} = 8\frac{2}{4} - 6\frac{3}{4}$$

$$l = 7\frac{6}{4} - 6\frac{3}{4}$$

$$l = 1\frac{3}{4}$$

This tells us that the total width of the margins will be $1\frac{3}{4}$ in. Since the margins are of equal width, we divide by 2 to find the width of each margin. Let m = this width, in inches. We have:

$$m = 1\frac{3}{4} \div 2$$

$$m = \frac{7}{4} \div 2$$

$$m = \frac{7}{4} \cdot \frac{1}{2} = \frac{7}{8}$$

Check. We repeat the calculations.

State. The table will fit on a standard piece of paper. Each margin will be $\frac{7}{8}$ in. wide.

49. *Familiarize*. We draw a picture.

$\frac{1}{3}$ lb	$\frac{1}{3}$ lb	$\cdots$	$\frac{1}{3}$ lb

$$\longleftarrow \quad 5\frac{1}{2} \text{ lb} \quad \longrightarrow$$

We let s = the number of servings that can be prepared from $5\frac{1}{2}$ lb of salmon fillet.

Translate. The situation corresponds to a division sentence.

$$s = 5\frac{1}{2} \div \frac{1}{3}$$

Solve. We carry out the division.

$$s = 5\frac{1}{2} \div \frac{1}{3} = \frac{11}{2} \div \frac{1}{3}$$

$$= \frac{11}{2} \cdot \frac{3}{1} = \frac{33}{2}$$

$$= 16\frac{1}{2}$$

Check. We check by multiplying. If $16\frac{1}{2}$ servings are prepared, then

$$16\frac{1}{2} \cdot \frac{1}{3} = \frac{33}{2} \cdot \frac{1}{3} = \frac{3 \cdot 11 \cdot 1}{2 \cdot 3} = \frac{3}{3} \cdot \frac{11 \cdot 1}{2} = \frac{11}{2} = 5\frac{1}{2} \text{ lb}$$

of flounder is used. Our answer checks.

State. $16\frac{1}{2}$ servings can be prepared from $5\frac{1}{2}$ lb of salmon fillet.

51. *Familiarize*. We let $w =$ the weight of $5\frac{1}{2}$ cubic feet of water.

***Translate*.** We write an equation.

$$
\underbrace{\text{Weight per}}_{} \cdot \underbrace{\text{Number of}}_{} = \underbrace{\text{Total}}_{}
$$
$$
\underbrace{\text{cubic foot}}_{} \qquad \underbrace{\text{cubic feet}}_{} \qquad \underbrace{\text{weight}}_{}
$$

$$
62\tfrac{1}{2} \quad \cdot \quad 5\tfrac{1}{2} \quad = \quad w
$$

***Solve*.** To solve the equation we carry out the multiplication.

$$
w = 62\tfrac{1}{2} \cdot 5\tfrac{1}{2}
$$
$$
= \frac{125}{2} \cdot \frac{11}{2} = \frac{125 \cdot 11}{2 \cdot 2}
$$
$$
= \frac{1375}{4} = 343\tfrac{3}{4}
$$

***Check*.** We repeat the calculation. We also note that $62\tfrac{1}{2} \approx 60$ and $5\tfrac{1}{2} \approx 5$. Then the product is about 300. Our answer seems reasonable.

***State*.** The weight of $5\tfrac{1}{2}$ cubic feet of water is $343\tfrac{3}{4}$ lb.

53. *Familiarize*. We let $t =$ the number of inches of tape used in 60 sec of recording.

***Translate*.** We write an equation.

$$
\underbrace{\text{Inches per}}_{} \cdot \underbrace{\text{Number of}}_{} = \underbrace{\text{Tape}}_{}
$$
$$
\underbrace{\text{second}}_{} \qquad \underbrace{\text{seconds}}_{} \qquad \underbrace{\text{used}}_{}
$$

$$
1\tfrac{3}{8} \quad \cdot \quad 60 \quad = \quad t
$$

***Solve*.** We carry out the multiplication.

$$
t = 1\tfrac{3}{8} \cdot 60 = \frac{11}{8} \cdot 60
$$
$$
= \frac{11 \cdot 4 \cdot 15}{2 \cdot 4} = \frac{11 \cdot 15}{2} \cdot \frac{4}{4}
$$
$$
= \frac{165}{2} = 82\tfrac{1}{2}
$$

***Check*.** We repeat the calculation.

***State*.** $82\tfrac{1}{2}$ in. of tape are used in 60 sec of recording in short-play mode.

55. *Familiarize*. We let $m =$ the number of miles per gallon the car got.

***Translate*.** We write an equation.

$$
\underbrace{\substack{\text{Total number} \\ \text{of} \\ \text{miles traveled}}}_{} \div \underbrace{\substack{\text{Number of} \\ \text{gallons of} \\ \text{gas used}}}_{} = \underbrace{\substack{\text{Miles} \\ \text{per} \\ \text{gallon}}}_{}
$$

$$
213 \quad \div \quad 14\tfrac{2}{10} \quad = \quad m
$$

***Solve*.** To solve the equation we carry out the division.

$$
m = 213 \div 14\tfrac{2}{10} = 213 \div \frac{142}{10}
$$
$$
= 213 \cdot \frac{10}{142} = \frac{3 \cdot 71 \cdot 2 \cdot 5}{2 \cdot 71 \cdot 1}
$$
$$
= \frac{2 \cdot 71}{2 \cdot 71} \cdot \frac{3 \cdot 5}{1} = 15
$$

***Check*.** We repeat the calculation.

***State*.** The car got 15 miles per gallon of gas.

57. *Familiarize*. To compute an average, we add the values and then divide the sum by the number of values. Let $w =$ the average birthweight, in pounds.

***Translate*.** We have

$$
w = \frac{2\frac{9}{16} + 2\frac{9}{32} + 2\frac{1}{8} + 2\frac{5}{16}}{4}.
$$

***Solve*.** First we add.

$$
2\tfrac{9}{16} + 2\tfrac{9}{32} + 2\tfrac{1}{8} + 2\tfrac{5}{16}
$$
$$
= 2\tfrac{18}{32} + 2\tfrac{9}{32} + 2\tfrac{4}{32} + 2\tfrac{10}{32}
$$
$$
= 8\tfrac{41}{32} = 9\tfrac{9}{32}
$$

Then we divide:

$$
9\tfrac{9}{32} \div 4 = \frac{297}{32} \div 4
$$
$$
= \frac{297}{32} \cdot \frac{1}{4}
$$
$$
= \frac{297}{128} = 2\tfrac{41}{128}
$$

***Check*.** As a partial check we note that the average weight is larger than the smallest weight and smaller than the largest weight. We could also repeat the calculations.

***State*.** The average birthweight of the quadruplets was $2\tfrac{41}{128}$ lb.

59. We add the numbers and divide by the number of addends.

$$
\frac{7\frac{3}{5} + 9\frac{1}{10} + 6\frac{1}{2} + 6\frac{9}{10} + 7\frac{1}{5}}{5}
$$
$$
= \frac{7\frac{6}{10} + 9\frac{1}{10} + 6\frac{5}{10} + 6\frac{9}{10} + 7\frac{2}{10}}{5}
$$
$$
= \frac{35\frac{23}{10}}{5} = \frac{37\frac{3}{10}}{5}
$$
$$
= \frac{\frac{373}{10}}{5} = \frac{373}{10} \cdot \frac{1}{5}
$$
$$
= \frac{373}{50} = 7\tfrac{23}{50}
$$

The average time was $7\tfrac{23}{50}$ sec.

61. *Familiarize*. We can refer to the drawing in the text. Let $a =$ the total area of the sod.

***Translate*.** The total area is the sum of the areas of the two rectangles.

$$a = 20 \cdot 15\frac{1}{2} + 12\frac{1}{2} \cdot 10\frac{1}{2}$$

Solve. We perform the multiplication and then add.

$$a = 20 \cdot 15\frac{1}{2} + 12\frac{1}{2} \cdot 10\frac{1}{2}$$

$$= 20 \cdot \frac{31}{2} + \frac{25}{2} \cdot \frac{21}{2}$$

$$= \frac{20 \cdot 31}{2} + \frac{25 \cdot 21}{4}$$

$$= \frac{\cancel{2} \cdot 10 \cdot 31}{\cancel{2} \cdot 1} + \frac{525}{4}$$

$$= 310 + 131\frac{1}{4}$$

$$= 441\frac{1}{4}$$

Check. We can perform a partial check by estimating the total area as $20 \cdot 16 + 13 \cdot 11 = 320 + 143 = 463 \approx 441\frac{1}{4}$. Our answer seems reasonable.

State. The total area of the sod is $441\frac{1}{4}$ ft^2.

63. Familiarize. The figure contains a square with sides of $10\frac{1}{2}$ ft and a rectangle with dimensions of $8\frac{1}{2}$ ft by 4 ft. The area of the shaded region consists of the area of the square less the area of the rectangle. Let A = the area of the shaded region, in square feet.

Translate. We write an equation.

$$A = 10\frac{1}{2} \cdot 10\frac{1}{2} - 8\frac{1}{2} \cdot 4$$

Solve. We multiply and then subtract.

$$A = 10\frac{1}{2} \cdot 10\frac{1}{2} - 8\frac{1}{2} \cdot 4$$

$$A = \frac{21}{2} \cdot \frac{21}{2} - \frac{17}{2} \cdot 4$$

$$A = \frac{441}{4} - \frac{68}{2}$$

$$A = \frac{441}{4} - \frac{68}{2} \cdot \frac{2}{2}$$

$$A = \frac{441}{4} - \frac{136}{4}$$

$$A = \frac{305}{4} = 76\frac{1}{4}$$

Check. We repeat the calculation.

State. The area of the shaded region is $76\frac{1}{4}$ ft^2.

65. Familiarize. The figure is a rectangle with dimensions $7\frac{1}{4}$ cm by $4\frac{1}{2}$ cm with a rectangular area cut out of it. One dimension of the cut-out area is $4\frac{1}{4}$ cm. We subtract to find the other dimension:

$$4\frac{1}{2} - \left(1\frac{1}{2} + 1\frac{3}{4}\right) = 4\frac{1}{2} - \left(1\frac{2}{4} + 1\frac{3}{4}\right) = 4\frac{1}{2} - 2\frac{5}{4} = 4\frac{1}{2} - 3\frac{1}{4} =$$

$$4\frac{2}{4} - 3\frac{1}{4} = 1\frac{1}{4}.$$

The area of the shaded region consists of the area of the larger rectangle less the area of the smaller rectangle. Let A = the area of the shaded region, in square centimeters.

Translate.

$$A = 7\frac{1}{4} \cdot 4\frac{1}{2} - 4\frac{1}{4} \cdot 1\frac{1}{4}$$

Solve.

$$A = 7\frac{1}{4} \cdot 4\frac{1}{2} - 4\frac{1}{4} \cdot 1\frac{1}{4}$$

$$A = \frac{29}{4} \cdot \frac{9}{2} - \frac{17}{4} \cdot \frac{5}{4}$$

$$A = \frac{29 \cdot 9}{4 \cdot 2} - \frac{17 \cdot 5}{4 \cdot 4}$$

$$A = \frac{261}{8} - \frac{85}{16}$$

$$A = \frac{261}{8} \cdot \frac{2}{2} - \frac{85}{16}$$

$$A = \frac{522}{16} - \frac{85}{16}$$

$$A = \frac{437}{16} = 27\frac{5}{16}$$

Check. We can perform a partial check by estimating the area as $7 \cdot 5 - 4 \cdot 1 = 35 - 4 = 31$. Our answer seems reasonable.

State. The area of the shaded region is $27\frac{5}{16}$ cm^2.

67. Discussion and Writing Exercise

69. The set $\{\ldots, -3, -2, -1, 0, 1, 2, 3, \ldots\}$ is the set of <u>integers</u>.

71. The numbers 91, 95, and 111 are examples of <u>composite</u> numbers.

73. To add fractions with different denominators, we must first find the <u>least common multiple</u> of the denominators.

75. In the expression $\frac{c}{d}$, we call c the <u>numerator</u>.

77. Discussion and Writing Exercise

79. $-8 \div \frac{1}{2} + \frac{3}{4} + \left(-5 - \frac{5}{8}\right)^2 = -8 \div \frac{1}{2} + \frac{3}{4} + \left(-\frac{40}{8} - \frac{5}{8}\right)^2 =$

$-8 \div \frac{1}{2} + \frac{3}{4} + \left(-\frac{45}{8}\right)^2 = -8 \div \frac{1}{2} + \frac{3}{4} + \frac{2025}{64} =$

$-8 \cdot 2 + \frac{3}{4} + \frac{2025}{64} = -16 + \frac{3}{4} + \frac{2025}{64} =$

$-\frac{1024}{64} + \frac{48}{64} + \frac{2025}{64} = \frac{1049}{64} = 16\frac{25}{64}$

81. $\frac{1}{3} \div \left(\frac{1}{2} - \frac{1}{5}\right) \times \frac{1}{4} + \frac{1}{6}$

$$= \frac{1}{3} \div \left(\frac{5}{10} - \frac{2}{10}\right) \times \frac{1}{4} + \frac{1}{6}$$

$$= \frac{1}{3} \div \frac{3}{10} \times \frac{1}{4} + \frac{1}{6}$$

$$= \frac{1}{3} \times \frac{10}{3} \times \frac{1}{4} + \frac{1}{6}$$

$$= \frac{10}{9} \times \frac{1}{4} + \frac{1}{6}$$

$$= \frac{2 \times 5 \times 1}{9 \times 2 \times 2} + \frac{1}{6} = \frac{2}{2} \times \frac{5 \times 1}{9 \times 2} + \frac{1}{6}$$

$$= \frac{5}{18} + \frac{1}{6} = \frac{5}{18} + \frac{3}{18} = \frac{8}{18} = \frac{4}{9}$$

83. $\dfrac{1}{r} = \dfrac{1}{40} + \dfrac{1}{60} + \dfrac{1}{80}$

$\dfrac{1}{r} = \dfrac{1}{40} \cdot \dfrac{6}{6} + \dfrac{1}{60} \cdot \dfrac{4}{4} + \dfrac{1}{80} \cdot \dfrac{3}{3}$

$\dfrac{1}{r} = \dfrac{6}{240} + \dfrac{4}{240} + \dfrac{3}{240}$

$\dfrac{1}{r} = \dfrac{13}{240}$

Then r is the reciprocal of $\dfrac{13}{240}$, so $r = \dfrac{240}{13}$, or $18\dfrac{6}{13}$.

85. *Familiarize.* Let $w =$ the amount of hot water required for two showers and two loads of wash. Note that washing one load of clothes requires $1\dfrac{3}{5} \cdot 20$ gallons of hot water.

Translate. We write an equation.

$$w = 2 \cdot 20 + 2 \cdot 1\dfrac{3}{5} \cdot 20$$

Solve. We perform the multiplications and then we add.

$w = 2 \cdot 20 + 2 \cdot 1\dfrac{3}{5} \cdot 20$

$w = 40 + 2 \cdot \dfrac{8}{5} \cdot 20$

$w = 40 + \dfrac{2 \cdot 8 \cdot 20}{5}$

$w = 40 + \dfrac{320}{5}$

$w = 40 + 64$

$w = 104$

Check. We repeat the calculations. The answer checks.

State. Two showers and two loads of wash require 104 gallons of hot water.

Chapter 4 Review Exercises

1. $16 = 2 \cdot 2 \cdot 2 \cdot 2 = 2^4$
$20 = 2 \cdot 2 \cdot 5 = 2^2 \cdot 5$

We form the LCM using the greatest power of each factor. The LCM is $2^4 \cdot 5$, or 80.

2. 1.) 45 is not a multiple of 18.

2.) Check multiples:

$2 \cdot 45 = 90$ A multiple of 18

The LCM is 90.

3. Note that 3 and 6 are factors of 30. Since the largest number, 30, has the other two numbers as factors, it is the LCM.

4. $\dfrac{2}{9} + \dfrac{5}{9} = \dfrac{2+5}{9} = \dfrac{7}{9}$

5. $\dfrac{7}{x} + \dfrac{2}{x} = \dfrac{7+2}{x} = \dfrac{9}{x}$

6. The LCM of 5 and 15 is 15.

$-\dfrac{6}{5} + \dfrac{11}{15} = -\dfrac{6}{5} \cdot \dfrac{3}{3} + \dfrac{11}{15} = -\dfrac{18}{15} + \dfrac{11}{15} = -\dfrac{7}{15}$

7. The LCM of 16 and 24 is 48.

$\dfrac{5}{16} + \dfrac{3}{24} = \dfrac{5}{16} \cdot \dfrac{3}{3} + \dfrac{3}{24} \cdot \dfrac{2}{2}$

$= \dfrac{15}{48} + \dfrac{6}{48} = \dfrac{21}{48}$

$= \dfrac{3 \cdot 7}{3 \cdot 16} = \dfrac{3}{3} \cdot \dfrac{7}{16} = \dfrac{7}{16}$

8. $\dfrac{7}{9} - \dfrac{5}{9} = \dfrac{7-5}{9} = \dfrac{2}{9}$

9. The LCM of 4 and 8 is 8.

$\dfrac{1}{4} - \dfrac{3}{8} = \dfrac{1}{4} \cdot \dfrac{2}{2} - \dfrac{3}{8} = \dfrac{2}{8} - \dfrac{3}{8} = -\dfrac{1}{8}$

10. The LCM of 27 and 9 is 27.

$\dfrac{10}{27} - \dfrac{2}{9} = \dfrac{10}{27} - \dfrac{2}{9} \cdot \dfrac{3}{3} = \dfrac{10}{27} - \dfrac{6}{27} = \dfrac{10-6}{27} = \dfrac{4}{27}$

11. The LCM of 6 and 9 is 18.

$\dfrac{5}{6} - \dfrac{2}{9} = \dfrac{5}{6} \cdot \dfrac{3}{3} - \dfrac{2}{9} \cdot \dfrac{2}{2}$

$= \dfrac{15}{18} - \dfrac{4}{18} = \dfrac{11}{18}$

12. The LCD is $7 \cdot 9$, or 63.

$\dfrac{4}{7} \cdot \dfrac{9}{9} = \dfrac{36}{63}$

$\dfrac{5}{9} \cdot \dfrac{7}{7} = \dfrac{35}{63}$

Since $36 > 35$, it follows that $\dfrac{36}{63} > \dfrac{35}{63}$, so $\dfrac{4}{7} > \dfrac{5}{9}$.

13. The LCD is $9 \cdot 13$, or 117.

$\dfrac{-8}{9} \cdot \dfrac{13}{13} = \dfrac{-104}{117}$

$\dfrac{-11}{13} \cdot \dfrac{9}{9} = \dfrac{-99}{117}$

Since $-104 < -99$, it follows that $\dfrac{-104}{117} < \dfrac{-99}{117}$, so $-\dfrac{8}{9} < -\dfrac{11}{13}$.

14. $x + \dfrac{2}{5} = \dfrac{7}{8}$

$x + \dfrac{2}{5} - \dfrac{2}{5} = \dfrac{7}{8} - \dfrac{2}{5}$

$x + 0 = \dfrac{7}{8} \cdot \dfrac{5}{5} - \dfrac{2}{5} \cdot \dfrac{8}{8}$

$x = \dfrac{35}{40} - \dfrac{16}{40}$

$x = \dfrac{19}{40}$

The solution is $\dfrac{19}{40}$.

15. $7a - 3 = 25$

$7a - 3 + 3 = 25 + 3$

$7a = 28$

$\dfrac{1}{7} \cdot 7a = \dfrac{1}{7} \cdot 28$

$a = 4$

The solution is 4.

16.
$$5 + \frac{16}{3}x = \frac{5}{9}$$
$$5 + \frac{16}{3}x - 5 = \frac{5}{9} - 5$$
$$\frac{16}{3}x = \frac{5}{9} - \frac{45}{9}$$
$$\frac{16}{3}x = -\frac{40}{9}$$
$$\frac{3}{16} \cdot \frac{16}{3}x = \frac{3}{16}\left(-\frac{40}{9}\right)$$
$$x = -\frac{3 \cdot 40}{16 \cdot 9} = -\frac{3 \cdot 5 \cdot 8}{2 \cdot 8 \cdot 3 \cdot 3}$$
$$x = -\frac{5}{2 \cdot 3} \cdot \frac{3 \cdot 8}{3 \cdot 8} = -\frac{5}{6}$$

The solution is $-\frac{5}{6}$.

17.
$$\frac{22}{5} = \frac{16}{5} + \frac{5}{2}x$$
$$\frac{22}{5} - \frac{16}{5} = \frac{16}{5} + \frac{5}{2}x - \frac{16}{5}$$
$$\frac{6}{5} = \frac{5}{2}x$$
$$\frac{2}{5} \cdot \frac{6}{5} = \frac{2}{5} \cdot \frac{5}{2}x$$
$$\frac{12}{25} = x$$

The solution is $\frac{12}{25}$.

18.
$$\frac{5}{3}x + \frac{5}{6} = \frac{3}{2}$$
$$6\left(\frac{5}{3}x + \frac{5}{6}\right) = 6 \cdot \frac{3}{2} \qquad \text{The LCD is 6.}$$
$$\frac{6 \cdot 5}{3}x + 6 \cdot \frac{5}{6} = \frac{18}{2}$$
$$\frac{2 \cdot 3 \cdot 5}{3 \cdot 1}x + 5 = 9$$
$$10x + 5 = 9$$
$$10x + 5 - 5 = 9 - 5$$
$$10x = 4$$
$$\frac{10x}{10} = \frac{4}{10}$$
$$x = \frac{2}{5}$$

The solution is $\frac{2}{5}$.

19. $7\frac{1}{2} = \frac{15}{2}$ ($7 \cdot 2 = 14$, $14 + 1 = 15$)

20. $30\frac{4}{9} = \frac{274}{9}$ ($30 \cdot 9 = 270$, $270 + 4 = 274$)

21. $-9\frac{2}{7} = -\left(9 + \frac{2}{7}\right) = -\frac{65}{7}$

($9 \cdot 7 = 63$; $63 + 2 = 65$; include the negative sign.)

22.
$$\begin{array}{r} 2 \\ 5\overline{\smash{)}13} \\ 10 \\ \hline 3 \end{array} \qquad \frac{13}{5} = 2\frac{3}{5}$$

23. First consider $\frac{27}{4}$.

$$\begin{array}{r} 6 \\ 4\overline{\smash{)}27} \\ 24 \\ \hline 3 \end{array} \qquad \frac{27}{4} = 6\frac{3}{4}$$

Since $\frac{27}{4} = 6\frac{3}{4}$, we have $\frac{-27}{4} = -6\frac{3}{4}$.

24.
$$\begin{array}{r} 7 \\ 8\overline{\smash{)}57} \\ 56 \\ \hline 1 \end{array} \qquad \frac{57}{8} = 7\frac{1}{8}$$

25.
$$\begin{array}{r} 3 \\ 2\overline{\smash{)}7} \\ 6 \\ \hline 1 \end{array} \qquad \frac{7}{2} = 3\frac{1}{2}$$

26. First we find $7896 \div 9$.

$$\begin{array}{r} 877 \\ 9\overline{\smash{)}7896} \\ 7200 \\ \hline 696 \\ 630 \\ \hline 66 \\ 63 \\ \hline 3 \end{array}$$

Since $877\frac{3}{9} = 877\frac{1}{3}$, we have $7896 \div (-9) = -877\frac{1}{3}$.

27. $\dfrac{80 + 82 + 85}{3} = \dfrac{247}{3} = 82\frac{1}{3}$

28.
$$\begin{array}{r} 7\frac{3}{5} \\ +2\frac{4}{5} \\ \hline 9\frac{7}{5} \end{array} = 9 + \frac{7}{5}$$
$$= 9 + 1\frac{2}{5}$$
$$= 10\frac{2}{5}$$

29.
$$\begin{array}{r} 6\,\boxed{\dfrac{1}{3} \cdot \dfrac{5}{5}} = 6\dfrac{5}{15} \\ +5\,\boxed{\dfrac{2}{5} \cdot \dfrac{3}{3}} = +5\dfrac{6}{15} \\ \hline 11\dfrac{11}{15} \end{array}$$

30. $-3\frac{5}{6} + \left(-5\frac{1}{6}\right)$

We add the absolute values and make the answer negative.

$$3\frac{5}{6}$$
$$+5\frac{1}{6}$$
$$\overline{8\frac{6}{6}} = 8 + 1 = 9$$

Thus, $-3\frac{5}{6} + \left(-5\frac{1}{6}\right) = -9$.

31. $-2\frac{3}{4} + 4\frac{1}{2} = 4\frac{1}{2} - 2\frac{3}{4}$

$$4\;\boxed{\frac{1}{2}\cdot\frac{2}{2}} = \;\;4\frac{2}{4} = \;\;3\frac{6}{4}$$
$$-2\;\frac{3}{4} \;\;= -2\frac{3}{4} = -2\frac{3}{4}$$
$$\overline{\phantom{-2\frac{3}{4} = -2\frac{3}{4} = }\;\;1\frac{3}{4}}$$

32.
$$14 \;\;= \;\;13\frac{9}{9}$$
$$-\;6\frac{2}{9} \;= -\;6\frac{2}{9}$$
$$\overline{\phantom{-\;6\frac{2}{9} = -\;}7\frac{7}{9}}$$

33.
$$9\;\boxed{\frac{3}{5}\cdot\frac{3}{3}} = \;\;9\frac{9}{15} = \;\;8\frac{24}{15}$$
$$-4\;\;\frac{13}{15} \;\;= -4\frac{13}{15} = -4\frac{13}{15}$$
$$\overline{\phantom{-4\frac{13}{15} = -4\frac{13}{15} = }\;\;4\frac{11}{15}}$$

34. $4\frac{5}{8} - 9\frac{3}{4} = 4\frac{5}{8} + \left(-9\frac{3}{4}\right)$

Since $9\frac{3}{4}$ is greater than $4\frac{5}{8}$, the answer will be negative. The difference of the absolute values is

$$9\;\boxed{\frac{3}{4}\cdot\frac{2}{2}} = \;\;9\frac{6}{8}$$
$$-4\;\frac{5}{8} \;\;= -4\frac{5}{8}$$
$$\overline{\phantom{-4\frac{5}{8} = -}\;\;5\frac{1}{8}}$$

so $4\frac{5}{8} - 9\frac{3}{4} = -5\frac{1}{8}$.

35. $-7\frac{1}{2} - 6\frac{3}{4} = -7\frac{1}{2} + \left(-6\frac{3}{4}\right)$

We add the absolute values and make the answer negative.

$$7\;\boxed{\frac{1}{2}\cdot\frac{2}{2}} = \;\;7\frac{2}{4}$$
$$+6\;\frac{3}{4} \;\;= +6\frac{3}{4}$$
$$\overline{\phantom{+6\frac{3}{4} = +}\;\;13\frac{5}{4} = 13 + \frac{5}{4}}$$
$$= 13 + 1\frac{1}{4}$$
$$= 14\frac{1}{4}$$

Thus, $-7\frac{1}{2} - 6\frac{3}{4} = -14\frac{1}{4}$.

36. $\dfrac{4}{9}x + \dfrac{1}{3}x = \dfrac{4}{9}x + \dfrac{1}{3}\cdot\dfrac{3}{3}x$
$$= \dfrac{4}{9}x + \dfrac{3}{9}x$$
$$= \dfrac{7}{9}x$$

37. $8\frac{3}{10}a - 5\frac{1}{8}a = \left(8\frac{3}{10} - 5\frac{1}{8}\right)a$
$$= \left(8\frac{12}{40} - 5\frac{5}{40}\right)a$$
$$= 3\frac{7}{40}a$$

38. $6\cdot 2\frac{2}{3} = 6\cdot\dfrac{8}{3} = \dfrac{6\cdot 8}{3} = \dfrac{2\cdot 3\cdot 8}{3\cdot 1} = \dfrac{3}{3}\cdot\dfrac{2\cdot 8}{1} = 16$

39. $-5\frac{1}{4}\cdot\dfrac{2}{3} = -\dfrac{21}{4}\cdot\dfrac{2}{3} = -\dfrac{21\cdot 2}{4\cdot 3} = -\dfrac{3\cdot 7\cdot 2}{2\cdot 2\cdot 3} =$
$$\dfrac{2\cdot 3}{2\cdot 3}\cdot\left(-\dfrac{7}{2}\right) = -\dfrac{7}{2} = -3\frac{1}{2}$$

40. $2\frac{1}{5}\cdot 1\frac{1}{10} = \dfrac{11}{5}\cdot\dfrac{11}{10} = \dfrac{11\cdot 11}{5\cdot 10} = \dfrac{121}{50} = 2\frac{21}{50}$

41. $2\frac{2}{5}\cdot 2\frac{1}{2} = \dfrac{12}{5}\cdot\dfrac{5}{2} = \dfrac{12\cdot 5}{5\cdot 2} = \dfrac{2\cdot 6\cdot 5}{5\cdot 2\cdot 1} = \dfrac{2\cdot 5}{2\cdot 5}\cdot\dfrac{6}{1} = 6$

42. $-54 \div 2\frac{1}{4} = -54 \div \dfrac{9}{4} = -54\cdot\dfrac{4}{9} = \dfrac{-54\cdot 4}{9} =$
$$\dfrac{-2\cdot 3\cdot 9\cdot 4}{9\cdot 1} = \dfrac{9}{9}\cdot\dfrac{-2\cdot 3\cdot 4}{1} = -24$$

43. $2\frac{2}{5} \div \left(-1\frac{7}{10}\right) = \dfrac{12}{5} \div \left(-\dfrac{17}{10}\right) = \dfrac{12}{5}\cdot\left(-\dfrac{10}{17}\right) =$
$$-\dfrac{12\cdot 10}{5\cdot 17} = -\dfrac{12\cdot 2\cdot 5}{5\cdot 17} = \dfrac{5}{5}\cdot\left(-\dfrac{12\cdot 2}{17}\right) = -\dfrac{24}{17} = -1\frac{7}{17}$$

44. $3\frac{1}{4} \div 26 = \dfrac{13}{4} \div 26 = \dfrac{13}{4}\cdot\dfrac{1}{26} = \dfrac{13\cdot 1}{4\cdot 26} = \dfrac{13\cdot 1}{4\cdot 2\cdot 13} =$
$$\dfrac{13}{13}\cdot\dfrac{1}{4\cdot 2} = \dfrac{1}{8}$$

45. $4\frac{1}{5} \div 4\frac{2}{3} = \dfrac{21}{5} \div \dfrac{14}{3} = \dfrac{21}{5}\cdot\dfrac{3}{14} = \dfrac{21\cdot 3}{5\cdot 14} = \dfrac{3\cdot 7\cdot 3}{5\cdot 2\cdot 7} =$
$$\dfrac{7}{7}\cdot\dfrac{3\cdot 3}{5\cdot 2} = \dfrac{9}{10}$$

46.
$$5x - y = 5 \cdot 3\frac{1}{5} - 2\frac{2}{7}$$
$$= 5 \cdot \frac{16}{5} - 2\frac{2}{7}$$
$$= \frac{\cancel{5} \cdot 16}{1 \cdot \cancel{5}} - 2\frac{2}{7}$$
$$= 16 - 2\frac{2}{7}$$
$$= 15\frac{7}{7} - 2\frac{2}{7}$$
$$= 13\frac{5}{7}$$

47.
$$2a \div b = 2 \cdot 5\frac{2}{11} \div 3\frac{4}{5}$$
$$= 2 \cdot \frac{57}{11} \div \frac{19}{5}$$
$$= \frac{2 \cdot 57}{11} \div \frac{19}{5}$$
$$= \frac{2 \cdot 57}{11} \cdot \frac{5}{19}$$
$$= \frac{2 \cdot 57 \cdot 5}{11 \cdot 19}$$
$$= \frac{2 \cdot 3 \cdot \cancel{19} \cdot 5}{11 \cdot \cancel{19}}$$
$$= \frac{30}{11}$$
$$= 2\frac{8}{11}$$

48. *Familiarize*. Let c = the number of cassettes that can be placed on each shelf.

Translate. We write a division sentence.
$$c = 27 \div 1\frac{1}{8}$$

Solve. We carry out the division.
$$c = 27 \div 1\frac{1}{8} = 27 \div \frac{9}{8}$$
$$= 27 \cdot \frac{8}{9} = \frac{27 \cdot 8}{9}$$
$$= \frac{3 \cdot \cancel{9} \cdot 8}{\cancel{9} \cdot 1}$$
$$= 24$$

Check. Since $24 \cdot 1\frac{1}{8} = 24 \cdot \frac{9}{8} = \frac{24 \cdot 9}{8} = \frac{3 \cdot 8 \cdot 9}{8} = 27$, the answer checks.

State. 24 cassettes can be placed on each shelf.

49. *Familiarize*. Let p = the number of pizzas that remained.

Translate.
$$p = \frac{3}{8} + 1\frac{1}{2} + 1\frac{1}{4}$$

Solve. We carry out the addition.

$$\frac{3}{8} = \frac{3}{8}$$
$$1\boxed{\frac{1}{2} \cdot \frac{4}{4}} = 1\frac{4}{8}$$
$$+1\boxed{\frac{1}{4} \cdot \frac{2}{2}} = +1\frac{2}{8}$$
$$\overline{} \quad 2\frac{9}{8} = 2 + \frac{9}{8}$$
$$= 2 + 1\frac{1}{8}$$
$$= 3\frac{1}{8}$$

Check. We repeat the calculation. The answer checks.

State. Altogether, $3\frac{1}{8}$ pizzas remained.

50. *Familiarize*. Let d = the distance Mica traveled, in miles.

Translate.
$$d = \frac{1}{10} + \frac{1}{2}$$

Solve. We carry out the addition.
$$d = \frac{1}{10} + \frac{1}{2} = \frac{1}{10} + \frac{1}{2} \cdot \frac{5}{5} = \frac{1}{10} + \frac{5}{10} = \frac{6}{10} = \frac{\cancel{2} \cdot 3}{\cancel{2} \cdot 5} = \frac{3}{5}$$

Check. We repeat the calculation. The answer checks.

State. Mica traveled $\frac{3}{5}$ mi.

51. *Familiarize*. Let c = how many fewer feet of cable Crew B can install per hour than Crew A.

Translate. This is a "how much more" situation.

Crew B's feet per hour	plus	How many more feet per hour	is	Crew A's feet per hour
↓	↓	↓	↓	↓
$31\frac{2}{3}$	$+$	c	$=$	$38\frac{1}{8}$

Solve.
$$31\frac{2}{3} + c = 38\frac{1}{8}$$
$$31\frac{2}{3} + c - 31\frac{2}{3} = 38\frac{1}{8} - 31\frac{2}{3}$$
$$c = 38\frac{3}{24} - 31\frac{16}{24}$$
$$c = 37\frac{27}{24} - 31\frac{16}{24}$$
$$c = 6\frac{11}{24}$$

Check. Since $31\frac{2}{3} + 6\frac{11}{24} = 31\frac{16}{24} + 6\frac{11}{24} = 37\frac{27}{24} = 37 + 1\frac{3}{24} = 38\frac{3}{24} = 38\frac{1}{8}$, the answer checks.

State. Crew B can install $6\frac{11}{24}$ fewer feet per hour than Crew A.

52. *Familiarize*. Let s = the number of cups of shortening in the lower calorie cake.

Translate.

New amount of shortening	plus	Amount of prune puree	is	Original amount of shortening
↓	↓	↓	↓	↓
s	$+$	$3\frac{5}{8}$	$=$	12

Solve. We subtract $3\frac{5}{8}$ on both sides of the equation.

$$12 \quad = 11\frac{8}{8}$$
$$-3\frac{5}{8} = -3\frac{5}{8}$$
$$\overline{\qquad\qquad 8\frac{3}{8}}$$

Thus, $s = 8\frac{3}{8}$.

Check. $8\frac{3}{8} + 3\frac{5}{8} = 11\frac{8}{8} = 12$, so the answer checks.

State. The lower calorie recipe uses $8\frac{3}{8}$ cups of shortening.

53. *Familiarize*. We draw a picture.

$8\frac{1}{2}$ in.

$9\frac{3}{4}$ in. ▭ $9\frac{3}{4}$ in.

$8\frac{1}{2}$ in.

Translate. We let D = the distance around the book.

Top distance	plus	Right-side distance	plus	Bottom distance	plus	Left-side distance	is	Total distance
↓	↓	↓	↓	↓	↓	↓	↓	↓
$8\frac{1}{2}$	$+$	$9\frac{3}{4}$	$+$	$8\frac{1}{2}$	$+$	$9\frac{3}{4}$	$=$	D

Solve. To solve we carry out the addition. The LCD is 4.

$$8\boxed{\frac{1}{2}\cdot\frac{2}{2}} = \quad 8\frac{2}{4}$$
$$9\frac{3}{4} = \quad 9\frac{3}{4}$$
$$8\boxed{\frac{1}{2}\cdot\frac{2}{2}} = \quad 8\frac{2}{4}$$
$$+9\frac{3}{4} = +\ 9\frac{3}{4}$$
$$\overline{\qquad\qquad\qquad 34\frac{10}{4} = 36\frac{2}{4} = 36\frac{1}{2}}$$

Check. We repeat the calculation.

State. The distance around the book is $36\frac{1}{2}$ in.

54. *Familiarize*. Let p = the population of Louisiana.

Translate.

Population of Louisiana	is	$2\frac{1}{2}$	times	Population of West Virginia
↓	↓	↓	↓	↓
p	$=$	$2\frac{1}{2}$	$\cdot$	$1,800,000$

Solve. We carry out the multiplication.

$$p = 2\frac{1}{2} \cdot 1,800,000 = \frac{5}{2} \cdot 1,800,000$$
$$= \frac{5 \cdot 1,800,000}{2} = \frac{5 \cdot \cancel{2} \cdot 900,000}{\cancel{2} \cdot 1}$$
$$= 4,500,000$$

Check. We repeat the calculation. The answer checks.

State. The population of Louisiana is about 4,500,000.

55. We find the area of each rectangle and then add to find the total area. Recall that the area of a rectangle is length × width.

Area of rectangle A:
$$12 \times 9\frac{1}{2} = 12 \times \frac{19}{2} = \frac{12 \times 19}{2} = \frac{2 \cdot 6 \cdot 19}{2 \cdot 1} = \frac{2}{2} \cdot \frac{6 \cdot 19}{1} =$$
114 in^2

Area of rectangle B:
$$8\frac{1}{2} \times 7\frac{1}{2} = \frac{17}{2} \times \frac{15}{2} = \frac{17 \times 15}{2 \times 2} = \frac{255}{4} = 63\frac{3}{4} \text{ in}^2$$

Sum of the areas:
$$114 \text{ in}^2 + 63\frac{3}{4} \text{ in}^2 = 177\frac{3}{4} \text{ in}^2$$

56. We subtract the area of rectangle B from the area of rectangle A.

$$114 \quad = 113\frac{4}{4}$$
$$-63\frac{3}{4} = -63\frac{3}{4}$$
$$\overline{\qquad\qquad 50\frac{1}{4}}$$

The area of rectangle A is $50\frac{1}{4}$ in^2 greater than the area of rectangle B.

57. *Discussion and Writing Exercise*. The student multiplied the whole numbers and multiplied the fractions. The mixed numerals should be converted to fraction notation before multiplying.

58. *Discussion and Writing Exercise*. Yes. We may need to find a common denominator before adding or subtracting. To find the least common denominator, we use the last common multiple of the denominators.

59. We use a calculator to find the prime factorization of each number.

$141 = 3 \cdot 47$

$2419 = 41 \cdot 59$

$1357 = 23 \cdot 59$

The LCM is $3 \cdot 47 \cdot 41 \cdot 59 \cdot 23$, or $7,844,817$.

60.
$$\frac{1}{100} + \frac{1}{150} + \frac{1}{200} = \frac{1}{100} \cdot \frac{6}{6} + \frac{1}{150} \cdot \frac{4}{4} + \frac{1}{200} \cdot \frac{3}{3}$$
$$= \frac{6}{600} + \frac{4}{600} + \frac{3}{600}$$
$$= \frac{13}{600}$$

Thus, $\frac{1}{r} = \frac{13}{600}$, so r is the reciprocal of $\frac{13}{600}$, or $\frac{600}{13}$.

61. a) $\dfrac{\square}{11}$ is greater than $\dfrac{1}{2}$ when the numerator is greater than $\dfrac{1}{2}$ of the denominator. Since $\dfrac{1}{2} \cdot 11 = \dfrac{11}{2} = 5\dfrac{1}{2}$, the smallest integer numerator possible is 6.

b) $\dfrac{\square}{8}$ is greater than $\dfrac{1}{2}$ when the numerator is greater than $\dfrac{1}{2}$ of the denominator. Since $\dfrac{1}{2} \cdot 8 = 4$, the smallest integer numerator possible is 5.

c) $\dfrac{\square}{23}$ is greater than $\dfrac{1}{2}$ when the numerator is greater than $\dfrac{1}{2}$ of the denominator. Since $\dfrac{1}{2} \cdot 23 = \dfrac{23}{2} = 11\dfrac{1}{2}$, the smallest integer numerator possible is 12.

d) $\dfrac{\square}{35}$ is greater than $\dfrac{1}{2}$ when the numerator is greater than $\dfrac{1}{2}$ of the denominator. Since $\dfrac{1}{2} \cdot 35 = \dfrac{35}{2} = 17\dfrac{1}{2}$, the smallest integer numerator possible is 18.

e) $\dfrac{-51}{\square}$ is greater than $\dfrac{1}{2}$ when the denominator is greater than twice he numerator. Since $2(-51) = -102$, the smallest integer denominator possible is -101.

f) $\dfrac{-78}{\square}$ is greater than $\dfrac{1}{2}$ when the denominator is greater than twice he numerator. Since $2(-78) = -156$, the smallest integer denominator possible is -155.

g) $\dfrac{-2}{\square}$ is greater than $\dfrac{1}{2}$ when the denominator is greater than twice he numerator. Since $2(-2) = -4$, the smallest integer denominator possible is -3.

h) $\dfrac{-1}{\square}$ is greater than $\dfrac{1}{2}$ when the denominator is greater than twice he numerator. Since $2(-1) = -2$, the smallest integer denominator possible is -1.

62. a) $\dfrac{7}{\square}$ is greater than 1 when the denominator is less than the numerator. Thus, for this fraction the largest integer denominator possible is 6.

b) $\dfrac{11}{\square}$ is greater than 1 when the denominator is less than the numerator. Thus, for this fraction the largest integer denominator possible is 10.

c) $\dfrac{47}{\square}$ is greater than 1 when the denominator is less than the numerator. Thus, for this fraction the largest integer denominator possible is 46.

d) $\dfrac{9}{\square}$ is greater than 1 when the denominator is less than the numerator. Since $\dfrac{9}{8} = 1\dfrac{1}{8}$, the largest integer denominator possible is 1.

e) $\dfrac{\square}{-13}$ is greater than 1 when the numerator is less than the denominator. Thus, for this fraction the largest integer denominator possible is -14.

f) $\dfrac{\square}{-27}$ is greater than 1 when the numerator is less than the denominator. Thus, for this fraction the largest integer denominator possible is -28.

g) $\dfrac{\square}{-1}$ is greater than 1 when the numerator is less than the denominator. Thus, for this fraction the largest integer denominator possible is -2.

h) $\dfrac{\square}{-\frac{1}{2}}$ is greater than 1 when the numerator is less than the denominator. Thus, for this fraction the largest integer denominator possible is -1.

Chapter 4 Test

1. $12 = 2 \cdot 2 \cdot 3 = 2^2 \cdot 3$

$16 = 2 \cdot 2 \cdot 2 \cdot 2 = 2^4$

We form the LCM using the greatest power of each factor. The LCM is $2^4 \cdot 3$, or 48.

2. $\dfrac{1}{2} + \dfrac{5}{2} = \dfrac{1+5}{2} = \dfrac{6}{2} = 3$

3.
$$-\frac{7}{8} + \frac{2}{3}$$
$$= \frac{-7}{8} + \frac{2}{3} \qquad \text{8 and 3 have no common factors,}$$
$$\qquad\qquad\qquad \text{so the LCD is } 8 \cdot 3, \text{ or } 24.$$
$$= \frac{-7}{8} \cdot \frac{3}{3} + \frac{2}{3} \cdot \frac{8}{8}$$
$$= \frac{-21}{24} + \frac{16}{24}$$
$$= \frac{-5}{24}$$

4. $\dfrac{5}{t} - \dfrac{3}{t} = \dfrac{5-3}{t} = \dfrac{2}{t}$

5. The LCM of 6 and 4 is 12.

$$\frac{5}{6} - \frac{3}{4} = \frac{5}{6} \cdot \frac{2}{2} - \frac{3}{4} \cdot \frac{3}{3}$$

$$= \frac{10}{12} - \frac{9}{12} = \frac{1}{12}$$

6. The LCM of 8 and 24 is 24.

$$\frac{5}{8} - \frac{17}{24} = \frac{5}{8} \cdot \frac{3}{3} - \frac{17}{24}$$

$$= \frac{15}{24} - \frac{17}{24} = \frac{15 - 17}{24}$$

$$= \frac{-2}{24} = \frac{-1 \cdot \cancel{2}}{\cancel{2} \cdot 12}$$

$$= \frac{-1}{12}, \text{ or } -\frac{1}{12}$$

7.

$$x + \frac{2}{3} = \frac{11}{12}$$

$$x + \frac{2}{3} - \frac{2}{3} = \frac{11}{12} - \frac{2}{3} \quad \text{Subtracting } \frac{2}{3} \text{ on both sides}$$

$$x + 0 = \frac{11}{12} - \frac{2}{3} \cdot \frac{4}{4} \quad \text{The LCD is 12.}$$

$$x = \frac{11}{12} - \frac{8}{12} = \frac{3}{12}$$

$$x = \frac{3 \cdot 1}{3 \cdot 4} = \frac{3}{3} \cdot \frac{1}{4}$$

$$x = \frac{1}{4}$$

8.

$$-5x - 3 = 9$$

$$-5x - 3 + 3 = 9 + 3$$

$$-5x = 12$$

$$\frac{-5x}{-5} = \frac{12}{-5}$$

$$x = \frac{12}{-5}$$

The solution is $\frac{12}{-5}$, or $\frac{-12}{5}$, or $-\frac{12}{5}$.

9. The LCM of the denominators is 12.

$$\frac{3}{4} = \frac{1}{2} + \frac{5}{3}x$$

$$12 \cdot \frac{3}{4} = 12\left(\frac{1}{2} + \frac{5}{3}x\right)$$

$$\frac{36}{4} = 12 \cdot \frac{1}{2} + \frac{12 \cdot 5}{3}x$$

$$9 = \frac{12}{2} + \frac{60}{3}x$$

$$9 = 6 + 20x$$

$$9 - 6 = 6 + 20x - 6$$

$$3 = 20x$$

$$\frac{3}{20} = \frac{20x}{20}$$

$$\frac{3}{20} = x$$

The solution is $\frac{3}{20}$.

10. The LCD is 175.

$$\frac{6}{7} \cdot \frac{25}{25} = \frac{150}{175}$$

$$\frac{21}{25} \cdot \frac{7}{7} = \frac{147}{175}$$

Since $150 > 147$, it follows that $\frac{150}{175} > \frac{147}{175}$, so $\frac{6}{7} > \frac{21}{25}$.

11. $3\frac{1}{2} = \frac{7}{2}$ $\quad (3 \cdot 2 = 6, \ 6 + 1 = 7)$

12. $-9\frac{3}{8} = -\left(9 + \frac{3}{8}\right) = -\frac{75}{8}$

$(9 \cdot 8 = 72; \ 72 + 3 = 75; \text{ include the negative sign.})$

13. First consider $\frac{74}{9}$.

$$9 \overline{)\begin{array}{c} 8 \\ 7\,4 \\ 7\,2 \\ \hline 2 \end{array}} \qquad \frac{74}{9} = 8\frac{2}{9}$$

Since $\frac{74}{9} = 8\frac{2}{9}$, we have $-\frac{74}{9} = -8\frac{2}{9}$.

14.

$$11 \overline{)\begin{array}{c} 1\,6\,2 \\ 1\,7\,8\,9 \\ 1\,1\,0\,0 \\ \hline 6\,8\,9 \\ 6\,6\,0 \\ \hline 2\,9 \\ 2\,2 \\ \hline 7 \end{array}}$$

The answer is $162\frac{7}{11}$.

15.

$$\begin{array}{r} 6\frac{2}{5} \\ +7\frac{4}{5} \\ \hline 13\frac{6}{5} \end{array} = 13 + \frac{6}{5}$$

$$= 13 + 1\frac{1}{5}$$

$$= 14\frac{1}{5}$$

16. The LCD is 12.

$$\begin{array}{r} 3\ \boxed{\dfrac{1}{4} \cdot \dfrac{3}{3}} = 3\dfrac{3}{12} \\ +9\ \boxed{\dfrac{1}{6} \cdot \dfrac{2}{2}} = +9\dfrac{2}{12} \\ \hline 12\dfrac{5}{12} \end{array}$$

17. The LCD is 24.

$$10\;\boxed{\frac{1}{6}\cdot\frac{4}{4}} = 10\,\frac{4}{24} = \;9\,\frac{28}{24}$$

$$-5\;\boxed{\frac{7}{8}\cdot\frac{3}{3}} = -5\,\frac{21}{24} = -5\,\frac{21}{24}$$

$$\phantom{-5\;\boxed{\frac{7}{8}\cdot\frac{3}{3}} = -5\,\frac{21}{24} = }\;4\,\frac{7}{24}$$

$\left(\text{Since } \dfrac{4}{24} \text{ is smaller than } \dfrac{21}{24}, \text{ we cannot subtract until we}\right.$
borrow: $10\dfrac{4}{24} = 9 + \dfrac{24}{24} + \dfrac{4}{24} = 9 + \dfrac{28}{24} = 9\dfrac{28}{24}.\Big)$

18. $14 + \left(-5\dfrac{3}{7}\right) = 13\dfrac{7}{7} + \left(-5\dfrac{3}{7}\right) = 8\dfrac{4}{7}$

19. $3\dfrac{4}{5} - 9\dfrac{1}{2} = 3\dfrac{4}{5} + \left(-9\dfrac{1}{2}\right)$

Since $-9\dfrac{1}{2}$ has the larger absolute value, the answer will be negative. We find the difference in absolute values.

$$9\;\boxed{\frac{1}{2}\cdot\frac{5}{5}} = \;9\,\frac{5}{10} = \;8\,\frac{15}{10}$$

$$-3\;\boxed{\frac{4}{5}\cdot\frac{2}{2}} = -3\,\frac{8}{10} = -3\,\frac{8}{10}$$

$$\phantom{-3\;\boxed{\frac{4}{5}\cdot\frac{2}{2}} = -3\,\frac{8}{10} = }\;5\,\frac{7}{10}$$

Thus, $3\dfrac{4}{5} - 9\dfrac{1}{2} = -5\dfrac{7}{10}$.

20. $\dfrac{3}{8}x - \dfrac{1}{2}x = \dfrac{3}{8}x - \dfrac{1}{2}\cdot\dfrac{4}{4}\cdot x$

$$= \dfrac{3}{8}x - \dfrac{4}{8}x$$

$$= -\dfrac{1}{8}x$$

21. $5\dfrac{2}{11}a - 3\dfrac{1}{5}a = \left(5\dfrac{2}{11} - 3\dfrac{1}{5}\right)a$

$$= \left(5\dfrac{10}{55} - 3\dfrac{11}{55}\right)a$$

$$= \left(4\dfrac{65}{55} - 3\dfrac{11}{55}\right)a$$

$$= 1\dfrac{54}{55}a$$

22. $9\cdot 4\dfrac{1}{3} = 9\cdot\dfrac{13}{3} = \dfrac{9\cdot 13}{3} = \dfrac{3\cdot 3\cdot 13}{3\cdot 1} = \dfrac{3}{3}\cdot\dfrac{3\cdot 13}{1} = 39$

23. $6\dfrac{3}{4}\cdot\left(-2\dfrac{2}{3}\right) = \dfrac{27}{4}\cdot\left(-\dfrac{8}{3}\right) = -\dfrac{27\cdot 8}{4\cdot 3} = -\dfrac{3\cdot 9\cdot 2\cdot 4}{4\cdot 3\cdot 1} =$

$$-\dfrac{9\cdot 2}{1}\cdot\dfrac{3\cdot 4}{3\cdot 4} = -18$$

24. $33 \div 5\dfrac{1}{2} = 33 \div \dfrac{11}{2} = 33\cdot\dfrac{2}{11} = \dfrac{33\cdot 2}{11} = \dfrac{3\cdot 11\cdot 2}{11\cdot 1} =$

$$\dfrac{11}{11}\cdot\dfrac{3\cdot 2}{1} = 6$$

25. $2\dfrac{1}{3} \div 1\dfrac{1}{6} = \dfrac{7}{3} \div \dfrac{7}{6} = \dfrac{7}{3}\cdot\dfrac{6}{7} = \dfrac{7\cdot 6}{3\cdot 7} = \dfrac{7\cdot 2\cdot 3}{3\cdot 7\cdot 1} =$

$$\dfrac{7\cdot 3}{7\cdot 3}\cdot\dfrac{2}{1} = 2$$

26. $\dfrac{2}{3}ab = \dfrac{2}{3}\cdot 7\cdot 4\dfrac{1}{5} = \dfrac{2\cdot 7}{3}\cdot 4\dfrac{1}{5} = \dfrac{2\cdot 7}{3}\cdot\dfrac{21}{5} = \dfrac{2\cdot 7\cdot 21}{3\cdot 5} =$

$$\dfrac{2\cdot 7\cdot 3\cdot 7}{3\cdot 5} = \dfrac{98}{5}, \text{ or } 19\dfrac{3}{5}$$

27. $4 + mn = 4 + 7\dfrac{2}{5}\cdot 3\dfrac{1}{4}$

$$= 4 + \dfrac{37}{5}\cdot\dfrac{13}{4}$$

$$= 4 + \dfrac{481}{20}$$

$$= 4 + 24\dfrac{1}{20}$$

$$= 28\dfrac{1}{20}$$

28. *Familiarize.* Let t = the number of pounds of turkey required for 5 batches of chili.

Translate. We write a multiplication sentence.

$$t = 5\cdot 1\dfrac{1}{2}$$

Solve. We carry out the multiplication.

$$t = 5\cdot 1\dfrac{1}{2} = 5\cdot\dfrac{3}{2} = \dfrac{15}{2} = 7\dfrac{1}{2}$$

Check. We can repeat the calculation. The answer checks.

State. $7\dfrac{1}{2}$ lb of turkey is required for 5 batches of chili.

29. *Familiarize.* Let b = the number of books in the order.

Translate.

Weight per book	times	Number of books	is	Total weight
↓	↓	↓	↓	↓
$2\dfrac{3}{4}$	$\cdot$	b	$=$	220

Solve

$$2\dfrac{3}{4}\cdot b = 220$$

$$\dfrac{11}{4}\cdot b = 220$$

$$\dfrac{4}{11}\cdot\dfrac{11}{4}\cdot b = \dfrac{4}{11}\cdot 220$$

$$b = \dfrac{4\cdot 220}{11} = \dfrac{4\cdot \cancel{11}\cdot 20}{\cancel{11}\cdot 1}$$

$$b = 80$$

Check. Since $2\dfrac{3}{4}\cdot 80 = \dfrac{11}{4}\cdot 80 = \dfrac{880}{4} = 220$, the answer checks.

State. There are 80 books in the order.

30. *Familiarize.* We add the three lengths across the top to find a and the three lengths across the bottom to find b.

Translate.

$$a = 1\frac{1}{8} + \frac{3}{4} + 1\frac{1}{8}$$

$$b = \frac{3}{4} + 3 + \frac{3}{4}$$

Solve. We carry out the additions.

$$a = 1\frac{1}{8} + \frac{6}{8} + 1\frac{1}{8} = 2\frac{8}{8} = 2 + 1 = 3$$

$$b = \frac{3}{4} + 3 + \frac{3}{4} = 3\frac{6}{4} = 3 + 1\frac{2}{4} = 3 + 1\frac{1}{2} = 4\frac{1}{2}$$

Check. We can repeat the calculations. The answer checks.

State. a) The short length a across the top is 3 in.

b) The length b across the bottom is $4\frac{1}{2}$ in.

31. *Familiarize.* Let t = the number of inches by which $\frac{3}{4}$ in. exceeds the actual thickness of the plywood.

Translate.

$$\underbrace{\text{Actual thickness}}_{} \ \text{plus} \ \underbrace{\text{Excess thickness}}_{} \ \text{is} \ \overbrace{\frac{3}{4}}^{} \ \text{in.}$$

$$\underset{\downarrow}{\frac{11}{16}} \quad \underset{\downarrow}{+} \quad \underset{\downarrow}{t} \quad \underset{\downarrow}{=} \quad \underset{\downarrow}{\frac{3}{4}}$$

Solve. We will subtract $\frac{11}{16}$ on both sides of the equation.

$$\frac{11}{16} + t = \frac{3}{4}$$

$$\frac{11}{16} + t - \frac{11}{16} = \frac{3}{4} - \frac{11}{16}$$

$$t = \frac{3}{4} \cdot \frac{4}{4} - \frac{11}{16}$$

$$t = \frac{12}{16} - \frac{11}{16}$$

$$t = \frac{1}{16}$$

Check. Since $\frac{11}{16} + \frac{1}{16} = \frac{12}{16} = \frac{3}{4}$, the answer checks.

State. A $\frac{3}{4}$-in. piece of plywood is actually $\frac{1}{16}$ in. thinner than its name indicates.

32. We add the heights and divide by the number of addends.

$$\frac{6\frac{5}{12} + 5\frac{11}{12} + 6\frac{7}{12}}{3} = \frac{17\frac{23}{12}}{3} = \frac{17 + 1\frac{11}{12}}{3} =$$

$$\frac{18\frac{11}{12}}{3} = \frac{227}{12} \div 3 = \frac{227}{12} \cdot \frac{1}{3} = \frac{227}{36} = 6\frac{11}{36}$$

The women's average height is $6\frac{11}{36}$ ft.

33. The length of the act can be expressed as a fraction with a denominator of 25 and a numerator that is the LCM of 6 and 8.

$$6 = 2 \cdot 3$$
$$8 = 2 \cdot 2 \cdot 2 = 2^3$$

The LCM is $2^3 \cdot 3$, or 24, so the act lasts $\frac{24}{25}$ min.

34. *Familiarize.* First compare $\frac{1}{7}$ mi and $\frac{1}{8}$ mi. The LCD is 56.

$$\frac{1}{7} = \frac{1}{7} \cdot \frac{8}{8} = \frac{8}{56}$$

$$\frac{1}{8} = \frac{1}{8} \cdot \frac{7}{7} = \frac{7}{56}$$

Since $8 > 7$, then $\frac{8}{56} > \frac{7}{56}$ so $\frac{1}{7} > \frac{1}{8}$.

This tells us that Cheri walks farther than Trent.

Next we will find how much farther Cheri walks on each lap and then multiply by 17 to find how much farther she walks in 17 laps. Let d represent how much farther Cheri walks on each lap, in miles.

Translate. An equation that fits this situation is

$$\frac{1}{8} + d = \frac{1}{7}, \ \text{or} \ \frac{7}{56} + d = \frac{8}{56}$$

Solve.

$$\frac{7}{56} + d = \frac{8}{56}$$

$$\frac{7}{56} + d - \frac{7}{56} = \frac{8}{56} - \frac{7}{56}$$

$$d = \frac{1}{56}$$

Now we multiply: $17 \cdot \frac{1}{56} = \frac{17}{56}$.

Check. We can think of the problem in a different way.

In 17 laps Cheri walks $17 \cdot \frac{1}{7}$, or $\frac{17}{7}$ mi, and Trent walks $17 \cdot \frac{1}{8}$, or $\frac{17}{8}$ mi. Then $\frac{17}{7} - \frac{17}{8} = \frac{17}{7} \cdot \frac{8}{8} - \frac{17}{8} \cdot \frac{7}{7} = \frac{136}{56} - \frac{119}{56} = \frac{17}{56}$, so Cheri walks $\frac{17}{56}$ mi farther and our answer checks.

State. Cheri walks $\frac{17}{56}$ mi farther than Trent.

35. a) We find some common multiples of 8 and 6.

Multiples of 8: $8, 16, 24, 32, 40, 48, 56, 64, 72, \ldots$

Multiples of 6: $6, 12, 18, 24, 30, 36, 42, 48, 54, 60, 66, 72, \ldots$

Some common multiples are 24, 48, and 72. These are some class sizes for which study groups of 8 students or of 6 students can be organized with no students left out.

b) The smallest such class size is the least common multiple, 24.

36. a) $\frac{1}{1 \cdot 2} = \frac{1}{2}$

b) $\frac{1}{1 \cdot 2} + \frac{1}{2 \cdot 3} = \frac{1}{2} + \frac{1}{6} = \frac{1}{2} \cdot \frac{3}{3} + \frac{1}{6} = \frac{3}{6} + \frac{1}{6} = \frac{4}{6} = \frac{2 \cdot 2}{2 \cdot 3} = \frac{2}{2} \cdot \frac{2}{3} = \frac{2}{3}$

c) $\frac{1}{1 \cdot 2} + \frac{1}{2 \cdot 3} + \frac{1}{3 \cdot 4} = \left(\frac{1}{1 \cdot 2} + \frac{1}{2 \cdot 3} \right) + \frac{1}{3 \cdot 4} = \frac{2}{3} + \frac{1}{12} = \frac{2}{3} \cdot \frac{4}{4} + \frac{1}{12} = \frac{8}{12} + \frac{1}{12} = \frac{9}{12} = \frac{3 \cdot 3}{3 \cdot 4} =$

$$\frac{3}{3} \cdot \frac{3}{4} = \frac{3}{4}$$

d) $\dfrac{1}{1 \cdot 2} + \dfrac{1}{2 \cdot 3} + \dfrac{1}{3 \cdot 4} + \dfrac{1}{4 \cdot 5} =$

$\left(\dfrac{1}{1 \cdot 2} + \dfrac{1}{2 \cdot 3} + \dfrac{1}{3 \cdot 4} \right) + \dfrac{1}{4 \cdot 5} = \dfrac{3}{4} + \dfrac{1}{20} = \dfrac{3}{4} \cdot \dfrac{5}{5} + \dfrac{1}{20} =$

$\dfrac{15}{20} + \dfrac{1}{20} = \dfrac{16}{20} = \dfrac{4 \cdot 4}{4 \cdot 5} = \dfrac{4}{4} \cdot \dfrac{4}{5} = \dfrac{4}{5}$

e) In each case, the sum is a fraction in which the numerator is the smaller factor in the denominator of the last addend and the denominator is the larger factor in the denominator of the last addend. Then the given sum is $\dfrac{9}{10}$.

Chapter 5

Decimal Notation

---

1. 63.05

  a) Write a word name for

    the whole number. | Sixty-three |

  b) Write "and" for the   Sixty-three

    decimal point. | and |

  c) Write a word name for

    the number to the right   Sixty-three

    of the decimal point,   and

    followed by the place | five hundredths |

    value of the last digit.

A word name for 63.05 is sixty-three and five hundredths.

3. 26.59

  a) Write a word name for

    the whole number. | Twenty-six |

  b) Write "and" for the   Twenty-six

    decimal point. | and |

  c) Write a word name for

    the number to the right   Twenty-six

    of the decimal point,   and

    followed by the place | fifty-nine |
    value of the last digit. | hundredths |

A word name for 26.59 is twenty-six and fifty-nine hundredths.

5. A word name for 8.35 is eight and thirty-five hundredths.

7. A word name for 24.6875 is twenty-four and six thousand eight hundred seventy-five ten-thousandths.

9. 5.63

  a) Write a word name for

    the whole number. | Five |

  b) Write "and" for the   Five

    decimal point. | and |

  c) Write a word name for

    the number to the right   Five

    of the decimal point,   and

    followed by the place | sixty-three |
    value of the last digit. | hundredths |

A word name for 5.63 is five and sixty-three hundredths.

11. Write "and 95 cents" as "and $\dfrac{95}{100}$ dollars." A word name for \$524.95 is five hundred twenty-four and $\dfrac{95}{100}$ dollars.

13. Write "and 72 cents" as "and $\dfrac{72}{100}$ dollars." A word name for \$36.72 is thirty-six and $\dfrac{72}{100}$ dollars.

15. 7.3   7.3. $\dfrac{73}{10}$

          └↑

     1 place   1 zero

$7.3 = \dfrac{73}{10}$

To write 7.3 as a mixed numeral, we rewrite the whole number part and express the rest in fraction form.

$$7.3 = 7\frac{3}{10}$$

17. 203.6   203.6. $\dfrac{2036}{10}$

            └↑

     1 place   1 zero

$203.6 = \dfrac{2036}{10}$

To write 203.6 as a mixed numeral, we rewrite the whole number part and express the rest in fraction form.

$$203.6 = 203\frac{6}{10}$$

19. −2.703   −2.703. $\dfrac{-2073}{1000}$

              └──↑

     3 places   3 zeros

$-2.703 = \dfrac{-2073}{1000}$, or $-\dfrac{2703}{1000}$

To write −2.703 as a mixed numeral, we rewrite the whole number part and express the rest in fraction form.

$$-2.703 = -2\frac{703}{1000}$$

21. 0.0109   0.0109. $\dfrac{109}{10,000}$

              └───↑

     4 places   4 zeros

$0.0109 = \dfrac{109}{10,000}$

Since the whole number part of 0.0109 is zero, we cannot express this number as a mixed numeral.

23. -4.0003 $-4.0003.$ $\dfrac{-40,003}{10,000}$

4 places   4 zeros

$$-4.0003 = -\dfrac{40,003}{10,000}$$

To write -4.0003 as a mixed numeral, we rewrite the whole number part and express the rest in fraction form.

$$-4.0003 = -4\dfrac{3}{10,000}$$

25. -0.0207 $-0.0207.$ $-\dfrac{207}{10,000}$

4 places   4 zeros

$$-0.0207 = -\dfrac{207}{10,000}$$

Since the whole number part of -0.0207 is zero, we cannot express this number as a mixed numeral.

27. 70.00105 $70.00105.$ $\dfrac{7,000,105}{100,000}$

5 places   5 zeros

$$70.00105 = -\dfrac{7,000,105}{100,000}$$

To write $70,00105$ as a mixed numeral, we rewrite the whole number part and express the rest in fraction form.

$$70.00105 = 70\dfrac{105}{100,000}$$

29. $\dfrac{3}{10}$ $0.3.$

1 zero   Move 1 place.

$$\dfrac{3}{10} = 0.3$$

31. $-\dfrac{59}{100}$ $-0.59.$

2 zeros   Move 2 places.

$$-\dfrac{59}{100} = -0.59$$

33. $\dfrac{3798}{1000}$ $3.798.$

3 zeros   Move 3 places.

$$\dfrac{3798}{1000} = 3.798$$

35. $\dfrac{78}{10,000}$ $0.0078.$

4 zeros   Move 4 places.

$$\dfrac{78}{10,000} = 0.0078$$

37. $\dfrac{-18}{100,000}$ $-0.00018.$

5 zeros   Move 5 places.

$$\dfrac{-18}{100,000} = -0.00018$$

39. $\dfrac{486,197}{1,000,000}$ $0.486197.$

6 zeros   Move 6 places.

$$\dfrac{486,197}{1,000,000} = 0.486197$$

41. $7\dfrac{13}{1000} = 7 + \dfrac{13}{1000} = 7 \text{ and } \dfrac{13}{1000} = 7.013$

43. $-8\dfrac{431}{1000}$

First consider $8\dfrac{431}{1000}$:

$$8\dfrac{431}{1000} = 8 + \dfrac{431}{1000} = 8 \text{ and } \dfrac{431}{1000} = 8.431$$

Since $8\dfrac{431}{1000} = 8.431$, we have $-8\dfrac{431}{1000} = -8.431$.

45. $2\dfrac{1739}{10,000} = 2 + \dfrac{1739}{10,000} = 2 \text{ and } \dfrac{1739}{10,000} = 2.1739$

47. $8\dfrac{953,073}{1,000,000} = 8 + \dfrac{953,073}{1,000,000} =$

$8 \text{ and } \dfrac{953,073}{1,000,000} = 8.953073$

49. To compare two numbers in decimal notation, start at the left and compare corresponding digits moving from left to right. When two digits differ, the number with the larger digit is the larger of the two numbers.

0.06

$\downarrow$ Different; 5 is larger than 0.

0.58

Thus, 0.58 is larger.

51. 0.403

$\updownarrow$ Starting at the left, these digits are the first to differ; 1 is larger than 0.

0.410

Thus, 0.410 is larger.

53. -5.046

$\updownarrow$ Starting at the left, these digits are the first to differ, and 3 is smaller than 6.

-5.043

Thus, -5.043 is larger.

55. 234.07

$\updownarrow$ Starting at the left, these digits are the first to differ, and 5 is larger than 4.

235.07

Thus, 235.07 is larger.

57. $\frac{7}{100} = 0.07$ so we compare 0.007 and 0.07.

0.007

↑ Starting at the left, these digits are the first to differ, and 7 is larger than 0.

0.07

Thus, 0.07 or $\frac{7}{100}$ is larger.

59. −0.872

↑ Starting at the left, these digits are the first to differ, and 2 is smaller than 3.

−0.873

Thus, −0.872 is larger.

61. 0.2│3│ Hundredths digit is 4 or lower. Round down.

0.2

63. −0.3│7│2 Hundredths digit is 5 or higher. Round from −0.372 to −0.4.

−0.4

65. 2.9│5│1 Hundredths digit is 5 or higher. Round up.

3.0

(When we make the tenths digit a 10, we carry 1 to the ones place.)

67. −327.2│3│47 Hundredths digit is 4 or lower. Round from −327.2347 to −327.2.

−327.2

69. 0.89│3│ Thousandths digit is 4 or lower. Round down.

0.89

71. −0.66│6│6 Thousandths digit is 5 or higher. Round from −0.6666 to −0.67.

−0.67

73. 0.99│5│2 Thousandths digit is 5 or higher. Round up.

1.00

(When we make the hundredths digit a 10, we carry 1 to the tenths place. This then requires us to carry 1 to the ones place.)

75. −0.03│4│88 Thousandths digit is 4 or lower. Round from −0.03488 to −0.03.

−0.03

77. 0.572│4│ Ten-thousandths digit is 4 or lower. Round down.

0.572

79. 17.001│5│ Ten-thousandths digit is 5 or higher. Round up.

17.002

81. −20.202│0│2 Ten-thousandths digit is 4 or lower. Round −20.20202 to −20.202.

−20.202

83. 9.984│8│ Ten-thousandths digit is 5 or higher. Round up.

9.985

85. 809.4│7│321 Hundredths digit is 5 or higher. Round up.

809.5

87. 809.47│3│21 Thousandths digit is 4 or lower. Round down.

809.47

89. Discussion and Writing Exercise

91.
$$\begin{array}{r} \overset{1\ \ 1}{6\ 8\ 1} \\ +\ 1\ 4\ 9 \\ \hline 8\ 3\ 0 \end{array}$$

93.
$$\begin{array}{r} \overset{1\ 16}{2\ \cancel{6}\ 7} \\ -\ \ \ 8\ 5 \\ \hline 1\ 8\ 2 \end{array}$$

95. $\frac{37}{55} - \frac{49}{55} = \frac{37-49}{55} = \frac{-12}{55}$, or $-\frac{12}{55}$

97.
$$\begin{array}{r} \overset{\ \ \ \ \ \ \ 18}{\ \ \ \ 3\ \overset{8}{\cancel{9}}\ 9\ 13} \\ 3\ 4,\cancel{9}\ \cancel{0}\ \cancel{3} \\ -\ \ \ 1\ 9\ 4\ 5 \\ \hline 3\ 2,\ 9\ 5\ 8 \end{array}$$

99. Discussion and Writing Exercise

101. All of the numbers except −1.09 have 3 decimal places, so we add a zero to −1.09 so that each number has 3 decimal places. Then we start at the left and compare corresponding digits moving from left to right. Keep in mind that, when we compare two negative numbers and when digits differ, the number with the smaller digit is the larger of the two numbers. The numbers, listed from smallest to largest, are −1.09, −1.009, −0.989, −0.898, and −0.098.

103. 6.78346│123│ ←Drop all decimal places past the fifth place.

The answer is 6.78346.

105. 99.99999 9999 ←Drop all decimal places
past the fifth place.

The answer is 99.99999.

107. From the graph we see that the years for which the vertical bars lie above +0.4° are 1983, 1988, 1990, 1991, 1992, 1995, 1997, 1998, 1999, 2000, 2001, 2002, 2003, and 2004.

109. From the graph we see that the last year for which the corresponding vertical bar line below 0.0° is 1985.

Exercise Set 5.2

1.
```
      1
   4 2 6.2 5      Add hundredths.
 +   3 8.1 2      Add tenths.
 ───────────      Write a decimal point in the answer.
   4 6 4.3 7      Add ones.
                  Add tens.
                  Add hundreds.
```

3.
```
        1 1
   6 5 9.4 0 3      Add thousandths.
 + 9 1 6.8 1 2      Add hundredths.
 ─────────────      Add tenths.
 1 5 7 6.2 1 5      Write a decimal point in the answer.
                    Add ones.
                    Add tens.
                    Add hundreds.
```

5.
```
       1     1
       9.1 0 4
 + 1 2 3.4 5 6
 ─────────────
 1 3 2.5 6 0
```

7. Line up the decimal points.
```
       1
   2.0 0 6
 + 5.8 1 7
 ─────────
   7.8 2 3
```

9. Line up the decimal points.
```
   2 0 0.1 2 4
 + 3 0 0.1 2 4
 ─────────────
   5 0 0.2 4 8
```

11. Line up the decimal points.
```
   0.8 3 0      Writing an extra zero
 + 0.0 0 5
 ─────────
   0.8 3 5
```

13. Line up the decimal points.
```
   1
     0.3 4 0      Writing an extra zero
     3.5 0 0      Writing 2 extra zeros
     0.1 2 7
 + 7 6 8.0 0 0    Writing in the decimal point
 ─────────────    and 3 extra zeros
   7 7 1.9 6 7    Adding
```

15.
```
      1   1 1
   1 7.0 0 0 0      Writing in the decimal point.
    3.2 4 0 0       You may find it helpful to
    0.2 5 6 0       write extra zeros.
 +  0.3 6 8 9
 ─────────────
   2 0.8 6 4 9
```

17.
```
      1 2 1     1
       2.7 0 3 0
      7 8.3 3 0 0
      2 8.0 0 0 9
 + 1 1 8.4 3 4 1
 ───────────────
   2 2 7.4 6 8 0
```

19.
```
   4 7.5 9 6      Subtract thousandths.
 −    6.2 1 5     Subtract hundredths.
 ───────────      Subtract tenths.
   4 1.3 8 1      Write a decimal point in the answer.
                  Subtract ones.
                  Subtract tens.
```

21.
```
   4 11 2 11
   5̶ 1̶.3̶ 1̶      Borrow tenths to subtract hundredths.
 −     2.2 9     Subtract hundredths.
 ───────────     Subtract tenths.
   4 9.0 2       Write a decimal point in the answer.
                 Borrow tens to subtract ones.
                 Subtract ones.
                 Subtract tens.
```

23.
```
        5 9 10
   3.6̶ 0̶ 0̶      Writing 2 extra zeros
 − 0.0 3 6
 ─────────
   3.5 6 4
```

25.
```
           11
      8 1̶ 13
   9̶ 2̶.3̶ 4 1
 −    6.4 2
 ───────────
   8 5.9 2 1
```

27.
```
      2 9 10 6 14
   3.0̶ 0̶ 7̶ 4̶
 − 1.3 4 0 8
 ───────────
   1.6 6 6 6
```

29.
```
        6 9 10
   6.0 7̶ 0̶ 0̶      Writing 2 extra zeros
 − 2.0 0 7 8
 ───────────
   4.0 6 2 2
```

31. Line up the decimal points. Write an extra zero if desired.
```
         11 13
    2 9 1̶ 3̶ 10
   3̶.0̶ 2̶ 4̶ 0̶
 −     0.2 4 1
 ─────────────
   2 9.9 9 9
```

33.
```
      3 10
   3 4.0̶ 7
 − 3 0.7
 ─────────
    3.3 7
```

35.
```
      4 10
   8.4 5̶ 0̶
 − 7.4 0 5
 ─────────
   1.0 4 5
```

37.
$$\begin{array}{r} \overset{5\ \ 10}{\cancel{6}.\cancel{0}\ 0\ 3} \\ -\ 2.\ 3 \\ \hline 3.7\ 0\ 3 \end{array}$$

39.
$$\begin{array}{r} \overset{1\ \ 9\ \ 9\ \ 9\ \ 10}{2.\cancel{0}\ \cancel{0}\ \cancel{0}\ \cancel{0}} \\ -1.0\ 9\ 0\ 8 \\ \hline 0.9\ 0\ 9\ 2 \end{array}$$
Writing in the decimal point and 4 extra zeros
Subtracting

41.
$$\begin{array}{r} \overset{\ \ \ \ \ \ \ \ \ 13}{\overset{1\ \ \cancel{3}\ \ 9\ \ 10}{6\ \cancel{2}\ \cancel{4}.\cancel{0}\ \cancel{0}}} \\ -\ \ \ 1\ 8.\ 7\ 9 \\ \hline 6\ 0\ 5.\ 2\ 1 \end{array}$$

43.
$$\begin{array}{r} 5\ 7.\ 8\ 0\ 3 \\ -\ \ \ 4.\ 6 \\ \hline 5\ 3.\ 2\ 0\ 3 \end{array}$$

45.
$$\begin{array}{r} \overset{\ \ \ \ \ \ \ 6\ \ 10}{2\ 6\ 3.\ \cancel{7}\ \cancel{0}} \\ -1\ 0\ 2.\ 0\ 8 \\ \hline 1\ 6\ 1.\ 6\ 2 \end{array}$$

47.
$$\begin{array}{r} \overset{4\ \ 9\ \ 9\ \ 10}{4\ \cancel{5}.\cancel{0}\ \cancel{0}\ \cancel{0}} \\ -\ \ \ 0.\ 9\ 9\ 9 \\ \hline 4\ 4.\ 0\ 0\ 1 \end{array}$$

49. $-5.02 + 1.73$ A positive and a negative number

a) $|-5.02| = 5.02$, $|1.73| = 1.73$, and $|-5.02| > |-1.73|$, so the answer is negative.

b)
$$\begin{array}{r} \overset{4\ \ 9\ \ 12}{\cancel{5}.\cancel{0}\ \cancel{2}} \\ -\ 1.\ 7\ 3 \\ \hline 3.\ 2\ 9 \end{array}$$
Find the difference in the absolute values.

c) $-5.02 + 1.73 = -3.29$

51. $12.9 - 15.4 = 12.9 + (-15.4)$
We add the opposite of 15.4. We have a positive and a negative number.

a) $|12.9| = 12.9$, $|-15.4| = 15.4$, and $|15.4| > |12.9|$, so the answer is negative.

b)
$$\begin{array}{r} \overset{4\ \ 14}{1\ \cancel{5}.\cancel{4}} \\ -\ 1\ 2.\ 9 \\ \hline 2.\ 5 \end{array}$$
Finding the difference in the absolute values

c) $12.9 - 15.4 = -2.5$

53. $-2.9 + (-4.3)$ Two negative numbers

a)
$$\begin{array}{r} \overset{1}{2}\ .9 \\ +\ 4\ .3 \\ \hline 7\ .2 \end{array}$$
Adding the absolute values

b) $-2.9 + (-4.3) = -7.2$ The sum of two negative numbers is negative.

55. $-4.301 + 7.68$ A negative and a positive number

a) $|-4.301| = 4.301$, $|7.68| = 7.68$, and $|7.68| > |-4.301|$, so the answer is positive.

b)
$$\begin{array}{r} \overset{\ \ \ \ 7\ \ 10}{7.6\ \cancel{8}\ \cancel{0}} \\ -\ 4.3\ 0\ 1 \\ \hline 3.3\ 7\ 9 \end{array}$$
Finding the difference in the absolute values

c) $-4.301 + 7.68 = 3.379$

57.
$-12.9 - 3.7$
$= -12.9 + (-3.7)$ Adding the opposite of 3.7
$= -16.6$ The sum of two negatives is negative.

59.
$-2.1 - (-4.6)$
$= -2.1 + 4.6$ Adding the opposite of -4.6
$= 2.5$ Subtracting absolute values. Since 4.6 has the larger absolute value, the answer is positive.

61.
$14.301 + (-17.82)$
$= -3.519$ Subtracting absolute values. Since -17.82 has the larger absolute value, the answer is negative.

63.
$7.201 - (-2.4)$
$= 7.201 + 2.4$ Adding the opposite of -2.4
$= 9.601$ Adding

65.
$96.9 + (-21.4)$
$= 75.5$ Subtracting absolute values. Since 96.9 has the larger absolute value, the answer is positive.

67.
$-8.9 - (-12.7)$
$= -8.9 + 12.7$ Adding the opposite of -12.7
$= 3.8$ Subtracting absolute values. Since 12.7 has the larger absolute value, the answer is positive.

69.
$-4.9 - 5.392$
$= -4.9 + (-5.392)$ Adding the opposite of 5.392
$= -10.292$ The sum of two negatives is negative.

71.
$14.7 - 23.5$
$= 14.7 + (-23.5)$ Adding the opposite of 23.5
$= -8.8$ Subtracting absolute values. Since -23.5 has the larger absolute value, the answer is negative.

73.
$1.8x + 3.9x$
$= (1.8 + 3.9)x$ Using the distributive law
$= 5.7x$ Adding

75.
$17.59a - 12.73a$
$= (17.59 - 12.73)a$
$= 4.86a$

77.
$15.2t + 7.9 + 5.9t$
$= 15.2t + 5.9t + 7.9$ Using the commutative law
$= (15.2 + 5.9)t + 7.9$ Using the distributive law
$= 21.1t + 7.9$

79. $5.217x - 8.134x$

= $(5.217 - 8.134)x$ Using the distributive law

= $(5.217 + (-8.134))x$ Adding the opposite of 8.134

= $-2.917x$ Subtracting absolute values. The coefficient is negative since -8.134 has the larger absolute value.

81. $4.906y - 7.1 + 3.2y$

= $4.906y + 3.2y - 7.1$

= $(4.906 + 3.2)y - 7.1$

= $8.106y - 7.1$

83. $4.8x + 1.9y - 5.7x + 1.2y$

= $4.8x + 1.9y + (-5.7x) + 1.2y$ Rewriting as addition

= $4.8x + (-5.7x) + 1.9y + 1.2y$ Using the commutative law

= $(4.8 + (-5.7))x + (1.9 + 1.2)y$

= $-0.9x + 3.1y$

85. $4.9 - 3.9t + 2.3 - 4.5t$

= $4.9 + (-3.9t) + 2.3 + (-4.5t)$

= $4.9 + 2.3 + (-3.9t) + (-4.5t)$

= $(4.9 + 2.3) + (-3.9 + (-4.5))t$

= $7.2 + (-8.4t)$

= $7.2 - 8.4t$

87. Discussion and Writing Exercise

89. $\dfrac{3}{5} \cdot \dfrac{4}{7} = \dfrac{3 \cdot 4}{5 \cdot 7} = \dfrac{12}{35}$

91. $\dfrac{3}{10} \cdot \dfrac{21}{100} = \dfrac{3 \cdot 21}{10 \cdot 100} = \dfrac{63}{1000}$

93. $5 - 3x^2$

= $5 - 3 \cdot 2^2$ Substituting 2 for x

= $5 - 3 \cdot 4$

= $5 - 12$

= -7

95. Discussion and Writing Exercise

97. $-3.928 - 4.39a + 7.4b - 8.073 + 2.0001a - 9.931b - 9.8799a + 12.897b$

= $-3.928 - 8.073 - 4.39a + 2.0001a - 9.8799a + 7.4b - 9.931b + 12.897b$

= $-12.001 - 12.2698a + 10.366b$

99. $39.123a - 42.458b - 72.457a + 31.462b - 59.491 + 37.927a$

= $39.123a - 72.457a + 37.927a - 42.458b + 31.462b - 59.491$

= $4.593a - 10.996b - 59.491$

101. First "undo" the incorrect subtraction by adding 349.2 to the incorrect answer:

$$-836.9 + 349.2 = -487.7$$

Now add 349.2 to -487.7:

$$-487.7 + 349.2 = -138.5$$

The correct answer is -138.5.

103.
$$\begin{array}{r} \overset{\scriptstyle 3\ 13}{9\,3.\,a\,\cancel{4}\,\cancel{3}} \\ -8\,7.\,9\,6\,9 \\ \hline 5.\,2\,7\,4 \end{array}$$

We need to borrow 1 tenth in order to subtract hundredths. Then when we subtract 9 tenths from $(a-1)$ tenths we get 2. Since $11 - 9 = 2$, we know that we had to borrow a one in order to subtract tenths and this was added to 1 tenth, or $a - 1$. Then if $a - 1 = 1$, we know that $a = 2$.

Exercise Set 5.3

1.
$$\begin{array}{r} 6.\,8 \\ \times \quad 7 \\ \hline 4\,7.\,6 \end{array}$$
(1 decimal place)
(0 decimal places)
(1 decimal place)

3.
$$\begin{array}{r} 0.\,8\,4 \\ \times \quad\quad 8 \\ \hline 6.\,7\,2 \end{array}$$
(2 decimal places)
(0 decimal places)
(2 decimal places)

5.
$$\begin{array}{r} 6.\,3 \\ \times\,0.\,0\,4 \\ \hline 0.\,2\,5\,2 \end{array}$$
(1 decimal place)
(2 decimal places)
(3 decimal places)

7.
$$\begin{array}{r} 2\,8.\,6 \\ \times\,0.\,0\,9 \\ \hline 2.\,5\,7\,4 \end{array}$$
(1 decimal place)
(2 decimal places)

9. 10×42.63 $42.6.3$

1 zero   Move 1 place to the right.

$10 \times 42.63 = 426.3$

11. $-1000 \times 783.686852 = -(1000 \times 783.686852)$

$-(1000 \times 783.686852)$ $-783.686.852$

3 zeros   Move 3 places to the right.

$-1000 \times 783.686852 = -783,686.852$

13. -7.8×100 $-7.80.$

2 zeros   Move 2 places to the right.

$-7.8 \times 100 = -780$

15. 0.1×79.18 $7.9.18$

1 decimal place   Move 1 place to the left.

$0.1 \times 79.18 = 7.918$

17. 0.001×97.68 $0.097.68$

3 decimal places   Move 3 places to the left.

$0.001 \times 97.68 = 0.09768$

19. $28.7 \times (-0.01) = -(28.7 \times 0.01)$

$-(28.7 \times 0.01)$ $-0.28.7$

2 decimal places   Move 2 places to the left.

$28.7 \times (-0.01) = -0.287$

21.

$$
\begin{array}{r}
2.\,7\,3 \\
\times\quad 1\,6 \\
\hline
1\,6\,3\,8 \\
2\,7\,3\,0 \\
\hline
4\,3.\,6\,8
\end{array}
$$

(2 decimal places)
(0 decimal places)

(2 decimal places)

23.

$$
\begin{array}{r}
0.\,9\,8\,4 \\
\times\quad 3.\,3 \\
\hline
2\,9\,5\,2 \\
2\,9\,5\,2\,0 \\
\hline
3.\,2\,4\,7\,2
\end{array}
$$

(3 decimal places)
(1 decimal place)

(4 decimal places)

25. We multiply the absolute values.

$$
\begin{array}{r}
3\,7.\,4 \\
\times\quad 2.\,4 \\
\hline
1\,4\,9\,6 \\
7\,4\,8\,0 \\
\hline
8\,9.\,7\,6
\end{array}
$$

(1 decimal place)
(1 decimal place)

(2 decimal places)

Since the product of two negative numbers is positive, the answer is 89.76.

27. We multiply the absolute values.

$$
\begin{array}{r}
7\,4\,9 \\
\times\quad 0.\,4\,3 \\
\hline
2\,2\,4\,7 \\
2\,9\,9\,6\,0 \\
\hline
3\,2\,2.\,0\,7
\end{array}
$$

(0 decimal places)
(2 decimal places)

(2 decimal places)

Since the product of a positive number and a negative number is negative, the answer is −322.07.

29.

$$
\begin{array}{r}
0.\,8\,7 \\
\times\quad 6\,4 \\
\hline
3\,4\,8 \\
5\,2\,2\,0 \\
\hline
5\,5.\,6\,8
\end{array}
$$

(2 decimal places)
(0 decimal places)

(2 decimal places)

31.

$$
\begin{array}{r}
4\,6.\,5\,0 \\
\times\quad 7\,5 \\
\hline
2\,3\,2\,5\,0 \\
3\,2\,5\,5\,0\,0 \\
\hline
3\,4\,8\,7.\,5\,0
\end{array}
$$

(2 decimal places)
(0 decimal places)

(2 decimal places)

Since the last decimal place is 0, we could also write this answer as 3487.5.

33. We multiply the absolute values.

$$
\begin{array}{r}
0.\,2\,3\,1 \\
\times\quad 0.\,5 \\
\hline
0.\,1\,1\,5\,5
\end{array}
$$

(3 decimal places)
(1 decimal place)
(4 decimal places)

Since the product of two negative numbers is positive, the answer is 0.1155.

35. $9.42 \times (-1000) = -(9.42 \times 1000)$

$-(9.42 \times 1\underline{000})$ $-9.420.$

3 zeros   Move 3 places to the right.

$9.42 \times (-1000) = -9420$

37. $-95.3 \times (-0.0001) = 95.3 \times 0.0001$

$95.3 \times 0.\underline{0001}$ $0.0095.3$

4 decimal places   Move 4 places to the left.

$-95.3 \times (-0.0001) = 0.00953$

39. Move 2 places to the right.

$\$57.06.\cancel{c}$

Change from $ sign in front to ¢ sign at end.

$\$57.06 = 5706\cancel{c}$

41. Move 2 places to the right.

$\$0.95.\cancel{c}$

Change from $ sign in front to ¢ sign at end.

$\$0.95 = 95\cancel{c}$

43. Move 2 places to the right.

$\$0.01.\cancel{c}$

Change from $ sign in front to ¢ sign at end.

$\$0.01 = 1\cancel{c}$

45. Move 2 places to the left.

$\$0.72.\cancel{c}$

Change from ¢ sign at end to $ sign in front.

$72\cancel{c} = \$0.72$

47. Move 2 places to the left.

$\$0.02.\cancel{c}$

Change from ¢ sign at end to $ sign in front.

$2\cancel{c} = \$0.02$

49. Move 2 places to the left.

$\$63.99.\cancel{c}$

Change from ¢ sign at end to $ sign in front.

$6399\cancel{c} = \$63.99$

51. 3.156 billion $= 3.156 \times 1,\underbrace{000,000,000}_{9 \text{ zeros}}$

$3.156000000.$

Move 9 places to the right.

3.156 billion $= 3,156,000,000$

53. 63.1 trillion $= 63.1 \times 1,\underbrace{000,000,000,000}_{12 \text{ zeros}}$

$63.1000000000000.$

Move 12 places to the right.

63.1 trillion $= 63,100,000,000,000$

55. 11.98 million $= 11.98 \times 1,\underbrace{000,000}_{6 \text{ zeros}}$

$\$11.980000.$

Move 6 places to the right.

11.98 million $= 11,980,000$

57. $\quad P + Prt$
$= 10,000 + 10,000(0.04)(2.5)$ Substituting
$= 10,000 + 400(2.5)$ Multiplying and dividing
$= 10,000 + 1000$ in order from left to right
$= 11,000$ Adding

59. $\quad vt + at^2$
$= 10(1.5) + 4.9(1.5)^2$
$= 10(1.5) + 4.9(1.5)(1.5)$
$= 10(1.5) + 4.9(2.25)$ Squaring first
$= 15 + 11.025$
$= 26.025$

61. a) $P = 2l + 2w$
$= 2(12.5) + 2(9.5)$
$= 25 + 19$
$= 44$

The perimeter is 44 ft.

b) $A = l \cdot w$
$= (12.5)(9.5)$
$= 118.75$

The area is 118.75 ft^2.

63. a) $P = 2l + 2w$
$= 2(10.5) + 2(8.4)$
$= 21 + 16.8$
$= 37.8$

The perimeter is 37.8 m.

b) $A = l \cdot w$
$= (10.5)(8.4)$
$= 88.2$

The area is 88.2 m^2.

65. In 2010, $t = 2010 - 2000 = 10$. Substitute 10 for t.

$5.06t + 9.7 = 5.06(10) + 9.7$
$= 50.6 + 9.7$
$= 60.3$

In 2010 there will be approximately 60.3 billion e-mails each day in North America.

67. Discussion and Writing Exercise

69. $-162 \div 6 = -27$

(The signs are different, so the quotient is negative.)

71. $-1035 \div (-15) = 69$

(The signs are the same, so the quotient is positive.)

73. $-525 \div 25 = -21$

(The signs are different, so the quotient is negative.)

75. $-7050 \div 50 = -141$

(The signs are different, so the quotient is negative.)

77. Discussion and Writing Exercise

79. $\quad 85 \times 9.46 \times 10^{12}$
$= 85 \times 9.46 \times 10 \times 10 \times 10 \times 10 \times 10 \times 10 \times$
$\quad\quad 10 \times 10 \times 10 \times 10 \times 10 \times 10$
$= 8.04.1 \times 1,000,000,000,000$
$= (804.1 \times 1000) \times 1,000,000,000$
$= 804,100 \times 1,000,000,000$
$= 804,100 \times 1$ billion

Regulus is 804,100 billion km from Earth.

81. Use a calculator.
$\quad d + vt + at^2$
$= 79.2 + 3.029(7.355) + 4.9(7.355)^2$
$= 79.2 + 3.029(7.355) + 4.9(54.096025)$
$= 79.2 + 22.278295 + 265.0705225$
$= 101.478295 + 265.0705225$
$= 366.5488175$

83. Use a calculator.
$0.5(b_1 + b_2)h = 0.5(9.7 \text{ cm} + 13.4 \text{ cm})(6.32 \text{ cm})$
$= 0.5(23.1 \text{ cm})(6.32 \text{ cm})$
$= 72.996 \text{ cm}^2$

85. $(1 \text{ trillion}) \cdot (1 \text{ billion})$
$= 1,\underbrace{000,000,000,000}_{12 \text{ zeros}} \times 1,\underbrace{000,000,000}_{9 \text{ zeros}}$
$= 1,\underbrace{000,000,000,000,000,000,000}_{21 \text{ zeros}}$
$= 10^{21}$

87. For the British number 6.6 billion, we have:

6.6 billion
$= 6.6 \times 1,000,000 \times 1,000,000$
$= 6.6 \times 1,000,000,000,000$
$= 6,600,000,000,000$ Moving the decimal point
                   12 places to the right

89. The period from April 20 to May 20 consists of 30 days, so the "customer charge" is 30 × $0.374, or $11.22. The "energy charge" for the first 250 kilowatt-hours is 250 × $0.1174, or $29.35.

We subtract to find the number of kilowatt-hours in excess of 250: 480 − 250 = 230. Then the "energy charge" for the 230 kilowatt-hours in excess of 250 kilowatt-hours is 230 × $0.09079, or $20.88 (rounding to the nearest cent).

Finally, we add to find the total bill:

$11.22 + $29.35 + $20.88 = $61.45.

Exercise Set 5.4

1.
```
      12.6
  5 ⟌63.0
      50
      ‾‾
      13
      10
      ‾‾
       30    Write an extra 0.
       30
       ‾‾
        0
```

3.
```
      23.78
  4 ⟌95.12    Divide as though dividing whole num-
      8000     bers. Place the decimal point directly
      ‾‾‾‾     above the decimal point in the divi-
      1512     dend.
      1200
      ‾‾‾‾
       312
       280
       ‾‾‾
        32
        32
        ‾‾
         0
```

5.
```
       7.48
  12 ⟌89.76
       8400
       ‾‾‾‾
        576
        480
        ‾‾‾
         96
         96
         ‾‾
          0
```

7.
```
       7.2
  33 ⟌237.6
       2310
       ‾‾‾‾
         66
         66
         ‾‾
          0
```

9. We first consider 5.4 ÷ 6.
```
      0.9
  6 ⟌5.4
     54
     ‾‾
      0
```

Since a positive number divided by a negative number is negative, the answer is −0.9.

11. We first consider 9.144 ÷ 8.
```
      1.143
  8 ⟌9.144
     8000
     ‾‾‾‾
     1144
      800
      ‾‾‾
      344
      320
      ‾‾‾
       24
       24
       ‾‾
        0
```

Since a negative number divided by a positive number is negative, the answer is −1.143.

13.
```
           140.
  0.06∧⟌8.40∧    Multiply the divisor by 100 (move the
         600      decimal point 2 places). Multiply the
         ‾‾‾      same way in the dividend (move 2
         240      places). Then divide.
         240
         ‾‾‾
           0
```

15.
```
          40.
  2.6∧⟌104.0∧    Put a decimal point at the end of
        1040      the whole number. Multiply the di-
        ‾‾‾‾      visor by 10 (move the decimal point
           0      1 place). Multiply the same way in
                  the dividend (move 1 place), adding
                  an extra 0. Then divide.
```

17. We first consider 1.8 ÷ 12.
```
       0.15        Divide as though dividing whole num-
  12 ⟌1.80         bers. Place the decimal point directly
       12          above the decimal point in the divi-
       ‾‾          dend.
       60
       60
       ‾‾
        0
```

Since a positive number divided by a negative number is negative, the answer is −0.15.

19.
```
           48.
  2.7∧⟌129.6∧
        1080
        ‾‾‾‾
         216
         216
         ‾‾‾
           0
```

21.
```
           3.2
  8.5∧⟌27.2∧0
        255
        ‾‾‾
        170    Write an extra zero.
        170
        ‾‾‾
          0
```

23. We first consider 5 ÷ 8.
```
       0.625
  8 ⟌5.000
     48
     ‾‾
      20    Write an extra 0.
      16
      ‾‾
       40   Write an extra 0.
       40
       ‾‾
        0
```

Since a negative number divided by a negative number is positive, the answer is 0.625.

25.

$$\begin{array}{r} 0.2\,6 \\ 0.4\,7_\wedge\!\overline{\big)\,0.1\,2_\wedge2\,2} \\ \underline{9\,4\,0} \\ 2\,8\,2 \\ \underline{2\,8\,2} \\ 0 \end{array}$$

27.

$$\begin{array}{r} 2.3\,4 \\ 0.0\,3\,2_\wedge\!\overline{\big)\,0.0\,7\,4_\wedge8\,8} \\ \underline{6\,4\,0\,0} \\ 1\,0\,8\,8 \\ \underline{9\,6\,0} \\ 1\,2\,8 \\ \underline{1\,2\,8} \\ 0 \end{array}$$

29. We first consider $24.969 \div 82$.

$$\begin{array}{r} 0.3\,0\,4\,5 \\ 8\,2\,\overline{\big)\,2\,4.9\,6\,9\,0} \\ \underline{2\,4\,6\,0\,0} \\ 3\,6\,9 \\ \underline{3\,2\,8} \\ 4\,1\,0 \quad \text{Write an extra 0.} \\ \underline{4\,1\,0} \\ 0 \end{array}$$

Since a negative number divided by a positive number is negative, the answer is -0.3045.

31. $\dfrac{-213.4567}{1\underline{00}}$ $-2\underset{\uparrow\_}{.13}.4567$

2 zeros   Move 2 places to the left.

$\dfrac{-213.4567}{100} = -2.134567$

33. $\dfrac{1.0237}{0.\underline{001}}$ $1.023\underset{\,\,\underline{}\uparrow}{.7}$

3 decimal places   Move 3 places to the right.

$\dfrac{1.0237}{0.001} = 1023.7$

35. $\dfrac{92.36}{-0.01} = \dfrac{-92.36}{0.01}$

$\dfrac{-92.36}{0.\underline{01}}$ $-92\underset{\,\,\underline{}\uparrow}{.36}$

2 decimal places   Move 2 places to the right.

$\dfrac{92.36}{-0.01} = \dfrac{-92.36}{0.01} = -9236$

37. $\dfrac{0.8172}{1\underline{0}}$ $0\underset{\underline{}\uparrow}{.0}.8172$

1 zero   Move 1 place to the left.

$\dfrac{0.8172}{10} = 0.08172$

39. $\dfrac{0.97}{0.\underline{1}}$ $0.9\underset{\underline{}\uparrow}{.7}$

1 decimal place   Move 1 place to the right.

$\dfrac{0.97}{0.1} = 9.7$

41. $\dfrac{52.7}{-1000} = \dfrac{-52.7}{1000}$

$\dfrac{-52.7}{1\underline{000}}$ $-0\underset{\uparrow\underline{}}{.052}.7$

3 zeros   Move 3 places to the left.

$\dfrac{52.7}{-1000} = \dfrac{-52.7}{1000} = -0.0527$

43. $\dfrac{75.3}{-0.001} = -\dfrac{75.3}{0.001}$

$-\dfrac{75.3}{0.\underline{001}}$ $-75.300\underset{\underline{}\uparrow}{.}$

3 decimal places   Move 3 places to the right.

$\dfrac{75.3}{-0.001} = -75,300$

45. $\dfrac{-75.3}{1\underline{000}}$ $-0\underset{\uparrow\underline{}}{.075}.3$

3 zeros   Move 3 places to the left.

$\dfrac{-75.3}{1000} = -0.0753$

47. $14 \times (82.6 + 67.9)$
$= 14 \times (150.5)$ Doing the calculation inside
                          the parentheses
$= 2107$ Multiplying

49. $0.003 + 3.03 \div (-0.01) = 0.003 - 303$ Dividing first
                                   $= -302.997$ Subtracting

51. $(4.9 - 18.6) \times 13$
$= -13.7 \times 13$ Doing the calculation inside
                        the parentheses
$= -178.1$ Multiplying

53. $210.3 - 4.24 \times 1.01$
$= 210.3 - 4.2824$ Multiplying
$= 206.0176$ Subtracting

55. $12 \div (-0.03) - 12 \times 0.03^2$
$= 12 \div (-0.03) - 12 \times 0.0009$ Evaluating the
                                           exponential expression
$= -400 - 0.0108$ Dividing and multiplying
                      in order from left to right
$= -400.0108$ Subtracting

57. $\quad (4 - 2.5)^2 \div 100 + 0.1 \times 6.5$

$= (1.5)^2 \div 100 + 0.1 \times 6.5 \qquad$ Doing the calculation inside the parentheses

$= 2.25 \div 100 + 0.1 \times 6.5 \qquad$ Evaluating the exponential expression

$= 0.0225 + 0.65 \qquad$ Dividing and multiplying in order from left to right

$= 0.6725 \qquad$ Adding

59. $\quad 6 \times 0.9 - 0.1 \div 4 + 0.2^3$

$= 6 \times 0.9 - 0.1 \div 4 + 0.008 \qquad$ Evaluating the exponential expression

$= 5.4 - 0.025 + 0.008 \qquad$ Multiplying and dividing in order from left to right

$= 5.383 \qquad$ Subtracting and adding in order from left to right

61. $\quad 12^2 \div (12 + 2.4) - [(2 - 2.4) \div 0.8]$

$= 12^2 \div (12 + 2.4) - [-0.4 \div 0.8] \qquad$ Doing the calculations in the innermost parentheses first

$= 12^2 \div 14.4 - [-0.5] \qquad$ Doing the calculations inside the parentheses

$= 12^2 \div 14.4 + 0.5 \qquad$ Simplifying

$= 144 \div 14.4 + 0.5 \qquad$ Evaluating the exponential expression

$= 10 + 0.5 \qquad$ Dividing

$= 10.5 \qquad$ Adding

63. We add the amounts and divide by the number of addends, 5.

$$\frac{131.8 + 168.7 + 230.2 + 250.0 + 251.0}{5}$$

$$= \frac{1031.7}{5} = 206.34$$

The average amount paid per year in individual income tax over the five-year period was $206.34 billion.

65. We add the amounts and divide by the number of addends, 5.

$$\frac{83.51 + 81.71 + 81.71 + 81.59 + 81.25}{5}$$

$$= \frac{409.77}{5} = 81.954$$

The average life expectancy for the given countries is 81.954 yr.

67. We add the amounts and divide by the number of addends, 5.

$$\frac{33.5 + 31.1 + 16.0 + 13.8 + 12.3}{5}$$

$$= \frac{106.7}{5} = 21.34$$

The average length of the tunnels is 21.34 mi.

69. Discussion and Writing Exercise

71. $\dfrac{33}{44} = \dfrac{3 \cdot 11}{4 \cdot 11} = \dfrac{3}{4} \cdot \dfrac{11}{11} = \dfrac{3}{4}$

73. $-\dfrac{27}{18} = -\dfrac{3 \cdot 9}{2 \cdot 9} = -\dfrac{3}{2} \cdot \dfrac{9}{9} = -\dfrac{3}{2}$

75. $\dfrac{9a}{27} = \dfrac{9 \cdot a}{9 \cdot 3} = \dfrac{9}{9} \cdot \dfrac{a}{3} = \dfrac{a}{3}$

77. $\dfrac{4r}{20r} = \dfrac{4 \cdot r \cdot 1}{5 \cdot 4 \cdot r} = \dfrac{4 \cdot r}{4 \cdot r} \cdot \dfrac{1}{5} = \dfrac{1}{5}$

79. Discussion and Writing Exercise

81. Use a calculator.

$\quad 7.434 \div (-1.2) \times 9.5 + 1.47^2$

$= 7.434 \div (-1.2) \times 9.5 + 2.1609$

$\qquad$ Evaluating the exponential expression

$= -6.195 \times 9.5 + 2.1609$

$\qquad$ Multiplying and dividing

$= -58.8525 + 2.1609 \qquad$ in order from left to right

$= -56.6916 \qquad$ Adding

83. Use a calculator.

$\quad 9.0534 - 2.041^2 \times 0.731 \div 1.043^2$

$= 9.0534 - 4.165681 \times 0.731 \div 1.087849$

$\qquad$ Evaluating the exponential expressions

$= 9.0534 - 3.045112811 \div 1.087849$

$\qquad$ Multiplying and dividing

$= 9.0534 - 2.799205415 \qquad$ in order from left to right

$= 6.254194585 \qquad$ Subtracting

85. $\quad 439.57 \times 0.01 \div 1000 \cdot x = 4.3957$

$\qquad\qquad 4.3957 \div 1000 \cdot x = 4.3957$

$\qquad\qquad\quad 0.0043957 \cdot x = 4.3957$

$$x = \frac{4.3957}{0.0043957}$$

$$x = 1000$$

The solution is 1000.

87. $\qquad 0.0329 \div 0.001 \times 10^4 \div x = 3290$

$\quad 0.0329 \div 0.001 \times 10{,}000 \div x = 3290$

$\qquad\qquad 32.9 \times 10{,}000 \div x = 3290$

$\qquad\qquad\qquad 329{,}000 \div x = 3290$

We need to divide 329,000 by a number that moves the decimal point 2 places to the left. Thus, we need to divide by 100. The solution is 100.

89. We divide. Note that 5.6 million $= 5.6 \times 1{,}000{,}000 = 5{,}600{,}000$.

```
            5.71
980,000 ) 5,600,000.00
          4 900 000
          ─────────
            700 0000
            686 0000
          ─────────
              14 00000
               9 80000
          ─────────
               4 20000
```

Rounding to the nearest tenth, we have 5.7 rating points.

91. The period from August 20 to September 20 consists of 31 days.

The "customer charge" is $31 \times \$0.374 = \$11.59$ (rounded to the nearest cent). The "energy charge" for the first 250 kilowatt-hours is $250 \times \$0.1174 = \$29.35$.

Subtract to find the "energy charge" for the kilowatt-hours in excess of 250:

$59.10 - $11.59 - $29.35 = $18.16

Divide to find the number of kilowatt-hours in excess of 250:

18.16 ÷ $0.09079 = 200 (rounded to the nearest hour)

The total number of kilowatt-hours of electricity used is 250 + 200 = 450 kwh.

Exercise Set 5.5

1. Since $\frac{3}{8}$ means $3 \div 8$ we have:

$$
\begin{array}{r}
0.375 \\
8\overline{)3.000} \\
\underline{24} \\
60 \\
\underline{56} \\
40 \\
\underline{40} \\
0
\end{array}
$$

$\frac{3}{8} = 0.375$

3. Since $\frac{-1}{2}$ is negative, we divide 1 by 2 and make the results negative.

$$
\begin{array}{r}
0.5 \\
2\overline{)1.0} \\
\underline{10} \\
0
\end{array}
$$

Thus, $\frac{-1}{2} = -0.5$.

5. Since $\frac{3}{25}$ means $3 \div 25$, we have:

$$
\begin{array}{r}
0.12 \\
25\overline{)3.00} \\
\underline{25} \\
50 \\
\underline{50} \\
0
\end{array}
$$

$\frac{3}{25} = 0.12$

7. Since $\frac{9}{40}$ means $9 \div 40$, we have:

$$
\begin{array}{r}
0.225 \\
40\overline{)9.000} \\
\underline{80} \\
100 \\
\underline{80} \\
200 \\
\underline{200} \\
0
\end{array}
$$

$\frac{9}{40} = 0.225$

9. Since $\frac{13}{25}$ means $13 \div 25$, we have:

$$
\begin{array}{r}
0.52 \\
25\overline{)13.00} \\
\underline{125} \\
50 \\
\underline{50} \\
0
\end{array}
$$

$\frac{13}{25} = 0.52$

11. Since $\frac{-17}{20}$ is negative, we divide 17 by 20 and make the result negative.

$$
\begin{array}{r}
0.85 \\
20\overline{)17.00} \\
\underline{160} \\
100 \\
\underline{100} \\
0
\end{array}
$$

Thus, $\frac{-17}{20} = -0.85$.

13. Since $-\frac{9}{16}$ is negative, we divide 9 by 16 and make the result negative.

$$
\begin{array}{r}
0.5625 \\
16\overline{)9.0000} \\
\underline{80} \\
100 \\
\underline{96} \\
40 \\
\underline{32} \\
80 \\
\underline{80} \\
0
\end{array}
$$

Thus, $-\frac{9}{16} = -0.5625$.

15. Since $\frac{7}{5}$ means $7 \div 5$, we have:

$$
\begin{array}{r}
1.4 \\
5\overline{)7.0} \\
\underline{5} \\
20 \\
\underline{20} \\
0
\end{array}
$$

$\frac{7}{5} = 1.4$

17. Since $\frac{28}{25}$ means $28 \div 25$, we have:

$$
\begin{array}{r}
1.12 \\
25\overline{)28.00} \\
\underline{25} \\
30 \\
\underline{25} \\
50 \\
\underline{50} \\
0
\end{array}
$$

$\frac{28}{25} = 1.12$

19. Since $\dfrac{11}{-8}$ is negative, we divide 11 by 8 and make the result negative.

```
        1. 3 7 5
    8 ⟌ 1 1. 0 0 0
        8
      ─────
        3 0
        2 4
      ─────
          6 0
          5 6
        ─────
            4 0
            4 0
          ─────
              0
```

Thus, $\dfrac{11}{-8} = -1.375$.

21. Since $-\dfrac{39}{40}$ is negative, we divide 39 by 40 and make the result negative.

```
          0. 9 7 5
    4 0 ⟌ 3 9. 0 0 0
          3 6 0
        ─────
            3 0 0
            2 8 0
          ─────
              2 0 0
              2 0 0
            ─────
                0
```

Thus, $-\dfrac{39}{40} = -0.975$.

23. Since $\dfrac{121}{200}$ means $121 \div 200$, we have:

```
              0. 6 0 5
    2 0 0 ⟌ 1 2 1. 0 0 0
            1 2 0 0
          ─────────
              1 0 0 0
              1 0 0 0
            ─────────
                    0
```

$\dfrac{121}{200} = 0.605$

25. Since $\dfrac{8}{15}$ means $8 \div 15$, we have:

```
          0. 5 3 3
    1 5 ⟌ 8. 0 0 0
          7 5
        ─────
          5 0
          4 5
        ─────
            5 0
            4 5
          ─────
              5
```

Since 5 keeps reappearing as a remainder, the digits repeat and

$\dfrac{8}{15} = 0.533\ldots$ or $0.5\overline{3}$.

27. Since $\dfrac{1}{3}$ means $1 \div 3$, we have:

```
        0. 3 3 3
    3 ⟌ 1. 0 0 0
        9
      ─────
        1 0
          9
        ─────
          1 0
            9
          ─────
            1
```

Since 1 keeps reappearing as a remainder, the digits repeat and

$\dfrac{1}{3} = 0.333\ldots$ or $0.\overline{3}$.

29. Since $\dfrac{-4}{3}$ is negative, we divide by 4 and 3 and make the result negative.

```
        1. 3 3
    3 ⟌ 4. 0 0
        3
      ─────
        1 0
          9
        ─────
          1 0
            9
          ─────
            1
```

Since 1 keeps reappearing as a remainder, the digits repeat and

$\dfrac{4}{3} = 1.333\ldots$ or $1.\overline{3}$.

Thus, $\dfrac{-4}{3} = -1.\overline{3}$.

31. Since $\dfrac{7}{6}$ means $7 \div 6$, we have:

```
        1. 1 6 6
    6 ⟌ 7. 0 0 0
        6
      ─────
        1 0
          6
        ─────
          4 0
          3 6
        ─────
            4 0
            3 6
          ─────
              4
```

Since 4 keeps reappearing as a remainder, the digits repeat and

$\dfrac{7}{6} = 1.166\ldots$ or $1.1\overline{6}$.

33. Since $-\dfrac{14}{11}$ is negative, we divide 14 by 11 and make the result negative.

$$
\begin{array}{r}
1.2727 \\
11\overline{\smash{\big)}\,14.0000} \\
\underline{11} \\
30 \\
\underline{22} \\
80 \\
\underline{77} \\
30 \\
\underline{22} \\
80 \\
\underline{77} \\
3
\end{array}
$$

Since 3 and 8 keep reappearing as remainders, the sequence of digits "27" repeats in the quotient and $\dfrac{14}{11} = 1.2727\ldots,$ or $1.\overline{27}.$

Thus, $-\dfrac{14}{11} = -1.\overline{27}.$

35. Since $-\dfrac{5}{12}$ is negative, we divide 5 by 12 and make the result negative.

$$
\begin{array}{r}
0.4166 \\
12\overline{\smash{\big)}\,5.0000} \\
\underline{48} \\
20 \\
\underline{12} \\
80 \\
\underline{72} \\
80 \\
\underline{72} \\
8
\end{array}
$$

Since 8 keeps reappearing as a remainder, the digits repeat and

$\dfrac{5}{12} = 0.4166\ldots,$ or $0.41\overline{6}.$

Thus, $\dfrac{-5}{12} = -0.41\overline{6}.$

37. Since $\dfrac{127}{500}$ means $127 \div 500$, we have:

$$
\begin{array}{r}
0.254 \\
500\overline{\smash{\big)}\,127.000} \\
\underline{1000} \\
2700 \\
\underline{2500} \\
2000 \\
\underline{2000} \\
0
\end{array}
$$

$\dfrac{127}{500} = 0.254$

39. Since $\dfrac{4}{33}$ means $4 \div 33$, we have:

$$
\begin{array}{r}
0.1212 \\
33\overline{\smash{\big)}\,4.0000} \\
\underline{33} \\
70 \\
\underline{66} \\
40 \\
\underline{33} \\
70 \\
\underline{66} \\
4
\end{array}
$$

Since 7 and 4 keep reappearing as remainders, the sequence of digits "12" repeats in the quotient and

$\dfrac{4}{33} = 0.1212\ldots,$ or $0.\overline{12}.$

41. Since $-\dfrac{12}{55}$ is negative, we divide 12 by 55 and make the result negative.

$$
\begin{array}{r}
0.21818 \\
55\overline{\smash{\big)}\,12.00000} \\
\underline{110} \\
100 \\
\underline{55} \\
450 \\
\underline{440} \\
100 \\
\underline{55} \\
450 \\
\underline{440} \\
10
\end{array}
$$

Since 10 and 45 keep reappearing as remainders, the sequence of digits "18" repeats in the quotient and $\dfrac{12}{55} = 0.21818\ldots,$ or $0.2\overline{18}.$

Thus, $\dfrac{-12}{55} = -0.2\overline{18}.$

43. Since $\dfrac{35}{111}$ means $35 \div 111$, we have:

$$
\begin{array}{r}
0.315315 \\
111\overline{\smash{\big)}\,35.000000} \\
\underline{333} \\
170 \\
\underline{111} \\
590 \\
\underline{555} \\
350 \\
\underline{333} \\
170 \\
\underline{111} \\
590 \\
\underline{555} \\
35
\end{array}
$$

Since 17, 59, and 35 keep reappearing as remainders, the sequence of digits "315" repeats in the quotient and

$\dfrac{35}{111} = 0.315315\ldots,$ or $0.\overline{315}.$

45. Since $\dfrac{4}{7}$ means $4 \div 7$, we have:

$$
\begin{array}{r}
0.5\,7\,1\,4\,2\,8 \\
7\,\overline{|\,4.0\,0\,0\,0\,0\,0} \\
\underline{3\,5} \\
5\,0 \\
\underline{4\,9} \\
1\,0 \\
\underline{7} \\
3\,0 \\
\underline{2\,8} \\
2\,0 \\
\underline{1\,4} \\
6\,0 \\
\underline{5\,6} \\
4
\end{array}
$$

Since we have already divided 7 into 4, the sequence of digits "571428" repeats in the quotient and $\dfrac{4}{7} = 0.571428571428\ldots$, or $0.\overline{571428}$.

47. Since $\dfrac{-37}{25}$ is negative, we divide 37 by 25 and make the result negative.

$$
\begin{array}{r}
1.\,4\,8 \\
2\,5\,\overline{|\,3\,7.0\,0} \\
\underline{2\,5} \\
1\,2\,0 \\
\underline{1\,0\,0} \\
2\,0\,0 \\
\underline{2\,0\,0} \\
0
\end{array}
$$

Thus, $\dfrac{-37}{25} = -1.48$.

49. In Example 4 we see that $\dfrac{4}{11} = 0.\overline{36}$.

Round $0.3\,\boxed{6}\,3\,6\ldots$ to the nearest tenth.

Hundredths digit is 6 or more.

0.4 Round up.

Round $0.3\,\underline{6}\,\boxed{3}\,6\ldots$ to the nearest hundredth.

Thousandths digit is 4 or less.

$0.3\,6$ Round down.

Round $0.3\,6\,\underline{3}\,\boxed{6}\ldots$ to the nearest thousandth.

Ten-thousandths digit is 5 or more.

$0.3\,6\,4$ Round up.

51. First we find decimal notation for $-\dfrac{5}{3}$.

$$
\begin{array}{r}
1.\,6\,6 \\
3\,\overline{|\,5.0\,0} \\
\underline{3} \\
2\,0 \\
\underline{1\,8} \\
2\,0 \\
\underline{1\,8} \\
2
\end{array}
$$

$\dfrac{5}{3} = 1.\overline{6}$, so $-\dfrac{5}{3} = -1.6$

Round $-1.\,6\,\boxed{6}\,6\,6\ldots$ to the nearest tenth.

Hundredths digit is 5 or more.

$-1.\,7$ Round to -1.7.

Round $-1.\,6\,6\,\underline{6}\,\boxed{6}\,6\ldots$ to the nearest hundredth.

Thousandths digit is 5 or more.

$-1.\,6\,7$ Round to -1.67.

Round $-1.\,6\,6\,6\,\underline{6}\,\boxed{6}\ldots$ to the nearest thousandth.

Ten-thousandths digit is 5 or more.

$-1.\,6\,6\,7$ Round to -1.667.

53. First we find decimal notation for $\dfrac{-8}{17}$.

$$
\begin{array}{r}
0.\,4\,7\,0\,5\,8\,8 \\
1\,7\,\overline{|\,8.0\,0\,0\,0\,0} \\
\underline{6\,8} \\
1\,2\,0 \\
\underline{1\,1\,9} \\
1\,0\,0 \\
\underline{8\,5} \\
1\,5\,0 \\
\underline{1\,3\,6} \\
1\,4\,0 \\
\underline{1\,3\,6} \\
4
\end{array}
$$

The digits repeat eventually but we have enough decimal places now to be able to round as instructed.

We have $\dfrac{8}{17} \approx 0.47059$, so $\dfrac{-8}{17} \approx -0.47059$.

Round $-0.\,4\,\boxed{7}\,0\,5\,9\ldots$ to the nearest tenth.

Hundredths digit is 5 or more.

$-0.\,5$ Round to -0.5.

Round $-0.\,4\,\underline{7}\,\boxed{0}\,5\,9\ldots$ to the nearest hundredth.

Thousandths digit is 4 or less.

$-0.\,4\,7$ Round to -0.47.

Round $-0.\,4\,7\,\underline{0}\,\boxed{5}\,9\ldots$ to the nearest thousandth.

Ten-thousandths digit is 5 or more.

$-0.\,4\,7\,1$ Round to -0.471.

55. First find decimal notation for $\dfrac{7}{12}$.

$$
\begin{array}{r}
0.\,5\,8\,3\,3 \\
1\,2\,\overline{|\,7.0\,0\,0\,0} \\
\underline{6\,0} \\
1\,0\,0 \\
\underline{9\,6} \\
4\,0 \\
\underline{3\,6} \\
4\,0 \\
\underline{3\,6} \\
4
\end{array}
$$

$\dfrac{7}{12} = 0.58\overline{3}$

Round $0.\,\underline{5}\,\boxed{8}\,3\,3\,\ldots$ to the nearest tenth.

$\quad\quad\downarrow\quad\underset{\textstyle\longleftarrow}{}$ Hundredths digit is 5 or more.

$\quad 0.\,6$ $\quad\quad\quad$ Round up.

Round $0.\,5\,\underline{8}\,\boxed{3}\,3\,\ldots$ to the nearest hundredth.

$\quad\quad\downarrow\quad\underset{\textstyle\longleftarrow}{}$ Thousandths digit is 4 or less.

$\quad 0.\,5\,8$ $\quad\quad\quad$ Round down.

Round $0.\,5\,8\,\underline{3}\,\boxed{3}\,\ldots$ to the nearest thousandth.

$\quad\quad\quad\uparrow\underset{\textstyle\longleftarrow}{}$ Ten-thousandths digit is 4 or less.

$\quad 0.\,5\,8\,3$ $\quad\quad\quad$ Round down.

57. First find decimal notation for $\dfrac{29}{-150}$.

$$
\begin{array}{r}
0.1\,9\,3\,3 \\
150\,\overline{)\,29.0\,0\,0\,0} \\
1\,5\,0 \\
\overline{1\,4\,0\,0} \\
1\,3\,5\,0 \\
\overline{5\,0\,0} \\
4\,5\,0 \\
\overline{5\,0\,0} \\
4\,5\,0 \\
\overline{5\,0}
\end{array}
$$

We have $\dfrac{29}{150} = 0.19\overline{3}$, so $\dfrac{29}{-150} = -0.19\overline{3}$.

Round $-0.\,\underline{1}\,\boxed{9}\,3\,3\,\ldots$ to the nearest tenth.

$\quad\quad\downarrow\quad\underset{\textstyle\longleftarrow}{}$ Hundredths digit is 5 or more.

$\quad -0.\,2$ $\quad\quad\quad$ Round to -0.2.

Round $-0.\,1\,\underline{9}\,\boxed{3}\,3\,\ldots$ to the nearest hundredth.

$\quad\quad\downarrow\quad\underset{\textstyle\longleftarrow}{}$ Thousandths digit is 3 or less.

$\quad -0.\,1\,9$ $\quad\quad\quad$ Round to -0.19.

Round $-0.\,1\,9\,\underline{3}\,\boxed{3}\,\ldots$ to the nearest thousandth.

$\quad\quad\quad\uparrow\underset{\textstyle\longleftarrow}{}$ Ten-thousandths digit is 3 or less.

$\quad -0.\,1\,9\,3$ $\quad\quad\quad$ Round to -0.193.

59. First find decimal notation for $\dfrac{7}{-9}$.

$$
\begin{array}{r}
0.7\,7 \\
9\,\overline{)\,7.0\,0} \\
6\,3 \\
\overline{7\,0} \\
6\,3 \\
\overline{7}
\end{array}
$$

We have $\dfrac{7}{9} = 0.\overline{7}$, so $\dfrac{7}{-9} = -0.\overline{7}$.

Round $-0.\,\underline{7}\,\boxed{7}\,7\,7\,\ldots$ to the nearest tenth.

$\quad\quad\downarrow\quad\underset{\textstyle\longleftarrow}{}$ Hundredths digit is 5 or more.

$\quad -0.\,8$ $\quad\quad\quad$ Round to -0.8.

Round $-0.\,7\,\underline{7}\,\boxed{7}\,7\,\ldots$ to the nearest hundredth.

$\quad\quad\downarrow\quad\underset{\textstyle\longleftarrow}{}$ Thousandths digit is 5 or more.

$\quad -0.\,7\,8$ $\quad\quad\quad$ Round to -0.78.

Round $-0.\,7\,7\,\underline{7}\,\boxed{7}\,\ldots$ to the nearest thousandth.

$\quad\quad\quad\uparrow\underset{\textstyle\longleftarrow}{}$ Ten-thousandths digit is 5 or more.

$\quad -0.\,7\,7\,8$ $\quad\quad\quad$ Round to -0.778.

61. We will use the first method discussed in the text.

$$\dfrac{7}{8}(10.84) = \dfrac{7}{8} \times \dfrac{10.84}{1} = \dfrac{7 \times 10.84}{8} = \dfrac{75.88}{8} = 9.485$$

63. We will use the third method discussed in the text.

$$
\begin{aligned}
\dfrac{47}{9}(-79.95) &= \dfrac{47}{9} \cdot \left(-\dfrac{7995}{100}\right) \\
&= -\dfrac{47 \cdot 7995}{9 \cdot 100} \\
&= -\dfrac{47 \cdot \cancel{3} \cdot \cancel{5} \cdot 533}{\cancel{3} \cdot 3 \cdot \cancel{5} \cdot 20} \\
&= -\dfrac{25,051}{60} \\
&= -417.51\overline{6}
\end{aligned}
$$

65. We will use the first method discussed in the text.

$$
\begin{aligned}
\left(\dfrac{1}{6}\right)0.0765 + \left(\dfrac{3}{4}\right)0.1124 &= \dfrac{1}{6} \times \dfrac{0.0765}{1} + \dfrac{3}{4} \times \dfrac{0.1124}{1} \\
&= \dfrac{0.0765}{6} + \dfrac{3 \times 0.1124}{4} \\
&= \dfrac{0.0765}{6} + \dfrac{0.3372}{4} \\
&= 0.01275 + 0.0843 \\
&= 0.09705
\end{aligned}
$$

67. We will use the third method discussed in the text.

$$
\begin{aligned}
&\dfrac{3}{4} \times 2.56 - \dfrac{7}{8} \times 3.94 \\
&= \dfrac{3}{4} \times \dfrac{256}{100} - \dfrac{7}{8} \times \dfrac{394}{100} \\
&= \dfrac{768}{400} - \dfrac{2758}{800} \\
&= \dfrac{768}{400} \cdot \dfrac{2}{2} - \dfrac{2758}{800} \\
&= \dfrac{1536}{800} - \dfrac{2758}{800} \\
&= \dfrac{-1222}{800} = -\dfrac{1222}{800} \\
&= -\dfrac{2 \cdot 611}{2 \cdot 400} = \dfrac{2}{2} \cdot \left(-\dfrac{611}{400}\right) \\
&= -\dfrac{611}{400}, \text{ or } -1.5275
\end{aligned}
$$

69. We will use the second method discussed in the text.

$$
\begin{aligned}
5.2 \times 1\dfrac{7}{8} \div 0.4 &= 5.2 \times 1.875 \div 0.4 \\
&= 9.75 \div 0.4 \\
&= 24.375
\end{aligned}
$$

71. Familiarize. We draw a picture and recall that the formula for the area A of a triangle with base b and height h is $A = \frac{1}{2} \times b \times h$.

Translate. We substitute 1.2 for b and 1.8 for h.
$$A = \frac{1}{2} \times b \times h = \frac{1}{2} \times 1.2 \times 1.8$$

Solve. We carry out the computation.

$$A = \frac{1}{2} \times 1.2 \times 1.8$$
$$= \frac{1.2}{2} \times 1.8 \qquad \text{Multiplying } \frac{1}{2} \text{ and } 1.2$$
$$= 0.6 \times 1.8 \qquad \text{Dividing}$$
$$= 1.08 \qquad \text{Multiplying}$$

Check. We repeat the calculations using a different method.
$$\frac{1}{2} \times 1.2 \times 1.8 = 0.5 \times (1.2 \times 1.8) = 0.5 \times 2.16 = 1.08$$
Our answer checks.

State. The area of the shawl is 1.08 m^2.

73. Familiarize. We draw a picture and recall that the formula for the area A of a triangle with base b and height h is $A = \frac{1}{2} \times b \times h$.

Translate. We substitute 3.4 for b and 3.4 for h.
$$A = \frac{1}{2} \times b \times h = \frac{1}{2} \times 3.4 \times 3.4$$

Solve. We carry out the computation.

$$A = \frac{1}{2} \times 3.4 \times 3.4$$
$$= \frac{3.4}{2} \times 3.4 \qquad \text{Multiplying } \frac{1}{2} \text{ and } 3.4$$
$$= 1.7 \times 3.4 \qquad \text{Dividing}$$
$$= 5.78 \qquad \text{Multiplying}$$

Check. We repeat the calculations using a different method.
$$\frac{1}{2} \times 3.4 \times 3.4 = 0.5 \times (3.4 \times 3.4) = 0.5 \times 11.56 = 5.78$$
Our answer checks.

State. The area of the stamp is 5.78 cm^2.

75. Familiarize. First combine the lengths of the two 19.5-in. segments: 19.5 in. + 19.5 in. = 39 in. Now we can think of the area as the sum of the areas of two triangles, one with base 39 in. and height 11.25 in. and the other with base 39 in. and height 29.31 in.

Translate. We use the formula $A = \frac{1}{2}bh$ twice and add.
$$A = \frac{1}{2} \times 39 \times 11.25 + \frac{1}{2} \times 39 \times 29.31$$

Solve. We carry out the computation.

$$A = \frac{1}{2} \times 39 \times 11.25 + \frac{1}{2} \times 39 \times 29.31$$
$$= \frac{39}{2} \times 11.25 + \frac{39}{2} \times 29.31 \qquad \begin{array}{l}\text{Multiplying } \frac{1}{2} \text{ and} \\ 39 \text{ twice}\end{array}$$
$$= 19.5 \times 11.25 + 19.5 \times 29.31 \quad \text{Dividing}$$
$$= 219.375 + 571.545$$
$$= 790.92$$

Check. We repeat the calculation using a different method.

$$\frac{1}{2} \times 39 \times 11.25 + \frac{1}{2} \times 39 \times 29.31$$
$$= 0.5 \times (39 \times 11.25) + 0.5 \times (39 \times 29.31)$$
$$= 0.5 \times 438.75 + 0.5 \times 1143.09$$
$$= 219.375 + 571.545$$
$$= 790.92$$
The answer checks.

State. The area of the kite is 790.92 in^2.

77. Discussion and Writing Exercise

79. 3 5 7 $\underline{7}$ $\boxed{2}$

The ones digit is 4 or less so we round down to 3570.

81. 7 8, $\underline{9}$ $\boxed{5}$ 1

The tens digit is 5 or more so we round up to 79,000.

83. $\frac{n}{1} = n$, for any integer n.

Thus, $\frac{95}{-1} = \frac{-95}{1} = -95$.

85.
$$9 - 4 + 2 \div (-1) \cdot 6$$
$$= 9 - 4 - 2 \cdot 6 \qquad \begin{array}{l}\text{Multiplying and dividing in} \\ \text{order from left to right}\end{array}$$
$$= 9 - 4 - 12$$
$$= 5 - 12 \qquad \begin{array}{l}\text{Adding and subtracting in} \\ \text{order from left to right}\end{array}$$
$$= -7$$

87. Discussion and Writing Exercise

89. Using a calculator we find that
$$\frac{1}{7} = 1 \div 7 = 0.\overline{142857}.$$

91. Using a calculator we find that
$$\frac{3}{7} = 3 \div 7 = 0.\overline{428571}.$$

93. Using a calculator we find that
$$\frac{5}{7} = 5 \div 7 = 0.\overline{714285}.$$

95. Using a calculator we find that
$$\frac{1}{9} = 1 \div 9 = 0.\overline{1}.$$

97. Using a calculator we find that
$$\frac{1}{999} = 0.\overline{001}.$$

99. We substitute $\frac{22}{7}$ for π and 2.1 for r.

$$A = \pi r^2$$
$$= \frac{22}{7}(2.1)^2$$
$$= \frac{22}{7}(4.41)$$
$$= \frac{22 \times 4.41}{7}$$
$$= \frac{97.02}{7}$$
$$= 13.86 \text{ cm}^2$$

101. We substitute 3.14 for π and $\frac{3}{4}$ for r.

$$A = \pi r^2$$
$$= 3.14\left(\frac{3}{4}\right)^2$$
$$= 3.14\left(\frac{9}{16}\right)$$
$$= \frac{3.14 \times 9}{16}$$
$$= \frac{28.26}{16}$$
$$= 1.76625 \text{ ft}^2$$

When the calculation is done using the π key on a calculator, the result is 1.767145868 ft^2.

103. Discussion and Writing Exercise

Exercise Set 5.6

1. We are estimating the sum

$$\$279 + \$149.99.$$

We round both numbers to the nearest ten. The estimate is

$$\$280 + \$150 = \$430.$$

3. We are estimating the difference

$$\$279 - \$149.99.$$

We round both numbers to the nearest ten. The estimate is

$$\$280 - \$150 = \$130.$$

5. We are estimating the product

$$6 \times \$79.95.$$

We round $79.95 to the nearest ten. The estimate is

$$6 \times \$80 = \$480.$$

7. We are estimating the quotient

$$\$830 \div \$79.95.$$

We round $830 to the nearest hundred and $79.95 to the nearest ten. The estimate is

$$\$800 \div \$80 = 10 \text{ sets.}$$

9. This is about $0.0 + 1.3 + 0.3$, so the answer is about 1.6.

11. This is about $6 + 0 + 0$, so the answer is about 6.

13. This is about $52 + 1 + 7$, so the answer is about 60.

15. This is about $2.7 - 0.4$, so the answer is about 2.3.

17. This is about $200 - 20$, so the answer is about 180.

19. This is about 50×8, rounding 49 to the nearest ten and 7.89 to the nearest one, so the answer is about 400. Answer (a) is correct.

21. This is about 100×0.08, rounding 98.4 to the nearest ten and 0.083 to the nearest hundredth, so the answer is about 8. Answer (c) is correct.

23. This is about $4 \div 4$, so the answer is about 1. Answer (b) is correct.

25. This is about $75 \div 25$, so the answer is about 3. Answer (b) is correct.

27. We estimate the quotient $1760 \div 8.625$.

$$1800 \div 9 = 200$$

We estimate that 200 posts will be needed. Answers may vary depending on how the rounding was done.

29. Discussion and Writing Exercise

31. The decimal $0.57\overline{3}$ is an example of a <u>repeating</u> decimal.

33. The sentence $5(3+8) = 5 \cdot 3 + 5 \cdot 8$ illustrates the <u>distributive</u> law.

35. The number 1 is the <u>multiplicative</u> identity.

37. The least common <u>denominator</u> of two or more fractions is the least common <u>multiple</u> of their denominators.

39. Discussion and Writing Exercise

41. We round each factor to the nearest ten. The estimate is $180 \times 60 = 10,800$. The estimate is close to the result given, so the decimal point was placed correctly.

43. We round each number on the left to the nearest one. The estimate is $19 - 1 \times 4 = 19 - 4 = 15$. The estimate is not close to the result given, so the decimal point was not placed correctly.

45. a) Observe that $2^{13} = 8192 \approx 8000$, $156,876.8 \approx 160,000$, and $8000 \times 20 = 160,000$. Thus, we want to find the product of 2^{13} and a number that is approximately 20. Since $0.37 + 18.78 = 19.15 \approx 20$, we add inside the parentheses and then multiply:

$$(0.37 + 18.78) \times 2^{13} = 156,876.8$$

We can use a calculator to confirm this result.

   b) Observe that $312.84 \approx 6 \cdot 50$. We start by multiplying 6.4 and 51.2, getting 327.68. Then we can use a calculator to find that if we add 2.56 to this product and then subtract 17.4, we have the desired result. Thus, we have

$$2.56 + 6.4 \times 51.2 - 17.4 = 312.84.$$

Exercise Set 5.7

1. $5x = 27$

$\dfrac{5x}{5} = \dfrac{27}{5}$ Dividing both sides by 5

$x = 5.4$

Check: $\dfrac{5x = 27}{}$

$5(5.4) \ ? \ 27$

$27 \ | \ 27$ TRUE

The solution is 5.4.

3. $x + 15.7 = 3.1$

$x + 15.7 - 15.7 = 3.1 - 15.7$ Adding -15.7 to (or subtracting 15.7 from) both sides

$x = -12.6$

Check: $\dfrac{x + 15.7 = 3.1}{}$

$-12.6 + 15.7 \ ? \ 3.1$

$3.1 \ | \ 3.1$ TRUE

The solution is -12.6.

5. $5x - 8 = 22$

$5x - 8 + 8 = 22 + 8$ Adding 8 to both sides

$5x = 30$

$\dfrac{5x}{5} = \dfrac{30}{5}$ Dividing both sides by 5

$x = 6$

The solution is 6.

7. $6.9x - 8.4 = 4.02$

$6.9x - 8.4 + 8.4 = 4.02 + 8.4$ Adding 8.4 to both sides

$6.9x = 12.42$

$\dfrac{6.9x}{6.9} = \dfrac{12.42}{6.9}$ Dividing both sides by 6.9

$x = 1.8$

The solution is 1.8.

9. $21.6 + 4.1t = 6.43$

$21.6 + 4.1t - 21.6 = 6.43 - 21.6$ Subtracting 21.6 from both sides

$4.1t = -15.17$

$\dfrac{4.1t}{4.1} = \dfrac{-15.17}{4.1}$ Dividing both sides by 4.1

$t = -3.7$

The solution is -3.7.

11. $-26.05 = 7.5x + 9.2$

$-26.05 - 9.2 = 7.5x + 9.2 - 9.2$

$-35.25 = 7.5x$

$-4.7 = x$

The solution is -4.7.

13. $-4.2x + 3.04 = -4.1$

$-4.2x + 3.04 - 3.04 = -4.1 - 3.04$

$-4.2x = -7.14$

$\dfrac{-4.2x}{-4.2} = \dfrac{-7.14}{-4.2}$

$x = 1.7$

The solution is 1.7.

15. $-3.05 = 7.24 - 3.5t$

$-3.05 - 7.24 = 7.24 - 3.5t - 7.24$

$-10.29 = -3.5t$

$\dfrac{-10.29}{-3.5} = \dfrac{-3.5t}{-3.5}$

$2.94 = t$

The solution is 2.94.

17. $9x - 2 = 5x + 34$

$9x - 2 + 2 = 5x + 34 + 2$ Adding 2 to both sides

$9x = 5x + 36$

$9x - 5x = 5x + 36 - 5x$ Subtracting $5x$ from both sides

$4x = 36$

$\dfrac{4x}{4} = \dfrac{36}{4}$ Dividing both sides by 4

$x = 9$

Check: $\dfrac{9x - 2 = 5x + 34}{}$

$9 \cdot 9 - 2 \ ? \ 5 \cdot 9 + 34$

$81 - 2 \ | \ 45 + 34$

$79 \ | \ 79$ TRUE

The solution is 9.

19. $2x + 6 = 7x - 10$

$2x + 6 - 6 = 7x - 10 - 6$ Subtracting 6 from both sides

$2x = 7x - 16$

$2x - 7x = 7x - 16 - 7x$ Subtracting $7x$ from both sides

$-5x = -16$

$\dfrac{-5x}{-5} = \dfrac{-16}{-5}$ Dividing both sides by -5

$x = 3.2$

Check: $\dfrac{2x + 6 = 7x - 10}{}$

$2(3.2) + 6 \ ? \ 7(3.2) - 10$

$6.4 + 6 \ | \ 22.4 - 10$

$12.4 \ | \ 12.4$ TRUE

The solution is 3.2.

21.
$$5y - 3 = 4 + 9y$$
$$5y - 3 + 3 = 4 + 9y + 3$$
$$5y = 9y + 7$$
$$5y - 9y = 9y + 7 - 9y$$
$$-4y = 7$$
$$\frac{-4y}{-4} = \frac{7}{-4}$$
$$y = -1.75$$
The solution is -1.75.

23.
$$5.9x + 67 = 7.6x + 16$$
$$5.9x + 67 - 16 = 7.6x + 16 - 16$$
$$5.9x + 51 = 7.6x$$
$$5.9x + 51 - 5.9x = 7.6x - 5.9x$$
$$51 = 1.7x$$
$$\frac{51}{1.7} = \frac{1.7x}{1.7}$$
$$30 = x$$
The solution is 30.

25.
$$7.8a + 2 = 2.4a + 19.28$$
$$7.8a + 2 - 2 = 2.4a + 19.28 - 2$$
$$7.8a = 2.4a + 17.28$$
$$7.8a - 2.4a = 2.4a + 17.28 - 2.4a$$
$$5.4a = 17.28$$
$$\frac{5.4a}{5.4} = \frac{17.28}{5.4}$$
$$a = 3.2$$
The solution is 3.2

27.
$$6(x + 2) = 4x + 30$$
$$6x + 12 = 4x + 30 \quad \text{Using the distributive law}$$
$$6x + 12 - 12 = 4x + 30 - 12$$
$$6x = 4x + 18$$
$$6x - 4x = 4x + 18 - 4x$$
$$2x = 18$$
$$\frac{2x}{2} = \frac{18}{2}$$
$$x = 9$$
Check:

$6(x + 2) = 4x + 30$	
$6(9 + 2)$?	$4 \cdot 9 + 30$
$6(11)$	$36 + 30$
66	66

TRUE

The solution is 9.

29.
$$5(x + 3) = 15x - 6$$
$$5x + 15 = 15x - 6 \quad \text{Using the distributive law}$$
$$5x + 15 - 15 = 15x - 6 - 15$$
$$5x = 15x - 21$$
$$5x - 15x = 15x - 21 - 15x$$
$$-10x = -21$$
$$\frac{-10x}{-10} = \frac{-21}{-10}$$
$$x = 2.1$$

Check:

$5(x + 3) = 15x - 6$	
$5(2.1 + 3)$?	$15(2.1) - 6$
$5(5.1)$	$31.5 - 6$
25.5	25.5

TRUE

The solution is 2.1.

31.
$$7a - 9 = 15(a - 3)$$
$$7a - 9 = 15a - 45 \quad \text{Using the distributive law}$$
$$7a - 9 + 9 = 15a - 45 + 9$$
$$7a = 15a - 36$$
$$7a - 15a = 15a - 36 - 15a$$
$$-8a = -36$$
$$\frac{-8a}{-8} = \frac{-36}{-8}$$
$$a = 4.5$$
The solution is 4.5.

33.
$$2.9(x + 8.1) = 7.8x - 3.95$$
$$2.9x + 23.49 = 7.8x - 3.95$$
$$2.9x + 23.49 - 23.49 = 7.8x - 3.95 - 23.49$$
$$2.9x = 7.8x - 27.44$$
$$2.9x - 7.8x = 7.8x - 27.44 - 7.8x$$
$$-4.9x = -27.44$$
$$\frac{-4.9x}{-4.9} = \frac{-27.44}{-4.9}$$
$$x = 5.6$$
The solution is 5.6.

35.
$$-6.21 - 4.3t = 9.8(t + 2.1)$$
$$-6.21 - 4.3t = 9.8t + 20.58$$
$$-6.21 - 4.3t + 6.21 = 9.8t + 20.58 + 6.21$$
$$-4.3t = 9.8t + 26.79$$
$$-4.3t - 9.8t = 26.79$$
$$-14.1t = 26.79$$
$$\frac{-14.1t}{-14.1} = \frac{26.79}{-14.1}$$
$$t = -1.9$$
The solution is -1.9.

37. $4(x-2)-9 = 2x+9$

$4x-8-9 = 2x+9$

$4x-17 = 2x+9$

$4x-17+17 = 2x+9+17$

$4x = 2x+26$

$4x-2x = 2x+26-2x$

$2x = 26$

$\dfrac{2x}{2} = \dfrac{26}{2}$

$x = 13$

The solution is 13.

39. $43(7-2x)+34 = 50(x-4.1)+744$

$301-86x+34 = 50x-205+744$

$-86x+335 = 50x+539$

$-86x+335-335 = 50x+539-335$

$-86x = 50x+204$

$-86x-50x = 50x+204-50x$

$-136x = 204$

$\dfrac{-136x}{-136} = \dfrac{204}{-136}$

$x = -1.5$

The solution is -1.5.

41. Discussion and Writing Exercise

43. We use the formula $A = \dfrac{1}{2}\cdot b\cdot h$ and substitute 7 m for b and 4 m for h.

$A = \dfrac{1}{2}\cdot b\cdot h$

$= \dfrac{1}{2}\cdot 7\text{ m}\cdot 4\text{ m}$

$= \dfrac{7\cdot 4}{2}\text{ m}^2$

$= 14\text{ m}^2$

45. We use the formula $A = \dfrac{1}{2}\cdot b\cdot h$ and substitute 5 in. for b and 5 in. for h.

$A = \dfrac{1}{2}\cdot b\cdot h$

$= \dfrac{1}{2}\cdot 5\text{ in.}\cdot 5\text{ in.}$

$= \dfrac{5\cdot 5}{2}\text{ in}^2$

$= \dfrac{25}{2}\text{ in}^2$, or 12.5 in^2

47. The area of the figure is the sum of the areas of two triangles, each with base 5 ft and height 1 ft. Then we have

$A = \dfrac{1}{2}\cdot b\cdot h + \dfrac{1}{2}\cdot b\cdot h$

$= \dfrac{1}{2}\cdot 5\text{ ft}\cdot 1\text{ ft} + \dfrac{1}{2}\cdot 5\text{ ft}\cdot 1\text{ ft}$

$= \dfrac{5\cdot 1}{2}\text{ ft}^2 + \dfrac{5\cdot 1}{2}\text{ ft}^2$

$= \dfrac{5}{2}\text{ ft}^2 + \dfrac{5}{2}\text{ ft}^2$

$= 5\text{ ft}^2$

49. $\dfrac{3}{25}-\dfrac{7}{10} = \dfrac{3}{25}\cdot\dfrac{2}{2}-\dfrac{7}{10}\cdot\dfrac{5}{5}$ The LCM is 50.

$= \dfrac{6}{50}-\dfrac{35}{50}$

$= -\dfrac{29}{50}$

51. We add in order from left to right.

$-17+24+(-9) = 7+(-9) = -2$

53. Discussion and Writing Exercise

55. $7.035(4.91x-8.21)+17.401 =$

$23.902x-7.372815$

$34.54185x-57.75735+17.401 =$

$23.902x-7.372815$

$34.54185x-40.35635 =$

$23.902x-7.372815$

$34.54185x-40.35635-23.902x =$

$23.902x-7.372815-23.902x$

$10.63985x-40.35635 = -7.372815$

$10.63985x-40.35635+40.35635 =$

$-7.372815+40.35635$

$10.63985x = 32.983535$

$\dfrac{10.63985x}{10.63985} = \dfrac{32.983535}{10.63985}$

$x = 3.1$

The solution is 3.1.

57. $5(x-4.2)+3[2x-5(x+7)] =$

$39+2(7.5-6x)+3x$

$5(x-4.2)+3[2x-5x-35] =$

$39+2(7.5-6x)+3x$

$5(x-4.2)+3[-3x-35] = 39+2(7.5-6x)+3x$

$5x-21-9x-105 = 39+15-12x+3x$

$-4x-126 = 54-9x$

$-4x-126+9x = 54-9x+9x$

$5x-126 = 54$

$5x-126+126 = 54+126$

$5x = 180$

$\dfrac{5x}{5} = \dfrac{180}{5}$

$x = 36$

The solution is 36.

59. $3.5(4.8x - 2.9) + 4.5 = 9.4x - 3.4(x - 1.9)$

$16.8x - 10.15 + 4.5 = 9.4x - 3.4x + 6.46$

$16.8x - 5.65 = 6x + 6.46$

$16.8x - 5.65 - 6x = 6x + 6.46 - 6x$

$10.8x - 5.65 = 6.46$

$10.8x - 5.65 + 5.65 = 6.46 + 5.65$

$10.8x = 12.11$

$\dfrac{10.8x}{10.8} = \dfrac{12.11}{10.8}$

$x \approx 1.1212963$

The solution is approximately 1.1212963.

Exercise Set 5.8

1. Familiarize. Repeated addition fits this situation. We let C = the cost of 7 jackets.

$$\underbrace{\boxed{\$32.98} + \boxed{\$32.98} + \cdots + \boxed{\$32.98}}_{\text{7 addends}}$$

Translate.

Price per jacket	times	Number of jackets	is	Total cost
↓	↓	↓	↓	↓
32.98	×	7	=	C

Solve. We carry out the multiplication.

$$\begin{array}{r} 3\,2.\,9\,8 \\ \times \qquad 7 \\ \hline 2\,3\,0.\,8\,6 \end{array}$$

Thus, $C = 230.86$.

Check. We obtain a partial check by rounding and estimating:

$$32.98 \times 7 \approx 30 \times 7 = 210 \approx 230.86.$$

State. Seven jackets cost $230.86.

3. Familiarize. Repeated addition fits this situation. We let c = the cost of 20.4 gal of gasoline, in dollars.

Translate.

Cost per gallon	times	Number of gallons	is	Total cost
↓	↓	↓	↓	↓
2.249	·	20.4	=	c

Solve. We carry out the multiplication.

$$\begin{array}{r} 2.\,2\,4\,9 \\ \times \qquad 2\,0.\,4 \\ \hline 8\,9\,9\,6 \\ 4\,4\,9\,8\,0\,0 \\ \hline 4\,5.\,8\,7\,9\,6 \end{array}$$

Thus, $c = 45.8796$.

Check. We obtain a partial check by rounding and estimating:

$$2.249 \times 20.4 \approx 2.25 \times 20 = 45 \approx 45.8796.$$

State. We round $45.8796 to the nearest cent and find that the cost of the gasoline is $45.88.

5. Familiarize. We visualize the situation. We let n = the new temperature.

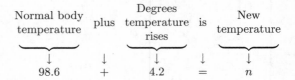

Translate. We are combining amounts.

Normal body temperature	plus	Degrees temperature rises	is	New temperature
↓	↓	↓	↓	↓
98.6	+	4.2	=	n

Solve. To solve the equation we carry out the addition.

$$\begin{array}{r} \overset{1}{9}\,8.6 \\ +\quad 4.2 \\ \hline 1\,0\,2.8 \end{array}$$

Thus, $n = 102.8$.

Check. We can check by repeating the addition. We can also check by rounding:

$$98.6 + 4.2 \approx 99 + 4 = 103 \approx 102.8$$

State. The new temperature was 102.8°F.

7. Familiarize. We visualize the situation. Let w = each winner's share.

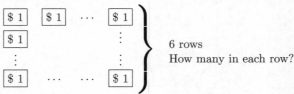

Translate.

Total prize	÷	Number of winners	=	Each winner's share
↓	↓	↓	↓	↓
127,315	÷	6	=	w

Solve. We carry out the division.

$$
\begin{array}{r}
2\,1,2\,1\,9.\,1\,6\,6 \\
6\,\overline{\smash{)}\,1\,2\,7,3\,1\,5.\,0\,0\,0} \\
\underline{1\,2\,0\,0\,0\,0} \\
7\,3\,1\,5 \\
\underline{6\,0\,0\,0} \\
1\,3\,1\,5 \\
\underline{1\,2\,0\,0} \\
1\,1\,5 \\
\underline{6\,0} \\
5\,5 \\
\underline{5\,4} \\
1\,0 \\
\underline{6} \\
4\,0 \\
\underline{3\,6} \\
4\,0 \\
\underline{3\,6} \\
4
\end{array}
$$

Rounding to the nearest cent, or hundredth, we get $w =$ 21,219.17.

Check. We can repeat the calculation. The answer checks.

State. Each winner's share is $21,219.17.

9. *Familiarize*. Let $A =$ the area, in sq cm, and $P =$ the perimeter, in cm.

Translate. We use the formulas $A = l \cdot w$ and $P = l + w + l + w$ and substitute 3.25 for l and 2.5 for w.

$$A = l \cdot w = (3.25) \cdot (2.5)$$

$$P = l + w + l + w = 3.25 + 2.5 + 3.25 + 2.5$$

Solve. To find the area we carry out the multiplication.

$$
\begin{array}{r}
3.\,2\,5 \\
\times\,2.\,5 \\
\hline
1\,6\,2\,5 \\
6\,5\,0\,0 \\
\hline
8.\,1\,2\,5
\end{array}
$$

Thus, $A = 8.125$

To find the perimeter we carry out the addition.

$$
\begin{array}{r}
3.\,2\,5 \\
2.\,5 \\
3.\,2\,5 \\
+\,2.\,5 \\
\hline
1\,1.\,5\,0
\end{array}
$$

Then $P = 11.5$.

Check. We can obtain partial checks by estimating.

$(3.25) \times (2.5) \approx 3 \times 3 \approx 9 \approx 8.125$

$3.25 + 2.5 + 3.25 + 2.5 \approx 3 + 3 + 3 + 3 = 12 \approx 11.5$

The answers check.

State. The area of the stamp is 8.125 sq cm, and the perimeter is 11.5 cm.

11. *Familiarize*. We visualize the situation. We let $m =$ the odometer reading at the end of the trip.

22,456.8 mi	234.7 mi
m	

Translate. We are combining amounts.

Reading before trip	plus	Miles driven	is	Reading at end of trip
↓	↓	↓	↓	↓
22,456.8	+	234.7	=	m

Solve. To solve the equation we carry out the addition.

$$
\begin{array}{r}
\overset{1\ \ 1}{2\,2,4\,5\,6.8} \\
+\quad 2\,3\,4.7 \\
\hline
2\,2,6\,9\,1.5
\end{array}
$$

Thus, $m = 22,691.5$.

Check. We can check by repeating the addition. We can also check by rounding:

$22,456.8 + 234.7 \approx 22,460 + 230 = 22,690 \approx 22,691.5$

State. The odometer reading at the end of the trip was 22,691.5.

13. *Familiarize*. Let $c =$ the amount of change.

Translate. We subtract the price of the DVD and the amount of the sales tax from $50.

$$c = 50 - 29.24 - 1.61$$

Solve. We carry out the subtractions in order from left to right.

$$
\begin{array}{r}
\overset{4\ \ 9\ \ 9\ 10}{5\,0.\,0\,\cancel{0}} \\
-\,2\,9.\,2\,4 \\
\hline
2\,0.\,7\,6
\end{array}
\qquad
\begin{array}{r}
\overset{1\ 10}{\cancel{2}\,\cancel{0}.\,7\,6} \\
-\quad 1.\,6\,1 \\
\hline
1\,9.\,1\,5
\end{array}
$$

Check. If we add the cost of the DVD and the amount of the sales tax to the change we should get $50.

$$\$19.15 + \$29.24 + \$1.61 = \$48.39 + \$1.61 = \$50$$

The answer checks.

State. Andrew received $19.15 in change.

15. *Familiarize*. We visualize the situation. We let $d =$ the number of degrees Wanda's temperature dropped.

103.2°F	
99.7°F	d

Translate. This is a "take-away" situation.

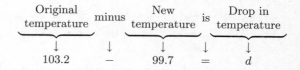

Original temperature	minus	New temperature	is	Drop in temperature
↓	↓	↓	↓	↓
103.2	−	99.7	=	d

Solve. We carry out the subtraction.

$$\begin{array}{r} \overset{9}{\cancel{1}}\,\overset{12}{\cancel{0}}\,\overset{12}{\cancel{3}}.\overset{}{\cancel{2}} \\ -\ \ 9\ 9.7 \\ \hline 3.5 \end{array}$$

Thus, $d = 3.5$.

Check. We check by adding 3.5 to 99.7 to get 103.2. The answer checks.

State. Wanda's temperature dropped 3.5°F.

17. Familiarize. Let c = the cost per serving.

Translate.

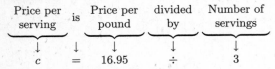

$$c\ \ =\ \ 16.95\ \ \div\ \ 3$$

Solve. We carry out the division.

$$\begin{array}{r} 5.6\,5 \\ 3\overline{\smash{\big)}\,16.9\,5} \\ 15\,0\,0 \\ \hline 1\,9\,5 \\ 1\,8\,0 \\ \hline 1\,5 \\ 1\,5 \\ \hline 0 \end{array}$$

Thus, $c = 5.65$.

Check. We can check by multiplying: $3 \times 5.65 = 16.95$. The answer checks.

State. The cost per serving is $5.65.

19. Familiarize. We are combining amounts. Let d = the total number of gallons of liquids the average U.S. citizen drinks each year.

Translate.

$$\underset{\text{drinks}}{\text{Soft}} + \text{Water} + \text{Milk} + \text{Coffee} + \underset{\text{juice}}{\text{Fruit}} = \text{Total}$$

$$49.0 + 41.2 + 25.3 + 24.8 + 7.8 = d$$

Solve. We carry out the addition.

$$\begin{array}{r} \overset{2\ 2}{} \\ 4\,9.0 \\ 4\,1.2 \\ 2\,5.3 \\ 2\,4.8 \\ +\ \ 7.8 \\ \hline 1\,4\,8.1 \end{array}$$

Thus, $d = 148.1$.

Check. We repeat the calculation. The answer checks.

State. The average U.S. citizen drinks 148.1 gallons of liquids each year.

21. Familiarize. This is a two-step problem. First, we find the number of miles that have been driven between fillups. This is a "how-much-more" situation. We let n = the number of miles driven.

Translate and Solve.

$$\underset{\text{reading}}{\underset{\text{odometer}}{\text{First}}} \text{ plus } \underset{\text{driven}}{\underset{\text{of miles}}{\text{Number}}} \text{ is } \underset{\text{reading}}{\underset{\text{odometer}}{\text{Second}}}$$

$$26{,}342.8\ +\ n\ =\ 26{,}736.7$$

To solve the equation we subtract 26,342.8 on both sides.

$$n = 26{,}736.7 - 26{,}342.8$$
$$n = 393.9$$

$$\begin{array}{r} 2\,6,7\,3\,6.7 \\ -\ 2\,6,3\,4\,2.8 \\ \hline 3\,9\,3.9 \end{array}$$

Second, we divide the total number of miles driven by the number of gallons. This gives us m = the number of miles per gallon.

$$393.9 \div 19.5 = m$$

To find the number m, we divide.

$$\begin{array}{r} 2\,0.2 \\ 19.5_{\wedge}\overline{\smash{\big)}\,393.9_{\wedge}0} \\ 3\,9\,0\,0 \\ \hline 3\,9\ \,0 \\ 3\,9\ \,0 \\ \hline 0 \end{array}$$

Thus, $m = 20.2$.

Check. To check, we first multiply the number of miles per gallon times the number of gallons:

$$19.5 \times 20.2 = 393.9$$

Then we add 393.9 to 26,342.8:

$$26{,}342.8 + 393.9 = 26{,}736.7$$

The number 20.2 checks.

State. The van gets 20.2 miles per gallon.

23. Familiarize. Let a = the amount of insulin consumed in a week, in cubic centimeters. Note that 38 units of insulin corresponds to 0.38 cc. (See Example 2.)

Translate.

$$\underset{\text{each day}}{\underset{\text{used}}{\text{Amount}}} \text{ times } \underset{\text{a week}}{\underset{\text{days in}}{\text{Number of}}} \text{ is } \underset{\text{used}}{\underset{\text{amount}}{\text{Total}}}$$

$$0.38 \times 7 = a$$

Solve. We carry out the multiplication.

$$\begin{array}{r} 0.3\,8 \\ \times\ \ \ 7 \\ \hline 2.6\,6 \end{array}$$

Thus, $a = 2.66$.

Check. We can approximate the product.

$$0.38 \times 7 \approx 0.4 \times 7 = 2.8 \approx 2.66$$

The answer checks.

State. Phil consumes 2.66 cc of insulin in a week.

25. Familiarize. This is a multi-step problem. We find the total area of the poster and the area devoted to the painting and then we subtract to find the area not devoted to the painting.

Translate and Solve. First we use the formula $A = l \times w$ to find the total area of the poster.

$A = l \times w$

$= 27.4 \text{ in.} \times 19.3 \text{ in.}$

$= 528.82 \text{ in}^2$

$$\begin{array}{r} 2\ 7.\ 4 \\ \times\ 1\ 9.\ 3 \\ \hline 8\ 2\ 2 \\ 2\ 4\ 6\ 6\ 0 \\ 2\ 7\ 4\ 0\ 0 \\ \hline 5\ 2\ 8.\ 8\ 2 \end{array}$$

Next we use the formula $A = l \times w$ again to find the area of the painting.

$A = l \times w$

$= 18.8 \text{ in.} \times 15.7 \text{ in.}$

$= 295.16 \text{ in}^2$

$$\begin{array}{r} 1\ 8.\ 8 \\ \times\ 1\ 5.\ 7 \\ \hline 1\ 3\ 1\ 6 \\ 9\ 4\ 0\ 0 \\ 1\ 8\ 8\ 0\ 0 \\ \hline 2\ 9\ 5.\ 1\ 6 \end{array}$$

Let a = the area not devoted to the painting. We subtract to find this area.

$a = 528.82 \text{ in}^2 - 295.16 \text{ in}^2$

$= 233.66 \text{ in}^2$

$$\begin{array}{r} {\scriptstyle 4\ 12\quad 7\ 12} \\ \cancel{5}\ \cancel{2}\ 8.\ \cancel{8}\ \cancel{2} \\ -\ 2\ 9\ 5.\ 1\ 6 \\ \hline 2\ 3\ 3.\ 6\ 6 \end{array}$$

Check. We repeat the calculations. The answer checks.

State. The area not devoted to the painting is 233.66 in^2.

27. ***Familiarize***. Let n = the area available for notes. The total area available is twice the area of one side of the card. Recall that the formula for the area of a rectangle with length l and width w is $A = l \times w$.

Translate.

$$\underbrace{\text{Total area}}_{\downarrow \atop n} \ \underset{\downarrow \atop =}{\text{is}} \ \underset{\downarrow \atop 2\times}{\text{twice}} \ \underbrace{\text{Area of one side}}_{\downarrow \atop 12.7 \times 7.6}$$

Solve. We find the product.

$n = 2 \times 12.7 \times 7.6$

$= 25.4 \times 7.6$

$= 193.04$

$$\begin{array}{r} 2\ 5.\ 4 \\ \times\ 7.\ 6 \\ \hline 1\ 5\ 2\ 4 \\ 1\ 7\ 7\ 8\ 0 \\ \hline 1\ 9\ 3.\ 0\ 4 \end{array}$$

Check. We can approximate the product.

$2 \times 12.7 \times 7.6 \approx 2 \times 13 \times 7.5 = 195 \approx 193.04$

The answer checks.

State. The area available for notes is 193.04 cm^2.

29. ***Familiarize.*** This is a two-step problem. First, we find the number of games that can be played in one hour. Think of an array containing 60 minutes (1 hour = 60 minutes) with 1.5 minutes in each row. We want to find how many rows there are. We let g represent this number.

Translate and Solve. We think (Number of minutes) $\div$ (Number of minutes per game) = (Number of games).

$60 \div 1.5 = g$

To solve the equation we carry out the division.

$$\begin{array}{r} 4\ 0. \\ 1.5_{\wedge}\overline{\smash{)}6\ 0.\ 0_{\wedge}} \\ \underline{6\ 0\ 0} \\ 0 \\ \underline{0} \\ 0 \end{array}$$

Thus, $g = 40$.

Second, we find the cost t of playing 40 video games. Repeated addition fits this situation. (We express 75¢ as \$0.75.)

$$\underbrace{\text{Cost of} \atop \text{one game}}_{\downarrow \atop 0.75} \ \underset{\downarrow \atop \times}{\text{times}} \ \underbrace{\text{Number of} \atop \text{games played}}_{\downarrow \atop 40} \ \underset{\downarrow \atop =}{\text{is}} \ \underbrace{\text{Total} \atop \text{cost}}_{\downarrow \atop t}$$

To solve the equation we carry out the multiplication.

$$\begin{array}{r} 0.\ 7\ 5 \\ \times\ 4\ 0 \\ \hline 3\ 0.\ 0\ 0 \end{array}$$

Thus, $t = 30$.

Check. To check, we first divide the total cost by the cost per game to find the number of games played:

$30 \div 0.75 = 40$

Then we multiply 40 by 1.5 to find the total time:

$1.5 \times 40 = 60$

The number 30 checks.

State. It costs \$30 to play video games for one hour.

31. ***Familiarize***. This is a multistep problem. First we find the sum s of the two 0.8 cm segments. Then we use this length to find d.

Translate and Solve.

$$\underbrace{\text{Length of one} \atop \text{small segment}}_{\downarrow \atop 0.8} \ \underset{\downarrow \atop +}{\text{plus}} \ \underbrace{\text{Length of other} \atop \text{small segment}}_{\downarrow \atop 0.8} \ \underset{\downarrow \atop =}{\text{is}} \ \underbrace{\text{Total} \atop \text{length}}_{\downarrow \atop s}$$

To solve we carry out the addition.

$$\begin{array}{r} {\scriptstyle 1} \\ 0.\ 8 \\ +\ 0.\ 8 \\ \hline 1.\ 6 \end{array}$$

Thus, $s = 1.6$.

Now we find d.

$$\underbrace{\text{Total length of} \atop \text{smaller segments}}_{\downarrow \atop 1.6} \ \underset{\downarrow \atop +}{\text{plus}} \ \underbrace{\text{length} \atop \text{of } d}_{\downarrow \atop d} \ \underset{\downarrow \atop =}{\text{is}} \ \underset{\downarrow \atop 3.91}{3.91 \text{ cm}}$$

To solve we subtract 1.6 on both sides of the equation.

$d = 3.91 - 1.6$

$d = 2.31$

$$\begin{array}{r} 3.\ 9\ 1 \\ -\ 1.\ 6\ 0 \\ \hline 2.\ 3\ 1 \end{array}$$

Check. We repeat the calculations.

State. The length d is 2.31 cm.

33. Familiarize. We make and label a drawing. The question deals with a rectangle and a circle, so we also list the relevant area formulas. We let $d =$ the amount of decking needed.

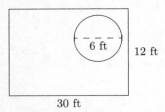

Area of a rectangle with length l and width w:
$A = l \times w$

Area of a circle with radius r: $A = \pi r^2$, where $\pi \approx 3.14$

Translate. We subtract the area of the circle from the area of the rectangle. Recall that a circle's radius is half of its diameter.

Area of rectangle	minus	Area of circle	is	Area covered by decking
$\downarrow$	$\downarrow$	$\downarrow$	$\downarrow$	$\downarrow$
30×12	$-$	$3.14\left(\dfrac{6}{2}\right)^2$	$=$	d

Solve. We carry out the computations.

$$30 \times 12 - 3.14\left(\frac{6}{2}\right)^2 = d$$
$$30 \times 12 - 3.14(3)^2 = d$$
$$30 \times 12 - 3.14 \times 9 + d$$
$$360 - 28.26 = d$$
$$331.74 = d$$

Check. We can repeat the calculations. Also note that 331.74 is less than the area of the yard but more than the area of the flower garden. This agrees with the impression given by our drawing.

State. The amount of decking needed is 331.74 ft^2.

35. Familiarize. This is a multistep problem. First we find the number of minutes in excess of 450. Then we find the charge for the excess minutes. Finally we add this charge to the monthly charge for 450 minutes to find the total cost for the month. Let $m =$ the number of minutes in excess of 450.

Translate and Solve. First we have a "how much more" situation.

First 450 minutes	plus	Excess minutes	is	Total minutes
$\downarrow$	$\downarrow$	$\downarrow$	$\downarrow$	$\downarrow$
450	$+$	m	$=$	479

We subtract 450 on both sides of the equation.

$$450 + m = 479$$
$$450 + m - 450 = 479 - 450$$
$$m = 29$$

We see that 29 minutes are charged at the rate of $0.45 per minute. We multiply to find c, the cost of these minutes.

$$\begin{array}{r} 2\,9 \\ \times\ 0.\,4\,5 \\ \hline 1\,4\,5 \\ 1\,1\,6\,0 \\ \hline 1\,3.\,0\,5 \end{array}$$

Thus, $c = 13.05$.

Finally we add the cost of the first 450 minutes and the cost of the additional 29 minutes to find t, the total cost for the month.

$$\begin{array}{r} 3\,9.\,9\,9 \\ +\ 1\,3.\,0\,5 \\ \hline 5\,3.\,0\,4 \end{array}$$

Thus, $t = 53.04$.

Check. We can repeat the calculations. The answer checks.

State. The total cost for the month was $53.04.

37. Familiarize. Let $m =$ the number of megabytes Nikki paid for. We will express 9 cents as $0.09.

Translate.

Processing fee	plus	$0.09	times	Number of megabytes	is	Total bill
$\downarrow$	$\downarrow$	$\downarrow$	$\downarrow$	$\downarrow$	$\downarrow$	$\downarrow$
10	$+$	0.09	$\cdot$	m	$=$	88.75

Solve.

$$10 + 0.09 \cdot m = 88.75$$
$$10 + 0.09 \cdot m - 10 = 88.75 - 10$$
$$0.09 \cdot m = 78.75$$
$$\frac{0.09 \cdot m}{0.09} = \frac{78.75}{0.09}$$
$$m = 875$$

Check. $0.09 \cdot 875 = \$78.75$ and $\$78.75 + \$10 = \$88.75$, so the answer checks.

State. Nikki paid for 875 megabytes.

39. Familiarize. This is a multistep problem. First we find the amount charged for minutes in excess of 900. Then we find the number of minutes in excess of 900 and finally we find the total number of minutes used. Let $c =$ the amount charged for the minutes in excess of 900.

Translate and Solve.

Access fee	plus	Additional fees	plus	Charge for excess minutes	is	Total bill
$\downarrow$	$\downarrow$	$\downarrow$	$\downarrow$	$\downarrow$	$\downarrow$	$\downarrow$
59.99	$+$	5.79	$+$	c	$=$	89.78

We solve this equation.

$$59.99 + 5.79 + c = 89.78$$
$$65.78 + c = 89.78 \qquad \text{Adding on the left side}$$
$$65.78 + c - 65.78 = 89.78 - 65.78$$
$$c = 24$$

We see that $24 was charged for the number of minutes used in excess of 900 minutes. Let $m =$ the number of minutes in excess of 900. We will express 40 cents as $0.40.

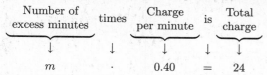

We solve this equation.

$$m \cdot 0.40 = 24$$
$$\frac{m \cdot 0.40}{0.40} = \frac{24}{0.40}$$
$$m = 60$$

Finally let $t =$ the total number of minutes Jeff used. We add.

Anytime minutes plus Excess minutes is Total number of minutes

$$900 \quad + \quad 60 \quad = \quad t$$

We solve the equation.

$$900 + 60 = t$$
$$960 = t$$

Check. We can repeat the calculations. The answer checks.

State. Jeff used 960 minutes.

41. Familiarize. Let $m =$ the number of minutes of video capture the sorority can buy. Then $m - 30 =$ the number of minutes in excess of 30 minutes. We will express 50 cents as $0.50.

Translate.

Charge for first 30 minutes plus Number of minutes in excess of 30 times Charge per minute is Total charge

$$50 \quad + \quad (m - 30) \quad \cdot \quad 0.50 \quad = \quad 95$$

Solve.

$$50 + (m - 30) \cdot 0.50 = 95$$
$$50 + 0.50m - 15 = 95$$
$$35 + 0.50m = 95$$
$$35 + 0.50m - 35 = 95 - 35$$
$$0.50m = 60$$
$$\frac{0.50m}{0.50} = \frac{60}{0.50}$$
$$m = 120$$

Check. If 120 min are captured, then the number of minutes in excess of 30 is $120 - 30$, or 90 minutes.

The charge for the excess minutes is $90 \cdot \$0.50$, or $45, and $45 + \$50 = \$95$. The answer checks.

State. The sorority can buy 120 min of video capture.

43. Familiarize. Let $c =$ the number of credit card transactions processed. Then $c - 500 =$ the number of transactions in excess of 500. We will express 10 cents as $0.10.

Translate.

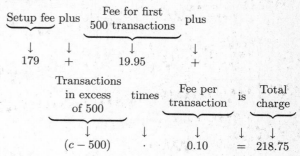

Solve.

$$179 + 19.95 + (c - 500) \cdot 0.10 = 218.75$$
$$179 + 19.95 + 0.10c - 50 = 218.75$$
$$148.95 + 0.10c = 218.75$$
$$148.95 + 0.10c - 193.95 = 218.75 - 148.95$$
$$0.10c = 69.8$$
$$\frac{0.10c}{0.10} = \frac{69.8}{0.10}$$
$$c = 698$$

Check. The number of transactions in excess of 500 is $698 - 500$, or 198. At $0.10 each, the fee for processing 198 transactions is $0.10(198)$, or $19.80. Then the total charge is

$$\$179 + \$19.95 + \$19.80, \text{ or } \$218.75.$$

The answer checks.

State. VeriSign processed 698 credit card transactions.

45. Familiarize. This is a three-step problem. We will find the area S of a standard soccer field and the area F of a standard football field using the formula Area $= l \cdot w$. Then we will find E, the amount by which the area of a soccer field exceeds the area of a football field.

Translate and Solve.

$$S = l \cdot w = 114.9 \times 74.4 = 8548.56$$
$$F = l \cdot w = 120 \times 53.3 = 6396$$

Area of football field plus Excess area of soccer field is Area of soccer field

$$6396 \quad + \quad E \quad = \quad 8548.56$$

To solve the equation we subtract 6396 on both sides.

$$E = 8548.56 - 6396$$
$$E = 2152.56$$

$$\begin{array}{r} {}^{4\ 14} \\ 8\,\cancel{5}\,\cancel{4}\,8.\,5\,6 \\ -\ 6\,3\,9\,6.\,0\,0 \\ \hline 2\,1\,5\,2.\,5\,6 \end{array}$$

Check. We can obtain a partial check by rounding and estimating:

$$114.9 \times 74.4 \approx 110 \times 75 = 8250 \approx 8548.56$$
$$120 \times 53.3 \approx 120 \times 50 = 6000 \approx 6396$$

$8250 - 6000 = 2250 \approx 2152.56$

State. The area of a soccer field is 2152.56 yd^2 greater than the area of a football field.

47. Familiarize. This is a multistep problem. First, we find the cost of the cheese. We let $c =$ the cost of the cheese.

Translate and Solve.

$$\underbrace{\text{Number}}_{6} \underbrace{\text{times}}_{\cdot} \underbrace{\text{Price per}}_{\$4.79} \underbrace{\text{is}}_{=} \underbrace{\text{Cost of}}_{c}$$

Number of pounds times Price per pound is Cost of cheese

To solve the equation we carry out the multiplication.

$$\begin{array}{r} \$4.\,7\,9 \\ \times \quad 6 \\ \hline \$2\,8.\,7\,4 \end{array}$$

Thus, $c = \$28.74$.

Next, we subtract to find how much money m is left to purchase seltzer.

$$m = \$40 - \$28.74 \qquad \begin{array}{r} {\scriptstyle 3\ 9\ 9\ 10} \\ \cancel{4\,0.\,0\,0} \\ -\,2\,8.\,7\,4 \\ \hline 1\,1.\,2\,6 \end{array}$$
$$m = \$11.26$$

Finally, we divide the amount of money left over by the cost of a bottle of seltzer to find how many bottles can be purchased. We let $b =$ the number of bottles of seltzer that can be purchased.

$$\$11.26 \div \$0.64 = b$$

To find b we carry out the division.

$$\begin{array}{r} 1\,7. \\ 0.6\,4_\wedge\overline{)\,1\,1.2\,6_\wedge} \\ 6\,4\,0 \\ \hline 4\,8\,6 \\ 4\,4\,8 \\ \hline 3\,8 \end{array}$$

We stop dividing at this point, because Frank cannot purchase a fraction of a bottle. Thus, $b = 17$ (rounded to the nearest 1).

Check. The cost of the seltzer is $17 \cdot \$0.64$ or $\$10.88$. The cost of the cheese is $6 \cdot \$4.79$, or $\$28.74$. Frank has spent a total of $\$10.88 + \$28.74$, or $\$39.62$. Frank has $\$40 - \$39.62$, or $\$0.38$ left over. This is not enough to purchase another bottle of seltzer, so our answer checks.

State. Frank should buy 17 bottles of seltzer.

49. Discussion and Writing Exercise

51. $\dfrac{0}{n} = 0$, for any integer n that is not 0.

Thus, $\dfrac{0}{-13} = 0$.

53. $\dfrac{8}{11} - \dfrac{4}{3} = \dfrac{8}{11} \cdot \dfrac{3}{3} - \dfrac{4}{3} \cdot \dfrac{11}{11}$ \qquad The LCM is 33.

$$= \dfrac{24}{33} - \dfrac{44}{33}$$
$$= \dfrac{-20}{33}, \text{ or } -\dfrac{20}{33}$$

55.
$$\begin{array}{r} 4\dfrac{1}{3} = \quad 4\dfrac{1}{3} \cdot \dfrac{2}{2} = \quad 4\dfrac{2}{6} \\ +\,2\dfrac{1}{2} = +\,2\dfrac{1}{2} \cdot \dfrac{3}{3} = +\,2\dfrac{3}{6} \\ \hline 6\dfrac{5}{6} \end{array}$$

57. Discussion and Writing Exercise

59. Discussion and Writing Exercise

61. Familiarize. We will use the formula Distance = Speed × Time in the form Time = Distance ÷ Speed. Let $t =$ the difference in the travel times for the two routes, in hours.

Translate.

Time for faster route minus Time for slower route is Difference in times

$$\underbrace{\dfrac{7.6}{65}}_{} \underbrace{-}_{} \underbrace{\dfrac{5.6}{50}}_{} \underbrace{=}_{} \underbrace{t}_{}$$

Solve. We carry out the subtraction. First we multiply each fraction by 1 in the form $\dfrac{10}{10}$ to eliminate the decimals in the numerators.

$$\dfrac{7.6}{65} - \dfrac{5.6}{50} = t$$
$$\dfrac{7.6}{65} \cdot \dfrac{10}{10} - \dfrac{5.6}{50} \cdot \dfrac{10}{10} = t$$
$$\dfrac{76}{650} - \dfrac{56}{500} = t$$
$$\dfrac{76}{650} \cdot \dfrac{10}{10} - \dfrac{56}{500} \cdot \dfrac{13}{13} = t \qquad \text{The LCD is 6500.}$$
$$\dfrac{760}{6500} - \dfrac{728}{6500} = t$$
$$\dfrac{32}{6500} = t$$
$$\dfrac{\cancel{4} \cdot 8}{\cancel{4} \cdot 1625} = t$$
$$\dfrac{8}{1625} = t$$

The difference in times is $\dfrac{8}{1625}$ hr. We can convert this time to minutes as follows.

$$\dfrac{8}{1625} \text{ hr} = \dfrac{8}{1625} \text{ hr} \cdot \dfrac{60 \text{ min}}{1 \text{ hr}}$$
$$= \dfrac{8 \cdot 60}{1625} \cdot \dfrac{\text{hr}}{\text{hr}} \cdot \text{min}$$
$$= \dfrac{8 \cdot \cancel{5} \cdot 12}{\cancel{5} \cdot 325} \text{ min} \qquad \left(\dfrac{\text{hr}}{\text{hr}} = 1\right)$$
$$= \dfrac{96}{325} \text{ min}$$

We can convert this to seconds.

$$\dfrac{96}{325} \text{ min} = \dfrac{96}{365} \text{ min} \cdot \dfrac{60 \text{ sec}}{1 \text{ min}}$$
$$= \dfrac{96 \cdot 60}{325} \cdot \dfrac{\text{min}}{\text{min}} \cdot \text{sec}$$
$$= \dfrac{5760}{325} \text{ sec}$$
$$\approx 18 \text{ sec}$$

Check. We can repeat the calculations. The answer checks.

State. You can save $\frac{96}{325}$ min, or about 18 sec, by taking the faster route.

63. We must make some assumptions. First we assume that the figures are nested squares formed by connecting the midpoints of consecutive sides of the next larger square. Next assume that the shaded area is the same as the area of the innermost square. (It appears that if we folded the shaded area into the innermost square, it would exactly fill the square.) Finally assume that the length of a side of the innermost square is 5 cm. (If we project the vertices of the innermost square onto the corresponding sides of the largest square, it appears that the distance between each projection and the nearest vertex of the largest square is one-fourth the length of a side of the largest square. Thus, the distance between projections on each side of the largest square is $\frac{1}{2} \cdot 10$ cm, or 5 cm and, hence, the length of a side of the innermost square is 5 cm.) Then the area of the innermost square is 5 cm $\cdot$ 5 cm, or 25 cm^2, so the shaded area is 25 cm^2.

65. **Familiarize**. This is a multistep problem. First we will subtract the amounts of the contracts in successive years to find the yearly increases. Let $x =$ the increase from 2003 to 2004, $y =$ the increase from 2004 to 2005, and $z =$ the increase from 2005 to 2006, in millions.

Translate and Solve. We have three "take away" situations.
$$x = 12.1 - 10.3 = 1.8$$
$$y = 17.2 - 12.1 = 5.1$$
$$z = 22.6 - 17.2 = 5.4$$

Now we add the increases and divide by the number of addends, 3, to find the average increase. Let $a =$ the average yearly increase.
$$a = \frac{1.84 + 5.1 + 5.4}{3} = \frac{12.3}{3} = 4.1$$

Check. We check the calculations. The answer checks.

State. The average yearly increase in a contract was $4.1 million.

Chapter 5 Review Exercises

1. 6.59 million $= 6.59 \times 1,\underbrace{000,000}_{\text{6 zeros}}$

6.590000.

Move 6 places to the right.

6.59 million $= 6,590,000$

2. 6.9 million $= 6.9 \times 1,\underbrace{000,000}_{\text{6 zeros}}$

6.900000.

Move 6 places to the right.

6.9 million $= 6,900,000$

3. A word name for 3.47 is three and forty-seven hundredths.

4. A word name for 0.031 is thirty-one thousandths.

5. $0.\underline{09}$ 0.09. $\frac{9}{100}$

2 places   Move 2 places.   2 zeros

$0.09 = \frac{9}{100}$

6. $-4.\underline{561}$ $-4.561.$ $-\frac{4561}{1000}$

3 places   Move 3 places.   3 zeros

$-4.561 = -\frac{4561}{1000}$, and $-4.561 = -4\frac{561}{1000}$.

7. $-0.\underline{089}$ $-0.089.$ $-\frac{89}{1000}$

3 places   Move 3 places.   3 zeros

$-0.089 = -\frac{89}{1000}$

8. $3.\underline{0227}$ 3.0227. $\frac{30,227}{10,000}$

4 places   Move 4 places.   4 zeros

$3.0227 = \frac{30,227}{10,000}$, and $3.0227 = 3\frac{227}{10,000}$.

9. $-\frac{34}{1000}$ $-0.034.$

3 zeros   Move 3 places.

$-\frac{34}{1000} = -0.034$

10. $\frac{42,603}{10,000}$ 4.2603.

4 zeros   Move 4 places.

$\frac{42,603}{10,000} = 4.2603$

11. $27\frac{91}{100} = 27 + \frac{91}{100} = 27$ and $\frac{91}{100} = 27.91$

12. $867\frac{6}{1000} = 867 + \frac{6}{1000} = 867$ and $\frac{6}{1000} = 867.006$

$867\frac{6}{1000} = 867.006$, so $-867\frac{6}{1000} = -867.006$.

13. 0.034

Starting at the left, these digits are the first to differ; 3 is larger than 1.

0.0185

Thus, 0.034 is larger.

14. -0.91

Starting at the left, these digits are the first to differ; 1 is smaller than 9.

-0.19

Thus, -0.19 is larger.

15.

17.4⟨2⟩87   Hundredths digit is 4 or lower.
↓   Round down.
17.4

16.

17.428⟨7⟩   Ten-thousandths digit is 5 or higher.
↓   Round up.
17.429

17.
```
          1
     2 3 6.2 3 1
     2 6 3.4
  +     0.1 9 8
  ─────────────
     4 9 9.8 2 9
```

18.
```
            13
    2  17 5  3  15
    3  7.6  4  5
  −    8.4  9  7
  ───────────────
    2  9.1  4  8
```

19.
```
        1 1
     2 1 9.3
         2.8
  +      7.0
  ──────────
     2 2 9.1
```

20.
```
       13 14   10
    6  3  4  9  0  10
    7  4  5.0  1  0  9
  −    5  9.9  5  9  0
  ───────────────────
    6  8  5.0  5  1  9
```

21. $-37.8 + (-19.5)$

Add the absolute values: $37.8 + 19.5 = 57.3$

Make the answer negative: $-37.8 + (-19.5) = -57.3$

22. $-7.52 - (-9.89) = -7.52 + 9.89 = 2.37$

23.
```
         4 8
    × 0. 2 7
  ──────────
       3 3 6
       9 6 0
  ──────────
    1 2. 9 6
```

24. $-3.7(0.29) = -1.073$

25.
```
      2 4. 6 8
    × 1 0 0 0
```
The number 1000 has 3 zeros so we move the decimal point in 24.68 three places to the right. The product is 24,680.

26.
```
          3.2
    2 5 ) 8 0.0
          7 5
         ────
          5 0
          5 0
         ────
            0
```

27. First we consider $11.52 \div 7.2$.
```
              1. 6
    7.2∧) 1 1 5.∧2
              7 2 0
            ───────
              4 3 2
              4 3 2
            ───────
                  0
```

Since we have a positive number divided by a negative number, the answer is -1.6.

28. $\dfrac{276.3}{1000}$

The number 1000 has 3 zeros, so we move the decimal point in the numerator 3 places to the left.

$$\frac{276.3}{1000} = 0.2763$$

29.
$$3.7x - 5.2y - 1.5x - 3.9y$$
$$= 3.7x + (-5.2y) + (-1.5x) + (-3.9y)$$
$$= 3.7x + (-1.5x) + (-5.2y) + (-3.9y)$$
$$= [3.7 + (-1.5)]x + [-5.2 + (-3.9)]y$$
$$= 2.2x - 9.1y$$

30.
$$7.94 - 3.89a + 4.63 + 1.05a$$
$$= 7.94 + (-3.89a) + 4.63 + 1.05a$$
$$= -3.89a + 1.05a + 7.94 + 4.63$$
$$= (-3.89 + 1.05)a + (7.94 + 4.63)$$
$$= -2.84a + 12.57$$

31. $P - Prt = 1000 - 1000(0.05)(1.5)$
$$= 1000 - 50(1.5)$$
$$= 1000 - 75$$
$$= 925$$

32.
$$9 - 3.2(-1.5) + 5.2^2$$
$$= 9 - 3.2(-1.5) + 5.2(5.2)$$
$$= 9 - 3.2(-1.5) + 27.04$$
$$= 9 + 4.8 + 27.04$$
$$= 13.8 + 27.04$$
$$= 40.84$$

33. This is about $7.3 + 4.0$, so the sum is about 11.3.

34. This is about $50.0 \div 2.5$, so about 20 videotapes can be purchased.

35. Move 2 places to the left.

$15.49.¢

Change from ¢ sign at end to \$ sign in front.
$$1549¢ = \$15.49$$

36. Round

2 4 8. 2 7 ⟨2⟩ 7 ... to the nearest hundredth.
↓   Thousandths digit is 4 or less.
2 4 8. 2 7   Round down.

37. $\dfrac{13}{5} = \dfrac{13}{5} \cdot \dfrac{2}{2} = \dfrac{26}{10} = 2.6$

38. $\dfrac{32}{25} = \dfrac{32}{25} \cdot \dfrac{4}{4} = \dfrac{128}{100} = 1.28$

39.

$$\begin{array}{r} 3.25 \\ 4\overline{\smash{)}13.00} \\ \underline{12} \\ 10 \\ \underline{8} \\ 20 \\ \underline{20} \\ 0 \end{array}$$

$$\frac{13}{4} = 3.25$$

40. Since $-\dfrac{7}{6}$ is negative, we divide 7 by 6 and make the result negative.

$$\begin{array}{r} 1.166 \\ 6\overline{\smash{)}7.000} \\ \underline{6} \\ 10 \\ \underline{6} \\ 40 \\ \underline{36} \\ 40 \\ \underline{36} \\ 4 \end{array}$$

Since 4 keeps reappearing as a remainder, the digits repeat and

$$\frac{7}{6} = 1.166\ldots, \text{ or } 1.1\overline{6}.$$

Thus, $-\dfrac{7}{6} = -1.1\overline{6}$.

41. $\dfrac{4}{15} \times 79.05 = \dfrac{4}{15} \times \dfrac{79.05}{1} = \dfrac{4 \times 79.05}{15 \times 1} = \dfrac{316.2}{15} = 21.08$

42.
$$t - 4.3 = -7.5$$
$$t - 4.3 + 4.3 = -7.5 + 4.3$$
$$t = -3.2$$
The solution is -3.2.

43.
$$4.1x + 5.6 = -6.7$$
$$4.1x + 5.6 - 5.6 = -6.7 - 5.6$$
$$4.1x = -12.3$$
$$\frac{4.1x}{4.1} = \frac{-12.3}{4.1}$$
$$x = -3$$
The solution is -3.

44.
$$6x - 11 = 8x + 4$$
$$6x - 11 + 11 = 8x + 4 + 11$$
$$6x = 8x + 15$$
$$6x - 8x = 8x + 15 - 8x$$
$$-2x = 15$$
$$\frac{-2x}{-2} = \frac{15}{-2}$$
$$x = -7.5$$
The solution is -7.5.

45.
$$3(x + 2) = 5x - 7$$
$$3x + 6 = 5x - 7$$
$$3x + 6 - 6 = 5x - 7 - 6$$
$$3x = 5x - 13$$
$$3x - 5x = 5x - 13 - 5x$$
$$-2x = -13$$
$$\frac{-2x}{-2} = \frac{-13}{-2}$$
$$x = 6.5$$
The solution is 6.5.

46. Familiarize. Let $t =$ the number by which the number of telephone poles for every 100 people in the U.S. exceeds the number in Canada.

Translate. We have a "how much more" situation.

Number of poles in Canada	plus	How many more poles	is	Number of poles in U.S.
↓	↓	↓	↓	↓
40.65	+	t	=	51.81

Solve.
$$40.65 + t = 51.81$$
$$40.65 + t - 40.65 = 51.81 - 40.65$$
$$t = 11.16$$

Check. Since $40.65 + 11.16 = 51.81$, the answer checks.

State. There are 11.16 more telephone poles for every 100 people in the U.S. than in Canada.

47. Familiarize. Let $h =$ Stacia's hourly wage.

Translate.

Hourly wage	times	Number of hours worked	is	Total earnings
↓	↓	↓	↓	↓
h	·	40	=	620.74

Solve.
$$h \cdot 40 = 620.74$$
$$\frac{h \cdot 40}{40} = \frac{620.74}{40}$$
$$h \approx 15.52$$

Check. $40 \cdot \$15.52 = \$620.80 \approx \$620.74$, so the answer checks. (Remember, we rounded the solution of the equation.)

State. Stacia earns $15.52 per hour.

48. Familiarize. We let $a =$ the area of grass in the yard. Recall that the area of a rectangle with length l and width w is $A = l \times w$ and the area of a circle with radius r is $A = \pi r^2$, where $\pi \approx 3.14$.

Translate. We subtract the area of the base of the fountain from the area of the yard. Recall that a circle's radius is half of its diameter, or width.

$$\underbrace{\begin{matrix}\text{Area of}\\\text{yard}\end{matrix}}\quad \text{minus}\quad \underbrace{\begin{matrix}\text{Area of}\\\text{fountain}\end{matrix}}\quad \text{is}\quad \underbrace{\begin{matrix}\text{Area to}\\\text{be seeded}\end{matrix}}$$

$$20 \times 15 \quad - \quad 3.14\left(\frac{8}{2}\right)^2 \quad = \quad a$$

Solve. We carry out the computations.

$$20 \times 15 - 3.14\left(\frac{8}{2}\right)^2 = a$$
$$20 \times 15 - 3.14(4)^2 = a$$
$$20 \times 15 - 3.14(16) = a$$
$$300 - 50.24 = a$$
$$249.76 = a$$

Check. We recheck the calculations. Our answer checks.

State. The area of grass in the yard is 249.76 ft^2.

49. Familiarize. Let a = the amount left in the account after the purchase was made.

Translate. We write a subtraction sentence.

$$a = 6274.35 - 485.79$$

Solve. We carry out the subtraction.

$$\begin{array}{r} {\scriptstyle 11\ 1613\ 12}\\ {\scriptstyle 5\ 1\ 6\ 3\ 2\ 15}\\ 6\ 2\ 7\ 4.\ 3\ 5\\ -\ \ \ 4\ 8\ 5.\ 7\ 9\\ \hline 5\ 7\ 8\ 8.\ 5\ 6 \end{array}$$

Thus, $a = 5788.56$.

Check. $5788.56 + 485.79 = 6274.35$, so the answer checks.

State. There is $5788.56 left in the account.

50. Familiarize. This is a multistep problem. First we find the number of minutes in excess of 900. Then we find the charge for the excess minutes. Finally we add this charge to the monthly charge for 900 minutes to find the total cost for the month. Let m = the number of minutes in excess of 900.

Translate and Solve. First we have a "how much more" situation.

$$\underbrace{\begin{matrix}\text{First 900}\\\text{minutes}\end{matrix}}\quad \text{plus}\quad \underbrace{\begin{matrix}\text{Excess}\\\text{minutes}\end{matrix}}\quad \text{is}\quad \underbrace{\begin{matrix}\text{Total}\\\text{minutes}\end{matrix}}$$

$$900 \quad + \quad m \quad = \quad 946$$

We subtract 900 on both sides of the equation.

$$900 + m = 946$$
$$900 + m - 900 = 946 - 900$$
$$m = 46$$

We see that 46 minutes are charged at the rate of $0.40 per minute. We multiply to find c, the cost of these minutes.

$$\begin{array}{r} 4\ 6\\ \times\ 0.\ 4\ 0\\ \hline 1\ 8.\ 4\ 0 \end{array}$$

Thus, $c = 18.40$.

Finally we add the cost of the first 900 minutes and the cost of the additional 46 minutes to find t, the total cost for the month.

$$\begin{array}{r} 5\ 9.\ 9\ 9\\ +\ 1\ 8.\ 4\ 0\\ \hline 7\ 8.\ 3\ 9 \end{array}$$

Thus, $t = 78.39$.

Check. We can repeat the calculations. The answer checks.

State. The total cost for the month was $78.39.

51. Familiarize. Let m = the number of megabytes Cody paid for. We will express 2 cents as $0.02.

Translate.

$$\underbrace{\begin{matrix}\text{Processing}\\\text{fee}\end{matrix}}\quad \text{plus}\quad \$0.02\quad \text{times}\quad \underbrace{\begin{matrix}\text{Number of}\\\text{megabytes}\end{matrix}}\quad \text{is}\quad \underbrace{\begin{matrix}\text{Total}\\\text{bill}\end{matrix}}$$

$$10 \quad + \quad 0.02 \quad \cdot \quad m \quad = \quad 46.60$$

Solve.

$$10 + 0.02 \cdot m = 46.60$$
$$10 + 0.02 \cdot m - 10 = 46.60 - 10$$
$$0.02 \cdot m = 36.60$$
$$\frac{0.02 \cdot m}{0.02} = \frac{36.60}{0.02}$$
$$m = 1830$$

Check. $0.02 \cdot 1830 = 36.60$ and $36.60 + 10 = 46.60$, so the answer checks.

State. Cody paid for 1830 megabytes of storage.

52. Familiarize. This is a two-step problem. First, we find the number of miles that have been driven between fillups. This is a "how-much-more" situation. We let n = the number of miles driven.

Translate and Solve.

$$\underbrace{\begin{matrix}\text{First}\\\text{odometer}\\\text{reading}\end{matrix}}\ \text{plus}\ \underbrace{\begin{matrix}\text{Number}\\\text{of miles}\\\text{driven}\end{matrix}}\ \text{is}\ \underbrace{\begin{matrix}\text{Second}\\\text{odometer}\\\text{reading}\end{matrix}}$$

$$36,057.1 \quad + \quad n \quad = \quad 36,217.6$$

To solve the equation we subtract 36,057.1 on both sides.

$$n = 36,217.6 - 36,057.1$$
$$n = 160.5$$

$$\begin{array}{r} 3\ 6,\ 2\ 1\ 7.6\\ -\ 3\ 6,\ 0\ 5\ 7.1\\ \hline 1\ 6\ 0.5 \end{array}$$

Second, we divide the total number of miles driven by the number of gallons. This gives us m = the number of miles per gallon.

$$160.5 \div 11.1 = m$$

To find the number m, we divide.

$$\begin{array}{r} 1\ 4.4\ 5\\ 1\ 1.1_{\wedge}\overline{)\ 1\ 6\ 0.\ 5_{\wedge}0\ 0}\\ \underline{1\ 1\ 1\ 0}\\ 4\ 9\ 5\\ \underline{4\ 4\ 4}\\ 5\ 1\ 0\\ \underline{4\ 4\ 4}\\ 6\ 6\ 0\\ \underline{5\ 5\ 5}\\ 1\ 0\ 5 \end{array}$$

Thus, $m \approx 14.5$.

Check. To check, we first multiply the number of miles per gallon times the number of gallons:

$$11.1 \times 14.5 = 160.95$$

Then we add 160.95 to 36,057.1:

$$36,057.1 + 160.95 = 36,218.05 \approx 36,217.6$$

The number 14.5 checks.

State. Inge gets 14.5 miles per gallon.

53. a) **Familiarize.** Let $s =$ the total consumption of seafood per person, in pounds, for the seven given years.

Translate. We add the seven amounts shown in the graph in the text.

$$s = 12.4 + 15.0 + 14.9 + 14.8 + 15.2 + 14.7 + 15.6$$

Solve. We carry out the addition.

```
    3 3
    1 2. 4
    1 5. 0
    1 4. 9
    1 4. 8
    1 5. 2
    1 4. 7
  + 1 5. 6
  1 0 2. 6
```

Check. We repeat the calculation. The answer checks.

State. The total consumption of seafood per person for the seven given years was 102.6 lb.

b) We add the amounts and divide by the number of addends. From part (a) we know that the sum of the seven numbers is 102.6, so we have $102.6 \div 7$:

```
      1 4. 6 5
  7 ) 1 0 2. 6 0
      7 0 0
      3 2 6
      2 8 0
        4 6
        4 2
          4 0
          3 5
            5
```

Rounding to the nearest tenth, we find that the average seafood consumption per person was about 14.7 lb.

54. Familiarize. Let $d =$ the number of miles that an out-of-towner can travel for $15.23. We will express 95¢ as $0.95.

Translate.

Initial charge	plus	$0.95	times	Distance traveled	is	Fare
↓	↓	↓	↓	↓	↓	↓
7.25	+	0.95	·	d	=	15.23

Solve.

$$7.25 + 0.95 \cdot d = 15.23$$
$$7.25 + 0.95 \cdot d - 7.25 = 15.23 - 7.25$$
$$0.95 \cdot d = 7.98$$
$$\frac{0.95 \cdot d}{0.95} = \frac{7.98}{0.95}$$
$$d = 8.4$$

Check. $0.95 \cdot 8.4 = \$7.98$ and $\$7.98 + \$7.25 = \$15.23$, so the answer checks.

State. An out-of-towner can travel 8.4 mi for $15.23.

55. Familiarize. Let $c =$ the cost per serving.

Translate.

Cost	divided by	Number of servings	is	Cost per serving
↓	↓	↓	↓	↓
5.99	÷	4.5	=	c

Solve. We carry out the division.

```
        1. 3 3 1
  4.5 ) 5.9 9 0 0
        4 5 0
        1 4 9
        1 3 5
          1 4 0
          1 3 5
            5 0
            4 5
              5
```

Rounding to the nearest cent, we have $c \approx 1.33$.

Check. We find the cost of 4.5 servings at $1.33 per serving.

$$4.5 \cdot \$1.33 = \$5.985 \approx \$5.99$$

The answer checks

State. The ham costs about $1.33 per serving.

56. Familiarize. We will find the perimeter P of the room to determine how much crown molding is needed, and we will find the area A of the floor to determine how many square feet of bamboo tiles are needed. We will use the formulas $P = 2l + 2w$ and $A = l \cdot w$.

Translate.

$$P = 2 \cdot 14.5 \text{ ft} + 2 \cdot 16.25 \text{ ft}$$
$$A = 16.25 \text{ ft} \cdot 14.5 \text{ ft}$$

Solve. We carry out the calculations.

$$P = 2 \cdot 14.5 \text{ ft} + 2 \cdot 16.25 \text{ ft}$$
$$= 29 \text{ ft} + 32.5 \text{ ft}$$
$$= 61.5 \text{ ft}$$
$$A = 16.25 \text{ ft} \cdot 14.5 \text{ ft} = 235.625 \text{ ft}^2$$

Check. We can repeat the calculations. The answer checks.

State. 61.5 ft of crown molding and 235.625 ft^2 of bamboo tiles are needed.

57. *Discussion and Writing Exercise.* Since there are 20 nickels to a dollar, $\frac{3}{20}$ corresponds to 3 nickels, or 15¢, which is 0.15 dollars.

58. *Discussion and Writing Exercise.* In decimal notation, $\frac{1}{3}$ and $\frac{1}{6}$ both must be rounded before they can be multiplied. The best way to express $\frac{1}{3} \cdot \frac{1}{6}$ as a decimal is to multiply the fractions and then convert the product $\frac{1}{18}$ to decimal notation.

59. a) By trial we find the following true sentence.

$$2.56 - 6.4 + 51.2 - 17.4 + 89.7 = 119.66$$

b) By trial we find the following true sentence.

$$(11.12 - 0.29)3^4 = 877.23$$

60. First we find decimal notation for each fraction. Then we compare these numbers in decimal notation.

$$-\frac{2}{3} = -0.\overline{6}, \quad -\frac{15}{19} \approx -0.789474, \quad -\frac{11}{13} = -0.\overline{846153},$$

$$\frac{-5}{7} = -0.\overline{714285}, \quad \frac{-13}{15} = -0.8\overline{6}, \quad \frac{-17}{20} = -0.85$$

Arranging these numbers from smallest to largest and writing them in fraction notation, we have

$$\frac{-13}{15}, \frac{-17}{20}, -\frac{11}{13}, \frac{15}{19}, \frac{-5}{7}, -\frac{2}{3}$$

61. *Familiarize.* Let $m =$ the number of miles Quentin drove the car in 2006. Then $m - 10,000 =$ the number of miles in excess of 10,000. At \$396 per month, the leasing cost for 1 year, or 12 months, is $12 \cdot \$396$. We will express 20 cents as \$0.20.

Translate.

Leasing cost	plus	\$0.20	times	Miles over 10,000	is	Total bill
↓	↓	↓	↓	↓	↓	↓
$12 \cdot 396$	$+$	0.20	$\cdot$	$(m - 10,000)$	$=$	5952

Solve.

$$12 \cdot 396 + 0.20 \cdot (m - 10,000) = 5952$$
$$4752 + 0.20 \cdot (m - 10,000) = 5952$$
$$4752 + 0.20m - 2000 = 5952$$
$$2752 + 0.20m = 5952$$
$$2752 + 0.20m - 2752 = 5952 - 2752$$
$$0.20m = 3200$$
$$\frac{0.20m}{0.20} = \frac{3200}{0.20}$$
$$m = 16,000$$

Check. If Quentin drives 16,000 mi, then he drives $16,000 - 10,000$, or 6000 mi, in excess of 10,000. The charge for the excess miles is $\$0.20 \cdot 6000$, or \$1200. The leasing fee is $12 \cdot \$396$, or \$4752, and $\$4752 + \$1200 = \$5952$, so the answer checks.

State. Quentin drove the car 16,000 mi in 2006.

62. *Discussion and Writing Exercise.* The Sicilian pizza, at $\frac{4.4¢}{\text{in}^2}$, is a better buy than the round pizza which costs $\frac{5.5¢}{\text{in}^2}$.

Chapter 5 Test

1.   8.9 billion

$= 8.9 \times 1$ billion

$= 8.9 \times 1,000,000,000$ 9 zeros

$= 8,900,000,000$ Moving the decimal point 9 places to the right

2.   3.756 million

$= 3.756 \times 1$ million

$= 3.756 \times 1,000,000$ 6 zeros

$= 3,756,000$ Moving the decimal point 6 places to the right

3. 2.34

a) Write a word name for the whole number. | Two |

b) Write "and" for the decimal point.   Two | and |

c) Write a word name for the number to the right   Two
of the decimal point,   and
followed by the place | thirty-four hundredths |
value of the last digit.

A word name for 2.34 is two and thirty-four hundredths.

4. 105.0005

a) Write a word name for the whole number. | One hundred five |

b) Write "and" for the decimal point.   One hundred five | and |

c) Write a word name for the number to the right   One hundred five
of the decimal point,   and
followed by the place | five ten-thousandths |
value of the last digit.

A word name for 105.0005 is one hundred five and five ten-thousandths.

5. -0.3 $-0.3.$ $-\dfrac{3}{10}$

                          1 place   1 zero

$$-0.3 = -\frac{3}{10}$$

6. $2.\underline{769}$ $2.769.$ $\dfrac{2769}{1000}$

3 places Move 3 places. 3 zeros

$2.769 = \dfrac{2769}{1000}$

7. $\dfrac{74}{1\underline{000}}$ $0.074.$

3 zeros Move 3 places.

$\dfrac{74}{1000} = 0.074$

8. $-\dfrac{37,047}{10,\underline{000}}$ $-3.7047.$

4 zeros Move 4 places.

$-\dfrac{37,047}{10,000} = -3.7047$

9. $756\dfrac{9}{100} = 756 + \dfrac{9}{100} = 756 \text{ and } \dfrac{9}{100} = 756.09$

10. $91\dfrac{703}{1000} = 91 + \dfrac{703}{1000} = 91 \text{ and } \dfrac{703}{1000} = 91.703$

11. To compare two positive numbers in decimal notation, start at the left and compare corresponding digits moving from left to right. When two digits differ, the number with the larger digit is the larger of the two numbers.

0.07

Different; 1 is larger than 0.

0.162

Thus, 0.162 is larger.

12. To compare two negative numbers in decimal notation, start at the left and compare corresponding digits moving from left to right. When two digits differ, the number with the smaller digit is the larger of the two numbers.

−0.173

Different; 1 is smaller than 2.

−0.25

Thus, −0.173 is larger.

13.

$9.4\boxed{5}23$ Hundredths digit is 5 or higher. Round up.

9.5

14.

$9.452\boxed{3}$ Ten-thousandths digit is 4 or lower. Round down.

9.452

15.
```
    1
  4 0 2. 3
     2. 8 1
+    0. 1 0 9
  4 0 5. 2 1 9
```

16.
```
    0. 1 2 5   (3 decimal places)
 ×  0. 2 4     (2 decimal places)
    5 0 0
  2 5 0 0
  0.0 3 0 0 0  (5 decimal places)
```

17. $0.\underline{001} \times 213.45$ $0.213.45$

3 decimal places   Move 3 places to the left.

$0.001 \times 213.45 = 0.21345$

18.
```
        11
    4 7 10 9 11
    5 2. 0 9 1
 −   7. 3 4 5
  4 4. 7 4 6
```

19.
```
    1 1
  3 4 2. 9
      8. 1
+     5. 3 7
  3 5 6. 3 7
```

20. $-9.5 + 7.3$

The absolute values are 9.5 and 7.3. The difference is 2.2. The negative number has the larger absolute value, so the answer is negative.

$-9.5 + 7.3 = -2.2$

21. We write extra zeros.
```
  1 9 9 9 10
  2 0 0 0 0
 −0. 0 0 5 4
  1. 9 9 4 6
```

22. $1\underline{000} \times 73.962$ $73.962.$

3 zeros   Move 3 places to the right.

$1000 \times 73.962 = 73,962$

23.
```
        4. 7 5
  4 ) 1 9. 0 0
      1 6
       3 0
       2 8
         2 0
         2 0
          0
```

24.
```
           3 0. 4
  3.3∧) 1 0 0.3∧2
        9 9 0 0
          1 3 2
          1 3 2
            0
```

25. $\dfrac{-346.82}{1\underline{000}}$ $-0.346.82$

3 zeros   Move 3 places to the left.

$\dfrac{-346.82}{1000} = -0.34682$

26. $\dfrac{346.82}{0.\underline{01}}$ 346.82.

2 decimal places   Move 2 places to the right.

$$\dfrac{346.82}{0.01} = 34,682$$

27. Move 2 places to the right.

$\$179.82.\cancel{c}$

Change from $\$$ sign in front to $\cancel{c}$ sign at end.

$$\$179.82 = 17,982\cancel{c}$$

28. $4.1x + 5.2 - 3.9y + 5.7x - 9.8$
$= 4.1x + 5.2 + (-3.9y) + 5.7x + (-9.8)$
$= 4.1x + 5.7x + (-3.9y) + 5.2 + (-9.8)$
$= (4.1 + 5.7)x + (-3.9y) + (5.2 + (-9.8))$
$= 9.8x - 3.9y - 4.6$

29. $2l + 4w + 2h = 2 \cdot 2.4 + 4 \cdot 1.3 + 2 \cdot 0.8$
$= 4.8 + 5.2 + 1.6$
$= 10.0 + 1.6$
$= 11.6$

30. $20 \div 5(-2)^2 - 8.4 = 20 \div 5 \cdot 4 - 8.4$
$= 4 \cdot 4 - 8.4$
$= 16 - 8.4$
$= 7.6$

31. *Familiarize*. Let $g =$ the number of gallons of gasoline that can be bought with $\$20$.

Translate.

Price per gallon	times	Number of gallons	is	Total cost
↓	↓	↓	↓	↓
2.749	·	g	=	20

Solve.
$$2.749 \cdot g = 20$$
$$\dfrac{2.749 \cdot g}{2.749} = \dfrac{20}{2.749}$$
$$g \approx 7 \quad \text{Rounding to the nearest gallon}$$

Check. $\$2.749 \cdot 7 \approx \$19.24 \approx \$20$, so the answer checks. Remember that we rounded to the nearest gallon.

State. About 7 gal of gasoline can be bought with $\$20$.

32. $48.7\boxed{4}74\ldots$

↓ ↑——— Hundredths digit is 4 or lower.

48.7   Round down.

33. $\dfrac{8}{5} = \dfrac{8}{5} \cdot \dfrac{2}{2} = \dfrac{16}{10} = 1.6$

34. $\dfrac{21}{4} = \dfrac{21}{4} \cdot \dfrac{25}{25} = \dfrac{525}{100} = 5.25$

35. First consider $\dfrac{7}{16}$.

```
      0. 4 3 7 5
 1 6 | 7. 0 0 0 0
      6 4
      ───
        6 0
        4 8
        ───
        1 2 0
        1 1 2
        ─────
            8 0
            8 0
            ───
              0
```

Since $\dfrac{7}{16} = 0.4375$, we have $-\dfrac{7}{16} = -0.4375$.

36.
```
       1. 5 5
  9 | 1 4. 0 0
      9
      ──
      5 0
      4 5
      ───
        5 0
        4 5
        ───
          5
```

Since 5 keeps reappearing as a remainder, the digit 5 repeats and

$$\dfrac{14}{9} = 1.55\ldots = 1.\overline{5}.$$

37. $1.55\boxed{5}5\ldots$

↓ ↑——— Thousandths digit is 5 or higher.

1.56   Round up.

38. $8.91 \times 22.457 \approx 9 \times 22 = 198$

39. $78.2209 \div 16.09 \approx 80 \div 20 = 4$

40. $\dfrac{3}{8} \times 45.6 - \dfrac{1}{5} \times 36.9$
$= \dfrac{3 \times 45.6}{8} - \dfrac{36.9}{5}$
$= \dfrac{136.8}{8} - \dfrac{36.9}{5}$
$= 17.1 - 7.38$
$= 9.72$

41. $17y - 3.12 = -58.2$
$17y - 3.12 + 3.12 = -58.2 + 3.12$
$17y = -55.08$
$\dfrac{17y}{17} = \dfrac{-55.08}{17}$
$y = -3.24$

The solution is -3.24.

42.
$$9t - 4 = 6t + 26$$
$$9t - 4 + 4 = 6t + 26 + 4$$
$$9t = 6t + 30$$
$$9t - 6t = 6t + 30 - 6t$$
$$3t = 30$$
$$\frac{3t}{3} = \frac{30}{3}$$
$$t = 10$$
The solution is 10.

43.
$$4 + 2(x - 3) = 7x - 9$$
$$4 + 2x - 6 = 7x - 9$$
$$2x - 2 = 7x - 9$$
$$2x - 2 + 2 = 7x - 9 + 2$$
$$2x = 7x - 7$$
$$2x - 7x = 7x - 7 - 7x$$
$$-5x = -7$$
$$\frac{-5x}{-5} = \frac{-7}{-5}$$
$$x = 1.4$$
The solution is 1.4.

44. We add the numbers and divide by the number of addends.
$$\frac{76.1 + 69.4 + 55.0 + 53.2 + 37.5}{5} = \frac{291.2}{5} = 58.24$$
The average number of passengers is 58.24 million.

45. *Familiarize*. This is a multistep problem. First we find the amount charged for minutes in excess of 2000. Then we find the number of minutes in excess of 2000 and finally we find the total number of minutes used. Let $c =$ the amount charged for the minutes in excess of 2000.

***Translate and Solve*.**

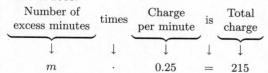

Access fee	plus	Charge for extra minutes	is	Total bill
↓	↓	↓	↓	↓
99.99	+	c	=	314.99

We solve this equation.
$$99.99 + c = 314.99$$
$$99.99 + c - 99.99 = 314.99 - 99.99$$
$$c = 215$$
We see that $215 was charged for the number of minutes used in excess of 2000. Let $m =$ the number of minutes in excess of 2000.

Number of excess minutes	times	Charge per minute	is	Total charge
↓	↓	↓	↓	↓
m	·	0.25	=	215

We solve this equation.
$$m \cdot 0.25 = 215$$
$$\frac{m \cdot 0.25}{0.25} = \frac{215}{0.25}$$
$$m = 860$$

Finally let $t =$ the total number of minutes Trey used. We add.

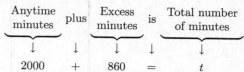

Anytime minutes	plus	Excess minutes	is	Total number of minutes
↓	↓	↓	↓	↓
2000	+	860	=	t

We solve the equation.
$$2000 + 860 = t$$
$$2860 = t$$

***Check*.** We can repeat the calculations. The answer checks.

***State*.** Trey used 2860 minutes.

46. *Familiarize*. This is a two-step problem. First we will find the number of miles that are driven between fillups. Then we find the gas mileage. Let $n =$ the number of miles driven between fillups.

***Translate and Solve*.**

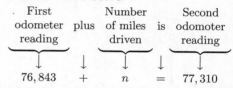

First odometer reading	plus	Number of miles driven	is	Second odometer reading
↓	↓	↓	↓	↓
76,843	+	n	=	77,310

To solve the equation, we subtract 76,843 on both sides.
$$n = 77,310 - 76,843 = 467$$

Now let $m =$ the number of miles driven per gallon.

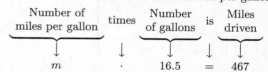

Number of miles per gallon	times	Number of gallons	is	Miles driven
↓	↓	↓	↓	↓
m	·	16.5	=	467

We divide by 16.5 on both sides to find m.
$$m = 467 \div 16.5$$
$$m = 28.\overline{30}$$
$$m \approx 28.3 \quad \text{Rounding to the nearest tenth}$$

***Check*.** First we multiply the number of miles per gallon by the number of gallons to find the number of miles driven:
$$16.5 \cdot 28.3 = 466.95 \approx 467$$
Then we add 467 mi to the first odometer reading:
$$76,843 + 467 = 77,310$$
This is the second odometer reading, so the answer checks.

***State*.** The gas mileage is about 28.3 miles per gallon.

47. *Familiarize*. Let $b =$ the balance after the purchases are made.

***Translate*.** We subtract the amounts of the three purchases from the original balance:
$$b = 10,200 - 123.89 - 56.68 - 3446.98$$

Solve. We carry out the calculations.

$$b = 10,200 - 123.89 - 56.68 - 3446.98$$
$$= 10,076.11 - 56.68 - 3446.98$$
$$= 10,019.43 - 3446.98$$
$$= 6572.45$$

Check. We can find the total amount of the purchases and then subtract to find the new balance.

$$\$123.89 + \$56.68 + \$3446.98 = \$3627.55$$
$$\$10,200 - \$3627.55 = \$6572.45$$

The answer checks.

State. After the purchases were made, the balance was $6572.45.

48. *Familiarize*. Let c = the total cost of the copy paper.

Translate.

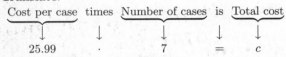

$$\underbrace{\text{Cost per case}}_{25.99} \quad \underbrace{\text{times}}_{\cdot} \quad \underbrace{\text{Number of cases}}_{7} \quad \underbrace{\text{is}}_{=} \quad \underbrace{\text{Total cost}}_{c}$$

Solve. We carry out the multiplication.

$$\begin{array}{r} 2\,5.9\,9 \quad \text{(2 decimal places)} \\ \times \quad\quad 7 \\ \hline 1\,8\,1.9\,3 \quad \text{(2 decimal places)} \end{array}$$

Thus, $c = 181.93$.

Check. We can obtain a partial check by rounding and estimating:

$$25.99 \times 7 \approx 25 \times 7 = 175 \approx 181.93$$

State. The total cost of the copy paper is $181.93.

49. a) The product of two numbers greater than 0 and less than 1 is <u>always</u> less than 1.

b) The product of two numbers greater than 1 is <u>never</u> less than 1.

c) The product of a number greater than 1 and a number less than 1 is <u>sometimes</u> equal to 1.

d) The product of a number greater than 1 and a number less than 1 is <u>sometimes</u> equal to 0.

50. *Familiarize*. This is a multistep problem. First we will subtract the amounts of bottled water consumed in successive years to find the yearly increases. Let x = the increase from 2002 to 2003, y = the increase from 2003 to 2004, and z = the increase from 2004 to 2005, in gallons per person.

Translate and Solve. We have three "take away" situations.

$$x = 22.1 - 20.7 = 1.4$$
$$y = 23.8 - 22.1 = 1.7$$
$$z = 25.0 - 23.8 = 1.2$$

Now we add the increases and divide by the number of addends, 3, to find the average increase. Let a = the average yearly increase.

$$a = \frac{1.4 + 1.7 + 1.2}{3} = \frac{4.3}{3} = 1.4\overline{3}$$

Check. We check the calculations. The answer checks.

State. The average yearly increase in bottled water consumption was $1.4\overline{3}$ gal per person.

51. The cost to drive roundtrip is $2 \cdot \$0.32 \cdot 320$, or $204.80.

a) Since $\$189 < \$204.80$, it is more economical for an individual to fly.

b) The airfare for a couple is $2 \cdot \$189$, or $378. Since $\$378 > \$204.80$, it is more economical for a couple to drive.

c) Since we found in part (b) that it is more economical for a couple to drive, it seems reasonable that it is also more economical for a family of 3 to drive. We can verify this by first finding the airfare for 3 people: $3 \cdot \$189 = \$567$. Since $\$567 > \$204.80$, we confirm that it is more economical for a family of 3 to drive.

Chapter 6

Percent Notation

Exercise Set 6.1

---

1. The ratio of 4 to 5 is $\dfrac{4}{5}$.

3. The ratio of 56.78 to 98.35 is $\dfrac{56.78}{98.35}$.

5. The ratio of physicians to residents in Connecticut was $\dfrac{356}{100,000}$.

The ratio of physicians to residents in Wyoming was $\dfrac{173}{100,000}$.

7. The ratio of 18 to 24 is $\dfrac{18}{24} = \dfrac{3 \cdot 6}{4 \cdot 6} = \dfrac{3}{4} \cdot \dfrac{6}{6} = \dfrac{3}{4}$.

9. The ratio of 2.8 to 3.6 is $\dfrac{2.8}{3.6} = \dfrac{2.8}{3.6} \cdot \dfrac{10}{10} = \dfrac{28}{36} = \dfrac{4 \cdot 7}{4 \cdot 9} = \dfrac{4}{4} \cdot \dfrac{7}{9} = \dfrac{7}{9}$.

11. The ratio of length to width is $\dfrac{478}{213}$.

The ratio of width to length is $\dfrac{213}{478}$.

13. $\dfrac{120 \text{ km}}{3 \text{ hr}} = 40 \dfrac{\text{km}}{\text{hr}}$

15. $\dfrac{217 \text{ mi}}{29 \text{ sec}} \approx 7.48 \dfrac{\text{mi}}{\text{sec}}$

17. $\dfrac{448.5 \text{ mi}}{19.5 \text{ gal}} = 23 \text{ mpg}$

19. $\dfrac{1465 \text{ points}}{80 \text{ games}} \approx 18.3 \dfrac{\text{points}}{\text{game}}$

21. $\dfrac{623 \text{ gal}}{1000 \text{ sq ft}} = 0.623 \text{ gal/ft}^2$

23. $\dfrac{1500 \text{ beats}}{60 \text{ min}} = 25 \dfrac{\text{beats}}{\text{min}}$

25. We can use cross-products:

$$5 \cdot 9 = 45 \qquad \begin{array}{c} 5 \quad 7 \\ \diagdown \\ 6 \quad 9 \end{array} \qquad 6 \cdot 7 = 42$$

Since the cross-products are not the same, $45 \neq 42$, we know that the numbers are not proportional.

27. We can use cross-products:

$$1 \cdot 20 = 20 \qquad \begin{array}{c} 1 \quad 10 \\ \diagdown \\ 2 \quad 20 \end{array} \qquad 2 \cdot 10 = 20$$

Since the cross-products are the same, $20 = 20$, we know that $\dfrac{1}{2} = \dfrac{10}{20}$, so the numbers are proportional.

29. We can use cross-products:

$$2.4 \cdot 2.7 = 6.48 \qquad \begin{array}{c} 2.4 \quad 1.8 \\ \diagdown \\ 3.6 \quad 2.7 \end{array} \qquad 3.6 \cdot 1.8 = 6.48$$

Since the cross-products are the same, $6.48 = 6.48$, we know that $\dfrac{2.4}{3.6} = \dfrac{1.8}{2.7}$, so the numbers are proportional.

31. We can use cross-products:

$$5\tfrac{1}{3} \cdot 9\tfrac{1}{2} = 50\tfrac{2}{3} \qquad \begin{array}{c} 5\tfrac{1}{3} \quad 2\tfrac{1}{5} \\ \diagdown \\ 8\tfrac{1}{4} \quad 9\tfrac{1}{2} \end{array} \qquad 8\tfrac{1}{4} \cdot 2\tfrac{1}{5} = 18\tfrac{3}{20}$$

Since the cross-products are not the same, $50\tfrac{2}{3} \neq 18\tfrac{3}{20}$, we know that the numbers are not proportional.

33.
$$\dfrac{18}{4} = \dfrac{x}{10}$$
$$18 \cdot 10 = 4 \cdot x \qquad \text{Equating cross-products}$$
$$\dfrac{18 \cdot 10}{4} = \dfrac{4 \cdot x}{4} \qquad \text{Dividing by 4}$$
$$\dfrac{18 \cdot 10}{4} = x$$
$$\dfrac{180}{4} = x \qquad \text{Multiplying}$$
$$45 = x \qquad \text{Dividing}$$

35.
$$\dfrac{t}{12} = \dfrac{5}{6}$$
$$6 \cdot t = 12 \cdot 5$$
$$\dfrac{6 \cdot t}{6} = \dfrac{12 \cdot 5}{6}$$
$$t = \dfrac{12 \cdot 5}{6}$$
$$t = \dfrac{60}{6}$$
$$t = 10$$

37.
$$\dfrac{2}{5} = \dfrac{8}{n}$$
$$2 \cdot n = 5 \cdot 8$$
$$\dfrac{2 \cdot n}{2} = \dfrac{5 \cdot 8}{2}$$
$$n = \dfrac{5 \cdot 8}{2}$$
$$n = \dfrac{40}{2}$$
$$n = 20$$

39.
$$\frac{16}{12} = \frac{24}{x}$$
$$16 \cdot x = 12 \cdot 24$$
$$\frac{16 \cdot x}{16} = \frac{12 \cdot 24}{6}$$
$$x = \frac{12 \cdot 24}{16}$$
$$x = \frac{288}{16}$$
$$x = 18$$

41.
$$\frac{t}{0.16} = \frac{0.15}{0.40}$$
$$0.40 \times t = 0.16 \times 0.15$$
$$\frac{0.40 \times t}{0.40} = \frac{0.16 \times 0.15}{0.40}$$
$$t = \frac{0.16 \times 0.15}{0.40}$$
$$t = \frac{0.024}{0.40}$$
$$t = 0.06$$

43.
$$\frac{100}{25} = \frac{20}{n}$$
$$100 \cdot n = 25 \cdot 20$$
$$\frac{100 \cdot n}{100} = \frac{25 \cdot 20}{100}$$
$$n = \frac{25 \cdot 20}{100}$$
$$n = \frac{500}{100}$$
$$n = 5$$

45.
$$\frac{\frac{1}{4}}{\frac{1}{2}} = \frac{\frac{1}{2}}{x}$$
$$\frac{1}{4} \cdot x = \frac{1}{2} \cdot \frac{1}{2}$$
$$\frac{\frac{1}{4} \cdot x}{\frac{1}{4}} = \frac{\frac{1}{2} \cdot \frac{1}{2}}{\frac{1}{4}}$$
$$x = \frac{\frac{1}{2} \cdot \frac{1}{2}}{\frac{1}{4}}$$
$$x = \frac{\frac{1}{4}}{\frac{1}{4}}$$
$$x = 1$$

47.
$$\frac{1.28}{3.76} = \frac{4.28}{y}$$
$$1.28 \times y = 3.76 \times 4.28$$
$$\frac{1.28 \times y}{1.28} = \frac{3.76 \times 4.28}{1.28}$$
$$y = \frac{3.76 \times 4.28}{1.28}$$
$$y = \frac{16.0928}{1.28}$$
$$y = 12.5725$$

49. *Familiarize*. Let c = the number of calories in 6 cups of cereal.

Translate. We translate to a proportion, keeping the number of calories in the numerators.

$$\text{Calories} \rightarrow \frac{110}{3/4} = \frac{c}{6} \leftarrow \text{Calories}$$
$$\text{Cups} \rightarrow \qquad \qquad \leftarrow \text{Cups}$$

Solve. We solve the proportion.

$$110 \cdot 6 = \frac{3}{4} \cdot c \quad \text{Equating cross products}$$
$$\frac{110 \cdot 6}{3/4} = \frac{\frac{3}{4} \cdot c}{3/4}$$
$$\frac{110 \cdot 6}{3/4} = c$$
$$110 \cdot 6 \cdot \frac{4}{3} = c$$
$$880 = c$$

Check. We substitute into the proportion and check cross products.

$$\frac{110}{3/4} = \frac{880}{6}$$
$$110 \cdot 6 = 660; \frac{3}{4} \cdot 880 = 660$$

The cross products are the same.

State. There are 880 calories in 6 cups of cereal.

51. *Familiarize*. Let n = the number of Americans who would be considered overweight.

Translate. We translate to a proportion.

$$\text{Overweight} \rightarrow \frac{60}{100} = \frac{n}{295,000,000} \leftarrow \text{Overweight}$$
$$\text{Total} \rightarrow \qquad \qquad \leftarrow \text{Total}$$

Solve. We solve the proportion.

$$60 \cdot 295,000,000 = 100 \cdot n \quad \text{Equating cross products}$$
$$\frac{60 \cdot 295,000,000}{100} = \frac{100 \cdot n}{100}$$
$$\frac{60 \cdot 295,000,000}{100} = n$$
$$177,000,000 = n$$

Check. We substitute in the proportion and check cross products.

$$\frac{60}{100} = \frac{177,000,000}{295,000,000}$$
$$60 \cdot 295,000,000 = 17,700,000,000$$
$$100 \cdot 177,700,000 = 17,700,000,000$$

The cross products are the same.

State. 177,000,000, or 177 million, Americans would be considered overweight.

53. *Familiarize*. Let m = the number of miles the car will be driven in 1 year. Note that 1 year = 12 months.

Translate.

$$\text{Months} \rightarrow \frac{8}{9000} = \frac{12}{m} \leftarrow \text{Months}$$
$$\text{Miles} \rightarrow \qquad \qquad \leftarrow \text{Miles}$$

Solve.

$$8 \cdot m = 9000 \cdot 12$$

$$m = \frac{9000 \cdot 12}{8}$$

$$m = \frac{2 \cdot 4500 \cdot 3 \cdot 4}{2 \cdot 4}$$

$$m = 4500 \cdot 3$$

$$m = 13,500$$

Check. We find the average number of miles driven in 1 month and then multiply to find the number of miles the car will be driven in 1 yr, or 12 months.

$$9000 \div 8 = 1125 \text{ and } 12 \cdot 1125 = 13,500$$

The answer checks.

State. At the given rate, the car will be driven 13,500 mi in one year.

55. Familiarize. Let $z =$ the number of pounds of zinc in the alloy.

Translate. We translate to a proportion.

$$\begin{array}{c} \text{Zinc} \rightarrow \\ \text{Copper} \rightarrow \end{array} \frac{3}{13} = \frac{z}{520} \begin{array}{c} \leftarrow \text{Zinc} \\ \leftarrow \text{Copper} \end{array}$$

Solve.

$$3 \cdot 520 = 13 \cdot z$$

$$\frac{3 \cdot 520}{13} = z$$

$$\frac{3 \cdot 13 \cdot 40}{13} = z$$

$$3 \cdot 40 = z$$

$$120 = z$$

Check. We substitute in the proportion and check cross products.

$$\frac{3}{13} = \frac{120}{520}$$

$$3 \cdot 520 = 1560; \ 13 \cdot 120 = 1560$$

The cross products are the same.

State. There are 120 lb of zinc in the alloy.

57. a) Familiarize. Let $g =$ the number of gallons of gasoline needed to drive 2690 mi.

Translate. We translate to a proportion.

$$\begin{array}{c} \text{Gallons} \rightarrow \\ \text{Miles} \rightarrow \end{array} \frac{15.5}{372} = \frac{g}{2690} \begin{array}{c} \leftarrow \text{Gallons} \\ \leftarrow \text{Miles} \end{array}$$

Solve.

$$15.5 \cdot 2690 = 372 \cdot g \quad \text{Equating cross products}$$

$$\frac{15.5 \cdot 2690}{372} = g$$

$$112 \approx g$$

Check. We find how far the car can be driven on 1 gallon of gasoline and then divide to find the number of gallons required for a 2690-mi trip.

$$372 \div 15.5 = 24 \text{ and } 2690 \div 24 \approx 112$$

The answer checks.

State. It will take about 112 gal of gasoline to drive 2690 mi.

b) Familiarize. Let $d =$ the number of miles the car can be driven on 140 gal of gasoline.

Translate. We translate to a proportion.

$$\begin{array}{c} \text{Gallons} \rightarrow \\ \text{Miles} \rightarrow \end{array} \frac{15.5}{372} = \frac{140}{d} \begin{array}{c} \leftarrow \text{Gallons} \\ \leftarrow \text{Miles} \end{array}$$

Solve.

$$15.5 \cdot d = 372 \cdot 140 \quad \text{Equating cross products}$$

$$d = \frac{372 \cdot 140}{15.5}$$

$$d = 3360$$

Check. From the check in part (a) we know that the car can be driven 24 mi on 1 gal of gasoline. We multiply to find how far it can be driven on 140 gal.

$$140 \cdot 24 = 3360$$

The answer checks.

State. The car can be driven 3360 mi on 140 gal of gasoline.

59. a) Familiarize. Let $a =$ the number of Mexican pesos equivalent to 120 U.S. dollars.

Translate. We translate to a proportion.

$$\begin{array}{c} \text{U.S.} \rightarrow \\ \text{Mexican} \rightarrow \end{array} \frac{1}{10.4555} = \frac{120}{a} \begin{array}{c} \leftarrow \text{U.S.} \\ \leftarrow \text{Mexican} \end{array}$$

Solve.

$$1 \cdot a = 10.4555 \cdot 120 \quad \text{Equating cross products}$$

$$a = 1254.66$$

Check. We substitute in the proportion and check cross products.

$$\frac{1}{10.4555} = \frac{120}{1254.66}$$

$$1 \cdot 1254.66 = 1254.66; \ 10.4555 \cdot 120 = 1254.66$$

The cross products are the same.

State. 120 U.S. dollars would be worth 1254.66 Mexican pesos.

b) Familiarize. Let $c =$ the cost of the watch in U.S. dollars.

Translate. We translate to a proportion.

$$\begin{array}{c} \text{U.S.} \rightarrow \\ \text{Mexican} \rightarrow \end{array} \frac{1}{10.4555} = \frac{c}{3600} \begin{array}{c} \leftarrow \text{U.S.} \\ \leftarrow \text{Mexican} \end{array}$$

Solve.

$$1 \cdot 3600 = 10.4555 \cdot c \quad \text{Equating cross products}$$

$$\frac{1 \cdot 3600}{10.4555} = c$$

$$344.32 \approx c$$

Check. We substitute in the proportion and check cross products.

$$\frac{1}{10.4555} = \frac{344.32}{3600}$$

$$1 \cdot 3600 = 3600; \ 10.4555 \cdot 344.32 = 3600.03776 \approx 3600$$

The cross products are about the same. Remember that we rounded the value of c.

State. The watch cost $344.32 in U.S. dollars.

61. Familiarize. Let d = the number of defective bulbs in a lot of 2500.

Translate. We translate to a proportion.

Defective bulbs $\rightarrow$ $\dfrac{7}{100} = \dfrac{d}{2500}$ $\leftarrow$ Defective bulbs
Bulbs in lot $\rightarrow$ $\phantom{\dfrac{7}{100}}$ $\phantom{\dfrac{d}{2500}}$ $\leftarrow$ Bulbs in lot

Solve.
$$7 \cdot 2500 = 100 \cdot d$$
$$\frac{7 \cdot 2500}{100} = d$$
$$\frac{7 \cdot 25 \cdot 100}{100} = d$$
$$7 \cdot 25 = d$$
$$175 = d$$

Check. We substitute in the proportion and check cross products.
$$\frac{7}{100} = \frac{175}{2500}$$
$$7 \cdot 2500 = 17{,}500; \; 100 \cdot 175 = 17{,}500$$

State. There will be 175 defective bulbs in a lot of 2500.

63. Familiarize. Let s = the number of square feet of siding that Fred can paint with 7 gal of paint.

Translate. We translate to a proportion.

Gallons $\rightarrow$ $\dfrac{3}{1275} = \dfrac{7}{s}$ $\leftarrow$ Gallons
Siding $\rightarrow$ $\phantom{\dfrac{3}{1275}}$ $\phantom{\dfrac{7}{s}}$ $\leftarrow$ Siding

Solve.
$$3 \cdot s = 1275 \cdot 7$$
$$s = \frac{1275 \cdot 7}{3}$$
$$s = \frac{3 \cdot 425 \cdot 7}{3}$$
$$s = 425 \cdot 7$$
$$s = 2975$$

Check. We find the number of square feet covered by 1 gallon of paint and then multiply that number by 7.
$$1275 \div 3 = 425 \text{ and } 425 \cdot 7 = 2975$$
The answer checks.

State. Fred can paint 2975 ft^2 of siding with 7 gal of paint.

65. Familiarize. Let s = the number of ounces of grass seed needed for 5000 ft^2 of lawn.

Translate. We translate to a proportion.

Seed $\rightarrow$ $\dfrac{60}{3000} = \dfrac{s}{5000}$ $\leftarrow$ Seed
Area $\rightarrow$ $\phantom{\dfrac{60}{3000}}$ $\phantom{\dfrac{s}{5000}}$ $\leftarrow$ Area

Solve.
$$60 \cdot 5000 = 3000 \cdot s$$
$$\frac{60 \cdot 5000}{3000} = s$$
$$100 = s$$

Check. We find the number of ounces of seed needed for 1 ft^2 of lawn and then multiply this number by 5000:
$$60 \div 3000 = 0.02 \text{ and } 5000(0.02) = 100$$
The answer checks.

State. 100 oz of grass seed would be needed to seed 5000 ft^2 of lawn.

67. Familiarize. Let D = the number of deer in the game preserve.

Translate. We translate to a proportion.

Deer tagged originally $\rightarrow$ $\dfrac{318}{D} = \dfrac{56}{168}$ $\leftarrow$ Tagged deer caught later
Deer in game preserve $\rightarrow$ $\phantom{\dfrac{318}{D}}$ $\phantom{\dfrac{56}{168}}$ $\leftarrow$ Deer caught later

Solve.
$$318 \cdot 168 = 56 \cdot D$$
$$\frac{318 \cdot 168}{56} = D$$
$$954 = D$$

Check. We substitute in the proportion and check cross products.
$$\frac{318}{954} = \frac{56}{168}; \; 318 \cdot 168 = 53{,}424; \; 954 \cdot 56 = 53{,}424$$
Since the cross products are the same, the answer checks.

State. We estimate that there are 954 deer in the game preserve.

69. Discussion and Writing Exercise

71.
$$\begin{array}{r} 50 \\ 4\overline{)200} \\ \underline{200} \\ 0 \\ \underline{0} \\ 0 \end{array}$$

The answer is 50.

73.
$$\begin{array}{r} 14.5 \\ 16\overline{)232.0} \\ \underline{160} \\ 72 \\ \underline{64} \\ 80 \\ \underline{80} \\ 0 \end{array}$$

The answer is 14.5.

75. Familiarize. We let h = Rocky's excess height.

Translate. We have a "how much more" situation.

Height of daughter $\underbrace{}$ plus How much more height $\underbrace{}$ is Rocky's height $\underbrace{}$

$$\begin{array}{ccccc} \downarrow & \downarrow & \downarrow & \downarrow & \downarrow \\ 180\frac{3}{4} & + & h & = & 187\frac{1}{10} \end{array}$$

Solve. We solve the equation as follows:
$$h = 187\frac{1}{10} - 180\frac{3}{4}$$
$$187 \; \boxed{\frac{1}{10} \cdot \frac{2}{2}} = 187 \, \frac{2}{20}$$
$$180 \; \boxed{\frac{3}{4} \cdot \frac{5}{5}} = 180 \, \frac{15}{20}$$

$$187\frac{1}{10} = \quad 187\frac{2}{20} = \quad 186\frac{22}{20}$$
$$-180\frac{3}{4} = -180\frac{15}{20} = -180\frac{15}{20}$$
$$\overline{\qquad\qquad\qquad\qquad\qquad\qquad 6\frac{7}{20}}$$

Thus, $h = 6\frac{7}{20}$.

Check. We add Rocky's excess height to his daughter's height:

$$180\frac{3}{4} + 6\frac{7}{20} = 180\frac{15}{20} + 6\frac{7}{20} = 186\frac{22}{20} = 187\frac{2}{20} = 187\frac{1}{10}$$

The answer checks.

State. Rocky is $6\frac{7}{20}$ cm taller.

77. Familiarize. Let r = the number of earned runs Cy Young gave up in his career.

Translate. We translate to a proportion.

$$\begin{array}{c} \text{Runs} \rightarrow \\ \text{Innings} \rightarrow \end{array} \frac{2.63}{9} = \frac{r}{7356} \begin{array}{c} \leftarrow \text{Runs} \\ \leftarrow \text{Innings} \end{array}$$

Solve.

$$2.63 \cdot 7356 = 9 \cdot r$$
$$\frac{2.63 \cdot 7356}{9} = r$$
$$2150 \approx r$$

Check. We substitute in the proportion and check cross products.

$$\frac{2.63}{9} = \frac{2150}{7356}$$

$2.63 \cdot 7356 = 19,346.28$ and $9 \cdot 2150 = 19,350 \approx 19,346.28$

State. Cy Young gave up 2150 earned runs in his career.

Exercise Set 6.2

1. $90\% = \frac{90}{100}$ $\qquad$ A ratio of 90 to 100

$90\% = 90 \times \frac{1}{100}$ $\qquad$ Replacing % with $\times \frac{1}{100}$

$90\% = 90 \times 0.01$ $\qquad$ Replacing % with $\times 0.01$

3. $12.5\% = \frac{12.5}{100}$ $\qquad$ A ratio of 12.5 to 100

$12.5\% = 12.5 \times \frac{1}{100}$ $\qquad$ Replacing % with $\times \frac{1}{100}$

$12.5\% = 12.5 \times 0.01$ $\qquad$ Replacing % with $\times 0.01$

5. 67%

a) Replace the percent symbol with $\times 0.01$.

67×0.01

b) Move the decimal point two places to the left.

$0.67.$

Thus, $67\% = 0.67$.

7. 45.6%

a) Replace the percent symbol with $\times 0.01$.

45.6×0.01

b) Move the decimal point two places to the left.

$0.45.6$

Thus, $45.6\% = 0.456$.

9. 59.01%

a) Replace the percent symbol with $\times 0.01$.

59.01×0.01

b) Move the decimal point two places to the left.

$0.59.01$

Thus, $59.01\% = 0.5901$.

11. 10%

a) Replace the percent symbol with $\times 0.01$.

10×0.01

b) Move the decimal point two places to the left.

$0.10.$

Thus, $10\% = 0.1$.

13. 1%

a) Replace the percent symbol with $\times 0.01$.

1×0.01

b) Move the decimal point two places to the left.

$0.01.$

Thus, $1\% = 0.01$.

15. 200%

a) Replace the percent symbol with $\times 0.01$.

200×0.01

b) Move the decimal point two places to the left.

$2.00.$

Thus, $200\% = 2$.

17. 0.1%

a) Replace the percent symbol with $\times 0.01$.

0.1×0.01

b) Move the decimal point two places to the left.

$0.00.1$

Thus, $0.1\% = 0.001$.

19. 0.09%

a) Replace the percent symbol with ×0.01.

0.09 × 0.01

b) Move the decimal point two places to the left.

0.00.09

Thus, 0.09% = 0.0009.

21. 0.18%

a) Replace the percent symbol with ×0.01.

0.18 × 0.01

b) Move the decimal point two places to the left.

0.00.18

Thus, 0.18% = 0.0018.

23. 23.19%

a) Replace the percent symbol with ×0.01.

23.19 × 0.01

b) Move the decimal point two places to the left.

0.23.19

Thus, 23.19% = 0.2319.

25. $14\frac{7}{8}$%

a) Convert $14\frac{7}{8}$ to decimal notation and replace the percent symbol with ×0.01.

14.875 × 0.01

b) Move the decimal point two places to the left.

0.14.875

Thus, $14\frac{7}{8}$% = 0.14875.

27. $56\frac{1}{2}$%

a) Convert $56\frac{1}{2}$ to decimal notation and replace the percent symbol with ×0.01.

56.5 × 0.01

b) Move the decimal point two places to the left.

0.56.5

Thus, $56\frac{1}{2}$% = 0.565.

29. 9%

a) Replace the percent symbol with ×0.01.

9 × 0.01

b) Move the decimal point two places to the left.

0.09.

Thus, 9% = 0.09.

58%

a) Replace the percent symbol with ×0.01.

58 × 0.01

b) Move the decimal point two places to the left.

0.58.

Thus, 58% = 0.58.

31. 44%

a) Replace the percent symbol with ×0.01.

44 × 0.01

b) Move the decimal point two places to the left.

0.44.

Thus, 44% = 0.44.

33. 36%

a) Replace the percent symbol with ×0.01.

36 × 0.01

b) Move the decimal point two places to the left.

0.36.

Thus, 36% = 0.36.

35. 0.47

a) Move the decimal point two places to the right.

0.47.

b) Write a percent symbol: 47%

Thus, 0.47 = 47%.

37. 0.03

a) Move the decimal point two places to the right.

0.03.

b) Write a percent symbol: 3%

Thus, 0.03 = 3%.

39. 8.7

a) Move the decimal point two places to the right.

8.70.

b) Write a percent symbol: 870%

Thus, 8.7 = 870%.

41. 0.334

a) Move the decimal point two places to the right.

0.33.4

b) Write a percent symbol: 33.4%

Thus, 0.334 = 33.4%.

43. 0.75

    a) Move the decimal point two places to the right.

      0.75.

      ⌞↑

    b) Write a percent symbol: 75%

    Thus, 0.75 = 75%.

45. 0.4

    a) Move the decimal point two places to the right.

      0.40.

      ⌞↑

    b) Write a percent symbol: 40%

    Thus, 0.4 = 40%.

47. 0.006

    a) Move the decimal point two places to the right.

      0.00.6

      ⌞↑

    b) Write a percent symbol: 0.6%

    Thus, 0.006 = 0.6%.

49. 0.017

    a) Move the decimal point two places to the right.

      0.01.7

      ⌞↑

    b) Write a percent symbol: 1.7%

    Thus, 0.017 = 1.7%.

51. 0.2718

    a) Move the decimal point two places to the right.

      0.27.18

      ⌞↑

    b) Write a percent symbol: 27.18%

    Thus, 0.2718 = 27.18%.

53. 0.0239

    a) Move the decimal point two places to the right.

      0.02.39

      ⌞↑

    b) Write a percent symbol: 2.39%

    Thus, 0.0239 = 2.39%.

55. 0.26

    a) Move the decimal point two places to the right.

      0.26.

      ⌞↑

    b) Write a percent symbol: 26%

    Thus, 0.26 = 26%.

0.38

    a) Move the decimal point two places to the right.

      0.38.

      ⌞↑

    b) Write a percent symbol: 38%

    Thus, 0.38 = 38%.

57. 0.177

    a) Move the decimal point two places to the right.

      0.17.7

      ⌞↑

    b) Write a percent symbol: 17.7%

    Thus, 0.117 = 17.7%.

59. 0.215

    a) Move the decimal point two places to the right.

      0.21.5

      ⌞↑

    b) Write a percent symbol: 21.5%

    Thus, 0.215 = 21.5%.

61. Discussion and Writing Exercise

63. To convert $\dfrac{100}{3}$ to a mixed numeral, we divide.

$$
\begin{array}{r}
3\,3 \\
3\,\overline{\smash{)}1\,0\,0} \\
9\,0 \\
\hline
1\,0 \\
9 \\
\hline
1
\end{array}
\qquad \frac{100}{3} = 33\frac{1}{3}
$$

65. To convert $\dfrac{75}{8}$ to a mixed numeral, we divide.

$$
\begin{array}{r}
9 \\
8\,\overline{\smash{)}7\,5} \\
7\,2 \\
\hline
3
\end{array}
\qquad \frac{75}{8} = 9\frac{3}{8}
$$

67. To convert $\dfrac{567}{98}$ to a mixed numeral, we divide.

$$
\begin{array}{r}
5 \\
9\,8\,\overline{\smash{)}5\,6\,7} \\
4\,9\,0 \\
\hline
7\,7
\end{array}
\qquad \frac{567}{98} = 5\frac{77}{98} = 5\frac{11}{14}
$$

69. To convert $\dfrac{2}{3}$ to decimal notation, we divide.

$$
\begin{array}{r}
0.6\,6 \\
3\,\overline{\smash{)}2.0\,0} \\
1\,8 \\
\hline
2\,0 \\
1\,8 \\
\hline
2
\end{array}
$$

Since 2 keeps reappearing as a remainder, the digits repeat and

$$
\frac{2}{3} = 0.66\ldots \quad \text{or} \quad 0.\overline{6}.
$$

71. To convert $\frac{5}{6}$ to decimal notation, we divide.

$$
\begin{array}{r}
0.8\,3 \\
6\,\overline{\smash{)}5.0\,0} \\
\underline{4\,8} \\
2\,0 \\
\underline{1\,8} \\
2
\end{array}
$$

Since 2 keeps reappearing as a remainder, the digits repeat and

$$\frac{5}{6} = 0.833\ldots \quad \text{or} \quad 0.8\overline{3}.$$

73. To convert $\frac{8}{3}$ to decimal notation, we divide.

$$
\begin{array}{r}
2.6\,6 \\
3\,\overline{\smash{)}8.0\,0} \\
\underline{6} \\
2\,0 \\
\underline{1\,8} \\
2\,0
\end{array}
$$

Since 2 keeps reappearing as a remainder, the digits repeat and

$$\frac{8}{3} = 2.66\ldots \text{ or } 2.\overline{6}.$$

Exercise Set 6.3

1. We use the definition of percent as a ratio.

$$\frac{41}{100} = 41\%$$

3. We use the definition of percent as a ratio.

$$\frac{5}{100} = 5\%$$

5. We multiply by 1 to get 100 in the denominator.

$$\frac{2}{10} = \frac{2}{10} \cdot \frac{10}{10} = \frac{20}{100} = 20\%$$

7. We multiply by 1 to get 100 in the denominator.

$$\frac{3}{10} = \frac{3}{10} \cdot \frac{10}{10} = \frac{30}{100} = 30\%$$

9. $\frac{1}{2} = \frac{1}{2} \cdot \frac{50}{50} = \frac{50}{100} = 50\%$

11. Find decimal notation by division.

$$
\begin{array}{r}
0.8\,7\,5 \\
8\,\overline{\smash{)}7.0\,0\,0} \\
\underline{6\,4} \\
6\,0 \\
\underline{5\,6} \\
4\,0 \\
\underline{4\,0} \\
0
\end{array}
$$

$$\frac{7}{8} = 0.875$$

Convert to percent notation.

0.87.5
└─↑

$$\frac{7}{8} = 87.5\%, \text{ or } 87\frac{1}{2}\%$$

13. $\frac{4}{5} = \frac{4}{5} \cdot \frac{20}{20} = \frac{80}{100} = 80\%$

15. Find decimal notation by division.

$$
\begin{array}{r}
0.6\,6\,6 \\
3\,\overline{\smash{)}2.0\,0\,0} \\
\underline{1\,8} \\
2\,0 \\
\underline{1\,8} \\
2\,0 \\
\underline{1\,8} \\
2
\end{array}
$$

We get a repeating decimal: $\frac{2}{3} = 0.66\overline{6}$

Convert to percent notation.

0.66.$\overline{6}$
└─↑

$$\frac{2}{3} = 66.\overline{6}\%, \text{ or } 66\frac{2}{3}\%$$

17.
$$
\begin{array}{r}
0.1\,6\,6 \\
6\,\overline{\smash{)}1.0\,0\,0} \\
\underline{6} \\
4\,0 \\
\underline{3\,6} \\
4\,0 \\
\underline{3\,6} \\
4
\end{array}
$$

We get a repeating decimal: $\frac{1}{6} = 0.16\overline{6}$

Convert to percent notation.

0.16.$\overline{6}$
└─↑

$$\frac{1}{6} = 16.\overline{6}\%, \text{ or } 16\frac{2}{3}\%$$

19.
$$
\begin{array}{r}
0.1\,8\,7\,5 \\
16\,\overline{\smash{)}3.0\,0\,0\,0} \\
\underline{1\,6} \\
1\,4\,0 \\
\underline{1\,2\,8} \\
1\,2\,0 \\
\underline{1\,1\,2} \\
8\,0 \\
\underline{8\,0} \\
0
\end{array}
$$

$$\frac{3}{16} = 0.1875$$

Convert to percent notation.

0.18.75
└─↑

$$\frac{3}{16} = 18.75\%, \text{ or } 18\frac{3}{4}\%$$

21.
$$\begin{array}{r} 0.8\,1\,2\,5 \\ 16\overline{)1\,3.0\,0\,0\,0} \\ \underline{1\,2\,8} \\ 2\,0 \\ \underline{1\,6} \\ 4\,0 \\ \underline{3\,2} \\ 8\,0 \\ \underline{8\,0} \\ 0 \end{array}$$

$$\frac{13}{16} = 0.8125$$

Convert to percent notation.

$$0.81.25$$

$$\frac{13}{16} = 81.25\%, \text{ or } 81\frac{1}{4}\%$$

23. $\dfrac{4}{25} = \dfrac{4}{25} \cdot \dfrac{4}{4} = \dfrac{16}{100} = 16\%$

25. $\dfrac{1}{20} = \dfrac{1}{20} \cdot \dfrac{5}{5} = \dfrac{5}{100} = 5\%$

27. $\dfrac{17}{50} = \dfrac{17}{50} \cdot \dfrac{2}{2} = \dfrac{34}{100} = 34\%$

29. $\dfrac{2}{5} = \dfrac{2}{5} \cdot \dfrac{20}{20} = \dfrac{40}{100} = 40\%;$

$\dfrac{9}{50} = \dfrac{9}{50} \cdot \dfrac{2}{2} = \dfrac{18}{100} = 18\%$

31. $\dfrac{11}{50} = \dfrac{11}{50} \cdot \dfrac{2}{2} = \dfrac{22}{100} = 22\%$

33. $\dfrac{1}{20} = \dfrac{1}{20} \cdot \dfrac{5}{5} = \dfrac{5}{100} = 5\%$

35. $\dfrac{9}{100} = 9\%$

37. $85\% = \dfrac{85}{100}$ Definition of percent

$$= \dfrac{5 \cdot 17}{5 \cdot 20}$$

$$= \dfrac{5}{5} \cdot \dfrac{17}{20} \quad\Big\} \text{ Simplifying}$$

$$= \dfrac{17}{20}$$

39. $62.5\% = \dfrac{62.5}{100}$ Definition of percent

$$= \dfrac{62.5}{100} \cdot \dfrac{10}{10} \quad \begin{array}{l}\text{Multiplying by 1 to elim-}\\\text{inate the decimal point}\\\text{in the numerator}\end{array}$$

$$= \dfrac{625}{1000}$$

$$= \dfrac{5 \cdot 125}{8 \cdot 125}$$

$$= \dfrac{5}{8} \cdot \dfrac{125}{125} \quad\Big\} \text{ Simplifying}$$

$$= \dfrac{5}{8}$$

41. $33\dfrac{1}{3}\% = \dfrac{100}{3}\%$ Converting from mixed numeral to fraction notation

$$= \dfrac{100}{3} \times \dfrac{1}{100} \quad \text{Definition of percent}$$

$$= \dfrac{100 \cdot 1}{3 \cdot 100} \quad \text{Multiplying}$$

$$= \dfrac{1}{3} \cdot \dfrac{100}{100} \quad\Big\}$$

$$= \dfrac{1}{3} \quad \text{Simplifying}$$

43. $16.\overline{6}\% = 16\dfrac{2}{3}\%$ $\left(16.\overline{6} = 16\dfrac{2}{3}\right)$

$$= \dfrac{50}{3}\% \quad \begin{array}{l}\text{Converting from mixed nu-}\\\text{meral to fractional notation}\end{array}$$

$$= \dfrac{50}{3} \times \dfrac{1}{100} \quad \text{Definition of percent}$$

$$= \dfrac{50 \cdot 1}{3 \cdot 50 \cdot 2} \quad \text{Multiplying}$$

$$= \dfrac{1}{2 \cdot 3} \cdot \dfrac{50}{50} \quad\Big\}$$

$$= \dfrac{1}{6} \quad \text{Simplifying}$$

45. $7.25\% = \dfrac{7.25}{100} = \dfrac{7.25}{100} \cdot \dfrac{100}{100}$

$$= \dfrac{725}{10,000} = \dfrac{29 \cdot 25}{400 \cdot 25} = \dfrac{29}{400} \cdot \dfrac{25}{25}$$

$$= \dfrac{29}{400}$$

47. $0.8\% = \dfrac{0.8}{100} = \dfrac{0.8}{100} \cdot \dfrac{10}{10}$

$$= \dfrac{8}{1000} = \dfrac{1 \cdot 8}{125 \cdot 8} = \dfrac{1}{125} \cdot \dfrac{8}{8}$$

$$= \dfrac{1}{125}$$

49. $25\dfrac{3}{8}\% = \dfrac{203}{8}\%$

$$= \dfrac{203}{8} \times \dfrac{1}{100} \quad \text{Definition of percent}$$

$$= \dfrac{203}{800}$$

51. $78\dfrac{2}{9}\% = \dfrac{704}{9}\%$

$$= \dfrac{704}{9} \times \dfrac{1}{100} \quad \text{Definition of percent}$$

$$= \dfrac{4 \cdot 176 \cdot 1}{9 \cdot 4 \cdot 25}$$

$$= \dfrac{4}{4} \cdot \dfrac{176 \cdot 1}{9 \cdot 25}$$

$$= \dfrac{176}{225}$$

53. $64\dfrac{7}{11}\% = \dfrac{711}{11}\%$

$\qquad = \dfrac{711}{11} \times \dfrac{1}{100}$

$\qquad = \dfrac{711}{1100}$

55. $150\% = \dfrac{150}{100} = \dfrac{3 \cdot 50}{2 \cdot 50} = \dfrac{3}{2} \cdot \dfrac{50}{50} = \dfrac{3}{2}$

57. $0.0325\% = \dfrac{0.0325}{100} = \dfrac{0.0325}{100} \cdot \dfrac{10,000}{10,000} = \dfrac{325}{1,000,000} =$

$\dfrac{25 \cdot 13}{25 \cdot 40,000} = \dfrac{25}{25} \cdot \dfrac{13}{40,000} = \dfrac{13}{40,000}$

59. Note that $33.\overline{3}\% = 33\dfrac{1}{3}\%$ and proceed as in Exercise 41;

$33.\overline{3}\% = \dfrac{1}{3}$.

61. $8\% = \dfrac{8}{100}$

$\qquad = \dfrac{4 \cdot 2}{4 \cdot 25} = \dfrac{4}{4} \cdot \dfrac{2}{25}$

$\qquad = \dfrac{2}{25}$

63. $60\% = \dfrac{60}{100}$

$\qquad = \dfrac{20 \cdot 3}{20 \cdot 5} = \dfrac{20}{20} \cdot \dfrac{3}{5}$

$\qquad = \dfrac{3}{5}$

65. $2\% = \dfrac{2}{100}$

$\qquad = \dfrac{1 \cdot 2}{50 \cdot 2} = \dfrac{1}{50} \cdot \dfrac{2}{2}$

$\qquad = \dfrac{1}{50}$

67. $35\% = \dfrac{35}{100}$

$\qquad = \dfrac{7 \cdot 5}{20 \cdot 5} = \dfrac{7}{20} \cdot \dfrac{5}{5}$

$\qquad = \dfrac{7}{20}$

69. $47\% = \dfrac{47}{100}$

71. $\dfrac{1}{8} = 1 \div 8$

$$\begin{array}{r} 0.1\,2\,5 \\ 8\,\overline{)\,1.0\,0\,0} \\ \underline{8} \\ 2\,0 \\ \underline{1\,6} \\ 4\,0 \\ \underline{4\,0} \\ 0 \end{array}$$

$\dfrac{1}{8} = 0.125 = 12\dfrac{1}{2}\%,\ \text{or}\ \textbf{12.5\%}$

$\dfrac{1}{6} = 1 \div 6$

$$\begin{array}{r} 0.1\,6\,6 \\ 6\,\overline{)\,1.0\,0\,0} \\ \underline{6} \\ 4\,0 \\ \underline{3\,6} \\ 4\,0 \\ \underline{3\,6} \\ 4 \end{array}$$

We get a repeating decimal: $0.1\overline{6}$

$0.16.\overline{6}\qquad\qquad 0.1\overline{6} = 16.\overline{6}\%$

$\dfrac{1}{6} = 0.1\overline{6} = 16.\overline{6}\%,\ \text{or}\ 16\dfrac{2}{3}\%$

$20\% = \dfrac{20}{100} = \dfrac{1}{5} \cdot \dfrac{20}{20} = \dfrac{1}{5}$

$0.20.\qquad\qquad 20\% = 0.2$

$\dfrac{1}{5} = \textbf{0.2} = \textbf{20\%}$

$0.25.\qquad\qquad 0.25 = 25\%$

$25\% = \dfrac{25}{100} = \dfrac{1}{4} \cdot \dfrac{25}{25} = \dfrac{1}{4}$

$\dfrac{1}{4} = \textbf{0.25} = \textbf{25\%}$

$33\dfrac{1}{3}\% = \dfrac{100}{3}\% = \dfrac{100}{3} \times \dfrac{1}{100} = \dfrac{100}{300} = \dfrac{1}{3} \cdot \dfrac{100}{100} = \dfrac{1}{3}$

$0.33.\overline{3}\qquad\qquad 33.\overline{3}\% = 0.33\overline{3},\ \text{or}\ 0.\overline{3}$

$\dfrac{1}{3} = \textbf{0.}\overline{\textbf{3}} = \textbf{33}\dfrac{1}{3}\%,\ \text{or}\ \textbf{33.}\overline{\textbf{3}}\%$

$37.5\% = \dfrac{37.5}{100} = \dfrac{37.5}{100} \cdot \dfrac{10}{10} = \dfrac{375}{1000} = \dfrac{3}{8} \cdot \dfrac{125}{125} = \dfrac{3}{8}$

$0.37.5\qquad\qquad 37.5\% = 0.375$

$\dfrac{3}{8} = \textbf{0.375} = \textbf{37}\dfrac{1}{2}\%,\ \text{or}\ \textbf{37.5\%}$

$40\% = \dfrac{40}{100} = \dfrac{2}{5} \cdot \dfrac{20}{20} = \dfrac{2}{5}$

$0.40.\qquad\qquad 40\% = 0.4$

$\dfrac{2}{5} = \textbf{0.4} = \textbf{40\%}$

$\dfrac{1}{2} = \dfrac{1}{2} \cdot \dfrac{5}{5} = \dfrac{5}{10} = 0.5$

$\dfrac{1}{2} = \dfrac{1}{2} \cdot \dfrac{50}{50} = \dfrac{50}{100} = 5\%$

$\dfrac{1}{2} = \textbf{0.5} = \textbf{50\%}$

73. 0.50. $0.5 = 50\%$

$50\% = \dfrac{50}{100} = \dfrac{1}{2} \cdot \dfrac{50}{50} = \dfrac{1}{2}$

$\dfrac{1}{2} = 0.5 = 50\%$

$\dfrac{1}{3} = 1 \div 3$

$\begin{array}{r} 0.3 \\ 3\overline{)1.0} \\ \underline{9} \\ 1 \end{array}$

We get a repeating decimal: $0.\overline{3}$

0.33.$\overline{3}$ $0.\overline{3} = 33.\overline{3}\%$

$\dfrac{1}{3} = 0.\overline{3} = 33.\overline{3}\%, \text{ or } 33\dfrac{1}{3}\%$

$25\% = \dfrac{25}{100} = \dfrac{25}{25} \cdot \dfrac{1}{4} = \dfrac{1}{4}$

0.25. $25\% = 0.25$

$\dfrac{1}{4} = 0.25 = 25\%$

$16\dfrac{2}{3}\% = \dfrac{50}{3}\% = \dfrac{50}{3} \times \dfrac{1}{100} = \dfrac{50 \cdot 1}{3 \cdot 2 \cdot 50} = \dfrac{50}{50} \cdot \dfrac{1}{6} = \dfrac{1}{6}$

$\dfrac{1}{6} = 1 \div 6$

$\begin{array}{r} 0.1\,6 \\ 6\overline{)1.0\,0} \\ \underline{6} \\ 4\,0 \\ \underline{3\,6} \\ 4 \end{array}$

We get a repeating decimal: $0.1\overline{6}$

$\dfrac{1}{6} = 0.1\overline{6} = 16\dfrac{2}{3}\%, \text{ or } 16.\overline{6}\%$

0.12.5 $0.125 = 12.5\%$

$12.5\% = \dfrac{12.5}{100} = \dfrac{12.5}{100} \cdot \dfrac{10}{10} = \dfrac{125}{1000} = \dfrac{125}{125} \cdot \dfrac{1}{8} = \dfrac{1}{8}$

$\dfrac{1}{8} = 0.125 = 12.5\%, \text{ or } 12\dfrac{1}{2}\%$

$\dfrac{3}{4} = \dfrac{3}{4} \cdot \dfrac{25}{25} = \dfrac{75}{100} = 75\%$

0.75. $75\% = 0.75$

$\dfrac{3}{4} = 0.75 = 75\%$

$0.8\overline{3} = 0.83.\overline{3}$ $0.8\overline{3} = 83.\overline{3}\%$

$83.\overline{3}\% = 83\dfrac{1}{3}\% = \dfrac{250}{3}\% = \dfrac{250}{3} \times \dfrac{1}{100} = \dfrac{5 \cdot 50}{3 \cdot 2 \cdot 50} =$
$\dfrac{5}{6} \cdot \dfrac{50}{50} = \dfrac{5}{6}$

$\dfrac{5}{6} = 0.8\overline{3} = 83.\overline{3}\%, \text{ or } 83\dfrac{1}{3}\%$

$\dfrac{3}{8} = 3 \div 8$

$\begin{array}{r} 0.3\,7\,5 \\ 8\overline{)3.0\,0\,0} \\ \underline{2\,4} \\ 6\,0 \\ \underline{5\,6} \\ 4\,0 \\ \underline{4\,0} \\ 0 \end{array}$

$\dfrac{3}{8} = 0.375$

0.37.5 $0.375 = 37.5\%$

$\dfrac{3}{8} = 0.375 = 37.5\%, \text{ or } 37\dfrac{1}{2}\%$

75. Discussion and Writing Exercise

77. $13 \cdot x = 910$

$\dfrac{13 \cdot x}{13} = \dfrac{910}{13}$

$x = 70$

79. $0.05 \times b = 20$

$\dfrac{0.05 \times b}{0.05} = \dfrac{20}{0.05}$

$b = 400$

81. $\dfrac{24}{37} = \dfrac{15}{x}$

$24 \cdot x = 37 \cdot 15$ Equating cross products

$x = \dfrac{37 \cdot 15}{24}$

$x = 23.125$

83. $\dfrac{9}{10} = \dfrac{x}{5}$

$9 \cdot 5 = 10 \cdot x$

$\dfrac{9 \cdot 5}{10} = x$

$\dfrac{45}{10} = x$

$\dfrac{9}{2} = x, \text{ or }$

$4.5 = x$

85.

$$3\overline{)\begin{array}{r}3\,3\\1\,0\,0\\\underline{9\,0}\\1\,0\\\underline{9}\\1\end{array}}$$

$$\frac{100}{3} = 33\frac{1}{3}$$

87.

$$3\overline{)\begin{array}{r}8\,3\\2\,5\,0\\\underline{2\,4\,0}\\1\,0\\\underline{9}\\1\end{array}}$$

$$\frac{250}{3} = 83\frac{1}{3}$$

89.

$$8\overline{)\begin{array}{r}4\,3\\3\,4\,5\\\underline{3\,2\,0}\\2\,5\\\underline{2\,4}\\1\end{array}}$$

$$\frac{345}{8} = 43\frac{1}{8}$$

91.

$$4\overline{)\begin{array}{r}1\,8\\7\,5\\\underline{4\,0}\\3\,5\\\underline{3\,2}\\3\end{array}}$$

$$\frac{75}{4} = 18\frac{3}{4}$$

93. $1\frac{1}{17} = \frac{18}{17}$ $(1 \cdot 17 = 17, \; 17 + 1 = 18)$

95. $101\frac{1}{2} = \frac{203}{2}$ $(101 \cdot 2 = 202, \; 202 + 1 = 203)$

97. Use a calculator.

$$\frac{41}{369} = 0.11.\overline{1} = 11.\overline{1}\%$$

99. $2.57\overline{4631} = 2.57.\overline{46317} = 257.\overline{46317}\%$

101. $\frac{14}{9}\% = \frac{14}{9} \times \frac{1}{100} = \frac{2 \cdot 7 \cdot 1}{9 \cdot 2 \cdot 50} = \frac{2}{2} \cdot \frac{7}{450} = \frac{7}{450}$

To find decimal notation for $\frac{7}{450}$ we divide.

$$450\overline{)\begin{array}{r}0.0\,1\,5\,5\\7.0\,0\,0\,0\\\underline{4\,5\,0}\\2\,5\,0\,0\\\underline{2\,2\,5\,0}\\2\,5\,0\,0\\\underline{2\,2\,5\,0}\\2\,5\,0\end{array}}$$

We get a repeating decimal: $\frac{14}{9}\% = 0.01\overline{5}$

103. $\frac{729}{7}\% = \frac{729}{7} \times \frac{1}{100} = \frac{729}{700}$

To find decimal notation for $\frac{729}{700}$ we divide.

$$700\overline{)\begin{array}{r}1.0\,4\,1\,4\,2\,8\,5\,7\\7\,2\,9.0\,0\,0\,0\,0\,0\,0\,0\\\underline{7\,0\,0}\\2\,9\,0\,0\\\underline{2\,8\,0\,0}\\1\,0\,0\,0\\\underline{7\,0\,0}\\3\,0\,0\,0\\\underline{2\,8\,0\,0}\\2\,0\,0\,0\\\underline{1\,4\,0\,0}\\6\,0\,0\,0\\\underline{5\,6\,0\,0}\\4\,0\,0\,0\\\underline{3\,5\,0\,0}\\5\,0\,0\,0\\\underline{4\,9\,0\,0}\\1\,0\,0\end{array}}$$

We get a repeating decimal: $\frac{729}{7}\% = 1.04\overline{142857}$.

Exercise Set 6.4

1. What is 32% of 78?

$$\begin{array}{ccccc}\downarrow&\downarrow&\downarrow&\downarrow&\downarrow\\a&=&32\%&\times&78\end{array}$$

3. 89 is what percent of 99?

$$\begin{array}{cccccc}\downarrow&\downarrow&&\downarrow&&\downarrow\;\downarrow\\89&=&&p&&\times\;99\end{array}$$

5. 13 is 25% of what?

$$\begin{array}{ccccc}\downarrow&\downarrow&\downarrow&\downarrow&\downarrow\\13&=&25\%&\times&b\end{array}$$

7. What is 85% of 276?

Translate: $a = 85\% \cdot 276$

Solve: The letter is by itself. To solve the equation we convert 85% to decimal notation and multiply.

$$\begin{array}{r}2\,7\,6\\\times\,0.\,8\,5\\\hline 1\,3\,8\,0\\2\,2\,0\,8\,0\\\hline a=2\,3\,4.\,6\,0\end{array}\quad (85\% = 0.85)$$

234.6 is 85% of 276. The answer is 234.6.

9. 150% of 30 is what?

Translate: $150\% \times 30 = a$

Solve: Convert 150% to decimal notation and multiply.

$$\begin{array}{r}3\,0\\\times\,1.\,5\\\hline 1\,5\,0\\3\,0\,0\\\hline a=4\,5.\,0\end{array}\quad (150\% = 1.5)$$

150% of 30 is 45. The answer is 45.

11. What is 6% of $300?

Translate: $a = 6\% \cdot \$300$

Solve: Convert 6% to decimal notation and multiply.

$$\begin{array}{r} \$\,3\,0\,0 \\ \times\,0.0\,6 \quad (6\% = 0.06) \\ \hline a = \$\,1\,8.0\,0 \end{array}$$

$18 is 6% of $300. The answer is $18.

13. 3.8% of 50 is what?

Translate: $3.8\% \cdot 50 = a$

Solve: Convert 3.8% to decimal notation and multiply.

$$\begin{array}{r} 5\,0 \\ \times\,0.0\,3\,8 \quad (3.8\% = 0.038) \\ \hline 4\,0\,0 \\ 1\,5\,0\,0 \\ \hline a = 1.9\,0\,0 \end{array}$$

3.8% of 50 is 1.9. The answer is 1.9.

15. $39 is what percent of $50?

Translate: $39 = n \times 50$

Solve: To solve the equation we divide on both sides by 50 and convert the answer to percent notation.

$$n \cdot 50 = 39$$

$$\frac{n \cdot 50}{50} = \frac{39}{50}$$

$$n = 0.78 = 78\%$$

$39 is 78% of $50. The answer is 78%.

17. 20 is what percent of 10?

Translate: $20 = n \times 10$

Solve: To solve the equation we divide on both sides by 10 and convert the answer to percent notation.

$$n \cdot 10 = 20$$

$$\frac{n \cdot 10}{10} = \frac{20}{10}$$

$$n = 2 = 200\%$$

20 is 200% of 10. The answer is 200%.

19. What percent of $300 is $150?

Translate: $n \times 300 = 150$

Solve: $n \cdot 300 = 150$

$$\frac{n \cdot 300}{300} = \frac{150}{300}$$

$$n = 0.5 = 50\%$$

50% of $300 is $150. The answer is 50%.

21. What percent of 80 is 100?

Translate: $n \times 80 = 100$

Solve: $n \cdot 80 = 100$

$$\frac{n \cdot 80}{80} = \frac{100}{80}$$

$$n = 1.25 = 125\%$$

125% of 80 is 100. The answer is 125%.

23. 20 is 50% of what?

Translate: $20 = 50\% \times b$

Solve: To solve the equation we divide on both sides by 50%:

$$\frac{20}{50\%} = \frac{50\% \times b}{50\%}$$

$$\frac{20}{0.5} = b \quad (50\% = 0.5)$$

$$40 = b$$

$$\begin{array}{r} 4\,0. \\ 0.5_\wedge\overline{)2\,0.0_\wedge} \\ 2\,0\,0 \\ \hline 0 \\ 0 \\ \hline 0 \end{array}$$

20 is 50% of 40. The answer is 40.

25. 40% of what is $16?

Translate: $40\% \times b = 16$

Solve: To solve the equation we divide on both sides by 40%:

$$\frac{40\% \times b}{40\%} = \frac{16}{40\%}$$

$$b = \frac{16}{0.4} \quad (40\% = 0.4)$$

$$b = 40$$

$$\begin{array}{r} 4\,0. \\ 0.4_\wedge\overline{)1\,6.0_\wedge} \\ 1\,6\,0 \\ \hline 0 \\ 0 \\ \hline 0 \end{array}$$

40% of $40 is $16. The answer is $40.

27. 56.32 is 64% of what?

Translate: $56.32 = 64\% \times b$

Solve: $\dfrac{56.32}{64\%} = \dfrac{64\% \times b}{64\%}$

$$\frac{56.32}{0.64} = b$$

$$88 = b$$

$$\begin{array}{r} 8\,8. \\ 0.6\,4_\wedge\overline{)5\,6.3\,2_\wedge} \\ 5\,1\,2\,0 \\ \hline 5\,1\,2 \\ 5\,1\,2 \\ \hline 0 \end{array}$$

56.32 is 64% of 88. The answer is 88.

29. 70% of what is 14?

Translate: $70\% \times b = 14$

Solve: $\dfrac{70\% \times b}{70\%} = \dfrac{14}{70\%}$

$$b = \frac{14}{0.7}$$

$$b = 20$$

$$\begin{array}{r} 2\,0. \\ 0.7_\wedge\overline{)1\,4.0_\wedge} \\ 1\,4\,0 \\ \hline 0 \\ 0 \\ \hline 0 \end{array}$$

70% of 20 is 14. The answer is 20.

31. What is $62\frac{1}{2}\%$ of 10?

Translate: $a = 62\frac{1}{2}\% \times 10$

Solve: $a = 0.625 \times 10$ $(62\frac{1}{2}\% = 0.625)$

$\quad\quad a = 6.25$ Multiplying

6.25 is $62\frac{1}{2}\%$ of 10. The answer is 6.25.

33. What is 8.3% of $10,200?

Translate: $a = 8.3\% \times 10,200$

Solve: $a = 8.3\% \times 10,200$
$\quad\quad\quad a = 0.083 \times 10,200$ $(8.3\% = 0.083)$
$\quad\quad\quad a = 846.6$ Multiplying

$846.60 is 8.3% of $10,200. The answer is $846.60.

35. Discussion and Writing Exercise

37. $0.\underline{09}$ $=$ $\dfrac{9}{100}$

   2 decimal   2 zeros
   places

39. $0.\underline{875}$ $=$ $\dfrac{875}{1000}$

   3 decimal   3 zeros
   places

$\dfrac{875}{1000} = \dfrac{7 \cdot 125}{8 \cdot 125} = \dfrac{7}{8} \cdot \dfrac{125}{125} = \dfrac{7}{8}$

Thus, $0.875 = \dfrac{875}{1000}$, or $\dfrac{7}{8}$.

41. $0.\underline{9375}$ $=$ $\dfrac{9375}{10,000}$

   4 decimal   4 zeros
   places

$\dfrac{9375}{10,000} = \dfrac{15 \cdot 625}{16 \cdot 625} = \dfrac{15}{16} \cdot \dfrac{625}{625} = \dfrac{15}{16}$

Thus, $0.9375 = \dfrac{9375}{10,000}$, or $\dfrac{15}{16}$

43. $\dfrac{89}{100}$ $0.89.$
                        $\uparrow\_\rfloor$

   2 zeros   Move 2 places

$\dfrac{89}{100} = 0.89$

45. $\dfrac{3}{10}$ $0.3.$
                    $\uparrow\rfloor$

   1 zero   Move 1 place

$\dfrac{3}{10} = 0.3$

47. Estimate: Round 7.75% to 8% and $10,880 to $11,000.
Then translate:
   What is 8% of $11,000?
   $\downarrow$ $\downarrow$ $\downarrow$ $\downarrow$ $\downarrow$
   a $= 8\% \times$ $11,000$

We convert 8% to decimal notation and multiply.

$\quad 1\,1,0\,0\,0$
$\underline{\times \quad 0.0\,8}$ $(8\% = 0.08)$
$\quad 8\,8\,0.0\,0$

$880 is about 7.75% of $10,880. (Answers may vary.)

Calculate: First we translate.
   What is 7.75% of $10,880?
   $\downarrow$ $\downarrow$ $\downarrow$ $\downarrow$ $\downarrow$
   a $= 7.75\% \times$ $10,880$

Use a calculator to multiply:

$\quad 0.0775 \times 10,880 = 843.2$

$843.20 is 7.75% of $10,880.

49. Estimate: Round $2496 to $2500 and 24% to 25%. Then
translate:
   $2500 is 25% of what?
   $\downarrow$ $\downarrow$ $\downarrow$ $\downarrow$ $\downarrow$
   2500 $= 25\% \times$ b

We convert 25% to decimal notation and divide.

$\dfrac{2500}{0.25} = \dfrac{0.25 \times b}{0.25}$
$10,000 = b$

$2496 is 24% of about $10,000. (Answers may vary.)

Calculate: First we translate.
   $2496 is 24% of what?
   $\downarrow$ $\downarrow$ $\downarrow$ $\downarrow$ $\downarrow$
   2496 $= 0.24\times$ b

Use a calculator to divide:
$\dfrac{2496}{0.24} = 10,400$

$2496 is 24% of $10,400.

51. We translate:
   40% of $18\frac{3}{4}\%$ of $25,000 is what?
   $\downarrow$ $\downarrow$ $\downarrow$ $\downarrow$ $\downarrow$ $\downarrow$ $\downarrow$
   $40\% \times 18\frac{3}{4}\% \times$ $25,000 =$ a

We convert 40% and $18\frac{3}{4}\%$ to decimal notation and multiply.

$\quad\quad 0.4 \times 0.1875 \times 25,000 = a$

$\quad 0.1\,8\,7\,5$
$\underline{\times \quad\quad 0.4}$
$\quad 0.0\,7\,5\,0\,0$

$\quad\quad 2\,5,0\,0\,0$
$\underline{\times \quad\quad 0.0\,7\,5}$
$\quad 1\,2\,5\,0\,0\,0$
$\underline{1\,7\,5\,0\,0\,0\,0}$
$\quad 1\,8\,7\,5.0\,0\,0$

40% of $18\frac{3}{4}\%$ of $25,000 is $1875.

Exercise Set 6.5

1. What is 37% of 74?

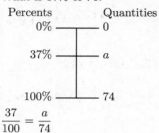

$$\frac{37}{100} = \frac{a}{74}$$

3. 4.3 is what percent of 5.9?

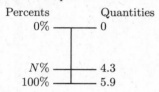

$$\frac{N}{100} = \frac{4.3}{5.9}$$

5. 14 is 25% of what?

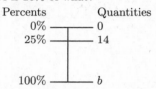

$$\frac{25}{100} = \frac{14}{b}$$

7. What is 76% of 90?

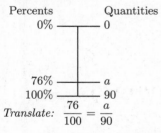

Translate: $\dfrac{76}{100} = \dfrac{a}{90}$

Solve: $76 \cdot 90 = 100 \cdot a$ Equating cross-products

$$\frac{76 \cdot 90}{100} = \frac{100 \cdot a}{100}$$ Dividing by 100

$$\frac{6840}{100} = a$$

$$68.4 = a$$ Simplifying

68.4 is 76% of 90. The answer is 68.4.

9. 70% of 660 is what?

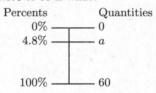

Translate: $\dfrac{70}{100} = \dfrac{a}{660}$

Solve: $70 \cdot 660 = 100 \cdot a$ Equating cross-products

$$\frac{70 \cdot 660}{100} = \frac{100 \cdot a}{100}$$ Dividing by 100

$$\frac{46,200}{100} = a$$

$$462 = a$$ Simplifying

70% of 660 is 462. The answer is 462.

11. What is 4% of 1000?

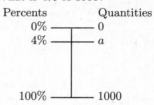

Translate: $\dfrac{4}{100} = \dfrac{a}{1000}$

Solve: $4 \cdot 1000 = 100 \cdot a$

$$\frac{4 \cdot 1000}{100} = \frac{100 \cdot a}{100}$$

$$\frac{4000}{100} = a$$

$$40 = a$$

40 is 4% of 1000. The answer is 40.

13. 4.8% of 60 is what?

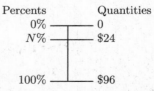

Translate: $\dfrac{4.8}{100} = \dfrac{a}{60}$

Solve: $4.8 \cdot 60 = 100 \cdot a$

$$\frac{4.8 \cdot 60}{100} = \frac{100 \cdot a}{100}$$

$$\frac{288}{100} = a$$

$$2.88 = a$$

4.8% of 60 is 2.88. The answer is 2.88.

15. $24 is what percent of $96?

Percents	Quantities
0%	0
N%	$24
100%	$96

Translate: $\dfrac{N}{100} = \dfrac{24}{96}$

Solve: $96 \cdot N = 100 \cdot 24$

$$\dfrac{96N}{96} = \dfrac{100 \cdot 24}{96}$$

$$N = \dfrac{100 \cdot 24}{96}$$

$$N = 25$$

$24 is 25% of $96. The answer is 25%.

17. 102 is what percent of 100?

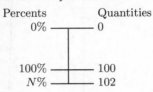

Translate: $\dfrac{N}{100} = \dfrac{102}{100}$

Solve: $100 \cdot N = 100 \cdot 102$

$$\dfrac{100 \cdot N}{100} = \dfrac{100 \cdot 102}{100}$$

$$N = \dfrac{100 \cdot 102}{100}$$

$$N = 102$$

102 is 102% of 100. The answer is 102%.

19. What percent of $480 is $120?

Percents	Quantities
0%	0
N%	$120
100%	$480

Translate: $\dfrac{N}{100} = \dfrac{120}{480}$

Solve: $480 \cdot N = 100 \cdot 120$

$$\dfrac{480 \cdot N}{480} = \dfrac{100 \cdot 120}{480}$$

$$N = \dfrac{100 \cdot 120}{480}$$

$$N = 25$$

25% of $480 is $120. The answer is 25%.

21. What percent of 160 is 150?

Percents	Quantities
0%	0
N%	150
100%	160

Translate: $\dfrac{N}{100} = \dfrac{150}{160}$

Solve: $160 \cdot N = 100 \cdot 150$

$$\dfrac{160 \cdot N}{160} = \dfrac{100 \cdot 150}{160}$$

$$N = \dfrac{100 \cdot 150}{160}$$

$$N = 93.75$$

93.75% of 160 is 150. The answer is 93.75%.

23. $18 is 25% of what?

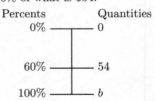

Translate: $\dfrac{25}{100} = \dfrac{18}{b}$

Solve: $25 \cdot b = 100 \cdot 18$

$$\dfrac{25 \cdot b}{b} = \dfrac{100 \cdot 18}{25}$$

$$b = \dfrac{100 \cdot 18}{25}$$

$$b = 72$$

$18 is 25% of $72. The answer is $72.

25. 60% of what is $54.

Percents	Quantities
0%	0
60%	54
100%	b

Translate: $\dfrac{60}{100} = \dfrac{54}{b}$

Solve: $60 \cdot b = 100 \cdot 54$

$$\dfrac{60 \cdot b}{b} = \dfrac{100 \cdot 54}{60}$$

$$b = \dfrac{100 \cdot 54}{60}$$

$$b = 90$$

60% of 90 is 54. The answer is 90.

27. 65.12 is 74% of what?

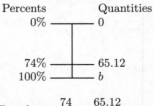

Translate: $\dfrac{74}{100} = \dfrac{65.12}{b}$

Solve: $74 \cdot b = 100 \cdot 65.12$

$$\frac{74 \cdot b}{74} = \frac{100 \cdot 65.12}{74}$$

$$b = \frac{100 \cdot 65.12}{74}$$

$$b = 88$$

65.12 is 74% of 88. The answer is 88.

29. 80% of what is 16?

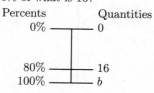

Translate: $\dfrac{80}{100} = \dfrac{16}{b}$

Solve: $80 \cdot b = 100 \cdot 16$

$$\frac{80 \cdot b}{80} = \frac{100 \cdot 16}{80}$$

$$b = \frac{100 \cdot 16}{80}$$

$$b = 20$$

80% of 20 is 16. The answer is 20.

31. What is $62\frac{1}{2}\%$ of 40?

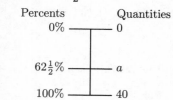

Translate: $\dfrac{62\frac{1}{2}}{100} = \dfrac{a}{40}$

Solve: $62\dfrac{1}{2} \cdot 40 = 100 \cdot a$

$$\frac{125}{2} \cdot \frac{40}{1} = 100 \cdot a$$

$$2500 = 100 \cdot a$$

$$\frac{2500}{100} = \frac{100 \cdot a}{100}$$

$$25 = a$$

25 is $62\frac{1}{2}\%$ of 40. The answer is 25.

33. What is 9.4% of $8300?

Percents   Quantities

0% —— 0

9.4% —— a

100% —— 8300

Translate: $\dfrac{9.4}{100} = \dfrac{a}{8300}$

Solve: $9.4 \cdot 8300 = 100 \cdot a$

$$\frac{9.4 \cdot 8300}{100} = \frac{100 \cdot a}{100}$$

$$\frac{78,020}{100} = a$$

$$780.2 = a$$

$780.20 is 9.4% of $8300. The answer is $780.20.

35. Discussion and Writing Exercise

37. $\dfrac{x}{188} = \dfrac{2}{47}$

$$47 \cdot x = 188 \cdot 2$$

$$x = \frac{188 \cdot 2}{47}$$

$$x = \frac{4 \cdot 47 \cdot 2}{47}$$

$$x = 8$$

39. $\dfrac{4}{7} = \dfrac{x}{14}$

$$4 \cdot 14 = 7 \cdot x$$

$$\frac{4 \cdot 14}{7} = x$$

$$\frac{4 \cdot 2 \cdot 7}{7} = x$$

$$8 = x$$

41. $\dfrac{5000}{t} = \dfrac{3000}{60}$

$$5000 \cdot 60 = 3000 \cdot t$$

$$\frac{5000 \cdot 60}{3000} = t$$

$$\frac{5 \cdot 1000 \cdot 3 \cdot 20}{3 \cdot 1000} = t$$

$$100 = t$$

43. $\dfrac{x}{1.2} = \dfrac{36.2}{5.4}$

$$5.4 \cdot x = 1.2(36.2)$$

$$x = \frac{1.2(36.2)}{5.4}$$

$$x = 8.0\overline{4}$$

45. *Familiarize*. Let $q =$ the number of quarts of liquid ingredients the recipe calls for.

***Translate*.**

Butter-milk plus Skim milk plus Oil is Total liquid ingredients

$$\frac{1}{2} + \frac{1}{3} + \frac{1}{16} = q$$

***Solve*.** We carry out the addition. The LCM of the denominators is 48, so the LCD is 48.

$$\frac{1}{2} \cdot \frac{24}{24} + \frac{1}{3} \cdot \frac{16}{16} + \frac{1}{16} \cdot \frac{3}{3} = q$$

$$\frac{24}{48} + \frac{16}{48} + \frac{3}{48} = q$$

$$\frac{43}{48} = q$$

Check. We repeat the calculation. The answer checks.

State. The recipe calls for $\frac{43}{48}$ qt of liquid ingredients.

47. Estimate: Round 8.85% to 9%, and $12,640 to $12,600.
What is 9% of $12,600?

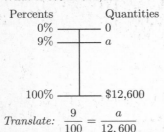

Translate: $\frac{9}{100} = \frac{a}{12,600}$

Solve: $9 \cdot 12,600 = 100 \cdot a$

$$\frac{9 \cdot 12,600}{100} = \frac{100 \cdot a}{100}$$

$$\frac{113,400}{100} = a$$

$$1134 = a$$

$1134 is about 8.85% of $12,640. (Answers may vary.)

Calculate:

What is 8.85% of $12,640?

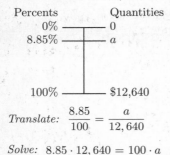

Translate: $\frac{8.85}{100} = \frac{a}{12,640}$

Solve: $8.85 \cdot 12,640 = 100 \cdot a$

$$\frac{8.85 \cdot 12,640}{100} = \frac{100 \cdot a}{100}$$

$$\frac{111,864}{100} = a \qquad \text{Use a calculator to}$$
$$\text{multiply and divide.}$$

$$1118.64 = a$$

$1118.64 is 8.85% of $12,640.

Exercise Set 6.6

1. *Familiarize*. Let w = the number of wild horses in Nevada.

Translate. We translate to a percent equation.

What number is 48.4% of 27,369?

$$\begin{array}{ccccc} \downarrow & & \downarrow & \downarrow & \downarrow & \downarrow \\ w & & = 48.4\% & \cdot & 27,369 \end{array}$$

Solve. We convert 48.4% to decimal notation and multiply.

$$w = 0.484 \cdot 27,369 = 13,246.596 \approx 13,247$$

Check. We can repeat the calculations. We can also do a partial check by estimating: $48.4\% \cdot 27,369 \approx 50\% \cdot 27,00 = 13,500$. Since 13,500 is close to 13,247, our answer is reasonable.

State. There are about 13,247 wild horses in Nevada.

3. *Familiarize*. Let t = the value of a Nissan 350Z after three years and f = the value after five years.

Translate. We translate to two proportions.

$$\frac{62}{100} = \frac{t}{34,000} \text{ and } \frac{52}{100} = \frac{f}{34,000}$$

Solve. We solve each proportion.

$$\frac{62}{100} = \frac{t}{34,000}$$
$$62 \cdot 34,000 = 100 \cdot t$$
$$\frac{62 \cdot 34,000}{100} = \frac{100 \cdot t}{100}$$
$$21,080 = t$$

$$\frac{52}{100} = \frac{f}{34,000}$$
$$52 \cdot 34,000 = 100 \cdot f$$
$$\frac{52 \cdot 34,000}{100} = \frac{100 \cdot f}{100}$$
$$17,680 = f$$

Check. We can repeat the calculations. We can also do a partial check by estimating.

$$62\% \cdot 34,000 \approx 60\% \cdot 35,000 = 21,000 \approx 21,080$$
$$52\% \cdot 34,000 \approx 50\% \cdot 34,000 = 17,000 \approx 17,680$$

The answers check.

State. The value of a Nissan 350Z will be $21,080 after three years; the value after five years will be $17,680.

5. *Familiarize*. Let x = the number of people in the U.S. who are overweight and y = the number who are obese, in millions.

Translate. We translate to percent equations.

What number is 60% of 294?

$$\begin{array}{ccccc} \downarrow & & \downarrow & \downarrow & \downarrow & \downarrow \\ x & & = 60\% & \cdot & 294 \end{array}$$

What number is 25% of 294?

$$\begin{array}{ccccc} \downarrow & & \downarrow & \downarrow & \downarrow & \downarrow \\ y & & = 25\% & \cdot & 294 \end{array}$$

Solve.

$$x = 0.6 \cdot 294 = 176.4$$
$$y = 0.25 \cdot 294 = 73.5$$

Check. We can repeat the calculations. Also note that $60\% \cdot 294 \approx 60\% \cdot 300 = 180$ and $25\% \cdot 294 \approx 25\% \cdot 300 = 75$. Since 176.4 is close to 180 and 73.5 is close to 75, the answers check.

State. 176.4 million, or 176,400,000 people in the U.S. are overweight and 73.5 million, or 73,500,000 are obese.

7. *Familiarize.* First we find the amount of the solution that is acid. We let $a =$ this amount.

Translate. We translate to a percent equation.

What is 3% of 680?

$$\downarrow \quad \downarrow \quad \downarrow \quad \downarrow \quad \downarrow$$
$$a \quad = 3\% \text{ of } 680$$

Solve. We convert 3% to decimal notation and multiply.

$$a = 3\% \times 680 = 0.03 \times 680 = 20.4$$

Now we find the amount that is water. We let $w =$ this amount.

Total amount	minus	Amount of acid	is	Amount of water
$\downarrow$	$\downarrow$	$\downarrow$	$\downarrow$	$\downarrow$
680	$-$	20.4	$=$	w

To solve the equation we carry out the subtraction.

$$w = 680 - 20.4 = 659.6$$

Check. We can repeat the calculations. Also, observe that, since 3% of the solution is acid, 97% is water. Because 97% of $680 = 0.97 \times 680 = 659.6$, our answer checks.

State. The solution contains 20.4 mL of acid and 659.6 mL of water.

9. *Familiarize.* Let $n =$ the number of miles of the Mississippi River that are navigable.

Translate. We translate to a proportion.

$$\frac{77}{100} = \frac{n}{2348}$$

Solve.

$$\frac{77}{100} = \frac{n}{2348}$$
$$77 \cdot 2348 = 100 \cdot n$$
$$\frac{77 \cdot 2348}{100} = \frac{100 \cdot n}{100}$$
$$1808 \approx n$$

Check. We can repeat the calculations. Also note that $\frac{1808}{2348} \approx \frac{1800}{2400} = 0.75 = 75\% \approx 77\%$. The answer checks.

State. About 1808 miles of the Mississippi River are navigable.

11. *Familiarize.* Let $h =$ the number of Hispanic people in the U.S. in 2003, in millions.

Translate. We translate to a proportion.

$$\frac{13.7}{100} = \frac{h}{291}$$

Solve.

$$\frac{13.7}{100} = \frac{h}{291}$$
$$13.7 \cdot 291 = 100 \cdot h$$
$$\frac{13.7 \cdot 291}{100} = \frac{100 \cdot h}{100}$$
$$39.867 = h$$

Check. We can repeat the calculation. Also note that $\frac{39.867}{291} \approx \frac{40}{300} = 13.\overline{3}\% \approx 13.7\%$. The answer checks.

State. In 2003 about 39.867 million, or 39,867,000, Hispanic people lived in the United States.

13. *Familiarize.* First we find the number of items Christina got correct. Let b represent this number.

Translate. We translate to a percent equation.

What number is 91% of 40?

$$\downarrow \qquad \downarrow \quad \downarrow \quad \downarrow \quad \downarrow$$
$$b \qquad = 91\% \cdot \quad 40$$

Solve. We convert 91% to decimal notation and multiply.

$$b = 0.91 \cdot 40 = 36.4$$

We subtract to find the number of items Christina got incorrect:

$$40 - 36.4 = 3.6$$

Check. We can repeat the calculation. Also note that $91\% \cdot 40 \approx 90\% \cdot 40 = 36 \approx 36.4$. The answer checks.

State. Christina got 36.4 items correct and 3.6 items incorrect.

15. *Familiarize.* Let $a =$ the number of items on the test.

Translate. We translate to a proportion.

$$\frac{86}{100} = \frac{81.7}{a}$$

Solve.

$$\frac{86}{100} = \frac{81.7}{a}$$
$$86 \cdot a = 100 \cdot 81.7$$
$$\frac{86 \cdot a}{86} = \frac{100 \cdot 81.7}{86}$$
$$a = 95$$

Check. We can repeat the calculation. Also note that $\frac{81.7}{95} \approx \frac{82}{100} = 82\% \approx 86\%$. The answer checks.

State. There were 95 items on the test.

17. *Familiarize.* We let $n =$ the percent of time that television sets are on.

Translate. We translate to a percent equation.

2190 is what percent of 8760?

$$\downarrow \quad \downarrow \qquad \downarrow \qquad \downarrow \quad \downarrow$$
$$2190 = \qquad n \qquad \times 8760$$

Solve. We divide on both sides by 8760 and convert the result to percent notation.

$$2190 = n \times 8760$$
$$\frac{2190}{8760} = \frac{n \times 8760}{8760}$$
$$0.25 = n$$
$$25\% = n$$

Check. To check we find 25% of 8760:

$$25\% \times 8760 = 0.25 \times 8760 = 2190.$$ The answer checks.

State. Television sets are on for 25% of the year.

19. First we find the maximum heart rate for a 25 year old person.

Familiarize. Note that $220 - 25 = 195$. We let $x =$ the maximum heart rate for a 25 year old person.

Translate. We translate to a percent equation.

What is 85% of 195?

$\downarrow \quad \downarrow \quad \downarrow \quad \downarrow \quad \downarrow$

$x \quad = 85\% \times \ 195$

Solve. We convert 85% to a decimal and simplify.

$x = 0.85 \times 195 = 165.75 \approx 166$

Check. We can repeat the calculations. Also, 85% of $195 \approx 0.85 \times 200 = 170 \approx 166$. The answer checks.

State. The maximum heart rate for a 25 year old person is 166 beats per minute.

Next we find the maximum heart rate for a 36 year old person.

Familiarize. Note that $220 - 36 = 184$. We let $x =$ the maximum heart rate for a 36 year old person.

Translate. We translate to a percent equation.

What is 85% of 184?

$\downarrow \quad \downarrow \quad \downarrow \quad \downarrow \quad \downarrow$

$x \quad = 85\% \times \ 184$

Solve. We convert 85% to a decimal and simplify.

$x = 0.85 \times 184 = 156.4 \approx 156$

Check. We can repeat the calculations. Also, 85% of $184 \approx 0.9 \times 180 = 162 \approx 156$. The answer checks.

State. The maximum heart rate for a 36 year old person is 156 beats per minute.

Next we find the maximum heart rate for a 48 year old person.

Familiarize. Note that $220 - 48 = 172$. We let $x =$ the maximum heart rate for a 48 year old person.

Translate. We translate to a percent equation.

What is 85% of 172?

$\downarrow \quad \downarrow \quad \downarrow \quad \downarrow \quad \downarrow$

$x \quad = 85\% \times \ 172$

Solve. We convert 85% to a decimal and simplify.

$x = 0.85 \times 172 = 146.2 \approx 146$

Check. We can repeat the calculations. Also, 85% of $172 \approx 0.9 \times 170 = 153 \approx 146$. The answer checks.

State. The maximum heart rate for a 48 year old person is 146 beats per minute.

We find the maximum heart rate for a 55 year old person.

Familiarize. Note that $220 - 55 = 165$. We let $x =$ the maximum heart rate for a 55 year old person.

Translate. We translate to a percent equation.

What is 85% of 165?

$\downarrow \quad \downarrow \quad \downarrow \quad \downarrow \quad \downarrow$

$x \quad = 85\% \times \ 165$

Solve. We convert 85% to a decimal and simplify.

$x = 0.85 \times 165 = 140.25 \approx 140$

Check. We can repeat the calculations. Also, 85% of $165 \approx 0.9 \times 160 = 144 \approx 140$. The answer checks.

State. The maximum heart rate for a 55 year old person is 140 beats per minute.

Finally we find the maximum heart rate for a 76 year old person.

Familiarize. Note that $220 - 76 = 144$. We let $x =$ the maximum heart rate for a 76 year old person.

Translate. We translate to a percent equation.

What is 85% of 144?

$\downarrow \quad \downarrow \quad \downarrow \quad \downarrow \quad \downarrow$

$x \quad = 85\% \times \ 144$

Solve. We convert 85% to a decimal and simplify.

$x = 0.85 \times 144 = 122.4 \approx 122$

Check. We can repeat the calculations. Also, 85% of $144 \approx 0.9 \times 140 = 126 \approx 122$. The answer checks.

State. The maximum heart rate for a 76 year old person is 122 beats per minute.

21. *Familiarize*. Use the drawing in the text to visualize the situation. Note that the increase in the amount was $16.

Let $n =$ the percent of increase.

Translate. We translate to a percent equation.

$16 is \underbrace{\text{what percent}}\ of \$200?$

$\downarrow \quad \downarrow \qquad\quad \downarrow \qquad\ \downarrow \quad \downarrow$

$16 = \qquad\quad n \quad \times \ 200$

Solve. We divide by 200 on both sides and convert the result to percent notation.

$16 = n \times 200$

$\dfrac{16}{200} = \dfrac{n \times 200}{200}$

$0.08 = n$

$8\% = n$

Check. Find 8% of 200: $8\% \times 200 = 0.08 \times 200 = 16$. Since this is the amount of the increase, the answer checks.

State. The percent of increase was 8%.

23. *Familiarize*. We use the drawing in the text to visualize the situation. Note that the reduction is $18.

We let $n =$ the percent of decrease.

Translate. We translate to a percent equation.

$18 is \underbrace{\text{what percent}}\ of \$90?$

$\downarrow \quad \downarrow \qquad\quad \downarrow \qquad \downarrow \quad \downarrow$

$18 = \qquad\quad n \quad \times \ 90$

Solve. To solve the equation, we divide on both sides by 90 and convert the result to percent notation.

$\dfrac{18}{90} = \dfrac{n \times 90}{90}$

$0.2 = n$

$20\% = n$

Check. We find 20% of 90: $20\% \times 90 = 0.2 \times 90 = 18$. Since this is the price decrease, the answer checks.

State. The percent of decrease was 20%.

25. *Familiarize*. First we find the amount of increase.

$$\begin{array}{r} 2,2\,4\,1,1\,5\,4 \\ -1,2\,0\,1,8\,3\,3 \\ \hline 1,0\,3\,9,3\,2\,1 \end{array}$$

Let N = the percent of increase.

Translate. We translate to a proportion.

$$\frac{N}{100} = \frac{1,039,321}{1,201,833}$$

Solve.

$$\frac{N}{100} = \frac{1,039,321}{1,201,833}$$

$$N \cdot 1,201,833 = 100 \cdot 1,039,321$$

$$\frac{N \cdot 1,201,833}{1,201,833} = \frac{100 \cdot 1,039,321}{1,201,833}$$

$$N \approx 86.5$$

Check. We can repeat the calculation. Also note that $86.5\% \cdot 1,201,833 \approx 90\% \cdot 1,200,000 = 1,080,000 \approx 1,039,321$. The answer checks.

State. The percent of increase was about 86.5%.

27. Familiarize. We note that the amount of the raise can be found and then added to the old salary. A drawing helps us visualize the situation.

$28,600	$?
100%	5%

We let x = the new salary.

Translate. We translate to a percent equation.

What is the old salary plus 5% of the old salary?

$$x = 28,600 + 5\% \times 28,600$$

Solve. We convert 5% to a decimal and simplify.

$$x = 28,600 + 0.05 \times 28,600$$
$$= 28,600 + 1430 \qquad \text{The raise is } \$1430.$$
$$= 30,030$$

Check. To check, we note that the new salary is 100% of the old salary plus 5% of the old salary, or 105% of the old salary. Since $1.05 \times 28,600 = 30,030$, our answer checks.

State. The new salary is $30,030.

29. Familiarize. Let d = the amount of depreciation the first year.

Translate. We translate to a proportion.

$$\frac{25}{100} = \frac{d}{21,566}$$

Solve.

$$\frac{25}{100} = \frac{d}{21,566}$$

$$25 \cdot 21,566 = 100 \cdot d$$

$$\frac{25 \cdot 21,566}{100} = d$$

$$5391.50 = d$$

Now we subtract to find the depreciated value after 1 year.

$$\begin{array}{r} 2\,1,5\,6\,6.0\,0 \\ -5\,3\,9\,1.5\,0 \\ \hline 1\,6,1\,7\,4.5\,0 \end{array}$$

The second year the car depreciates 25% of the value after 1 year. We use a proportion to find this amount, a.

$$\frac{25}{100} = \frac{a}{16,174.50}$$

$$25 \cdot 16,174.50 = 100 \cdot a$$

$$\frac{25 \cdot 16,174.50}{100} = a$$

$$4043.63 \approx a$$

Now we subtract to find the value of the car after 2 years.

$$\begin{array}{r} 1\,6,1\,7\,4.5\,0 \\ -4\,0\,4\,3.6\,3 \\ \hline 1\,2,1\,3\,0.8\,7 \end{array}$$

Check. We can repeat the calculations. Also note that after 1 year the value of the car will be $100\% - 25\%$, or 75%, of the original value:

$$75\% \times \$21,566 = \$16,174.50$$

After 2 years the value of the car will be $100\% - 25\%$, or 75%, of the value after 1 year:

$$75\% \times \$16,174.50 \approx \$12,130.88$$

The slight discrepancy in this amount is due to rounding. The answers check.

State. After 1 year the value of the car will be $16,174.50. After 2 years, its value will be $12,130.87.

31. Familiarize. First we find the amount of the decrease.

$$\begin{array}{r} 8\,9.9\,5 \\ -6\,5.4\,9 \\ \hline 2\,4.4\,6 \end{array}$$

Let p = the percent of decrease.

Translate. We translate to a percent equation.

$24.46 is what percent of $89.95

$$24.46 = p \cdot 89.95$$

Solve. We divide by 89.95 on both sides and convert to percent notation.

$$\frac{24.46}{89.95} = \frac{p \cdot 89.95}{89.95}$$

$$0.27 \approx p$$

$$27\% \approx p$$

Check. We find 27% of 89.95: $27\% \cdot 89.95 = 0.27 \cdot 89.95 \approx 24.29$. This is approximately the amount of the decrease, so the answer checks. (Remember that we rounded the percent.)

State. The percent of decrease is about 27%.

33. Familiarize. This is a multistep problem. First we find the area of a cross-section of a finished board and of a rough board using the formula $A = l \cdot w$. Then we find the amount of wood removed in planing and drying and finally we find the percent of wood removed. Let f = the area of a cross-section of a finished board and let r = the area of a cross-section of a rough board.

Translate. We find the areas.

$$f = 3\frac{1}{2} \cdot 1\frac{1}{2}$$

$$r = 4 \cdot 2$$

Solve. We carry out the multiplications.

$$f = 3\frac{1}{2} \cdot 1\frac{1}{2} = \frac{7}{2} \cdot \frac{3}{2} = \frac{21}{4}$$
$$r = 4 \cdot 2 = 8$$

Now we subtract to find the amount of wood removed in planing and drying.

$$8 - \frac{21}{4} = \frac{32}{4} - \frac{21}{4} = \frac{11}{4}$$

Finally we find p, the percent of wood removed in planing and drying.

$$\frac{11}{4} \text{ is } \underbrace{\text{what percent}} \text{ of } 8?$$
$$\downarrow \quad \downarrow \qquad \downarrow \qquad \downarrow \quad \downarrow$$
$$\frac{11}{4} = \qquad p \qquad \cdot \quad 8$$

We solve the equation.

$$\frac{11}{4} = p \cdot 8$$
$$\frac{1}{8} \cdot \frac{11}{4} = p$$
$$\frac{11}{32} = p$$
$$0.34375 = p$$
$$34.375\% = p, \text{ or}$$
$$34\frac{3}{8}\% = p$$

Check. We repeat the calculations. The answer checks.

State. 34.375%, or $34\frac{3}{8}\%$, of the wood is removed in planing and drying.

35. a) *Familiarize*. First we find the amount of the decrease.

$$\begin{array}{r} 1,0\,2\,8,0\,0\,0 \\ -\ 9\,5\,1,0\,0\,0 \\ \hline 7\,7,0\,0\,0 \end{array}$$

Let $N =$ the percent of decrease.

Translate. We translate to a proportion.

$$\frac{N}{100} = \frac{77,000}{1,028,000}$$

Solve.

$$\frac{N}{100} = \frac{77,000}{1,028,000}$$
$$N \cdot 1,028,000 = 100 \cdot 77,000$$
$$\frac{N \cdot 1,028,000}{1,028,000} = \frac{100 \cdot 77,000}{1,028,000}$$
$$N \approx 7.5$$

Check. We can repeat the calculations. Also note that $7.5\% \cdot 1,028,000 \approx 7.5\% \cdot 1,000,000 = 75,000 \approx 77,000$. The answer checks.

State. The percent of decrease was about 7.5%

b) *Familiarize*. First we find the amount of the decrease in the next decade. Let b represent this number.

Translate. We translate to a proportion.

$$\frac{7.5}{100} = \frac{b}{951,000}$$

Solve.

$$\frac{7.5}{100} = \frac{b}{951,000}$$
$$7.5 \cdot 951,000 = 100 \cdot b$$
$$\frac{7.5 \cdot 951,000}{100} = \frac{100 \cdot b}{100}$$
$$71,325 \approx b$$

We subtract to find the population in 2010:

$$951,000 - 71,325 = 879,675$$

Check. We can repeat the calculations. Also note that the population in 2010 will be $100\% - 7.5\%$, or 92.5%, of the 2000 population and $92.5\% \cdot 951,000 \approx 879,675$. The answer checks.

State. In 2010 the population will be 879,675.

37. *Familiarize*. First we subtract to find the amount of the increase.

$$\begin{array}{r} 7\,3\,5 \\ -4\,3\,0 \\ \hline 3\,0\,5 \end{array}$$

Now let $p =$ the percent of increase.

Translate. We translate to an equation.

$$305 \text{ is } \underbrace{\text{what percent}} \text{ of } 430?$$
$$\downarrow \quad \downarrow \qquad \downarrow \qquad \downarrow \quad \downarrow$$
$$305 = \qquad p \qquad \cdot \quad 430$$

Solve.

$$305 = p \cdot 430$$
$$\frac{305}{430} = \frac{p \cdot 430}{430}$$
$$0.71 \approx p$$
$$71\% \approx p$$

Check. We can repeat the calculations. Also note that $171\% \cdot 430 = 735.3 \approx 735$. The answer checks.

State. The percent of increase is about 71%.

39. *Familiarize*. Let $a =$ the amount of the increase.

Translate. We translate to a proportion.

$$\frac{100}{100} = \frac{a}{780}$$

Solve.

$$\frac{100}{100} = \frac{a}{780}$$
$$100 \cdot 780 = 100 \cdot a$$
$$\frac{100 \cdot 780}{100} = a$$
$$780 = a$$

Now we add to find the higher rate:

$$\begin{array}{r} 7\,8\,0 \\ +\ 7\,8\,0 \\ \hline 1\,5\,6\,0 \end{array}$$

Check. We can repeat the calculations. Also note that $200\% \cdot \$780 = \$1560$. The answer checks.

State. The rate for smokers is $1560.

41. *Familiarize*. First we subtract to find the amount of the increase.

$$\begin{array}{r} 2\,9\,5\,5 \\ -1\,6\,4\,5 \\ \hline 1\,3\,1\,0 \end{array}$$

Now let $p =$ the percent of increase.

***Translate*.** We translate to an equation.

1310 is what percent of 1645?

$$\underset{1310}{\downarrow} \; \underset{=}{\downarrow} \qquad \underset{p}{\downarrow} \qquad \underset{\cdot}{\downarrow} \; \underset{1645}{\downarrow}$$

***Solve*.**

$$1310 = p \cdot 1645$$
$$\frac{1310}{1645} = p$$
$$0.80 \approx p$$
$$80\% \approx p$$

***Check*.** We can repeat the calculations. Also note that $180\% \cdot 1645 = 2961 \approx 2955$. The answer checks.

***State*.** The percent of increase is about 80%.

43. *Familiarize*. First we subtract to find the amount of change.

$$\begin{array}{r} 6\,4\,8,8\,1\,8 \\ -5\,5\,0,0\,4\,3 \\ \hline 9\,8,7\,7\,5 \end{array}$$

Now let $p =$ the percent of change.

***Translate*.** We translate to a proportion.

$$\frac{p}{100} = \frac{98,775}{550,043}$$

***Solve*.**

$$\frac{p}{100} = \frac{98,775}{550,043}$$
$$p \cdot 550,043 = 100 \cdot 98,775$$
$$p = \frac{100 \cdot 98,775}{550,043}$$
$$p \approx 18.0$$

***Check*.** We can repeat the calculations. Also note that $118\% \cdot 550,043 \approx 649,051 \approx 648,818$. The answer checks.

***State*.** The population of Alaska increased by 98,775. This was an 18% increase.

45. *Familiarize*. First we subtract to find the population in 1990.

$$\begin{array}{r} 9\,1\,7,6\,2\,1 \\ -1\,1\,8,5\,5\,6 \\ \hline 7\,9\,9,0\,6\,5 \end{array}$$

Now let $p =$ the percent of change.

***Translate*.** We translate to an equation.

118,556 is what percent of 799,065?

$$\underset{118,556}{\downarrow} \; \underset{=}{\downarrow} \qquad \underset{p}{\downarrow} \qquad \underset{\cdot}{\downarrow} \; \underset{799,065}{\downarrow}$$

***Solve*.**

$$118,556 = p \cdot 799,065$$
$$\frac{118,556}{799,065} = p$$
$$0.148 \approx p$$
$$14.8\% \approx p$$

***Check*.** We can repeat the calculations. Also note that $114.8\% \cdot 799,065 \approx 917,327 \approx 917,621$. The answer checks.

***State*.** The population of Montana was 799,065 in 1990. The population had increased by about 14.8% in 2003.

47. *Familiarize*. First we add to find the population in 2003.

$$\begin{array}{r} 3,2\,9\,4,3\,9\,4 \\ +1,2\,5\,6,2\,9\,4 \\ \hline 4,5\,5\,0,6\,8\,8 \end{array}$$

Now let $p =$ the percent of change.

***Translate*.** We translate to a proportion.

$$\frac{p}{100} = \frac{1,256,294}{3,294,394}$$

***Solve*.**

$$\frac{p}{100} = \frac{1,256,294}{3,294,394}$$
$$p \cdot 3,294,394 = 100 \cdot 1,256,294$$
$$p = \frac{100 \cdot 1,256,294}{3,294,394}$$
$$p \approx 38.1$$

***Check*.** We can repeat the calculations. Also note that $138.1\% \cdot 3,294,394 \approx 4,549,558 \approx 4,550,688$. The answer checks.

***State*.** The population of Colorado in 2003 was 4,550,688. The population had increased by about 38.1% in 2003.

49. *Familiarize*. Since the car depreciates 25% in the first year, its value after the first year is $100\% - 25\%$, or 75%, of the original value. To find the decrease in value, we ask:

$27,300 is 75% of what?

Let $b =$ the original cost.

***Translate*.** We translate to an equation.

$27,300 is 75% of what?

$$\underset{\$27,300}{\downarrow} \; \underset{=}{\downarrow} \; \underset{75\%}{\downarrow} \; \underset{\times}{\downarrow} \; \underset{b}{\downarrow}$$

***Solve*.**

$$27,300 = 75\% \times b$$
$$\frac{27,300}{75\%} = \frac{75\% \times b}{75\%}$$
$$\frac{27,300}{0.75} = b$$
$$36,400 = b$$

***Check*.** We find 25% of 36,400 and then subtract this amount from 36,400:

$$0.25 \times 36,400 = 9100 \text{ and}$$
$$36,400 - 9100 = 27,300$$

The answer checks.

State. The original cost was \$36,400.

51. *Familiarize*. First we use the formula $A = l \times w$ to find the area of the strike zone:

$$A = 30 \times 17 = 510 \text{ in}^2$$

When a 2-in. border is added to the outside of the strike zone, the dimensions of the larger zone are 19 in. by 34 in. The area of this zone is

$$A = 34 \times 21 = 714 \text{ in}^2$$

We subtract to find the increase in area:

$$714 \text{ in}^2 - 510 \text{ in}^2 = 204 \text{ in}^2$$

We let $p =$ the percent of increase in the area.

Translate. We translate to a proportion.

$$\underbrace{204 \text{ is}}_{} \underbrace{\text{what percent}}_{} \text{ of } 510?$$
$$\begin{array}{ccccc} \downarrow \ \downarrow & & \downarrow & & \downarrow \ \downarrow \\ 204 = & & P & & \times \ 510 \end{array}$$

Solve. We divide by 510 on both sides and convert to percent notation.

$$\frac{204}{510} = \frac{p \times 510}{510}$$
$$0.4 = p$$
$$40\% = p$$

Check. We repeat the calculations.

State. The area of the strike zone is increased by 40%.

53. Discussion and Writing Exercise

55. $\dfrac{25}{11} = 25 \div 11$

$$\begin{array}{r} 2.\,2\,7 \\ 1\,1\,\overline{\smash{)}\,2\,5.\,0\,0} \\ \underline{2\,2} \\ 3\,0 \\ \underline{2\,2} \\ 8\,0 \\ \underline{7\,7} \\ 3 \end{array}$$

Since the remainders begin to repeat, we have a repeating decimal.

$$\frac{25}{11} = 2.\overline{27}$$

57. $\dfrac{27}{8} = 27 \div 8$

$$\begin{array}{r} 3.\,3\,7\,5 \\ 8\,\overline{\smash{)}\,2\,7.\,0\,0\,0} \\ \underline{2\,4} \\ 3\,0 \\ \underline{2\,4} \\ 6\,0 \\ \underline{5\,6} \\ 4\,0 \\ \underline{4\,0} \\ 0 \end{array}$$

$$\frac{27}{8} = 3.375$$

We could also do this conversion as follows:

$$\frac{27}{8} = \frac{27}{8} \cdot \frac{125}{125} = \frac{3375}{1000} = 3.375$$

59. $\dfrac{23}{25} = \dfrac{23}{25} \cdot \dfrac{4}{4} = \dfrac{92}{100} = 0.92$

61. $\dfrac{14}{32} = 14 \div 32$

$$\begin{array}{r} 0.\,4\,3\,7\,5 \\ 3\,2\,\overline{\smash{)}\,1\,4.\,0\,0\,0\,0} \\ \underline{1\,2\,8} \\ 1\,2\,0 \\ \underline{9\,6} \\ 2\,4\,0 \\ \underline{2\,2\,4} \\ 1\,6\,0 \\ \underline{1\,6\,0} \\ 0 \end{array}$$

$$\frac{14}{32} = 0.4375$$

(Note that we could have simplified the fraction first, getting $\dfrac{7}{16}$ and then found the quotient $7 \div 16$.)

63. Since 10,000 has 4 zeros, we move the decimal point in the number in the numerator 4 places to the left.

$$\frac{34,809}{10,000} = 3.4809$$

65. *Familiarize*. We will express 4 ft, 8 in. as 56 in. (4 ft $+$ 8 in. $= 4 \cdot 12$ in. $+ 8$ in. $= 48$ in. $+ 8$ in. $= 56$ in.) We let $h =$ Cynthia's final adult height.

Translate. We translate to an equation.

$$\underbrace{56 \text{ in.}}_{} \text{ is } 84.4\% \text{ of what?}$$
$$\begin{array}{ccccc} \downarrow & \downarrow & \downarrow & \downarrow & \downarrow \\ 56 & = & 84.4\% & \times & h \end{array}$$

Solve. First we convert 84.4% to a decimal.

$$56 = 0.844 \times h$$
$$\frac{56}{0.844} = \frac{0.844 \times h}{0.844}$$
$$66 \approx h$$

Check. We find 84.4% of 66: $0.844 \times 66 \approx 56$. The answer checks.

State. Cynthia's final adult height will be about 66 in., or 5 ft 6 in.

67. *Familiarize*. If p is 120% of q, then $p = 1.2q$. Let $n =$ the percent of p that q represents.

Translate. We translate to an equation. We use $1.2q$ for p.

$$q \text{ is } \underbrace{\text{what percent}}_{} \text{ of } p?$$
$$\begin{array}{ccccc} \downarrow \ \downarrow & & \downarrow & & \downarrow \ \downarrow \\ q = & & n & & \times \ 1.2q \end{array}$$

Solve.

$$q = n \times 1.2q$$

$$\frac{q}{1.2q} = \frac{n \times 1.2q}{1.2q}$$

$$\frac{1}{1.2} = n$$

$$0.8\overline{3} = n$$

$$83.\overline{3}\%, \text{ or } 83\frac{1}{3}\% = n$$

Check. We find $83\frac{1}{3}\%$ of $1.2q$:

$$0.8\overline{3} \times 1.2q = q$$

The answer checks.

State. q is $83.\overline{3}\%$, or $83\frac{1}{3}\%$, of p.

Exercise Set 6.7

1. The sales tax on an item costing $49.99 is

$$\underbrace{\text{Sales tax rate}} \times \underbrace{\text{Purchase price}}$$
$$\downarrow \qquad\qquad \downarrow \qquad\qquad \downarrow$$
$$5.3\% \qquad\quad \times \qquad \$49.99,$$

or 0.053×49.99, or about 2.65. Thus the tax is $2.65.

3. The sales tax on an item costing $279 is

$$\underbrace{\text{Sales tax rate}} \times \underbrace{\text{Purchase price}}$$
$$\downarrow \qquad\qquad \downarrow \qquad\qquad \downarrow$$
$$7\% \qquad\quad \times \qquad \$279,$$

or 0.07×279, or 19.53. Thus the tax is $19.53.

5. a) We first find the cost of the telephones. It is

$$5 \times \$69 = \$345.$$

b) The sales tax on items costing $345 is

$$\underbrace{\text{Sales tax rate}} \times \underbrace{\text{Purchase price}}$$
$$\downarrow \qquad\qquad \downarrow \qquad\qquad \downarrow$$
$$4.75\% \qquad\quad \times \qquad \$345,$$

or 0.0475×345, or about 16.39. Thus the tax is $16.39.

c) The total price is given by the purchase price plus the sales tax:

$$\$345 + \$16.39 = \$361.39.$$

To check, note that the total price is the purchase price plus 4.75% of the purchase price. Thus the total price is 104.75% of the purchase price. Since $1.0475 \times \$345 = \$361.3875 \approx \$361.39$, we have a check. The total price is $361.39.

7. *Rephrase:*

$$\underbrace{\substack{\text{Sales} \\ \text{tax}}} \quad \text{is} \quad \underbrace{\substack{\text{what} \\ \text{percent}}} \quad \text{of} \quad \underbrace{\substack{\text{purchase} \\ \text{price?}}}$$

Translate: $\quad 48 \quad = \quad r \quad \times \quad 960$

To solve the equation, we divide on both sides by 960.

$$\frac{48}{960} = \frac{r \times 960}{960}$$

$$0.05 = r$$

$$5\% = r$$

The sales tax rate is 5%.

9. *Rephrase:* $\quad$ Sales tax is 5% of what?

$$\downarrow \qquad \downarrow \;\; \downarrow \;\; \downarrow \qquad \downarrow$$

Translate: $\quad 100 \quad = \; 5\% \; \times \quad b, \quad$ or
$$\qquad\qquad\quad 100 \quad = 0.05 \times \quad b$$

To solve the equation, we divide on both sides by 0.05.

$$\frac{100}{0.05} = \frac{0.05 \times b}{0.05}$$

$$2000 = b$$

$$\begin{array}{r} 2\,0\,0\,0. \\ 0.\,0\,5_{\wedge}\overline{)\,1\,0\,0.0\,0_{\wedge}} \\ \underline{1\,0\,0\,0\,0} \\ 0 \end{array}$$

The purchase price is $2000.

11. *Rephrase:* $\quad$ Sales tax is 3.5% of what?

$$\downarrow \qquad \downarrow \;\; \downarrow \;\; \downarrow \qquad \downarrow$$

Translate: $\quad 28 \quad = \; 3.5\% \; \times \quad b, \quad$ or
$$\qquad\qquad\quad 28 \quad = 0.035 \times \quad b$$

To solve the equation, we divide on both sides by 0.035.

$$\frac{28}{0.035} = \frac{0.035 \times b}{0.035}$$

$$800 = b$$

$$\begin{array}{r} 8\,0\,0. \\ 0.\,0\,3\,5_{\wedge}\overline{)\,2\,8.0\,0\,0_{\wedge}} \\ \underline{2\,8\,0\,0\,0} \\ 0 \end{array}$$

The purchase price is $800.

13. a) We first find the cost of the shower units. It is

$$2 \times \$332.50 = \$665.$$

b) The total tax rate is the city tax rate plus the state tax rate, or $2\% + 6.25\% = 8.25\%$. The sales tax paid on items costing $665 is

$$\underbrace{\text{Sales tax rate}} \times \underbrace{\text{Purchase price}}$$
$$\downarrow \qquad\qquad \downarrow \qquad\qquad \downarrow$$
$$8.25\% \qquad\quad \times \qquad \$665,$$

or 0.0825×665, or about 54.86. Thus the tax is $54.86.

c) The total price is given by the purchase price plus the sales tax:

$$\$665 + \$54.86 = \$719.86.$$

To check, note that the total price is the purchase price plus 8.25% of the purchase price. Thus the total price is 108.25% of the purchase price. Since $1.0825 \times 665 \approx 719.86$, we have a check. The total amount paid for the 2 shower units is $719.86.

15. *Rephrase:*

$$\underbrace{\substack{\text{Sales} \\ \text{tax}}} \quad \text{is} \quad \underbrace{\substack{\text{what} \\ \text{percent}}} \quad \text{of} \quad \underbrace{\substack{\text{purchase} \\ \text{price?}}}$$

Translate: $\quad 1030.40 = \quad r \quad \times \quad 18,400$

To solve the equation, we divide on both sides by 18,400.

$$\frac{1030.40}{18,400} = \frac{r \times 18,400}{18,400}$$
$$0.056 = r$$
$$5.6\% = r$$

The sales tax rate is 5.6%.

17. Commission = Commission rate × Sales
    C = 6% × 45,000

    This tells us what to do. We multiply.

    $$\begin{array}{r} 4\,5,0\,0\,0 \\ \times\quad 0.0\,6 \\ \hline 2\,7\,0\,0.0\,0 \end{array} \qquad (6\% = 0.06)$$

    The commission is $2700.

19. Commission = Commission rate × Sales
    120 = r × 2400

    To solve this equation we divide on both sides by 2400:

    $$\frac{120}{2400} = \frac{r \times 2400}{2400}$$

    We can divide, but this time we simplify by removing a factor of 1:

    $$r = \frac{120}{2400} = \frac{1}{20} \cdot \frac{120}{120} = \frac{1}{20} = 0.05 = 5\%$$

    The commission rate is 5%.

21. Commission = Commission rate × Sales
    392 = 40% × S

    To solve this equation we divide on both sides by 0.4:

    $$\frac{392}{0.4} = \frac{0.4 \times S}{0.4}$$
    $$980 = S$$

    $$\begin{array}{r} 9\,8\,0. \\ 0.4\,{\scriptstyle\wedge}\overline{\smash{)}3\,9\,2.0\,{\scriptstyle\wedge}} \\ \underline{3\,6\,0\,0} \\ 3\,2\,0 \\ \underline{3\,2\,0} \\ 0 \\ \underline{0} \\ 0 \end{array}$$

    $980 worth of artwork was sold.

23. Commission = Commission rate × Sales
    C = 6% × 98,000

    This tells us what to do. We multiply.

    $$\begin{array}{r} 9\,8,0\,0\,0 \\ \times\quad 0.0\,6 \\ \hline 5\,8\,8\,0.0\,0 \end{array} \qquad (6\% = 0.06)$$

    The commission is $5880.

25. First we find the commission on the first $2000 of sales.
    Commission = Commission rate × Sales
    C = 5% × 2000

    This tells us what to do. We multiply.

    $$\begin{array}{r} 2\,0\,0\,0 \\ \times\quad 0.0\,5 \\ \hline 1\,0\,0.0\,0 \end{array}$$

    The commission on the first $2000 of sales is $100.

Next we subtract to find the amount of sales over $2000.

$6000 - $2000 = $4000

Miguel had $4000 in sales over $2000.

Then we find the commission on the sales over $2000.

Commission = Commission rate × Sales
C = 8% × 4000

This tells us what to do. We multiply.

$$\begin{array}{r} 4\,0\,0\,0 \\ \times\quad 0.0\,8 \\ \hline 3\,2\,0.0\,0 \end{array}$$

The commission on the sales over $2000 is $320.

Finally we add to find the total commission.

$100 + $320 = $420

The total commission is $420.

27. Discount = Rate of discount × Marked price
    D = 10% × $300

    Convert 10% to decimal notation and multiply.

    $$\begin{array}{r} 3\,0\,0 \\ \times\quad 0.1 \\ \hline 3\,0.0 \end{array} \qquad (10\% = 0.10 = 0.1)$$

    The discount is $30.

    Sale price = Marked price − Discount
    S = 300 − 30

    We subtract: $\begin{array}{r} 3\,0\,0 \\ -\quad 3\,0 \\ \hline 2\,7\,0 \end{array}$

    To check, note that the sale price is 90% of the marked price: $0.9 \times 300 = 270$.

    The sale price is $270.

29. Discount = Rate of discount × Marked price
    D = 15% × $17

    Convert 15% to decimal notation and multiply.

    $$\begin{array}{r} 1\,7 \\ \times\,0.1\,5 \\ \hline 8\,5 \\ 1\,7\,0 \\ \hline 2.5\,5 \end{array} \qquad (15\% = 0.15)$$

    The discount is $2.55.

    Sale price = Marked price − Discount
    S = 17 − 2.55

    We subtract: $\begin{array}{r} 1\,7.0\,0 \\ -\quad 2.5\,5 \\ \hline 1\,4.4\,5 \end{array}$

    To check, note that the sale price is 85% of the marked price: $0.85 \times 17 = 14.45$.

    The sale price is $14.45.

31. Discount = Rate of discount × Marked price
    12.50 = 10% × M

    To solve the equation we divide on both sides by 0.1.

$$\frac{12.50}{0.1} = \frac{0.1 \times M}{0.1}$$
$$125 = M$$

The marked price is $125.

Sale price = Marked price − Discount
$$S \quad = \quad 125.00 \quad - \quad 12.50$$

We subtract:
```
  125.00
−  12.50
  112.50
```

To check, note that the sale price is 90% of the marked price: $0.9 \times 125 = 112.50$.

The sale price is $112.50.

33. Discount = Rate of discount × Marked price
$$240 \quad = \quad r \quad \times \quad 600$$

To solve the equation we divide on both sides by 600.
$$\frac{240}{600} = \frac{r \times 600}{600}$$

We can simplify by removing a factor of 1:
$$r = \frac{240}{600} = \frac{2}{5} \cdot \frac{120}{120} = \frac{2}{5} = 0.4 = 40\%$$

The rate of discount is 40%.

Sale price = Marked price − Discount
$$S \quad = \quad 600 \quad - \quad 240$$

We subtract:
```
  600
− 240
  360
```

To check, note that a 40% discount rate means that 60% of the marked price is paid. Since $\frac{360}{600} = 0.6$, or 60%, we have a check.

The sale price is $360.

35. Discount = Marked price − Sale price
$$D \quad = \quad 179.99 \quad - \quad 149.99$$

We subtract:
```
  179.99
− 149.99
   30.00
```

The discount is $30.

Discount = Rate of discount × Marked price
$$30 \quad = \quad R \quad \times \quad 179.99$$

To solve the equation we divide on both sides by 179.99.
$$\frac{30}{179.99} = \frac{R \times 179.99}{179.99}$$
$$0.167 \approx R$$
$$16.7\% \approx R$$

To check, note that a discount rate of 16.7% means that 83.3% of the marked price is paid: $0.833 \times 179.99 = 149.93167 \approx 149.99$. Since that is the sale price, the answer checks.

The rate of discount is 16.7%.

37. $I = P \cdot r \cdot t$
$$= \$200 \times 4\% \times 1$$
$$= \$200 \times 0.04$$
$$= \$8$$

39. $I = P \cdot r \cdot t$
$$= \$2000 \times 8.4\% \times \frac{1}{2}$$
$$= \frac{\$2000 \times 0.084}{2}$$
$$= \$84$$

41. $I = P \cdot r \cdot t$
$$= \$4300 \times 10.56\% \times \frac{1}{4}$$
$$= \frac{\$4300 \times 0.1056}{4}$$
$$= \$113.52$$

43. $I = P \cdot r \cdot t$
$$= \$20,000 \times 4\frac{5}{8}\% \times 1$$
$$= \$20,000 \times 0.04625$$
$$= \$925$$

45. $I = P \cdot r \cdot t$
$$= \$50,000 \times 5\frac{3}{8}\% \times \frac{1}{4}$$
$$= \frac{\$50,000 \times 0.05375}{4}$$
$$\approx \$671.88$$

47. a) We express 90 days as a fractional part of a year and find the interest.
$$I = P \cdot r \cdot t$$
$$= \$6500 \times 5\% \times \frac{90}{365}$$
$$= \$6500 \times 0.05 \times \frac{90}{365}$$
$$\approx \$80.14 \quad \text{Using a calculator}$$

The interest due for 90 days is $80.14.

b) The total amount that must be paid after 90 days is the principal plus the interest.
$$6500 + 80.14 = 6580.14$$

The total amount due is $6580.14.

49. a) We express 60 days as a fractional part of a year and find the interest.
$$I = P \cdot r \cdot t$$
$$= \$10,000 \times 9\% \times \frac{60}{365}$$
$$= \$10,000 \times 0.09 \times \frac{60}{365}$$
$$\approx \$147.95 \quad \text{Using a calculator}$$

The interest due for 60 days is $147.95.

b) The total amount that must be paid after 60 days is the principal plus the interest.
$$10,000 + 147.95 = 10,147.95$$

The total amount due is $10,147.95.

51. a) We express 30 days as a fractional part of a year
and find the interest.

$$I = P \cdot r \cdot t$$
$$= \$5600 \times 10\% \times \frac{30}{365}$$
$$= \$5600 \times 0.1 \times \frac{30}{365}$$
$$\approx \$46.03 \quad \text{Using a calculator}$$

The interest due for 30 days is $46.03.

b) The total amount that must be paid after 30 days
is the principal plus the interest.

$$5600 + 46.03 = 5646.03$$

The total amount due is $5646.03.

53. a) After 1 year, the account will contain 105% of $400.

$$1.05 \times \$400 = \$420$$

$$\begin{array}{r} 4\,0\,0 \\ \times \quad 1.\,0\,5 \\ \hline 2\,0\,0\,0 \\ 4\,0\,0\,0\,0 \\ \hline 4\,2\,0.0\,0 \end{array}$$

b) At the end of the second year, the account will contain 1.05% of $420.

$$1.05 \times \$420 = \$441$$

$$\begin{array}{r} 4\,2\,0 \\ \times \quad 1.\,0\,5 \\ \hline 2\,1\,0\,0 \\ 4\,2\,0\,0\,0 \\ \hline 4\,4\,1.0\,0 \end{array}$$

The amount in the account after 2 years is $441.
(Note that we could have used the formula
$A = P \cdot \left(1 + \dfrac{r}{n}\right)^{n \cdot t}$, substituting $400 for P, 5% for r,
1 for n, and 2 for t.)

55. We use the compound interest formula, substituting $2000
for P, 8.8% for r, 1 for n, and 4 for t.

$$A = P \cdot \left(1 + \frac{r}{n}\right)^{n \cdot t}$$
$$= \$2000 \cdot \left(1 + \frac{8.8\%}{1}\right)^{1 \cdot 4}$$
$$= \$2000 \cdot (1 + 0.088)^4$$
$$= \$2000 \cdot (1.088)^4$$
$$\approx \$2802.50$$

The amount in the account after 4 years is $2802.50.

57. We use the compound interest formula, substituting $4300
for P, 10.56% for r, 1 for n, and 6 for t.

$$A = P \cdot \left(1 + \frac{r}{n}\right)^{n \cdot t}$$
$$= \$4300 \cdot \left(1 + \frac{10.56\%}{1}\right)^{1 \cdot 6}$$
$$= \$4300 \cdot (1 + 0.1056)^6$$
$$= \$4300 \cdot (1.1056)^6$$
$$\approx \$7853.38$$

The amount in the account after 4 years is $7853.38.

59. We use the compound interest formula, substituting
$20,000 for P, $6\frac{5}{8}\%$ for r, 1 for n, and 25 for t.

$$A = P \cdot \left(1 + \frac{r}{n}\right)^{n \cdot t}$$
$$= \$20,000 \cdot \left(1 + \frac{6\frac{5}{8}\%}{1}\right)^{1 \cdot 25}$$
$$= \$20,000 \cdot (1 + 0.06625)^{25}$$
$$= \$20,000 \cdot (1.06625)^{25}$$
$$\approx \$99,427.40$$

The amount in the account after 25 years is $99,427.40.

61. We use the compound interest formula, substituting $4000
for P, 6% for r, 2 for n, and 1 for t.

$$A = P \cdot \left(1 + \frac{r}{n}\right)^{n \cdot t}$$
$$= \$4000 \cdot \left(1 + \frac{6\%}{2}\right)^{2 \cdot 1}$$
$$= \$4000 \cdot \left(1 + \frac{0.06}{2}\right)^2$$
$$= \$4000 \cdot (1.03)^2$$
$$= \$4243.60$$

The amount in the account after 1 year is $4243.60.

63. We use the compound interest formula, substituting
$20,000 for P, 8.8% for r, 2 for n, and 4 for t.

$$A = P \cdot \left(1 + \frac{r}{n}\right)^{n \cdot t}$$
$$= \$20,000 \cdot \left(1 + \frac{8.8\%}{2}\right)^{2 \cdot 4}$$
$$= \$20,000 \cdot \left(1 + \frac{0.088}{2}\right)^8$$
$$= \$20,000 \cdot (1.044)^8$$
$$\approx \$28,225.00$$

The amount in the account after 4 years is $28,225.00.

65. We use the compound interest formula, substituting $5000
for P, 10.56% for r, 2 for n, and 6 for t.

$$A = P \cdot \left(1 + \frac{r}{n}\right)^{n \cdot t}$$
$$= \$5000 \cdot \left(1 + \frac{10.56\%}{2}\right)^{2 \cdot 6}$$
$$= \$5000 \cdot \left(1 + \frac{0.1056}{2}\right)^{12}$$
$$= \$5000 \cdot (1.0528)^{12}$$
$$\approx \$9270.87$$

The amount in the account after 6 years is $9270.87.

67. We use the compound interest formula, substituting
$20,000 for P, $7\frac{5}{8}\%$ for r, 2 for n, and 25 for t.

$$A = P \cdot \left(1 + \frac{r}{n}\right)^{n \cdot t}$$

$$= \$20,000 \cdot \left(1 + \frac{7\frac{5}{8}\%}{2}\right)^{2 \cdot 25}$$

$$= \$20,000 \cdot \left(1 + \frac{0.07625}{2}\right)^{50}$$

$$= \$20,000 \cdot (1.038125)^{50}$$

$$\approx \$129,871.09$$

The amount in the account after 25 years is \$129,871.09.

69. We use the compound interest formula, substituting \$4000 for P, 6% for r, 12 for n, and $\frac{5}{12}$ for t.

$$A = P \cdot \left(1 + \frac{r}{n}\right)^{n \cdot t}$$

$$= \$4000 \cdot \left(1 + \frac{6\%}{2}\right)^{12 \cdot \frac{5}{12}}$$

$$= \$4000 \cdot \left(1 + \frac{0.06}{12}\right)^{5}$$

$$= \$4000 \cdot (1.005)^{5}$$

$$\approx \$4101.01$$

The amount in the account after 5 months is \$4101.01.

71. Discussion and Writing Exercise

73. If the product of two numbers is 1, they are <u>reciprocals</u> of each other.

75. The number 0 is the <u>additive</u> identity.

77. The distance around an object is its <u>perimeter</u>.

79. A natural number that has exactly two different factors, only itself and 1, is called a <u>prime</u> number.

81. *Familiarize.* Let x = the original price of the plaque that was sold for a profit. Then the plaque was sold for 100% of x plus 20% of x, or 120% of x. Let y = the original price of the plaque that was sold for a loss. This plaque was sold for 100% of y less 20% of y, or 80% of y.

Translate. First we consider the plaque that was sold for a profit.

Selling price	is	120%	of	Original price
↓	↓	↓	↓	↓
200	=	120%	·	x

Next we consider the plaque that was sold for a loss.

Selling price	is	80%	of	Original price
↓	↓	↓	↓	↓
200	=	80%	·	y

Solve. We solve each equation.

$$200 = 120\% \cdot x$$
$$200 = 1.2 \cdot x$$
$$\frac{200}{1.2} = \frac{1.2 \cdot x}{1.2}$$
$$166.\overline{6} = x, \text{ or}$$
$$166\frac{2}{3} = x$$

$$200 = 80\% \cdot y$$
$$200 = 0.8 \cdot y$$
$$\frac{200}{0.8} = \frac{0.8 \cdot y}{0.8}$$
$$250 = y$$

Then the plaques were bought for $\$166\frac{2}{3} + \$250$, or $\$416\frac{2}{3}$ and were sold for $\$200 + \$200$, or \$400, so Herb lost money on the sale.

Check. 20% of $166\frac{2}{3} = 0.2 \times 166\frac{2}{3} = 33\frac{1}{3}$ and $166\frac{2}{3} + 33\frac{1}{3} = 200$. Also, 20% of $250 = 0.2 \times 250 = 50$ and $250 - 50 = 200$. Since we get the selling price in each case, the answer checks.

State. Herb lost money on the sale.

83. For a principle P invested at 9% compounded monthly, to find the amount in the account at the end of 1 year we would multiply P by $(1 + 0.09/12)^{12}$. Since $(1 + 0.09/12)^{12} = 1.0075^{12} \approx 1.0938$, the effective yield is approximately 9.38%.

Exercise Set 6.8

1. a) We multiply the balance by 2%:

$$0.02 \times \$4876.54 = \$97.5308.$$

Antonio's minimum payment, rounded to the nearest dollar, is \$98.

b) We find the amount of interest on \$4876.54 at 21.3% for one month.

$$I = P \cdot r \cdot t$$
$$= \$4876.54 \times 0.213 \times \frac{1}{12}$$
$$\approx \$86.56$$

We subtract to find the amount applied to decrease the principal in the first payment.

$$\$98 - \$86.56 = \$11.44$$

The principal is decreased by \$11.44 with the first payment.

c) We find the amount of interest on \$4876.54 at 12.6% for one month.

$$I = P \cdot r \cdot t$$
$$= \$4876.54 \times 0.126 \times \frac{1}{12}$$
$$\approx \$51.20$$

We subtract to find the amount applied to decrease the principal in the first payment.

$98 − $51.20 = $46.80.

The principal is decreased by $46.80 with the first payment.

d) With the 12.6% rate the principal was decreased by $46.80 − $11.44, or $35.36 more than at the 21.3% rate. This also means that the interest at 12.6% is $35.36 less than at 21.3%.

3. a) We find the interest on $44,560 at 3.37% for one month.
$$I = P \cdot r \cdot t$$
$$= \$44,560 \times 0.0337 \times \frac{1}{12}$$
$$\approx \$125.14$$

The amount of interest in the first payment is $125.14.

We subtract to find amount applied to the principal.

$437.93 − $125.14 = $312.79

With the first payment the principal will decrease by $312.79.

b) We find the interest on $44,560 at 4.75% for one month.
$$I = P \cdot r \cdot t$$
$$= \$44,560 \times 0.0475 \times \frac{1}{12}$$
$$\approx \$176.38$$

At 4.75% the additional interest in the first payment is $176.38 − $125.14 = $51.24.

c) For the 3.37% loan there will be 120 payments of $437.93:

$$120 \times \$437.93 = \$52,551.60$$

The total interest at this rate is

$$\$52,551.60 − \$44,560 = \$7991.60.$$

For the 4.75% loan there will be 120 payments of $467.20:

$$120 \times \$467.20 = \$56,064$$

The total interest at this rate is

$$\$56,064 − \$44,560 = \$11,504$$

At 4.75% Grace would pay

$$\$11,504 − \$7991.60 = \$3521.40$$

more in interest than at 3.37%.

5. a) We find the interest on $164,000 at $6\frac{1}{4}$%, or 6.25% for one month.
$$I = P \cdot r \cdot t$$
$$= \$164,000 \times 0.0625 \times \frac{1}{12}$$
$$\approx \$854.17$$

The amount applied to the principal is

$$\$1009.78 − \$854.17 = \$155.61.$$

b) The total paid will be

$$360 \times \$1009.78 = \$363,520.80.$$

Then the total amount of interest paid is

$$\$363,520.80 − \$164,000 = \$199,520.80.$$

c) We subtract to find the new principal after the first payment.

$$\$164,000 − \$155.61 = \$163,844.39$$

Now we find the interest on $163,844.39 at $6\frac{1}{4}$% for one month.
$$I = P \cdot r \cdot t$$
$$= \$163,844.39 \times 0.0625 \times \frac{1}{12}$$
$$\approx \$853.36$$

We subtract to find the amount applied to the principal.

$$\$1009.78 − \$853.36 = \$156.42$$

7. a) From Exercise 5(a) we know that the amount of interest in the first payment is $854.17. The amount applied to the principal is

$$\$1406.17 − \$854.17 = \$552$$

b) The total paid will be

$$180 \times \$1406.17 = \$253,110.60.$$

Then the total amount of interest paid is

$$\$253,110.60 − \$164,000 = \$89,110.60.$$

c) On the 15-yr loan the Martinez family will pay

$$\$199,520.80 − \$89,110.60 = \$110,410.20$$

less in interest than on the 30-yr loan.

9. Interest in first payment:
$$I = P \cdot r \cdot t$$
$$= \$100,000 \times 0.0698 \times \frac{1}{12}$$
$$\approx \$581.67$$

Amount of principal in first payment:

$$\$663.96 − \$581.67 = \$82.29$$

Principal after first payment:

$$\$100,000 − \$82.29 = \$99,917.71$$

Interest in second payment:
$$I = P \cdot r \cdot t$$
$$= \$99,917.71 \times 0.0698 \times \frac{1}{12}$$
$$\approx \$581.19$$

Amount of principal in second payment:

$$\$663.96 − \$581.19 = \$82.77$$

Principal after second payment:

$$\$99,917.71 − \$82.77 = \$99,834.94$$

11. Interest in first payment:

$$I = P \cdot r \cdot t$$

$$= \$100,000 \times 0.0804 \times \frac{1}{12}$$

$$\approx \$670.00$$

Amount of principal in first payment:

$$\$957.96 - \$670.00 = \$287.96$$

Principal after first payment:

$$\$100,000 - \$287.96 = \$99,712.04$$

Interest in second payment:

$$I = P \cdot r \cdot t$$

$$= \$99,712.04 \times 0.0804 \times \frac{1}{12}$$

$$\approx \$668.07$$

Amount of principal in second payment:

$$\$957.96 - \$668.07 = \$289.89$$

Principal after second payment:

$$\$99,712.04 - \$289.89 = \$99,422.15$$

13. Interest in first payment:

$$I = P \cdot r \cdot t$$

$$= \$150,000 \times 0.0724 \times \frac{1}{12}$$

$$= \$905.00$$

Amount of principal in first payment:

$$\$1022.25 - \$905.00 = \$117.25$$

Principal after first payment:

$$\$150,000 - \$117.25 = \$149,882.75$$

Interest in second payment:

$$I = P \cdot r \cdot t$$

$$= \$149,882.75 \times 0.0724 \times \frac{1}{12}$$

$$\approx \$904.29$$

Amount of principal in second payment:

$$\$1022.25 - \$904.29 = \$117.96$$

Principal after second payment:

$$\$149,882.75 - \$117.96 = \$149,764.79$$

15. Interest in first payment:

$$I = P \cdot r \cdot t$$

$$= \$200,000 \times 0.0724 \times \frac{1}{12}$$

$$\approx \$1206.67$$

Amount of principal in first payment:

$$\$1824.60 - \$1206.67 = \$617.93$$

Principal after first payment:

$$\$200,000 - \$617.93 = \$199,382.07$$

Interest in second payment:

$$I = P \cdot r \cdot t$$

$$= \$199,382.07 \times 0.0724 \times \frac{1}{12}$$

$$\approx \$1202.94$$

Amount of principal in second payment:

$$\$1824.60 - \$1202.94 = \$621.66$$

Principal after second payment:

$$\$199,382.07 - \$621.66 = \$198,760.41$$

17. a) The down payment is 10% of $23,950, or

$$0.1 \times \$23,950 = \$2395.$$

The amount borrowed is

$$\$23,950 - \$2395 = \$21,555$$

b) Interest in first payment:

$$I = P \cdot r \cdot t$$

$$= \$21,555 \times 0.029 \times \frac{1}{12}$$

$$\approx \$52.09$$

The amount of the first payment that is applied to reduce the principal is

$$\$454.06 - \$52.09 = \$401.97.$$

c) The total amount paid is

$$48 \cdot \$454.06 = \$21,794.88$$

Then the total interest paid is

$$\$21,794.88 - \$21,555 = \$239.88.$$

19. a) The down payment is 5% of $11,900:

$$0.05 \times \$11,900 = \$595$$

We subtract to find the amount borrowed:

$$\$11,900 - \$595 = \$11,305$$

b) We find the interest on $11,305 at 9.3% for one month.

$$I = P \cdot r \cdot t$$

$$= \$11,305 \times 0.093 \times \frac{1}{12}$$

$$\approx \$87.61$$

We subtract to find the amount applied to reduce the principal:

$$\$361.08 - \$87.61 = \$273.47$$

c) There will be 36 payments of $361.08:

$$36 \times \$361.08 = \$12,998.88$$

The total interest paid will be

$$\$12,998.88 - \$11,305 = \$1693.88$$

21. Discussion and Writing Exercise

23. Discussion and Writing Exercise

25.

$$\frac{x}{12} = \frac{24}{16}$$

$$16 \cdot x = 12 \cdot 24 \qquad \text{Equating cross products}$$

$$x = \frac{12 \cdot 24}{16} \qquad \text{Dividing by 16 on both sides}$$

$$x = \frac{288}{16}$$

$$x = 18$$

The solution is 18.

27. $0.64 \times x = 170$

$$\frac{0.64 \cdot x}{0.64} = \frac{170}{0.64} \qquad \text{Dividing by 0.64 on both sides}$$

$$x = 265.625$$

The solution is 265.625.

29. $\dfrac{5}{9} = 5 \div 9$

$$\begin{array}{r} 0.5\,5 \\ 9\,\overline{)\,5.0\,0} \\ 4\,5 \\ \hline 5\,0 \\ 4\,5 \\ \hline 5 \end{array}$$

We get a repeating decimal.

$$\frac{5}{9} = 0.\overline{5}$$

31. $\dfrac{11}{12} = 11 \div 12$

$$\begin{array}{r} 0.9\,1\,6\,6 \\ 1\,2\,\overline{)\,1\,1.0\,0\,0\,0} \\ 1\,0\,8 \\ \hline 2\,0 \\ 1\,2 \\ \hline 8\,0 \\ 7\,2 \\ \hline 8\,0 \\ 7\,2 \\ \hline 8 \end{array}$$

We get a repeating decimal.

$$\frac{11}{12} = 0.91\overline{6}$$

33. $\dfrac{15}{7} = 15 \div 7$

$$\begin{array}{r} 2.1\,4\,2\,8\,5\,7 \\ 7\,\overline{)\,1\,5.0\,0\,0\,0\,0\,0} \\ 1\,4 \\ \hline 1\,0 \\ 7 \\ \hline 3\,0 \\ 2\,8 \\ \hline 2\,0 \\ 1\,4 \\ \hline 6\,0 \\ 5\,6 \\ \hline 4\,0 \\ 3\,5 \\ \hline 5\,0 \\ 4\,9 \\ \hline 1 \end{array}$$

We get a repeating decimal.

$$\frac{15}{7} = 2.\overline{142857}$$

35. $4.03 trillion = 4.03×1 trillion
$\qquad\qquad\qquad = \$4.03 \times 1,000,000,000,000$
$\qquad\qquad\qquad = \$4,030,000,000,000$

37. 42.7 million $= 42.7 \times 1$ million
$\qquad\qquad\qquad = 42.7 \times 1,000,000$
$\qquad\qquad\qquad = 42,700,000$

Chapter 6 Review Exercises

1. The ratio of 47 to 84 is $\dfrac{47}{84}$.

2. The ratio of 46 to 1.27 is $\dfrac{46}{1.27}$.

3. The ratio of 83 to 100 is $\dfrac{83}{100}$.

4. The ratio of 0.72 to 197 is $\dfrac{0.72}{197}$.

5. $\dfrac{9}{12} = \dfrac{3 \cdot 3}{4 \cdot 4} = \dfrac{3}{3} \cdot \dfrac{3}{4} = \dfrac{3}{4}$

6. $\dfrac{3.6}{6.4} = \dfrac{3.6}{6.4} \cdot \dfrac{10}{10} = \dfrac{36}{64} = \dfrac{4 \cdot 9}{4 \cdot 16} = \dfrac{4}{4} \cdot \dfrac{9}{16} = \dfrac{9}{16}$

7. $\dfrac{377 \text{ mi}}{14.5 \text{ gal}} = 26 \dfrac{\text{mi}}{\text{gal}}$, or 26 mpg

8. $\dfrac{472,500 \text{ revolutions}}{75 \text{ min}} = 6300 \dfrac{\text{revolutions}}{\text{min}}$, or 6300 rpm

9. $\dfrac{319 \text{ gal}}{500 \text{ ft}^2} = 0.638 \text{ gal/ft}^2$

10. $\dfrac{18 \text{ servings}}{25 \text{ lb}} = 0.72 \text{ serving/lb}$

11. We can use cross products:

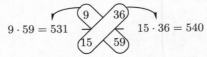

$9 \cdot 59 = 531 \qquad\qquad 15 \cdot 36 = 540$

Since the cross products are not the same, $531 \neq 540$, we know that the numbers are not proportional.

12. We can use cross products:

$24 \cdot 46.25 = 1110 \qquad\qquad 37 \cdot 40 = 1480$

Since the cross products are not the same, $1110 \neq 1480$, we know that the numbers are not proportional.

13. $\dfrac{8}{9} = \dfrac{x}{36}$

$8 \cdot 36 = 9 \cdot x \qquad \text{Equating cross products}$

$\dfrac{8 \cdot 36}{9} = \dfrac{9 \cdot x}{9}$

$\dfrac{288}{9} = x$

$32 = x$

14.
$$\frac{6}{x} = \frac{48}{56}$$
$$6 \cdot 56 = x \cdot 48$$
$$\frac{6 \cdot 56}{48} = \frac{x \cdot 48}{48}$$
$$\frac{336}{48} = x$$
$$7 = x$$

15.
$$\frac{120}{\frac{3}{7}} = \frac{7}{x}$$
$$120 \cdot x = \frac{3}{7} \cdot 7$$
$$120 \cdot x = 3$$
$$\frac{120 \cdot x}{120} = \frac{3}{120}$$
$$x = \frac{1}{40}$$

16.
$$\frac{4.5}{120} = \frac{0.9}{x}$$
$$4.5 \cdot x = 120 \cdot 0.9$$
$$\frac{4.5 \cdot x}{4.5} = \frac{120 \cdot 0.9}{4.5}$$
$$x = \frac{108}{4.5}$$
$$x = 24$$

17. *Familiarize.* Let $p =$ the price of 5 dozen eggs.

Translate. We translate to a proportion.

$$\text{Eggs} \to \quad \frac{3}{2.67} = \frac{5}{p} \quad \leftarrow \text{Eggs}$$
$$\text{Price} \to \qquad\qquad\qquad \leftarrow \text{Price}$$

Solve. We solve the proportion.
$$\frac{3}{2.67} = \frac{5}{p}$$
$$3 \cdot p = 2.67 \cdot 5$$
$$\frac{3 \cdot p}{3} = \frac{2.67 \cdot 5}{3}$$
$$p = 4.45$$

Check. We substitute in the proportion and check cross products.
$$\frac{3}{2.67} = \frac{5}{4.45}$$
$$3 \cdot 4.45 = 13.35; \; 2.67 \cdot 5 = 13.35$$

The cross products are the same, so the answer checks.

State. 5 dozen eggs would cost \$4.45.

18. *Familiarize.* Let $d =$ the number of defective circuits in a lot of 585.

Translate. We translate to a proportion.

$$\text{Defective} \to \quad \frac{3}{65} = \frac{d}{585} \quad \leftarrow \text{Defective}$$
$$\text{Total circuits} \to \qquad\qquad\qquad \leftarrow \text{Total circuits}$$

Solve. We solve the proportion.
$$\frac{3}{65} = \frac{d}{585}$$
$$3 \cdot 585 = 65 \cdot d$$
$$\frac{3 \cdot 585}{65} = d$$
$$27 = d$$

Check. We substitute in the proportion and check cross products.
$$\frac{3}{65} = \frac{27}{585}$$
$$3 \cdot 585 = 1755; \; 65 \cdot 27 = 1755$$

The cross products are the same, so the answer checks.

State. It would be expected that 27 defective circuits would occur in a lot of 585 circuits.

19. *Familiarize.* Let $d =$ the number of miles the train will travel in 13 hr.

Translate. We translate to a proportion.

$$\text{Miles} \to \quad \frac{448}{7} = \frac{d}{13} \quad \leftarrow \text{Miles}$$
$$\text{Hours} \to \qquad\qquad\qquad \leftarrow \text{Hours}$$

Solve.
$$448 \cdot 13 = 7 \cdot d \quad \text{Equating cross products}$$
$$\frac{448 \cdot 13}{7} = \frac{7 \cdot d}{7}$$
$$832 = d$$

Check. We find how far the train travels in 1 hr and then multiply by 13:
$$448 \div 7 = 64 \text{ and } 64 \cdot 13 = 832$$

The answer checks.

State. The train will travel 832 mi in 13 hr.

20. *Familiarize.* Let $g =$ the number of kilograms of garbage produced in San Diego in one day.

Translate. We translate to a proportion.

$$\text{Garbage} \to \quad \frac{13}{5} = \frac{g}{1,266,753} \quad \leftarrow \text{Garbage}$$
$$\text{People} \to \qquad\qquad\qquad \leftarrow \text{People}$$

Solve.
$$13 \cdot 1,266,753 = 5 \cdot g \quad \text{Equating cross products}$$
$$\frac{13 \cdot 1,266,753}{5} = \frac{5 \cdot g}{5}$$
$$3,293,558 \approx g$$

Check. We can divide to find the amount of garbage produced by one person and then multiply to find the amount produced by 1,266,753 people.

$13 \div 5 = 2.6$ and $2.6 \cdot 1,266,753 = 3,293,557.8 \approx 3,293,558$. The answer checks.

State. About 3,293,558 kg of garbage is produced in San Diego in one day.

21. a) *Familiarize*. Let a = the number of Euros equivalent to 250 U.S. dollars.

Translate. We translate to a proportion.

$$\text{U.S. dollars} \rightarrow \quad \frac{1}{0.8244} = \frac{250}{a} \quad \begin{array}{l} \leftarrow \text{ U.S. dollars} \\ \leftarrow \quad \text{Euros} \end{array}$$

Solve.

$1 \cdot a = 0.8244 \cdot 250$ Equating cross products

$a = 206.1$

Check. We substitute in the proportion and check cross products.

$$\frac{1}{0.8244} = \frac{250}{206.1}$$

$1 \cdot 206.1 = 206.1; \; 0.8244 \cdot 250 = 206.1$

The cross products are the same, so the answer checks.

State. 250 U.S. dollars would be worth 206.1 Euros.

b) *Familiarize*. Let c = the cost of the sweatshirt in U.S. dollars.

Translate. We translate to a proportion.

$$\text{U.S. dollars} \rightarrow \quad \frac{1}{0.8244} = \frac{c}{50} \quad \begin{array}{l} \leftarrow \text{ U.S. dollars} \\ \leftarrow \quad \text{Euros} \end{array}$$

Solve.

$1 \cdot 50 = 0.8244 \cdot c$ Equating cross products

$\dfrac{1 \cdot 50}{0.8244} = c$

$60.65 \approx c$

Check. We substitute in the proportion and check cross products.

$$\frac{1}{0.8244} = \frac{60.65}{50}$$

$1 \cdot 50 = 50; \; 0.8244 \cdot 60.65 \approx 50$

The cross products are about the same. Remember that we rounded the value of c.

State. The sweatshirt cost \$60.65 in U.S. dollars.

22. *Familiarize*. Let a = the number of acres required to produce 97.2 bushels of tomatoes.

Translate. We translate to a proportion.

$$\begin{array}{l} \text{Acres} \rightarrow \\ \text{Bushels} \rightarrow \end{array} \frac{15}{54} = \frac{a}{97.2}$$

Solve.

$15 \cdot 97.2 = 54 \cdot a$ Equating cross products

$\dfrac{15 \cdot 97.2}{54} = \dfrac{54 \cdot a}{54}$

$27 = a$

Check. We substitute in the proportion and check cross products.

$$\frac{15}{54} = \frac{27}{97.2}$$

$15 \cdot 97.2 = 1458; \; 54 \cdot 27 = 1458$

The answer checks.

State. 27 acres are required to produce 97.2 bushels of tomatoes.

23. *Familiarize*. Let w = the number of inches of water to which $4\frac{1}{2}$ ft of snow melts.

Translate. We translate to a proportion.

$$\begin{array}{l} \text{Snow} \rightarrow \\ \text{Water} \rightarrow \end{array} \frac{1\frac{1}{2}}{2} = \frac{4\frac{1}{2}}{w} \begin{array}{l} \leftarrow \text{ Snow} \\ \leftarrow \text{ Water} \end{array}$$

Solve.

$1\frac{1}{2} \cdot w = 2 \cdot 4\frac{1}{2}$ Equating cross products

$\dfrac{3}{2} \cdot w = 2 \cdot \dfrac{9}{2}$

$\dfrac{3}{2} \cdot w = 9$

$w = 9 \div \dfrac{3}{2}$ Dividing by $\dfrac{3}{2}$ on both sides

$w = 9 \cdot \dfrac{2}{3}$

$w = \dfrac{9 \cdot 2}{3}$

$w = 6$

Check. We substitute in the proportion and check cross products.

$$\frac{1\frac{1}{2}}{2} = \frac{4\frac{1}{2}}{6}$$

$1\frac{1}{2} \cdot 6 = \dfrac{3}{2} \cdot 6 = \dfrac{3 \cdot 6}{2} = 9; \; 2 \cdot 4\frac{1}{2} = 2 \cdot \dfrac{9}{2} = \dfrac{2 \cdot 9}{2} = 9$

The cross products are the same, so the answer checks.

State. $4\frac{1}{2}$ ft of snow will melt to 6 in. of water.

24. *Familiarize*. Let l = the number of lawyers we would expect to find in Detroit.

Translate. We translate to a proportion.

$$\begin{array}{l} \text{Lawyers} \rightarrow \\ \text{Population} \rightarrow \end{array} \frac{2.3}{1000} = \frac{l}{911,402} \begin{array}{l} \leftarrow \text{ Lawyers} \\ \leftarrow \text{ Population} \end{array}$$

Solve.

$2.3 \cdot 911,402 = 1000 \cdot l$ Equating cross products

$\dfrac{2.3 \cdot 911,402}{1000} = \dfrac{1000 \cdot l}{1000}$

$2096 \approx l$

Check. We substitute in the proportion and check cross products.

$$\frac{2.3}{1000} = \frac{2096}{911,402}$$

$2.3 \cdot 911,402 = 2,096,224.6; \; 1000 \cdot 2096 = 2,096,000 \approx 2,096,224.6$

The answer checks.

State. We would expect that there would be about 2096 lawyers in Detroit.

25. Move the decimal point two places to the right and write a percent symbol.

$0.56 = 56\%$

26. Move the decimal point two places to the right and write a percent symbol.

$$0.017 = 1.7\%$$

27. First we divide to find decimal notation.

```
    0.3 7 5
8 | 3.0 0 0
    2 4
    ───
      6 0
      5 6
      ───
        4 0
        4 0
        ───
          0
```

$$\frac{3}{8} = 0.375$$

Now convert 0.375 to percent notation by moving the decimal point two places to the right and writing a percent symbol.

$$\frac{3}{8} = 37.5\%$$

28. First we divide to find decimal notation.

```
    0.3 3 3
3 | 1.0 0 0
    9
    ───
    1 0
      9
      ───
      1 0
        9
        ───
        1
```

We get a repeating decimal: $\frac{1}{3} = 0.33\overline{3}$. We convert $0.33\overline{3}$ to percent notation by moving the decimal point two places to the right and writing a percent symbol.

$$\frac{1}{3} = 33.\overline{3}\%, \text{ or } 33\frac{1}{3}\%$$

29. 73.5%

a) Replace the percent symbol with $\times 0.01$.

$$73.5 \times 0.01$$

b) Move the decimal point two places to the left.

$$0.73.5$$

Thus, $73.5\% = 0.735$.

30. $6\frac{1}{2}\% = 6.5\%$

a) Replace the percent symbol with $\times 0.01$.

$$6.5 \times 0.01$$

b) Move the decimal point two places to the left.

$$0.06.5$$

Thus, $6.5\% = 0.065$.

31. $24\% = \frac{24}{100} = \frac{4 \cdot 6}{4 \cdot 25} = \frac{4}{4} \cdot \frac{6}{25} = \frac{6}{25}$

32. $6.3\% = \frac{6.3}{100} = \frac{6.3}{100} \cdot \frac{10}{10} = \frac{63}{1000}$

33. *Translate.* $30.6 = p \times 90$

Solve. We divide by 90 on both sides and convert to percent notation.

$$30.6 = p \times 90$$
$$\frac{30.6}{90} = \frac{p \times 90}{90}$$
$$0.34 = p$$
$$34\% = p$$

30.6 is 34% of 90.

34. *Translate.* $63 = 84\% \times n$

Solve. We divide by 84% on both sides.

$$63 = 84\% \times n$$
$$\frac{63}{84\%} = \frac{84\% \times n}{84\%}$$
$$\frac{63}{0.84} = n$$
$$75 = n$$

63 is 84% of 75.

35. *Translate.* $y = 38\frac{1}{2}\% \times 168$

Solve. Convert $38\frac{1}{2}\%$ to decimal notation and multiply.

```
        1 6 8
    × 0. 3 8 5
    ─────────
        8 4 0
    1 3 4 4 0
    5 0 4 0 0
    ─────────
    6 4.6 8 0
```

64.68 is $38\frac{1}{2}\%$ of 168.

36. 24 percent of what is 16.8?

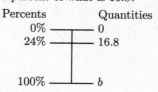

Translate: $\dfrac{24}{100} = \dfrac{16.8}{b}$

Solve: $24 \cdot b = 100 \cdot 16.8$

$$\frac{24 \cdot b}{24} = \frac{100 \cdot 16.8}{24}$$
$$b = \frac{100 \cdot 16.8}{24}$$
$$b = 70$$

24% of 70 is 16.8. The answer is 70.

37. 42 is what percent of 30?

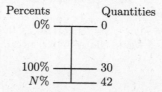

Translate: $\dfrac{N}{100} = \dfrac{42}{30}$

Solve: $30 \cdot N = 100 \cdot 42$

$$\frac{30 \cdot N}{30} = \frac{100 \cdot 42}{30}$$

$$N = \frac{4200}{30}$$

$$N = 140$$

42 is 140% of 30. The answer is 140%.

38. What is 10.5% of 84?

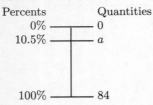

Percents	Quantities
0%	0
10.5%	a
100%	84

Translate: $\dfrac{10.5}{100} = \dfrac{a}{84}$

Solve: $10.5 \cdot 84 = 100 \cdot a$

$$\frac{10.5 \cdot 84}{100} = \frac{100 \cdot a}{100}$$

$$\frac{882}{100} = a$$

$$8.82 = a$$

8.82 is 10.5% of 84. The answer is 8.82.

39. ***Familiarize***. Let c = the number of students who would choose chocolate as their favorite ice cream and b = the number who would choose butter pecan.

Translate. We translate to two equations.

$$\underbrace{\text{What number}}_{\downarrow \atop c} \text{ is 8.9\% of 2500?}$$
$$ = 8.9\% \cdot 2500$$

$$\underbrace{\text{What number}}_{\downarrow \atop b} \text{ is 4.2\% of 2500?}$$
$$ = 4.2\% \cdot 2500$$

Solve. We convert percent notation to decimal notation and multiply.

$$c = 0.089 \cdot 2500 = 222.5 \approx 223$$

$$b = 0.042 \cdot 2500 = 105$$

Check. We can repeat the calculation. We can also do partial checks by estimating.

$$8.9\% \cdot 2500 \approx 10\% \cdot 2500 = 250;$$

$$4.2\% \cdot 2500 \approx 4\% \cdot 2500 = 100$$

Since 250 is close to 223 and 100 is closer to 105, our answers seem reasonable.

State. 223 students would choose chocolate as their favorite ice cream and 105 would choose butter pecan.

40. ***Familiarize***. Let p = the percent of people in the U.S. who take at least one kind of prescription drug per day.

Translate. We translate to a proportion.

$$\frac{p}{100} = \frac{123.64}{295}$$

Solve. We equate cross products.

$$p \cdot 295 = 100 \cdot 123.64$$

$$\frac{p \cdot 295}{295} = \frac{100 \cdot 123.64}{295}$$

$$p \approx 0.42$$

$$p \approx 42\%$$

Check. $42\% \cdot 295$ million $= 0.42 \times 295$ million $=$ 123.9 million $\approx$ 123.64 million. The answer seems reasonable.

State. In the U.S. about 42% of the people take at least one kind of prescription drug per day.

41. ***Familiarize***. Let w = the total output of water from the body per day.

Translate.

$$\underset{200}{200 \text{ mL}} \underset{=}{\text{ is }} \underset{8\%}{8\%} \underset{\cdot}{\text{ of }} \underbrace{\text{what number?}}_{w}$$

Solve.

$$200 = 8\% \cdot w$$

$$200 = 0.08 \cdot w$$

$$\frac{200}{0.08} = \frac{0.08 \cdot w}{0.08}$$

$$2500 = w$$

Check. $8\% \cdot 2500 = 0.08 \cdot 2500 = 200$, so the answer checks.

State. The total output of water from the body is 2500 mL per day.

42. ***Familiarize***. First we subtract to find the amount of the increase.

$$\begin{array}{r} {\scriptstyle 7\ 14} \\ 8\!\!\!/\ 4\!\!\!/ \\ -\ 7\ 5 \\ \hline 9 \end{array}$$

Now let p = the percent of increase.

Translate. We translate to a proportion.

$$\frac{p}{100} = \frac{9}{75}$$

Solve. We equate cross products.

$$p \cdot 75 = 100 \cdot 9$$

$$\frac{p \cdot 75}{75} = \frac{100 \cdot 9}{75}$$

$$p = 12$$

Check. $12\% \cdot 75 = 0.12 \cdot 75 = 9$, the amount of the increase, so the answer checks.

State. Jason's score increased 12%.

43. *Familiarize*. Let s = the new score. Note that the new score is the original score plus 15% of the original score.

$$\underbrace{\text{New score}}_{\downarrow} \quad \underbrace{\text{is}}_{\downarrow} \quad \underbrace{\text{Original score}}_{\downarrow} \quad \underbrace{\text{plus 15% of}}_{\downarrow \quad \downarrow \quad \downarrow} \quad \underbrace{\text{Original score}}_{\downarrow}$$
$$s \quad = \quad 80 \quad + \quad 15\% \cdot \quad 80$$

Solve. We convert 15% to decimal notation and carry out the computation.

$$s = 80 + 0.15 \cdot 80 = 80 + 12 = 92$$

Check. We repeat the calculation. The answer checks.

State. Jenny's new score was 92.

44. The meals tax is

$$\underbrace{\text{Meal tax rate}}_{\downarrow} \times \underbrace{\text{Cost of meal}}_{\downarrow}$$
$$4\frac{1}{2}\% \quad \times \quad \$320,$$

or $0.045 \times \$320$, or $14.40.

45.

$$\underbrace{\text{Sales tax}}_{378} \quad \underbrace{\text{is}}_{} \quad \underbrace{\text{what percent}}_{r} \quad \underbrace{\text{of}}_{\times} \quad \underbrace{\text{purchase price?}}_{7560}$$

To solve the equation, we divide on both sides by 7560.

$$\frac{378}{7560} = \frac{r \times 7560}{7560}$$
$$0.05 = r$$
$$5\% = r$$

The sales tax rate is 5%.

46.
$$\text{Commission} = \text{Commission rate} \times \text{Sales}$$
$$753.50 \quad = \quad r \quad \times 6850$$

To solve this equation, we divide on both sides by 6850.

$$\frac{753.50}{6850} = \frac{r \times 6850}{6850}$$
$$0.11 = r$$
$$11\% = r$$

The commission rate is 11%.

47.
$$\text{Discount} = \text{Rate of discount} \times \text{Marked price}$$
$$D \quad = \quad 12\% \quad \times \quad \$350$$

Convert 12% to decimal notation and multiply.

$$\begin{array}{r} 350 \\ \times\, 0.12 \\ \hline 700 \\ 3500 \\ \hline 42.00 \end{array}$$

The discount is $42.

$$\text{Sale price} = \text{Marked price} - \text{Discount}$$
$$S \quad = \quad \$350 \quad - \quad \$42$$

We subtract:

$$\begin{array}{r} \overset{4\ \ 10}{3\,\cancel{5}\,\cancel{0}} \\ -\quad 42 \\ \hline 308 \end{array}$$

The sale price is $308.

48.
$$\text{Discount} = \text{Marked price} - \text{Sale price}$$
$$D \quad = \quad 305 \quad - \quad 262.30$$

We subtract:

$$\begin{array}{r} \overset{2\ \ 10\ \ 4\ \ 10}{\cancel{3}\,\cancel{0}\,\cancel{5}.\,\cancel{0}\,0} \\ -\ 262.30 \\ \hline 42.70 \end{array}$$

The discount is $42.70.

$$\text{Discount} = \text{Rate of discount} \times \text{Marked price}$$
$$42.70 \quad = \quad R \quad \times \quad 305$$

To solve the equation we divide on both sides by 305.

$$\frac{42.70}{305} = \frac{R \times 305}{305}$$
$$0.14 = R$$
$$14\% = R$$

The rate of discount is 14%.

49.
$$\text{Commission} = \text{Commission rate} \times \text{Sales}$$
$$C \quad = \quad 7\% \quad \times 42,000$$

We convert 7% to decimal notation and multiply.

$$\begin{array}{r} 42,000 \\ \times\quad 0.07 \\ \hline 2940.00 \end{array}$$

The commission is $2940.

50. First we subtract to find the discount.

$$\begin{array}{r} \overset{3\ \ 18}{\cancel{4}\,\cancel{8}\,9.99} \\ -\ 399.69 \\ \hline 90.30 \end{array}$$

$$\text{Discount} = \text{Rate of discount} \times \text{Marked price}$$
$$90.30 \quad = \quad r \quad \times \quad 489.99$$

We divide on both sides by 489.99.

$$\frac{90.30}{489.99} = \frac{r \times 489.99}{489.99}$$
$$0.184 \approx r$$
$$18.4\% \approx r$$

The rate of discount is about 18.4%.

51. $I = P \cdot r \cdot t$

$$= \$1800 \times 6\% \times \frac{1}{3}$$
$$= \$1800 \times 0.06 \times \frac{1}{3}$$
$$= \$36$$

52. a) $I = P \cdot r \cdot t$

$$= \$24,000 \times 10\% \times \frac{60}{365}$$
$$= \$24,000 \times 0.1 \times \frac{60}{365}$$
$$\approx \$394.52$$

b) $\$24,000 + \$394.52 = \$24,394.52$

53. $I = P \cdot r \cdot t$

$= \$2200 \times 5.5\% \times 1$

$= \$2200 \times 0.055 \times 1$

$= \$121$

54. $A = P \cdot \left(1 + \dfrac{r}{n}\right)^{n \cdot t}$

$= \$7500 \cdot \left(1 + \dfrac{12\%}{12}\right)^{12 \cdot \frac{1}{4}}$

$= \$7500 \cdot \left(1 + \dfrac{0.12}{12}\right)^{3}$

$= \$7500 \cdot (1 + 0.01)^3$

$= \$7500 \cdot (1.01)^3$

$\approx \$7727.26$

55. $A = P \cdot \left(1 + \dfrac{r}{n}\right)^{n \cdot t}$

$= \$8000 \cdot \left(1 + \dfrac{9\%}{1}\right)^{1 \cdot 2}$

$= \$8000 \cdot (1 + 0.09)^2$

$= \$8000 \cdot (1.09)^2$

$= \$9504.80$

56. a) 2% of $\$6428.74 = 0.02 \times \$6428.74 \approx \$129$

b) $I = P \cdot r \cdot t$

$= \$6428.74 \times 0.187 \times \dfrac{1}{12}$

$\approx \$100.18$

The amount of interest is $\$100.18$.

$\$129 - \$100.18 = \$28.82$, so the principal is reduced by $\$28.82$.

c) $I = P \cdot r \cdot t$

$= \$6428.74 \times 0.132 \times \dfrac{1}{12}$

$\approx \$70.72$

The amount of interest is $\$70.72$.

$\$129 - \$70.72 = \$58.28$, so the principal is reduced by $\$58.28$ with the lower interest rate.

d) With the 13.2% rate the principal was decreased by $\$58.28 - \$28.82$, or $\$29.46$, more than at the 18.7% rate. This also means that the interest at 13.2% is $\$29.46$ less than at 18.7%.

57. *Discussion and Writing Exercise.* No; the 10% discount was based on the original price rather than on the sale price.

58. *Discussion and Writing Exercise.* A 40% discount is better. When successive discounts are taken, each is based on the previous discounted price rather than on the original price. A 20% discount followed by a 22% discount is the same as a 37.6% discount off the original price.

59. First we divide to find how many gallons of finishing paint are needed.

$4950 \div 450 = 11$ gal

Next we write and solve a proportion to find how many gallons of primer are needed. Let $p =$ the amount of primer needed.

Finishing paint $\rightarrow$ $\dfrac{2}{}$ $= \dfrac{11}{p}$ $\leftarrow$ Finishing paint

Primer $\rightarrow$ $\dfrac{}{3}$ $\phantom{= \dfrac{11}{p}}$ $\leftarrow$ Primer

$2 \cdot p = 3 \cdot 11$

$p = \dfrac{3 \cdot 11}{2}$

$p = \dfrac{33}{2}$, or 16.5

Thus, 11 gal of finishing paint and 16.5 gal of primer should be purchased.

60. *Familiarize.* Let $d =$ the original price of the dress. After the 40% discount, the sale price is 60% of d, or $0.6d$. Let $p =$ the percent by which the sale price must be increased to return to the original price.

Translate.

Sale price	plus	what percent	of	sale price	is	original price?
$\downarrow$	$\downarrow$	$\downarrow$	$\downarrow$	$\downarrow$	$\downarrow$	$\downarrow$
$0.6d$	$+$	p	$\cdot$	$0.6d$	$=$	d

Solve.

$0.6d + p \cdot 0.6d = d$

$(1 + p)(0.6d) = d$ Factoring on the left

$\dfrac{(1 + p)(0.6d)}{0.6d} = \dfrac{d}{0.6d}$

$1 + p = \dfrac{1}{0.6} \cdot \dfrac{d}{d}$

$1 + p = 1.66\overline{6}$

$p = 1.66\overline{6} - 1$

$p = 0.66\overline{6}$

$p = 66.\overline{6}\%$, or $66\dfrac{2}{3}\%$

Check. Suppose the dress cost $\$100$. Then the sale price is 60% of $\$100$, or $\$60$. Now $66\dfrac{2}{3}\% \cdot \$60 = \$40$ and $\$60 + \$40 = \$100$, the original price. Since the answer checks for this specific price, it seems to be reasonable.

State. The sale price must be increased $66\dfrac{2}{3}\%$ after the sale to return to the original price.

Chapter 6 Test

---

1. The ratio of 85 to 97 is $\dfrac{85}{97}$.

2. The ratio of 0.34 to 124 is $\dfrac{0.34}{124}$.

3. $\dfrac{18}{20} = \dfrac{2 \cdot 9}{2 \cdot 10} = \dfrac{2}{2} \cdot \dfrac{9}{10} = \dfrac{9}{10}$

4. $\dfrac{0.75}{0.96} = \dfrac{0.75}{0.96} \cdot \dfrac{100}{100}$ Clearing the decimals

$= \dfrac{75}{96}$

$= \dfrac{3 \cdot 25}{3 \cdot 32} = \dfrac{3}{3} \cdot \dfrac{25}{32}$

$= \dfrac{25}{32}$

5. $\dfrac{319 \text{ mi}}{14.5 \text{ gal}} = \dfrac{319}{14.5} \dfrac{\text{mi}}{\text{gal}} = 22$ mpg

6. $\dfrac{16 \text{ servings}}{12 \text{ lb}} = \dfrac{16}{12} \dfrac{\text{servings}}{\text{lb}} = \dfrac{4}{3}$ servings/lb, or

$1\dfrac{1}{3}$ servings/lb

7. We can use cross products:

$7 \cdot 72 = 504 \qquad 8 \cdot 63 = 504$

Since the cross products are the same, $504 = 504$, we know that $\dfrac{7}{8} = \dfrac{63}{72}$, so the numbers are proportional.

8. We can use cross products:

$1.3 \cdot 15.2 = 19.76 \qquad 3.4 \cdot 5.6 = 19.04$

Since the cross products are not the same, $19.76 \neq 19.04$, we know that $\dfrac{1.3}{3.4} \neq \dfrac{5.6}{15.2}$, so the numbers are not proportional.

9. $\dfrac{9}{4} = \dfrac{27}{x}$

$9 \cdot x = 4 \cdot 27$ Equating cross products

$\dfrac{9 \cdot x}{9} = \dfrac{4 \cdot 27}{9}$

$x = \dfrac{4 \cdot \cancel{9} \cdot 3}{\cancel{9} \cdot 1}$

$x = 12$

10. $\dfrac{150}{2.5} = \dfrac{x}{6}$

$150 \cdot 6 = 2.5 \cdot x$ Equating cross products

$\dfrac{150 \cdot 6}{2.5} = \dfrac{2.5 \cdot x}{2.5}$

$\dfrac{900}{2.5} = x$

$360 = x$

11. Familiarize. Let $m =$ the number of minutes the watch will lose in 24 hr.

Translate. We translate to a proportion.

$$\begin{array}{c} \text{Minutes lost} \rightarrow \\ \text{Hours} \rightarrow \end{array} \dfrac{2}{10} = \dfrac{m}{24} \begin{array}{c} \leftarrow \text{Minutes lost} \\ \leftarrow \text{Hours} \end{array}$$

Solve.

$2 \cdot 24 = 10 \cdot m$ Equating cross products

$\dfrac{2 \cdot 24}{10} = \dfrac{10 \cdot m}{10}$

$4.8 = m$ Multiplying and dividing

Check. We substitute in the proportion and check cross products.

$$\dfrac{2}{10} = \dfrac{4.8}{24}$$

$2 \cdot 24 = 48; \; 10 \cdot 4.8 = 48$

The cross products are the same, so the answer checks.

State. The watch will lose 4.8 min in 24 hr.

12. Familiarize. Let $d =$ the actual distance between the cities.

Translate. We translate to a proportion.

$$\begin{array}{c} \text{Map distance} \rightarrow \\ \text{Actual distance} \rightarrow \end{array} \dfrac{3}{225} = \dfrac{7}{d} \begin{array}{c} \leftarrow \text{Map distance} \\ \leftarrow \text{Actual distance} \end{array}$$

Solve.

$3 \cdot d = 225 \cdot 7$ Equating cross products

$\dfrac{3 \cdot d}{3} = \dfrac{225 \cdot 7}{3}$

$d = 525$ Multiplying and dividing

Check. We substitute in the proportion and check cross products.

$$\dfrac{3}{225} = \dfrac{7}{525}$$

$3 \cdot 525 = 1575; \; 225 \cdot 7 = 1575$

The cross products are the same, so the answer checks.

State. The cities are 525 mi apart.

13. a) **Familiarize.** Let $c =$ the value of 450 U.S. dollars in Hong Kong dollars.

Translate. We translate to a proportion.

$$\begin{array}{c} \text{U.S. dollars} \rightarrow \\ \text{Hong Kong} \rightarrow \\ \text{dollars} \end{array} \dfrac{1}{7.7565} = \dfrac{450}{c} \begin{array}{c} \leftarrow \text{U.S. dollars} \\ \leftarrow \text{Hong Kong} \\ \text{dollars} \end{array}$$

Solve.

$1 \cdot c = 7.7565 \cdot 450$ Equating cross products

$c = 3490.425$

Check. We substitute in the proportion and check cross products.

$$\dfrac{1}{7.7565} = \dfrac{450}{3490.425}$$

$1 \cdot 3490.425 = 3490.425; \; 7.7565 \cdot 450 = 3490.425$

The cross products are the same, so the answer checks.

State. 450 U.S. dollars would be worth 3490.425 Hong Kong dollars.

b) *Familiarize*. Let d = the price of the DVD player in U.S. dollars.

Translate. We translate to a proportion.

U.S. dollars → $\dfrac{1}{7.7565} = \dfrac{d}{795}$ ← U.S. dollars

Hong Kong → $\quad 7.7565 \quad 795$ ← Hong Kong
  dollars   dollars

Solve.

$$1 \cdot 795 = 7.7565 \cdot d \quad \text{Equating cross products}$$

$$\frac{1 \cdot 795}{7.7565} = \frac{7.7565 \cdot d}{7.7565}$$

$$102.49 \approx d$$

Check. We use a different approach. Since 1 U.S. dollar is worth 7.7565 Hong Kong dollars, we multiply 102.49 by 7.7565:

$$102.49(7.7565) \approx 795.$$

This is the price in Hong Kong dollars, so the answer checks.

State. The DVD player would cost $102.49 in U.S. dollars.

14. *Familiarize*. Let a = the number of arrests that could be made if the number of officers were increased to 2500.

Translate. We translate to a proportion.

Officers → $\dfrac{1088}{37,493} = \dfrac{2500}{a}$ ← Officers

Arrests → $\qquad\qquad$ ← Arrests

Solve.

$$1088 \cdot a = 37,493 \cdot 2500$$

$$\frac{1088 \cdot a}{1088} = \frac{37,493 \cdot 2500}{1088}$$

$$a \approx 86,151$$

Check. We substitute in the proportion and check cross products.

$$\frac{1088}{37,493} = \frac{2500}{86,151}$$

$1088 \cdot 86,151 = 93,732,288$; $37,493 \cdot 2500 = 93,732,500$

Since $93,732,288 \approx 93,732,500$, the answer checks.

State. If the number of officers were increased to 2500, about 86,151 arrests would be made.

15. 6.4%

a) Replace the percent symbol with ×0.01.

$$6.4 \times 0.01$$

b) Move the decimal point two places to the left.

0.06.4

Thus, 6.4% = 0.064.

16. 0.38

a) Move the decimal point two places to the right.

0.38.

b) Write a percent symbol: 38%

Thus, 0.38 = 38%.

17.

$$\begin{array}{r} 1.3\,7\,5 \\ 8\,\overline{)\,1\,1.0\,0\,0} \\ \underline{8} \\ 3\,0 \\ \underline{2\,4} \\ 6\,0 \\ \underline{5\,6} \\ 4\,0 \\ \underline{4\,0} \\ 0 \end{array}$$

$$\frac{11}{8} = 1.375$$

Convert to percent notation.

1.37.5

$$\frac{11}{8} = 137.5\%, \text{ or } 137\frac{1}{2}\%$$

18. $65\% = \dfrac{65}{100}$ $\qquad$ Definition of percent

$$\left. \begin{array}{l} = \dfrac{5 \cdot 13}{5 \cdot 20} \\[2mm] = \dfrac{5}{5} \cdot \dfrac{13}{20} \\[2mm] = \dfrac{13}{20} \end{array} \right\} \quad \text{Simplifying}$$

19. Translate: What is 40% of 55?

$$\begin{array}{ccccc} \downarrow & \downarrow & \downarrow & \downarrow & \downarrow \\ a & = & 40\% & \cdot & 55 \end{array}$$

Solve: We convert 40% to decimal notation and multiply.

$$a = 40\% \cdot 55$$

$$= 0.4 \cdot 55 = 22$$

The answer is 22.

20. What percent of 80 is 65?

Percents	Quantities
0% ────	── 0
$N\%$ ────	── 65
100% ────	── 80

Translate: $\dfrac{N}{100} = \dfrac{65}{80}$

Solve: $80 \cdot N = 100 \cdot 65$

$$\frac{80 \cdot N}{80} = \frac{100 \cdot 65}{80}$$

$$N = \frac{6500}{80}$$

$$N = 81.25$$

The answer is 81.25%.

21. *Familiarize*. Let x = the number of passengers in the 25-34 age group. Let y = the number of passengers in the 35-44 age group.

Translate. We will translate to two equations.

What number is 16% of 2500?

$$x = 16\% \cdot 2500$$

What number is 23% of 2500?

$$y = 23\% \cdot 2500$$

Solve. To solve each equation we convert percent notation to decimal notation and multiply.

$$x = 16\% \cdot 2500 = 0.16 \cdot 2500 = 400$$
$$y = 23\% \cdot 2500 = 0.23 \cdot 2500 = 575$$

Check. We repeat the calculations. The answers check.

State. There are 400 passengers in the 25-34 age group and 575 passengers in the 35-44 age group.

22. *Familiarize*. Let $b =$ the number of at-bats.

Translate. We translate to a proportion. We are asking "202 is 30.9% of what?"

$$\frac{30.9}{100} = \frac{202}{b}$$

Solve.

$$30.9 \cdot b = 100 \cdot 202 \quad \text{Equating cross products}$$
$$\frac{30.9 \cdot b}{30.9} = \frac{100 \cdot 202}{30.9}$$
$$b \approx 654$$

Check. We can repeat the calculation. The answer checks.

State. Derek Jeter had about 654 at-bats.

23. *Familiarize*. We first find the amount of decrease, in billions of dollars.

$$\begin{array}{r} \overset{4}{\cancel{5}}.\overset{15}{\cancel{5}} \\ - \, 2.7 \\ \hline 2.8 \end{array}$$

Let $p =$ the percent of decrease.

Translate. We translate to an equation.

2.8 is what percent of 5.5?

$$2.8 = p \cdot 5.5$$

Solve.

$$2.8 = p \cdot 5.5$$
$$\frac{2.8}{5.5} = \frac{p \cdot 5.5}{5.5}$$
$$0.50\overline{90} = p$$
$$50.\overline{90}\% = p$$

Check. Note that $50.\overline{90}\% \approx 51\%$. With a decrease of approximately 51%, the profit in 2000 should be about $100\% - 51\%$, or 49%, of the profit in 1999. Since $49\% \cdot 5.5 = 0.49 \cdot 5.5 = 2.695 \approx 2.7$, the answer checks.

State. The percent of decrease was $50.\overline{90}\%$.

24. *Familiarize*. Let $p =$ the percent of people who have ever lived who are alive today. Note that the population numbers are given in billions.

Translate. We translate to an equation.

6.6 is what percent of 120?

$$6.6 = p \cdot 120$$

Solve.

$$6.6 = p \cdot 120$$
$$\frac{6.6}{120} = \frac{p \cdot 120}{120}$$
$$0.055 = p$$
$$5.5\% = p$$

Check. We find 5.5% of 120:

$$5.5\% \cdot 120 = 0.055 \cdot 120 = 6.6$$

The answer checks.

State. 5.5% of the people who have ever lived are alive today.

25. The sales tax on an item costing $324 is

Sales tax rate $\times$ Purchase price

$$5\% \quad \times \quad \$324,$$

or 0.05×324, or 16.2. Thus the tax is $16.20.

The total price is given by the purchase price plus the sales tax:

$$\$324 + \$16.20 = \$340.20$$

26. Commission = Commission rate $\times$ Sales

$$\begin{aligned} C &= 15\% &\times 4200 \\ C &= 0.15 &\times 4200 \\ C &= 630 \end{aligned}$$

The commission is $630.

27. Discount = Rate of discount $\times$ Marked price

$$D \quad = \quad 20\% \quad \times \quad \$200$$

Convert 20% to decimal notation and multiply.

$$\begin{array}{r} 2\,0\,0 \\ \times \quad 0.\,2 \\ \hline 4\,0.\,0 \end{array} \quad (20\% = 0.20 = 0.2)$$

The discount is $40.

Sale price = Marked price − Discount

$$S \quad = \quad 200 \quad - \quad 40$$

We subtract:

$$\begin{array}{r} 2\,0\,0 \\ - \quad 4\,0 \\ \hline 1\,6\,0 \end{array}$$

To check, note that the sale price is 80% of the marked price: $0.8 \times 200 = 160$.

The sale price is $160.

28. $I = P \cdot r \cdot t = \$120 \times 7.1\% \times 1$

$$= \$120 \times 0.071 \times 1$$
$$= \$8.52$$

29. $I = P \cdot r \cdot t = \$5200 \times 6\% \times \dfrac{1}{2}$

$= \$5200 \times 0.06 \times \dfrac{1}{2}$

$= \$312 \times \dfrac{1}{2}$

$= \$156$

The interest earned is \$156. The amount in the account is the principal plus the interest: $\$5200 + \$156 = \$5356$.

30. $A = P \cdot \left(1 + \dfrac{r}{n}\right)^{n \cdot t}$

$= \$1000 \cdot \left(1 + \dfrac{5\frac{3}{8}\%}{1}\right)^{1 \cdot 2}$

$= \$1000 \cdot \left(1 + \dfrac{0.05375}{1}\right)^{2}$

$= \$1000(1.05375)^2$

$\approx \$1110.39$

31. $A = P \cdot \left(1 + \dfrac{r}{n}\right)^{n \cdot t}$

$= \$10,000 \cdot \left(1 + \dfrac{4.9\%}{12}\right)^{12 \cdot 3}$

$= \$10,000 \cdot \left(1 + \dfrac{0.049}{12}\right)^{36}$

$\approx \$11,580.07$

32. Registered nurses: We add to find the number of jobs in 2012; $2.3 + 0.6 = 2.9$, so we project that there will be 2.9 million jobs for registered nurses in 2012. We solve an equation to find the percent of increase, p.

$\underset{\downarrow}{0.6} \ \underset{\downarrow}{\text{is}} \ \underbrace{\text{what percent}}_{\downarrow} \ \underset{\downarrow}{\text{of}} \ \underset{\downarrow}{2.3?}$

$0.6 \ = \ \quad\ p \quad\ \cdot \ 2.3$

Solve.

$\dfrac{0.6}{2.3} = \dfrac{p \cdot 2.3}{2.3}$

$0.261 \approx p$

$26.1\% \approx p$

Post-secondary teachers: We subtract to find the change; $2.2 - 1.6 = 0.6$, so we project that the change will be 0.6 million. We solve an equation to find the percent of increase, p.

$\underset{\downarrow}{0.6} \ \underset{\downarrow}{\text{is}} \ \underbrace{\text{what percent}}_{\downarrow} \ \underset{\downarrow}{\text{of}} \ \underset{\downarrow}{1.6?}$

$0.6 \ = \ \quad\ p \quad\ \cdot \ 1.6$

Solve.

$\dfrac{0.6}{1.6} = \dfrac{p \cdot 1.6}{1.6}$

$0.375 = p$

$37.5\% = p$

Food preparation and service workers: We subtract to find the number of jobs in 2002; $2.4 - 0.4 = 2.0$, so there were 2.0 million jobs for food preparation and service

workers in 2002. We solve an equation to find the percent of increase, p.

$\underset{\downarrow}{0.4} \ \underset{\downarrow}{\text{is}} \ \underbrace{\text{what percent}}_{\downarrow} \ \underset{\downarrow}{\text{of}} \ \underset{\downarrow}{2.0?}$

$0.4 \ = \ \quad\ p \quad\ \cdot \ 2.0$

Solve.

$\dfrac{0.4}{2.0} = \dfrac{p \cdot 2.0}{2.0}$

$0.2 = p$

$20\% = p$

Restaurant servers: Let $n =$ the number of jobs in 2002, in millions. If the number of jobs increases by 19.0%, then the new number of jobs is 100% of $n + 19\%$ of n, or 119% of n, or $1.19 \cdot n$. We solve an equation to find n.

$\underbrace{\text{The number of jobs in 2002 increased by 19\%}}_{\downarrow} \ \underset{\downarrow}{\text{is}} \ \underset{\downarrow}{\underline{\text{2.5 million}}}$

$1.19 \cdot n \quad = \quad\ 2.5$

Solve.

$\dfrac{1.19 \cdot n}{1.19} = \dfrac{2.5}{1.19}$

$n \approx 2.1$

There were 2.1 million jobs for restaurant servers in 2002. We subtract to find the change; $2.5 - 2.1 = 0.4$, so we project that the change will be 0.4 million.

33. Discount = Marked price $-$ Sale price

$\ \ D \quad = \quad\ \ 1950 \quad\ - \quad\ 1675$

We subtract: $\begin{array}{r} 1\,9\,5\,0 \\ -\,1\,6\,7\,5 \\ \hline 2\,7\,5 \end{array}$

The discount is \$275.

Discount = Rate of discount $\times$ Marked price

$\ \ 275 \quad = \quad\quad\ R \quad\ \times \quad\ 1950$

To solve the equation we divide on both sides by 1950.

$\dfrac{275}{1950} = \dfrac{R \times 1950}{1950}$

$0.141 \approx R$

$14.1\% \approx R$

To check, note that a discount rate of 14.1% means that 85.9% of the marked price is paid: $0.859 \times 1950 = 1675.05 \approx 1675$. Since that is the sale price, the answer checks.

The rate of discount is about 14.1%.

34. To find the principal after the first payment, we first use the formula $I = P \cdot r \cdot t$ to find the amount of interest paid in the first payment.

$I = P \cdot r \cdot t = \$120,000 \cdot 0.074 \cdot \dfrac{1}{12} \approx \$740.$

Then the amount of the principal applied to the first payment is

$\$830.86 - \$740 = \$90.86.$

Finally, we find that the principal after the first payment is

$$\$120,000 - \$90.86 = \$119,909.14.$$

To find the principal after the second payment, we first use the formula $I = P \cdot r \cdot t$ to find the amount of interest paid in the second payment.

$$I = \$119,909.14 \cdot 0.074 \cdot \frac{1}{12} \approx \$739.44$$

Then the amount of the principal applied to the second payment is

$$\$830.86 - \$739.44 = \$91.42.$$

Finally, we find that the principal after the second payment is

$$\$119,909.14 - \$91.42 = \$119,817.72.$$

35. *Familiarize.* Let $p = $ the price for which a realtor would have to sell the house in order for Juan and Marie to receive $180,000 from the sale. The realtor's commission would be $7.5\% \cdot p$, or $0.075 \cdot p$, and Juan and Marie would receive 100% of $p - 7.5\%$ of p, or 92.5% of p, or $0.925 \cdot p$.

Translate.

$$\underbrace{\text{Amount Juan and Marie receive}}_{\downarrow} \quad \text{is} \quad \$180,000$$
$$\quad\quad\quad\quad\quad\quad\quad\downarrow \quad\quad\quad\quad\quad\quad \downarrow \quad\quad \downarrow$$
$$\quad\quad\quad\quad\quad 0.925 \cdot p \quad\quad\quad = \quad 180,000$$

Solve.

$$\frac{0.925 \cdot p}{0.925} = \frac{180,000}{0.925}$$
$$p \approx 194,600 \quad \text{Rounding to the}$$
$$\quad\quad\quad\quad\quad\quad\quad\quad\quad \text{nearest hundred}$$

Check. 7.5% of $\$194,600 = 0.075 \cdot \$194,600 = \$14,595$ and $\$194,600 - \$14,595 = \$180,005 \approx \$180,000$. The answer checks.

State. A realtor would need to sell the house for about $194,600.

36. First we find the commission.

Commission	=	Commission rate	×	Sales
C	=	16%	×	$\$15,000$
C	=	0.16	×	$\$15,000$
C	=	$\$2400$		

Now we find the amount in the account after 6 months.

$$A = P \cdot \left(1 + \frac{r}{n}\right)^{n \cdot t}$$
$$= \$2400 \cdot \left(1 + \frac{12\%}{4}\right)^{4 \cdot \frac{1}{2}}$$
$$= \$2400 \cdot \left(1 + \frac{0.12}{4}\right)^{2}$$
$$= \$2400 \cdot (1 + 0.03)^2$$
$$= \$2400 \cdot (1.03)^2$$
$$= \$2400(1.0609)$$
$$= \$2546.16$$

Chapter 7
Data, Graphs, and Statistics

Exercise Set 7.1

1. To find the average, add the numbers. Then divide by the number of addends.
$$\frac{17 + 19 + 29 + 18 + 14 + 29}{6} = \frac{126}{6} = 21$$
The average is 21.

To find the median, first list the numbers in order from smallest to largest. Then locate the middle number.
$$14, 17, 18, 19, 29, 29$$
$$\uparrow$$
Middle number

The median is halfway between 18 and 19. It is the average of the two middle numbers:
$$\frac{18 + 19}{2} = \frac{37}{2} = 18.5$$

Find the mode:
The number that occurs most often is 29. The mode is 29.

3. To find the average, add the numbers. Then divide by the number of addends.
$$\frac{5 + 37 + 20 + 20 + 35 + 5 + 25}{7} = \frac{147}{7} = 21$$
The average is 21.

To find the median, first list the numbers in order from smallest to largest. Then locate the middle number.
$$5, 5, 20, 20, 25, 35, 37$$
$$\uparrow$$
Middle number

The median is 20.

Find the mode:
There are two numbers that occur most often, 5 and 20. Thus the modes are 5 and 20.

5. Find the average:
$$\frac{4.3 + 7.4 + 1.2 + 5.7 + 7.4}{5} = \frac{26}{5} = 5.2$$
The average is 5.2.

Find the median:
$$1.2, 4.3, 5.7, 7.4, 7.4$$
$$\uparrow$$
Middle number

The median is 5.7.

Find the mode:
The number that occurs most often is 7.4. The mode is 7.4.

7. Find the average:
$$\frac{234 + 228 + 234 + 229 + 234 + 278}{6} = \frac{1437}{6} = 239.5$$
The average is 239.5.

Find the median:
$$228, 229, 234, 234, 234, 278$$
$$\uparrow$$
Middle number

The median is halfway between 234 and 234. Although it seems clear that this is 234, we can compute it as follows:
$$\frac{234 + 234}{2} = \frac{468}{2} = 234$$
The median is 234.

Find the mode:
The number that occurs most often is 234. The mode is 234.

9. Find the average:
$$\frac{1 + 1 + 11 + 25 + 60 + 72 + 29 + 15 + 1}{9} = \frac{215}{9} = 23.\overline{8}$$
The average is $23.\overline{8}$.

Find the median:
$$1, 1, 1, 11, 15, 25, 29, 60, 72$$
$$\uparrow$$
Middle number

The median is 15.

Find the mode:
The number that occurs most often is 1. The mode is 1.

11. We divide the total number of miles, 279, by the number of gallons, 9.
$$\frac{279}{9} = 31$$
The average was 31 miles per gallon.

13. To find the GPA we first add the grade point values for each hour taken. This is done by first multiplying the grade point value by the number of hours in the course and then adding as follows:

$$
\begin{array}{lll}
\text{B} & 3.0 \cdot 4 = & 12 \\
\text{A} & 4.0 \cdot 5 = & 20 \\
\text{D} & 1.0 \cdot 3 = & 3 \\
\text{C} & 2.0 \cdot 4 = & \underline{8} \\
& & 43 \text{ (Total)}
\end{array}
$$

The total number of hours taken is
$$4 + 5 + 3 + 4, \text{ or } 16.$$
We divide 43 by 16 and round to the nearest tenth.
$$\frac{43}{16} = 2.6875 \approx 2.7$$
The student's grade point average is 2.7.

15. Find the average price per pound:

$$\frac{\$6.99 + \$8.49 + \$8.99 + \$6.99 + \$9.49}{5} = \frac{\$40.95}{5} = \$8.19$$

The average price per pound of Atlantic salmon was $8.19.

Find the median price per pound:

List the prices in order:

$$\$6.99, \$6.99, \$8.49, \$8.99, \$9.49$$
$$\uparrow$$
Middle number

The median is $8.49.

Find the mode:

The number that occurs most often is $6.99. The mode is $6.99.

17. We can find the total of the five scores needed as follows:

$$80 + 80 + 80 + 80 + 80 = 400.$$

The total of the scores on the first four tests is

$$80 + 74 + 81 + 75 = 310.$$

Thus Rich needs to get at least

$$400 - 310, \text{ or } 90$$

to get a B. We can check this as follows:

$$\frac{80 + 74 + 81 + 75 + 90}{5} = \frac{400}{5} = 80.$$

19. We can find the total number of days needed as follows:

$$266 + 266 + 266 + 266 = 1064.$$

The total number of days for Marta's first three pregnancies is

$$270 + 259 + 272 = 801.$$

Thus, Marta's fourth pregnancy must last

$$1064 - 801 = 263 \text{ days}$$

in order to equal the worldwide average. We can check this as follows:

$$\frac{270 + 259 + 272 + 263}{4} = \frac{1064}{4} = 266.$$

21. Compare the averages of the two sets of data.

Bulb A: Average $= (983 + 964 + 1214 + 1417 + 1211 + 1521 + 1084 + 1075 + 892 + 1423 + 949 + 1322)/12 = 1171.25$

Bulb B: Average $= (979 + 1083 + 1344 + 984 + 1445 + 975 + 1492 + 1325 + 1283 + 1325 + 1352 + 1432)/12 \approx 1251.58$

Since the average life of Bulb A is 1171.25 hr and of Bulb B is about 1251.58 hr, Bulb B is better.

23. Discussion and Writing Exercise

25.
$$
\begin{array}{r}
1\,4 \\
\times\,1\,4 \\
\hline
5\,6 \\
1\,4\,0 \\
\hline
1\,9\,6 \\
\end{array}
$$

27.
$$
\begin{array}{rl}
1.\,4 & \text{(1 decimal place)} \\
\times\,1.\,4 & \text{(1 decimal place)} \\
\hline
5\,6 & \\
1\,4\,0 & \\
\hline
1.\,9\,6 & \text{(2 decimal places)} \\
\end{array}
$$

29.
$$
\begin{array}{rl}
1\,2.\,8\,6 & \text{(2 decimal places)} \\
\times\,1\,7.\,5 & \text{(1 decimal place)} \\
\hline
6\,4\,3\,0 & \\
9\,0\,0\,2\,0 & \\
1\,2\,8\,6\,0\,0 & \\
\hline
2\,2\,5.\,0\,5\,0 & \text{(3 decimal places)} \\
\end{array}
$$

31.
$$\frac{4}{5} \cdot \frac{3}{28} = \frac{4 \cdot 3}{5 \cdot 28}$$
$$= \frac{4 \cdot 3}{5 \cdot 4 \cdot 7}$$
$$= \frac{4}{4} \cdot \frac{3}{5 \cdot 7}$$
$$= \frac{3}{35}$$

33. First we divide to find the decimal notation.

$$
\begin{array}{r}
1.\,1\,8\,7\,5 \\
16\,\overline{)\,1\,9.\,0\,0\,0\,0} \\
\underline{1\,6} \\
3\,0 \\
\underline{1\,6} \\
1\,4\,0 \\
\underline{1\,2\,8} \\
1\,2\,0 \\
\underline{1\,1\,2} \\
8\,0 \\
\underline{8\,0} \\
0 \\
\end{array}
$$

Then we move the decimal point two places to the right and write a percent symbol.

$$\frac{19}{16} = 1.1875 = 118.75\%$$

35. First we divide to find the decimal notation.

$$
\begin{array}{r}
0.\,5\,1\,2 \\
125\,\overline{)\,6\,4.\,0\,0\,0} \\
\underline{6\,2\,5} \\
1\,5\,0 \\
\underline{1\,2\,5} \\
2\,5\,0 \\
\underline{2\,5\,0} \\
0 \\
\end{array}
$$

Then we move the decimal point two places to the right and write a percent symbol.

$$\frac{64}{125} = 0.512 = 51.2\%$$

37. Divide the total by the number of games. Use a calculator.

$$\frac{547}{3} \approx 182.33$$

Drop the amount to the right of the decimal point.

$$\underline{182} \cdot \boxed{33}$$

This is the $\uparrow$ $\uparrow$ Drop this
average.   amount.

The bowler's average is 182.

39. We can find the total number of home runs needed over Aaron's 22-yr career as follows:

$$22 \cdot 34\frac{7}{22} = 22 \cdot \frac{755}{22} = \frac{22 \cdot 755}{22} = \frac{22}{22} \cdot \frac{755}{1} = 755.$$

The total number of home runs during the first 21 years of Aaron's career was

$$21 \cdot 35\frac{10}{21} = 21 \cdot \frac{745}{21} = \frac{21 \cdot 745}{21} = \frac{21}{21} \cdot \frac{745}{1} = 745.$$

Then Aaron hit

$$755 - 745 = 10 \text{ home runs}$$

in his final year.

41. The total of the scores on the first four tests was

$$90.5 + 90.5 + 90.5 + 90.5 = 362.$$

The total of the scores on all five tests was

$$84.0 + 84.0 + 84.0 + 84.0 + 84.0 = 420.$$

We subtract to find the score on the fifth test:

$$420 - 362 = 58$$

Exercise Set 7.2

1. Go down the Planet column to Jupiter. Then go across to the column headed Average Distance from Sun (in miles) and read the entry, 483,612,200. The average distance from the sun to Jupiter is 483,612,200 miles.

3. Go down the column headed Time of Revolution in Earth Time (in years) to 164.78. Then go across the Planet column. The entry there is Neptune, so Neptune has a time of revolution of 164.78 days.

5. All of the entries in the column headed Average Distance from Sun (in miles) are greater than 1,000,000. Thus, all of the planets have an average distance from the sun that is greater than 1,000,000 mi.

7. Go down the Planet column to earth and then across to the Diameter (in miles) column to find that the diameter of Earth is 7926 mi. Similarly, find that the diameter of Jupiter is 88,846 mi. Then divide:

$$\frac{88,846}{7926} \approx 11$$

It would take about 11 Earth diameters to equal one Jupiter diameter.

9. Find the average of all the numbers in the column headed Diameter (in miles):

$$(3031 + 7520 + 7926 + 4221 + 88,846 + 74,898 + 31,763 + 31,329 + 1423)/9 = 27,884.\overline{1}$$

The average of the diameters of the planets is $27,884.\overline{1}$ mi.

To find the median of the diameters of the planets we first list the diameters in order from smallest to largest:

1423, 3031, 4221, 7520, 7926, 31,329, 31,763, 74,898, 88,846.

The middle number is 7926, so the median of the diameters is 7926 mi.

Since no number appears more than once in the Diameter (in miles) column, there is no mode.

11. Go down the column headed Actual Temperature (°F) to 80°. Then go across to the Relative Humidity column headed 60%. The entry is 92, so the apparent temperature is 92°F.

13. Go down the column headed Actual Temperature (°F) to 85°. Then go across the Relative Humidity column headed 90%. The entry is 108, so the apparent temperature is 108°F.

15. The number 100 appears in the columns headed Apparent Temperature (°F) 3 times, so there are 3 temperature-humidity combinations that given an apparent temperature of 100°.

17. Go down the Relative Humidity column headed 50% and find all the entries greater than 100. The last 4 entries are greater than 100. Then go across to the column headed Actual Temperature (°F) and read the temperatures that correspond to these entries. At 50% humidity, the actual temperatures 90° and higher give an apparent temperature above 100°.

19. Go down the column headed Actual Temperature (°F) to 95°. Then read across to locate the entries greater than 100. All of the entries except the first two are greater than 100. Go up from each entry to find the corresponding relative humidity. At an actual temperature of 95°, relative humidities of 30% and higher give an apparent temperature above 100°.

21. Go down the column headed Actual Temperature (°F) to 85°, then across to 94, and up to find that the corresponding relative humidity is 40%. Similarly, go down to 85°, across to 108, and up to 90%. At an actual temperature of 85°, the humidity would have to increase by

$$90\% - 40\%, \text{ or } 50\%$$

to raise the apparent temperature from 94° to 108°.

23. The number 1976 lies below the heading "1940," and the number 3849 lies below "1980." Thus, the cigarette consumption in 1940 was 1976 cigarettes per capita and in 1980 it was 3849 cigarettes per capita.

To find the percent of increase we first find the amount of increase.

$$\begin{array}{r} \overset{17}{} \\ \overset{2\ \cancel{7}\ 14}{\cancel{3}\ \cancel{8}\ \cancel{4}\ 9} \\ -\ 1\ 9\ 7\ 6 \\ \hline 1\ 8\ 7\ 3 \end{array}$$

Let $p =$ the percent of increase.

Translate: 1873 is what percent of 1976?

$$1873 = p \cdot 1976$$

Solve.

$$1873 = p \cdot 1976$$
$$\frac{1873}{1976} = \frac{p \cdot 1976}{1976}$$
$$0.948 \approx p$$
$$94.8\% \approx p$$

Cigarette consumption increased about 94.8% from 1940 to 1980.

25. For 1920 to 1950 the average consumption is:
$$\frac{665 + 1485 + 1976 + 3552}{4} = \frac{7678}{4} \approx 1920$$

For 1970 to 2000 the average consumption is:
$$\frac{3985 + 3849 + 2817 + 2092}{4} = \frac{12,743}{4} \approx 3186$$

We subtract to find by how many cigarettes the second average exceeds the first:

$$\begin{array}{r} \overset{2\ 11}{\cancel{3}\ \cancel{1}\ 8\ 6} \\ -\ 1\ 9\ 2\ 0 \\ \hline 1\ 2\ 6\ 6 \end{array}$$

The latter average exceeds the former by 1266 cigarettes per capita.

27. The world population in 1850 is represented by 1 symbol, so the population was 1 billion.

29. The 2070 (projected) population is represented by the most symbols, so the population will be largest in 2070.

31. The smallest increase in the number of symbols is represented by $\frac{1}{2}$ symbol from 1650 to 1850 (as opposed to 1 or more symbols for each of the other pairs). Then the growth was the least between these two years.

33. The world population in 1975 is represented by 4 symbols so it was 4×1 billion, or 4 billion people. The population in 2012 is represented by 7 symbols so it will be 7×1 billion, or 7 billion people. We subtract to find the difference:

$$7 \text{ billion} - 4 \text{ billion} = 3 \text{ billion}$$

The world population in 2012 will be 3 billion more than in 1975.

To find the percent of increase from 1975 to 2012, we divide the amount of increase by the population in 1975:

$$\frac{3 \text{ billion}}{4 \text{ billion}} = 0.75 = 75\%$$

The percent of increase in the world population from 1975 to 2012 will be 75%.

35. The smallest portion of a symbol represents Africa, so the smallest amount of water, per person, is consumed in Africa.

37. North America is represented by $4\frac{3}{4}$ symbols so the water consumption, per person, in North America is $4\frac{3}{4} \times 10,000 = \frac{19}{4} \times 100,000 = \frac{1,900,000}{4} = 475,000$ gal.

39. From Exercise 37, we know that 475,000 gal of water are consumed, per person, in North America. Asia is represented by $1\frac{1}{2}$ symbols so the water consumption, per person, in Asia is $1\frac{1}{2} \times 100,000 = \frac{3}{2} \times 100,000 = \frac{300,000}{2} = 150,000$ gal.

We subtract to find how many more gallons are consumed, per person, in North America than in Asia:

$$475,000 - 150,000 = 325,000 \text{ gal}$$

41. Discussion and Writing Exercise

43. Cabinets: 50% of $26,888 = 0.5(\$26,888) = \$13,444$

Countertops: 15% of $26,888 = 0.15(\$26,888) = \$4033.20$

Appliances: 8% of $26,888 = 0.08(\$26,888) = \$2151.04$

Fixtures: 3% of $26,888 = 0.03(\$26,888) = \$806.64$

45. $24\% = \frac{24}{100} = \frac{4 \cdot 6}{4 \cdot 25} = \frac{4}{4} \cdot \frac{6}{25} = \frac{6}{25}$

47. $4.8\% = \frac{4.8}{100} = \frac{4.8}{100} \cdot \frac{10}{10} = \frac{48}{1000} = \frac{8 \cdot 6}{8 \cdot 125} = \frac{8}{8} \cdot \frac{6}{125} = \frac{6}{125}$

49. $53.1\% = \frac{53.1}{100} = \frac{53.1}{100} \cdot \frac{10}{10} = \frac{531}{1000}$

51. $100\% = \frac{100}{100} = 1$

53. Find the coffee consumption, per person, for each country in Example 4.

Germany: $11 \times 100 = 1100$ cups

United States: $6\frac{1}{10} \times 100 = 610$ cups

Switzerland: $12\frac{1}{5} \times 100 = 1220$ cups

France: $7\frac{9}{10} \times 100 = 790$ cups

Italy: $7\frac{1}{2} \times 100 = 750$ cups

Now divide to determine the number of symbols required when each symbol represents 150 cups of coffee.

Germany: $1100 \div 150 = 7.\overline{3}$, or $7\frac{1}{3}$

United States: $610 \div 150 = 4.0\overline{6} \approx 4$

Switzerland: $1220 \div 150 = 8.1\overline{3} \approx 8\frac{1}{10}$

France: $790 \div 150 = 5.2\overline{6} \approx 5\frac{1}{4}$

Italy: $750 \div 150 = 5$

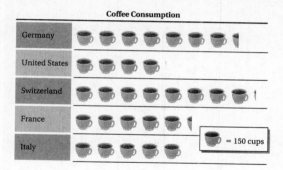

Coffee Consumption

Germany	
United States	
Switzerland	
France	
Italy	

= 150 cups

Exercise Set 7.3

1. Move to the right along the bar representing 1 cup of hot cocoa with skim milk. We read that there are about 190 calories in the cup of cocoa.

3. The longest bar is for 1 slice of chocolate cake with fudge frosting. Thus, it has the highest caloric content.

5. We locate 460 calories at the bottom of the graph and then go up until we reach a bar that ends at approximately 460 calories. Now go across to the left and read the dessert, 1 cup of premium chocolate ice cream.

7. From the graph we see that 1 cup of hot cocoa made with whole milk has about 310 calories and 1 cup of hot cocoa made with skim milk has about 190 calories. We subtract to find the difference:

$$310 - 190 = 120$$

The cocoa made with whole milk has about 120 more calories than the cocoa made with skim milk.

9. From Exercise 5 we know that 1 cup of premium ice cream has about 460 calories. We multiply to find the caloric content of 2 cups:

$$2 \times 460 = 920$$

Kristin consumes about 920 calories.

11. From the graph we see that a 2-oz chocolate bar with peanuts contains about 270 calories. We multiply to find the number of extra calories Paul adds to his diet in 1 year:

$$365 \times 270 \text{ calories} = 98,550 \text{ calories}$$

Then we divide to determine the number of pounds he will gain:

$$\frac{98,550}{3500} \approx 28$$

Paul will gain about 28 pounds.

13. Find the bar representing men with bachelor's degrees in 1970 and read $11,000 on the vertical scale. Do the same for the bar representing men with bachelor's degrees in 2002 and read $58,000.

Subtract to find the amount of increase:

$$\$58,000 - \$11,000 = \$47,000$$

Let p = the percent of increase. We write and solve an equation to find p.

$$47,000 = p \cdot 11,000$$
$$\frac{47,000}{11,000} = p$$
$$4.27 \approx p$$
$$427\% \approx p$$

The percent of increase is approximately 427%.

15. Find the bar representing women with a high school diploma in 1970 and read $6000. Do the same for the bar representing women with a high school diploma in 2002 and read $25,000.

Subtract to find the increase:

$$\$25,000 - \$6000 = \$19,000$$

Let p = the percent of increase. We write and solve an equation to find p.

$$19,000 = p \cdot 6000$$
$$\frac{19,000}{6000} = p$$
$$3.17 \approx p$$
$$317\% \approx p$$

The percent of increase is approximately 317%.

17. From Exercise 13 we know that men with bachelor's degrees earned $11,000 in 1970. Find the bar representing men with a high school diploma in 1970 and read $7000.

Subtract to find the increase:

$$\$11,000 - \$7000 = \$4000$$

19. Find the bar representing women with bachelor's degrees in 2002 and read $46,000. Do the same for the bar representing men with high school diplomas in 2002 and read $35,000.

Subtract to find how much more the women earned:

$$\$46,000 - \$35,000 = \$11,000$$

21. On the horizontal scale in six equally spaced intervals indicate the names of the cities. Label this scale "City." Then label the vertical scale "Commuting Time (in minutes)." Note that the smallest time is 21.6 minutes and the largest is 39.0 minutes. We could start the vertical scale at 0 or we could start it at 20, using a jagged line to indicate the missing numbers. We choose the second option. Label the marks on the vertical scale by 5's. Finally, draw vertical bars above the cities to show the commuting times.

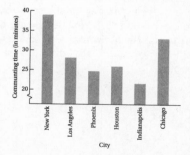

23. The shortest bar represents Indianapolis, so it has the least commuting time.

25. We add the commuting times and divide by the number of addends, 6:
$$\frac{39.0+28.1+24.7+25.9+21.6+33.1}{6} = \frac{172.4}{6} = 28.7\overline{3} \text{ min}$$

27. From the table or the bar graph we see that there was an increase in profit between the following pairs of years: 1995 and 1996, 1996 and 1997, 1998 and 1999.

29. First we subtract to find the amount of decrease:
$$5.5 - 2.7 = 2.8$$
Then let $p =$ the percent of decrease. We write and solve a proportion to find p.
$$\frac{p}{100} = \frac{2.8}{5.5}$$
$$5.5 \cdot p = 100 \cdot 2.8$$
$$p = \frac{100 \cdot 2.8}{5.5}$$
$$p \approx 51$$
The percent of decrease is 51%.

31. We add the net profits, in billions, and divide by the number of addends, 6:
$$\frac{\$2.3 + \$2.8 + \$5.2 + \$4.9 + \$5.5 + \$2.7}{6} = \frac{\$23.4}{6} = \$3.9$$
The average net profit was $3.9 billion.

33. From the graph we read that the average driving distance was 256.9 yd in 1980 and 287.3 yd in 2004. We subtract to find the increase:
$$287.3 - 256.9 = 30.4$$
The driving distance in 2004 was 30.4 yd farther than in 1980.

35. Find 264 on the vertical scale and observe that the horizontal line representing 264 intersects the graph at the points corresponding to 1988 and 1995 on the horizontal scale. Thus, the average driving distance was about 264 yd in 1988 and in 1995.

37. First indicate the years on the horizontal scale and label it "Year." The years range from 1980 to 2030 and increase by 10's. We could start the vertical scale at 0, but the graph will be more compact if we start at a higher number. The years lived beyond age 65 range from 14 to 17.5 so we choose to label the vertical scale from 13 to 18. We use a jagged line to indicate that we are not starting at 0. Label the vertical scale "Average number of years men are estimated to live beyond 65." Next, at the appropriate level above each year on the horizontal scale, mark the corresponding number of years. Finally, draw line segments connecting the points.

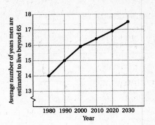

39. First we subtract to find the amount of the increase:
$$17.5 - 14 = 3.5$$
Let $p =$ the percent of increase. We write and solve an equation to find p.
$$3.5 = p \cdot 14$$
$$\frac{3.5}{14} = p$$
$$0.25 = p$$
$$25\% = p$$
Longevity is estimated to increase 25% between 1980 and 2030.

41. First we subtract to find the amount of increase:
$$17.5 - 15.9 = 1.6$$
Let $p =$ the percent of increase. We write and solve an equation to find p.
$$1.6 = p \cdot 15.9$$
$$\frac{1.6}{15.9} = p$$
$$0.101 \approx p$$
$$10.1\% \approx p$$
Longevity is estimated to increase about 10.1% between 2000 and 2030.

43. From the table or the graph we see that the increase in murders was greatest between 1995 and 1996.

45. We add the number of murders committed and divide by the number of addends, 6:
$$\frac{325 + 331 + 311 + 313 + 305 + 262}{6} = \frac{1847}{6} \approx 308 \text{ murders}$$

47. First we subtract to find the amount of the decrease:
$$305 - 262 = 43$$
Let $p =$ the percent of decrease. We write and solve a proportion to find p.
$$\frac{p}{100} = \frac{43}{305}$$
$$p \cdot 305 = 100 \cdot 43$$
$$p = \frac{100 \cdot 43}{305}$$
$$p \approx 14.1$$
Murders decreased about 14.1% between 1999 and 2000.

49. Discussion and Writing Exercise

51. The set of numbers $1, 2, 3, 4, 5, \ldots$ is call the set of <u>natural</u> numbers.

53. The simple interest I on principal P, invested for t years at interest rate r, is given by $I = P \cdot r \cdot t$.

55. When interest is paid on interest, it is called compound interest.

57. In the equation $103 - 13 = 90$, the subtrahend is 13.

59. Using the bar graph we estimate that there were about 27,000 indoor movie screens in 1995, about 28,500 in 1996, about 30,500 in 1997, about 33,000 in 1998, and 37,185 in 1999. We draw a line graph representing these data, putting years on the horizontal scale and the number of indoor movie screens on the vertical scale. We extend the horizontal axis to 2003 since we are interested in making estimates for 2000, 2001, and 2003.

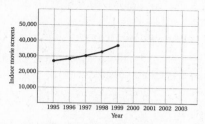

We extend the line to make predictions. We can use any pair of points to determine the extended line. It appears that we will get a good estimate if we use the line containing the points for 1996 and 1998. Using this line we estimate that the number of indoor movie screens was about 37,500 in 2000, about 40,000 in 2001, an about 44,500 in 2003. Answers will vary depending on the points used to extend the line. Actual data could be collected to determine if the estimates are accurate.

Exercise Set 7.4

1. We see from the graph that 17.1% of freshmen major in engineering.

3. We see from the graph that 5.8% of freshmen major in education. Find 5.8% of 10,562:

$$0.058 \times 10,562 \approx 613 \text{ students.}$$

5. First we add the percents corresponding to biological science and social science.

$$6.2\% + 7.7\% = 13.9\%$$

Then we subtract this percent from 100% to find the percent of all freshman who do not major in biological science or social science.

$$100\% - 13.9\% = 86.1\%$$

7. The section of the graph representing food is the largest, so food accounts for the greatest expense.

9. We add percents:

12% (medical care) + 2% (personal care) = 14%

11. Using a circle with 100 equally-spaced tick marks we first draw a line from the center to any tick mark. From that tick mark we count off 28.7 tick marks to graph 28.7% and

label the wedge "Pre-dawn, before 6 AM." We continue in this manner with the other preferences. Finally we title the graph "Holiday Baking: When Is It Done?"

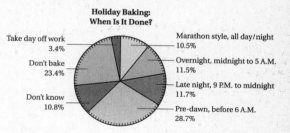

13. Using a circle with 100 equally-spaced tick marks, we first draw a line from the center to any tick mark. From that tick mark we count off 20 tick marks to graph 20% and label it "Less than 20." We continue in this manner with the other categories. Finally we title the graph "Weight Gain During Pregnancy."

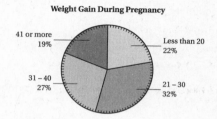

15. Using a circle with 100 equally-spaced tick marks, we first draw a line from the center to any tick mark. From that tick mark we count off 44 tick marks to graph 44% and label it "Motor Vehicle Accidents." We continue in this manner with the other causes of injury. Finally we title the graph "Causes of Spinal Injuries."

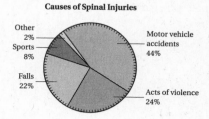

17. Discussion and Writing Exercise

Chapter 7 Review Exercises

1. Go down the FedEx Letter column to 3. Then go across to the column headed FedEx Priority Overnight and read the entry, $32.25. Thus the cost of a 3-lb FedEx Priority Overnight delivery is $32.25.

2. Go down the FedEx Letter Column to 10. Then go across to the column headed FedEx Standard Overnight and read the entry, $47.00. Thus the cost of a 10-lb FedEx Standard Overnight delivery is $47.00.

3. From the table we see that it costs $11.30 to send a 3-lb letter by FedEx 2Day delivery. Now we subtract to find the amount saved by using 2Day delivery:

$$\$32.25 - \$11.30 = \$20.95$$

4. From the table we see that it costs $21.60 to send a 10-lb letter by FedEx 2Day delivery. Now we subtract to find the amount saved by using 2Day delivery:

$$\$47.00 - \$21.60 = \$25.40$$

5. Within each category the price is the same for all packages up to 8 oz, so there is no difference in price between sending a 5-oz package FedEx Priority Overnight and sending an 8-oz package in the same way.

6. Cost for 4-lb package: $31.50

   Cost for 5-lb package: $33.25

   Total cost for a 4-lb package and a 5-lb package:

   $$\$31.50 + \$33.25 = \$64.75$$

   Weight of combined packages: 4 lb + 5 lb = 9 lb

   Cost for 9-lb package: $45.00

   Amount saved by sending both packages as one:

   $$\$64.75 - \$45.00 = \$19.75$$

7. The Chicago police force is represented by 7 symbols, so there are 7×2000, or 14,000, officers.

8. $9000 \div 2000 = 4.5$, so we look for a city represented by about 4.5 symbols. It is Los Angeles.

9. Houston is represented by the smallest number of symbols, so it has the smallest police force.

10. First we find the number of symbols representing each police force. Answers may vary slightly depending on how partial symbols are counted.

    New York: 17

    Chicago: 7

    Los Angeles: 4.8

    Philadelphia: 3.6

    Washington, D.C.: 2.6

    Houston: 2.5

    Now we find the average of these numbers.

    $$\frac{17 + 7 + 4.8 + 3.6 + 2.6 + 2.5}{6} = \frac{37.5}{6} = 6.25$$

    Finally we multiply to find the number of officers represented by 6.25 symbols.

    $$6.25 \times 2000 = 12,500$$

    The average size of the six police forces is 12,500 officers.

11. The number that occurs most often is 26, so 26 is the mode.

12. The numbers that occur most often are 11 and 17. They are the modes.

13. The number that occurs most often is 0.2, so 0.2 is the mode.

14. The numbers that occur most often are 700 and 800. They are the modes.

15. The number that occurs most often is $17, so $17 is the mode.

16. The number that occurs most often is 20, so 20 is the mode.

17. To find the average we add the amounts and divide by the number of addends.

    $$\frac{\$102 + \$112 + \$130 + \$98}{4} = \frac{\$442}{4} = \$110.50$$

    To find the median we first write the numbers in order from smallest to largest. Then locate the middle number.

    $$\$98, \$102, \$112, \$130$$
    $$\uparrow$$
    Middle number

    The median is halfway between $102 and $112. It is the average of the two middle numbers.

    $$\frac{\$102 + \$112}{2} = \frac{\$214}{2} = \$107$$

18. We divide the number of miles, 528, by the number of gallons, 16.

    $$\frac{528}{16} = 33$$

    The average was 33 miles per gallon.

19. We can find the total of the four scores needed as follows:

    $$90 + 90 + 90 + 90 = 360.$$

    The total of the scores on the first three tests is

    $$94 + 78 + 92 = 264.$$

    Thus Marcus needs to get at least

    $$360 - 264 = 96$$

    to get an A. We can check this as follows:

    $$\frac{90 + 78 + 92 + 96}{4} = \frac{360}{4} = 90.$$

20. Move to the right along the bar representing a Single with Everything. We read that there are about 420 calories in this sandwich.

21. Move to the right along the bar representing a Breaded Chicken sandwich. We read that there are about 440 calories in this sandwich.

22. The longest bar represents a Big Bacon Classic. This is the sandwich with the highest caloric content.

23. The shortest bar represents a Plain Single. This is the sandwich with the lowest caloric content.

24. Locate 360 on the horizontal axis and then go up to a bar that ends above this point. That bar represents a Plain Single, so this is the sandwich that contains about 360 calories.

25. Locate 470 on the horizontal axis and then go up to a bar that ends above this point. That bar represents a Chicken Club, so this is the sandwich that contains about 470 calories.

26. Using the information from Exercises 20 and 25, we subtract to find how many more calories are in a Chicken Club than in a Single with Everything.

$$470 - 420 = 50 \text{ calories}$$

27. From the graph we see that a Big Bacon Classic contains about 580 calories. In Exercise 24 we found that a Plain Single contains about 360 calories. We subtract to find how many more calories the Big Bacon Classic contains.

$$580 - 360 = 220 \text{ calories}$$

28. The highest point on the graph lies above the Under 20 label on the horizontal scale, so the under 20 age group has the most accidents per 100 drivers.

29. Find the lowest point on the graph and then move across to the vertical scale to read that 12 accidents is the fewest number of accidents per 100 drivers in any age group.

30. From the graph we see that people 75 and over have 25 accidents per 100 drivers and those in the 65-74 age range have about 12 accidents per 100 drivers. We subtract to find the difference.

$$25 - 12 = 13 \text{ accidents per 100 drivers}$$

31. We see that the line is nearly horizontal (it rises and falls only slightly) from the 45-54 age group to the 65-74 age group. Thus the number of accidents stays basically the same from ages 45 to 74.

32. From the graph we see that people in the 25-34 age group have about 23 accidents per 100 drivers and those in the 20-24 age group have about 34. We subtract to find the difference.

$$34 - 23 = 11 \text{ accidents per 100 drivers}$$

33. From the graph we see that people in the 55-64 age group have about 12 accidents per 100 drivers. Then $3 \cdot 12 = 36$ and we see that people under 20 have about 36 accidents per 100 drivers, so people in this age group have about three times as many accidents as those in the 55-64 age group.

34. From the graph we see that 22% of travelers prefer a first-class hotel.

35. From the graph we see that 11% of travelers prefer an economy hotel.

36. From the graph we see that 64% of travelers prefer a moderate hotel. We find 64% of 2500 travelers: $0.64 \times 2500 = 1600$ travelers.

37. 22% of travelers prefer a first class hotel and 3% prefer a deluxe hotel. Then $22\% + 3\% = 25\%$ prefer either a first-class or deluxe hotel.

38. On the horizontal scale in seven equally spaced intervals indicate the years. Label this scale "Year." Then label the vertical scale "Cost of first-class postage." The smallest cost is 20¢ and the largest is 39¢, so we start the vertical scale at 0 and extend it to 40¢, labeling it by 5's. Finally, draw vertical bars above the years to show the cost of the postage.

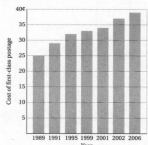

39. Prepare horizontal and vertical scales as described in Exercise 38. Then, at the appropriate level above each year, mark the corresponding postage. Finally, draw line segments connecting the points.

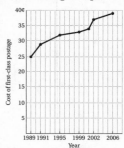

40. Battery A:

$$(38.9 + 39.3 + 40.4 + 53.1 + 41.7 + 38.0 + 36.8 + 47.7 +$$
$$48.1 + 38.2 + 46.9 + 47.4) \div 12 = \frac{516.5}{12} \approx 43.04$$

Battery B:

$$(39.3 + 38.6 + 38.8 + 37.4 + 47.6 + 37.9 + 46.9 + 37.8 +$$
$$38.1 + 47.9 + 50.1 + 38.2) \div 12 = \frac{498.6}{12} \approx 41.55$$

Because the average time for Battery A is longer, it is the better battery.

41. $\dfrac{26 + 34 + 43 + 51}{4} = \dfrac{154}{4} = 38.5$

42. $\dfrac{11 + 14 + 17 + 18 + 7}{5} = \dfrac{67}{5} = 13.4$

43. $\dfrac{0.2 + 1.7 + 1.9 + 2.4}{4} = \dfrac{6.2}{4} = 1.55$

44. $\dfrac{700 + 2700 + 3000 + 900 + 1900}{5} = \dfrac{9200}{5} = 1840$

45. $\dfrac{\$2 + \$14 + \$17 + \$17 + \$21 + \$29}{6} = \dfrac{\$100}{6} = \$16.\overline{6}$

46. $\dfrac{20 + 190 + 280 + 470 + 470 + 500}{6} = \dfrac{1930}{6} = 321.\overline{6}$

47. $26, 34, 43, 51$
              ↑
          Middle number

The median is halfway between 34 and 43. It is the average of the two middle numbers.

$$\frac{34 + 43}{2} = \frac{77}{2} = 38.5$$

The median is 38.5.

48. $7, 11, 14, 17, 18$
                 ↑
          Middle number

The median is 14.

49. $0.2, 1.7, 1.9, 2.4$
                 ↑
          Middle number

The median is halfway between 1.7 and 1.9. It is the average of the two middle numbers.

$$\frac{1.7 + 1.9}{2} = \frac{3.6}{2} = 1.8$$

The median is 1.8.

50. $700, 900, 1900, 2700, 3000$
                       ↑
              Middle number

The median is 1900.

51. We arrange the numbers from smallest to largest.

$$\$2, \$14, \$17, \$17, \$21, \$29$$
                 ↑
          Middle number

The median is halfway between $17 and $17. Although it seems clear that this is $17, we can compute it as follows:

$$\frac{\$17 + \$17}{2} = \frac{\$34}{2} = \$17$$

The median is $17.

52. We arrange the numbers from smallest to largest.

$$20, 190, 280, 470, 470, 500$$
                     ↑
              Middle number

The median is halfway between 280 and 470. It is the average of the two middle numbers.

$$\frac{280 + 470}{2} = \frac{750}{2} = 375$$

The median is 375.

53. To find the GPA we first add the grade point values for each hour taken. This is done by first multiplying the grade point value by the number of hours in the course and then adding as follows:

$$
\begin{array}{lll}
\text{A} & 4.0 \cdot 5 = & 20 \\
\text{B} & 3.0 \cdot 3 = & 9 \\
\text{C} & 2.0 \cdot 4 = & 8 \\
\text{B} & 3.0 \cdot 3 = & 9 \\
\text{B} & 3.0 \cdot 1 = & \underline{3} \\
 & & 49 \ \text{(Total)}
\end{array}
$$

The total number of hours taken is

$5 + 3 + 4 + 3 + 1$, or 16.

We divide 49 by 16 and round to the nearest tenth.

$$\frac{49}{16} = 3.0625 \approx 3.1$$

The student's grade point average is 3.1.

54. *Discussion and Writing Exercise.* The average, the median, and the mode are "center points" that characterize a set of data. You might use the average to find a center point that is midway between the extreme values of the data. The median is a center point that is in the middle of all the data. That is, there are as many values less than the median as there are values greater than the median. The mode is a center point that represents the value or values that occur most frequently.

55. *Discussion and Writing Exercise.* The equation could represent a person's average income during a 4-yr period. Answers may vary.

56. a is the middle number and the median is 316, so $a = 316$.

The average is 326 so the data must add to $326 + 326 + 326 + 326 + 326 + 326 + 326$, or 2282.

The sum of the known data items, including a, is $298 + 301 + 305 + 316 + 323 + 390$, or 1933.

We subtract to find b:

$$b = 2282 - 1933 = 349$$

Chapter 7 Test

1. Go down the column in the first table labeled "Height" to the entry "6 ft, 1 in." Then go to the right and read the entry in the column headed "Medium Frame." We see that the desirable weight is 179 lb.

2. Go down the column in the second table labeled "Height" to the entry "5 ft, 3 in." Then go to the right and read the entry in the column headed "Small Frame." We see that the desirable weight is 111 lb.

3. Locate the number 120 in the second table and observe that it is in the column headed "Medium Frame." Then go to the left and observe that the corresponding entry in the "Height" column is 5 ft, 3 in. Thus a 5 ft, 3 in. woman with a medium frame has a desirable weight of 120 lb.

4. Locate the number 169 in the first table and observe that it is in the column headed "Medium Frame." Then go to the left and observe that the corresponding entry in the "Height" column is 5 ft, 11 in. Thus a 5 ft, 11 in. man with a medium frame has a desirable weight of 169 lb.

5. Since $600 \div 100 = 6$, we look for a country represented by 6 symbols. We find that it is Japan.

6. Since $1000 \div 100 = 10$, we look for a country represented by 10 symbols. We find that it is the United States.

7. The amount of waste generated per person per year in France is represented by 8 symbols, so each person generates $8 \cdot 100$, or 800 lb, of waste per year.

8. The amount of waste generated per person per year in Finland is represented by 4 symbols, so each person generates $4 \cdot 100$, or 400 lb, of waste per year.

9. We add the numbers and then divide by the number of items of data.
$$\frac{45 + 49 + 52 + 52}{4} = \frac{198}{4} = 49.5$$

10. We add the numbers and then divide by the number of items of data.
$$\frac{1 + 1 + 3 + 5 + 3}{5} = \frac{13}{5} = 2.6$$

11. We add the numbers and then divide by the number of items of data.
$$\frac{3 + 17 + 17 + 18 + 18 + 20}{6} = \frac{93}{6} = 15.5$$

12. 45, 49, 52, 52

Find the median: There is an even number of numbers. The median is the average of the two middle numbers:
$$\frac{49 + 52}{2} = \frac{101}{2} = 50.5$$

Find the mode: The number that occurs most often is 52. It is the mode.

13. Find the median: First we rearrange the numbers from the smallest to largest.

$$1, 1, 3, 3, 5$$
$$\uparrow$$
Middle number

The median is 3.

Find the mode: There are two numbers that occur most often, 1 and 3. They are the modes.

14. 3, 17, 17, 18, 18, 20

Find the median: There is an even number of numbers. The median is the average of the two middle numbers:
$$\frac{17 + 18}{2} = \frac{35}{2} = 17.5$$

Find the mode: There are two numbers that occur most often, 17 and 18. They are the modes.

15. We divide the number of miles by the number of gallons.
$$\frac{432}{16} = 27 \text{ mpg}$$

16. The total of the four scores needed is
$$70 + 70 + 70 + 70 = 4 \cdot 70, \text{ or } 280.$$

The total of the scores on the first three tests is
$$68 + 71 + 65 = 204.$$

Thus the student needs to get at least
$$280 - 204, \text{ or } 76$$

on the fourth test.

17. Find 2010 on the bottom scale and move up from there to the line. The line is labeled 53% at that point, so 53% of meals will be eaten away from home in 2010.

18. Find 1985 halfway between 1980 and 1990 on the bottom scale and move up from that point to the line. Then go straight across to the left and find that about 41% of meals were eaten away from home in 1985.

19. Locate 30% on the vertical scale. Then move to the right to the line. Look down to the bottom scale and observe that the year 1967 corresponds to this point.

20. Locate 50% on the vertical scale. Then move to the right to the line. Look down to the bottom scale and observe that the year 2006 corresponds to this point.

21. First indicate the names of the animals in seven equally spaced intervals on the horizontal scale. Title this scale "Animals." Now note that the lowest speed is 28 mph and the highest is 225 mph. We start the vertical scaling at 0 and label the marks on the scale by 50's from 0 to 300. Title this scale "Maximum speed (in miles per hour)." Finally, draw vertical bars above the names of the animals to show the speeds.

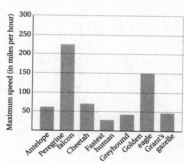

22. From the table or the bar graph, we see that the slowest speed is 28 mph and the fastest is 225 mph. Then the fastest speed exceeds the slowest by
$$225 - 28, \text{ or } 197 \text{ mph}.$$

23. The fastest human's maximum speed is 28 mph and a greyhound's maximum speed is 42 mph. Thus a human cannot outrun a greyhound because a greyhound can run $42 - 28$, or 14 mph, faster than a human.

24. We add the speeds and then divide by the number of speeds.
$$\frac{61 + 225 + 70 + 28 + 42 + 150 + 47}{7} = \frac{623}{7} = 89 \text{ mph}$$

25. First we write the numbers from smallest to largest.
$$28, 42, 47, 61, 70, 150, 225$$
$$\uparrow$$
Middle number

The median speed is 61 mph.

26. Using a circle with 100 equally spaced tick marks, we first draw a line from the center to any tick mark. From that tick mark, count off 44 tick marks and draw another line to graph 44%. Label this wedge "Employee theft." Continue in this manner with the other types of losses. Finally, title the graph.

Retailing Losses

Other 0.7%
Vendor fraud 5.1%
Administrative error 17.5%
Employee theft 44%
Shoplifting 32.7%

27. Employee theft: 44% of $23 billion = 0.44($23 billion) = $10.12 billion

Shoplifting: 32.7% of $23 billion = 0.327($23 billion) = $7.521 billion

Administrative error: 17.5% of $23 billion = 0.175($23 billion) = $4.025 billion

Vendor fraud: 5.1% of $23 billion = 0.051($23 billion) = $1.173 billion

Other: 0.7% of $23 billion = 0.007($23 billion) = $0.161 billion

28. We will make a vertical bar graph. First indicate the years in five equally spaced intervals on the horizontal scale and title this scale "Year." Now note that the sales range from 7152 to 23,000. We start the vertical scaling with 0 and label the marks by 5000's from 0 to 30,000. Title this scale "U.S. Porsche sales." Finally, draw vertical bars to show the sales numbers.

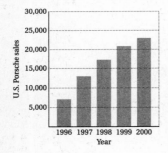

29. First indicate the years on the horizontal scale and title this scale "Year." We scale the vertical axis by 5000's and title it "U.S. Porsche sales." Next mark the number of sales at the appropriate level above each year. Then draw line segments connecting adjacent points.

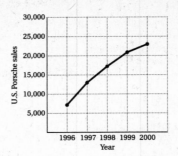

30. We find the average of each set of ratings.

Pecan:
$$\frac{9+10+8+10+9+7+6+9+10+7+8+8}{12} =$$
$$\frac{101}{12} \approx 8.417$$

Hazelnut:
$$\frac{10+6+8+9+10+10+8+7+6+9+10+8}{12} =$$
$$\frac{101}{12} \approx 8.417$$

Since the averages are equal, the chocolate bars are of equal quality.

31. To find the GPA we first add the grade point values for each class taken. This is done by first multiplying the grade point value by the number of hours in the course and then adding as follows:

$$
\begin{array}{lll}
B & 3.0 \cdot 3 = & 9 \\
A & 4.0 \cdot 3 = & 12 \\
C & 2.0 \cdot 4 = & 8 \\
B & 3.0 \cdot 3 = & 9 \\
B & 3.0 \cdot 2 = & 6 \\
\hline
& & 44 \ \text{(Total)}
\end{array}
$$

The total number of hours taken is

   $3 + 3 + 4 + 3 + 2$, or 15.

We divide 44 by 15 and round to the nearest tenth.

   $\frac{44}{15} = 2.9\overline{3} \approx 2.9$

The grade point average is 2.9.

32. a is the middle number in the ordered set of data, so a is the median, 74.

Since the mean, or average, is 82, the total of the seven numbers is $7 \cdot 82$, or 574.

The total of the known numbers is

   $69 + 71 + 73 + 74 + 78 + 98$, or 463.

Then $b = 574 - 463 = 111$.

Chapter 8

Geometry

Exercise Set 8.1

1. The segment consists of the endpoints G and H and all points between them.

It can be named $\overline{GH}$ or $\overline{HG}$.

3. The ray with endpoint Q extends forever in the direction of point D.

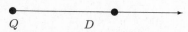

In naming a ray, the endpoint is always given first. This ray is named $\overrightarrow{QD}$.

5.

The line can be named with the small letter l, or it can be named by any two points on it. This line can be named

$$l, \overleftrightarrow{DE}, \overleftrightarrow{ED}, \overleftrightarrow{DF}, \overleftrightarrow{FD}, \overleftrightarrow{EF}, \text{ or } \overleftrightarrow{FE}.$$

7. The angle can be named in five different ways:

angle GHI, angle IHG, $\angle GHI$, $\angle IHG$, or $\angle H$.

9. Place the $\triangle$ of the protractor at the vertex of the angle, and line up one of the sides at 0°. We choose the horizontal side. Since 0° is on the inside scale, we check where the other side of the angle crosses the inside scale. It crosses at 10°. Thus, the measure of the angle is 10°.

11. Place the $\triangle$ of the protractor at the vertex of the angle, point B. Line up one of the sides at 0°. We choose the side that contains point A. Since 0° is on the outside scale, we check where the other side crosses the outside scale. It crosses at 180°. Thus, the measure of the angle is 180°.

13. Place the $\triangle$ of the protractor at the vertex of the angle, and line up one of the sides at 0°. We choose the horizontal side. Since 0° is on the inside scale, we check where the other side crosses the inside scale. It crosses at 130°. Thus, the measure of the angle is 130°.

15. Using a protractor, we find that the measure of the angle in Exercise 7 is 148°. Since its measure is greater than 90°and less than 180°, it is an obtuse angle.

17. The measure of the angle in Exercise 9 is 10°. Since its measure is greater than 0°and less than 90°, it is an acute angle.

19. The measure of the angle in Exercise 11 is 180°. It is a straight angle.

21. The measure of the angle in Exercise 13 is 130°. Since its measure is greater than 90°and less than 180°, it is an obtuse angle.

23. The measure of the angle in Margin Exercise 12 is 30°. Since its measure is greater than 0°and less than 90°, it is an acute angle.

25. The measure of the angle in Margin Exercise 14 is 126°. Since its measure is greater than 90°and less than 180°, it is an obtuse angle.

27. Using a protractor, we find that the lines do not intersect to form a right angle. They are not perpendicular.

29. Using a protractor, we find that the lines intersect to form a right angle. They are perpendicular.

31. All the sides are of different lengths. The triangle is a scalene triangle.

One angle is an obtuse angle. The triangle is an obtuse triangle.

33. All the sides are of different lengths. The triangle is a scalene triangle.

One angle is a right angle. The triangle is a right triangle.

35. All the sides are the same length. The triangle is an equilateral triangle.

All three angles are acute. The triangle is an acute triangle.

37. All the sides are of different lengths. The triangle is a scalene triangle.

One angle is an obtuse angle. The triangle is an obtuse triangle.

39. The polygon has 4 sides. It is a quadrilateral.

41. The polygon has 5 sides. It is a pentagon.

43. The polygon has 3 sides. It is a triangle.

45. The polygon has 5 sides. It is a pentagon.

47. The polygon has 6 sides. It is a hexagon.

49. If a polygon has n sides, the sum of its angle measures is $(n-2) \cdot 180°$. A decagon has 10 sides. Substituting 10 for n in the formula, we get

$$(n-2) \cdot 180° = (10-2) \cdot 180°$$
$$= 8 \cdot 180°$$
$$= 1440°.$$

51. If a polygon has n sides, the sum of its angle measures is $(n-2) \cdot 180°$. A heptagon has 7 sides. Substituting 7 for n in the formula, we get

$$(n-2) \cdot 180° = (7-2) \cdot 180°$$
$$= 5 \cdot 180°$$
$$= 900°.$$

53. If a polygon has n sides, the sum of its angle measures is $(n-2) \cdot 180°$. To find the sum of the angle measures for a 14-sided polygon, substitute 14 for n in the formula.

$$(n-2) \cdot 180° = (14-2) \cdot 180°$$
$$= 12 \cdot 180°$$
$$= 2160°$$

55. If a polygon has n sides, the sum of its angle measures is $(n-2) \cdot 180°$. To find the sum of the angle measures for a 20-sided polygon, substitute 20 for n in the formula.

$$(n-2) \cdot 180° = (20-2) \cdot 180°$$
$$= 18 \cdot 180°$$
$$= 3240°$$

57.
$$m(\angle A) + m(\angle B) + m(\angle C) = 180°$$
$$42° + 92° + x = 180°$$
$$134° + x = 180°$$
$$x = 180° - 134°$$
$$x = 46°$$

59.
$$31° + 29° + x = 180°$$
$$60° + x = 180°$$
$$x = 180° - 60°$$
$$x = 120°$$

61.
$$m(\angle R) + m(\angle S) + m(\angle T) = 180°$$
$$x + 58° + 79° = 180°$$
$$x + 137° = 180°$$
$$x = 180° - 137°$$
$$x = 43°$$

63. Discussion and Writing Exercise

65.
$$I = P \cdot r \cdot t$$
$$= \$2000 \cdot 8\% \cdot 1$$
$$= \$2000 \cdot 0.08 \cdot 1$$
$$= \$160$$

67.
$$I = P \cdot r \cdot t$$
$$= \$4000 \cdot 7.4\% \cdot \frac{1}{2}$$
$$= \$4000 \cdot 0.074 \cdot \frac{1}{2}$$
$$= \$148$$

69.
$$A = P \cdot \left(1 + \frac{r}{n}\right)^{n \cdot t}$$
$$= \$25,000 \cdot \left(1 + \frac{6\%}{2}\right)^{2 \cdot 5}$$
$$= \$25,000 \cdot \left(1 + \frac{0.06}{2}\right)^{10}$$
$$= \$25,000(1.03)^{10}$$
$$\approx \$33,597.91$$

71.
$$A = P \cdot \left(1 + \frac{r}{n}\right)^{n \cdot t}$$
$$= \$150,000 \cdot \left(1 + \frac{7.4\%}{2}\right)^{2 \cdot 20}$$
$$= \$150,000 \cdot \left(1 + \frac{0.074}{2}\right)^{40}$$
$$= \$150,000(1.037)^{40}$$
$$\approx \$641,566.26$$

73. We find $m \angle 2$:
$$m \angle 6 + m \angle 1 + m \angle 2 = 180°$$
$$33.07° + 79.8° + m \angle 2 = 180°$$
$$112.87° + m \angle 2 = 180°$$
$$m \angle 2 = 180° - 112.87°$$
$$m \angle 2 = 67.13°$$

The measure of angle 2 is 67.13°.

We find $m \angle 3$:
$$m \angle 1 + m \angle 2 + m \angle 3 = 180°$$
$$79.8° + 67.13° + m \angle 3 = 180°$$
$$146.93° + m \angle 3 = 180°$$
$$m \angle 3 = 180° - 146.93°$$
$$m \angle 3 = 33.07°$$

The measure of angle 3 is 33.07°.

We find $m \angle 4$:
$$m \angle 2 + m \angle 3 + m \angle 4 = 180°$$
$$67.13° + 33.07° + m \angle 4 = 180°$$
$$100.2° + m \angle 4 = 180°$$
$$m \angle 4 = 180° - 100.2°$$
$$m \angle 4 = 79.8°$$

The measure of angle 4 is 79.8°.

To find $m \angle 5$, note that $m \angle 6 + m \angle 1 + m \angle 5 = 180°$. Then to find $m \angle 5$ we follow the same procedure we used to find $m \angle 2$. Thus, the measure of angle 5 is 67.13°.

75. $\angle ACB$ and $\angle ACD$ are complementary angles. Since $m \angle ACD = 40°$ and $90° - 40° = 50°$, we have $m \angle ACB = 50°$.

Now consider triangle ABC. We know that the sum of the measures of the angles is 180°. Then

$$m \angle ABC + m \angle BCA + m \angle CAB = 180°$$
$$50° + 90° + m \angle CAB = 180°$$
$$140° + m \angle CAB = 180°$$
$$m \angle CAB = 180° - 140°$$
$$m \angle CAB = 40°,$$

so $m \angle CAB = 40°$.

To find $m\angle EBC$ we first find $m\angle CEB$. We note that $\angle DEC$ and $\angle CEB$ are supplementary angles. Since $m\angle DEC = 100°$ and $180° - 100° = 80°$, we have $m\angle CEB = 80°$. Now consider triangle BCE. We know that the sum of the measures of the angles is $180°$. Note that $\angle ACB$ can also be named $\angle BCE$. Then

$$m\angle BCE + m\angle CEB + m\angle EBC = 180°$$
$$50° + 80° + m\angle EBC = 180°$$
$$130° + m\angle EBC = 180°$$
$$m\angle EBC = 180° - 130°$$
$$m\angle EBC = 50°,$$

so $m\angle EBC = 50°$.

$\angle EBA$ and $\angle EBC$ are complementary angles. Since $m\angle EBC = 50°$ and $90° - 50° = 40°$, we have $m\angle EBA = 40°$.

Now consider triangle ABE. We know that the sum of the measures of the angles is $180°$. Then

$$m\angle CAB + m\angle EBA + m\angle AEB = 180°$$
$$40° + 40° + m\angle AEB = 180°$$
$$80° + m\angle AEB = 180°$$
$$m\angle AEB = 180° - 80°$$
$$m\angle AEB = 100°,$$

so $m\angle AEB = 100°$.

To find $m\angle ADB$ we first find $m\angle EDC$. Consider triangle CDE. We know that the sum of the measures of the angles is $180°$. Then

$$m\angle DEC + m\angle ECD + m\angle EDC = 180°$$
$$100° + 40° + m\angle EDC = 180°$$
$$140° + m\angle EDC = 180°$$
$$m\angle EDC = 180° - 140°$$
$$m\angle EDC = 40°,$$

so $m\angle EDC = 40°$. We now note that $\angle ADB$ and $\angle EDC$ are complementary angles. Since $m\angle EDC = 40°$ and $90° - 40° = 50°$, we have $m\angle ADB = 50°$.

Exercise Set 8.2

1. Perimeter $= 4\text{ mm} + 6\text{ mm} + 7\text{ mm}$
$= (4 + 6 + 7)\text{ mm}$
$= 17\text{ mm}$

3. Perimeter $= 3.5\text{ in.} + 3.5\text{ in.} + 4.25\text{ in.} +$
$\qquad\qquad 0.5\text{ in.} + 3.5\text{ in.}$
$= (3.5 + 3.5 + 4.25 + 0.5 + 3.5)\text{ in.}$
$= 15.25\text{ in.}$

5. $P = 2 \cdot (l + w)$ \qquad Perimeter of a rectangle
$P = 2 \cdot (5.6\text{ km} + 3.4\text{ km})$
$P = 2 \cdot (9\text{ km})$
$P = 18\text{ km}$

7. $P = 2 \cdot (l + w)$ \qquad Perimeter of a rectangle
$P = 2 \cdot (5\text{ ft} + 10\text{ ft})$
$P = 2 \cdot (15\text{ ft})$
$P = 30\text{ ft}$

9. $P = 2 \cdot (l + w)$ \qquad Perimeter of a rectangle
$P = 2 \cdot (34.67\text{ cm} + 4.9\text{ cm})$
$P = 2 \cdot (39.57\text{ cm})$
$P = 79.14\text{ cm}$

11. $P = 4 \cdot s$ \qquad Perimeter of a square
$P = 4 \cdot 22\text{ ft}$
$P = 88\text{ ft}$

13. $P = 4 \cdot s$ \qquad Perimeter of a square
$P = 4 \cdot 45.5\text{ mm}$
$P = 182\text{ mm}$

15. *Familiarize.* First we find the perimeter of the field. Then we multiply to find the cost of the fence wire. We make a drawing.

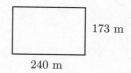

173 m

240 m

Translate. The perimeter of the field is given by

$$P = 2 \cdot (l + w) = 2 \cdot (240\text{ m} + 173\text{ m}).$$

Solve. We calculate the perimeter.

$$P = 2 \cdot (240\text{ m} + 173\text{ m}) = 2 \cdot (413\text{ m}) = 826\text{ m}$$

Then we multiply to find the cost of the fence wire.

$$\begin{aligned}\text{Cost} &= \$7.29/\text{m} \times \text{Perimeter}\\ &= \$7.29/\text{m} \times 826\text{ m}\\ &= \$6021.54\end{aligned}$$

Check. Repeat the calculations.

State. The perimeter of the field is 826 m. The fencing will cost \$6021.54.

17. *Familiarize.* We make a drawing and let $P =$ the perimeter.

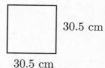

30.5 cm

30.5 cm

Translate. The perimeter of the square is given by

$$P = 4 \cdot s = 4 \cdot (30.5\text{ cm}).$$

Solve. We do the calculation.

$$P = 4 \cdot (30.5\text{ cm}) = 122\text{ cm}.$$

Check. Repeat the calculation.

State. The perimeter of the tile is 122 cm.

19. Familiarize. We label the missing lengths on the drawing and let P = the perimeter.

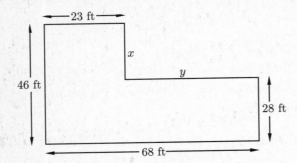

Translate. First we find the missing lengths x and y.

28 ft	plus	how many more ft	is	46 ft
↓	↓	↓	↓	↓
28	+	x	=	46

23 ft	plus	how many more ft	is	68 ft
↓	↓	↓	↓	↓
23	+	y	=	68

Solve. We solve for x and y.

$$28 + x = 46 \qquad 23 + y = 68$$
$$x = 46 - 28 \qquad y = 68 - 23$$
$$x = 18 \qquad\qquad y = 45$$

a) To find the perimeter we add the lengths of the sides of the house.

$$P = 23 \text{ ft} + 18 \text{ ft} + 45 \text{ ft} + 28 \text{ ft} + 68 \text{ ft} + 46 \text{ ft}$$
$$= (23 + 18 + 45 + 28 + 68 + 46) \text{ ft}$$
$$= 228 \text{ ft}$$

b) Next we find t, the total cost of the gutter.

Cost per foot	times	Number of feet	is	Total cost
↓	↓	↓	↓	↓
4.59	×	228	=	t

We carry out the multiplication.

```
      2 2 8
   ×  4 .5 9
   ---------
    2 0 5 2
  1 1 4 0 0
  9 1 2 0 0
  ---------
1 0 4 6 .5 2
```

Thus, $t = 1046.52$.

Check. We can repeat the calculations.

State. (a) The perimeter of the house is 228 ft. (b) The total cost of the gutter is \$1046.52.

21. Discussion and Writing Exercise

23. Interest $= P \cdot r \cdot t$
$$= \$600 \times 6.4\% \times \frac{1}{2}$$
$$= \frac{\$600 \times 0.064}{2}$$
$$= \$19.20$$

The interest is \$19.20.

25. $10^3 = 10 \cdot 10 \cdot 10 = 1000$

27. $15^2 = 15 \cdot 15 = 225$

29. $7^2 = 7 \cdot 7 = 49$

31. Rephrase:

Sales tax	is	what percent	of	purchase price?
↓	↓	↓	↓	↓

Translate: $\qquad 878 \quad = \quad r \quad \times \quad 17,560$

To solve the equation we divide on both sides by 17,560.

$$\frac{878}{17,560} = \frac{r \times 17,560}{17,560}$$
$$0.05 = r$$
$$5\% = r$$

The sales tax rate is 5%.

33. $18 \text{ in.} = 18 \text{ in.} \times \dfrac{1 \text{ ft}}{12 \text{ in.}} = \dfrac{18}{12} \times 1 \text{ ft} = \dfrac{3}{2} \text{ ft}$

$$P = 2 \cdot (l + w)$$
$$P = 2 \cdot \left(3 \text{ ft} + \frac{3}{2} \text{ ft}\right)$$
$$P = 2 \cdot \left(\frac{9}{2} \text{ ft}\right)$$
$$P = 9 \text{ ft}$$

Exercise Set 8.3

1. $A = l \cdot w$ Area of a rectangular region
$A = (5 \text{ km}) \cdot (3 \text{ km})$
$A = 5 \cdot 3 \cdot \text{ km} \cdot \text{ km}$
$A = 15 \text{ km}^2$

3. $A = l \cdot w$ Area of a rectangular region
$A = (2 \text{ in.}) \cdot (0.7 \text{ in.})$
$A = 2 \cdot 0.7 \cdot \text{ in.} \cdot \text{ in.}$
$A = 1.4 \text{ in}^2$

5. $A = s \cdot s$ Area of a square
$A = \left(2\frac{1}{2} \text{ yd}\right) \cdot \left(2\frac{1}{2} \text{ yd}\right)$
$A = \left(\frac{5}{2} \text{ yd}\right) \cdot \left(\frac{5}{2} \text{ yd}\right)$
$A = \frac{5}{2} \cdot \frac{5}{2} \cdot \text{ yd} \cdot \text{ yd}$
$A = \frac{25}{4} \text{ yd}^2$, or $6\frac{1}{4} \text{ yd}^2$

7. $A = s \cdot s$ Area of a square
$A = (90 \text{ ft}) \cdot (90 \text{ ft})$
$A = 90 \cdot 90 \cdot \text{ ft} \cdot \text{ ft}$
$A = 8100 \text{ ft}^2$

9. $A = l \cdot w$ Area of a rectangular region
$A = (10 \text{ ft}) \cdot (5 \text{ ft})$
$A = 10 \cdot 5 \cdot \text{ ft} \cdot \text{ ft}$
$A = 50 \text{ ft}^2$

11. $A = l \cdot w$ Area of a rectangular region
$A = (34.67 \text{ cm}) \cdot (4.9 \text{ cm})$
$A = 34.67 \cdot 4.9 \cdot \text{ cm} \cdot \text{ cm}$
$A = 169.883 \text{ cm}^2$

13. $A = l \cdot w$ Area of a rectangular region
$A = \left(4\frac{2}{3} \text{ in.}\right) \cdot \left(8\frac{5}{6} \text{ in.}\right)$
$A = \left(\frac{14}{3} \text{ in.}\right) \cdot \left(\frac{53}{6} \text{ in.}\right)$
$A = \frac{14}{3} \cdot \frac{53}{6} \cdot \text{ in.} \cdot \text{ in.}$
$A = \frac{2 \cdot 7 \cdot 53}{3 \cdot 2 \cdot 3} \text{ in}^2$
$A = \frac{2}{2} \cdot \frac{7 \cdot 53}{3 \cdot 3} \text{ in}^2$
$A = \frac{371}{9} \text{ in}^2, \text{ or } 41\frac{2}{9} \text{ in}^2$

15. $A = s \cdot s$ Area of a square
$A = (22 \text{ ft}) \cdot (22 \text{ ft})$
$A = 22 \cdot 22 \cdot \text{ ft} \cdot \text{ ft}$
$A = 484 \text{ ft}^2$

17. $A = s \cdot s$ Area of a square
$A = (56.9 \text{ km}) \cdot (56.9 \text{ km})$
$A = 56.9 \cdot 56.9 \cdot \text{ km} \cdot \text{ km}$
$A = 3237.61 \text{ km}^2$

19. $A = s \cdot s$ Area of a square
$A = \left(5\frac{3}{8} \text{ yd}\right) \cdot \left(5\frac{3}{8} \text{ yd}\right)$
$A = \left(\frac{43}{8} \text{ yd}\right) \cdot \left(\frac{43}{8} \text{ yd}\right)$
$A = \frac{43}{8} \cdot \frac{43}{8} \cdot \text{ yd} \cdot \text{ yd}$
$A = \frac{1849}{64} \text{ yd}^2, \text{ or } 28\frac{57}{64} \text{ yd}^2$

21. $A = b \cdot h$ Area of a parallelogram
$A = 8 \text{ cm} \cdot 4 \text{ cm}$ Substituting 8 cm for b and
                4 cm for h
$A = 32 \text{ cm}^2$

23. $A = \frac{1}{2} \cdot b \cdot h$ Area of a triangle
$A = \frac{1}{2} \cdot 15 \text{ in.} \cdot 8 \text{ in.}$ Substituting 15 in. for b and
                     8 in. for h
$A = 60 \text{ in}^2$

25. $A = \frac{1}{2} \cdot h \cdot (a + b)$ Area of a trapezoid
$A = \frac{1}{2} \cdot 8 \text{ ft} \cdot (6 + 20) \text{ ft}$ Substituting 8 ft for h, 6 ft
                        for a, and 20 ft for b
$A = \frac{8 \cdot 26}{2} \text{ ft}^2$
$A = 104 \text{ ft}^2$

27. $A = \frac{1}{2} \cdot h \cdot (a + b)$ Area of a trapezoid
$A = \frac{1}{2} \cdot 7 \text{ in.} \cdot (4.5 + 8.5) \text{ in.}$ Substituting 7 in. for h,
                            4.5 in. for a, and 8.5 in.
                            for b
$A = \frac{7 \cdot 13}{2} \text{ in}^2$
$A = \frac{91}{2} \text{ in}^2$
$A = 45.5 \text{ in}^2$

29. $A = b \cdot h$ Area of a parallelogram
$A = 2.3 \text{ cm} \cdot 3.5 \text{ cm}$ Substituting 2.3 cm for b
                     and 3.5 cm for h
$A = 8.05 \text{ cm}^2$

31. $A = \frac{1}{2} \cdot h \cdot (a + b)$ Area of a trapezoid
$A = \frac{1}{2} \cdot 18 \text{ cm} \cdot (9 + 24) \text{ cm}$ Substituting 18 cm for
                           h, 9 cm for a, and 24
                           cm for b
$A = \frac{18 \cdot 33}{2} \text{ cm}^2$
$A = 297 \text{ cm}^2$

33. $A = \frac{1}{2} \cdot b \cdot h$ Area of a triangle
$A = \frac{1}{2} \cdot 4 \text{ m} \cdot 3.5 \text{ m}$ Substituting 4 m for b and
                     3.5 m for h
$A = \frac{4 \cdot 3.5}{2} \text{ m}^2$
$A = 7 \text{ m}^2$

35. *Familiarize.* We draw a picture.

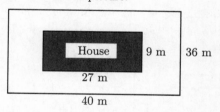

Translate. We let $A =$ the area left over.

Area left over	is	Area of lot	minus	Area of house
↓	↓	↓	↓	↓
A	$=$	$(40 \text{ m}) \cdot (36 \text{ m})$	$-$	$(27 \text{ m}) \cdot (9 \text{ m})$

Solve. The area of the lot is

$(40 \text{ m}) \cdot (36 \text{ m}) = 40 \cdot 36 \cdot \text{ m} \cdot \text{ m} = 1440 \text{ m}^2.$

The area of the house is

$(27 \text{ m}) \cdot (9 \text{ m}) = 27 \cdot 9 \cdot \text{ m} \cdot \text{ m} = 243 \text{ m}^2.$

The area left over is

$A = 1440 \text{ m}^2 - 243 \text{ m}^2 = 1197 \text{ m}^2.$

Check. Repeat the calculations.

State. The area left over for the lawn is 1197 m^2.

37. a) First find the area of the entire yard, including the basketball court:

$$A = l \cdot w = \left(110\frac{2}{3} \text{ ft}\right) \cdot (80 \text{ ft})$$

$$= \left(\frac{332}{3} \text{ ft}\right) \cdot (80 \text{ ft})$$

$$= \frac{26,560}{3} \text{ ft}^2$$

$$= 8853\frac{1}{3} \text{ ft}^2$$

Now find the area of the basketball court:

$$A = s \cdot s = \left(19\frac{1}{2} \text{ ft}\right) \cdot \left(19\frac{1}{2} \text{ ft}\right) =$$

$$\frac{39}{2} \cdot \frac{39}{2} \text{ ft}^2 = \frac{1521}{4} \text{ ft}^2 = 380\frac{1}{4} \text{ ft}^2$$

Finally, subtract to find the area of the lawn:

$$8853\frac{1}{3} \text{ ft}^2 - 380\frac{1}{4} \text{ ft}^2 = 8853\frac{4}{12} \text{ ft}^2 - 380\frac{3}{12} \text{ ft}^2 =$$

$$8473\frac{1}{12} \text{ ft}^2 \approx 8473 \text{ ft}^2$$

b) Let c = the cost of mowing the law. We translate to an equation.

$$\underbrace{\text{The cost of mowing}}_{c} \quad \underbrace{\text{is \$0.012 times}}_{= \ 0.012 \ \cdot} \quad \underbrace{\text{the area of the lawn.}}_{8473}$$

We multiply to solve the equation.

$$c = 0.012 \cdot 8473 \approx \$102$$

The total cost of the mowing is about \$102.

39. *Familiarize.* We use the drawing in the text.

Translate. We let A = the area of the sidewalk, in square feet.

$$\underbrace{\text{Area of sidewalk}}_{A} \ \underbrace{\text{is}}_{=} \ \underbrace{\text{Total area}}_{(113.4 \text{ ft}) \times (75.4 \text{ ft})} \ \underbrace{\text{minus}}_{-} \ \underbrace{\text{Area of building}}_{(110 \text{ ft}) \times (72 \text{ ft})}$$

Solve. The total area is

$$(113.4 \text{ ft}) \times (75.4 \text{ ft}) = 113.4 \times 75.4 \times \text{ ft} \times \text{ ft} = 8550.36 \text{ ft}^2.$$

The area of the building is

$$(110 \text{ ft}) \times (72 \text{ ft}) = 110 \times 72 \times \text{ ft} \times \text{ ft} = 7920 \text{ ft}^2.$$

The area of the sidewalk is

$$A = 8550.36 \text{ ft}^2 - 7920 \text{ ft}^2 = 630.36 \text{ ft}^2.$$

Check. Repeat the calculations.

State. The area of the sidewalk is 630.36 ft^2.

41. *Familiarize.* The dimensions are as follows:

Two walls are 15 ft by 8 ft.

Two walls are 20 ft by 8 ft.

The ceiling is 15 ft by 20 ft.

The total area of the walls and ceiling is the total area of the rectangles described above less the area of the windows and the door.

Translate. a) We let A = the total area of the walls and ceiling. The total area of the two 15 ft by 8 ft walls is

$$2 \cdot (15 \text{ ft}) \cdot (8 \text{ ft}) = 2 \cdot 15 \cdot 8 \cdot \text{ ft} \cdot \text{ ft} = 240 \text{ ft}^2$$

The total area of the two 20 ft by 8 ft walls is

$$2 \cdot (20 \text{ ft}) \cdot (8 \text{ ft}) = 2 \cdot 20 \cdot 8 \cdot \text{ ft} \cdot \text{ ft} = 320 \text{ ft}^2$$

The area of the ceiling is

$$(15 \text{ ft}) \cdot (20 \text{ ft}) = 15 \cdot 20 \cdot \text{ ft} \cdot \text{ ft} = 300 \text{ ft}^2$$

The area of the two windows is

$$2 \cdot (3 \text{ ft}) \cdot (4 \text{ ft}) = 2 \cdot 3 \cdot 4 \cdot \text{ ft} \cdot \text{ ft} = 24 \text{ ft}^2$$

The area of the door is

$$\left(2\frac{1}{2} \text{ ft}\right) \cdot \left(6\frac{1}{2} \text{ ft}\right) = \left(\frac{5}{2} \text{ ft}\right) \cdot \left(\frac{13}{2} \text{ ft}\right)$$

$$= \frac{5}{2} \cdot \frac{13}{2} \cdot \text{ ft} \cdot \text{ ft}$$

$$= \frac{65}{4} \text{ ft}^2, \text{ or } 16\frac{1}{4} \text{ ft}^2$$

Thus

$$A = 240 \text{ ft}^2 + 320 \text{ ft}^2 + 300 \text{ ft}^2 - 24 \text{ ft}^2 - 16\frac{1}{4} \text{ ft}^2$$

$$= 819\frac{3}{4} \text{ ft}^2, \text{ or } 819.75 \text{ ft}^2$$

b) We divide to find how many gallons of paint are needed.

$$819.75 \div 86.625 \approx 9.46$$

It will be necessary to buy 10 gallons of paint in order to have the required 9.46 gallons.

c) We multiply to find the cost of the paint.

$$10 \times \$17.95 = \$179.50$$

Check. We repeat the calculations.

State. (a) The total area of the walls and ceiling is 819.75 ft^2. (b) 10 gallons of paint are needed. (c) It will cost \$179.50 to paint the room.

43.

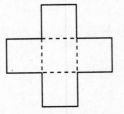

Each side is 4 cm.

The region is composed of 5 squares, each with sides of length 4 cm. The area is

$$A = 5 \cdot (s \cdot s) = 5 \cdot (4 \text{ cm} \cdot 4 \text{ cm}) = 5 \cdot 4 \cdot 4 \text{ cm} \cdot \text{ cm} = 80 \text{ cm}^2$$

45. Familiarize. We look for the kinds of figures whose areas we can calculate using area formulas that we already know.

Translate. The shaded region consists of a square region with a triangular region removed from it. The sides of the square are 30 cm, and the triangle has base 30 cm and height 15 cm. We find the area of the square using the formula $A = s \cdot s$, and the area of the triangle using $A = \frac{1}{2} \cdot b \cdot h$. Then we subtract.

Solve. Area of the square: $A = 30 \text{ cm} \cdot 30 \text{ cm} = 900 \text{ cm}^2$.

Area of the triangle: $A = \frac{1}{2} \cdot 30 \text{ cm} \cdot 15 \text{ cm} = 225 \text{ cm}^2$.

Area of the shaded region: $A = 900 \text{ cm}^2 - 225 \text{ cm}^2 = 675 \text{ cm}^2$.

Check. We repeat the calculations.

State. The area of the shaded region is 675 cm².

47. Familiarize. We have one large triangle with height and base each 6 cm. We also have 6 small triangles, each with height and base 1 cm.

Translate. We will find the area of each type of triangle using the formula $A = \frac{1}{2} \cdot b \cdot h$. Next we will multiply the area of the smaller triangle by 6. And, finally, we will add this product to the area of the larger triangle to find the total area.

Solve.

For the large triangle: $A = \frac{1}{2} \cdot 6 \text{ cm} \cdot 6 \text{ cm} = 18 \text{ cm}^2$

For one small triangle: $A = \frac{1}{2} \cdot 1 \text{ cm} \cdot 1 \text{ cm} = \frac{1}{2} \text{ cm}^2$

Find the area of the 6 small triangles: $6 \cdot \frac{1}{2} \text{ cm}^2 = 3 \text{ cm}^2$

Add to find the total area: $18 \text{ cm}^2 + 3 \text{ cm}^2 = 21 \text{ cm}^2$

Check. We repeat the calculations.

State. The area of the shaded region is 21 cm².

49. Familiarize. We make a drawing, shading the area left over after the triangular piece is cut from the sailcloth.

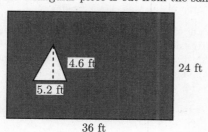

Translate. The shaded region consists of a rectangular region with a triangular region removed from it. The rectangular region has dimensions 36 ft by 24 ft, and the triangular region has base 5.2 ft and height 4.6 ft. We will find the area of the rectangular region using the formula $A = b \cdot h$, and the area of the triangle using $A = \frac{1}{2} \cdot b \cdot h$. Then we will subtract to find the area of the shaded region.

Solve. Area of the rectangle: $A = 36 \text{ ft} \cdot 24 \text{ ft} = 864 \text{ ft}^2$.

Area of the triangle: $A = \frac{1}{2} \cdot 5.2 \text{ ft} \cdot 4.6 \text{ ft} = 11.96 \text{ ft}^2$.

Area of the shaded region: $A = 864 \text{ ft}^2 - 11.96 \text{ ft}^2 = 852.04 \text{ ft}^2$.

Check. We repeat the calculation.

State. The area left over is 852.04 ft².

51. Discussion and Writing Exercise

53. A number is divisible by 8 if the number named by the last <u>three</u> digits is divisible by 8.

55. Two lines are <u>perpendicular</u> if they intersect to form a right angle.

57. An <u>angle</u> is a set of points consisting of two rays.

59. The <u>perimeter</u> of a polygon is the sum of the lengths of its sides.

61.

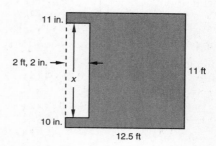

$2 \text{ ft} = 2 \times 1 \text{ ft} = 2 \times 12 \text{ in.} = 24 \text{ in.}$, so 2 ft, 2 in. $= 2 \text{ ft} + 2 \text{ in.} = 24 \text{ in.} + 2 \text{ in.} = 26 \text{ in.}$

$11 \text{ ft} = 11 \times 1 \text{ ft} = 11 \times 12 \text{ in.} = 132 \text{ in.}$

$12.5 \text{ ft} = 12.5 \times 1 \text{ ft} = 12.5 \times 12 \text{ in.} = 150 \text{ in.}$

We solve an equation to find x, in inches:

$$11 + x + 10 = 132$$
$$21 + x = 132$$
$$21 + x - 21 = 132 - 21$$
$$x = 111$$

Then the area of the shaded region is the area of a 150 in. by 132 in. rectangle less the area of a 111 in. by 26 in. rectangle.

$$A = (150 \text{ in.}) \cdot (132 \text{ in.}) - (111 \text{ in.}) \cdot (26 \text{ in.})$$
$$A = 19,800 \text{ in}^2 - 2886 \text{ in}^2$$
$$A = 16,914 \text{ in}^2$$

Exercise Set 8.4

1. $d = 2 \cdot r$

$d = 2 \cdot 7 \text{ cm} = 14 \text{ cm}$

$C = 2 \cdot \pi \cdot r$

$C \approx 2 \cdot \frac{22}{7} \cdot 7 \text{ cm} = \frac{2 \cdot 22 \cdot 7}{7} \text{ cm} = 44 \text{ cm}$

$A = \pi \cdot r \cdot r$

$A \approx \frac{22}{7} \cdot 7 \text{ cm} \cdot 7 \text{ cm} = \frac{22}{7} \cdot 49 \text{ cm}^2 = 154 \text{ cm}^2$

3. $d = 2 \cdot r$

$d = 2 \cdot \dfrac{3}{4}$ in. $= \dfrac{6}{4}$ in. $= \dfrac{3}{2}$ in., or $1\dfrac{1}{2}$ in.

$C = 2 \cdot \pi \cdot r$

$C \approx 2 \cdot \dfrac{22}{7} \cdot \dfrac{3}{4}$ in. $= \dfrac{2 \cdot 22 \cdot 3}{7 \cdot 4}$ in. $= \dfrac{132}{28}$ in. $= \dfrac{33}{7}$ in.,

or $4\dfrac{5}{7}$ in.

$A = \pi \cdot r \cdot r$

$A \approx \dfrac{22}{7} \cdot \dfrac{3}{4}$ in. $\cdot \dfrac{3}{4}$ in. $= \dfrac{22 \cdot 3 \cdot 3}{7 \cdot 4 \cdot 4}$ in^2 $= \dfrac{99}{56}$ in^2, or $1\dfrac{43}{56}$ in^2

5. $r = \dfrac{d}{2}$

$r = \dfrac{32 \text{ ft}}{2} = 16$ ft

$C = \pi \cdot d$

$C \approx 3.14 \cdot 32 \text{ ft} = 100.48$ ft

$A = \pi \cdot r \cdot r$

$A \approx 3.14 \cdot 16 \text{ ft} \cdot 16 \text{ ft} \quad \left(r = \dfrac{d}{2}; r = \dfrac{32 \text{ ft}}{2} = 16 \text{ ft}\right)$

$A = 3.14 \cdot 256 \text{ ft}^2$

$A = 803.84 \text{ ft}^2$

7. $r = \dfrac{d}{2}$

$r = \dfrac{1.4 \text{ cm}}{2} = 0.7$ cm

$C = \pi \cdot d$

$C \approx 3.14 \cdot 1.4 \text{ cm} = 4.396$ cm

$A = \pi \cdot r \cdot r$

$A \approx 3.14 \cdot 0.7 \text{ cm} \cdot 0.7 \text{ cm}$

$\quad\quad \left(r = \dfrac{d}{2}; r = \dfrac{1.4 \text{ cm}}{2} = 0.7 \text{ cm}\right)$

$A = 3.14 \cdot 0.49 \text{ cm}^2 = 1.5386 \text{ cm}^2$

9. $r = \dfrac{d}{2}$

$r = \dfrac{6 \text{ cm}}{2} = 3$ cm

The radius is 3 cm.

$C = \pi \cdot d$

$C \approx 3.14 \cdot 6 \text{ cm} = 18.84$ cm

The circumference is about 18.84 cm.

$A = \pi \cdot r \cdot r$

$A \approx 3.14 \cdot 3 \text{ cm} \cdot 3 \text{ cm} = 28.26 \text{ cm}^2$

The area is about 28.26 cm^2.

11. $r = \dfrac{d}{2}$

$r = \dfrac{14 \text{ ft}}{2} = 7$ ft

$A = \pi \cdot r \cdot r$

$A \approx 3.14 \cdot 7 \text{ ft} \cdot 7 \text{ ft} = 153.86 \text{ ft}^2$

The area of the trampoline is about 153.86 ft^2.

13. $\quad\quad C = \pi \cdot d$

$7.85 \text{ cm} \approx 3.14 \cdot d \quad$ Substituting 7.85 cm for C and 3.14 for π

$\dfrac{7.85 \text{ cm}}{3.14} = d \quad$ Dividing on both sides by 3.14

$2.5 \text{ cm} = d$

The diameter is about 2.5 cm.

$r = \dfrac{d}{2}$

$r = \dfrac{2.5 \text{ cm}}{2} = 1.25$ cm

The radius is about 1.25 cm.

$A = \pi \cdot r \cdot r$

$A \approx 3.14 \cdot 1.25 \text{ cm} \cdot 1.25 \text{ cm} = 4.90625 \text{ cm}^2$

The area is about 4.90625 cm^2.

15. $C = \pi \cdot d$

$C \approx 3.14 \cdot 1.1 \text{ ft} = 3.454$ ft

The circumference of the elm tree is about 3.454 ft.

17. Find the area of the larger circle (pool plus walk). Its diameter is 1 yd + 20 yd + 1 yd, or 22 yd. Thus its radius is $\dfrac{22}{2}$ yd, or 11 yd.

$$A = \pi \cdot r \cdot r$$
$$A \approx 3.14 \cdot 11 \text{ yd} \cdot 11 \text{ yd} = 379.94 \text{ yd}^2$$

Find the area of the pool. Its diameter is 20 yd. Thus its radius is $\dfrac{20}{2}$ yd, or 10 yd.

$$A = \pi \cdot r \cdot r$$
$$A \approx 3.14 \cdot 10 \text{ yd} \cdot 10 \text{ yd} = 314 \text{ yd}^2$$

We subtract to find the area of the walk:

$$A = 379.94 \text{ yd}^2 - 314 \text{ yd}^2$$
$$A = 65.94 \text{ yd}^2$$

The area of the walk is 65.94 yd^2.

19. The perimeter consists of the circumferences of three semicircles, each with diameter 8 ft, and one side of a square of length 8 ft. We first find the circumference of one semicircle. This is one-half the circumference of a circle with diameter 8 ft:

$$\frac{1}{2} \cdot \pi \cdot d \approx \frac{1}{2} \cdot 3.14 \cdot 8 \text{ ft} = 12.56 \text{ ft}$$

Then we multiply by 3:

$$3 \cdot (12.56 \text{ ft}) = 37.68 \text{ ft}$$

Finally we add the circumferences of the semicircles and the length of the side of the square:

$$37.68 \text{ ft} + 8 \text{ ft} = 45.68 \text{ ft}$$

The perimeter is 45.68 ft.

21. The perimeter consists of three-fourths of the circumference of a circle with radius 4 yd and two sides of a square with sides of length 4 yd. We first find three-fourths of the circumference of the circle:

$$\frac{3}{4} \cdot 2 \cdot \pi \cdot r \approx 0.75 \cdot 2 \cdot 3.14 \cdot 4 \text{ yd} = 18.84 \text{ yd}$$

Then we add this length to the lengths of two sides of the square:

$$18.84 \text{ yd} + 4 \text{ yd} + 4 \text{ yd} = 26.84 \text{ yd}$$

The perimeter is 26.84 yd.

23. The perimeter consists of three-fourths of the perimeter of a square with side of length 10 yd and the circumference of a semicircle with diameter 10 yd. First we find three-fourths of the perimeter of the square:

$$\frac{3}{4} \cdot 4 \cdot s = \frac{3}{4} \cdot 4 \cdot 10 \text{ yd} = 30 \text{ yd}$$

Then we find one-half of the circumference of a circle with diameter 10 yd:

$$\frac{1}{2} \cdot \pi \cdot d \approx \frac{1}{2} \cdot 3.14 \cdot 10 \text{ yd} = 15.7 \text{ yd}$$

Then we add:

$$30 \text{ yd} + 15.7 \text{ yd} = 45.7 \text{ yd}$$

The perimeter is 45.7 yd.

25. The shaded region consists of a circle of radius 8 m, with two circles each of diameter 8 m, removed. First we find the area of the large circle:

$$A = \pi \cdot r \cdot r \approx 3.14 \cdot 8 \text{ m} \cdot 8 \text{ m} = 200.96 \text{ m}^2$$

Then we find the area of one of the small circles:
The radius is $\frac{8 \text{ m}}{2} = 4 \text{ m}$.

$$A = \pi \cdot r \cdot r \approx 3.14 \cdot 4 \text{ m} \cdot 4 \text{ m} = 50.24 \text{ m}^2$$

We multiply this area by 2 to find the area of the two small circles:

$$2 \cdot 50.24 \text{ m}^2 = 100.48 \text{ m}^2$$

Finally we subtract to find the area of the shaded region:

$$200.96 \text{ m}^2 - 100.48 \text{ m}^2 = 100.48 \text{ m}^2$$

The area of the shaded region is 100.48 m².

27. The shaded region consists of one-half of a circle with diameter 2.8 cm and a triangle with base 2.8 cm and height 2.8 cm. First we find the area of the semicircle. The radius is $\frac{2.8 \text{ cm}}{2} = 1.4 \text{ cm}$.

$$A = \frac{1}{2} \cdot \pi \cdot r \cdot r \approx \frac{1}{2} \cdot 3.14 \cdot 1.4 \text{ cm} \cdot 1.4 \text{ cm} = 3.0772 \text{ cm}^2$$

Then we find the area of the triangle:

$$A = \frac{1}{2} \cdot b \cdot h = \frac{1}{2} \cdot 2.8 \text{ cm} \cdot 2.8 \text{ cm} = 3.92 \text{ cm}^2$$

Finally we add to find the area of the shaded region:

$$3.0772 \text{ cm}^2 + 3.92 \text{ cm}^2 = 6.9972 \text{ cm}^2$$

The area of the shaded region is 6.9972 cm².

29. The shaded area consists of a rectangle of dimensions 11.4 in. by 14.6 in., with the area of two semicircles, each of diameter 11.4 in., removed. This is equivalent to removing one circle with diameter 11.4 in. from the rectangle. First we find the area of the rectangle:

$$l \cdot w = (11.4 \text{ in.}) \cdot (14.6 \text{ in.}) = 166.44 \text{ in}^2$$

Then we find the area of the circle. The radius is $\frac{11.4 \text{ in.}}{2} = 5.7 \text{ in.}$

$$\pi \cdot r \cdot r \approx 3.14 \cdot 5.7 \text{ in.} \cdot 5.7 \text{ in.} = 102.0186 \text{ in}^2$$

Finally we subtract to find the area of the shaded region:

$$166.44 \text{ in}^2 - 102.0186 \text{ in}^2 = 64.4214 \text{ in}^2$$

31. Discussion and Writing Exercise

33. $2^4 = 2 \cdot 2 \cdot 2 \cdot 2 = 16$

35. $5^3 = 5 \cdot 5 \cdot 5 = 125$

37. a) Find decimal notation using long division.

$$
\begin{array}{r}
0.375 \\
8\overline{\smash{)}3.000} \\
\underline{2\ 4} \\
6\ 0 \\
\underline{5\ 6} \\
4\ 0 \\
\underline{4\ 0} \\
0
\end{array}
$$

$$\frac{3}{8} = 0.375$$

b) Convert the decimal notation to percent notation. Move the decimal point two places to the right, and write a % symbol.

0.37.5

$$\frac{3}{8} = 37.5\%$$

39. a) Find decimal notation using long division.

$$
\begin{array}{r}
0.66 \\
3\overline{\smash{)}2.00} \\
\underline{1\ 8} \\
2\ 0 \\
\underline{1\ 8} \\
2
\end{array}
$$

$$\frac{2}{3} = 0.\overline{6}$$

b) Convert the decimal notation to percent notation. Move the decimal point two places to the right, and write a % symbol.

$$0.66.\overline{6}$$

$$\frac{2}{3} = 66.\overline{6}\%$$

41. First we find the discount.

Discount = Marked price − Sale price

$$D \quad = \quad 100 \quad - \quad 58.99$$

We subtract:
$$\begin{array}{r} 1\,0\,0.\,0\,0 \\ -\quad 5\,8.\,9\,9 \\ \hline 4\,1.\,0\,1 \end{array}$$

The discount is $41.01.

Now we find the rate of discount.

Discount = Rate of discount × Marked price

$$41.01 \quad = \quad R \quad \times \quad 100$$

To solve the equation we divide on both sides by 100.

$$\frac{41.01}{100} = \frac{R \times 100}{100}$$

$$0.4101 \approx R$$

$$41.01\% \approx R$$

To check, note that a discount rate of 41.01% means that 58.99% of the marked price is paid: $0.5899 \times 100 = 58.99$. Since this is the sale price, the answer checks. The rate of discount is about 41.01%.

43. Let c represent the number of cans that can be bought for $7.45. We translate to a proportion and solve.

$$\begin{array}{c} \text{Cans} \rightarrow \\ \text{Cost} \rightarrow \end{array} \frac{2}{\$1.49} = \frac{c}{\$7.45} \begin{array}{c} \leftarrow \text{Cans} \\ \leftarrow \text{Cost} \end{array}$$

$$2 \cdot \$7.45 = \$1.49 \cdot c \quad \text{Equating cross-products}$$

$$\frac{2 \cdot \$7.45}{\$1.49} = c$$

$$\frac{\$14.90}{\$1.49} = c$$

$$10 = c$$

You can buy 10 cans for $7.45.

45. Find $3927 \div 1250$ using a calculator.

$$\frac{3927}{1250} = 3.1416$$

47. The height of the stack of tennis balls is three times the diameter of one ball, or $3 \cdot d$.

The circumference of one ball is given by $\pi \cdot d$.

The circumference of one ball is greater than the height of the stack of balls, because $\pi > 3$.

Exercise Set 8.5

1. $V = l \cdot w \cdot h$

$V = 12 \text{ cm} \cdot 8 \text{ cm} \cdot 8 \text{ cm}$

$V = 12 \cdot 64 \text{ cm}^3$

$V = 768 \text{ cm}^3$

$SA = 2lw + 2lh + 2wh$

$\quad = 2 \cdot 12 \text{ cm} \cdot 8 \text{ cm} + 2 \cdot 12 \text{ cm} \cdot 8 \text{ cm} +$

$\qquad 2 \cdot 8 \text{ cm} \cdot 8 \text{ cm}$

$\quad = 192 \text{ cm}^2 + 192 \text{ cm}^2 + 128 \text{ cm}^2$

$\quad = 512 \text{ cm}^2$

3. $V = l \cdot w \cdot h$

$V = 7.5 \text{ in.} \cdot 2 \text{ in.} \cdot 3 \text{ in.}$

$V = 7.5 \cdot 6 \text{ in}^3$

$V = 45 \text{ in}^3$

$SA = 2lw + 2lh + 2wh$

$\quad = 2 \cdot 7.5 \text{ in.} \cdot 2 \text{ in.} + 2 \cdot 7.5 \text{ in.} \cdot 3 \text{ in.} +$

$\qquad 2 \cdot 2 \text{ in.} \cdot 3 \text{ in.}$

$\quad = 30 \text{ in}^2 + 45 \text{ in}^2 + 12 \text{ in}^2$

$\quad = 87 \text{ in}^2$

5. $V = l \cdot w \cdot h$

$V = 10 \text{ m} \cdot 5 \text{ m} \cdot 1.5 \text{ m}$

$V = 10 \cdot 7.5 \text{ m}^3$

$V = 75 \text{ m}^3$

$SA = 2lw + 2lh + 2wh$

$\quad = 2 \cdot 10 \text{ m} \cdot 5 \text{ m} + 2 \cdot 10 \text{ m} \cdot 1.5 \text{ m} +$

$\qquad 2 \cdot 5 \text{ m} \cdot 1.5 \text{ m}$

$\quad = 100 \text{ m}^2 + 30 \text{ m}^2 + 15 \text{ m}^2$

$\quad = 145 \text{ m}^2$

7. $V = l \cdot w \cdot h$

$V = 6\frac{1}{2} \text{ yd} \cdot 5\frac{1}{2} \text{ yd} \cdot 10 \text{ yd}$

$V = \frac{13}{2} \cdot \frac{11}{2} \cdot 10 \text{ yd}^3$

$V = \frac{715}{2} \text{ yd}^3$

$V = 357\frac{1}{2} \text{ yd}^3$

$$SA = 2lw + 2lh + 2wh$$
$$= 2 \cdot 6\frac{1}{2} \text{ yd} \cdot 5\frac{1}{2} \text{ yd} + 2 \cdot 6\frac{1}{2} \text{ yd} \cdot 10 \text{ yd} +$$
$$2 \cdot 5\frac{1}{2} \text{ yd} \cdot 10 \text{ yd}$$
$$= 2 \cdot \frac{13}{2} \cdot \frac{11}{2} \text{ yd}^2 + 2 \cdot \frac{13}{2} \cdot 10 \text{ yd}^2 +$$
$$2 \cdot \frac{11}{2} \cdot 10 \text{ yd}^2$$
$$= \frac{143}{2} \text{ yd}^2 + 130 \text{ yd}^2 + 110 \text{ yd}^2$$
$$= 311\frac{1}{2} \text{ yd}^2$$

9. $V = Bh = \pi \cdot r^2 \cdot h$
$\approx 3.14 \times 8 \text{ in.} \times 8 \text{ in.} \times 4 \text{ in.}$
$= 803.84 \text{ in}^3$

11. $V = Bh = \pi \cdot r^2 \cdot h$
$\approx 3.14 \times 5 \text{ cm} \times 5 \text{ cm} \times 4.5 \text{ cm}$
$= 353.25 \text{ cm}^3$

13. $V = Bh = \pi \cdot r^2 \cdot h$
$\approx \frac{22}{7} \times 210 \text{ yd} \times 210 \text{ yd} \times 300 \text{ yd}$
$= 41,580,000 \text{ yd}^3$

15. $V = \frac{4}{3} \cdot \pi \cdot r^3$
$\approx \frac{4}{3} \times 3.14 \times (100 \text{ in.})^3$
$= \frac{4 \times 3.14 \times 1,000,000 \text{ in}^3}{3}$
$= 4,186,666\frac{2}{3} \text{ in}^3$

17. $V = \frac{4}{3} \cdot \pi \cdot r^3$
$\approx \frac{4}{3} \times 3.14 \times (3.1 \text{ m})^3$
$= \frac{4 \times 3.14 \times 29.791 \text{ m}^3}{3}$
$\approx 124.72 \text{ m}^3$

19. $V = \frac{4}{3} \cdot \pi \cdot r^3$
$\approx \frac{4}{3} \times \frac{22}{7} \times \left(7\frac{3}{4} \text{ ft}\right)^3$
$= \frac{4}{3} \times \frac{22}{7} \times \left(\frac{31}{4} \text{ ft}\right)^3$
$= \frac{4 \times 22 \times 29,791 \text{ ft}^3}{3 \times 7 \times 64}$
$\approx 1950\frac{101}{168} \text{ ft}^3$

21. $V = \frac{1}{3} \cdot \pi \cdot r^2 \cdot h$
$\approx \frac{1}{3} \times 3.14 \times 33 \text{ ft} \times 33 \text{ ft} \times 100 \text{ ft}$
$\approx 113,982 \text{ ft}^3$

23. $V = \frac{1}{3} \cdot \pi \cdot r^2 \cdot h$
$\approx \frac{1}{3} \times \frac{22}{7} \times 1.4 \text{ cm} \times 1.4 \text{ cm} \times 12 \text{ cm}$
$\approx 24.64 \text{ cm}^3$

25. We must find the radius of the base in order to use the formula for the volume of a circular cylinder.
$$r = \frac{d}{2} = \frac{12 \text{ cm}}{2} = 6 \text{ cm}$$
$$V = Bh = \pi \cdot r^2 \cdot h$$
$$\approx 3.14 \times 6 \text{ cm} \times 6 \text{ cm} \times 42 \text{ cm}$$
$$\approx 4747.68 \text{ cm}^3$$

27. We must find the radius of the silo in order to use the formula for the volume of a circular cylinder.
$$r = \frac{d}{2} = \frac{6 \text{ m}}{2} = 3 \text{ m}$$
$$V = Bh = \pi \cdot r^2 \cdot h$$
$$\approx 3.14 \times 3 \text{ m} \times 3 \text{ m} \times 13 \text{ m}$$
$$= 367.38 \text{ m}^3$$

29. First we find the radius of the ball:
$$r = \frac{d}{2} = \frac{6.5 \text{ cm}}{2} = 3.25 \text{ cm}$$
Then we find the volume, using the formula for the volume of a sphere.
$$V = \frac{4}{3} \cdot \pi \cdot r^3$$
$$\approx \frac{4}{3} \cdot 3.14 \cdot (3.25 \text{ cm})^3$$
$$\approx 143.72 \text{ cm}^3$$

31. First we find the radius of the earth:
$$\frac{3980 \text{ mi}}{2} = 1990 \text{ mi}$$
Then we find the volume, using the formula for the volume of a sphere.
$$V = \frac{4}{3} \cdot \pi \cdot r^3$$
$$\approx \frac{4}{3} \cdot 3.14 \cdot (1990 \text{ mi})^3$$
$$\approx 32,993,441,150 \text{ mi}^3$$

33. First we find the radius of the can.
$$r = \frac{d}{2} = \frac{6.5 \text{ cm}}{2} = 3.25 \text{ cm}$$
The height of the can is the length of the diameters of 3 tennis balls.
$$h = 3(6.5 \text{ cm}) = 19.5 \text{ cm}$$
Now we find the volume.
$$V = Bh = \pi \cdot r^2 \cdot h$$
$$\approx 3.14 \times 3.25 \text{ cm} \times 3.25 \text{ cm} \times 19.5 \text{ cm}$$
$$\approx 646.74 \text{ cm}^3$$

35. $V = Bh = \pi \cdot r^2 \cdot h$

$\approx \dfrac{22}{7} \cdot 14 \text{ cm} \cdot 14 \text{ cm} \cdot 100 \text{ cm}$

$= 61,600 \text{ cm}^3$

37. A cube is a rectangular solid.

$V = l \cdot w \cdot h$

$= 18 \text{ yd} \cdot 18 \text{ yd} \cdot 18 \text{ yd}$

$= 5832 \text{ yd}^3$

39. Discussion and Writing Exercise

41. $35\% = \dfrac{35}{100} = \dfrac{5 \cdot 7}{5 \cdot 20} = \dfrac{5}{5} \cdot \dfrac{7}{20} = \dfrac{7}{20}$

43. $37\dfrac{1}{2}\% = \dfrac{75}{2}\% = \dfrac{75}{2} \times \dfrac{1}{100} = \dfrac{75}{2 \cdot 100} =$

$\dfrac{25 \cdot 3}{2 \cdot 25 \cdot 4} = \dfrac{25}{25} \cdot \dfrac{3}{2 \cdot 4} = \dfrac{3}{8}$

45. $83.\overline{3}\% = 83\dfrac{1}{3}\% = \dfrac{250}{3}\% = \dfrac{250}{3} \times \dfrac{1}{100} =$

$\dfrac{250 \cdot 1}{3 \cdot 100} = \dfrac{5 \cdot 50 \cdot 1}{3 \cdot 2 \cdot 50} = \dfrac{50}{50} \cdot \dfrac{5 \cdot 1}{3 \cdot 2} = \dfrac{5}{6}$

47. **Familiarize.** Let s = the number of sheets in 15 reams of paper. Repeated addition works well here.

$$\underbrace{\boxed{500} + \boxed{500} + \cdots + \boxed{500}}_{\text{15 addends}}$$

Translate.

Sheets in one ream	times	Number of reams	is	Total number of sheets
↓	↓	↓	↓	↓
500	×	15	=	s

Solve. We multiply.

$500 \times 15 = 7500$, so $7500 = s$, or $s = 7500$.

Check. We can repeat the calculation. The answer checks.

State. There are 7500 sheets in 15 reams of paper.

49. First find the volume of one one-dollar bill in cubic inches:

$V = l \cdot w \cdot h$

$V = 6.0625 \text{ in.} \times 2.3125 \text{ in.} \times 0.0041 \text{ in.}$

$V = 0.05748 \text{ in}^3 \quad \text{Rounding}$

Then multiply to find the volume of one million one-dollar bills in cubic inches:

$$1,000,000 \times 0.05748 \text{ in}^3 = 57,480 \text{ in}^3$$

Thus the volume of one million one-dollar bills is about $57,480 \text{ in}^3$.

51. Radius of water stream: $\dfrac{2 \text{ cm}}{2} = 1 \text{ cm}$

To convert 30 m to centimeters, think: 1 meter is 100 times as large as 1 centimeter. Thus, we move the decimal point 2 places to the right:

$$30 \text{ m} = 3000 \text{ cm}$$

$V = Bh = \pi \cdot r^2 \cdot h$

$\approx 3.141593 \cdot 1 \text{ cm} \cdot 1 \text{ cm} \cdot 3000 \text{ cm}$

$\approx 9425 \text{ cm}^3$

Now we convert 9425 cm^3 to liters:

$9425 \text{ cm}^3 = 9425 \text{ cm}^3 \cdot \dfrac{1 \text{ L}}{1000 \text{ cm}^3}$

$= 9.425 \text{ L}$

There is about 9.425 L of water in the hose.

53. Find the diameter of the earth at the equator:

$C = \pi \cdot d$

$24,901.55 \text{ mi} \approx 3.14 \cdot d$

$7930 \text{ mi} \approx d$

Find the diameter of the earth through the north and south poles:

$C = \pi \cdot d$

$24,859.82 \text{ mi} \approx 3.14 \cdot d$

$7917 \text{ mi} \approx d$

Find the average of these two diameters:

$$\dfrac{7930 \text{ mi} + 7917 \text{ mi}}{2} = 7923.5 \text{ mi}$$

Use this average to estimate the volume of the earth:

$r = \dfrac{d}{2} = \dfrac{7923.5 \text{ mi}}{2} = 3961.75 \text{ mi}$

$V = \dfrac{4}{3} \cdot \pi \cdot r^3$

$\approx \dfrac{4 \times 3.14 \times (3961.75 \text{ mi})^3}{3}$

$\approx 260,000,000,000 \text{ mi}^3$

55. The length of a diagonal of the cube is the length of the diameter of the sphere, 1 m. Visualize a triangle whose hypotenuse is a diagonal of the cube and with one leg a side s of the cube and the other leg a diagonal c of a side of the cube.

We want to find the length of a side s of the cube in order to find the volume. We begin by using the Pythagorean theorem to find c.

$s^2 + s^2 = c^2$

$2s^2 = c^2$

$\sqrt{2s^2} = c$

Now use the Pythagorean theorem again to find s.

$c^2 + s^2 = 1^2$

$(\sqrt{2s^2})^2 + s^2 = 1 \quad \text{Subtracting } \sqrt{2s^2} \text{ for } c$

$2s^2 + s^2 = 1$

$3s^2 = 1$

$s^2 = \dfrac{1}{3}$

$s = \sqrt{\dfrac{1}{3}} \approx 0.577$

Next we find the volume of the cube.
$$V = l \cdot w \cdot h$$
$$= 0.577 \text{ m} \cdot 0.577 \text{ m} \cdot 0.577 \text{ m}$$
$$= 0.192 \text{ m}^3$$

Find the volume of the sphere. The radius is $\frac{1 \text{ m}}{2}$, or 0.5 m.

$$V = \frac{4}{3} \cdot \pi \cdot r^3$$
$$\approx \frac{4}{3} \times 3.14 \times (0.5 \text{ m})^3$$
$$\approx \frac{4 \times 3.14 \times 0.125 \text{ m}^3}{3}$$
$$\approx 0.523 \text{ m}^3$$

Finally we subtract to find how much more volume is in the sphere.

$$0.523 \text{ m}^3 - 0.192 \text{ m}^3 = 0.331 \text{ m}^3$$

There is 0.331 m³ more volume in the sphere.

Exercise Set 8.6

1. Two angles are complementary if the sum of their measures is 90°.

$$90° - 11° = 79°.$$

The measure of a complement is 79°.

3. Two angles are complementary if the sum of their measures is 90°.

$$90° - 67° = 23°.$$

The measure of a complement is 23°.

5. Two angles are complementary if the sum of their measures is 90°.

$$90° - 58° = 32°.$$

The measure of a complement is 32°.

7. Two angles are complementary if the sum of their measures is 90°.

$$90° - 29° = 61°.$$

The measure of a complement is 61°.

9. Two angles are supplementary if the sum of their measures is 180°.

$$180° - 3° = 177°.$$

The measure of a supplement is 177°.

11. Two angles are supplementary if the sum of their measures is 180°.

$$180° - 139° = 41°.$$

The measure of a supplement is 41°.

13. Two angles are supplementary if the sum of their measures is 180°.

$$180° - 85° = 95°.$$

The measure of a supplement is 95°.

15. Two angles are supplementary if the sum of their measures is 180°.

$$180° - 102° = 78°.$$

The measure of a supplement is 78°.

17. The segments have different lengths. They are not congruent.

19. $m\angle G = m\angle R$, so $\angle G \cong \angle R$.

21. Since $\angle 2$ and $\angle 5$ are vertical angles, $m\angle 2 = 67°$. Likewise, $\angle 1$ and $\angle 4$ are vertical angles, so $m\angle 4 = 80°$.

$$m\angle 1 + m\angle 2 + m\angle 3 = 180°$$
$$80° + 67° + m\angle 3 = 180° \qquad \text{Substituting}$$
$$147° + m\angle 3 = 180°$$
$$m\angle 3 = 180° - 147°$$
$$m\angle 3 = 33°$$

Since $\angle 3$ and $\angle 6$ are vertical angles, $m\angle 6 = 33°$.

23. a) The pairs of corresponding angles are

$\angle 1$ and $\angle 3$,

$\angle 2$ and $\angle 4$,

$\angle 8$ and $\angle 6$,

$\angle 7$ and $\angle 5$.

b) The interior angles are $\angle 2$, $\angle 3$, $\angle 6$, and $\angle 7$.

c) The pairs of alternate interior angles are

$\angle 2$ and $\angle 6$,

$\angle 3$ and $\angle 7$.

25. $\angle 4$ and $\angle 6$ are vertical angles, so $m\angle 6 = 125°$.

$\angle 4$ and $\angle 2$ are corresponding angles. By Property 1, $m\angle 2 = 125°$.

$\angle 6$ and $\angle 8$ are corresponding angles. By Property 1, $m\angle 8 = 125°$.

$\angle 2$ and $\angle 3$ are interior angles on the same side of the transversal. Using Property 4 and $m\angle 2 = 125°$, $m\angle 3 = 55°$.

$\angle 6$ and $\angle 7$ are interior angles on the same side of the transversal. Using Property 4 and $m\angle 6 = 125°$, $m\angle 7 = 55°$.

$\angle 3$ and $\angle 5$ are vertical angles, so $m\angle 5 = 55°$.

$\angle 7$ and $\angle 1$ are vertical angles, so $m\angle 1 = 55°$.

27. Considering the transversal $\overleftrightarrow{BC}$, $\angle ABE$ and $\angle DCE$ are alternate interior angles. By Property 2, $\angle ABE \cong \angle DCE$. Then $m\angle ABE = m\angle DCE = 95°$.

Considering the transversal $\overleftrightarrow{AD}$, $\angle BAE$ and $\angle CDE$ are alternate interior angles. By Property 2, $\angle BAE \cong \angle CDE$. We cannot determine the measure of these angles.

$\angle AEB$ and $\angle DEC$ are vertical angles, so $\angle AEB \cong \angle DEC$. We cannot determine the measure of these angles.

$\angle BED$ and $\angle AEC$ are also vertical angles, so $\angle BED \cong \angle AEC$. We cannot determine their measures.

29. Considering the transversal $\overleftrightarrow{CE}$, $\angle AEC$ and $\angle DCE$ are alternate interior angles. By Property 2, $\angle AEC \cong \angle DCE$. Then $m\angle AEC = m\angle DCE = 50°$.

Considering the transversal $\overleftrightarrow{DE}$, $\angle BED$ and $\angle EDC$ are alternate interior angles. By Property 2, $\angle BED \cong \angle EDC$. Then $m\angle BED = m\angle EDC = 41°$.

31.
$$6 \times 1\frac{7}{8} = 6 \times \frac{15}{8}$$
$$= \frac{6 \times 15}{8}$$
$$= \frac{2 \times 3 \times 15}{2 \times 4}$$
$$= \frac{2}{2} \times \frac{3 \times 15}{4}$$
$$= \frac{45}{4}, \text{ or } 11\frac{1}{4}$$

33.
$$8\frac{3}{7} \times 14 = \frac{59}{7} \times 14$$
$$= \frac{59 \times 14}{7}$$
$$= \frac{59 \times 2 \times 7}{7 \times 1}$$
$$= \frac{7}{7} \times \frac{59 \times 2}{1}$$
$$= 118$$

Exercise Set 8.7

1. The notation tells us the way in which the vertices of the two triangles are matched.

$$\triangle ABC \cong \triangle RST$$

$\triangle ABC \cong \triangle RST$ means

$\angle A \cong \angle R$ and $\overline{AB} \cong \overline{RS}$
$\angle B \cong \angle S$ $\overline{AC} \cong \overline{RT}$
$\angle C \cong \angle T$ $\overline{BC} \cong \overline{ST}$

3. The notation tells us the way in which the vertices of the two triangles are matched.

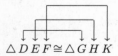

$$\triangle DEF \cong \triangle GHK$$

$\triangle DEF \cong \triangle GHK$ means

$\angle D \cong \angle G$ and $\overline{DE} \cong \overline{GH}$
$\angle E \cong \angle H$ $\overline{DF} \cong \overline{GK}$
$\angle F \cong \angle K$ $\overline{EF} \cong \overline{HK}$

5. The notation tells us the way in which the vertices of the two triangles are matched.

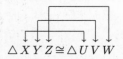

$$\triangle XYZ \cong \triangle UVW$$

$\triangle XYZ \cong \triangle UVW$ means

$\angle X \cong \angle U$ and $\overline{XY} \cong \overline{UV}$
$\angle Y \cong \angle V$ $\overline{XZ} \cong \overline{UW}$
$\angle Z \cong \angle W$ $\overline{YZ} \cong \overline{VW}$

7. The notation tells us the way in which the vertices of the two triangles are matched.

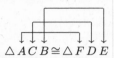

$$\triangle ACB \cong \triangle FDE$$

$\triangle ACB \cong \triangle FDE$ means

$\angle A \cong \angle F$ and $\overline{AC} \cong \overline{FD}$
$\angle C \cong \angle D$ $\overline{AB} \cong \overline{FE}$
$\angle B \cong \angle E$ $\overline{CB} \cong \overline{DE}$

9. The notation tells us the way in which the vertices of the two triangles are matched.

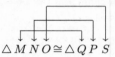

$$\triangle MNO \cong \triangle QPS$$

$\triangle MNO \cong \triangle QPS$ means

$\angle M \cong \angle Q$ and $\overline{MN} \cong \overline{QP}$
$\angle N \cong \angle P$ $\overline{MO} \cong \overline{QS}$
$\angle O \cong \angle S$ $\overline{NO} \cong \overline{PS}$

11. We cannot determine from the information given that two sides of one triangle and the included angle are congruent to two sides and the included angle of the other triangle. Therefore, we cannot use the SAS Property.

13. Two sides of one triangle and the included angle are congruent to two sides and the included angle of the other triangle. They are congruent by the SAS Property.

15. Two sides of one triangle and the included angle are congruent to two sides and the included angle of the other triangle. They are congruent by the SAS Property.

17. We cannot determine from the information given that three sides of one triangle are congruent to three sides of the other triangle. Therefore, we cannot use the SSS Property.

19. Three sides of one triangle are congruent to three sides of the other triangle. They are congruent by the SSS Property.

21. Three sides of one triangle are congruent to three sides of the other triangle. They are congruent by the SSS Property.

23. Two angles and the included side are of one triangle are congruent to two angles and the included side of the other triangle. They are congruent by the ASA Property.

25. Two angles and the included side are of one triangle are congruent to two angles and the included side of the other triangle. They are congruent by the ASA Property.

27. The vertical angles are congruent so two angles and the included side are of one triangle are congruent to two angles and the included side of the other triangle. They are congruent by the ASA Property.

29. Two angles and the included side are of one triangle are congruent to two angles and the included side of the other triangle. They are congruent by the ASA Property.

31. Two sides of one triangle and the included angle are congruent to two sides and the included angle of the other triangle. They are congruent by the SAS Property.

33. Three sides of one triangle are congruent to three sides of the other triangle. In addition, two sides of one triangle and the included angle are congruent to two sides and the included angle of the other triangle. Therefore, we can use either the SSS Property or the SAS Property to show that they are congruent.

35. Since R is the midpoint of $\overline{PT}$, $\overline{PR} \cong \overline{TR}$.

Since R is the midpoint of $\overline{QS}$, $\overline{RQ} \cong \overline{RS}$.

$\angle PRQ$ and $\angle TRS$ are vertical angles, so $\angle PRQ \cong \angle TRS$.

Two sides and the included angle of $\triangle PRQ$ are congruent to two sides and the included angle of $\triangle TRS$, so $\triangle PRQ \cong \triangle TRS$ by the SAS Property.

37. Since $GL \perp KM$, $m\angle GLK = m\angle GLM = 90°$. Then $\angle GLK \cong \angle GLM$.

Since L is the midpoint of $\overline{KM}$, $\overline{KL} \cong \overline{LM}$.

$\overline{GL} \cong \overline{GL}$.

Two sides and the included angle of $\triangle KLG$ are congruent to two sides and the included angle of $\triangle MLG$, so $\triangle KLG \cong \triangle MLG$ by the SAS Property.

39. The information given tells us that $\overline{AE} \cong \overline{CB}$ and $\overline{AB} \cong \overline{CD}$.

Since B is the midpoint of $\overline{ED}$, $\overline{EB} \cong \overline{BD}$.

Three sides of $\triangle AEB$ are congruent to three sides of $\triangle CDB$, so $\triangle AEB \cong \triangle CDB$ by the SSS Property.

41. The information given tells us that $\overline{HK} \cong \overline{KJ}$ and $\overline{GK} \cong \overline{LK}$.

Since $\overline{GK} \perp \overline{LJ}$, $m\angle HKL = m\angle GKJ = 90°$.

Then $\angle HKL \cong \angle GKJ$.

Two sides and the included angle of $\triangle LKH$ are congruent to two sides and the included angle of $\triangle GKJ$, so $\triangle LKH \cong \triangle GKJ$ by the SAS Property. This means that the remaining corresponding parts of the two triangles are congruent. That is, $\angle HLK \cong \angle JGK$, $\angle LHK \cong \angle GJK$, and $\overline{LH} \cong \overline{GJ}$.

43. Two angles and the included side of $\triangle PED$ are congruent to two angles and the included side of $\triangle PFG$, so $\triangle PED \cong \triangle PFG$ by the ASA Property. Then corresponding parts of the two triangles are congruent, so $\overline{EP} \cong \overline{FP}$. Therefore, P is the midpoint of $\overline{EF}$.

45. $\angle A$ and $\angle C$ are opposite angles, so $m\angle A = 70°$ by Property 2.

$\angle C$ and $\angle B$ are consecutive angles, so the are supplementary by Property 4. Then

$$m\angle B = 180° - m\angle C$$

$$m\angle B = 180° - 70°$$

$$m\angle B = 110°.$$

$\angle B$ and $\angle D$ are opposite angles, so $m\angle D = 110°$ by Property 2.

47. $\angle M$ and $\angle K$ are opposite angles, so $m\angle M = 71°$ by Property 2.

$\angle K$ and $\angle L$ are consecutive angles, so the are supplementary by Property 4. Then

$$m\angle L = 180° - m\angle K$$

$$m\angle L = 180° - 71°$$

$$m\angle L = 109°.$$

$\angle J$ and $\angle L$ are opposite angles, so $m\angle J = 109°$ by Property 2.

49. $\overline{ON}$ and $\overline{TU}$ are opposite sides of the parallelogram. So are $\overline{OT}$ and $\overline{NU}$. The opposite sides of a parallelogram are congruent (Property 3), so $TU = 9$ and $NU = 15$.

51. $\overline{JM}$ and $\overline{KL}$ are opposite sides of the parallelogram. So are $\overline{JK}$ and $\overline{ML}$. The opposite sides of a parallelogram are congruent (Property 3). Then $KL = 3\frac{1}{2}$ and $JK + LM = 22 - 3\frac{1}{2} - 3\frac{1}{2} = 15$. Thus, $JK = LM = \frac{1}{2} \cdot 15 = 7\frac{1}{2}$.

53. The diagonals of a parallelogram bisect each other (Property 5). Then

$$AC = 2 \cdot AB = 2 \cdot 14 = 28,$$

$$ED = 2 \cdot BD = 2 \cdot 19 = 38.$$

55. 0.452   0.45.2   Move the decimal point 2 places to the right.

Write a % symbol: 45.2%

$0.452 = 45.2\%$

57. We multiply by 1 to get 100 in the denominator.

$$\frac{11}{20} = \frac{11}{20} \cdot \frac{5}{5} = \frac{55}{100} = 55\%$$

59. The ratio of the amount spent in Florida to the total amount spent is $\dfrac{2.7}{13.1}$. This can also be expressed as follows:

$$\frac{2.7}{13.1} = \frac{2.7}{13.1} \cdot \frac{10}{10} = \frac{27}{131}$$

The ratio of the total amount spent to the amount spent in Florida is $\dfrac{13.1}{2.7}$, or $\dfrac{131}{27}$.

61.

$$\begin{array}{r} 1.7\,5 \\ 12\,\overline{\smash{)}2\,1.0\,0} \\ \underline{1\,2\,0\,0} \\ 9\,0\,0 \\ \underline{8\,4\,0} \\ 6\,0 \\ \underline{6\,0} \\ 0 \end{array}$$

The answer is 1.75.

63. To divide by 100, move the decimal point 2 places to the left.

$$23.4 \qquad .23.4$$

$$23.4 \div 100 = 0.234$$

65.

$$\begin{array}{r} 3.\,1\,4 \quad \text{(2 decimal places)} \\ \times\,4.\,4\,1 \quad \text{(2 decimal places)} \\ \hline 3\,1\,4 \\ 1\,2\,5\,6\,0 \\ 1\,2\,5\,6\,0\,0 \\ \hline 1\,3.\,8\,4\,7\,4 \quad \text{(4 decimal places)} \end{array}$$

Round

13. 8 4 $\boxed{7}$ 4   to the nearest hundredth.

Thousandths digit is 5 or higher.

13. 8 5   Round up.

Exercise Set 8.8

1. Vertex R is matched with vertex A, vertex S is matched with vertex B, and vertex T is matched with vertex C. Then

$$\begin{array}{lll} \overline{RS} \longleftrightarrow \overline{AB} & \text{and} & \angle R \longleftrightarrow \angle A \\ \overline{ST} \longleftrightarrow \overline{BC} & & \angle S \longleftrightarrow \angle B \\ \overline{TR} \longleftrightarrow \overline{CA} & & \angle T \longleftrightarrow \angle C \end{array}$$

3. Vertex C is matched with vertex W, vertex B is matched with vertex J, and vertex S is matched with vertex Z. Then

$$\begin{array}{lll} \overline{CB} \longleftrightarrow \overline{WJ} & \text{and} & \angle C \longleftrightarrow \angle W \\ \overline{BS} \longleftrightarrow \overline{JZ} & & \angle B \longleftrightarrow \angle J \\ \overline{SC} \longleftrightarrow \overline{ZW} & & \angle S \longleftrightarrow \angle Z \end{array}$$

5. The notation tells us the way in which the vertices are matched.

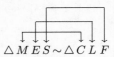

$$\triangle A\,B\,C \sim \triangle R\,S\,T$$

$\triangle ABC \sim \triangle RST$ means

$$\begin{array}{l} \angle A \cong \angle R \\ \angle B \cong \angle S \quad \text{and} \quad \dfrac{AB}{RS} = \dfrac{AC}{RT} = \dfrac{BC}{ST}. \\ \angle C \cong \angle T \end{array}$$

7. The notation tells us the way in which the vertices are matched.

$$\triangle M\,E\,S \sim \triangle C\,L\,F$$

$\triangle MES \sim \triangle CLF$ means

$$\begin{array}{l} \angle M \cong \angle C \\ \angle E \cong \angle L \quad \text{and} \quad \dfrac{ME}{CL} = \dfrac{MS}{CF} = \dfrac{ES}{LF}. \\ \angle S \cong \angle F \end{array}$$

9. If we match P with N, S with D, and Q with M, the corresponding angles will be congruent. That is, $\triangle PSQ \sim \triangle NDM$. Then

$$\frac{PS}{ND} = \frac{PQ}{NM} = \frac{SQ}{DM}.$$

11. If we match T with G, A with F, and W with C, the corresponding angles will be congruent. That is, $\triangle TAW \sim \triangle GFC$. Then

$$\frac{TA}{GF} = \frac{TW}{GC} = \frac{AW}{FC}.$$

13. Since $\triangle ABC \sim \triangle PQR$, the corresponding sides are proportional. Then

$$\begin{array}{ll} \dfrac{3}{6} = \dfrac{4}{PR} \quad \text{and} & \dfrac{3}{6} = \dfrac{5}{QR} \\ 3(PR) = 6 \cdot 4 & 3(QR) = 6 \cdot 5 \\ 3(PR) = 24 & 3(QR) = 30 \\ PR = 8 & QR = 10 \end{array}$$

15. Recall that if a transversal intersects two parallel lines, then the alternate interior angles are congruent. Thus,

$$\angle A \cong \angle B \text{ and } \angle D \cong \angle C.$$

Since $\angle AED$ and $\angle CEB$ are vertical angles, they are congruent. Thus,

$$\angle AED \cong \angle CEB.$$

Then $\triangle AED \sim \triangle CEB$, and the lengths of the corresponding sides are proportional.

$$\begin{array}{l} \dfrac{AD}{CB} = \dfrac{ED}{EC} \\ \dfrac{7}{21} = \dfrac{6}{EC} \\ 7 \cdot EC = 126 \\ EC = 18 \end{array}$$

17. If we use the sun's rays to represent the third side of a triangle in a drawing of the situation, we see that we have similar triangles. We let $h =$ the height of the tree.

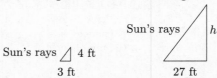

The ratio of h to 4 is the same as the ratio of 27 to 3. We have the proportion

$$\frac{h}{4} = \frac{27}{3}.$$

Solve: $3 \cdot h = 4 \cdot 27$

$$h = \frac{4 \cdot 27}{3}$$

$$h = 36$$

The tree is 36 ft tall.

19. Since the ratio of d to 25 ft is the same as the ratio of 40 ft to 10 ft, we have the proportion

$$\frac{d}{25} = \frac{40}{10}.$$

Solve: $10 \cdot d = 25 \cdot 40$

$$d = \frac{25 \cdot 40}{10}$$

$$d = 100$$

The distance across the river is 100 ft.

21. Discussion and Writing Exercise

23. $2\frac{4}{5} \times 10\frac{1}{2} = \frac{14}{5} \times \frac{21}{2} = \frac{14 \times 21}{5 \times 2} =$

$\frac{2 \times 7 \times 21}{5 \times 2} = \frac{2 \times 7 \times 21}{5 \times 2} = \frac{147}{5} = 29\frac{2}{5}$

25. $8 \times 9\frac{3}{4} = \frac{8}{1} \times \frac{39}{4} = \frac{8 \times 39}{1 \times 4} = \frac{2 \times 4 \times 39}{1 \times 4} =$

$\frac{2 \times 4 \times 39}{1 \times 4} = \frac{78}{1} = 78$

Chapter 8 Review Exercises

1. Place the $\triangle$ of the protractor at the vertex of the angle, and line up one of the sides at 0°. We choose the nearly horizontal side. Since 0° is on the inside scale, we check where the other side of the angle crosses the inside scale. It crosses at 54°. Thus, the measure of the angle is 54°.

2. Place the $\triangle$ of the protractor at the vertex of the angle, point B. Line up one of the sides at 0°. We choose the side that contains point P. Since 0° is on the outside scale, we check where the other side crosses the outside scale. It crosses at 180°. Thus, the measure of the angle is 180°.

3. Place the $\triangle$ of the protractor at the vertex of the angle, and line up one of the sides at 0°. We choose the horizontal side. Since 0° is on the inside scale, we check where the other side crosses the inside scale. It crosses at 140°. Thus, the measure of the angle is 140°.

4. Place the $\triangle$ of the protractor at the vertex of the angle, and line up one of the sides at 0°. We choose the horizontal side. Since 0° is on the inside scale, we check where the other side crosses the inside scale. It crosses at 90°. Thus, the measure of the angle is 90°.

5. The measure of the angle in Exercise 1 is 54°. Since its measure is greater than 0° and less than 90°, it is an acute angle.

6. The measure of the angle in Exercise 2 is 180°. It is a straight angle.

7. The measure of the angle in Exercise 3 is 140°. Since its measure is greater than 90° and less than 180°, it is an obtuse angle.

8. The measure of the angle in Exercise 4 is 90°. It is a right angle.

9. $30° + 90° + x = 180°$

$120° + x = 180°$

$x = 180° - 120°$

$x = 60°$

10. All the sides are of different lengths. The triangle is a scalene triangle.

11. One angle is a right angle. The triangle is a right triangle.

12. $(n - 2) \cdot 180°$

$= (6 - 2) \cdot 180°$ Substituting 6 for n

$= 4 \cdot 180°$

$= 720°$

13. Perimeter $= 5\,\text{m} + 7\,\text{m} + 4\,\text{m} + 4\,\text{m} + 3\,\text{m}$

$= (5 + 7 + 4 + 4 + 3)\,\text{m}$

$= 23\,\text{m}$

14. Perimeter $= 0.5\,\text{m} + 1.9\,\text{m} + 1.2\,\text{m} + 0.8\,\text{m}$

$= (0.5 + 1.9 + 1.2 + 0.8)\,\text{m}$

$= 4.4\,\text{m}$

15. $P = 2 \cdot l + 2 \cdot w$

$P = 2 \cdot 78\,\text{ft} + 2 \cdot 36\,\text{ft}$

$P = 156\,\text{ft} + 72\,\text{ft}$

$P = 228\,\text{ft}$

$A = l \cdot w$

$A = 78\,\text{ft} \cdot 36\,\text{ft}$

$A = 2808\,\text{ft}^2$

16. $P = 4 \cdot s$

$P = 4 \cdot 9\,\text{ft}$

$P = 36\,\text{ft}$

$A = s \cdot s$

$A = 9\,\text{ft} \cdot 9\,\text{ft}$

$A = 81\,\text{ft}^2$

17. $P = 2 \cdot (l + w)$

$P = 2 \cdot (7\,\text{cm} + 1.8\,\text{cm})$

$P = 2 \cdot (8.8\,\text{cm})$

$P = 17.6\,\text{cm}$

$A = l \cdot w$

$A = 7 \text{ cm} \cdot 1.8 \text{ cm}$

$A = 12.6 \text{ cm}^2$

18. $A = b \cdot h$

$A = 12 \text{ cm} \cdot 5 \text{ cm}$

$A = 60 \text{ cm}^2$

19. $A = \dfrac{1}{2} \cdot h \cdot (a + b)$

$A = \dfrac{1}{2} \cdot 5 \text{ mm} \cdot (4 + 10) \text{ mm}$

$A = \dfrac{5 \cdot 14}{2} \text{ mm}^2$

$A = 35 \text{ mm}^2$

20. $A = \dfrac{1}{2} \cdot b \cdot h$

$A = \dfrac{1}{2} \cdot 15 \text{ m} \cdot 3 \text{ m}$

$A = \dfrac{15 \cdot 3}{2} \text{ m}^2$

$A = 22.5 \text{ m}^2$

21. $A = \dfrac{1}{2} \cdot b \cdot h$

$A = \dfrac{1}{2} \cdot 11.4 \text{ cm} \cdot 5.2 \text{ cm}$

$A = \dfrac{11.4 \cdot 5.2}{2} \text{ cm}^2$

$A = 29.64 \text{ cm}^2$

22. $A = \dfrac{1}{2} \cdot h \cdot (a + b)$

$A = \dfrac{1}{2} \cdot 8 \text{ m} \cdot (5 + 17) \text{ m}$

$A = \dfrac{8 \cdot 22}{2} \text{ m}^2$

$A = 88 \text{ m}^2$

23. $A = b \cdot h$

$A = 21\dfrac{5}{6} \text{ in.} \cdot 6\dfrac{2}{3} \text{ in.}$

$A = \dfrac{131}{6} \cdot \dfrac{20}{3} \text{ in}^2$

$A = \dfrac{131 \cdot 20}{6 \cdot 3} \text{ in}^2$

$A = \dfrac{1310}{9} \text{ in}^2$

$A = 145\dfrac{5}{9} \text{ in}^2$

24. *Familiarize.* The seeded area is the total area of the house and the seeded area less the area of the house. From the drawing in the text we see that the total area is the area of a rectangle with length 70 ft and width 25 ft + 7 ft, or 32 ft. The length of the rectangular house is 70 ft − 7 ft − 7 ft, or 56 ft, and its width is 25 ft. We let $A =$ the seeded area.

Translate.

$$\underbrace{\text{Seeded area}}_{\downarrow} \ \ \text{is} \ \ \underbrace{\text{Total area}}_{\downarrow} \ \ \text{minus} \ \ \underbrace{\text{Area of house}}_{\downarrow}$$

$$A \qquad = \ 70 \text{ ft} \cdot 32 \text{ ft} \ - \ \quad 56 \text{ ft} \cdot 25 \text{ ft}$$

Solve.

$A = 70 \text{ ft} \cdot 32 \text{ ft} - 56 \text{ ft} \cdot 25 \text{ ft}$

$A = 2240 \text{ ft}^2 - 1400 \text{ ft}^2$

$A = 840 \text{ ft}^2$

Check. We can repeat the calculations. The answer checks.

State. The seeded area is 840 ft^2.

25. $r = \dfrac{d}{2} = \dfrac{16 \text{ m}}{2} = 8 \text{ m}$

26. $r = \dfrac{d}{2} = \dfrac{\frac{28}{11} \text{ in.}}{2} = \dfrac{28}{11} \text{ in.} \cdot \dfrac{1}{2}$

$= \dfrac{28}{11 \cdot 2} \text{ in.} = \dfrac{2 \cdot 14}{11 \cdot 2} \text{ in.}$

$= \dfrac{14}{11} \text{ in., or } 1\dfrac{3}{11} \text{ in.}$

27. $d = 2 \cdot r = 2 \cdot 7 \text{ ft} = 14 \text{ ft}$

28. $d = 2 \cdot r = 2 \cdot 10 \text{ cm} = 20 \text{ cm}$

29. $C = \pi \cdot d$

$C \approx 3.14 \cdot 16 \text{ m}$

$= 50.24 \text{ m}$

30. $C = \pi \cdot d$

$C \approx \dfrac{22}{7} \cdot \dfrac{28}{11} \text{ in.}$

$= \dfrac{22 \cdot 28}{7 \cdot 11} \text{ in.} = \dfrac{2 \cdot 11 \cdot 4 \cdot 7}{7 \cdot 11 \cdot 1} \text{ in.}$

$= 8 \text{ in.}$

31. In Exercise 25 we found that the radius of the circle is 8 m.

$A = \pi \cdot r \cdot r$

$A \approx 3.14 \cdot 8 \text{ m} \cdot 8 \text{ m}$

$A = 200.96 \text{ m}^2$

32. In Exercise 26 we found that the radius of the circle is $\dfrac{14}{11}$ in.

$A = \pi \cdot r \cdot r$

$A \approx \dfrac{22}{7} \cdot \dfrac{14}{11} \text{ in.} \cdot \dfrac{14}{11} \text{ in.}$

$A = \dfrac{22 \cdot 14 \cdot 14}{7 \cdot 11 \cdot 11} \text{ in}^2 = \dfrac{2 \cdot 11 \cdot 2 \cdot 7 \cdot 14}{7 \cdot 11 \cdot 11}$

$A = \dfrac{56}{11} \text{ in}^2, \text{ or } 5\dfrac{1}{11} \text{ in}^2$

33. The shaded area is the area of a circle with radius of 21 ft less the area of a circle with a diameter of 21 ft. The radius of the smaller circle is $\dfrac{21 \text{ ft}}{2}$, or 10.5 ft.

$A = \pi \cdot 21 \text{ ft} \cdot 21 \text{ ft} - \pi \cdot 10.5 \text{ ft} \cdot 10.5 \text{ ft}$

$A \approx 3.14 \cdot 21 \text{ ft} \cdot 21 \text{ ft} - 3.14 \cdot 10.5 \text{ ft} \cdot 10.5 \text{ ft}$

$A = 1384.74 \text{ ft}^2 - 346.185 \text{ ft}^2$

$A = 1038.555 \text{ ft}^2$

34. $V = l \cdot w \cdot h$

$V = 12 \text{ m} \cdot 3 \text{ m} \cdot 2.6 \text{ m}$

$V = 36 \cdot 2.6 \text{ m}^3$

$V = 93.6 \text{ m}^3$

35. $V = l \cdot w \cdot h$

$V = 4.6 \text{ cm} \cdot 3 \text{ cm} \cdot 14 \text{ cm}$

$V = 13.8 \cdot 14 \text{ cm}^3$

$V = 193.2 \text{ cm}^3$

36. $r = \dfrac{20 \text{ ft}}{2} = 10 \text{ ft}$

$V = B \cdot h = \pi \cdot r^2 \cdot h$

$\approx 3.14 \times 10 \text{ ft} \times 10 \text{ ft} \times 100 \text{ ft}$

$= 31,400 \text{ ft}^3$

37. $V = \dfrac{1}{3} \cdot \pi \cdot r^2 \cdot h$

$\approx \dfrac{1}{3} \times 3.14 \times 1 \text{ in.} \times 1 \text{ in.} \times 4.5 \text{ in.}$

$= 4.71 \text{ in}^3$

38. $V = \dfrac{4}{3} \cdot \pi \cdot r^3$

$\approx \dfrac{4}{3} \times 3.14 \times (2 \text{ cm})^3$

$= \dfrac{4 \times 3.14 \times 8 \text{ cm}^3}{3}$

$= 33.49\overline{3} \text{ cm}^3$

39. $V = B \cdot h = \pi \cdot r^2 \cdot h$

$\approx 3.14 \times 5 \text{ cm} \times 5 \text{ cm} \times 12 \text{ cm}$

$= 942 \text{ cm}^3$

40. $90° - 82° = 9°$

41. $90° - 5° = 85°$

42. $180° - 33° = 147°$

43. $180° - 133° = 47°$

44. $\angle 1$ and $\angle 4$ are vertical angles, so $m \angle 4 = 38°$. Likewise, $\angle 5$ and $\angle 2$ and vertical angles, so $m \angle 2 = 105°$.

$m \angle 1 + m \angle 2 + m \angle 3 = 180°$

$38° + 105° + m \angle 3 = 180°$

$143° + m \angle 3 = 180°$

$m \angle 3 = 37°$

Since $\angle 3$ and $\angle 6$ are vertical angles, $m \angle 6 = 37°$.

45. a) The pairs of corresponding angles are

$\angle 1$ and $\angle 5$,

$\angle 2$ and $\angle 6$,

$\angle 3$ and $\angle 7$,

$\angle 4$ and $\angle 8$.

b) The interior angles are $\angle 4$, $\angle 2$, $\angle 5$, and $\angle 7$.

c) The pairs of alternate interior angles are

$\angle 2$ and $\angle 5$,

$\angle 4$ and $\angle 7$.

46. $\angle 4$ and $\angle 6$ are vertical angles, so $m\angle 6 = 135°$.

$\angle 4$ and $\angle 2$ are corresponding angles. By Property 1, $m\angle 2 = 135°$.

$\angle 6$ and $\angle 8$ are corresponding angles. By Property 1, $m\angle 8 = 135°$.

$\angle 2$ and $\angle 3$ are interior angles on the same side of the transversal. Using Property 4 and $m\angle 2 = 135°$, $m\angle 3 = 45°$.

$\angle 6$ and $\angle 7$ are interior angles on the same side of the transversal. Using Property 4 and $m\angle 6 = 135°$, $m\angle 7 = 45°$.

$\angle 3$ and $\angle 5$ are vertical angles, so $m\angle 5 = 45°$.

$\angle 7$ and $\angle 1$ are vertical angles, so $m\angle 1 = 45°$.

47. The notation tells us the way in which the vertices of the two triangles are matched.

$\triangle DHJ \cong \triangle RZK$ means

$\angle D \cong \angle R$ and $\overline{DH} \cong \overline{RZ}$

$\angle H \cong \angle Z$ $\overline{DJ} \cong \overline{RK}$

$\angle J \cong \angle K$ $\overline{HJ} \cong \overline{ZK}$.

48. $\triangle ABC \cong \triangle GDF$

The notation tells us the way in which the vertices of the two triangles are matched.

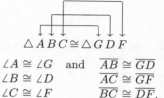

$\angle A \cong \angle G$ and $\overline{AB} \cong \overline{GD}$

$\angle B \cong \angle D$ $\overline{AC} \cong \overline{GF}$

$\angle C \cong \angle F$ $\overline{BC} \cong \overline{DF}$.

49. Two angles and the included side are of one triangle are congruent to two angles and the included side of the other triangle. They are congruent by the ASA Property.

50. Three sides of one triangle are congruent to three sides of the other triangle. They are congruent by the SSS Property.

51. Since we know only that three angles of one triangle are congruent to three angles of the other triangle, none of the properties can be used to show that the triangles are congruent.

52. Since J is the midpoint of $\overline{JK}$, $IJ \cong KJ$.

$\angle HJI \cong \angle LJK$ because they are vertical angles.

$\angle HIJ \cong \angle LKJ$ because they are opposite interior angles.

Two angles and the included side of $\triangle JIH$ are congruent to two angles and the included side of $\triangle JKL$, so $\triangle JIH \cong \triangle JKL$ by the ASA Property.

53. $\angle A$ and $\angle C$ are opposite angles, so $m\angle A = 63°$ by Property 2.

$\angle C$ and $\angle B$ are consecutive angles, so the are supplementary by Property 4. Then

$$m\angle B = 180° - m\angle C$$
$$m\angle B = 180° - 63°$$
$$m\angle B = 117°.$$

$\angle B$ and $\angle D$ are opposite angles, so $m\angle D = 117°$ by Property 2.

The opposite sides of a parallelogram are congruent (Property 3), so $CD = 13$ and $BC = 23$.

54. The notation tells us the way in which the vertices are matched.

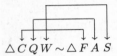

$$\triangle CQW \sim \triangle FAS$$

$\triangle CQW \sim \triangle FAS$ means

$$\angle C \cong \angle F$$
$$\angle Q \cong \angle A \quad \text{and} \quad \frac{CQ}{FA} = \frac{CW}{FS} = \frac{QW}{AS}.$$
$$\angle W \cong \angle S$$

55. Since $\triangle NMO \sim \triangle STR$, the corresponding sides are proportional. Thus

$$\frac{10}{15} = \frac{MO}{21}$$
$$10 \cdot 21 = 15(MO)$$
$$210 = 15(MO)$$
$$14 = MO.$$

56. *Discussion and Writing Exercise.* Linear measure is one-dimensional, area is two-dimensional, and volume is three-dimensional.

57. *Discussion and Writing Exercise.* Volume of two spheres, each with radius r: $2\left(\frac{4}{3}\pi r^3\right) = \frac{8}{3}\pi r^3$; volume of one sphere with radius $2r$: $\frac{4}{3}\pi(2r)^3 = \frac{32}{3}\pi r^2$. The volume of the sphere with radius $2r$ is four times the volume of the two spheres, each with radius r: $\frac{32}{3}\pi r^3 = 4 \cdot \frac{8}{3}\pi r^3$.

58. *Familiarize.* Let $s =$ the length of a side of the square, in feet. When the square is cut in half the resulting rectangle has length s and width $s/2$.

Translate.

$$\underbrace{\text{Perimeter of rectangle}}_{\downarrow} \quad \text{is} \quad \underbrace{30 \text{ ft.}}_{\downarrow \quad \downarrow}$$
$$2 \cdot s + 2 \cdot \frac{s}{2} \qquad = \qquad 30$$

Solve.

$$2 \cdot s + 2 \cdot \frac{s}{2} = 30$$
$$2 \cdot s + s = 30$$
$$3 \cdot s = 30$$
$$s = 10$$

If $s = 10$, then the area of the square is 10 ft $\cdot$ 10 ft, or 100 ft^2.

Check. If $s = 10$, then $s/2 = 10/2 = 5$ and the perimeter of a rectangle with length 10 ft and width 5 ft is $2 \cdot 10$ ft $+ 2 \cdot 5$ ft $= 20$ ft $+ 10$ ft $= 30$ ft. We can also recheck the calculation for the area of the square. The answer checks.

State. The area of the square is 100 ft^2.

59. The area A of the shaded region is the area of a square with sides 2.8 m less the areas of the four small squares cut out at each corner. Each of the small squares has sides of 1.8 mm, or 0.0018 m.

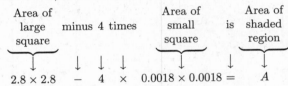

$$\begin{array}{ccccccc} \text{Area of} & & & \text{Area of} & & \text{Area of} \\ \text{large} & \text{minus 4 times} & & \text{small} & \text{is} & \text{shaded} \\ \text{square} & & & \text{square} & & \text{region} \\ \downarrow & \downarrow\ \downarrow\ \downarrow & & \downarrow & \downarrow & \downarrow \\ 2.8 \times 2.8 & - \quad 4 \quad \times & & 0.0018 \times 0.0018 & = & A \end{array}$$

We carry out the calculations.

$$7.84 - 4 \times 0.00000324 = 7.84 - 0.00001296 = 7.83998704$$

The area of the shaded region is 7.83998704 m^2.

60. The shaded region consists of one large triangle with base 8.4 cm and height 10 and 7 small triangles, each with height and base 1.3 mm, or 0.13 cm. Let $A =$ the area of the shaded region.

$$\begin{array}{ccccccc} \text{Area of} & & \text{Area of} & & & \text{Area of} \\ \text{shaded} & \text{is} & \text{large} & \text{plus 7 times} & & \text{small} \\ \text{region} & & \text{triangle} & & & \text{triangle} \\ \downarrow & \downarrow & \downarrow & \downarrow\ \downarrow\ \downarrow & & \downarrow \\ A & = & \frac{1}{2} \cdot 8.4 \cdot 10 & + \quad 7 \quad \cdot & & \frac{1}{2} \cdot 0.13 \cdot 0.13 \end{array}$$

We carry out the computations.

$$A = \frac{1}{2} \cdot 8.4 \cdot 10 + 7 \cdot \frac{1}{2} \cdot 0.13 \cdot 0.13$$
$$A = 42 + 0.05915$$
$$A = 42.05915$$

The area of the shaded region is 42.05915 cm^2.

Chapter 8 Test

1. Using a protractor, we find that the measure of the angle is 90°.

2. Using a protractor, we find that the measure of the angle is 35°.

3. Using a protractor, we find that the measure of the angle is 180°.

4. Using a protractor, we find that the measure of the angle is 113°.

5. The measure of the angle in Exercise 1 is 90°. It is a right angle.

6. The measure of the angle in Exercise 2 is 35°. Since its measure is greater than 0°and less than 90°, it is an acute angle.

7. The measure of the angle in Exercise 3 is 180°. It is a straight angle.

8. The measure of the angle in Exercise 4 is 113°. Since its measure is greater than 90°and less than 180°, it is an obtuse angle.

9.
$$m(\angle A) + m(\angle H) + m(\angle F) = 180°$$
$$35° + 110° + x = 180°$$
$$145° + x = 180°$$
$$x = 180° - 145°$$
$$x = 35°$$

10. From the labels on the triangle, we see that two sides are the same length. By measuring we find that the third side is a different length. Thus, this is an isosceles triangle.

11. One angle is an obtuse angle, so this is an obtuse triangle.

12.
$$(n - 2) \cdot 180°$$
$$= (5 - 2) \cdot 180° \quad \text{Substituting 5 for } n$$
$$= 3 \cdot 180°$$
$$= 540°$$

13.
$$P = 2 \cdot (l + w)$$
$$= 2 \cdot (9.4 \text{ cm} + 7.01 \text{ cm})$$
$$= 2 \cdot (16.41 \text{ cm})$$
$$= 32.82 \text{ cm}$$

$$A = l \cdot w$$
$$= (9.4 \text{ cm}) \cdot (7.01 \text{ cm})$$
$$= 9.4 \cdot 7.01 \cdot \text{cm} \cdot \text{cm}$$
$$= 65.894 \text{ cm}^2$$

14.
$$P = 4 \cdot s$$
$$= 4 \cdot 4\frac{7}{8} \text{ in.}$$
$$= 4 \cdot \frac{39}{8} \text{ in.}$$
$$= \frac{4 \cdot 39}{8} \text{ in.}$$
$$= \frac{4 \cdot 39}{2 \cdot 4} \text{ in.}$$
$$= \frac{39}{2} \text{ in., or } 19\frac{1}{2} \text{ in.}$$

$$A = s \cdot s$$
$$= \left(4\frac{7}{8} \text{ in.}\right) \cdot \left(4\frac{7}{8} \text{ in.}\right)$$
$$= 4\frac{7}{8} \cdot 4\frac{7}{8} \cdot \text{in.} \cdot \text{in.}$$
$$= \frac{39}{8} \cdot \frac{39}{8} \cdot \text{in}^2$$
$$= \frac{1521}{64} \text{ in}^2, \text{ or } 23\frac{49}{64} \text{ in}^2$$

15.
$$A = b \cdot h$$
$$= 10 \text{ cm} \cdot 2.5 \text{ cm}$$
$$= 25 \text{ cm}^2$$

16.
$$A = \frac{1}{2} \cdot b \cdot h$$
$$= \frac{1}{2} \cdot 8 \text{ m} \cdot 3 \text{ m}$$
$$= \frac{8 \cdot 3}{2} \text{ m}^2$$
$$= 12 \text{ m}^2$$

17.
$$A = \frac{1}{2} \cdot h \cdot (a + b)$$
$$= \frac{1}{2} \cdot 3 \text{ ft} \cdot (8 \text{ ft} + 4 \text{ ft})$$
$$= \frac{1}{2} \cdot 3 \text{ ft} \cdot 12 \text{ ft}$$
$$= \frac{3 \cdot 12}{2} \text{ ft}^2$$
$$= 18 \text{ ft}^2$$

18. $d = 2 \cdot r = 2 \cdot \frac{1}{8}$ in. $= \frac{1}{4}$ in.

19. $r = \dfrac{d}{2} = \dfrac{18 \text{ cm}}{2} = 9$ cm

20.
$$C = 2 \cdot \pi \cdot r$$
$$\approx 2 \cdot \frac{22}{7} \cdot \frac{1}{8} \text{ in.}$$
$$= \frac{2 \cdot 22 \cdot 1}{7 \cdot 8} \text{ in.}$$
$$= \frac{\cancel{2} \cdot \cancel{2} \cdot 11 \cdot 1}{7 \cdot \cancel{2} \cdot \cancel{2} \cdot 2} \text{ in.}$$
$$= \frac{11}{14} \text{ in.}$$

21. In Exercise 19 we found that the radius of the circle is 9 cm.
$$A = \pi \cdot r \cdot r$$
$$\approx 3.14 \cdot 9 \text{ cm} \cdot 9 \text{ cm}$$
$$= 3.14 \cdot 81 \text{ cm}^2$$
$$= 254.34 \text{ cm}^2$$

22. The perimeter of the shaded region consists of 2 sides of length 18.6 km and the circumferences of two semicircles with diameter 9.0 km. Note that the sum of the circumferences of the two semicircles is the same as the circumference of one circle with diameter 9.0 km.

The total length of the 2 sides of length 18.6 km is

$2 \cdot 18.6$ km $= 37.2$ km.

Next we find the perimeter, or circumference, of the circle.

$$C = \pi \cdot d$$
$$\approx 3.14 \cdot 9.0 \text{ km}$$
$$= 28.26 \text{ km}$$

Finally we add to find the perimeter of the shaded region.

$$37.2 \text{ km} + 28.26 \text{ km} = 65.46 \text{ km}$$

The shaded region is the area of a rectangle that is 18.6 km by 9.0 km less the area of two semicircles, each with diameter 9.0 km. Note that the two semicircles have the same area as one circle with diameter 9.0 km.

First we find the area of the rectangle.

$$A = l \cdot w$$
$$= 18.6 \text{ km} \cdot 9.0 \text{ km}$$
$$= 167.4 \text{ km}^2$$

Now find the area of the circle. The radius is $\dfrac{9.0 \text{ km}}{2}$, or 4.5 km.

$$A = \pi \cdot r \cdot r$$
$$\approx 3.14 \cdot 4.5 \text{ km} \cdot 4.5 \text{ km}$$
$$= 3.14 \cdot 20.25 \text{ km}^2$$
$$= 63.585 \text{ km}^2$$

Finally, we subtract to find the area of the shaded region.

$$167.4 \text{ km}^2 - 63.585 \text{ km}^2 = 103.815 \text{ km}^2$$

23. $V = l \cdot w \cdot h$
$$= 4 \text{ cm} \cdot 2 \text{ cm} \cdot 10.5 \text{ cm}$$
$$= 8 \cdot 10.5 \text{ cm}^3$$
$$= 84 \text{ cm}^3$$

24. $V = l \cdot w \cdot h$
$$= 10\frac{1}{2} \text{ in.} \cdot 8 \text{ in.} \cdot 5 \text{ in.}$$
$$= \frac{21}{2} \text{ in.} \cdot 8 \text{ in.} \cdot 5 \text{ in.}$$
$$= \frac{21 \cdot 8 \cdot 5}{2} \text{ in}^3$$
$$= \frac{21 \cdot \cancel{2} \cdot 4 \cdot 5}{\cancel{2} \cdot 1} \text{ in}^3$$
$$= 420 \text{ in}^3$$

25. $V = \pi \cdot r^2 \cdot h$
$$\approx 3.14 \times 5 \text{ ft} \times 5 \text{ ft} \times 15 \text{ ft}$$
$$= 1177.5 \text{ ft}^3$$

26. $r = \dfrac{d}{2} = \dfrac{20 \text{ yd}}{2} = 10 \text{ yd}$
$$V = \frac{4}{3} \cdot \pi \cdot r^3$$
$$\approx \frac{4}{3} \times 3.14 \times (10 \text{ yd})^3$$
$$= 4186.\overline{6} \text{ yd}^3$$

27. $V = \dfrac{1}{3} \pi \cdot r^2 \cdot h$
$$\approx \frac{1}{3} \times 3.14 \times 3 \text{ cm} \times 3 \text{ cm} \times 12 \text{ cm}$$
$$= 113.04 \text{ cm}^3$$

28. $180° - 31° = 149°$

29. $90° - 79° = 11°$

30. $\angle 1$ and $\angle 4$ are vertical angles, so $m \angle 4 = 62°$. Likewise, $\angle 5$ and $\angle 2$ are vertical angles, so $m \angle 2 = 110°$.

$$m \angle 1 + m \angle 2 + m \angle 3 = 180°$$
$$62° + 110° + m \angle 3 = 180°$$
$$172° + m \angle 3 = 180°$$
$$m \angle 3 = 8°$$

Since $\angle 3$ and $\angle 6$ are vertical angles, $m \angle 6 = 8°$.

31. $\angle 4$ and $\angle 6$ are vertical angles, so $m\angle 6 = 120°$.

$\angle 4$ and $\angle 2$ are corresponding angles. By Property 1, $m\angle 2 = 120°$.

$\angle 6$ and $\angle 8$ are corresponding angles. By Property 1, $m\angle 8 = 120°$.

$\angle 2$ and $\angle 3$ are interior angles on the same side of the transversal. Using Property 4 and $m\angle 2 = 120°$, $m\angle 3 = 60°$.

$\angle 6$ and $\angle 7$ are interior angles on the same side of the transversal. Using Property 4 and $m\angle 6 = 120°$, $m\angle 7 = 60°$.

$\angle 3$ and $\angle 5$ are vertical angles, so $m\angle 5 = 60°$.

$\angle 7$ and $\angle 1$ are vertical angles, so $m\angle 1 = 60°$.

32. The notation tells us the way in which the vertices of the two triangles are matched.

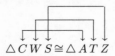

$$\triangle C W S \cong \triangle A T Z$$

$\triangle CWS \sim \triangle ATZ$ means

$$\angle C \cong \angle A \quad \text{and} \quad CW \cong AT$$
$$\angle W \cong \angle T \qquad\quad CS \cong AZ$$
$$\angle S \cong \angle Z \qquad\quad WS \cong TZ.$$

33. Two sides of one triangle and the included angle are congruent to two sides and the included angle of the other triangle. They are congruent by the SAS Property.

34. Since we know only that three angles of one triangle are congruent to three angles of the other triangle, none of the properties can be used to show that the triangles are congruent.

35. Two angles and the included side are of one triangle are congruent to two angles and the included side of the other triangle. They are congruent by the ASA Property.

36. We know only that two sides of one triangle as well as an angle that is not the included angle are congruent to the corresponding sides and angle of the other triangle. Thus, none of the properties can be used to show that the triangles are congruent.

37. $\angle E$ and $\angle G$ are opposite angles, so $m \angle G = 105°$ by Property 2.

$\angle D$ and $\angle E$ are consecutive angles, so they are supplementary by Property 4. Then

$$m \angle D = 180° - m \angle E$$
$$m \angle D = 180° - 105°$$
$$m \angle D = 75°$$

$\angle D$ and $\angle F$ are opposite angles, so $m \angle F = 75°$ by Property 2.

The opposite sides of a parallelogram are congruent (Property 3), so $EF = 11$. Similarly, $DE = GF$. Let $l =$ the length of DE. Then $l =$ the length of GF also. The perimeter of the parallelogram is 62, so we have:

$$11 + l + 11 + l = 62$$
$$22 + 2 \cdot l = 62$$
$$2 \cdot l = 40$$
$$l = 20$$

Then $DE = GF = 20$.

38. The diagonals of a parallelogram bisect each other (Property 5). Then

$$LJ = 2 \cdot JN = 2 \cdot 3.2 = 6.4,$$
$$KM = 2 \cdot KN = 2 \cdot 3 = 6.$$

39. The notation tells us the way in which the vertices are matched.

$$\triangle ERS \sim \triangle TGF$$

$$\angle E \cong \angle T$$
$$\angle R \cong \angle G \quad \text{and} \quad \frac{ER}{TG} = \frac{ES}{TF} = \frac{RS}{GF}.$$
$$\angle S \cong \angle F$$

40. The corresponding sides of the triangle are proportional.

$$\frac{24}{8} = \frac{EK}{6} \qquad \frac{24}{8} = \frac{ZK}{9}$$
$$24 \cdot 6 = 8(EK) \qquad 24 \cdot 9 = 8(ZK)$$
$$144 = 8(EK) \qquad 216 = 8(ZK)$$
$$18 = EK \qquad 27 = ZK$$

41. First we convert 3 in. to feet.

$$3 \text{ in.} = 3 \text{ in.} \cdot \frac{1 \text{ ft}}{12 \text{ in.}}$$
$$= \frac{3}{12} \cdot \frac{\text{in.}}{\text{in.}} \cdot 1 \text{ ft}$$
$$= \frac{1}{4} \text{ ft}$$

Now we find the area of the rectangle.

$$A = l \cdot w$$
$$= 8 \text{ ft} \cdot \frac{1}{4} \text{ ft}$$
$$= \frac{8}{4} \text{ ft}^2$$
$$= 2 \text{ ft}^2$$

42. We convert both units of measure to feet. From Exercise 41 we know that $3 \text{ in.} = \frac{1}{4}$ ft. We also have

$$5 \text{ yd} = 5 \times 1 \text{ yd}$$
$$= 5 \times 3 \text{ ft}$$
$$= 15 \text{ ft.}$$

Now we find the area.

$$A = \frac{1}{2} \cdot b \cdot h$$
$$= \frac{1}{2} \cdot 15 \text{ ft} \cdot \frac{1}{4} \text{ ft}$$
$$= \frac{15}{2 \cdot 4} \text{ ft}^2$$
$$= \frac{15}{8} \text{ ft}^2, \text{ or } 1.875 \text{ ft}^2$$

43. First we convert 2.6 in. and 3 in. to feet.

$$2.6 \text{ in.} = 2.6 \text{ in.} \cdot \frac{1 \text{ ft}}{12 \text{ in.}}$$
$$= \frac{2.6}{12} \cdot \frac{\text{in.}}{\text{in.}} \cdot 1 \text{ ft}$$
$$= \frac{2.6}{12} \text{ ft}$$

From Exercise 41 we know that $3 \text{ in.} = \frac{1}{4}$ ft. Now we find the volume.

$$V = l \cdot w \cdot h$$
$$= 12 \text{ ft} \cdot \frac{1}{4} \text{ ft} \cdot \frac{2.6}{12} \text{ ft}$$
$$= \frac{12 \cdot 2.6}{4 \cdot 12} \text{ ft}^3$$
$$= 0.65 \text{ ft}^3$$

44. First we convert 1 in. to feet.

$$1 \text{ in.} = 1 \text{ in.} \cdot \frac{1 \text{ ft}}{12 \text{ in.}}$$
$$= \frac{1}{12} \cdot \frac{\text{in.}}{\text{in.}} \cdot 1 \text{ ft}$$
$$= \frac{1}{12} \text{ ft}$$

Now we find the volume.

$$V = \frac{1}{3} \pi \cdot r^2 \cdot h$$
$$\approx \frac{1}{3} \cdot 3.14 \cdot \frac{1}{12} \text{ ft} \cdot \frac{1}{12} \text{ ft} \cdot 4.5 \text{ ft}$$
$$= \frac{3.14 \cdot 4.5}{3 \cdot 12 \cdot 12} \text{ ft}^3$$
$$\approx 0.033 \text{ ft}^3$$

45. First we find the radius of the cylinder.

$$r = \frac{d}{2} = \frac{\frac{3}{4}\text{ in.}}{2} = \frac{3}{4}\text{ in.} \cdot \frac{1}{2} = \frac{3}{8}\text{ in.}$$

Now we convert $\frac{3}{8}$ in. to feet.

$$\frac{3}{8}\text{ in.} = \frac{3}{8}\text{ in.} \cdot \frac{1\text{ ft}}{12\text{ in.}}$$

$$= \frac{3}{8 \cdot 12} \cdot \frac{\text{in.}}{\text{in.}} \cdot 1\text{ ft}$$

$$= \frac{1}{32}\text{ ft}$$

Finally, we find the volume.

$$V = B \cdot h = \pi \cdot r^2 \cdot h$$

$$\approx 3.14 \cdot \frac{1}{32}\text{ ft} \cdot \frac{1}{32}\text{ ft} \cdot 18\text{ ft}$$

$$= \frac{3.14 \cdot 18}{32 \cdot 32}\text{ ft}^3$$

$$\approx 0.055\text{ ft}^3$$

Chapter 9

Introduction to Real Numbers and Algebraic Expressions

Exercise Set 9.1

1. Substitute 56 for x: $56 - 24 = 32$, so it takes Erin 32 min to get to work if it takes George 56 min.

   Substitute 93 for x: $93 - 24 = 69$, so it takes Erin 69 min to get to work if it takes George 93 min.

   Substitute 105 for x: $105 - 24 = 81$, so it takes Erin 81 min to get to work if it takes George 105 min.

3. Substitute 45 m for b and 86 m for h, and carry out the multiplication:

$$A = \frac{1}{2}bh = \frac{1}{2}(45 \text{ m})(86 \text{ m})$$
$$= \frac{1}{2}(45)(86)(\text{m})(\text{m})$$
$$= 1935 \text{ m}^2$$

5. Substitute 65 for r and 4 for t, and carry out the multiplication:
$$d = rt = 65 \cdot 4 = 260 \text{ mi}$$

7. We substitute 6 ft for l and 4 ft for w in the formula for the area of a rectangle.
$$A = lw = (6 \text{ ft})(4 \text{ ft})$$
$$= (6)(4)(\text{ft})(\text{ft})$$
$$= 24 \text{ ft}^2$$

9. $8x = 8 \cdot 7 = 56$

11. $\dfrac{a}{b} = \dfrac{24}{3} = 8$

13. $\dfrac{3p}{q} = \dfrac{3 \cdot 2}{6} = \dfrac{6}{6} = 1$

15. $\dfrac{x + y}{5} = \dfrac{10 + 20}{5} = \dfrac{30}{5} = 6$

17. $\dfrac{x - y}{8} = \dfrac{20 - 4}{8} = \dfrac{16}{8} = 2$

19. $b + 7$, or $7 + b$

21. $c - 12$

23. $4 + q$, or $q + 4$

25. $a + b$, or $b + a$

27. $x \div y$, or $\dfrac{x}{y}$, or x/y, or $x \cdot \dfrac{1}{y}$

29. $x + w$, or $w + x$

31. $n - m$

33. $x + y$, or $y + x$

35. $2z$

37. $3m$

39. $4a + 6$, or $6 + 4a$

41. $xy - 8$

43. $2t - 5$

45. $3n + 11$, or $11 + 3n$

47. $4x + 3y$, or $3y + 4x$

49. Let s represent your salary. Then we have $89\%s$, or $0.89s$.

51. A 5% increase in s is represented by $0.05s$, so we have $s + 0.05s$.

53. The distance traveled is the product of the speed and the time. Thus, Danielle traveled $65t$ miles.

55. $\$50 - x$

57. Discussion and Writing Exercise

59. We use a factor tree.

   The prime factorization is $2 \cdot 3 \cdot 3 \cdot 3$.

61. We use the list of primes. The first prime that is a factor of 108 is 2.
$$108 = 2 \cdot 54$$

   We keep dividing by 2 until it is no longer possible to do so.
$$108 = 2 \cdot 2 \cdot 27$$

   Now we do the same thing for the next prime, 3.
$$108 = 2 \cdot 2 \cdot 3 \cdot 3 \cdot 3$$

   This is the prime factorization of 108.

63. We use the list of primes. The first prime number that is a factor of 1023 is 3.
$$1023 = 3 \cdot 341$$

   We continue through the list of prime numbers until we have

$1023 = 3 \cdot 11 \cdot 31.$

Since 3, 11, and 31 are prime numbers, the prime factorization of 1023 is $3 \cdot 11 \cdot 31$.

65. $6 = 2 \cdot 3$

$24 = 2 \cdot 2 \cdot 2 \cdot 3$

$32 = 2 \cdot 2 \cdot 2 \cdot 2 \cdot 2$

The LCM is $2 \cdot 2 \cdot 2 \cdot 2 \cdot 2 \cdot 3$, or 96.

67. $16 = 2 \cdot 2 \cdot 2 \cdot 2$

$24 = 2 \cdot 2 \cdot 2 \cdot 3$

$32 = 2 \cdot 2 \cdot 2 \cdot 2 \cdot 2$

The LCM is $2 \cdot 2 \cdot 2 \cdot 2 \cdot 2 \cdot 3$, or 96.

69.
$$\frac{a - 2b + c}{4b - a} = \frac{20 - 2 \cdot 10 + 5}{4 \cdot 10 - 20}$$
$$= \frac{20 - 20 + 5}{40 - 20}$$
$$= \frac{0 + 5}{20}$$
$$= \frac{5}{20} = \frac{\cancel{5} \cdot 1}{\cancel{5} \cdot 4}$$
$$= \frac{1}{4}$$

71. $\dfrac{12 - c}{c + 12b} = \dfrac{12 - 12}{12 + 12 \cdot 1} = \dfrac{0}{12 + 12} = \dfrac{0}{24} = 0$

Exercise Set 9.2

1. The integer $-34{,}000{,}000$ corresponds to paying a fine of \$34 million.

3. The integer 24 corresponds to 24° above zero; the integer -2 corresponds to 2° below zero.

5. The integer 950,000,000 corresponds to a temperature of 950,000,000°F; the integer -460 corresponds to a temperature of 460°F below zero.

7. The integer -34 describes the situation from the Alley Cats' point of view. The integer 34 describes the situation from the Strikers' point of view.

9. The number $\dfrac{10}{3}$ can be named $3\dfrac{1}{3}$, or $3.3\overline{3}$. The graph is $\dfrac{1}{3}$ of the way from 3 to 4.

11. The graph of -5.2 is $\dfrac{2}{10}$ of the way from -5 to -6.

13. The graph of $-4\dfrac{2}{5}$ is $\dfrac{2}{5}$ of the way from -4 to -5.

15. We first find decimal notation for $\dfrac{7}{8}$. Since $\dfrac{7}{8}$ means $7 \div 8$, we divide.

```
    0.8 7 5
8 ) 7.0 0 0
    6 4
    ─────
      6 0
      5 6
      ─────
        4 0
        4 0
        ───
          0
```

Thus $\dfrac{7}{8} = 0.875$, so $-\dfrac{7}{8} = -0.875$.

17. $\dfrac{5}{6}$ means $5 \div 6$, so we divide.

```
    0.8 3 3 ...
6 ) 5.0 0 0
    4 8
    ─────
      2 0
      1 8
      ─────
        2 0
        1 8
        ───
          0
```

We have $\dfrac{5}{6} = 0.8\overline{3}$.

19. First we find decimal notation for $\dfrac{7}{6}$. Since $\dfrac{7}{6}$ means $7 \div 6$, we divide.

```
    1.1 6 6 ...
6 ) 7.0 0 0
    6
    ───
    1 0
      6
    ───
      4 0
      3 6
      ─────
        4 0
        3 6
        ───
          4
```

Thus $\dfrac{7}{6} = 1.1\overline{6}$, so $-\dfrac{7}{6} = -1.1\overline{6}$.

21. $\dfrac{2}{3}$ means $2 \div 3$, so we divide.

```
    0.6 6 6 ...
3 ) 2.0 0 0
    1 8
    ─────
      2 0
      1 8
      ─────
        2 0
        1 8
        ───
          2
```

We have $\dfrac{2}{3} = 0.\overline{6}$.

23. $\frac{1}{10}$ means $1 \div 10$, so we divide.

$$
\begin{array}{r}
0.1 \\
1\,0\,\overline{\smash{)}1.0} \\
\underline{1\,0} \\
0
\end{array}
$$

We have $\frac{1}{10} = 0.1$

25. We first find decimal notation for $\frac{1}{2}$. Since $\frac{1}{2}$ means $1 \div 2$, we divide.

$$
\begin{array}{r}
0.5 \\
2\,\overline{\smash{)}1.0} \\
\underline{1\,0} \\
0
\end{array}
$$

Thus $\frac{1}{2} = 0.5$, so $-\frac{1}{2} = -0.5$

27. $\frac{4}{25}$ means $4 \div 25$, so we divide.

$$
\begin{array}{r}
0.1\,6 \\
2\,5\,\overline{\smash{)}4.0\,0} \\
\underline{2\,5} \\
1\,5\,0 \\
\underline{1\,5\,0} \\
0
\end{array}
$$

We have $\frac{4}{25} = 0.16$.

29. Since 8 is to the right of 0, we have $8 > 0$.

31. Since -8 is to the left of 3, we have $-8 < 3$.

33. Since -8 is to the left of 8, we have $-8 < 8$.

35. Since -8 is to the left of -5, we have $-8 < -5$.

37. Since -5 is to the right of -11, we have $-5 > -11$.

39. Since -6 is to the left of -5, we have $-6 < -5$.

41. Since 2.14 is to the right of 1.24, we have $2.14 > 1.24$.

43. Since -14.5 is to the left of 0.011, we have $-14.5 < 0.011$.

45. Since -12.88 is to the left of -6.45, we have $-12.88 < -6.45$.

47. $-\frac{1}{2} = -\frac{1}{2} \cdot \frac{3}{3} = -\frac{3}{6}$

$-\frac{2}{3} = -\frac{2}{3} \cdot \frac{2}{2} = -\frac{4}{6}$

Since $-\frac{3}{6}$ is to the right of $-\frac{4}{6}$, then $-\frac{1}{2}$ is to the right of $-\frac{2}{3}$, and we have $-\frac{1}{2} > -\frac{2}{3}$.

49. Since $-\frac{2}{3}$ is to the left of $\frac{1}{3}$, we have $-\frac{2}{3} < \frac{1}{3}$.

51. Convert to decimal notation $\frac{5}{12} = 0.4166\ldots$ and $\frac{11}{25} = 0.44$. Since $0.4166\ldots$ is to the left of 0.44, $\frac{5}{12} < \frac{11}{25}$.

53. $-3 \geq -11$ is true since $-3 > -11$ is true.

55. $0 \geq 8$ is false since neither $0 > 8$ nor $0 = 8$ is true.

57. $x < -6$ has the same meaning as $-6 > x$.

59. $y \geq -10$ has the same meaning as $-10 \leq y$.

61. The distance of -3 from 0 is 3, so $|-3| = 3$.

63. The distance of 10 from 0 is 10, so $|10| = 10$.

65. The distance of 0 from 0 is 0, so $|0| = 0$.

67. The distance of -30.4 from 0 is 30.4, so $|-30.4| = 30.4$.

69. The distance of $-\frac{2}{3}$ from 0 is $\frac{2}{3}$, so $\left|-\frac{2}{3}\right| = \frac{2}{3}$.

71. The distance of $\frac{0}{4}$ from 0 is $\frac{0}{4}$, or 0, so $\left|\frac{0}{4}\right| = 0$.

73. The distance of $-3\frac{5}{8}$ from 0 is $3\frac{5}{8}$, so $\left|-3\frac{5}{8}\right| = 3\frac{5}{8}$.

75. Discussion and Writing Exercise

77. 63% $\quad$ $0.63.$

Move the decimal point 2 places to the left.

$63\% = 0.63$

79. 110% $\quad$ $1.10.$

Move the decimal point 2 places to the left.

$110\% = 1.1$

81. $\frac{13}{25} = 0.52 = 52\%$

83. From Exercise 17 we know that $\frac{5}{6} = 0.8\overline{3}$, or $0.83\overline{3}$, so $\frac{5}{6} = 83.\overline{3}\%$, or $83\frac{1}{3}\%$.

85. $-\frac{2}{3}, \frac{1}{2}, -\frac{3}{4}, -\frac{5}{6}, \frac{3}{8}, \frac{1}{6}$ can be written in decimal notation as $-0.\overline{6}$, 0.5, -0.75, $-0.8\overline{3}$, 0.375, $0.1\overline{6}$, respectively. Listing from least to greatest, we have

$$-\frac{5}{6}, -\frac{3}{4}, -\frac{2}{3}, \frac{1}{6}, \frac{3}{8}, \frac{1}{2}.$$

87. $-8.76, -5.16, -4.24, -2.13, 1.85, 5.23$

89. $0.\overline{1} = \frac{0.\overline{3}}{3} = \frac{\frac{1}{3}}{3} = \frac{1}{3} \cdot \frac{1}{3} = \frac{1}{9}$

91. First consider $0.\overline{5}$.

$0.\overline{5} = 0.\overline{3} \cdot \frac{5}{3} = \frac{1}{3} \cdot \frac{5}{3} = \frac{5}{9}$

Then, $5.\overline{5} = 5 + 0.\overline{5} = 5 + \frac{5}{9} = 5\frac{5}{9}$, or $\frac{50}{9}$.

Exercise Set 9.3

1. $2 + (-9)$ The absolute values are 2 and 9. The difference is $9 - 2$, or 7. The negative number has the larger absolute value, so the answer is negative. $2 + (-9) = -7$

3. $-11 + 5$ The absolute values are 11 and 5. The difference is $11 - 5$, or 6. The negative number has the larger absolute value, so the answer is negative. $-11 + 5 = -6$

5. $-8 + 8$ A negative and a positive number. The numbers have the same absolute value. The sum is 0. $-8 + 8 = 0$

7. $-3 + (-5)$ Two negatives. Add the absolute values, getting 8. Make the answer negative. $-3 + (-5) = -8$

9. $-7 + 0$ One number is 0. The answer is the other number. $-7 + 0 = -7$

11. $0 + (-27)$ One number is 0. The answer is the other number. $0 + (-27) = -27$

13. $17 + (-17)$ A negative and a positive number. The numbers have the same absolute value. The sum is 0. $17 + (-17) = 0$

15. $-17 + (-25)$ Two negatives. Add the absolute values, getting 42. Make the answer negative. $-17 + (-25) = -42$

17. $18 + (-18)$ A positive and a negative number. The numbers have the same absolute value. The sum is 0. $18 + (-18) = 0$

19. $-28 + 28$ A negative and a positive number. The numbers have the same absolute value. The sum is 0. $-28 + 28 = 0$

21. $8 + (-5)$ The absolute values are 8 and 5. The difference is $8 - 5$, or 3. The positive number has the larger absolute value, so the answer is positive. $8 + (-5) = 3$

23. $-4 + (-5)$ Two negatives. Add the absolute values, getting 9. Make the answer negative. $-4 + (-5) = -9$

25. $13 + (-6)$ The absolute values are 13 and 6. The difference is $13 - 6$, or 7. The positive number has the larger absolute value, so the answer is positive. $13 + (-6) = 7$

27. $-25 + 25$ A negative and a positive number. The numbers have the same absolute value. The sum is 0. $-25 + 25 = 0$

29. $53 + (-18)$ The absolute values are 53 and 18. The difference is $53 - 18$, or 35. The positive number has the larger absolute value, so the answer is positive. $53 + (-18) = 35$

31. $-8.5 + 4.7$ The absolute values are 8.5 and 4.7. The difference is $8.5 - 4.7$, or 3.8. The negative number has the larger absolute value, so the answer is negative. $-8.5 + 4.7 = -3.8$

33. $-2.8 + (-5.3)$ Two negatives. Add the absolute values, getting 8.1. Make the answer negative. $-2.8 + (-5.3) = -8.1$

35. $-\frac{3}{5} + \frac{2}{5}$ The absolute values are $\frac{3}{5}$ and $\frac{2}{5}$. The difference is $\frac{3}{5} - \frac{2}{5}$, or $\frac{1}{5}$. The negative number has the larger absolute value, so the answer is negative. $-\frac{3}{5} + \frac{2}{5} = -\frac{1}{5}$

37. $-\frac{2}{9} + \left(-\frac{5}{9}\right)$ Two negatives. Add the absolute values, getting $\frac{7}{9}$. Make the answer negative. $-\frac{2}{9} + \left(-\frac{5}{9}\right) = -\frac{7}{9}$

39. $-\frac{5}{8} + \frac{1}{4}$ The absolute values are $\frac{5}{8}$ and $\frac{1}{4}$. The difference is $\frac{5}{8} - \frac{2}{8}$, or $\frac{3}{8}$. The negative number has the larger absolute value, so the answer is negative. $-\frac{5}{8} + \frac{1}{4} = -\frac{3}{8}$

41. $-\frac{5}{8} + \left(-\frac{1}{6}\right)$ Two negatives. Add the absolute values, getting $\frac{15}{24} + \frac{4}{24}$, or $\frac{19}{24}$. Make the answer negative. $-\frac{5}{8} + \left(-\frac{1}{6}\right) = -\frac{19}{24}$

43. $-\frac{3}{8} + \frac{5}{12}$ The absolute values are $\frac{3}{8}$ and $\frac{5}{12}$. The difference is $\frac{10}{24} - \frac{9}{24}$, or $\frac{1}{24}$. The positive number has the larger absolute value, so the answer is positive. $-\frac{3}{8} + \frac{5}{12} = \frac{1}{24}$

45. $-\frac{1}{6} + \frac{7}{10}$ The absolute values are $\frac{1}{6}$ and $\frac{7}{10}$. The difference is $\frac{21}{30} - \frac{5}{30} = \frac{16}{30} = \frac{2 \cdot 8}{2 \cdot 15} = \frac{8}{15}$. The positive number has the larger absolute value, so the answer is positive. $-\frac{1}{6} + \frac{7}{10} = \frac{8}{15}$

47. $\frac{7}{15} + \left(-\frac{1}{9}\right)$ The absolute values are $\frac{7}{15}$ and $\frac{1}{9}$. The difference is $\frac{21}{45} - \frac{5}{45} = \frac{16}{45}$. The positive number has the larger absolute value, so the answer is positive. $\frac{7}{15} + \left(-\frac{1}{9}\right) = \frac{16}{45}$

49. $76 + (-15) + (-18) + (-6)$

a) Add the negative numbers: $-15 + (-18) + (-6) = -39$

b) Add the results: $76 + (-39) = 37$

51. $-44 + \left(-\frac{3}{8}\right) + 95 + \left(-\frac{5}{8}\right)$

a) Add the negative numbers: $-44 + \left(-\frac{3}{8}\right) + \left(-\frac{5}{8}\right) = -45$

b) Add the results: $-45 + 95 = 50$

53. We add from left to right.

$$
\begin{aligned}
& 98 + (-54) + 113 + (-998) + 44 + (-612) \\
&= \quad 44 + 113 + (-998) + 44 + (-612) \\
&= \quad\quad 157 + (-998) + 44 + (-612) \\
&= \quad\quad\quad -841 + 44 + (-612) \\
&= \quad\quad\quad\quad -797 + (-612) \\
&= \quad\quad\quad\quad\quad -1409
\end{aligned}
$$

55. The additive inverse of 24 is -24 because $24 + (-24) = 0$.

57. The additive inverse of -26.9 is 26.9 because $-26.9 + 26.9 = 0$.

59. If $x = 8$, then $-x = -8$. (The opposite of 8 is -8.)

61. If $x = -\dfrac{13}{8}$ then $-x = -\left(-\dfrac{13}{8}\right) = \dfrac{13}{8}$. (The opposite of $-\dfrac{13}{8}$ is $\dfrac{13}{8}$.)

63. If $x = -43$ then $-(-x) = -(-(-43)) = -43$. (The opposite of the opposite of -43 is -43.)

65. If $x = \dfrac{4}{3}$ then $-(-x) = -\left(-\dfrac{4}{3}\right) = \dfrac{4}{3}$. (The opposite of the opposite of $\dfrac{4}{3}$ is $\dfrac{4}{3}$.)

67. $-(-24) = 24$ (The opposite of -24 is 24.)

69. $-\left(-\dfrac{3}{8}\right) = \dfrac{3}{8}$ (The opposite of $-\dfrac{3}{8}$ is $\dfrac{3}{8}$.)

71. Let $E =$ the elevation of Mauna Kea above sea level.

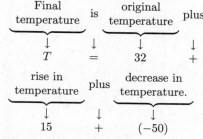

$$E = 33,480 + (-19,684)$$

We carry out the addition.

$$E = 33,480 + (-19,684) = 13,796$$

The elevation of Mauna Kea is 13,796 ft above sea level.

73. Let $T =$ the final temperature. We will express the rise in temperature as a positive number and a decrease in the temperature as a negative number.

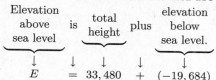

We add from left to right.

$$
\begin{aligned}
T &= 32 + 15 + (-50) \\
&= 47 + (-50) \\
&= -3
\end{aligned}
$$

The final temperature was $-3°$F.

75. Let $S =$ the sum of the profits and losses. We add the five numbers in the bar graph to find S.

$S = \$10,500 + (-\$16,600) + (-\$12,800) + (-\$9600) + \$8200 = -\$20,300$

The sum of the profits and losses is $-\$20,300$.

77. Let $B =$ the new balance in the account at the end of August. We will express the payments as positive numbers and the original balance and the amount of the new charge as negative numbers.

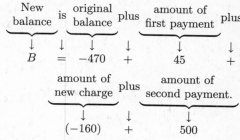

We add from left to right.

$$
\begin{aligned}
B &= -470 + 45 + (-160) + 500 \\
&= -425 + (-160) + 500 \\
&= -585 + 500 \\
&= -85
\end{aligned}
$$

The balance in the account is $-\$85$. Lyle owes $\$85$.

79. Discussion and Writing Exercise

81. $57\% \qquad 0.57.$

Move the decimal point two places to the left.

$57\% = 0.57$.

83. $23\dfrac{4}{5}\% = 23.8\% \qquad 0.23.8$

Move the decimal point two places to the left.

$23\dfrac{4}{5}\% = 0.238$

85. $\dfrac{5}{4} = 1.25 = 125\%$

87. $\dfrac{13}{25} = 0.52 = 52\%$

89. When x is positive, the opposite of x, $-x$, is negative, so $-x$ is negative for all positive numbers x.

91. If a is positive, $-a$ is negative. Thus $-a + b$, the sum of two negative numbers, is negative. The correct answer is (b).

Exercise Set 9.4

1. $2 - 9 = 2 + (-9) = -7$

3. $-8 - (-2) = -8 + 2 = -6$

5. $-11 - (-11) = -11 + 11 = 0$

7. $12 - 16 = 12 + (-16) = -4$

9. $20 - 27 = 20 + (-27) = -7$

11. $-9 - (-3) = -9 + 3 = -6$

13. $-40 - (-40) = -40 + 40 = 0$

15. $7 - (-7) = 7 + 7 = 14$

17. $8 - (-3) = 8 + 3 = 11$

19. $-6 - 8 = -6 + (-8) = -14$

21. $-4 - (-9) = -4 + 9 = 5$

23. $-6 - (-5) = -6 + 5 = -1$

25. $8 - (-10) = 8 + 10 = 18$

27. $-5 - (-2) = -5 + 2 = -3$

29. $-7 - 14 = -7 + (-14) = -21$

31. $0 - (-5) = 0 + 5 = 5$

33. $-8 - 0 = -8 + 0 = -8$

35. $7 - (-5) = 7 + 5 = 12$

37. $2 - 25 = 2 + (-25) = -23$

39. $-42 - 26 = -42 + (-26) = -68$

41. $-71 - 2 = -71 + (-2) = -73$

43. $24 - (-92) = 24 + 92 = 116$

45. $-50 - (-50) = -50 + 50 = 0$

47. $-\dfrac{3}{8} - \dfrac{5}{8} = -\dfrac{3}{8} + \left(-\dfrac{5}{8}\right) = -\dfrac{8}{8} = -1$

49. $\dfrac{3}{4} - \dfrac{2}{3} = \dfrac{3}{4} + \left(-\dfrac{2}{3}\right) = \dfrac{9}{12} + \left(-\dfrac{8}{12}\right) = \dfrac{1}{12}$

51. $-\dfrac{3}{4} - \dfrac{2}{3} = -\dfrac{3}{4} + \left(-\dfrac{2}{3}\right) = -\dfrac{9}{12} + \left(-\dfrac{8}{12}\right) = -\dfrac{17}{12}$

53. $-\dfrac{5}{8} - \left(-\dfrac{3}{4}\right) = -\dfrac{5}{8} + \dfrac{3}{4} = -\dfrac{5}{8} + \dfrac{6}{8} = \dfrac{1}{8}$

55. $6.1 - (-13.8) = 6.1 + 13.8 = 19.9$

57. $-2.7 - 5.9 = -2.7 + (-5.9) = -8.6$

59. $0.99 - 1 = 0.99 + (-1) = -0.01$

61. $-79 - 114 = -79 + (-114) = -193$

63. $0 - (-500) = 0 + 500 = 500$

65. $-2.8 - 0 = -2.8 + 0 = -2.8$

67. $7 - 10.53 = 7 + (-10.53) = -3.53$

69. $\dfrac{1}{6} - \dfrac{2}{3} = \dfrac{1}{6} + \left(-\dfrac{2}{3}\right) = \dfrac{1}{6} + \left(-\dfrac{4}{6}\right) = -\dfrac{3}{6}$, or $-\dfrac{1}{2}$

71. $-\dfrac{4}{7} - \left(-\dfrac{10}{7}\right) = -\dfrac{4}{7} + \dfrac{10}{7} = \dfrac{6}{7}$

73. $-\dfrac{7}{10} - \dfrac{10}{15} = -\dfrac{7}{10} + \left(-\dfrac{10}{15}\right) = -\dfrac{21}{30} + \left(-\dfrac{20}{30}\right) = -\dfrac{41}{30}$

75. $\dfrac{1}{5} - \dfrac{1}{3} = \dfrac{1}{5} + \left(-\dfrac{1}{3}\right) = \dfrac{3}{15} + \left(-\dfrac{5}{15}\right) = -\dfrac{2}{15}$

77. $\dfrac{5}{12} - \dfrac{7}{16} = \dfrac{5}{12} + \left(-\dfrac{7}{16}\right) = \dfrac{20}{48} + \left(-\dfrac{21}{48}\right) = -\dfrac{1}{48}$

79. $-\dfrac{2}{15} - \dfrac{7}{12} = -\dfrac{2}{15} + \left(-\dfrac{7}{12}\right) = -\dfrac{8}{60} + \left(-\dfrac{35}{60}\right) = -\dfrac{43}{60}$

81. $18 - (-15) - 3 - (-5) + 2 = 18 + 15 + (-3) + 5 + 2 = 37$

83. $-31 + (-28) - (-14) - 17 = (-31) + (-28) + 14 + (-17) = -62$

85. $-34 - 28 + (-33) - 44 = (-34) + (-28) + (-33) + (-44) = -139$

87. $-93 - (-84) - 41 - (-56) = (-93) + 84 + (-41) + 56 = 6$

89. $-5.4 - (-30.9) + 30.8 + 40.2 - (-12) = -5.4 + 30.9 + 30.8 + 40.2 + 12 = 108.5$

91. $-\dfrac{7}{12} + \dfrac{3}{4} - \left(-\dfrac{5}{8}\right) - \dfrac{13}{24} = -\dfrac{7}{12} + \dfrac{3}{4} + \dfrac{5}{8} + \left(-\dfrac{13}{24}\right) = -\dfrac{28}{48} + \dfrac{36}{48} + \dfrac{30}{48} + \left(-\dfrac{26}{48}\right) = \dfrac{12}{48} = \dfrac{\cancel{12} \cdot 1}{4 \cdot \cancel{12}} = \dfrac{1}{4}$

93. Let $D =$ the difference in elevation.

$$\underbrace{\text{Difference in elevation}}_{\downarrow \atop D} \; \underbrace{\text{is}}_{\downarrow \atop =} \; \underbrace{\substack{\text{larger} \\ \text{depth}}}_{\downarrow \atop 10{,}924} \; \underbrace{\text{minus}}_{\downarrow \atop -} \; \underbrace{\substack{\text{smaller} \\ \text{depth.}}}_{\downarrow \atop 8605}$$

We carry out the subtraction.

$D = 10{,}924 - 8605 = 2319$

The difference in elevation is 2319 m.

95. Let $A =$ the amount owed.

$$\underbrace{\substack{\text{Amount} \\ \text{owed}}}_{\downarrow \atop A} \; \underbrace{\text{is}}_{\downarrow \atop =} \; \underbrace{\substack{\text{amount} \\ \text{of charge}}}_{\downarrow \atop 476.89} \; \underbrace{\text{minus}}_{\downarrow \atop -} \; \underbrace{\substack{\text{amount} \\ \text{of return.}}}_{\downarrow \atop 128.95}$$

We subtract.

$A = 476.89 - 128.95 = 347.94$

Claire owes $347.94.

97. a) We subtract the number of home runs allowed from the number of home runs hit.

Home run differential $= 197 - 120 = 77$

b) We subtract the number of home runs allowed from the number of home runs hit.

Home run differential $= 153 - 194 = -41$

99. Let $D =$ the difference in elevation.

$$\underbrace{\substack{\text{Difference} \\ \text{in} \\ \text{elevation}}}_{\downarrow \atop D} \; \underbrace{\text{is}}_{\downarrow \atop =} \; \underbrace{\substack{\text{higher} \\ \text{elevation}}}_{\downarrow \atop -131} \; \underbrace{\text{minus}}_{\downarrow \atop -} \; \underbrace{\substack{\text{lower} \\ \text{elevation.}}}_{\downarrow \atop (-512)}$$

We carry out the subtraction.

$D = -131 - (-512) = -131 + 512 = 381$

Lake Assal is 381 ft lower than the Valdes Peninsula.

101. Discussion and Writing Exercise

103. $256 \div 64 \div 2^3 + 100 = 256 \div 64 \div 8 + 100$
$$= 4 \div 8 + 100$$
$$= \frac{1}{2} + 100$$
$$= 100\frac{1}{2}, \text{ or } 100.5$$

105. $2^5 \div 4 + 20 \div 2^2 = 32 \div 4 + 20 \div 4 = 8 + 5 = 13$

107. $\frac{1}{8} + \frac{7}{12} + \frac{5}{24} = \frac{3}{24} + \frac{14}{24} + \frac{5}{24} = \frac{22}{24} = \frac{2 \cdot 11}{2 \cdot 12} = \frac{11}{12}$

109. False. $3 - 0 = 3, 0 - 3 = -3, 3 - 0 \neq 0 - 3$

111. True

113. True by definition of opposites.

Exercise Set 9.5

1. -8

3. -48

5. -24

7. -72

9. 16

11. 42

13. -120

15. -238

17. 1200

19. 98

21. -72

23. -12.4

25. 30

27. 21.7

29. $\frac{2}{3} \cdot \left(-\frac{3}{5}\right) = -\left(\frac{2 \cdot 3}{3 \cdot 5}\right) = -\left(\frac{2}{5} \cdot \frac{3}{3}\right) = -\frac{2}{5}$

31. $-\frac{3}{8} \cdot \left(-\frac{2}{9}\right) = \frac{3 \cdot 2}{8 \cdot 9} = \frac{3 \cdot 2 \cdot 1}{4 \cdot 2 \cdot 3 \cdot 3} = \frac{3 \cdot 2}{3 \cdot 2} \cdot \frac{1}{4 \cdot 3} = \frac{1}{12}$

33. -17.01

35. $-\frac{5}{9} \cdot \frac{3}{4} = -\left(\frac{5 \cdot 3}{9 \cdot 4}\right) = -\frac{5 \cdot 3}{3 \cdot 3 \cdot 4} = -\frac{5}{3 \cdot 4} \cdot \frac{3}{3} = -\frac{5}{12}$

37. $7 \cdot (-4) \cdot (-3) \cdot 5 = 7 \cdot 12 \cdot 5 = 7 \cdot 60 = 420$

39. $-\frac{2}{3} \cdot \frac{1}{2} \cdot \left(-\frac{6}{7}\right) = -\frac{2}{6} \cdot \left(-\frac{6}{7}\right) = \frac{2 \cdot 6}{7 \cdot 6} = \frac{2}{7} \cdot \frac{6}{6} = \frac{2}{7}$

41. $-3 \cdot (-4) \cdot (-5) = 12 \cdot (-5) = -60$

43. $-2 \cdot (-5) \cdot (-3) \cdot (-5) = 10 \cdot 15 = 150$

45. $-\frac{2}{45}$

47. $-7 \cdot (-21) \cdot 13 = 147 \cdot 13 = 1911$

49. $-4 \cdot (-1.8) \cdot 7 = (7.2) \cdot 7 = 50.4$

51. $-\frac{1}{9} \cdot \left(-\frac{2}{3}\right) \cdot \left(\frac{5}{7}\right) = \frac{2}{27} \cdot \frac{5}{7} = \frac{10}{189}$

53. $4 \cdot (-4) \cdot (-5) \cdot (-12) = -16 \cdot (60) = -960$

55. $0.07 \cdot (-7) \cdot 6 \cdot (-6) = 0.07 \cdot 6 \cdot (-7) \cdot (-6) = 0.42 \cdot (42) = 17.64$

57. $\left(-\frac{5}{6}\right)\left(\frac{1}{8}\right)\left(-\frac{3}{7}\right)\left(-\frac{1}{7}\right) = \left(-\frac{5}{48}\right)\left(\frac{3}{49}\right) = -\frac{5 \cdot 3}{16 \cdot 3 \cdot 49} =$
$$-\frac{5}{16 \cdot 49} \cdot \frac{3}{3} = -\frac{5}{784}$$

59. 0, The product of 0 and any real number is 0.

61. $(-8)(-9)(-10) = 72(-10) = -720$

63. $(-6)(-7)(-8)(-9)(-10) = 42 \cdot 72 \cdot (-10) = 3024 \cdot (-10) = -30,240$

65. $\quad (-1)^{12}$
$$= (-1)(-1)(-1)(-1)(-1)(-1)(-1)(-1)(-1)(-1)(-1)(-1)$$
$$= 1 \cdot 1 \cdot 1 \cdot 1 \cdot 1 \cdot 1 = 1$$

67. For $x = 4$:
$$(-x)^2 = (-4)^2 = 16$$
$$-x^2 = -(4)^2 = -(16) = -16$$
For $x = -4$:
$$(-x)^2 = [-(-4)]^2 = [4]^2 = 16$$
$$-x^2 = -(-4)^2 = -(16) = -16$$

69. $(-3x)^2 = (-3 \cdot 7)^2$ Substituting
$$= (-21)^2 \qquad \text{Multiplying inside the parentheses}$$
$$= (-21)(-21) \quad \text{Evaluating the power}$$
$$= 441$$

    $-3x^2 = -3(7)^2$ Substituting
$$= -3 \cdot 49 \quad \text{Evaluating the power}$$
$$= -147$$

71. When $x = 2$: $5x^2 = 5(2)^2$ Substituting
$$= 5 \cdot 4 \quad \text{Evaluating the power}$$
$$= 20$$

When $x = -2$: $5x^2 = 5(-2)^2$ Substituting
$$= 5 \cdot 4 \quad \text{Evaluating the power}$$
$$= 20$$

73. When $x = 1$, $-2x^3 = -2 \cdot 1^3 = -2 \cdot 1 = -2$.
When $x = -1$, $-2x^3 = -2(-1)^3 = -2(-1) = 2$.

75. Let $w = $ the total weight change. Since Dave's weight decreases 2 lb each week for 10 weeks we have
$$w = 10 \cdot (-2) = -20.$$
Thus, the total weight change is -20 lb.

77. This is a multistep problem. First we find the number of degrees the temperature dropped. Since it dropped 3°C each minute for 18 minutes we have a drop d given by
$$d = 18 \cdot (-3) = -54.$$
Now let $T = $ the temperature at 10:18 AM.
$$T = 0 + (-54) = -54$$
The temperature was -54°C at 10:18 AM.

79. This is a multistep problem. First we find the total decrease in price. Since it decreased $1.38 each hour for 8 hours we have a decrease in price d given by

$$d = 8(-\$1.38) = -\$11.04.$$

Now let $P = $ the price of the stock after 8 hours.

$$P = \$23.75 + (-\$11.04) = \$12.71$$

After 8 hours the price of the stock was $12.71.

81. This is a multistep problem. First we find the total distance the diver rises. Since the diver rises 7 meters each minute for 9 minutes, the total distance d the diver rises is given by

$$d = 9 \cdot 7 = 63.$$

Now let $E = $ the diver's elevation after 9 minutes.

$$E = -95 + 63 = -32$$

The diver's elevation is -32 m, or 32 m below the surface.

83. Discussion and Writing Exercise

85. $36 = 2 \cdot 2 \cdot 3 \cdot 3$
$60 = 2 \cdot 2 \cdot 3 \cdot 5$
LCM $= 2 \cdot 2 \cdot 3 \cdot 3 \cdot 5$, or 180

87. $\dfrac{26}{39} = \dfrac{2 \cdot \cancel{13}}{3 \cdot \cancel{13}} = \dfrac{2}{3}$

89. $\dfrac{264}{484} = \dfrac{\cancel{2} \cdot \cancel{2} \cdot 2 \cdot 3 \cdot \cancel{11}}{\cancel{2} \cdot \cancel{2} \cdot \cancel{11} \cdot 11} = \dfrac{6}{11}$

91. $\dfrac{275}{800} = \dfrac{\cancel{25} \cdot 11}{\cancel{25} \cdot 32} = \dfrac{11}{32}$

93. $\dfrac{11}{264} = \dfrac{\cancel{11} \cdot 1}{\cancel{11} \cdot 24} = \dfrac{1}{24}$

95. If a is positive and b is negative, then ab is negative and thus $-ab$ is positive. The correct answer is (a).

97. To locate $2x$, start at 0 and measure off two adjacent lengths of x to the right of 0.

To locate $3x$, start at 0 and measure off three adjacent lengths of x to the right of 0.

To locate $2y$, start at 0 and measure off two adjacent lengths of y to the right of 0.

To locate $-x$, start at 0 and measure off the length x to the left of 0.

To locate $-y$, start at 0 and measure off the length y to the left of 0.

To locate $x + y$, start at 0 and measure off the length x to the right of 0 followed by the length y immediately to the right of x. (We could also measure off y followed by x.)

To locate $x - y$, start at 0 and measure off the length x to the right of 0. Then, from that point, measure off the length y going to the left.

To locate $x - 2y$, first locate $x - y$ as described above. Then, from that point, measure off another length y going to the left.

Exercise Set 9.6

1. $48 \div (-6) = -8$ Check: $-8(-6) = 48$

3. $\dfrac{28}{-2} = -14$ Check: $-14(-2) = 28$

5. $\dfrac{-24}{8} = -3$ Check: $-3 \cdot 8 = -24$

7. $\dfrac{-36}{-12} = 3$ Check: $3(-12) = -36$

9. $\dfrac{-72}{9} = -8$ Check: $-8 \cdot 9 = -72$

11. $-100 \div (-50) = 2$ Check: $2(-50) = -100$

13. $-108 \div 9 = -12$ Check: $9(-12) = -108$

15. $\dfrac{200}{-25} = -8$ Check: $-8(-25) = 200$

17. Not defined

19. $\dfrac{0}{-2.6} = 0$ Check: $0(-2.6) = 0$

21. The reciprocal of $\dfrac{15}{7}$ is $\dfrac{7}{15}$ because $\dfrac{15}{7} \cdot \dfrac{7}{15} = 1$.

23. The reciprocal of $-\dfrac{47}{13}$ is $-\dfrac{13}{47}$ because $\left(-\dfrac{47}{13}\right)\left(-\dfrac{13}{47}\right) = 1$.

25. The reciprocal of 13 is $\dfrac{1}{13}$ because $13 \cdot \dfrac{1}{13} = 1$.

27. The reciprocal of 4.3 is $\dfrac{1}{4.3}$ because $4.3 \cdot \dfrac{1}{4.3} = 1$.

29. The reciprocal of $-\dfrac{1}{7.1}$ is -7.1 because $\left(-\dfrac{1}{7.1}\right)(-7.1) = 1$.

31. The reciprocal of $\dfrac{p}{q}$ is $\dfrac{q}{p}$ because $\dfrac{p}{q} \cdot \dfrac{q}{p} = 1$.

33. The reciprocal of $\dfrac{1}{4y}$ is $4y$ because $\dfrac{1}{4y} \cdot 4y = 1$.

35. The reciprocal of $\dfrac{2a}{3b}$ is $\dfrac{3b}{2a}$ because $\dfrac{2a}{3b} \cdot \dfrac{3b}{2a} = 1$.

37. $4 \cdot \dfrac{1}{17}$

39. $8 \cdot \left(-\dfrac{1}{13}\right)$

41. $13.9 \cdot \left(-\dfrac{1}{1.5}\right)$

43. $x \cdot y$

45. $(3x + 4)\left(\dfrac{1}{5}\right)$

47. $(5a - b)\left(\dfrac{1}{5a + b}\right)$

49. $\dfrac{3}{4} \div \left(-\dfrac{2}{3}\right) = \dfrac{3}{4} \cdot \left(-\dfrac{3}{2}\right) = -\dfrac{9}{8}$

51. $-\dfrac{5}{4} \div \left(-\dfrac{3}{4}\right) = -\dfrac{5}{4} \cdot \left(-\dfrac{4}{3}\right) = \dfrac{20}{12} = \dfrac{5 \cdot 4}{3 \cdot 4} = \dfrac{5}{3}$

53. $-\dfrac{2}{7} \div \left(-\dfrac{4}{9}\right) = -\dfrac{2}{7} \cdot \left(-\dfrac{9}{4}\right) = \dfrac{18}{28} = \dfrac{9 \cdot 2}{14 \cdot 2} = \dfrac{9}{14}$

55. $-\dfrac{3}{8} \div \left(-\dfrac{8}{3}\right) = -\dfrac{3}{8} \cdot \left(-\dfrac{3}{8}\right) = \dfrac{9}{64}$

57. $-6.6 \div 3.3 = -2$ Do the long division. Make the answer negative.

59. $\dfrac{-11}{-13} = \dfrac{11}{13}$ The opposite of a number divided by the opposite of another number is the quotient of the two numbers.

61. $\dfrac{48.6}{-3} = -16.2$ Do the long division. Make the answer negative.

63. $\dfrac{-9}{17 - 17} = \dfrac{-9}{0}$ Division by 0 is not defined.

65. $\dfrac{81}{183} \approx 0.443 \approx 44.3\%$

67. $\dfrac{-22}{428} \approx -0.051 \approx -5.1\%$

69. Discussion and Writing Exercise

71. $2^3 - 5 \cdot 3 + 8 \cdot 10 \div 2$

$\begin{aligned} &= 8 - 5 \cdot 3 + 8 \cdot 10 \div 2 &&\text{Evaluating the power} \\ &= 8 - 15 + 80 \div 2 &&\text{Multiplying and dividing} \\ &= 8 - 15 + 40 &&\text{in order from left to right} \\ &= -7 + 40 &&\text{Adding and subtracting} \\ &= 33 &&\text{in order from left to right} \end{aligned}$

73. $\quad 1000 \div 100 \div 10$

$\begin{aligned} &= 10 \div 10 &&\text{Dividing in order from} \\ &= 1 &&\text{left to right} \end{aligned}$

75. $\dfrac{264}{468} = \dfrac{4 \cdot 66}{4 \cdot 117} = \dfrac{\cancel{4} \cdot \cancel{3} \cdot 22}{\cancel{4} \cdot \cancel{3} \cdot 39} = \dfrac{22}{39}$

77. $\dfrac{7}{8} = 0.875 = 87.5\%$

79. $\dfrac{12}{25} \div \dfrac{32}{75} = \dfrac{12}{25} \cdot \dfrac{75}{32}$

$\begin{aligned} &= \dfrac{12 \cdot 75}{25 \cdot 32} \\ &= \dfrac{3 \cdot \cancel{4} \cdot 3 \cdot \cancel{25}}{\cancel{25} \cdot \cancel{4} \cdot 8} \\ &= \dfrac{9}{8} \end{aligned}$

81. The reciprocal of -10.5 is $\dfrac{1}{-10.5}$.

The reciprocal of $\dfrac{1}{-10.5} = -10.5$.

We see that the reciprocal of the reciprocal is the original number.

83. $-a$ is positive and b is negative, so $\dfrac{-a}{b}$ is the quotient of a positive and a negative number and, thus, is negative.

85. a is negative and $-b$ is positive, so $\dfrac{a}{-b}$ is the quotient of a negative number and a positive number and, thus, is negative. Then $-\left(\dfrac{a}{-b}\right)$ is the opposite of a negative number and, thus, is positive.

87. $-a$ and $-b$ are both positive, so $\dfrac{-a}{-b}$ is the quotient of two positive numbers and, thus, is positive. Then $-\left(\dfrac{-a}{-b}\right)$ is the opposite of a positive number and, thus, is negative.

Exercise Set 9.7

1. Note that $5y = 5 \cdot y$. We multiply by 1, using y/y as an equivalent expression for 1:

$$\dfrac{3}{5} = \dfrac{3}{5} \cdot 1 = \dfrac{3}{5} \cdot \dfrac{y}{y} = \dfrac{3y}{5y}$$

3. Note that $15x = 3 \cdot 5x$. We multiply by 1, using $5x/5x$ as an equivalent expression for 1:

$$\dfrac{2}{3} = \dfrac{2}{3} \cdot 1 = \dfrac{2}{3} \cdot \dfrac{5x}{5x} = \dfrac{10x}{15x}$$

5. Note that $x^2 = x \cdot x$. We multiply by 1, using x/x as an equivalent expression for 1.

$$\dfrac{2}{x} = \dfrac{2}{x} \cdot 1 = \dfrac{2}{x} \cdot \dfrac{x}{x} = \dfrac{2x}{x^2}$$

7. $\begin{aligned} -\dfrac{24a}{16a} &= -\dfrac{3 \cdot 8a}{2 \cdot 8a} \\ &= -\dfrac{3}{2} \cdot \dfrac{8a}{8a} \\ &= -\dfrac{3}{2} \cdot 1 &&\left(\dfrac{8a}{8a} = 1\right) \\ &= -\dfrac{3}{2} &&\text{Identity property of 1} \end{aligned}$

9. $\begin{aligned} -\dfrac{42ab}{36ab} &= -\dfrac{7 \cdot 6ab}{6 \cdot 6ab} \\ &= -\dfrac{7}{6} \cdot \dfrac{6ab}{6ab} \\ &= -\dfrac{7}{6} \cdot 1 &&\left(\dfrac{6ab}{6ab} = 1\right) \\ &= -\dfrac{7}{6} &&\text{Identity property of 1} \end{aligned}$

11. $\begin{aligned} \dfrac{20st}{15t} &= \dfrac{4s \cdot 5t}{3 \cdot 5t} \\ &= \dfrac{4s}{3} \cdot \dfrac{5t}{5t} \\ &= \dfrac{4s}{3} \cdot 1 &&\left(\dfrac{5t}{5t} = 1\right) \\ &= \dfrac{4s}{3} &&\text{Identity property of 1} \end{aligned}$

13. $8 + y$, commutative law of addition

15. nm, commutative law of multiplication

17. $xy + 9$, commutative law of addition

$9 + yx$, commutative law of multiplication

19. $c + ab$, commutative law of addition

$ba + c$, commutative law of multiplication

21. $(a + b) + 2$, associative law of addition

23. $8(xy)$, associative law of multiplication

25. $a + (b + 3)$, associative law of addition

27. $(3a)b$, associative law of multiplication

29. a) $(a + b) + 2 = a + (b + 2)$, associative law of addition

b) $(a + b) + 2 = (b + a) + 2$, commutative law of addition

c) $(a + b) + 2 = (b + a) + 2$ Using the commutative law first,

$\qquad = b + (a + 2)$ then the associative law

There are other correct answers.

31. a) $5 + (v + w) = (5 + v) + w$, associative law of addition

b) $5 + (v + w) = 5 + (w + v)$, commutative law of addition

c) $5 + (v + w) = 5 + (w + v)$ Using the commutative law first,

$\qquad = (5 + w) + v$ then the associative law

There are other correct answers.

33. a) $(xy)3 = x(y3)$, associative law of multiplication

b) $(xy)3 = (yx)3$, commutative law of multiplication

c) $(xy)3 = (yx)3$ Using the commutative law first,

$\qquad = y(x3)$ then the associative law

There are other correct answers.

35. a) $7(ab) = (7a)b$

b) $7(ab) = (7a)b = b(7a)$

c) $7(ab) = 7(ba) = (7b)a$

There are other correct answers.

37. $2(b + 5) = 2 \cdot b + 2 \cdot 5 = 2b + 10$

39. $7(1 + t) = 7 \cdot 1 + 7 \cdot t = 7 + 7t$

41. $6(5x + 2) = 6 \cdot 5x + 6 \cdot 2 = 30x + 12$

43. $7(x + 4 + 6y) = 7 \cdot x + 7 \cdot 4 + 7 \cdot 6y = 7x + 28 + 42y$

45. $7(x - 3) = 7 \cdot x - 7 \cdot 3 = 7x - 21$

47. $-3(x - 7) = -3 \cdot x - (-3) \cdot 7 = -3x - (-21) = -3x + 21$

49. $\frac{2}{3}(b - 6) = \frac{2}{3} \cdot b - \frac{2}{3} \cdot 6 = \frac{2}{3}b - 4$

51. $7.3(x - 2) = 7.3 \cdot x - 7.3 \cdot 2 = 7.3x - 14.6$

53. $-\frac{3}{5}(x - y + 10) = -\frac{3}{5} \cdot x - \left(-\frac{3}{5}\right) \cdot y + \left(-\frac{3}{5}\right) \cdot 10 =$

$-\frac{3}{5}x - \left(-\frac{3}{5}y\right) + (-6) = -\frac{3}{5}x + \frac{3}{5}y - 6$

55. $-9(-5x - 6y + 8) = -9(-5x) - (-9)6y + (-9)8$

$= 45x - (-54y) + (-72) = 45x + 54y - 72$

57. $-4(x - 3y - 2z) = -4 \cdot x - (-4)3y - (-4)2z$

$= -4x - (-12y) - (-8z) = -4x + 12y + 8z$

59. $3.1(-1.2x + 3.2y - 1.1) = 3.1(-1.2x) + (3.1)3.2y - 3.1(1.1)$

$= -3.72x + 9.92y - 3.41$

61. $4x + 3z$ Parts are separated by plus signs. The terms are $4x$ and $3z$.

63. $7x + 8y - 9z = 7x + 8y + (-9z)$ Separating parts with plus signs

The terms are $7x$, $8y$, and $-9z$.

65. $2x + 4 = 2 \cdot x + 2 \cdot 2 = 2(x + 2)$

67. $30 + 5y = 5 \cdot 6 + 5 \cdot y = 5(6 + y)$

69. $14x + 21y = 7 \cdot 2x + 7 \cdot 3y = 7(2x + 3y)$

71. $5x + 10 + 15y = 5 \cdot x + 5 \cdot 2 + 5 \cdot 3y = 5(x + 2 + 3y)$

73. $8x - 24 = 8 \cdot x - 8 \cdot 3 = 8(x - 3)$

75. $-4y + 32 = -4 \cdot y - 4(-8) = -4(y - 8)$

We could also factor this expression as follows:

$-4y + 32 = 4(-y) + 4 \cdot 8 = 4(-y + 8)$

77. $8x + 10y - 22 = 2 \cdot 4x + 2 \cdot 5y - 2 \cdot 11 = 2(4x + 5y - 11)$

79. $ax - a = a \cdot x - a \cdot 1 = a(x - 1)$

81. $ax - ay - az = a \cdot x - a \cdot y - a \cdot z = a(x - y - z)$

83. $-18x + 12y + 6 = -6 \cdot 3x - 6(-2y) - 6(-1) = -6(3x - 2y - 1)$

We could also factor this expression as follows:

$-18x + 12y + 6 = 6(-3x) + 6 \cdot 2y + 6 \cdot 1 = 6(-3x + 2y + 1)$

85. $\frac{2}{3}x - \frac{5}{3}y + \frac{1}{3} = \frac{1}{3} \cdot 2x - \frac{1}{3} \cdot 5y + \frac{1}{3} \cdot 1 =$

$\frac{1}{3}(2x - 5y + 1)$

87. $9a + 10a = (9 + 10)a = 19a$

89. $10a - a = 10a - 1 \cdot a = (10 - 1)a = 9a$

91. $2x + 9z + 6x = 2x + 6x + 9z = (2 + 6)x + 9z = 8x + 9z$

93. $7x + 6y^2 + 9y^2 = 7x + (6 + 9)y^2 = 7x + 15y^2$

95. $41a + 90 - 60a - 2 = 41a - 60a + 90 - 2$

$= (41 - 60)a + (90 - 2)$

$= -19a + 88$

97. $23 + 5t + 7y - t - y - 27$

$= 23 - 27 + 5t - 1 \cdot t + 7y - 1 \cdot y$

$= (23 - 27) + (5 - 1)t + (7 - 1)y$

$= -4 + 4t + 6y$, or $4t + 6y - 4$

99. $\frac{1}{2}b + \frac{1}{2}b = \left(\frac{1}{2} + \frac{1}{2}\right)b = 1b = b$

101. $2y + \dfrac{1}{4}y + y = 2y + \dfrac{1}{4}y + 1 \cdot y = \left(2 + \dfrac{1}{4} + 1\right)y = 3\dfrac{1}{4}y$, or $\dfrac{13}{4}y$

103. $11x - 3x = (11 - 3)x = 8x$

105. $6n - n = (6 - 1)n = 5n$

107. $y - 17y = (1 - 17)y = -16y$

109. $\quad -8 + 11a - 5b + 6a - 7b + 7$
$= 11a + 6a - 5b - 7b - 8 + 7$
$= (11 + 6)a + (-5 - 7)b + (-8 + 7)$
$= 17a - 12b - 1$

111. $9x + 2y - 5x = (9 - 5)x + 2y = 4x + 2y$

113. $11x + 2y - 4x - y = (11 - 4)x + (2 - 1)y = 7x + y$

115. $2.7x + 2.3y - 1.9x - 1.8y = (2.7 - 1.9)x + (2.3 - 1.8)y = 0.8x + 0.5y$

117. $\quad \dfrac{13}{2}a + \dfrac{9}{5}b - \dfrac{2}{3}a - \dfrac{3}{10}b - 42$
$= \left(\dfrac{13}{2} - \dfrac{2}{3}\right)a + \left(\dfrac{9}{5} - \dfrac{3}{10}\right)b - 42$
$= \left(\dfrac{39}{6} - \dfrac{4}{6}\right)a + \left(\dfrac{18}{10} - \dfrac{3}{10}\right)b - 42$
$= \dfrac{35}{6}a + \dfrac{15}{10}b - 42$
$= \dfrac{35}{6}a + \dfrac{3}{2}b - 42$

119. Discussion and Writing Exercise

121. $16 = 2 \cdot 2 \cdot 2 \cdot 2$
$18 = 2 \cdot 3 \cdot 3$
The LCM is $2 \cdot 2 \cdot 2 \cdot 2 \cdot 3 \cdot 3$, or 144.

123. $16 = 2 \cdot 2 \cdot 2 \cdot 2$
$18 = 2 \cdot 3 \cdot 3$
$24 = 2 \cdot 2 \cdot 2 \cdot 3$
The LCM is $2 \cdot 2 \cdot 2 \cdot 2 \cdot 3 \cdot 3$, or 144.

125. $16 = 2 \cdot 2 \cdot 2 \cdot 2$
$32 = 2 \cdot 2 \cdot 2 \cdot 2 \cdot 2$
The LCM is $2 \cdot 2 \cdot 2 \cdot 2 \cdot 2$, or 32.

127. $15 = 3 \cdot 5$
$45 = 3 \cdot 3 \cdot 5$
$90 = 2 \cdot 3 \cdot 3 \cdot 5$
The LCM is $2 \cdot 3 \cdot 3 \cdot 5$, or 90.

129. $\dfrac{11}{12} + \dfrac{15}{16} = \dfrac{11}{12} \cdot \dfrac{4}{4} + \dfrac{15}{16} \cdot \dfrac{3}{3}$ LCD is 48
$= \dfrac{44}{48} + \dfrac{45}{48}$
$= \dfrac{89}{48}$

131. $\dfrac{1}{8} - \dfrac{1}{3} = \dfrac{1}{8} + \left(-\dfrac{1}{3}\right) = \dfrac{3}{24} + \left(-\dfrac{8}{24}\right) = -\dfrac{5}{24}$

133. No; for any replacement other than 5 the two expressions do not have the same value. For example, let $t = 2$. Then $3 \cdot 2 + 5 = 6 + 5 = 11$, but $3 \cdot 5 + 2 = 15 + 2 = 17$.

135. Yes; commutative law of addition

137. $\quad q + qr + qrs + qrst$ There are no like terms.
$= q \cdot 1 + q \cdot r + q \cdot rs + q \cdot rst$
$= q(1 + r + rs + rst)$ Factoring

Exercise Set 9.8

1. $-(2x + 7) = -2x - 7$ Changing the sign of each term

3. $-(8 - x) = -8 + x$ Changing the sign of each term

5. $-4a + 3b - 7c$

7. $-6x + 8y - 5$

9. $-3x + 5y + 6$

11. $8x + 6y + 43$

13. $9x - (4x + 3) = 9x - 4x - 3$ Removing parentheses by changing the sign of every term
$\qquad\qquad\qquad = 5x - 3$ Collecting like terms

15. $2a - (5a - 9) = 2a - 5a + 9 = -3a + 9$

17. $2x + 7x - (4x + 6) = 2x + 7x - 4x - 6 = 5x - 6$

19. $2x - 4y - 3(7x - 2y) = 2x - 4y - 21x + 6y = -19x + 2y$

21. $\quad 15x - y - 5(3x - 2y + 5z)$
$= 15x - y - 15x + 10y - 25z$ Multiplying each term in parentheses by -5
$= 9y - 25z$

23. $(3x + 2y) - 2(5x - 4y) = 3x + 2y - 10x + 8y = -7x + 10y$

25. $\quad (12a - 3b + 5c) - 5(-5a + 4b - 6c)$
$= 12a - 3b + 5c + 25a - 20b + 30c$
$= 37a - 23b + 35c$

27. $[9 - 2(5 - 4)] = [9 - 2 \cdot 1]$ Computing $5 - 4$
$\qquad\qquad\quad = [9 - 2]$ Computing $2 \cdot 1$
$\qquad\qquad\quad = 7$

29. $8[7 - 6(4 - 2)] = 8[7 - 6(2)] = 8[7 - 12] = 8[-5] = -40$

31. $\quad [4(9 - 6) + 11] - [14 - (6 + 4)]$
$= [4(3) + 11] - [14 - 10]$
$= [12 + 11] - [14 - 10]$
$= 23 - 4$
$= 19$

33. $\quad [10(x + 3) - 4] + [2(x - 1) + 6]$
$= [10x + 30 - 4] + [2x - 2 + 6]$
$= [10x + 26] + [2x + 4]$
$= 10x + 26 + 2x + 4$
$= 12x + 30$

35. $\quad [7(x+5)-19]-[4(x-6)+10]$
$= [7x+35-19]-[4x-24+10]$
$= [7x+16]-[4x-14]$
$= 7x+16-4x+14$
$= 3x+30$

37. $\quad 3\{[7(x-2)+4]-[2(2x-5)+6]\}$
$= 3\{[7x-14+4]-[4x-10+6]\}$
$= 3\{[7x-10]-[4x-4]\}$
$= 3\{7x-10-4x+4\}$
$= 3\{3x-6\}$
$= 9x-18$

39. $\quad 4\{[5(x-3)+2]-3[2(x+5)-9]\}$
$= 4\{[5x-15+2]-3[2x+10-9]\}$
$= 4\{[5x-13]-3[2x+1]\}$
$= 4\{5x-13-6x-3\}$
$= 4\{-x-16\}$
$= -4x-64$

41. $8-2\cdot3-9 = 8-6-9$ Multiplying
$\qquad\qquad = 2-9$ Doing all additions and subtractions in order from
$\qquad\qquad = -7$ left to right

43. $(8-2\cdot3)-9 = (8-6)-9$ Multiplying inside the parentheses
$\qquad\qquad = 2-9$ Subtracting inside the parentheses
$\qquad\qquad = -7$

45. $[(-24)\div(-3)]\div\left(-\dfrac{1}{2}\right) = 8\div\left(-\dfrac{1}{2}\right) = 8\cdot(-2) = -16$

47. $16\cdot(-24)+50 = -384+50 = -334$

49. $2^4+2^3-10 = 16+8-10 = 24-10 = 14$

51. $5^3+26\cdot71-(16+25\cdot3) = 5^3+26\cdot71-(16+75) =$
$5^3+26\cdot71-91 = 125+26\cdot71-91 = 125+1846-91 =$
$1971-91 = 1880$

53. $4\cdot5-2\cdot6+4 = 20-12+4 = 8+4 = 12$

55. $4^3/8 = 64/8 = 8$

57. $8(-7)+6(-5) = -56-30 = -86$

59. $19-5(-3)+3 = 19+15+3 = 34+3 = 37$

61. $9\div(-3)+16\div8 = -3+2 = -1$

63. $-4^2+6 = -16+6 = -10$

65. $-8^2-3 = -64-3 = -67$

67. $12-20^3 = 12-8000 = -7988$

69. $2\cdot10^3-5000 = 2\cdot1000-5000 = 2000-5000 = -3000$

71. $6[9-(3-4)] = 6[9-(-1)] = 6[9+1] = 6[10] = 60$

73. $-1000\div(-100)\div10 = 10\div10 = 1$

75. $8-(7-9) = 8-(-2) = 8+2 = 10$

77. $\dfrac{10-6^2}{9^2+3^2} = \dfrac{10-36}{81+9} = \dfrac{-26}{90} = -\dfrac{13}{45}$

79. $\dfrac{3(6-7)-5\cdot4}{6\cdot7-8(4-1)} = \dfrac{3(-1)-5\cdot4}{42-8\cdot3} = \dfrac{-3-20}{42-24} = -\dfrac{23}{18}$

81. $\dfrac{|2^3-3^2|+|12\cdot5|}{-32\div(-16)\div(-4)} = \dfrac{|8-9|+|12\cdot5|}{-32\div(-16)\div(-4)} =$
$\dfrac{|-1|+|60|}{2\div(-4)} = \dfrac{1+60}{-\frac{1}{2}} = \dfrac{61}{-\frac{1}{2}} = 61(-2) = -122$

83. Discussion and Writing Exercise

85. The set of <u>integers</u> is
$\{\ldots,-5,\overline{-4,-3},-2,-1,0,1,2,3,\ldots\}$.

87. The <u>commutative law</u> of addition says that $a+b=b+a$ for any real numbers a and b.

89. The <u>associative law</u> of addition says that $a+(b+c)=(a+b)+c$ for any real numbers a, b, and c.

91. Two numbers whose product is 1 are called <u>multiplicative inverses</u> of each other.

93. $6y+2x-3a+c = 6y-(-2x)-3a-(-c) = 6y-(-2x+3a-c)$

95. $6m+3n-5m+4b = 6m-(-3n)-5m-(-4b) =$
$6m-(-3n+5m-4b)$

97. $\quad \{x-[f-(f-x)]+[x-f]\}-3x$
$= \{x-[f-f+x]+[x-f]\}-3x$
$= \{x-[x]+[x-f]\}-3x$
$= \{x-x+x-f\}-3x = x-f-3x = -2x-f$

99. a) $\quad x^2+3 = 7^2+3 = 49+3 = 52$;
$\qquad x^2+3 = (-7)^2+3 = 49+3 = 52$;
$\qquad x^2+3 = (-5.013)^2+3 = 25.130169+3 = 28.130169$
    b) $\quad 1-x^2 = 1-5^2 = 1-25 = -24$;
$\qquad 1-x^2 = 1-(-5)^2 = 1-25 = -24$;
$\qquad 1-x^2 = 1-(-10.455)^2 = 1-109.307025 =$
$\qquad -108.307025$

101. $\dfrac{-15+20+50+(-82)+(-7)+(-2)}{6} = \dfrac{-36}{6} = -6$

Chapter 9 Review Exercises

1. Substitute 17 for x and 5 for y and carry out the computation.
$$\frac{x-y}{3} = \frac{17-5}{3} = \frac{12}{3} = 4$$

2. $19\%x$, $0.19x$

3. The integer -45 corresponds to a debt of \$45; the integer 72 corresponds to having \$72 in a savings account.

4. The distance of -38 from 0 is 38, so $|-38| = 38$.

5. The graph of -2.5 is halfway between -3 and -2.

6. The graph of $\frac{8}{9}$ is $\frac{8}{9}$ of the way from 0 to 1.

7. Since -3 is to the left of 10, we have $-3 < 10$.

8. Since -1 is to the right of -6, we have $-1 > -6$.

9. Since 0.126 is to the right of -12.6, we have $0.126 > -12.6$.

10. $-\dfrac{2}{3} = -\dfrac{2}{3} \cdot \dfrac{10}{10} = -\dfrac{20}{30}$

$-\dfrac{1}{10} = -\dfrac{1}{10} \cdot \dfrac{3}{3} = -\dfrac{3}{30}$

Since $-\dfrac{20}{30}$ is to the left of $-\dfrac{3}{30}$, then $-\dfrac{2}{3}$ is to the left of $-\dfrac{1}{10}$ and we have $-\dfrac{2}{3} < -\dfrac{1}{10}$.

11. The opposite of 3.8 is -3.8 because $3.8 + (-3.8) = 0$.

12. The opposite of $-\dfrac{3}{4}$ is $\dfrac{3}{4}$ because $-\dfrac{3}{4} + \dfrac{3}{4} = 0$.

13. The reciprocal of $\dfrac{3}{8}$ is $\dfrac{8}{3}$ because $\dfrac{3}{8} \cdot \dfrac{8}{3} = 1$.

14. The reciprocal of -7 is $-\dfrac{1}{7}$ because $-7 \cdot \left(-\dfrac{1}{7}\right) = 1$.

15. If $x = -34$, then $-x = -(-34) = 34$.

16. If $x = 5$, then $-(-x) = -(-5) = 5$.

17. $4 + (-7)$

The absolute values are 4 and 7. The difference is $7 - 4$, or 3. The negative number has the larger absolute value, so the answer is negative. $4 + (-7) = -3$

18. $6 + (-9) + (-8) + 7$

a) Add the negative numbers: $-9 + (-8) = -17$

b) Add the positive numbers: $6 + 7 = 13$

c) Add the results: $-17 + 13 = -4$

19. $-3.8 + 5.1 + (-12) + (-4.3) + 10$

a) Add the negative numbers: $-3.8 + (-12) + (-4.3) = -20.1$

b) Add the positive numbers: $5.1 + 10 = 15.1$

c) Add the results: $-20.1 + 15.1 = -5$

20. $-3 - (-7) = -3 + 7 = 4$

21. $-\dfrac{9}{10} - \dfrac{1}{2} = -\dfrac{9}{10} - \dfrac{5}{10} = -\dfrac{9}{10} + \left(-\dfrac{5}{10}\right) = -\dfrac{14}{10} = -\dfrac{7 \cdot 2}{5 \cdot 2} = -\dfrac{7}{5} \cdot \dfrac{2}{2} = -\dfrac{7}{5}$

22. $-3.8 - 4.1 = -3.8 + (-4.1) = -7.9$

23. $-9 \cdot (-6) = 54$

24. $-2.7(3.4) = -9.18$

25. $\dfrac{2}{3} \cdot \left(-\dfrac{3}{7}\right) = -\left(\dfrac{2 \cdot 3}{3 \cdot 7}\right) = -\left(\dfrac{2}{7} \cdot \dfrac{3}{3}\right) = -\dfrac{2}{7}$

26. $3 \cdot (-7) \cdot (-2) \cdot (-5) = -21 \cdot 10 = -210$

27. $35 \div (-5) = -7$ Check: $-7 \cdot (-5) = 35$

28. $-5.1 \div 1.7 = -3$ Check: $-3 \cdot (1.7) = -5.1$

29. $-\dfrac{3}{11} \div -\dfrac{4}{11} = -\dfrac{3}{11} \cdot \left(-\dfrac{11}{4}\right) = \dfrac{3 \cdot 11}{11 \cdot 4} = \dfrac{3}{4} \cdot \dfrac{11}{11} = \dfrac{3}{4}$

30. $(-3.4 - 12.2) - 8(-7) = -15.6 - 8(-7)$
$$= -15.6 + 56$$
$$= 40.4$$

31. $\dfrac{-12(-3) - 2^3 - (-9)(-10)}{3 \cdot 10 + 1} = \dfrac{-12(-3) - 8 - (-9)(-10)}{30 + 1}$

$$= \dfrac{36 - 8 - 90}{31}$$

$$= \dfrac{28 - 90}{31}$$

$$= \dfrac{-62}{31}$$

$$= -2$$

32. $-16 \div 4 - 30 \div (-5) = -4 - (-6)$
$$= -4 + 6$$
$$= 2$$

33. $\dfrac{9[(7 - 14) - 13]}{|-2(8) - 4|} = \dfrac{9[-7 - 13]}{|-16 - 4|} = \dfrac{9[-20]}{|-20|} = \dfrac{-180}{20} = -9$

34. Let $t =$ the total gain or loss. We represent the gains as positive numbers and the loss as a negative number. We add the gains and the loss to find t.

$$t = 5 + (-12) + 15 = -7 + 15 = 8$$

There is a total gain of 8 yd.

35. Let $a =$ Kaleb's total assets after he borrows \$300.

Total assets	is	Initial assets	minus	Amount of loan
↓	↓	↓	↓	↓
a	$=$	170	$-$	300

We carry out the subtraction.

$$a = 170 - 300 = -130$$

Kaleb's total assets were $-\$130$.

36. First we multiply to find the total drop d in the price:

$$d = 8(-\$1.63) = -\$13.04$$

Now we add this number to the opening price to find the price p after 8 hr:

$$p = \$17.68 + (-\$13.04) = \$4.64$$

After 8 hr the price of the stock was \$4.64 per share.

37. Yuri spent the \$68 in his account plus an additional \$64.65, so he spent a total of $\$68 + \$64.65$, or \$132.65, on seven equally-priced DVDs. Then each DVD cost $\dfrac{\$132.65}{7}$, or \$18.95.

38. $5(3x - 7) = 5 \cdot 3x - 5 \cdot 7 = 15x - 35$

39. $-2(4x - 5) = -2 \cdot 4x - (-2)(5) = -8x - (-10) = -8x + 10$

40. $10(0.4x + 1.5) = 10 \cdot 0.4x + 10 \cdot 1.5 = 4x + 15$

41. $-8(3 - 6x) = -8 \cdot 3 - (-8)(6x) = -24 - (-48x) = -24 + 48x$

42. $2x - 14 = 2 \cdot x - 2 \cdot 7 = 2(x - 7)$

43. $-6x + 6 = -6 \cdot x - 6(-1) = -6(x - 1)$

The expression can also be factored as follows:

$-6x + 6 = 6(-x) + 6 \cdot 1 = 6(-x + 1)$

44. $5x + 10 = 5 \cdot x + 5 \cdot 2 = 5(x + 2)$

45. $-3x + 12y - 12 = -3 \cdot x - 3(-4y) - 3 \cdot 4 = -3(x - 4y + 4)$

We could also factor this expression as follows:

$-3x + 12y - 12 = 3(-x) + 3 \cdot 4y + 3(-4) = 3(-x + 4y - 4)$

46. $\begin{aligned} 11a + 2b - 4a - 5b &= 11a - 4a + 2b - 5b \\ &= (11 - 4)a + (2 - 5)b \\ &= 7a - 3b \end{aligned}$

47. $\begin{aligned} 7x - 3y - 9x + 8y &= 7x - 9x - 3y + 8y \\ &= (7 - 9)x + (-3 + 8)y \\ &= -2x + 5y \end{aligned}$

48. $\begin{aligned} 6x + 3y - x - 4y &= 6x - x + 3y - 4y \\ &= (6 - 1)x + (3 - 4)y \\ &= 5x - y \end{aligned}$

49. $\begin{aligned} -3a + 9b + 2a - b &= -3a + 2a + 9b - b \\ &= (-3 + 2)a + (9 - 1)b \\ &= -a + 8b \end{aligned}$

50. $2a - (5a - 9) = 2a - 5a + 9 = -3a + 9$

51. $3(b + 7) - 5b = 3b + 21 - 5b = -2b + 21$

52. $3[11 - 3(4 - 1)] = 3[11 - 3 \cdot 3] = 3[11 - 9] = 3 \cdot 2 = 6$

53. $2[6(y - 4) + 7] = 2[6y - 24 + 7] = 2[6y - 17] = 12y - 34$

54. $\begin{aligned} &[8(x + 4) - 10] - [3(x - 2) + 4] \\ &= [8x + 32 - 10] - [3x - 6 + 4] \\ &= 8x + 22 - [3x - 2] \\ &= 8x + 22 - 3x + 2 \\ &= 5x + 24 \end{aligned}$

55. $\begin{aligned} &5\{[6(x - 1) + 7] - [3(3x - 4) + 8]\} \\ &= 5\{[6x - 6 + 7] - [9x - 12 + 8]\} \\ &= 5\{6x + 1 - [9x - 4]\} \\ &= 5\{6x + 1 - 9x + 4\} \\ &= 5\{-3x + 5\} \\ &= -15x + 25 \end{aligned}$

56. $-9 \leq 11$ is true since $-9 < 11$ is true.

57. $-11 \geq -3$ is false since neither $-11 > -3$ nor $-11 = -3$ is true.

58. $x > -3$ has the same meaning as $-3 < x$.

59. *Discussion and Writing Exercise.* If the sum of two numbers is 0, they are opposites, or additive inverses of each other. For every real number a, the opposite of a can be named $-a$, and $a + (-a) = (-a) + a = 0$.

60. *Discussion and Writing Exercise.* No; $|0| = 0$, and 0 is not positive.

61. $\begin{aligned} -\left|\frac{7}{8} - \left(-\frac{1}{2}\right) - \frac{3}{4}\right| &= -\left|\frac{7}{8} + \frac{1}{2} - \frac{3}{4}\right| \\ &= -\left|\frac{7}{8} + \frac{4}{8} - \frac{6}{8}\right| \\ &= -\left|\frac{11}{8} - \frac{6}{8}\right| \\ &= -\left|\frac{5}{8}\right| \\ &= -\frac{5}{8} \end{aligned}$

62. $\begin{aligned} &(|2.7 - 3| + 3^2 - |-3|) \div (-3) \\ &= (|2.7 - 3| + 9 - |-3|) \div (-3) \\ &= (|-0.3| + 9 - |-3|) \div (-3) \\ &= (0.3 + 9 - 3) \div (-3) \\ &= (9.3 - 3) \div (-3) \\ &= 6.3 \div (-3) \\ &= -2.1 \end{aligned}$

63. $\underbrace{2000 - 1990}_{\downarrow \atop 10} + \underbrace{1980 - 1970}_{\downarrow \atop 10} + \ldots + \underbrace{20 - 10}_{\downarrow \atop 10}$

$10 \quad + \quad 10 \quad + \ldots + \quad 10$

Counting by 10's from 10 through 2000 gives us $2000/10$, or 200, numbers in the expression. There are $200/2$, or 100, pairs of numbers in the expression. Each pair is a difference that is equivalent to 10. Thus, the expression is equal to $100 \cdot 10$, or 1000.

64. Note that the sum of the lengths of the three horizontal segments at the top of the figure, two of which are not labeled and one of which is labeled b, is equivalent to the length of the horizontal segment at the bottom of the figure, a. Then the perimeter is $a + b + b + a + a + a$, or $4a + 2b$.

Chapter 9 Test

1. Substitute 10 for x and 5 for y and carry out the computations.

$$\frac{3x}{y} = \frac{3 \cdot 10}{5} = \frac{30}{5} = 6$$

2. Using x for "some number," we have $x - 9$.

3. Substitute 16 ft for b and 30 ft for h in the formula for the area of a triangle and then carry out the multiplication.

$$A = \frac{1}{2}bh = \frac{1}{2}(16\text{ ft})(30\text{ ft})$$
$$= \frac{1}{2}(16)(30)(\text{ft})(\text{ft})$$
$$= 240\text{ ft}^2$$

4. Since -4 is to the left of 0 on the number line, we have $-4 < 0$.

5. Since -3 is to the right of -8 on the number line, we have $-3 > -8$.

6. Since -0.78 is to the right of -0.87 on the number line, we have $-0.78 > -0.87$.

7. Since $-\frac{1}{8}$ is to the left of $\frac{1}{2}$ on the number line, we have $-\frac{1}{8} < \frac{1}{2}$.

8. The distance of -7 from 0 is 7, so $|-7| = 7$.

9. The distance of $\frac{9}{4}$ from 0 is $\frac{9}{4}$, so $\left|\frac{9}{4}\right| = \frac{9}{4}$.

10. The distance of -2.7 from 0 is 2.7, so $|-2.7| = 2.7$.

11. The opposite of $\frac{2}{3}$ is $-\frac{2}{3}$ because $\frac{2}{3} + \left(-\frac{2}{3}\right) = 0$.

12. The opposite of -1.4 is 1.4 because $-1.4 + 1.4 = 0$.

13. If $x = -8$, then $-x = -(-8) = 8$.

14. The reciprocal of -2 is $-\frac{1}{2}$ because $-2\left(-\frac{1}{2}\right) = 1$.

15. The reciprocal of $\frac{4}{7}$ is $\frac{7}{4}$ because $\frac{4}{7} \cdot \frac{7}{4} = 1$.

16. $3.1 - (-4.7) = 3.1 + 4.7 = 7.8$

17.
$$-8 + 4 + (-7) + 3 = -4 + (-7) + 3$$
$$= -11 + 3$$
$$= -8$$

18.
$$-\frac{1}{5} + \frac{3}{8} = -\frac{1}{5} \cdot \frac{8}{8} + \frac{3}{8} \cdot \frac{5}{5}$$
$$= -\frac{8}{40} + \frac{15}{40}$$
$$= \frac{7}{40}$$

19. $2 - (-8) = 2 + 8 = 10$

20. $3.2 - 5.7 = 3.2 + (-5.7) = -2.5$

21.
$$\frac{1}{8} - \left(-\frac{3}{4}\right) = \frac{1}{8} + \frac{3}{4}$$
$$= \frac{1}{8} + \frac{3}{4} \cdot \frac{2}{2}$$
$$= \frac{1}{8} + \frac{6}{8}$$
$$= \frac{7}{8}$$

22. $4 \cdot (-12) = -48$

23. $-\frac{1}{2} \cdot \left(-\frac{3}{8}\right) = \frac{3}{16}$

24. $-45 \div 5 = -9$ Check: $-9 \cdot 5 = -45$

25. $-\frac{3}{5} \div \left(-\frac{4}{5}\right) = -\frac{3}{5} \cdot \left(-\frac{5}{4}\right) = \frac{3 \cdot 5}{5 \cdot 4} = \frac{3 \cdot \cancel{5}}{\cancel{5} \cdot 4} = \frac{3}{4}$

26. $4.864 \div (-0.5) = -9.728$

27.
$$-2(16) - [2(-8) - 5^3] = -2(16) - [2(-8) - 125]$$
$$= -2(16) - [-16 - 125]$$
$$= -2(16) - [-141]$$
$$= -2(16) + 141$$
$$= -32 + 141$$
$$= 109$$

28. $-20 \div (-5) + 36 \div (-4) = 4 + (-9) = -5$

29. Let $D = $ the difference in the temperatures.

Difference in temperature	is	Higher temperature	minus	Lower temperature
↓	↓	↓	↓	↓
D	$=$	-67	$-$	(-81)

We carry out the subtraction.
$$D = -67 - (-81) = -67 + 81 = 14$$

The average high temperature is 14°F higher than the average low temperature.

30. Let $P = $ the number of points by which the market has changed over the five week period.

Total change	$=$	Week 1 change	$+$	Week 2 change	$+$	Week 3 change	$+$
↓	↓	↓	↓	↓	↓	↓	
P	$=$	-13	$+$	(-16)	$+$	36	$+$

Week 4 change	$+$	Week 5 change
↓	↓	↓
(-11)	$+$	19

We carry out the computation.
$$P = -13 + (-16) + 36 + (-11) + 19$$
$$= -29 + 36 + (-11) + 19$$
$$= 7 + (-11) + 19$$
$$= -4 + 19$$
$$= 15$$

The market rose 15 points.

31. First we multiply to find the total decrease d in the population.
$$d = 6 \cdot 420 = 2520$$

The population decreased by 2520 over the six year period.

Now we subtract to find the new population p.
$$18,600 - 2520 = 16,080$$

After 6 yr the population was 16,080.

32. First we subtract to find the total drop in temperature t.

$$t = 16°C - (-17°C) = 16°C + 17°C = 33°C$$

Then we divide to find by how many degrees d the temperature dropped each minute in the 35 minutes from 11:08 A.M. to 11:43 A.M.

$$d = 33 \div 35 = \frac{33}{35}$$

The temperature dropped about $\frac{33}{35}°C$ each minute.

33. $3(6 - x) = 3 \cdot 6 - 3 \cdot x = 18 - 3x$

34. $-5(y - 1) = -5 \cdot y - (-5)(1) = -5y - (-5) = -5y + 5$

35. $12 - 22x = 2 \cdot 6 - 2 \cdot 11x = 2(6 - 11x)$

36. $7x + 21 + 14y = 7 \cdot x + 7 \cdot 3 + 7 \cdot 2y = 7(x + 3 + 2y)$

37. $\begin{aligned} 6 + 7 - 4 - (-3) &= 6 + 7 + (-4) + 3 \\ &= 13 + (-4) + 3 \\ &= 9 + 3 \\ &= 12 \end{aligned}$

38. $5x - (3x - 7) = 5x - 3x + 7 = 2x + 7$

39. $\begin{aligned} 4(2a - 3b) + a - 7 &= 8a - 12b + a - 7 \\ &= 9a - 12b - 7 \end{aligned}$

40. $\begin{aligned} &4\{3[5(y - 3) + 9] + 2(y + 8)\} \\ &= 4\{3[5y - 15 + 9] + 2y + 16\} \\ &= 4\{3[5y - 6] + 2y + 16\} \\ &= 4\{15y - 18 + 2y + 16\} \\ &= 4\{17y - 2\} \\ &= 68y - 8 \end{aligned}$

41. $256 \div (-16) \div 4 = -16 \div 4 = -4$

42. $\begin{aligned} 2^3 - 10[4 - (-2 + 18)3] &= 2^3 - 10[4 - (16)3] \\ &= 2^3 - 10[4 - 48] \\ &= 2^3 - 10[-44] \\ &= 8 - 10[-44] \\ &= 8 + 440 \\ &= 448 \end{aligned}$

43. $-2 \geq x$ has the same meaning as $x \leq -2$.

44. $\begin{aligned} &|-27 - 3(4)| - |-36| + |-12| \\ &= |-27 - 12| - |-36| + |-12| \\ &= |-39| - |-36| + |-12| \\ &= 39 - 36 + 12 \\ &= 3 + 12 \\ &= 15 \end{aligned}$

45. $\begin{aligned} &a - \{3a - [4a - (2a - 4a)]\} \\ &= a - \{3a - [4a - (-2a)]\} \\ &= a - \{3a - [4a + 2a]\} \\ &= a - \{3a - 6a\} \\ &= a - \{-3a\} \\ &= a + 3a \\ &= 4a \end{aligned}$

46. The perimeter is equivalent to the perimeter of a square with sides x along with four additional segments of length y. We have $x + x + x + x + y + y + y + y = 4x + 4y$.

Chapter 10

Solving Equations and Inequalities

Exercise Set 10.1

1. $\underline{x + 17 = 32}$ Writing the equation

$15 + 17$? 32 Substituting 15 for x

32 | TRUE

Since the left-hand and right-hand sides are the same, 15 is a solution of the equation.

3. $\underline{x - 7 = 12}$ Writing the equation

$21 - 7$? 12 Substituting 21 for x

14 | FALSE

Since the left-hand and right-hand sides are not the same, 21 is not a solution of the equation.

5. $\underline{6x = 54}$ Writing the equation

$6(-7)$? 54 Substituting

-42 | FALSE

-7 is not a solution of the equation.

7. $\dfrac{x}{6} = 5$ Writing the equation

$\dfrac{30}{6}$? 5 Substituting

5 | TRUE

5 is a solution of the equation.

9. $\underline{5x + 7 = 107}$

$5 \cdot 19 + 7$? 107 Substituting

$95 + 7$ |

102 | FALSE

19 is not a solution of the equation.

11. $\underline{7(y - 1) = 63}$

$7(-11 - 1)$? 63 Substituting

$7(-12)$ |

-84 | FALSE

-11 is not a solution of the equation.

13. $x + 2 = 6$

$x + 2 - 2 = 6 - 2$ Subtracting 2 on both sides

$x = 4$ Simplifying

Check: $\underline{x + 2 = 6}$

$4 + 2$? 6

6 | TRUE

The solution is 4.

15. $x + 15 = -5$

$x + 15 - 15 = -5 - 15$ Subtracting 15 on both sides

$x = -20$

Check: $\underline{x + 15 = -5}$

$-20 + 15$? -5

-5 | TRUE

The solution is -20.

17. $x + 6 = -8$

$x + 6 - 6 = -8 - 6$

$x = -14$

Check: $\underline{x + 6 = -8}$

$-14 + 6$? -8

-8 | TRUE

The solution is -14.

19. $x + 16 = -2$

$x + 16 - 16 = -2 - 16$

$x = -18$

Check: $\underline{x + 16 = -2}$

$-18 + 16$? -2

-2 | TRUE

The solution is -18.

21. $x - 9 = 6$

$x - 9 + 9 = 6 + 9$

$x = 15$

Check: $\underline{x - 9 = 6}$

$15 - 9$? 6

6 | TRUE

The solution is 15.

23. $x - 7 = -21$

$x - 7 + 7 = -21 + 7$

$x = -14$

Check: $\underline{x - 7 = -21}$

$-14 - 7$? -21

-21 | TRUE

The solution is -14.

25. $5 + t = 7$

$-5 + 5 + t = -5 + 7$

$t = 2$

Check: $\underline{5 + t = 7}$

$5 + 2$? ? 7

7 | TRUE

The solution is 2.

27.
$$-7 + y = 13$$
$$7 + (-7) + y = 7 + 13$$
$$y = 20$$

Check:
$$-7 + y = 13$$
$$\overline{}$$
$$-7 + 20 \;?\; 13$$
$$13 \;\big|\; \qquad \text{TRUE}$$

The solution is 20.

29.
$$-3 + t = -9$$
$$3 + (-3) + t = 3 + (-9)$$
$$t = -6$$

Check:
$$-3 + t = -9$$
$$\overline{}$$
$$-3 + (-6) \;?\; -9$$
$$-9 \;\big|\; \qquad \text{TRUE}$$

The solution is -6.

31.
$$x + \frac{1}{2} = 7$$
$$x + \frac{1}{2} - \frac{1}{2} = 7 - \frac{1}{2}$$
$$x = 6\frac{1}{2}$$

Check:
$$x + \frac{1}{2} = 7$$
$$\overline{\phantom{6\frac{1}{2} + \frac{1}{2}}}$$
$$6\frac{1}{2} + \frac{1}{2} \;?\; 7$$
$$7 \;\big|\; \qquad \text{TRUE}$$

The solution is $6\frac{1}{2}$.

33.
$$12 = a - 7.9$$
$$12 + 7.9 = a - 7.9 + 7.9$$
$$19.9 = a$$

Check:
$$12 = a - 7.9$$
$$\overline{}$$
$$12 \;?\; 19.9 - 7.9$$
$$\big|\; 12 \qquad \text{TRUE}$$

The solution is 19.9.

35.
$$r + \frac{1}{3} = \frac{8}{3}$$
$$r + \frac{1}{3} - \frac{1}{3} = \frac{8}{3} - \frac{1}{3}$$
$$r = \frac{7}{3}$$

Check:
$$r + \frac{1}{3} = \frac{8}{3}$$
$$\overline{\phantom{\frac{7}{3} + \frac{1}{3}}}$$
$$\frac{7}{3} + \frac{1}{3} \;?\; \frac{8}{3}$$
$$\frac{8}{3} \;\big|\; \qquad \text{TRUE}$$

The solution is $\frac{7}{3}$.

37.
$$m + \frac{5}{6} = -\frac{11}{12}$$
$$m + \frac{5}{6} - \frac{5}{6} = -\frac{11}{12} - \frac{5}{6}$$
$$m = -\frac{11}{12} - \frac{5}{6} \cdot \frac{2}{2}$$
$$m = -\frac{11}{12} - \frac{10}{12}$$
$$m = -\frac{21}{12} = -\frac{\cancel{3} \cdot 7}{\cancel{3} \cdot 4}$$
$$m = -\frac{7}{4}$$

Check:
$$m + \frac{5}{6} = -\frac{11}{12}$$
$$\overline{\phantom{-\frac{7}{4} + \frac{5}{6}}}$$
$$-\frac{7}{4} + \frac{5}{6} \;?\; -\frac{11}{12}$$
$$-\frac{21}{12} + \frac{10}{12} \;\Big|$$
$$-\frac{11}{12} \;\Big| \qquad \text{TRUE}$$

The solution is $-\frac{7}{4}$.

39.
$$x - \frac{5}{6} = \frac{7}{8}$$
$$x - \frac{5}{6} + \frac{5}{6} = \frac{7}{8} + \frac{5}{6}$$
$$x = \frac{7}{8} \cdot \frac{3}{3} + \frac{5}{6} \cdot \frac{4}{4}$$
$$x = \frac{21}{24} + \frac{20}{24}$$
$$x = \frac{41}{24}$$

Check:
$$x - \frac{5}{6} = \frac{7}{8}$$
$$\overline{\phantom{\frac{41}{24} - \frac{5}{6}}}$$
$$\frac{41}{24} - \frac{5}{6} \;?\; \frac{7}{8}$$
$$\frac{41}{24} - \frac{20}{24} \;\Big|\; \frac{21}{24}$$
$$\frac{21}{24} \;\Big| \qquad \text{TRUE}$$

The solution is $\frac{41}{24}$.

41.
$$-\frac{1}{5} + z = -\frac{1}{4}$$
$$\frac{1}{5} - \frac{1}{5} + z = \frac{1}{5} - \frac{1}{4}$$
$$z = \frac{1}{5} \cdot \frac{4}{4} - \frac{1}{4} \cdot \frac{5}{5}$$
$$z = \frac{4}{20} - \frac{5}{20}$$
$$z = -\frac{1}{20}$$

Check:
$$-\frac{1}{5} + z = -\frac{1}{4}$$

$$-\frac{1}{5} + \left(-\frac{1}{20}\right) \; ? \; -\frac{1}{4}$$
$$-\frac{4}{20} + \left(-\frac{1}{20}\right) \; \bigg| \; -\frac{5}{20}$$
$$-\frac{5}{20} \; \bigg| \qquad \text{TRUE}$$

The solution is $-\frac{1}{20}$.

43.
$$x + 2.3 = 7.4$$
$$x + 2.3 - 2.3 = 7.4 - 2.3$$
$$x = 5.1$$

Check:
$$x + 2.3 = 7.4$$
$$5.1 + 2.3 \; ? \; 7.4$$
$$7.4 \; \bigg| \qquad \text{TRUE}$$

The solution is 5.1.

45.
$$7.6 = x - 4.8$$
$$7.6 + 4.8 = x - 4.8 + 4.8$$
$$12.4 = x$$

Check:
$$7.6 = x - 4.8$$
$$7.6 \; ? \; 12.4 - 4.8$$
$$\bigg| \; 7.6 \qquad \text{TRUE}$$

The solution is 12.4.

47.
$$-9.7 = -4.7 + y$$
$$4.7 + (-9.7) = 4.7 + (-4.7) + y$$
$$-5 = y$$

Check:
$$-9.7 = -4.7 + y$$
$$-9.7 \; ? \; -4.7 + (-5)$$
$$\bigg| \; -9.7 \qquad \text{TRUE}$$

The solution is -5.

49.
$$5\frac{1}{6} + x = 7$$
$$-5\frac{1}{6} + 5\frac{1}{6} + x = -5\frac{1}{6} + 7$$
$$x = -\frac{31}{6} + \frac{42}{6}$$
$$x = \frac{11}{6}, \text{ or } 1\frac{5}{6}$$

Check:
$$5\frac{1}{6} + x = 7$$
$$5\frac{1}{6} + 1\frac{5}{6} \; ? \; 7$$
$$7 \; \bigg| \qquad \text{TRUE}$$

The solution is $\frac{11}{6}$, or $1\frac{5}{6}$.

51.
$$q + \frac{1}{3} = -\frac{1}{7}$$
$$q + \frac{1}{3} - \frac{1}{3} = -\frac{1}{7} - \frac{1}{3}$$
$$q = -\frac{1}{7} \cdot \frac{3}{3} - \frac{1}{3} \cdot \frac{7}{7}$$
$$q = -\frac{3}{21} - \frac{7}{21}$$
$$q = -\frac{10}{21}$$

Check:
$$q + \frac{1}{3} = -\frac{1}{7}$$
$$-\frac{10}{21} + \frac{1}{3} \; ? \; -\frac{1}{7}$$
$$-\frac{10}{21} + \frac{7}{21} \; \bigg| \; -\frac{3}{21}$$
$$-\frac{3}{21} \; \bigg| \qquad \text{TRUE}$$

The solution is $-\frac{10}{21}$.

53. Discussion and Writing Exercise

55. $-3 + (-8)$ Two negative numbers. We add the absolute values, getting 11, and make the answer negative.
$$-3 + (-8) = -11$$

57. $-\frac{2}{3} \cdot \frac{5}{8} = -\frac{2 \cdot 5}{3 \cdot 8} = -\frac{\not{2} \cdot 5}{3 \cdot \not{2} \cdot 4} = -\frac{5}{12}$

59. $\frac{2}{3} \div \left(-\frac{4}{9}\right) = \frac{2}{3} \cdot \left(-\frac{9}{4}\right) = -\frac{2 \cdot 9}{3 \cdot 4} = -\frac{\not{2} \cdot \not{3} \cdot 3}{\not{3} \cdot \not{2} \cdot 2} = -\frac{3}{2}$

61. $-\frac{2}{3} - \left(-\frac{5}{8}\right) = -\frac{2}{3} + \frac{5}{8}$
$$= -\frac{2}{3} \cdot \frac{8}{8} + \frac{5}{8} \cdot \frac{3}{3}$$
$$= -\frac{16}{24} + \frac{15}{24}$$
$$= -\frac{1}{24}$$

63. The translation is $\$83 - x$.

65.
$$-356.788 = -699.034 + t$$
$$699.034 + (-356.788) = 699.034 + (-699.034) + t$$
$$342.246 = t$$

The solution is 342.246.

67.
$$x + \frac{4}{5} = -\frac{2}{3} - \frac{4}{15}$$
$$x + \frac{4}{5} = -\frac{2}{3} \cdot \frac{5}{5} - \frac{4}{15} \qquad \text{Adding on the right side}$$
$$x + \frac{4}{5} = -\frac{10}{15} - \frac{4}{15}$$
$$x + \frac{4}{5} = -\frac{14}{15}$$
$$x + \frac{4}{5} - \frac{4}{5} = -\frac{14}{15} - \frac{4}{5}$$
$$x = -\frac{14}{15} - \frac{4}{5} \cdot \frac{3}{3}$$
$$x = -\frac{14}{15} - \frac{12}{15}$$
$$x = -\frac{26}{15}$$

The solution is $-\frac{26}{15}$.

69. $16 + x - 22 = -16$
$$x - 6 = -16 \qquad \text{Adding on the left side}$$
$$x - 6 + 6 = -16 + 6$$
$$x = -10$$
The solution is -10.

71.
$$x + 3 = 3 + x$$
$$x + 3 - 3 = 3 + x - 3$$
$$x = x$$

$x = x$ is true for all real numbers. Thus the solution is all real numbers.

73.
$$-\frac{3}{2} + x = -\frac{5}{17} - \frac{3}{2}$$
$$\frac{3}{2} - \frac{3}{2} + x = \frac{3}{2} - \frac{5}{17} - \frac{3}{2}$$
$$x = \left(\frac{3}{2} - \frac{3}{2}\right) - \frac{5}{17}$$
$$x = -\frac{5}{17}$$

The solution is $-\frac{5}{17}$.

75.
$$|x| + 6 = 19$$
$$|x| + 6 - 6 = 19 - 6$$
$$|x| = 13$$

x represents a number whose distance from 0 is 13. Thus $x = -13$ or $x = 13$.

The solutions are -13 and 13.

Exercise Set 10.2

1. $6x = 36$
$$\frac{6x}{6} = \frac{36}{6} \qquad \text{Dividing by 6 on both sides}$$
$$1 \cdot x = 6 \qquad \text{Simplifying}$$
$$x = 6 \qquad \text{Identity property of 1}$$
Check: $\quad \dfrac{6x = 36}{6 \cdot 6 \; ? \; 36}$
$$36 \mid \qquad \text{TRUE}$$
The solution is 6.

3. $5x = 45$
$$\frac{5x}{5} = \frac{45}{5} \qquad \text{Dividing by 5 on both sides}$$
$$1 \cdot x = 9 \qquad \text{Simplifying}$$
$$x = 9 \qquad \text{Identity property of 1}$$
Check: $\quad \dfrac{5x = 45}{5 \cdot 9 \; ? \; 45}$
$$45 \mid \qquad \text{TRUE}$$
The solution is 9.

5. $84 = 7x$
$$\frac{84}{7} = \frac{7x}{7} \qquad \text{Dividing by 7 on both sides}$$
$$12 = 1 \cdot x$$
$$12 = x$$
Check: $\quad \dfrac{84 = 7x}{84 \; ? \; 7 \cdot 12}$
$$\mid 84 \qquad \text{TRUE}$$
The solution is 12.

7. $-x = 40$
$$-1 \cdot x = 40$$
$$\frac{-1 \cdot x}{-1} = \frac{40}{-1}$$
$$1 \cdot x = -40$$
$$x = -40$$
Check: $\quad \dfrac{-x = 40}{-(-40) \; ? \; 40}$
$$40 \mid \qquad \text{TRUE}$$
The solution is -40.

9. $-x = -1$
$$-1 \cdot x = -1$$
$$\frac{-1 \cdot x}{-1} = \frac{-1}{-1}$$
$$1 \cdot x = 1$$
$$x = 1$$
Check: $\quad \dfrac{-x = -1}{-(1) \; ? \; -1}$
$$-1 \mid \qquad \text{TRUE}$$
The solution is 1.

11. $7x = -49$
$$\frac{7x}{7} = \frac{-49}{7}$$
$$1 \cdot x = -7$$
$$x = -7$$
Check: $\quad \dfrac{7x = -49}{7(-7) \; ? \; -49}$
$$-49 \mid \qquad \text{TRUE}$$
The solution is -7.

13. $-12x = 72$

$$\frac{-12x}{-12} = \frac{72}{-12}$$

$$1 \cdot x = -6$$

$$x = -6$$

Check: $\dfrac{-12x = 72}{}$

$$-12(-6) \; ? \; 72$$

$$72 \;\Big|\qquad \text{TRUE}$$

The solution is -6.

15. $-21x = -126$

$$\frac{-21x}{-21} = \frac{-126}{-21}$$

$$1 \cdot x = 6$$

$$x = 6$$

Check: $\dfrac{-21x = -126}{}$

$$-21 \cdot 6 \; ? \; -126$$

$$-126 \;\Big|\qquad \text{TRUE}$$

The solution is 6.

17. $\dfrac{t}{7} = -9$

$$7 \cdot \frac{1}{7}t = 7 \cdot (-9)$$

$$1 \cdot t = -63$$

$$t = -63$$

Check: $\dfrac{\dfrac{t}{7} = -9}{}$

$$\frac{-63}{7} \; ? \; -9$$

$$-9 \;\Big|\qquad \text{TRUE}$$

The solution is -63.

19. $\dfrac{3}{4}x = 27$

$$\frac{4}{3} \cdot \frac{3}{4}x = \frac{4}{3} \cdot 27$$

$$1 \cdot x = \frac{4 \cdot \cancel{3} \cdot 3 \cdot 3}{\cancel{3} \cdot 1}$$

$$x = 36$$

Check: $\dfrac{\dfrac{3}{4}x = 27}{}$

$$\frac{3}{4} \cdot 36 \; ? \; 27$$

$$27 \;\Big|\qquad \text{TRUE}$$

The solution is 36.

21. $\dfrac{-t}{3} = 7$

$$3 \cdot \frac{1}{3} \cdot (-t) = 3 \cdot 7$$

$$-t = 21$$

$$-1 \cdot (-1 \cdot t) = -1 \cdot 21$$

$$1 \cdot t = -21$$

$$t = -21$$

Check: $\dfrac{\dfrac{-t}{3} = 7}{}$

$$\frac{-(-21)}{3} \; ? \; 7$$

$$\frac{21}{3}$$

$$7 \;\Big|\qquad \text{TRUE}$$

The solution is -21.

23. $-\dfrac{m}{3} = \dfrac{1}{5}$

$$-\frac{1}{3} \cdot m = \frac{1}{5}$$

$$-3 \cdot \left(-\frac{1}{3} \cdot m\right) = -3 \cdot \frac{1}{5}$$

$$m = -\frac{3}{5}$$

Check: $\dfrac{-\dfrac{m}{3} = \dfrac{1}{5}}{}$

$$-\frac{-\dfrac{3}{5}}{3} \; ? \; \frac{1}{5}$$

$$-\left(-\frac{3}{5} \div 3\right)\;\Big|$$

$$-\left(-\frac{3}{5} \cdot \frac{1}{3}\right)$$

$$-\left(-\frac{1}{5}\right)$$

$$\frac{1}{5} \;\Big|\qquad \text{TRUE}$$

The solution is $-\dfrac{3}{5}$.

25. $-\dfrac{3}{5}r = \dfrac{9}{10}$

$$-\frac{5}{3} \cdot \left(-\frac{3}{5}r\right) = -\frac{5}{3} \cdot \frac{9}{10}$$

$$1 \cdot r = -\frac{\cancel{5} \cdot \cancel{3} \cdot 3}{\cancel{3} \cdot \cancel{5} \cdot 2}$$

$$r = -\frac{3}{2}$$

Check: $\dfrac{-\dfrac{3}{5}r = \dfrac{9}{10}}{}$

$$-\frac{3}{5} \cdot \left(-\frac{3}{2}\right) \; ? \; \frac{9}{10}$$

$$\frac{9}{10} \;\Big|\qquad \text{TRUE}$$

The solution is $-\dfrac{3}{2}$.

27.
$$-\frac{3}{2}r = -\frac{27}{4}$$
$$-\frac{2}{3}\cdot\left(-\frac{3}{2}r\right) = -\frac{2}{3}\cdot\left(-\frac{27}{4}\right)$$
$$1\cdot r = \frac{2\cdot 3\cdot 3\cdot 3}{3\cdot 2\cdot 2}$$
$$r = \frac{9}{2}$$

Check:
$$-\frac{3}{2}r = -\frac{27}{4}$$
$$-\frac{3}{2}\cdot\frac{9}{2} \;?\; -\frac{27}{4}$$
$$-\frac{27}{4} \;\Big|\; \qquad \text{TRUE}$$

The solution is $\frac{9}{2}$.

29. $6.3x = 44.1$
$$\frac{6.3x}{6.3} = \frac{44.1}{6.3}$$
$$1\cdot x = 7$$
$$x = 7$$

Check:
$$6.3x = 44.1$$
$$6.3\cdot 7 \;?\; 44.1$$
$$44.1 \;\Big|\; \qquad \text{TRUE}$$

The solution is 7.

31. $-3.1y = 21.7$
$$\frac{-3.1y}{-3.1} = \frac{21.7}{-3.1}$$
$$1\cdot y = -7$$
$$y = -7$$

Check:
$$3.1y = 21.7$$
$$-3.1(-7) \;?\; 21.7$$
$$21.7 \;\Big|\; \qquad \text{TRUE}$$

The solution is -7.

33. $38.7m = 309.6$
$$\frac{38.7m}{38.7} = \frac{309.6}{38.7}$$
$$1\cdot m = 8$$
$$m = 8$$

Check:
$$38.7m = 309.6$$
$$38.7\cdot 8 \;?\; 309.6$$
$$309.6 \;\Big|\; \qquad \text{TRUE}$$

The solution is 8.

35.
$$-\frac{2}{3}y = -10.6$$
$$-\frac{3}{2}\cdot\left(-\frac{2}{3}y\right) = -\frac{3}{2}\cdot(-10.6)$$
$$1\cdot y = \frac{31.8}{2}$$
$$y = 15.9$$

Check:
$$-\frac{2}{3}y = -10.6$$
$$-\frac{2}{3}\cdot(15.9) \;?\; -10.6$$
$$-\frac{31.8}{3}$$
$$-10.6 \;\Big|\; \qquad \text{TRUE}$$

The solution is 15.9.

37.
$$\frac{-x}{5} = 10$$
$$5\cdot\frac{-x}{5} = 5\cdot 10$$
$$-x = 50$$
$$-1\cdot(-x) = -1\cdot 50$$
$$x = -50$$

Check:
$$\frac{-x}{5} = 10$$
$$\frac{-(-50)}{5} \;?\; 10$$
$$\frac{50}{5}$$
$$10 \;\Big|\; \qquad \text{TRUE}$$

The solution is -50.

39.
$$-\frac{t}{2} = 7$$
$$2\cdot\left(-\frac{t}{2}\right) = 2\cdot 7$$
$$-t = 14$$
$$-1\cdot(-t) = -1\cdot 14$$
$$t = -14$$

Check:
$$-\frac{t}{2} = 7$$
$$-\frac{-14}{2} \;?\; 7$$
$$-(-7)$$
$$7 \;\Big|\; \qquad \text{TRUE}$$

The solution is -14.

41. Discussion and Writing Exercise

43. $3x + 4x = (3+4)x = 7x$

45. $-4x + 11 - 6x + 18x = (-4 - 6 + 18)x + 11 = 8x + 11$

47. $3x - (4 + 2x) = 3x - 4 - 2x = x - 4$

49. $8y - 6(3y + 7) = 8y - 18y - 42 = -10y - 42$

51. The translation is $8r$ miles.

53.
$$-0.2344m = 2028.732$$
$$\frac{-0.2344m}{-0.2344} = \frac{2028.732}{-0.2344}$$
$$1\cdot m = -8655$$
$$m = -8655$$

The solution is -8655.

55. For all x, $0 \cdot x = 0$. There is no solution to $0 \cdot x = 9$.

57.
$$2|x| = -12$$
$$\frac{2|x|}{2} = \frac{-12}{2}$$
$$1 \cdot |x| = -6$$
$$|x| = -6$$

Absolute value cannot be negative. The equation has no solution.

59.
$$3x = \frac{b}{a}$$
$$\frac{1}{3} \cdot 3x = \frac{1}{3} \cdot \frac{b}{a}$$
$$x = \frac{b}{3a}$$

The solution is $\frac{b}{3a}$.

61.
$$\frac{a}{b}x = 4$$
$$\frac{b}{a} \cdot \frac{a}{b}x = \frac{b}{a} \cdot 4$$
$$x = \frac{4b}{a}$$

The solution is $\frac{4b}{a}$.

Exercise Set 10.3

1.
$$5x + 6 = 31$$
$$5x + 6 - 6 = 31 - 6 \qquad \text{Subtracting 6 on both sides}$$
$$5x = 25 \qquad \text{Simplifying}$$
$$\frac{5x}{5} = \frac{25}{5} \qquad \text{Dividing by 5 on both sides}$$
$$x = 5 \qquad \text{Simplifying}$$

Check:
$$\begin{array}{c|c} 5x + 6 = 31 \\ \hline 5 \cdot 5 + 6 \ ? \ 31 \\ 25 + 6 \\ 31 & \text{TRUE} \end{array}$$

The solution is 5.

3.
$$8x + 4 = 68$$
$$8x + 4 - 4 = 68 - 4 \qquad \text{Subtracting 4 on both sides}$$
$$8x = 64 \qquad \text{Simplifying}$$
$$\frac{8x}{8} = \frac{64}{8} \qquad \text{Dividing by 8 on both sides}$$
$$x = 8 \qquad \text{Simplifying}$$

Check:
$$\begin{array}{c|c} 8x + 4 = 68 \\ \hline 8 \cdot 8 + 4 \ ? \ 68 \\ 64 + 4 \\ 68 & \text{TRUE} \end{array}$$

The solution is 8.

5.
$$4x - 6 = 34$$
$$4x - 6 + 6 = 34 + 6 \qquad \text{Adding 6 on both sides}$$
$$4x = 40$$
$$\frac{4x}{4} = \frac{40}{4} \qquad \text{Dividing by 4 on both sides}$$
$$x = 10$$

Check:
$$\begin{array}{c|c} 4x - 6 = 34 \\ \hline 4 \cdot 10 - 6 \ ? \ 34 \\ 40 - 6 \\ 34 & \text{TRUE} \end{array}$$

The solution is 10.

7.
$$3x - 9 = 33$$
$$3x - 9 + 9 = 33 + 9$$
$$3x = 42$$
$$\frac{3x}{3} = \frac{42}{3}$$
$$x = 14$$

Check:
$$\begin{array}{c|c} 3x - 9 = 33 \\ \hline 3 \cdot 14 - 9 \ ? \ 33 \\ 42 - 9 \\ 33 & \text{TRUE} \end{array}$$

The solution is 14.

9.
$$7x + 2 = -54$$
$$7x + 2 - 2 = -54 - 2$$
$$7x = -56$$
$$\frac{7x}{7} = \frac{-56}{7}$$
$$x = -8$$

Check:
$$\begin{array}{c|c} 7x + 2 = -54 \\ \hline 7(-8) + 2 \ ? \ -54 \\ -56 + 2 \\ -54 & \text{TRUE} \end{array}$$

The solution is -8.

11.
$$-45 = 6y + 3$$
$$-45 - 3 = 6y + 3 - 3$$
$$-48 = 6y$$
$$\frac{-48}{6} = \frac{6y}{6}$$
$$-8 = y$$

Check:
$$\begin{array}{c|c} -45 = 6y + 3 \\ \hline -45 \ ? \ 6(-8) + 3 \\ -48 + 3 \\ -45 & \text{TRUE} \end{array}$$

The solution is -8.

13.
$$-4x + 7 = 35$$
$$-4x + 7 - 7 = 35 - 7$$
$$-4x = 28$$
$$\frac{-4x}{-4} = \frac{28}{-4}$$
$$x = -7$$

Check: $-4x + 7 = 35$

$$-4(-7) + 7 \;?\; 35$$
$$28 + 7 \;\Big|$$
$$35 \;\Big|\qquad \text{TRUE}$$

The solution is -7.

15. $-8x - 24 = -29\frac{1}{3}$

$$-8x - 24 + 24 = -29\frac{1}{3} + 24$$

$$-8x = -5\frac{1}{3}$$

$$-8x = -\frac{16}{3} \qquad \left(-5\frac{1}{3} = -\frac{16}{3}\right)$$

$$-\frac{1}{8}(-8x) = -\frac{1}{8}\left(-\frac{16}{3}\right)$$

$$x = \frac{16}{8 \cdot 3} = \frac{2 \cdot \cancel{8}}{\cancel{8} \cdot 3}$$

$$x = \frac{2}{3}$$

Check: $-8x - 24 = -29\frac{1}{3}$

$$-8 \cdot \frac{2}{3} - 24 \;?\; -29\frac{1}{3}$$

$$-\frac{16}{3} - 24 \;\Big|$$

$$-5\frac{1}{3} - 24 \;\Big|$$

$$-29\frac{1}{3} \;\Big|\qquad \text{TRUE}$$

The solution is $\frac{2}{3}$.

17. $5x + 7x = 72$

$$12x = 72 \qquad \text{Collecting like terms}$$
$$\frac{12x}{12} = \frac{72}{12} \qquad \text{Dividing by 12 on both sides}$$
$$x = 6$$

Check: $5x + 7x = 72$

$$5 \cdot 6 + 7 \cdot 6 \;?\; 72$$
$$30 + 42 \;\Big|$$
$$72 \;\Big|\qquad \text{TRUE}$$

The solution is 6.

19. $8x + 7x = 60$

$$15x = 60 \qquad \text{Collecting like terms}$$
$$\frac{15x}{15} = \frac{60}{15} \qquad \text{Dividing by 15 on both sides}$$
$$x = 4$$

Check: $8x + 7x = 60$

$$8 \cdot 4 + 7 \cdot 4 \;?\; 60$$
$$32 + 28 \;\Big|$$
$$60 \;\Big|\qquad \text{TRUE}$$

The solution is 4.

21. $4x + 3x = 42$

$$7x = 42$$
$$\frac{7x}{7} = \frac{42}{7}$$
$$x = 6$$

Check: $4x + 3x = 42$

$$4 \cdot 6 + 3 \cdot 6 \;?\; 42$$
$$24 + 18 \;\Big|$$
$$42 \;\Big|\qquad \text{TRUE}$$

The solution is 6.

23. $-6y - 3y = 27$

$$-9y = 27$$
$$\frac{-9y}{-9} = \frac{27}{-9}$$
$$y = -3$$

Check: $-6y - 3y = 27$

$$-6(-3) - 3(-3) \;?\; 27$$
$$18 + 9 \;\Big|$$
$$27 \;\Big|\qquad \text{TRUE}$$

The solution is -3.

25. $-7y - 8y = -15$

$$-15y = -15$$
$$\frac{-15y}{-15} = \frac{-15}{-15}$$
$$y = 1$$

Check: $-7y - 8y = -15$

$$-7 \cdot 1 - 8 \cdot 1 \;?\; -15$$
$$-7 - 8 \;\Big|$$
$$-15 \;\Big|\qquad \text{TRUE}$$

The solution is 1.

27. $x + \frac{1}{3}x = 8$

$$\left(1 + \frac{1}{3}\right)x = 8$$

$$\frac{4}{3}x = 8$$

$$\frac{3}{4} \cdot \frac{4}{3}x = \frac{3}{4} \cdot 8$$

$$x = 6$$

Check: $x + \frac{1}{3}x = 8$

$$6 + \frac{1}{3} \cdot 6 \;?\; 8$$

$$6 + 2 \;\Big|$$

$$8 \;\Big|\qquad \text{TRUE}$$

The solution is 6.

29. $10.2y - 7.3y = -58$

$\qquad 2.9y = -58$

$\qquad \dfrac{2.9y}{2.9} = \dfrac{-58}{2.9}$

$\qquad y = -20$

Check: $\qquad \dfrac{10.2y - 7.3y = -58}{}$

$\qquad \dfrac{10.2(-20) - 7.3(-20) \ ? \ -58}{\begin{array}{c|c} -204 + 146 & \\ -58 & \end{array}}$ TRUE

The solution is -20.

31. $8y - 35 = 3y$

$\qquad 8y = 3y + 35 \qquad$ Adding 35 and simplifying

$8y - 3y = 35 \qquad$ Subtracting $3y$ and simplifying

$\qquad 5y = 35 \qquad$ Collecting like terms

$\qquad \dfrac{5y}{5} = \dfrac{35}{5} \qquad$ Dividing by 5

$\qquad y = 7$

Check: $\qquad \dfrac{8y - 35 = 3y}{}$

$\qquad \dfrac{8 \cdot 7 - 35 \ ? \ 3 \cdot 7}{\begin{array}{c|c} 56 - 35 & 21 \\ 21 & \end{array}}$ TRUE

The solution is 7.

33. $8x - 1 = 23 - 4x$

$8x + 4x = 23 + 1 \qquad$ Adding 1 and $4x$ and simplifying

$\qquad 12x = 24 \qquad$ Collecting like terms

$\qquad \dfrac{12x}{12} = \dfrac{24}{12} \qquad$ Dividing by 12

$\qquad x = 2$

Check: $\qquad \dfrac{8x - 1 = 23 - 4x}{}$

$\qquad \dfrac{8 \cdot 2 - 1 \ ? \ 23 - 4 \cdot 2}{\begin{array}{c|c} 16 - 1 & 23 - 8 \\ 15 & 15 \end{array}}$ TRUE

The solution is 2.

35. $2x - 1 = 4 + x$

$2x - x = 4 + 1 \qquad$ Adding 1 and $-x$

$\qquad x = 5 \qquad$ Collecting like terms

Check: $\qquad \dfrac{2x - 1 = 4 + x}{}$

$\qquad \dfrac{2 \cdot 5 - 1 \ ? \ 4 + 5}{\begin{array}{c|c} 10 - 1 & 9 \\ 9 & \end{array}}$ TRUE

The solution is 5.

37. $6x + 3 = 2x + 11$

$6x - 2x = 11 - 3$

$\qquad 4x = 8$

$\qquad \dfrac{4x}{4} = \dfrac{8}{4}$

$\qquad x = 2$

Check: $\qquad \dfrac{6x + 3 = 2x + 11}{}$

$\qquad \dfrac{6 \cdot 2 + 3 \ ? \ 2 \cdot 2 + 11}{\begin{array}{c|c} 12 + 3 & 4 + 11 \\ 15 & 15 \end{array}}$ TRUE

The solution is 2.

39. $5 - 2x = 3x - 7x + 25$

$\qquad 5 - 2x = -4x + 25$

$4x - 2x = 25 - 5$

$\qquad 2x = 20$

$\qquad \dfrac{2x}{2} = \dfrac{20}{2}$

$\qquad x = 10$

Check: $\qquad \dfrac{5 - 2x = 3x - 7x + 25}{}$

$\qquad \dfrac{5 - 2 \cdot 10 \ ? \ 3 \cdot 10 - 7 \cdot 10 + 25}{\begin{array}{c|c} 5 - 20 & 30 - 70 + 25 \\ -15 & -40 + 25 \\ & -15 \end{array}}$ TRUE

The solution is 10.

41. $4 + 3x - 6 = 3x + 2 - x$

$\qquad 3x - 2 = 2x + 2 \qquad$ Collecting like terms on each side

$3x - 2x = 2 + 2$

$\qquad x = 4$

Check: $\qquad \dfrac{4 + 3x - 6 = 3x + 2 - x}{}$

$\qquad \dfrac{4 + 3 \cdot 4 - 6 \ ? \ 3 \cdot 4 + 2 - 4}{\begin{array}{c|c} 4 + 12 - 6 & 12 + 2 - 4 \\ 16 - 6 & 14 - 4 \\ 10 & 10 \end{array}}$ TRUE

The solution is 4.

43. $4y - 4 + y + 24 = 6y + 20 - 4y$

$\qquad 5y + 20 = 2y + 20$

$\qquad 5y - 2y = 20 - 20$

$\qquad 3y = 0$

$\qquad y = 0$

Check: $\qquad \dfrac{4y - 4 + y + 24 = 6y + 20 - 4y}{}$

$\qquad \dfrac{4 \cdot 0 - 4 + 0 + 24 \ ? \ 6 \cdot 0 + 20 - 4 \cdot 0}{\begin{array}{c|c} 0 - 4 + 0 + 24 & 0 + 20 - 0 \\ 20 & 20 \end{array}}$ TRUE

The solution is 0.

45. $\frac{7}{2}x + \frac{1}{2}x = 3x + \frac{3}{2} + \frac{5}{2}x$

The least common multiple of all the denominators is 2. We multiply by 2 on both sides.

$$2\left(\frac{7}{2}x + \frac{1}{2}x\right) = 2\left(3x + \frac{3}{2} + \frac{5}{2}x\right)$$

$$2 \cdot \frac{7}{2}x + 2 \cdot \frac{1}{2}x = 2 \cdot 3x + 2 \cdot \frac{3}{2} + 2 \cdot \frac{5}{2}x$$

$$7x + x = 6x + 3 + 5x$$

$$8x = 11x + 3$$

$$8x - 11x = 3$$

$$-3x = 3$$

$$\frac{-3x}{-3} = \frac{3}{-3}$$

$$x = -1$$

Check: $\quad \frac{7}{2}x + \frac{1}{2}x = 3x + \frac{3}{2} + \frac{5}{2}x$

$$\begin{array}{c|c} \frac{7}{2}(-1) + \frac{1}{2}(-1) \ ? \ 3(-1) + \frac{3}{2} + \frac{5}{2}(-1) \\ \hline -\frac{7}{2} - \frac{1}{2} & -3 + \frac{3}{2} - \frac{5}{2} \\ -4 & -\frac{8}{2} \\ & -4 \qquad \text{TRUE} \end{array}$$

The solution is -1.

47. $\frac{2}{3} + \frac{1}{4}t = \frac{1}{3}$

The least common multiple of all the denominators is 12. We multiply by 12 on both sides.

$$12\left(\frac{2}{3} + \frac{1}{4}t\right) = 12 \cdot \frac{1}{3}$$

$$12 \cdot \frac{2}{3} + 12 \cdot \frac{1}{4}t = 12 \cdot \frac{1}{3}$$

$$8 + 3t = 4$$

$$3t = 4 - 8$$

$$3t = -4$$

$$\frac{3t}{3} = \frac{-4}{3}$$

$$t = -\frac{4}{3}$$

Check: $\quad \frac{2}{3} + \frac{1}{4}t = \frac{1}{3}$

$$\begin{array}{c|c} \frac{2}{3} + \frac{1}{4}\left(-\frac{4}{3}\right) \ ? \ \frac{1}{3} \\ \hline \frac{2}{3} - \frac{1}{3} & \\ \frac{1}{3} & \text{TRUE} \end{array}$$

The solution is $-\frac{4}{3}$.

49. $\frac{2}{3} + 3y = 5y - \frac{2}{15}$,   LCM is 15

$$15\left(\frac{2}{3} + 3y\right) = 15\left(5y - \frac{2}{15}\right)$$

$$15 \cdot \frac{2}{3} + 15 \cdot 3y = 15 \cdot 5y - 15 \cdot \frac{2}{15}$$

$$10 + 45y = 75y - 2$$

$$10 + 2 = 75y - 45y$$

$$12 = 30y$$

$$\frac{12}{30} = \frac{30y}{30}$$

$$\frac{2}{5} = y$$

Check: $\quad \frac{2}{3} + 3y = 5y - \frac{2}{15}$

$$\begin{array}{c|c} \frac{2}{3} + 3 \cdot \frac{2}{5} \ ? \ 5 \cdot \frac{2}{5} - \frac{2}{15} \\ \hline \frac{2}{3} + \frac{6}{5} & 2 - \frac{2}{15} \\ \frac{10}{15} + \frac{18}{15} & \frac{30}{15} - \frac{2}{15} \\ \frac{28}{15} & \frac{28}{15} \qquad \text{TRUE} \end{array}$$

The solution is $\frac{2}{5}$.

51. $\frac{5}{3} + \frac{2}{3}x = \frac{25}{12} + \frac{5}{4}x + \frac{3}{4}$,   LCM is 12

$$12\left(\frac{5}{3} + \frac{2}{3}x\right) = 12\left(\frac{25}{12} + \frac{5}{4}x + \frac{3}{4}\right)$$

$$12 \cdot \frac{5}{3} + 12 \cdot \frac{2}{3}x = 12 \cdot \frac{25}{12} + 12 \cdot \frac{5}{4}x + 12 \cdot \frac{3}{4}$$

$$20 + 8x = 25 + 15x + 9$$

$$20 + 8x = 15x + 34$$

$$20 - 34 = 15x - 8x$$

$$-14x = 7x$$

$$\frac{-14}{7} = \frac{7x}{7}$$

$$-2 = x$$

Check: $\quad \frac{5}{3} + \frac{2}{3}x = \frac{25}{12} + \frac{5}{4}x + \frac{3}{4}$

$$\begin{array}{c|c} \frac{5}{3} + \frac{2}{3}(-2) \ ? \ \frac{25}{12} + \frac{5}{4}(-2) + \frac{3}{4} \\ \hline \frac{5}{3} - \frac{4}{3} & \frac{25}{12} - \frac{5}{2} + \frac{3}{4} \\ \frac{1}{3} & \frac{25}{12} - \frac{30}{12} + \frac{9}{12} \\ & \frac{4}{12} \\ & \frac{1}{3} \qquad \text{TRUE} \end{array}$$

The solution is -2.

53.
$$2.1x + 45.2 = 3.2 - 8.4x$$

Greatest number of decimal places is 1

$$10(2.1x + 45.2) = 10(3.2 - 8.4x)$$

Multiplying by 10 to clear decimals

$$10(2.1x) + 10(45.2) = 10(3.2) - 10(8.4x)$$

$$21x + 452 = 32 - 84x$$

$$21x + 84x = 32 - 452$$

$$105x = -420$$

$$\frac{105x}{105} = \frac{-420}{105}$$

$$x = -4$$

Check: $\dfrac{2.1x + 45.2 = 3.2 - 8.4x}{}$

$$\begin{array}{c|c} 2.1(-4) + 45.2 \ ? \ 3.2 - 8.4(-4) \\ -8.4 + 45.2 & 3.2 + 33.6 \\ 36.8 & 36.8 \end{array} \quad \text{TRUE}$$

The solution is -4.

55.
$$1.03 - 0.62x = 0.71 - 0.22x$$

Greatest number of decimal places is 2

$$100(1.03 - 0.62x) = 100(0.71 - 0.22x)$$

Multiplying by 100 to clear decimals

$$100(1.03) - 100(0.62x) = 100(0.71) - 100(0.22x)$$

$$103 - 62x = 71 - 22x$$

$$32 = 40x$$

$$\frac{32}{40} = \frac{40x}{40}$$

$$\frac{4}{5} = x, \text{ or}$$

$$0.8 = x$$

Check: $\dfrac{1.03 - 0.62x = 0.71 - 0.22x}{}$

$$\begin{array}{c|c} 1.03 - 0.62(0.8) \ ? \ 0.71 - 0.22(0.8) \\ 1.03 - 0.496 & 0.71 - 0.176 \\ 0.534 & 0.534 \end{array} \quad \text{TRUE}$$

The solution is $\dfrac{4}{5}$, or 0.8.

57.
$$\frac{2}{7}x - \frac{1}{2}x = \frac{3}{4}x + 1, \text{ LCM is 28}$$

$$28\left(\frac{2}{7}x - \frac{1}{2}x\right) = 28\left(\frac{3}{4}x + 1\right)$$

$$28 \cdot \frac{2}{7}x - 28 \cdot \frac{1}{2}x = 28 \cdot \frac{3}{4}x + 28 \cdot 1$$

$$8x - 14x = 21x + 28$$

$$-6x = 21x + 28$$

$$-6x - 21x = 28$$

$$-27x = 28$$

$$x = -\frac{28}{27}$$

Check: $\dfrac{\frac{2}{7}x - \frac{1}{2}x = \frac{3}{4}x + 1}{}$

$$\begin{array}{c|c} \frac{2}{7}\left(-\frac{28}{27}\right) - \frac{1}{2}\left(-\frac{28}{27}\right) \ ? \ \frac{3}{4}\left(-\frac{28}{27}\right) + 1 \\ -\frac{8}{27} + \frac{14}{27} & -\frac{21}{27} + 1 \\ \frac{6}{27} & \frac{6}{27} \end{array} \quad \text{TRUE}$$

The solution is $-\dfrac{28}{27}$.

59.
$$3(2y - 3) = 27$$

$$6y - 9 = 27 \qquad \text{Using a distributive law}$$

$$6y = 27 + 9 \qquad \text{Adding 9}$$

$$6y = 36$$

$$y = 6 \qquad \text{Dividing by 6}$$

Check: $\dfrac{3(2y - 3) = 27}{}$

$$\begin{array}{c|c} 3(2 \cdot 6 - 3) \ ? \ 27 \\ 3(12 - 3) & \\ 3 \cdot 9 & \\ 27 & \end{array} \quad \text{TRUE}$$

The solution is 6.

61.
$$40 = 5(3x + 2)$$

$$40 = 15x + 10 \qquad \text{Using a distributive law}$$

$$40 - 10 = 15x$$

$$30 = 15x$$

$$2 = x$$

Check: $\dfrac{40 = 5(3x + 2)}{}$

$$\begin{array}{c|c} 40 \ ? \ 5(3 \cdot 2 + 2) \\ & 5(6 + 2) \\ & 5 \cdot 8 \\ & 40 \end{array} \quad \text{TRUE}$$

The solution is 2.

63.
$$-23 + y = y + 25$$

$$-y - 23 + y = -y + y + 25$$

$$-23 = 25 \qquad \text{FALSE}$$

The equation has no solution.

65.
$$-23 + x = x - 23$$

$$-x - 23 + x = -x + x - 23$$

$$-23 = -23 \qquad \text{TRUE}$$

All real numbers are solutions.

67.
$$2(3 + 4m) - 9 = 45$$

$$6 + 8m - 9 = 45 \qquad \text{Collecting like terms}$$

$$8m - 3 = 45$$

$$8m = 45 + 3$$

$$8m = 48$$

$$m = 6$$

Check: $\dfrac{2(3+4m)-9=45}{}$

$2(3+4\cdot 6)-9 \ ? \ 45$

$2(3+24)-9 \quad \Big|$

$2\cdot 27-9 \quad \Big|$

$54-9 \quad \Big|$

$45 \quad \Big| \qquad$ TRUE

The solution is 6.

69. $5r-(2r+8)=16$

$5r-2r-8=16$

$3r-8=16 \qquad$ Collecting like terms

$3r=16+8$

$3r=24$

$r=8$

Check: $\dfrac{5r-(2r+8)=16}{}$

$5\cdot 8-(2\cdot 8+8) \ ? \ 16$

$40-(16+8) \quad \Big|$

$40-24 \quad \Big|$

$16 \quad \Big| \qquad$ TRUE

The solution is 8.

71. $6-2(3x-1)=2$

$6-6x+2=2$

$8-6x=2$

$8-2=6x$

$6=6x$

$1=x$

Check: $\dfrac{6-2(3x-1)=2}{}$

$6-2(3\cdot 1-1) \ ? \ 2$

$6-2(3-1) \quad \Big|$

$6-2\cdot 2 \quad \Big|$

$6-4 \quad \Big|$

$2 \quad \Big| \qquad$ TRUE

The solution is 1.

73. $5x+5-7x=15-12x+10x-10$

$-2x+5=5-2x \quad$ Collecting like terms

$2x-2x+5=2x+5-2x \quad$ Adding $2x$

$5=5 \qquad$ TRUE

All real numbers are solutions.

75. $22x-5-15x+3=10x-4-3x+11$

$7x-2=7x+7 \quad$ Collecting like terms

$-7x+7x-2=-7x+7x+7$

$-2=7 \qquad$ FALSE

The equation has no solution.

77. $5(d+4)=7(d-2)$

$5d+20=7d-14$

$20+14=7d-5d$

$34=2d$

$17=d$

Check: $\dfrac{5(d+4)=7(d-2)}{}$

$5(17+4) \ ? \ 7(17-2)$

$5\cdot 21 \quad \Big| \quad 7\cdot 15$

$105 \quad \Big| \quad 105 \qquad$ TRUE

The solution is 17.

79. $8(2t+1)=4(7t+7)$

$16t+8=28t+28$

$16t-28t=28-8$

$-12t=20$

$t=-\dfrac{20}{12}$

$t=-\dfrac{5}{3}$

Check: $\dfrac{8(2t+1)=4(7t+7)}{}$

$8\Big(2\Big(-\dfrac{5}{3}\Big)+1\Big) \ ? \ 4\Big(7\Big(-\dfrac{5}{3}\Big)+7\Big)$

$8\Big(-\dfrac{10}{3}+1\Big) \quad \Big| \quad 4\Big(-\dfrac{35}{3}+7\Big)$

$8\Big(-\dfrac{7}{3}\Big) \quad \Big| \quad 4\Big(-\dfrac{14}{3}\Big)$

$-\dfrac{56}{3} \quad \Big| \quad -\dfrac{56}{3} \qquad$ TRUE

The solution is $-\dfrac{5}{3}$.

81. $3(r-6)+2=4(r+2)-21$

$3r-18+2=4r+8-21$

$3r-16=4r-13$

$13-16=4r-3r$

$-3=r$

Check: $\dfrac{3(r-6)+2=4(r+2)-21}{}$

$3(-3-6)+2 \ ? \ 4(-3+2)-21$

$3(-9)+2 \quad \Big| \quad 4(-1)-21$

$-27+2 \quad \Big| \quad -4-21$

$-25 \quad \Big| \quad -25 \qquad$ TRUE

The solution is -3.

83. $19-(2x+3)=2(x+3)+x$

$19-2x-3=2x+6+x$

$16-2x=3x+6$

$16-6=3x+2x$

$10=5x$

$2=x$

Check: $\dfrac{19-(2x+3)=2(x+3)+x}{}$

$19-(2\cdot 2+3) \ ? \ 2(2+3)+2$

$19-(4+3) \quad \Big| \quad 2\cdot 5+2$

$19-7 \quad \Big| \quad 10+2$

$12 \quad \Big| \quad 12 \qquad$ TRUE

The solution is 2.

85. $2[4 - 2(3 - x)] - 1 = 4[2(4x - 3) + 7] - 25$

$2[4 - 6 + 2x] - 1 = 4[8x - 6 + 7] - 25$

$2[-2 + 2x] - 1 = 4[8x + 1] - 25$

$-4 + 4x - 1 = 32x + 4 - 25$

$4x - 5 = 32x - 21$

$-5 + 21 = 32x - 4x$

$16 = 28x$

$\dfrac{16}{28} = x$

$\dfrac{4}{7} = x$

The check is left to the student.

The solution is $\dfrac{4}{7}$.

87. $11 - 4(x + 1) - 3 = 11 + 2(4 - 2x) - 16$

$11 - 4x - 4 - 3 = 11 + 8 - 4x - 16$

$4 - 4x = 3 - 4x$

$4x + 4 - 4x = 4x + 3 - 4x$

$4 = 3 \quad$ FALSE

The equation has no solution.

89. $22x - 1 - 12x = 5(2x - 1) + 4$

$22x - 1 - 12x = 10x - 5 + 4$

$10x - 1 = 10x - 1$

$-10x + 10x - 1 = -10x + 10x - 1$

$-1 = -1 \quad$ TRUE

All real numbers are solutions.

91. $0.7(3x + 6) = 1.1 - (x + 2)$

$2.1x + 4.2 = 1.1 - x - 2$

$10(2.1x + 4.2) = 10(1.1 - x - 2) \quad$ Clearing decimals

$21x + 42 = 11 - 10x - 20$

$21x + 42 = -10x - 9$

$21x + 10x = -9 - 42$

$31x = -51$

$x = -\dfrac{51}{31}$

The check is left to the student.

The solution is $-\dfrac{51}{31}$.

93. Discussion and Writing Exercise

95. Do the long division. The answer is negative.

```
        6 . 5
3. 4∧⟌ 2 2.1∧0
      2 0 4
      ─────
        1 7 0
        1 7 0
        ─────
            0
```

$-22.1 \div 3.4 = -6.5$

97. $7x - 21 - 14y = 7 \cdot x - 7 \cdot 3 - 7 \cdot 2y = 7(x - 3 - 2y)$

99. $-3 + 2(-5)^2(-3) - 7 = -3 + 2(25)(-3) - 7$

$= -3 + 50(-3) - 7$

$= -3 - 150 - 7$

$= -153 - 7$

$= -160$

101. $23(2x - 4) - 15(10 - 3x) = 46x - 92 - 150 + 45x = 91x - 242$

103. First we multiply to remove the parentheses.

$\dfrac{2}{3}\left(\dfrac{7}{8} - 4x\right) - \dfrac{5}{8} = \dfrac{3}{8}$

$\dfrac{7}{12} - \dfrac{8}{3}x - \dfrac{5}{8} = \dfrac{3}{8}, \text{ LCM is } 24$

$24\left(\dfrac{7}{12} - \dfrac{8}{3}x - \dfrac{5}{8}\right) = 24 \cdot \dfrac{3}{8}$

$24 \cdot \dfrac{7}{12} - 24 \cdot \dfrac{8}{3}x - 24 \cdot \dfrac{5}{8} = 9$

$14 - 64x - 15 = 9$

$-1 - 64x = 9$

$-64x = 10$

$x = -\dfrac{10}{64}$

$x = -\dfrac{5}{32}$

The solution is $-\dfrac{5}{32}$.

105. $\dfrac{4 - 3x}{7} = \dfrac{2 + 5x}{49} - \dfrac{x}{14}$

$98\left(\dfrac{4 - 3x}{7}\right) = 98\left(\dfrac{2 + 5x}{49} - \dfrac{x}{14}\right), \text{ LCM is } 98$

$\dfrac{98(4 - 3x)}{7} = 98\left(\dfrac{2 + 5x}{49}\right) - 98 \cdot \dfrac{x}{14}$

$14(4 - 3x) = 2(2 + 5x) - 7x$

$56 - 42x = 4 + 10x - 7x$

$56 - 42x = 4 + 3x$

$56 - 42x + 42x = 4 + 3x + 42x$

$56 = 4 + 45x$

$56 - 4 = 4 + 45x - 4$

$52 = 45x$

$\dfrac{52}{45} = x$

The solution is $\dfrac{52}{45}$.

Exercise Set 10.4

1. a) We substitute 1900 for a and calculate B.

$B = 30a = 30 \cdot 1900 = 57{,}000$

The minimum furnace output is 57,000 Btu's.

b) $B = 30a$

$\dfrac{B}{30} = \dfrac{30a}{30} \quad$ Dividing by 30

$\dfrac{B}{30} = a$

3. a) We substitute 8 for t and calculate M.
$$M = \frac{1}{5} \cdot 8 = \frac{8}{5}, \text{ or } 1\frac{3}{5}$$

The storm is $1\frac{3}{5}$ miles away.

b) $M = \frac{1}{5}t$

$5 \cdot M = 5 \cdot \frac{1}{5}t$

$5M = t$

5. a) We substitute 21,345 for n and calculate f.
$$f = \frac{21,345}{15} = 1423$$

There are 1423 full-time equivalent students.

b) $f = \frac{n}{15}$

$15 \cdot f = 15 \cdot \frac{n}{15}$

$15f = n$

7. We substitute 84 for c and 8 for w and calculate D.
$$D = \frac{c}{w} = \frac{84}{8} = 10.5$$

The calorie density is 10.5 calories per oz.

9. We substitute 7 for n and calculate N.
$$N = n^2 - n = 7^2 - 7 = 49 - 7 = 42$$

42 games are played.

11. $y = 5x$

$\frac{y}{5} = \frac{5x}{5}$

$\frac{y}{5} = x$

13. $a = bc$

$\frac{a}{b} = \frac{bc}{b}$

$\frac{a}{b} = c$

15. $y = 13 + x$

$y - 13 = 13 + x - 13$

$y - 13 = x$

17. $y = x + b$

$y - b = x + b - b$

$y - b = x$

19. $y = 5 - x$

$y - 5 = 5 - x - 5$

$y - 5 = -x$

$-1 \cdot (y - 5) = -1 \cdot (-x)$

$-y + 5 = x, \text{ or }$

$5 - y = x$

21. $y = a - x$

$y - a = a - x - a$

$y - a = -x$

$-1 \cdot (y - a) = -1 \cdot (-x)$

$-y + a = x, \text{ or }$

$a - y = x$

23. $8y = 5x$

$\frac{8y}{8} = \frac{5x}{8}$

$y = \frac{5x}{8}, \text{ or } \frac{5}{8}x$

25. $By = Ax$

$\frac{By}{A} = \frac{Ax}{A}$

$\frac{By}{A} = x$

27. $W = mt + b$

$W - b = mt + b - b$

$W - b = mt$

$\frac{W - b}{m} = \frac{mt}{m}$

$\frac{W - b}{m} = t$

29. $y = bx + c$

$y - c = bx + c - c$

$y - c = bx$

$\frac{y - c}{b} = \frac{bx}{b}$

$\frac{y - c}{b} = x$

31. $A = \dfrac{a + b + c}{3}$

$3A = a + b + c$ Multiplying by 3

$3A - a - c = b$ Subtracting a and c

33. $A = at + b$

$A - b = at$ Subtracting b

$\dfrac{A - b}{a} = t$ Dividing by a

35. $A = bh$

$\dfrac{A}{b} = \dfrac{bh}{b}$ Dividing by b

$\dfrac{A}{b} = h$

37. $P = 2l + 2w$

$P - 2l = 2l + 2w - 2l$ Subtracting $2l$

$P - 2l = 2w$

$\dfrac{P - 2l}{2} = \dfrac{2w}{2}$ Dividing by 2

$\dfrac{P - 2l}{2} = w, \text{ or }$

$\dfrac{1}{2}P - l = w$

39.
$$A = \frac{a+b}{2}$$
$$2A = a + b \quad \text{Multiplying by 2}$$
$$2A - b = a \qquad \text{Subtracting } b$$

41. $F = ma$
$$\frac{F}{m} = \frac{ma}{m} \quad \text{Dividing by } m$$
$$\frac{F}{m} = a$$

43. $E = mc^2$
$$\frac{E}{m} = \frac{mc^2}{m} \quad \text{Dividing by } m$$
$$\frac{E}{m} = c^2$$

45. $Ax + By = c$
$$Ax = c - By \quad \text{Subtracting } By$$
$$\frac{Ax}{A} = \frac{c - By}{A} \quad \text{Dividing by } A$$
$$x = \frac{c - By}{A}$$

47. $v = \frac{3k}{t}$
$$tv = t \cdot \frac{3k}{t} \quad \text{Multiplying by } t$$
$$tv = 3k$$
$$\frac{tv}{v} = \frac{3k}{v} \quad \text{Dividing by } v$$
$$t = \frac{3k}{v}$$

49. Discussion and Writing Exercise

51. We divide:

```
       0.9 2
 2 5 ⟌ 2 3.0 0
       2 2 5
       ─────
         5 0
         5 0
       ─────
           0
```

Decimal notation for $\frac{23}{25}$ is 0.92.

53. $0.082 + (-9.407) = -9.325$

55. $-45.8 - (-32.6) = -45.8 + 32.6 = -13.2$

57. $3.1\% \qquad 0.03.1$

Move the decimal point 2 places to the left.

$3.1\% = 0.031$

59. $-\dfrac{2}{3} + \dfrac{5}{6} = -\dfrac{2}{3} \cdot \dfrac{2}{2} + \dfrac{5}{6}$
$$= -\frac{4}{6} + \frac{5}{6}$$
$$= \frac{1}{6}$$

61. a) We substitute 120 for w, 67 for h, and 23 for a and calculate K.
$$K = 917 + 6(w + h - a)$$
$$K = 917 + 6(120 + 67 - 23)$$
$$K = 917 + 6(164)$$
$$K = 917 + 984$$
$$K = 1901 \text{ calories}$$

b) Solve for a:
$$K = 917 + 6(w + h - a)$$
$$K = 917 + 6w + 6h - 6a$$
$$K + 6a = 917 + 6w + 6h$$
$$6a = 917 + 6w + 6h - K$$
$$a = \frac{917 + 6w + 6h - K}{6}$$

Solve for h:
$$K = 917 + 6(w + h - a)$$
$$K = 917 + 6w + 6h - 6a$$
$$K - 917 - 6w + 6a = 6h$$
$$\frac{K - 917 - 6w + 6a}{6} = h$$

Solve for w:
$$K = 917 + 6(w + h - a)$$
$$K = 917 + 6w + 6h - 6a$$
$$K - 917 - 6h + 6a = 6w$$
$$\frac{K - 917 - 6h + 6a}{6} = w$$

63.
$$H = \frac{2}{a - b}$$
$$(a - b)H = (a - b)\left(\frac{2}{a - b}\right)$$
$$Ha - Hb = 2$$
$$Ha - Hb - Ha = 2 - Ha$$
$$-Hb = 2 - Ha$$
$$-1(-Hb) = -1(2 - Ha)$$
$$Hb = -2 + Ha$$
$$\frac{Hb}{H} = \frac{-2 + Ha}{H}$$
$$b = \frac{-2 + Ha}{H}, \text{ or } \frac{Ha - 2}{H}, \text{ or } a - \frac{2}{H}$$

$$H = \frac{2}{a - b}$$
$$(a - b)H = (a - b) \cdot \frac{2}{a - b}$$
$$Ha - Hb = 2$$
$$Ha - Hb + Hb = 2 + Hb$$
$$Ha = 2 + Hb$$
$$\frac{Ha}{H} = \frac{2 + Hb}{H}, \text{ or } \frac{2}{H} + b$$

65. $A = lw$

When l and w both double, we have

$$2l \cdot 2w = 4lw = 4A,$$

so A quadruples.

67. $A = \frac{1}{2}bh$

When b increases by 4 units we have

$$\frac{1}{2}(b+4)h = \frac{1}{2}bh + 2h = A + 2h,$$

so A increases by $2h$ units.

Exercise Set 10.5

1. *Translate*.

$$\underbrace{\text{What percent}}_{\downarrow} \text{ of } \underbrace{180}_{\downarrow} \text{ is } \underbrace{36?}_{\downarrow}$$
$$p \quad\quad \cdot \quad 180 \,\, = \,\, 36$$

Solve. We divide by 36 on both sides and convert the answer to percent notation.

$$p \cdot 180 = 36$$
$$\frac{p \cdot 180}{180} = \frac{36}{180}$$
$$p = 0.2$$
$$p = 20\%$$

Thus, 36 is 20% of 180. The answer is 20%.

3. *Translate*.

$$\begin{array}{ccccc} 45 & \text{is} & 30\% & \text{of} & \text{what?} \\ \downarrow & \downarrow & \downarrow & \downarrow & \downarrow \\ 45 & = & 30\% & \cdot & b \end{array}$$

Solve. We solve the equation.

$$45 = 30\% \cdot b$$
$$45 = 0.3b \quad\quad \text{Converting to decimal notation}$$
$$\frac{45}{0.3} = \frac{b}{0.3}$$
$$150 = b$$

Thus, 45 is 30% of 150. The answer is 150.

5. *Translate*.

$$\begin{array}{ccccc} \text{What} & \text{is} & 65\% & \text{of} & 840? \\ \downarrow & \downarrow & \downarrow & \downarrow & \downarrow \\ a & = & 65\% & \cdot & 840 \end{array}$$

Solve. We convert 65% to decimal notation and multiply.

$$a = 65\% \cdot 840$$
$$a = 0.65 \times 840$$
$$a = 546$$

Thus, 546 is 65% of 840. The answer is 546.

7. *Translate*.

$$\begin{array}{ccccc} 30 & \text{is} & \underbrace{\text{what percent}} & \text{of} & 125? \\ \downarrow & \downarrow & \downarrow & \downarrow & \downarrow \\ 30 & = & p & \cdot & 125 \end{array}$$

Solve. We solve the equation.

$$30 = p \cdot 125$$
$$\frac{30}{125} = \frac{p \cdot 125}{125}$$
$$0.24 = p$$
$$24\% = p$$

Thus, 30 is 24% of 125. The answer is 24%.

9. *Translate*.

$$\begin{array}{ccccc} 12\% & \text{of} & \underbrace{\text{what number}} & \text{is} & 0.3? \\ \downarrow & \downarrow & \downarrow & \downarrow & \downarrow \\ 12\% & \cdot & b & = & 0.3 \end{array}$$

Solve. We solve the equation.

$$12\% \cdot b = 0.3$$
$$0.12b = 0.3 \quad\quad \text{Converting to decimal notation}$$
$$\frac{b}{0.12} = \frac{0.3}{0.12}$$
$$b = 2.5$$

Thus, 12% of 2.5 is 0.3. The answer is 2.5.

11. *Translate*.

$$\begin{array}{ccccc} \underbrace{2} & \text{is} & \underbrace{\text{what percent}} & \text{of} & \underbrace{40?} \\ \downarrow & \downarrow & \downarrow & \downarrow & \downarrow \\ 2 & = & p & \cdot & 40 \end{array}$$

Solve. We divide by 40 on both sides and convert the answer to percent notation.

$$2 = p \cdot 40$$
$$\frac{2}{40} = \frac{p \cdot 40}{40}$$
$$0.05 = p$$
$$5\% = p$$

Thus, 2 is 5% of 40. The answer is 5%.

13. *Translate*.

$$\underbrace{\text{What percent}}_{\downarrow} \text{ of } 68 \text{ is } 17?$$
$$p \quad\quad \cdot \quad 68 \,\, = \,\, 17$$

Solve. We divide by 68 on both sides and then convert to percent notation.

$$p \cdot 68 = 17$$
$$p = \frac{17}{68}$$
$$p = 0.25 = 25\%$$

The answer is 25%.

15. *Translate.*

What is 35% of 240?
$$a = 35\% \cdot 240$$

Solve. We convert 35% to decimal notation and multiply.
$$a = 35\% \cdot 240$$
$$a = 0.35 \cdot 240$$
$$a = 84$$
The answer is 84.

17. *Translate.*

What percent of 125 is 30?
$$p \cdot 125 = 30$$

Solve. We divide by 125 on both sides and then convert to percent notation.
$$p \cdot 125 = 30$$
$$p = \frac{30}{125}$$
$$p = 0.24 = 24\%$$
The answer is 24%.

19. *Translate.*

What percent of 300 is 48?
$$p \cdot 300 = 48$$

Solve. We divide by 300 on both sides and then convert to percent notation.
$$p \cdot 300 = 48$$
$$p = \frac{48}{300}$$
$$p = 0.16 = 16\%$$
The answer is 16%.

21. *Translate.*

14 is 30% of what number?
$$14 = 30\% \cdot b$$

Solve. We solve the equation.
$$14 = 0.3b \quad (30\% = 0.3)$$
$$\frac{14}{0.3} = b$$
$$46.\overline{6} = b$$

The answer is $46.\overline{6}$, or $46\frac{2}{3}$, or $\frac{140}{3}$.

23. *Translate.*

What is 2% of 40?
$$a = 2\% \cdot 40$$

Solve. We convert 2% to decimal notation and multiply.
$$a = 2\% \cdot 40$$
$$a = 0.02 \cdot 40$$
$$a = 0.8$$
The answer is 0.8.

25. *Translate.*

0.8 is 16% of what number?
$$0.8 = 16\% \cdot b$$

Solve. We solve the equation.
$$0.8 = 0.16b \quad (16\% = 0.16)$$
$$\frac{0.8}{0.16} = b$$
$$5 = b$$
The answer is 5.

27. *Translate.*

54 is 135% of what number?
$$54 = 135\% \cdot b$$

Solve. We solve the equation.
$$54 = 1.35b \quad (135\% = 1.35)$$
$$\frac{54}{1.35} = b$$
$$40 = b$$
The answer is 40.

29. First we reword and translate.

What is 3% of $6600?
$$a = 3\% \cdot 6600$$

Solve. We convert 3% to decimal notation and multiply.
$$a = 3\% \cdot 6600 = 0.03 \cdot 6600 = 198$$
The price of the dog is $198.

31. First we reword and translate.

What is 24% of $6600?
$$a = 24\% \cdot 6600$$

Solve. We convert 24% to decimal notation and multiply.
$$a = 24\% \cdot 6600 = 0.24 \cdot 6600 = 1584$$
Veterinarian expenses are $1584.

33. First we reword and translate.

What is 8% of $6600?
$$a = 8\% \cdot 6600$$

Solve. We convert 8% to decimal notation and multiply.
$$a = 8\% \cdot 6600 = 0.08 \cdot 6600 = 528$$
The cost of supplies is $528.

35. To find the percent of the imported cars that were manufactured in Japan, we first reword and translate.

1,003,745 is what percent of 2,268,093?
$$1{,}003{,}745 = p \cdot 2{,}268{,}093$$

Solve. We divide by 2,268,093 on both sides and convert to percent notation.

$$1,003,745 = p \cdot 2,268,093$$

$$\frac{1,003,745}{2,268,093} = p$$

$$0.443 \approx p$$

$$44.3\% \approx p$$

About 44.3% of the imported cars were manufactured in Japan.

To find the percent of imported cars that were manufactured in Germany, we first reword and translate.

$$\underbrace{564,910}\;\; \text{is}\;\; \underbrace{\text{what percent}}\;\; \text{of}\;\; \underbrace{2,268,093}?$$
$$\downarrow \qquad \quad \downarrow \qquad\quad \downarrow \qquad\quad \downarrow \qquad \downarrow$$
$$564,910 \;=\; \qquad\quad p \qquad\quad \cdot \quad 2,268,093$$

Solve. We divide by 2,268,093 on both sides and convert to percent notation.

$$564,910 = p \cdot 2,268,093$$

$$\frac{564,910}{2,268,093} = p$$

$$0.249 \approx p$$

$$24.9\% \approx p$$

About 24.9% of the imported cars were manufactured in Germany.

37. First we reword and translate.

$$\begin{array}{ccccc} 193 & \text{is} & 32\% & \text{of} & \underbrace{\text{what number}}? \\ \downarrow & \downarrow & \downarrow & \downarrow & \downarrow \\ 193 & = & 32\% & \cdot & b \end{array}$$

Solve. We solve the equation.

$$193 = 0.32 \cdot b \quad (32\% = 0.32)$$

$$\frac{193}{0.32} = b$$

$$603 \approx b$$

Sammy Sosa had 603 at-bats.

39. First we reword and translate.

$$\begin{array}{ccccc} \text{What} & \text{is} & 3\% & \text{of} & \$6500? \\ \downarrow & \downarrow & \downarrow & \downarrow & \downarrow \\ a & = & 3\% & \cdot & 6500 \end{array}$$

Solve. We convert 3% to decimal notation and multiply.

$$a = 3\% \cdot 6500 = 0.03 \cdot 6500 = 195$$

Sarah will pay $195 in interest.

41. a) First we reword and translate.

$$\begin{array}{cccc} \underbrace{\text{What percent}} & \text{of}\;\$25 & \text{is} & \$4? \\ \downarrow & & \downarrow\; \downarrow\; \downarrow \\ p & \cdot & 25 & = & 4 \end{array}$$

Solve. We divide by 25 on both sides and convert to percent notation.

$$p \cdot 25 = 4$$

$$\frac{p \cdot 25}{25} = \frac{4}{25}$$

$$p = 0.16$$

$$p = 16\%$$

The tip was 16% of the cost of the meal.

b) We add to find the total cost of the meal, including tip:

$$\$25 + \$4 = \$29$$

43. a) First we reword and translate.

$$\begin{array}{ccccc} \text{What} & \text{is} & 15\% & \text{of} & \$25? \\ \downarrow & \downarrow & \downarrow & \downarrow & \downarrow \\ a & = & 15\% & \cdot & 25 \end{array}$$

Solve. We convert 15% to decimal notation and multiply.

$$a = 15\% \cdot 25$$

$$a = 0.15 \times 25$$

$$a = 3.75$$

The tip was $3.75.

b) We add to find the total cost of the meal, including tip:

$$\$25 + \$3.75 = \$28.75$$

45. a) First we reword and translate.

$$\begin{array}{ccccc} 15\% & \text{of} & \text{what} & \text{is} & \$4.32? \\ \downarrow & \downarrow & \downarrow & \downarrow & \downarrow \\ 15\% & \cdot & b & = & 4.32 \end{array}$$

Solve. We solve the equation.

$$15\% \cdot b = 4.32$$

$$0.15 \cdot b = 4.32$$

$$\frac{0.15 \cdot b}{0.15} = \frac{4.32}{0.15}$$

$$b = 28.8$$

The cost of the meal before the tip was $28.80.

b) We add to find the total cost of the meal, including tip:

$$\$28.80 + \$4.32 = \$33.12$$

47. First we reword and translate.

$$\begin{array}{ccccc} 8\% & \text{of} & \text{what} & \text{is} & 16? \\ \downarrow & \downarrow & \downarrow & \downarrow & \downarrow \\ 8\% & \cdot & b & = & 16 \end{array}$$

Solve. We solve the equation.

$$8\% \cdot b = 16$$

$$0.08 \cdot b = 16$$

$$\frac{0.08 \cdot b}{0.08} = \frac{16}{0.08}$$

$$b = 200$$

There were 200 women in the original study.

49. First we reword and translate.

$$\begin{array}{ccccc} \text{What} & \text{is} & 16.5\% & \text{of} & 191? \\ \downarrow & \downarrow & \downarrow & \downarrow & \downarrow \\ a & = & 16.5\% & \cdot & 191 \end{array}$$

Solve. We convert 16.5% to decimal notation and multiply.

$$a = 16.5\% \cdot 191$$
$$a = 0.165 \cdot 191$$
$$a = 31.515 \approx 31.5$$

About 31.5 lb of the author's body weight is fat.

51. We subtract to find the increase.

$$\$990 - \$335 = \$655$$

The increase is $655.

Now we find the percent of increase.

$$\$655 \text{ is } \underbrace{\text{what percent}} \text{ of } \$335?$$
$$\downarrow \quad \downarrow \qquad \downarrow \qquad \downarrow \quad \downarrow$$
$$655 = \qquad p \qquad \cdot \quad 335$$

We divide by 335 on both sides and then convert to percent notation.

$$655 = p \cdot 335$$
$$\frac{655}{335} = p$$
$$1.96 \approx p$$
$$196\% \approx p$$

The percent of increase is about 196%.

53. First we find the increase in the rate for smokers.

$$\underbrace{\text{Rate increase}} \text{ is } 198\% \text{ of } \$735.$$
$$\downarrow \qquad \downarrow \quad \downarrow \quad \downarrow \quad \downarrow$$
$$a \qquad = 198\% \quad \cdot \quad 735$$

We convert 198% to decimal notation and multiply.

$$a = 198\% \cdot 735 = 1.98 \times 735 \approx 1455$$

The rate increase is $1455.

Now we add the rate increase to the rate for nonsmokers to find the rate for smokers.

$$\$735 + \$1455 = \$2190$$

55. We subtract to find the increase.

$$\$5445 - \$1510 = \$3935$$

The increase is $3935.

Now we find the percent of increase.

$$\$3935 \text{ is } \underbrace{\text{what percent}} \text{ of } \$1510?$$
$$\downarrow \quad \downarrow \qquad \downarrow \qquad \downarrow \quad \downarrow$$
$$3935 = \qquad p \qquad \cdot \quad 1510$$

We divide by 1510 on both sides and then convert to percent notation.

$$3935 = p \cdot 1510$$
$$\frac{3935}{1510} = p$$
$$2.61 \approx p$$
$$261\% \approx p$$

The percent of increase is about 261%.

57. Discussion and Writing Exercise

59.
$$\begin{array}{r} 1\,8\,1\,.5\,2 \\ 0.0\,5_{\wedge}\overline{)\,9.0\,7_{\wedge}6\,0\,} \\ 5 \\ \overline{4\,0} \\ 4\,0 \\ \overline{7} \\ 5 \\ \overline{2\,6} \\ 2\,5 \\ \overline{1\,0} \\ 1\,0 \\ \overline{0} \end{array}$$

The answer is 181.52.

61.
$$\begin{array}{r} {\scriptstyle 1\ 1\ 1} \\ 1.0\,8\,9\,0 \\ 1\,0.8\,9\,0\,0 \\ +\ \ 0.1\,0\,8\,9 \\ \hline 1\,2.0\,8\,7\,9 \end{array}$$

63.
$$-5a + 3c - 2(c - 3a)$$
$$= -5a + 3c - 2 \cdot c - 2(-3a)$$
$$= -5a + 3c - 2c + 6a$$
$$= (-5 + 6)a + (3 - 2)c$$
$$= 1 \cdot a + 1 \cdot c$$
$$= a + c$$

65. $-6.5 + 2.6 = -3.9$ The absolute values are 6.5 and 2.6. The difference is 3.9. The negative number has the larger absolute value, so the answer is negative, -3.9.

67. To simplify the calculation $18 - 24 \div 3 - 48 \div (-4)$, do all the <u>division</u> calculations first, and then the <u>subtraction</u> calculations.

69. Since 6 ft $= 6 \times 1$ ft $= 6 \times 12$ in. $= 72$ in., we can express 6 ft 4 in. as 72 in. $+ 4$ in., or 76 in.

Translate. We reword the problem.

$$96.1\% \text{ of what is } \underbrace{76 \text{ in.}}?$$
$$\downarrow \quad \downarrow \quad \downarrow \quad \downarrow \qquad \downarrow$$
$$96.1\% \ \cdot \quad b \quad = \quad 76$$

Solve. We solve the equation.

$$96.1\% \cdot b = 76$$
$$0.961 \cdot b = 76$$
$$\frac{0.961 \cdot b}{0.961} = \frac{76}{0.961}$$
$$b \approx 79$$

Note that 79 in. $= 72$ in. $+ 7$ in. $= 6$ ft 7 in.

Jaraan's final adult height will be about 6 ft 7 in.

Exercise Set 10.6

1. *Familiarize.* Using the labels on the drawing in the text, we let $x =$ the length of the shorter piece, in inches, and $3x =$ the length of the longer piece, in inches.

Translate. We reword the problem.

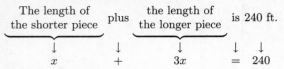

$$x + 3x = 240$$

Solve. We solve the equation.

$$x + 3x = 240$$
$$4x = 240 \qquad \text{Collecting like terms}$$
$$\frac{4x}{4} = \frac{240}{4}$$
$$x = 60$$

If x is 60, then $3x = 3 \cdot 60$, or 180.

Check. 180 is three times 60, and $60 + 180 = 240$. The answer checks.

State. The lengths of the pieces are 60 in. and 180 in.

3. *Familiarize*. Let $c = $ the cost of one box of Cinnamon Life cereal.

Translate.

Total cost	is	Number of boxes	times	Price of one box
↓	↓	↓	↓	↓
17.16	=	4	·	c

Solve. We solve the equation.

$$17.16 = 4 \cdot c$$
$$\frac{17.16}{4} = c \qquad \text{Dividing by 4}$$
$$4.29 = c$$

Check. If one box of Cinnamon Life costs \$4.29, then 4 boxes cost 4(\$4.29), or \$17.16. The answer checks.

State. One box of Cinnamon Life costs \$4.29.

5. *Familiarize*. Let $d = $ the amount spent on women's dresses, in billions of dollars.

Translate.

Amount spent on blouses	was	\$0.2 billion	more than	amount spent on dresses
↓	↓	↓	↓	↓
6.5	=	0.2	+	d

Solve. We solve the equation.

$$6.5 = 0.2 + d$$
$$6.5 - 0.2 = 0.2 + d - 0.2 \qquad \text{Subtracting 0.2}$$
$$6.3 = d$$

Check. If we add \$0.2 billion to \$6.3 billion, we get \$6.5 billion. The answer checks.

State. \$6.3 billion was spent on dresses.

7. *Familiarize*. Let $d = $ the musher's distance from Nome, in miles. Then $2d = $ the distance from Anchorage, in miles. This is the number of miles the musher has completed. The sum of the two distances is the length of the race, 1049 miles.

Translate.

Distance from Nome	plus	distance from Anchorage	is 1049 mi.
↓	↓	↓	↓ ↓
d	+	$2d$	= 1049

Solve. We solve the equation.

$$d + 2d = 1049$$
$$3d = 1049 \qquad \text{Collecting like terms}$$
$$\frac{3d}{3} = \frac{1049}{3}$$
$$d = \frac{1049}{3}$$

If $d = \dfrac{1049}{3}$, then $2d = 2 \cdot \dfrac{1049}{3} = \dfrac{2098}{3} = 699\dfrac{1}{3}$.

Check. $\dfrac{2098}{3}$ is twice $\dfrac{1049}{3}$, and $\dfrac{1049}{3} + \dfrac{2098}{3} = \dfrac{3147}{3} = 1049$. The result checks.

State. The musher has traveled $699\dfrac{1}{3}$ miles.

9. *Familiarize*. Let $x = $ the smaller number and $x + 1 = $ the larger number.

Translate. We reword the problem.

First number	+	second number	is 2409
↓	↓	↓	↓ ↓
x	+	$(x + 1)$	= 2409

Solve. We solve the equation.

$$x + (x + 1) = 2409$$
$$2x + 1 = 2409 \qquad \text{Collecting like terms}$$
$$2x + 1 - 1 = 2409 - 1 \qquad \text{Subtracting 1}$$
$$2x = 2408$$
$$\frac{2x}{2} = \frac{2408}{2} \qquad \text{Dividing by 2}$$
$$x = 1204$$

If x is 1204, then $x + 1$ is 1205.

Check. 1204 and 1205 are consecutive integers, and their sum is 2409. The answer checks.

State. The apartment numbers are 1204 and 1205.

11. *Familiarize*. Let $a = $ the first number. Then $a + 1 = $ the second number, and $a + 2 = $ the third number.

Translate. We reword the problem.

First number	+	second number	+	third number	is 126
↓	↓	↓	↓	↓	↓ ↓
a	+	$(a + 1)$	+	$(a + 2)$	= 114

Solve. We solve the equation.

$$a + (a + 1) + (a + 2) = 126$$
$$3a + 3 = 126 \qquad \text{Collecting like terms}$$
$$3a + 3 - 3 = 126 - 3$$
$$3a = 123$$
$$\frac{3a}{3} = \frac{123}{3}$$
$$a = 41$$

If a is 41, then $a + 1$ is 42 and $a + 2$ is 43.

Check. 41, 42, and 43 are consecutive integers, and their sum is 126. The answer checks.

State. The numbers are 41, 42, and 43.

13. Familiarize. Let $x =$ the first odd integer. Then $x + 2 =$ the next odd integer and $(x + 2) + 2$, or $x + 4 =$ the third odd integer.

Translate. We reword the problem.

$$\underbrace{\text{First odd integer}}_{x} + \underbrace{\text{second odd integer}}_{(x+2)} + \underbrace{\text{third odd integer}}_{(x+4)} \underbrace{\text{is}}_{=} \underbrace{189}_{189}$$

Solve. We solve the equation.

$$x + (x + 2) + (x + 4) = 189$$
$$3x + 6 = 189 \qquad \text{Collecting like terms}$$
$$3x + 6 - 6 = 189 - 6$$
$$3x = 183$$
$$\frac{3x}{3} = \frac{183}{3}$$
$$x = 61$$

If x is 61, then $x + 2$ is 63 and $x + 4$ is 65.

Check. 61, 63, and 65 are consecutive odd integers, and their sum is 189. The answer checks.

State. The integers are 61, 63, and 65.

15. Familiarize. Using the labels on the drawing in the text, we let $w =$ the width and $3w + 6 =$ the length. The perimeter P of a rectangle is given by the formula $2l + 2w = P$, where $l =$ the length and $w =$ the width.

Translate. Substitute $3w + 6$ for l and 124 for P:

$$2l + 2w = P$$
$$2(3w + 6) + 2w = 124$$

Solve. We solve the equation.

$$2(3w + 6) + 2w = 124$$
$$6w + 12 + 2w = 124$$
$$8w + 12 = 124$$
$$8w + 12 - 12 = 124 - 12$$
$$8w = 112$$
$$\frac{8w}{8} = \frac{112}{8}$$
$$w = 14$$

The possible dimensions are $w = 14$ ft and $l = 3w + 6 = 3(14) + 6$, or 48 ft.

Check. The length, 48 ft, is 6 ft more than three times the width, 14 ft. The perimeter is $2(48 \text{ ft}) + 2(14 \text{ ft}) = 96 \text{ ft} + 28 \text{ ft} = 124$ ft. The answer checks.

State. The width is 14 ft, and the length is 48 ft.

17. Familiarize. Let $p =$ the regular price of the shoes. At 15% off, Amy paid 85% of the regular price.

Translate.

$$\underbrace{\$63.75}_{63.75} \underbrace{\text{is}}_{=} \underbrace{85\%}_{0.85} \underbrace{\text{of}}_{\cdot} \underbrace{\text{the regular price.}}_{p}$$

Solve. We solve the equation.

$$63.75 = 0.85p$$
$$\frac{63.75}{0.08} = p \qquad \text{Dividing both sides by 0.85}$$
$$75 = p$$

Check. 85% of \$75, or 0.85(\$75), is \$63.75. The answer checks.

State. The regular price was \$75.

19. Familiarize. Let $b =$ the price of the book itself. When the sales tax rate is 5%, the tax paid on the book is 5% of b, or $0.05b$.

Translate.

$$\underbrace{\text{Price of book}}_{b} \underbrace{\text{plus}}_{+} \underbrace{\text{sales tax}}_{0.05b} \underbrace{\text{is}}_{=} \underbrace{\$89.25.}_{89.25}$$

Solve. We solve the equation.

$$b + 0.05b = 89.25$$
$$1.05b = 89.25$$
$$b = \frac{89.25}{1.05}$$
$$b = 85$$

Check. 5% of \$85, or 0.05(\$85), is \$4.25 and \$85 + \$4.25 is \$89.25, the total cost. The answer checks.

State. The book itself cost \$85.

21. Familiarize. Let $n =$ the number of visits required for a total parking cost of \$27.00. The parking cost for each $1\frac{1}{2}$ hour visit is \$1.50 for the first hour plus \$1.00 for part of a second hour, or \$2.50. Then the total parking cost for n visits is $2.50n$ dollars.

Translate. We reword the problem.

$$\underbrace{\text{Total parking cost}}_{2.50n} \underbrace{\text{is}}_{=} \underbrace{\$27.00.}_{27.00}$$

Solve. We solve the equation.

$$2.5n = 27$$
$$10(2.5n) = 10(27) \qquad \text{Clearing the decimal}$$
$$25n = 270$$
$$\frac{25n}{25} = \frac{270}{25}$$
$$n = 10.8$$

If the total parking cost is \$27.00 for 10.8 visits, then the cost will be more than \$27.00 for 11 or more visits.

Check. The parking cost for 10 visits is \$2.50(10), or \$25, and the parking cost for 11 visits is \$2.50(11), or \$27.50. Since 11 is the smallest number for which the parking cost exceeds \$27.00, the answer checks.

State. The minimum number of weekly visits for which it is worthwhile to buy a parking pass is 11.

23. Familiarize. Let x = the measure of the first angle. Then $3x$ = the measure of the second angle, and $x + 40$ = the measure of the third angle. Recall that the sum of measures of the angles of a triangle is $180°$.

Translate.

$$\underbrace{\text{Measure of first angle}} + \underbrace{\text{measure of second angle}} + \underbrace{\text{measure of third angle}} \text{ is } 180.$$
$$x + 3x + (x + 40) = 180$$

Solve. We solve the equation.
$$x + 3x + (x + 40) = 180$$
$$5x + 40 = 180$$
$$5x + 40 - 40 = 180 - 40$$
$$5x = 140$$
$$\frac{5x}{5} = \frac{140}{5}$$
$$x = 28$$

Possible answers for the angle measures are as follows:

First angle: $x = 28°$

Second angle: $3x = 3(28) = 84°$

Third angle: $x + 40 = 28 + 40 = 68°$

Check. Consider $28°$, $84°$, and $68°$. The second angle is three times the first, and the third is $40°$ more than the first. The sum, $28° + 84° + 68°$, is $180°$. These numbers check.

State. The measures of the angles are $28°$, $84°$, and $68°$.

25. Familiarize. Using the labels on the drawing in the text, we let x = the measure of the first angle, $x + 5$ = the measure of the second angle, and $3x + 10$ = the measure of the third angle. Recall that the sum of measures of the angles of a triangle is $180°$.

Translate.

$$\underbrace{\text{Measure of first angle}} + \underbrace{\text{measure of second angle}} + \underbrace{\text{measure of third angle}} \text{ is } 180.$$
$$x + (x + 5) + (3x + 10) = 180$$

Solve. We solve the equation.
$$x + (x + 5) + (3x + 10) = 180$$
$$5x + 15 = 180$$
$$5x + 15 - 15 = 180 - 15$$
$$5x = 165$$
$$\frac{5x}{5} = \frac{165}{5}$$
$$x = 33$$

Possible answers for the angle measures are as follows:

First angle: $x = 33°$

Second angle: $x + 5 = 33 + 5 = 38°$

Third angle: $3x + 10 = 3(33) + 10 = 109°$

Check. The second angle is $5°$ more than the first, and the third is $10°$ more than 3 times the first. The sum, $33° + 38° + 109°$, is $180°$. The numbers check.

State. The measures of the angles are $33°$, $38°$, and $109°$.

27. Familiarize. Let a = the amount Sarah invested. The investment grew by 28% of a, or $0.28a$.

Translate.

$$\underbrace{\text{Amount invested}} \text{ plus } \underbrace{\text{amount of growth}} \text{ is } \$448.$$
$$a + 0.28a = 448$$

Solve. We solve the equation.
$$a + 0.28a = 448$$
$$1.28a = 448$$
$$a = 350$$

Check. 28% of \$350 is 0.28(\$350), or \$98, and \$350 + \$98 = \$448. The answer checks.

State. Sarah invested \$350.

29. Familiarize. Let b = the balance in the account at the beginning of the month. The balance grew by 2% of b, or $0.02b$.

Translate.

$$\underbrace{\text{Original balance}} \text{ plus } \underbrace{\text{amount of growth}} \text{ is } \$870.$$
$$b + 0.02b = 870$$

Solve. We solve the equation.
$$b + 0.02b = 870$$
$$1.02b = 870$$
$$b \approx \$852.94$$

Check. 2% of \$852.94 is 0.02(\$852.94), or \$17.06, and \$852.94 + \$17.06 = \$870. The answer checks.

State. The balance at the beginning of the month was \$852.94.

31. Familiarize. The total cost is the initial charge plus the mileage charge. Let d = the distance, in miles, that Courtney can travel for \$12. The mileage charge is the cost per mile times the number of miles traveled or $0.75d$.

Translate.

$$\underbrace{\text{Initial charge}} \text{ plus } \underbrace{\text{mileage charge}} \text{ is } \$12.$$
$$3 + 0.75d = 12$$

Solve. We solve the equation.
$$3 + 0.75d = 12$$
$$0.75d = 9$$
$$d = 12$$

Check. A 12-mi taxi ride from the airport would cost $3 + 12(\$0.75)$, or \$3 + \$9, or \$12. The answer checks.

State. Courtney can travel 12 mi from the airport for \$12.

33. *Familiarize.* Let c = the cost of the meal before the tip. We know that the cost of the meal before the tip plus the tip, 15% of the cost, is the total cost, $41.40.

Translate.

$$\underbrace{\text{Cost of meal}}_{\downarrow} \; \underset{\downarrow}{\text{plus}} \; \underset{\downarrow}{\text{tip}} \; \underset{\downarrow}{\text{is}} \; \underset{\downarrow}{\$41.40}$$
$$c \quad + \quad 15\%c = 41.40$$

Solve. We solve the equation.

$$c + 15\%c = 41.40$$
$$c + 0.15c = 41.40$$
$$1c + 0.15c = 41.40$$
$$1.15c = 41.40$$
$$\frac{1.15c}{1.15} = \frac{41.40}{1.15}$$
$$c = 36$$

Check. We find 15% of $36 and add it to $36:

$15\% \times \$36 = 0.15 \times \$36 = \$5.40$ and $\$36 + \$5.40 = \$41.40$. The answer checks.

State. The cost of the meal before the tip was added was $36.

35. *Familiarize.* Tom paid a total of $3 \cdot \$34$, or $102, for the three ties. Let t = the price of one of the ties. Then $2t$ = the price of another and we are told that the remaining tie cost $27.

Translate.

$$\underbrace{\text{Total cost of the ties}}_{\downarrow} \; \underset{\downarrow}{\text{is}} \; \underset{\downarrow}{\$102.}$$
$$t + 2t + 27 \quad = \quad 102$$

Solve. We solve the equation.

$$t + 2t + 27 = 102$$
$$3t + 27 = 102$$
$$3t + 27 - 27 = 102 - 27$$
$$3t = 75$$
$$\frac{3t}{3} = \frac{75}{3}$$
$$t = 25$$

If $t = 25$, then $2t = 2 \cdot 25 = 50$.

Check. The $50 tie costs twice as much as the $25 tie, and the total cost of the ties is $\$25 + \$50 + \$27$, or $102. The answer checks.

State. One tie cost $25 and another cost $50.

37. Discussion and Writing Exercise

39.
$$-\frac{4}{5} - \frac{3}{8} = -\frac{4}{5} + \left(-\frac{3}{8}\right)$$
$$= -\frac{32}{40} + \left(-\frac{15}{40}\right)$$
$$= -\frac{47}{40}$$

41.
$$-\frac{4}{5} \cdot \frac{3}{8} = -\frac{4 \cdot 3}{5 \cdot 8}$$
$$= -\frac{4 \cdot 3}{5 \cdot 2 \cdot 4}$$
$$= -\frac{\cancel{4} \cdot 3}{5 \cdot 2 \cdot \cancel{4}}$$
$$= -\frac{3}{10}$$

43.
$$\frac{1}{10} \div \left(-\frac{1}{100}\right) = \frac{1}{10} \cdot \left(-\frac{100}{1}\right) = -\frac{1 \cdot 100}{10 \cdot 1} =$$
$$-\frac{\cancel{1} \cdot \cancel{10} \cdot 10}{\cancel{10} \cdot \cancel{1} \cdot 1} = -\frac{10}{1} = -10$$

45. $-25.6(-16) = 409.6$

47. $-25.6 + (-16) = -41.6$

49. *Familiarize.* Let a = the original number of apples. Then $\frac{1}{3}a$, $\frac{1}{4}a$, $\frac{1}{8}a$, and $\frac{1}{5}a$ are given to four people, respectively. The fifth and sixth people get 10 apples and 1 apple, respectively.

Translate. We reword the problem.

$$\underbrace{\text{The total number of apples}}_{\downarrow} \; \underset{\downarrow}{\text{is}} \; \underbrace{a}_{\downarrow}$$
$$\frac{1}{3}a + \frac{1}{4}a + \frac{1}{8}a + \frac{1}{5}a + 10 + 1 = \quad a$$

Solve. We solve the equation.

$$\frac{1}{3}a + \frac{1}{4}a + \frac{1}{8}a + \frac{1}{5}a + 10 + 1 = a, \text{ LCD is } 120$$
$$120\left(\frac{1}{3}a + \frac{1}{4}a + \frac{1}{8}a + \frac{1}{5}a + 11\right) = 120 \cdot a$$
$$40a + 30a + 15a + 24a + 1320 = 120a$$
$$109a + 1320 = 120a$$
$$1320 = 11a$$
$$120 = a$$

Check. If the original number of apples was 120, then the first four people got $\frac{1}{3} \cdot 120$, $\frac{1}{4} \cdot 120$, $\frac{1}{8} \cdot 120$, and $\frac{1}{5} \cdot 120$, or 40, 30, 15, and 24 apples, respectively. Adding all the apples we get $40 + 30 + 15 + 24 + 10 + 1$, or 120. The result checks.

State. There were originally 120 apples in the basket.

51. Divide the largest triangle into three triangles, each with a vertex at the center of the circle and with height x as shown.

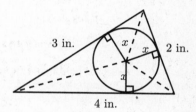

Then the sum of the areas of the three smaller triangles is the area of the original triangle. We have:

$$\frac{1}{2} \cdot 3x + \frac{1}{2} \cdot 2x + \frac{1}{2} \cdot 4x = 2.9047$$

$$2\left(\frac{1}{2} \cdot 3x + \frac{1}{2} \cdot 2x + \frac{1}{2} \cdot 4x\right) = 2(2.9047)$$

$$3x + 2x + 4x = 5.8094$$

$$9x = 5.8094$$

$$x \approx 0.65$$

Thus, x is about 0.65 in.

53. *Familiarize.* Let p = the price of the gasoline as registered on the pump. Then the sales tax will be $9\%p$.

Translate. We reword the problem.

$$\underbrace{\text{Price on pump}}_{p} \text{ plus } \underbrace{\text{sales tax}}_{9\%p} \text{ is } \underbrace{\$10}_{10}$$

$$\text{plus} \quad \downarrow \qquad \downarrow \quad \downarrow$$
$$+ \qquad 9\%p \quad = \quad 10$$

Solve. We solve the equation.

$$p + 9\%p = 10$$

$$1p + 0.09p = 10$$

$$1.09p = 10$$

$$\frac{1.09p}{1.09} = \frac{10}{1.09}$$

$$p \approx 9.17$$

Check. We find 9% of $9.17 and add it to $9.17:

$$9\% \times \$9.17 = 0.09 \times \$9.17 \approx \$0.83$$

Then $9.17 + $0.83 = $10, so $9.17 checks.

State. The attendant should have filled the tank until the pump read $9.17, not $9.10.

Exercise Set 10.7

1. $x > -4$

   a) Since $4 > -4$ is true, 4 is a solution.

   b) Since $0 > -4$ is true, 0 is a solution.

   c) Since $-4 > -4$ is false, -4 is not a solution.

   d) Since $6 > -4$ is true, 6 is a solution.

   e) Since $5.6 > -4$ is true, 5.6 is a solution.

3. $x \geq 6.8$

   a) Since $-6 \geq 6.8$ is false, -6 is not a solution.

   b) Since $0 \geq 6.8$ is false, 0 is not a solution.

   c) Since $6 \geq 6.8$ is false, 6 is not a solution.

   d) Since $8 \geq 6.8$ is true, 8 is a solution.

   e) Since $-3\frac{1}{2} \geq 6.8$ is false, $-3\frac{1}{2}$ is not a solution.

5. The solutions of $x > 4$ are those numbers greater than 4. They are shown on the graph by shading all points to the right of 4. The open circle at 4 indicates that 4 is not part of the graph.

7. The solutions of $t < -3$ are those numbers less than -3. They are shown on the graph by shading all points to the left of -3. The open circle at -3 indicates that -3 is not part of the graph.

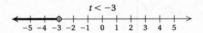

9. The solutions of $m \geq -1$ are are shown by shading the point for -1 and all points to the right of -1. The closed circle at -1 indicates that -1 is part of the graph.

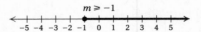

11. In order to be a solution of the inequality $-3 < x \leq 4$, a number must be a solution of both $-3 < x$ and $x \leq 4$. The solution set is graphed as follows:

The open circle at -3 means that -3 is not part of the graph. The closed circle at 4 means that 4 is part of the graph.

13. In order to be a solution of the inequality $0 < x < 3$, a number must be a solution of both $0 < x$ and $x < 3$. The solution set is graphed as follows:

The open circles at 0 and at 3 mean that 0 and 3 are not part of the graph.

15.
$$x + 7 > 2$$
$$x + 7 - 7 > 2 - 7 \quad \text{Subtracting 7}$$
$$x > -5 \qquad \text{Simplifying}$$

The solution set is $\{x | x > -5\}$.

The graph is as follows:

17.
$$x + 8 \leq -10$$
$$x + 8 - 8 \leq -10 - 8 \quad \text{Subtracting 8}$$
$$x \leq -18 \qquad \text{Simplifying}$$

The solution set is $\{x | x \leq -18\}$.

The graph is as follows:

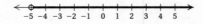

19.
$$y - 7 > -12$$
$$y - 7 + 7 > -12 + 7 \quad \text{Adding 7}$$
$$y > -5 \qquad \text{Simplifying}$$

The solution set is $\{y | y > -5\}$.

21.
$$2x + 3 > x + 5$$
$$2x + 3 - 3 > x + 5 - 3 \quad \text{Subtracting 3}$$
$$2x > x + 2 \quad \text{Simplifying}$$
$$2x - x > x + 2 - x \quad \text{Subtracting } x$$
$$x > 2 \quad \text{Simplifying}$$

The solution set is $\{x | x > 2\}$.

23.
$$3x + 9 \leq 2x + 6$$
$$3x + 9 - 9 \leq 2x + 6 - 9 \quad \text{Subtracting 9}$$
$$3x \leq 2x - 3 \quad \text{Simplifying}$$
$$3x - 2x \leq 2x - 3 - 2x \quad \text{Subtracting } 2x$$
$$x \leq -3 \quad \text{Simplifying}$$

The solution set is $\{x | x \leq -3\}$.

25.
$$5x - 6 < 4x - 2$$
$$5x - 6 + 6 < 4x - 2 + 6$$
$$5x < 4x + 4$$
$$5x - 4x < 4x + 4 - 4x$$
$$x < 4$$

The solution set is $\{x | x < 4\}$.

27.
$$-9 + t > 5$$
$$-9 + t + 9 > 5 + 9$$
$$t > 14$$

The solution set is $\{t | t > 14\}$.

29.
$$y + \frac{1}{4} \leq \frac{1}{2}$$
$$y + \frac{1}{4} - \frac{1}{4} \leq \frac{1}{2} - \frac{1}{4}$$
$$y \leq \frac{2}{4} - \frac{1}{4} \quad \text{Obtaining a common denominator}$$
$$y \leq \frac{1}{4}$$

The solution set is $\left\{y \middle| y \leq \frac{1}{4}\right\}$.

31.
$$x - \frac{1}{3} > \frac{1}{4}$$
$$x - \frac{1}{3} + \frac{1}{3} > \frac{1}{4} + \frac{1}{3}$$
$$x > \frac{3}{12} + \frac{4}{12} \quad \text{Obtaining a common denominator}$$
$$x > \frac{7}{12}$$

The solution set is $\left\{x \middle| x > \frac{7}{12}\right\}$.

33.
$$5x < 35$$
$$\frac{5x}{5} < \frac{35}{5} \quad \text{Dividing by 5}$$
$$x < 7$$

The solution set is $\{x | x < 7\}$. The graph is as follows:

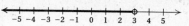

35.
$$-12x > -36$$
$$\frac{-12x}{-12} < \frac{-36}{-12} \quad \text{Dividing by } -12$$
$$\quad\quad\quad \text{The symbol has to be reversed.}$$
$$x < 3 \quad \text{Simplifying}$$

The solution set is $\{x | x < 3\}$. The graph is as follows:

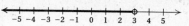

37.
$$5y \geq -2$$
$$\frac{5y}{5} \geq \frac{-2}{5} \quad \text{Dividing by 5}$$
$$y \geq -\frac{2}{5}$$

The solution set is $\left\{y \middle| y \geq -\frac{2}{5}\right\}$.

39.
$$-2x \leq 12$$
$$\frac{-2x}{-2} \geq \frac{12}{-2} \quad \text{Dividing by } -2$$
$$\quad\quad\quad \text{The symbol has to be reversed.}$$
$$x \geq -6 \quad \text{Simplifying}$$

The solution set is $\{x | x \geq -6\}$.

41.
$$-4y \geq -16$$
$$\frac{-4y}{-4} \leq \frac{-16}{-4} \quad \text{Dividing by } -4$$
$$\quad\quad\quad \text{The symbol has to be reversed.}$$
$$y \leq 4 \quad \text{Simplifying}$$

The solution set is $\{y | y \leq 4\}$.

43.
$$-3x < -17$$
$$\frac{-3x}{-3} > \frac{-17}{-3} \quad \text{Dividing by } -3$$
$$\quad\quad\quad \text{The symbol has to be reversed.}$$
$$x > \frac{17}{3} \quad \text{Simplifying}$$

The solution set is $\left\{x \middle| x > \frac{17}{3}\right\}$.

45.
$$-2y > \frac{1}{7}$$
$$-\frac{1}{2} \cdot (-2y) < -\frac{1}{2} \cdot \frac{1}{7}$$
$$\quad\quad\quad \text{The symbol has to be reversed.}$$
$$y < -\frac{1}{14}$$

The solution set is $\left\{y \middle| y < -\frac{1}{14}\right\}$.

47.
$$-\frac{6}{5} \leq -4x$$
$$-\frac{1}{4} \cdot \left(-\frac{6}{5}\right) \geq -\frac{1}{4} \cdot (-4x)$$
$$\frac{6}{20} \geq x$$
$$\frac{3}{10} \geq x, \text{ or } x \leq \frac{3}{10}$$

The solution set is $\left\{x \middle| \frac{3}{10} \geq x\right\}$, or $\left\{x \middle| x \leq \frac{3}{10}\right\}$.

49.
$$4 + 3x < 28$$
$$-4 + 4 + 3x < -4 + 28 \quad \text{Adding } -4$$
$$3x < 24 \quad \text{Simplifying}$$
$$\frac{3x}{3} < \frac{24}{3} \quad \text{Dividing by 3}$$
$$x < 8$$

The solution set is $\{x | x < 8\}$.

51. $3x - 5 \le 13$
$3x - 5 + 5 \le 13 + 5$ Adding 5
$3x \le 18$
$\dfrac{3x}{3} \le \dfrac{18}{3}$ Dividing by 3
$x \le 6$

The solution set is $\{x | x \le 6\}$.

53. $13x - 7 < -46$
$13x - 7 + 7 < -46 + 7$
$13x < -39$
$\dfrac{13x}{13} < \dfrac{-39}{13}$
$x < -3$

The solution set is $\{x | x < -3\}$.

55. $30 > 3 - 9x$
$30 - 3 > 3 - 9x - 3$ Subtracting 3
$27 > -9x$
$\dfrac{27}{-9} < \dfrac{-9x}{-9}$ Dividing by -9
$\uparrow$___ The symbol has to be reversed.
$-3 < x$

The solution set is $\{x | -3 < x\}$, or $\{x | x > -3\}$.

57. $4x + 2 - 3x \le 9$
$x + 2 \le 9$ Collecting like terms
$x + 2 - 2 \le 9 - 2$
$x \le 7$

The solution set is $\{x | x \le 7\}$.

59. $-3 < 8x + 7 - 7x$
$-3 < x + 7$ Collecting like terms
$-3 - 7 < x + 7 - 7$
$-10 < x$

The solution set is $\{x | -10 < x\}$, or $\{x | x > -10\}$.

61. $6 - 4y > 4 - 3y$
$6 - 4y + 4y > 4 - 3y + 4y$ Adding $4y$
$6 > 4 + y$
$-4 + 6 > -4 + 4 + y$ Adding -4
$2 > y$, or $y < 2$

The solution set is $\{y | 2 > y\}$, or $\{y | y < 2\}$.

63. $5 - 9y \le 2 - 8y$
$5 - 9y + 9y \le 2 - 8y + 9y$
$5 \le 2 + y$
$-2 + 5 \le -2 + 2 + y$
$3 \le y$, or $y \ge 3$

The solution set is $\{y | 3 \le y\}$, or $\{y | y \ge 3\}$.

65. $19 - 7y - 3y < 39$
$19 - 10y < 39$ Collecting like terms
$-19 + 19 - 10y < -19 + 39$
$-10y < 20$
$\dfrac{-10y}{-10} > \dfrac{20}{-10}$
$\uparrow$___ The symbol has to be reversed.
$y > -2$

The solution set is $\{y | y > -2\}$.

67. $2.1x + 45.2 > 3.2 - 8.4x$
$10(2.1x + 45.2) > 10(3.2 - 8.4x)$ Multiplying by 10 to clear decimals
$21x + 452 > 32 - 84x$
$21x + 84x > 32 - 452$ Adding $84x$ and subtracting 452
$105x > -420$
$x > -4$ Dividing by 105

The solution set is $\{x | x > -4\}$.

69. $\dfrac{x}{3} - 2 \le 1$
$3\left(\dfrac{x}{3} - 2\right) \le 3 \cdot 1$ Multiplying by 3 to to clear the fraction
$x - 6 \le 3$ Simplifying
$x \le 9$ Adding 6

The solution set is $\{x | x \le 9\}$.

71. $\dfrac{y}{5} + 1 \le \dfrac{2}{5}$
$5\left(\dfrac{y}{5} + 1\right) \le 5 \cdot \dfrac{2}{5}$ Clearing fractions
$y + 5 \le 2$
$y \le -3$ Subtracting 5

The solution set is $\{y | y \le -3\}$.

73. $3(2y - 3) < 27$
$6y - 9 < 27$ Removing parentheses
$6y < 36$ Adding 9
$y < 6$ Dividing by 6

The solution set is $\{y | y < 6\}$.

75. $2(3 + 4m) - 9 \ge 45$
$6 + 8m - 9 \ge 45$ Removing parentheses
$8m - 3 \ge 45$ Collecting like terms
$8m \ge 48$ Adding 3
$m \ge 6$ Dividing by 8

The solution set is $\{m | m \ge 6\}$.

77. $8(2t + 1) > 4(7t + 7)$
$16t + 8 > 28t + 28$
$16t - 28t > 28 - 8$
$-12t > 20$
$t < -\dfrac{20}{12}$ Dividing by -12 and reversing the symbol
$t < -\dfrac{5}{3}$

The solution set is $\left\{t \,\middle|\, t < -\dfrac{5}{3}\right\}$.

79. $3(r - 6) + 2 < 4(r + 2) - 21$
$3r - 18 + 2 < 4r + 8 - 21$
$3r - 16 < 4r - 13$
$-16 + 13 < 4r - 3r$
$-3 < r$, or $r > -3$

The solution set is $\{r | r > -3\}$.

81.
$$0.8(3x + 6) \geq 1.1 - (x + 2)$$
$$2.4x + 4.8 \geq 1.1 - x - 2$$
$$10(2.4x + 4.8) \geq 10(1.1 - x - 2) \quad \text{Clearing decimals}$$
$$24x + 48 \geq 11 - 10x - 20$$
$$24x + 48 \geq -10x - 9 \quad \text{Collecting like terms}$$
$$24x + 10x \geq -9 - 48$$
$$34x \geq -57$$
$$x \geq -\frac{57}{34}$$

The solution set is $\left\{ x \middle| x \geq -\frac{57}{34} \right\}$.

83. $\dfrac{5}{3} + \dfrac{2}{3}x < \dfrac{25}{12} + \dfrac{5}{4}x + \dfrac{3}{4}$

The number 12 is the least common multiple of all the denominators. We multiply by 12 on both sides.

$$12\left(\frac{5}{3} + \frac{2}{3}x\right) < 12\left(\frac{25}{12} + \frac{5}{4}x + \frac{3}{4}\right)$$
$$12 \cdot \frac{5}{3} + 12 \cdot \frac{2}{3}x < 12 \cdot \frac{25}{12} + 12 \cdot \frac{5}{4}x + 12 \cdot \frac{3}{4}$$
$$20 + 8x < 25 + 15x + 9$$
$$20 + 8x < 34 + 15x$$
$$20 - 34 < 15x - 8x$$
$$-14 < 7x$$
$$-2 < x, \text{ or } x > -2$$

The solution set is $\{x | x > -2\}$.

85. Discussion and Writing Exercise

87. $-56 + (-18)$ Two negative numbers. Add the absolute values and make the answer negative.

$$-56 + (-18) = -74$$

89. $-\dfrac{3}{4} + \dfrac{1}{8}$ One negative and one positive number. Find the difference of the absolute values. Then make the answer negative, since the negative number has the larger absolute value.

$$-\frac{3}{4} + \frac{1}{8} = -\frac{6}{8} + \frac{1}{8} = -\frac{5}{8}$$

91. $-56 - (-18) = -56 + 18 = -38$

93. $-2.3 - 7.1 = -2.3 + (-7.1) = -9.4$

95. $5 - 3^2 + (8 - 2)^2 \cdot 4 = 5 - 3^2 + 6^2 \cdot 4$
$$= 5 - 9 + 36 \cdot 4$$
$$= 5 - 9 + 144$$
$$= -4 + 144$$
$$= 140$$

97. $5(2x - 4) - 3(4x + 1) = 10x - 20 - 12x - 3 =$
$-2x - 23$

99. $|x| < 3$

a) Since $|0| = 0$ and $0 < 3$ is true, 0 is a solution.

b) Since $|-2| = 2$ and $2 < 3$ is true, -2 is a solution.

c) Since $|-3| = 3$ and $3 < 3$ is false, -3 is not a solution.

d) Since $|4| = 4$ and $4 < 3$ is false, 4 is not a solution.

e) Since $|3| = 3$ and $3 < 3$ is false, 3 is not a solution.

f) Since $|1.7| = 1.7$ and $1.7 < 3$ is true, 1.7 is a solution.

g) Since $|-2.8| = 2.8$ and $2.8 < 3$ is true, -2.8 is a solution.

101. $x + 3 \leq 3 + x$
$$x - x \leq 3 - 3 \quad \text{Subtracting } x \text{ and 3}$$
$$0 \leq 0$$

We get an inequality that is true for all values of x, so the inequality is true for all real numbers.

Exercise Set 10.8

1. $n \geq 7$

3. $w > 2$ kg

5. 90 mph $< s <$ 110 mph

7. $a \leq 1,200,000$

9. $c \leq \$1.50$

11. $x > 8$

13. $y \leq -4$

15. $n \geq 1300$

17. $a \leq 500$ L

19. $3x + 2 < 13$, or $2 + 3x < 13$

21. *Familiarize*. Let s represent the score on the fourth test.

Translate.

$$\underbrace{\text{The average score}}_{\frac{82 + 76 + 78 + s}{4}} \quad \underbrace{\text{is at least}}_{\geq} \quad \underbrace{80.}_{80}$$

Solve.

$$\frac{82 + 76 + 78 + s}{4} \geq 80$$
$$4\left(\frac{82 + 76 + 78 + s}{4}\right) \geq 4 \cdot 80$$
$$82 + 76 + 78 + s \geq 320$$
$$236 + s \geq 320$$
$$s \geq 84$$

Check. As a partial check we show that the average is at least 80 when the fourth test score is 84.

$$\frac{82 + 76 + 78 + 84}{4} = \frac{320}{4} = 80$$

State. The student will get at least a B if the score on the fourth test is at least 84. The solution set is $\{s | s \geq 84\}$.

23. *Familiarize*. We use the formula for converting Celsius temperatures to Fahrenheit temperatures, $F = \dfrac{9}{5}C + 32$.

Translate.

$$\underbrace{\text{Fahrenheit temperature}}_{\frac{9}{5}C + 32} \quad \underbrace{\text{is less than}}_{\leq} \quad \underbrace{1945.4.}_{1945.4}$$

Solve.

$$\frac{9}{5}C + 32 < 1945.4$$

$$\frac{9}{5}C < 1913.4$$

$$\frac{5}{9} \cdot \frac{9}{5}C < \frac{5}{9}(1913.4)$$

$$C < 1063$$

Check. As a partial check we can show that the Fahrenheit temperature is less than 1945.4° for a Celsius temperature less than 1063° and is greater than 1945.4° for a Celsius temperature greater than 1063°.

$$F = \frac{9}{5} \cdot 1062 + 32 = 1943.6 < 1945.4$$

$$F = \frac{9}{5} \cdot 1064 + 32 = 1947.2 > 1945.4$$

State. Gold stays solid for temperatures less than 1063°C. The solution set is $\{C|C < 1063°\}$.

25. Familiarize. $R = -0.075t + 3.85$

In the formula R represents the world record and t represents the years since 1930. When $t = 0$ (1930), the record was $-0.075 \cdot 0 + 3.85$, or 3.85 minutes. When $t = 2$ (1932), the record was $-0.075(2) + 3.85$, or 3.7 minutes. For what values of t will $-0.075t + 3.85$ be less than 3.5?

Translate. The record is to be less than 3.5. We have the inequality

$$R < 3.5.$$

To find the t values which satisfy this condition we substitute $-0.075t + 3.85$ for R.

$$-0.075t + 3.85 < 3.5$$

Solve.

$$-0.075t + 3.85 < 3.5$$
$$-0.075t < 3.5 - 3.85$$
$$-0.075t < -0.35$$
$$t > \frac{-0.35}{-0.075}$$
$$t > 4\frac{2}{3}$$

Check. With inequalities it is impossible to check each solution. But we can check to see if the solution set we obtained seems reasonable.

When $t = 4\frac{1}{2}$, $R = -0.075(4.5) + 3.85$, or 3.5125.

When $t = 4\frac{2}{3}$, $R = -0.075\left(\frac{14}{3}\right) + 3.85$, or 3.5.

When $t = 4\frac{3}{4}$, $R = -0.075(4.75) + 3.85$, or 3.49375.

Since $r = 3.5$ when $t = 4\frac{2}{3}$ and R decreases as t increases, R will be less than 3.5 when t is greater than $4\frac{2}{3}$.

State. The world record will be less than 3.5 minutes more than $4\frac{2}{3}$ years after 1930. If we let $Y = $ the year, then the solution set is $\{Y|Y \geq 1935\}$.

27. Familiarize. As in the drawing in the text, we let $L = $ the length of the envelope. Recall that the area of a rectangle is the product of the length and the width.

Translate.

Length	times	width	is at least	$17\frac{1}{2}$ in^2
↓	↓	↓	↓	↓
L	$\cdot$	$3\frac{1}{2}$	$\geq$	$17\frac{1}{2}$

Solve.

$$L \cdot 3\frac{1}{2} \geq 17\frac{1}{2}$$

$$L \cdot \frac{7}{2} \geq \frac{35}{2}$$

$$L \cdot \frac{7}{2} \cdot \frac{2}{7} \geq \frac{35}{2} \cdot \frac{2}{7}$$

$$L \geq 5$$

The solution set is $\{L|L \geq 5\}$.

Check. We can obtain a partial check by substituting a number greater than or equal to 5 in the inequality. For example, when $L = 6$:

$$L \cdot 3\frac{1}{2} = 6 \cdot 3\frac{1}{2} = 6 \cdot \frac{7}{2} = 21 \geq 17\frac{1}{2}$$

The result appears to be correct.

State. Lengths of 5 in. or more will satisfy the constraints. The solution set is $\{L|L \geq 5 \text{ in.}\}$.

29. Familiarize. Let $c = $ the number of copies Myra has made. The total cost of the copies is the setup fee of \$5 plus \$4 times the number of copies, or \$4 $\cdot c$.

Translate.

Setup fee	plus	copying cost	cannot exceed	\$65.
↓	↓	↓	↓	↓
5	$+$	$4c$	$\leq$	65

Solve. We solve the inequality.

$$5 + 4c \leq 65$$
$$4c \leq 60$$
$$c \leq 15$$

Check. As a partial check, we show that Myra can have 15 copies made and not exceed her \$65 budget.

$$\$5 + \$4 \cdot 15 = 5 + 60 = \$65$$

State. Myra can have 15 or fewer copies made and stay within her budget.

31. Familiarize. Let m represent the length of a telephone call, in minutes.

Translate.

\$0.75 charge	plus	charge for time used	is at least	\$3.00.
↓	↓	↓	↓	↓
0.75	$+$	$0.45m$	$\geq$	3

Solve. We solve the inequality.

$$0.75 + 0.45m \geq 3$$
$$0.45m \geq 2.25$$
$$m \geq 5$$

Check. As a partial check, we can show that if a call lasts 5 minutes it costs at least $3.00:

$$\$0.75 + \$0.45(5) = \$0.75 + \$2.25 = \$3.00.$$

State. Simon's calls last at least 5 minutes each.

33. *Familiarize*. Let c = the number of courses for which Angelica registers. Her total tuition is the $35 registration fee plus $375 times the number of courses for which she registers, or $\$375 \cdot c$.

Translate.

Registration fee	plus	fee for courses	cannot exceed	$1000.
↓	↓	↓	↓	↓
35	+	$375 \cdot c$	≤	1000

Solve. We solve the inequality.

$$35 + 375c \leq 1000$$
$$375c \leq 965$$
$$c \leq 2.57\overline{3}$$

Check. Although the solution set of the inequality is all numbers less than or equal to $2.57\overline{3}$, since c represents the number of courses for which Angelica registers, we round down to 2. If she registers for 2 courses, her tuition is $\$35 + \$375 \cdot 2$, or $785 which does not exceed $1000. If she registers for 3 courses, her tuition is $\$35 + \$375 \cdot 3$, or $1160 which exceeds $1000.

State. Angelica can register for at most 2 courses.

35. *Familiarize*. Let s = the number of servings of fruits or vegetables Dale eats on Saturday.

Translate.

Average number of fruit or vegetable servings	is at least	5.
↓	↓	↓
$\dfrac{4+6+7+4+6+4+s}{7}$	≥	5

Solve. We first multiply by 7 to clear the fraction.

$$7\left(\frac{4+6+7+4+6+4+s}{7}\right) \geq 7 \cdot 5$$
$$4+6+7+4+6+4+s \geq 35$$
$$31 + s \geq 35$$
$$s \geq 4$$

Check. As a partial check, we show that Dale can eat 4 servings of fruits or vegetables on Saturday and average at least 5 servings per day for the week:

$$\frac{4+6+7+4+6+4+4}{7} = \frac{35}{7} = 5.$$

State. Dale should eat at least 4 servings of fruits or vegetables on Saturday.

37. *Familiarize*. We first make a drawing. We let l represent the length, in feet.

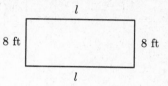

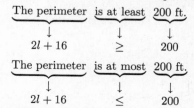

The perimeter is $P = 2l + 2w$, or $2l + 2 \cdot 8$, or $2l + 16$.

Translate. We translate to 2 inequalities.

The perimeter	is at least	200 ft.
↓	↓	↓
$2l + 16$	≥	200

The perimeter	is at most	200 ft.
↓	↓	↓
$2l + 16$	≤	200

Solve. We solve each inequality.

$$2l + 16 \geq 200 \qquad 2l + 16 \leq 200$$
$$2l \geq 184 \qquad\qquad 2l \leq 184$$
$$l \geq 92 \qquad\qquad\quad l \leq 92$$

Check. We check to see if the solutions seem reasonable.

When $l = 91$ ft, $P = 2 \cdot 91 + 16$, or 198 ft.

When $l = 92$ ft, $P = 2 \cdot 92 + 16$, or 200 ft.

When $l = 93$ ft, $P = 2 \cdot 93 + 16$, or 202 ft.

From these calculations, it appears that the solutions are correct.

State. Lengths greater than or equal to 92 ft will make the perimeter at least 200 ft. Lengths less than or equal to 92 ft will make the perimeter at most 200 ft.

39. *Familiarize*. Using the label on the drawing in the text, we let L represent the length.

The area is the length times the width, or $4L$.

Translate.

Area	is less than	86 cm².
↓	↓	↓
$4L$	<	86

Solve.

$$4L < 86$$
$$L < 21.5$$

Check. We check to see if the solution seems reasonable.

When $L = 22$, the area is $22 \cdot 4$, or 88 cm².

When $L = 21.5$, the area is $21.5(4)$, or 86 cm².

When $L = 21$, the area is $21 \cdot 4$, or 84 cm².

From these calculations, it would appear that the solution is correct.

State. The area will be less than 86 cm² for lengths less than 21.5 cm.

41. *Familiarize*. Let v = the blue book value of the car. Since the car was repaired, we know that $8500 does not exceed $0.8v$ or, in other words, $0.8v$ is at least $8500.

Translate.

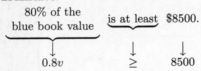

$$0.8v \qquad\qquad \geq \qquad 8500$$

Solve.

$$0.8v \geq 8500$$
$$v \geq \frac{8500}{0.8}$$
$$v \geq 10,625$$

Check. As a partial check, we show that 80% of $10,625 is at least $8500:

$$0.8(\$10,625) = \$8500$$

State. The blue book value of the car was at least $10,625.

43. **Familiarize.** Let $r =$ the amount of fat in a serving of the regular peanut butter, in grams. If reduced fat peanut butter has at least 25% less fat than regular peanut butter, then it has at most 75% as much fat as the regular peanut butter.

Translate.

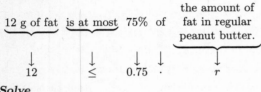

$$12 \qquad\qquad \leq \qquad 0.75 \;\cdot\qquad\qquad r$$

Solve.

$$12 \leq 0.75r$$
$$16 \leq r$$

Check. As a partial check, we show that 12 g of fat does not exceed 75% of 16 g of fat:

$$0.75(16) = 12$$

State. Regular peanut butter contains at least 16 g of fat per serving.

45. **Familiarize.** Let $w =$ the number of weeks after July 1. After w weeks the water level has dropped $\frac{2}{3}w$ ft.

Translate.

Original depth	minus	drop in water level	does not exceed	21 ft.
↓	↓	↓	↓	↓
25	−	$\frac{2}{3}w$	≤	21

Solve. We solve the inequality.

$$25 - \frac{2}{3}w \leq 21$$
$$-\frac{2}{3}w \leq -4$$
$$w \geq -\frac{3}{2}(-4)$$
$$w \geq 6$$

Check. As a partial check we show that the water level is 21 ft 6 weeks after July 1.

$$25 - \frac{2}{3} \cdot 6 = 25 - 4 = 21 \text{ ft}$$

Since the water level continues to drop during the weeks after July 1, the answer seems reasonable.

State. The water level will not exceed 21 ft for dates at least 6 weeks after July 1.

47. **Familiarize.** Let $h =$ the height of the triangle, in ft. Recall that the formula for the area of a triangle with base b and height h is $A = \frac{1}{2}bh$.

Translate.

Area	is at least	3 ft².
↓	↓	↓
$\frac{1}{2}\left(1\frac{1}{2}\right)h$	≥	3

Solve. We solve the inequality.

$$\frac{1}{2}\left(1\frac{1}{2}\right)h \geq 3$$
$$\frac{1}{2} \cdot \frac{3}{2} \cdot h \geq 3$$
$$\frac{3}{4}h \geq 3$$
$$h \geq \frac{4}{3} \cdot 3$$
$$h \geq 3$$

Check. As a partial check, we show that the area of the triangle is 3 ft² when the height is 4 ft.

$$\frac{1}{2}\left(1\frac{1}{2}\right)(4) = \frac{1}{2} \cdot \frac{3}{2} \cdot \frac{4}{1} = 3$$

State. The height should be at least 4 ft.

49. **Familiarize.** The average number of calls per week is the sum of the calls for the three weeks divided by the number of weeks, 3. We let c represent the number of calls made during the third week.

Translate. The average of the three weeks is given by

$$\frac{17 + 22 + c}{3}.$$

Since the average must be at least 20, this means that it must be greater than or equal to 20. Thus, we can translate the problem to the inequality

$$\frac{17 + 22 + c}{3} \geq 20.$$

Solve. We first multiply by 3 to clear the fraction.

$$3\left(\frac{17 + 22 + c}{3}\right) \geq 3 \cdot 20$$
$$17 + 22 + c \geq 60$$
$$39 + c \geq 60$$
$$c \geq 21$$

Check. Suppose c is a number greater than or equal to 21. Then by adding 17 and 22 on both sides of the inequality we get

$$17 + 22 + c \geq 17 + 22 + 21$$
$$17 + 22 + c \geq 60$$

so

$$\frac{17 + 22 + c}{3} \geq \frac{60}{3}, \text{ or } 20.$$

State. 21 calls or more will maintain an average of at least 20 for the three-week period.

51. Discussion and Writing Exercise.

53. The product of an <u>even</u> number of negative numbers is always positive.

55. The <u>additive</u> inverse of a negative number is always positive.

57. Equations with the same solutions are called <u>equivalent</u> equations.

59. The <u>multiplication principle</u> for inequalities asserts that when we multiply or divide by a negative number on both sides of an inequality, the direction of the inequality symbol <u>is reversed</u>.

61. *Familiarize*. We use the formula $F = \frac{9}{5}C + 32$.

Translate. We are interested in temperatures such that $5° < F < 15°$. Substituting for F, we have:

$$5 < \frac{9}{5}C + 32 < 15$$

Solve.

$$5 < \frac{9}{5}C + 32 < 15$$

$$5 \cdot 5 < 5\left(\frac{9}{5}C + 32\right) < 5 \cdot 15$$

$$25 < 9C + 160 < 75$$

$$-135 < 9C < -85$$

$$-15 < C < -9\frac{4}{9}$$

Check. The check is left to the student.

State. Green ski wax works best for temperatures between $-15°C$ and $-9\frac{4}{9}°C$.

63. *Familiarize*. Let f = the fat content of a serving of regular tortilla chips, in grams. A product that contains 60% less fat than another product has 40% of the fat content of that product. If Reduced Fat Tortilla Pops cannot be labeled lowfat, then they contain at least 3 g of fat.

Translate.

40% of	the fat content of regular tortilla chips	is at least	3 grams of fat
↓ ↓	↓	↓	↓
0.4 ·	f	≥	3

Solve.

$$0.4f \geq 3$$

$$f \geq 7.5$$

Check. As a partial check, we show that 40% of 7.5 g is not less than 3 g.

$$0.4(7.5) = 3$$

State. A serving of regular tortilla chips contains at least 7.5 g of fat.

Chapter 10 Review Exercises

1.
$$x + 5 = -17$$
$$x + 5 - 5 = -17 - 5$$
$$x = -22$$
The solution is -22.

2.
$$n - 7 = -6$$
$$n - 7 + 7 = -6 + 7$$
$$n = 1$$
The solution is 1.

3.
$$x - 11 = 14$$
$$x - 11 + 11 = 14 + 11$$
$$x = 25$$
The solution is 25.

4.
$$y - 0.9 = 9.09$$
$$y - 0.9 + 0.9 = 9.09 + 0.9$$
$$y = 9.99$$
The solution is 9.99.

5.
$$-\frac{2}{3}x = -\frac{1}{6}$$
$$-\frac{3}{2} \cdot \left(-\frac{2}{3}x\right) = -\frac{3}{2} \cdot \left(-\frac{1}{6}\right)$$
$$1 \cdot x = \frac{\cancel{3} \cdot 1}{2 \cdot 2 \cdot \cancel{3}}$$
$$x = \frac{1}{4}$$
The solution is $\frac{1}{4}$.

6.
$$-8x = -56$$
$$\frac{-8x}{-8} = \frac{-56}{-8}$$
$$x = 7$$
The solution is 7.

7.
$$-\frac{x}{4} = 48$$
$$4 \cdot \frac{1}{4} \cdot (-x) = 4 \cdot 48$$
$$-x = 192$$
$$-1 \cdot (-1 \cdot x) = -1 \cdot 192$$
$$x = -192$$
The solution is -192.

8.
$$15x = -35$$
$$\frac{15x}{15} = \frac{-35}{15}$$
$$x = -\frac{\cancel{5} \cdot 7}{3 \cdot \cancel{5}}$$
$$x = -\frac{7}{3}$$
The solution is $-\frac{7}{3}$.

9. $\frac{4}{5}y = -\frac{3}{16}$

$\frac{5}{4} \cdot \frac{4}{5}y = \frac{5}{4} \cdot \left(-\frac{3}{16}\right)$

$y = -\frac{15}{64}$

The solution is $-\frac{15}{64}$.

10. $5 - x = 13$

$5 - x - 5 = 13 - 5$

$-x = 8$

$-1 \cdot (-1 \cdot x) = -1 \cdot 8$

$x = -8$

The solution is -8.

11. $\frac{1}{4}x - \frac{5}{8} = \frac{3}{8}$

$\frac{1}{4}x - \frac{5}{8} + \frac{5}{8} = \frac{3}{8} + \frac{5}{8}$

$\frac{1}{4}x = 1$

$4 \cdot \frac{1}{4}x = 4 \cdot 1$

$x = 4$

The solution is 4.

12. $5t + 9 = 3t - 1$

$5t + 9 - 3t = 3t - 1 - 3t$

$2t + 9 = -1$

$2t + 9 - 9 = -1 - 9$

$2t = -10$

$\frac{2t}{2} = \frac{-10}{2}$

$t = -5$

The solution is -5.

13. $7x - 6 = 25x$

$7x - 6 - 7x = 25x - 7x$

$-6 = 18x$

$\frac{-6}{18} = \frac{18x}{18}$

$-\frac{\cancel{6} \cdot 1}{3 \cdot \cancel{6}} = x$

$-\frac{1}{3} = x$

The solution is $-\frac{1}{3}$.

14. $14y = 23y - 17 - 10$

$14y = 23y - 27$ Collecting like terms

$14y - 23y = 23y - 27 - 23y$

$-9y = -27$

$\frac{-9y}{-9} = \frac{-27}{-9}$

$y = 3$

The solution is 3.

15. $0.22y - 0.6 = 0.12y + 3 - 0.8y$

$0.22y - 0.6 = -0.68y + 3$ Collecting like terms

$0.22y - 0.6 + 0.68y = -0.68y + 3 + 0.68y$

$0.9y - 0.6 = 3$

$0.9y - 0.6 + 0.6 = 3 + 0.6$

$0.9y = 3.6$

$\frac{0.9y}{0.9} = \frac{3.6}{0.9}$

$y = 4$

The solution is 4.

16. $\frac{1}{4}x - \frac{1}{8}x = 3 - \frac{1}{16}x$

$\frac{2}{8}x - \frac{1}{8}x = 3 - \frac{1}{16}x$

$\frac{1}{8}x = 3 - \frac{1}{16}x$

$\frac{1}{8}x + \frac{1}{16}x = 3 - \frac{1}{16}x + \frac{1}{16}x$

$\frac{2}{16}x + \frac{1}{16}x = 3$

$\frac{3}{16}x = 3$

$\frac{16}{3} \cdot \frac{3}{16}x = \frac{16}{3} \cdot 3$

$x = \frac{16 \cdot \cancel{3}}{\cancel{3} \cdot 1}$

$x = 16$

The solution is 16.

17. $14y + 17 + 7y = 9 + 21y + 8$

$21y + 17 = 21y + 17$

$21y + 17 - 21y = 21y + 17 - 21y$

$17 = 17$ TRUE

All real numbers are solutions.

18. $4(x + 3) = 36$

$4x + 12 = 36$

$4x + 12 - 12 = 36 - 12$

$4x = 24$

$\frac{4x}{4} = \frac{24}{4}$

$x = 6$

The solution is 6.

19. $3(5x - 7) = -66$

$15x - 21 = -66$

$15x - 21 + 21 = -66 + 21$

$15x = -45$

$\frac{15x}{15} = \frac{-45}{15}$

$x = -3$

The solution is -3.

20. $8(x-2) - 5(x+4) = 20 + x$

$8x - 16 - 5x - 20 = 20 + x$

$3x - 36 = 20 + x$

$3x - 36 - x = 20 + x - x$

$2x - 36 = 20$

$2x - 36 + 36 = 20 + 36$

$2x = 56$

$\dfrac{2x}{2} = \dfrac{56}{2}$

$x = 28$

The solution is 28.

21. $-5x + 3(x+8) = 16$

$-5x + 3x + 24 = 16$

$-2x + 24 = 16$

$-2x + 24 - 24 = 16 - 24$

$-2x = -8$

$\dfrac{-2x}{-2} = \dfrac{-8}{-2}$

$x = 4$

The solution is 4.

22. $6(x-2) - 16 = 3(2x-5) + 11$

$6x - 12 - 16 = 6x - 15 + 11$

$6x - 28 = 6x - 4$

$6x - 28 - 6x = 6x - 4 - 6x$

$-28 = -4 \qquad$ False

There are no solutions.

23. Since $-3 \le 4$ is true, -3 is a solution.

24. Since $7 \le 4$ is false, 7 is not a solution.

25. Since $4 \le 4$ is true, 4 is a solution.

26. $y + \dfrac{2}{3} \ge \dfrac{1}{6}$

$y + \dfrac{2}{3} - \dfrac{2}{3} \ge \dfrac{1}{6} - \dfrac{2}{3}$

$y \ge \dfrac{1}{6} - \dfrac{4}{6}$

$y \ge -\dfrac{3}{6}$

$y \ge -\dfrac{1}{2}$

The solution set is $\left\{ y \middle| y \ge -\dfrac{1}{2} \right\}$.

27. $9x \ge 63$

$\dfrac{9x}{9} \ge \dfrac{63}{9}$

$x \ge 7$

The solution set is $\{x | x \ge 7\}$.

28. $2 + 6y > 14$

$2 + 6y - 2 > 14 - 2$

$6y > 12$

$\dfrac{6y}{6} > \dfrac{12}{6}$

$y > 2$

The solution set is $\{y | y > 2\}$.

29. $7 - 3y \ge 27 + 2y$

$7 - 3y - 2y \ge 27 + 2y - 2y$

$7 - 5y \ge 27$

$7 - 5y - 7 \ge 27 - 7$

$-5y \ge 20$

$\dfrac{-5y}{-5} \le \dfrac{20}{-5} \qquad$ Reversing the inequality symbol

$y \le -4$

The solution set is $\{y | y \le -4\}$.

30. $3x + 5 < 2x - 6$

$3x + 5 - 2x < 2x - 6 - 2x$

$x + 5 < -6$

$x + 5 - 5 < -6 - 5$

$x < -11$

The solution set is $\{x | x < -11\}$.

31. $-4y < 28$

$\dfrac{-4y}{-4} > \dfrac{28}{-4} \qquad$ Reversing the inequality symbol

$y > -7$

The solution set is $\{y | y > -7\}$.

32. $4 - 8x < 13 + 3x$

$4 - 8x - 3x < 13 + 3x - 3x$

$4 - 11x < 13$

$4 - 11x - 4 < 13 - 4$

$-11x < 9$

$\dfrac{-11x}{-11} > \dfrac{9}{-11} \qquad$ Reversing the inequality symbol

$x > -\dfrac{9}{11}$

The solution set is $\left\{ x \middle| x > -\dfrac{9}{11} \right\}$.

33. $-4x \le \dfrac{1}{3}$

$-\dfrac{1}{4} \cdot (-4x) \ge -\dfrac{1}{4} \cdot \dfrac{1}{3} \qquad$ Reversing the inequality symbol

$x \ge -\dfrac{1}{12}$

The solution set is $\left\{ x \middle| x \ge -\dfrac{1}{12} \right\}$.

34.

$$4x - 6 < x + 3$$
$$4x - 6 - x < x + 3 - x$$
$$3x - 6 < 3$$
$$3x - 6 + 6 < 3 + 6$$
$$3x < 9$$
$$\frac{3x}{3} < \frac{9}{3}$$
$$x < 3$$

The solution set is $\{x | x < 3\}$. The graph is as follows:

35. In order to be a solution of $-2 < x \le 5$, a number must be a solution of both $-2 < x$ and $x \le 5$. The solution set is graphed as follows:

36. The solutions of $y > 0$ are those numbers greater than 0. The graph is as follows:

37.

$$C = \pi d$$
$$\frac{C}{\pi} = \frac{\pi d}{\pi}$$
$$\frac{C}{\pi} = d$$

38.

$$V = \frac{1}{3}Bh$$
$$3 \cdot V = 3 \cdot \frac{1}{3}Bh$$
$$3V = Bh$$
$$\frac{3V}{h} = \frac{Bh}{h}$$
$$\frac{3V}{h} = B$$

39.

$$A = \frac{a+b}{2}$$
$$2 \cdot A = 2 \cdot \left(\frac{a+b}{2}\right)$$
$$2A = a + b$$
$$2A - b = a + b - b$$
$$2A - b = a$$

40.

$$y = mx + b$$
$$y - b = mx + b - b$$
$$y - b = mx$$
$$\frac{y-b}{m} = \frac{mx}{m}$$
$$\frac{y-b}{m} = x$$

41. Familiarize. Let w = the width, in miles. Then $w + 90$ = the length. Recall that the perimeter P of a rectangle with length l and width w is given by $P = 2l + 2w$.

Translate. Substitute 1280 for P and $w + 90$ for l in the formula above.

$$P = 2l + 2w$$
$$1280 = 2(w + 90) + 2w$$

Solve. We solve the equation.

$$1280 = 2(w + 90) + 2w$$
$$1280 = 2w + 180 + 2w$$
$$1280 = 4w + 180$$
$$1280 - 180 = 4w + 180 - 180$$
$$1100 = 4w$$
$$\frac{1100}{4} = \frac{4w}{4}$$
$$275 = w$$

If $w = 275$, then $w + 90 = 275 + 90 = 365$.

Check. The length, 365 mi, is 90 mi more than the width, 275 mi. The perimeter is $2 \cdot 365$ mi $+ 2 \cdot 275$ mi $= 730$ mi $+ 550$ mi $= 1280$ mi. The answer checks.

State. The length is 365 mi, and the width is 275 mi.

42. Familiarize. Let x = the number on the first marker. Then $x + 1$ = the number on the second marker.

Translate.

First number plus second number is 691.
$$x + (x + 1) = 691$$

Solve. We solve the equation.

$$x + (x + 1) = 691$$
$$2x + 1 = 691$$
$$2x + 1 - 1 = 691 - 1$$
$$2x = 690$$
$$\frac{2x}{2} = \frac{690}{2}$$
$$x = 345$$

If $x = 345$, then $x + 1 = 345 + 1 = 346$.

Check. 345 and 346 are consecutive integers and $345 + 346 = 691$. The answer checks.

State. The numbers on the markers are 345 and 346.

43. Familiarize. Let c = the cost of the entertainment center in February.

Translate.

Cost in February plus \$332 is Cost in June
$$c + 332 = 2449$$

Solve. We solve the equation.

$$c + 332 = 2449$$
$$c + 332 - 332 = 2449 - 332$$
$$c = 2117$$

Check. \$2117 + \$332 = \$2449, so the answer checks.

State. The entertainment center cost \$2117 in February.

44. **Familiarize.** Let a = the number of appliances Ty sold.

Translate.

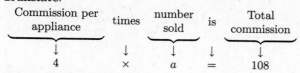

Commission per appliance	times	number sold	is	Total commission
↓	↓	↓	↓	↓
4	×	a	=	108

Solve. We solve the equation.

$$4 \cdot a = 108$$
$$\frac{4 \cdot a}{4} = \frac{108}{4}$$
$$a = 27$$

Check. $\$4 \cdot 27 = \$108$, so the answer checks.

State. Ty sold 27 appliances.

45. **Familiarize.** Let x = the measure of the first angle. Then $x + 50$ = the measure of the second angle, and $2x - 10$ = the measure of the third angle. Recall that the sum of measures of the angles of a triangle is $180°$.

Translate.

Measure of first angle	+	measure of second angle	+	measure of third angle	is $180°$.	
↓	↓	↓	↓	↓	↓	↓
x	+	$(x + 50)$	+	$(2x - 10)$	=	180

Solve. We solve the equation.

$$x + (x + 50) + (2x - 10) = 180$$
$$4x + 40 = 180$$
$$4x + 40 - 40 = 180 - 40$$
$$4x = 140$$
$$\frac{4x}{4} = \frac{140}{4}$$
$$x = 35$$

If $x = 35$, then $x + 50 = 35 + 50 = 85$ and $2x - 10 = 2 \cdot 35 - 10 = 70 - 10 = 60$.

Check. The measure of the second angle is $50°$ more than the measure of the first angle, and the measure of the third angle is $10°$ less than twice the measure of the first angle. The sum of the measure is $35° + 85° + 60° = 180°$. The answer checks.

State. The measures of the angles are $35°$, $85°$, and $60°$.

46. **Translate.**

What number	is	20%	of	75?
↓	↓	↓	↓	↓
a	=	20%	·	75

Solve. We convert 20% to decimal notation and multiply.

$$a = 20\% \cdot 75$$
$$a = 0.2 \cdot 75$$
$$a = 15$$

Thus, 15 is 20% of 75.

47. **Translate.**

15	is	what percent	of	80?
↓	↓	↓	↓	↓
15	=	p	·	80

Solve. We solve the equation.

$$15 = p \cdot 80$$
$$\frac{15}{80} = \frac{p \cdot 80}{80}$$
$$0.1875 = p$$
$$18.75\% = p$$

Thus, 15 is 18.75% of 80.

48. **Translate.**

18	is	3%	of	what number?
↓	↓	↓	↓	↓
18	=	3%	·	b

Solve. We solve the equation.

$$18 = 3\% \cdot b$$
$$18 = 0.03 \cdot b$$
$$\frac{18}{0.03} = \frac{0.03 \cdot b}{0.03}$$
$$600 = b$$

Thus, 18 is 3% of 600.

49. We subtract to find the increase, in thousands.

$$1141 - 905 = 236$$

The increase is 236 thousand.

Now we find the percent of increase.

236	is	what percent	of	905?
↓	↓	↓	↓	↓
236	=	p	·	905

We divide by 905 on both sides and then convert to percent notation.

$$236 = p \cdot 905$$
$$\frac{236}{905} = \frac{p \cdot 905}{905}$$
$$0.26 \approx p$$
$$26\% \approx p$$

The percent of increase is about 26%.

50. **Familiarize.** Let p = the price before the reduction.

Translate.

Price before reduction	minus	30%	of	price	is	$154.
↓	↓	↓	↓	↓	↓	↓
p	−	30%	·	p	=	154

Solve. We solve the equation.

$$p - 30\% \cdot p = 154$$
$$p - 0.3p = 154$$
$$0.7p = 154$$
$$\frac{0.7p}{0.7} = \frac{154}{0.7}$$
$$p = 220$$

Check. 30% of $220 is $0.3 \cdot \$220 = \$66$ and $\$220 - \$66 = \$154$, so the answer checks.

State. The price before the reduction was $220.

51. Familiarize. Let s = the previous salary.

Translate.

$$\underbrace{\text{Previous salary}}_{\downarrow \atop s} \text{ plus } 15\% \text{ of } \underbrace{\text{previous salary}}_{\downarrow \atop s} \text{ is } \$61,410.$$

$$s + 15\% \cdot s = 61,410$$

Solve. We solve the equation.

$$s + 15\% \cdot s = 61,410$$
$$s + 0.15s = 61,410$$
$$1.15s = 61,410$$
$$\frac{1.15s}{1.15} = \frac{61,410}{1.15}$$
$$s = 53,400$$

Check. 15% of $53,400 = 0.15 \cdot \$53,400 = \$8010$ and $\$53,400 + \$8010 = \$61,410$, so the answer checks.

State. The previous salary was $53,400.

52. Familiarize. Let a = the amount the charity actually owes. This is the price of the pump without sales tax added. Then the incorrect amount is $a + 5\%$ of a, or $a + 0.05a$, or $1.05a$.

Translate.

$$\underbrace{\text{Incorrect amount}}_{\downarrow \atop 1.05a} \text{ is } \$145.90.$$

$$1.05a = 145.90$$

Solve. We solve the equation.

$$1.05a = 145.90$$
$$\frac{1.05a}{1.05} = \frac{145.90}{1.05}$$
$$a \approx 138.95$$

Check. 5% of $138.95 is $0.05 \cdot \$138.95 \approx \$6.95$, and $\$138.95 + \$6.95 = \$145.90$, so the answer checks.

State. The charity actually owes $138.95.

53. Familiarize. Let s represent the score on the next test.

Translate.

$$\underbrace{\text{The average score}}_{\downarrow \atop \frac{71+75+82+86+s}{5}} \underbrace{\text{is at least}}_{\downarrow \atop \geq} \underbrace{80.}_{\downarrow \atop 80}$$

Solve.

$$\frac{71 + 75 + 82 + 86 + s}{5} \geq 80$$
$$5\left(\frac{71 + 75 + 82 + 86 + s}{5}\right) \geq 5 \cdot 80$$
$$71 + 75 + 82 + 86 + s \geq 400$$
$$314 + s \geq 400$$
$$s \geq 86$$

Check. As a partial check we show that the average is at least 80 when the next test score is 86.

$$\frac{71 + 75 + 82 + 86 + 86}{5} = \frac{400}{5} = 80$$

State. The lowest grade you can get on the next test and have an average test score of 80 is 86.

54. Familiarize. Let w represent the width of the rectangle, in cm. The perimeter is given by $P = 2l + 2w$, or $2 \cdot 43 + 2w$, or $86 + 2w$.

Translate.

$$\underbrace{\text{The perimeter}}_{\downarrow \atop 86 + 2w} \underbrace{\text{is greater than}}_{\downarrow \atop >} \underbrace{120 \text{ cm}}_{\downarrow \atop 120}.$$

Solve.

$$86 + 2w > 120$$
$$2w > 34$$
$$w > 17$$

Check. We check to see if the solution seems reasonable.

When $w = 16$ cm, $P = 2 \cdot 43 + 2 \cdot 16$, or 118 cm.

When $w = 17$ cm, $P = 2 \cdot 43 + 2 \cdot 17$, or 120 cm.

When $w = 18$ cm, $P = 2 \cdot 43 + 2 \cdot 18$, or 122 cm.

It appears that the solution is correct.

State. The solution set is $\{w | w > 17 \text{ cm}\}$.

55. Discussion and Writing Exercise. The end result is the same either way. If s is the original salary, the new salary after a 5% raise followed by an 8% raise is $1.08(1.05s)$. If the raises occur in the opposite order, the new salary is $1.05(1.08s)$. By the commutative and associate laws of multiplication, we see that these are equal. However, it would be better to receive the 8% raise first, because this increase yields a higher salary initially than a 5% raise.

56. Discussion and Writing Exercise. The inequalities are equivalent by the multiplication principle for inequalities. If we multiply both sides of one inequality by -1, the other inequality results.

57.
$$2|x| + 4 = 50$$
$$2|x| = 46$$
$$|x| = 23$$

The solutions are the numbers whose distance from 0 is 23. Those numbers are -23 and 23.

58. $|3x| = 60$

The solutions are the values of x for which the distance of $3 \cdot x$ from 0 is 60. Then we have:

$$3x = -60 \quad or \quad 3x = 60$$
$$x = -20 \quad or \quad x = 20$$

The solutions are -20 and 20.

59.
$$y = 2a - ab + 3$$
$$y - 3 = 2a - ab$$
$$y - 3 = a(2 - b)$$
$$\frac{y - 3}{2 - b} = a$$

Chapter 10 Test

1.
$$x + 7 = 15$$
$$x + 7 - 7 = 15 - 7$$
$$x = 8$$
The solution is 8.

2.
$$t - 9 = 17$$
$$t - 9 + 9 = 17 + 9$$
$$t = 26$$
The solution is 26.

3.
$$3x = -18$$
$$\frac{3x}{3} = \frac{-18}{3}$$
$$x = -6$$
The solution is -6.

4.
$$-\frac{4}{7}x = -28$$
$$-\frac{7}{4} \cdot \left(-\frac{4}{7}x\right) = -\frac{7}{4} \cdot (-28)$$
$$x = \frac{7 \cdot 4 \cdot 7}{4 \cdot 1}$$
$$x = 49$$
The solution is 49.

5.
$$3t + 7 = 2t - 5$$
$$3t + 7 - 2t = 2t - 5 - 2t$$
$$t + 7 = -5$$
$$t + 7 - 7 = -5 - 7$$
$$t = -12$$
The solution is -12.

6.
$$\frac{1}{2}x - \frac{3}{5} = \frac{2}{5}$$
$$\frac{1}{2}x - \frac{3}{5} + \frac{3}{5} = \frac{2}{5} + \frac{3}{5}$$
$$\frac{1}{2}x = 1$$
$$2 \cdot \frac{1}{2}x = 2 \cdot 1$$
$$x = 2$$
The solution is 2.

7.
$$8 - y = 16$$
$$8 - y - 8 = 16 - 8$$
$$-y = 8$$
$$-1 \cdot (-1 \cdot y) = -1 \cdot 8$$
$$y = -8$$
The solution is -8.

8.
$$-\frac{2}{5} + x = -\frac{3}{4}$$
$$-\frac{2}{5} + x + \frac{2}{5} = -\frac{3}{4} + \frac{2}{5}$$
$$x = -\frac{15}{20} + \frac{8}{20}$$
$$x = -\frac{7}{20}$$
The solution is $-\frac{7}{20}$.

9.
$$3(x + 2) = 27$$
$$3x + 6 = 27$$
$$3x + 6 - 6 = 27 - 6$$
$$3x = 21$$
$$\frac{3x}{3} = \frac{21}{3}$$
$$x = 7$$
The solution is 7.

10.
$$-3x - 6(x - 4) = 9$$
$$-3x - 6x + 24 = 9$$
$$-9x + 24 = 9$$
$$-9x + 24 - 24 = 9 - 24$$
$$-9x = -15$$
$$\frac{-9x}{-9} = \frac{-15}{-9}$$
$$x = \frac{3 \cdot 5}{3 \cdot 3}$$
$$x = \frac{5}{3}$$
The solution is $\frac{5}{3}$.

11. We multiply by 10 to clear the decimals.
$$0.4p + 0.2 = 4.2p - 7.8 - 0.6p$$
$$10(0.4p + 0.2) = 10(4.2p - 7.8 - 0.6p)$$
$$4p + 2 = 42p - 78 - 6p$$
$$4p + 2 = 36p - 78$$
$$4p + 2 - 36p = 36p - 78 - 36p$$
$$-32p + 2 = -78$$
$$-32p + 2 - 2 = -78 - 2$$
$$-32p = -80$$
$$\frac{-32p}{-32} = \frac{-80}{-32}$$
$$p = \frac{5 \cdot 16}{2 \cdot 16}$$
$$p = \frac{5}{2}$$
The solution is $\frac{5}{2}$.

12. $4(3x - 1) + 11 = 2(6x + 5) - 8$

$12x - 4 + 11 = 12x + 10 - 8$

$12x + 7 = 12x + 2$

$12x + 7 - 12x = 12x + 2 - 12x$

$7 = 2 \qquad \text{FALSE}$

There are no solutions.

13. $-2 + 7x + 6 = 5x + 4 + 2x$

$7x + 4 = 7x + 4$

$7x + 4 - 7x = 7x + 4 - 7x$

$4 = 4 \qquad \text{TRUE}$

All real numbers are solutions.

14. $x + 6 \leq 2$

$x + 6 - 6 \leq 2 - 6$

$x \leq -4$

The solution set is $\{x | x \leq -4\}$.

15. $14x + 9 > 13x - 4$

$14x + 9 - 13x > 13x - 4 - 13x$

$x + 9 > -4$

$x + 9 - 9 > -4 - 9$

$x > -13$

The solution set is $\{x | x > -13\}$.

16. $12x \leq 60$

$\dfrac{12x}{12} \leq \dfrac{60}{12}$

$x \leq 5$

The solution set is $\{x | x \leq 5\}$.

17. $-2y \geq 26$

$\dfrac{-2y}{-2} \leq \dfrac{26}{-2}$ \quad Reversing the inequality symbol

$y \leq -13$

The solution set is $\{y | y \leq -13\}$.

18. $-4y \leq -32$

$\dfrac{-4y}{-4} \geq \dfrac{-32}{-4}$ \quad Reversing the inequality symbol

$y \geq 8$

The solution set is $\{y | y \geq 8\}$.

19. $-5x \geq \dfrac{1}{4}$

$-\dfrac{1}{5} \cdot (-5x) \leq -\dfrac{1}{5} \cdot \dfrac{1}{4}$ \quad Reversing the inequality symbol

$x \leq -\dfrac{1}{20}$

The solution set is $\left\{ x \Big| x \leq -\dfrac{1}{20} \right\}$.

20. $4 - 6x > 40$

$4 - 6x - 4 > 40 - 4$

$-6x > 36$

$\dfrac{-6x}{-6} < \dfrac{36}{-6}$ \quad Reversing the inequality symbol

$x < -6$

The solution set is $\{x | x < -6\}$.

21. $5 - 9x \geq 19 + 5x$

$5 - 9x - 5x \geq 19 + 5x - 5x$

$5 - 14x \geq 19$

$5 - 14x - 5 \geq 19 - 5$

$-14x \geq 14$

$\dfrac{-14x}{-14} \leq \dfrac{14}{-14}$ \quad Reversing the inequality symbol

$x \leq -1$

The solution set is $\{x | x \leq -1\}$.

22. The solutions of $y \leq 9$ are shown by shading the point for 9 and all points to the left of 9. The closed circle at 9 indicates that 9 is part of the graph.

23. $6x - 3 < x + 2$

$6x - 3 - x < x + 2 - x$

$5x - 3 < 2$

$5x - 3 + 3 < 2 + 3$

$5x < 5$

$\dfrac{5x}{5} < \dfrac{5}{5}$

$x < 1$

The solution set is $\{x | x < 1\}$. The graph is as follows:

24. In order to be a solution of the inequality $-2 \leq x \leq 2$, a number must be a solution of both $-2 \leq x$ and $x \leq 2$. The solution set is graphed as follows:

25. *Translate.*

<u>What number</u> is 24% of 75?

$\quad \downarrow \qquad\qquad \downarrow \quad \downarrow \quad \downarrow \quad \downarrow$

$\quad a \qquad\quad = 24\% \cdot 75$

Solve. We convert 24% to decimal notation and multiply.

$a = 24\% \cdot 75$

$a = 0.24 \cdot 75$

$a = 18$

Thus, 18 is 24% of 75.

26. *Translate.*

15.84 is $\underbrace{\text{what percent}}$ of 96?

$\downarrow$ $\downarrow$ $\downarrow$ $\downarrow$ $\downarrow$
15.84 $=$ p $\cdot$ 96

Solve.

$$15.84 = p \cdot 96$$
$$\frac{15.84}{96} = \frac{p \cdot 96}{96}$$
$$0.165 = p$$
$$16.5\% = p$$

Thus, 15.84 is 16.5% of 96.

27. *Translate.*

800 is 2% of $\underbrace{\text{what number?}}$

$\downarrow$ $\downarrow$ $\downarrow$ $\downarrow$ $\downarrow$
800 $=$ 2% $\cdot$ b

Solve.

$$800 = 2\% \cdot b$$
$$800 = 0.02 \cdot b$$
$$\frac{800}{0.02} = \frac{0.02 \cdot b}{0.02}$$
$$40,000 = b$$

Thus, 800 is 2% of 40,000.

28. We subtract to find the increase.

$$89,000 - 58,000 = 31,000$$

Now we find the percent of increase.

31,000 is $\underbrace{\text{what percent}}$ of 58,000?

$\downarrow$ $\downarrow$ $\downarrow$ $\downarrow$ $\downarrow$
31,000 $=$ p $\cdot$ 58,000

We divide by 58,000 on both sides and then convert to percent notation.

$$31,000 = p \cdot 58,000$$
$$\frac{31,000}{58,000} = \frac{p \cdot 58,000}{58,000}$$
$$0.534 \approx p$$
$$53.4\% \approx p$$

The percent of increase is about 53.4%.

29. *Familiarize.* Let w = the width of the photograph, in cm. Then $w + 4$ = the length. Recall that the perimeter P of a rectangle with length l and width w is given by $P = 2l + 2w$.

Translate. We substitute 36 for P and $w + 4$ for l in the formula above.

$$P = 2l + 2w$$
$$36 = 2(w + 4) + 2w$$

Solve. We solve the equation.

$$36 = 2(w + 4) + 2w$$
$$36 = 2w + 8 + 2w$$
$$36 = 4w + 8$$
$$36 - 8 = 4w + 8 - 8$$
$$28 = 4w$$
$$\frac{28}{4} = \frac{4w}{4}$$
$$7 = w$$

If $w = 7$, then $w + 4 = 7 + 4 = 11$.

Check. The length, 11 cm, is 4 cm more than the width, 7 cm. The perimeter is $2 \cdot 11$ cm $+ 2 \cdot 7$ cm $= 22$ cm $+ 14$ cm $= 36$ cm. The answer checks.

State. The width is 7 cm, and the length is 11 cm.

30. *Familiarize.* Let c = the amount that was given to charities in general in 2003, in billions of dollars.

Translate.

$\underbrace{\$86.4 \text{ billion}}$ is 35.9% of what?

$\downarrow$ $\downarrow$ $\downarrow$ $\downarrow$ $\downarrow$
86.4 $=$ 35.9% $\cdot$ c

Solve. We solve the equation.

$$86.4 = 35.9\% \cdot c$$
$$86.4 = 0.359 \cdot c$$
$$\frac{86.4}{0.359} = \frac{0.359 \cdot c}{0.359}$$
$$240.7 \approx c$$

Check. 35.9% of \$240.7 billion is 0.359(\$240.7 billion) $\approx$ \$86.4 billion, so the answer checks.

State. In 2003 about \$240.7 billion was given to charities.

31. *Familiarize.* Let x = the first integer. Then $x + 1$ = the second and $x + 2$ = the third.

Translate.

$\underbrace{\text{First integer}}$ plus $\underbrace{\text{second integer}}$ plus $\underbrace{\text{third integer}}$ is 7530.

$\downarrow$ $\downarrow$ $\downarrow$ $\downarrow$ $\downarrow$ $\downarrow$ $\downarrow$
x $+$ $(x+1)$ $+$ $(x+2)$ $=$ 7530

Solve.

$$x + (x + 1) + (x + 2) = 7530$$
$$3x + 3 = 7530$$
$$3x + 3 - 3 = 7530 - 3$$
$$3x = 7527$$
$$\frac{3x}{3} = \frac{7527}{3}$$
$$x = 2509$$

If $x = 2509$, then $x + 1 = 2510$ and $x + 2 = 2511$.

Check. The numbers 2509, 2510, and 2511 are consecutive integers and $2509 + 2510 + 2511 = 7530$. The answer checks.

State. The integers are 2509, 2510, and 2511.

32. *Familiarize.* Let $x =$ the amount originally invested. Using the formula for simple interest, $I = Prt$, the interest earned in one year will be $x \cdot 5\% \cdot 1$, or $5\%x$.

Translate.

$$\underbrace{\text{Amount invested}}_{\downarrow \atop x} \underbrace{\text{plus}}_{\downarrow \atop +} \underbrace{\text{interest}}_{\downarrow \atop 5\%x} \underbrace{\text{is}}_{\downarrow \atop =} \underbrace{\text{amount after 1 year.}}_{\downarrow \atop 924}$$

Solve. We solve the equation.

$$x + 5\%x = 924$$
$$x + 0.05x = 924$$
$$1.05x = 924$$
$$\frac{1.05x}{1.05} = \frac{924}{1.05}$$
$$x = 880$$

Check. 5% of $880 is $0.05 \cdot \$880 = \$44$ and $\$880 + \$44 = \$924$, so the answer checks.

State. $880 was originally invested.

33. *Familiarize.* Using the labels on the drawing in the text, we let $x =$ the length of the shorter piece, in meters, and $x + 2 =$ the length of the longer piece.

Translate.

$$\underbrace{\text{Length of shorter piece}}_{\downarrow \atop x} \underbrace{\text{plus}}_{\downarrow \atop +} \underbrace{\text{length of longer piece}}_{\downarrow \atop (x+2)} \underbrace{\text{is}}_{\downarrow \atop =} \underbrace{8\text{ m}}_{\downarrow \atop 8}.$$

Solve. We solve the equation.

$$x + (x + 2) = 8$$
$$2x + 2 = 8$$
$$2x + 2 - 2 = 8 - 2$$
$$2x = 6$$
$$\frac{2x}{2} = \frac{6}{2}$$
$$x = 3$$

If $x = 3$, then $x + 2 = 3 + 2 = 5$.

Check. One piece is 2 m longer than the other and the sum of the lengths is 3 m+5 m, or 8 m. The answer checks.

State. The lengths of the pieces are 3 m and 5 m.

34. *Familiarize.* Let $l =$ the length of the rectangle, in yd. The perimeter is given by $P = 2l + 2w$, or $2l + 2 \cdot 96$, or $2l + 192$.

Translate.

$$\underbrace{\text{The perimeter}}_{\downarrow \atop 2l + 192} \underbrace{\text{is at least}}_{\downarrow \atop \geq} \underbrace{540\text{ yd}}_{\downarrow \atop 540}.$$

Solve.

$$2l + 192 \geq 540$$
$$2l \geq 348$$
$$l \geq 174$$

Check. We check to see if the solution seems reasonable.

When $l = 174$ yd, $P = 2 \cdot 174 + 2 \cdot 96$, or 540 yd.

When $l = 175$ yd, $P = 2 \cdot 175 + 2 \cdot 96$, or 542 yd.

It appears that the solution is correct.

State. For lengths that are at least 174 yd, the perimeter will be at least 540 yd. The solution set can be expressed as $\{l|l \geq 174$ yd$\}$.

35. *Familiarize.* Let $s =$ the amount Jason spends in the sixth month.

Translate.

$$\underbrace{\text{Average spending}}_{\downarrow \atop \frac{98 + 89 + 110 + 85 + 83 + s}{6}} \underbrace{\text{is no more than}}_{\downarrow \atop \leq} \underbrace{\$95.}_{\downarrow \atop 95}$$

Solve.

$$\frac{98 + 89 + 110 + 85 + 83 + s}{6} \leq 95$$
$$6\left(\frac{98 + 89 + 110 + 85 + 83 + s}{6}\right) \leq 6 \cdot 95$$
$$98 + 89 + 110 + 85 + 83 + s \leq 570$$
$$465 + s \leq 570$$
$$s \leq 105$$

Check. As a partial check we show that the average spending is $95 when Jason spends $105 in the sixth month.

$$\frac{98 + 89 + 110 + 85 + 83 + 105}{6} = \frac{570}{6} = 95$$

State. Jason can spend no more than $105 in the sixth month. The solution set can be expressed as $\{s|s \leq \$105\}$.

36. *Familiarize.* Let $c =$ the number of copies made. For 3 months, the rental charge is $3 \cdot \$225$, or $675. Expressing 1.2¢ as $0.012, the charge for the copies is given by $\$0.012 \cdot c$.

Translate.

$$\underbrace{\text{Rental charge}}_{\downarrow \atop 675} \underbrace{\text{plus}}_{\downarrow \atop +} \underbrace{\text{copy charge}}_{\downarrow \atop 0.012c} \underbrace{\text{is no more than}}_{\downarrow \atop \leq} \underbrace{\$2400.}_{\downarrow \atop 2400}$$

Solve.

$$675 + 0.012c \leq 2400$$
$$0.012c \leq 1725$$
$$c \leq 143,750$$

Check. We check to see if the solution seems reasonable.

When $c = 143,749$, the total cost is $\$675 + \$0.012(143,749)$, or about $2399.99.

When $c = 143,750$, the total cost is $\$675 + \$0.012(143,750)$, or about $2400.

It appears that the solution is correct.

State. No more than 143,750 copies can be made. The solution set can be expressed as $\{c|c \leq 143,750\}$.

37.
$$A = 2\pi rh$$
$$\frac{A}{2\pi h} = \frac{2\pi rh}{2\pi h}$$
$$\frac{A}{2\pi h} = r$$

38.
$$y = 8x + b$$
$$y - b = 8x + b - b$$
$$y - b = 8x$$
$$\frac{y - b}{8} = \frac{8x}{8}$$
$$\frac{y - b}{8} = x$$

39.
$$c = \frac{1}{a - d}$$
$$(a - d) \cdot c = a - d \cdot \left(\frac{1}{a - d}\right)$$
$$ac - dc = 1$$
$$ac - dc - ac = 1 - ac$$
$$-dc = 1 - ac$$
$$\frac{-dc}{-c} = \frac{1 - ac}{-c}$$
$$d = \frac{1 - ac}{-c}$$

Since $\dfrac{1 - ac}{-c} = \dfrac{-1}{-1} \cdot \dfrac{1 - ac}{-c} = \dfrac{-1(1 - ac)}{-1(-c)} = \dfrac{-1 + ac}{c}$, or $\dfrac{ac - 1}{c}$, we can also express the result as $d = \dfrac{ac - 1}{c}$.

40. $3|w| - 8 = 37$
$$3|w| = 45$$
$$|w| = 15$$

The solutions are the numbers whose distance from 0 is 15. They are -15 and 15.

41. Familiarize. Let $t =$ the number of tickets given away.

Translate. We add the number of tickets given to the five people.

$$\frac{1}{3}t + \frac{1}{4}t + \frac{1}{5}t + 8 + 5 = t$$

Solve.

$$\frac{1}{3}t + \frac{1}{4}t + \frac{1}{5}t + 8 + 5 = t$$
$$\frac{20}{60}t + \frac{15}{60}t + \frac{12}{60}t + 8 + 5 = t$$
$$\frac{47}{60}t + 13 = t$$
$$13 = t - \frac{47}{60}t$$
$$13 = \frac{60}{60}t - \frac{47}{60}t$$
$$13 = \frac{13}{60}t$$
$$\frac{60}{13} \cdot 13 = \frac{60}{13} \cdot \frac{13}{60}t$$
$$60 = t$$

Check. $\dfrac{1}{3} \cdot 60 = 20$, $\dfrac{1}{4} \cdot 60 = 15$, $\dfrac{1}{5} \cdot 60 = 12$; then $20 + 15 + 12 + 8 + 5 = 60$. The answer checks.

State. 60 tickets were given away.

Chapter 11

Graphs of Linear Equations

Exercise Set 11.1

1. $(2, 5)$ is 2 units right and 5 units up.

   $(-1, 3)$ is 1 unit left and 3 units up.

   $(3, -2)$ is 3 units right and 2 units down.

   $(-2, -4)$ is 2 units left and 4 units down.

   $(0, 4)$ is 0 units left or right and 4 units up.

   $(0, -5)$ is 0 units left or right and 5 units down.

   $(5, 0)$ is 5 units right and 0 units up or down.

   $(-5, 0)$ is 5 units left and 0 units up or down.

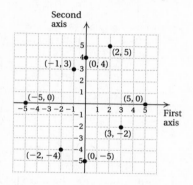

3. Since the first coordinate is negative and the second coordinate positive, the point $(-5, 3)$ is located in quadrant II.

5. Since the first coordinate is positive and the second coordinate negative, the point $(100, -1)$ is in quadrant IV.

7. Since both coordinates are negative, the point $(-6, -29)$ is in quadrant III.

9. Since one of the coordinates is 0, the point $(3.8, 0)$ lies on an axis.

11. Since the first coordinate is negative and the second coordinate is positive, the point $\left(-\frac{1}{3}, \frac{15}{7}\right)$ is in quadrant II.

13. Since the first coordinate is positive and the second coordinate is negative, the point $\left(12\frac{7}{8}, -1\frac{1}{2}\right)$ is in quadrant IV.

15.

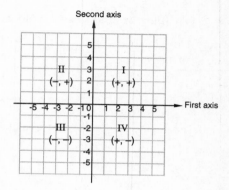

If the first coordinate is negative and the second coordinate is positive, the point is in quadrant II.

17. See the figure in Exercise 15.

   If the first coordinate is positive, then the point must be in either quadrant I or quadrant IV.

19. If the first and second coordinates are equal, they must either be both positive or both negative. The point must be in either quadrant I (both positive) or quadrant III (both negative).

21.

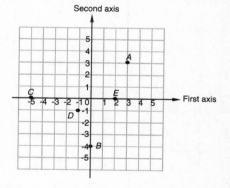

Point A is 3 units right and 3 units up. The coordinates of A are $(3, 3)$.

Point B is 0 units left or right and 4 units down. The coordinates of B are $(0, -4)$.

Point C is 5 units left and 0 units up or down. The coordinates of C are $(-5, 0)$.

Point D is 1 unit left and 1 unit down. The coordinates of D are $(-1, -1)$.

Point E is 2 units right and 0 units up or down. The coordinates of E are $(2, 0)$.

23. We substitute 2 for x and 9 for y (alphabetical order of variables).

$$\frac{y = 3x - 1}{\begin{array}{l} 9 \; ? \; 3 \cdot 2 - 1 \\ \quad | \quad 6 - 1 \\ \quad | \quad 5 \qquad \text{FALSE} \end{array}}$$

Since $9 = 5$ is false, the pair $(2, 9)$ is not a solution.

25. We substitute 4 for x and 2 for y.

$$\frac{2x + 3y = 12}{\begin{array}{l} 2 \cdot 4 + 3 \cdot 2 \; ? \; 12 \\ \quad 8 + 6 \quad | \\ \quad\quad 14 \quad | \qquad \text{FALSE} \end{array}}$$

Since $14 = 12$ is false, the pair $(4, 2)$ is not a solution.

27. We substitute 3 for a and -1 for b.

$$\frac{3a - 4b = 13}{\begin{array}{l} 3 \cdot 3 - 4(-1) \; ? \; 13 \\ \quad 9 + 4 \quad | \\ \quad\quad 13 \quad | \qquad \text{TRUE} \end{array}}$$

Since $13 = 13$ is true, the pair $(3, -1)$ is a solution.

29. To show that a pair is a solution, we substitute, replacing x with the first coordinate and y with the second coordinate in each pair.

$$\frac{y = x - 5}{\begin{array}{l} -1 \; ? \; 4 - 5 \\ \quad | \; -1 \quad \text{TRUE} \end{array}} \qquad \frac{y = x - 5}{\begin{array}{l} -4 \; ? \; 1 - 5 \\ \quad | \; -4 \quad \text{TRUE} \end{array}}$$

In each case the substitution results in a true equation. Thus, $(4, -1)$ and $(1, -4)$ are both solutions of $y = x - 5$. We graph these points and sketch the line passing through them.

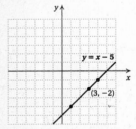

The line appears to pass through $(3, -2)$ also. We check to determine if $(3, -2)$ is a solution of $y = x - 5$.

$$\frac{y = x - 5}{\begin{array}{l} -2 \; ? \; 3 - 5 \\ \quad | \; -2 \quad \text{TRUE} \end{array}}$$

Thus, $(3, -2)$ is another solution. There are other correct answers, including $(-1, -6)$, $(2, -3)$, $(0, -5)$, $(5, 0)$, and $(6, 1)$.

31. To show that a pair is a solution, we substitute, replacing x with the first coordinate and y with the second coordinate in each pair.

$$\frac{y = \frac{1}{2}x + 3}{\begin{array}{l} 5 \; ? \; \frac{1}{2} \cdot 4 + 3 \\ \quad | \quad 2 + 3 \\ \quad | \quad 5 \qquad \text{TRUE} \end{array}} \qquad \frac{y = \frac{1}{2}x + 3}{\begin{array}{l} 2 \; ? \; \frac{1}{2}(-2) + 3 \\ \quad | \quad -1 + 3 \\ \quad | \quad 2 \qquad \text{TRUE} \end{array}}$$

In each case the substitution results in a true equation. Thus, $(4, 5)$ and $(-2, 2)$ are both solutions of $y = \frac{1}{2}x + 3$. We graph these points and sketch the line passing through them.

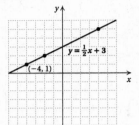

The line appears to pass through $(-4, 1)$ also. We check to determine if $(-4, 1)$ is a solution of $y = \frac{1}{2}x + 3$.

$$\frac{y = \frac{1}{2}x + 3}{\begin{array}{l} 1 \; ? \; \frac{1}{2}(-4) + 3 \\ \quad | \quad -2 + 3 \\ \quad | \quad 1 \qquad \text{TRUE} \end{array}}$$

Thus, $(-4, 1)$ is another solution. There are other correct answers, including $(-6, 0)$, $(0, 3)$, $(2, 4)$, and $(6, 6)$.

33. To show that a pair is a solution, we substitute, replacing x with the first coordinate and y with the second coordinate in each pair.

$$\frac{4x - 2y = 10}{\begin{array}{l} 4 \cdot 0 - 2(-5) \; ? \; 10 \\ \quad\quad\quad 10 \quad | \qquad \text{TRUE} \end{array}}$$

$$\frac{4x - 2y = 10}{\begin{array}{l} 4 \cdot 4 - 2 \cdot 3 \; ? \; 10 \\ \quad 16 - 6 \quad | \\ \quad\quad 10 \quad | \qquad \text{TRUE} \end{array}}$$

In each case the substitution results in a true equation. Thus, $(0, -5)$ and $(4, 3)$ are both solutions of $4x - 2y = 10$. We graph these points and sketch the line passing through them.

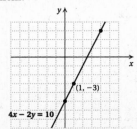

The line appears to pass through $(1, -3)$ also. We check to determine if $(1, -3)$ is a solution of $4x - 2y = 10$.

$$\begin{array}{c|c} 4x - 2y = 10 \\ \hline 4 \cdot 1 - 2(-3) \ ? \ 10 \\ 4 + 6 \\ 10 & \text{TRUE} \end{array}$$

Thus, $(1, -3)$ is another solution. There are other correct answers, including $(2, -1)$, $(3, 1)$, and $(5, 5)$.

35. $y = x + 1$

The equation is in the form $y = mx + b$. The y-intercept is $(0, 1)$. We find five other pairs.

When $x = -2$, $y = -2 + 1 = -1$.

When $x = -1$, $y = -1 + 1 = 0$.

When $x = 1$, $y = 1 + 1 = 2$.

When $x = 2$, $y = 2 + 1 = 3$.

When $x = 3$, $y = 3 + 1 = 4$.

x	y
-2	-1
-1	0
0	1
1	2
2	3
3	4

Plot these points, draw the line they determine, and label the graph $y = x + 1$.

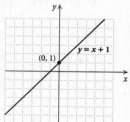

37. $y = x$

The equation is equivalent to $y = x + 0$. The y-intercept is $(0, 0)$. We find five other points.

When $x = -2$, $y = -2$.

When $x = -1$, $y = -1$.

When $x = 1$, $y = 1$.

When $x = 2$, $y = 2$.

When $x = 3$, $y = 3$.

x	y
-2	-2
-1	-1
0	0
1	1
2	2
3	3

Plot these points, draw the line they determine, and label the graph $y = x$.

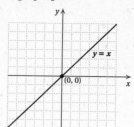

39. $y = \frac{1}{2}x$

The equation is equivalent to $y = \frac{1}{2}x + 0$. The y-intercept is $(0, 0)$. We find two other points.

When $x = -2$, $y = \frac{1}{2}(-2) = -1$.

When $x = 4$, $y = \frac{1}{2} \cdot 4 = 2$.

x	y
-2	-1
0	0
4	2

Plot these points, draw the line they determine, and label the graph $y = \frac{1}{2}x$.

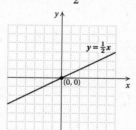

41. $y = x - 3$

The equation is equivalent to $y = x + (-3)$. The y-intercept is $(0, -3)$. We find two other points.

When $x = -2$, $y = -2 - 3 = -5$.

When $x = 4$, $y = 4 - 3 = 1$.

x	y
-2	-5
0	-3
4	1

Plot these points, draw the line they determine, and label the graph $y = x - 3$.

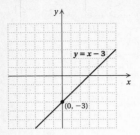

43. $y = 3x - 2 = 3x + (-2)$

The y-intercept is $(0, -2)$. We find two other points.

When $x = -2$, $y = 3(-2) + 2 = -6 + 2 = -4$.

When $x = 1$, $y = 3 \cdot 1 + 2 = 3 + 2 = 5$.

x	y
-2	-4
0	-2
1	5

Plot these points, draw the line they determine, and label the graph $y = 3x + 2$.

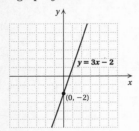

45. $y = \frac{1}{2}x + 1$

The y-intercept is $(0, 1)$. We find two other points using multiples of 2 for x to avoid fractions.

When $x = -4$, $y = \frac{1}{2}(-4) + 1 = -2 + 1 = -1$.

When $x = 4$, $y = \frac{1}{2} \cdot 4 + 1 = 2 + 1 = 3$.

x	y
-4	-1
0	1
4	3

Plot these points, draw the line they determine, and label the graph $y = \frac{1}{2}x + 1$.

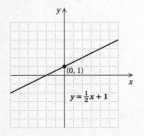

47. $x + y = -5$

$\qquad y = -x - 5$

$\qquad y = -x + (-5)$

The y-intercept is $(0, -5)$. We find two other points.

When $x = -4$, $y = -(-4) - 5 = 4 - 5 = -1$.

When $x = -1$, $y = -(-1) - 5 = 1 - 5 = -4$.

x	y
-4	-1
0	-5
-1	-4

Plot these points, draw the line they determine, and label the graph $x + y = -5$.

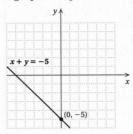

49. $y = \frac{5}{3}x - 2 = \frac{5}{3}x + (-2)$

The y-intercept is $(0, -2)$. We find two other points using multiples of 3 for x to avoid fractions.

When $x = -3$, $y = \frac{5}{3}(-3) - 2 = -5 - 2 = -7$.

When $x = 3$, $y = \frac{5}{3} \cdot 3 - 2 = 5 - 2 = 3$.

x	y
-3	-7
0	-2
3	3

Plot these points, draw the line they determine, and label the graph $y = \frac{5}{3}x - 2$.

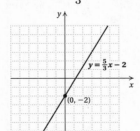

51. $x + 2y = 8$

$\qquad 2y = -x + 8$

$\qquad y = -\frac{1}{2}x + 4$

The y-intercept is $(0, 4)$. We find two other points using multiples of 2 for x to avoid fractions.

When $x = -2$, $y = -\dfrac{1}{2}(-2) + 4 = 1 + 4 = 5$.

When $x = 4$, $y = -\dfrac{1}{2} \cdot 4 + 4 = -2 + 4 = 2$.

x	y
-2	5
0	4
4	2

Plot these points, draw the line they determine, and label the graph $x + 2y = 8$.

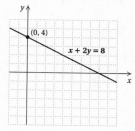

53. $y = \dfrac{3}{2}x + 1$

The y-intercept is $(0, 1)$. We find two other points using multiples of 2 for x to avoid fractions.

When $x = -4$, $y = \dfrac{3}{2}(-4) + 1 = -6 + 1 = -5$.

When $x = 2$, $y = \dfrac{3}{2} \cdot 2 + 1 = 3 + 1 = 4$.

x	y
-4	-5
0	1
2	4

Plot these points, draw the line they determine, and label the graph $y = \dfrac{3}{2}x + 1$.

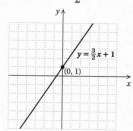

55. $8x - 2y = -10$
$-2y = -8x - 10$
$y = 4x + 5$

The y-intercept is $(0, 5)$. We find two other points.

When $x = -2$, $y = 4(-2) + 5 = -8 + 5 = -3$.

When $x = -1$, $y = 4(-1) + 5 = -4 + 5 = 1$.

x	y
-2	-3
-1	1
0	5

Plot these points, draw the line they determine, and label the graph $8x - 2y = -10$.

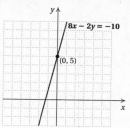

57. $8y + 2x = -4$
$8y = -2x - 4$
$y = -\dfrac{1}{4}x - \dfrac{1}{2}$
$y = -\dfrac{1}{4}x + \left(-\dfrac{1}{2} \right)$

The y-intercept is $\left(0, -\dfrac{1}{2} \right)$. We find two other points.

When $x = -2$, $y = -\dfrac{1}{4}(-2) - \dfrac{1}{2} = \dfrac{1}{2} - \dfrac{1}{2} = 0$.

When $x = 2$, $y = -\dfrac{1}{4} \cdot 2 - \dfrac{1}{2} = -\dfrac{1}{2} - \dfrac{1}{2} = -1$.

x	y
-2	0
0	$-\dfrac{1}{2}$
2	-1

Plot these points, draw the line they determine, and label the graph $8y + 2x = -4$.

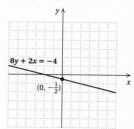

59. a) We substitute 0, 4, and 6 for t and then calculate V.

If $t = 0$, then $V = -50 \cdot 0 + 300 = \$300$.

If $t = 4$, then $V = -50 \cdot 4 + 300 = -200 + 300 = \$100$.

If $t = 6$, then $V = -50 \cdot 6 + 300 = -300 + 300 = \$0$.

b) We plot the three ordered pairs we found in part (a). Note the negative t- and V-values have no meaning in this problem.

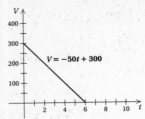

To use the graph to estimate the value of the software after 5 years we must determine which V-value is paired with $t = 5$. We locate 5 on the t-axis, go up to the graph, and then find the value on the V-axis that corresponds to that point. It appears that after 5 years the value of the software is \$50.

c) Substitute 150 for V and then solve for t.
$$V = -50t + 300$$
$$150 = -50t + 300$$
$$-150 = -50t$$
$$3 = t$$

The value of the software is \$150 after 3 years.

61. a) When $d = 1$, $N = 0.8(1)+21.2 = 0.8+21.2 = 22$ gal.

In 2000, $d = 2000 - 1995 = 5$. When $d = 5$, $N = 0.8(5) + 21.2 = 4 + 21.2 = 25.2$ gal.

In 2006, $d = 2006 - 1995 = 11$. When $d = 11$, $N = 0.8(11) + 21.2 = 8.8 + 21.2 = 30$ gal.

In 2010, $d = 2010 - 1995 = 15$. When $d = 15$, $N = 0.8(15) + 21.2 = 12 + 21.2 = 33.2$ gal.

b) Plot the four ordered pairs we found in part (a). Note that negative d- and N-values have no meaning in this problem.

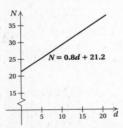

To use the graph to estimate what tea consumption was in 2002 we must determine which N-value is paired with 2002, or with $d = 7$. We locate 7 on the d-axis, go up to the graph, and then find the value on the N-axis that corresponds to that point. It appears that tea consumption was about 27 gallons in 2002.

c) Substitute 31.6 for N and then solve for d.
$$N = 0.8d + 21.2$$
$$31.6 = 0.8d + 21.2$$
$$10.4 = 0.8d$$
$$13 = d$$

Tea consumption will be about 31.6 gallons 13 years after 1995, or in 2008.

63. Discussion and Writing Exercise

65. The distance of -12 from 0 is 12, so $|-12| = 12$.

67. The distance of 0 from 0 is 0, so $|0| = 0$.

69. The distance of -3.4 from 0 is 3.4, so $|-3.4| = 3.4$.

71. The distance of $\frac{2}{3}$ from 0 is $\frac{2}{3}$, so $\left|\frac{2}{3}\right| = \frac{2}{3}$.

73. **Familiarize.** Let $p =$ the average price of a ticket to a major-league baseball game in 2000. The price in 2004 was $p + 17.9\%p$, or $p + 0.179p$, or $1.179p$.

Translate.

$$\underbrace{\text{The price in 2004}}_{1.179p} \quad \underset{=}{\text{was}} \quad \underset{19.82}{\$19.82}.$$

Solve.
$$1.179p = 19.82$$
$$p = \frac{19.82}{1.179}$$
$$p \approx 16.81$$

Check. 17.9% of \$16.81 is $0.179(\$16.81) \approx \$3.01$ and $\$16.81 + \$3.01 = \$19.82$. The answer checks.

State. In 2000 the average price of a major-league baseball ticket was \$16.81.

75.

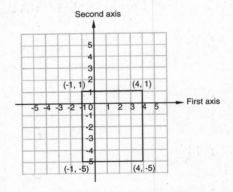

The coordinates of the fourth vertex are $(-1, -5)$.

77. Answers may vary.

We select eight points such that the sum of the coordinates for each point is 6.

$(-1, 7)$	$-1 + 7 = 6$
$(0, 6)$	$0 + 6 = 6$
$(1, 5)$	$1 + 5 = 6$
$(2, 4)$	$2 + 4 = 6$
$(3, 3)$	$3 + 3 = 6$
$(4, 2)$	$4 + 2 = 6$
$(5, 1)$	$5 + 1 = 6$
$(6, 0)$	$6 + 0 = 6$

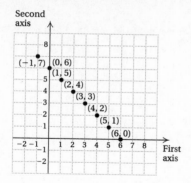

79.

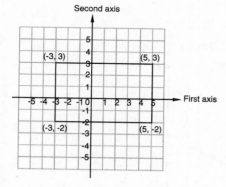

The length is 8 linear units, and the width is 5 linear units.

$P = 2l + 2w$

$P = 2 \cdot 8 + 2 \cdot 5 = 16 + 10 = 26$ linear units

81. Point A is 10 units right and 18 units up. The coordinates of A are $(10, 18)$.

Point B is 8 units right and 15 units up. The coordinates of B are $(8, 15)$.

Point C is 3 units right and 13 units up. The coordinates of C are $(3, 13)$.

Point D is 8 units right and 11 units up. The coordinates of D are $(8, 11)$.

Point E is 3 units right and 8 units up. The coordinates of E are $(3, 8)$.

Point F is 8 units right and 5 units up. The coordinates of F are $(8, 5)$.

Point G is 12 units right and 5 units up. The coordinates of G are $(12, 5)$.

Point H is 17 units right and 8 units up. The coordinates of H are $(17, 8)$.

Point I is 12 units right and 11 units up. The coordinates of I are $(12, 11)$.

Point J is 17 units right and 13 units up. The coordinates of J are $(17, 13)$.

Point K is 12 units right and 15 units up. The coordinates of K are $(12, 15)$.

83. Subtracting 3 from each of the y-coordinates found in Exercise 81 gives us the points $(10, 15)$, $(8, 12)$, $(3, 10)$, $(8, 8)$, $(3, 5)$, $(8, 2)$, $(12, 2)$, $(17, 5)$, $(12, 8)$, $(17, 10)$, and $(12, 12)$.

When we plot these points and connect them in the same order as in Exercise 81, we get a figure that has the same shape as the figure in Exercise 81 and that is translated 3 units down.

Exercise Set 11.2

1. (a) The graph crosses the y-axis at $(0, 5)$, so the y-intercept is $(0, 5)$.

(b) The graph crosses the x- axis at $(2, 0)$, so the x-intercept is $(2, 0)$.

3. (a) The graph crosses the y-axis at $(0, -4)$, so the y-intercept is $(0, -4)$.

(b) The graph crosses the x-axis at $(3, 0)$, so the x-intercept is $(3, 0)$.

5. $3x + 5y = 15$

(a) To find the y-intercept, let $x = 0$. This is the same as covering up the x-term and then solving.

$$5y = 15$$
$$y = 3$$

The y-intercept is $(0, 3)$.

(b) To find the x-intercept, let $y = 0$. This is the same as covering up the y-term and then solving.

$$3x = 15$$
$$x = 5$$

The x-intercept is $(5, 0)$.

7. $7x - 2y = 28$

(a) To find the y-intercept, let $x = 0$. This is the same as covering up the x-term and then solving.

$$-2y = 28$$
$$y = -14$$

The $y-$intercept is $(0, -14)$.

(b) To find the x-intercept, let $y = 0$. This is the same as covering up the y-term and then solving.

$$7x = 28$$
$$x = 4$$

The x-intercept is $(4, 0)$.

9. $-4x + 3y = 10$

(a) To find the y-intercept, let $x = 0$. This is the same as covering up the x-term and then solving.

$$3y = 10$$
$$y = \frac{10}{3}$$

The y-intercept is $\left(0, \frac{10}{3}\right)$.

(b) To find the x-intercept, let $y = 0$. This is the same as covering up the y-term and then solving.

$$-4x = 10$$
$$x = -\frac{5}{2}$$

The x-intercept is $\left(-\frac{5}{2}, 0\right)$.

11. $6x - 3 = 9y$

$6x - 9y = 3$ Writing the equation in the
form $Ax + By = C$

(a) To find the y-intercept, let $x = 0$. This is the same
as covering up the x-term and then solving.

$$-9y = 3$$
$$y = -\frac{1}{3}$$

The y-intercept is $\left(0, -\frac{1}{3}\right)$.

(b) To find the x-intercept, let $y = 0$. This is the same
as covering up the y-term and then solving.

$$6x = 3$$
$$x = \frac{1}{2}$$

The x-intercept is $\left(\frac{1}{2}, 0\right)$.

13. $x + 3y = 6$

To find the x-intercept, let $y = 0$. Then solve for x.

$$x + 3y = 6$$
$$x + 3 \cdot 0 = 6$$
$$x = 6$$

Thus, $(6, 0)$ is the x-intercept.

To find the y-intercept, let $x = 0$. Then solve for y.

$$x + 3y = 6$$
$$0 + 3y = 6$$
$$3y = 6$$
$$y = 2$$

Thus, $(0, 2)$ is the y-intercept.

Plot these points and draw the line.

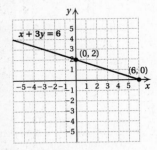

A third point should be used as a check. We substitute
any value for x and solve for y.

We let $x = 3$. Then

$$x + 3y = 6$$
$$3 + 3y = 6$$
$$3y = 3$$
$$y = 1$$

The point $(3, 1)$ is on the graph, so the graph is probably
correct.

15. $-x + 2y = 4$

To find the x-intercept, let $y = 0$. Then solve for x.

$$-x + 2y = 4$$
$$-x + 2 \cdot 0 = 4$$
$$-x = 4$$
$$x = -4$$

Thus, $(-4, 0)$ is the x-intercept.

To find the y-intercept, let $x = 0$. Then solve for y.

$$-x + 2y = 4$$
$$-0 + 2y = 4$$
$$2y = 4$$
$$y = 2$$

Thus, $(0, 2)$ is the y-intercept.

Plot these points and draw the line.

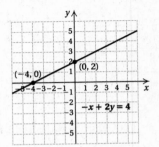

A third point should be used as a check. We substitute
any value for x and solve for y.

We let $x = 4$. Then

$$-x + 2y = 4$$
$$-4 + 2y = 4$$
$$2y = 8$$
$$y = 4$$

The point $(4, 4)$ is on the graph, so the graph is probably
correct.

17. $3x + y = 6$

To find the x-intercept, let $y = 0$. Then solve for x.

$$3x + y = 6$$
$$3x + 0 = 6$$
$$3x = 6$$
$$x = 2$$

Thus, $(2, 0)$ is the x-intercept.

To find the y-intercept, let $x = 0$. Then solve for y.

$$3x + y = 6$$
$$3 \cdot 0 + y = 6$$
$$y = 6$$

Thus, $(0, 6)$ is the y-intercept.

Plot these points and draw the line.

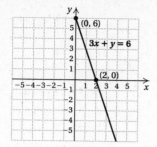

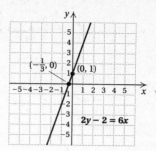

A third point should be used as a check. We substitute any value for x and solve for y.

We let $x = 1$. Then

$$3x + y = 6$$
$$3 \cdot 1 + y = 6$$
$$3 + y = 6$$
$$y = 3$$

The point $(1, 3)$ is on the graph, so the graph is probably correct.

19. $2y - 2 = 6x$

To find the x-intercept, let $y = 0$. Then solve for x.

$$2y - 2 = 6x$$
$$2 \cdot 0 - 2 = 6x$$
$$-2 = 6x$$
$$-\frac{1}{3} = x$$

Thus, $\left(-\frac{1}{3}, 0\right)$ is the x-intercept.

To find the y-intercept, let $x = 0$. Then solve for y.

$$2y - 2 = 6x$$
$$2y - 2 = 6 \cdot 0$$
$$2y - 2 = 0$$
$$2y = 2$$
$$y = 1$$

Thus, $(0, 1)$ is the y-intercept.

It is helpful to plot another point since the intercepts are so close together. This point can also serve as a check.

We let $x = 1$. Then

$$2y - 2 = 6x$$
$$2y - 2 = 6 \cdot 1$$
$$2y - 2 = 6$$
$$2y = 8$$
$$y = 4$$

Plot the point $(1, 4)$ and the intercepts and draw the line.

21. $3x - 9 = 3y$

To find the x-intercept, let $y = 0$. Then solve for x.

$$3x - 9 = 3y$$
$$3x - 9 = 3 \cdot 0$$
$$3x - 9 = 0$$
$$3x = 9$$
$$x = 3$$

Thus, $(3, 0)$ is the x-intercept.

To find the y-intercept, let $x = 0$. Then solve for y.

$$3x - 9 = 3y$$
$$3 \cdot 0 - 9 = 3y$$
$$-9 = 3y$$
$$-3 = y$$

Thus, $(0, -3)$ is the y-intercept.

Plot these points and draw the line.

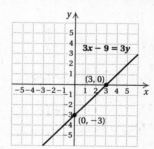

A third point should be used as a check. We substitute any value for x and solve for y.

We let $x = 1$. Then

$$3x - 9 = 3y$$
$$3 \cdot 1 - 9 = 3y$$
$$3 - 9 = 3y$$
$$-6 = 3y$$
$$-2 = y$$

The point $(1, -2)$ is on the graph, so the graph is probably correct.

23. $2x - 3y = 6$

To find the x-intercept, let $y = 0$. Then solve for x.

$$2x - 3y = 6$$
$$2x - 3 \cdot 0 = 6$$
$$2x = 6$$
$$x = 3$$

Thus, $(3, 0)$ is the x-intercept.

To find the y-intercept, let $x = 0$. Then solve for y.

$$2x - 3y = 6$$
$$2 \cdot 0 - 3y = 6$$
$$-3y = 6$$
$$y = -2$$

Thus, $(0, -2)$ is the y-intercept.

Plot these points and draw the line.

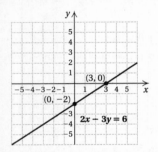

A third point should be used as a check. We substitute any value for x and solve for y.

We let $x = -3$.

$$2x - 3y = 6$$
$$2(-3) - 3y = 6$$
$$-6 - 3y = 6$$
$$-3y = 12$$
$$y = -4$$

The point $(-3, -4)$ is on the graph, so the graph is probably correct.

25. $4x + 5y = 20$

To find the x-intercept, let $y = 0$. Then solve for x.

$$4x + 5y = 20$$
$$4x + 5 \cdot 0 = 20$$
$$4x = 20$$
$$x = 5$$

Thus, $(5, 0)$ is the x-intercept.

To find the y-intercept, let $x = 0$. Then solve for y.

$$4x + 5y = 20$$
$$4 \cdot 0 + 5y = 20$$
$$5y = 20$$
$$y = 4$$

Thus, $(0, 4)$ is the y-intercept.

Plot these points and draw the graph.

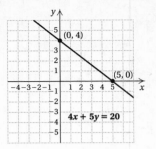

A third point should be used as a check. We substitute any value for x and solve for y.

We let $x = 4$. Then

$$4x + 5y = 20$$
$$4 \cdot 4 + 5y = 20$$
$$16 + 5y = 20$$
$$5y = 4$$
$$y = \frac{4}{5}$$

The point $\left(4, \dfrac{4}{5}\right)$ is on the graph, so the graph is probably correct.

27. $2x + 3y = 8$

To find the x-intercept, let $y = 0$. Then solve for x.

$$2x + 3y = 8$$
$$2x + 3 \cdot 0 = 8$$
$$2x = 8$$
$$x = 4$$

Thus, $(4, 0)$ is the x-intercept.

To find the y-intercept, let $x = 0$. Then solve for y.

$$2x + 3y = 8$$
$$2 \cdot 0 + 3y = 8$$
$$3y = 8$$
$$y = \frac{8}{3}$$

Thus, $\left(0, \dfrac{8}{3}\right)$ is the y-intercept.

Plot these points and draw the graph.

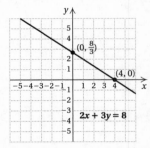

A third point should be used as a check.

We let $x = 1$. Then

$$2x + 3y = 8$$
$$2 \cdot 1 + 3y = 8$$
$$2 + 3y = 8$$
$$3y = 6$$
$$y = 2$$

The point $(1, 2)$ is on the graph, so the graph is probably correct.

29. $x - 3 = y$

To find the x-intercept, let $y = 0$. Then solve for x.

$$x - 3 = y$$
$$x - 3 = 0$$
$$x = 3$$

Thus, $(3, 0)$ is the x-intercept.

To find the y-intercept, let $x = 0$. Then solve for y.

$$x - 3 = y$$
$$0 - 3 = y$$
$$-3 = y$$

Thus, $(0, -3)$ is the y-intercept.

Plot these points and draw the line.

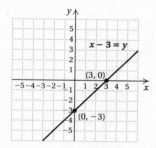

A third point should be used as a check.

We let $x = -2$. Then

$$x - 3 = y$$
$$-2 - 3 = y$$
$$-5 = y$$

The point $(-2, -5)$ is on the graph, so the graph is probably correct.

31. $3x - 2 = y$

To find the x-intercept, let $y = 0$. Then solve for x.

$$3x - 2 = y$$
$$3x - 2 = 0$$
$$3x = 2$$
$$x = \frac{2}{3}$$

Thus, $\left(\frac{2}{3}, 0\right)$ is the x-intercept.

To find the y-intercept, let $x = 0$. Then solve for y.

$$3x - 2 = y$$
$$3 \cdot 0 - 2 = y$$
$$-2 = y$$

Thus, $(0, -2)$ is the y-intercept.

Plot these points and draw the line.

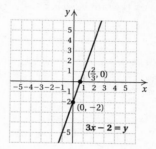

A third point should be used as a check.

We let $x = 2$. Then

$$3x - 2 = y$$
$$3 \cdot 2 - 2 = y$$
$$6 - 2 = y$$
$$4 = y$$

The point $(2, 4)$ is on the graph, so the graph is probably correct.

33. $6x - 2y = 12$

To find the x-intercept, let $y = 0$. Then solve for x.

$$6x - 2y = 12$$
$$6x - 2 \cdot 0 = 12$$
$$6x = 12$$
$$x = 2$$

Thus, $(2, 0)$ is the x-intercept.

To find the y-intercept, let $x = 0$. Then solve for y.

$$6x - 2y = 12$$
$$6 \cdot 0 - 2y = 12$$
$$-2y = 12$$
$$y = -6$$

Thus, $(0, -6)$ is the y-intercept.

Plot these points and draw the line.

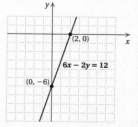

We use a third point as a check.

We let $x = 1$. Then

$$6x - 2y = 12$$
$$6 \cdot 1 - 2y = 12$$
$$6 - 2y = 12$$
$$-2y = 6$$
$$y = -3$$

The point $(1, -3)$ is on the graph, so the graph is probably correct.

35. $3x + 4y = 5$

To find the x-intercept, let $y = 0$. Then solve for x.

$$3x + 4y = 5$$
$$3x + 4 \cdot 0 = 5$$
$$3x = 5$$
$$x = \frac{5}{3}$$

Thus, $\left(\frac{5}{3}, 0\right)$ is the x-intercept.

To find the y-intercept, let $x = 0$. Then solve for y.

$$3x + 4y = 5$$
$$3 \cdot 0 + 4y = 5$$
$$4y = 5$$
$$y = \frac{5}{4}$$

Thus, $\left(0, \frac{5}{4}\right)$ is the y-intercept.

It is helpful to plot another point since the intercepts are so close together. This point can also serve as a check.

We let $x = 3$. Then

$$3x + 4y = 5$$
$$3 \cdot 3 + 4y = 5$$
$$9 + 4y = 5$$
$$4y = -4$$
$$y = -1$$

Plot the point $(3, -1)$ and the intercepts and draw the line.

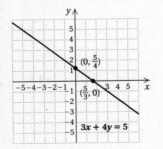

37. $y = -3 - 3x$

To find the x-intercept, let $y = 0$. Then solve for x.

$$y = -3 - 3x$$
$$0 = -3 - 3x$$
$$3x = -3$$
$$x = -1$$

Thus, $(-1, 0)$ is the x-intercept.

To find the y-intercept, let $x = 0$. Then solve for y.

$$y = -3 - 3x$$
$$y = -3 - 3 \cdot 0$$
$$y = -3$$

Thus, $(0, -3)$ is the y-intercept.

Plot these points and draw the graph.

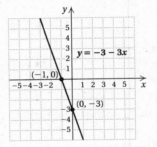

We use a third point as a check.

We let $x = -2$. Then

$$y = -3 - 3x$$
$$y = -3 - 3 \cdot (-2)$$
$$y = -3 + 6$$
$$y = 3$$

The point $(-2, 3)$ is on the graph, so the graph is probably correct.

39. $y - 3x = 0$

To find the x-intercept, let $y = 0$. Then solve for x.

$$0 - 3x = 0$$
$$-3x = 0$$
$$x = 0$$

Thus, $(0, 0)$ is the x-intercept. Note that this is also the y-intercept.

In order to graph the line, we will find a second point.

$$\text{When } x = 1, \quad y - 3 \cdot 1 = 0$$
$$y - 3 = 0$$
$$y = 3$$

Plot the points and draw the graph.

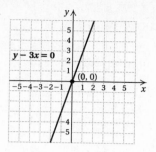

We use a third point as a check.

We let $x = -1$. Then

$$y - 3(-1) = 0$$
$$y + 3 = 0$$
$$y = -3$$

The point $(-1, -3)$ is on the graph, so the graph is probably correct.

41. $x = -2$

Any ordered pair $(-2, y)$ is a solution. The variable x must be -2, but y can be any number we choose. A few solutions are listed below. Plot these points and draw the line.

x	y
-2	-2
-2	0
-2	4

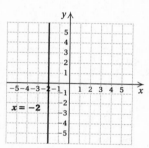

43. $y = 2$

Any ordered pair $(x, 2)$ is a solution. The variable y must be 2, but x can be any number we choose. A few solutions are listed below. Plot these points and draw the line.

x	y
-3	2
0	2
2	2

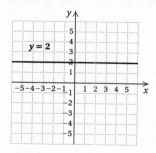

45. $x = 2$

Any ordered pair $(2, y)$ is a solution. The variable x must be 2, but y can be any number we choose. A few solutions are listed below. Plot these points and draw the line.

x	y
2	-1
2	4
2	5

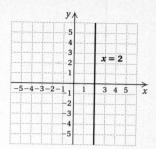

47. $y = 0$

Any ordered pair $(x, 0)$ is a solution. The variable y must be 0, but x can be any number we choose. A few solutions are listed below. Plot these points and draw the line.

x	y
-5	0
-1	0
3	0

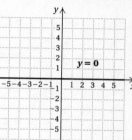

49. $x = \dfrac{3}{2}$

Any ordered pair $\left(\dfrac{3}{2}, y\right)$ is a solution. The variable x must be $\dfrac{3}{2}$, but y can be any number we choose. A few solutions are listed below. Plot these points and draw the line.

x	y
$\dfrac{3}{2}$	-2
$\dfrac{3}{2}$	0
$\dfrac{3}{2}$	4

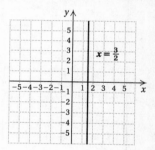

51. $3y = -5$

$$y = -\frac{5}{3} \qquad \text{Solving for } y$$

Any ordered pair $\left(x, -\dfrac{5}{3}\right)$ is a solution. A few solutions are listed below. Plot these points and draw the line.

x	y
-3	$-\dfrac{5}{3}$
0	$-\dfrac{5}{3}$
2	$-\dfrac{5}{3}$

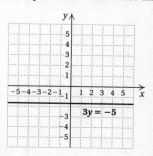

53. $4x + 3 = 0$

$\qquad 4x = -3$

$\qquad x = -\dfrac{3}{4}$ Solving for x

Any ordered pair $\left(-\dfrac{3}{4}, y\right)$ is a solution. A few solutions are listed below. Plot these points and draw the line.

x	y
$-\dfrac{3}{4}$	-2
$-\dfrac{3}{4}$	0
$-\dfrac{3}{4}$	3

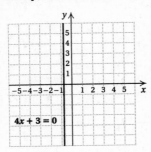

55. $48 - 3y = 0$

$\qquad -3y = -48$

$\qquad\quad y = 16$ Solving for y

Any ordered pair $(x, 16)$ is a solution. A few solutions are listed below. Plot these points and draw the line.

x	y
-4	16
0	16
2	16

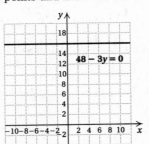

57. Note that every point on the horizontal line passing through $(0, -1)$ has -1 as the y-coordinate. Thus, the equation of the line is $y = -1$.

59. Note that every point on the vertical line passing through $(4, 0)$ has 4 as the x-coordinate. Thus, the equation of the line is $x = 4$.

61. Discussion and Writing Exercise

63. *Familiarize.* Let $p =$ the percent of desserts sold that will be pie.

Translate. We reword the problem.

40 is what percent of 250?

$\downarrow\ \downarrow \qquad\quad \downarrow \qquad\ \downarrow\ \downarrow$

$40 = \qquad p \qquad \cdot\ 250$

Solve. We solve the equation.

$\qquad 40 = p \cdot 250$

$\qquad \dfrac{40}{250} = \dfrac{p \cdot 250}{250}$

$\qquad 0.16 = p$

$\qquad 16\% = p$

Check. We can find 16% of 250:

$\qquad 16\% \cdot 250 = 0.16 \cdot 250 = 40$

The answer checks.

State. 16% of the desserts sold will be pie.

65. $-1.6x < 64$

$\qquad \dfrac{-1.6x}{-1.6} > \dfrac{64}{-1.6}$ Dividing by -1.6 and reversing the inequality symbol

$\qquad\quad x > -40$

The solution set is $\{x | x > -40\}$.

67. $x + (x - 1) < (x + 2) - (x + 1)$

$\qquad 2x - 1 < x + 2 - x - 1$

$\qquad 2x - 1 < 1$

$\qquad\quad 2x < 2$

$\qquad\quad\ x < 1$

The solution set is $\{x | x < 1\}$.

69. A line parallel to the x-axis has an equation of the form $y = b$. Since the y-coordinate of one point on the line is -4, then $b = -4$ and the equation is $y = -4$.

71. Substitute -4 for x and 0 for y.

$\qquad 3(-4) + k = 5 \cdot 0$

$\qquad\quad -12 + k = 0$

$\qquad\qquad\quad k = 12$

Exercise Set 11.3

1. We consider (x_1, y_1) to be $(-3, 5)$ and (x_2, y_2) to be $(4, 2)$.

$$m = \frac{y_2 - y_1}{x_2 - x_1} = \frac{2 - 5}{4 - (-3)} = \frac{-3}{7} = -\frac{3}{7}$$

3. We can choose any two points. We consider (x_1, y_1) to be $(-3, -1)$ and (x_2, y_2) to be $(0, 1)$.

$$m = \frac{y_2 - y_1}{x_2 - x_1} = \frac{1 - (-1)}{0 - (-3)} = \frac{2}{3}$$

5. We can choose any two points. We consider (x_1, y_1) to be $(-4, -2)$ and (x_2, y_2) to be $(4, 4)$.

$$m = \frac{y_2 - y_1}{x_2 - x_1} = \frac{4 - (-2)}{4 - (-4)} = \frac{6}{8} = \frac{3}{4}$$

7. We consider (x_1, y_1) to be $(-4, -2)$ and (x_2, y_2) to be $(3, -2)$.

$$m = \frac{y_2 - y_1}{x_2 - x_1} = \frac{-2 - (-2)}{3 - (-4)} = \frac{0}{7} = 0$$

9. We plot $(-2, 4)$ and $(3, 0)$ and draw the line containing these points.

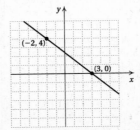

To find the slope, consider (x_1, y_1) to be $(-2, 4)$ and (x_2, y_2) to be $(3, 0)$.

$$m = \frac{y_2 - y_1}{x_2 - x_1} = \frac{0 - 4}{3 - (-2)} = \frac{-4}{5} = -\frac{4}{5}$$

11. We plot $(-4, 0)$ and $(-5, -3)$ and draw the line containing these points.

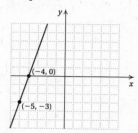

To find the slope, consider (x_1, y_1) to be $(-4, 0)$ and (x_2, y_2) to be $(-5, -3)$.

$$m = \frac{y_2 - y_1}{x_2 - x_1} = \frac{-3 - 0}{-5 - (-4)} = \frac{-3}{-1} = 3$$

13. We plot $(-4, 2)$ and $(2, -3)$ and draw the line containing these points.

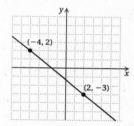

To find the slope, consider (x_1, y_1) to be $(-4, 2)$ and (x_2, y_2) to be $(2, -3)$.

$$m = \frac{y_2 - y_1}{x_2 - x_1} = \frac{-3 - 2}{2 - (-4)} = \frac{-5}{6} = -\frac{5}{6}$$

15. We plot $(5, 3)$ and $(-3, -4)$ and draw the line containing these points.

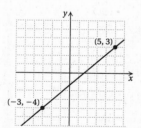

To find the slope, consider (x_1, y_1) to be $(5, 3)$ and (x_2, y_2) to be $(-3, -4)$.

$$m = \frac{y_2 - y_1}{x_2 - x_1} = \frac{-4 - 3}{-3 - 5} = \frac{-7}{-8} = \frac{7}{8}$$

17. $m = \dfrac{-\dfrac{1}{2} - \dfrac{3}{2}}{2 - 5} = \dfrac{-2}{-3} = \dfrac{2}{3}$

19. $m = \dfrac{-2 - 3}{4 - 4} = \dfrac{-5}{0}$

Since division by 0 is not defined, the slope is not defined.

21. $m = \dfrac{-3 - 7}{15 - (-11)} = \dfrac{-10}{26} = -\dfrac{5}{13}$

23. $m = \dfrac{\dfrac{3}{11} - \dfrac{3}{11}}{\dfrac{5}{4} - \left(-\dfrac{1}{2}\right)} = \dfrac{0}{\dfrac{7}{4}} = 0$

25. $m = \dfrac{\text{rise}}{\text{run}} = \dfrac{2.4}{8.2} = \dfrac{2.4}{8.2} \cdot \dfrac{10}{10} = \dfrac{24}{82}$

$= \dfrac{\cancel{2} \cdot 12}{\cancel{2} \cdot 41} = \dfrac{12}{41}$

27. $m = \dfrac{\text{rise}}{\text{run}} = \dfrac{56}{258} = \dfrac{\cancel{2} \cdot 28}{\cancel{2} \cdot 129} = \dfrac{28}{129}$

29. Long's Peak rises 14,255 ft $-$ 9600 ft $=$ 4655 ft.

Grade $= \dfrac{4655}{15,840} \approx 0.294 \approx 29.4\%$

31. The rate of change is the slope of the line. We can use any two ordered pairs to find the slope. We choose $(2, 50)$ and $(8, 200)$.

Rate of change $= \dfrac{200 \text{ mi} - 50 \text{ mi}}{8 \text{ gal} - 2 \text{ gal}} = \dfrac{150 \text{ mi}}{6 \text{ gal}} =$ 25 miles per gallon

33. The rate of change is the slope of the line. We can use any two ordered pairs to find the slope. We choose $(2, 2000)$ and $(4, 1000)$. (Note that units on the vertical axis are given in thousands.)

Rate of change $= \dfrac{\$1000 - \$2000}{4 \text{ yr} - 2 \text{ yr}} = \dfrac{-\$1000}{2 \text{ yr}} = -\$500$ per year

35. The rate of change is the slope of the line. We can use any two ordered pairs to find the slope. We choose $(1990, 550,000)$ and $(2003, 649,000)$.

Rate of change $= \dfrac{649,000 - 550,000}{2003 - 1990} = \dfrac{99,000}{13} \approx$ 7600 people per year

37. $y = -10x + 7$

The equation is in the form $y = mx + b$, where $m = -10$. Thus, the slope is -10.

39. $y = 3.78x - 4$

The equation is in the form $y = mx + b$, where $m = 3.78$. Thus, the slope is 3.78.

41. We solve for y, obtaining an equation of the form $y = mx + b$.

$$3x - y = 4$$
$$-y = -3x + 4$$
$$-1(-y) = -1(-3x + 4)$$
$$y = 3x - 4$$

The slope is 3.

43. We solve for y, obtaining an equation of the form $y = mx + b$.

$$x + 5y = 10$$
$$5y = -x + 10$$
$$y = \frac{1}{5}(-x + 10)$$
$$y = -\frac{1}{5}x + 2$$

The slope is $-\frac{1}{5}$.

45. We solve for y, obtaining an equation of the form $y = mx + b$.

$$3x + 2y = 6$$
$$2y = -3x + 6$$
$$y = \frac{1}{2}(-3x + 6)$$
$$y = -\frac{3}{2}x + 3$$

The slope is $-\frac{3}{2}$.

47. The graph of $x = \frac{2}{15}$ is a vertical line, so the slope is not defined.

49. $y = -2.74x$

The equation is in the form $y = mx + b$, where $m = -2.74$. Thus, the slope is -2.74.

51. We solve for y, obtaining an equation of the form $y = mx + b$.

$$9x = 3y + 5$$
$$9x - 5 = 3y$$
$$\frac{1}{3}(9x - 5) = y$$
$$3x - \frac{5}{3} = y$$

The slope is 3.

53. We solve for y, obtaining an equation of the form $y = mx + b$.

$$5x - 4y + 12 = 0$$
$$5x + 12 = 4y$$
$$\frac{1}{4}(5x + 12) = y$$
$$\frac{5}{4}x + 3 = y$$

The slope is $\frac{5}{4}$.

55. $y = 4$

The equation can be thought of as $y = 0 \cdot x + 4$, so the slope is 0.

57. Discussion and Writing Exercise

59. $16\% = \frac{16}{100} = \frac{4 \cdot 4}{4 \cdot 25} = \frac{4}{25}$

61. $37.5\% = \frac{37.5}{100} = \frac{37.5}{100} \cdot \frac{10}{10} = \frac{375}{1000} = \frac{3 \cdot 125}{8 \cdot 125} = \frac{3}{8}$

63. Translate.

What is 15% of \$23.80?
$$\downarrow \quad \downarrow \quad \downarrow \quad \downarrow \quad \downarrow$$
$$a \quad = \quad 15\% \quad \cdot \quad 23.80$$

Solve. We convert to decimal notation and multiply.

$$a = 15\% \cdot 23.80 = 0.15 \cdot 23.80 = 3.57$$

The answer is \$3.57.

65. Familiarize. Let $p =$ the percent of the cost of the meal represented by the tip.

Translate. We reword the problem.

\$8.50 is what percent of \$42.50?
$$\downarrow \quad \downarrow \qquad \downarrow \qquad \downarrow \quad \downarrow$$
$$8.50 = \qquad p \qquad \cdot \quad 42.50$$

Solve. We solve the equation.

$$8.50 = p \cdot 42.50$$
$$0.2 = p$$
$$20\% = p$$

Check. We can find 20% of 42.50.

$$20\% \cdot 42.50 = 0.2 \cdot 42.50 = 8.50$$

The answer checks.

State. The tip was 20% of the cost of the meal.

67. Familiarize. Let $c =$ the cost of the meal before the tip was added. Then the tip is $15\% \cdot c$.

Translate. We reword the problem.

Cost of meal plus tip is total cost
$$\downarrow \qquad \qquad \downarrow \quad \downarrow \quad \downarrow \qquad \downarrow$$
$$c \qquad + \quad 15\% \cdot c = \quad 51.92$$

Solve. We solve the equation.

$$c + 15\% \cdot c = 51.92$$
$$1 \cdot c + 0.15c = 51.92$$
$$1.15c = 51.92$$
$$c \approx 45.15$$

Check. We can find 15% of 45.15 and then add this to 45.15.

$$15\% \cdot 45.15 = 0.15 \cdot 45.15 \approx 6.77 \text{ and } 45.15 + 6.77 = 51.92$$

The answer checks.

State. Before the tip the meal cost \$45.15.

69. Note that the sum of the coordinates of each point on the graph is 5. Thus, we have $x + y = 5$, or $y = -x + 5$.

71. Note that each y-coordinate is 2 more than the corresponding x-coordinate. Thus, we have $y = x + 2$.

73.

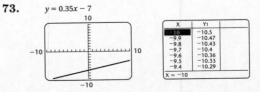

75.

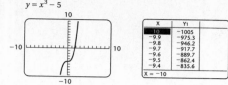

$y = x^3 - 5$

X	Y₁
-10	-1005
-9.9	-975.3
-9.8	-946.2
-9.7	-917.7
-9.6	-889.7
-9.5	-862.4
-9.4	-835.6

X = -10

Exercise Set 11.4

1. $y = -4x - 9$

The equation is already in the form $y = mx + b$. The slope is -4 and the y-intercept is $(0, -9)$.

3. $y = 1.8x$

We can think of $y = 1.8x$ as $y = 1.8x + 0$. The slope is 1.8 and the y-intercept is $(0, 0)$.

5. We solve for y.
$$-8x - 7y = 21$$
$$-7y = 8x + 21$$
$$y = -\frac{1}{7}(8x + 21)$$
$$y = -\frac{8}{7}x - 3$$

The slope is $-\frac{8}{7}$ and the y-intercept is $(0, -3)$.

7. We solve for y.
$$4x = 9y + 7$$
$$4x - 7 = 9y$$
$$\frac{1}{9}(4x - 7) = y$$
$$\frac{4}{9}x - \frac{7}{9} = y$$

The slope is $\frac{4}{9}$ and the y-intercept is $\left(0, -\frac{7}{9}\right)$.

9. We solve for y.
$$-6x = 4y + 2$$
$$-6x - 2 = 4y$$
$$\frac{1}{4}(-6x - 2) = y$$
$$-\frac{3}{2}x - \frac{1}{2} = y$$

The slope is $-\frac{3}{2}$ and the y-intercept is $\left(0, -\frac{1}{2}\right)$.

11. $y = -17$

We can think of $y = -17$ as $y = 0x - 17$. The slope is 0 and the y-intercept is $(0, -17)$.

13. We substitute -7 for m and -13 for b in the equation $y = mx + b$.
$$y = -7x - 13$$

15. We substitute 1.01 for m and -2.6 for b in the equation $y = mx + b$.
$$y = 1.01x - 2.6$$

17. We know the slope is -2, so the equation is $y = -2x + b$. Using the point $(-3, 0)$, we substitute -3 for x and 0 for y in $y = -2x + b$. Then we solve for b.
$$y = -2x + b$$
$$0 = -2(-3) + b$$
$$0 = 6 + b$$
$$-6 = b$$

Thus, we have the equation $y = -2x - 6$.

19. We know the slope is $\frac{3}{4}$, so the equation is $y = \frac{3}{4}x + b$. Using the point $(2, 4)$, we substitute 2 for x and 4 for y in $y = \frac{3}{4}x + b$. Then we solve for b.
$$y = \frac{3}{4}x + b$$
$$4 = \frac{3}{4} \cdot 2 + b$$
$$4 = \frac{3}{2} + b$$
$$\frac{5}{2} = b$$

Thus, we have the equation $y = \frac{3}{4}x + \frac{5}{2}$.

21. We know the slope is 1, so the equation is $y = 1 \cdot x + b$, or $y = x + b$. Using the point $(2, -6)$, we substitute 2 for x and -6 for y in $y = x + b$. Then we solve for y.
$$y = x + b$$
$$-6 = 2 + b$$
$$-8 = b$$

Thus, we have the equation $y = x - 8$.

23. We substitute -3 for m and 3 for b in the equation $y = mx + b$.
$$y = -3x + 3$$

25. $(12, 16)$ and $(1, 5)$

First we find the slope.
$$m = \frac{16 - 5}{12 - 1} = \frac{11}{11} = 1$$

Thus, $y = 1 \cdot x + b$, or $y = x + b$. We can use either point to find b. We choose $(1, 5)$. Substitute 1 for x and 5 for y in $y = x + b$.
$$y = x + b$$
$$5 = 1 + b$$
$$4 = b$$

Thus, the equation is $y = x + 4$.

27. $(0, 4)$ and $(4, 2)$

First we find the slope.
$$m = \frac{4 - 2}{0 - 4} = \frac{2}{-4} = -\frac{1}{2}$$

Thus, $y = -\frac{1}{2}x + b$. One of the given points is the y-intercept $(0, 4)$. Thus, we substitute 4 for b in $y = -\frac{1}{2}x + b$. The equation is $y = -\frac{1}{2}x + 4$.

29. $(3, 2)$ and $(1, 5)$

First we find the slope.

$$m = \frac{2 - 5}{3 - 1} = \frac{-3}{2} = -\frac{3}{2}$$

Thus, $y = -\frac{3}{2}x + b$.   We can use either point to find b. We choose $(3, 2)$. Substitute 3 for x and 2 for y in $y = -\frac{3}{2}x + b$.

$$y = -\frac{3}{2}x + b$$
$$2 = -\frac{3}{2} \cdot 3 + b$$
$$2 = -\frac{9}{2} + b$$
$$\frac{13}{2} = b$$

Thus, the equation is $y = -\frac{3}{2}x + \frac{13}{2}$.

31. $(-4, 5)$ and $(-2, -3)$

First we find the slope.

$$m = \frac{5 - (-3)}{-4 - (-2)} = \frac{8}{-2} = -4$$

Thus, $y = -4x + b$. We can use either point to find b. We choose $(-4, 5)$. Substitute -4 for x and 5 for y in $y = -4x + b$.

$$y = -4x + b$$
$$5 = -4(-4) + b$$
$$5 = 16 + b$$
$$-11 = b$$

Thus, the equation is $y = -4x - 11$.

33. a) First we find the slope.

$$m = \frac{150 - 105}{20 - 80} = \frac{45}{-60} = -0.75$$

Thus, $T = -0.75a + b$. We can use either point to find b. We choose $(20, 150)$. Substitute 20 for a and 150 for T in $T = -0.75a + b$.

$$150 = -0.75(20) + b$$
$$150 = -15 + b$$
$$165 = b$$

Thus, the equation is $T = -0.75a + 165$.

b) The rate of change is the slope, -0.75 beats per minute per year.

c) Substitute 50 for a and calculate T.

$$T = -0.75a + 165$$
$$T = -0.75(50) + 165$$
$$T = -37.5 + 165$$
$$T = 127.5$$

The target heart rate for a 50 year old person is 127.5 beats per minute.

35. Discussion and Writing Exercise

37.
$$3x - 4(9 - x) = 17$$
$$3x - 36 + 4x = 17$$
$$7x - 36 = 17$$
$$7x = 53$$
$$x = \frac{53}{7}$$

The solution is $\frac{53}{7}$.

39.
$$40(2x - 7) = 50(4 - 6x)$$
$$80x - 280 = 200 - 300x$$
$$380x - 280 = 200$$
$$380x = 480$$
$$x = \frac{480}{380}$$
$$x = \frac{24}{19}$$

The solution is $\frac{24}{19}$.

41.
$$3x - 9x + 21x - 15x = 6x - 12 - 24x + 18$$
$$0 = -18x + 6$$
$$18x = 6$$
$$x = \frac{6}{18}$$
$$x = \frac{1}{3}$$

The solution is $\frac{1}{3}$.

43.
$$3(x - 9x) + 21(x - 15x) = 6(x - 12) - 24(x + 18)$$
$$3(-8x) + 21(-14x) = 6x - 72 - 24x - 432$$
$$-24x - 294x = -18x - 504$$
$$-318x = -18x - 504$$
$$-300x = -504$$
$$x = \frac{504}{300} = \frac{2 \cdot 6 \cdot 6 \cdot 7}{2 \cdot 6 \cdot 5 \cdot 5}$$
$$x = \frac{42}{25}$$

The solution is $\frac{42}{25}$.

45. First find the slope of $3x - y + 4 = 0$.

$$3x - y + 4 = 0$$
$$3x + 4 = y$$

The slope is 3.

Thus, $y = 3x + b$. Using the point $(2, -3)$, we substitute 2 for x and -3 for y in $y = 3x + b$. Then we solve for b.

$$y = 3x + b$$
$$-3 = 3 \cdot 2 + b$$
$$-3 = 6 + b$$
$$-9 = b$$

Thus, the equation is $y = 3x - 9$.

47. First find the slope of $3x - 2y = 8$.

$$3x - 2y = 8$$
$$-2y = -3x + 8$$
$$y = \frac{3}{2}x - 4$$

The slope is $\frac{3}{2}$.

Then find the y-intercept of $2y + 3x = -4$.

$$2y + 3x = -4$$
$$2y = -3x - 4$$
$$y = -\frac{3}{2}x - 2$$

The y-intercept is $(0, -2)$.

Finally, write the equation of the line with slope $\frac{3}{2}$ and y-intercept $(0, -2)$.

$$y = mx + b$$
$$y = \frac{3}{2}x + (-2)$$
$$y = \frac{3}{2}x - 2$$

Exercise Set 11.5

1. Slope $\frac{2}{5}$; y-intercept $(0, 1)$

We plot $(0, 1)$ and from there move up 2 units and right 5 units. This locates the point $(5, 3)$. We plot $(5, 3)$ and draw a line passing through $(0, 1)$ and $(5, 3)$.

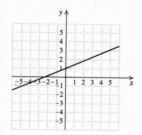

3. Slope $\frac{5}{3}$; y-intercept $(0, -2)$

We plot $(0, -2)$ and from there move up 5 units and right 3 units. This locates the point $(3, 3)$. We plot $(3, 3)$ and draw a line passing through $(0, -2)$ and $(3, 3)$.

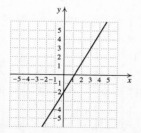

5. Slope $-\frac{3}{4}$; y-intercept $(0, 5)$

We plot $(0, 5)$. We can think of the slope as $\frac{-3}{4}$, so from $(0, 5)$ we move down 3 units and right 4 units. This locates the point $(4, 2)$. We plot $(4, 2)$ and draw a line passing through $(0, 5)$ and $(4, 2)$.

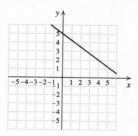

7. Slope $-\frac{1}{2}$; y-intercept $(0, 3)$

We plot $(0, 3)$. We can think of the slope as $\frac{-1}{2}$, so from $(0, 3)$ we move down 1 unit and right 2 units. This locates the point $(2, 2)$. We plot $(2, 2)$ and draw a line passing through $(0, 3)$ and $(2, 2)$

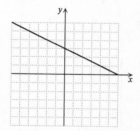

9. Slope 2; y-intercept $(0, -4)$

We plot $(0, -4)$. We can think of the slope as $\frac{2}{1}$, so from $(0, -4)$ we move up 2 units and right 1 unit. This locates the point $(1, -2)$. We plot $(1, -2)$ and draw a line passing through $(0, -4)$ and $(1, -2)$.

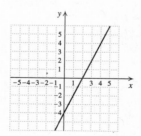

11. Slope -3; y-intercept $(0, 2)$

We plot $(0, 2)$. We can think of the slope as $\frac{-3}{1}$, so from $(0, 2)$ we move down 3 units and right 1 unit. This locates the point $(1, -1)$. We plot $(1, -1)$ and draw a line passing through $(0, 2)$ and $(1, -1)$.

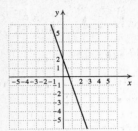

13. $y = \dfrac{3}{5}x + 2$

First we plot the y-intercept $(0, 2)$. We can start at the y-intercept and use the slope, $\dfrac{3}{5}$, to find another point. We move up 3 units and right 5 units to get a new point $(5, 5)$. Thinking of the slope as $\dfrac{-3}{-5}$ we can start at $(0, 2)$ and move down 3 units and left 5 units to get another point $(-5, -1)$.

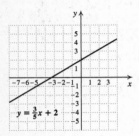

15. $y = -\dfrac{3}{5}x + 1$

First we plot the y-intercept $(0, 1)$. We can start at the y-intercept and, thinking of the slope as $\dfrac{-3}{5}$, find another point by moving down 3 units and right 5 units to the point $(5, -2)$. Thinking of the slope as $\dfrac{3}{-5}$ we can start at $(0, 1)$ and move up 3 units and left 5 units to get another point $(-5, 4)$.

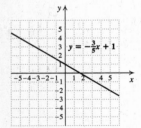

17. $y = \dfrac{5}{3}x + 3$

First we plot the y-intercept $(0, 3)$. We can start at the y-intercept and use the slope, $\dfrac{5}{3}$, to find another point. We move up 5 units and right 3 units to get a new point $(3, 8)$. Thinking of the slope as $\dfrac{-5}{-3}$ we can start at $(0, 3)$ and move down 5 units and left 3 units to get another point $(-3, -2)$.

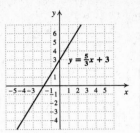

19. $y = -\dfrac{3}{2}x - 2$

First we plot the y-intercept $(0, -2)$. We can start at the y-intercept and, thinking of the slope as $\dfrac{-3}{2}$, find another point by moving down 3 units and right 2 units to the point $(2, -5)$. Thinking of the slope as $\dfrac{3}{-2}$ we can start at $(0, -2)$ and move up 3 units and left 2 units to get another point $(-2, 1)$.

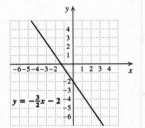

21. We first rewrite the equation in slope-intercept form.
$$2x + y = 1$$
$$y = -2x + 1$$

Now we plot the y-intercept $(0, 1)$. We can start at the y-intercept and, thinking of the slope as $\dfrac{-2}{1}$, find another point by moving down 2 units and right 1 unit to the point $(1, -1)$. In a similar manner, we can move from the point $(1, -1)$ to find a third point $(2, -3)$.

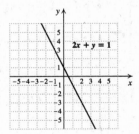

23. We first rewrite the equation in slope-intercept form.
$$3x - y = 4$$
$$-y = -3x + 4$$
$$y = 3x - 4 \quad \text{Multiplying by } -1$$

Now we plot the y-intercept $(0, -4)$. We can start at the y-intercept and, thinking of the slope as $\dfrac{3}{1}$, find another point by moving up 3 units and right 1 unit to the point $(1, -1)$. In a similar manner, we can move from the point $(1, -1)$ to find a third point $(2, 2)$.

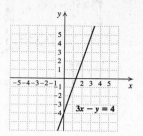

3x − y = 4

25. We first rewrite the equation in slope-intercept form.

$$2x + 3y = 9$$
$$3y = -2x + 9$$
$$y = \frac{1}{3}(-2x + 9)$$
$$y = -\frac{2}{3}x + 3$$

Now we plot the y-intercept $(0, 3)$. We can start at the y-intercept and, thinking of the slope as $\frac{-2}{3}$, find another point by moving down 2 units and right 3 units to the point $(3, 1)$. Thinking of the slope as $\frac{2}{-3}$ we can start at $(0, 3)$ and move up 2 units and left 3 units to get another point $(-3, 5)$.

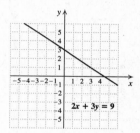

2x + 3y = 9

27. We first rewrite the equation in slope-intercept form.

$$x - 4y = 12$$
$$-4y = -x + 12$$
$$y = -\frac{1}{4}(-x + 12)$$
$$y = \frac{1}{4}x - 3$$

Now we plot the y-intercept $(0, -3)$. We can start at the y-intercept and use the slope, $\frac{1}{4}$, to find another point. We move up 1 unit and right 4 units to the point $(4, -2)$. Thinking of the slope as $\frac{-1}{-4}$ we can start at $(0, -3)$ and move down 1 unit and left 4 units to get another point $(-4, -4)$.

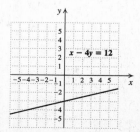

x − 4y = 12

29. We first rewrite the equation in slope-intercept form.

$$x + 2y = 6$$
$$2y = -x + 6$$
$$y = \frac{1}{2}(-x + 6)$$
$$y = -\frac{1}{2}x + 3$$

Now we plot the y-intercept $(0, 3)$. We can start at the y-intercept and, thinking of the slope as $\frac{-1}{2}$, find another point by moving down 1 unit and right 2 units to the point $(2, 2)$. Thinking of the slope as $\frac{1}{-2}$ we can start at $(0, 3)$ and move up 1 unit and left 2 units to get another point $(-2, 4)$.

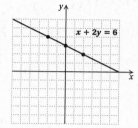

x + 2y = 6

31. Discussion and Writing Exercise

33. $m = \dfrac{y_2 - y_1}{x_2 - x_1} = \dfrac{7 - (-6)}{8 - (-2)} = \dfrac{13}{10}$

35. $m = \dfrac{y_2 - y_1}{x_2 - x_1} = \dfrac{4.6 - (-2.3)}{14.5 - 4.5} = \dfrac{6.9}{10} = \dfrac{69}{100}$, or 0.69

37. $m = \dfrac{y_2 - y_1}{x_2 - x_1} = \dfrac{-6 - (-6)}{8 - (-2)} = \dfrac{0}{10} = 0$

39. $m = \dfrac{y_2 - y_1}{x_2 - x_1} = \dfrac{-4 - (-1)}{11 - 11} = \dfrac{-3}{0}$

Since division by 0 is not defined, the slope is not defined.

41. Rate of change $= \dfrac{\text{Change in number of transplants}}{\text{Change in years}} =$

$\dfrac{15,120 - 8873}{2003 - 1988} = \dfrac{6247}{15} \approx 416$

Kidney transplants are increasing at a rate of about 416 per year. The slope of the line is 416.

43. For residents in excess of 2, the rate of change is 1.5 ft³ per person. For 1-2 people, the number of residents in excess of 2 is 0, so the x-intercept is $(0, 16)$. Then the equation is $y = 1.5x + 16$.

45. First we plot $(-3, 1)$. Then, thinking of the slope as $\frac{2}{1}$, from $(-3, 1)$ we move up 2 units and right 1 unit to locate the point $(-2, 3)$. We plot $(-2, 3)$ and draw a line passing through $(-3, 1)$ and $(-2, 3)$.

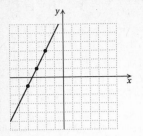

Exercise Set 11.6

1. 1. The first equation is already solved for y:

$$y = x + 4$$

2. We solve the second equation for y:

$$y - x = -3$$
$$y = x - 3$$

The slope of each line is 1. The y-intercepts, $(0, 4)$ and $(0, -3)$, are different. The lines are parallel.

3. We solve each equation for y:

1. $y + 3 = 6x$ 2. $-6x - y = 2$

    $y = 6x - 3$ $-y = 6x + 2$

                       $y = -6x - 2$

The slope of the first line is 6 and of the second is -6. Since the slopes are different, the lines are not parallel.

5. We solve each equation for y:

1. $10y + 32x = 16.4$ 2. $y + 3.5 = 0.3125x$

   $10y = -32x + 16.4$ $y = 0.3125x - 3.5$

     $y = -3.2x + 1.64$

The slope of the first line is -3.2 and of the second is 0.3125. Since the slopes are different, the lines are not parallel.

7. 1. The first equation is already solved for y:

$$y = 2x + 7$$

2. We solve the second equation for y:

$$5y + 10x = 20$$
$$5y = -10x + 20$$
$$y = -2x + 4$$

The slope of the first line is 2 and of the second is -2. Since the slopes are different, the lines are not parallel.

9. We solve each equation for y:

1. $3x - y = -9$ 2. $2y - 6x = -2$

   $3x + 9 = y$ $2y = 6x - 2$

                  $y = 3x - 1$

The slope of each line is 3. The y-intercepts, $(0, 9)$ and $(0, -1)$ are different. The lines are parallel.

11. $x = 3$,

$x = 4$

These are vertical lines with equations of the form $x = p$ and $x = q$, where $p \neq q$. Thus, they are parallel.

13. 1. The first equation is already solved for y:

$$y = -4x + 3$$

2. We solve the second equation for y:

$$4y + x = -1$$
$$4y = -x - 1$$
$$y = -\frac{1}{4}x - \frac{1}{4}$$

The slopes are -4 and $-\frac{1}{4}$. Their product is $-4\left(-\frac{1}{4}\right) = 1$. Since the product of the slopes is not -1, the lines are not perpendicular.

15. We solve each equation for y:

1. $x + y = 6$ 2. $4y - 4x = 12$

   $y = -x + 6$ $4y = 4x + 12$

                    $y = x + 3$

The slopes are -1 and 1. Their product is $-1 \cdot 1 = -1$. The lines are perpendicular.

17. 1. The first equation is already solved for y:

$$y = -0.3125x + 11$$

2. We solve the second equation for y:

$$y - 3.2x = -14$$
$$y = 3.2x - 14$$

The slopes are -0.3125 and 3.2. Their product is $-0.3125(3.2) = -1$. The lines are perpendicular.

19. 1. The first equation is already solved for y:

$$y = -x + 8$$

2. We solve the second equation for y:

$$x - y = -1$$
$$x + 1 = y$$

The slopes are -1 and 1. Their product is $-1 \cdot 1 = -1$. The lines are perpendicular.

21. We solve each equation for y:

1. $\frac{3}{8}x - \frac{y}{2} = 1$

$$8\left(\frac{3}{8}x - \frac{y}{2}\right) = 8 \cdot 1$$

$$8 \cdot \frac{3}{8}x - 8 \cdot \frac{y}{2} = 8$$

$$3x - 4y = 8$$

$$-4y = -3x + 8$$

$$y = \frac{3}{4}x - 2$$

2. $\frac{4}{3}x - y + 1 = 0$

$$\frac{4}{3}x + 1 = y$$

The slopes are $\frac{3}{4}$ and $\frac{4}{3}$. Their product is $\frac{3}{4}\left(\frac{4}{3}\right) = 1$. Since the product of the slopes is not -1, the lines are not perpendicular.

23. $x = 0$,

$y = -2$

The first line is vertical and the second is horizontal, so the lines are perpendicular.

25. We solve each equation for y:

1. $3y + 21 = 2x$ 2. $3y = 2x + 24$

$3y = 2x - 21$ $y = \dfrac{2}{3}x + 8$

$y = \dfrac{2}{3}x - 7$

The slope of each line is $\dfrac{2}{3}$. The y-intercepts, $(0, -7)$ and $(0, 8)$, are different. The lines are parallel.

27. We solve each equation for y:

1. $3y = 2x - 21$ 2. $2y - 16 = 3x$

$y = \dfrac{2}{3}x - 7$ $2y = 3x + 16$

$y = \dfrac{3}{2}x + 8$

The slopes, $\dfrac{2}{3}$ and $\dfrac{3}{2}$, are different so the lines are not parallel. The product of the slopes is $\dfrac{2}{3} \cdot \dfrac{3}{2} = 1 \neq -1$, so the lines are not perpendicular. Thus, the lines are neither parallel nor perpendicular.

29. Discussion and Writing Exercise

31. Equations with the same solutions are called <u>equivalent equations</u>.

33. The <u>multiplication principle</u> for equations asserts that when we multiply or divide by the same non-zero number on both sides of an equation, we get equivalent equations.

35. <u>Vertical</u> lines are graphs of equations of the $x = a$.

37. The <u>x-intercept</u> of a line, if it exists, indicates where the line crosses the x-axis.

39. First we find the slope of the given line:

$y - 3x = 4$

$y = 3x + 4$

The slope is 3.

Then we use the slope-intercept equation to write the equation of a line with slope 3 and y-intercept $(0, 6)$:

$y = mx + b$

$y = 3x + 6$ Substituting 3 for m and 6 for b

41. First we find the slope of the given line:

$3y - x = 0$

$3y = x$

$y = \dfrac{1}{3}x$

The slope is $\dfrac{1}{3}$.

We can find the slope of the line perpendicular to the given line by taking the reciprocal of $\dfrac{1}{3}$ and changing the sign. We get -3.

Then we use the slope-intercept equation to write the equation of a line with slope -3 and y-intercept $(0, 2)$:

$y = mx + b$

$y = -3x + 2$ Substituting -3 for m and 2 for b

43. First we find the slope of the given line:

$4x - 8y = 12$

$-8y = -4x + 12$

$y = \dfrac{1}{2}x - \dfrac{3}{2}$

The slope is $\dfrac{1}{2}$, so the equation is $y = \dfrac{1}{2}x + b$. Substitute -2 for x and 0 for y and solve for b.

$y = \dfrac{1}{2}x + b$

$0 = \dfrac{1}{2}(-2) + b$

$0 = -1 + b$

$1 = b$

Thus, the equation is $y = \dfrac{1}{2}x + 1$.

45. We find the slope of each line:

1. $4y = kx - 6$ 2. $5x + 20y = 12$

$y = \dfrac{k}{4}x - \dfrac{3}{2}$ $20y = -5x + 12$

$y = -\dfrac{1}{4}x + \dfrac{3}{5}$

The slopes are $\dfrac{k}{4}$ and $-\dfrac{1}{4}$. If the lines are perpendicular, the product of their slopes is -1.

$\dfrac{k}{4}\left(-\dfrac{1}{4}\right) = -1$

$-\dfrac{k}{16} = -1$

$k = 16$

47. First we find the equation of A, a line containing the points $(1, -1)$ and $(4, 3)$:

The slope is $\dfrac{3 - (-1)}{4 - 1} = \dfrac{4}{3}$, so the equation is $y = \dfrac{4}{3}x + b$. Use either point to find b. We choose $(1, -1)$.

$y = \dfrac{4}{3}x + b$

$-1 = \dfrac{4}{3} \cdot 1 + b$

$-1 = \dfrac{4}{3} + b$

$-\dfrac{7}{3} = b$

Thus, the equation of line A is $y = \dfrac{4}{3}x - \dfrac{7}{3}$.

The slope of A is $\frac{4}{3}$. Since A and B are perpendicular we find the slope of B by taking the reciprocal of $\frac{4}{3}$ and changing the sign. We get $-\frac{3}{4}$, so the equation is $y = -\frac{3}{4}x + b$. We use the point $(1, -1)$ to find b.

$$y = -\frac{3}{4}x + b$$

$$-1 = -\frac{3}{4} \cdot 1 + b$$

$$-1 = -\frac{3}{4} + b$$

$$-\frac{1}{4} = b$$

Thus, the equation of line B is $y = -\frac{3}{4}x - \frac{1}{4}$.

Exercise Set 11.7

1. We use alphabetical order to replace x by -3 and y by -5.

$$\begin{array}{c|c}
\multicolumn{2}{c}{-x - 3y < 18} \\
\hline
-(-3) - 3(-5) \ ? \ 18 & \\
3 + 15 & \\
18 & \text{FALSE}
\end{array}$$

Since $18 < 18$ is false, $(-3, -5)$ is not a solution.

3. We use alphabetical order to replace x by $\frac{1}{2}$ and y by $-\frac{1}{4}$.

$$\begin{array}{c|c}
\multicolumn{2}{c}{7y - 9x \leq -3} \\
\hline
7\left(-\frac{1}{4}\right) - 9 \cdot \frac{1}{2} \ ? \ -3 & \\
-\frac{7}{4} - \frac{9}{2} & \\
-\frac{7}{4} - \frac{18}{4} & \\
-\frac{25}{4} & \\
-6\frac{1}{4} & \text{TRUE}
\end{array}$$

Since $-6\frac{1}{4} \leq -3$ is true, $\left(\frac{1}{2}, -\frac{1}{4}\right)$ is a solution.

5. Graph $x > 2y$.

First graph the line $x = 2y$, or $y = \frac{1}{2}x$. Two points on the line are $(0, 0)$ and $(4, 2)$. We draw a dashed line since the inequality symbol is $>$. Then we pick a test point that is not on the line. We try $(-2, 1)$.

$$\begin{array}{c|c}
\multicolumn{2}{c}{x > 2y} \\
\hline
-2 \ ? \ 2 \cdot 1 & \\
2 & \text{FALSE}
\end{array}$$

We see that $(-2, 1)$ is not a solution of the inequality, so we shade the points in the region that does not contain $(-2, 1)$.

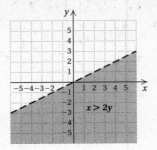

7. Graph $y \leq x - 3$.

First graph the line $y = x - 3$. The intercepts are $(0, -3)$ and $(3, 0)$. We draw a solid line since the inequality symbol is $\leq$. Then we pick a test point that is not on the line. We try $(0, 0)$.

$$\begin{array}{c|c}
\multicolumn{2}{c}{y \leq x - 3} \\
\hline
0 \ ? \ 0 - 3 & \\
-3 & \text{FALSE}
\end{array}$$

We see that $(0, 0)$ is not a solution of the inequality, so we shade the region that does not contain $(0, 0)$.

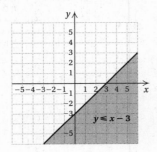

9. Graph $y < x + 1$.

First graph the line $y = x + 1$. The intercepts are $(0, 1)$ and $(-1, 0)$. We draw a dashed line since the inequality symbol is $<$. Then we pick a test point that is not on the line. We try $(0, 0)$.

$$\begin{array}{c|c}
\multicolumn{2}{c}{y < x + 1} \\
\hline
0 \ ? \ 0 + 1 & \\
1 & \text{TRUE}
\end{array}$$

Since $(0, 0)$ is a solution of the inequality, we shade the region that contains $(0, 0)$.

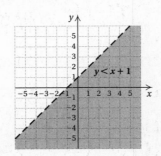

11. Graph $y \geq x - 2$.

First graph the line $y = x - 2$. The intercepts are $(0, -2)$ and $(2, 0)$. We draw a solid line since the inequality symbol is $\geq$. Then we test the point $(0, 0)$.

$$\frac{y \geq x - 2}{0 \ ? \ 0 - 2}$$
$$\bigg| \ {-2} \quad \text{TRUE}$$

Since $(0, 0)$ is a solution of the inequality, we shade the region containing $(0, 0)$.

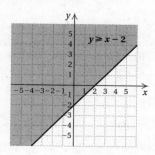

13. Graph $y \leq 2x - 1$.

First graph the line $y = 2x - 1$. The intercepts are $(0, -1)$ and $\left(\frac{1}{2}, 0\right)$. We draw a solid line since the inequality symbol is $\leq$. Then we test the point $(0, 0)$.

$$\frac{y \leq 2x - 1}{0 \ ? \ 2 \cdot 0 - 1}$$
$$\bigg| \ {-1} \quad \text{FALSE}$$

Since $(0, 0)$ is not a solution of the inequality, we shade the region that does not contain $(0, 0)$.

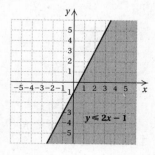

15. Graph $x + y \leq 3$.

First graph the line $x + y = 3$. The intercepts are $(0, 3)$ and $(3, 0)$. We draw a solid line since the inequality symbol is $\leq$. Then we test the point $(0, 0)$.

$$\frac{x + y \leq 3}{0 + 0 \ ? \ 3}$$
$$0 \ \bigg| \quad \text{TRUE}$$

Since $(0, 0)$ is a solution of the inequality, we shade the region that contains $(0, 0)$.

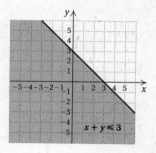

17. Graph $x - y > 7$.

First graph the line $x - y = 7$. The intercepts are $(0, -7)$ and $(7, 0)$. We draw a dashed line since the inequality symbol is $>$. Then we test the point $(0, 0)$.

$$\frac{x - y > 7}{0 - 0 \ ? \ 7}$$
$$0 \ \bigg| \quad \text{FALSE}$$

Since $(0, 0)$ is not a solution of the inequality, we shade the region that does not contain $(0, 0)$.

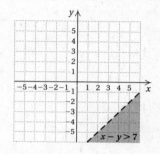

19. Graph $2x + 3y \leq 12$.

First graph the line $2x + 3y = 12$. The intercepts are $(0, 4)$ and $(6, 0)$. We draw a solid line since the inequality symbol is $\leq$. Then we test the point $(0, 0)$.

$$\frac{2x + 3y \leq 12}{2 \cdot 0 + 3 \cdot 0 \ ? \ 12}$$
$$0 \ \bigg| \quad \text{TRUE}$$

Since $(0, 0)$ is a solution of the inequality, we shade the region containing $(0, 0)$.

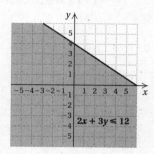

21. Graph $y \geq 1 - 2x$.

First graph the line $y = 1 - 2x$. The intercepts are $(0, 1)$ and $\left(\frac{1}{2}, 0\right)$. We draw a solid line since the inequality symbol is $\geq$. Then we test the point $(0, 0)$.

$$\frac{y \geq 1 - 2x}{\begin{array}{c} 0 \; ? \; 1 - 2 \cdot 0 \\ \quad \mid \quad 1 \qquad \text{FALSE} \end{array}}$$

Since $(0, 0)$ is not a solution of the inequality, we shade the region that does not contain $(0, 0)$.

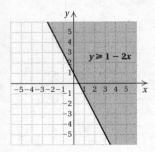

23. Graph $2x - 3y > 6$.

First graph the line $2x - 3y = 6$. The intercepts are $(0, -2)$ and $(3, 0)$. We draw a dashed line since the inequality symbol is $>$. Then we test the point $(0, 0)$.

$$\frac{2x - 3y > 6}{\begin{array}{c} 2 \cdot 0 - 3 \cdot 0 \; ? \; 6 \\ \quad 0 \quad \mid \quad \text{FALSE} \end{array}}$$

Since $(0, 0)$ is not a solution of the inequality, we shade the region that does not contain $(0, 0)$.

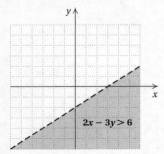

25. Graph $y \leq 3$.

First graph the line $y = 3$ using a solid line since the inequality symbol is $\leq$. Then pick a test point that is not on the line. We choose $(1, -2)$. We can write the inequality as $0x + y \leq 3$.

$$\frac{0x + y \leq 3}{\begin{array}{c} 0 \cdot 1 + (-2) \; ? \; 3 \\ \quad -2 \quad \mid \quad \text{TRUE} \end{array}}$$

Since $(1, -2)$ is a solution of the inequality, we shade the region containing $(1, -2)$.

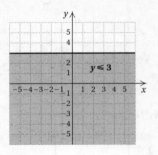

27. Graph $x \geq -1$.

Graph the line $x = 1$ using a solid line since the inequality symbol is $\geq$. Then pick a test point that is not on the line. We choose $(2, 3)$. We can write the inequality as $x + 0y \geq -1$.

$$\frac{x + 0y \geq -1}{\begin{array}{c} 2 + 0 \cdot 3 \; ? \; -1 \\ \quad 2 \quad \mid \quad \text{TRUE} \end{array}}$$

Since $(2, 3)$ is a solution of the inequality, we shade the region containing $(2, 3)$.

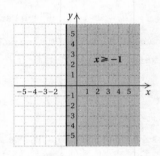

29. Discussion and Writing Exercise

31. First we solve each equation for y:

1. $5y + 50 = 4x$ 2. $5y = 4x + 15$

   $5y = 4x - 50$ $y = \dfrac{4}{5}x + 3$

   $y = \dfrac{4}{5}x - 10$

The slope of each line is $\dfrac{4}{5}$. The y-intercepts, $(0, -10)$ and $(0, 3)$, are different. The lines are parallel.

33. First we solve each equation for y:

1. $5y + 50 = 4x$ 2. $4y = 5x + 12$

   $5y = 4x - 50$ $y = \dfrac{5}{4}x + 3$

   $y = \dfrac{4}{5}x - 10$

The slope, $\dfrac{4}{5}$ and $\dfrac{5}{4}$, are different, so the lines are not parallel. The product of the slopes is $\dfrac{4}{5} \cdot \dfrac{5}{4} = 1 \neq -1$, so the lines are not perpendicular. Thus, the lines are neither parallel nor perpendicular.

35. The c children weigh $35c$ kg, and the a adults weigh $75a$ kg. Together, the children and adults weigh $35c + 75a$ kg.

When this total is more than 1000 kg the elevator is over-loaded, so we have $35c + 75a > 1000$. (Of course, c and a would also have to be nonnegative, but we will not deal with nonnegativity constraints here.)

To graph $35c + 75a > 1000$, we first graph $35c + 75a = 1000$ using a dashed line. Two points on the line are $(4, 20)$ and $(11, 5)$. (We are using alphabetical order of variables.) Then we test the point $(0, 0)$.

$$\frac{35c + 75a > 1000}{35 \cdot 0 + 75 \cdot 0 \ ? \ 1000}$$
$$0 \ \Big| \quad \text{FALSE}$$

Since $(0, 0)$ is not a solution of the inequality, we shade the region that does not contain $(0, 0)$.

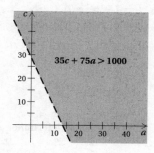

Chapter 11 Review Exercises

1. Point A is 5 units left and 1 unit down. The coordinates of A are $(-5, -1)$.

2. Point B is 2 units left and 5 units up. The coordinates of B are $(-2, 5)$.

3. Point C is 3 units right and 0 units up or down. The coordinates of C are $(3, 0)$.

4. $(2, 5)$ is 2 units right and 5 units up. See the graph following Exercise 6 below.

5. $(0, -3)$ is 0 units right or left and 3 units down. See the graph following Exercise 6 below.

6. $(-4, -2)$ is 4 units left and 2 units down. See the graph below.

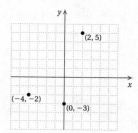

7. Since the first coordinate is positive and the second coordinate is negative, the point $(3, -8)$ is in quadrant IV.

8. Since both coordinates are negative, the point $(-20, -14)$ is in quadrant III.

9. Since both coordinates are positive, the point $(4.9, 1.3)$ is in quadrant I.

10. We substitute 2 for x and -6 for y.

$$\frac{2y - x = 10}{2(-6) - 2 \ ? \ 10}$$
$$\frac{-12 - 2}{-14} \ \Big| \quad \text{FALSE}$$

Since $-14 = 10$ is false, the pair $(2, -6)$ is not a solution.

11. We substitute 0 for x and 5 for y.

$$\frac{2y - x = 10}{2 \cdot 5 - 0 \ ? \ 10}$$
$$\frac{10 - 0}{10} \ \Big| \quad \text{TRUE}$$

Since $10 = 10$ is true, the pair $(0, 5)$ is a solution.

12. To show that a pair is a solution, we substitute, replacing x with the first coordinate and y with the second coordinate in each pair.

$$\frac{2x - y = 3}{2 \cdot 0 - (-3) \ ? \ 3}$$
$$\frac{0 + 3}{3} \ \Big| \quad \text{TRUE}$$

$$\frac{2x - y = 3}{2 \cdot 2 - 1 \ ? \ 3}$$
$$\frac{4 - 1}{3} \ \Big| \quad \text{TRUE}$$

In each case the substitution results in a true equation. Thus, $(0, -3)$ and $(2, 1)$ are both solutions of $2x - y = 3$. We graph these points and sketch the line passing through them.

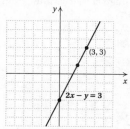

The line appears to pass through $(3, 3)$ also. We check to determine if $(3, 3)$ is a solution of $2x - y = 3$.

$$\frac{2x - y = 3}{2 \cdot 3 - 3 \ ? \ 3}$$
$$\frac{6 - 3}{3} \ \Big| \quad \text{TRUE}$$

Thus, $(3, 3)$ is another solution. There are other correct answers, including $(-1, -5)$ and $(4, 5)$.

13. $y = 2x - 5$

The y-intercept is $(0, -5)$. We find two other points.

When $x = 2$, $y = 2 \cdot 2 - 5 = 4 - 5 = -1$.

When $x = 4$, $y = 2 \cdot 4 - 5 = 8 - 5 = 3$.

x	y
0	−5
2	−1
4	3

Plot these points, draw the line they determine, and label the graph $y = 2x - 5$.

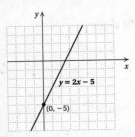

14. $y = -\dfrac{3}{4}x$

The equation is equivalent to $y = -\dfrac{3}{4}x + 0$. The y-intercept is $(0,0)$. We find two other points.

When $x = -4$, $y = -\dfrac{3}{4}(-4) = 3$.

When $x = 4$, $y = -\dfrac{3}{4} \cdot 4 = -3$.

x	y
−4	3
0	0
4	−3

Plot these points, draw the line they determine, and label the graph $y = -\dfrac{3}{4}x$.

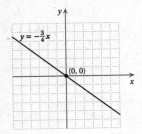

15. $y = -x + 4$

The y-intercept is $(0,4)$. We find two other points.

When $x = -1$, $y = -(-1) + 4 = 1 + 4 = 5$.

When $x = 4$, $y = -4 + 4 = 0$.

x	y
−1	5
0	4
4	0

Plot these points, draw the line they determine, and label the graph $y = -x + 4$.

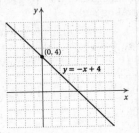

16. $y = 3 - 4x$, or $y = -4x + 3$

The y-intercept is $(0,3)$. We find two other points.

When $x = 1$, $y = -4 \cdot 1 + 3 = -4 + 3 = -1$.

When $x = 2$, $y = -4 \cdot 2 + 3 = -8 + 3 = -5$.

x	y
0	3
1	−1
2	−5

Plot these points, draw the line they determine, and label the graph $y = 3 - 4x$.

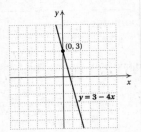

17. $y = 3$

Any ordered pair $(x, 3)$ is a solution. The variable y must be 3, but x can be any number we choose. A few solutions are listed below. Plot these points and draw the line.

x	y
−3	3
0	3
2	3

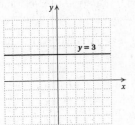

18. $5x - 4 = 0$

$5x = 4$

$x = \dfrac{4}{5}$ Solving for x

Any ordered pair $\left(\dfrac{4}{5}, y\right)$ is a solution. A few solutions are listed below. Plot these points and draw the graph.

x	y
$\frac{4}{5}$	-3
$\frac{4}{5}$	0
$\frac{4}{5}$	2

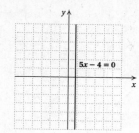

19. $x - 2y = 6$

To find the x-intercept, let $y = 0$. Then solve for x.

$$x - 2y = 6$$
$$x - 2 \cdot 0 = 6$$
$$x = 6$$

Thus, $(6, 0)$ is the x-intercept.

To find the y-intercept, let $x = 0$. Then solve for y.

$$x - 2y = 6$$
$$0 - 2y = 6$$
$$-2y = 6$$
$$y = -3$$

Thus, $(0, -3)$ is the y-intercept.

Plot these points and draw the graph.

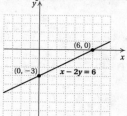

We use a third point as a check.

We let $x = -2$. Then

$$x - 2y = 6$$
$$-2 - 2y = 6$$
$$-2y = 8$$
$$y = -4.$$

The point $(-2, -4)$ is on the graph, so the graph is probably correct.

20. $5x - 2y = 10$

To find the x-intercept, let $y = 0$. Then solve for x.

$$5x - 2y = 10$$
$$5x - 2 \cdot 0 = 10$$
$$5x = 10$$
$$x = 2$$

Thus, $(2, 0)$ is the x-intercept.

To find the y-intercept, let $x = 0$. Then solve for y.

$$5x - 2y = 10$$
$$5 \cdot 0 - 2y = 10$$
$$-2y = 10$$
$$y = -5$$

Thus, $(0, -5)$ is the y-intercept.

Plot these points and draw the graph.

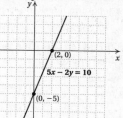

We use a third point as a check.

We let $x = 4$. Then

$$5x - 2y = 10$$
$$5 \cdot 4 - 2y = 10$$
$$20 - 2y = 10$$
$$-2y = -10$$
$$y = 5.$$

The point $(4, 5)$ is on the graph, so the graph is probably correct.

21. a) When $n = 1$, $S = \frac{3}{2} \cdot 1 + 13 = \frac{3}{2} + 13 = 1\frac{1}{2} + 13 = 14\frac{1}{2}$ ft^3.

When $n = 2$, $S = \frac{3}{2} \cdot 2 + 13 = 3 + 13 = 16$ ft^3.

When $n = 5$, $S = \frac{3}{2} \cdot 5 + 13 = \frac{15}{2} + 13 = 7\frac{1}{2} + 13 = 20\frac{1}{2}$ ft^3.

When $n = 10$, $S = \frac{3}{2} \cdot 10 + 13 = 15 + 13 = 28$ ft^3.

b) We plot the points found in part (a): $\left(1, 14\frac{1}{2}\right)$, $(2, 16)$, $\left(5, 20\frac{1}{2}\right)$ and $(10, 28)$. Then we draw the graph.

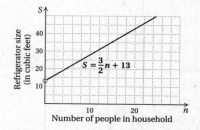

From the graph, it appears that an x-value of 3 corresponds to the S-value of $17\frac{1}{2}$, so the recommended size is $17\frac{1}{2}$ ft^3.

c) We substitute 22 for S and solve for n.

$$22 = \frac{3}{2}n + 13$$

$$9 = \frac{3}{2}n$$

$$\frac{2}{3} \cdot 9 = \frac{2}{3} \cdot \frac{3}{2}n$$

$$6 = n$$

A 22-ft^3 refrigerator is recommended for a household of 6 residents.

22. 5:30 P.M. is 2.5 hr after 3:00 P.M. In this time the number of driveways plowed was $13 - 7$, or 6.

a) Rate of change $= \dfrac{6 \text{ driveways}}{2.5 \text{ hr}} = 2.4$ driveways per hour

b) 2.5 hr $= 2.5 \times 1$ hr $= 2.5 \times 60$ min $= 150$ min

Rate of change $= \dfrac{150 \text{ min}}{6 \text{ driveways}} = 25$ minutes per driveway

23. We will use the points (11:00 A.M., 6 manicures) and (1:00 P.M., 14 manicures) to find the rate of change. Note that 1:00 P.M. is 2 hr after 11:00 A.M.

Rate of change $= \dfrac{14 \text{ manicures} - 6 \text{ manicures}}{2 \text{ hr}} =$

$\dfrac{8 \text{ manicures}}{2 \text{ hr}} = 4$ manicures per hour

24. We can choose any two points. We consider (x_1, y_1) to be $(-3, 1)$ and (x_2, y_2) to be $(3, 3)$.

$$m = \frac{y_2 - y_1}{x_2 - x_1} = \frac{3 - 1}{3 - (-3)} = \frac{2}{6} = \frac{1}{3}$$

25. We can choose any two points. We consider (x_1, y_1) to be $(3, 1)$ and (x_2, y_2) to be $(-3, 3)$.

$$m = \frac{3 - 1}{-3 - 3} = \frac{2}{-6} = -\frac{1}{3}$$

26. We plot $(-5, -2)$ and $(5, 4)$ and draw the line containing those points.

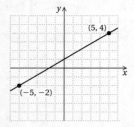

To find the slope, consider (x_1, y_1) to be $(-5, -2)$ and (x_2, y_2) to be $(5, 4)$.

$$m = \frac{y_2 - y_1}{x_2 - x_1} = \frac{4 - (-2)}{5 - (-5)} = \frac{6}{10} = \frac{3}{5}$$

27. We plot $(-5, 5)$ and $(4, -4)$ and draw the line containing those points.

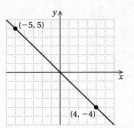

To find the slope, consider (x_1, y_1) to be $(4, -4)$ and (x_2, y_2) to be $(-5, 5)$.

$$m = \frac{y_2 - y_1}{x_2 - x_1} = \frac{5 - (-4)}{-5 - 4} = \frac{9}{-9} = -1$$

28. Grade $= \dfrac{315}{4500} = 0.07 = 7\%$

29. $y = -\dfrac{5}{8}x - 3$

The equation is in the form $y = mx + b$, where $m = -\dfrac{5}{8}$. Thus, the slope is $-\dfrac{5}{8}$.

30. We solve for y, obtaining an equation of the form $y = mx + b$.

$$2x - 4y = 8$$

$$-4y = -2x + 8$$

$$-\frac{1}{4}(-4y) = -\frac{1}{4}(-2x + 8)$$

$$y = \frac{1}{2}x - 2$$

The slope is $\dfrac{1}{2}$.

31. The graph of $x = -2$ is a vertical line, so the slope is not defined.

32. $y = 9$, or $y = 0 \cdot x + 9$

The slope is 0.

33. $y = -9x + 46$

The equation is in the form $y = mx + b$. The slope is -9 and the y-intercept is $(0, 46)$.

34. We solve for y.

$$x + y = 9$$

$$y = -x + 9$$

The slope is -1 and the y-intercept is $(0, 9)$.

35. We solve for y.

$$3x - 5y = 4$$

$$-5y = -3x + 4$$

$$-\frac{1}{5}(-5y) = -\frac{1}{5}(-3x + 4)$$

$$y = \frac{3}{5}x - \frac{4}{5}$$

The slope is $\dfrac{3}{5}$ and the y-intercept is $\left(0, -\dfrac{4}{5}\right)$.

36. We substitute -2.8 for m and 19 for b in the equation $y = mx + b$.

$$y = -2.8x + 19$$

37. We substitute $\dfrac{5}{8}$ for m and $-\dfrac{7}{8}$ for b in the equation $y = mx + b$.

$$y = \frac{5}{8}x - \frac{7}{8}$$

38. We know the slope is 3, so the equation is $y = 3x + b$. Using the point $(1, 2)$, we substitute 1 for x and 2 for y in $y = 3x + b$. Then we solve for b.

$$y = 3x + b$$
$$2 = 3 \cdot 1 + b$$
$$2 = 3 + b$$
$$-1 = b$$

Thus, we have the equation $y = 3x - 1$.

39. We know the slope is $\dfrac{2}{3}$, so the equation is $y = \dfrac{2}{3}x + b$. Using the point $(-2, -5)$, we substitute -2 for x and -5 for y in $y = \dfrac{2}{3}x + b$. Then we solve for b.

$$y = \frac{2}{3}x + b$$
$$-5 = \frac{2}{3}(-2) + b$$
$$-5 = -\frac{4}{3} + b$$
$$-\frac{11}{3} = b$$

Thus, we have the equation $y = \dfrac{2}{3}x - \dfrac{11}{3}$.

40. The slope is -2 and the y-intercept is $(0, -4)$, so we have the equation $y = -2x - 4$.

41. First we find the slope.

$$m = \frac{1 - 7}{-1 - 5} = \frac{-6}{-6} = 1$$

Thus, $y = 1 \cdot x + b$, or $y = x + b$. We can use either point to find b. We choose $(5, 7)$. Substitute 5 for x and 7 for y in $y = x + b$.

$$y = x + b$$
$$7 = 5 + b$$
$$2 = b$$

Thus, the equation is $y = x + 2$.

42. First we find the slope.

$$m = \frac{-3 - 0}{-4 - 2} = \frac{-3}{-6} = \frac{1}{2}$$

Thus, $y = \dfrac{1}{2}x + b$. We can use either point to find b. We choose $(2, 0)$. Substitute 2 for x and 0 for y in $y = \dfrac{1}{2}x + b$.

$$y = \frac{1}{2}x + b$$
$$0 = \frac{1}{2} \cdot 2 + b$$
$$0 = 1 + b$$
$$-1 = b$$

Thus, the equation is $y = \dfrac{1}{2}x - 1$.

43. a) First we find the slope.

$$m = \frac{59.5 - 33.9}{53 - 0} = \frac{25.6}{53} \approx 0.48$$

The y-intercept is $(0, 33.9)$, so we have the equation $y = 0.48x + 33.9$.

b) The rate of change is the slope, 0.48 percent per year.

c) In 2005, $x = 2005 - 1950 = 55$.

$$y = 0.48x + 33.9$$
$$y = 0.48(55) + 33.9$$
$$y = 26.4 + 33.9$$
$$y = 60.3$$

The percent of female workers in the labor force in 2005 will be 60.3%.

44. Slope -1, y-intercept $(0, 4)$

We plot $(0, 4)$. We can think of the slope as $\dfrac{-1}{1}$, so from $(0, 4)$ we move down 1 unit and right 1 unit. This locates the point $(1, 3)$. We plot $(1, 3)$ and draw a line passing through $(0, 4)$ and $(1, 3)$.

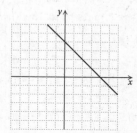

45. Slope $\dfrac{5}{3}$, y-intercept $(0, -3)$.

Plot $(0, -3)$ and from there move up 5 units and right 3 units. This locates the point $(3, 2)$. We plot $(3, 2)$ and draw a line passing through $(0, -3)$ and $(3, 2)$.

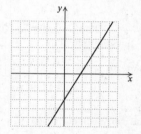

46. $y = -\frac{3}{5}x + 2$

First we plot the y-intercept $(0, 2)$. We can start at the y-intercept and, thinking of the slope as $\frac{-3}{5}$, find another point by moving down 3 units and right 5 units to the point $(5, -1)$. Thinking of the slope as $\frac{3}{-5}$ we can start at $(0, 2)$ and move up 3 units and left 5 units to get another point $(-5, 5)$.

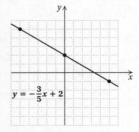

47. First we rewrite the equation in slope-intercept form.
$$2y - 3x = 6$$
$$2y = 3x + 6$$
$$y = \frac{1}{2}(3x + 6)$$
$$y = \frac{3}{2}x + 3$$

Now we plot the y-intercept $(0, 3)$. We can start at the y-intercept and use the slope, $\frac{3}{2}$, to find another point. We move up 3 units and right 2 units to the point $(2, 6)$. Thinking of the slope as $\frac{-3}{-2}$ we can start at $(0, 3)$ and move down 3 units and left 2 units to get another point, $(-2, 0)$.

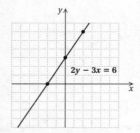

48. First we solve each equation for y:

1. $4x + y = 6$ 2. $4x + y = 8$
     $y = -4x + 6$ $y = -4x + 8$

The slope of each line is -4. The y-intercepts, $(0, 6)$ and $(0, 8)$, are different. The lines are parallel.

49. We solve the first equation for y.
$$2x + y = 10$$
$$y = -2x + 10$$
The second equation is already solved for y.
$$y = \frac{1}{2}x - 4$$

The slopes, -2 and $\frac{1}{2}$, are not the same so the lines are not parallel. The product of the slopes is $-2 \cdot \frac{1}{2} = -1$, so the lines are perpendicular.

50. First we solve each equation for y:

1. $x + 4y = 8$ 2. $x = -4y - 10$
    $4y = -x + 8$ $x + 10 = -4y$
     $y = \frac{1}{4}(-x + 8)$ $-\frac{1}{4}(x + 10) = y$
     $y = -\frac{1}{4}x + 2$ $-\frac{1}{4}x - \frac{5}{2} = y$

The slope of each line is $-\frac{1}{4}$. The y-intercepts, $(0, 2)$ and $\left(0, -\frac{5}{2}\right)$, are different. The lines are parallel.

51. First we solve each equation for y:

1. $3x - y = 6$ 2. $3x + y = 8$
   $-y = -3x + 6$ $y = -3x + 8$
     $y = -1(-3x + 6)$
     $y = 3x - 6$

The slopes, 3 and -3, are not the same so the lines are not parallel. The product of the slopes is $3(-3) = -9 \neq -1$, so the lines are not perpendicular. Thus, the lines are neither parallel nor perpendicular.

52.
$$\frac{x - 2y > 1}{0 - 2 \cdot 0 \; ? \; 1}$$
         $0 \quad | \quad$ FALSE

Since $0 > 1$ is false, $(0, 0)$ is not a solution.

53.
$$\frac{x - 2y > 1}{1 - 2 \cdot 3 \; ? \; 1}$$
       $1 - 6 \quad |$
         $-5 \quad |$ FALSE

Since $-5 > 1$ is false, $(1, 3)$ is not a solution.

54.
$$\frac{x - 2y > 1}{4 - 2(-1) \; ? \; 1}$$
       $4 + 2 \quad |$
         $6 \quad |$ TRUE

Since $6 > 1$ is true, $(4, -1)$ is a solution.

55. Graph $x < y$.

First graph the line $x = y$, or $y = x$. Two points on the line are $(0, 0)$ and $(3, 3)$. We draw a dashed line since the inequality symbol is $<$. Then we pick a test point that is not on the line. We try $(1, 2)$.

$$\frac{x < y}{1 \; ? \; 2 \; \text{TRUE}}$$

We see that $(1, 2)$ is a solution of the inequality, so we shade the region that contains $(1, 2)$.

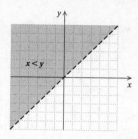

56. Graph $x + 2y \geq 4$.

First graph the line $x + 2y = 4$. The intercepts are $(0, 2)$ and $(4, 0)$. We draw a solid line since the inequality symbol is $\geq$. Then we test the point $(0, 0)$.

$$\begin{array}{c|c} x + 2y \geq 4 \\ \hline 0 + 2 \cdot 0 \; ? \; 1 \\ 0 & \text{FALSE} \end{array}$$

Since $(0, 0)$ is not a solution of the inequality, we shade the region that does not contain $(0, 0)$.

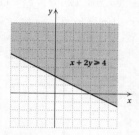

57. Graph $x > -2$.

Graph the line $x = -2$ using a dashed line since the inequality symbol is $>$. Then pick a test point that is not on the line. We choose $(0, 0)$. We can write the inequality as $x + 0y > -2$.

$$\begin{array}{c|c} x + 0y > -2 \\ \hline 0 + 0 \cdot 0 \; ? \; -2 \\ 0 & \text{TRUE} \end{array}$$

Since $(0, 0)$ is a solution of the inequality, we shade the region containing $(0, 0)$.

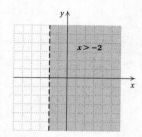

58. *Discussion and Writing Exercise.* The y-intercept is the point at which the graph crosses the y-axis. Since a point on the y-axis is neither left nor right of the origin, the first or x-coordinate of the point is 0.

59. *Discussion and Writing Exercise.* The graph of $x < 1$ on a number line consists of the points in the set $\{x | x < 1\}$.

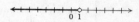

The graph of $x < 1$ on a plane consists of the points, or ordered pairs, in the set $\{(x, y) | x + 0 \cdot y < 1\}$. This is the set of ordered pairs with first coordinate less than 1.

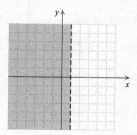

60. *Discussion and Writing Exercise.* First plot the y-intercept, $(0, 2458)$. Then, thinking of the slope as $\dfrac{37}{100}$, plot a second point on the line by moving up 37 units and right 100 units from the y-intercept and plot a third point by moving down 37 units and left 100 units. Finally, draw a line through the three points.

61. Substitute -2 for x and 5 for y and then solve for m.

$$\begin{aligned} y &= mx + 3 \\ 5 &= m(-2) + 3 \\ 5 &= -2m + 3 \\ 2 &= -2m \\ -1 &= m \end{aligned}$$

62. We plot the given points. We see that the fourth vertex is $(-2, -3)$.

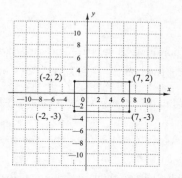

The length of the rectangle is 9 units and the width is 5 units.

$$\begin{aligned} A &= l \cdot w \\ A &= (9 \text{ units})(5 \text{ units}) = 45 \text{ square units} \end{aligned}$$

$$\begin{aligned} P &= 2l + 2w \\ P &= 2 \cdot 9 \text{ units} + 2 \cdot 5 \text{ units} \\ P &= 18 \text{ units} + 10 \text{ units} = 28 \text{ units} \end{aligned}$$

63. From 7:00 A.M. to 1:25 P.M., the elapsed time is 6 hours, 25 minutes.

$$\begin{aligned} 6 \text{ hours} &= 6 \times 1 \text{ hour} \\ &= 6 \times 60 \text{ minutes} \\ &= 360 \text{ minutes} \end{aligned}$$

Then 6 hours, 25 minutes is 360 minutes + 25 minutes = 385 minutes.

The distance climbed is 29,028 ft − 27,600 ft = 1428 ft.

a) Rate = $\dfrac{1428 \text{ ft}}{385 \text{ min}} \approx 3.709$ feet per minute

b) Rate = $\dfrac{385 \text{ min}}{1428 \text{ ft}} \approx 0.2696$ minute per foot

64. Consider a move to a, b, c, or d to be a move up; consider a move to e, f, g, or h to be a move down. Also consider a move to c, d, e, or f to be a move to the right; consider a move to a, b, g, or h to be a move to the left.

A move to a is up 1 and 2 to the left, or $\dfrac{1}{-2}$, or $-\dfrac{1}{2}$.

A move to b is up 2 and 1 to the left, or $\dfrac{2}{-1}$, or -2.

A move to c is up 2 and 1 to the right, or $\dfrac{2}{1}$, or 2.

A move to d is up 1 and 2 to the right, or $\dfrac{1}{2}$.

A move to e is down 1 and 2 to the right, or $\dfrac{-1}{2}$, or $-\dfrac{1}{2}$.

A move to f is down 2 and 1 to the right, or $\dfrac{-2}{1}$, or -2.

A move to g is down 2 and 1 to the left, or $\dfrac{-2}{-1}$, or 2.

A move to h is down 1 and 2 to the left, or $\dfrac{-1}{-2}$, or $\dfrac{1}{2}$.

Thus the slopes that are possible are $-\dfrac{1}{2}$, -2, 2, and $\dfrac{1}{2}$.

Chapter 11 Test

---

1. Since the first coordinate is negative and the second coordinate is positive, the point $\left(-\dfrac{1}{2}, 7\right)$ is in quadrant II.

2. Since both coordinates are negative, the point $(-5, -6)$ is in quadrant III.

3. Point A is 3 units right and 4 units up. The coordinates of A are $(3, 4)$.

4. Point B is 0 units left or right and 4 units down. The coordinates of B are $(0, -4)$.

5.

$$\begin{array}{c|c} y - 2x = 5 & \\ \hline -3 - 2(-4) \ ? \ 5 & \\ -3 + 8 & \\ 5 & \text{TRUE} \end{array}$$

$$\begin{array}{c|c} y - 2x = 5 & \\ \hline 3 - 2(-1) \ ? \ 5 & \\ 3 + 2 & \\ 5 & \text{TRUE} \end{array}$$

In each case we get a true equation, so $(-4, -3)$ and $(-1, 3)$ are solutions of $y - 2x = 5$. We plot these points and draw the line passing through them.

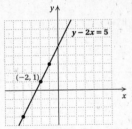

The line appears to pass through $(-2, 1)$. We check to determine if $(-2, 1)$ is a solution of $y - 2x = 5$.

$$\begin{array}{c|c} y - 2x = 5 & \\ \hline 1 - 2(-2) \ ? \ 5 & \\ 1 + 4 & \\ 5 & \text{TRUE} \end{array}$$

Thus, $(-2, 1)$ is another solution. There are other correct answers, including $(-5, -5)$, $(-3, -1)$, and $(0, 5)$.

6. $y = 2x - 1$

The y-intercept is $(0, -1)$. We find two other points.

When $x = -2$, $y = 2(-2) - 1 = -4 - 1 = -5$.

When $x = 3$, $y = 2 \cdot 3 - 1 = 6 - 1 = 5$.

x	y
-2	-5
0	-1
3	5

Plot these points, draw the line they determine, and label the graph $y = 2x - 1$.

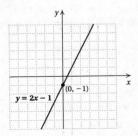

7. $y = -\dfrac{3}{2}x$, or $y = -\dfrac{3}{2}x + 0$

The y-intercept is $(0, 0)$. We find two other points.

When $x = -2$, $y = -\dfrac{3}{2}(-2) = 3$.

When $x = 2$, $y = -\dfrac{3}{2} \cdot 2 = -3$.

x	y
-2	3
0	0
2	-3

Plot these points, draw the line they determine, and label the graph $y = -\dfrac{3}{2}x$.

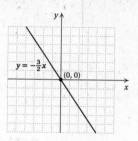

8. $2x + 8 = 0$

$\qquad 2x = -8$

$\qquad x = -4 \quad$ Solving for x

Any ordered pair $(-4, y)$ is a solution. A few solutions are listed below. Plot these points and draw the graph.

x	y
-4	-3
-4	0
-4	2

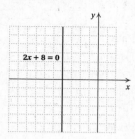

9. $y = 5$

Any ordered pair $(x, 5)$ is a solution. A few solutions are listed below. Plot these points and draw the graph.

x	y
-2	5
0	5
3	5

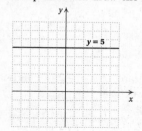

10. $2x - 4y = -8$

To find the x-intercept, let $y = 0$. Then solve for x.

$\qquad 2x - 4y = -8$

$\qquad 2x - 4 \cdot 0 = -8$

$\qquad\qquad 2x = -8$

$\qquad\qquad x = -4$

Thus, $(-4, 0)$ is the x-intercept.

To find the y-intercept, let $x = 0$. Then solve for y.

$\qquad 2x - 4y = -8$

$\qquad 2 \cdot 0 - 4y = -8$

$\qquad\qquad -4y = -8$

$\qquad\qquad y = 2$

Thus, $(0, 2)$ is the y-intercept.

Plot these points and draw the graph.

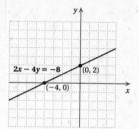

We use a third point as a check.

We let $x = 4$. Then

$\qquad 2x - 4y = -8$

$\qquad 2 \cdot 4 - 4y = -8$

$\qquad\qquad 8 - 4y = -8$

$\qquad\qquad -4y = -16$

$\qquad\qquad y = 4$

The point $(4, 4)$ is on the graph, so the graph is probably correct.

11. $2x - y = 3$

To find the x-intercept, let $y = 0$. Then solve for x.

$\qquad 2x - y = 3$

$\qquad 2x - 0 = 3$

$\qquad\qquad 2x = 3$

$\qquad\qquad x = \dfrac{3}{2}$

Thus, $\left(\dfrac{3}{2}, 0\right)$ is the x-intercept.

To find the y-intercept, let $x = 0$. Then solve for y.

$\qquad 2x - y = 3$

$\qquad 2 \cdot 0 - y = 3$

$\qquad\qquad -y = 3$

$\qquad\qquad y = -3$

Thus, $(0, -3)$ is the y-intercept.

Plot these points and draw the graph.

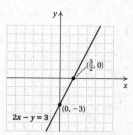

We use a third point as a check.

We let $x = 3$. Then

$\qquad 2x - y = 3$

$\qquad 2 \cdot 3 - y = 3$

$\qquad\qquad 6 - y = 3$

$\qquad\qquad -y = -3$

$\qquad\qquad y = 3$

The point $(3,3)$ is on the line, so the graph is probably correct.

12. a) In 1985, $n = 0$, and $T = \dfrac{3}{5} \cdot 0 + 5 = 5$ so the cost of tuition was $5 thousand, or $5000.

In 1996, $n = 1996 - 1985 = 11$, and $T = \dfrac{3}{5} \cdot 11 + 5 = \dfrac{33}{5} + 5 = 6.6 + 5 = 11.6$ so the cost of tuition was $11.6 thousand, or $11,600.

In 2000, $n = 2000 - 1985 = 15$, and $T = \dfrac{3}{5} \cdot 15 + 5 = 9 + 5 = 14$, so the cost of tuition was $14 thousand, or $14,000.

In 2004, $n = 2004 - 1985 = 19$, and $T = \dfrac{3}{5} \cdot 19 + 5 = \dfrac{57}{5} + 5 = 11.4 + 5 = 16.4$, so the cost of tuition was $16.4 thousand, or $16,400.

b) We plot the points found in part (a), $(0, 5)$, $(11, 11.6)$, $(15, 14)$, and $(19, 16.4)$. Then we draw the line passing through these points.

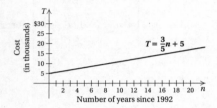

In 2005, $n = 2005 - 1985 = 20$. From the graph it appears that an n-value of 20 corresponds to the T-value of 17, so we estimate that the cost of tuition was $17 thousand, or $17,000, in 2005.

c) We substitute 23 for T and solve for n.
$$T = \frac{3}{5}n + 5$$
$$23 = \frac{3}{5}n + 5$$
$$18 = \frac{3}{5}n$$
$$\frac{5}{3} \cdot 18 = n$$
$$30 = n$$

A tuition cost of $23,000 will occur 30 years after 1985, or in 2015.

13. The time that elapses from 2:38 to 2:40 is 2 minutes. In that time the elevator travels $34 - 5$, or 29 floors.

a) Rate $= \dfrac{29 \text{ floors}}{2 \text{ minutes}} = 14.5$ floors per minute

b) 2 min $= 2 \times 1$ min $= 2 \times 60$ sec $= 120$ sec

Rate $= \dfrac{120 \text{ seconds}}{29 \text{ floors}} = \dfrac{120}{29}$ seconds per floor $= 4\dfrac{4}{29}$ seconds per floor

14. The time that elapses from 1:00 P.M. to 5:00 P.M. is 4 hours.

Rate $= \dfrac{450 \text{ miles} - 100 \text{ miles}}{4 \text{ hours}} = \dfrac{350 \text{ miles}}{4 \text{ hours}} =$ 87.5 miles per hour

15. We can choose any two points. We consider (x_1, y_1) to be $(2, 4)$ and (x_2, y_2) to be $(5, -2)$.
$$m = \frac{y_2 - y_1}{x_2 - x_1} = \frac{-2 - 4}{5 - 2} = \frac{-6}{3} = -2$$

16. We plot $(-3, 1)$ and $(5, 4)$ and draw the line containing these points.

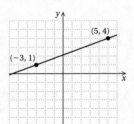

To find the slope, consider (x_1, y_1) to be $(5, 4)$ and (x_2, y_2) to be $(-3, 1)$.
$$m = \frac{y_2 - y_1}{x_2 - x_1} = \frac{1 - 4}{-3 - 5} = \frac{-3}{-8} = \frac{3}{8}$$

17. Slope $= \dfrac{-54}{1080} = -\dfrac{1}{20}$

18. a) We solve for y.
$$2x - 5y = 10$$
$$-5y = -2x + 10$$
$$y = -\frac{1}{5}(-2x + 10)$$
$$y = \frac{2}{5}x - 2$$
The slope is $\dfrac{2}{5}$.

b) The graph of $x = -2$ is a vertical line, so the slope is not defined.

19. Slope $-\dfrac{3}{2}$, y-intercept $(0, 1)$

We plot $(0, 1)$. We can think of the slope as $\dfrac{-3}{2}$, so from $(0, 1)$ we move down 3 units and right 2 units. This locates the point $(2, -2)$. We plot $(2, -2)$ and draw a line passing through $(0, 1)$ and $(2, -2)$.

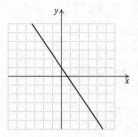

20. $y = 2x - 3$

The slope is 2 and the y-intercept is $(0, -3)$. We plot $(0, -3)$. We can start at the y-intercept and, thinking of the slope as $\dfrac{2}{1}$, find another point by moving up 2 units and right 1 unit to the point $(1, -1)$. Thinking of the slope as $\dfrac{-2}{-1}$ we can start at $(0, -3)$ and move down 2 units and left 1 unit to the point $(-1, -5)$.

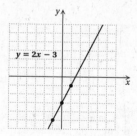

21. $y = 2x - \dfrac{1}{4}$

The equation is in the form $y = mx + b$. The slope is 2 and the y-intercept is $\left(0, -\dfrac{1}{4}\right)$.

22. We solve for y.

$$-4x + 3y = -6$$
$$3y = 4x - 6$$
$$y = \dfrac{1}{3}(4x - 6)$$
$$y = \dfrac{4}{3}x - 2$$

The slope is $\dfrac{4}{3}$ and the y-intercept is $(0, -2)$.

23. We substitute 1.8 for m and -7 for b in the equation $y = mx + b$.

$$y = 1.8x - 7$$

24. We substitute $-\dfrac{3}{8}$ for m and $-\dfrac{1}{8}$ for b in the equation $y = mx + b$.

$$y = -\dfrac{3}{8}x - \dfrac{1}{8}$$

25. We know the slope is 1, so the equation is $y = 1 \cdot x + b$, or $y = x + b$. Using the point $(3, 5)$, we substitute 3 for x and 5 for y in $y = x + b$. Then we solve for b.

$$y = x + b$$
$$5 = 3 + b$$
$$2 = b$$

Thus, the equation is $y = x + 2$.

26. We know the slope is -3, so the equation is $y = -3x + b$. Using the point $(-2, 0)$, we substitute -2 for x and 0 for y in $y = -3x + b$. Then we solve for b.

$$y = -3x + b$$
$$0 = -3(-2) + b$$
$$0 = 6 + b$$
$$-6 = b$$

Thus, the equation is $y = -3x - 6$.

27. First we find the slope.

$$m = \dfrac{-2 - 1}{2 - 1} = \dfrac{-3}{1} = -3$$

Thus, $y = -3x + b$. We can use either point to find b. We choose $(1, 1)$. Substitute 1 for both x and y in $y = -3x + b$.

$$y = -3x + b$$
$$1 = -3 \cdot 1 + b$$
$$1 = -3 + b$$
$$4 = b$$

Thus, the equation is $y = -3x + 4$.

28. First we find the slope.

$$m = \dfrac{-1 - (-3)}{4 - (-4)} = \dfrac{2}{8} = \dfrac{1}{4}$$

Thus, $y = \dfrac{1}{4}x + b$. We can use either point to find b. We choose $(4, -1)$. Substitute 4 for x and -1 for y in $y = \dfrac{1}{4}x + b$.

$$y = \dfrac{1}{4}x + b$$
$$-1 = \dfrac{1}{4} \cdot 4 + b$$
$$-1 = 1 + b$$
$$-2 = b$$

Thus, the equation is $y = \dfrac{1}{4}x - 2$.

29. a) First we find the slope.

$$m = \dfrac{956 - 203}{12 - 2} = \dfrac{753}{10} = 75.3$$

The slope is 75.3, so the equation is $y = 75.3x + b$.

Using the point $(2, 203)$, we substitute 2 for x and 203 for y in $y = 75.3x + b$.

$$y = 75.3x + b$$
$$203 = 75.3(2) + b$$
$$203 = 150.6 + b$$
$$52.4 = b$$

Thus, the equation is $y = 75.3x + 52.4$.

b) The rate of change is the slope, 75.3 lung transplants per year.

c) In 2005, $x = 2005 - 1988 = 17$.

$$y = 75.3x + 52.4$$
$$y = 75.3(17) + 52.4$$
$$y = 1280.1 + 52.4$$
$$y = 1332.5 \approx 1333$$

There were about 1333 lung transplants in 2005.

30. First we solve each equation for y:

  1. $2x + y = 8$ 2. $2x + y = 4$

      $y = -2x + 8$ $y = -2x + 4$

The slope of each line is -2. The y-intercepts, $(0, 8)$ and $(0, 4)$ are different. The lines are parallel.

31. We solve the first equation for y.

$$2x + 5y = 2$$
$$5y = -2x + 2$$
$$y = \frac{1}{5}(-2x + 2)$$
$$y = -\frac{2}{5}x + \frac{2}{5}$$

The second equation is in the form $y = mx + b$:

$$y = 2x + 4$$

The slopes, $-\frac{2}{5}$ and 2, are not the same, so the lines are not parallel. The product of the slopes is $-\frac{2}{5} \cdot 2 = -\frac{4}{5} \neq -1$, so the lines are not perpendicular. Thus, the lines are neither parallel nor perpendicular.

32. First we solve each equation for y:

1. $x + 2y = 8$ 2. $-2x + y = 8$
$$2y = -x + 8 \qquad\qquad y = 2x + 8$$
$$y = \frac{1}{2}(-x + 8)$$
$$y = -\frac{1}{2}x + 4$$

The slopes, $-\frac{1}{2}$ and 2, are not the same, so the lines are not parallel. The product of the slope is $-\frac{1}{2} \cdot 2 = -1$, so the lines are perpendicular.

33.
$$\frac{3y - 2x < -2}{3 \cdot 0 - 2 \cdot 0 \ ? \ -2}$$
$$0 \ | \qquad \text{FALSE}$$

Since $0 < -2$ is false, $(0, 0)$ is not a solution.

34.
$$\frac{3y - 2x < -2}{3(-10) - 2(-4) \ ? \ -2}$$
$$-30 + 8 \ |$$
$$-22 \ | \qquad \text{TRUE}$$

Since $-22 < -2$ is true, $(-4, -10)$ is a solution.

35. Graph $y > x - 1$.

First graph the line $y = x - 1$. Two points on the line are $(0, -1)$ and $(4, 3)$. We draw a dashed line since the inequality symbol is $>$. Then we test a point that is not on the line. We try $(0, 0)$.

$$\frac{y > x - 1}{0 \ ? \ 0 - 1}$$
$$| \ -1 \qquad \text{TRUE}$$

Since $(0, 0)$ is a solution of the inequality, we shade the region containing $(0, 0)$.

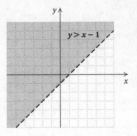

36. Graph $2x - y \leq 4$.

First graph the line $2x - y = 4$. The intercepts are $(0, -4)$ and $(2, 0)$. We draw a solid line since the inequality symbol is $\leq$. Then we test a point that is not on the line. We try $(0, 0)$.

$$\frac{2x - y \leq 4}{2 \cdot 0 - 0 \ ? \ 4}$$
$$0 \ | \qquad \text{TRUE}$$

Since $(0, 0)$ is a solution of the inequality, we shade the region containing $(0, 0)$.

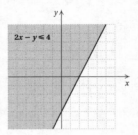

37. We plot the given points. We see that the other vertices of the square are $(-3, 4)$ and $(2, -1)$.

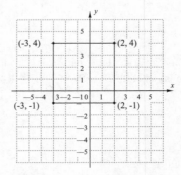

The length of the sides of the square is 5 units.

$$A = s^2 = (5 \text{ units})^2 = 25 \text{ square units}$$
$$P = 4s = 4 \cdot 5 \text{ units} = 20 \text{ units}$$

38. First solve each equation for y.

$$3x + 7y = 14$$
$$7y = -3x + 14$$
$$y = \frac{1}{7}(-3x + 14)$$
$$y = -\frac{3}{7}x + 2$$

$$ky - 7x = -3$$
$$ky = 7x - 3$$
$$y = \frac{1}{k}(7x - 3)$$
$$y = \frac{7}{k}x - \frac{3}{k}$$

If the lines are perpendicular, the product of their slopes is -1.

$$-\frac{3}{7} \cdot \frac{7}{k} = -1$$
$$-\frac{3 \cdot 7}{7 \cdot k} = -1$$
$$-\frac{3}{k} = -1$$
$$k\left(-\frac{3}{k}\right) = k(-1)$$
$$-3 = -k$$
$$3 = k$$

Chapter 12

Polynomials: Operations

Exercise Set 12.1

1. 3^4 means $3 \cdot 3 \cdot 3 \cdot 3$.

3. $(-1.1)^5$ means $(-1.1)(-1.1)(-1.1)(-1.1)(-1.1)$.

5. $\left(\dfrac{2}{3}\right)^4$ means $\left(\dfrac{2}{3}\right)\left(\dfrac{2}{3}\right)\left(\dfrac{2}{3}\right)\left(\dfrac{2}{3}\right)$.

7. $(7p)^2$ means $(7p)(7p)$.

9. $8k^3$ means $8 \cdot k \cdot k \cdot k$.

11. $-6y^4$ means $-6 \cdot y \cdot y \cdot y \cdot y$.

13. $a^0 = 1,\ a \neq 0$

15. $b^1 = b$

17. $\left(\dfrac{2}{3}\right)^0 = 1$

19. $(-7.03)^1 = -7.03$

21. $8.38^0 = 1$

23. $(ab)^1 = ab$

25. $ab^0 = a \cdot b^0 = a \cdot 1 = a$

27. $m^3 = 3^3 = 3 \cdot 3 \cdot 3 = 27$

29. $p^1 = 19^1 = 19$

31. $-x^4 = -(-3)^4 = -(-3)(-3)(-3)(-3) = -81$

33. $x^4 = 4^4 = 4 \cdot 4 \cdot 4 \cdot 4 = 256$

35. $y^2 - 7 = 10^2 - 7$
$\qquad = 100 - 7 \quad$ Evaluating the power
$\qquad = 93 \qquad\quad$ Subtracting

37. $161 - b^2 = 161 - 5^2$
$\qquad\quad = 161 - 25 \quad$ Evaluating the power
$\qquad\quad = 136 \qquad\quad$ Subtracting

39. $x^1 + 3 = 7^1 + 3$
$\qquad\quad = 7 + 3 \quad (7^1 = 7)$
$\qquad\quad = 10$

$x^0 + 3 = 7^0 + 3$
$\qquad\quad = 1 + 3 \quad (7^0 = 1)$
$\qquad\quad = 4$

41. $A = \pi r^2 \approx 3.14 \times (34 \text{ ft})^2$
$\qquad \approx 3.14 \times 1156 \text{ ft}^2 \quad$ Evaluating the power
$\qquad \approx 3629.84 \text{ ft}^2$

43. $3^{-2} = \dfrac{1}{3^2} = \dfrac{1}{9}$

45. $10^{-3} = \dfrac{1}{10^3} = \dfrac{1}{1000}$

47. $7^{-3} = \dfrac{1}{7^3} = \dfrac{1}{343}$

49. $a^{-3} = \dfrac{1}{a^3}$

51. $\dfrac{1}{8^{-2}} = 8^2 = 64$

53. $\dfrac{1}{y^{-4}} = y^4$

55. $\dfrac{1}{z^{-n}} = z^n$

57. $\dfrac{1}{4^3} = 4^{-3}$

59. $\dfrac{1}{x^3} = x^{-3}$

61. $\dfrac{1}{a^5} = a^{-5}$

63. $2^4 \cdot 2^3 = 2^{4+3} = 2^7$

65. $8^5 \cdot 8^9 = 8^{5+9} = 8^{14}$

67. $x^4 \cdot x^3 = x^{4+3} = x^7$

69. $9^{17} \cdot 9^{21} = 9^{17+21} = 9^{38}$

71. $(3y)^4 (3y)^8 = (3y)^{4+8} = (3y)^{12}$

73. $(7y)^1 (7y)^{16} = (7y)^{1+16} = (7y)^{17}$

75. $3^{-5} \cdot 3^8 = 3^{-5+8} = 3^3$

77. $x^{-2} \cdot x = x^{-2+1} = x^{-1} = \dfrac{1}{x}$

79. $x^{14} \cdot x^3 = x^{14+3} = x^{17}$

81. $x^{-7} \cdot x^{-6} = x^{-7+(-6)} = x^{-13} = \dfrac{1}{x^{13}}$

83. $a^{11} \cdot a^{-3} \cdot a^{-18} = a^{11+(-3)+(-18)} = a^{-10} = \dfrac{1}{a^{10}}$

85. $t^8 \cdot t^{-8} = t^{8+(-8)} = t^0 = 1$

87. $\dfrac{7^5}{7^2} = 7^{5-2} = 7^3$

89. $\dfrac{8^{12}}{8^6} = 8^{12-6} = 8^6$

91. $\dfrac{y^9}{y^5} = y^{9-5} = y^4$

93. $\dfrac{16^2}{16^8} = 16^{2-8} = 16^{-6} = \dfrac{1}{16^6}$

95. $\dfrac{m^6}{m^{12}} = m^{6-12} = m^{-6} = \dfrac{1}{m^6}$

97. $\dfrac{(8x)^6}{(8x)^{10}} = (8x)^{6-10} = (8x)^{-4} = \dfrac{1}{(8x)^4}$

99. $\dfrac{(2y)^9}{(2y)^9} = (2y)^{9-9} = (2y)^0 = 1$

101. $\dfrac{x}{x^{-1}} = x^{1-(-1)} = x^2$

103. $\dfrac{x^7}{x^{-2}} = x^{7-(-2)} = x^9$

105. $\dfrac{z^{-6}}{z^{-2}} = z^{-6-(-2)} = z^{-4} = \dfrac{1}{z^4}$

107. $\dfrac{x^{-5}}{x^{-8}} = x^{-5-(-8)} = x^3$

109. $\dfrac{m^{-9}}{m^{-9}} = m^{-9-(-9)} = m^0 = 1$

111. $5^2 = 5 \cdot 5 = 25$

$5^{-2} = \dfrac{1}{5^2} = \dfrac{1}{25}$

$\left(\dfrac{1}{5}\right)^2 = \dfrac{1}{5} \cdot \dfrac{1}{5} = \dfrac{1}{25}$

$\left(\dfrac{1}{5}\right)^{-2} = \dfrac{1}{\left(\dfrac{1}{5}\right)^2} = \dfrac{1}{\dfrac{1}{25}} = 1 \cdot \dfrac{25}{1} = 25$

$-5^2 = -(5)(5) = -25$

$(-5)^2 = (-5)(-5) = 25$

$-\left(-\dfrac{1}{5}\right)^2 = -\left(-\dfrac{1}{5}\right)\left(-\dfrac{1}{5}\right) = -\dfrac{1}{25}$

$\left(-\dfrac{1}{5}\right)^{-2} = \dfrac{1}{\left(-\dfrac{1}{5}\right)^2} = \dfrac{1}{\dfrac{1}{25}} = 1 \cdot \dfrac{25}{1} = 25$

113. Discussion and Writing Exercise

115. *Familiarize*. Let $x =$ the length of the shorter piece. Then $2x =$ the length of the longer piece.

Translate.

$$\underbrace{\text{Length of shorter piece}}_{x} \ \underbrace{\text{plus}}_{+} \ \underbrace{\text{length of longer piece}}_{2x} \ \underbrace{\text{is}}_{=} \ \underbrace{\text{12 in.}}_{12}$$

Solve.

$x + 2x = 12$

$3x = 12$

$\dfrac{3x}{3} = \dfrac{12}{3}$

$x = 4$

If $x = 4$, $2x = 2 \cdot 4 = 8$.

Check. The longer piece, 8 in., is twice as long as the shorter piece, 4 in. Also, 4 in. + 8 in. = 12 in., the total length of the sandwich. The answer checks.

State. The lengths of the pieces are 4 in. and 8 in.

117. *Familiarize*. Let $w =$ the width. Then $w + 15 =$ the length. We draw a picture.

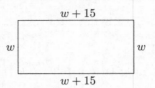

We will use the fact that the perimeter is 640 ft to find w (the width). Then we can find $w + 15$ (the length) and multiply the length and the width to find the area.

Translate.

Width+Width+ Length + Length =Perimeter

$w \ \ + \ \ w \ \ +(w+15)+(w+15)= \ \ 640$

Solve.

$w + w + (w + 15) + (w + 15) = 640$

$4w + 30 = 640$

$4w = 610$

$w = 152.5$

If the width is 152.5, then the length is $152.5+15$, or 167.5. The area is $(167.5)(152.5)$, or 25,543.75 ft^2.

Check. The length, 167.5 ft, is 15 ft greater than the width, 152.5 ft. The perimeter is $152.5 + 152.5 + 167.5 + 167.5$, or 640 ft. We should also recheck the computation we used to find the area. The answer checks.

State. The area is 25,543.75 ft^2.

119.

$-6(2 - x) + 10(5x - 7) = 10$

$-12 + 6x + 50x - 70 = 10$

$56x - 82 = 10 \quad \text{Collecting like terms}$

$56x - 82 + 82 = 10 + 82 \quad \text{Adding 82}$

$56x = 92$

$\dfrac{56x}{56} = \dfrac{92}{56} \quad \text{Dividing by 56}$

$x = \dfrac{23}{14}$

The solution is $\dfrac{23}{14}$.

121. $4x - 12 + 24y = 4 \cdot x - 4 \cdot 3 + 4 \cdot 6y = 4(x - 3 + 6y)$

123. Let $y_1 = (x+1)^2$ and $y_2 = x^2+1$. A graph of the equations or a table of values shows that $(x + 1)^2 = x^2 + 1$ is not correct.

125. Let $y_1 = (5x)^0$ and $y_2 = 5x^0$. A graph of the equations or a table of values shows that $(5x)^0 = 5x^0$ is not correct.

127. $(y^{2x})(y^{3x}) = y^{2x+3x} = y^{5x}$

129. $\dfrac{a^{6t}(a^{7t})}{a^{9t}} = \dfrac{a^{6t+7t}}{a^{9t}} = \dfrac{a^{13t}}{a^{9t}} = a^{13t-9t} = a^{4t}$

131. $\dfrac{(0.8)^5}{(0.8)^3(0.8)^2} = \dfrac{(0.8)^5}{(0.8)^{3+2}} = \dfrac{(0.8)^5}{(0.8)^5} = 1$

133. Since the bases are the same, the expression with the larger exponent is larger. Thus, $3^5 > 3^4$.

135. Since the exponents are the same, the expression with the larger base is larger. Thus, $4^3 < 5^3$.

137. $\dfrac{1}{-z^4} = \dfrac{1}{-(-10)^4} = \dfrac{1}{-(-10)(-10)(-10)(-10)} = \dfrac{1}{-10,000} = -\dfrac{1}{10,000}$

Exercise Set 12.2

1. $(2^3)^2 = 2^{3\cdot2} = 2^6$

3. $(5^2)^{-3} = 5^{2(-3)} = 5^{-6} = \dfrac{1}{5^6}$

5. $(x^{-3})^{-4} = x^{(-3)(-4)} = x^{12}$

7. $(a^{-2})^9 = a^{-2\cdot9} = a^{-18} = \dfrac{1}{a^{18}}$

9. $(t^{-3})^{-6} = t^{(-3)(-6)} = t^{18}$

11. $(t^4)^{-3} = t^{4(-3)} = t^{-12} = \dfrac{1}{t^{12}}$

13. $(x^{-2})^{-4} = x^{-2)(-4)} = x^8$

15. $(ab)^3 = a^3b^3$ Raising each factor to the third power

17. $(ab)^{-3} = a^{-3}b^{-3} = \dfrac{1}{a^3b^3}$

19. $(mn^2)^{-3} = m^{-3}(n^2)^{-3} = m^{-3}n^{2(-3)} = m^{-3}n^{-6} = \dfrac{1}{m^3n^6}$

21. $(4x^3)^2 = 4^2(x^3)^2$ Raising each factor to the second power
$= 16x^6$

23. $(3x^{-4})^2 = 3^2(x^{-4})^2 = 3^2x^{-4\cdot2} = 9x^{-8} = \dfrac{9}{x^8}$

25. $(x^4y^5)^{-3} = (x^4)^{-3}(y^5)^{-3} = x^{4(-3)}y^{5(-3)} = x^{-12}y^{-15} = \dfrac{1}{x^{12}y^{15}}$

27. $(x^{-6}y^{-2})^{-4} = (x^{-6})^{-4}(y^{-2})^{-4} = x^{(-6)(-4)}y^{(-2)(-4)} = x^{24}y^8$

29. $(a^{-2}b^7)^{-5} = (a^{-2})^{-5}(b^7)^{-5} = a^{10}b^{-35} = \dfrac{a^{10}}{b^{35}}$

31. $(5r^{-4}t^3)^2 = 5^2(r^{-4})^2(t^3)^2 = 25r^{-4\cdot2}t^{3\cdot2} = 25r^{-8}t^6 = \dfrac{25t^6}{r^8}$

33. $(a^{-5}b^7c^{-2})^3 = (a^{-5})^3(b^7)^3(c^{-2})^3 = a^{-5\cdot3}b^{7\cdot3}c^{-2\cdot3} = a^{-15}b^{21}c^{-6} = \dfrac{b^{21}}{a^{15}c^6}$

35. $(3x^3y^{-8}z^{-3})^2 = 3^2(x^3)^2(y^{-8})^2(z^{-3})^2 = 9x^6y^{-16}z^{-6} = \dfrac{9x^6}{y^{16}z^6}$

37. $(-4x^3y^{-2})^2 = (-4)^2(x^3)^2(y^{-2})^2 = 16x^6y^{-4} = \dfrac{16x^6}{y^4}$

39. $(-a^{-3}b^{-2})^{-4} = (-1\cdot a^{-3}b^{-2})^{-4} = (-1)^{-4}(a^{-3})^{-4}(b^{-2})^{-4} = \dfrac{1}{(-1)^4}\cdot a^{12}b^8 = \dfrac{a^{12}b^8}{1} = a^{12}b^8$

41. $\left(\dfrac{y^3}{2}\right)^2 = \dfrac{(y^3)^2}{2^2} = \dfrac{y^6}{4}$

43. $\left(\dfrac{a^2}{b^3}\right)^4 = \dfrac{(a^2)^4}{(b^3)^4} = \dfrac{a^8}{b^{12}}$

45. $\left(\dfrac{y^2}{2}\right)^{-3} = \dfrac{(y^2)^{-3}}{2^{-3}} = \dfrac{y^{-6}}{2^{-3}} = \dfrac{\frac{1}{y^6}}{\frac{1}{2^3}} = \dfrac{1}{y^6}\cdot\dfrac{2^3}{1} = \dfrac{8}{y^6}$

47. $\left(\dfrac{7}{x^{-3}}\right)^2 = \dfrac{7^2}{(x^{-3})^2} = \dfrac{49}{x^{-6}} = 49x^6$

49. $\left(\dfrac{x^2y}{z}\right)^3 = \dfrac{(x^2)^3y^3}{z^3} = \dfrac{x^6y^3}{z^3}$

51. $\left(\dfrac{a^2b}{cd^3}\right)^{-2} = \dfrac{(a^2)^{-2}b^{-2}}{c^{-2}(d^3)^{-2}} = \dfrac{a^{-4}b^{-2}}{c^{-2}d^{-6}} = \dfrac{\frac{1}{a^4}\cdot\frac{1}{b^2}}{\frac{1}{c^2}\cdot\frac{1}{d^6}} = \dfrac{\frac{1}{a^4b^2}}{\frac{1}{c^2d^6}} = \dfrac{1}{a^4b^2}\cdot\dfrac{c^2d^6}{1} = \dfrac{c^2d^6}{a^4b^2}$

53. $2.8,000,000,000.$
$\underset{\text{10 places}}{\underline{\qquad\qquad}}$
Large number, so the exponent is positive.
$28,000,000,000 = 2.8\times10^{10}$

55. $9.07,000,000,000,000,000.$
$\underset{\text{17 places}}{\underline{\qquad\qquad}}$
Large number, so the exponent is positive.
$907,000,000,000,000,000 = 9.07\times10^{17}$

57. $0.000003.04$
$\underset{\text{6 places}}{\underline{\qquad\qquad}}$
Small number, so the exponent is negative.
$0.00000304 = 3.04\times10^{-6}$

59. $0.00000001.8$
$\underset{\text{8 places}}{\underline{\qquad\qquad}}$
Small number, so the exponent is negative.
$0.000000018 = 1.8\times10^{-8}$

61. $1.00,000,000,000.$
$\underset{\text{11 places}}{\underline{\qquad\qquad}}$
Large number, so the exponent is positive.
$100,000,000,000 = 1.0\times10^{11} = 10^{11}$

63. 296 million $= 296{,}000{,}000$

2.96,000,000.

↑_____| 8 places

Large number, so the exponent is positive.

296 million $= 2.96 \times 10^8$

65. $\dfrac{1}{10{,}000{,}000} = 0.0000001$

0.0000001.

|_____↑ 7 places

Small number, so the exponent is negative.

$\dfrac{1}{10{,}000{,}000} = 1 \times 10^{-7}$, or 10^{-7}

67. 8.74×10^7

Positive exponent, so the answer is a large number.

8.7400000.

|_____↑ 7 places

$8.74 \times 10^7 = 87{,}400{,}000$

69. 5.704×10^{-8}

Negative exponent, so the answer is a small number.

0.00000005.704

↑_____| 8 places

$5.704 \times 10^{-8} = 0.00000005704$

71. $10^7 = 1 \times 10^7$

Positive exponent, so the answer is a large number.

1.0000000.

|_____↑ 7 places

$10^7 = 10{,}000{,}000$

73. $10^{-5} = 1 \times 10^{-5}$

Negative exponent, so the answer is a small number.

0.00001.

↑____| 5 places

$10^{-5} = 0.00001$

75. $(3 \times 10^4)(2 \times 10^5) = (3 \cdot 2) \times (10^4 \cdot 10^5)$
$$= 6 \times 10^9$$

77. $(5.2 \times 10^5)(6.5 \times 10^{-2}) = (5.2 \cdot 6.5) \times (10^5 \cdot 10^{-2})$
$$= 33.8 \times 10^3$$

The answer at this stage is 33.8×10^3 but this is not scientific notation since 33.8 is not a number between 1 and 10. We convert 33.8 to scientific notation and simplify.

$33.8 \times 10^3 = (3.38 \times 10^1) \times 10^3 = 3.38 \times (10^1 \times 10^3) = 3.38 \times 10^4$

The answer is 3.38×10^4.

79. $(9.9 \times 10^{-6})(8.23 \times 10^{-8}) = (9.9 \cdot 8.23) \times (10^{-6} \cdot 10^{-8})$
$$= 81.477 \times 10^{-14}$$

The answer at this stage is 81.477×10^{-14}. We convert 81.477 to scientific notation and simplify.

$81.477 \times 10^{-14} = (8.1477 \times 10^1) \times 10^{-14} =$
$8.1477 \times (10^1 \times 10^{-14}) = 8.1477 \times 10^{-13}$.

The answer is 8.1477×10^{-13}.

81. $\dfrac{8.5 \times 10^8}{3.4 \times 10^{-5}} = \dfrac{8.5}{3.4} \times \dfrac{10^8}{10^{-5}}$
$$= 2.5 \times 10^{8-(-5)}$$
$$= 2.5 \times 10^{13}$$

83. $(3.0 \times 10^6) \div (6.0 \times 10^9) = \dfrac{3.0 \times 10^6}{6.0 \times 10^9}$
$$= \dfrac{3.0}{6.0} \times \dfrac{10^6}{10^9}$$
$$= 0.5 \times 10^{6-9}$$
$$= 0.5 \times 10^{-3}$$

The answer at this stage is 0.5×10^{-3}. We convert 0.5 to scientific notation and simplify.

$0.5 \times 10^{-3} = (5.0 \times 10^{-1}) \times 10^{-3} =$
$5.0 \times (10^{-1} \times 10^{-3}) = 5.0 \times 10^{-4}$

85. $\dfrac{7.5 \times 10^{-9}}{2.5 \times 10^{12}} = \dfrac{7.5}{2.5} \times \dfrac{10^{-9}}{10^{12}}$
$$= 3.0 \times 10^{-9-12}$$
$$= 3.0 \times 10^{-21}$$

87. There are 60 seconds in one minute and 60 minutes in one hour, so there are 60(60), or 3600 seconds in one hour. There are 24 hours in one day and 365 days in one year, so there are 3600(24)(365), or 31,536,000 seconds in one year.

$4{,}200{,}000 \times 31{,}536{,}000$
$= (4.2 \times 10^6) \times (3.1536 \times 10^7)$
$= (4.2 \times 3.1536) \times (10^6 \times 10^7)$
$\approx 13.25 \times 10^{13}$
$\approx (1.325 \times 10) \times 10^{13}$
$\approx 1.325 \times (10 \times 10^{13})$
$\approx 1.325 \times 10^{14}$

About 1.325×10^{14} cubic feet of water is discharged from the Amazon River in 1 yr.

89. $\dfrac{1.908 \times 10^{24}}{6 \times 10^{21}} = \dfrac{1.908}{6} \times \dfrac{10^{24}}{10^{21}}$
$$= 0.318 \times 10^3$$
$$= (3.18 \times 10^{-1}) \times 10^3$$
$$= 3.18 \times (10^{-1} \times 10^3)$$
$$= 3.18 \times 10^2$$

The mass of Jupiter is 3.18×10^2 times the mass of Earth.

91. 10 billion trillion $= 1 \times 10 \times 10^9 \times 10^{12}$
$$= 1 \times 10^{22}$$

There are 1×10^{22} stars in the known universe.

93. We divide the mass of the sun by the mass of earth.

$\dfrac{1.998 \times 10^{27}}{6 \times 10^{21}} = 0.333 \times 10^6$
$$= (3.33 \times 10^{-1}) \times 10^6$$
$$= 3.33 \times 10^5$$

The mass of the sun is 3.33×10^5 times the mass of Earth.

95. First we divide the distance from the earth to the moon by 3 days to find the number of miles per day the space vehicle travels. Note that $240,000 = 2.4 \times 10^5$.

$$\frac{2.4 \times 10^5}{3} = 0.8 \times 10^5 = 8 \times 10^4$$

The space vehicle travels 8×10^4 miles per day. Now divide the distance from the earth to Mars by 8×10^4 to find how long it will take the space vehicle to reach Mars. Note that $35,000,000 = 3.5 \times 10^7$.

$$\frac{3.5 \times 10^7}{8 \times 10^4} = 0.4375 \times 10^3 = 4.375 \times 10^2$$

It takes 4.375×10^2 days for the space vehicle to travel from the earth to Mars.

97. Discussion and Writing Exercise

99. $9x - 36 = 9 \cdot x - 9 \cdot 4 = 9(x - 4)$

101. $3s + 3t + 24 = 3 \cdot s + 3 \cdot t + 3 \cdot 8 = 3(s + t + 8)$

103.
$$\begin{aligned}
2x - 4 - 5x + 8 &= x - 3 \\
-3x + 4 &= x - 3 && \text{Collecting like terms} \\
-3x + 4 - 4 &= x - 3 - 4 && \text{Subtracting 4} \\
-3x &= x - 7 \\
-3x - x &= x - 7 - x && \text{Subtracting } x \\
-4x &= -7 \\
\frac{-4x}{-4} &= \frac{-7}{-4} && \text{Dividing by } -4 \\
x &= \frac{7}{4}
\end{aligned}$$

The solution is $\dfrac{7}{4}$.

105.
$$\begin{aligned}
8(2x + 3) - 2(x - 5) &= 10 \\
16x + 24 - 2x + 10 &= 10 && \text{Removing parentheses} \\
14x + 34 &= 10 && \text{Collecting like terms} \\
14x + 34 - 34 &= 10 - 34 && \text{Subtracting 34} \\
14x &= -24 \\
\frac{14x}{14} &= \frac{-24}{14} && \text{Dividing by 14} \\
x &= -\frac{12}{7} && \text{Simplifying}
\end{aligned}$$

The solution is $-\dfrac{12}{7}$.

107. $y = x - 5$

The equation is equivalent to $y = x + (-5)$. The y-intercept is $(0, -5)$. We find two other points.

When $x = 2$, $y = 2 - 5 = -3$.

When $x = 4$, $y = 4 - 5 = -1$.

x	y
0	-5
2	-3
4	-1

Plot these points, draw the line they determine, and label the graph $y = x - 5$.

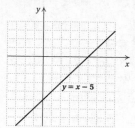

109.
$$\begin{aligned}
\frac{(5.2 \times 10^6)(6.1 \times 10^{-11})}{1.28 \times 10^{-3}} &= \frac{(5.2 \cdot 6.1)}{1.28} \times \frac{(10^6 \cdot 10^{-11})}{10^{-3}} \\
&= 24.78125 \times 10^{-2} \\
&= (2.478125 \times 10^1) \times 10^{-2} \\
&= 2.478125 \times 10^{-1}
\end{aligned}$$

111. $\dfrac{(5^{12})^2}{5^{25}} = \dfrac{5^{24}}{5^{25}} = 5^{24-25} = 5^{-1} = \dfrac{1}{5}$

113. $\dfrac{(3^5)^4}{3^5 \cdot 3^4} = \dfrac{3^{5 \cdot 4}}{3^{5+4}} = \dfrac{3^{20}}{3^9} = 3^{20-9} = 3^{11}$

115. $\dfrac{49^{18}}{7^{35}} = \dfrac{(7^2)^{18}}{7^{35}} = \dfrac{7^{36}}{7^{35}} = 7$

117. $\dfrac{(0.4)^5}{\left((0.4)^3\right)^2} = \dfrac{(0.4)^5}{(0.4)^6} = (0.4)^{-1} = \dfrac{1}{0.4}$, or 2.5

119. False; let $x = 2$, $y = 3$, $m = 4$, and $n = 2$:
$$2^4 \cdot 3^2 = 16 \cdot 9 = 144, \text{ but}$$
$$(2 \cdot 3)^{4 \cdot 2} = 6^8 = 1,679,616$$

121. False; let $x = 5$, $y = 3$, and $m = 2$:
$$(5 - 3)^2 = 2^2 = 4, \text{ but}$$
$$5^2 - 3^2 = 25 - 9 = 16$$

123. True; $(-x)^{2m} = (-1 \cdot x)^{2m} = (-1)^{2m} \cdot x^{2m} = [(-1)^2]^m \cdot x^{2m} = 1^m \cdot x^{2m} = x^{2m}$

Exercise Set 12.3

1. $-5x + 2 = -5 \cdot 4 + 2 = -20 + 2 = -18$;

$-5x + 2 = -5(-1) + 2 = 5 + 2 = 7$

3. $2x^2 - 5x + 7 = 2 \cdot 4^2 - 5 \cdot 4 + 7 = 2 \cdot 16 - 20 + 7 = 32 - 20 + 7 = 19$;

$2x^2 - 5x + 7 = 2(-1)^2 - 5(-1) + 7 = 2 \cdot 1 + 5 + 7 = 2 + 5 + 7 = 14$

5. $x^3 - 5x^2 + x = 4^3 - 5 \cdot 4^2 + 4 = 64 - 5 \cdot 16 + 4 = 64 - 80 + 4 = -12$;

$x^3 - 5x^2 + x = (-1)^3 - 5(-1)^2 + (-1) = -1 - 5 \cdot 1 - 1 = -1 - 5 - 1 = -7$

7. $\dfrac{1}{3}x + 5 = \dfrac{1}{3}(-2) + 5 = -\dfrac{2}{3} + 5 = -\dfrac{2}{3} + \dfrac{15}{3} = \dfrac{13}{3}$;

$\dfrac{1}{3}x + 5 = \dfrac{1}{3} \cdot 0 + 5 = 0 + 5 = 5$

9. $x^2 - 2x + 1 = (-2)^2 - 2(-2) + 1 = 4 + 4 + 1 = 9;$

   $x^2 - 2x + 1 = 0^2 - 2 \cdot 0 + 1 = 0 - 0 + 1 = 1$

11. $-3x^3 + 7x^2 - 3x - 2 = -3(-2)^3 + 7(-2)^2 - 3(-2) - 2 =$
    $-3(-8) + 7(4) - 3(-2) - 2 = 24 + 28 + 6 - 2 = 56;$

    $-3x^3 + 7x^2 - 3x - 2 = -3 \cdot 0^3 + 7 \cdot 0^2 - 3 \cdot 0 - 2 =$
    $-3 \cdot 0 + 7 \cdot 0 - 0 - 2 = 0 + 0 - 0 - 2 = -2$

13. We evaluate the polynomial for $t = 10$:

    $S = 11.12t^2 = 11.12(10)^2 = 11.12(100) = 1112$

    The skydiver has fallen approximately 1112 ft.

15. a) In 2001, $t = 0$.

    $E = 90.28(0) + 1138.34 = 0 + 1138.34 = 1138.34.$

    The consumption of electricity in 2001 was 1138.34 billion kilowatt-hours.

    In 2005, $t = 2005 - 2001 = 4$.

    $E = 90.28(4) + 1138.34 = 361.12 + 1138.34 = 1499.46$

    The consumption of electricity in 2005 was 1499.46 billion kilowatt-hours.

    In 2010, $t = 2010 - 2001 = 9$.

    $E = 90.28(9) + 1138.34 = 812.52 + 1138.34 = 1950.86$

    The consumption of electricity in 2010 will be 1950.86 billion kilowatt-hours.

    In 2015, $t = 2015 - 2001 = 14$.

    $E = 90.28(14) + 1138.34 = 1263.92 + 1138.34 = 2402.26$

    The consumption of electricity in 2015 will be 2402.26 billion kilowatt-hours.

    In 2025, $t = 2025 - 2001 = 24$.

    $E = 90.28(24) + 1138.34 = 2166.72 + 1138.34 = 3305.06$

    The consumption of electricity in 2025 will be 3305.06 billion kilowatt-hours.

    b) It appears that the points $(0, 1138.34)$, $(4, 1499.46)$, $(9, 1950.86)$, $(14, 2402.26)$, and $(24, 3305.06)$ are on the graph, so the results check.

17. We evaluate the polynomial for $x = 75$:

    $$\begin{aligned} R = 280x - 0.4x^2 &= 280(75) - 0.4(75)^2 \\ &= 280(75) - 0.4(5625) \\ &= 21,000 - 2250 \\ &= 18,750 \end{aligned}$$

    The total revenue from the sale of 75 TVs is $18,750.

    We evaluate the polynomial for $x = 100$:

    $$\begin{aligned} R = 280x - 0.4x^2 &= 280(100) - 0.4(100)^2 \\ &= 280(100) - 0.4(10,000) \\ &= 28,000 - 4000 \\ &= 24,000 \end{aligned}$$

    The total revenue from the sale of 100 TVs is $24,000.

19. Locate -3 on the x-axis. Then move vertically to the graph and horizontally to the y-axis. It appears that the y-value that is paired with -3 is -4. Thus, the value of $y = 5 - x^2$ is -4 when $x = -3$.

    Locate -1 on the x-axis. Then move vertically to the graph and horizontally to the y-axis. It appears that the y-value that is paired with -1 is 4. Thus, the value of $y = 5 - x^2$ is 4 when $x = -1$.

    Locate 0 on the x-axis. Then move vertically to the graph. We arrive at a point on the y-axis with the y-value 5. Thus, the value of $5 - x^2$ is 5 when $x = 0$.

    Locate 1.5 on the x-axis. Then move vertically to the graph and horizontally to the y-axis. It appears that the y-value that is paired with 1.5 is 2.75. Thus, the value of $y = 5 - x^2$ is 2.75 when $x = 1.5$.

    Locate 2 on the x-axis. Then move vertically to the graph and horizontally to the y-axis. It appears that the y-value that is paired with 2 is 1. Thus, the value of $y = 5 - x^2$ is 1 when $x = 2$.

21. We evaluate the polynomial for $x = 20$:

    $$\begin{aligned} N &= -0.00006(20)^3 + 0.006(20)^2 - 0.1(20) + 1.9 \\ &= -0.00006(8000) + 0.006(400) - 0.1(20) + 1.9 \\ &= -0.48 + 2.4 - 2.0 + 1.9 \\ &= 1.82 \end{aligned}$$

    There are about 1.82 million or 1,820,000 hearing-impaired Americans of age 20.

    We evaluate the polynomial for $x = 40$:

    $$\begin{aligned} N &= -0.00006(40)^3 + 0.006(40)^2 - 0.1(40) + 1.9 \\ &= -0.00006(64,000) + 0.006(1600) - 0.1(40) + 1.9 \\ &= -3.84 + 9.6 - 4.0 + 1.9 \\ &= 3.66 \end{aligned}$$

    There are about 3.66 million, or 3,660,000, hearing-impaired Americans of age 40.

23. Locate 10 on the horizontal axis. From there move vertically to the graph and then horizontally to the M-axis. This locates an M-value of about 9. Thus, about 9 words were memorized in 10 minutes.

25. Locate 8 on the horizontal axis. From there move vertically to the graph and then horizontally to the M-axis. This locates an M-value of about 6. Thus, the value of $-0.001t^3 + 0.1t^2$ for $t = 8$ is approximately 6.

27. Locate 13 on the horizontal axis. It is halfway between 12 and 14. From there move vertically to the graph and then horizontally to the M-axis. This locates an M-value of about 15. Thus, the value of $-0.001t^3 + 0.1t^2$ when t is 13 is approximately 15.

29. $2 - 3x + x^2 = 2 + (-3x) + x^2$

    The terms are 2, $-3x$, and x^2.

31. $-2x^4 + \frac{1}{3}x^3 - x + 3 = -2x^4 + \frac{1}{3}x^3 + (-x) + 3$

    The terms are $-2x^4$, $\frac{1}{3}x^3$, $-x$, and 3.

33. $5x^3 + 6x^2 - 3x^2$

Like terms: $6x^2$ and $-3x^2$ Same variable and exponent

35. $2x^4 + 5x - 7x - 3x^4$

Like terms: $2x^4$ and $-3x^4$ Same variable and

Like terms: $5x$ and $-7x$ exponent

37. $3x^5 - 7x + 8 + 14x^5 - 2x - 9$

Like terms: $3x^5$ and $14x^5$

Like terms: $-7x$ and $-2x$

Like terms: 8 and -9 Constant terms are like terms.

39. $-3x + 6$

The coefficient of $-3x$, the first term, is -3.

The coefficient of 6, the second term, is 6.

41. $5x^2 + \dfrac{3}{4}x + 3$

The coefficient of $5x^2$, the first term, is 5.

The coefficient of $\dfrac{3}{4}x$, the second term, is $\dfrac{3}{4}$.

The coefficient of 3, the third term, is 3.

43. $-5x^4 + 6x^3 - 2.7x^2 + 8x - 2$

The coefficient of $-5x^4$, the first term, is -5.

The coefficient of $6x^3$, the second term, is 6.

The coefficient of $-2.7x^2$, the third term, is -2.7.

The coefficient of $8x$, the fourth term, is 8.

The coefficient of -2, the fifth term, is -2.

45. $2x - 5x = (2 - 5)x = -3x$

47. $x - 9x = 1x - 9x = (1 - 9)x = -8x$

49. $5x^3 + 6x^3 + 4 = (5 + 6)x^3 + 4 = 11x^3 + 4$

51. $5x^3 + 6x - 4x^3 - 7x = (5 - 4)x^3 + (6 - 7)x =$

$1x^3 + (-1)x = x^3 - x$

53. $6b^5 + 3b^2 - 2b^5 - 3b^2 = (6 - 2)b^5 + (3 - 3)b^2 =$

$4b^5 + 0b^2 = 4b^5$

55. $\dfrac{1}{4}x^5 - 5 + \dfrac{1}{2}x^5 - 2x - 37 =$

$\left(\dfrac{1}{4} + \dfrac{1}{2}\right)x^5 - 2x + (-5 - 37) = \dfrac{3}{4}x^5 - 2x - 42$

57. $6x^2 + 2x^4 - 2x^2 - x^4 - 4x^2 =$

$6x^2 + 2x^4 - 2x^2 - 1x^4 - 4x^2 =$

$(6 - 2 - 4)x^2 + (2 - 1)x^4 = 0x^2 + 1x^4 =$

$0 + x^4 = x^4$

59. $\dfrac{1}{4}x^3 - x^2 - \dfrac{1}{6}x^2 + \dfrac{3}{8}x^3 + \dfrac{5}{16}x^3 =$

$\dfrac{1}{4}x^3 - 1x^2 - \dfrac{1}{6}x^2 + \dfrac{3}{8}x^3 + \dfrac{5}{16}x^3 =$

$\left(\dfrac{1}{4} + \dfrac{3}{8} + \dfrac{5}{16}\right)x^3 + \left(-1 - \dfrac{1}{6}\right)x^2 =$

$\left(\dfrac{4}{16} + \dfrac{6}{16} + \dfrac{5}{16}\right)x^3 + \left(-\dfrac{6}{6} - \dfrac{1}{6}\right)x^2 = \dfrac{15}{16}x^3 - \dfrac{7}{6}x^2$

61. $x^5 + x + 6x^3 + 1 + 2x^2 = x^5 + 6x^3 + 2x^2 + x + 1$

63. $5y^3 + 15y^9 + y - y^2 + 7y^8 =$

$15y^9 + 7y^8 + 5y^3 - y^2 + y$

65. $3x^4 - 5x^6 - 2x^4 + 6x^6 = x^4 + x^6 = x^6 + x^4$

67. $-2x + 4x^3 - 7x + 9x^3 + 8 = -9x + 13x^3 + 8 =$

$13x^3 - 9x + 8$

69. $3x + 3x + 3x - x^2 - 4x^2 = 9x - 5x^2 = -5x^2 + 9x$

71. $-x + \dfrac{3}{4} + 15x^4 - x - \dfrac{1}{2} - 3x^4 = -2x + \dfrac{1}{4} + 12x^4 =$

$12x^4 - 2x + \dfrac{1}{4}$

73. $2x - 4 = 2x^1 - 4x^0$

The degree of $2x$ is 1.

The degree of -4 is 0.

The degree of the polynomial is 1, the largest exponent.

75. $3x^2 - 5x + 2 = 3x^2 - 5x^1 + 2x^0$

The degree of $3x^2$ is 2.

The degree of $-5x$ is 1.

The degree of 2 is 0.

The degree of the polynomial is 2, the largest exponent.

77. $-7x^3 + 6x^2 + \dfrac{3}{5}x + 7 = -7x^3 + 6x^2 + \dfrac{3}{5}x^1 + 7x^0$

The degree of $-7x^3$ is 3.

The degree of $6x^2$ is 2.

The degree of $\dfrac{3}{5}x$ is 1.

The degree of 7 is 0.

The degree of the polynomial is 3, the largest exponent.

79. $x^2 - 3x + x^6 - 9x^4 = x^2 - 3x^1 + x^6 - 9x^4$

The degree of x^2 is 2.

The degree of $-3x$ is 1.

The degree of x^6 is 6.

The degree of $-9x^4$ is 4.

The degree of the polynomial is 6, the largest exponent.

81. See the answer section in the text.

83. In the polynomial $x^3 - 27$, there are no x^2 or x terms. The x^2 term (or second-degree term) and the x term (or first-degree term) are missing.

85. In the polynomial $x^4 - x$, there are no x^3, x^2, or x^0 terms. The x^3 term (or third-degree term), the x^2 term (or second-degree term), and the x^0 term (or zero-degree term) are missing.

87. No terms are missing in the polynomial

$2x^3 - 5x^2 + x - 3$.

89. $x^3 - 27 = x^3 + 0x^2 + 0x - 27$

$x^3 - 27 = x^3 \qquad\qquad - 27$

91. $x^4 - x = x^4 + 0x^3 + 0x^2 - x + 0x^0$

$ x^4 - x = x^4 - x$

93. There are no missing terms.

95. The polynomial $x^2 - 10x + 25$ is a *trinomial* because it has just three terms.

97. The polynomial $x^3 - 7x^2 + 2x - 4$ is *none of these* because it has more than three terms.

99. The polynomial $4x^2 - 25$ is a *binomial* because it has just two terms.

101. The polynomial $40x$ is a *monomial* because it has just one term.

103. Discussion and Writing Exercise

105. *Familiarize*. Let $a =$ the number of apples the campers had to begin with. Then the first camper ate $\frac{1}{3}a$ apples and $a - \frac{1}{3}a$, or $\frac{2}{3}a$, apples were left. The second camper ate $\frac{1}{3}\left(\frac{2}{3}a\right)$, or $\frac{2}{9}a$, apples, and $\frac{2}{3}a - \frac{2}{9}a$, or $\frac{4}{9}a$, apples were left. The third camper ate $\frac{1}{3}\left(\frac{4}{9}a\right)$, or $\frac{4}{27}a$, apples, and $\frac{4}{9}a - \frac{4}{27}a$, or $\frac{8}{27}a$, apples were left.

Translate. We write an equation for the number of apples left after the third camper eats.

$$\underbrace{\text{Number of apples left}}_{\displaystyle \frac{8}{27}a} \ \underset{\displaystyle =}{\overset{\downarrow\ \downarrow}{\text{is}}} \ \underset{\displaystyle 8}{8.}$$

Solve. We solve the equation.

$$\frac{8}{27}a = 8$$
$$a = \frac{27}{8} \cdot 8$$
$$a = 27$$

Check. If the campers begin with 27 apples, then the first camper eats $\frac{1}{3} \cdot 27$, or 9, and $27 - 9$, or 18, are left. The second camper then eats $\frac{1}{3} \cdot 18$, or 6 apples and $18 - 6$, or 12, are left. Finally, the third camper eats $\frac{1}{3} \cdot 12$, or 4 apples and $12 - 4$, or 8, are left. The answer checks.

State. The campers had 27 apples to begin with.

107. $\frac{1}{8} - \frac{5}{6} = \frac{1}{8} + \left(-\frac{5}{6}\right)$, LCM is 24

$$= \frac{1}{8} \cdot \frac{3}{3} + \left(-\frac{5}{6}\right)\left(\frac{4}{4}\right)$$
$$= \frac{3}{24} + \left(-\frac{20}{24}\right)$$
$$= -\frac{17}{24}$$

109. $5.6 - 8.2 = 5.6 + (-8.2) = -2.6$

111. $\qquad C = ab - r$

$\qquad C + r = ab \qquad$ Adding r

$\qquad \dfrac{C+r}{a} = \dfrac{ab}{a} \qquad$ Dividing by a

$\qquad \dfrac{C+r}{a} = b \qquad$ Simplifying

113. $3x - 15y + 63 = 3 \cdot x - 3 \cdot 5y + 3 \cdot 21 = 3(x - 5y + 21)$

115. $\quad (3x^2)^3 + 4x^2 \cdot 4x^4 - x^4(2x)^2 + [(2x)^2]^3 - 100x^2(x^2)^2$

$= 27x^6 + 4x^2 \cdot 4x^4 - x^4 \cdot 4x^2 + (2x)^6 - 100x^2 \cdot x^4$

$= 27x^6 + 16x^6 - 4x^6 + 64x^6 - 100x^6$

$= 3x^6$

117. $(5m^5)^2 = 5^2 m^{5 \cdot 2} = 25m^{10}$

The degree is 10.

119. Graph $y = 5 - x^2$. Then use VALUE from the CALC menu to find the y-values that correspond to $x = -3$, $x = -1$, x=0, $x = 1.5$, and $x = 2$. As before, we find that these values are -4, 4, 5, 2.75, and 1, respectively.

121. Graph $y = -0.00006x^3 + 0.006x^2 - 0.1x + 1.9$. Then use VALUE from the CALC menu to find the y-values that correspond to $x = 20$ and $x = 40$. As before, we find that these values are 1.82 and 3.66, respectively. These results represent 1,820,000 and 3,660,000 hearing-impaired Americans.

Exercise Set 12.4

1. $(3x + 2) + (-4x + 3) = (3 - 4)x + (2 + 3) = -x + 5$

3. $(-6x + 2) + \left(x^2 + \frac{1}{2}x - 3\right) = x^2 + \left(-6 + \frac{1}{2}\right)x + (2-3) =$

$x^2 + \left(-\frac{12}{2} + \frac{1}{2}\right)x + (2 - 3) = x^2 - \frac{11}{2}x - 1$

5. $(x^2 - 9) + (x^2 + 9) = (1 + 1)x^2 + (-9 + 9) = 2x^2$

7. $(3x^2 - 5x + 10) + (2x^2 + 8x - 40) =$

$(3 + 2)x^2 + (-5 + 8)x + (10 - 40) = 5x^2 + 3x - 30$

9. $(1.2x^3 + 4.5x^2 - 3.8x) + (-3.4x^3 - 4.7x^2 + 23) =$

$(1.2 - 3.4)x^3 + (4.5 - 4.7)x^2 - 3.8x + 23 =$

$-2.2x^3 - 0.2x^2 - 3.8x + 23$

11. $(1 + 4x + 6x^2 + 7x^3) + (5 - 4x + 6x^2 - 7x^3) =$

$(1 + 5) + (4 - 4)x + (6 + 6)x^2 + (7 - 7)x^3 =$

$6 + 0x + 12x^2 + 0x^3 = 6 + 12x^2$, or $12x^2 + 6$

13. $\left(\frac{1}{4}x^4 + \frac{2}{3}x^3 + \frac{5}{8}x^2 + 7\right) + \left(-\frac{3}{4}x^4 + \frac{3}{8}x^2 - 7\right) =$

$\left(\frac{1}{4} - \frac{3}{4}\right)x^4 + \frac{2}{3}x^3 + \left(\frac{5}{8} + \frac{3}{8}\right)x^2 + (7 - 7) =$

$-\frac{2}{4}x^4 + \frac{2}{3}x^3 + \frac{8}{8}x^2 + 0 =$

$-\frac{1}{2}x^4 + \frac{2}{3}x^3 + x^2$

15. $(0.02x^5 - 0.2x^3 + x + 0.08) + (-0.01x^5 + x^4 - 0.8x - 0.02) =$
$(0.02 - 0.01)x^5 + x^4 - 0.2x^3 + (1 - 0.8)x + (0.08 - 0.02) =$
$0.01x^5 + x^4 - 0.2x^3 + 0.2x + 0.06$

17. $9x^8 - 7x^4 + 2x^2 + 5) + (8x^7 + 4x^4 - 2x) +$
$(-3x^4 + 6x^2 + 2x - 1) = 9x^8 + 8x^7 + (-7 + 4 - 3)x^4 +$
$(2 + 6)x^2 + (-2 + 2)x + (5 - 1) =$
$9x^8 + 8x^7 - 6x^4 + 8x^2 + 4$

19. Rewrite the problem so the coefficients of like terms have the same number of decimal places.

$$
\begin{array}{r}
0.15x^4 + 0.10x^3 - 0.90x^2 \\
- 0.01x^3 + 0.01x^2 + x \\
1.25x^4 + 0.11x^2 + 0.01 \\
0.27x^3 + 0.99 \\
-0.35x^4 + 15.00x^2 - 0.03 \\
\hline
1.05x^4 + 0.36x^3 + 14.22x^2 + x + 0.97
\end{array}
$$

21. We change the sign of the term inside the parentheses.
$-(-5x) = 5x$

23. We change the sign of every term inside the parentheses.
$-\left(-x^2 + \dfrac{3}{2}x - 2\right) = x^2 - \dfrac{3}{2}x + 2$

25. We change the sign of every term inside the parentheses.
$-(12x^4 - 3x^3 + 3) = -12x^4 + 3x^3 - 3$

27. We change the sign of every term inside parentheses.
$-(3x - 7) = -3x + 7$

29. We change the sign of every term inside parentheses.
$-(4x^2 - 3x + 2) = -4x^2 + 3x - 2$

31. We change the sign of every term inside parentheses.
$-\left(-4x^4 + 6x^2 + \dfrac{3}{4}x - 8\right) = 4x^4 - 6x^2 - \dfrac{3}{4}x + 8$

33. $(3x + 2) - (-4x + 3) = 3x + 2 + 4x - 3$
Changing the sign of every term inside parentheses
$= 7x - 1$

35. $(-6x + 2) - (x^2 + x - 3) = -6x + 2 - x^2 - x + 3$
$= -x^2 - 7x + 5$

37. $(x^2 - 9) - (x^2 + 9) = x^2 - 9 - x^2 - 9 = -18$

39. $(6x^4 + 3x^3 - 1) - (4x^2 - 3x + 3)$
$= 6x^4 + 3x^3 - 1 - 4x^2 + 3x - 3$
$= 6x^4 + 3x^3 - 4x^2 + 3x - 4$

41. $(1.2x^3 + 4.5x^2 - 3.8x) - (-3.4x^3 - 4.7x^2 + 23)$
$= 1.2x^3 + 4.5x^2 - 3.8x + 3.4x^3 + 4.7x^2 - 23$
$= 4.6x^3 + 9.2x^2 - 3.8x - 23$

43. $\dfrac{5}{8}x^3 - \dfrac{1}{4}x - \dfrac{1}{3} - \left(-\dfrac{1}{8}x^3 + \dfrac{1}{4}x - \dfrac{1}{3}\right)$
$= \dfrac{5}{8}x^3 - \dfrac{1}{4}x - \dfrac{1}{3} + \dfrac{1}{8}x^3 - \dfrac{1}{4}x + \dfrac{1}{3}$
$= \dfrac{6}{8}x^3 - \dfrac{2}{4}x$
$= \dfrac{3}{4}x^3 - \dfrac{1}{2}x$

45. $(0.08x^3 - 0.02x^2 + 0.01x) - (0.02x^3 + 0.03x^2 - 1)$
$= 0.08x^3 - 0.02x^2 + 0.01x - 0.02x^3 - 0.03x^2 + 1$
$= 0.06x^3 - 0.05x^2 + 0.01x + 1$

47.
$$
\begin{array}{l}
x^2 + 5x + 6 \\
x^2 + 2x \\
\hline
\end{array}
$$
$$
\begin{array}{ll}
x^2 + 5x + 6 & \\
-x^2 - 2x & \text{Changing signs} \\
\hline
 3x + 6 & \text{Adding}
\end{array}
$$

49.
$$
\begin{array}{l}
5x^4 + 6x^3 - 9x^2 \\
-6x^4 - 6x^3 + 8x + 9 \\
\hline
\end{array}
$$
$$
\begin{array}{ll}
5x^4 + 6x^3 - 9x^2 & \\
6x^4 + 6x^3 - 8x - 9 & \text{Changing signs} \\
\hline
11x^4 + 12x^3 - 9x^2 - 8x - 9 & \text{Adding}
\end{array}
$$

51.
$$
\begin{array}{l}
x^5 - 1 \\
x^5 - x^4 + x^3 - x^2 + x - 1 \\
\hline
\end{array}
$$
$$
\begin{array}{ll}
x^5 - 1 & \\
-x^5 + x^4 - x^3 + x^2 - x + 1 & \text{Changing signs} \\
\hline
 x^4 - x^3 + x^2 - x & \text{Adding}
\end{array}
$$

53. We add the lengths of the sides:
$$4a + 7 + a + \dfrac{1}{2}a + 3 + a + 2a + 3a$$
$$= \left(4 + 1 + \dfrac{1}{2} + 1 + 2 + 3\right)a + (7 + 3)$$
$$= 11\dfrac{1}{2}a + 10, \text{ or } \dfrac{23}{2}a + 10$$

55.

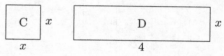

The area of a rectangle is the product of the length and width. The sum of the areas is found as follows:

$$\begin{array}{ccccccccc} \text{Area} & & \text{Area} & & \text{Area} & & \text{Area} \\ \text{of } A & + & \text{of } B & + & \text{of } C & + & \text{of } D \\ = 3x \cdot x & + & x \cdot x & + & x \cdot x & + & 4 \cdot x \\ = 3x^2 & + & x^2 & + & x^2 & + & 4x \\ = 5x^2 & + & 4x & & & & \end{array}$$

A polynomial for the sum of the areas is $5x^2 + 4x$.

57.

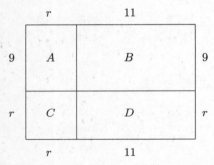

The length and width of the figure can be expressed as $r + 11$ and $r + 9$, respectively. The area of this figure (a rectangle) is the product of the length and width. An algebraic expression for the area is $(r + 11) \cdot (r + 9)$.

The area of the figure can also be found by adding the areas of the four rectangles A, B, C, and D. The area of a rectangle is the product of the length and the width.

$$\begin{array}{ccccccccc} \text{Area} & & \text{Area} & & \text{Area} & & \text{Area} \\ \text{of } A & + & \text{of } B & + & \text{of } C & + & \text{of } D \\ = 9 \cdot r & + & 11 \cdot 9 & + & r \cdot r & + & 11 \cdot r \\ = 9r & + & 99 & + & r^2 & + & 11r \end{array}$$

A second algebraic expression for the area of the figure is $9r + 99 + r^2 + 11r$, or $r^2 + 20r + 99$.

59.

The length and width of the figure can each be expressed as $x + 3$. The area can be expressed as $(x+3) \cdot (x+3)$, or $(x+3)^2$.

Another way to express the area is to find an expression for the sum of the areas of the four rectangles A, B, C, and D. The area of each rectangle is the product of its length and width.

$$\begin{array}{ccccccccc} \text{Area} & & \text{Area} & & \text{Area} & & \text{Area} \\ \text{of } A & + & \text{of } B & + & \text{of } C & + & \text{of } D \\ = x \cdot x & + & 3 \cdot x & + & 3 \cdot x & + & 3 \cdot 3 \\ = x^2 & + & 3x & + & 3x & + & 9 \end{array}$$

Then a second algebraic expression for the area of the figure is $x^2 + 3x + 3x + 9$, or $x^2 + 6x + 9$.

61.

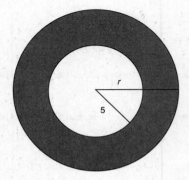

Familiarize. Recall that the area of a circle is the product of π and the square of the radius, r^2.

$$A = \pi r^2$$

Translate.

$$\begin{array}{ccc} \text{Area of circle} & & \text{Area of circle} & & \text{Shaded} \\ \text{with radius } r & - & \text{with radius } 5 & = & \text{area} \\ \pi \cdot r^2 & - & \pi \cdot 5^2 & = \text{Shaded area} \end{array}$$

Carry out. We simplify the expression.

$$\pi \cdot r^2 - \pi \cdot 5^2 = \pi r^2 - 25\pi$$

Check. We can go over our calculations. We can also assign some value to r, say 7, and carry out the computation in two ways.

Difference of areas: $\pi \cdot 7^2 - \pi \cdot 5^2 = 49\pi - 25\pi = 24\pi$

Substituting in the polynomial: $\pi \cdot 7^2 - 25\pi = 49\pi - 25\pi = 24\pi$

Since the results are the same, our solution is probably correct.

State. A polynomial for the shaded area is $\pi r^2 - 25\pi$.

63. Familiarize. We label the figure with additional information.

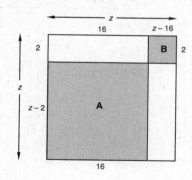

Translate.

Area of shaded sections = Area of A + Area of B

Area of shaded sections = $16(z - 2) + 2(z - 16)$

Carry out. We simplify the expression.

$16(z - 2) + 2(z - 16) = 16z - 32 + 2z - 32 = 18z - 64$

Check. We can go over the calculations. We can also assign some value to z, say 30, and carry out the computation in two ways.

Sum of areas:

$16 \cdot 28 + 2 \cdot 14 = 448 + 28 = 476$

Substituting in the polynomial:

$18 \cdot 30 - 64 = 540 - 64 = 476$

Since the results are the same, our solution is probably correct.

State. A polynomial for the shaded area is $18z - 64$.

65. Discussion and Writing Exercise

67.
$$8x + 3x = 66$$
$$11x = 66 \quad \text{Collecting like terms}$$
$$\frac{11x}{11} = \frac{66}{11} \quad \text{Dividing by 11}$$
$$x = 6$$

The solution is 6.

69.
$$\frac{3}{8}x + \frac{1}{4} - \frac{3}{4}x = \frac{11}{16} + x, \quad \text{LCM is 16}$$
$$16\left(\frac{3}{8}x + \frac{1}{4} - \frac{3}{4}x\right) = 16\left(\frac{11}{16} + x\right) \quad \text{Clearing fractions}$$
$$6x + 4 - 12x = 11 + 16x$$
$$-6x + 4 = 11 + 16x \quad \text{Collecting like terms}$$
$$-6x + 4 - 4 = 11 + 16x - 4 \quad \text{Subtracting 4}$$
$$-6x = 7 + 16x$$
$$-6x - 16x = 7 + 16x - 16x \quad \text{Subtracting } 16x$$
$$-22x = 7$$
$$\frac{-22x}{-22} = \frac{7}{-22} \quad \text{Dividing by } -22$$
$$x = -\frac{7}{22}$$

The solution is $-\frac{7}{22}$.

71.
$$1.5x - 2.7x = 22 - 5.6x$$
$$10(1.5x - 2.7x) = 10(22 - 5.6x) \quad \text{Clearing decimals}$$
$$15x - 27x = 220 - 56x$$
$$-12x = 220 - 56x \quad \text{Collecting like terms}$$
$$44x = 220 \quad \text{Adding } 56x$$
$$x = \frac{220}{44} \quad \text{Dividing by 44}$$
$$x = 5 \quad \text{Simplifying}$$

The solution is 5.

73.
$$6(y - 3) - 8 = 4(y + 2) + 5$$
$$6y - 18 - 8 = 4y + 8 + 5 \quad \text{Removing parentheses}$$
$$6y - 26 = 4y + 13 \quad \text{Collecting like terms}$$
$$6y - 26 + 26 = 4y + 13 + 26 \quad \text{Adding 26}$$
$$6y = 4y + 39$$
$$6y - 4y = 4y + 39 - 4y \quad \text{Subtracting } 4y$$
$$2y = 39$$
$$\frac{2y}{2} = \frac{39}{2} \quad \text{Dividing by 2}$$
$$y = \frac{39}{2}$$

The solution is $\frac{39}{2}$.

75.
$$3x - 7 \le 5x + 13$$
$$-2x - 7 \le 13 \quad \text{Subtracting } 5x$$
$$-2x \le 20 \quad \text{Adding 7}$$
$$x \ge -10 \quad \text{Dividing by } -2 \text{ and reversing the inequality symbol}$$

The solution set is $\{x | x \ge -10\}$.

77. Familiarize. The surface area is $2lw + 2lh + 2wh$, where l = length, w = width, and h = height of the rectangular solid. Here we have $l = 3$, $w = w$, and $h = 7$.

Translate. We substitute in the formula above.

$2 \cdot 3 \cdot w + 2 \cdot 3 \cdot 7 + 2 \cdot w \cdot 7$

Carry out. We simplify the expression.

$2 \cdot 3 \cdot w + 2 \cdot 3 \cdot 7 + 2 \cdot w \cdot 7$
$= 6w + 42 + 14w$
$= 20w + 42$

Check. We can go over the calculations. We can also assign some value to w, say 6, and carry out the computation in two ways.

Using the formula: $2 \cdot 3 \cdot 6 + 2 \cdot 3 \cdot 7 + 2 \cdot 6 \cdot 7 = 36 + 42 + 84 = 162$

Substituting in the polynomial: $20 \cdot 6 + 42 = 120 + 42 = 162$

Since the results are the same, our solution is probably correct.

State. A polynomial for the surface area is $20w + 42$.

79. *Familiarize*. The surface area is $2lw + 2lh + 2wh$, where $l = $ length, $w = $ width, and $h = $ height of the rectangular solid. Here we have $l = x$, $w = x$, and $h = 5$.

Translate. We substitute in the formula above.

$$2 \cdot x \cdot x + 2 \cdot x \cdot 5 + 2 \cdot x \cdot 5$$

Carry out. We simplify the expression.

$$\begin{aligned} & 2 \cdot x \cdot x + 2 \cdot x \cdot 5 + 2 \cdot x \cdot 5 \\ &= 2x^2 + 10x + 10x \\ &= 2x^2 + 20x \end{aligned}$$

Check. We can go over the calculations. We can also assign some value to x, say 3, and carry out the computation in two ways.

Using the formula: $2 \cdot 3 \cdot 3 + 2 \cdot 3 \cdot 5 + 2 \cdot 3 \cdot 5 = 18 + 30 + 30 = 78$

Substituting in the polynomial: $2 \cdot 3^2 + 20 \cdot 3 = 2 \cdot 9 + 60 = 18 + 60 = 78$

Since the results are the same, our solution is probably correct.

State. A polynomial for the surface area is $2x^2 + 20x$.

81.

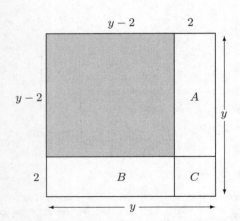

The shaded area is $(y-2)^2$. We find it as follows:

$$\begin{array}{ccccccc} \text{Shaded} \\ \text{area} & = & \text{Area of} \\ & & \text{square} & - & \text{Area} \\ & & & & \text{of } A & - & \text{Area} \\ & & & & & & \text{of } B & - & \text{Area} \\ & & & & & & & & \text{of } C \end{array}$$

$\text{Shaded area} = \text{Area of square} - \text{Area of } A - \text{Area of } B - \text{Area of } C$

$$(y-2)^2 = y^2 - 2(y-2) - 2(y-2) - 2 \cdot 2$$
$$(y-2)^2 = y^2 - 2y + 4 - 2y + 4 - 4$$
$$(y-2)^2 = y^2 - 4y + 4$$

83. $(7y^2 - 5y + 6) - (3y^2 + 8y - 12) + (8y^2 - 10y + 3)$
$= 7y^2 - 5y + 6 - 3y^2 - 8y + 12 + 8y^2 - 10y + 3$
$= 12y^2 - 23y + 21$

85. $(-y^4 - 7y^3 + y^2) + (-2y^4 + 5y - 2) - (-6y^3 + y^2)$
$= -y^4 - 7y^3 + y^2 - 2y^4 + 5y - 2 + 6y^3 - y^2$
$= -3y^4 - y^3 + 5y - 2$

Exercise Set 12.5

1. $(8x^2)(5) = (8 \cdot 5)x^2 = 40x^2$

3. $(-x^2)(-x) = (-1x^2)(-1x) = (-1)(-1)(x^2 \cdot x) = x^3$

5. $(8x^5)(4x^3) = (8 \cdot 4)(x^5 \cdot x^3) = 32x^8$

7. $(0.1x^6)(0.3x^5) = (0.1)(0.3)(x^6 \cdot x^5) = 0.03x^{11}$

9. $\left(-\dfrac{1}{5}x^3\right)\left(-\dfrac{1}{3}x\right) = \left(-\dfrac{1}{5}\right)\left(-\dfrac{1}{3}\right)(x^3 \cdot x) = \dfrac{1}{15}x^4$

11. $(-4x^2)(0) = 0$ Any number multiplied by 0 is 0.

13. $(3x^2)(-4x^3)(2x^6) = (3)(-4)(2)(x^2 \cdot x^3 \cdot x^6) = -24x^{11}$

15. $\begin{aligned} 2x(-x + 5) &= 2x(-x) + 2x(5) \\ &= -2x^2 + 10x \end{aligned}$

17. $\begin{aligned} -5x(x - 1) &= -5x(x) - 5x(-1) \\ &= -5x^2 + 5x \end{aligned}$

19. $\begin{aligned} x^2(x^3 + 1) &= x^2(x^3) + x^2(1) \\ &= x^5 + x^2 \end{aligned}$

21. $\begin{aligned} 3x(2x^2 - 6x + 1) &= 3x(2x^2) + 3x(-6x) + 3x(1) \\ &= 6x^3 - 18x^2 + 3x \end{aligned}$

23. $\begin{aligned} -6x^2(x^2 + x) &= -6x^2(x^2) - 6x^2(x) \\ &= -6x^4 - 6x^3 \end{aligned}$

25. $\begin{aligned} 3y^2(6y^4 + 8y^3) &= 3y^2(6y^4) + 3y^2(8y^3) \\ &= 18y^6 + 24y^5 \end{aligned}$

27. $\begin{aligned} (x + 6)(x + 3) &= (x + 6)x + (x + 6)3 \\ &= x \cdot x + 6 \cdot x + x \cdot 3 + 6 \cdot 3 \\ &= x^2 + 6x + 3x + 18 \\ &= x^2 + 9x + 18 \end{aligned}$

29. $\begin{aligned} (x + 5)(x - 2) &= (x + 5)x + (x + 5)(-2) \\ &= x \cdot x + 5 \cdot x + x(-2) + 5(-2) \\ &= x^2 + 5x - 2x - 10 \\ &= x^2 + 3x - 10 \end{aligned}$

31. $\begin{aligned} (x - 4)(x - 3) &= (x - 4)x + (x - 4)(-3) \\ &= x \cdot x - 4 \cdot x + x(-3) - 4(-3) \\ &= x^2 - 4x - 3x + 12 \\ &= x^2 - 7x + 12 \end{aligned}$

33. $\begin{aligned} (x + 3)(x - 3) &= (x + 3)x + (x + 3)(-3) \\ &= x \cdot x + 3 \cdot x + x(-3) + 3(-3) \\ &= x^2 + 3x - 3x - 9 \\ &= x^2 - 9 \end{aligned}$

35. $\begin{aligned} (5 - x)(5 - 2x) &= (5 - x)5 + (5 - x)(-2x) \\ &= 5 \cdot 5 - x \cdot 5 + 5(-2x) - x(-2x) \\ &= 25 - 5x - 10x + 2x^2 \\ &= 25 - 15x + 2x^2 \end{aligned}$

37. $(2x+5)(2x+5) = (2x+5)2x + (2x+5)5$
$= 2x \cdot 2x + 5 \cdot 2x + 2x \cdot 5 + 5 \cdot 5$
$= 4x^2 + 10x + 10x + 25$
$= 4x^2 + 20x + 25$

39. $\left(x - \dfrac{5}{2}\right)\left(x + \dfrac{2}{5}\right) = \left(x - \dfrac{5}{2}\right)x + \left(x - \dfrac{5}{2}\right)\dfrac{2}{5}$
$= x \cdot x - \dfrac{5}{2} \cdot x + x \cdot \dfrac{2}{5} - \dfrac{5}{2} \cdot \dfrac{2}{5}$
$= x^2 - \dfrac{5}{2}x + \dfrac{2}{5}x - 1$
$= x^2 - \dfrac{25}{10}x + \dfrac{4}{10}x - 1$
$= x^2 - \dfrac{21}{10}x - 1$

41. $(x - 2.3)(x + 4.7) = (x - 2.3)x + (x - 2.3)4.7$
$= x \cdot x - 2.3 \cdot x + x \cdot 4.7 - 2.3(4.7)$
$= x^2 - 2.3x + 4.7x - 10.81$
$= x^2 + 2.4x - 10.81$

43. The length of the rectangle is $x+6$ and the width is $x+2$, so the area is $(x+6)(x+2)$. If we carry out the multiplication, we have $x^2 + 8x + 12$.

45. The length of the rectangle is $x+6$ and the width is $x+1$, so the area is $(x+6)(x+1)$. If we carry out the multiplication, we have $x^2 + 7x + 6$.

47. Illustrate $x(x + 5)$ as the area of a rectangle with width x and length $x + 5$.

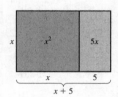

49. Illustrate $(x + 1)(x + 2)$ as the area of a rectangle with width $x + 1$ and length $x + 2$.

51. Illustrate $(x + 5)(x + 3)$ as the area of a rectangle with length $x + 5$ and width $x + 3$.

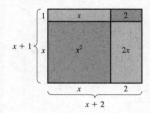

53. $(x^2 + x + 1)(x - 1)$
$= (x^2 + x + 1)x + (x^2 + x + 1)(-1)$
$= x^2 \cdot x + x \cdot x + 1 \cdot x + x^2(-1) + x(-1) + 1(-1)$
$= x^3 + x^2 + x - x^2 - x - 1$
$= x^3 - 1$

55. $(2x + 1)(2x^2 + 6x + 1)$
$= 2x(2x^2 + 6x + 1) + 1(2x^2 + 6x + 1)$
$= 2x \cdot 2x^2 + 2x \cdot 6x + 2x \cdot 1 + 1 \cdot 2x^2 + 1 \cdot 6x + 1 \cdot 1$
$= 4x^3 + 12x^2 + 2x + 2x^2 + 6x + 1$
$= 4x^3 + 14x^2 + 8x + 1$

57. $(y^2 - 3)(3y^2 - 6y + 2)$
$= y^2(3y^2 - 6y + 2) - 3(3y^2 - 6y + 2)$
$= y^2 \cdot 3y^2 + y^2(-6y) + y^2 \cdot 2 - 3 \cdot 3y^2 - 3(-6y) - 3 \cdot 2$
$= 3y^4 - 6y^3 + 2y^2 - 9y^2 + 18y - 6$
$= 3y^4 - 6y^3 - 7y^2 + 18y - 6$

59. $(x^3 + x^2)(x^3 + x^2 - x)$
$= x^3(x^3 + x^2 - x) + x^2(x^3 + x^2 - x)$
$= x^3 \cdot x^3 + x^3 \cdot x^2 + x^3(-x) + x^2 \cdot x^3 + x^2 \cdot x^2 + x^2(-x)$
$= x^6 + x^5 - x^4 + x^5 + x^4 - x^3$
$= x^6 + 2x^5 - x^3$

61. $(-5x^3 - 7x^2 + 1)(2x^2 - x)$
$= (-5x^3 - 7x^2 + 1)2x^2 + (-5x^3 - 7x^2 + 1)(-x)$
$= -5x^3 \cdot 2x^2 - 7x^2 \cdot 2x^2 + 1 \cdot 2x^2 - 5x^3(-x) - 7x^2(-x) + 1(-x)$
$= -10x^5 - 14x^4 + 2x^2 + 5x^4 + 7x^3 - x$
$= -10x^5 - 9x^4 + 7x^3 + 2x^2 - x$

63.
$$
\begin{array}{ll}
\quad 1 + x + x^2 & \text{Line up like terms} \\
\underline{\;-1 - x + x^2} & \text{in columns} \\
\quad\; x^2 + x^3 + x^4 & \text{Multiplying the top row by } x^2 \\
-\;\; x - x^2 - x^3 & \text{Multiplying by } -x \\
\underline{-1 - \; x - x^2 \qquad} & \text{Multiplying by } -1 \\
-1 - 2x - x^2 \qquad + x^4 &
\end{array}
$$

65.
$$
\begin{array}{ll}
\quad 2t^2 - \; t - 4 & \\
\underline{\quad 3t^2 + 2t - 1} & \\
\;-\; 2t^2 + \; t + 4 & \text{Multiplying by } -1 \\
\;4t^3 - \; 2t^2 - 8t & \text{Multiplying by } 2t \\
\underline{6t^4 - 3t^3 - 12t^2} & \text{Multiplying by } 3t^2 \\
6t^4 + \; t^3 - 16t^2 - 7t + 4 &
\end{array}
$$

67.
$$
\begin{array}{ll}
\quad x \quad\; - x^3 \qquad + x^5 & \\
\underline{-1 + x^2 \qquad + x^4} & \text{Rewriting in ascending order} \\
\qquad\quad x^5 - x^7 + x^9 & \text{Multiplying by } x^4 \\
\quad x^3 - x^5 + x^7 & \text{Multiplying by } x^2 \\
\underline{-x + \; x^3 - x^5} & \text{Multiplying by } -1 \\
-x + 2x^3 - x^5 \qquad + x^9 &
\end{array}
$$

69.
$$
\begin{array}{l}
\quad x^3 + x^2 + x + 1 \\
\underline{\qquad\qquad\quad x - 1} \\
\;-x^3 - x^2 - x - 1 \\
\underline{x^4 + x^3 + x^2 + x} \\
\;x^4 \qquad\qquad\quad -1
\end{array}
$$

71. We will multiply horizontally while still aligning like terms.

$$(x+1)(x^3 + 7x^2 + 5x + 4)$$

$$
\begin{array}{ll}
= x^4 + 7x^3 + 5x^2 + 4x & \text{Multiplying by } x \\
\underline{ + x^3 + 7x^2 + 5x + 4} & \text{Multiplying by } 1 \\
= x^4 + 8x^3 + 12x^2 + 9x + 4 &
\end{array}
$$

73. We will multiply horizontally while still aligning like terms.

$$\left(x - \frac{1}{2}\right)\left(2x^3 - 4x^2 + 3x - \frac{2}{5}\right)$$

$$
\begin{array}{l}
= 2x^4 - 4x^3 + 3x^2 - \dfrac{2}{5}x \\[2mm]
\underline{ - x^3 + 2x^2 - \dfrac{3}{2}x + \dfrac{1}{5}} \\[2mm]
2x^4 - 5x^3 + 5x^2 - \dfrac{19}{10}x + \dfrac{1}{5}
\end{array}
$$

75. Discussion and Writing Exercise

77. $-\dfrac{1}{4} - \dfrac{1}{2} = -\dfrac{1}{4} - \dfrac{1}{2} \cdot \dfrac{2}{2} = -\dfrac{1}{4} - \dfrac{2}{4} = -\dfrac{3}{4}$

79. $(10 - 2)(10 + 2) = 8 \cdot 12 = 96$

81. $15x - 18y + 12 = 3 \cdot 5x - 3 \cdot 6y + 3 \cdot 4 =$
$3(5x - 6y + 4)$

83. $-9x - 45y + 15 = -3 \cdot 3x - 3 \cdot 15y - 3(-5) =$
$-3(3x + 15y - 5)$

85. $y = \dfrac{1}{2}x - 3$

The equation is equivalent to $y = \dfrac{1}{2}x + (-3)$. The y-intercept is $(0, -3)$. We find two other points, using multiples of 2 for x to avoid fractions.

When $x = -2$, $y = \dfrac{1}{2}(-2) - 3 = -1 - 3 = -4$.

When $x = 4$, $y = \dfrac{1}{2} \cdot 4 - 3 = 2 - 3 = -1$.

x	y
0	-3
-2	-4
4	-1

Plot these points, draw the line they determine, and label the graph $y = \dfrac{1}{2}x - 3$.

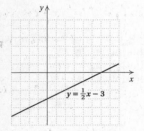

87. The shaded area is the area of the large rectangle, $6y(14y - 5)$ less the area of the unshaded rectangle, $3y(3y + 5)$. We have:

$$6y(14y - 5) - 3y(3y + 5)$$
$$= 84y^2 - 30y - 9y^2 - 15y$$
$$= 75y^2 - 45y$$

89.

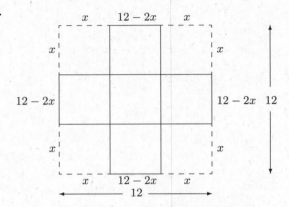

The dimensions, in inches, of the box are $12 - 2x$ by $12 - 2x$ by x. The volume is the product of the dimensions (volume = length × width × height):

$$
\begin{aligned}
\text{Volume} &= (12 - 2x)(12 - 2x)x \\
&= (144 - 48x + 4x^2)x \\
&= (144x - 48x^2 + 4x^3) \text{ in}^3, \text{ or} \\
&\quad (4x^3 - 48x^2 + 144x) \text{ in}^3
\end{aligned}
$$

The outside surface area is the sum of the area of the bottom and the areas of the four sides. The dimensions, in inches, of the bottom are $12 - 2x$ by $12 - 2x$, and the dimensions, in inches, of each side are x by $12 - 2x$.

$$
\begin{aligned}
\text{Surface area} &= \text{Area of bottom} + \\
&\quad\quad 4 \cdot \text{Area of each side} \\
&= (12 - 2x)(12 - 2x) + 4 \cdot x(12 - 2x) \\
&= 144 - 24x - 24x + 4x^2 + 48x - 8x^2 \\
&= 144 - 48x + 4x^2 + 48x - 8x^2 \\
&= (144 - 4x^2) \text{ in}^2, \text{ or } (-4x^2 + 144) \text{ in}^2
\end{aligned}
$$

91. Let $n =$ the missing number.

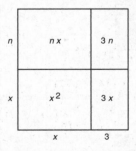

The area of the figure is $x^2 + 3x + nx + 3n$. This is equivalent to $x^2 + 8x + 15$, so we have $3x + nx = 8x$ and $3n = 15$. Solving either equation for n, we find that the missing number is 5.

93. We have a rectangular solid with dimensions x m by x m by $x+2$ m with a rectangular solid piece with dimensions 6 m by 5 m by 7 m cut out of it.

$$\text{Volume} = \begin{matrix}\text{Volume of}\\\text{large solid}\end{matrix} - \begin{matrix}\text{Volume of}\\\text{small solid}\end{matrix}$$

$$= (x\text{ m})(x\text{ m})(x+2\text{ m}) - (6\text{ m})(5\text{ m})(7\text{ m})$$
$$= x^2(x+2)\text{ m}^3 - 210\text{ m}^3$$
$$= (x^3 + 2x^2 - 210)\text{ m}^3$$

95. $(x-2)(x-7) - (x-7)(x-2)$

First observe that, by the commutative law of multiplication, $(x-2)(x-7)$ and $(x-7)(x-2)$ are equivalent expressions. Then when we subtract $(x-7)(x-2)$ from $(x-2)(x-7)$, the result is 0.

97.
$$(x-a)(x-b)\cdots(x-x)(x-y)(x-z)$$
$$= (x-a)(x-b)\cdots 0 \cdot (x-y)(x-z)$$
$$= 0$$

Exercise Set 12.6

1. $(x+1)(x^2+3)$

$\qquad$ F $\qquad$ O $\qquad$ I $\qquad$ L
$$= x \cdot x^2 + x \cdot 3 + 1 \cdot x^2 + 1 \cdot 3$$
$$= x^3 + 3x + x^2 + 3$$

3. $(x^3+2)(x+1)$

$\qquad$ F $\qquad$ O $\qquad$ I $\qquad$ L
$$= x^3 \cdot x + x^3 \cdot 1 + 2 \cdot x + 2 \cdot 1$$
$$= x^4 + x^3 + 2x + 2$$

5. $(y+2)(y-3)$

$\qquad$ F $\qquad$ O $\qquad$ I $\qquad$ L
$$= y \cdot y + y \cdot (-3) + 2 \cdot y + 2 \cdot (-3)$$
$$= y^2 - 3y + 2y - 6$$
$$= y^2 - y - 6$$

7. $(3x+2)(3x+2)$

$\qquad$ F $\qquad$ O $\qquad$ I $\qquad$ L
$$= 3x \cdot 3x + 3x \cdot 2 + 2 \cdot 3x + 2 \cdot 2$$
$$= 9x^2 + 6x + 6x + 4$$
$$= 9x^2 + 12x + 4$$

9. $(5x-6)(x+2)$

$\qquad$ F $\qquad$ O $\qquad$ I $\qquad$ L
$$= 5x \cdot x + 5x \cdot 2 + (-6) \cdot x + (-6) \cdot 2$$
$$= 5x^2 + 10x - 6x - 12$$
$$= 5x^2 + 4x - 12$$

11. $(3t-1)(3t+1)$

$\qquad$ F $\qquad$ O $\qquad$ I $\qquad$ L
$$= 3t \cdot 3t + 3t \cdot 1 + (-1) \cdot 3t + (-1) \cdot 1$$
$$= 9t^2 + 3t - 3t - 1$$
$$= 9t^2 - 1$$

13. $(4x-2)(x-1)$

$\qquad$ F $\qquad$ O $\qquad$ I $\qquad$ L
$$= 4x \cdot x + 4x \cdot (-1) + (-2) \cdot x + (-2) \cdot (-1)$$
$$= 4x^2 - 4x - 2x + 2$$
$$= 4x^2 - 6x + 2$$

15. $\left(p - \dfrac{1}{4}\right)\left(p + \dfrac{1}{4}\right)$

$\qquad$ F $\qquad$ O $\qquad$ I $\qquad$ L
$$= p \cdot p + p \cdot \frac{1}{4} + \left(-\frac{1}{4}\right) \cdot p + \left(-\frac{1}{4}\right) \cdot \frac{1}{4}$$
$$= p^2 + \frac{1}{4}p - \frac{1}{4}p - \frac{1}{16}$$
$$= p^2 - \frac{1}{16}$$

17. $(x-0.1)(x+0.1)$

$\qquad$ F $\qquad$ O $\qquad$ I $\qquad$ L
$$= x \cdot x + x \cdot (0.1) + (-0.1) \cdot x + (-0.1)(0.1)$$
$$= x^2 + 0.1x - 0.1x - 0.01$$
$$= x^2 - 0.01$$

19. $(2x^2+6)(x+1)$

$\qquad$ F $\qquad$ O $\qquad$ I $\qquad$ L
$$= 2x^3 + 2x^2 + 6x + 6$$

21. $(-2x+1)(x+6)$

$\qquad$ F $\qquad$ O $\qquad$ I $\qquad$ L
$$= -2x^2 - 12x + x + 6$$
$$= -2x^2 - 11x + 6$$

23. $(a+7)(a+7)$

$\qquad$ F $\qquad$ O $\qquad$ I $\qquad$ L
$$= a^2 + 7a + 7a + 49$$
$$= a^2 + 14a + 49$$

25. $(1+2x)(1-3x)$

$\qquad$ F $\qquad$ O $\qquad$ I $\qquad$ L
$$= 1 - 3x + 2x - 6x^2$$
$$= 1 - x - 6x^2$$

27. $\left(\dfrac{3}{8}y - \dfrac{5}{6}\right)\left(\dfrac{3}{8}y - \dfrac{5}{6}\right)$

$\qquad$ F $\qquad$ O $\qquad$ I $\qquad$ L
$$= \frac{9}{64}y^2 - \frac{15}{48}y - \frac{15}{48}y + \frac{25}{36}$$
$$= \frac{9}{64}y^2 - \frac{30}{48}y + \frac{25}{36}$$
$$= \frac{9}{64}y^2 - \frac{5}{8}y + \frac{25}{36}$$

29. $(x^2+3)(x^3-1)$

$\qquad$ F $\qquad$ O $\qquad$ I $\qquad$ L
$$= x^5 - x^2 + 3x^3 - 3$$

31. $(3x^2-2)(x^4-2)$

$\qquad$ F $\qquad$ O $\qquad$ I $\qquad$ L
$$= 3x^6 - 6x^2 - 2x^4 + 4$$

33. $(2.8x - 1.5)(4.7x + 9.3)$

$\qquad$ F $\qquad$ O $\qquad$ I $\qquad$ L

$= 2.8x(4.7x) + 2.8x(9.3) - 1.5(4.7x) - 1.5(9.3)$

$= 13.16x^2 + 26.04x - 7.05x - 13.95$

$= 13.16x^2 + 18.99x - 13.95$

35. $(3x^5 + 2)(2x^2 + 6)$

$\qquad$ F $\qquad$ O $\qquad$ I $\qquad$ L

$= 6x^7 + 18x^5 + 4x^2 + 12$

37. $(8x^3 + 1)(x^3 + 8)$

$\qquad$ F $\qquad$ O $\qquad$ I $\qquad$ L

$= 8x^6 + 64x^3 + x^3 + 8$

$= 8x^6 + 65x^3 + 8$

39. $(4x^2 + 3)(x - 3)$

$\qquad$ F $\qquad$ O $\qquad$ I $\qquad$ L

$= 4x^3 - 12x^2 + 3x - 9$

41. $(4y^4 + y^2)(y^2 + y)$

$\qquad$ F $\qquad$ O $\qquad$ I $\qquad$ L

$= 4y^6 + 4y^5 + y^4 + y^3$

43. $(x + 4)(x - 4)$ Product of sum and difference of two terms

$= x^2 - 4^2$

$= x^2 - 16$

45. $(2x + 1)(2x - 1)$ Product of sum and difference of two terms

$= (2x)^2 - 1^2$

$= 4x^2 - 1$

47. $(5m - 2)(5m + 2)$ Product of sum and difference of two terms

$= (5m)^2 - 2^2$

$= 25m^2 - 4$

49. $(2x^2 + 3)(2x^2 - 3)$ Product of sum and difference of two terms

$= (2x^2)^2 - 3^2$

$= 4x^4 - 9$

51. $(3x^4 - 4)(3x^4 + 4)$

$= (3x^4)^2 - 4^2$

$= 9x^8 - 16$

53. $(x^6 - x^2)(x^6 + x^2)$

$= (x^6)^2 - (x^2)^2$

$= x^{12} - x^4$

55. $(x^4 + 3x)(x^4 - 3x)$

$= (x^4)^2 - (3x)^2$

$= x^8 - 9x^2$

57. $(x^{12} - 3)(x^{12} + 3)$

$= (x^{12})^2 - 3^2$

$= x^{24} - 9$

59. $(2y^8 + 3)(2y^8 - 3)$

$= (2y^8)^2 - 3^2$

$= 4y^{16} - 9$

61. $\left(\dfrac{5}{8}x - 4.3\right)\left(\dfrac{5}{8}x + 4.3\right)$

$= \left(\dfrac{5}{8}x\right)^2 - (4.3)^2$

$= \dfrac{25}{64}x^2 - 18.49$

63. $(x + 2)^2 = x^2 + 2 \cdot x \cdot 2 + 2^2$ Square of a binomial sum

$\qquad\qquad = x^2 + 4x + 4$

65. $(3x^2 + 1)$ Square of a binomial sum

$= (3x^2)^2 + 2 \cdot 3x^2 \cdot 1 + 1^2$

$= 9x^4 + 6x^2 + 1$

67. $\left(a - \dfrac{1}{2}\right)^2$ Square of a binomial sum

$= a^2 - 2 \cdot a \cdot \dfrac{1}{2} + \left(\dfrac{1}{2}\right)^2$

$= a^2 - a + \dfrac{1}{4}$

69. $(3 + x)^2 = 3^2 + 2 \cdot 3 \cdot x + x^2$

$\qquad\qquad = 9 + 6x + x^2$

71. $(x^2 + 1)^2 = (x^2)^2 + 2 \cdot x^2 \cdot 1 + 1^2$

$\qquad\qquad = x^4 + 2x^2 + 1$

73. $(2 - 3x^4)^2 = 2^2 - 2 \cdot 2 \cdot 3x^4 + (3x^4)^2$

$\qquad\qquad = 4 - 12x^4 + 9x^8$

75. $(5 + 6t^2)^2 = 5^2 + 2 \cdot 5 \cdot 6t^2 + (6t^2)^2$

$\qquad\qquad = 25 + 60t^2 + 36t^4$

77. $\left(x - \dfrac{5}{8}\right)^2 = x^2 - 2 \cdot x \cdot \dfrac{5}{8} + \left(\dfrac{5}{8}\right)^2$

$\qquad\qquad = x^2 - \dfrac{5}{4}x + \dfrac{25}{64}$

79. $(3 - 2x^3)^2 = 3^2 - 2 \cdot 3 \cdot 2x^3 + (2x^3)^2$

$\qquad\qquad = 9 - 12x^3 + 4x^6$

81. $4x(x^2 + 6x - 3)$ Product of a monomial and a trinomial

$= 4x \cdot x^2 + 4x \cdot 6x + 4x(-3)$

$= 4x^3 + 24x^2 - 12x$

83. $\left(2x^2 - \frac{1}{2}\right)\left(2x^2 - \frac{1}{2}\right)$ Square of a binomial difference

$= (2x^2)^2 - 2 \cdot 2x^2 \cdot \frac{1}{2} + \left(\frac{1}{2}\right)^2$

$= 4x^4 - 2x^2 + \frac{1}{4}$

85. $(-1 + 3p)(1 + 3p)$

$= (3p - 1)(3p + 1)$ Product of the sum and difference of two terms

$= (3p)^2 - 1^2$

$= 9p^2 - 1$

87. $3t^2(5t^3 - t^2 + t)$ Product of a monomial and a trinomial

$= 3t^2 \cdot 5t^3 + 3t^2(-t^2) + 3t^2 \cdot t$

$= 15t^5 - 3t^4 + 3t^3$

89. $(6x^4 + 4)^2$ Square of a binomial sum

$= (6x^4)^2 + 2 \cdot 6x^4 \cdot 4 + 4^2$

$= 36x^8 + 48x^4 + 16$

91. $(3x + 2)(4x^2 + 5)$ Product of two binomials; use FOIL

$= 3x \cdot 4x^2 + 3x \cdot 5 + 2 \cdot 4x^2 + 2 \cdot 5$

$= 12x^3 + 15x + 8x^2 + 10$

93. $(8 - 6x^4)^2$ Square of a binomial difference

$= 8^2 - 2 \cdot 8 \cdot 6x^4 + (6x^4)^2$

$= 64 - 96x^4 + 36x^8$

95.
$$
\begin{array}{r}
t^2 + t + 1 \\
t - 1 \\
\hline
-t^2 - t - 1 \\
t^3 + t^2 + t \\
\hline
t^3 \qquad\quad -1
\end{array}
$$

97. $3^2 + 4^2 = 9 + 16 = 25$

$(3 + 4)^2 = 7^2 = 49$

99. $9^2 - 5^2 = 81 - 25 = 56$

$(9 - 5)^2 = 4^2 = 16$

101.

We can find the shaded area in two ways.

Method 1: The figure is a square with side $a + 1$, so the area is $(a + 1)^2 = a^2 + 2a + 1$.

Method 2: We add the areas of A, B, C, and D.

$1 \cdot a + 1 \cdot 1 + 1 \cdot a + a \cdot a = a + 1 + a + a^2 = a^2 + 2a + 1$.

Either way we find that the total shaded area is $a^2 + 2a + 1$.

103.

We can find the shaded area in two ways.

Method 1: The figure is a rectangle with dimensions $t + 6$ by $t + 4$, so the area is $(t + 6)(t + 4) = t^2 + 4t + 6t + 24 = t^2 + 10t + 24$.

Method 2: We add the areas of A, B, C, and D.

$t \cdot t + t \cdot 6 + 6 \cdot 4 + 4 \cdot t = t^2 + 6t + 24 + 4t = t^2 + 10t + 24$.

Either way, we find that the total shaded area is $t^2 + 10t + 24$.

105. Discussion and Writing Exercise

107. *Familiarize*. Let $t =$ the number of watts used by the television set. Then $10t =$ the number of watts used by the lamps, and $40t =$ the number of watts used by the air conditioner.

Translate.

Lamp watts	+	Air conditioner watts	+	Television watts	=	Total watts
↓	↓	↓	↓	↓	↓	↓
$10t$	+	$40t$	+	t	=	2550

Solve. We solve the equation.

$$10t + 40t + t = 2550$$
$$51t = 2550$$
$$t = 50$$

The possible solution is:

Television, t: 50 watts

Lamps, $10t$: $10 \cdot 50$, or 500 watts

Air conditioner, $40t$: $40 \cdot 50$, or 2000 watts

Check. The number of watts used by the lamps, 500, is 10 times 50, the number used by the television. The number of watts used by the air conditioner, 2000, is 40 times 50, the number used by the television. Also, $50 + 500 + 2000 = 2550$, the total wattage used.

State. The television uses 50 watts, the lamps use 500 watts, and the air conditioner uses 2000 watts.

109. $3(x-2) = 5(2x+7)$

$3x - 6 = 10x + 35$ Removing parentheses

$3x - 6 + 6 = 10x + 35 + 6$ Adding 6

$3x = 10x + 41$

$3x - 10x = 10x + 41 - 10x$ Subtracting $10x$

$-7x = 41$

$\dfrac{-7x}{-7} = \dfrac{41}{-7}$ Dividing by -7

$x = -\dfrac{41}{7}$

The solution is $-\dfrac{41}{7}$.

111. $3x - 2y = 12$

$-2y = -3x + 12$ Subtracting $3x$

$\dfrac{-2y}{-2} = \dfrac{-3x+12}{-2}$ Dividing by -2

$y = \dfrac{3x-12}{2}$, or

$y = \dfrac{3}{2}x - 6$

113. $5x(3x-1)(2x+3)$

$= 5x(6x^2 + 7x - 3)$ Using FOIL

$= 30x^3 + 35x^2 - 15x$

115. $[(a-5)(a+5)]^2$

$= (a^2 - 25)^2$ Finding the product of a sum
and difference of same two terms

$= a^4 - 50a^2 + 625$ Squaring a binomial

117. $(3t^4 - 2)^2 1(3t^4 + 2)^2$

$= [(3t^4 - 2)(3t^4 + 2)]^2$

$= (9t^8 - 4)^2$

$= 81t^{16} - 72t^8 + 16$

119. $(x+2)(x-5) = (x+1)(x-3)$

$x^2 - 5x + 2x - 10 = x^2 - 3x + x - 3$

$x^2 - 3x - 10 = x^2 - 2x - 3$

$-3x - 10 = -2x - 3$ Adding $-x^2$

$-3x + 2x = 10 - 3$ Adding $2x$ and 10

$-x = 7$

$x = -7$

The solution is -7.

121. See the answer section in the text.

123. Enter $y_1 = (x-1)^2$ and $y_2 = x^2 - 2x + 1$. Then compare the graphs or the y_1-and y_2-values in a table. It appears that the graphs are the same and that the y_1-and y_2-values are the same, so $(x-1)^2 = x^2 - 2x + 1$ is correct.

125. Enter $y_1 = (x-3)(x+3)$ and $y_2 = x^2 - 6$. Then compare the graphs or the y_1-and y_2-values in a table. The graphs are not the same nor are the y_1-and y_2-values, so $(x-3)(x+3) = x^2 - 6$ is not correct.

Exercise Set 12.7

1. We replace x by 3 and y by -2.

$x^2 - y^2 + xy = 3^2 - (-2)^2 + 3(-2) = 9 - 4 - 6 = -1$

3. We replace x by 3 and y by -2.

$x^2 - 3y^2 + 2xy = 3^2 - 3(-2)^2 + 2 \cdot 3(-2) =$
$9 - 3 \cdot 4 + 2 \cdot 3(-2) = 9 - 12 - 12 = -15$

5. We replace x by 3, y by -2, and z by -5.

$8xyz = 8 \cdot 3 \cdot (-2) \cdot (-5) = 240$

7. We replace x by 3, y by -2, and z by -5.

$xyz^2 - z = 3(-2)(-5)^2 - (-5) = 3(-2)(25) - (-5) =$
$-150 + 5 = -145$

9. We replace h by 165 and A by 20.

$C = 0.041h - 0.018A - 2.69$
$= 0.041(165) - 0.018(20) - 2.69$
$= 6.765 - 0.36 - 2.69$
$= 6.405 - 2.69$
$= 3.715$

The lung capacity of a 20-year-old woman who is 165 cm tall is 3.715 liters.

11. Evaluate the polynomial for $h = 32$, $v = 40$, and $t = 2$.

$h = h_0 + vt - 4.9t^2$
$= 32 + 40 \cdot 2 - 4.9(2)^2$
$= 32 + 80 - 19.6$
$= 92.4$

The rocket will be 92.4 m above the ground 2 seconds after blast off.

13. Replace h by 4.7, r by 1.2, and π by 3.14.

$S = 2\pi rh + 2\pi r^2$
$\approx 2(3.14)(1.2)(4.7) + 2(3.14)(1.2)^2$
$\approx 2(3.14)(1.2)(4.7) + 2(3.14)(1.44)$
$\approx 35.4192 + 9.0432$
≈ 44.46

The surface area of the can is about 44.46 in^2.

15. Evaluate the polynomial for $h = 7\dfrac{1}{2}$, or $\dfrac{15}{2}$, $r = 1\dfrac{1}{4}$, or $\dfrac{5}{4}$, and $\pi \approx 3.14$.

$S = 2\pi rh + \pi r^2$

$\approx 2(3.14)\left(\dfrac{5}{4}\right)\left(\dfrac{15}{2}\right) + (3.14)\left(\dfrac{5}{4}\right)^2$

$\approx 2(3.14)\left(\dfrac{5}{4}\right)\left(\dfrac{15}{2}\right) + (3.14)\left(\dfrac{25}{16}\right)$

$\approx 58.875 + 4.90625$

≈ 63.78125

The surface area is about 63.78125 in^2.

17. $x^3y - 2xy + 3x^2 - 5$

Term	Coefficient	Degree	
x^3y	1	4	(Think: $x^3y = x^3y^1$)
$-2xy$	-2	2	(Think: $-2xy = -2x^1y^1$)
$3x^2$	3	2	
-5	-5	0	(Think: $-5 = -5x^0$)

The degree of the polynomial is the degree of the term of highest degree. The term of highest degree is x^3y. Its degree is 4. The degree of the polynomial is 4.

19. $17x^2y^3 - 3x^3yz - 7$

Term	Coefficient	Degree	
$17x^2y^3$	17	5	
$-3x^3yz$	-3	5	(Think: $-3x^3yz =$
			$-3x^3y^1z^1$)
-7	-7	0	(Think: $-7 = -7x^0$)

The terms of highest degree are $17x^2y^3$ and $-3x^3yz$. Each has degree 5. The degree of the polynomial is 5.

21. $a + b - 2a - 3b = (1-2)a + (1-3)b = -a - 2b$

23. $3x^2y - 2xy^2 + x^2$

There are *no* like terms, so none of the terms can be collected.

25.
$$6au + 3av + 14au + 7av$$
$$= (6 + 14)au + (3 + 7)av$$
$$= 20au + 10av$$

27.
$$2u^2v - 3uv^2 + 6u^2v - 2uv^2$$
$$= (2 + 6)u^2v + (-3 - 2)uv^2$$
$$= 8u^2v - 5uv^2$$

29.
$$(2x^2 - xy + y^2) + (-x^2 - 3xy + 2y^2)$$
$$= (2 - 1)x^2 + (-1 - 3)xy + (1 + 2)y^2$$
$$= x^2 - 4xy + 3y^2$$

31.
$$(r - 2s + 3) + (2r + s) + (s + 4)$$
$$= (1 + 2)r + (-2 + 1 + 1)s + (3 + 4)$$
$$= 3r + 0s + 7$$
$$= 3r + 7$$

33.
$$(b^3a^2 - 2b^2a^3 + 3ba + 4) + (b^2a^3 - 4b^3a^2 + 2ba - 1)$$
$$= (1 - 4)b^3a^2 + (-2 + 1)b^2a^3 + (3 + 2)ba + (4 - 1)$$
$$= -3b^3a^2 - b^2a^3 + 5ba + 3$$

35.
$$(a^3 + b^3) - (a^2b - ab^2 + b^3 + a^3)$$
$$= a^3 + b^3 - a^2b + ab^2 - b^3 - a^3$$
$$= (1 - 1)a^3 - a^2b + ab^2 + (1 - 1)b^3$$
$$= -a^2b + ab^2, \text{ or } ab^2 - a^2b$$

37.
$$(xy - ab - 8) - (xy - 3ab - 6)$$
$$= xy - ab - 8 - xy + 3ab + 6$$
$$= (1 - 1)xy + (-1 + 3)ab + (-8 + 6)$$
$$= 2ab - 2$$

39.
$$(-2a + 7b - c) - (-3b + 4c - 8d)$$
$$= -2a + 7b - c + 3b - 4c + 8d$$
$$= -2a + (7 + 3)b + (-1 - 4)c + 8d$$
$$= -2a + 10b - 5c + 8d$$

41.
$$\overset{\text{F \quad O \quad I \quad L}}{(3z - u)(2z + 3u)} = 6z^2 + 9zu - 2uz - 3u^2$$
$$= 6z^2 + 7zu - 3u^2$$

43.
$$\overset{\text{F \quad O \quad I \quad L}}{(a^2b - 2)(a^2b - 5)} = a^4b^2 - 5a^2b - 2a^2b + 10$$
$$= a^4b^2 - 7a^2b + 10$$

45.
$$(a^3 + bc)(a^3 - bc) = (a^3)^2 - (bc)^2$$
$$[(A + B)(A - B) = A^2 - B^2]$$
$$= a^6 - b^2c^2$$

47.
$$\begin{array}{r} y^4x + y^2 + 1 \\ y^2 + 1 \\ \hline y^4x + y^2 + 1 \\ y^6x + y^4 \quad\quad + y^2 \\ \hline y^6x + y^4 + y^4x + 2y^2 + 1 \end{array}$$

49. $(3xy - 1)(4xy + 2)$
$$\overset{\text{F \quad O \quad I \quad L}}{}$$
$$= 12x^2y^2 + 6xy - 4xy - 2$$
$$= 12x^2y^2 + 2xy - 2$$

51. $(3 - c^2d^2)(4 + c^2d^2)$
$$\overset{\text{F \quad O \quad I \quad L}}{}$$
$$= 12 + 3c^2d^2 - 4c^2d^2 - c^4d^4$$
$$= 12 - c^2d^2 - c^4d^4$$

53. $(m^2 - n^2)(m + n)$
$$\overset{\text{F \quad O \quad I \quad L}}{}$$
$$= m^3 + m^2n - mn^2 - n^3$$

55. $(xy + x^5y^5)(x^4y^4 - xy)$
$$\overset{\text{F \quad O \quad I \quad L}}{}$$
$$= x^5y^5 - x^2y^2 + x^9y^9 - x^6y^6$$
$$= x^9y^9 - x^6y^6 + x^5y^5 - x^2y^2$$

57. $(x + h)^2$
$$= x^2 + 2xh + h^2 \quad [(A + B)^2 = A^2 + 2AB + B^2]$$

59. $(r^3t^2 - 4)^2$
$$= (r^3t^2)^2 - 2 \cdot r^3t^2 \cdot 4 + 4^2$$
$$[(A - B)^2 = A^2 - 2AB + B^2]$$
$$= r^6t^4 - 8r^3t^2 + 16$$

61. $(p^4 + m^2n^2)^2$
$$= (p^4)^2 + 2 \cdot p^4 \cdot m^2n^2 + (m^2n^2)^2$$
$$[(A + B)^2 = A^2 + 2AB + B^2]$$
$$= p^8 + 2p^4m^2n^2 + m^4n^4$$

63. $\left(2a^3 - \frac{1}{2}b^3\right)^2$
$$= (2a^3)^2 - 2 \cdot 2a^3 \cdot \frac{1}{2}b^3 + \left(\frac{1}{2}b^3\right)^2$$
$$[(A - B)^2 = A^2 - 2AB + B^2]$$
$$= 4a^6 - 2a^3b^3 + \frac{1}{4}b^6$$

65. $3a(a - 2b)^2 = 3a(a^2 - 4ab + 4b^2)$
$$= 3a^3 - 12a^2b + 12ab^2$$

67. $(2a - b)(2a + b) = (2a)^2 - b^2 = 4a^2 - b^2$

69. $(c^2 - d)(c^2 + d) = (c^2)^2 - d^2$
$$= c^4 - d^2$$

71. $(ab + cd^2)(ab - cd^2) = (ab)^2 - (cd^2)^2$
$$= a^2b^2 - c^2d^4$$

73. $\quad(x + y - 3)(x + y + 3)$
$= [(x + y) - 3][(x + y) + 3]$
$= (x + y)^2 - 3^2$
$= x^2 + 2xy + y^2 - 9$

75. $\quad[x + y + z][x - (y + z)]$
$= [x + (y + z)][x - (y + z)]$
$= x^2 - (y + z)^2$
$= x^2 - (y^2 + 2yz + z^2)$
$= x^2 - y^2 - 2yz - z^2$

77. $\quad(a + b + c)(a - b - c)$
$= [a + (b + c)][a - (b + c)]$
$= a^2 - (b + c)^2$
$= a^2 - (b^2 + 2bc + c^2)$
$= a^2 - b^2 - 2bc - c^2$

79.
$$
\begin{array}{r}
x^2 \quad - \ 4y \ + \ 2 \\
3x^2 \ + \ 5y \ - \ 3 \\
\hline
-3x^2 + 12y - 6 \\
+ \ 10y \quad - \ 20y^2 + \ 5x^2y \\
6x^2 \qquad\qquad - \ 12x^2y + 3x^4 \\
\hline
3x^2 \ + 22y - 6 - 20y^2 - 7x^2y + 3x^4
\end{array}
$$

We could also write the result as
$3x^4 - 7x^2y + 3x^2 - 20y^2 + 22y - 6$.

81. Discussion and Writing Exercise

83. The first coordinate is positive and the second coordinate is negative, so $(2, -5)$ is in quadrant IV.

85. Both coordinates are positive, so $(16, 23)$ is in quadrant I.

87. $2x = -10$
$\quad\ x = -5$

Any ordered pair $(-5, y)$ is a solution. The variable x must be -5, but y can be any number we choose. A few solutions are listed below. Plot these points and draw the line.

x	y
-5	-3
-5	0
-5	4

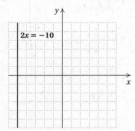

89. $8y - 16 = 0$
$\qquad 8y = 16$
$\qquad\ \ y = 2$

Any ordered pair $(x, 2)$ is a solution. The variable y must be 2, but x can be any number we choose. A few solutions are listed below. Plot these points and draw the line.

x	y
-4	2
0	2
3	2

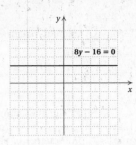

91. It is helpful to add additional labels to the figure.

The area of the large square is $x \cdot x$, or x^2. The area of the small square is $(x - 2y)(x - 2y)$, or $(x - 2y)^2$.

$$
\begin{array}{ccc}
\text{Area of shaded} & = & \text{Area of large} \quad - \quad \text{Area of small} \\
\text{region} & & \text{square} \qquad\qquad \text{square}
\end{array}
$$

$$
\begin{array}{ccc}
\text{Area of shaded} & = & x^2 \qquad - \qquad (x - 2y)^2 \\
\text{region} & &
\end{array}
$$

$$= x^2 - (x^2 - 4xy + 4y^2)$$
$$= x^2 - x^2 + 4xy - 4y^2$$
$$= 4xy - 4y^2$$

93. It is helpful to add additional labels to the figure.

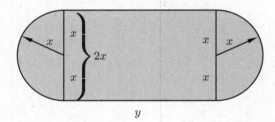

The two semicircles make a circle with radius x. The area of that circle is πx^2. The area of the rectangle is $2x \cdot y$. The sum of the two regions, $\pi x^2 + 2xy$, is the area of the shaded region.

95. The lateral surface area of the outer portion of the solid is the lateral surface area of a right circular cylinder with radius n and height h. The lateral surface area of the inner portion is the lateral surface area of a right circular cylinder with radius m and height h. Recall that the formula for the lateral surface area of a right circular cylinder with radius r and height h is $2\pi rh$.

The surface area of the top is the area of a circle with radius n less the area of a circle with radius m. The surface area of the bottom is the same as the surface area of the top.

Thus, the surface area of the solid is
$$2\pi nh + 2\pi mh + 2\pi n^2 - 2\pi m^2.$$

97. In the formula for the surface area of a silo, $S = 2\pi rh + \pi r^2$, the term πr^2 represents the area of the base. Since the base of the observatory rests on the ground, it will not need to be painted. Thus, we will subtract this term from the formula and find the remaining surface area, $2\pi rh$.

The height of the observatory is 40 ft and its radius is 30/2, or 15 ft, so the surface area is $2\pi rh \approx 2(3.14)(15)(40) \approx 3768$ ft^2. Since 3768 ft^2/250 ft^2 = 15.072, 16 gallons of paint should be purchased.

99. Substitute \$10,400 for P, 8.5% or 0.085 for r, and 5 for t.

$$P(1+r)^t$$
$$= \$10,400(1 + 0.085)^5$$
$$= \$10,400(1.085)^5$$
$$\approx \$15,638.03$$

Exercise Set 12.8

1. $\dfrac{24x^4}{8} = \dfrac{24}{8} \cdot x^4 = 3x^4$

Check: We multiply.
$$3x^4 \cdot 8 = 24x^4$$

3. $\dfrac{25x^3}{5x^2} = \dfrac{25}{5} \cdot \dfrac{x^3}{x^2} = 5x^{3-2} = 5x$

Check: We multiply.
$$5x \cdot 5x^2 = 25x^3$$

5. $\dfrac{-54x^{11}}{-3x^8} = \dfrac{-54}{-3} \cdot \dfrac{x^{11}}{x^8} = 18x^{11-8} = 18x^3$

Check: We multiply.
$$18x^3(-3x^8) = -54x^{11}$$

7. $\dfrac{64a^5b^4}{16a^2b^3} = \dfrac{64}{16} \cdot \dfrac{a^5}{a^2} \cdot \dfrac{b^4}{b^3} = 4a^{5-2}b^{4-3} = 4a^3b$

Check: We multiply.
$$(4a^3b)(16a^2b^3) = 64a^5b^4$$

9. $\dfrac{24x^4 - 4x^3 + x^2 - 16}{8}$

$$= \dfrac{24x^4}{8} - \dfrac{4x^3}{8} + \dfrac{x^2}{8} - \dfrac{16}{8}$$
$$= 3x^4 - \dfrac{1}{2}x^3 + \dfrac{1}{8}x^2 - 2$$

Check: We multiply.
$$\begin{array}{r} 3x^4 - \dfrac{1}{2}x^3 + \dfrac{1}{8}x^2 - 2 \\ \underline{8} \\ 24x^4 - 4x^3 + x^2 - 16 \end{array}$$

11. $\dfrac{u - 2u^2 - u^5}{u}$

$$= \dfrac{u}{u} - \dfrac{2u^2}{u} - \dfrac{u^5}{u}$$
$$= 1 - 2u - u^4$$

Check: We multiply.

$$\begin{array}{r} 1 - 2u - u^4 \\ \underline{u} \\ u - 2u^2 - u^5 \end{array}$$

13. $(15t^3 + 24t^2 - 6t) \div (3t)$

$$= \dfrac{15t^3 + 24t^2 - 6t}{3t}$$
$$= \dfrac{15t^3}{3t} + \dfrac{24t^2}{3t} - \dfrac{6t}{3t}$$
$$= 5t^2 + 8t - 2$$

Check: We multiply.
$$\begin{array}{r} 5t^2 + 8t - 2 \\ \underline{3t} \\ 15t^3 + 24t^2 - 6t \end{array}$$

15. $(20x^6 - 20x^4 - 5x^2) \div (-5x^2)$

$$= \dfrac{20x^6 - 20x^4 - 5x^2}{-5x^2}$$
$$= \dfrac{20x^6}{-5x^2} - \dfrac{20x^4}{-5x^2} - \dfrac{5x^2}{-5x^2}$$
$$= -4x^4 - (-4x^2) - (-1)$$
$$= -4x^4 + 4x^2 + 1$$

Check: We multiply.
$$\begin{array}{r} -4x^4 + 4x^2 + 1 \\ \underline{-5x^2} \\ 20x^6 - 20x^4 - 5x^2 \end{array}$$

17. $(24x^5 - 40x^4 + 6x^3) \div (4x^3)$

$$= \dfrac{24x^5 - 40x^4 + 6x^3}{4x^3}$$
$$= \dfrac{24x^5}{4x^3} - \dfrac{40x^4}{4x^3} + \dfrac{6x^3}{4x^3}$$
$$= 6x^2 - 10x + \dfrac{3}{2}$$

Check: We multiply.
$$\begin{array}{r} 6x^2 - 10x + \dfrac{3}{2} \\ \underline{4x^3} \\ 24x^5 - 40x^4 + 6x^3 \end{array}$$

19. $\dfrac{18x^2 - 5x + 2}{2}$

$$= \dfrac{18x^2}{2} - \dfrac{5x}{2} + \dfrac{2}{2}$$
$$= 9x^2 - \dfrac{5}{2}x + 1$$

Check: We multiply.
$$\begin{array}{r} 9x^2 - \dfrac{5}{2}x + 1 \\ \underline{2} \\ 18x^2 - 5x + 2 \end{array}$$

21. $\dfrac{12x^3 + 26x^2 + 8x}{2x}$

$= \dfrac{12x^3}{2x} + \dfrac{26x^2}{2x} + \dfrac{8x}{2x}$

$= 6x^2 + 13x + 4$

Check: We multiply.

$$\begin{array}{r} 6x^2 + 13x + 4 \\ 2x \\ \hline 12x^3 + 26x^2 + 8x \end{array}$$

23. $\dfrac{9r^2s^2 + 3r^2s - 6rs^2}{3rs}$

$= \dfrac{9r^2s^2}{3rs} + \dfrac{3r^2s}{3rs} - \dfrac{6rs^2}{3rs}$

$= 3rs + r - 2s$

Check: We multiply.

$$\begin{array}{r} 3rs + r - 2s \\ 3rs \\ \hline 9r^2s^2 + 3r^2s - 6rs^2 \end{array}$$

25.
$$\begin{array}{r} x + 2 \\ x+2 \enclose{longdiv}{x^2+4x+4} \\ \underline{x^2+2x} \\ 2x+4 \leftarrow (x^2+4x)-(x^2+2x) \\ \underline{2x+4} \\ 0 \leftarrow (2x+4)-(2x+4) \end{array}$$

The answer is $x + 2$.

27.
$$\begin{array}{r} x - 5 \\ x-5 \enclose{longdiv}{x^2-10x-25} \\ \underline{x^2-5x} \\ -5x-25 \leftarrow (x^2-10x)-(x^2-5x) \\ \underline{-5x+25} \\ -50 \leftarrow (-5x-25)-(-5x+25) \end{array}$$

The answer is $x - 5 + \dfrac{-50}{x-5}$.

29.
$$\begin{array}{r} x - 2 \\ x+6 \enclose{longdiv}{x^2+4x-14} \\ \underline{x^2+6x} \\ -2x-14 \leftarrow (x^2+4x)-(x^2+6x) \\ \underline{-2x-12} \\ -2 \leftarrow (-2x-14)-(-2x-12) \end{array}$$

The answer is $x - 2 + \dfrac{-2}{x+6}$.

31.
$$\begin{array}{r} x - 3 \\ x+3 \enclose{longdiv}{x^2+0x-9} \leftarrow \text{Filling in the missing term} \\ \underline{x^2+3x} \\ -3x-9 \leftarrow x^2-(x^2+3x) \\ \underline{-3x-9} \\ 0 \leftarrow (-3x-9)-(-3x-9) \end{array}$$

The answer is $x - 3$.

33.
$$\begin{array}{r} x^4 - x^3 + x^2 - x + 1 \\ x+1 \enclose{longdiv}{x^5+0x^4+0x^3+0x^2+0x+1} \leftarrow \text{Filling in missing} \\ \text{terms} \\ \underline{x^5+x^4} \\ -x^4 \quad \leftarrow x^5-(x^5+x^4) \\ \underline{-x^4-x^3} \\ x^3 \quad \leftarrow -x^4-(-x^4-x^3) \\ \underline{x^3+x^2} \\ -x^2 \quad \leftarrow x^3-(x^3+x^2) \\ \underline{-x^2-x} \\ x+1 \leftarrow -x^2-(-x^2-x) \\ \underline{x+1} \\ 0 \leftarrow (x+1)-(x+1) \end{array}$$

The answer is $x^4 - x^3 + x^2 - x + 1$.

35.
$$\begin{array}{r} 2x^2 - 7x + 4 \\ 4x+3 \enclose{longdiv}{8x^3-22x^2-5x+12} \\ \underline{8x^3+6x^2} \\ -28x^2-5x \leftarrow (8x^3-22x^2)-(8x^3+6x^2) \\ \underline{-28x^2-21x} \\ 16x+12 \leftarrow (-28x^2-5x)- \\ (-28x^2-21x) \\ \underline{16x+12} \\ 0 \leftarrow (16x+12)-(16x+12) \end{array}$$

The answer is $2x^2 - 7x + 4$.

37.
$$\begin{array}{r} x^3 - 6 \\ x^3-7 \enclose{longdiv}{x^6-13x^3+42} \\ \underline{x^6-7x^3} \\ -6x^3+42 \leftarrow (x^6-13x^3)-(x^6-7x^3) \\ \underline{-6x^3+42} \\ 0 \leftarrow (-6x^3+42)-(-6x^3+42) \end{array}$$

The answer is $x^3 - 6$.

39.
$$\begin{array}{r} 3x^2 + x + 2 \\ 5x+1 \enclose{longdiv}{15x^3+8x^2+11x+12} \\ \underline{15x^3+3x^2} \\ 5x^2+11x \\ \underline{5x^2+x} \\ 10x+12 \\ \underline{10x+2} \\ 10 \end{array}$$

The answer is $3x^2 + x + 2 + \dfrac{10}{5x+1}$.

41.
$$\begin{array}{r} t^2 + 1 \\ t-1 \enclose{longdiv}{t^3-t^2+t-1} \\ \underline{t^3-t^2} \quad \leftarrow (t^3-t^2)-(t^3-t^2) \\ 0+t-1 \\ \underline{t-1} \leftarrow (t-1)-(t-1) \\ 0 \end{array}$$

The answer is $t^2 + 1$.

43. Discussion and Writing Exercise

45. The product rule asserts that when multiplying with exponential notation, if the bases are the same, keep the base and add the exponents.

47. The <u>multiplication</u> principle asserts that when we multiply or divide by the same nonzero number on each side of an equation, we get equivalent equations.

49. A <u>trinomial</u> is a polynomial with three terms, such as $5x^4 - 7x^2 + 4$.

51. The <u>absolute value</u> of a number is its distance from zero on a number line.

53.
$$
\begin{array}{r}
x^2 + 5 \\
x^2 + 4 \overline{\smash{\big)}\ x^4 + 9x^2 + 20} \\
\underline{x^4 + 4x^2} \\
5x^2 + 20 \\
\underline{5x^2 + 20} \\
0
\end{array}
$$

The answer is $x^2 + 5$.

55.
$$
\begin{array}{r}
a + 3 \\
5a^2 - 7a - 2 \overline{\smash{\big)}\ 5a^3 + 8a^2 - 23a - 1} \\
\underline{5a^3 - 7a^2 - 2a} \\
15a^2 - 21a - 1 \\
\underline{15a^2 - 21a - 6} \\
5
\end{array}
$$

The answer is $a + 3 + \dfrac{5}{5a^2 - 7a - 2}$.

57. We rewrite the dividend in descending order.
$$
\begin{array}{r}
2x^2 + x - 3 \\
3x^3 - 2x - 1 \overline{\smash{\big)}\ 6x^5 + 3x^4 - 13x^3 - 4x^2 + 5x + 3} \\
\underline{6x^5 \qquad\ - 4x^3 - 2x^2} \\
3x^4 - 9x^3 - 2x^2 + 5x \\
\underline{3x^4 \qquad\ - 2x^2 - x} \\
-9x^3 \qquad\ + 6x + 3 \\
\underline{-9x^3 \qquad\ + 6x + 3} \\
0
\end{array}
$$

The answer is $2x^2 + x - 3$.

59.
$$
\begin{array}{r}
a^5 + a^4b + a^3b^2 + a^2b^3 + ab^4 + b^5 \\
a - b \overline{\smash{\big)}\ a^6 + 0a^5b + 0a^4b^2 + 0a^3b^3 + 0a^2b^4 + 0ab^5 - b^6} \\
\underline{a^6 - a^5b} \\
a^5b \\
\underline{a^5b - a^4b^2} \\
a^4b^2 \\
\underline{a^4b^2 - a^3b^3} \\
a^3b^3 \\
\underline{a^3b^3 - a^2b^4} \\
a^2b^4 \\
\underline{a^2b^4 - ab^5} \\
ab^5 - b^6 \\
\underline{ab^5 - b^6} \\
0
\end{array}
$$

The answer is $a^5 + a^4b + a^3b^2 + a^2b^3 + ab^4 + b^5$.

61.
$$
\begin{array}{r}
x + 5 \\
x - 1 \overline{\smash{\big)}\ x^2 + 4x + c} \\
\underline{x^2 - x} \\
5x + c \\
\underline{5x - 5} \\
c + 5
\end{array}
$$

We set the remainder equal to 0.
$$
c + 5 = 0
$$
$$
c = -5
$$

Thus, c must be -5.

63.
$$
\begin{array}{r}
c^2 x + (-2c + c^2) \\
x - 1 \overline{\smash{\big)}\ c^2 x^2 - 2cx + 1} \\
\underline{c^2 x^2 - c^2 x} \\
(-2c + c^2)x + 1 \\
\underline{(-2c + c^2)x - (-2c + c^2)} \\
1 + (-2c + c^2)
\end{array}
$$

We set the remainder equal to 0.
$$
c^2 - 2c + 1 = 0
$$
$$
(c - 1)^2 = 0
$$
$$
c = 1
$$

Thus, c must be 1.

Chapter 12 Review Exercises

1. $7^2 \cdot 7^{-4} = 7^{2+(-4)} = 7^{-2} = \dfrac{1}{7^2}$

2. $y^7 \cdot y^3 \cdot y = y^{7+3+1} = y^{11}$

3. $(3x)^5(3x)^9 = (3x)^{5+9} = (3x)^{14}$

4. $t^8 \cdot t^0 = t^8 \cdot 1 = t^8$, or
$t^8 \cdot t^0 = t^{8+0} = t^8$

5. $\dfrac{4^5}{4^2} = 4^{5-2} = 4^3$

6. $\dfrac{a^5}{a^8} = a^{5-8} = a^{-3} = \dfrac{1}{a^3}$

7. $\dfrac{(7x)^4}{(7x)^4} = 1$

8. $(3t^4)^2 = 3^2 \cdot (t^4)^2 = 9 \cdot t^{4 \cdot 2} = 9t^8$

9. $(2x^3)^2(-3x)^2 = 2^2 \cdot (x^3)^2(-3)^2 x^2 = 4 \cdot x^6 \cdot 9 \cdot x^2 = 36x^8$

10. $\left(\dfrac{2x}{y}\right)^{-3} = \left(\dfrac{y}{2x}\right)^3 = \dfrac{y^3}{2^3 \cdot x^3} = \dfrac{y^3}{8x^3}$

11. $\dfrac{1}{t^5} = t^{-5}$

12. $y^{-4} = \dfrac{1}{y^4}$

13. $0.000003.28$

$\underline{\qquad\qquad}\uparrow$ 6 places

Small number, so the exponent is negative.

$0.00000328 = 3.28 \times 10^{-6}$

14. 8.3×10^6

$8.300000.$

$\underline{\qquad\uparrow}$ 6 places

Positive exponent, so the answer is a large number.

$8.3 \times 10^6 = 8,300,000$

15. $(3.8 \times 10^4)(5.5 \times 10^{-1}) = (3.8 \cdot 5.5) \times (10^4 \cdot 10^{-1})$
$$= 20.9 \times 10^3$$
$$= (2.09 \times 10) \times 10^3$$
$$= 2.09 \times 10^4$$

16. $\dfrac{1.28 \times 10^{-8}}{2.5 \times 10^{-4}} = \dfrac{1.28}{2.5} \times \dfrac{10^{-8}}{10^{-4}}$
$$= 0.512 \times 10^{-4}$$
$$= (5.12 \times 10^{-1}) \times 10^{-4}$$
$$= 5.12 \times 10^{-5}$$

17. 292 million $= 292 \times 10^6$
$$= (2.92 \times 10^2) \times 10^6$$
$$= 2.92 \times 10^8$$

Also, $15.3 = 1.53 \times 10$. Then we have

$(2.92 \times 10^8)(1.53 \times 10) = (2.92 \cdot 1.53) \times (10^8 \cdot 10)$
$$= 4.4676 \times 10^9$$

In 2005, 4.4676×10^9 gal of diet drinks were consumed in the United States.

18. $x^2 - 3x + 6 = (-1)^2 - 3(-1) + 6 = 1 + 3 + 6 = 10$

19. $-4y^5 + 7y^2 - 3y - 2 = -4y^5 + 7y^2 + (-3y) + (-2)$

The terms are $-4y^5$, $7y^2$, $-3y$, and -2.

20. In the polynomial $x^3 + x$ there are no x^2 or x^0 terms. Thus, the x^2 term (or second-degree term) and the x^0 term (or zero-degree term) are missing.

21. $4x^3 + 6x^2 - 5x + \dfrac{5}{3} = 4x^3 + 6x^2 - 5x^1 + \dfrac{5}{3}x^0$

The degree of $4x^3$ is 3.

The degree of $6x^2$ is 2.

The degree of $-5x$ is 1.

The degree of $\dfrac{5}{3}$ is 0.

The degree of the polynomial is 3, the largest exponent.

22. The polynomial $4x^3 - 1$ is a binomial because it has just two terms.

23. The polynomial $4 - 9t^3 - 7t^4 + 10t^2$ is none of these because it has more than three terms.

24. The polynomial $7y^2$ is a monomial because it has just one term.

25. $3x^2 - 2x + 3 - 5x^2 - 1 - x$
$$= (3-5)x^2 + (-2-1)x + (3-1)$$
$$= -2x^2 - 3x + 2$$

26. $-x + \dfrac{1}{2} + 14x^4 - 7x^2 - 1 - 4x^4$
$$= (14 - 4)x^4 - 7x^2 - x + \left(\dfrac{1}{2} - 1\right)$$
$$= 10x^4 - 7x^2 - x - \dfrac{1}{2}$$

27. $(3x^4 - x^3 + x - 4) + (x^5 + 7x^3 - 3x^2 - 5) + (-5x^4 + 6x^2 - x) = (3-5)x^4 + (-1+7)x^3 + (1-1)x + (-4-5) + x^5 + (-3+6)x^2 = -2x^4 + 6x^3 - 9 + x^5 + 3x^2$, or $x^5 - 2x^4 + 6x^3 + 3x^2 - 9$

28. $(3x^5 - 4x^4 + x^3 - 3) + (3x^4 - 5x^3 + 3x^2) + (-5x^5 - 5x^2) + (-5x^4 + 2x^3 + 5) = (3-5)x^5 + (-4+3-5)x^4 + (1-5+2)x^3 + (-3+5) + (3-5)x^2 = -2x^5 - 6x^4 - 2x^3 + 2 - 2x^2$, or $-2x^5 - 6x^4 - 2x^3 - 2x^2 + 2$

29. $(5x^2 - 4x + 1) - (3x^2 + 1) = 5x^2 - 4x + 1 - 3x^2 - 1$
$$= 2x^2 - 4x$$

30. $(3x^5 - 4x^4 + 3x^2 + 3) - (2x^5 - 4x^4 + 3x^3 + 4x^2 - 5)$
$$= 3x^5 - 4x^4 + 3x^2 + 3 - 2x^5 + 4x^4 - 3x^3 - 4x^2 + 5$$
$$= x^5 - 3x^3 - x^2 + 8$$

31. $P = 2(w + 3) + 2w = 2w + 6 + 2w = 4w + 6$

$A = w(w + 3) = w^2 + 3w$

32. Regarding the figure as one large rectangle with length $t + 4$ and width $t + 3$, we have $(t + 4)(t + 3)$. We can also add the areas of the four smaller rectangles:

$3 \cdot t + 4 \cdot 3 + 4 \cdot t + t \cdot t$, or $3t + 12 + 4t + t^2$, or

$t^2 + 7t + 12$

33. $\left(x + \dfrac{2}{3}\right)\left(x + \dfrac{1}{2}\right) = x^2 + \dfrac{1}{2}x + \dfrac{2}{3}x + \dfrac{2}{6}$
$$= x^2 + \dfrac{3}{6}x + \dfrac{4}{6}x + \dfrac{1}{3}$$
$$= x^2 + \dfrac{7}{6}x + \dfrac{1}{3}$$

34. $(7x + 1)^2 = (7x)^2 + 2 \cdot 7x \cdot 1 + 1^2$
$$= 49x^2 + 14x + 1$$

35.
$$\begin{array}{r} 4x^2 - 5x + 1 \\ 3x - 2 \\ \hline -8x^2 + 10x - 2 \\ 12x^3 - 15x^2 + 3x \\ \hline 12x^3 - 23x^2 + 13x - 2 \end{array}$$

36. $(3x^2 + 4)(3x^2 - 4) = (3x^2)^2 - 4^2 = 9x^4 - 16$

37. $5x^4(3x^3 - 8x^2 + 10x + 2)$
$$= 5x^4 \cdot 3x^3 - 5x^4 \cdot 8x^2 + 5x^4 \cdot 10x + 5x^4 \cdot 2$$
$$= 15x^7 - 40x^6 + 50x^5 + 10x^4$$

38. $(x + 4)(x - 7) = x^2 - 7x + 4x - 28 = x^2 - 3x - 28$

39. $(3y^2 - 2y)^2 = (3y^2)^2 - 2 \cdot 3y^2 \cdot 2y + (2y)^2 = 9y^4 - 12y^3 + 4y^2$

40. $(2t^2 + 3)(t^2 - 7) = 4t^4 - 14t^2 + 3t^2 - 21 = 4t^4 - 11t^2 - 21$

41. $\quad 2 - 5xy + y^2 - 4xy^3 + x^6$

$= 2 - 5(-1)(2) + 2^2 - 4(-1)(2)^3 + (-1)^6$

$= 2 - 5(-1)(2) + 4 - 4(-1)(8) + 1$

$= 2 + 10 + 4 + 32 + 1$

$= 49$

42. $x^5y - 7xy + 9x^2 - 8$

Term	Coefficient	Degree	
x^5y	1	6	$(x^5y = 1 \cdot x^5y^1)$
$-7xy$	-7	2	$(-7xy = -7x^1y^1)$
$9x^2$	9	2	
-8	-8	0	$(-8 = -8x^0)$

The degree of the polynomial is the degree of the term of highest degree. The term of highest degree is x^5y. Its degree is 6, so the degree of the polynomial is 6.

43. $y + w - 2y + 8w - 5 = (1 - 2)y + (1 + 8)w - 5$

$\qquad\qquad\qquad\qquad = -y + 9w - 5$

44. $\quad m^6 - 2m^2n + m^2n^2 + n^2m - 6m^3 + m^2n^2 + 7n^2m$

$= m^6 - 2m^2n + (1+1)m^2n^2 + (1+7)n^2m - 6m^3$

$= m^6 - 2m^2n + 2m^2n^2 + 8n^2m - 6m^3$

45. $\quad (5x^2 - 7xy + y^2) + (-6x^2 - 3xy - y^2) + (x^2 + xy - 2y^2)$

$= (5 - 6 + 1)x^2 + (-7 - 3 + 1)xy + (1 - 1 - 2)y^2$

$= -9xy - 2y^2$

46. $\quad (6x^3y^2 - 4x^2y - 6x) - (-5x^3y^2 + 4x^2y + 6x^2 - 6)$

$= 6x^3y^2 - 4x^2y - 6x + 5x^3y^2 - 4x^2y - 6x^2 + 6$

$= (6 + 5)x^3y^2 + (-4 - 4)x^2y - 6x - 6x^2 + 6$

$= 11x^3y^2 - 8x^2y - 6x - 6x^2 + 6$

47.
$$
\begin{array}{r}
p^2 \;+\; pq \;+\; q^2 \\
p \;-\; q \\
\hline
-\,p^2q - pq^2 - q^3 \\
p^3 + p^2q + pq^2 \\
\hline
p^3 \qquad\qquad -\; q^3
\end{array}
$$

48. $\left(3a^4 - \dfrac{1}{3}b^3\right)^2 = (3a^4)^2 - 2 \cdot 3a^4 \cdot \dfrac{1}{3}b^3 + \left(\dfrac{1}{3}b^3\right)^2$

$\qquad\qquad\qquad\quad = 9a^8 - 2a^4b^3 + \dfrac{1}{9}b^6$

49. $\dfrac{10x^3 - x^2 + 6x}{2x} = \dfrac{10x^3}{2x} - \dfrac{x^2}{2x} + \dfrac{6x}{2x}$

$\qquad\qquad\qquad\quad = 5x^2 - \dfrac{1}{2}x + 3$

50.
$$
\begin{array}{r}
3x^2 \;-\; 7x \;+\; 4 \\
2x+3\,\overline{)\,6x^3 \;-\; 5x^2 \;-\; 13x \;+\; 13} \\
\underline{6x^3 \;+\; 9x^2 } \\
-\,14x^2 \;-\; 13x \\
\underline{-\,14x^2 \;-\; 21x } \\
8x \;+\; 13 \\
\underline{8x \;+\; 12} \\
1
\end{array}
$$

The answer is $3x^2 - 7x + 4 + \dfrac{1}{2x + 3}$.

51. Locate -1 on the x-axis. Then move vertically to the graph and horizontally to the y-axis. It appears that the y-value that is paired with -1 is 0. Thus, the value of $y = 10x^3 - 10x$ is 0 when $x = -1$.

Locate -0.5 on the x-axis. Then move vertically to the graph and horizontally to the y-axis. It appears that the y-value that is paired with -0.5 is about 3.75. Thus, the value of $y = 10x^3 - 10x$ is 3.75 when $x = -0.5$.

Locate 0.5 on the x-axis. Then move vertically to the graph and horizontally to the y-axis. It appears that the y-value that is paired with 0.5 is about -3.75. Thus, the value of $y = 10x^3 - 10x$ is -3.75 when $x = 0.5$.

Locate 1 on the x-axis. Then move vertically to the graph and horizontally to the y-axis. It appears that the y-value that is paired with 1 is 0. Thus, the value of $y = 10x^3 - 10x$ is 0 when $x = 1$.

52. *Discussion and Writing Exercise.* 578.6×10^{-7} is not in scientific notation because 578.6 is larger than 10.

53. *Discussion and Writing Exercise.* A monomial is an expression of the type ax^n, where n is a whole number and a is a real number. A binomial is a sum of two monomials and has two terms. A trinomial is a sum of three monomials and has three terms. A general polynomial is a monomial or a sum of monomials and has one or more terms.

54. $A = \dfrac{1}{2}bh$

$\quad A = \dfrac{1}{2}(x + y)(x - y) = \dfrac{1}{2}(x^2 - y^2) = \dfrac{1}{2}x^2 - \dfrac{1}{2}y^2$

55. The shaded area is the area of a square with side 20 minus the area of 4 small squares, each with side a.

$A = 20^2 - 4 \cdot a^2 = 400 - 4a^2$

56. $\quad -3x^5 \cdot 3x^3 - x^6(2x)^2 + (3x^4)^2 + (2x^2)^4 - 40x^2(x^3)^2$

$= -3x^5 \cdot 3x^3 - x^6(4x^2) + 9x^8 + 16x^8 - 40x^2(x^6)$

$= -9x^8 - 4x^8 + 9x^2 + 16x^8 - 40x^8$

$= -28x^8$

57.
$$(x - 7)(x + 10) = (x - 4)(x - 6)$$

$$x^2 + 10x - 7x - 70 = x^2 - 6x - 4x + 24$$

$$x^2 + 3x - 70 = x^2 - 10x + 24$$

$$3x - 70 = -10x + 24 \quad \text{Subtracting } x^2$$

$$13x - 70 = 24 \quad\quad\quad\text{Adding } 10x$$

$$13x = 94 \quad\quad\quad\quad\text{Adding } 70$$

$$x = \frac{94}{13}$$

The solution is $\dfrac{94}{13}$.

58. Let P represent the other polynomial. Then we have

$(x-1)P = x^5 - 1$, or $P = \dfrac{x^5 - 1}{x - 1}$. We divide to find P.

$$
\begin{array}{r}
x^4 + x^3 + x^2 + x + 1 \\
x-1 \overline{\smash{\big)}\, x^5 + 0x^4 + 0x^3 + 0x^2 + 0x - 1} \\
\underline{x^5 - x^4} \\
x^4 \\
\underline{x^4 - x^3} \\
x^3 \\
\underline{x^3 - x^2} \\
x^2 \\
\underline{x^2 - x} \\
x - 1 \\
\underline{x - 1} \\
0
\end{array}
$$

The other polynomial is $x^4 + x^3 + x^2 + x + 1$.

59. *Familiarize*. Let $w =$ the width of the garden, in feet. Then $2w =$ the length. From the drawing in the text we see that the width of the garden and the sidewalk together is $w + 4 + 4$, or $w + 8$, and the length of the garden and the sidewalk together is $2w + 4 + 4$, or $2w + 8$.

Translate. The area of the sidewalk is the area of the garden and sidewalk together minus the area of the garden. Recall that the formula for the area of a rectangle is $A = l \cdot w$. Thus, we have

$$256 = (2w + 8)(w + 8) - 2w \cdot w.$$

Solve. We solve the equation.

$$
\begin{aligned}
256 &= (2w + 8)(w + 8) - 2w \cdot w \\
256 &= 2w^2 + 16w + 8w + 64 - 2w^2 \\
256 &= 24w + 64 \\
192 &= 24w \\
8 &= w
\end{aligned}
$$

If $w = 8$, then $2w = 2 \cdot 8 = 16$.

Check. The dimensions of the garden and the sidewalk together are $2 \cdot 8 + 8$ by $8 + 8$, or 24 by 16. Then the area of the garden and sidewalk together is $24 \text{ ft} \cdot 16 \text{ ft}$, or 384 ft^2 and the area of the garden is $16 \text{ ft} \cdot 8 \text{ ft}$, or 128 ft^2. Subtracting to find the area of the sidewalk, we get $384 \text{ ft}^2 - 128 \text{ ft}^2$, or 256 ft^2, so the answer checks.

State. The dimensions of the garden are 16 ft by 8 ft.

Chapter 12 Test

1. $6^{-2} \cdot 6^{-3} = 6^{-2+(-3)} = 6^{-5} = \dfrac{1}{6^5}$

2. $x^6 \cdot x^2 \cdot x = x^{6+2+1} = x^9$

3. $(4a)^3 \cdot (4a)^8 = (4a)^{3+8} = (4a)^{11}$

4. $\dfrac{3^5}{3^2} = 3^{5-2} = 3^3$

5. $\dfrac{x^3}{x^8} = x^{3-8} = x^{-5} = \dfrac{1}{x^5}$

6. $\dfrac{(2x)^5}{(2x)^5} = 1$

7. $(x^3)^2 = x^{3 \cdot 2} = x^6$

8. $(-3y^2)^3 = (-3)^3(y^2)^3 = -27y^{2 \cdot 3} = -27y^6$

9. $(2a^3b)^4 = 2^4(a^3)^4 \cdot b^4 = 16a^{12}b^4$

10. $\left(\dfrac{ab}{c}\right)^3 = \dfrac{(ab)^3}{c^3} = \dfrac{a^3b^3}{c^3}$

11. $(3x^2)^3(-2x^5)^3 = 3^3(x^2)^3(-2)^3(x^5)^3 = 27x^6(-8)x^{15} = -216x^{21}$

12. $3(x^2)^3(-2x^5)^3 = 3x^6(-2)^3(x^5)^3 = 3x^6(-8)x^{15} = -24x^{21}$

13. $2x^2(-3x^2)^4 = 2x^2(-3)^4(x^2)^4 = 2x^2 \cdot 81x^8 = 162x^{10}$

14. $(2x)^2(-3x^2)^4 = 2^2x^2(-3)^4(x^2)^4 = 4x^2 \cdot 81x^8 = 324x^{10}$

15. $5^{-3} = \dfrac{1}{5^3}$

16. $\dfrac{1}{y^8} = y^{-8}$

17. $3,900,000,000$

$$3 \, . \, 900,000,000.$$
$$\underset{\text{9 places}}{\underrightarrow{}}$$

Large number, so the exponent is positive.

$3,900,000,000 = 3.9 \times 10^9$

18. 5×10^{-8}

Negative exponent, so the answer is a small number.

$$0 \, . \, 00000005.$$
$$\underset{\text{8 places}}{\underleftarrow{}}$$

$5 \times 10^{-8} = 0.00000005$

19. $\dfrac{5.6 \times 10^6}{3.2 \times 10^{-11}} = \dfrac{5.6}{3.2} \times \dfrac{10^6}{10^{-11}} = 1.75 \times 10^{6-(-11)} = 1.75 \times 10^{17}$

20. $(2.4 \times 10^5)(5.4 \times 10^{16}) = (2.4 \cdot 5.4) \times (10^5 \cdot 10^{16}) = 12.96 \times 10^{21} = (1.296 \times 10) \times 10^{21} = 1.296 \times 10^{22}$

21. $600 \text{ million} = 600 \times 1 \text{ million} = 600 \times 1,000,000 = 600,000,000 = 6 \times 10^8$

$40,000 = 4 \times 10^4$

We divide:

$$\dfrac{6 \times 10^8}{4 \times 10^4} = 1.5 \times 10^4$$

A CD-ROM can hold 1.5×10^4 sound files.

22. $x^5 + 5x - 1 = (-2)^5 + 5(-2) - 1 = -32 - 10 - 1 = -43$

23. $\dfrac{1}{3}x^5 - x + 7$

The coefficient of $\dfrac{1}{3}x^5$ is $\dfrac{1}{3}$.

The coefficient of $-x$, or $-1 \cdot x$, is -1.

The coefficient of 7 is 7.

24. $2x^3 - 4 + 5x + 3x^6$

The degree of $2x^3$ is 3.

The degree of -4, or $-4x^0$, is 0.

The degree of $5x$, or $5x^1$, is 1.

The degree of $3x^6$ is 6.

The degree of the polynomial is 6, the largest exponent.

25. $7 - x$ is a binomial because it has just 2 terms.

26. $4a^2 - 6 + a^2 = (4 + 1)a^2 - 6 = 5a^2 - 6$

27. $y^2 - 3y - y + \dfrac{3}{4}y^2 = \left(1 + \dfrac{3}{4}\right)y^2 + (-3 - 1)y =$

$\left(\dfrac{4}{4} + \dfrac{3}{4}\right)y^2 + (-3 - 1)y = \dfrac{7}{4}y^2 - 4y$

28. $3 - x^2 + 2x^3 + 5x^2 - 6x - 2x + x^5$

$= 3 + (-1 + 5)x^2 + 2x^3 + (-6 - 2)x + x^5$

$= 3 + 4x^2 + 2x^3 - 8x + x^5$

$= x^5 + 2x^3 + 4x^2 - 8x + 3$

29. $(3x^5 + 5x^3 - 5x^2 - 3) + (x^5 + x^4 - 3x^3 - 3x^2 + 2x - 4)$

$= (3 + 1)x^5 + x^4 + (5 - 3)x^3 + (-5 - 3)x^2 + 2x + (-3 - 4)$

$= 4x^5 + x^4 + 2x^3 - 8x^2 + 2x - 7$

30. $\left(x^4 + \dfrac{2}{3}x + 5\right) + \left(4x^4 + 5x^2 + \dfrac{1}{3}x\right)$

$= (1 + 4)x^4 + 5x^2 + \left(\dfrac{2}{3} + \dfrac{1}{3}\right)x + 5$

$= 5x^4 + 5x^2 + x + 5$

31. $(2x^4 + x^3 - 8x^2 - 6x - 3) - (6x^4 - 8x^2 + 2x)$

$= 2x^4 + x^3 - 8x^2 - 6x - 3 - 6x^4 + 8x^2 - 2x$

$= (2 - 6)x^4 + x^3 + (-8 + 8)x^2 + (-6 - 2)x - 3$

$= -4x^4 + x^3 - 8x - 3$

32. $(x^3 - 0.4x^2 - 12) - (x^5 + 0.3x^3 + 0.4x^2 + 9)$

$= x^3 - 0.4x^2 - 12 - x^5 - 0.3x^3 - 0.4x^2 - 9$

$= -x^5 + (1 - 0.3)x^3 + (-0.4 - 0.4)x^2 + (-12 - 9)$

$= -x^5 + 0.7x^3 - 0.8x^2 - 21$

33. $-3x^2(4x^2 - 3x - 5) = -3x^2 \cdot 4x^2 - 3x^2(-3x) - 3x^2(-5) =$

$-12x^4 + 9x^3 + 15x^2$

34. $\left(x - \dfrac{1}{3}\right)^2 = x^2 - 2 \cdot x \cdot \dfrac{1}{3} + \left(\dfrac{1}{3}\right)^2 = x^2 - \dfrac{2}{3}x + \dfrac{1}{9}$

35. $(3x + 10)(3x - 10) = (3x)^2 - 10^2 = 9x^2 - 100$

36. $(3b + 5)(b - 3) = 3b^2 - 9b + 5b - 15 = 3b^2 - 4b - 15$

37. $(x^6 - 4)(x^8 + 4) = x^{14} + 4x^6 - 4x^8 - 16$, or

$x^{14} - 4x^8 + 4x^6 - 16$

38. $(8 - y)(6 + 5y) = 48 + 40y - 6y - 5y^2 = 48 + 34y - 5y^2$

39.
$$
\begin{array}{r}
3x^2 \ - \ 5x \ - \ 3 \\
2x \ + \ 1 \\
\hline
3x^2 \ - \ 5x \ - \ 3 \\
6x^3 \ - \ 10x^2 \ - \ 6x \\
\hline
6x^3 \ - \ 7x^2 \ - \ 11x \ - \ 3
\end{array}
$$

40. $(5t + 2)^2 = (5t)^2 + 2 \cdot 5t \cdot 2 + 2^2 = 25t^2 + 20t + 4$

41. $x^3y - y^3 + xy^3 + 8 - 6x^3y - x^2y^2 + 11$

$= (1 - 6)x^3y - y^3 + xy^3 + (8 + 11) - x^2y^2$

$= -5x^3y - y^3 + xy^3 + 19 - x^2y^2$

42. $(8a^2b^2 - ab + b^3) - (-6ab^2 - 7ab - ab^3 + 5b^3)$

$= 8a^2b^2 - ab + b^3 + 6ab^2 + 7ab + ab^3 - 5b^3$

$= 8a^2b^2 + (-1 + 7)ab + (1 - 5)b^3 + 6ab^2 + ab^3$

$= 8a^2b^2 + 6ab - 4b^3 + 6ab^2 + ab^3$

43. $(3x^5 - 4y^5)(3x^5 + 4y^5) = (3x^5)^2 - (4y^5)^2 =$

$9x^{10} - 16y^{10}$

44. $(12x^4 + 9x^3 - 15x^2) \div (3x^2)$

$= \dfrac{12x^4 + 9x^3 - 15x^2}{3x^2}$

$= \dfrac{12x^4}{3x^2} + \dfrac{9x^3}{3x^2} - \dfrac{15x^2}{3x^2}$

$= 4x^2 + 3x - 5$

45.
$$
\begin{array}{r}
2x^2 \ - \ 4x \ - \ 2 \\
3x + 2 \overline{)\ 6x^3 \ - \ 8x^2 \ - \ 14x \ + \ 13} \\
\underline{6x^3 \ + \ 4x^2 } \\
-12x^2 \ - \ 14x \\
\underline{-12x^2 \ - \ 8x } \\
-6x \ + \ 13 \\
\underline{-6x \ - \ 4} \\
17
\end{array}
$$

The answer is $2x^2 - 4x - 2 + \dfrac{17}{3x + 2}$.

46. Locate -1 on the x-axis. Then move vertically to the graph and horizontally to the y-axis. It appears that the y-value that is paired with -1 is 3. Thus, the value of $y = x^3 - 5x - 1$ is 3 when $x = -1$.

Locate -0.5 on the x-axis. Then move vertically to the graph and horizontally to the y-axis. It appears that the y-value that is paired with -0.5 is 1.5. Thus, the value of $y = x^3 - 5x - 1$ is 1.5 when $x = -0.5$.

Locate 0.5 on the x-axis. Then move vertically to the graph and horizontally to the y-axis. It appears that the y-value that is paired with 0.5 is -3.5. Thus, the value of $y = x^3 - 5x - 1$ is -3.5 when $x = 0.5$.

Locate 1 on the x-axis. Then move vertically to the graph and horizontally to the y-axis. It appears that the y-value that is paired with 1 is -5. Thus, the value of $y = x^3 - 5x - 1$ is -5 when $x = 1$.

Locate 1.1 on the x-axis. Then move vertically to the graph and horizontally to the y-axis. It appears that the y-value that is paired with 1.1 is -5.25. Thus, the value of $y = x^3 - 5x - 1$ is -5.25 when $x = 1.1$.

47. Two sides have dimensions 9 by 5, two other sides have dimensions a by 5, and the two remaining sides have dimensions a by 9. Then the surface area is $2 \cdot 9 \cdot 5 + 2 \cdot a \cdot 5 + 2 \cdot a \cdot 9$, or $90 + 10a + 18a$, or $90 + 28a$.

48. When we regard the figure as one large rectangle with dimensions $t + 2$ by $t + 2$, we can express the area as $(t + 2)(t + 2)$.

Next we will regard the figure as the sum of four smaller rectangles with dimensions t by t, 2 by t, 2 by t, and 2 by 2. The sum of the areas of these rectangles is $t\cdot t + 2\cdot t + 2\cdot t + 2\cdot 2$, or $t^2 + 2t + 2t + 4$, or $t^2 + 4t + 4$.

49. Let $l =$ the length of the box. Then the height is $l - 1$ and the width is $l - 2$. The volume of the box is length $\times$ width $\times$ height.

$$V = l(l - 2)(l - 1)$$
$$V = (l^2 - 2l)(l - 1)$$
$$V = l^3 - l^2 - 2l^2 + 2l$$
$$V = l^3 - 3l^2 + 2l$$

50.

$$
\begin{aligned}
(x - 5)(x + 5) &= (x + 6)^2 \\
x^2 - 25 &= x^2 + 12x + 36 \\
-25 &= 12x + 36 \qquad \text{Subtracting } x^2 \\
-61 &= 12x \\
-\frac{61}{12} &= x
\end{aligned}
$$

The solution is $-\dfrac{61}{12}$.

Chapter 13

Polynomials: Factoring

Exercise Set 13.1

1. $x^2 = x^2$

$-6x = -1 \cdot 2 \cdot 3 \cdot x$

The coefficients have no common prime factor. The GCF of the powers of x is x because 1 is the smallest exponent of x. Thus the GCF is x.

3. $3x^4 = 3 \cdot x^4$

$x^2 = x^2$

The coefficients have no common prime factor. The GCF of the powers of x is x^2 because 2 is the smallest exponent of x. Thus the GCF is x^2.

5. $2x^2 = 2 \cdot x^2$

$2x = 2 \cdot x$

$-8 = -1 \cdot 2 \cdot 2 \cdot 2$

Each coefficient has a factor of 2. There are no other common prime factors. The GCF of the powers of x is 1 since -8 has no x-factor. Thus the GCF is 2.

7. $-17x^5y^3 = -1 \cdot 17 \cdot x^5 \cdot y^3$

$34x^3y^2 = 2 \cdot 17 \cdot x^3 \cdot y^2$

$51xy = 3 \cdot 17 \cdot x \cdot y$

Each coefficient has a factor of 17. There are no other common prime factors. The GCF of the powers of x is x because 1 is the smallest exponent of x. Similarly, the GCF of the powers of y is y because 1 is the smallest exponent of y. Thus the GCF is $17xy$.

9. $-x^2 = -1 \cdot x^2$

$-5x = -1 \cdot 5 \cdot x$

$-20x^3 = -1 \cdot 2 \cdot 2 \cdot 5 \cdot x^3$

The coefficients have no common prime factor. (Note that -1 is not a prime number.) The GCF of the powers of x is x because 1 is the smallest exponent of x. Thus the GCF is x.

11. $x^5y^5 = x^5 \cdot y^5$

$x^4y^3 = x^4 \cdot y^3$

$x^3y^3 = x^3 \cdot y^3$

$-x^2y^2 = -1 \cdot x^2 \cdot y^2$

There is no common prime factor. The GCF of the powers of x is x^2 because 2 is the smallest exponent of x. Similarly, the GCF of the powers of y is y^2 because 2 is the smallest exponent of y. Thus the GCF is x^2y^2.

13. $x^2 - 6x = x \cdot x - x \cdot 6$ Factoring each term

$\qquad = x(x-6)$ Factoring out the common factor x

15. $2x^2 + 6x = 2x \cdot x + 2x \cdot 3$ Factoring each term

$\qquad = 2x(x+3)$ Factoring out the common factor $2x$

17. $x^3 + 6x^2 = x^2 \cdot x + x^2 \cdot 6$ Factoring each term

$\qquad = x^2(x+6)$ Factoring out x^2

19. $8x^4 - 24x^2 = 8x^2 \cdot x^2 - 8x^2 \cdot 3$

$\qquad = 8x^2(x^2 - 3)$ Factoring out $8x^2$

21. $2x^2 + 2x - 8 = 2 \cdot x^2 + 2 \cdot x - 2 \cdot 4$

$\qquad = 2(x^2 + x - 4)$ Factoring out 2

23. $17x^5y^3 + 34x^3y^2 + 51xy$

$= 17xy \cdot x^4y^2 + 17xy \cdot 2x^2y + 17xy \cdot 3$

$= 17xy(x^4y^2 + 2x^2y + 3)$

25. $6x^4 - 10x^3 + 3x^2 = x^2 \cdot 6x^2 - x^2 \cdot 10x + x^2 \cdot 3$

$\qquad = x^2(6x^2 - 10x + 3)$

27. $x^5y^5 + x^4y^3 + x^3y^3 - x^2y^2$

$= x^2y^2 \cdot x^3y^3 + x^2y^2 \cdot x^2y + x^2y^2 \cdot xy + x^2y^2(-1)$

$= x^2y^2(x^3y^3 + x^2y + xy - 1)$

29. $2x^7 - 2x^6 - 64x^5 + 4x^3$

$= 2x^3 \cdot x^4 - 2x^3 \cdot x^3 - 2x^3 \cdot 32x^2 + 2x^3 \cdot 2$

$= 2x^3(x^4 - x^3 - 32x^2 + 2)$

31. $1.6x^4 - 2.4x^3 + 3.2x^2 + 6.4x$

$= 0.8x(2x^3) - 0.8x(3x^2) + 0.8x(4x) + 0.8x(8)$

$= 0.8x(2x^3 - 3x^2 + 4x + 8)$

33. $\dfrac{5}{3}x^6 + \dfrac{4}{3}x^5 + \dfrac{1}{3}x^4 + \dfrac{1}{3}x^3$

$= \dfrac{1}{3}x^3(5x^3) + \dfrac{1}{3}x^3(4x^2) + \dfrac{1}{3}x^3(x) + \dfrac{1}{3}x^3(1)$

$= \dfrac{1}{3}x^3(5x^3 + 4x^2 + x + 1)$

35. Factor: $x^2(x+3) + 2(x+3)$

The binomial $x+3$ is common to both terms:

$x^2(x+3) + 2(x+3) = (x^2 + 2)(x+3)$

37. $5a^3(2a-7) - (2a-7)$

$= 5a^3(2a-7) - 1(2a-7)$

$= (5a^3 - 1)(2a - 7)$

39. $x^3 + 3x^2 + 2x + 6$

$= (x^3 + 3x^2) + (2x + 6)$

$= x^2(x+3) + 2(x+3)$ Factoring each binomial

$= (x^2 + 2)(x+3)$ Factoring out the common factor $x+3$

41. $2x^3 + 6x^2 + x + 3$

$= (2x^3 + 6x^2) + (x + 3)$

$= 2x^2(x + 3) + 1(x + 3)$ Factoring each binomial

$= (2x^2 + 1)(x + 3)$

43. $8x^3 - 12x^2 + 6x - 9 = 4x^2(2x - 3) + 3(2x - 3)$

$= (4x^2 + 3)(2x - 3)$

45. $12p^3 - 16p^2 + 3p - 4$

$= 4p^2(3p - 4) + 1(3p - 4)$ Factoring 1 out of the second binomial

$= (4p^2 + 1)(3p - 4)$

47. $5x^3 - 5x^2 - x + 1$

$= (5x^3 - 5x^2) + (-x + 1)$

$= 5x^2(x - 1) - 1(x - 1)$ Check: $-1(x-1) = -x+1$

$= (5x^2 - 1)(x - 1)$

49. $x^3 + 8x^2 - 3x - 24 = x^2(x + 8) - 3(x + 8)$

$= (x^2 - 3)(x + 8)$

51. $2x^3 - 8x^2 - 9x + 36 = 2x^2(x - 4) - 9(x - 4)$

$= (2x^2 - 9)(x - 4)$

53.   Discussion and Writing Exercise

55. $-2x < 48$

$x > -24$ Dividing by -2 and reversing the inequality symbol

The solution set is $\{x | x > -24\}$.

57. $\dfrac{-108}{-4} = 27$ (The quotient of two negative numbers is positive.)

59. $(y + 5)(y + 7) = y^2 + 7y + 5y + 35$ Using FOIL

$= y^2 + 12y + 35$

61. $(y + 7)(y - 7) = y^2 - 7^2 = y^2 - 49$

$[(A + B))(A - B) = A^2 - B^2]$

63. $x + y = 4$

To find the x-intercept, let $y = 0$. Then solve for x.

$x + y = 4$

$x + 0 = 4$

$x = 4$

The x-intercept is $(4, 0)$.

To find the y-intercept, let $x = 0$. Then solve for y.

$x + y = 4$

$0 + y = 4$

$y = 4$

The y-intercept is $(0, 4)$.

Plot these points and draw the line.

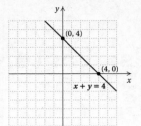

A third point should be used as a check. We substitute any value for x and solve for y. We let $x = 2$. Then

$x + y = 4$

$2 + y = 4$

$y = 2$

The point $(2, 2)$ is on the graph, so the graph is probably correct.

65. $5x - 3y = 15$

To find the x-intercept, let $y = 0$. Then solve for x.

$5x - 3y = 15$

$5x - 3 \cdot 0 = 15$

$5x = 15$

$x = 3$

The x-intercept is $(3, 0)$.

To find the y-intercept, let $x = 0$. Then solve for y.

$5x - 3y = 15$

$5 \cdot 0 - 3y = 15$

$-3y = 15$

$y = -5$

The y-intercept is $(0, -5)$.

Plot these points and draw the line.

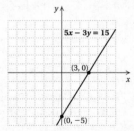

A third point should be used as a check. We substitute any value for x and solve for y. We let $x = 6$. Then

$5x - 3y = 15$

$5 \cdot 6 - 3y = 15$

$30 - 3y = 15$

$-3y = -15$

$y = 5$

The point $(6, 5)$ is on the graph, so the graph is probably correct.

67. $4x^5 + 6x^3 + 6x^2 + 9 = 2x^3(2x^2 + 3) + 3(2x^2 + 3)$

$= (2x^3 + 3)(2x^2 + 3)$

69. $x^{12} + x^7 + x^5 + 1 = x^7(x^5 + 1) + (x^5 + 1)$
$$= (x^7 + 1)(x^5 + 1)$$

71. $p^3 + p^2 - 3p + 10 = p^2(p + 1) - (3p - 10)$

This polynomial is not factorable using factoring by grouping.

Exercise Set 13.2

1. $x^2 + 8x + 15$

Since the constant term and coefficient of the middle term are both positive, we look for a factorization of 15 in which both factors are positive. Their sum must be 8.

Pairs of factors	Sums of factors
1, 15	16
3, 5	8

The numbers we want are 3 and 5.
$x^2 + 8x + 15 = (x + 3)(x + 5)$.

3. $x^2 + 7x + 12$

Since the constant term is positive and the coefficient of the middle term is positive, we look for a factorization of 12 in which both factors are positive. Their sum must be 7.

Pairs of factors	Sums of factors
1, 12	13
2, 6	8
3, 4	7

The numbers we want are 3 and 4.
$x^2 + 7x + 12 = (x + 3)(x + 4)$.

5. $x^2 - 6x + 9$

Since the constant term is positive and the coefficient of the middle term is negative, we look for a factorization of 9 in which both factors are negative. Their sum must be -6.

Pairs of factors	Sums of factors
$-1, -9$	-10
$-3, -3$	-6

The numbers we want are -3 and -3.
$x^2 - 6x + 9 = (x - 3)(x - 3)$, or $(x - 3)^2$.

7. $x^2 - 5x - 14$

Since the constant term is negative, we look for a factorization of -14 in which one factor is positive and one factor is negative. Their sum must be -5, the coefficient of the middle term.

Pairs of factors	Sums of factors
$-1, 14$	13
1, -14	-13
$-2, 7$	5
2, -7	-5

The numbers we want are 2 and -7.
$x^2 - 5x - 14 = (x + 2)(x - 7)$.

9. $b^2 + 5b + 4$

Since the constant term is positive and the coefficient of the middle term is positive, we look for a factorization of 4 in which both factors are positive. Their sum must be 5.

Pairs of factors	Sums of factors
1, 4	5
2, 2	4

The numbers we want are 1 and 4.
$b^2 + 5b + 4 = (b + 1)(b + 4)$.

11. $x^2 + \dfrac{2}{3}x + \dfrac{1}{9}$

Since the constant term is positive and the coefficient of the middle term is positive, we look for a factorization of $\dfrac{1}{9}$ in which both factors are positive. Their sum must be $\dfrac{2}{3}$.

Pairs of factors	Sums of factors
$1, \dfrac{1}{9}$	$\dfrac{10}{9}$
$\dfrac{1}{3}, \dfrac{1}{3}$	$\dfrac{2}{3}$

The numbers we want are $\dfrac{1}{3}$ and $\dfrac{1}{3}$.
$x^2 + \dfrac{2}{3}x + \dfrac{1}{9} = \left(x + \dfrac{1}{3}\right)\left(x + \dfrac{1}{3}\right)$, or $\left(x + \dfrac{1}{3}\right)^2$.

13. $d^2 - 7d + 10$

Since the constant term is positive and the coefficient of the middle term is negative, we look for a factorization of 10 in which both factors are negative. Their sum must be -7.

Pairs of factors	Sums of factors
$-1, -10$	-11
$-2, -5$	-7

The numbers we want are -2 and -5.
$d^2 - 7d + 10 = (d - 2)(d - 5)$.

15. $y^2 - 11y + 10$

Since the constant term is positive and the coefficient of the middle term is negative, we look for a factorization of 10 in which both factors are negative. Their sum must be -11.

Pairs of factors	Sums of factors
$-1, -10$	-11
$-2, -5$	-7

The numbers we want are -1 and -10.
$y^2 - 11y + 10 = (y - 1)(y - 10)$.

17. $x^2 + x + 1$

Since the constant term and the coefficient of the middle term are both positive, we look for a factorization of 1 in which both factors are positive. The sum must be 1. The only possible pair of factors is 1 and 1, but their sum is not 1. Thus, this polynomial is not factorable into binomials. It is prime.

19. $x^2 - 7x - 18$

Since the constant term is negative, we look for a factorization of -18 in which one factor is positive and one factor is negative. Their sum must be -7, the coefficient of the middle term.

Pairs of factors	Sums of factors
$-1,\quad 18$	17
$1,\quad -18$	-17
$-2,\quad 9$	7
$2,\quad -9$	-7
$-3,\quad 6$	3
$3,\quad -6$	-3

The numbers we want are 2 and -9.

$x^2 - 7x - 18 = (x + 2)(x - 9)$.

21. $x^3 - 6x^2 - 16x = x(x^2 - 6x - 16)$

After factoring out the common factor, x, we consider $x^2 - 6x - 16$. Since the constant term is negative, we look for a factorization of -16 in which one factor is positive and one factor is negative. Their sum must be -6, the coefficient of the middle term.

Pairs of factors	Sums of factors
$-1,\quad 16$	15
$1,\quad -16$	-15
$-2,\quad 8$	6
$2,\quad -8$	-6
$-4,\quad 4$	0

The numbers we want are 2 and -8.

Then $x^2 - 6x - 16 = (x + 2)(x - 8)$, so $x^3 - 6x^2 - 16x = x(x + 2)(x - 8)$.

23. $y^3 - 4y^2 - 45y = y(y^2 - 4y - 45)$

After factoring out the common factor, y, we consider $y^2 - 4y - 45$. Since the constant term is negative, we look for a factorization of -45 in which one factor is positive and one factor is negative. Their sum must be -4, the coefficient of the middle term.

Pairs of factors	Sums of factors
$-1,\quad 45$	44
$1,\quad -45$	-44
$-3,\quad 15$	12
$3,\quad -15$	-12
$-5,\quad 9$	4
$5,\quad -9$	-4

The numbers we want are 5 and -9.

Then $y^2 - 4y - 45 = (y + 5)(y - 9)$, so $y^3 - 4y^2 - 45y = y(y + 5)(y - 9)$.

25. $-2x - 99 + x^2 = x^2 - 2x - 99$

Since the constant term is negative, we look for a factorization of -99 in which one factor is positive and one factor is negative. Their sum must be -2, the coefficient of the middle term.

Pairs of factors	Sums of factors
$-1,\quad 99$	98
$1,\quad -99$	-98
$-3,\quad 33$	30
$3,\quad -33$	-30
$-9,\quad 11$	2
$9,\quad -11$	-2

The numbers we want are 9 and -11.

$-2x - 99 + x^2 = (x + 9)(x - 11)$.

27. $c^4 + c^2 - 56$

Consider this trinomial as $(c^2)^2 + c^2 - 56$. We look for numbers p and q such that $c^4 + c^2 - 56 = (c^2 + p)(c^2 + q)$. Since the constant term is negative, we look for a factorization of -56 in which one factor is positive and one factor is negative. Their sum must be 1.

Pairs of factors	Sums of factors
$-1,\quad 56$	55
$1,\quad -56$	-55
$-2,\quad 28$	26
$2,\quad -28$	-26
$-4,\quad 14$	12
$4,\quad -14$	-12
$-7,\quad 8$	1
$7,\quad -8$	-1

The numbers we want are -7 and 8.

$c^4 + c^2 - 56 = (c^2 - 7)(c^2 + 8)$.

29. $a^4 + 2a^2 - 35$

Consider this trinomial as $(a^2)^2 + 2a^2 - 35$. We look for numbers p and q such that $a^4 + 2a^2 - 35 = (a^2 + p)(a^2 + q)$. Since the constant term is negative, we look for a factorization of -35 in which one factor is positive and one factor is negative. Their sum must be 2.

Pairs of factors	Sums of factors
$-1,\quad 35$	34
$1,\quad -35$	-34
$-5,\quad 7$	2
$5,\quad -7$	-2

The numbers we want are -5 and 7.

$a^4 + 2a^2 - 35 = (a^2 - 5)(a^2 + 7)$.

31. $x^2 + x - 42$

Since the constant term is negative, we look for a factorization of -42 in which one factor is positive and one factor is negative. Their sum must be 1, the coefficient of the middle term.

Pairs of factors	Sums of factors
−1, 42	41
1, −42	−41
−2, 21	19
2, −21	−19
−3, 14	11
3, −14	−11
−6, 7	1
6, −7	−1

The numbers we want are −6 and 7.

$x^2 + x - 42 = (x - 6)(x + 7)$.

33. $7 - 2p + p^2 = p^2 - 2p + 7$

Since the constant term is positive and the coefficient of the middle term is negative, we look for a factorization of 7 in which both factors are negative. The sum must be −2. The only possible pair of factors is −1 and −7, but their sum is not −2. Thus, this polynomial is not factorable into binomials. It is prime.

35. $x^2 + 20x + 100$

We look for two factors, both positive, whose product is 100 and whose sum is 20.

They are 10 and 10. $10 \cdot 10 = 100$ and $10 + 10 = 20$.

$x^2 + 20x + 100 = (x + 10)(x + 10)$, or $(x + 10)^2$.

37. $30 + 7x - x^2 = -x^2 + 7x + 30 = -1(x^2 - 7x - 30)$

Now we factor $x^2 - 7x - 30$. Since the constant term is negative, we look for a factorization of −30 in which one factor is positive and one factor is negative. Their sum must be −7, the coefficient of the middle term.

Pairs of factors	Sums of factors
−1, 30	29
1, −30	−29
−2, 15	13
2, −15	−13
−3, 10	7
3, −10	−7
−5, 6	1
5, −6	−1

The numbers we want are 3 and −10. Then

$x^2 - 7x - 30 = (x + 3)(x - 10)$, so we have:

$-x^2 + 7x + 30$

$= -1(x + 3)(x - 10)$

$= (-x - 3)(x - 10)$ Multiplying $x + 3$ by -1

$= (x + 3)(-x + 10)$ Multiplying $x - 10$ by -1

39. $24 - a^2 - 10a = -a^2 - 10a + 24 = -1(a^2 + 10a - 24)$

Now we factor $a^2 + 10a - 24$. Since the constant term is negative, we look for a factorization of −24 in which one factor is positive and one factor is negative. Their sum must be 10, the coefficient of the middle term.

Pairs of factors	Sums of factors
−1, 24	23
1, −24	−23
−2, 12	10
2, −12	−10
−3, 8	5
3, −8	−5
−4, 6	2
4, −6	−2

The numbers we want are −2 and 12. Then

$a^2 + 10a - 24 = (a - 2)(a + 12)$, so we have:

$-a^2 - 10a + 24$

$= -1(a - 2)(a + 12)$

$= (-a + 2)(a + 12)$ Multiplying $a - 2$ by -1

$= (a - 2)(-a - 12)$ Multiplying $a + 12$ by -1

41. $x^4 - 21x^3 - 100x^2 = x^2(x^2 - 21x - 100)$

After factoring out the common factor, x^2, we consider $x^2 - 21x - 100$. We look for two factors, one positive and one negative, whose product is −100 and whose sum is −21. They are 4 and −25. $4 \cdot (-25) = -100$ and $4 + (-25) = -21$.

Then $x^2 - 21x - 100 = (x + 4)(x - 25)$, so $x^4 - 21x^3 - 100x^2 = x^2(x + 4)(x - 25)$.

43. $x^2 - 21x - 72$

We look for two factors, one positive and one negative, whose product is −72 and whose sum is −21. They are 3 and −24.

$x^2 - 21x - 72 = (x + 3)(x - 24)$.

45. $x^2 - 25x + 144$

We look for two factors, both negative, whose product is 144 and whose sum is −25. They are −9 and −16.

$x^2 - 25x + 144 = (x - 9)(x - 16)$.

47. $a^2 + a - 132$

We look for two factors, one positive and one negative, whose product is −132 and whose sum is 1. They are −11 and 12.

$a^2 + a - 132 = (a - 11)(a + 12)$.

49. $120 - 23x + x^2 = x^2 - 23x + 120$

We look for two factors, both negative, whose product is 120 and whose sum is −23. They are −8 and −15.

$x^2 - 23x + 120 = (x - 8)(x - 15)$.

51. First write the polynomial in descending order and factor out −1.

$108 - 3x - x^2 = -x^2 - 3x + 108 = -1(x^2 + 3x - 108)$

Now we factor the polynomial $x^2 + 3x - 108$. We look for two factors, one positive and one negative, whose product is −108 and whose sum is 3. They are −9 and 12.

$x^2 + 3x - 108 = (x - 9)(x + 12)$

The final answer must include -1 which was factored out above.

$$-x^2 - 3x + 108$$
$$= -1(x-9)(x+12)$$
$$= (-x+9)(x+12) \qquad \text{Multiplying } x-9 \text{ by } -1$$
$$= (x-9)(-x-12) \qquad \text{Multiplying } x+12 \text{ by } -1$$

53. $y^2 - 0.2y - 0.08$

We look for two factors, one positive and one negative, whose product is -0.08 and whose sum is -0.2. They are -0.4 and 0.2.

$$y^2 - 0.2y - 0.08 = (y - 0.4)(y + 0.2).$$

55. $p^2 + 3pq - 10q^2 = p^2 + 3pq - 10q^2$

Think of $3q$ as a "coefficient" of p. Then we look for factors of $-10q^2$ whose sum is $3q$. They are $5q$ and $-2q$.

$$p^2 + 3pq - 10q^2 = (p + 5q)(p - 2q).$$

57. $84 - 8t - t^2 = -t^2 - 8t + 84 = -1(t^2 + 8t - 84)$

Now we factor $t^2 + 8t - 84$. We look for two factors, one positive and one negative, whose product is -84 and whose sum is 8. They are 14 and -6.

Then $t^2 + 8t - 84 = (t + 14)(t - 6)$, so we have:

$$-t^2 - 8t + 84$$
$$= -1(t+14)(t-6)$$
$$= (-t-14)(t-6) \qquad \text{Multiplying } t+14 \text{ by } -1$$
$$= (t+14)(-t+6) \qquad \text{Multiplying } t-6 \text{ by } -1$$

59. $m^2 + 5mn + 4n^2 = m^2 + 5nm + 4n^2$

We look for factors of $4n^2$ whose sum is $5n$. They are $4n$ and n.

$$m^2 + 5mn + 4n^2 = (m + 4n)(m + n)$$

61. $s^2 - 2st - 15t^2 = s^2 - 2ts - 15t^2$

We look for factors of $-15t^2$ whose sum is $-2t$. They are $-5t$ and $3t$.

$$s^2 - 2st - 15t^2 = (s - 5t)(s + 3t)$$

63. $6a^{10} - 30a^9 - 84a^8 = 6a^8(a^2 - 5a - 14)$

After factoring out the common factor, $6a^8$, we consider $a^2 - 5a - 14$. We look for two factors, one positive and one negative, whose product is -14 and whose sum is -5. They are 2 and -7.

$a^2 - 5a - 14 = (a+2)(a-7)$, so $6a^{10} - 30a^9 - 84a^8 = 6a^8(a+2)(a-7)$.

65. Discussion and Writing Exercise

67. Discussion and Writing Exercise

69. $8x(2x^2 - 6x + 1) = 8x \cdot 2x^2 - 8x \cdot 6x + 8x \cdot 1 = 16x^3 - 48x^2 + 8x$

71. $(7w+6)^2 = (7w)^2 + 2 \cdot 7w \cdot 6 + 6^2 = 49w^2 + 84w + 36$

73. $(4w-11)(4w+11) = (4w)^2 - (11)^2 = 16w^2 - 121$

75. $(3x-5y)(2x+7y) = 3x \cdot 2x + 3x \cdot 7y - 5y \cdot 2x - 5y \cdot 7y = 6x^2 + 21xy - 10xy - 35y^2 = 6x^2 + 11xy - 35y^2$

77. $3x - 8 = 0$
$$3x = 8 \qquad \text{Adding 8 on both sides}$$
$$x = \frac{8}{3} \qquad \text{Dividing by 3 on both sides}$$

The solution is $\frac{8}{3}$.

79. *Familiarize*. Let $n =$ the number of people arrested the year before.

Translate. We reword the problem.

$$\underbrace{\text{Number arrested the year before}} \quad \text{less} \quad 1.2\% \text{ of} \quad \underbrace{\text{that number}} \quad \text{is } 29,200.$$

$$n \quad - \quad 1.2\% \cdot \quad n \quad = 29,200$$

Carry out. We solve the equation.

$$n - 1.2\% \cdot n = 29,200$$
$$1 \cdot n - 0.012n = 29,200$$
$$0.988n = 29,200$$
$$n \approx 29,555 \qquad \text{Rounding}$$

Check. 1.2% of $29,555$ is $0.012(29,555) \approx 355$ and $29,555 - 355 = 29,200$. The answer checks.

State. Approximately $29,555$ people were arrested the year before.

81. $y^2 + my + 50$

We look for pairs of factors whose product is 50. The sum of each pair is represented by m.

Pairs of factors whose product is -50	Sums of factors
$1, \quad 50$	51
$-1, -50$	-51
$2, \quad 25$	27
$-2, -25$	-27
$5, \quad 10$	15
$-5, -10$	-15

The polynomial $y^2 + my + 50$ can be factored if m is 51, -51, 27, -27, 15, or -15.

83. $x^2 - \frac{1}{2}x - \frac{3}{16}$

We look for two factors, one positive and one negative, whose product is $-\frac{3}{16}$ and whose sum is $-\frac{1}{2}$.

They are $-\frac{3}{4}$ and $\frac{1}{4}$.

$$-\frac{3}{4} \cdot \frac{1}{4} = -\frac{3}{16} \text{ and } -\frac{3}{4} + \frac{1}{4} = -\frac{2}{4} = -\frac{1}{2}.$$

$$x^2 - \frac{1}{2}x - \frac{3}{16} = \left(x - \frac{3}{4}\right)\left(x + \frac{1}{4}\right)$$

85. $x^2 + \dfrac{30}{7}x - \dfrac{25}{7}$

We look for two factors, one positive and one negative, whose product is $-\dfrac{25}{7}$ and whose sum is $\dfrac{30}{7}$.

They are 5 and $-\dfrac{5}{7}$.

$5 \cdot \left(-\dfrac{5}{7}\right) = -\dfrac{25}{7}$ and $5 + \left(-\dfrac{5}{7}\right) = \dfrac{35}{7} + \left(-\dfrac{5}{7}\right) = \dfrac{30}{7}$.

$x^2 + \dfrac{30}{7}x - \dfrac{25}{7} = (x + 5)\left(x - \dfrac{5}{7}\right)$

87. $b^{2n} + 7b^n + 10$

Consider this trinomial as $(b^n)^2 + 7b^n + 10$. We look for numbers p and q such that $b^{2n} + 7b^n + 10 = (b^n + p)(b^n + q)$. We find two factors, both positive, whose product is 10 and whose sum is 7. They are 5 and 2.

$b^{2n} + 7b^n + 10 = (b^n + 5)(b^n + 2)$

89. We first label the drawing with additional information.

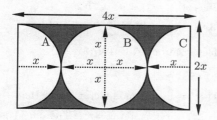

$4x$ represents the length of the rectangle and $2x$ the width. The area of the rectangle is $4x \cdot 2x$, or $8x^2$.

The area of semicircle A is $\dfrac{1}{2}\pi x^2$.

The area of circle B is πx^2.

The area of semicircle C is $\dfrac{1}{2}\pi x^2$.

$$\begin{array}{l} \text{Area of} \\ \text{shaded region} \end{array} = \begin{array}{l} \text{Area of} \\ \text{rectangle} \end{array} - \begin{array}{c} \text{Area} \\ \text{of} \\ A \end{array} - \begin{array}{c} \text{Area} \\ \text{of} \\ B \end{array} - \begin{array}{c} \text{Area} \\ \text{of} \\ C \end{array}$$

$$\begin{array}{l} \text{Area of} \\ \text{shaded region} \end{array} = \quad 8x^2 \quad - \dfrac{1}{2}\pi x^2 - \pi x^2 - \dfrac{1}{2}\pi x^2$$

$$= 8x^2 - 2\pi x^2$$

$$= 2x^2(4 - \pi)$$

The shaded area can be represented by $2x^2(4 - \pi)$.

Exercise Set 13.3

1. $2x^2 - 7x - 4$

(1) Look for a common factor. There is none (other than 1 or -1).

(2) Factor the first term, $2x^2$. The only possibility is $2x$, x. The desired factorization is of the form:

$$(2x + \quad)(x + \quad)$$

(3) Factor the last term, -4, which is negative. The possibilities are -4, 1 and 4, -1 and 2, -2.

These factors can also be written as 1, -4 and -1, 4 and -2, 2.

(4) Look for combinations of factors from steps (2) and (3) such that the sum of their products is the middle term, $-7x$. We try some possibilities:

$$(2x - 4)(x + 1) = 2x^2 - 2x - 4$$
$$(2x + 4)(x - 1) = 2x^2 + 2x - 4$$
$$(2x + 2)(x - 2) = 2x^2 - 2x - 4$$
$$(2x + 1)(x - 4) = 2x^2 - 7x - 4$$

The factorization is $(2x + 1)(x - 4)$.

3. $5x^2 - x - 18$

(1) There is no common factor (other than 1 or -1).

(2) Factor the first term, $5x^2$. The only possibility is $5x$, x. The desired factorization is of the form:

$$(5x + \quad)(x + \quad)$$

(3) Factor the last term, -18. The possibilities are -18, 1 and 18, -1 and -9, 2 and 9, -2 and -6, 3 and 6, -3.

These factors can also be written as 1, -18 and -1, 18 and 2, -9 and -2, 9 and 3, -6 and -3, 6.

(4) Look for combinations of factors from steps (2) and (3) such that the sum of their products is the middle term, x. We try some possibilities:

$$(5x - 18)(x + 1) = 5x^2 - 13x - 18$$
$$(5x + 18)(x - 1) = 5x^2 + 13x - 18$$
$$(5x + 9)(x - 2) = 5x^2 - x - 18$$

The factorization is $(5x + 9)(x - 2)$.

5. $6x^2 + 23x + 7$

(1) There is no common factor (other than 1 or -1).

(2) Factor the first term, $6x^2$. The possibilities are $6x$, x and $3x$, $2x$. We have these as possibilities for factorizations:

$$(6x + \quad)(x + \quad) \text{ and } (3x + \quad)(2x + \quad)$$

(3) Factor the last term, 7. The possibilities are 7, 1 and -7, -1.

These factors can also be written as 1, 7 and -1, -7.

(4) Look for combinations of factors from steps (2) and (3) such that the sum of their products is the middle term, $23x$. Since all signs are positive, we need consider only plus signs. We try some possibilities:

$$(6x + 7)(x + 1) = 6x^2 + 13x + 7$$
$$(3x + 7)(2x + 1) = 6x^2 + 17x + 7$$
$$(6x + 1)(x + 7) = 6x^2 + 43x + 7$$
$$(3x + 1)(2x + 7) = 6x^2 + 23x + 7$$

The factorization is $(3x + 1)(2x + 7)$.

7. $3x^2 + 4x + 1$

(1) There is no common factor (other than 1 or -1).

(2) Factor the first term, $3x^2$. The only possibility is $3x$, x. The desired factorization is of the form:

$$(3x+\quad)(x+\quad)$$

(3) Factor the last term, 1. The possibilities are 1, 1 and -1, -1.

(4) Look for combinations of factors from steps (2) and (3) such that the sum of their products is the middle term, $4x$. Since all signs are positive, we need consider only plus signs. There is only one such possibility:

$$(3x + 1)(x + 1) = 3x^2 + 4x + 1$$

The factorization is $(3x + 1)(x + 1)$.

9. $4x^2 + 4x - 15$

(1) There is no common factor (other than 1 or -1).

(2) Factor the first term, $4x^2$. The possibilities are $4x$, x and $2x$, $2x$. We have these as possibilities for factorizations:

$$(4x+\quad)(x+\quad) \text{ and } (2x+\quad)(2x+\quad)$$

(3) Factor the last term, -15. The possibilities are 15, -1 and -15, 1 and 5, -3 and -5, 3.

These factors can also be written as -1, 15 and 1, -15 and -3, 5 and 3, -5.

(4) We try some possibilities:

$$(4x + 15)(x - 1) = 4x^2 + 11x - 15$$
$$(2x + 15)(2x - 1) = 4x^2 + 28x - 15$$
$$(4x - 15)(x + 1) = 4x^2 - 11x - 15$$
$$(2x - 15)(2x + 1) = 4x^2 - 28x - 15$$
$$(4x + 5)(x - 3) = 4x^2 - 7x - 15$$
$$(2x + 5)(2x - 3) = 4x^2 + 4x - 15$$

The factorization is $(2x + 5)(2x - 3)$.

11. $2x^2 - x - 1$

(1) There is no common factor (other than 1 or -1).

(2) Factor the first term, $2x^2$. The only possibility is $2x$, x. The desired factorization is of the form:

$$(2x+\quad)(x+\quad)$$

(3) Factor the last term, -1. The only possibility is -1, 1.

These factors can also be written as 1, -1.

(4) We try the possibilities:

$$(2x - 1)(x + 1) = 2x^2 + x - 1$$
$$(2x + 1)(x - 1) = 2x^2 - x - 1$$

The factorization is $(2x + 1)(x - 1)$.

13. $9x^2 + 18x - 16$

(1) There is no common factor (other than 1 or -1).

(2) Factor the first term, $9x^2$. The possibilities are $9x$, x and $3x$, $3x$. We have these as possibilities for factorizations:

$$(9x+\quad)(x+\quad) \text{ and } (3x+\quad)(3x+\quad)$$

(3) Factor the last term, -16. The possibilities are 16, -1 and -16, 1 and 8, -2 and -8, 2 and 4, -4.

These factors can also be written as -1, 16 and 1, -16 and -2, 8 and 2, -8 and -4, 4.

(4) We try some possibilities:

$$(9x + 16)(x - 1) = 9x^2 + 7x - 16$$
$$(3x + 16)(3x - 1) = 9x^2 + 45x - 16$$
$$(9x - 16)(x + 1) = 9x^2 - 7x - 16$$
$$(3x - 16)(3x + 1) = 9x^2 - 45x - 16$$
$$(9x + 8)(x - 2) = 9x^2 - 10x - 16$$
$$(3x + 8)(3x - 2) = 9x^2 + 18x - 16$$

The factorization is $(3x + 8)(3x - 2)$.

15. $3x^2 - 5x - 2$

(1) There is no common factor (other than 1 or -1).

(2) Factor the first term, $3x^2$. The only possibility is $3x$, x. The desired factorization is of the form:

$$(3x+\quad)(x+\quad)$$

(3) Factor the last term, -2. The possibilities are 2, -1 and -2 and 1.

These factors can also be written as -1, 2 and 1, -2.

(4) We try some possibilities:

$$(3x + 2)(x - 1) = 3x^2 - x - 2$$
$$(3x - 2)(x + 1) = 3x^2 + x - 2$$
$$(3x - 1)(x + 2) = 3x^2 + 5x - 2$$
$$(3x + 1)(x - 2) = 3x^2 - 5x - 2$$

The factorization is $(3x + 1)(x - 2)$.

17. $12x^2 + 31x + 20$

(1) There is no common factor (other than 1 or -1).

(2) Factor the first term, $12x^2$. The possibilities are $12x$, x and $6x$, $2x$ and $4x$, $3x$. We have these as possibilities for factorizations:

$$(12x+\quad)(x+\quad) \text{ and } (6x+\quad)(2x+\quad) \text{ and }$$
$$(4x+\quad)(3x+\quad)$$

(3) Factor the last term, 20. Since all signs are positive, we need consider only positive pairs of factors. Those factor pairs are 20, 1 and 10, 2 and 5, 4.

These factors can also be written as 1, 20 and 2, 10 and 4, 5.

(4) We can immediately reject all possibilities in which either factor has a common factor, such as $(12x+20)$ or $(6x+4)$, because we determined at the outset that there are no common factors. We try some of the remaining possibilities:

$$(12x + 1)(x + 20) = 12x^2 + 241x + 20$$
$$(12x + 5)(x + 4) = 12x^2 + 53x + 20$$
$$(6x + 1)(2x + 20) = 12x^2 + 122x + 20$$
$$(4x + 5)(3x + 4) = 12x^2 + 31x + 20$$

The factorization is $(4x + 5)(3x + 4)$.

19. $14x^2 + 19x - 3$

(1) There is no common factor (other than 1 or −1).

(2) Factor the first term, $14x^2$. The possibilities are $14x$, x and $7x$, $2x$. We have these as possibilities for factorizations:

$$(14x+\quad)(x+\quad) \text{ and } (7x+\quad)(2x+\quad)$$

(3) Factor the last term, −3. The possibilities are −1, 3 and −3, 1.

These factors can also be written as 3, −1 and 1, −3.

(4) We try some possibilities:

$$(14x - 1)(x + 3) = 14x^2 + 41x - 3$$
$$(7x - 1)(2x + 3) = 7x^2 + 19x - 3$$

The factorization is $(7x - 1)(2x + 3)$.

21. $9x^2 + 18x + 8$

(1) There is no common factor (other than 1 or −1).

(2) Factor the first term, $9x^2$. The possibilities are $9x$, x and $3x$, $3x$. We have these as possibilities for factorizations:

$$(9x+\quad)(x+\quad) \text{ and } (3x+\quad)(3x+\quad)$$

(3) Factor the last term, 8. Since all signs are positive, we need consider only positive pairs of factors. Those factor pairs are 8, 1 and 4, 2.

These factors can also be written as 1, 8 and 2, 4.

(4) We try some possibilities:

$$(9x + 8)(x + 1) = 9x^2 + 17x + 8$$
$$(3x + 8)(3x + 1) = 9x^2 + 27x + 8$$
$$(9x + 4)(x + 2) = 9x^2 + 22x + 8$$
$$(3x + 4)(3x + 2) = 9x^2 + 18x + 8$$

The factorization is $(3x + 4)(3x + 2)$.

23. $49 - 42x + 9x^2 = 9x^2 - 42x + 49$

(1) There is no common factor (other than 1 or −1).

(2) Factor the first term, $9x^2$. The possibilities are $9x$, x and $3x$, $3x$. We have these as possibilities for factorizations:

$$(9x+\quad)(x+\quad) \text{ and } (3x+\quad)(3x+\quad)$$

(3) Factor 49. Since 49 is positive and the middle term is negative, we need consider only negative pairs of factors. Those factor pairs are −49, −1 and −7, −7.

The first pair of factors can also be written as −1, −49.

(4) We try some possibilities:

$$(9x - 49)(x - 1) = 9x^2 - 58x + 49$$
$$(3x - 49)(3x - 1) = 9x^2 - 150x + 49$$
$$(9x - 7)(x - 7) = 9x^2 - 70x + 49$$
$$(3x - 7)(3x - 7) = 9x^2 - 42x + 49$$

The factorization is $(3x - 7)(3x - 7)$, or $(3x - 7)^2$. This can also be expressed as follows:

$$(3x - 7)^2 = (-1)^2(3x - 7)^2 = [-1 \cdot (3x - 7)]^2 =$$
$$(-3x + 7)^2, \text{ or } (7 - 3x)^2$$

25. $24x^2 + 47x - 2$

(1) There is no common factor (other than 1 or −1).

(2) Factor the first term, $24x^2$. The possibilities are $24x$, x and $12x$, $2x$ and $6x$, $4x$ and $3x$, $8x$. We have these as possibilities for factorizations:

$$(24x+\quad)(x+\quad) \text{ and } (12x+\quad)(2x+\quad) \text{ and }$$
$$(6x+\quad)(4x+\quad) \text{ and } (3x+\quad)(8x+\quad)$$

(3) Factor the last term, −2. The possibilities are 2, −1 and −2, 1.

These factors can also be written as −1, 2 and 1, −2.

(4) We can immediately reject all possibilities in which either factor has a common factor, such as $(24x+2)$ or $(12x - 2)$, because we determined at the outset that there are no common factors. We try some of the remaining possibilities:

$$(24x - 1)(x + 2) = 24x^2 + 47x - 2$$

The factorization is $(24x - 1)(x + 2)$.

27. $35x^2 - 57x - 44$

(1) There is no common factor (other than 1 or −1).

(2) Factor the first term, $35x^2$. The possibilities are $35x$, x and $7x$, $5x$. We have these as possibilities for factorizations:

$$(35x+\quad)(x+\quad) \text{ and } (7x+\quad)(5x+\quad)$$

(3) Factor the last term, −44. The possibilities are 1, −44 and −1, 44 and 2, −22 and −2, 22 and 4, −11, and −4, 11.

These factors can also be written as −44, 1 and 44, −1 and −22, 2 and 22, −2 and −11, 4 and 11, −4.

(4) We try some possibilities:

$$(35x + 1)(x - 44) = 35x^2 - 1539x - 44$$
$$(7x + 1)(5x - 44) = 35x^2 - 303x - 44$$
$$(35x + 2)(x - 22) = 35x^2 - 768x - 44$$
$$(7x + 2)(5x - 22) = 35x^2 - 144x - 44$$

$$(35x + 4)(x - 11) = 35x^2 - 381x - 44$$
$$(7x + 4)(5x - 11) = 35x^2 - 57x - 44$$

The factorization is $(7x + 4)(5x - 11)$.

29. $20 + 6x - 2x^2 = -2x^2 + 6x + 20$

We factor out the common factor, -2. Factoring out -2 rather than 2 gives us a positive leading coefficient.

$-2(x^2 - 3x - 10)$

Then we factor the trinomial $x^2 - 3x - 10$. We look for a pair of factors whose product is -10 and whose sum is -3. The numbers are -5 and 2. The factorization of $x^2 - 3x - 10$ is $(x - 5)(x + 2)$. Then $20 + 6x - 2x^2 = -2(x - 5)(x + 2)$. If we think of -2 and $-1 \cdot 2$ then we can write other correct factorizations:

$$20 + 6x - 2x^2$$
$$= 2(-x + 5)(x + 2) \qquad \text{Multiplying } x - 5 \text{ by } -1$$
$$= 2(x - 5)(-x - 2) \qquad \text{Multplying } x + 2 \text{ by } -1$$

Note that we can also express $2(-x + 5)(x + 2)$ as $2(5 - x)(x + 2)$ since $-x + 5 = 5 - x$ by the commutative law of addition.

31. $12x^2 + 28x - 24$

(1) We factor out the common factor, 4:
$4(3x^2 + 7x - 6)$

Then we factor the trinomial $3x^2 + 7x - 6$.

(2) Factor $3x^2$. The only possibility is $3x$, x. The desired factorization is of the form:

$$(3x + \quad)(x + \quad)$$

(3) Factor -6. The possibilities are 6, -1 and -6, 1 and 3, -2 and -3, 2.

These factors can also be written as -1, 6 and 1, -6 and -2, 3 and 2, -3.

(4) We can immediately reject all possibilities in which either factor has a common factor, such as $(3x + 6)$ or $(3x - 3)$, because we factored out the largest common factor at the outset. We try some of the remaining possibilities:

$$(3x - 1)(x + 6) = 3x^2 + 17x - 6$$
$$(3x - 2)(x + 3) = 3x^2 + 7x - 6$$

The factorization of $3x^2 + 7x - 6$ is $(3x - 2)(x + 3)$. We must include the common factor in order to get a factorization of the original trinomial.

$12x^2 + 28x - 24 = 4(3x - 2)(x + 3)$

33. $30x^2 - 24x - 54$

(1) We factor out the common factor, 6:
$6(5x^2 - 4x - 9)$

Then we factor the trinomial $5x^2 - 4x - 9$.

(2) Factor $5x^2$. The only possibility is $5x$, x. The desired factorization is of the form:

$$(5x + \quad)(x + \quad)$$

(3) Factor -9. The possibilities are 9, -1 and -9, 1 and -3, 3.

These factors can also be written as -1, 9 and 1, -9 and 3, -3.

(4) We try some possibilities:

$$(5x + 9)(x - 1) = 5x^2 + 4x - 9$$
$$(5x - 9)(x + 1) = 5x^2 - 4x - 9$$

The factorization of $5x^2 - 4x - 9$ is $(5x - 9)(x + 1)$. We must include the common factor in order to get a factorization of the original trinomial.

$30x^2 - 24x - 54 = 6(5x - 9)(x + 1)$

35. $4y + 6y^2 - 10 = 6y^2 + 4y - 10$

(1) We factor out the common factor, 2:
$2(3y^2 + 2y - 5)$

Then we factor the trinomial $3y^2 + 2y - 5$.

(2) Factor $3y^2$. The only possibility is $3y$, y. The desired factorization is of the form:

$$(3y + \quad)(y + \quad)$$

(3) Factor -5. The possibilities are 5, -1 and -5, 1.

These factors can also be written as -1, 5 and 1, -5.

(4) We try some possibilities:

$$(3y + 5)(y - 1) = 3y^2 + 2y - 5$$

Then $3y^2 + 2y - 5 = (3y + 5)(y - 1)$, so $6y^2 + 4y - 10 = 2(3y + 5)(y - 1)$.

37. $3x^2 - 4x + 1$

(1) There is no common factor (other than 1 or -1).

(2) Factor the first term, $3x^2$. The only possibility is $3x$, x. The desired factorization is of the form:

$$(3x + \quad)(x + \quad)$$

(3) Factor the last term, 1. Since 1 is positive and the middle term is negative, we need consider only negative factor pairs. The only such pair is -1, -1.

(4) There is only one possibility:

$$(3x - 1)(x - 1) = 3x^2 - 4x + 1$$

The factorization is $(3x - 1)(x - 1)$.

39. $12x^2 - 28x - 24$

(1) We factor out the common factor, 4:
$4(3x^2 - 7x - 6)$

Then we factor the trinomial $3x^2 - 7x - 6$.

(2) Factor $3x^2$. The only possibility is $3x$, x. The desired factorization is of the form:

$$(3x + \quad)(x + \quad)$$

(3) Factor -6. The possibilities are 6, -1 and -6, 1 and 3, -2 and -3, 2.

These factors can also be written as -1, 6 and 1, -6 and -2, 3 and 2, -3.

(4) We can immediately reject all possibilities in which either factor has a common factor, such as $(3x - 6)$ or $(3x + 3)$, because we factored out the largest common factor at the outset. We try some of the remaining possibilities:

$$(3x - 1)(x + 6) = 3x^2 + 17x - 6$$
$$(3x - 2)(x + 3) = 3x^2 + 7x - 6$$
$$(3x + 2)(x - 3) = 3x^2 - 7x - 6$$

Then $3x^2 - 7x - 6 = (3x + 2)(x - 3)$, so $12x^2 - 28x - 24 = 4(3x + 2)(x - 3)$.

41. $-1 + 2x^2 - x = 2x^2 - x - 1$

(1) There is no common factor (other than 1 or -1).

(2) Factor the first term, $2x^2$. The only possibility is $2x$, x. The desired factorization is of the form:

$$(2x + \quad)(x + \quad)$$

(3) Factor -1. The only possibility is 1, -1.

(4) We try some possibilities:

$$(2x + 1)(x - 1) = 2x^2 - x - 1$$

The factorization is $(2x + 1)(x - 1)$.

43. $9x^2 - 18x - 16$

(1) There is no common factor (other than 1 or -1).

(2) Factor the first term, $9x^2$. The possibilities are $9x$, x and $3x$, $3x$. We have these as possibilities for factorizations:

$$(9x + \quad)(x + \quad) \text{ and } (3x + \quad)(3x + \quad)$$

(3) Factor the last term, -16. The possibilities are 16, -1 and -16, 1 and 8, -2 and -8, 2 and 4, -4.

These factors can also be written as -1, 16 and 1, -16 and -2, 8 and and 2, -8 and -4, 4.

(4) We try some possibilities:

$$(9x + 16)(x - 1) = 9x^2 + 7x - 16$$
$$(3x + 16)(3x - 1) = 9x^2 + 45x - 16$$
$$(9x + 8)(x - 2) = 9x^2 - 10x - 16$$
$$(3x + 8)(3x - 2) = 9x^2 + 18x - 16$$
$$(3x - 8)(3x + 2) = 9x^2 - 18x - 16$$

The factorization is $(3x - 8)(3x + 2)$.

45. $15x^2 - 25x - 10$

(1) Factor out the common factor, 5:

$$5(3x^2 - 5x - 2)$$

Then we factor the trinomial $3x^2 - 5x - 2$. This was done in Exercise 15. We know that $3x^2 - 5x - 2 = (3x + 1)(x - 2)$, so $15x^2 - 25x - 10 = 5(3x + 1)(x - 2)$.

47. $12p^3 + 31p^2 + 20p$

(1) We factor out the common factor, p:

$$p(12p^2 + 31p + 20)$$

Then we factor the trinomial $12p^2 + 31p + 20$. This was done in Exercise 17 although the variable is x in that exercise. We know that $12p^2 + 31p + 20 = (3p + 4)(4p + 5)$, so $12p^3 + 31p^2 + 20p = p(3p + 4)(4p + 5)$.

49.
$$16 + 18x - 9x^2 = -9x^2 + 18x + 16$$
$$= -1(9x^2 - 18x - 16)$$
$$= -1(3x - 8)(3x + 2) \quad \text{Using the result from Exercise 43}$$

Other correct factorizations are:
$$16 + 18x - 9x^2$$
$$= (-3x + 8)(3x + 2) \quad \text{Multiplying } 3x - 8 \text{ by } -1$$
$$= (3x - 8)(-3x - 2) \quad \text{Multiplying } 3x + 2 \text{ by } -1$$

We can also express $(-3x + 8)(3x + 2)$ as $(8 - 3x)(3x + 2)$ since $-3x + 8 = 8 - 3x$ by the commutative law of addition.

51. $-15x^2 + 19x - 6 = -1(15x^2 - 19x + 6)$

Now we factor $15x^2 - 19x + 6$.

(1) There is no common factor (other than 1 or -1).

(2) Factor the first term, $15x^2$. The possibilities are $15x$, x and $5x$, $3x$. We have these as possibilities for factorizations:

$$(15x + \quad)(x + \quad) \text{ and } (5x + \quad)(3x + \quad)$$

(3) Factor the last term, 6. The possibilities are 6, 1 and -6, -1 and 3, 2 and -3, -2.

These factors can also be written as 1, 6 and -1, -6 and 2, 3 and -2, -3.

(4) We try some possibilities:

$$(15x + 1)(x + 6) = 15x^2 + 91x + 6$$
$$(5x + 3)(3x + 2) = 15x^2 - 19x + 6$$
$$(5x - 3)(3x - 2) = 15x^2 - 19x + 6$$

The factorization of $15x^2 - 19x + 6$ is $(5x - 3)(3x - 2)$.

Then $-15x^2 + 19x - 6 = -1(5x - 3)(3x - 2)$. Other correct factorizations are:
$$-15x^2 + 19x - 6$$
$$= (-5x + 3)(3x - 2) \quad \text{Multiplying } 5x - 3 \text{ by } -1$$
$$= (5x - 3)(-3x + 2) \quad \text{Multiplying } 3x - 2 \text{ by } -1$$

Note that we can also express $(-5x + 3)(3x - 2)$ as $(3 - 5x)(3x - 2)$ since $-5x + 3 = 3 - 5x$ by the commutative law of addition. Similarly, we can express $(5x - 3)(-3x + 2)$ as $(5x - 3)(2 - 3x)$.

53. $14x^4 + 19x^3 - 3x^2$

(1) Factor out the common factor, x^2: $x^2(14x^2 + 19x - 3)$

Then we factor the trinomial $14x^2 + 19x - 3$. This was done in Exercise 19. We know that $14x^2 + 19x - 3 = (7x - 1)(2x + 3)$, so $14x^4 + 19x^3 - 3x^2 = x^2(7x - 1)(2x + 3)$.

55. $168x^3 - 45x^2 + 3x$

(1) Factor out the common factor, $3x$:

$3x(56x^2 - 15x + 1)$

Then we factor the trinomial $56x^2 - 15x + 1$.

(2) Factor $56x^2$. The possibilities are $56x$, x and $28x$, $2x$ and $14x$, $4x$ and $7x$, $8x$. We have these as possibilities for factorizations:

$(56x+\quad)(x+\quad)$ and $(28x+\quad)(2x+\quad)$ and $(14x+\quad)(4x+\quad)$ and $(7x+\quad)(8x+\quad)$

(3) Factor 1. Since 1 is positive and the middle term is negative we need consider only the negative factor pair $-1, -1$.

(4) We try some possibilities:

$(56x - 1)(x - 1) = 56x^2 - 57x + 1$
$(28x - 1)(2x - 1) = 56x^2 - 30x + 1$
$(14x - 1)(4x - 1) = 56x^2 - 18x + 1$
$(7x - 1)(8x - 1) = 56x^2 - 15x + 1$

Then $56x^2 - 15x + 1 = (7x - 1)(8x - 1)$, so
$168x^3 - 45x^2 + 3x = 3x(7x - 1)(8x - 1)$.

57. $15x^4 - 19x^2 + 6 = 15(x^2)^2 - 19x^2 + 6$

(1) There is no common factor (other than 1 or -1).

(2) Factor the first term, $15x^4$. The possibilities are $15x^2$, x^2 and $5x^2$, $3x^2$. We have these as possibilities for factorizations:

$(15x^2+\quad)(x^2+\quad)$ and $(5x^2+\quad)(3x^2+\quad)$

(3) Factor 6. Since 6 is positive and the middle term is negative, we need consider only negative factor pairs. Those pairs are $-6, -1$ and $-3, -2$.

These factors can also be written as $-1, -6$ and $-2, -3$.

(4) We can immediately reject all possibilities in which either factor has a common factor, such as $(15x^2 - 6)$ or $(3x^2 - 3)$, because we determined at the outset that there is no common factor. We try some of the remaining possibilities:

$(15x^2 - 1)(x^2 - 6) = 15x^4 - 91x^2 + 6$
$(15x^2 - 2)(x^2 - 3) = 15x^4 - 47x^2 + 6$
$(5x^2 - 6)(3x^2 - 1) = 15x^4 - 23x^2 + 6$
$(5x^2 - 3)(3x^2 - 2) = 15x^4 - 19x^2 + 6$

The factorization is $(5x^2 - 3)(3x^2 - 2)$.

59. $25t^2 + 80t + 64$

(1) There is no common factor (other than 1 or -1).

(2) Factor the first term, $25t^2$. The possibilities are $25t$, t and $5t$, $5t$. We have these as possibilities for factorizations:

$(25t+\quad)(t+\quad)$ and $(5t+\quad)(5t+\quad)$

(3) Factor the last term, 64. Since all signs are positive, we need consider only positive pairs of factors. Those factor pairs are 64, 1 and 32, 2 and 16, 4 and 8, 8.

These first three pairs can also be written as 1, 64 and 2, 32 and 4, 16.

(4) We try some possibilities:

$(25t + 64)(t + 1) = 25t^2 + 89t + 64$
$(5t + 32)(5t + 2) = 25t^2 + 170t + 64$
$(25t + 16)(t + 4) = 25t^2 + 116t + 64$
$(5t + 8)(5t + 8) = 25t^2 + 80t + 64$

The factorization is $(5t + 8)(5t + 8)$ or $(5t + 8)^2$.

61. $6x^3 + 4x^2 - 10x$

(1) Factor out the common factor, $2x$: $2x(3x^2 + 2x - 5)$

Then we factor the trinomial $3x^2 + 2x - 5$. We did this in Exercise 35 (after we factored 2 out of the original trinomial). We know that $3x^2 + 2x - 5 = (3x + 5)(x - 1)$, so $6x^3 + 4x^2 - 10x = 2x(3x + 5)(x - 1)$.

63. $25x^2 + 79x + 64$

We follow the same procedure as in Exercise 59. None of the possibilities works. Thus, $25x^2 + 79x + 64$ is not factorable. It is prime.

65. $6x^2 - 19x - 5$

(1) There is no common factor (other than 1 or -1).

(2) Factor the first term, $6x^2$. The possibilities are $6x$, x and $3x$, $2x$. We have these as possibilities for factorizations:

$(6x+\quad)(x+\quad)$ and $(3x+\quad)(2x+\quad)$

(3) Factor the last term, -5. The possibilities are -5, 1 and 5, -1.

These factors can also be written as 1, -5 and -1, 5.

(4) We try some possibilities:

$(6x - 5)(x + 1) = 6x^2 + x - 5$
$(6x + 5)(x - 1) = 6x^2 - x - 5$
$(6x + 1)(x - 5) = 6x^2 - 29x - 5$
$(6x - 1)(x + 5) = 6x^2 + 29x - 5$
$(3x - 5)(2x + 1) = 6x^2 - 7x - 5$
$(3x + 5)(2x - 1) = 6x^2 + 7x - 5$
$(3x + 1)(2x - 5) = 6x^2 - 13x - 5$
$(3x - 1)(2x + 5) = 6x^2 + 13x - 5$

None of the possibilities works. Thus, $6x^2 - 19x - 5$ is not factorable. It is prime.

67. $12m^2 - mn - 20n^2$

(1) There is no common factor (other than 1 or -1).

(2) Factor the first term, $12m^2$. The possibilities are $12m$, m and $6m$, $2m$ and $3m$, $4m$. We have these as possibilities for factorizations:

$$(12m+ \quad)(m+ \quad) \text{ and } (6m+ \quad)(2m+ \quad)$$
$$\text{and } (3m+ \quad)(4m+ \quad)$$

(3) Factor the last term, $-20n^2$. The possibilities are $20n$, $-n$ and $-20n$, n and $10n$, $-2n$ and $-10n$, $2n$ and $5n$, $-4n$ and $-5n$, $4n$.

These factors can also be written as $-n$, $20n$ and n, $-20n$ and $-2n$, $10n$ and $2n$, $-10n$ and $-4n$, $5n$ and $4n$, $-5n$.

(4) We can immediately reject all possibilities in which either factor has a common factor, such as $(12m + 20n)$ or $(4m - 2n)$, because we determined at the outset that there is no common factor. We try some of the remaining possibilities:

$$(12m - n)(m + 20n) = 12m^2 + 239mn - 20n^2$$
$$(12m + 5n)(m - 4n) = 12m^2 - 43mn - 20n^2$$
$$(3m - 20n)(4m + n) = 12m^2 - 77mn - 20n^2$$
$$(3m - 4n)(4m + 5n) = 12m^2 - mn - 20n^2$$

The factorization is $(3m - 4n)(4m + 5n)$.

69. $6a^2 - ab - 15b^2$

(1) There is no common factor (other than 1 or -1).

(2) Factor the first term, $6a^2$. The possibilities are $6a$, a and $3a$, $2a$. We have these as possibilities for factorizations:

$$(6a+ \quad)(a+ \quad) \text{ and } (3a+ \quad)(2a+ \quad)$$

(3) Factor the last term, $-15b^2$. The possibilities are $15b$, $-b$ and $-15b$, b and $5b$, $-3b$ and $-5b$, $3b$.

These factors can also be written as $-b$, $15b$ and b, $-15b$, and $-3b$, $5b$ and $3b$, $-5b$.

(4) We can immediately reject all possibilities in which either factor has a common factor, such as $(6a+15b)$ or $(3a - 3b)$, because we determined at the outset that there is no common factor. We try some of the remaining possibilities:

$$(6a - b)(a + 15b) = 6a^2 + 89ab - 15b^2$$
$$(3a - b)(2a + 15b) = 6a^2 + 43ab - 15b^2$$
$$(6a + 5b)(a - 3b) = 6a^2 - 13ab - 15b^2$$
$$(3a + 5b)(2a - 3b) = 6a^2 + ab - 15b^2$$
$$(3a - 5b)(2a + 3b) = 6a^2 - ab - 15b^2$$

The factorization is $(3a - 5b)(2a + 3b)$.

71. $9a^2 + 18ab + 8b^2$

(1) There is no common factor (other than 1 or -1).

(2) Factor the first term, $9a^2$. The possibilities are $9a$, a and $3a$, $3a$. We have these as possibilities for factorizations:

$$(9a+ \quad)(a+ \quad) \text{ and } (3a+ \quad)(3a+ \quad)$$

(3) Factor $8b^2$. Since all signs are positive, we need consider only pairs of factors with positive coefficients. Those factor pairs are $8b$, b and $4b$, $2b$.

These factors can also be written as b, $8b$ and $2b$, $4b$.

(4) We try some possibilities:

$$(9a + 8b)(a + b) = 9a^2 + 17ab + 8b^2$$
$$(3a + 8b)(3a + b) = 9a^2 + 27ab + 8b^2$$
$$(9a + 4b)(a + 2b) = 9a^2 + 22ab + 8b^2$$
$$(3a + 4b)(3a + 2b) = 9a^2 + 18ab + 8b^2$$

The factorization is $(3a + 4b)(3a + 2b)$.

73. $35p^2 + 34pq + 8q^2$

(1) There is no common factor (other than 1 or -1).

(2) Factor the first term, $35p^2$. The possibilities are $35p$, p and $7p$, $5p$. We have these as possibilities for factorizations:

$$(35p+ \quad)(p+ \quad) \text{ and } (7p+ \quad)(5p+ \quad)$$

(3) Factor $8q^2$. Since all signs are positive, we need consider only pairs of factors with positive coefficients. Those factor pairs are $8q$, q and $4q$, $2q$.

These factors can also be written as q, $8q$ and $2q$, $4q$.

(4) We try some possibilities:

$$(35p + 8q)(p + q) = 35p^2 + 43pq + 8q^2$$
$$(7p + 8q)(5p + q) = 35p^2 + 47pq + 8q^2$$
$$(35p + 4q)(p + 2q) = 35p^2 + 74pq + 8q^2$$
$$(7p + 4q)(5p + 2q) = 35p^2 + 34pq + 8p^2$$

The factorization is $(7p + 4q)(5p + 2q)$.

75. $18x^2 - 6xy - 24y^2$

(1) Factor out the common factor, 6:

$$6(3x^2 - xy - 4y^2)$$

Then we factor the trinomial $3x^2 - xy - 4y^2$.

(2) Factor $3x^2$. The only possibility is $3x$, x. The desired factorization is of the form:

$$(3x+ \quad)(x+ \quad)$$

(3) Factor $-4y^2$. The possibilities are $4y$, $-y$ and $-4y$, y and $2y$, $-2y$.

These factors can also be written as $-y$, $4y$ and y, $-4y$ and $-2y$, $2y$.

(4) We try some possibilities:

$$(3x + 4y)(x - y) = 3x^2 + xy - 4y^2$$
$$(3x - 4y)(x + y) = 3x^2 - xy - 4y^2$$

Then $3x^2 - xy - 4y^2 = (3x - 4y)(x + y)$, so $18x^2 - 6xy - 24y^2 = 6(3x - 4y)(x + y)$.

77. Discussion and Writing Exercise

79.
$$A = pq - 7$$
$$A + 7 = pq \qquad \text{Adding 7}$$
$$\frac{A + 7}{p} = q \qquad \text{Dividing by } p$$

81. $3x + 2y = 6$
$$2y = 6 - 3x \quad \text{Subtracting } 3x$$
$$y = \frac{6 - 3x}{2} \quad \text{Dividing by 2}$$

83. $5 - 4x < -11$
$$-4x < -16 \quad \text{Subtracting 5}$$
$$x > 4 \qquad \text{Dividing by } -4 \text{ and reversing the}$$
$$\text{inequality symbol}$$

The solution set is $\{x | x > 4\}$.

85. Graph: $y = \dfrac{2}{5}x - 1$

Because the equation is in the form $y = mx + b$, we know the y-intercept is $(0, -1)$. We find two other points on the line, substituting multiples of 5 for x to avoid fractions.

When $x = -5$, $y = \dfrac{2}{5}(-5) - 1 = -2 - 1 = -3$.

When $x = 5$, $y = \dfrac{2}{5}(5) - 1 = 2 - 1 = 1$.

x	y
0	-1
-5	-3
5	1

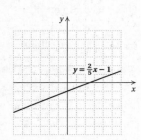

87. $4x - 16y = 64$

To find the x-intercept, let $y = 0$ and solve for x.
$$4x - 16y = 64$$
$$4x - 16 \cdot 0 = 64$$
$$4x = 64$$
$$x = 16$$

The x-intercept is $(16, 0)$.

To find the y-intercept, let $x = 0$ and solve for y.
$$4x - 16y = 64$$
$$4 \cdot 0 - 16y = 64$$
$$-16y = 64$$
$$y = -4$$

The y-intercept is $(0, -4)$.

89. $x - 1.3y = 6.5$

To find the x-intercept, let $y = 0$ and solve for x.
$$x - 1.3y = 6.5$$
$$x - 1.3(0) = 6.5$$
$$x = 6.5$$

The x-intercept is $(6.5, 0)$.

To find the y-intercept, let $x = 0$ and solve for y.
$$x - 1.3y = 6.5$$
$$0 - 1.3y = 6.5$$
$$-1.3y = 6.5$$
$$y = -5$$

The y-intercept is $(0, -5)$.

91. $y = 4 - 5x$

To find the x-intercept, let $y = 0$ and solve for x.
$$y = 4 - 5x$$
$$0 = 4 - 5x$$
$$5x = 4$$
$$x = \frac{4}{5}$$

The x-intercept is $\left(\dfrac{4}{5}, 0\right)$.

To find the y-intercept, let $x = 0$ and solve for y.
$$y = 4 - 5x$$
$$y = 4 - 5 \cdot 0$$
$$y = 4$$

The y-intercept is $(0, 4)$.

93. $20x^{2n} + 16x^n + 3 = 20(x^n)^2 + 16x^n + 3$

(1) There is no common factor (other than 1 and -1).

(2) Factor the first term, $20x^{2n}$. The possibilities are $20x^n$, x^n and $10x^n$, $2x^n$ and $5x^n$, $4x^n$. We have these as possibilities for factorizations:
$$(20x^n + \quad)(x^n + \quad) \text{ and } (10x^n + \quad)(2x^n + \quad)$$
$$\text{and } (5x^n + \quad)(4x^n + \quad)$$

(3) Factor the last term, 3. Since all signs are positive, we need consider only the positive factor pair 3, 1.

(4) We try some possibilities:
$$(20x^n + 3)(x^n + 1) = 20x^{2n} + 23x^n + 3$$
$$(10x^n + 3)(2x^n + 1) = 20x^{2n} + 16x^n + 3$$

The factorization is $(10x^n + 3)(2x^n + 1)$.

95. $3x^{6a} - 2x^{3a} - 1 = 3(x^{3a})^2 - 2x^{3a} - 1$

(1) There is no common factor (other than 1 or -1).

(2) Factor the first term, $3x^{6a}$. The only possibility is $3x^{3a}$, x^{3a}. The desired factorization is of the form:
$$(3x^{3a} + \quad)(x^{3a} + \quad)$$

(3) Factor the last term, -1. The only possibility is -1, 1.

(4) We try the possibilities:
$$(3x^{3a} - 1)(x^{3a} + 1) = 3x^{6a} + 2x^{3a} - 1$$
$$(3x^{3a} + 1)(x^{3a} - 1) = 3x^{6a} - 2x^{3a} - 1$$

The factorization is $(3x^{3a} + 1)(x^{3a} - 1)$.

97.-105. Left to the student

Exercise Set 13.4

1. $x^2 + 2x + 7x + 14 = (x^2 + 2x) + (7x + 14)$
$$= x(x+2) + 7(x+2)$$
$$= (x+7)(x+2)$$

3. $x^2 - 4x - x + 4 = (x^2 - 4x) + (-x + 4)$
$$= x(x-4) - 1(x-4)$$
$$= (x-1)(x-4)$$

5. $6x^2 + 4x + 9x + 6 = (6x^2 + 4x) + (9x + 6)$
$$= 2x(3x+2) + 3(3x+2)$$
$$= (2x+3)(3x+2)$$

7. $3x^2 - 4x - 12x + 16 = (3x^2 - 4x) + (-12x + 16)$
$$= x(3x-4) - 4(3x-4)$$
$$= (x-4)(3x-4)$$

9. $35x^2 - 40x + 21x - 24 = (35x^2 - 40x) + (21x - 24)$
$$= 5x(7x-8) + 3(7x-8)$$
$$= (5x+3)(7x-8)$$

11. $4x^2 + 6x - 6x - 9 = (4x^2 + 6x) + (-6x - 9)$
$$= 2x(2x+3) - 3(2x+3)$$
$$= (2x-3)(2x+3)$$

13. $2x^4 + 6x^2 + 5x^2 + 15 = (2x^4 + 6x^2) + (5x^2 + 15)$
$$= 2x^2(x^2+3) + 5(x^2+3)$$
$$= (2x^2+5)(x^2+3)$$

15. $2x^2 + 7x - 4$

(1) First factor out a common factor, if any. There is none (other than 1 or -1).

(2) Multiply the leading coefficient, 2 and the constant, -4: $2(-4) = -8$.

(3) Look for a factorization of -8 in which the sum of the factors is the coefficient of the middle term, 7.

Pairs of factors	Sums of factors
$-1,\ 8$	7
$1,\ -8$	-7
$-2,\ 4$	2
$2,\ -4$	-2

(4) Split the middle term: $7x = -1x + 8x$

(5) Factor by grouping:
$$2x^2 + 7x - 4 = 2x^2 - x + 8x - 4$$
$$= (2x^2 - x) + (8x - 4)$$
$$= x(2x-1) + 4(2x-1)$$
$$= (x+4)(2x-1)$$

17. $3x^2 - 4x - 15$

(1) First factor out a common factor, if any. There is none (other than 1 or -1).

(2) Multiply the leading coefficient, 3, and the constant, -15: $3(-15) = -45$.

(3) Look for a factorization of -45 in which the sum of the factors is the coefficient of the middle term, -4.

Pairs of factors	Sums of factors
$-1,\ 45$	44
$1,\ -45$	-44
$-3,\ 15$	12
$3,\ -15$	-12
$-5,\ 9$	4
$5,\ -9$	-4

(4) Split the middle term: $-4x = 5x - 9x$

(5) Factor by grouping:
$$3x^2 - 4x - 15 = 3x^2 + 5x - 9x - 15$$
$$= (3x^2 + 5x) + (-9x - 15)$$
$$= x(3x+5) - 3(3x+5)$$
$$= (x-3)(3x+5)$$

19. $6x^2 + 23x + 7$

(1) First factor out a common factor, if any. There is none (other than 1 or -1).

(2) Multiply the leading coefficient, 6, and the constant, 7: $6 \cdot 7 = 42$.

(3) Look for a factorization of 42 in which the sum of the factors is the coefficient of the middle term, 23. We only need to consider positive factors.

Pairs of factors	Sums of factors
$1,\ 42$	43
$2,\ 21$	23
$3,\ 14$	17
$6,\ 7$	13

(4) Split the middle term: $23x = 2x + 21x$

(5) Factor by grouping:
$$6x^2 + 23x + 7 = 6x^2 + 2x + 21x + 7$$
$$= (6x^2 + 2x) + (21x + 7)$$
$$= 2x(3x+1) + 7(3x+1)$$
$$= (2x+7)(3x+1)$$

21. $3x^2 - 4x + 1$

(1) First factor out a common factor, if any. There is none (other than 1 or -1).

(2) Multiply the leading coefficient, 3, and the constant, 1: $3 \cdot 1 = 3$.

(3) Look for a factorization of 3 in which the sum of the factors is the coefficient of the middle term, -4. The numbers we want are -1 and -3: $-1 \cdot (-3) = 3$ and $-1 + (-3) = -4$.

(4) Split the middle term: $-4x = -1x - 3x$

(5) Factor by grouping:

$$\begin{aligned} 3x^2 - 4x + 1 &= 3x^2 - x - 3x + 1 \\ &= (3x^2 - x) + (-3x + 1) \\ &= x(3x - 1) - 1(3x - 1) \\ &= (x - 1)(3x - 1) \end{aligned}$$

23. $4x^2 - 4x - 15$

(1) First factor out a common factor, if any. There is none (other than 1 or -1).

(2) Multiply the leading coefficient, 4, and the constant, -15: $4(-15) = -60$.

(3) Look for a factorization of -60 in which the sum of the factors is the coefficient of the middle term, -4.

Pairs of factors	Sums of factors
-1, 60	59
1, -60	-59
-2, 30	28
2, -30	-28
-3, 20	17
3, -20	-17
-4, 15	11
4, -15	-11
-5, 12	7
5, -12	-7
-6, 10	4
6, -10	-4

(4) Split the middle term: $-4x = 6x - 10x$

(5) Factor by grouping:

$$\begin{aligned} 4x^2 - 4x - 15 &= 4x^2 + 6x - 10x - 15 \\ &= (4x^2 + 6x) + (-10x - 15) \\ &= 2x(2x + 3) - 5(2x + 3) \\ &= (2x - 5)(2x + 3) \end{aligned}$$

25. $2x^2 + x - 1$

(1) First factor out a common factor, if any. There is none (other than 1 or -1).

(2) Multiply the leading coefficient, 2, and the constant, -1: $2(-1) = -2$.

(3) Look for a factorization of -2 in which the sum of the factors is the coefficient of the middle term, 1. The numbers we want are 2 and -1: $2(-1) = -2$ and $2 - 1 = 1$.

(4) Split the middle term: $x = 2x - 1x$

(5) Factor by grouping:

$$\begin{aligned} 2x^2 + x - 1 &= 2x^2 + 2x - x - 1 \\ &= (2x^2 + 2x) + (-x - 1) \\ &= 2x(x + 1) - 1(x + 1) \\ &= (2x - 1)(x + 1) \end{aligned}$$

27. $9x^2 - 18x - 16$

(1) First factor out a common factor, if any. There is none (other than 1 or -1).

(2) Multiply the leading coefficient, 9, and the constant, -16: $9(-16) = -144$.

(3) Look for a factorization of -144, so the sum of the factors is the coefficient of the middle term, -18.

Pairs of factors	Sums of factors
-1, 144	143
1, -144	-143
-2, 72	70
2, -72	-70
-3, 48	45
3, -48	-45
-4, 36	32
4, -36	-32
-6, 24	18
6, -24	-18
-8, 18	10
8, -18	-10
-9, 16	7
9, -16	-7
-12, 12	0

(4) Split the middle term: $-18x = 6x - 24x$

(5) Factor by grouping:

$$\begin{aligned} 9x^2 - 18x - 16 &= 9x^2 + 6x - 24x - 16 \\ &= (9x^2 + 6x) + (-24x - 16) \\ &= 3x(3x + 2) - 8(3x + 2) \\ &= (3x - 8)(3x + 2) \end{aligned}$$

29. $3x^2 + 5x - 2$

(1) First factor out a common factor, if any. There is none (other than 1 or -1).

(2) Multiply the leading coefficient, 3, and the constant, -2: $3(-2) = -6$.

(3) Look for a factorization of -6 in which the sum of the factors is the coefficient of the middle term, 5. The numbers we want are -1 and 6: $-1(6) = -6$ and $-1 + 6 = 5$.

(4) Split the middle term: $5x = -1x + 6x$

(5) Factor by grouping:

$$\begin{aligned} 3x^2 + 5x - 2 &= 3x^2 - x + 6x - 2 \\ &= (3x^2 - x) + (6x - 2) \\ &= x(3x - 1) + 2(3x - 1) \\ &= (x + 2)(3x - 1) \end{aligned}$$

31. $12x^2 - 31x + 20$

(1) First factor out a common factor, if any. There is none (other than 1 or -1).

(2) Multiply the leading coefficient, 12, and the constant, 20: $12 \cdot 20 = 240$.

(3) Look for a factorization of 240 in which the sum of the factors is the coefficient of the middle term, -31. We only need to consider negative factors.

Pairs of factors	Sums of factors
$-1, -240$	-241
$-2, -120$	-122
$-3, -8$	-83
$-4, -60$	-64
$-5, -48$	-53
$-6, -40$	-46
$-8, -30$	-38
$-10, -24$	-34
$-12, -20$	-32
$-15, -16$	-31

(4) Split the middle term: $-31x = -15x - 16x$

(5) Factor by grouping:
$$12x^2 - 31x + 20 = 12x^2 - 15x - 16x + 20$$
$$= (12x^2 - 15x) + (-16x + 20)$$
$$= 3x(4x - 5) - 4(4x - 5)$$
$$= (3x - 4)(4x - 5)$$

33. $14x^2 - 19x - 3$

(1) First factor out a common factor, if any. There is none (other than 1 or -1).

(2) Multiply the leading coefficient, 14, and the constant, -3: $14(-3) = -42$.

(3) Look for a factorization of -42 so that the sum of the factors is the coefficient of the middle term, -19.

Pairs of factors	Sums of factors
$-1, \ 42$	41
$1, -42$	-41
$-2, \ 21$	19
$2, -21$	-19
$-3, \ 14$	11
$3, -14$	-11
$-6, \ 7$	1
$6, \ -7$	-1

(4) Split the middle term: $-19x = 2x - 21x$

(5) Factor by grouping:
$$14x^2 - 19x - 3 = 14x^2 + 2x - 21x - 3$$
$$= (14x^2 + 2x) + (-21x - 3)$$
$$= 2x(7x + 1) - 3(7x + 1)$$
$$= (2x - 3)(7x + 1)$$

35. $9x^2 + 18x + 8$

(1) First factor out a common factor, if any. There is none (other than 1 or -1).

(2) Multiply the leading coefficient, 9, and the constant, 8: $9 \cdot 8 = 72$.

(3) Look for a factorization of 72 in which the sum of the factors is the coefficient of the middle term, 18. We only need to consider positive factors.

Pairs of factors	Sums of factors
$1, \quad 72$	73
$2, \quad 36$	38
$3, \quad 24$	27
$4, \quad 18$	22
$6, \quad 12$	18
$8, \quad 9$	17

(4) Split the middle term: $18x = 6x + 12x$

(5) Factor by grouping:
$$9x^2 + 18x + 8 = 9x^2 + 6x + 12x + 8$$
$$= (9x^2 + 6x) + (12x + 8)$$
$$= 3x(3x + 2) + 4(3x + 2)$$
$$= (3x + 4)(3x + 2)$$

37. $49 - 42x + 9x^2 = 9x^2 - 42x + 49$

(1) First factor out a common factor, if any. There is none (other than 1 or -1).

(2) Multiply the leading coefficient, 9, and the constant, 49: $9 \cdot 49 = 441$.

(3) Look for a factorization of 441 in which the sum of the factors is the coefficient of the middle term, -42. We only need to consider negative factors.

Pairs of factors	Sums of factors
$-1, -441$	-442
$-3, -147$	-150
$-7, -63$	-70
$-9, -49$	-58
$-21, -21$	-42

(4) Split the middle term: $-42x = -21x - 21x$

(5) Factor by grouping:
$$9x^2 - 42x + 49 = 9x^2 - 21x - 21x + 49$$
$$= (9x^2 - 21x) + (-21x + 49)$$
$$= 3x(3x - 7) - 7(3x - 7)$$
$$= (3x - 7)(3x - 7), \text{ or}$$
$$(3x - 7)^2$$

39. $24x^2 - 47x - 2$

(1) First factor out a common factor, if any. There is none (other than 1 or -1).

(2) Multiply the leading coefficient, 24, and the constant, -2: $24(-2) = -48$.

(3) Look for a factorization of -48 in which the sum of the factors is the coefficient of the middle term, -47. The numbers we want are -48 and 1: $-48 \cdot 1 = -48$ and $-48 + 1 = -47$.

(4) Split the middle term: $-47x = -48x + 1x$

(5) Factor by grouping:
$$24x^2 - 47x - 2 = 24x^2 - 48x + x - 2$$
$$= (24x^2 - 48x) + (x - 2)$$
$$= 24x(x - 2) + 1(x - 2)$$
$$= (24x + 1)(x - 2)$$

41. $5 - 9a^2 - 12a = -9a^2 - 12a + 5 = -1(9a^2 + 12a - 5)$

Now we factor $9a^2 + 12a - 5$.

(1) We have already factored out the common factor, -1, to make the leading coefficient positive.

(2) Multiply the leading coefficient, 9, and the constant, -5: $9(-5) = -45$.

(3) Look for a factorization of -45 in which the sum of the factors is the coefficient of the middle term, 12. The numbers we want are 15 and -3: $15(-3) = -45$ and $15 + (-3) = 12$.

(4) Split the middle term: $12a = 15a - 3a$

(5) Factor by grouping:
$$9a^2 + 12a - 5 = 9a^2 + 15a - 3a - 5$$
$$= (9a^2 + 15a) + (-3a - 5)$$
$$= 3a(3a + 5) - (3a + 5)$$
$$= (3a - 1)(3a + 5)$$

Then we have
$$5 - 9a^2 - 12a$$
$$= -1(3a - 1)(3a + 5)$$
$$= (-3a + 1)(3a + 5) \qquad \text{Multiplying } 3a-1 \text{ by } -1$$
$$= (3a - 1)(-3a - 5) \qquad \text{Multiplying } 3a+5 \text{ by } -1$$

Note that we can also express $(-3a + 1)(3a + 5)$ as $(1 - 3a)(3a + 5)$ since $-3a + 1 = 1 - 3a$ by the commutative law of addition.

43. $20 + 6x - 2x^2 = -2x^2 + 6x + 20$

(1) Factor out the common factor -2. We factor out -2 rather than 2 in order to make the leading coefficient of the trinomial factor positive.
$$-2x^2 + 6x + 20 = -2(x^2 - 3x - 10)$$

To factor $x^2 - 3x - 10$, we look for two factors of -10 whose sum is -3. The numbers we want are -5 and 2. Then $x^2 - 3x - 10 = (x - 5)(x + 2)$, so we have:
$$20 + 6x - 2x^2$$
$$= -2(x - 5)(x + 2)$$
$$= 2(-x + 5)(x + 2) \qquad \text{Multiplying } x - 5 \text{ by } -1$$
$$= 2(x - 5)(-x - 2) \qquad \text{Multiplying } x + 2 \text{ by } -1$$

Note that we can also express $2(-x + 5)(x + 2)$ as $2(5 - x)(x + 2)$ since $-x + 5 = 5 - x$ by the commutative law of addition.

45. $12x^2 + 28x - 24$

(1) Factor out the common factor, 4:
$$12x^2 + 28x - 24 = 4(3x^2 + 7x - 6)$$

(2) Now we factor the trinomial $3x^2 + 7x - 6$. Multiply the leading coefficient, 3, and the constant, -6: $3(-6) = -18$.

(3) Look for a factorization of -18 in which the sum of the factors is the coefficient of the middle term, 7. The numbers we want are 9 and -2: $9(-2) = -18$ and $9 + (-2) = 7$.

(4) Split the middle term: $7x = 9x - 2x$

(5) Factor by grouping:
$$3x^2 + 7x - 6 = 3x^2 + 9x - 2x - 6$$
$$= (3x^2 + 9x) + (-2x - 6)$$
$$= 3x(x + 3) - 2(x + 3)$$
$$= (3x - 2)(x + 3)$$

We must include the common factor to get a factorization of the original trinomial.
$$12x^2 + 28x - 24 = 4(3x - 2)(x + 3)$$

47. $30x^2 - 24x - 54$

(1) Factor out the common factor, 6.
$$30x^2 - 24x - 54 = 6(5x^2 - 4x - 9)$$

(2) Now we factor the trinomial $5x^2 - 4x - 9$. Multiply the leading coefficient, 5, and the constant, -9: $5(-9) = -45$.

(3) Look for a factorization of -45 in which the sum of the factors is the coefficient of the middle term, -4. The numbers we want are -9 and 5: $-9 \cdot 5 = -45$ and $-9 + 5 = -4$.

(4) Split the middle term: $-4x = -9x + 5x$

(5) Factor by grouping:
$$5x^2 - 4x - 9 = 5x^2 - 9x + 5x - 9$$
$$= (5x^2 - 9x) + (5x - 9)$$
$$= x(5x - 9) + (5x - 9)$$
$$= (x + 1)(5x - 9)$$

We must include the common factor to get a factorization of the original trinomial.
$$30x^2 - 24x - 54 = 6(x + 1)(5x - 9)$$

49. $4y + 6y^2 - 10 = 6y^2 + 4y - 10$

(1) Factor out the common factor, 2.
$$6y^2 + 4y - 10 = 2(3y^2 + 2y - 5)$$

(2) Now we factor the trinomial $3y^2 + 2y - 5$. Multiply the leading coefficient, 3, and the constant, -5: $3(-5) = -15$.

(3) Look for a factorization of -15 in which the sum of the factors is the coefficient of the middle term, 2. The numbers we want are 5 and -3: $5(-3) = -15$ and $5 + (-3) = 2$.

(4) Split the middle term: $2y = 5y - 3y$

(5) Factor by grouping:

$$3y^2 + 2y - 5 = 3y^2 + 5y - 3y - 5$$
$$= (3y^2 + 5y) + (-3y - 5)$$
$$= y(3y + 5) - (3y + 5)$$
$$= (y - 1)(3y + 5)$$

We must include the common factor to get a factorization of the original trinomial.

$$4y + 6y^2 - 10 = 2(y - 1)(3y + 5)$$

51. $3x^2 - 4x + 1$

(1) There is no common factor (other than 1 or -1).

(2) Multiply the leading coefficient, 3, and the constant, 1: $3 \cdot 1 = 3$.

(3) Look for a factorization of 3 in which the sum of the factors is the coefficient of the middle term, -4. The numbers we want are -1 and -3: $-1(-3) = 3$ and $-1 + (-3) = -4$.

(4) Split the middle term: $-4x = -1x - 3x$

(5) Factor by grouping:

$$3x^2 - 4x + 1 = 3x^2 - x - 3x + 1$$
$$= (3x^2 - x) + (-3x + 1)$$
$$= x(3x - 1) - (3x - 1)$$
$$= (x - 1)(3x - 1)$$

53. $12x^2 - 28x - 24$

(1) Factor out the common factor, 4:

$$12x^2 - 28x - 24 = 4(3x^2 - 7x - 6)$$

(2) Now we factor the trinomial $3x^2 - 7x - 6$. Multiply the leading coefficient, 3, and the constant, -6: $3(-6) = -18$.

(3) Look for a factorization of -18 in which the sum of the factors is the coefficient of the middle term, -7. The numbers we want are -9 and 2: $-9 \cdot 2 = -18$ and $-9 + 2 = -7$.

(4) Split the middle term: $-7x = -9x + 2x$

(5) Factor by grouping:

$$3x^2 - 7x - 6 = 3x^2 - 9x + 2x - 6$$
$$= (3x^2 - 9x) + (2x - 6)$$
$$= 3x(x - 3) + 2(x - 3)$$
$$= (3x + 2)(x - 3)$$

We must include the common factor to get a factorization of the original trinomial.

$$12x^2 - 28x - 24 = 4(3x + 2)(x - 3)$$

55. $-1 + 2x^2 - x = 2x^2 - x - 1$

(1) There is no common factor (other than 1 or -1).

(2) Multiply the leading coefficient, 2, and the constant, -1: $2(-1) = -2$.

(3) Look for a factorization of -2 in which the sum of the factors is the coefficient of the middle term, -1. The numbers we want are -2 and 1: $-2 \cdot 1 = -2$ and $-2 + 1 = -1$.

(4) Split the middle term: $-x = -2x + 1x$

(5) Factor by grouping:

$$2x^2 - x - 1 = 2x^2 - 2x + x - 1$$
$$= (2x^2 - 2x) + (x - 1)$$
$$= 2x(x - 1) + (x - 1)$$
$$= (2x + 1)(x - 1)$$

57. $9x^2 + 18x - 16$

(1) There is no common factor (other than 1 or -1).

(2) Multiply the leading coefficient, 9, and the constant, -16: $9(-16) = -144$.

(3) Look for a factorization of -144 in which the sum of the factors is the coefficient of the middle term, 18. The numbers we want are 24 and -6: $24(-6) = -144$ and $24 + (-6) = 18$.

(4) Split the middle term: $18x = 24x - 6x$

(5) Factor by grouping:

$$9x^2 + 18x - 16 = 9x^2 + 24x - 6x - 16$$
$$= (9x^2 + 24x) + (-6x - 16)$$
$$= 3x(3x + 8) - 2(3x + 8)$$
$$= (3x - 2)(3x + 8)$$

59. $15x^2 - 25x - 10$

(1) Factor out the common factor, 5:

$$15x^2 - 25x - 10 = 5(3x^2 - 5x - 2)$$

(2) Now we factor the trinomial $3x^2 - 5x - 2$. Multiply the leading coefficient, 3, and the constant, -2: $3(-2) = -6$.

(3) Look for a factorization of -6 in which the sum of the factors is the coefficient of the middle term, -5. The numbers we want are -6 and 1: $-6 \cdot 1 = -6$ and $-6 + 1 = -5$.

(4) Split the middle term: $-5x = -6x + 1x$

(5) Factor by grouping:

$$3x^2 - 5x - 2 = 3x^2 - 6x + x - 2$$
$$= (3x^2 - 6x) + (x - 2)$$
$$= 3x(x - 2) + (x - 2)$$
$$= (3x + 1)(x - 2)$$

We must include the common factor to get a factorization of the original trinomial.

$$15x^2 - 25x - 10 = 5(3x + 1)(x - 2)$$

61. $12p^3 + 31p^2 + 20p$

(1) Factor out the common factor, p:

$$12p^3 + 31p^2 + 20p = p(12p^2 + 31p + 20)$$

(2) Now we factor the trinomial $12p^2 + 31p + 20$. Multiply the leading coefficient, 12, and the constant, 20: $12 \cdot 20 = 240$.

(3) Look for a factorization of 240 in which the sum of the factors is the coefficient of the middle term, 31. The numbers we want are 15 and 16: $15 \cdot 16 = 240$ and $15 + 16 = 31$.

(4) Split the middle term: $31p = 15p + 16p$

(5) Factor by grouping:
$$12p^2 + 31p + 20 = 12p^2 + 15p + 16p + 20$$
$$= (12p^2 + 15p) + (16p + 20)$$
$$= 3p(4p + 5) + 4(4p + 5)$$
$$= (3p + 4)(4p + 5)$$

We must include the common factor to get a factorization of the original trinomial.
$$12p^3 + 31p^2 + 20p = p(3p + 4)(4p + 5)$$

63. $4 - x - 5x^2 = -5x^2 - x + 4$

(1) Factor out -1 to make the leading coefficient positive:
$$-5x^2 - x + 4 = -1(5x^2 + x - 4)$$

(2) Now we factor the trinomial $5x^2 + x - 4$. Multiply the leading coefficient, 5, and the constant, -4: $5(-4) = -20$.

(3) Look for a factorization of -20 in which the sum of the factors is the coefficient of the middle term, 1. The numbers we want are 5 and -4: $5(-4) = -20$ and $5 + (-4) = 1$.

(4) Split the middle term: $x = 5x - 4x$

(5) Factor by grouping:
$$5x^2 + x - 4 = 5x^2 + 5x - 4x - 4$$
$$= (5x^2 + 5x) + (-4x - 4)$$
$$= 5x(x + 1) - 4(x + 1)$$
$$= (5x - 4)(x + 1)$$

We must include the common factor to get a factorization of the original trinomial.
$$4 - x - 5x^2$$
$$= -1(5x - 4)(x + 1)$$
$$= (-5x + 4)(x + 1) \quad \text{Multiplying } 5x - 4 \text{ by } -1$$
$$= (5x - 4)(-x - 1) \quad \text{Multiplying } x + 1 \text{ by } -1$$

Note that we can also express $(-5x + 4)(x + 1)$ as $(4 - 5x)(x + 1)$ since $-5x + 4 = 4 - 5x$ by the commutative law of addition.

65. $33t - 15 - 6t^2 = -6t^2 + 33t - 15$

(1) Factor out the common factor, -3. We factor out -3 rather than 3 in order to make the leading coefficient of the trinomial factor positive.
$$-6t^2 + 33t - 15 = -3(2t^2 - 11t + 5)$$

(2) Now we factor the trinomial $2t^2 - 11t + 5$. Multiply the leading coefficient, 2, and the constant, 5: $2 \cdot 5 = 10$.

(3) Look for a factorization of 10 in which the sum of the factors is the coefficient of the middle term, -11. The numbers we want are -1 and -10: $-1(-10) = 10$ and $-1 + (-10) = -11$.

(4) Split the middle term: $-11t = -1t - 10t$

(5) Factor by grouping:
$$2t^2 - 11t + 5 = 2t^2 - t - 10t + 5$$
$$= (2t^2 - t) + (-10t + 5)$$
$$= t(2t - 1) - 5(2t - 1)$$
$$= (t - 5)(2t - 1)$$

We must include the common factor to get a factorization of the original trinomial.
$$33t - 15 - 6t^2$$
$$= -3(t - 5)(2t - 1)$$
$$= 3(-t + 5)(2t - 1) \quad \text{Multiplying } t - 5 \text{ by } -1$$
$$= 3(t - 5)(-2t + 1) \quad \text{Multiplying } 2t - 1 \text{ by } -1$$

Note that we can also express $3(-t + 5)(2t - 1)$ as $3(5 - t)(2t - 1)$ since $-t + 5 = 5 - t$ by the commutative law of addition. Similarly, we can express $3(t - 5)(-2t + 1)$ as $3(t - 5)(1 - 2t)$.

67. $14x^4 + 19x^3 - 3x^2$

(1) Factor out the common factor, x^2:
$$14x^4 + 19x^3 - 3x^2 = x^2(14x^2 + 19x - 3)$$

(2) Now we factor the trinomial $14x^2 + 19x - 3$. Multiply the leading coefficient, 14, and the constant, -3: $14(-3) = -42$.

(3) Look for a factorization of -42 in which the sum of the factors is the coefficient of the middle term, 19. The numbers we want are 21 and -2: $21(-2) = -42$ and $21 + (-2) = 19$.

(4) Split the middle term: $19x = 21x - 2x$

(5) Factor by grouping:
$$14x^2 + 19x - 3 = 14x^2 + 21x - 2x - 3$$
$$= (14x^2 + 21x) + (-2x - 3)$$
$$= 7x(2x + 3) - (2x + 3)$$
$$= (7x - 1)(2x + 3)$$

We must include the common factor to get a factorization of the original trinomial.
$$14x^4 + 19x^3 - 3x^2 = x^2(7x - 1)(2x + 3)$$

69. $168x^3 - 45x^2 + 3x$

(1) Factor out the common factor, $3x$:
$$168x^3 - 45x^2 + 3x = 3x(56x^2 - 15x + 1)$$

(2) Now we factor the trinomial $56x^2 - 15x + 1$. Multiply the leading coefficient, 56, and the constant, 1: $56 \cdot 1 = 56$.

(3) Look for a factorization of 56 in which the sum of the factors is the coefficient of the middle term, -15. The numbers we want are -7 and -8: $-7(-8) = 56$ and $-7 + (-8) = -15$.

(4) Split the middle term: $-15x = -7x - 8x$

(5) Factor by grouping:
$$56x^2 - 15x + 1 = 56x^2 - 7x - 8x + 1$$
$$= (56x^2 - 7x) + (-8x + 1)$$
$$= 7x(8x - 1) - (8x - 1)$$
$$= (7x - 1)(8x - 1)$$

We must include the common factor to get a factorization of the original trinomial.
$$168x^3 - 45x^2 + 3x = 3x(7x - 1)(8x - 1)$$

71. $15x^4 - 19x^2 + 6$

(1) There are no common factors (other than 1 or -1).

(2) Multiply the leading coefficient, 15, and the constant, 6: $15 \cdot 6 = 90$.

(3) Look for a factorization of 90 in which the sum of the factors is the coefficient of the middle term, -19. The numbers we want are -9 and -10: $-9(-10) = 90$ and $-9 + (-10) = -19$.

(4) Split the middle term: $-19x^2 = -9x^2 - 10x^2$

(5) Factor by grouping:
$$15x^4 - 19x^2 + 6 = 15x^4 - 9x^2 - 10x^2 + 6$$
$$= (15x^4 - 9x^2) + (-10x^2 + 6)$$
$$= 3x^2(5x^2 - 3) - 2(5x^2 - 3)$$
$$= (3x^2 - 2)(5x^2 - 3)$$

73. $25t^2 + 80t + 64$

(1) There are no common factors (other than 1 or -1).

(2) Multiply the leading coefficient, 25, and the constant, 64: $25 \cdot 64 = 1600$.

(3) Look for a factorization of 1600 in which the sum of the factors is the coefficient of the middle term, 80. The numbers we want are 40 and 40: $40 \cdot 40 = 1600$ and $40 + 40 = 80$.

(4) Split the middle term: $80t = 40t + 40t$

(5) Factor by grouping:
$$25t^2 + 80t + 64 = 25t^2 + 40t + 40t + 64$$
$$= (25t^2 + 40t) + (40t + 64)$$
$$= 5t(5t + 8) + 8(5t + 8)$$
$$= (5t + 8)(5t + 8), \text{ or}$$
$$(5t + 8)^2$$

75. $6x^3 + 4x^2 - 10x$

(1) Factor out the common factor, $2x$:
$$6x^3 + 4x^2 - 10x = 2x(3x^2 + 2x - 5)$$

(2) - (5) Now we factor the trinomial $3x^2 + 2x - 5$. We did this in Exercise 49, using the variable y rather than x. We found that $3x^2 + 2x - 5 = (x-1)(3x+5)$. We must include the common factor to get a factorization of the original trinomial.
$$6x^3 + 4x^2 - 10x = 2x(x - 1)(3x + 5)$$

77. $25x^2 + 79x + 64$

(1) There are no common factors (other than 1 or -1).

(2) Multiply the leading coefficient, 25, and the constant, 64: $25 \cdot 64 = 1600$.

(3) Look for a factorization of 1600 in which the sum of the factors is the coefficient of the middle term, 79. It is not possible to find such a pair of numbers. Thus, $25x^2 + 79x + 64$ cannot be factored into a product of binomial factors. It is prime.

79. $6x^2 - 19x - 5$

(1) There are no common factors (other than 1 or -1).

(2) Multiply the leading coefficient, 6, and the constant, -5: $6(-5) = -30$.

(3) Look for a factorization of -30 in which the sum of the factors is the coefficient of the middle term, -19. There is no such pair of numbers. Thus, $6x^2 - 19x - 5$ cannot be factored into a product of binomial factors. It is prime.

81. $12m^2 - mn - 20n^2$

(1) There are no common factors (other than 1 or -1).

(2) Multiply the leading coefficient, 12, and the constant, -20: $12(-20) = -240$.

(3) Look for a factorization of -240 in which the sum of the factors is the coefficient of the middle term, -1. The numbers we want are 15 and -16: $15(-16) = -240$ and $15 + (-16) = -1$.

(4) Split the middle term: $-mn = 15mn - 16mn$

(5) Factor by grouping:
$$12m^2 - mn - 20n^2$$
$$= 12m^2 + 15mn - 16mn - 20n^2$$
$$= (12m^2 + 15mn) + (-16mn - 20n^2)$$
$$= 3m(4m + 5n) - 4n(4m + 5n)$$
$$= (3m - 4n)(4m + 5n)$$

83. $6a^2 - ab - 15b^2$

(1) There are no common factors (other than 1 or -1).

(2) Multiply the leading coefficient, 6, and the constant, -15: $6(-15) = -90$.

(3) Look for a factorization of -90 in which the sum of the factors is the coefficient of the middle term, -1. The numbers we want are -10 and 9: $-10 \cdot 9 = -90$ and $-10 + 9 = -1$.

(4) Split the middle term: $-ab = -10ab + 9ab$

(5) Factor by grouping:
$$6a^2 - ab - 15b^2 = 6a^2 - 10ab + 9ab - 15b^2$$
$$= (6a^2 - 10ab) + (9ab - 15b^2)$$
$$= 2a(3a - 5b) + 3b(3a - 5b)$$
$$= (2a + 3b)(3a - 5b)$$

85. $9a^2 - 18ab + 8b^2$

(1) There are no common factors (other than 1 or -1).

(2) Multiply the leading coefficient, 9, and the constant, 8: $9 \cdot 8 = 72$.

(3) Look for a factorization of 72 in which the sum of the factors is the coefficient of the middle term, -18. The numbers we want are -6 and -12: $-6(-12) = 72$ and $-6 + (-12) = -18$.

(4) Split the middle term: $-18ab = -6ab - 12ab$

(5) Factor by grouping:
$$9a^2 - 18ab + 8b^2 = 9a^2 - 6ab - 12ab + 8b^2$$
$$= (9a^2 - 6ab) + (-12ab + 8b^2)$$
$$= 3a(3a - 2b) - 4b(3a - 2b)$$
$$= (3a - 4b)(3a - 2b)$$

87. $35p^2 + 34pq + 8q^2$

(1) There are no common factors (other than 1 or -1).

(2) Multiply the leading coefficient, 35, and the constant, 8: $35 \cdot 8 = 280$.

(3) Look for a factorization of 280 in which the sum of the factors is the coefficient of the middle term, 34. The numbers we want are 14 and 20: $14 \cdot 20 = 280$ and $14 + 20 = 34$.

(4) Split the middle term: $34pq = 14pq + 20pq$

(5) Factor by grouping:
$$35p^2 + 34pq + 8q^2 = 35p^2 + 14pq + 20pq + 8q^2$$
$$= (35p^2 + 14pq) + (20pq + 8q^2)$$
$$= 7p(5p + 2q) + 4q(5p + 2q)$$
$$= (7p + 4q)(5p + 2q)$$

89. $18x^2 - 6xy - 24y^2$

(1) Factor out the common factor, 6.
$$18x^2 - 6xy - 24y^2 = 6(3x^2 - xy - 4y^2)$$

(2) Now we factor the trinomial $3x^2 - xy - 4y^2$. Multiply the leading coefficient, 3, and the constant, -4: $3(-4) = -12$.

(3) Look for a factorization of -12 in which the sum of the factors is the coefficient of the middle term, -1. The numbers we want are -4 and 3: $-4 \cdot 3 = -12$ and $-4 + 3 = -1$.

(4) Split the middle term: $-xy = -4xy + 3xy$

(5) Factor by grouping:
$$3x^2 - xy - 4y^2 = 3x^2 - 4xy + 3xy - 4y^2$$
$$= (3x^2 - 4xy) + (3xy - 4y^2)$$
$$= x(3x - 4y) + y(3x - 4y)$$
$$= (x + y)(3x - 4y)$$

We must include the common factor to get a factorization of the original trinomial.
$$18x^2 - 6xy - 24y^2 = 6(x + y)(3x - 4y)$$

91. $60x + 18x^2 - 6x^3 = -6x^3 + 18x^2 + 60x$

(1) Factor out the common factor, $-6x$. We factor out $-6x$ rather than $6x$ in order to have a positive leading coefficient in the trinomial factor.
$$-6x^3 + 18x^2 + 60x = -6x(x^2 - 3x - 10)$$

(2) - (5) We factor $x^2 - 3x - 10$ as we did in Exercise 43, getting the $(x - 5)(x + 2)$. Then we have:
$$60x + 18x^2 - 6x^3$$
$$= -6x(x - 5)(x + 2)$$
$$= 6x(-x + 5)(x + 2)$$
$$\qquad \text{Multiplying } x - 5 \text{ by } -1$$
$$= 6x(x - 5)(-x - 2)$$
$$\qquad \text{Multiplying } x + 2 \text{ by } -1$$

Note that we can express $6x(-x + 5)(x + 2)$ as $6x(5 - x)(x + 2)$ since $-x + 5 = 5 - x$ by the commutative law of addition.

93. $35x^5 - 57x^4 - 44x^3$

(1) We first factor out the common factor, x^3.
$$x^3(35x^2 - 57x - 44)$$

(2) Now we factor the trinomial $35x^2 - 57x - 44$. Multiply the leading coefficient, 35, and the constant, -44: $35(-44) = -1540$.

(3) Look for a factorization of -1540 in which the sum of the factors is the coefficient of the middle term, -57.

Pairs of factors	Sums of factors
$7, -220$	-213
$10, -154$	-144
$11, -140$	-129
$14, -110$	-96
$20, -77$	-57

(4) Split the middle term: $-57x = 20x - 77x$

(5) Factor by grouping:
$$35x^2 - 57x - 44 = 35x^2 + 20x - 77x - 44$$
$$= (35x^2 + 20x) + (-77x - 44)$$
$$= 5x(7x + 4) - 11(7x + 4)$$
$$= (5x - 11)(7x + 4)$$

We must include the common factor to get a factorization of the original trinomial.
$$35x^5 - 57x^4 - 44x^3 = x^3(5x - 11)(7x + 4)$$

95. Discussion and Writing Exercise

97.
$$-10x > 1000$$
$$\frac{-10x}{-10} < \frac{1000}{-10} \qquad \text{Dividing by } -10 \text{ and reversing the inequality symbol}$$
$$x < -100$$

The solution set is $\{x | x < -100\}$.

99. $6 - 3x \geq -18$

$\quad -3x \geq -24 \qquad$ Subtracting 6

$\quad\quad x \leq 8 \qquad$ Dividing by -3 and reversing the inequality symbol

The solution set is $\{x | x \leq 8\}$.

101. $\quad \dfrac{1}{2}x - 6x + 10 \leq x - 5x$

$\quad 2\left(\dfrac{1}{2}x - 6x + 10\right) \leq 2(x - 5x) \qquad$ Multiplying by 2 to clear the fraction

$\quad\quad x - 12x + 20 \leq 2x - 10x$

$\quad\quad -11x + 20 \leq -8x \qquad$ Collecting like terms

$\quad\quad\quad\quad 20 \leq 3x \qquad$ Adding $11x$

$\quad\quad\quad\quad \dfrac{20}{3} \leq x \qquad$ Dividing by 3

The solution set is $\left\{x \Big| x \geq \dfrac{20}{3}\right\}$.

103. $3x - 6x + 2(x - 4) > 2(9 - 4x)$

$\quad 3x - 6x + 2x - 8 > 18 - 8x \qquad$ Removing parentheses

$\quad\quad -x - 8 > 18 - 8x \qquad$ Collecting like terms

$\quad\quad\quad 7x > 26 \qquad$ Adding $8x$ and 8

$\quad\quad\quad x > \dfrac{26}{7} \qquad$ Dividing by 7

The solution set is $\left\{x \Big| x > \dfrac{26}{7}\right\}$.

105. *Familiarize*. We will use the formula $C = 2\pi r$, where C is circumference and r is radius, to find the radius in kilometers. Then we will multiply that number by 0.62 to find the radius in miles.

Translate.

$$\underbrace{\text{Circumference}}_{40,000} = \underbrace{2 \cdot \pi \cdot \text{radius}}_{\approx\ 2(3.14)r}$$

Solve. First we solve the equation.

$\quad 40,000 \approx 2(3.14)r$

$\quad 40,000 \approx 6.28r$

$\quad 6369 \approx r$

Then we multiply to find the radius in miles:

$6369(0.62) \approx 3949$

Check. If $r = 6369$, then $2\pi r = 2(3.14)(6369) \approx 40,000$. We should also recheck the multiplication we did to find the radius in miles. Both values check.

State. The radius of the earth is about 6369 km or 3949 mi. (These values may differ slightly if a different approximation is used for π.)

107. $9x^{10} - 12x^5 + 4$

(a) First factor out a common factor, if any. There is none (other than 1 or -1).

(b) Multiply the leading coefficient, 9, and the constant, 4: $9 \cdot 4 = 36$.

(c) Look for a factorization of 36 in which the sum of the factors is the coefficient of the middle term, -12. The factors we want are -6 and -6.

(d) Split the middle term: $-12x^5 = -6x^5 - 6x^5$

(e) Factor by grouping:

$9x^{10} - 12x^5 + 4 = 9x^{10} - 6x^5 - 6x^5 + 4$

$\quad\quad = (9x^{10} - 6x^5) + (-6x^5 + 4)$

$\quad\quad = 3x^5(3x^5 - 2) - 2(3x^5 - 2)$

$\quad\quad = (3x^5 - 2)(3x^5 - 2), \text{ or}$

$\quad\quad = (3x^5 - 2)^2$

109. $16x^{10} + 8x^5 + 1$

(a) First factor out a common factor, if any. There is none (other than 1 or -1).

(b) Multiply the leading coefficient, 16, and the constant, 1: $16 \cdot 1 = 16$.

(c) Look for a factorization of 16 in which the sum of the factors is the coefficient of the middle term, 8. The factors we want are 4 and 4.

(d) Split the middle term: $8x^5 = 4x^5 + 4x^5$

(e) Factor by grouping:

$16x^{10} + 8x^5 + 1 = 16x^{10} + 4x^5 + 4x^5 + 1$

$\quad\quad = (16x^{10} + 4x^5) + (4x^5 + 1)$

$\quad\quad = 4x^5(4x^5 + 1) + 1(4x^5 + 1)$

$\quad\quad = (4x^5 + 1)(4x^5 + 1), \text{ or}$

$\quad\quad = (4x^5 + 1)^2$

111.–119. Left to the student

Exercise Set 13.5

1. $x^2 - 14x + 49$

(a) We know that x^2 and 49 are squares.

(b) There is no minus sign before either x^2 or 49.

(c) If we multiply the square roots, x and 7, and double the product, we get $2 \cdot x \cdot 7 = 14x$. This is the opposite of the remaining term, $-14x$.

Thus, $x^2 - 14x + 49$ is a trinomial square.

3. $x^2 + 16x - 64$

Both x^2 and 64 are squares, but there is a minus sign before 64. Thus, $x^2 + 16x - 64$ is not a trinomial square.

5. $x^2 - 2x + 4$

(a) Both x^2 and 4 are squares.

(b) There is no minus sign before either x^2 or 4.

(c) If we multiply the square roots, x and 2, and double the product, we get $2 \cdot x \cdot 2 = 4x$. This is neither the remaining term nor its opposite.

Thus, $x^2 - 2x + 4$ is not a trinomial square.

7. $9x^2 - 36x + 24$

Only one term is a square. Thus, $9x^2 - 36x + 24$ is not a trinomial square.

9. $x^2 - 14x + 49 = x^2 - 2 \cdot x \cdot 7 + 7^2 = (x - 7)^2$
$$\qquad\qquad\quad \uparrow \quad \uparrow \;\; \uparrow \;\; \uparrow \quad\; \uparrow$$
$$\qquad\quad = A^2 - 2 \quad A \quad B + B^2 = (A - B)^2$$

11. $x^2 + 16x + 64 = x^2 + 2 \cdot x \cdot 8 + 8^2 = (x + 8)^2$
$$\qquad\qquad\quad \uparrow \quad \uparrow \;\; \uparrow \;\; \uparrow \quad\; \uparrow$$
$$\qquad\quad = A^2 + 2 \quad A \quad B + B^2 = (A + B)^2$$

13. $x^2 - 2x + 1 = x^2 - 2 \cdot x \cdot 1 + 1^2 = (x - 1)^2$

15. $4 + 4x + x^2 = x^2 + 4x + 4 \qquad$ Changing the order
$$= x^2 + 2 \cdot x \cdot 2 + 2^2$$
$$= (x + 2)^2$$

17. $q^4 - 6q^2 + 9 = (q^2)^2 - 2 \cdot q^2 \cdot 3 + 3^2 = (q^2 - 3)^2$

19. $49 + 56y + 16y^2 = 16y^2 + 56y + 49$
$$= (4y)^2 + 2 \cdot 4y \cdot 7 + 7^2$$
$$= (4y + 7)^2$$

21. $2x^2 - 4x + 2 = 2(x^2 - 2x + 1)$
$$= 2(x^2 - 2 \cdot x \cdot 1 + 1^2)$$
$$= 2(x - 1)^2$$

23. $x^3 - 18x^2 + 81x = x(x^2 - 18x + 81)$
$$= x(x^2 - 2 \cdot x \cdot 9 + 9^2)$$
$$= x(x - 9)^2$$

25. $12q^2 - 36q + 27 = 3(4q^2 - 12q + 9)$
$$= 3[(2q)^2 - 2 \cdot 2q \cdot 3 + 3^2]$$
$$= 3(2q - 3)^2$$

27. $49 - 42x + 9x^2 = 7^2 - 2 \cdot 7 \cdot 3x + (3x)^2$
$$= (7 - 3x)^2$$

29. $5y^4 + 10y^2 + 5 = 5(y^4 + 2y^2 + 1)$
$$= 5[(y^2)^2 + 2 \cdot y^2 \cdot 1 + 1^2]$$
$$= 5(y^2 + 1)^2$$

31. $1 + 4x^4 + 4x^2 = 1^2 + 2 \cdot 1 \cdot 2x^2 + (2x^2)^2$
$$= (1 + 2x^2)^2$$

33. $4p^2 + 12pq + 9q^2 = (2p)^2 + 2 \cdot 2p \cdot 3q + (3q)^2$
$$= (2p + 3q)^2$$

35. $a^2 - 6ab + 9b^2 = a^2 - 2 \cdot a \cdot 3b + (3b)^2$
$$= (a - 3b)^2$$

37. $81a^2 - 18ab + b^2 = (9a)^2 - 2 \cdot 9a \cdot b + b^2$
$$= (9a - b)^2$$

39. $36a^2 + 96ab + 64b^2 = 4(9a^2 + 24ab + 16b^2)$
$$= 4[(3a)^2 + 2 \cdot 3a \cdot 4b + (4b)^2]$$
$$= 4(3a + 4b)^2$$

41. $x^2 - 4$

(a) The first expression is a square: x^2

The second expression is a square: $4 = 2^2$

(b) The terms have different signs.

$x^2 - 4$ is a difference of squares.

43. $x^2 + 25$

The terms do not have different signs.

$x^2 + 25$ is not a difference of squares.

45. $x^2 - 45$

The number 45 is not a square.

$x^2 - 45$ is not a difference of squares.

47. $16x^2 - 25y^2$

(a) The first expression is a square: $16x^2 = (4x)^2$

The second expression is a square: $25y^2 = (5y)^2$

(b) The terms have different signs.

$16x^2 - 25y^2$ is a difference of squares.

49. $y^2 - 4 = y^2 - 2^2 = (y + 2)(y - 2)$

51. $p^2 - 9 = p^2 - 3^2 = (p + 3)(p - 3)$

53. $-49 + t^2 = t^2 - 49 = t^2 - 7^2 = (t + 7)(t - 7)$

55. $a^2 - b^2 = (a + b)(a - b)$

57. $25t^2 - m^2 = (5t)^2 - m^2 = (5t + m)(5t - m)$

59. $100 - k^2 = 10^2 - k^2 = (10 + k)(10 - k)$

61. $16a^2 - 9 = (4a)^2 - 3^2 = (4a + 3)(4a - 3)$

63. $4x^2 - 25y^2 = (2x)^2 - (5y)^2 = (2x + 5y)(2x - 5y)$

65. $8x^2 - 98 = 2(4x^2 - 49) = 2[(2x)^2 - 7^2] =$
$\quad 2(2x + 7)(2x - 7)$

67. $36x - 49x^3 = x(36 - 49x^2) = x[6^2 - (7x)^2] =$
$\quad x(6 + 7x)(6 - 7x)$

69. $\dfrac{1}{16} - 49x^8 = \left(\dfrac{1}{4}\right)^2 - (7x^4)^2 = \left(\dfrac{1}{4} + 7x^4\right)\left(\dfrac{1}{4} - 7x^4\right)$

71. $0.09y^2 - 0.0004 = (0.3y)^2 - (0.02)^2 =$
$\quad (0.3y + 0.02)(0.3y - 0.02)$

73. $49a^4 - 81 = (7a^2)^2 - 9^2 = (7a^2 + 9)(7a^2 - 9)$

75. $\quad a^4 - 16$
$$= (a^2)^2 - 4^2$$
$$= (a^2 + 4)(a^2 - 4) \qquad \text{Factoring a difference of squares}$$
$$= (a^2 + 4)(a + 2)(a - 2) \qquad \text{Factoring further: } a^2 - 4 \text{ is a difference of squares.}$$

77. $5x^4 - 405$

$5(x^4 - 81)$

$= 5[(x^2)^2 - 9^2]$

$= 5(x^2 + 9)(x^2 - 9)$

$= 5(x^2 + 9)(x + 3)(x - 3)$ Factoring $x^2 - 9$

79. $1 - y^8$

$= 1^2 - (y^4)^2$

$= (1 + y^4)(1 - y^4)$

$= (1 + y^4)(1 + y^2)(1 - y^2)$ Factoring $1 - y^4$

$= (1 + y^4)(1 + y^2)(1 + y)(1 - y)$ Factoring $1 - y^2$

81. $x^{12} - 16$

$= (x^6)^2 - 4^2$

$= (x^6 + 4)(x^6 - 4)$

$= (x^6 + 4)(x^3 + 2)(x^3 - 2)$ Factoring $x^6 - 4$

83. $y^2 - \dfrac{1}{16} = y^2 - \left(\dfrac{1}{4}\right)^2$

$= \left(y + \dfrac{1}{4}\right)\left(y - \dfrac{1}{4}\right)$

85. $25 - \dfrac{1}{49}x^2 = 5^2 - \left(\dfrac{1}{7}x\right)^2$

$= \left(5 + \dfrac{1}{7}x\right)\left(5 - \dfrac{1}{7}x\right)$

87. $16m^4 - t^4$

$= (4m^2)^2 - (t^2)^2$

$= (4m^2 + t^2)(4m^2 - t^2)$

$= (4m^2 + t^2)(2m + t)(2m - t)$ Factoring $4m^2 - t^2$

89. Discussion and Writing exercise

91. $-110 \div 10$ The quotient of a negative number and a positive number is negative.

$-110 \div 10 = -11$

93. $-\dfrac{2}{3} \div \dfrac{4}{5} = -\dfrac{2}{3} \cdot \dfrac{5}{4} = -\dfrac{10}{12} = -\dfrac{2 \cdot 5}{2 \cdot 6} = -\dfrac{\cancel{2} \cdot 5}{\cancel{2} \cdot 6} = -\dfrac{5}{6}$

95. $-64 \div (-32)$ The quotient of two negative numbers is a positive number.

$-64 \div (-32) = 2$

97. The shaded region is a square with sides of length $x - y - y$, or $x - 2y$. Its area is $(x - 2y)(x - 2y)$, or $(x - 2y)^2$. Multiplying, we get the polynomial $x^2 - 4xy + 4y^2$.

99. $y^5 \cdot y^7 = y^{5+7} = y^{12}$

101. $y - 6x = 6$

To find the x-intercept, let $y = 0$. Then solve for x.

$y - 6x = 6$

$0 - 6x = 6$

$-6x = 6$

$x = -1$

The x-intercept is $(-1, 0)$.

To find the y-intercept, let $x = 0$. Then solve for y.

$y - 6x = 6$

$y - 6 \cdot 0 = 6$

$y = 6$

The y-intercept is $(0, 6)$.

Plot these points and draw the line.

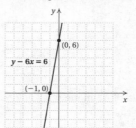

A third point should be used as a check. We substitute any value for x and solve for y. We let $x = -2$. Then

$y - 6x = 6$

$y - 6(-2) = 6$

$y + 12 = 6$

$y = -6$

The point $(-2, -6)$ is on the graph, so the graph is probably correct.

103. $49x^2 - 216$

There is no common factor. Also, $49x^2$ is a square, but 216 is not so this expression is not a difference of squares. It is not factorable. It is prime.

105. $x^2 + 22x + 121 = x^2 + 2 \cdot x \cdot 11 + 11^2$

$= (x + 11)^2$

107. $18x^3 + 12x^2 + 2x = 2x(9x^2 + 6x + 1)$

$= 2x[(3x)^2 + 2 \cdot 3x \cdot 1 + 1^2]$

$= 2x(3x + 1)^2$

109. $x^8 - 2^8$

$= (x^4 + 2^4)(x^4 - 2^4)$

$= (x^4 + 2^4)(x^2 + 2^2)(x^2 - 2^2)$

$= (x^4 + 2^4)(x^2 + 2^2)(x + 2)(x - 2)$, or

$= (x^4 + 16)(x^2 + 4)(x + 2)(x - 2)$

111. $3x^5 - 12x^3 = 3x^3(x^2 - 4) = 3x^3(x + 2)(x - 2)$

113. $18x^3 - \dfrac{8}{25}x = 2x\left(9x^2 - \dfrac{4}{25}\right) = 2x\left(3x + \dfrac{2}{5}\right)\left(3x - \dfrac{2}{5}\right)$

115. $0.49p - p^3 = p(0.49 - p^2) = p(0.7 + p)(0.7 - p)$

117. $0.64x^2 - 1.21 = (0.8x)^2 - (1.1)^2 = (0.8x + 1.1)(0.8x - 1.1)$

119. $(x+3)^2 - 9 = [(x+3)+3][(x+3)-3] = (x+6)x$, or $x(x+6)$

121. $x^2 - \left(\dfrac{1}{x}\right)^2 = \left(x + \dfrac{1}{x}\right)\left(x - \dfrac{1}{x}\right)$

123. $81 - b^{4k} = 9^2 - (b^{2k})^2$

$\quad\quad = (9 + b^{2k})(9 - b^{2k})$

$\quad\quad = (9 + b^{2k})[3^2 - (b^k)^2]$

$\quad\quad = (9 + b^{2k})(3 + b^k)(3 - b^k)$

125. $9b^{2n} + 12b^n + 4 = (3b^n)^2 + 2 \cdot 3b^n \cdot 2 + 2^2 =$

$(3b^n + 2)^2$

127. $(y + 3)^2 + 2(y + 3) + 1$

$\quad\quad = (y + 3)^2 + 2 \cdot (y + 3) \cdot 1 + 1^2$

$\quad\quad = [(y + 3) + 1]^2$

$\quad\quad = (y + 4)^2$

129. If $cy^2 + 6y + 1$ is the square of a binomial, then $2 \cdot a \cdot 1 = 6$ where $a^2 = c$. Then $a = 3$, so $c = a^2 = 3^2 = 9$. (The polynomial is $9y^2 + 6y + 1$.)

131. Enter $y_1 = x^2 + 9$ and $y_2 = (x + 3)(x + 3)$ and look at a table of values. The y_1-and y_2-values are not the same, so the factorization is not correct.

133. Enter $y_1 = x^2 + 9$ and $y_2 = (x + 3)^2$ and look at a table of values. The y_1-and y_2-values are not the same, so the factorization is not correct.

Exercise Set 13.6

1. $3x^2 - 192 = 3(x^2 - 64)$ $\quad$ 3 is a common factor.

$\quad\quad = 3(x^2 - 8^2)$ $\quad\quad$ Difference of squares

$\quad\quad = 3(x + 8)(x - 8)$

3. $a^2 + 25 - 10a = a^2 - 10a + 25$

$\quad\quad\quad\quad\quad = a^2 - 2 \cdot a \cdot 5 + 5^2$ $\quad$ Trinomial square

$\quad\quad\quad\quad\quad = (a - 5)^2$

5. $2x^2 - 11x + 12$

There is no common factor (other than 1). This polynomial has three terms, but it is not a trinomial square. Multiply the leading coefficient and the constant, 2 and 12: $2 \cdot 12 = 24$. Try to factor 24 so that the sum of the factors is -11. The numbers we want are -3 and -8: $-3(-8) = 24$ and $-3 + (-8) = -11$. Split the middle term and factor by grouping.

$2x^2 - 11x + 12 = 2x^2 - 3x - 8x + 12$

$\quad\quad\quad\quad\quad = (2x^2 - 3x) + (-8x + 12)$

$\quad\quad\quad\quad\quad = x(2x - 3) - 4(2x - 3)$

$\quad\quad\quad\quad\quad = (x - 4)(2x - 3)$

7. $x^3 + 24x^2 + 144x$

$\quad = x(x^2 + 24x + 144)$ $\quad\quad$ x is a common factor.

$\quad = x(x^2 + 2 \cdot x \cdot 12 + 12^2)$ Trinomial square

$\quad = x(x + 12)^2$

9. $x^3 + 3x^2 - 4x - 12$

$\quad = x^2(x + 3) - 4(x + 3)$ Factoring by grouping

$\quad = (x^2 - 4)(x + 3)$

$\quad = (x + 2)(x - 2)(x + 3)$ Factoring the difference of squares

11. $48x^2 - 3 = 3(16x^2 - 1)$ $\quad$ 3 is a common factor.

$\quad\quad = 3[(4x)^2 - 1^2]$ $\quad\quad$ Difference of squares

$\quad\quad = 3(4x + 1)(4x - 1)$

13. $9x^3 + 12x^2 - 45x$

$\quad = 3x(3x^2 + 4x - 15)$ $3x$ is a common factor.

$\quad = 3x(3x - 5)(x + 3)$ Factoring the trinomial

15. $x^2 + 4$ is a *sum* of squares with no common factor. It cannot be factored. It is prime.

17. $x^4 + 7x^2 - 3x^3 - 21x = x(x^3 + 7x - 3x^2 - 21)$

$\quad\quad\quad\quad\quad\quad\quad = x[x(x^2 + 7) - 3(x^2 + 7)]$

$\quad\quad\quad\quad\quad\quad\quad = x[(x - 3)(x^2 + 7)]$

$\quad\quad\quad\quad\quad\quad\quad = x(x - 3)(x^2 + 7)$

19. $x^5 - 14x^4 + 49x^3$

$\quad = x^3(x^2 - 14x + 49)$ $\quad\quad$ x^3 is a common factor.

$\quad = x^3(x^2 - 2 \cdot x \cdot 7 + 7^2)$ Trinomial square

$\quad = x^3(x - 7)^2$

21. $20 - 6x - 2x^2$

$\quad = -2(10 + 3x + x^2)$ $\quad$ -2 is a common factor.

$\quad = -2(x^2 + 3x - 10)$ $\quad$ Writing in descending order

$\quad = -2(x + 5)(x - 2),$ $\quad$ Using trial and error

$\quad\quad$ or $2(-x - 5)(x - 2),$

$\quad\quad$ or $2(x + 5)(-x + 2)$

23. $x^2 - 6x + 1$

There is no common factor (other than 1 or -1). This is not a trinomial square, because $-6x \neq 2 \cdot x \cdot 1$ and $-6x \neq -2 \cdot x \cdot 1$. We try factoring using the refined trial and error procedure. We look for two factors of 1 whose sum is -6. There are none. The polynomial cannot be factored. It is prime.

25. $4x^4 - 64$

$\quad = 4(x^4 - 16)$ 4 is a common factor.

$\quad = 4[(x^2)^2 - 4^2]$ $\quad$ Difference of squares

$\quad = 4(x^2 + 4)(x^2 - 4)$ $\quad$ Difference of squares

$\quad = 4(x^2 + 4)(x + 2)(x - 2)$

27. $1 - y^8$ $\quad\quad\quad\quad$ Difference of squares

$\quad = (1 + y^4)(1 - y^4)$ $\quad\quad$ Difference of squares

$\quad = (1 + y^4)(1 + y^2)(1 - y^2)$ Difference of squares

$\quad = (1 + y^4)(1 + y^2)(1 + y)(1 - y)$

29. $x^5 - 4x^4 + 3x^3$

$\quad = x^3(x^2 - 4x + 3)$ $\quad$ x^3 is a common factor.

$\quad = x^3(x - 3)(x - 1)$ Factoring the trinomial using trial and error

31. $\dfrac{1}{81}x^6 - \dfrac{8}{27}x^3 + \dfrac{16}{9}$

$= \dfrac{1}{9}\left(\dfrac{1}{9}x^6 - \dfrac{8}{3}x^3 + 16\right)$ $\quad \dfrac{1}{9}$ is a common factor.

$= \dfrac{1}{9}\left[\left(\dfrac{1}{3}x^3\right)^2 - 2 \cdot \dfrac{1}{3}x^3 \cdot 4 + 4^2\right]$ Trinomial square

$= \dfrac{1}{9}\left(\dfrac{1}{3}x^3 - 4\right)^2$

33. $mx^2 + my^2$

$= m(x^2 + y^2)$ $\quad m$ is a common factor.

The factor with more than one term cannot be factored further, so we have factored completely.

35. $9x^2y^2 - 36xy = 9xy(xy - 4)$

37. $2\pi rh + 2\pi r^2 = 2\pi r(h + r)$

39. $(a+b)(x-3) + (a+b)(x+4)$

$= (a+b)[(x-3) + (x+4)]$ $\quad (a+b)$ is a common factor.

$= (a+b)(2x+1)$

41. $(x-1)(x+1) - y(x+1) = (x+1)(x-1-y)$

$(x+1)$ is a common factor.

43. $n^2 + 2n + np + 2p$

$= n(n+2) + p(n+2)$ $\quad$ Factoring by grouping

$= (n+p)(n+2)$

45. $6q^2 - 3q + 2pq - p$

$= (6q^2 - 3q) + (2pq - p)$

$= 3q(2q-1) + p(2q-1)$ $\quad$ Factoring by grouping

$= (3q+p)(2q-1)$

47. $4b^2 + a^2 - 4ab$

$= a^2 - 4ab + 4b^2$ $\quad$ Rearranging

$= a^2 - 2 \cdot a \cdot 2b + (2b)^2$ Trinomial square

$= (a-2b)^2$

(Note that if we had rewritten the polynomial as $4b^2 - 4ab + a^2$, we might have written the result as $(2b-a)^2$. The two factorizations are equivalent.)

49. $16x^2 + 24xy + 9y^2$

$= (4x)^2 + 2 \cdot 4x \cdot 3y + (3y)^2$ Trinomial square

$= (4x + 3y)^2$

51. $49m^4 - 112m^2n + 64n^2$

$= (7m^2)^2 - 2 \cdot 7m^2 \cdot 8n + (8n)^2$ Trinomial square

$= (7m^2 - 8n)^2$

53. $y^4 + 10y^2z^2 + 25z^4$

$= (y^2)^2 + 2 \cdot y^2 \cdot 5z^2 + (5z^2)^2$ Trinomial square

$= (y^2 + 5z^2)^2$

55. $\dfrac{1}{4}a^2 + \dfrac{1}{3}ab + \dfrac{1}{9}b^2$

$= \left(\dfrac{1}{2}a\right)^2 + 2 \cdot \dfrac{1}{2}a \cdot \dfrac{1}{3}b + \left(\dfrac{1}{3}b\right)^2$

$= \left(\dfrac{1}{2}a + \dfrac{1}{3}b\right)^2$

57. $a^2 - ab - 2b^2 = (a-2b)(a+b)$ Using trial and error

59. $2mn - 360n^2 + m^2$

$= m^2 + 2mn - 360n^2$ $\quad$ Rewriting

$= (m+20n)(m-18n)$ Using trial and error

61. $m^2n^2 - 4mn - 32 = (mn-8)(mn+4)$ Using trial and error

63. $r^5s^2 - 10r^4s + 16r^3$

$= r^3(r^2s^2 - 10rs + 16)$ $\quad r^3$ is a common factor.

$= r^3(rs-2)(rs-8)$ $\quad$ Using trial and error

65. $a^5 + 4a^4b - 5a^3b^2$

$= a^3(a^2 + 4ab - 5b^2)$ $\quad a^3$ is a common factor.

$= a^3(a+5b)(a-b)$ $\quad$ Factoring the trinomial

67. $a^2 - \dfrac{1}{25}b^2$

$= a^2 - \left(\dfrac{1}{5}b\right)^2$ $\quad$ Difference of squares

$= \left(a + \dfrac{1}{5}b\right)\left(a - \dfrac{1}{5}b\right)$

69. $x^2 - y^2 = (x+y)(x-y)$ $\quad$ Difference of squares

71. $16 - p^4q^4$

$= 4^2 - (p^2q^2)^2$ $\quad$ Difference of squares

$= (4 + p^2q^2)(4 - p^2q^2)$ $\quad 4 - p^2q^2$ is a difference of squares.

$= (4 + p^2q^2)(2 + pq)(2 - pq)$

73. $1 - 16x^{12}y^{12}$

$= 1^2 - (4x^6y^6)^2$ $\quad$ Difference of squares

$= (1 + 4x^6y^6)(1 - 4x^6y^6)$ $\quad 1 - 4x^6y^6$ is a difference of squares.

$= (1 + 4x^6y^6)(1 + 2x^3y^3)(1 - 2x^3y^3)$

75. $q^3 + 8q^2 - q - 8$

$= q^2(q+8) - (q+8)$ $\quad$ Factoring by grouping

$= (q^2 - 1)(q+8)$

$= (q+1)(q-1)(q+8)$ $\quad$ Factoring the difference of squares

77. $112xy + 49x^2 + 64y^2$

$= 49x^2 + 112xy + 64y^2$ $\quad$ Rearranging

$= (7x)^2 + 2 \cdot 7x \cdot 8y + (8y)^2$ Trinomial square

$= (7x + 8y)^2$

79. Discussion and Writing Exercise

81. $(-4, 0)$; $m = -3$

The slope is -3, so the equation is $y = -3x + b$. Using the point $(-4, 0)$, we substitute -4 for x and 0 for y in $y = -3x + b$ and then solve for b.

$$y = -3x + b$$
$$0 = -3(-4) + b$$
$$0 = 12 + b$$
$$-12 = b$$

Then the equation is $y = -3x - 12$.

83. $(-4, 5)$; $m = -\dfrac{2}{3}$

The slope is $-\dfrac{2}{3}$, so the equation is $y = -\dfrac{2}{3}x + b$. Using the point $(-4, 5)$, we substitute -4 for x and 5 for y in $y = -\dfrac{2}{3}x + b$ and then solve for b.

$$y = -\frac{2}{3}x + b$$
$$5 = -\frac{2}{3}(-4) + b$$
$$5 = \frac{8}{3} + b$$
$$\frac{7}{3} = b$$

Then the equation is $y = -\dfrac{2}{3}x + \dfrac{7}{3}$.

85. $\dfrac{7}{5} \div \left(-\dfrac{11}{10} \right)$

$= \dfrac{7}{5} \cdot \left(-\dfrac{10}{11} \right)$ Multiplying by the reciprocal of the divisor

$= -\dfrac{7 \cdot 10}{5 \cdot 11}$

$= -\dfrac{7 \cdot 5 \cdot 2}{5 \cdot 11} = -\dfrac{7 \cdot 2}{11} \cdot \dfrac{5}{5}$

$= -\dfrac{14}{11}$

87.
$$A = aX + bX - 7$$
$$A + 7 = aX + bX$$
$$A + 7 = X(a + b)$$
$$\frac{A + 7}{a + b} = X$$

89. $a^4 - 2a^2 + 1 = (a^2)^2 - 2 \cdot a^2 \cdot 1 + 1^2$
$$= (a^2 - 1)^2$$
$$= [(a + 1)(a - 1)]^2$$
$$= (a + 1)^2(a - 1)^2$$

91. $12.25x^2 - 7x + 1 = (3.5x)^2 - 2 \cdot (3.5x) \cdot 1 + 1^2$
$$= (3.5x - 1)^2$$

93. $5x^2 + 13x + 7.2$

Multiply the leading coefficient and the constant, 5 and 7.2: $5(7.2) = 36$. Try to factor 36 so that the sum of the factors is 13. The numbers we want are 9 and 4. Split the middle term and factor by grouping:

$5x^2 + 13x + 7.2 = 5x^2 + 9x + 4x + 7.2$
$$= (5x^2 + 9x) + (4x + 7.2)$$
$$= 5x(x + 1.8) + 4(x + 1.8)$$
$$= (5x + 4)(x + 1.8)$$

95. $\quad 18 + y^3 - 9y - 2y^2$
$$= y^3 - 2y^2 - 9y + 18$$
$$= y^2(y - 2) - 9(y - 2)$$
$$= (y^2 - 9)(y - 2)$$
$$= (y + 3)(y - 3)(y - 2)$$

97. $a^3 + 4a^2 + a + 4 = a^2(a + 4) + 1(a + 4)$
$$= (a^2 + 1)(a + 4)$$

99. $x^3 - x^2 - 4x + 4 = x^2(x - 1) - 4(x - 1)$
$$= (x^2 - 4)(x - 1)$$
$$= (x + 2)(x - 2)(x - 1)$$

101. $\quad y^2(y - 1) - 2y(y - 1) + (y - 1)$
$$= (y - 1)(y^2 - 2y + 1)$$
$$= (y - 1)(y - 1)^2$$
$$= (y - 1)^3$$

103. $\quad (y + 4)^2 + 2x(y + 4) + x^2$
$$= (y + 4)^2 + 2 \cdot (y + 4) \cdot x + x^2 \quad \text{Trinomial square}$$
$$= (y + 4 + x)^2$$

Exercise Set 13.7

1. $(x + 4)(x + 9) = 0$

$x + 4 = 0 \quad or \quad x + 9 = 0$ Using the principle of zero products

$x = -4 \quad or \qquad\quad x = -9$ Solving the two equations separately

Check:

For -4
$$\begin{array}{c} \underline{(x + 4)(x + 9) = 0} \\ (-4 + 4)(-4 + 9) \ ? \ 0 \\ 0 \cdot 5 \\ 0 \ \big| \ \text{TRUE} \end{array}$$

For -9
$$\begin{array}{c} \underline{(x + 4)(x + 9) = 0} \\ (9 + 4)(-9 + 9) \ ? \ 0 \\ 13 \cdot 0 \\ 0 \ \big| \ \text{TRUE} \end{array}$$

The solutions are -4 and -9.

3. $(x + 3)(x - 8) = 0$

$x + 3 = 0 \quad or \quad x - 8 = 0$ Using the principle of zero products

$x = -3 \quad or \qquad\quad x = 8$

Check:

For -3

$$\frac{(x+3)(x-8)=0}{(-3+3)(-3-8) \; ? \; 0}$$
$$0(-11) \quad \Big|$$
$$0 \quad \Big| \quad \text{TRUE}$$

For 8

$$\frac{(x+3)(x-8)=0}{(8+3)(8-8) \; ? \; 0}$$
$$11 \cdot 0 \quad \Big|$$
$$0 \quad \Big| \quad \text{TRUE}$$

The solutions are -3 and 8.

5. $(x+12)(x-11)=0$

$x+12=0 \quad or \quad x-11=0$

$x=-12 \quad or \quad x=11$

The solutions are -12 and 11.

7. $x(x+3)=0$

$x=0 \quad or \quad x+3=0$

$x=0 \quad or \quad x=-3$

The solutions are 0 and -3.

9. $0=y(y+18)$

$y=0 \quad or \quad y+18=0$

$y=0 \quad or \quad y=-18$

The solutions are 0 and -18.

11. $(2x+5)(x+4)=0$

$2x+5=0 \quad or \quad x+4=0$

$2x=-5 \quad or \quad x=-4$

$x=-\dfrac{5}{2} \quad or \quad x=-4$

The solutions are $-\dfrac{5}{2}$ and -4.

13. $(5x+1)(4x-12)=0$

$5x+1=0 \quad or \quad 4x-12=0$

$5x=-1 \quad or \quad 4x=12$

$x=-\dfrac{1}{5} \quad or \quad x=3$

The solutions are $-\dfrac{1}{5}$ and 3.

15. $(7x-28)(28x-7)=0$

$7x-28=0 \quad or \quad 28x-7=0$

$7x=28 \quad or \quad 28x=7$

$x=4 \quad or \quad x=\dfrac{7}{28}=\dfrac{1}{4}$

The solutions are 4 and $\dfrac{1}{4}$.

17. $2x(3x-2)=0$

$2x=0 \quad or \quad 3x-2=0$

$x=0 \quad or \quad 3x=2$

$x=0 \quad or \quad x=\dfrac{2}{3}$

The solutions are 0 and $\dfrac{2}{3}$.

19. $\left(\dfrac{1}{5}+2x\right)\left(\dfrac{1}{9}-3x\right)=0$

$\dfrac{1}{5}+2x=0 \quad or \quad \dfrac{1}{9}-3x=0$

$2x=-\dfrac{1}{5} \quad or \quad -3x=-\dfrac{1}{9}$

$x=-\dfrac{1}{10} \quad or \quad x=\dfrac{1}{27}$

The solutions are $-\dfrac{1}{10}$ and $\dfrac{1}{27}$.

21. $(0.3x-0.1)(0.05x+1)=0$

$0.3x-0.1=0 \quad or \quad 0.05x+1=0$

$0.3x=0.1 \quad or \quad 0.05x=-1$

$x=\dfrac{0.1}{0.3} \quad or \quad x=-\dfrac{1}{0.05}$

$x=\dfrac{1}{3} \quad or \quad x=-20$

The solutions are $\dfrac{1}{3}$ and -20.

23. $9x(3x-2)(2x-1)=0$

$9x=0 \quad or \quad 3x-2=0 \quad or \quad 2x-1=0$

$x=0 \quad or \quad 3x=2 \quad or \quad 2x=1$

$x=0 \quad or \quad x=\dfrac{2}{3} \quad or \quad x=\dfrac{1}{2}$

The solutions are 0, $\dfrac{2}{3}$, and $\dfrac{1}{2}$.

25. $\quad x^2+6x+5=0$

$(x+5)(x+1)=0 \qquad \text{Factoring}$

$x+5=0 \quad or \quad x+1=0 \qquad$ Using the principle of zero products

$x=-5 \quad or \quad x=-1$

The solutions are -5 and -1.

27. $\quad x^2+7x-18=0$

$(x+9)(x-2)=0 \qquad \text{Factoring}$

$x+9=0 \quad or \quad x-2=0 \qquad$ Using the principle of zero products

$x=-9 \quad or \quad x=2$

The solutions are -9 and 2.

29. $\quad x^2-8x+15=0$

$(x-5)(x-3)=0$

$x-5=0 \quad or \quad x-3=0$

$x=5 \quad or \quad x=3$

The solutions are 5 and 3.

31. $x^2 - 8x = 0$

$x(x - 8) = 0$

$x = 0 \quad or \quad x - 8 = 0$

$x = 0 \quad or \qquad x = 8$

The solutions are 0 and 8.

33. $x^2 + 18x = 0$

$x(x + 18) = 0$

$x = 0 \quad or \quad x + 18 = 0$

$x = 0 \quad or \qquad x = -18$

The solutions are 0 and -18.

35. $x^2 = 16$

$x^2 - 16 = 0 \quad$ Subtracting 16

$(x - 4)(x + 4) = 0$

$x - 4 = 0 \quad or \quad x + 4 = 0$

$x = 4 \quad or \qquad x = -4$

The solutions are 4 and -4.

37. $9x^2 - 4 = 0$

$(3x - 2)(3x + 2) = 0$

$3x - 2 = 0 \quad or \quad 3x + 2 = 0$

$3x = 2 \quad or \qquad 3x = -2$

$x = \dfrac{2}{3} \quad or \qquad x = -\dfrac{2}{3}$

The solutions are $\dfrac{2}{3}$ and $-\dfrac{2}{3}$.

39. $0 = 6x + x^2 + 9$

$0 = x^2 + 6x + 9 \quad$ Writing in descending order

$0 = (x + 3)(x + 3)$

$x + 3 = 0 \quad or \quad x + 3 = 0$

$x = -3 \quad or \qquad x = -3$

There is only one solution, -3.

41. $x^2 + 16 = 8x$

$x^2 - 8x + 16 = 0 \quad$ Subtracting $8x$

$(x - 4)(x - 4) = 0$

$x - 4 = 0 \quad or \quad x - 4 = 0$

$x = 4 \quad or \qquad x = 4$

There is only one solution, 4.

43. $5x^2 = 6x$

$5x^2 - 6x = 0$

$x(5x - 6) = 0$

$x = 0 \quad or \quad 5x - 6 = 0$

$x = 0 \quad or \qquad 5x = 6$

$x = 0 \quad or \qquad x = \dfrac{6}{5}$

The solutions are 0 and $\dfrac{6}{5}$.

45. $6x^2 - 4x = 10$

$6x^2 - 4x - 10 = 0$

$2(3x^2 - 2x - 5) = 0$

$2(3x - 5)(x + 1) = 0$

$3x - 5 = 0 \quad or \quad x + 1 = 0$

$3x = 5 \quad or \qquad x = -1$

$x = \dfrac{5}{3} \quad or \qquad x = -1$

The solutions are $\dfrac{5}{3}$ and -1.

47. $12y^2 - 5y = 2$

$12y^2 - 5y - 2 = 0$

$(4y + 1)(3y - 2) = 0$

$4y + 1 = 0 \quad or \quad 3y - 2 = 0$

$4y = -1 \quad or \qquad 3y = 2$

$y = -\dfrac{1}{4} \quad or \qquad y = \dfrac{2}{3}$

The solutions are $-\dfrac{1}{4}$ and $\dfrac{2}{3}$.

49. $t(3t + 1) = 2$

$3t^2 + t = 2 \qquad$ Multiplying on the left

$3t^2 + t - 2 = 0 \qquad$ Subtracting 2

$(3t - 2)(t + 1) = 0$

$3t - 2 = 0 \quad or \quad t + 1 = 0$

$3t = 2 \quad or \qquad t = -1$

$t = \dfrac{2}{3} \quad or \qquad t = -1$

The solutions are $\dfrac{2}{3}$ and -1.

51. $100y^2 = 49$

$100y^2 - 49 = 0$

$(10y + 7)(10y - 7) = 0$

$10y + 7 = 0 \qquad or \quad 10y - 7 = 0$

$10y = -7 \quad or \qquad 10y = 7$

$y = -\dfrac{7}{10} \quad or \qquad y = \dfrac{7}{10}$

The solutions are $-\dfrac{7}{10}$ and $\dfrac{7}{10}$.

53. $x^2 - 5x = 18 + 2x$

$x^2 - 5x - 18 - 2x = 0 \qquad$ Subtracting 18 and $2x$

$x^2 - 7x - 18 = 0$

$(x - 9)(x + 2) = 0$

$x - 9 = 0 \quad or \quad x + 2 = 0$

$x = 9 \quad or \qquad x = -2$

The solutions are 9 and -2.

55. $10x^2 - 23x + 12 = 0$

$(5x - 4)(2x - 3) = 0$

$5x - 4 = 0 \quad or \quad 2x - 3 = 0$

$5x = 4 \quad or \quad 2x = 3$

$x = \dfrac{4}{5} \quad or \quad x = \dfrac{3}{2}$

The solutions are $\dfrac{4}{5}$ and $\dfrac{3}{2}$.

57. We let $y = 0$ and solve for x.

$0 = x^2 + 3x - 4$

$0 = (x + 4)(x - 1)$

$x + 4 = 0 \quad or \quad x - 1 = 0$

$x = -4 \quad or \quad x = 1$

The x-intercepts are $(-4, 0)$ and $(1, 0)$.

59. We let $y = 0$ and solve for x

$0 = 2x^2 + x - 10$

$0 = (2x + 5)(x - 2)$

$2x + 5 = 0 \quad or \quad x - 2 = 0$

$2x = -5 \quad or \quad x = 2$

$x = -\dfrac{5}{2} \quad or \quad x = 2$

The x-intercepts are $\left(-\dfrac{5}{2}, 0\right)$ and $(2, 0)$.

61. We let $y = 0$ and solve for x.

$0 = x^2 - 2x - 15$

$0 = (x - 5)(x + 3)$

$x - 5 = 0 \quad or \quad x + 3 = 0$

$x = 5 \quad or \quad x = -3$

The x-intercepts are $(5, 0)$ and $(-3, 0)$.

63. The solutions of the equation are the first coordinates of the x-intercepts of the graph. From the graph we see that the x-intercepts are $(-1, 0)$ and $(4, 0)$, so the solutions of the equation are -1 and 4.

65. The solutions of the equation are the first coordinates of the x-intercepts of the graph. From the graph we see that the x-intercepts are $(-1, 0)$ and $(3, 0)$, so the solutions of the equation are -1 and 3.

67. Discussion and Writing Exercise

69. $(a + b)^2$

71. The two numbers have different signs, so their quotient is negative.

$144 \div -9 = -16$

73. $-\dfrac{5}{8} \div \dfrac{3}{16} = -\dfrac{5}{8} \cdot \dfrac{16}{3}$

$= -\dfrac{5 \cdot 16}{8 \cdot 3}$

$= -\dfrac{5 \cdot 8 \cdot 2}{8 \cdot 3}$

$= -\dfrac{10}{3}$

75. $b(b + 9) = 4(5 + 2b)$

$b^2 + 9b = 20 + 8b$

$b^2 + 9b - 8b - 20 = 0$

$b^2 + b - 20 = 0$

$(b + 5)(b - 4) = 0$

$b + 5 = 0 \quad or \quad b - 4 = 0$

$b = -5 \quad or \quad b = 4$

The solutions are -5 and 4.

77. $(t - 3)^2 = 36$

$t^2 - 6t + 9 = 36$

$t^2 - 6t - 27 = 0$

$(t - 9)(t + 3) = 0$

$t - 9 = 0 \quad or \quad t + 3 = 0$

$t = 9 \quad or \quad t = -3$

The solutions are 9 and -3.

79. $x^2 - \dfrac{1}{64} = 0$

$\left(x - \dfrac{1}{8}\right)\left(x + \dfrac{1}{8}\right) = 0$

$x - \dfrac{1}{8} = 0 \quad or \quad x + \dfrac{1}{8} = 0$

$x = \dfrac{1}{8} \quad or \quad x = -\dfrac{1}{8}$

The solutions are $\dfrac{1}{8}$ and $-\dfrac{1}{8}$.

81. $\dfrac{5}{16}x^2 = 5$

$\dfrac{5}{16}x^2 - 5 = 0$

$5\left(\dfrac{1}{16}x^2 - 1\right) = 0$

$5\left(\dfrac{1}{4}x - 1\right)\left(\dfrac{1}{4}x + 1\right) = 0$

$\dfrac{1}{4}x - 1 = 0 \quad or \quad \dfrac{1}{4}x + 1 = 0$

$\dfrac{1}{4}x = 1 \quad or \quad \dfrac{1}{4}x = -1$

$x = 4 \quad or \quad x = -4$

The solutions are 4 and -4.

83. (a) $\quad x = -3 \quad or \quad x = 4$

$x + 3 = 0 \quad or \quad x - 4 = 0$

$(x + 3)(x - 4) = 0 \quad$ Principle of zero products

$x^2 - x - 12 = 0 \quad$ Multiplying

(b) $\quad x = -3 \quad or \quad x = -4$

$x + 3 = 0 \quad or \quad x + 4 = 0$

$(x + 3)(x + 4) = 0$

$x^2 + 7x + 12 = 0$

(c) $\qquad x = \dfrac{1}{2} \quad or \quad\quad x = \dfrac{1}{2}$

$$x - \dfrac{1}{2} = 0 \quad or \quad x - \dfrac{1}{2} = 0$$

$$\left(x - \dfrac{1}{2}\right)\left(x - \dfrac{1}{2}\right) = 0$$

$$x^2 - x + \dfrac{1}{4} = 0, \quad or$$

$$4x^2 - 4x + 1 = 0 \quad \text{Multiplying by 4}$$

(d) $\qquad (x - 5)(x + 5) = 0$

$$x^2 - 25 = 0$$

(e) $\quad (x - 0)(x - 0.1)\left(x - \dfrac{1}{4}\right) = 0$

$$x\left(x - \dfrac{1}{10}\right)\left(x - \dfrac{1}{4}\right) = 0$$

$$x\left(x^2 - \dfrac{7}{20}x + \dfrac{1}{40}\right) = 0$$

$$x^3 - \dfrac{7}{20}x^2 + \dfrac{1}{40}x = 0, \quad or$$

$$40x^3 - 14x^2 + x = 0 \quad \text{Multiplying by 40}$$

85. 2.33, 6.77

87. 0, 2.74

Exercise Set 13.8

1. *Familiarize*. We make a drawing. Let $w = $ the width, in cm. Then $w + 2 = $ the length, in cm.

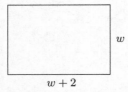

$$w + 2$$

Recall that the area of a rectangle is length times width.

Translate. We reword the problem.

Length $\;$ times $\;$ width $\;$ is $\;$ $\underbrace{144 \text{ cm}^2}$.

$\quad\downarrow\qquad\quad\downarrow\qquad\;\downarrow\quad\;\;\downarrow\qquad\downarrow$

$(w + 2) \qquad \cdot \qquad w \;\;=\;\; 24$

Solve. We solve the equation.

$$(w + 2)w = 24$$

$$w^2 + 2w = 24$$

$$w^2 + 2w - 24 = 0$$

$$(w + 6)(w - 4) = 0$$

$$w + 6 = 0 \quad or \quad w - 4 = 0$$

$$w = -6 \quad or \quad\quad w = 4$$

Check. Since the width must be positive, -6 cannot be a solution. If the width is 4 cm, then the length is $4 + 2$, or 6 cm, and the area is $6 \cdot 4$, or 24 cm^2. Thus, 4 checks.

State. The width is 4 cm, and the length is 6 cm.

3. *Familiarize*. Let $w = $ the width of the table, in feet. Then $6w = $ the length, in feet. Recall that the area of a rectangle is Length $\cdot$ Width.

Translate.

$\underbrace{\text{The area of the table}}$ $\;$ is $\;$ $\underbrace{24 \text{ ft}^2}$.

$\qquad\quad\downarrow\qquad\qquad\qquad\;\downarrow\qquad\;\;\downarrow$

$\qquad 6w \cdot w \qquad\qquad = \qquad 24$

Solve. We solve the equation.

$$6w \cdot w = 24$$

$$6w^2 = 24$$

$$6w^2 - 24 = 0$$

$$6(w^2 - 4) = 0$$

$$6(w + 2)(w - 2) = 0$$

$$w + 2 = 0 \quad or \quad w - 2 = 0$$

$$w = -2 \quad or \quad\quad w = 2$$

Check. Since the width must be positive, -2 cannot be a solution. If the width is 2 ft, then the length is $6 \cdot 2$ ft, or 12 ft, and the area is 12 ft $\cdot$ 2 ft $= 24$ ft^2. These numbers check.

State. The table is 12 ft long and 2 ft wide.

5. *Familiarize*. Using the labels shown on the drawing in the text, we let $h = $ the height, in cm, and $h + 10 = $ the base, in cm. Recall that the formula for the area of a triangle is $\dfrac{1}{2} \cdot \text{(base)} \cdot \text{(height)}$.

Translate.

$\dfrac{1}{2}$ $\;$ times $\;$ base $\;$ times $\;$ height $\;$ is $\;$ $\underbrace{28 \text{ cm}^2}$.

$\downarrow\qquad\downarrow\qquad\;\downarrow\qquad\;\downarrow\qquad\;\downarrow\quad\;\downarrow\qquad\downarrow$

$\dfrac{1}{2} \quad\cdot\quad (h + 10) \quad\cdot\quad h \;\;=\;\; 28$

Solve. We solve the equation.

$$\dfrac{1}{2}(h + 10)h = 28$$

$$(h + 10)h = 56 \quad \text{Multiplying by 2}$$

$$h^2 + 10h = 56$$

$$h^2 + 10h - 56 = 0$$

$$(h + 14)(h - 4) = 0$$

$$h + 14 = 0 \quad or \quad h - 4 = 0$$

$$h = -14 \quad or \quad\quad h = 4$$

Check. Since the height of the triangle must be positive, -14 cannot be a solution. If the height is 4 cm, then the base is $4 + 10$, or 14 cm, and the area is $\dfrac{1}{2} \cdot 14 \cdot 4$, or 28 cm^2. Thus, 4 checks.

State. The height of the triangle is 4 cm, and the base is 14 cm.

7. *Familiarize*. Using the labels shown on the drawing in the text, we let $h = $ the height of the triangle, in meters, and $\dfrac{1}{2}h = $ the length of the base, in meters. Recall that the formula for the area of a triangle is $\dfrac{1}{2} \cdot \text{(base)} \cdot \text{(height)}$.

Translate.

$\frac{1}{2}$ times base times height is $\underline{64\ \text{m}^2}$.

$\downarrow$ $\downarrow$ $\downarrow$ $\downarrow$ $\downarrow$ $\downarrow$ $\downarrow$

$\frac{1}{2}$ $\cdot$ $\frac{1}{2}h$ $\cdot$ h $=$ 64

Solve. We solve the equation.

$$\frac{1}{2} \cdot \frac{1}{2}h \cdot h = 64$$

$$\frac{1}{4}h^2 = 64$$

$$h^2 = 256 \quad \text{Multiplying by 4}$$

$$h^2 - 256 = 0$$

$$(h + 16)(h - 16) = 0$$

$$h + 16 = 0 \quad or \quad h - 16 = 0$$

$$h = -16 \quad or \quad h = 16$$

Check. The height of the triangle cannot be negative, so -16 cannot be a solution. If the height is 16 m, then the length of the base is $\frac{1}{2} \cdot 16$ m, or 8 m, and the area is $\frac{1}{2} \cdot 8$ m $\cdot 16$ m $= 64$ m^2. These numbers check.

State. The length of the base is 8 m, and the height is 16 m.

9. *Familiarize.* Reread Example 4 in Section 12.3.

Translate. Substitute 14 for n.

$$14^2 - 14 = N$$

Solve. We do the computation on the left.

$$14^2 - 14 = N$$

$$196 - 14 = N$$

$$182 = N$$

Check. We can redo the computation, or we can solve the equation $n^2 - n = 182$. The answer checks.

State. 182 games will be played.

11. *Familiarize.* Reread Example 4 in Section 12.3.

Translate. Substitute 132 for N.

$$n^2 - n = 132$$

Solve.

$$n^2 - n = 132$$

$$n^2 - n - 132 = 0$$

$$(n - 12)(n + 11) = 0$$

$$n - 12 = 0 \quad or \quad n + 11 = 0$$

$$n = 12 \quad or \quad n = -11$$

Check. The solutions of the equation are 12 and -11. Since the number of teams cannot be negative, -11 cannot be a solution. But 12 checks since $12^2 - 12 = 144 - 12 = 132$.

State. There are 12 teams in the league.

13. *Familiarize.* We will use the formula $N = \frac{1}{2}(n^2 - n)$.

Translate. Substitute 100 for n.

$$N = \frac{1}{2}(100^2 - 100)$$

Solve. We do the computation on the right.

$$N = \frac{1}{2}(10,000 - 100)$$

$$N = \frac{1}{2}(9900)$$

$$N = 4950$$

Check. We can redo the computation, or we can solve the equation $4950 = \frac{1}{2}(n^2 - n)$. The answer checks.

State. 4950 handshakes are possible.

15. *Familiarize.* We will use the formula $N = \frac{1}{2}(n^2 - n)$.

Translate. Substitute 300 for N.

$$300 = \frac{1}{2}(n^2 - n)$$

Solve. We solve the equation.

$$2 \cdot 300 = 2 \cdot \frac{1}{2}(n^2 - n) \quad \text{Multiplying by 2}$$

$$600 = n^2 - n$$

$$0 = n^2 - n - 600$$

$$0 = (n + 24)(n - 25)$$

$$n + 24 = 0 \quad or \quad n - 25 = 0$$

$$n = -24 \quad or \quad n = 25$$

Check. The number of people at a meeting cannot be negative, so -24 cannot be a solution. But 25 checks since $\frac{1}{2}(25^2 - 25) = \frac{1}{2}(625 - 25) = \frac{1}{2} \cdot 600 = 300$.

State. There were 25 people at the party.

17. *Familiarize.* We will use the formula $N = \frac{1}{2}(n^2 - n)$, since toasts can be substituted for handshakes.

Translate. Substitute 190 for N.

$$190 = \frac{1}{2}(n^2 - n)$$

Solve.

$$190 = \frac{1}{2}(n^2 - n)$$

$$380 = n^2 - n \quad \text{Multiplying by 2}$$

$$0 = n^2 - n - 380$$

$$0 = (n - 20)(n + 19)$$

$$n - 20 = 0 \quad or \quad n + 19 = 0$$

$$n = 20 \quad or \quad n = -19$$

Check. The solutions of the equation are 20 and -19. Since the number of people cannot be negative, -19 cannot be a solution. However, 20 checks since $\frac{1}{2}(20^2 - 20) = \frac{1}{2}(400 - 20) = \frac{1}{2}(380) = 190$.

State. 20 people took part in the toast.

19. *Familiarize.* The page numbers on facing pages are consecutive integers. Let $x =$ the smaller integer. Then $x + 1 =$ the larger integer.

Translate. We reword the problem.

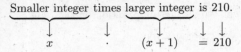

$$x \cdot (x+1) = 210$$

Solve. We solve the equation.

$$x(x+1) = 210$$
$$x^2 + x = 210$$
$$x^2 + x - 210 = 0$$
$$(x+15)(x-14) = 0$$
$$x + 15 = 0 \quad or \quad x - 14 = 0$$
$$x = -15 \quad or \quad x = 14$$

Check. The solutions of the equation are -15 and 14. Since a page number cannot be negative, -15 cannot be a solution of the original problem. We only need to check 14. When $x = 14$, then $x + 1 = 15$, and $14 \cdot 15 = 210$. This checks.

State. The page numbers are 14 and 15.

21. *Familiarize*. Let $x =$ the smaller even integer. Then $x + 2 =$ the larger even integer.

Translate. We reword the problem.

Smaller even integer times larger even integer is 168.

$$x \cdot (x+2) = 168$$

Solve.

$$x(x+2) = 168$$
$$x^2 + 2x = 168$$
$$x^2 + 2x - 168 = 0$$
$$(x+14)(x-12) = 0$$
$$x + 14 = 0 \quad or \quad x - 12 = 0$$
$$x = -14 \quad or \quad x = 12$$

Check. The solutions of the equation are -14 and 12. When x is -14, then $x + 2$ is -12 and $-14(-12) = 168$. The numbers -14 and -12 are consecutive even integers which are solutions of the problem. When x is 12, then $x + 2$ is 14 and $12 \cdot 14 = 168$. The numbers 12 and 14 are also consecutive even integers which are solutions of the problem.

State. We have two solutions, each of which consists of a pair of numbers: -14 and -12, and 12 and 14.

23. *Familiarize*. Let $x =$ the smaller odd integer. Then $x + 2 =$ the larger odd integer.

Translate. We reword the problem.

Smaller odd integer times larger odd integer is 255.

$$x \cdot (x+2) = 255$$

Solve.

$$x(x+2) = 255$$
$$x^2 + 2x = 255$$
$$x^2 + 2x - 255 = 0$$
$$(x-15)(x+17) = 0$$
$$x - 15 = 0 \quad or \quad x + 17 = 0$$
$$x = 15 \quad or \quad x = -17$$

Check. The solutions of the equation are 15 and -17. When x is 15, then $x + 2$ is 17 and $15 \cdot 17 = 255$. The numbers 15 and 17 are consecutive odd integers which are solutions to the problem. When x is -17, then $x + 2$ is -15 and $-17(-15) = 255$. The numbers -17 and -15 are also consecutive odd integers which are solutions to the problem.

State. We have two solutions, each of which consists of a pair of numbers: 15 and 17, and -17 and -15.

25. *Familiarize*. We make a drawing. Let $x =$ the length of the unknown leg. Then $x + 2 =$ the length of the hypotenuse.

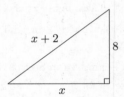

Translate. Use the Pythagorean theorem.

$$a^2 + b^2 = c^2$$
$$8^2 + x^2 = (x+2)^2$$

Solve. We solve the equation.

$$8^2 + x^2 = (x+2)^2$$
$$64 + x^2 = x^2 + 4x + 4$$
$$60 = 4x \qquad \text{Subtracting } x^2 \text{ and } 4$$
$$15 = x$$

Check. When $x = 15$, then $x + 2 = 17$ and $8^2 + 15^2 = 17^2$. Thus, 15 and 17 check.

State. The lengths of the hypotenuse and the other leg are 17 ft and 15 ft, respectively.

27. *Familiarize*. Consider the drawing in the text. We let $w =$ the width of Main Street, in feet.

Translate. Use the Pythagorean theorem.

$$a^2 + b^2 = c^2$$
$$24^2 + w^2 = 40^2$$

Solve. We solve the equation.

$$24^2 + w^2 = 40^2$$
$$576 + w^2 = 1600$$
$$w^2 - 1024 = 0$$
$$(w+32)(w-32) = 0$$
$$w + 32 = 0 \quad or \quad w - 32 = 0$$
$$w = -32 \quad or \quad w = 32$$

Check. The width of the street cannot be negative, so −32 cannot be a solution. If Main Street is 32 ft wide, we have $24^2 + 32^2 = 576 + 1024 = 1600$, which is 40^2. Thus, 32 ft checks.

State. Main Street is 32 ft wide.

29. Familiarize. Using the labels on the drawing in the text, we let h = the height of a brace, in feet. Note that we have a right triangle with hypotenuse 15 ft and legs of 12 ft and h.

Translate. We use the Pythagorean theorem.
$$a^2 + b^2 = c^2$$
$$12^2 + h^2 = 15^2 \quad \text{Substituting}$$

Solve.
$$12^2 + h^2 = 15^2$$
$$144 + h^2 = 225$$
$$h^2 - 81 = 0$$
$$(h + 9)(h - 9) = 0$$
$$h + 9 = 0 \quad or \quad h - 9 = 0$$
$$h = -9 \quad or \qquad h = 9$$

Check. The height of a brace cannot be negative, so −9 cannot be a solution. When $h = 9$, we have $12^2 + 9^2 = 144 + 81 = 225 = 15^2$. This checks.

State. The brace is 9 ft high.

31. Familiarize. We label the drawing. Let x = the length of a side of the dining room, in ft. Then the dining room has dimensions x by x and the kitchen has dimensions x by 10. The entire rectangular space has dimension x by $x + 10$. Recall that we multiply these dimensions to find the area of the rectangle.

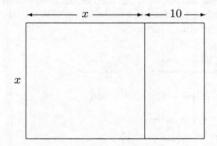

Translate.

$$\underbrace{\text{The area of the rectangular space}}_{x(x + 10)} \text{ is } \underbrace{264 \text{ ft}^2}_{264}.$$

Solve. We solve the equation.
$$x(x + 10) = 264$$
$$x^2 + 10x = 264$$
$$x^2 + 10x - 264 = 0$$
$$(x + 22)(x - 12) = 0$$
$$x + 22 = 0 \quad or \quad x - 12 = 0$$
$$x = -22 \; or \qquad x = 12$$

Check. Since the length of a side of the dining room must be positive, −22 cannot be a solution. If x is 12 ft, then $x + 10$ is 22 ft, and the area of the space is $12 \cdot 22$, or 264 ft^2. The number 12 checks.

State. The dining room is 12 ft by 12 ft, and the kitchen is 12 ft by 10 ft.

33. Familiarize. We will use the formula $h = 180t - 16t^2$.

Translate. Substitute 464 for h.
$$464 = 180t - 16t^2$$

Solve. We solve the equation.
$$464 = 180t - 16t^2$$
$$16t^2 - 180t + 464 = 0$$
$$4(4t^2 - 45t + 116) = 0$$
$$4(4t - 29)(t - 4) = 0$$
$$4t - 29 = 0 \quad or \quad t - 4 = 0$$
$$4t = 29 \quad or \qquad t = 4$$
$$t = \frac{29}{4} \quad or \qquad t = 4$$

Check. The solutions of the equation are $\frac{29}{4}$, or $7\frac{1}{4}$, and 4. Since we want to find how many seconds it takes the rocket to *first* reach a height of 464 ft, we check the smaller number, 4. We substitute 4 for t in the formula.
$$h = 180t - 16t^2$$
$$h = 180 \cdot 4 - 16(4)^2$$
$$h = 180 \cdot 4 - 16 \cdot 16$$
$$h = 720 - 256$$
$$h = 464$$

The answer checks.

State. The rocket will first reach a height of 464 ft after 4 seconds.

35. Familiarize. Let x = the smaller odd positive integer. Then $x + 2$ = the larger odd positive integer.

Translate.

$$\underbrace{\text{Square of the smaller odd positive integer}}_{x^2} + \underbrace{\text{Square of the larger odd positive integer}}_{(x + 2)^2} \text{ is } \underbrace{74}_{= 74}$$

Solve.
$$x^2 + (x + 2)^2 = 74$$
$$x^2 + x^2 + 4x + 4 = 74$$
$$2x^2 + 4x - 70 = 0$$
$$2(x^2 + 2x - 35) = 0$$
$$2(x + 7)(x - 5) = 0$$
$$x + 7 = 0 \quad or \quad x - 5 = 0$$
$$x = -7 \quad or \qquad x = 5$$

Check. The solutions of the equation are −7 and 5. The problem asks for odd positive integers, so −7 cannot be a solution. When x is 5, $x + 2$ is 7. The numbers 5 and

7 are consecutive odd positive integers. The sum of their squares, $25 + 49$, is 74. The numbers check.

State. The integers are 5 and 7.

37. Discussion and Writing Exercise

39. To <u>factor</u> a polynomial is to express it as a <u>product</u>.

41. A factorization of a polynomial is an expression that names that polynomial as a <u>product</u>.

43. The expression $-5x^2 + 8x - 7$ is an example of a <u>trinomial</u>.

45. For the graph of the equation $4x - 3y = 12$, the pair $(0, -4)$ is known as the <u>y-intercept</u>.

47. *Familiarize.* First we can use the Pythagorean theorem to find x, in ft. Then the height of the telephone pole is $x + 5$.

Translate. We use the Pythagorean theorem.
$$a^2 + b^2 = c^2$$
$$\left(\frac{1}{2}x + 1\right)^2 + x^2 = 34^2$$

Solve. We solve the equation.
$$\left(\frac{1}{2}x + 1\right)^2 + x^2 = 34^2$$
$$\frac{1}{4}x^2 + x + 1 + x^2 = 1156$$
$$x^2 + 4x + 4 + 4x^2 = 4624 \quad \text{Multiplying by 4}$$
$$5x^2 + 4 + 4 = 4624$$
$$5x^2 + 4x - 4620 = 0$$
$$(5x + 154)(x - 30) = 0$$
$$5x + 154 = 0 \quad or \quad x - 30 = 0$$
$$5x = -154 \quad or \quad x = 30$$
$$x = -30.8 \quad or \quad x = 30$$

Check. Since the length x must be positive, -30.8 cannot be a solution. If x is 30 ft, then $\frac{1}{2}x + 1$ is $\frac{1}{2} \cdot 30 + 1$, or 16 ft. Since $16^2 + 30^2 = 1156 = 34^2$, the number 30 checks. When x is 30 ft, then $x + 5$ is 35 ft.

State. The height of the telephone pole is 35 ft.

49. *Familiarize.* Using the labels shown on the drawing in the text, we let $x =$ the width of the walk. Then the length and width of the rectangle formed by the pool and walk together are $40 + 2x$ and $20 + 2x$, respectively.

Translate.

Area is length times width.

$$1500 = (40 + 2x) \cdot (20 + 2x)$$

Solve. We solve the equation.
$$1500 = (40 + 2x)(20 + 2x)$$
$$1500 = 2(20 + x) \cdot 2(10 + x) \quad \text{Factoring 2 out of each factor on the right}$$
$$1500 = 4 \cdot (20 + x)(10 + x)$$
$$375 = (20 + x)(10 + x) \quad \text{Dividing by 4}$$
$$375 = 200 + 30x + x^2$$
$$0 = x^2 + 30x - 175$$
$$0 = (x + 35)(x - 5)$$
$$x + 35 = 0 \quad or \quad x - 5 = 0$$
$$x = -35 \quad or \quad x = 5$$

Check. The solutions of the equation are -35 and 5. Since the width of the walk cannot be negative, -35 is not a solution. When $x = 5$, $40 + 2x = 40 + 2 \cdot 5$, or 50 and $20 + 2x = 20 + 2 \cdot 5$, or 30. The total area of the pool and walk is $50 \cdot 30$, or 1500 ft^2. This checks.

State. The width of the walk is 5 ft.

51. *Familiarize.* We make a drawing. Let $w =$ the width of the piece of cardboard. Then $2w =$ the length.

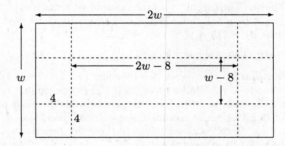

The box will have length $2w - 8$, width $w - 8$, and height 4. Recall that the formula for volume is $V = \text{length} \times \text{width} \times \text{height}$.

Translate.

The volume   is   616cm^3.

$$(2w - 8)(w - 8)(4) = 616$$

Solve. We solve the equation.
$$(2w - 8)(w - 8)(4) = 616$$
$$(2w^2 - 24w + 64)(4) = 616$$
$$8w^2 - 96w + 256 = 616$$
$$8w^2 - 96w - 360 = 0$$
$$8(w^2 - 12w - 45) = 0$$
$$w^2 - 12w - 45 = 0 \quad \text{Dividing by 8}$$
$$(w - 15)(w + 3) = 0$$
$$w - 15 = 0 \quad or \quad w + 3 = 0$$
$$w = 15 \quad or \quad w = -3$$

Check. The width cannot be negative, so we only need to check 15. When $w = 15$, then $2w = 30$ and the dimensions of the box are $30 - 8$ by $15 - 8$ by 4, or 22 by 7 by 4. The volume is $22 \cdot 7 \cdot 4$, or 616.

State. The cardboard is 30 cm by 15 cm.

53. *Familiarize.* Let x = the length of a side of the base of the box, in feet. Then each of the four sides of the box has dimensions 9 ft by x and the top and bottom each have dimensions x by x.

Translate. We add the areas of the four sides of the box and of the top and bottom.

$$4 \cdot 9 \cdot x + 2 \cdot x^2 = 350$$

Solve. We solve the equation.

$$4 \cdot 9 \cdot x + 2 \cdot x^2 = 350$$
$$36x + 2x^2 = 350$$
$$2x^2 + 36x - 350 = 0$$
$$2(x^2 + 18x - 175) = 0$$
$$x^2 + 18x - 175 = 0 \qquad \text{Dividing by 2}$$
$$(x + 25)(x - 7) = 0$$
$$x + 25 = 0 \quad or \quad x - 7 = 0$$
$$x = -25 \quad or \qquad x = 7$$

Check. The length of a side cannot be negative, so -25 cannot be a solution. If $x = 7$, then the surface area of the box is $4 \cdot 9 \text{ ft} \cdot 7 \text{ ft} + 2 \cdot (7 \text{ ft})^2 = 252 \text{ ft}^2 + 98 \text{ ft}^2 = 350 \text{ ft}^2$. The number 7 checks.

State. The length of a side of the base of the box is 7 ft.

Chapter 13 Review Exercises

1. $-15y^2 = -1 \cdot 3 \cdot 5 \cdot y^2$
$25y^6 = 5 \cdot 5 \cdot y^6$

Each coefficient has a factor of 5. There are no other common prime factors. The GCF of the powers of y is y^2 because 2 is the smallest exponent of y. Thus the GCF is $5y^2$.

2. $12x^3 = 2 \cdot 2 \cdot 3 \cdot x^3$
$-60x^2 y = -1 \cdot 2 \cdot 2 \cdot 3 \cdot 5 \cdot x^2 \cdot y$
$36xy = 2 \cdot 2 \cdot 3 \cdot 3 \cdot x \cdot y$

Each coefficient has two factors of 2 and one factor of 3. There are no other common prime factors. The GCF of the powers of x is x because 1 is the smallest exponent of x. The GCF of the powers of y is 1 because $12x^3$ has no y-factor. Thus the GCF is $2 \cdot 2 \cdot 3 \cdot x \cdot 1$, or $12x$.

3. $5 - 20x^6$
$= 5(1 - 4x^6) \qquad$ 5 is a common factor.
$= 5(1 - 2x^3)(1 + 2x^3) \quad$ Factoring the difference of squares

4. $x^2 - 3x = x(x - 3)$

5. $9x^2 - 4 = (3x + 2)(3x - 2) \quad$ Factoring a difference of squares

6. $x^2 + 4x - 12$

We look for a pair of factors of -12 whose sum is 4. The numbers we need are 6 and -2.

$x^2 + 4x - 12 = (x + 6)(x - 2)$

7. $x^2 + 14x + 49 = x^2 + 2 \cdot x \cdot 7 + 7^2 = (x + 7)^2$

8. $6x^3 + 12x^2 + 3x = 3x(2x^2 + 4x + 1)$

The trinomial $2x^2 + 4x + 1$ cannot be factored, so the factorization is complete.

9. $\quad x^3 + x^2 + 3x + 3$
$= (x^3 + x^2) + (3x + 3)$
$= x^2(x + 1) + 3(x + 1) \quad$ Factoring by grouping
$= (x^2 + 3)(x + 1)$

10. $6x^2 - 5x + 1$

There is no common factor (other than 1). This polynomial has three terms, but it is not a trinomial square. Multiply the leading coefficient and the constant, 6 and 1: $6 \cdot 1 = 6$. Try to factor 6 so that the sum of the factors is -5. The numbers we want are -2 and -3: $-2(-3) = 6$ and $-2 + (-3) = -5$. Split the middle term and factor by grouping.

$$6x^2 - 5x + 1 = 6x^2 - 2x - 3x + 1$$
$$= (6x^2 - 2x) + (-3x + 1)$$
$$= 2x(3x - 1) - 1(3x - 1)$$
$$= (2x - 1)(3x - 1)$$

11. $x^4 - 81 = (x^2 + 9)(x^2 - 9) = (x^2 + 9)(x + 3)(x - 3)$

12. $\quad 9x^3 + 12x^2 - 45x$
$= 3x(3x^2 + 4x - 15) \quad$ $3x$ is a common factor.
$= 3x(3x - 5)(x + 3) \quad$ Using trial and error

13. $2x^2 - 50 = 2(x^2 - 25) = 2(x + 5)(x - 5)$

14. $\quad x^4 + 4x^3 - 2x - 8 = (x^4 + 4x^3) + (-2x - 8)$
$= x^3(x + 4) - 2(x + 4)$
$= (x^3 - 2)(x + 4)$

15. $16x^4 - 1 = (4x^2 + 1)(4x^2 - 1) = (4x^2 + 1)(2x + 1)(2x - 1)$

16. $8x^6 - 32x^5 + 4x^4 = 4x^4(2x^2 - 8x + 1)$

The trinomial $2x^2 - 8x + 1$ cannot be factored, so the factorization is complete.

17. $\quad 75 + 12x^2 + 60x = 12x^2 + 60x + 75$
$= 3(4x^2 + 20x + 25)$
$= 3(2x + 5)^2$

18. $x^2 + 9$ is a sum of squares with no common factor, so it is prime.

19. $x^3 - x^2 - 30x = x(x^2 - x - 30) = x(x - 6)(x + 5)$

20. $4x^2 - 25 = (2x + 5)(2x - 5)$

21. $9x^2 + 25 - 30x = 9x^2 - 30x + 25 = (3x - 5)^2$

22. $\quad 6x^2 - 28x - 48 = 2(3x^2 - 14x - 24)$
$= 2(3x + 4)(x - 6)$

23. $x^2 - 6x + 9 = (x - 3)^2$

24. $2x^2 - 7x - 4 = (2x + 1)(x - 4)$

25. $18x^2 - 12x + 2 = 2(9x^2 - 6x + 1) = 2(3x - 1)^2$

26. $3x^2 - 27 = 3(x^2 - 9) = 3(x + 3)(x - 3)$

27. $15 - 8x + x^2 = x^2 - 8x + 15 = (x - 3)(x - 5)$

28. $25x^2 - 20x + 4 = (5x - 2)^2$

29. $49b^{10} + 4a^8 - 28a^4b^5 = 49b^{10} - 28a^4b^5 + 4a^8$
$$= (7b^5)^2 - 2 \cdot 7b^5 \cdot 2a^4 + (2a^4)^2$$
$$= (7b^5 - 2a^4)^2$$

30. $x^2y^2 + xy - 12 = (xy + 4)(xy - 3)$

31. $12a^2 + 84ab + 147b^2 = 3(4a^2 + 28ab + 49b^2) = 3(2a + 7b)^2$

32. $m^2 + 5m + mt + 5t = (m^2 + 5m) + (mt + 5t)$
$$= m(m + 5) + t(m + 5)$$
$$= (m + t)(m + 5)$$

33. $32x^4 - 128y^4z^4 = 32(x^4 - 4y^4z^4) =$
$32(x^2 + 2y^2z^2)(x^2 - 2y^2z^2)$

34. $(x - 1)(x + 3) = 0$
$x - 1 = 0 \quad or \quad x + 3 = 0$
$x = 1 \quad or \quad x = -3$
The solutions are 1 and -3.

35. $x^2 + 2x - 35 = 0$
$(x + 7)(x - 5) = 0$
$x + 7 = 0 \quad or \quad x - 5 = 0$
$x = -7 \quad or \quad x = 5$
The solutions are -7 and 5.

36. $x^2 + x - 12 = 0$
$(x + 4)(x - 3) = 0$
$x + 4 = 0 \quad or \quad x - 3 = 0$
$x = -4 \quad or \quad x = 3$
The solutions are -4 and 3.

37. $3x^2 + 2 = 5x$
$3x^2 - 5x + 2 = 0$
$(3x - 2)(x - 1) = 0$
$3x - 2 = 0 \quad or \quad x - 1 = 0$
$3x = 2 \quad or \quad x = 1$
$x = \frac{2}{3} \quad or \quad x = 1$
The solutions are $\frac{2}{3}$ and 1.

38. $2x^2 + 5x = 12$
$2x^2 + 5x - 12 = 0$
$(2x - 3)(x + 4) = 0$
$2x - 3 = 0 \quad or \quad x + 4 = 0$
$2x = 3 \quad or \quad x = -4$
$x = \frac{3}{2} \quad or \quad x = -4$
The solutions are $\frac{3}{2}$ and -4.

39. $16 = x(x - 6)$
$16 = x^2 - 6x$
$0 = x^2 - 6x - 16$
$0 = (x - 8)(x + 2)$
$x - 8 = 0 \quad or \quad x + 2 = 0$
$x = 8 \quad or \quad x = -2$
The solutions are 8 and -2.

40. *Familiarize*. Let $b =$ the length of the base, in cm. Then $b + 1 =$ the height.

Translate. We use the formula for the area of a triangle.

$$A = \frac{1}{2}bh$$
$$15 = \frac{1}{2}b(b + 1) \quad \text{Substituting}$$

Solve.
$$15 = \frac{1}{2}b(b + 1)$$
$$15 = \frac{1}{2}b^2 + \frac{1}{2}b$$
$$0 = \frac{1}{2}b^2 + \frac{1}{2}b - 15$$
$$2 \cdot 0 = 2\left(\frac{1}{2}b^2 + \frac{1}{2}b - 15\right) \quad \text{Clearing fractions}$$
$$0 = b^2 + b - 30$$
$$0 = (b + 6)(b - 5)$$
$$b + 6 = 0 \quad or \quad b - 5 = 0$$
$$b = -6 \quad or \quad b = 5$$

Check. The length of the base cannot be negative, so -6 cannot be a solution. If $b = 5$, then $b + 1 = 5 + 1 = 6$ and the area is $\frac{1}{2} \cdot 5 \cdot 6 = 15$ cm^2. The number 5 checks.

State. The base is 5 cm, and the height is 6 cm.

41. *Familiarize*. Let $x =$ the smaller integer. Then $x + 2$ is the other integer.

Translate.

Smaller even integer	times	larger even integer	is	288.
↓	↓	↓	↓	↓
x	$\cdot$	$(x + 2)$	$=$	288

Solve.
$$x(x + 2) = 288$$
$$x^2 + 2x = 288$$
$$x^2 + 2x - 288 = 0$$
$$(x + 18)(x - 16) = 0$$
$$x + 18 = 0 \quad or \quad x - 16 = 0$$
$$x = -18 \quad or \quad x = 16$$

If $x = -18$, then $x + 2 = -18 + 2 = -16$.

If $x = 16$, then $x + 2 = 16 + 2 = 18$.

Check. -18 and -16 are consecutive even integers and $-18(-16) = 288$. Similarly, 16 and 18 are consecutive even integers and $16 \cdot 18 = 288$. Both pairs of integers check.

State. The integers are -18 and -16 or 16 and 18.

42. Familiarize. Let $x =$ the smaller integer. Then $x + 2$ is the other integer.

Translate.

$$\underbrace{\text{Smaller odd integer}}_{\downarrow} \quad \underbrace{\text{times}}_{\downarrow} \quad \underbrace{\text{larger odd integer}}_{\downarrow} \quad \underbrace{\text{is}}_{\downarrow} \quad \underbrace{323.}_{\downarrow}$$

$$x \qquad \cdot \qquad (x+2) \qquad = \qquad 323$$

Solve.

$$x(x+2) = 323$$
$$x^2 + 2x = 323$$
$$x^2 + 2x - 323 = 0$$
$$(x+19)(x-17) = 0$$
$$x + 19 = 0 \quad or \quad x - 17 = 0$$
$$x = -19 \quad or \qquad x = 17$$

If $x = -19$, then $x + 2 = -19 + 2 = -17$.

If $x = 17$, then $x + 2 = 17 + 2 = 19$.

Check. -19 and -17 are consecutive odd integers and $-19(-17) = 323$. Similarly, 17 and 19 are consecutive odd integers and $17 \cdot 19 = 323$. Both pairs of integers check.

State. The integers are -19 and -17 or 17 and 19.

43. Familiarize. We make a drawing. Let $d =$ the height above the ground at which the cables are attached to the tree. Then $d + 2 =$ the length of the cable.

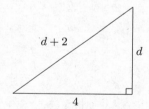

Translate. We use the Pythagorean theorem.

$$a^2 + b^2 = c^2$$
$$4^2 + d^2 = (d+2)^2 \quad \text{Substituting}$$

Solve.

$$4^2 + d^2 = (d+2)^2$$
$$16 + d^2 = d^2 + 4d + 4$$
$$16 = 4d + 4 \qquad \text{Subtracting } d^2$$
$$12 = 4d$$
$$3 = d$$

Check. If $d = 3$, then $d + 2 = 3 + 2 = 5$ and we have $4^2 + 3^2 = 16 + 9 = 25 = 5^2$. The answer checks.

State. The cables are attached to the tree 3 ft above the ground.

44. Familiarize. Let $s =$ the length of a side of the original square, in km. Then $s + 3 =$ the length of a side of the enlarged square.

Translate.

$$\underbrace{\text{Area of enlarged square}}_{\downarrow} \quad \underbrace{\text{is}}_{\downarrow} \quad \underbrace{81 \text{ km}^2.}_{\downarrow}$$

$$(s+3)^2 \qquad = \qquad 81$$

Solve.

$$(s+3)^2 = 81$$
$$s^2 + 6s + 9 = 81$$
$$s^2 + 6s - 72 = 0$$
$$(s+12)(s-6) = 0$$
$$s + 12 = 0 \quad or \quad s - 6 = 0$$
$$s = -12 \quad or \qquad s = 6$$

Check. The length of a side of the square cannot be negative, so -12 cannot be a solution. If the length of a side of the original square is 6 km, then the length of a side of the enlarged square is $6 + 3$, or 9 km. The area of the enlarged square is $(9 \text{ km})^2$, or 81 km^2, so the answer checks.

State. The length of a side of the original square is 6 km.

45. Let $y = 0$ and solve for x.

$$0 = x^2 + 9x + 20$$
$$0 = (x+5)(x+4)$$
$$x + 5 = 0 \quad or \quad x + 4 = 0$$
$$x = -5 \quad or \qquad x = -4$$

The x-intercepts are $(-5, 0)$ and $(-4, 0)$.

46. Let $y = 0$ and solve for x.

$$0 = 2x^2 - 7x - 15$$
$$0 = (2x+3)(x-5)$$
$$2x + 3 = 0 \quad or \quad x - 5 = 0$$
$$2x = -3 \quad or \qquad x = 5$$
$$x = -\frac{3}{2} \quad or \qquad x = 5$$

The x-intercepts are $\left(-\frac{3}{2}, 0\right)$ and $(5, 0)$.

47. Discussion and Writing Exercise. Answers may vary. The area of a rectangle is 90 m². The length is 1 m greater than the width. Find the length and the width.

48. Discussion and Writing Exercise. Because Sheri did not first factor out the largest common factor, 4, her factorization will not be "complete" until she removes a common factor of 2 from each binomial.

49. Familiarize. Let $w =$ the width of the margins, in cm. Then the printed area on each page has dimensions $20 - 2w$ by $15 - 2w$. The area of the margins constitutes one-half the area of each page, so the printed area also constitutes one-half of the area.

Translate.

$$\underbrace{\text{Printed area}}_{\downarrow} \quad \underbrace{\text{is}}_{\downarrow} \quad \underbrace{\text{one-half}}_{\downarrow} \quad \underbrace{\text{of}}_{\downarrow} \quad \underbrace{\text{total area.}}_{\downarrow}$$

$$(20-2w)(15-2w) \quad = \qquad \frac{1}{2} \qquad \cdot \qquad 20 \cdot 15$$

Solve.

$$(20 - 2w)(15 - 2w) = \frac{1}{2} \cdot 20 \cdot 15$$
$$300 - 70w + 4w^2 = 150$$
$$150 - 70w + 4w^2 = 0$$
$$4w^2 - 70w + 150 = 0$$
$$2(2w^2 - 35w + 75) = 0$$
$$2w^2 - 35w + 75 = 0 \quad \text{Dividing by 2}$$
$$(2w - 5)(w - 15) = 0$$
$$2w - 5 = 0 \quad or \quad w - 15 = 0$$
$$2w = 5 \quad or \quad w = 15$$
$$w = 2.5 \quad or \quad w = 15$$

Check. If $w = 15$, then $20 - 2w$ and $15 - 2w$ are both negative. Since the dimensions of the printed area cannot be negative, 15 cannot be a solution. If $w = 2.5$, then $20 - 2w = 20 - 2(2.5) = 20 - 5 = 15$ and $15 - 2w = 15 - 2(2.5) = 15 - 5 = 10$. Thus the printed area is $15 \cdot 10$, or 150 cm². This is one-half of the total area of the page, $20 \cdot 15$, or 300 cm². The number 2.5 checks.

State. The width of the margins is 2.5 cm.

50. Familiarize. Let $n =$ the number.

Translate.

The cube of a number | is | twice | the square of the number.
$\downarrow$ | $\downarrow$ | $\downarrow$ | $\downarrow$
n^3 | $=$ | $2 \cdot$ | n^2

Solve.

$$n^3 = 2n^2$$
$$n^3 - 2n^2 = 0$$
$$n^2(n - 2) = 0$$
$$n \cdot n(n - 2) = 0$$
$$n = 0 \quad or \quad n = 0 \quad or \quad n - 2 = 0$$
$$n = 0 \quad or \quad n = 0 \quad or \quad n = 2$$

Check. If $n = 0$, then $n^3 = 0^3 = 0$, $n^2 = 0^2 = 0$, and $0 = 2 \cdot 0$. If $n = 2$, then $n^3 = 2^3 = 8$, $n^2 = 2^2 = 4$, and $8 = 2 \cdot 4$. Both numbers check.

State. The number is 0 or 2.

51. Familiarize. Let $w =$ the width of the original rectangle. Then $2w =$ the length. The new length and width are $2w + 20$ and $w - 1$, respectively.

Translate.

The new area | is | 160.
$\downarrow$ | $\downarrow$ | $\downarrow$
$(2w + 20)(w - 1)$ | $=$ | 160

Solve.

$$(2w + 20)(w - 1) = 160$$
$$2w^2 + 18w - 20 = 160$$
$$2w^2 + 18w - 180 = 0$$
$$2(w^2 + 9w - 90) = 0$$
$$w^2 + 9w - 90 = 0$$
$$(w + 15)(w - 6) = 0$$

$$w + 15 = 0 \quad or \quad w - 6 = 0$$
$$w = -15 \quad or \quad w = 6$$

Check. The dimensions of the rectangle cannot be negative, so -15 cannot be a solution. If $w = 6$, then $2w = 2 \cdot 6 = 12$, $2w + 20 = 12 + 20 = 32$ and $w - 1 = 6 - 1 = 5$. The area of a rectangle with dimensions 32 by 5 is $32 \cdot 5$, or 160, so the answer checks.

State. The length of the original rectangle is 12 and the width is 6.

52. $x^2 + 25 = 0$

Since $x^2 + 25$ cannot be factored, the equation has no solution.

53. $(x - 2)(x + 3)(2x - 5) = 0$

$$x - 2 = 0 \quad or \quad x + 3 = 0 \quad or \quad 2x - 5 = 0$$
$$x = 2 \quad or \quad x = -3 \quad or \quad 2x = 5$$
$$x = 2 \quad or \quad x = -3 \quad or \quad x = \frac{5}{2}$$

The solutions are 2, -3, and $\frac{5}{2}$.

54. $(x - 3)4x^2 + 3x(x - 3) - (x - 3)10 = 0$

$$(4x^2 + 3x - 10)(x - 3) = 0 \quad \text{Factoring out } x - 3$$
$$(4x - 5)(x + 2)(x - 3) = 0 \quad \text{Factoring } 4x^2 + 3x - 10$$

$$4x - 5 = 0 \quad or \quad x + 2 = 0 \quad or \quad x - 3 = 0$$
$$4x = 5 \quad or \quad x = -2 \quad or \quad x = 3$$
$$x = \frac{5}{4} \quad or \quad x = -2 \quad or \quad x = 3$$

The solutions are $\frac{5}{4}$, -2, and 3.

55. The shaded area is the area of a circle with radius x less the area of a square with a diagonal of length $x + x$, or $2x$. The area of the circle is πx^2. The square can be thought of as two triangles, each with base $2x$ and height x. Then the area of the square is $2 \cdot \frac{1}{2} \cdot 2x \cdot x$, or $2x^2$. We subtract to find the shaded area.

$$\pi x^2 - 2x^2 = (\pi - 2)x^2$$

Chapter 13 Test

1. $28x^3 = 2 \cdot 2 \cdot 7 \cdot x^3$
$48x^7 = 2 \cdot 2 \cdot 2 \cdot 2 \cdot 3 \cdot x^7$

The coefficients each have two factors of 2. There are no other common prime factors. The GCF of the powers of x is x^3 because 3 is the smallest exponent of x. Thus the GCF is $2 \cdot 2 \cdot x^3$, or $4x^3$.

2. $x^2 - 7x + 10$

We look for a pair of factors of 10 whose sum is -7. The numbers we need are -2 and -5.

$$x^2 - 7x + 10 = (x - 2)(x - 5)$$

3. $x^2 + 25 - 10x = x^2 - 10x + 25$
$$= x^2 - 2 \cdot x \cdot 5 + 5^2$$
$$= (x - 5)^2$$

4. $6y^2 - 8y^3 + 4y^4 = 4y^4 - 8y^3 + 6y^2 =$
$2y^2 \cdot 2y^2 - 2y^2 \cdot 4y + 2y^3 \cdot 3 = 2y^2(2y^2 - 4y + 3)$

Since $2y^2 - 4y + 3$ cannot be factored, the factorization is complete.

5. $\quad x^3 + x^2 + 2x + 2$
$$= (x^3 + x^2) + (2x + 2)$$
$$= x^2(x + 1) + 2(x + 1) \quad \text{Factoring by grouping}$$
$$= (x^2 + 2)(x + 1)$$

6. $x^2 - 5x = x \cdot x - 5 \cdot x = x(x - 5)$

7. $\quad x^3 + 2x^2 - 3x$
$$= x(x^2 + 2x - 3) \quad x \text{ is a common factor.}$$
$$= x(x + 3)(x - 1) \quad \text{Factoring the trinomial}$$

8. $\quad 28x - 48 + 10x^2$
$$= 10x^2 + 28x - 48$$
$$= 2(5x^2 + 14x - 24) \quad 2 \text{ is a common factor.}$$
$$= 2(5x - 6)(x + 4) \quad \text{Factoring the trinomial}$$

9. $4x^2 - 9 = (2x)^2 - 3^2 \quad \text{Difference of squares}$
$$= (2x + 3)(2x - 3)$$

10. $x^2 - x - 12$

We look for a pair of factors of -12 whose sum is -1. The numbers we need are -4 and 3.
$$x^2 - x - 12 = (x - 4)(x + 3)$$

11. $\quad 6m^3 + 9m^2 + 3m$
$$= 3m(2m^2 + 3m + 1) \quad 3m \text{ is a common factor.}$$
$$= 3m(2m + 1)(m + 1) \quad \text{Factoring the trinomial}$$

12. $3w^2 - 75 = 3(w^2 - 25) \quad 3 \text{ is a common factor.}$
$$= 3(w^2 - 5^2) \quad \text{Difference of squares}$$
$$= 3(w + 5)(w - 5)$$

13. $\quad 60x + 45x^2 + 20$
$$= 45x^2 + 60x + 20$$
$$= 5(9x^2 + 12x + 4) \quad 5 \text{ is a common factor.}$$
$$= 5[(3x)^2 + 2 \cdot 3x \cdot 2 + 2^2] \quad \text{Trinomial square}$$
$$= 5(3x + 2)^2$$

14. $\quad 3x^4 - 48$
$$= 3(x^4 - 16) \quad 3 \text{ is a common factor.}$$
$$= 3[(x^2)^2 - 4^2] \quad \text{Difference of squares}$$
$$= 3(x^2 + 4)(x^2 - 4)$$
$$= 3(x^2 + 4)(x^2 - 2^2) \quad \text{Difference of squares}$$
$$= 3(x^2 + 4)(x + 2)(x - 2)$$

15. $\quad 49x^2 - 84x + 36$
$$= (7x)^2 - 2 \cdot 7x \cdot 6 + 6^2 \quad \text{Trinomial square}$$
$$= (7x - 6)^2$$

16. $5x^2 - 26x + 5$

There is no common factor (other than 1). This polynomial has 3 terms, but it is not a trinomial square. Using the ac-method we first multiply the leading coefficient and the constant term: $5 \cdot 5 = 25$. Try to factor 25 so that the sum of the factors is -26. The numbers we want are -1 and -25: $-1(-25) = 25$ and $-1 + (-25) = -26$. Split the middle term and factor by grouping:
$$5x^2 - 26x + 5 = 5x^2 - x - 25x + 5$$
$$= (5x^2 - x) + (-25x + 5)$$
$$= x(5x - 1) - 5(5x - 1)$$
$$= (x - 5)(5x - 1)$$

17. $\quad x^4 + 2x^3 - 3x - 6$
$$= (x^4 + 2x^3) + (-3x - 6)$$
$$= x^2(x + 2) - 3(x + 2) \quad \text{Factoring by grouping}$$
$$= (x^3 - 3)(x + 2)$$

18. $\quad 80 - 5x^4$
$$= 5(16 - x^4) \quad 5 \text{ is a common factor.}$$
$$= 5[4^2 - (x^2)^2] \quad \text{Difference of squares}$$
$$= 5(4 + x^2)(4 - x^2)$$
$$= 5(4 + x^2)(2^2 - x^2) \quad \text{Difference of squares}$$
$$= 5(4 + x^2)(2 + x)(2 - x)$$

19. $4x^2 - 4x - 15$

(1) There is no common factor (other than 1 or -1).

(2) Factor the first term, $4x^2$. The possibilities are x, $4x$ and $2x$, $2x$. We have these as possibilities for factorizations:
$$(x + \quad)(4x + \quad) \text{ and } (2x + \quad)(2x + \quad)$$

(3) Factor the last term, -15. The possibilities are 1, -15 and -1, 15 and 3, -5 and -3, 5. These factors can also be written as -15, 1 and 15, -1 and -5, 3 and 5, -3.

(4) We try some possibilities and find that the factorization is $(2x + 3)(2x - 5)$.

20. $\quad 6t^3 + 9t^2 - 15t$
$$= 3t(2t^2 + 3t - 5) \quad 3t \text{ is a common factor.}$$
$$= 3t(2t + 5)(t - 1) \quad \text{Factoring the trinomial}$$

21. $\quad 3m^2 - 9mn - 30n^2$
$$= 3(m^2 - 3mn - 10n^2) \quad 3 \text{ is a common factor.}$$
$$= 3(m - 5n)(m + 2n) \quad \text{Factoring the trinomial}$$

22. $\quad x^2 - x - 20 = 0$
$$(x - 5)(x + 4) = 0$$
$$x - 5 = 0 \quad or \quad x + 4 = 0$$
$$x = 5 \quad or \quad x = -4$$

The solutions are 5 and -4.

23.
$$2x^2 + 7x = 15$$
$$2x^2 + 7x - 15 = 0$$
$$(2x - 3)(x + 5) = 0$$
$$2x - 3 = 0 \quad or \quad x + 5 = 0$$
$$2x = 3 \quad or \qquad x = -5$$
$$x = \frac{3}{2} \quad or \qquad x = -5$$

The solutions are $\frac{3}{2}$ and -5.

24.
$$x(x - 3) = 28$$
$$x^2 - 3x = 28$$
$$x^2 - 3x - 28 = 0$$
$$(x - 7)(x + 4) = 0$$
$$x - 7 = 0 \quad or \quad x + 4 = 0$$
$$x = 7 \quad or \qquad x = -4$$

The solutions are 7 and -4.

25. *Familiarize*. Let w = the width, in meters. Then $w+2$ = the length. Recall that the area of a rectangle is (length) · (width).

Translate. We use the formula for the area of a rectangle.
$$48 = (w + 2)w$$

Solve.
$$48 = (w + 2)w$$
$$48 = w^2 + 2w$$
$$0 = w^2 + 2w - 48$$
$$0 = (w + 8)(w - 6)$$
$$w + 8 = 0 \quad or \quad w - 6 = 0$$
$$w = -8 \quad or \qquad w = 6$$

Check. The width cannot be negative, so -8 cannot be a solution. If $w = 6$, then $w + 2 = 8$ and the area is $(8 \text{ m}) \cdot (6 \text{ m})$, or 48 m^2. The number 6 checks.

State. The length is 8 m and the width is 6 m.

26. *Familiarize*. Using the labels on the drawing in the text, we let h = the height of the triangle, in cm, and $2h + 6$ = the base. Recall that the area of a triangle is $\frac{1}{2} \cdot$ (base) · (height).

Translate. We use the formula for the area of a triangle.
$$28 = \frac{1}{2} \cdot (2h + 6) \cdot h$$

Solve.
$$28 = \frac{1}{2}(2h + 6)h$$
$$28 = h^2 + 3h$$
$$0 = h^2 + 3h - 28$$
$$0 = (h + 7)(h - 4)$$
$$h + 7 = 0 \quad or \quad h - 4 = 0$$
$$h = -7 \quad or \qquad h = 4$$

Check. The height cannot be negative, so -7 cannot be a solution. If $h = 4$, then $2h + 6 = 2 \cdot 4 + 6 = 8 + 6 = 14$ and

the area is $\frac{1}{2} \cdot (14 \text{ cm}) \cdot (4 \text{ cm})$, or 28 cm^2. The number 4 checks.

State. The height is 4 cm and the base is 14 cm.

27. *Familiarize*. Using the labels on the drawing in the text, we let x = the distance between the two marked points, in feet. If the corner is a right angle, the lengths 3 ft, 4 ft, and x will satisfy the Pythagorean equation.

Translate.
$$a^2 + b^2 = c^2$$
$$3^2 + 4^2 = x^2 \quad \text{Substituting}$$

Solve.
$$3^2 + 4^2 = x^2$$
$$9 + 16 = x^2$$
$$25 = x^2$$
$$0 = x^2 - 25$$
$$0 = (x + 5)(x - 5)$$
$$x + 5 = 0 \quad or \quad x - 5 = 0$$
$$x = -5 \quad or \qquad x = 5$$

Check. The distance cannot be negative, so -5 cannot be a solution. If $x = 5$, then we have $3^2 + 4^2 = 9 + 16 = 25 = 5^2$, so the answer checks.

State. The distance between the marked points should be 5 ft.

28. We let $y = 0$ and solve for x.
$$0 = x^2 - 2x - 35$$
$$0 = (x + 5)(x - 7)$$
$$x + 5 = 0 \quad or \quad x - 7 = 0$$
$$x = -5 \quad or \qquad x = 7$$

The x-intercepts are $(-5, 0)$ and $(7, 0)$.

29. We let $y = 0$ and solve for x.
$$0 = 3x^2 - 5x + 2$$
$$0 = (3x - 2)(x - 1)$$
$$3x - 2 = 0 \quad or \quad x - 1 = 0$$
$$3x = 2 \quad or \qquad x = 1$$
$$x = \frac{2}{3} \quad or \qquad x = 1$$

The x-intercepts are $\left(\frac{2}{3}, 0\right)$ and $(1, 0)$.

30. *Familiarize*. Let w = the width of the original rectangle. Then $5w$ = the length. The new width and length are $w + 2$ and $5w - 3$, respectively.

Translate. We will use the formula for the area of a rectangle, Area = (length) · (width).
$$60 = (5w - 3)(w + 2)$$

Solve.
$$60 = (5w - 3)(w + 2)$$
$$60 = 5w^2 + 7w - 6$$
$$0 = 5w^2 + 7w - 66$$
$$0 = (5w + 22)(w - 3)$$

$$5w + 22 = 0 \quad or \quad w - 3 = 0$$
$$5w = -22 \quad or \quad w = 3$$
$$w = -\frac{22}{5} \quad or \quad w = 3$$

Check. The width cannot be negative, so $-\frac{22}{5}$ cannot be a solution. If $w = 3$, then $5w = 5 \cdot 3 = 15$. The new dimensions are $w + 2$, or $3 + 2$, or 5 and $5w - 3$, or $15 - 3$, or 12, and the new area is $12 \cdot 5$, or 60. The number 3 checks.

State. The original length is 15 and the width is 3.

31. $(a + 3)^2 - 2(a + 3) - 35$

We can think of $a + 3$ as the variable in this expression. Then we find a pair of factors of -35 whose sum is -2. The numbers we want are -7 and 5.

$(a + 3)^2 - 2(a + 3) - 35 = [(a + 3) - 7][(a + 3) + 5] = (a - 4)(a + 8)$

We could also do this exercise as follows:
$$(a + 3)^2 - 2(a + 3) - 35 = a^2 + 6a + 9 - 2a - 6 - 35$$
$$= a^2 + 4a - 32$$
$$= (a - 4)(a + 8)$$

32.
$$20x(x + 2)(x - 1) = 5x^3 - 24x - 14x^2$$
$$(20x^2 + 40x)(x - 1) = 5x^3 - 24x - 14x^2$$
$$20x^3 + 20x^2 - 40x = 5x^3 - 24x - 14x^2$$
$$15x^3 + 34x^2 - 16x = 0$$
$$x(15x^2 + 34x - 16) = 0$$
$$x(3x + 8)(5x - 2) = 0$$
$$x = 0 \quad or \quad 3x + 8 = 0 \quad or \quad 5x - 2 = 0$$
$$x = 0 \quad or \quad 3x = -8 \quad or \quad 5x = 2$$
$$x = 0 \quad or \quad x = -\frac{8}{3} \quad or \quad x = \frac{2}{5}$$

33. $x^2 - y^2 = (x + y)(x - y) = 4 \cdot 6 = 24$, so choice (d) is correct.

Chapter 14

Rational Expressions and Equations

Exercise Set 14.1

1. $\dfrac{-3}{2x}$

To determine the numbers for which the rational expression is not defined, we set the denominator equal to 0 and solve:

$$2x = 0 \quad = \dfrac{0}{2} = 0$$
$$x = 0$$

The expression is not defined for the replacement number 0.

3. $\dfrac{5}{x-8}$

To determine the numbers for which the rational expression is not defined, we set the denominator equal to 0 and solve:

$$x - 8 = 0$$
$$x = 8$$

The expression is not defined for the replacement number 8.

5. $\dfrac{3}{2y+5}$ $=0 \quad -\dfrac{5}{2}$

Set the denominator equal to 0 and solve:

$$2y + 5 = 0$$
$$2y = -5$$
$$y = -\dfrac{5}{2}$$

The expression is not defined for the replacement number $-\dfrac{5}{2}$.

7. $\dfrac{x^2 + 11}{x^2 - 3x - 28}$

Set the denominator equal to 0 and solve:

$$x^2 - 3x - 28 = 0$$
$$(x - 7)(x + 4) = 0$$
$$x - 7 = 0 \quad \text{or} \quad x + 4 = 0$$
$$x = 7 \quad \text{or} \quad x = -4$$

The expression is not defined for the replacement numbers 7 and -4.

9. $\dfrac{m^3 - 2m}{m^2 - 25}$

Set the denominator equal to 0 and solve:

$$m^2 - 25 = 0$$
$$(m + 5)(m - 5) = 0$$
$$m + 5 = 0 \quad \text{or} \quad m - 5 = 0$$
$$m = -5 \quad \text{or} \quad m = 5$$

The expression is not defined for the replacement numbers -5 and 5.

11. $\dfrac{x - 4}{3}$

Since the denominator is the constant 3, there are no replacement numbers for which the expression is not defined.

13. $\dfrac{4x}{4x} \cdot \dfrac{3x^2}{5y} = \dfrac{(4x)(3x^2)}{(4x)(5y)}$ Multiplying the numerators and the denominators

15. $\dfrac{2x}{2x} \cdot \dfrac{x-1}{x+4} = \dfrac{2x(x-1)}{2x(x+4)}$ Multiplying the numerators and the denominators

17. $\dfrac{3-x}{4-x} \cdot \dfrac{-1}{-1} = \dfrac{(3-x)(-1)}{(4-x)(-1)}$, or $\dfrac{-1(3-x)}{-1(4-x)}$

19. $\dfrac{y+6}{y+6} \cdot \dfrac{y-7}{y+2} = \dfrac{(y+6)(y-7)}{(y+6)(y+2)}$

21. $\dfrac{8x^3}{32x} = \dfrac{8 \cdot x \cdot x^2}{8 \cdot 4 \cdot x}$ Factoring numerator and denominator

$\quad = \dfrac{8x}{8x} \cdot \dfrac{x^2}{4}$ Factoring the rational expression

$\quad = 1 \cdot \dfrac{x^2}{4} \quad \left(\dfrac{8x}{8x} = 1\right)$

$\quad = \dfrac{x^2}{4}$ We removed a factor of 1.

23. $\dfrac{48p^7q^5}{18p^5q^4} = \dfrac{8 \cdot 6 \cdot p^5 \cdot p^2 \cdot q^4 \cdot q}{6 \cdot 3 \cdot p^5 \cdot q^4}$ Factoring numerator and denominator

$\quad = \dfrac{6p^5q^4}{6p^5q^4} \cdot \dfrac{8p^2q}{3}$ Factoring the rational expression

$\quad = 1 \cdot \dfrac{8p^2q}{3} \quad \left(\dfrac{6p^5q^4}{6p^5q^4} = 1\right)$

$\quad = \dfrac{8p^2q}{3}$ Removing a factor of 1

25. $\dfrac{4x - 12}{4x} = \dfrac{4(x-3)}{4 \cdot x}$

$\quad = \dfrac{4}{4} \cdot \dfrac{x-3}{x}$

$\quad = 1 \cdot \dfrac{x-3}{x}$

$\quad = \dfrac{x-3}{x}$

27. $\dfrac{3m^2 + 3m}{6m^2 + 9m} = \dfrac{3m(m+1)}{3m(2m+3)}$

$\quad = \dfrac{3m}{3m} \cdot \dfrac{m+1}{2m+3}$

$\quad = 1 \cdot \dfrac{m+1}{2m+3}$

$\quad = \dfrac{m+1}{2m+3}$

29. $\dfrac{a^2 - 9}{a^2 + 5a + 6} = \dfrac{(a-3)(a+3)}{(a+2)(a+3)}$

$\phantom{\dfrac{a^2 - 9}{a^2 + 5a + 6}} = \dfrac{a-3}{a+2} \cdot \dfrac{a+3}{a+3}$

$\phantom{\dfrac{a^2 - 9}{a^2 + 5a + 6}} = \dfrac{a-3}{a+2} \cdot 1$

$\phantom{\dfrac{a^2 - 9}{a^2 + 5a + 6}} = \dfrac{a-3}{a+2}$

31. $\dfrac{a^2 - 10a + 21}{a^2 - 11a + 28} = \dfrac{(a-7)(a-3)}{(a-7)(a-4)}$

$\phantom{\dfrac{a^2 - 10a + 21}{a^2 - 11a + 28}} = \dfrac{a-7}{a-7} \cdot \dfrac{a-3}{a-4}$

$\phantom{\dfrac{a^2 - 10a + 21}{a^2 - 11a + 28}} = 1 \cdot \dfrac{a-3}{a-4}$

$\phantom{\dfrac{a^2 - 10a + 21}{a^2 - 11a + 28}} = \dfrac{a-3}{a-4}$

33. $\dfrac{x^2 - 25}{x^2 - 10x + 25} = \dfrac{(x-5)(x+5)}{(x-5)(x-5)}$

$\phantom{\dfrac{x^2 - 25}{x^2 - 10x + 25}} = \dfrac{x-5}{x-5} \cdot \dfrac{x+5}{x-5}$

$\phantom{\dfrac{x^2 - 25}{x^2 - 10x + 25}} = 1 \cdot \dfrac{x+5}{x-5}$

$\phantom{\dfrac{x^2 - 25}{x^2 - 10x + 25}} = \dfrac{x+5}{x-5}$

35. $\dfrac{a^2 - 1}{a - 1} = \dfrac{(a-1)(a+1)}{a-1}$

$\phantom{\dfrac{a^2 - 1}{a - 1}} = \dfrac{a-1}{a-1} \cdot \dfrac{a+1}{1}$

$\phantom{\dfrac{a^2 - 1}{a - 1}} = 1 \cdot \dfrac{a+1}{1}$

$\phantom{\dfrac{a^2 - 1}{a - 1}} = a+1$

37. $\dfrac{x^2 + 1}{x + 1}$ cannot be simplified.

Neither the numerator nor the denominator can be factored.

39. $\dfrac{6x^2 - 54}{4x^2 - 36} = \dfrac{2 \cdot 3(x^2 - 9)}{2 \cdot 2(x^2 - 9)}$

$\phantom{\dfrac{6x^2 - 54}{4x^2 - 36}} = \dfrac{2(x^2-9)}{2(x^2-9)} \cdot \dfrac{3}{2}$

$\phantom{\dfrac{6x^2 - 54}{4x^2 - 36}} = 1 \cdot \dfrac{3}{2}$

$\phantom{\dfrac{6x^2 - 54}{4x^2 - 36}} = \dfrac{3}{2}$

41. $\dfrac{6t + 12}{t^2 - t - 6} = \dfrac{6(t+2)}{(t-3)(t+2)}$

$\phantom{\dfrac{6t + 12}{t^2 - t - 6}} = \dfrac{6}{t-3} \cdot \dfrac{t+2}{t+2}$

$\phantom{\dfrac{6t + 12}{t^2 - t - 6}} = \dfrac{6}{t-3} \cdot 1$

$\phantom{\dfrac{6t + 12}{t^2 - t - 6}} = \dfrac{6}{t-3}$

43. $\dfrac{2t^2 + 6t + 4}{4t^2 - 12t - 16} = \dfrac{2(t^2 + 3t + 2)}{4(t^2 - 3t - 4)}$

$\phantom{\dfrac{2t^2 + 6t + 4}{4t^2 - 12t - 16}} = \dfrac{2(t+2)(t+1)}{2 \cdot 2(t-4)(t+1)}$

$\phantom{\dfrac{2t^2 + 6t + 4}{4t^2 - 12t - 16}} = \dfrac{2(t+1)}{2(t+1)} \cdot \dfrac{t+2}{2(t-4)}$

$\phantom{\dfrac{2t^2 + 6t + 4}{4t^2 - 12t - 16}} = 1 \cdot \dfrac{t+2}{2(t-4)}$

$\phantom{\dfrac{2t^2 + 6t + 4}{4t^2 - 12t - 16}} = \dfrac{t+2}{2(t-4)}$

45. $\dfrac{t^2 - 4}{(t+2)^2} = \dfrac{(t-2)(t+2)}{(t+2)(t+2)}$

$\phantom{\dfrac{t^2 - 4}{(t+2)^2}} = \dfrac{t-2}{t+2} \cdot \dfrac{t+2}{t+2}$

$\phantom{\dfrac{t^2 - 4}{(t+2)^2}} = \dfrac{t-2}{t+2} \cdot 1$

$\phantom{\dfrac{t^2 - 4}{(t+2)^2}} = \dfrac{t-2}{t+2}$

47. $\dfrac{6 - x}{x - 6} = \dfrac{-(-6 + x)}{x - 6}$

$\phantom{\dfrac{6 - x}{x - 6}} = \dfrac{-1(x - 6)}{x - 6}$

$\phantom{\dfrac{6 - x}{x - 6}} = -1 \cdot \dfrac{x - 6}{x - 6}$

$\phantom{\dfrac{6 - x}{x - 6}} = -1 \cdot 1$

$\phantom{\dfrac{6 - x}{x - 6}} = -1$

49. $\dfrac{a - b}{b - a} = \dfrac{-(-a + b)}{b - a}$

$\phantom{\dfrac{a - b}{b - a}} = \dfrac{-1(b - a)}{b - a}$

$\phantom{\dfrac{a - b}{b - a}} = -1 \cdot \dfrac{b - a}{b - a}$

$\phantom{\dfrac{a - b}{b - a}} = -1 \cdot 1$

$\phantom{\dfrac{a - b}{b - a}} = -1$

51. $\dfrac{6t - 12}{2 - t} = \dfrac{-6(-t + 2)}{2 - t}$

$\phantom{\dfrac{6t - 12}{2 - t}} = \dfrac{-6(2 - t)}{2 - t}$

$\phantom{\dfrac{6t - 12}{2 - t}} = \dfrac{-6(2 - t)}{2 - t}$

$\phantom{\dfrac{6t - 12}{2 - t}} = -6$

53. $\dfrac{x^2 - 1}{1 - x} = \dfrac{(x+1)(x-1)}{-1(-1 + x)}$

$\phantom{\dfrac{x^2 - 1}{1 - x}} = \dfrac{(x+1)(x-1)}{-1(x-1)}$

$\phantom{\dfrac{x^2 - 1}{1 - x}} = \dfrac{(x+1)(x-1)}{-1(x-1)}$

$\phantom{\dfrac{x^2 - 1}{1 - x}} = -(x+1)$

$\phantom{\dfrac{x^2 - 1}{1 - x}} = -x - 1$

55. $\dfrac{4x^3}{3x} \cdot \dfrac{14}{x} = \dfrac{4x^3 \cdot 14}{3x \cdot x}$ Multiplying the numerators and the denominators

$= \dfrac{4 \cdot x \cdot x \cdot x \cdot 14}{3 \cdot x \cdot x}$ Factoring the numerator and the denominator

$= \dfrac{4 \cdot \cancel{x} \cdot \cancel{x} \cdot x \cdot 14}{3 \cdot \cancel{x} \cdot \cancel{x}}$ Removing a factor of 1

$= \dfrac{56x}{3}$ Simplifying

57. $\dfrac{3c}{d^2} \cdot \dfrac{4d}{6c^3} = \dfrac{3c \cdot 4d}{d^2 \cdot 6c^3}$ Multiplying the numerators and the denominators

$= \dfrac{3 \cdot c \cdot 2 \cdot 2 \cdot d}{d \cdot d \cdot 3 \cdot 2 \cdot c \cdot c \cdot c}$ Factoring the numerator and the denominator

$= \dfrac{\cancel{3} \cdot \cancel{c} \cdot \cancel{2} \cdot 2 \cdot \cancel{d}}{\cancel{d} \cdot d \cdot \cancel{3} \cdot \cancel{2} \cdot \cancel{c} \cdot c \cdot c}$

$= \dfrac{2}{dc^2}$

59. $\dfrac{x^2 - 3x - 10}{(x-2)^2} \cdot \dfrac{x-2}{x-5} = \dfrac{(x^2 - 3x - 10)(x-2)}{(x-2)^2(x-5)}$

$= \dfrac{(x-5)(x+2)(x-2)}{(x-2)(x-2)(x-5)}$

$= \dfrac{\cancel{(x-5)}(x+2)\cancel{(x-2)}}{\cancel{(x-2)}(x-2)\cancel{(x-5)}}$

$= \dfrac{x+2}{x-2}$

61. $\dfrac{a^2 - 9}{a^2} \cdot \dfrac{a^2 - 3a}{a^2 + a - 12} = \dfrac{(a-3)(a+3)(a)(a-3)}{a \cdot a(a+4)(a-3)}$

$= \dfrac{(a-3)(a+3)\cancel{(a)}(a-3)}{\cancel{a} \cdot a(a+4)\cancel{(a-3)}}$

$= \dfrac{(a-3)(a+3)}{a(a+4)}$

63. $\dfrac{4a^2}{3a^2 - 12a + 12} \cdot \dfrac{3a-6}{2a} = \dfrac{4a^2(3a-6)}{(3a^2 - 12a + 12)2a}$

$= \dfrac{2 \cdot 2 \cdot a \cdot a \cdot 3 \cdot (a-2)}{3 \cdot (a-2) \cdot (a-2) \cdot 2 \cdot a}$

$= \dfrac{\cancel{2} \cdot 2 \cdot \cancel{a} \cdot a \cdot \cancel{3} \cdot \cancel{(a-2)}}{\cancel{3} \cdot \cancel{(a-2)} \cdot (a-2) \cdot \cancel{2} \cdot \cancel{a}}$

$= \dfrac{2a}{a-2}$

65. $\dfrac{t^4 - 16}{t^4 - 1} \cdot \dfrac{t^2 + 1}{t^2 + 4}$

$= \dfrac{(t^4 - 16)(t^2 + 1)}{(t^4 - 1)(t^2 + 4)}$

$= \dfrac{(t^2 + 4)(t+2)(t-2)(t^2 + 1)}{(t^2 + 1)(t+1)(t-1)(t^2 + 4)}$

$= \dfrac{\cancel{(t^2 + 4)}(t+2)(t-2)\cancel{(t^2 + 1)}}{\cancel{(t^2 + 1)}(t+1)(t-1)\cancel{(t^2 + 4)}}$

$= \dfrac{(t+2)(t-2)}{(t+1)(t-1)}$

67. $\dfrac{(x+4)^3}{(x+2)^3} \cdot \dfrac{x^2 + 4x + 4}{x^2 + 8x + 16}$

$= \dfrac{(x+4)^3(x^2 + 4x + 4)}{(x+2)^3(x^2 + 8x + 16)}$

$= \dfrac{(x+4)(x+4)(x+4)(x+2)(x+2)}{(x+2)(x+2)(x+2)(x+4)(x+4)}$

$= \dfrac{\cancel{(x+4)}\cancel{(x+4)}(x+4)\cancel{(x+2)}\cancel{(x+2)}}{\cancel{(x+2)}\cancel{(x+2)}(x+2)\cancel{(x+4)}\cancel{(x+4)}}$

$= \dfrac{x+4}{x+2}$

69. $\dfrac{5a^2 - 180}{10a^2 - 10} \cdot \dfrac{20a + 20}{2a - 12} = \dfrac{(5a^2 - 180)(20a + 20)}{(10a^2 - 10)(2a - 12)}$

$= \dfrac{5(a+6)(a-6)(2)(10)(a+1)}{10(a+1)(a-1)(2)(a-6)}$

$= \dfrac{5(a+6)\cancel{(a-6)}\cancel{(2)}\cancel{(10)}\cancel{(a+1)}}{\cancel{10}\cancel{(a+1)}(a-1)\cancel{(2)}\cancel{(a-6)}}$

$= \dfrac{5(a+6)}{a-1}$

71. Discussion and Writing Exercise

73. *Familiarize.* Let x = the smaller even integer. Then $x + 2$ = the larger even integer.

Translate. We reword the problem.

$$\underbrace{\text{Smaller even integer}}_{x} \quad \underset{\cdot}{\text{times}} \quad \underbrace{\text{larger even integer}}_{(x+2)} \quad \underset{=\ 360}{\text{is 360.}}$$

Solve.

$$x(x+2) = 360$$
$$x^2 + 2x = 360$$
$$x^2 + 2x - 360 = 0$$
$$(x+20)(x-18) = 0$$
$$x + 20 = 0 \quad \text{or} \quad x - 18 = 0$$
$$x = -20 \quad \text{or} \qquad x = 18$$

Check. The solutions of the equation are -20 and 18. When $x = -20$, then $x + 2 = -18$ and $-20(-18) = 360$. The numbers -20 and -18 are consecutive even integers which are solutions to the problem. When $x = 18$, then $x + 2 = 20$ and $18 \cdot 20 = 360$. The numbers 18 and 20 are also consecutive even integers which are solutions to the problem.

State. We have two solutions, each of which consists of a pair of numbers: -20 and -18, and 18 and 20.

75. $x^2 - x - 56$

We look for a pair of numbers whose product is -56 and whose sum is -1. The numbers are -8 and 7.

$$x^2 - x - 56 = (x - 8)(x + 7)$$

77. $x^5 - 2x^4 - 35x^3 = x^3(x^2 - 2x - 35) = x^3(x-7)(x+5)$

79. $16 - t^4 = 4^2 - (t^2)^2$ Difference of squares

$= (4 + t^2)(4 - t^2)$

$= (4 + t^2)(2^2 - t^2)$ Difference of squares

$= (4 + t^2)(2 + t)(2 - t)$

81. $x^2 - 9x + 14$

We look for a pair of numbers whose product is 14 and whose sum is -9. The numbers are -2 and -7.

$$x^2 - 9x + 14 = (x-2)(x-7)$$

83. $\quad 16x^2 - 40xy + 25y^2$

$= (4x)^2 - 2 \cdot 4x \cdot 5y + (5y)^2 \quad$ Trinomial square

$= (4x - 5y)^2$

85. $\quad \dfrac{x^4 - 16y^2}{(x^2 + 4y^2)(x - 2y)}$

$= \dfrac{(x^2 + 4y^2)(x + 2y)(x - 2y)}{(x^2 + 4y^2)(x - 2y)}$

$= \dfrac{(x^2 + 4y^2)\,(x + 2y)(x - 2y)}{(x^2 + 4y^2)\,(x - 2y)(1)}$

$= x + 2y$

87. $\quad \dfrac{t^4 - 1}{t^4 - 81} \cdot \dfrac{t^2 - 9}{t^2 + 1} \cdot \dfrac{(t - 9)^2}{(t + 1)^2}$

$= \dfrac{(t^2 + 1)(t + 1)(t - 1)(t + 3)(t - 3)(t - 9)(t - 9)}{(t^2 + 9)(t + 3)(t - 3)(t^2 + 1)(t + 1)(t + 1)}$

$= \dfrac{(t^2 + 1)(t + 1)(t - 1)(t + 3)(t - 3)(t - 9)(t - 9)}{(t^2 + 9)(t + 3)(t - 3)(t^2 + 1)(t + 1)(t + 1)}$

$= \dfrac{(t - 1)(t - 9)(t - 9)}{(t^2 + 9)(t + 1)}, \text{ or } \dfrac{(t - 1)(t - 9)^2}{(t^2 + 9)(t + 1)}$

89. $\quad \dfrac{x^2 - y^2}{(x - y)^2} \cdot \dfrac{x^2 - 2xy + y^2}{x^2 - 4xy - 5y^2}$

$= \dfrac{(x + y)(x - y)(x - y)(x - y)}{(x - y)(x - y)(x - 5y)(x + y)}$

$= \dfrac{(x + y)(x - y)(x - y)(x - y)}{(x - y)(x - y)(x - 5y)(x + y)}$

$= \dfrac{x - y}{x - 5y}$

91. $\quad \dfrac{5(2x + 5) - 25}{10} = \dfrac{10x + 25 - 25}{10}$

$= \dfrac{10x}{10}$

$= x$

You get the same number you selected.

To do a number trick, ask someone to select a number and then perform these operations. The person will probably be surprised that the result is the original number.

Exercise Set 14.2

1. The reciprocal of $\dfrac{4}{x}$ is $\dfrac{x}{4}$ because $\dfrac{4}{x} \cdot \dfrac{x}{4} = 1$.

3. The reciprocal of $x^2 - y^2$ is $\dfrac{1}{x^2 - y^2}$ because

$$\dfrac{x^2 - y^2}{1} \cdot \dfrac{1}{x^2 - y^2} = 1.$$

5. The reciprocal of $\dfrac{1}{a + b}$ is $a + b$ because $\dfrac{1}{a + b} \cdot (a + b) = 1$.

7. The reciprocal of $\dfrac{x^2 + 2x - 5}{x^2 - 4x + 7}$ is $\dfrac{x^2 - 4x + 7}{x^2 + 2x - 5}$ because

$\dfrac{x^2 + 2x - 5}{x^2 - 4x + 7} \cdot \dfrac{x^2 - 4x + 7}{x^2 + 2x - 5} = 1.$

9. $\quad \dfrac{2}{5} \div \dfrac{4}{3} = \dfrac{2}{3} \cdot \dfrac{3}{4} \quad$ Multiplying by the reciprocal of the divisor

$= \dfrac{2 \cdot 3}{5 \cdot 4}$

$= \dfrac{2 \cdot 3}{5 \cdot 2 \cdot 2} \quad$ Factoring the denominator

$= \dfrac{2 \cdot 3}{5 \cdot 2 \cdot 2} \quad$ Removing a factor of 1

$= \dfrac{3}{10} \quad$ Simplifying

11. $\quad \dfrac{2}{x} \div \dfrac{8}{x} = \dfrac{2}{x} \cdot \dfrac{x}{8} \quad$ Multiplying by the reciprocal of the divisor

$= \dfrac{2 \cdot x}{x \cdot 8}$

$= \dfrac{2 \cdot x \cdot 1}{x \cdot 2 \cdot 4} \quad$ Factoring the numerator and the denominator

$= \dfrac{2 \cdot x \cdot 1}{x \cdot 2 \cdot 4} \quad$ Removing a factor of 1

$= \dfrac{1}{4} \quad$ Simplifying

13. $\quad \dfrac{a}{b^2} \div \dfrac{a^2}{b^3} = \dfrac{a}{b^2} \cdot \dfrac{b^3}{a^2} \quad$ Multiplying by the reciprocal of the divisor

$= \dfrac{a \cdot b^3}{b^2 \cdot a^2}$

$= \dfrac{a \cdot b^2 \cdot b}{b^2 \cdot a \cdot a}$

$= \dfrac{a \cdot b^2 \cdot b}{b^2 \cdot a \cdot a}$

$= \dfrac{b}{a}$

15. $\quad \dfrac{a + 2}{a - 3} \div \dfrac{a - 1}{a + 3} = \dfrac{a + 2}{a - 3} \cdot \dfrac{a + 3}{a - 1}$

$= \dfrac{(a + 2)(a + 3)}{(a - 3)(a - 1)}$

17. $\quad \dfrac{x^2 - 1}{x} \div \dfrac{x + 1}{x - 1} = \dfrac{x^2 - 1}{x} \cdot \dfrac{x - 1}{x + 1}$

$= \dfrac{(x^2 - 1)(x - 1)}{x(x + 1)}$

$= \dfrac{(x - 1)(x + 1)(x - 1)}{x(x + 1)}$

$= \dfrac{(x - 1)(x + 1)(x - 1)}{x(x + 1)}$

$= \dfrac{(x - 1)^2}{x}$

19. $\dfrac{x+1}{6} \div \dfrac{x+1}{3} = \dfrac{x+1}{6} \cdot \dfrac{3}{x+1}$

$\qquad = \dfrac{(x+1) \cdot 3}{6(x+1)}$

$\qquad = \dfrac{3(x+1)}{2 \cdot 3(x+1)}$

$\qquad = \dfrac{1 \cdot \cancel{3}(\cancel{x+1})}{2 \cdot \cancel{3}(\cancel{x+1})}$

$\qquad = \dfrac{1}{2}$

21. $\dfrac{5x-5}{16} \div \dfrac{x-1}{6} = \dfrac{5x-5}{16} \cdot \dfrac{6}{x-1}$

$\qquad = \dfrac{(5x-5) \cdot 6}{16(x-1)}$

$\qquad = \dfrac{5(x-1) \cdot 2 \cdot 3}{2 \cdot 8(x-1)}$

$\qquad = \dfrac{5(\cancel{x-1}) \cdot \cancel{2} \cdot 3}{\cancel{2} \cdot 8(\cancel{x-1})}$

$\qquad = \dfrac{15}{8}$

23. $\dfrac{-6+3x}{5} \div \dfrac{4x-8}{25} = \dfrac{-6+3x}{5} \cdot \dfrac{25}{4x-8}$

$\qquad = \dfrac{(-6+3x) \cdot 25}{5(4x-8)}$

$\qquad = \dfrac{3(x-2) \cdot 5 \cdot 5}{5 \cdot 4(x-2)}$

$\qquad = \dfrac{3(\cancel{x-2}) \cdot \cancel{5} \cdot 5}{\cancel{5} \cdot 4(\cancel{x-2})}$

$\qquad = \dfrac{15}{4}$

25. $\dfrac{a+2}{a-1} \div \dfrac{3a+6}{a-5} = \dfrac{a+2}{a-1} \cdot \dfrac{a-5}{3a+6}$

$\qquad = \dfrac{(a+2)(a-5)}{(a-1)(3a+6)}$

$\qquad = \dfrac{(a+2)(a-5)}{(a-1) \cdot 3 \cdot (a+2)}$

$\qquad = \dfrac{(\cancel{a+2})(a-5)}{(a-1) \cdot 3 \cdot (\cancel{a+2})}$

$\qquad = \dfrac{a-5}{3(a-1)}$

27. $\dfrac{x^2-4}{x} \div \dfrac{x-2}{x+2} = \dfrac{x^2-4}{x} \cdot \dfrac{x+2}{x-2}$

$\qquad = \dfrac{(x^2-4)(x+2)}{x(x-2)}$

$\qquad = \dfrac{(x-2)(x+2)(x+2)}{x(x-2)}$

$\qquad = \dfrac{(\cancel{x-2})(x+2)(x+2)}{x(\cancel{x-2})}$

$\qquad = \dfrac{(x+2)^2}{x}$

29. $\dfrac{x^2-9}{4x+12} \div \dfrac{x-3}{6} = \dfrac{x^2-9}{4x+12} \cdot \dfrac{6}{x-3}$

$\qquad = \dfrac{(x^2-9) \cdot 6}{(4x+12)(x-3)}$

$\qquad = \dfrac{(x-3)(x+3) \cdot 3 \cdot 2}{2 \cdot 2(x+3)(x-3)}$

$\qquad = \dfrac{(\cancel{x-3})(\cancel{x+3}) \cdot 3 \cdot \cancel{2}}{\cancel{2} \cdot 2(\cancel{x+3})(\cancel{x-3})}$

$\qquad = \dfrac{3}{2}$

31. $\dfrac{c^2+3c}{c^2+2c-3} \div \dfrac{c}{c+1} = \dfrac{c^2+3c}{c^2+2c-3} \cdot \dfrac{c+1}{c}$

$\qquad = \dfrac{(c^2+3c)(c+1)}{(c^2+2c-3)c}$

$\qquad = \dfrac{c(c+3)(c+1)}{(c+3)(c-1)c}$

$\qquad = \dfrac{\cancel{c}(\cancel{c+3})(c+1)}{(\cancel{c+3})(c-1)\cancel{c}}$

$\qquad = \dfrac{c+1}{c-1}$

33. $\dfrac{2y^2-7y+3}{2y^2+3y-2} \div \dfrac{6y^2-5y+1}{3y^2+5y-2}$

$= \dfrac{2y^2-7y+3}{2y^2+3y-2} \cdot \dfrac{3y^2+5y-2}{6y^2-5y+1}$

$= \dfrac{(2y^2-7y+3)(3y^2+5y-2)}{(2y^2+3y-2)(6y^2-5y+1)}$

$= \dfrac{(2y-1)(y-3)(3y-1)(y+2)}{(2y-1)(y+2)(3y-1)(2y-1)}$

$= \dfrac{(\cancel{2y-1})(y-3)(\cancel{3y-1})(\cancel{y+2})}{(\cancel{2y-1})(\cancel{y+2})(\cancel{3y-1})(2y-1)}$

$= \dfrac{y-3}{2y-1}$

35. $\dfrac{x^2-1}{4x+4} \div \dfrac{2x^2-4x+2}{8x+8} = \dfrac{x^2-1}{4x+4} \cdot \dfrac{8x+8}{2x^2-4x+2}$

$\qquad = \dfrac{(x^2-1)(8x+8)}{(4x+4)(2x^2-4x+2)}$

$\qquad = \dfrac{(x+1)(x-1)(2)(4)(x+1)}{4(x+1)(2)(x-1)(x-1)}$

$\qquad = \dfrac{(\cancel{x+1})(\cancel{x-1})(\cancel{2})(\cancel{4})(x+1)}{\cancel{4}(\cancel{x+1})(\cancel{2})(x-1)(\cancel{x-1})}$

$\qquad = \dfrac{x+1}{x-1}$

37. Discussion and Writing Exercise

39. *Familiarize.* Let $s =$ Bonnie's score on the last test.

Translate. The average of the four scores must be at least 90. This means it must be greater than or equal to 90. We translate.

$$\dfrac{96+98+89+s}{4} \geq 90$$

Solve. We solve the inequality. First we multiply by 4 to clear the fraction.

$$4\left(\frac{96+98+89+s}{4}\right) \geq 4\cdot 90$$
$$96+98+89+s \geq 360$$
$$283+s \geq 360$$
$$s \geq 77 \qquad \text{Subtracting 283}$$

Check. We can do a partial check by substituting a value for s less than 77 and a value for s greater than 77.

For $s = 76$: $\dfrac{96+98+89+76}{4} = 89.75 < 90$

For $s = 78$: $\dfrac{96+98+89+78}{4} = 90.25 \leq 90$

Since the average is less than 90 for a value of s less than 77 and greater than or equal to 90 for a value greater than or equal to 77, the answer is probably correct.

State. The scores on the last test that will earn Bonnie an A are $\{s|s \geq 77\}$.

41. $(8x^3 - 3x^2 + 7) - (8x^2 + 3x - 5) =$
$8x^3 - 3x^2 + 7 - 8x^2 - 3x + 5 =$
$8x^3 - 11x^2 - 3x + 12$

43. $(2x^{-3}y^4)^2 = 2^2(x^{-3})^2(y^4)^2$
$= 2^2 x^{-6} y^8 \qquad$ Multiplying exponents
$= 4x^{-6}y^8 \qquad (2^2 = 4)$
$= \dfrac{4y^8}{x^6} \qquad \left(x^{-6} = \dfrac{1}{x^6}\right)$

45. $\left(\dfrac{2x^3}{y^5}\right)^2 = \dfrac{2^2(x^3)^2}{(y^5)^2}$
$= \dfrac{2^2 x^6}{y^{10}} \qquad$ Multiplying exponents
$= \dfrac{4x^6}{y^{10}} \qquad (2^2 = 4)$

47. $\dfrac{3a^2 - 5ab - 12b^2}{3ab + 4b^2} \div (3b^2 - ab)$
$= \dfrac{3a^2 - 5ab - 12b^2}{3ab + 4b^2} \cdot \dfrac{1}{3b^2 - ab}$
$= \dfrac{(3a + 4b)(a - 3b)}{b(3a + 4b) \cdot b(3b - a)}$
$= \dfrac{(3a + 4b)(-1)(3b - a)}{b(3a + 4b) \cdot b(3b - a)}$
$= \dfrac{(3a + 4b)(-1)(3b - a)}{b(3a + 4b) \cdot b(3b - a)}$
$= -\dfrac{1}{b^2}$

49. $\dfrac{a^2b^2 + 3ab^2 + 2b^2}{a^2b^4 + 4b^4} \div (5a^2 + 10a)$
$= \dfrac{a^2b^2 + 3ab^2 + 2b^2}{a^2b^4 + 4b^4} \cdot \dfrac{1}{5a^2 + 10a}$
$= \dfrac{a^2b^2 + 3ab^2 + 2b^2}{(a^2b^4 + 4b^4)(5a^2 + 10a)}$
$= \dfrac{b^2(a^2 + 3a + 2)}{b^4(a^2 + 4)(5a)(a + 2)}$
$= \dfrac{b^2(a + 1)(a + 2)}{b^2 \cdot b^2(a^2 + 4)(5a)(a + 2)}$
$= \dfrac{b^2(a + 1)(a + 2)}{b^2 \cdot b^2(a^2 + 4)(5a)(a + 2)}$
$= \dfrac{a + 1}{5ab^2(a^2 + 4)}$

Exercise Set 14.3

1. $12 = 2 \cdot 2 \cdot 3$
$27 = 3 \cdot 3 \cdot 3$
LCM $= 2 \cdot 2 \cdot 3 \cdot 3 \cdot 3$, or 108

3. $8 = 2 \cdot 2 \cdot 2$
$9 = 3 \cdot 3$
LCM $= 2 \cdot 2 \cdot 2 \cdot 3 \cdot 3$, or 72

5. $6 = 2 \cdot 3$
$9 = 3 \cdot 3$
$21 = 3 \cdot 7$
LCM $= 2 \cdot 3 \cdot 3 \cdot 7$, or 126

7. $24 = 2 \cdot 2 \cdot 2 \cdot 3$
$36 = 2 \cdot 2 \cdot 3 \cdot 3$
$40 = 2 \cdot 2 \cdot 2 \cdot 5$
LCM $= 2 \cdot 2 \cdot 2 \cdot 3 \cdot 3 \cdot 5$, or 360

9. $10 = 2 \cdot 5$
$100 = 2 \cdot 2 \cdot 5 \cdot 5$
$500 = 2 \cdot 2 \cdot 5 \cdot 5 \cdot 5$
LCM $= 2 \cdot 2 \cdot 5 \cdot 5 \cdot 5$, or 500

(We might have observed at the outset that both 10 and 100 are factors of 500, so the LCM is 500.)

11. $24 = 2 \cdot 2 \cdot 2 \cdot 3$
$18 = 2 \cdot 3 \cdot 3$
LCD $= 2 \cdot 2 \cdot 2 \cdot 3 \cdot 3$, or 72
$\dfrac{7}{24} + \dfrac{11}{18} = \dfrac{7}{2 \cdot 2 \cdot 2 \cdot 3} \cdot \dfrac{3}{3} + \dfrac{11}{2 \cdot 3 \cdot 3} \cdot \dfrac{2 \cdot 2}{2 \cdot 2}$
$= \dfrac{21}{2 \cdot 2 \cdot 2 \cdot 3 \cdot 3} + \dfrac{44}{2 \cdot 2 \cdot 2 \cdot 3 \cdot 3}$
$= \dfrac{65}{72}$

13. $\dfrac{1}{6} + \dfrac{3}{40}$

$= \dfrac{1}{2 \cdot 3} + \dfrac{3}{2 \cdot 2 \cdot 2 \cdot 5}$

LCD is $2 \cdot 2 \cdot 2 \cdot 3 \cdot 5$, or 120

$= \dfrac{1}{2 \cdot 3} \cdot \dfrac{2 \cdot 2 \cdot 5}{2 \cdot 2 \cdot 5} + \dfrac{3}{2 \cdot 2 \cdot 2 \cdot 5} \cdot \dfrac{3}{3}$

$= \dfrac{20 + 9}{2 \cdot 2 \cdot 2 \cdot 3 \cdot 5}$

$= \dfrac{29}{120}$

15. $\dfrac{1}{20} + \dfrac{1}{30} + \dfrac{2}{45}$

$= \dfrac{1}{2 \cdot 2 \cdot 5} + \dfrac{1}{2 \cdot 3 \cdot 5} + \dfrac{2}{3 \cdot 3 \cdot 5}$

LCD is $2 \cdot 2 \cdot 3 \cdot 3 \cdot 5$, or 180

$= \dfrac{1}{2 \cdot 2 \cdot 5} \cdot \dfrac{3 \cdot 3}{3 \cdot 3} + \dfrac{1}{2 \cdot 3 \cdot 5} \cdot \dfrac{2 \cdot 3}{2 \cdot 3} + \dfrac{2}{3 \cdot 3 \cdot 5} \cdot \dfrac{2 \cdot 2}{2 \cdot 2}$

$= \dfrac{9 + 6 + 8}{2 \cdot 2 \cdot 3 \cdot 3 \cdot 5}$

$= \dfrac{23}{180}$

17. $6x^2 = 2 \cdot 3 \cdot x \cdot x$

$12x^3 = 2 \cdot 2 \cdot 3 \cdot x \cdot x \cdot x$

LCM $= 2 \cdot 2 \cdot 3 \cdot x \cdot x \cdot x$, or $12x^3$

19. $2x^2 = 2 \cdot x \cdot x$

$6xy = 2 \cdot 3 \cdot x \cdot y$

$18y^2 = 2 \cdot 3 \cdot 3 \cdot y \cdot y$

LCM $= 2 \cdot 3 \cdot 3 \cdot x \cdot x \cdot y \cdot y$, or $18x^2y^2$

21. $2(y - 3) = 2 \cdot (y - 3)$

$6(y - 3) = 2 \cdot 3 \cdot (y - 3)$

LCM $= 2 \cdot 3 \cdot (y - 3)$, or $6(y - 3)$

23. $t, t + 2, t - 2$

The expressions are not factorable, so the LCM is their product:

LCM $= t(t + 2)(t - 2)$

25. $x^2 - 4 = (x + 2)(x - 2)$

$x^2 + 5x + 6 = (x + 3)(x + 2)$

LCM $= (x + 2)(x - 2)(x + 3)$

27. $t^3 + 4t^2 + 4t = t(t^2 + 4t + 4) = t(t + 2)(t + 2)$

$t^2 - 4t = t(t - 4)$

LCM $= t(t + 2)(t + 2)(t - 4) = t(t + 2)^2(t - 4)$

29. $a + 1 = a + 1$

$(a - 1)^2 = (a - 1)(a - 1)$

$a^2 - 1 = (a + 1)(a - 1)$

LCM $= (a + 1)(a - 1)(a - 1) = (a + 1)(a - 1)^2$

31. $m^2 - 5m + 6 = (m - 3)(m - 2)$

$m^2 - 4m + 4 = (m - 2)(m - 2)$

LCM $= (m - 3)(m - 2)(m - 2) = (m - 3)(m - 2)^2$

33. $2 + 3x = 2 + 3x$

$4 - 9x^2 = (2 + 3x)(2 - 3x)$

$2 - 3x = 2 - 3x$

LCM $= (2 + 3x)(2 - 3x)$

35. $10v^2 + 30v = 10v(v + 3) = 2 \cdot 5 \cdot v(v + 3)$

$5v^2 + 35v + 60 = 5(v^2 + 7v + 12)$

$\qquad\qquad = 5(v + 4)(v + 3)$

LCM $= 2 \cdot 5 \cdot v(v + 3)(v + 4) = 10v(v + 3)(v + 4)$

37. $9x^3 - 9x^2 - 18x = 9x(x^2 - x - 2)$

$\qquad\qquad = 3 \cdot 3 \cdot x(x - 2)(x + 1)$

$6x^5 - 24x^4 + 24x^3 = 6x^3(x^2 - 4x + 4)$

$\qquad\qquad = 2 \cdot 3 \cdot x \cdot x \cdot x(x - 2)(x - 2)$

LCM $= 2 \cdot 3 \cdot 3 \cdot x \cdot x \cdot x(x - 2)(x - 2)(x + 1) =$
$18x^3(x - 2)^2(x + 1)$

39. $x^5 + 4x^4 + 4x^3 = x^3(x^2 + 4x + 4)$

$\qquad\qquad = x \cdot x \cdot x(x + 2)(x + 2)$

$3x^2 - 12 = 3(x^2 - 4) = 3(x + 2)(x - 2)$

$2x + 4 = 2(x + 2)$

LCM $= 2 \cdot 3 \cdot x \cdot x \cdot x(x + 2)(x + 2)(x - 2)$

$\qquad\qquad = 6x^3(x + 2)^2(x - 2)$

41. Discussion and Writing Exercise

43. $x^2 - 6x + 9 = x^2 - 2 \cdot x \cdot 3 + 3^2 \qquad$ Trinomial square

$\qquad\qquad = (x - 3)^2$

45. $x^2 - 9 = x^2 - 3^2 \qquad$ Difference of squares

$\qquad\qquad = (x + 3)(x - 3)$

47. $x^2 + 6x + 9 = x^2 + 2 \cdot x \cdot 3 + 3^2 \qquad$ Trinomial square

$\qquad\qquad = (x + 3)^2$

49. $40x^3 = 2 \cdot 2 \cdot 2 \cdot 5 \cdot x \cdot x \cdot x$

$24x^4 = 2 \cdot 2 \cdot 2 \cdot 3 \cdot x \cdot x \cdot x \cdot x$

LCM $= 2 \cdot 2 \cdot 2 \cdot 3 \cdot 5 \cdot x \cdot x \cdot x \cdot x = 120x^4$

GCF $= 2 \cdot 2 \cdot 2 \cdot x \cdot x \cdot x = 8x^3$

$120x^4(8x^3) = 960x^7$

51. $20x^2 = 2 \cdot 2 \cdot 5 \cdot x \cdot x$

$10x = 2 \cdot 5 \cdot x$

LCM $= 2 \cdot 2 \cdot 5 \cdot x \cdot x = 20x^2$

GCF $= 2 \cdot 5 \cdot x = 10x$

$20x^2(10x) = 200x^3$

53. $10x^2 = 2 \cdot 5 \cdot x \cdot x$

$24x^3 = 2 \cdot 2 \cdot 2 \cdot 3 \cdot x \cdot x \cdot x$

LCM $= 2 \cdot 2 \cdot 2 \cdot 3 \cdot 5 \cdot x \cdot x \cdot x = 120x^3$

GCF $= 2 \cdot x \cdot x = 2x^2$

$120x^3(2x^2) = 240x^5$

55. The time it takes Pedro and Maria to meet again at the starting place is the LCM of the times it takes them to complete one round of the course.

$6 = 2 \cdot 3$

$8 = 2 \cdot 2 \cdot 2$

$\text{LCM} = 2 \cdot 2 \cdot 2 \cdot 3$, or 24

It takes 24 min.

Exercise Set 14.4

1. $\dfrac{5}{8} + \dfrac{3}{8} = \dfrac{5+3}{8} = \dfrac{8}{8} = 1$

3. $\dfrac{1}{3+x} + \dfrac{5}{3+x} = \dfrac{1+5}{3+x} = \dfrac{6}{3+x}$

5. $\dfrac{x^2 + 7x}{x^2 - 5x} + \dfrac{x^2 - 4x}{x^2 - 5x} = \dfrac{(x^2 + 7x) + (x^2 - 4x)}{x^2 - 5x}$

$\qquad = \dfrac{2x^2 + 3x}{x^2 - 5x}$

$\qquad = \dfrac{x(2x + 3)}{x(x - 5)}$

$\qquad = \dfrac{\not{x}(2x + 3)}{\not{x}(x - 5)}$

$\qquad = \dfrac{2x + 3}{x - 5}$

7. $\dfrac{2}{x} + \dfrac{5}{x^2} = \dfrac{2}{x} + \dfrac{5}{x \cdot x} \qquad \text{LCD} = x \cdot x, \text{ or } x^2$

$\qquad = \dfrac{2}{x} \cdot \dfrac{x}{x} + \dfrac{5}{x \cdot x}$

$\qquad = \dfrac{2x + 5}{x^2}$

9. $\left. \begin{array}{l} 6r = 2 \cdot 3 \cdot r \\ 8r = 2 \cdot 2 \cdot 2 \cdot r \end{array} \right\} \text{LCD} = 2 \cdot 2 \cdot 2 \cdot 3 \cdot r, \text{ or } 24r$

$\dfrac{5}{6r} + \dfrac{7}{8r} = \dfrac{5}{6r} \cdot \dfrac{4}{4} + \dfrac{7}{8r} \cdot \dfrac{3}{3}$

$\qquad = \dfrac{20 + 21}{24r}$

$\qquad = \dfrac{41}{24r}$

11. $\left. \begin{array}{l} xy^2 = x \cdot y \cdot y \\ x^2 y = x \cdot x \cdot y \end{array} \right\} \text{LCD} = x \cdot x \cdot y \cdot y, \text{ or } x^2 y^2$

$\dfrac{4}{xy^2} + \dfrac{6}{x^2 y} = \dfrac{4}{xy^2} \cdot \dfrac{x}{x} + \dfrac{6}{x^2 y} \cdot \dfrac{y}{y}$

$\qquad = \dfrac{4x + 6y}{x^2 y^2}$

13. $\left. \begin{array}{l} 9t^3 = 3 \cdot 3 \cdot t \cdot t \cdot t \\ 6t^2 = 2 \cdot 3 \cdot t \cdot t \end{array} \right\} \text{LCD} = 2 \cdot 3 \cdot 3 \cdot t \cdot t \cdot t, \text{ or } 18t^3$

$\dfrac{2}{9t^3} + \dfrac{1}{6t^2} = \dfrac{2}{9t^3} \cdot \dfrac{2}{2} + \dfrac{1}{6t^2} \cdot \dfrac{3t}{3t}$

$\qquad = \dfrac{4 + 3t}{18t^3}$

15. $\text{LCD} = x^2 y^2$ (See Exercise 11.)

$\dfrac{x + y}{xy^2} + \dfrac{3x + y}{x^2 y} = \dfrac{x + y}{xy^2} \cdot \dfrac{x}{x} + \dfrac{3x + y}{x^2 y} \cdot \dfrac{y}{y}$

$\qquad = \dfrac{x(x + y) + y(3x + y)}{x^2 y^2}$

$\qquad = \dfrac{x^2 + xy + 3xy + y^2}{x^2 y^2}$

$\qquad = \dfrac{x^2 + 4xy + y^2}{x^2 y^2}$

17. The denominators do not factor, so the LCD is their product, $(x - 2)(x + 2)$.

$\dfrac{3}{x - 2} + \dfrac{3}{x + 2} = \dfrac{3}{x - 2} \cdot \dfrac{x + 2}{x + 2} + \dfrac{3}{x + 2} \cdot \dfrac{x - 2}{x - 2}$

$\qquad = \dfrac{3(x + 2) + 3(x - 2)}{(x - 2)(x + 2)}$

$\qquad = \dfrac{3x + 6 + 3x - 6}{(x - 2)(x + 2)}$

$\qquad = \dfrac{6x}{(x - 2)(x + 2)}$

19. $\left. \begin{array}{l} 3x = 3 \cdot x \\ x + 1 = x + 1 \end{array} \right\} \text{LCD} = 3x(x + 1)$

$\dfrac{3}{x + 1} + \dfrac{2}{3x} = \dfrac{3}{x + 1} \cdot \dfrac{3x}{3x} + \dfrac{2}{3x} \cdot \dfrac{x + 1}{x + 1}$

$\qquad = \dfrac{9x + 2(x + 1)}{3x(x + 1)}$

$\qquad = \dfrac{9x + 2x + 2}{3x(x + 1)}$

$\qquad = \dfrac{11x + 2}{3x(x + 1)}$

21. $\left. \begin{array}{l} x^2 - 16 = (x + 4)(x - 4) \\ x - 4 = x - 4 \end{array} \right\} \text{LCD} = (x + 4)(x - 4)$

$\dfrac{2x}{x^2 - 16} + \dfrac{x}{x - 4} = \dfrac{2x}{(x + 4)(x - 4)} + \dfrac{x}{x - 4} \cdot \dfrac{x + 4}{x + 4}$

$\qquad = \dfrac{2x + x(x + 4)}{(x + 4)(x - 4)}$

$\qquad = \dfrac{2x + x^2 + 4x}{(x + 4)(x - 4)}$

$\qquad = \dfrac{x^2 + 6x}{(x + 4)(x - 4)}$

23. $\dfrac{5}{z + 4} + \dfrac{3}{3z + 12} = \dfrac{5}{z + 4} + \dfrac{3}{3(z + 4)} \qquad \text{LCD} = 3(z + 4)$

$\qquad = \dfrac{5}{z + 4} \cdot \dfrac{3}{3} + \dfrac{3}{3(z + 4)}$

$\qquad = \dfrac{15 + 3}{3(z + 4)} = \dfrac{18}{3(z + 4)}$

$\qquad = \dfrac{3 \cdot 6}{3(z + 4)} = \dfrac{\not{3} \cdot 6}{\not{3}(z + 4)}$

$\qquad = \dfrac{6}{z + 4}$

25. $\dfrac{3}{x-1} + \dfrac{2}{(x-1)^2}$ LCD $= (x-1)^2$

$= \dfrac{3}{x-1} \cdot \dfrac{x-1}{x-1} + \dfrac{2}{(x-1)^2}$

$= \dfrac{3(x-1)+2}{(x-1)^2}$

$= \dfrac{3x-3+2}{(x-1)^2}$

$= \dfrac{3x-1}{(x-1)^2}$

27. $\dfrac{4a}{5a-10} + \dfrac{3a}{10a-20} = \dfrac{4a}{5(a-2)} + \dfrac{3a}{2 \cdot 5(a-2)}$

$\qquad\qquad\qquad\qquad$ LCD $= 2 \cdot 5(a-2)$

$\qquad\qquad = \dfrac{4a}{5(a-2)} \cdot \dfrac{2}{2} + \dfrac{3a}{2 \cdot 5(a-2)}$

$\qquad\qquad = \dfrac{8a+3a}{10(a-2)}$

$\qquad\qquad = \dfrac{11a}{10(a-2)}$

29. $\dfrac{x+4}{x} + \dfrac{x}{x+4}$ LCD $= x(x+4)$

$= \dfrac{x+4}{x} \cdot \dfrac{x+4}{x+4} + \dfrac{x}{x+4} \cdot \dfrac{x}{x}$

$= \dfrac{(x+4)^2 + x^2}{x(x+4)}$

$= \dfrac{x^2 + 8x + 16 + x^2}{x(x+4)}$

$= \dfrac{2x^2 + 8x + 16}{x(x+4)}$

31. $\dfrac{4}{a^2-a-2} + \dfrac{3}{a^2+4a+3}$

$= \dfrac{4}{(a-2)(a+1)} + \dfrac{3}{(a+3)(a+1)}$

$\qquad\qquad$ LCD $= (a-2)(a+1)(a+3)$

$= \dfrac{4}{(a-2)(a+1)} \cdot \dfrac{a+3}{a+3} + \dfrac{3}{(a+3)(a+1)} \cdot \dfrac{a-2}{a-2}$

$= \dfrac{4(a+3) + 3(a-2)}{(a-2)(a+1)(a+3)}$

$= \dfrac{4a+12+3a-6}{(a-2)(a+1)(a+3)}$

$= \dfrac{7a+6}{(a-2)(a+1)(a+3)}$

33. $\dfrac{x+3}{x-5} + \dfrac{x-5}{x+3}$ LCD $= (x-5)(x+3)$

$= \dfrac{x+3}{x-5} \cdot \dfrac{x+3}{x+3} + \dfrac{x-5}{x+3} \cdot \dfrac{x-5}{x-5}$

$= \dfrac{(x+3)^2 + (x-5)^2}{(x-5)(x+3)}$

$= \dfrac{x^2 + 6x + 9 + x^2 - 10x + 25}{(x-5)(x+3)}$

$= \dfrac{2x^2 - 4x + 34}{(x-5)(x+3)}$

35. $\dfrac{a}{a^2-1} + \dfrac{2a}{a^2-a}$

$= \dfrac{a}{(a+1)(a-1)} + \dfrac{2a}{a(a-1)}$

$\qquad\qquad$ LCD $= a(a+1)(a-1)$

$= \dfrac{a}{(a+1)(a-1)} \cdot \dfrac{a}{a} + \dfrac{2a}{a(a-1)} \cdot \dfrac{a+1}{a+1}$

$= \dfrac{a^2 + 2a(a+1)}{a(a+1)(a-1)} = \dfrac{a^2 + 2a^2 + 2a}{a(a+1)(a-1)}$

$= \dfrac{3a^2 + 2a}{a(a+1)(a-1)} = \dfrac{a(3a+2)}{a(a+1)(a-1)}$

$= \dfrac{\not{a}(3a+2)}{\not{a}(a+1)(a-1)} = \dfrac{3a+2}{(a+1)(a-1)}$

37. $\dfrac{7}{8} + \dfrac{5}{-8} = \dfrac{7}{8} + \dfrac{5}{-8} \cdot \dfrac{-1}{-1}$

$\qquad = \dfrac{7}{8} + \dfrac{-5}{8}$

$\qquad = \dfrac{7+(-5)}{8}$

$\qquad = \dfrac{2}{8} = \dfrac{\not{2} \cdot 1}{4 \cdot \not{2}}$

$\qquad = \dfrac{1}{4}$

39. $\dfrac{3}{t} + \dfrac{4}{-t} = \dfrac{3}{t} + \dfrac{4}{-t} \cdot \dfrac{-1}{-1}$

$\qquad = \dfrac{3}{t} + \dfrac{-4}{t}$

$\qquad = \dfrac{3+(-4)}{t}$

$\qquad = \dfrac{-1}{t}$

$\qquad = -\dfrac{1}{t}$

41. $\dfrac{2x+7}{x-6} + \dfrac{3x}{6-x} = \dfrac{2x+7}{x-6} + \dfrac{3x}{6-x} \cdot \dfrac{-1}{-1}$

$\qquad = \dfrac{2x+7}{x-6} + \dfrac{-3x}{x-6}$

$\qquad = \dfrac{(2x+7) + (-3x)}{x-6}$

$\qquad = \dfrac{-x+7}{x-6}$

43. $\dfrac{y^2}{y-3} + \dfrac{9}{3-y} = \dfrac{y^2}{y-3} + \dfrac{9}{3-y} \cdot \dfrac{-1}{-1}$

$= \dfrac{y^2}{y-3} + \dfrac{-9}{y-3}$

$= \dfrac{y^2 + (-9)}{y-3}$

$= \dfrac{y^2 - 9}{y-3}$

$= \dfrac{(y+3)(y-3)}{y-3}$

$= \dfrac{(y+3)\cancel{(y-3)}}{1\cancel{(y-3)}}$

$= y+3$

45. $\dfrac{b-7}{b^2-16} + \dfrac{7-b}{16-b^2} = \dfrac{b-7}{b^2-16} + \dfrac{7-b}{16-b^2} \cdot \dfrac{-1}{-1}$

$= \dfrac{b-7}{b^2-16} + \dfrac{b-7}{b^2-16}$

$= \dfrac{(b-7)+(b-7)}{b^2-16}$

$= \dfrac{2b-14}{b^2-16}$

47. $\dfrac{a^2}{a-b} + \dfrac{b^2}{b-a} = \dfrac{a^2}{a-b} + \dfrac{b^2}{b-a} \cdot \dfrac{-1}{-1}$

$= \dfrac{a^2}{a-b} + \dfrac{-b^2}{a-b}$

$= \dfrac{a^2 + (-b^2)}{a-b}$

$= \dfrac{a^2 - b^2}{a-b}$

$= \dfrac{(a+b)(a-b)}{a-b}$

$= \dfrac{(a+b)\cancel{(a-b)}}{1\cancel{(a-b)}}$

$= a+b$

49. $\dfrac{x+3}{x-5} + \dfrac{2x-1}{5-x} + \dfrac{2(3x-1)}{x-5}$

$= \dfrac{x+3}{x-5} + \dfrac{2x-1}{5-x} \cdot \dfrac{-1}{-1} + \dfrac{2(3x-1)}{x-5}$

$= \dfrac{x+3}{x-5} + \dfrac{1-2x}{x-5} + \dfrac{2(3x-1)}{x-5}$

$= \dfrac{(x+3)+(1-2x)+(6x-2)}{x-5}$

$= \dfrac{5x+2}{x-5}$

51. $\dfrac{2(4x+1)}{5x-7} + \dfrac{3(x-2)}{7-5x} + \dfrac{-10x-1}{5x-7}$

$= \dfrac{2(4x+1)}{5x-7} + \dfrac{3(x-2)}{7-5x} \cdot \dfrac{-1}{-1} + \dfrac{-10x-1}{5x-7}$

$= \dfrac{2(4x+1)}{5x-7} + \dfrac{-3(x-2)}{5x-7} + \dfrac{-10x-1}{5x-7}$

$= \dfrac{(8x+2)+(-3x+6)+(-10x-1)}{5x-7}$

$= \dfrac{-5x+7}{5x-7}$

$= \dfrac{-1(5x-7)}{5x-7}$

$= \dfrac{-1\cancel{(5x-7)}}{\cancel{5x-7}}$

$= -1$

53. $\dfrac{x+1}{(x+3)(x-3)} + \dfrac{4(x-3)}{(x-3)(x+3)} + \dfrac{(x-1)(x-3)}{(3-x)(x+3)}$

$= \dfrac{x+1}{(x+3)(x-3)} + \dfrac{4(x-3)}{(x-3)(x+3)} + \dfrac{(x-1)(x-3)}{(3-x)(x+3)} \cdot \dfrac{-1}{-1}$

$= \dfrac{x+1}{(x+3)(x-3)} + \dfrac{4(x-3)}{(x-3)(x+3)} + \dfrac{-1(x^2-4x+3)}{(x-3)(x+3)}$

$= \dfrac{(x+1)+(4x-12)+(-x^2+4x-3)}{(x+3)(x-3)}$

$= \dfrac{-x^2+9x-14}{(x+3)(x-3)}$

55. $\dfrac{6}{x-y} + \dfrac{4x}{y^2-x^2}$

$= \dfrac{6}{x-y} + \dfrac{4x}{(y-x)(y+x)}$

$= \dfrac{6}{x-y} + \dfrac{4x}{(y-x)(y+x)} \cdot \dfrac{-1}{-1}$

$= \dfrac{6}{x-y} + \dfrac{-4x}{(x-y)(x+y)}$

$[-1(y-x) = x-y; \; y+x = x+y]$

$\text{LCD} = (x-y)(x+y)$

$= \dfrac{6}{x-y} \cdot \dfrac{x+y}{x+y} + \dfrac{-4x}{(x-y)(x+y)}$

$= \dfrac{6(x+y)-4x}{(x-y)(x+y)}$

$= \dfrac{6x+6y-4x}{(x-y)(x+y)}$

$= \dfrac{2x+6y}{(x-y)(x+y)}$

57.
$$\frac{4-a}{25-a^2} + \frac{a+1}{a-5}$$

$$= \frac{4-a}{25-a^2} \cdot \frac{-1}{-1} + \frac{a+1}{a-5}$$

$$= \frac{a-4}{a^2-25} + \frac{a+1}{a-5}$$

$$= \frac{a-4}{(a+5)(a-5)} + \frac{a+1}{a-5}$$

$$\text{LCD} = (a+5)(a-5)$$

$$= \frac{a-4}{(a+5)(a-5)} + \frac{a+1}{a-5} \cdot \frac{a+5}{a+5}$$

$$= \frac{a-4}{(a+5)(a-5)} + \frac{(a+1)(a+5)}{(a+5)(a-5)}$$

$$= \frac{(a-4) + (a+1)(a+5)}{(a+5)(a-5)}$$

$$= \frac{a-4+a^2+6a+5}{(a+5)(a-5)}$$

$$= \frac{a^2+7a+1}{(a+5)(a-5)}$$

59.
$$\frac{2}{t^2+t-6} + \frac{3}{t^2-9}$$

$$= \frac{2}{(t+3)(t-2)} + \frac{3}{(t+3)(t-3)}$$

$$\text{LCD} = (t+3)(t-2)(t-3)$$

$$= \frac{2}{(t+3)(t-2)} \cdot \frac{t-3}{t-3} + \frac{3}{(t+3)(t-3)} \cdot \frac{t-2}{t-2}$$

$$= \frac{2(t-3) + 3(t-2)}{(t+3)(t-2)(t-3)}$$

$$= \frac{2t-6+3t-6}{(t+3)(t-2)(t-3)}$$

$$= \frac{5t-12}{(t+3)(t-2)(t-3)}$$

61. Discussion and Writing Exercise

63. $(x^2+x) - (x+1) = x^2+x-x-1 = x^2-1$

65. $(2x^4y^3)^{-3} = \dfrac{1}{(2x^4y^3)^3} = \dfrac{1}{2^3(x^4)^3(y^3)^3} = \dfrac{1}{8x^{12}y^9}$

67. $\left(\dfrac{x^{-4}}{y^7}\right)^3 = \dfrac{(x^{-4})^3}{(y^7)^3} = \dfrac{x^{-12}}{y^{21}} = \dfrac{1}{x^{12}y^{21}}$

69. $y = \dfrac{1}{2}x - 5 = \dfrac{1}{2}x + (-5)$

The y-intercept is $(0, -5)$. We find two other pairs.

When $x = 2$, $y = \dfrac{1}{2} \cdot 2 - 5 = 1 - 5 = -4$.

When $x = 4$, $y = \dfrac{1}{2} \cdot 4 - 5 = 2 - 5 = -3$.

x	y
0	-5
2	-4
4	-3

Plot these points, draw the line they determine, and label the graph $y = \dfrac{1}{2}x - 5$.

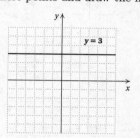

71. $y = 3$

Any ordered pair $(x, 3)$ is a solution. The variable y must be 3, but x can be any number we choose. A few solutions are listed below. Plot these points and draw the line.

x	y
-4	3
0	3
3	3

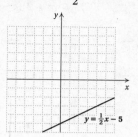

73.
$$3x - 7 = 5x + 9$$
$$-2x - 7 = 9 \qquad \text{Subtracting } 5x$$
$$-2x = 16 \qquad \text{Adding } 7$$
$$x = -8 \qquad \text{Dividing by } -2$$

The solution is -8.

75.
$$x^2 - 8x + 15 = 0$$
$$(x-3)(x-5) = 0$$
$$x - 3 = 0 \quad \text{or} \quad x - 5 = 0 \quad \text{Principle of zero products}$$
$$x = 3 \quad \text{or} \qquad x = 5$$

The solutions are 3 and 5.

77. To find the perimeter we add the lengths of the sides:

$$\frac{y+4}{3} + \frac{y+4}{3} + \frac{y-2}{5} + \frac{y-2}{5} \qquad \text{LCD} = 3 \cdot 5$$

$$= \frac{y+4}{3} \cdot \frac{5}{5} + \frac{y+4}{3} \cdot \frac{5}{5} + \frac{y-2}{5} \cdot \frac{3}{3} + \frac{y-2}{5} \cdot \frac{3}{3}$$

$$= \frac{5y+20+5y+20+3y-6+3y-6}{3 \cdot 5}$$

$$= \frac{16y+28}{15}$$

To find the area we multiply the length and the width:

$$\left(\frac{y+4}{3}\right)\left(\frac{y-2}{5}\right) = \frac{(y+4)(y-2)}{3 \cdot 5} = \frac{y^2+2y-8}{15}$$

79.
$$\frac{5}{z+2} + \frac{4z}{z^2-4} + 2$$

$$= \frac{5}{z+2} + \frac{4z}{(z+2)(z-2)} + \frac{2}{1} \quad \text{LCD} = (z+2)(z-2)$$

$$= \frac{5}{z+2} \cdot \frac{z-2}{z-2} + \frac{4z}{(z+2)(z-2)} + \frac{2}{1} \cdot \frac{(z+2)(z-2)}{(z+2)(z-2)}$$

$$= \frac{5z-10+4z+2(z^2-4)}{(z+2)(z-2)}$$

$$= \frac{5z-10+4z+2z^2-8}{(z+2)(z-2)} = \frac{2z^2+9z-18}{(z+2)(z-2)}$$

$$= \frac{(2z-3)(z+6)}{(z+2)(z-2)}$$

81.
$$\frac{3z^2}{z^4-4} + \frac{5z^2-3}{2z^4+z^2-6}$$

$$= \frac{3z^2}{(z^2+2)(z^2-2)} + \frac{5z^2-3}{(2z^2-3)(z^2+2)}$$

$$\text{LCD} = (z^2+2)(z^2-2)(2z^2-3)$$

$$= \frac{3z^2}{(z^2+2)(z^2-2)} \cdot \frac{2z^2-3}{2z^2-3} +$$

$$\qquad \frac{5z^2-3}{(2z^2-3)(z^2+2)} \cdot \frac{z^2-2}{z^2-2}$$

$$= \frac{6z^4-9z^2+5z^4-13z^2+6}{(z^2+2)(z^2-2)(2z^2-3)}$$

$$= \frac{11z^4-22z^2+6}{(z^2+2)(z^2-2)(2z^2-3)}$$

83.-85. Left to the student

Exercise Set 14.5

1. $\dfrac{7}{x} - \dfrac{3}{x} = \dfrac{7-3}{x} = \dfrac{4}{x}$

3. $\dfrac{y}{y-4} - \dfrac{4}{y-4} = \dfrac{y-4}{y-4} = 1$

5.
$$\frac{2x-3}{x^2+3x-4} - \frac{x-7}{x^2+3x-4}$$

$$= \frac{2x-3-(x-7)}{x^2+3x-4}$$

$$= \frac{2x-3-x+7}{x^2+3x-4}$$

$$= \frac{x+4}{x^2+3x-4}$$

$$= \frac{x+4}{(x+4)(x-1)}$$

$$= \frac{\cancel{(x+4)} \cdot 1}{\cancel{(x+4)}(x-1)}$$

$$= \frac{1}{x-1}$$

7. $\dfrac{a-2}{10} - \dfrac{a+1}{5} = \dfrac{a-2}{10} - \dfrac{a+1}{5} \cdot \dfrac{2}{2} \qquad \text{LCD} = 10$

$$= \frac{a-2}{10} - \frac{2(a+1)}{10}$$

$$= \frac{(a-2)-2(a+1)}{10}$$

$$= \frac{a-2-2a-2}{10}$$

$$= \frac{-a-4}{10}$$

9. $\dfrac{4z-9}{3z} - \dfrac{3z-8}{4z} = \dfrac{4z-9}{3z} \cdot \dfrac{4}{4} - \dfrac{3z-8}{4z} \cdot \dfrac{3}{3}$

$$\text{LCD} = 3 \cdot 4 \cdot z, \text{ or } 12z$$

$$= \frac{16z-36}{12z} - \frac{9z-24}{12z}$$

$$= \frac{16z-36-(9z-24)}{12z}$$

$$= \frac{16z-36-9z+24}{12z}$$

$$= \frac{7z-12}{12z}$$

11. $\dfrac{4x+2t}{3xt^2} - \dfrac{5x-3t}{x^2t} \qquad \text{LCD} = 3x^2t^2$

$$= \frac{4x+2t}{3xt^2} \cdot \frac{x}{x} - \frac{5x-3t}{x^2t} \cdot \frac{3t}{3t}$$

$$= \frac{4x^2+2tx}{3x^2t^2} - \frac{15xt-9t^2}{3x^2t^2}$$

$$= \frac{4x^2+2tx-(15xt-9t^2)}{3x^2t^2}$$

$$= \frac{4x^2+2tx-15xt+9t^2}{3x^2t^2}$$

$$= \frac{4x^2-13xt+9t^2}{3x^2t^2}$$

13. $\dfrac{5}{x+5} - \dfrac{3}{x-5} \qquad \text{LCD} = (x+5)(x-5)$

$$= \frac{5}{x+5} \cdot \frac{x-5}{x-5} - \frac{3}{x-5} \cdot \frac{x+5}{x+5}$$

$$= \frac{5x-25}{(x+5)(x-5)} - \frac{3x+15}{(x+5)(x-5)}$$

$$= \frac{5x-25-(3x+15)}{(x+5)(x-5)}$$

$$= \frac{5x-25-3x-15}{(x+5)(x-5)}$$

$$= \frac{2x-40}{(x+5)(x-5)}$$

15. $\dfrac{3}{2t^2 - 2t} - \dfrac{5}{2t - 2}$

$= \dfrac{3}{2t(t-1)} - \dfrac{5}{2(t-1)} \qquad \text{LCD} = 2t(t-1)$

$= \dfrac{3}{2t(t-1)} - \dfrac{5}{2(t-1)} \cdot \dfrac{t}{t}$

$= \dfrac{3}{2t(t-1)} - \dfrac{5t}{2t(t-1)}$

$= \dfrac{3 - 5t}{2t(t-1)}$

17. $\dfrac{2s}{t^2 - s^2} - \dfrac{s}{t - s} \qquad \text{LCD} = (t-s)(t+s)$

$= \dfrac{2s}{(t-s)(t+s)} - \dfrac{s}{t-s} \cdot \dfrac{t+s}{t+s}$

$= \dfrac{2s}{(t-s)(t+s)} - \dfrac{st + s^2}{(t-s)(t+s)}$

$= \dfrac{2s - (st + s^2)}{(t-s)(t+s)}$

$= \dfrac{2s - st - s^2}{(t-s)(t+s)}$

19. $\dfrac{y-5}{y} - \dfrac{3y-1}{4y} = \dfrac{y-5}{y} \cdot \dfrac{4}{4} - \dfrac{3y-1}{4y} \qquad \text{LCD} = 4y$

$= \dfrac{4y - 20}{4y} - \dfrac{3y-1}{4y}$

$= \dfrac{4y - 20 - (3y - 1)}{4y}$

$= \dfrac{4y - 20 - 3y + 1}{4y}$

$= \dfrac{y - 19}{4y}$

21. $\dfrac{a}{x+a} - \dfrac{a}{x-a} \qquad \text{LCD} = (x+a)(x-a)$

$= \dfrac{a}{x+a} \cdot \dfrac{x-a}{x-a} - \dfrac{a}{x-a} \cdot \dfrac{x+a}{x+a}$

$= \dfrac{ax - a^2}{(x+a)(x-a)} - \dfrac{ax + a^2}{(x+a)(x-a)}$

$= \dfrac{ax - a^2 - (ax + a^2)}{(x+a)(x-a)}$

$= \dfrac{ax - a^2 - ax - a^2}{(x+a)(x-a)}$

$= \dfrac{-2a^2}{(x+a)(x-a)}$

23. $\dfrac{11}{6} - \dfrac{5}{-6} = \dfrac{11}{6} - \dfrac{5}{-6} \cdot \dfrac{-1}{-1}$

$= \dfrac{11}{6} - \dfrac{-5}{6}$

$= \dfrac{11 - (-5)}{6}$

$= \dfrac{11 + 5}{6}$

$= \dfrac{16}{6}$

$= \dfrac{8}{3}$

25. $\dfrac{5}{a} - \dfrac{8}{-a} = \dfrac{5}{a} - \dfrac{8}{-a} \cdot \dfrac{-1}{-1}$

$= \dfrac{5}{a} - \dfrac{-8}{a}$

$= \dfrac{5 - (-8)}{a}$

$= \dfrac{5 + 8}{a}$

$= \dfrac{13}{a}$

27. $\dfrac{4}{y-1} - \dfrac{4}{1-y} = \dfrac{4}{y-1} - \dfrac{4}{1-y} \cdot \dfrac{-1}{-1}$

$= \dfrac{4}{y-1} - \dfrac{4(-1)}{(1-y)(-1)}$

$= \dfrac{4}{y-1} - \dfrac{-4}{y-1}$

$= \dfrac{4 - (-4)}{y-1}$

$= \dfrac{4 + 4}{y-1}$

$= \dfrac{8}{y-1}$

29. $\dfrac{3-x}{x-7} - \dfrac{2x-5}{7-x} = \dfrac{3-x}{x-7} - \dfrac{2x-5}{7-x} \cdot \dfrac{-1}{-1}$

$= \dfrac{3-x}{x-7} - \dfrac{(2x-5)(-1)}{(7-x)(-1)}$

$= \dfrac{3-x}{x-7} - \dfrac{5-2x}{x-7}$

$= \dfrac{(3-x) - (5-2x)}{x-7}$

$= \dfrac{3 - x - 5 + 2x}{x-7}$

$= \dfrac{x-2}{x-7}$

31. $\dfrac{a-2}{a^2-25} - \dfrac{6-a}{25-a^2} = \dfrac{a-2}{a^2-25} - \dfrac{6-a}{25-a^2} \cdot \dfrac{-1}{-1}$

$= \dfrac{a-2}{a^2-25} - \dfrac{(6-a)(-1)}{(25-a^2)(-1)}$

$= \dfrac{a-2}{a^2-25} - \dfrac{a-6}{a^2-25}$

$= \dfrac{(a-2)-(a-6)}{a^2-25}$

$= \dfrac{a-2-a+6}{a^2-25}$

$= \dfrac{4}{a^2-25}$

33. $\dfrac{4-x}{x-9} - \dfrac{3x-8}{9-x} = \dfrac{4-x}{x-9} - \dfrac{3x-8}{9-x} \cdot \dfrac{-1}{-1}$

$= \dfrac{4-x}{x-9} - \dfrac{8-3x}{x-9}$

$= \dfrac{(4-x)-(8-3x)}{x-9}$

$= \dfrac{4-x-8+3x}{x-9}$

$= \dfrac{2x-4}{x-9}$

35. $\dfrac{5x}{x^2-9} - \dfrac{4}{3-x}$

$= \dfrac{5x}{(x+3)(x-3)} - \dfrac{4}{3-x}$ $\quad$ $x-3$ and $3-x$ are opposites

$= \dfrac{5x}{(x+3)(x-3)} - \dfrac{4}{3-x} \cdot \dfrac{-1}{-1}$

$= \dfrac{5x}{(x+3)(x-3)} - \dfrac{-4}{x-3}$ $\quad$ LCD $= (x+3)(x-3)$

$= \dfrac{5x}{(x+3)(x-3)} - \dfrac{-4}{x-3} \cdot \dfrac{x+3}{x+3}$

$= \dfrac{5x}{(x+3)(x-3)} - \dfrac{-4x-12}{(x+3)(x-3)}$

$= \dfrac{5x-(-4x-12)}{(x+3)(x-3)}$

$= \dfrac{5x+4x+12}{(x+3)(x-3)}$

$= \dfrac{9x+12}{(x+3)(x-3)}$

37. $\dfrac{t^2}{2t^2-2t} - \dfrac{1}{2t-2}$

$= \dfrac{t^2}{2t(t-1)} - \dfrac{1}{2(t-1)}$ $\quad$ LCD $= 2t(t-1)$

$= \dfrac{t^2}{2t(t-1)} - \dfrac{1}{2(t-1)} \cdot \dfrac{t}{t}$

$= \dfrac{t^2}{2t(t-1)} - \dfrac{t}{2t(t-1)}$

$= \dfrac{t^2-t}{2t(t-1)}$

$= \dfrac{t(t-1)}{2t(t-1)}$

$= \dfrac{\cancel{t(t-1)}(1)}{2\cancel{t(t-1)}}$

$= \dfrac{1}{2}$

39. $\dfrac{x}{x^2+5x+6} - \dfrac{2}{x^2+3x+2}$

$= \dfrac{x}{(x+3)(x+2)} - \dfrac{2}{(x+2)(x+1)}$

$\quad$ LCD $= (x+3)(x+2)(x+1)$

$= \dfrac{x}{(x+3)(x+2)} \cdot \dfrac{x+1}{x+1} - \dfrac{2}{(x+2)(x+1)} \cdot \dfrac{x+3}{x+3}$

$= \dfrac{x^2+x}{(x+3)(x+2)(x+1)} - \dfrac{2x+6}{(x+3)(x+2)(x+1)}$

$= \dfrac{x^2+x-(2x+6)}{(x+3)(x+2)(x+1)}$

$= \dfrac{x^2+x-2x-6}{(x+3)(x+2)(x+1)}$

$= \dfrac{x^2-x-6}{(x+3)(x+2)(x+1)}$

$= \dfrac{(x-3)(x+2)}{(x+3)(x+2)(x+1)}$

$= \dfrac{(x-3)\cancel{(x+2)}}{(x+3)\cancel{(x+2)}(x+1)}$

$= \dfrac{x-3}{(x+3)(x+1)}$

41. $\dfrac{3(2x+5)}{x-1} - \dfrac{3(2x-3)}{1-x} + \dfrac{6x+1}{x-1}$

$= \dfrac{3(2x+5)}{x-1} - \dfrac{3(2x-3)}{1-x} \cdot \dfrac{-1}{-1} + \dfrac{6x-1}{x-1}$

$= \dfrac{3(2x+5)}{x-1} - \dfrac{-3(2x-3)}{x-1} + \dfrac{6x-1}{x-1}$

$= \dfrac{(6x+15)-(-6x+9)+(6x-1)}{x-1}$

$= \dfrac{6x+15+6x-9+6x-1}{x-1}$

$= \dfrac{18x+5}{x-1}$

43. $\dfrac{x-y}{x^2-y^2} + \dfrac{x+y}{x^2-y^2} - \dfrac{2x}{x^2-y^2}$

$= \dfrac{x-y+x+y-2x}{x^2-y^2}$

$= \dfrac{0}{x^2-y^2}$

$= 0$

45. $\dfrac{2(x-1)}{2x-3} - \dfrac{3(x+2)}{2x-3} - \dfrac{x-1}{3-2x}$

$= \dfrac{2(x-1)}{2x-3} - \dfrac{3(x+2)}{2x-3} - \dfrac{x-1}{3-2x} \cdot \dfrac{-1}{-1}$

$= \dfrac{2(x-1)}{2x-3} - \dfrac{3(x+2)}{2x-3} - \dfrac{1-x}{2x-3}$

$= \dfrac{(2x-2)-(3x+6)-(1-x)}{2x-3}$

$= \dfrac{2x-2-3x-6-1+x}{2x-3}$

$= \dfrac{-9}{2x-3}$

47. $\dfrac{10}{2y-1} - \dfrac{6}{1-2y} + \dfrac{y}{2y-1} + \dfrac{y-4}{1-2y}$

$= \dfrac{10}{2y-1} - \dfrac{6}{1-2y} \cdot \dfrac{-1}{-1} + \dfrac{y}{2y-1} + \dfrac{y-4}{1-2y} \cdot \dfrac{-1}{-1}$

$= \dfrac{10}{2y-1} - \dfrac{-6}{2y-1} + \dfrac{y}{2y-1} + \dfrac{4-y}{2y-1}$

$= \dfrac{10-(-6)+y+4-y}{2y-1}$

$= \dfrac{10+6+y+4-y}{2y-1}$

$= \dfrac{20}{2y-1}$

49. $\dfrac{a+6}{4-a^2} - \dfrac{a+3}{a+2} + \dfrac{a-3}{2-a}$

$= \dfrac{a+6}{(2+a)(2-a)} - \dfrac{a+3}{2+a} + \dfrac{a-3}{2-a}$

$\qquad a+2 = 2+a; \text{ LCD} = (2+a)(2-a)$

$= \dfrac{a+6}{(2+a)(2-a)} - \dfrac{a+3}{2+a} \cdot \dfrac{2-a}{2-a} + \dfrac{a-3}{2-a} \cdot \dfrac{2+a}{2+a}$

$= \dfrac{(a+6)-(a+3)(2-a)+(a-3)(2+a)}{(2+a)(2-a)}$

$= \dfrac{a+6-(-a^2-a+6)+(a^2-a-6)}{(2+a)(2-a)}$

$= \dfrac{a+6+a^2+a-6+a^2-a-6}{(2+a)(2-a)}$

$= \dfrac{2a^2+a-6}{(2+a)(2-a)}$

$= \dfrac{(2a-3)(a+2)}{(2+a)(2-a)}$

$= \dfrac{(2a-3)(2+a)}{(2+a)(2-a)}$

$= \dfrac{2a-3}{2-a}$

51. $\dfrac{2z}{1-2z} + \dfrac{3z}{2z+1} - \dfrac{3}{4z^2-1}$

$= \dfrac{2z}{1-2z} \cdot \dfrac{-1}{-1} + \dfrac{3z}{2z+1} - \dfrac{3}{4z^2-1}$

$= \dfrac{-2z}{2z-1} + \dfrac{3z}{2z+1} - \dfrac{3}{(2z-1)(2z+1)}$

$\qquad\qquad \text{LCD} = (2z-1)(2z+1)$

$= \dfrac{-2z}{2z-1} \cdot \dfrac{2z+1}{2z+1} + \dfrac{3z}{2z+1} \cdot \dfrac{2z-1}{2z-1} -$

$\qquad\qquad\qquad \dfrac{3}{(2z-1)(2z+1)}$

$= \dfrac{(-4z^2-2z)+(6z^2-3z)-3}{(2z-1)(2z+1)}$

$= \dfrac{2z^2-5z-3}{(2z-1)(2z+1)}$

$= \dfrac{(z-3)(2z+1)}{(2z-1)(2z+1)}$

$= \dfrac{(z-3)(2z+1)}{(2z-1)(2z+1)}$

$= \dfrac{z-3}{2z-1}$

53.
$$\frac{1}{x+y} - \frac{1}{x-y} + \frac{2x}{x^2 - y^2}$$
$$= \frac{1}{x+y} - \frac{1}{x-y} + \frac{2x}{(x+y)(x-y)}$$
$$\text{LCD} = (x+y)(x-y)$$
$$= \frac{1}{x+y} \cdot \frac{x-y}{x-y} - \frac{1}{x-y} \cdot \frac{x+y}{x+y} \cdot \frac{x+y}{x+y} +$$
$$\frac{2x}{(x+y)(x-y)}$$
$$= \frac{x-y-(x+y)+2x}{(x+y)(x-y)}$$
$$= \frac{x-y-x-y+2x}{(x+y)(x-y)}$$
$$= \frac{2x-2y}{(x+y)(x-y)}$$
$$= \frac{2(x-y)}{(x+y)(x-y)}$$
$$= \frac{2(\cancel{x-y})}{(x+y)(\cancel{x-y})}$$
$$= \frac{2}{x+y}$$

55. Discussion and Writing Exercise

57. $\dfrac{x^8}{x^3} = x^{8-3} = x^5$

59. $(a^2 b^{-5})^{-4} = a^{2(-4)} b^{-5(-4)} = a^{-8} b^{20} = \dfrac{b^{20}}{a^8}$

61. $\dfrac{66x^2}{11x^5} = \dfrac{6 \cdot \cancel{11} \cdot \cancel{x^2}}{\cancel{11} \cdot \cancel{x^2} \cdot x^3} = \dfrac{6}{x^3}$

63. The shaded area has dimensions $x-6$ by $x-3$. Then the area is $(x-6)(x-3)$, or $x^2 - 9x + 18$.

65.
$$\frac{2x+11}{x-3} \cdot \frac{3}{x+4} + \frac{2x+1}{4+x} \cdot \frac{3}{3-x}$$
$$= \frac{6x+33}{(x-3)(x+4)} + \frac{6x+3}{(4+x)(3-x)}$$
$$= \frac{6x+33}{(x-3)(x+4)} + \frac{6x+3}{(4+x)(3-x)} \cdot \frac{-1}{-1}$$
$$= \frac{6x+33}{(x-3)(x+4)} + \frac{-6x-3}{(x+4)(x-3)}$$
$$= \frac{6x+33-6x-3}{(x-3)(x+4)}$$
$$= \frac{30}{(x-3)(x+4)}$$

67.
$$\frac{x}{x^4 - y^4} - \left(\frac{1}{x+y}\right)^2$$
$$= \frac{x}{(x^2+y^2)(x+y)(x-y)} - \frac{1}{(x+y)^2}$$
$$\text{LCD} = (x^2+y^2)(x+y)^2(x-y)$$
$$= \frac{x}{(x^2+y^2)(x+y)(x-y)} \cdot \frac{x+y}{x+y} -$$
$$\frac{1}{(x+y)^2} \cdot \frac{(x^2+y^2)(x-y)}{(x^2+y^2)(x-y)}$$
$$= \frac{x(x+y) - (x^2+y^2)(x-y)}{(x^2+y^2)(x+y)^2(x-y)}$$
$$= \frac{x^2 + xy - (x^3 - x^2 y + xy^2 - y^3)}{(x^2+y^2)(x+y)^2(x-y)}$$
$$= \frac{x^2 + xy - x^3 + x^2 y - xy^2 + y^3}{(x^2+y^2)(x+y)^2(x-y)}$$

69. Let $l =$ the length of the missing side.
$$\frac{a^2 - 5a - 9}{a-6} + \frac{a^2 - 6}{a-6} + l = 2a + 5$$
$$\frac{2a^2 - 5a - 15}{a-6} + l = 2a + 5$$
$$l = 2a + 5 - \frac{2a^2 - 5a - 15}{a-6}$$
$$l = \left(2a+5\right) \cdot \frac{a-6}{a-6} - \frac{2a^2 - 5a - 15}{a-6}$$
$$l = \frac{2a^2 - 7a - 30}{a-6} - \frac{2a^2 - 5a - 15}{a-6}$$
$$l = \frac{2a^2 - 7a - 30 - (2a^2 - 5a - 15)}{a-6}$$
$$l = \frac{2a^2 - 7a - 30 - 2a^2 + 5a + 15}{a-6}$$
$$l = \frac{-2a - 15}{a-6}$$

The length of the missing side is $\dfrac{-2a - 15}{a-6}$.

Now find the area.
$$A = \frac{1}{2} \cdot b \cdot h$$
$$A = \frac{1}{2} \left(\frac{-2a - 15}{a-6}\right) \left(\frac{a^2 - 6}{a-6}\right)$$
$$A = \frac{(-2a - 15)(a^2 - 6)}{2(a-6)^2}, \text{ or}$$
$$A = \frac{-2a^3 - 15a^2 + 12a + 90}{2a^2 - 24a + 72}$$

71.-73. Left to the student

Exercise Set 14.6

1.
$$\frac{4}{5} - \frac{2}{3} = \frac{x}{9}, \text{ LCM} = 45$$

$$45\left(\frac{4}{5} - \frac{2}{3}\right) = 45 \cdot \frac{x}{9}$$

$$45 \cdot \frac{4}{5} - 45 \cdot \frac{2}{3} = 45 \cdot \frac{x}{9}$$

$$36 - 30 = 5x$$

$$6 = 5x$$

$$\frac{6}{5} = x$$

Check:
$$\frac{4}{5} - \frac{2}{3} = \frac{x}{9}$$

$$\frac{4}{5} - \frac{2}{3} \ ? \ \frac{\frac{6}{5}}{9}$$

$$\frac{12}{15} - \frac{10}{15} \ \Big| \ \frac{6}{5} \cdot \frac{1}{9}$$

$$\frac{2}{15} \ \Big| \ \frac{2}{15} \quad \text{TRUE}$$

This checks, so the solution is $\frac{6}{5}$.

3.
$$\frac{3}{5} + \frac{1}{8} = \frac{1}{x}, \text{ LCM} = 40x$$

$$40x\left(\frac{3}{5} + \frac{1}{8}\right) = 40x \cdot \frac{1}{x}$$

$$40x \cdot \frac{3}{5} + 40x \cdot \frac{1}{8} = 40x \cdot \frac{1}{x}$$

$$24x + 5x = 40$$

$$29x = 40$$

$$x = \frac{40}{29}$$

Check:
$$\frac{3}{5} + \frac{1}{8} = \frac{1}{x}$$

$$\frac{3}{5} + \frac{1}{8} \ ? \ \frac{1}{\frac{40}{29}}$$

$$\frac{24}{40} + \frac{5}{40} \ \Big| \ 1 \cdot \frac{29}{40}$$

$$\frac{29}{40} \ \Big| \ \frac{29}{40} \quad \text{TRUE}$$

This checks, so the solution is $\frac{40}{29}$.

5.
$$\frac{3}{8} + \frac{4}{5} = \frac{x}{20}, \text{ LCM} = 40$$

$$40\left(\frac{3}{8} + \frac{4}{5}\right) = 40 \cdot \frac{x}{20}$$

$$40 \cdot \frac{3}{8} + 40 \cdot \frac{4}{5} = 40 \cdot \frac{x}{20}$$

$$15 + 32 = 2x$$

$$47 = 2x$$

$$\frac{47}{2} = x$$

Check:
$$\frac{3}{8} + \frac{4}{5} = \frac{x}{20}$$

$$\frac{3}{8} + \frac{4}{5} \ ? \ \frac{\frac{47}{2}}{20}$$

$$\frac{15}{40} + \frac{32}{40} \ \Big| \ \frac{47}{2} \cdot \frac{1}{20}$$

$$\frac{47}{40} \ \Big| \ \frac{47}{40} \quad \text{TRUE}$$

This checks, so the solution is $\frac{47}{2}$.

7.
$$\frac{1}{x} = \frac{2}{3} - \frac{5}{6}, \text{ LCM} = 6x$$

$$6x \cdot \frac{1}{x} = 6x\left(\frac{2}{3} - \frac{5}{6}\right)$$

$$6x \cdot \frac{1}{x} = 6x \cdot \frac{2}{3} - 6x \cdot \frac{5}{6}$$

$$6 = 4x - 5x$$

$$6 = -x$$

$$-6 = x$$

Check:
$$\frac{1}{x} = \frac{2}{3} - \frac{5}{6}$$

$$\frac{1}{-6} \ ? \ \frac{2}{3} - \frac{5}{6}$$

$$-\frac{1}{6} \ \Big| \ \frac{4}{6} - \frac{5}{6}$$

$$\Big| \ -\frac{1}{6} \quad \text{TRUE}$$

This checks, so the solution is -6.

9.
$$\frac{1}{6} + \frac{1}{8} = \frac{1}{t}, \text{ LCM} = 24t$$

$$24t\left(\frac{1}{6} + \frac{1}{8}\right) = 24t \cdot \frac{1}{t}$$

$$24t \cdot \frac{1}{6} + 24t \cdot \frac{1}{8} = 24t \cdot \frac{1}{t}$$

$$4t + 3t = 24$$

$$7t = 24$$

$$t = \frac{24}{7}$$

Check:

$$\frac{1}{6} + \frac{1}{8} = \frac{1}{t}$$

$$\begin{array}{c|c} \frac{1}{6} + \frac{1}{8} \ ? \ \frac{1}{24/7} & \\ \frac{4}{24} + \frac{3}{24} & 1 \cdot \frac{7}{24} \\ \frac{7}{24} & \frac{7}{24} \quad \text{TRUE} \end{array}$$

This checks, so the solution is $\frac{24}{7}$.

11. $\qquad x + \frac{4}{x} = -5$, LCM $= x$

$$x\left(x + \frac{4}{x}\right) = x(-5)$$

$$x \cdot x + x \cdot \frac{4}{x} = x(-5)$$

$$x^2 + 4 = -5x$$

$$x^2 + 5x + 4 = 0$$

$$(x + 4)(x + 1) = 0$$

$$x + 4 = 0 \quad \text{or} \quad x + 1 = 0$$

$$x = -4 \quad \text{or} \qquad x = -1$$

Check:

$$\begin{array}{c|c} x + \frac{4}{x} = -5 & \\ \hline -4 + \frac{4}{-4} \ ? \ -5 & \\ -4 - 1 & \\ -5 & \text{TRUE} \end{array} \qquad \begin{array}{c|c} x + \frac{4}{x} = -5 & \\ \hline -1 + \frac{4}{-1} \ ? \ -5 & \\ -1 - 4 & \\ -5 & \text{TRUE} \end{array}$$

Both of these check, so the two solutions are -4 and -1.

13. $\qquad \frac{x}{4} - \frac{4}{x} = 0$, LCM $= 4x$

$$4x\left(\frac{x}{4} - \frac{4}{x}\right) = 4x \cdot 0$$

$$4x \cdot \frac{x}{4} - 4x \cdot \frac{4}{x} = 4x \cdot 0$$

$$x^2 - 16 = 0$$

$$(x + 4)(x - 4) = 0$$

$$x + 4 = 0 \quad \text{or} \quad x - 4 = 0$$

$$x = -4 \quad \text{or} \qquad x = 4$$

Check:

$$\begin{array}{c|c} \frac{x}{4} - \frac{4}{x} = 0 & \\ \hline \frac{-4}{4} - \frac{4}{-4} \ ? \ 0 & \\ -1 - (-1) & \\ -1 + 1 & \\ 0 & \text{TRUE} \end{array} \qquad \begin{array}{c|c} \frac{x}{4} - \frac{4}{x} = 0 & \\ \hline \frac{4}{4} - \frac{4}{4} \ ? \ 0 & \\ 1 - 1 & \\ 0 & \text{TRUE} \end{array}$$

Both of these check, so the two solutions are -4 and 4.

15. $\qquad \frac{5}{x} = \frac{6}{x} - \frac{1}{3}$, LCM $= 3x$

$$3x \cdot \frac{5}{x} = 3x\left(\frac{6}{x} - \frac{1}{3}\right)$$

$$3x \cdot \frac{5}{x} = 3x \cdot \frac{6}{x} - 3x \cdot \frac{1}{3}$$

$$15 = 18 - x$$

$$-3 = -x$$

$$3 = x$$

Check:

$$\frac{5}{x} = \frac{6}{x} - \frac{1}{3}$$

$$\begin{array}{c|c} \frac{5}{3} \ ? \ \frac{6}{3} - \frac{1}{3} & \\ & \frac{5}{3} \quad \text{TRUE} \end{array}$$

This checks, so the solution is 3.

17. $\qquad \frac{5}{3x} + \frac{3}{x} = 1$, LCM $= 3x$

$$3x\left(\frac{5}{3x} + \frac{3}{x}\right) = 3x \cdot 1$$

$$3x \cdot \frac{5}{3x} + 3x \cdot \frac{3}{x} = 3x \cdot 1$$

$$5 + 9 = 3x$$

$$14 = 3x$$

$$\frac{14}{3} = x$$

Check:

$$\frac{5}{3x} + \frac{3}{x} = 1$$

$$\begin{array}{c|c} \frac{5}{3 \cdot (14/3)} + \frac{3}{(14/3)} \ ? \ 1 & \\ \frac{5}{14} + \frac{9}{14} & \\ \frac{14}{14} & \\ 1 & \text{TRUE} \end{array}$$

This checks, so the solution is $\frac{14}{3}$.

19. $\qquad \frac{t-2}{t+3} = \frac{3}{8}$, LCM $= 8(t+3)$

$$8(t+3)\left(\frac{t-2}{t+3}\right) = 8(t+3)\left(\frac{3}{8}\right)$$

$$8(t-2) = 3(t+3)$$

$$8t - 16 = 3t + 9$$

$$5t = 25$$

$$t = 5$$

Check:

$$\frac{t-2}{t+3} = \frac{3}{8}$$

$$\frac{5-2}{5+3} \;\overset{?}{\mid}\; \frac{3}{8}$$

$$\frac{3}{8} \;\Big|\;\; \text{TRUE}$$

This checks, so the solution is 5.

21.
$$\frac{2}{x+1} = \frac{1}{x-2}, \text{ LCM} = (x+1)(x-2)$$

$$(x+1)(x-2) \cdot \frac{2}{x+1} = (x+1)(x-2) \cdot \frac{1}{x-2}$$

$$2(x-2) = x+1$$

$$2x - 4 = x + 1$$

$$x = 5$$

This checks, so the solution is 5.

23.
$$\frac{x}{6} - \frac{x}{10} = \frac{1}{6}, \text{ LCM} = 30$$

$$30\left(\frac{x}{6} - \frac{x}{10}\right) = 30 \cdot \frac{1}{6}$$

$$30 \cdot \frac{x}{6} - 30 \cdot \frac{x}{10} = 30 \cdot \frac{1}{6}$$

$$5x - 3x = 5$$

$$2x = 5$$

$$x = \frac{5}{2}$$

This checks, so the solution is $\frac{5}{2}$.

25.
$$\frac{t+2}{5} - \frac{t-2}{4} = 1, \text{ LCM} = 20$$

$$20\left(\frac{t+2}{5} - \frac{t-2}{4}\right) = 20 \cdot 1$$

$$20\left(\frac{t+2}{5}\right) - 20\left(\frac{t-2}{4}\right) = 20 \cdot 1$$

$$4(t+2) - 5(t-2) = 20$$

$$4t + 8 - 5t + 10 = 20$$

$$-t + 18 = 20$$

$$-t = 2$$

$$t = -2$$

This checks, so the solution is -2.

27.
$$\frac{5}{x-1} = \frac{3}{x+2},$$

$$\text{LCD} = (x-1)(x+2)$$

$$(x-1)(x+2) \cdot \frac{5}{x-1} = (x-1)(x+2) \cdot \frac{3}{x+2}$$

$$5(x+2) = 3(x-1)$$

$$5x + 10 = 3x - 3$$

$$2x = -13$$

$$x = -\frac{13}{2}$$

This checks, so the solution is $-\frac{13}{2}$.

29.
$$\frac{a-3}{3a+2} = \frac{1}{5}, \text{ LCM} = 5(3a+2)$$

$$5(3a+2) \cdot \frac{a-3}{3a+2} = 5(3a+2) \cdot \frac{1}{5}$$

$$5(a-3) = 3a+2$$

$$5a - 15 = 3a + 2$$

$$2a = 17$$

$$a = \frac{17}{2}$$

This checks, so the solution is $\frac{17}{2}$.

31.
$$\frac{x-1}{x-5} = \frac{4}{x-5}, \text{ LCM} = x-5$$

$$(x-5) \cdot \frac{x-1}{x-5} = (x-5) \cdot \frac{4}{x-5}$$

$$x - 1 = 4$$

$$x = 5$$

The number 5 is not a solution because it makes a denominator zero. Thus, there is no solution.

33.
$$\frac{2}{x+3} = \frac{5}{x}, \text{ LCM} = x(x+3)$$

$$x(x+3) \cdot \frac{2}{x+3} = x(x+3) \cdot \frac{5}{x}$$

$$2x = 5(x+3)$$

$$2x = 5x + 15$$

$$-15 = 3x$$

$$-5 = x$$

This checks, so the solution is -5.

35. $\qquad \dfrac{x-2}{x-3} = \dfrac{x-1}{x+1}$, LCM $= (x-3)(x+1)$

$$(x-3)(x+1) \cdot \frac{x-2}{x-3} = (x-3)(x+1) \cdot \frac{x-1}{x+1}$$

$$(x+1)(x-2) = (x-3)(x-1)$$

$$x^2 - x - 2 = x^2 - 4x + 3$$

$$-x - 2 = -4x + 3$$

$$3x = 5$$

$$x = \frac{5}{3}$$

This checks, so the solution is $\dfrac{5}{3}$.

37. $\qquad \dfrac{1}{x+3} + \dfrac{1}{x-3} = \dfrac{1}{x^2-9}$,

$$\text{LCM} = (x+3)(x-3)$$

$$(x+3)(x-3)\left(\frac{1}{x+3} + \frac{1}{x-3}\right) = (x+3)(x-3) \cdot \frac{1}{(x+3)(x-3)}$$

$$(x-3) + (x+3) = 1$$

$$2x = 1$$

$$x = \frac{1}{2}$$

This checks, so the solution is $\dfrac{1}{2}$.

39. $\qquad \dfrac{x}{x+4} - \dfrac{4}{x-4} = \dfrac{x^2+16}{x^2-16}$,

$$\text{LCM} = (x+4)(x-4)$$

$$(x+4)(x-4)\left(\frac{x}{x+4} - \frac{x}{x-4}\right) = (x+4)(x-4) \cdot \frac{x^2+16}{(x+4)(x-4)}$$

$$x(x-4) - 4(x+4) = x^2 + 16$$

$$x^2 - 4x - 4x - 16 = x^2 + 16$$

$$x^2 - 8x - 16 = x^2 + 16$$

$$-8x - 16 = 16$$

$$-8x = 32$$

$$x = -4$$

The number -4 is not a solution because it makes a denominator zero. Thus, there is no solution.

41. $\qquad \dfrac{4-a}{8-a} = \dfrac{4}{a-8} \qquad$ $8-a$ and $a-8$ are opposites

$$\frac{4-a}{8-a} \cdot \frac{-1}{-1} = \frac{4}{a-8}$$

$$\frac{a-4}{a-8} = \frac{4}{a-8}, \text{ LCM} = a-8$$

$$(a-8)\left(\frac{a-4}{a-8}\right) = (a-8)\left(\frac{4}{a-8}\right)$$

$$a - 4 = 4$$

$$a = 8$$

The number 8 is not a solution because it makes a denominator zero. Thus, there is no solution.

43. $\qquad 2 - \dfrac{a-2}{a+3} = \dfrac{a^2-4}{a+3}$, LCM $= a+3$

$$(a+3)\left(2 - \frac{a-2}{a+3}\right) = (a+3) \cdot \frac{a^2-4}{a+3}$$

$$2(a+3) - (a-2) = a^2 - 4$$

$$2a + 6 - a + 2 = a^2 - 4$$

$$0 = a^2 - a - 12$$

$$0 = (a-4)(a+3)$$

$$a - 4 = 0 \quad \text{or} \quad a + 3 = 0$$

$$a = 4 \quad \text{or} \qquad a = -3$$

Only 4 checks, so the solution is 4.

45. $\qquad \dfrac{x+1}{x+2} = \dfrac{x+3}{x+4}$,

$$\text{LCM} = (x+2)(x+4)$$

$$(x+2)(x+4)\left(\frac{x+1}{x+2}\right) = (x+2)(x+4)\left(\frac{x+3}{x+4}\right)$$

$$(x+4)(x+1) = (x+2)(x+3)$$

$$x^2 + 5x + 4 = x^2 + 5x + 6$$

$$4 = 6 \quad \text{Subtracting } x^2 \text{ and } 5x$$

We get a false equation, so the original equation has no solution.

47. $\qquad 4a - 3 = \dfrac{a+13}{a+1}$, LCM $= a+1$

$$(a+1)(4a-3) = (a+1) \cdot \frac{a+13}{a+1}$$

$$4a^2 + a - 3 = a + 13$$

$$4a^2 - 16 = 0$$

$$4(a+2)(a-2) = 0$$

$$a + 2 = 0 \quad \text{or} \quad a - 2 = 0$$

$$a = -2 \quad \text{or} \qquad a = 2$$

Both of these check, so the two solutions are -2 and 2.

49. $\qquad \dfrac{4}{y-2} - \dfrac{2y-3}{y^2-4} = \dfrac{5}{y+2}$,

$$\text{LCM} = (y+2)(y-2)$$

$$(y+2)(y-2)\left(\frac{4}{y-2} - \frac{2y-3}{(y+2)(y-2)}\right) =$$

$$(y+2)(y-2) \cdot \frac{5}{y+2}$$

$$4(y+2) - (2y-3) = 5(y-2)$$

$$4y + 8 - 2y + 3 = 5y - 10$$

$$2y + 11 = 5y - 10$$

$$21 = 3y$$

$$7 = y$$

This checks, so the solution is 7.

51. Discussion and Writing Exercise

53. A rational expression is a _quotient_ of two polynomials.

55. Two expressions are _reciprocals_ of each other if their product is 1.

57. To find the LCM, use each factor the greatest number of times that it appears in any one factorization.

59. The quotient rule asserts that when dividing with exponential notation, if the bases are the same, keep the base and subtract the exponent of the denominator from the exponent of the numerator.

61.
$$\frac{x}{x^2 + 3x - 4} + \frac{x+1}{x^2 + 6x + 8} = \frac{2x}{x^2 + x - 2}$$
$$\frac{x}{(x+4)(x-1)} + \frac{x+1}{(x+4)(x+2)} = \frac{2x}{(x+2)(x-1)}$$
$$x(x+2) + (x+1)(x-1) = 2x(x+4)$$

Multiplying by the LCM, $(x+4)(x-1)(x+2)$
$$x^2 + 2x + x^2 - 1 = 2x^2 + 8x$$
$$2x^2 + 2x - 1 = 2x^2 + 8x$$
$$2x - 1 = 8x$$
$$-1 = 6x$$
$$-\frac{1}{6} = x$$

This checks, so the solution is $-\frac{1}{6}$.

63. Left to the student

Exercise Set 14.7

1. *Familiarize*. The job takes Mandy 4 hours working alone and Omar 5 hours working alone. Then in 1 hour Mandy does $\frac{1}{4}$ of the job and Omar does $\frac{1}{5}$ of the job. Working together, they can do $\frac{1}{4} + \frac{1}{5}$, or $\frac{9}{20}$ of the job in 1 hour. In two hours, Mandy does $2\left(\frac{1}{4}\right)$ of the job and Omar does $2\left(\frac{1}{5}\right)$ of the job. Working together they can do $2\left(\frac{1}{4}\right) + 2\left(\frac{1}{5}\right)$, or $\frac{9}{10}$ of the job in 2 hours. In 3 hours they can do $3\left(\frac{1}{4}\right) + 3\left(\frac{1}{5}\right)$, or $1\frac{7}{20}$ of the job which is more of the job then needs to be done. The answer is somewhere between 2 hr and 3 hr.

Translate. If they work together t hours, then Mandy does $t\left(\frac{1}{4}\right)$ of the job and Omar does $t\left(\frac{1}{5}\right)$ of the job. We want some number t such that
$$t\left(\frac{1}{4}\right) + t\left(\frac{1}{5}\right) = 1, \text{ or } \frac{t}{4} + \frac{t}{5} = 1.$$

Solve. We solve the equation.
$$\frac{t}{4} + \frac{t}{5} = 1, \text{ LCM} = 20$$
$$20\left(\frac{t}{4} + \frac{t}{5}\right) = 20 \cdot 1$$
$$20 \cdot \frac{t}{4} + 20 \cdot \frac{t}{5} = 20$$
$$5t + 4t = 20$$
$$9t = 20$$
$$t = \frac{20}{9}, \text{ or } 2\frac{2}{9}$$

Check. The check can be done by repeating the computations. We also have a partial check in that we expected from our familiarization step that the answer would be between 2 hr and 3 hr.

State. Working together, it takes them $2\frac{2}{9}$ hr to complete the job.

3. *Familiarize*. The job takes Vern 45 min working alone and Nina 60 min working alone. Then in 1 minute Vern does $\frac{1}{45}$ of the job and Nina does $\frac{1}{60}$ of the job. Working together, they can do $\frac{1}{45} + \frac{1}{60}$, or $\frac{7}{180}$ of the job in 1 minute. In 20 minutes, Vern does $\frac{20}{45}$ of the job and Nina does $\frac{20}{60}$ of the job. Working together, they can do $\frac{20}{45} + \frac{20}{60}$, or $\frac{7}{9}$ of the job. In 30 minutes, they can do $\frac{30}{45} + \frac{30}{60}$, or $\frac{7}{6}$ of the job which is more of the job than needs to be done. The answer is somewhere between 20 minutes and 30 minutes.

Translate. If they work together t minutes, then Vern does $t\left(\frac{1}{45}\right)$ of the job and Nina does $t\left(\frac{1}{60}\right)$ of the job. We want some number t such that
$$t\left(\frac{1}{45}\right) + t\left(\frac{1}{60}\right) = 1, \text{ or } \frac{t}{45} + \frac{t}{60} = 1.$$

Solve. We solve the equation.
$$\frac{t}{45} + \frac{t}{60} = 1, \text{ LCM} = 180$$
$$180\left(\frac{t}{45} + \frac{t}{60}\right) = 180 \cdot 1$$
$$180 \cdot \frac{t}{45} + 180 \cdot \frac{t}{60} = 180$$
$$4t + 3t = 180$$
$$7t = 180$$
$$t = \frac{180}{7}, \text{ or } 25\frac{5}{7}$$

Check. The check can be done by repeating the computations. We also have a partial check in that we expected from our familiarization step that the answer would be between 20 minutes and 30 minutes.

State. It would take them $25\frac{5}{7}$ minutes to complete the job working together.

5. Familiarize. The job takes Kenny Dewitt 9 hours working alone and Betty Wohat 7 hours working alone. Then in 1 hour Kenny does $\frac{1}{9}$ of the job and Betty does $\frac{1}{7}$ of the job. Working together they can do $\frac{1}{9} + \frac{1}{7}$, or $\frac{16}{63}$ of the job in 1 hour. In two hours, Kenny does $2\left(\frac{1}{9}\right)$ of the job and Betty does $2\left(\frac{1}{7}\right)$ of the job. Working together they can do $2\left(\frac{1}{9}\right) + 2\left(\frac{1}{7}\right)$, or $\frac{32}{63}$ of the job in two hours. In five hours they can do $5\left(\frac{1}{9}\right) + 5\left(\frac{1}{7}\right)$, or $\frac{80}{63}$, or $1\frac{17}{63}$ of the job which is more of the job than needs to be done. The answer is somewhere between 2 hr and 5 hr.

Translate. If they work together t hours, Kenny does $t\left(\frac{1}{9}\right)$ of the job and Betty does $t\left(\frac{1}{7}\right)$ of the job. We want some number t such that

$$t\left(\frac{1}{9}\right) + t\left(\frac{1}{7}\right) = 1, \text{ or } \frac{t}{9} + \frac{t}{7} = 1.$$

Solve. We solve the equation.

$$\frac{t}{9} + \frac{t}{7} = 1, \text{ LCM} = 63$$

$$63\left(\frac{t}{9} + \frac{t}{7}\right) = 63 \cdot 1$$

$$63 \cdot \frac{t}{9} + 63 \cdot \frac{t}{7} = 63$$

$$7t + 9t = 63$$

$$16t = 63$$

$$t = \frac{63}{16}, \text{ or } 3\frac{15}{16}$$

Check. The check can be done by repeating the computations. We also have a partial check in that we expected from our familiarization step that the answer would be between 2 hr and 5 hr.

State. Working together, it takes them $3\frac{15}{16}$ hr to complete the job.

7. Familiarize. Let t = the number of minutes it takes Nicole and Glen to weed the garden, working together.

Translate. We use the work principle.

$$t\left(\frac{1}{50}\right) + t\left(\frac{1}{40}\right) = 1, \text{ or } \frac{t}{50} + \frac{t}{40} = 1$$

Solve. We solve the equation.

$$\frac{t}{50} + \frac{t}{40} = 1, \text{ LCM} = 200$$

$$200\left(\frac{t}{50} + \frac{t}{40}\right) = 200 \cdot 1$$

$$200 \cdot \frac{t}{50} + 200 \cdot \frac{t}{40} = 200$$

$$4t + 5t = 200$$

$$9t = 200$$

$$t = \frac{200}{9}, \text{ or } 22\frac{2}{9}$$

Check. In $\frac{200}{9}$ min, the portion of the job done is $\frac{1}{50} \cdot \frac{200}{9} + \frac{1}{40} \cdot \frac{200}{9} = \frac{4}{9} + \frac{5}{9} = 1$. The answer checks.

State. It would take $22\frac{2}{9}$ min to weed the garden if Nicole and Glen worked together.

9. Familiarize. Let t = the number of minutes it would take the two machines to make one copy of the report, working together.

Translate. We use the work principle.

$$t\left(\frac{1}{10}\right) + t\left(\frac{1}{6}\right) = 1, \text{ or } \frac{t}{10} + \frac{t}{6} = 1$$

Solve. We solve the equation.

$$\frac{t}{10} + \frac{t}{6} = 1, \text{ LCM} = 30$$

$$30\left(\frac{t}{10} + \frac{t}{6}\right) = 30 \cdot 1$$

$$30 \cdot \frac{t}{10} + 30 \cdot \frac{t}{6} = 30$$

$$3t + 5t = 30$$

$$8t = 30$$

$$t = \frac{15}{4}, \text{ or } 3\frac{3}{4}$$

Check. In $\frac{15}{4}$ min, the portion of the job done is $\frac{1}{10} \cdot \frac{15}{4} + \frac{1}{6} \cdot \frac{15}{4} = \frac{3}{8} + \frac{5}{8} = 1$. The answer checks.

State. It would take the two machines $3\frac{3}{4}$ min to make one copy of the report, working together.

11. Familiarize. We complete the table shown in the text.

$$d = r \cdot t$$

	Distance	Speed	Time	
Car	150	r	t	$\rightarrow 150 = r(t)$
Truck	350	$r+40$	t	$\rightarrow 350 = (r+40)t$

Translate. We apply the formula $d = rt$ along the rows of the table to obtain two equations:

$$150 = rt,$$

$$350 = (r+40)t$$

Then we solve each equation for t and set the results equal:

Solving $150 = rt$ for t: $t = \dfrac{150}{r}$

Solving $350 = (r+40)t$ for t: $t = \dfrac{350}{r+40}$

Thus, we have

$$\frac{150}{r} = \frac{350}{r+40}.$$

Solve. We multiply by the LCM, $r(r + 40)$.

$$r(r + 40) \cdot \frac{150}{r} = r(r + 40) \cdot \frac{350}{r + 40}$$

$$150(r + 40) = 350r$$

$$150r + 6000 = 350r$$

$$6000 = 200r$$

$$30 = r$$

Check. If r is 30 km/h, then $r + 40$ is 70 km/h. The time for the car is 150/30, or 5 hr. The time for the truck is 350/70, or 5 hr. The times are the same. The values check.

State. The speed of Sarah's car is 30 km/h, and the speed of Rick's truck is 70 km/h.

13. *Familiarize*. We complete the table shown in the text.

$$d = r \cdot t$$

	Distance	Speed	Time
Freight	330	$r - 14$	t
Passenger	400	r	t

Translate. From the rows of the table we have two equations:

$$330 = (r - 14)t,$$

$$400 = rt$$

We solve each equation for t and set the results equal:

Solving $330 = (r - 14)t$ for t: $t = \dfrac{330}{r - 14}$

Solving $400 = rt$ for t: $t = \dfrac{400}{r}$

Thus, we have

$$\frac{330}{r - 14} = \frac{400}{r}.$$

Solve. We multiply by the LCM, $r(r - 14)$.

$$r(r - 14) \cdot \frac{330}{r - 14} = r(r - 14) \cdot \frac{400}{r}$$

$$330r = 400(r - 14)$$

$$330r = 400r - 5600$$

$$-70r = -5600$$

$$r = 80$$

Then substitute 80 for r in either equation to find t:

$$t = \frac{400}{r}$$

$$t = \frac{400}{80} \quad \text{Substituting 80 for } r$$

$$t = 5$$

Check. If $r = 80$, then $r - 14 = 66$. In 5 hr the freight train travels $66 \cdot 5$, or 330 mi, and the passenger train travels $80 \cdot 5$, or 400 mi. The values check.

State. The speed of the passenger train is 80 mph. The speed of the freight train is 66 mph.

15. *Familiarize*. We let r represent the speed going. Then $2r$ is the speed returning. We let t represent the time going. Then $t - 3$ represents the time returning. We organize the information in a table.

$$d = r \cdot t$$

	Distance	Speed	Time
Going	120	r	t
Returning	120	$2r$	$t - 3$

Translate. The rows of the table give us two equations:

$$120 = rt,$$

$$120 = 2r(t - 3)$$

We can solve each equation for r and set the results equal:

Solving $120 = rt$ for r: $r = \dfrac{120}{t}$

Solving $120 = 2r(t - 3)$ for r: $r = \dfrac{120}{2(t - 3)}$, or

$$r = \frac{60}{t - 3}$$

Then $\dfrac{120}{t} = \dfrac{60}{t - 3}$.

Solve. We multiply on both sides by the LCM, $t(t - 3)$.

$$t(t - 3) \cdot \frac{120}{t} = t(t - 3) \cdot \frac{60}{t - 3}$$

$$120(t - 3) = 60t$$

$$120t - 360 = 60t$$

$$-360 = -60t$$

$$6 = t$$

Then substitute 6 for t in either equation to find r, the speed going:

$$r = \frac{120}{t}$$

$$r = \frac{120}{6} \quad \text{Substituting 6 for } t$$

$$r = 20$$

Check. If $r = 20$ and $t = 6$, then $2r = 2 \cdot 20$, or 40 mph and $t - 3 = 6 - 3$, or 3 hr. The distance going is $6 \cdot 20$, or 120 mi. The distance returning is $40 \cdot 3$, or 120 mi. The numbers check.

State. The speed going is 20 mph.

17. *Familiarize*. Let r = Kelly's speed, in km/h, and t = the time the bicyclists travel, in hours. Organize the information in a table.

	Distance	Speed	Time
Hank	42	$r - 5$	t
Kelly	57	r	t

Translate. We can replace the t's in the table above using the formula $t = d/r$.

	Distance	Speed	Time
Hank	42	$r-5$	$\dfrac{42}{r-5}$
Kelly	57	r	$\dfrac{57}{r}$

Since the times are the same for both bicyclists, we have the equation

$$\frac{42}{r-5}=\frac{57}{r}.$$

Solve. We first multiply by the LCD, $r(r-5)$.

$$r(r-5)\cdot\frac{42}{r-5}=r(r-5)\cdot\frac{57}{r}$$
$$42r=57(r-5)$$
$$42r=57r-285$$
$$-15r=-285$$
$$r=19$$

If $r=19$, then $r-5=14$.

Check. If Hank's speed is 14 km/h and Kelly's speed is 19 km/h, then Hank bicycles 5 km/h slower than Kelly. Hank's time is 42/14, or 3 hr. Kelly's time is 57/19, or 3 hr. Since the times are the same, the answer checks.

State. Hank travels at 14 km/h, and Kelly travels at 19 km/h.

19. **Familiarize.** Let r = Ralph's speed, in km/h. Then Bonnie's speed is $r+3$. Also set t = the time, in hours, that Ralph and Bonnie walk. We organize the information in a table.

	Distance	Speed	Time
Ralph	7.5	r	t
Bonnie	12	$r+3$	t

Translate. We can replace the t's in the table shown above using the formula $t=d/r$.

	Distance	Speed	Time
Ralph	7.5	r	$\dfrac{7.5}{r}$
Bonnie	12	$r+3$	$\dfrac{12}{r+3}$

Since the times are the same for both walkers, we have the equation

$$\frac{7.5}{r}=\frac{12}{r+3}.$$

Solve. We first multiply by the LCD, $r(r+3)$.

$$r(r+3)\cdot\frac{7.5}{r}=r(r+3)\cdot\frac{12}{r+3}$$
$$7.5(r+3)=12r$$
$$7.5r+22.5=12r$$
$$22.5=4.5r$$
$$5=r$$

If $r=5$, then $r+3=8$.

Check. If Ralph's speed is 5 km/h and Bonnie's speed is 8 km/h, then Bonnie walks 3 km/h faster than Ralph. Ralph's time is 7.5/5, or 1.5 hr. Bonnie's time is 12/8, or 1.5 hr. Since the times are the same, the answer checks.

State. Ralph's speed is 5 km/h, and Bonnie's speed is 8 km/h.

21. **Familiarize.** Let t = the time it takes Caledonia to drive to town and organize the given information in a table.

	Distance	Speed	Time
Caledonia	15	r	t
Manley	20	r	$t+1$

Translate. We can replace the r's in the table above using the formula $r=d/t$.

	Distance	Speed	Time
Caledonia	15	$\dfrac{15}{t}$	t
Manley	20	$\dfrac{20}{t+1}$	$t+1$

Since the speeds are the same for both riders, we have the equation

$$\frac{15}{t}=\frac{20}{t+1}.$$

Solve. We multiply by the LCD, $t(t+1)$.

$$t(t+1)\cdot\frac{15}{t}=t(t+1)\cdot\frac{20}{t+1}$$
$$15(t+1)=20t$$
$$15t+15=20t$$
$$15=5t$$
$$3=t$$

If $t=3$, then $t+1=3+1$, or 4.

Check. If Caledonia's time is 3 hr and Manley's time is 4 hr, then Manley's time is 1 hr more than Caledonia's. Caledonia's speed is 15/3, or 5 mph. Manley's speed is 20/4, or 5 mph. Since the speeds are the same, the answer checks.

State. It takes Caledonia 3 hr to drive to town.

23. $\dfrac{10\text{ divorces}}{18\text{ marriages}}=\dfrac{10}{18}$ divorce/marriage $=$ $\dfrac{5}{9}$ divorce/marriage

25. $\dfrac{4.6\text{ km}}{2\text{ hr}}=2.3$ km/h

27. **Familiarize.** A 120-lb person should eat at least 44 g of protein each day, and we wish to find the minimum protein required for a 180-lb person. We can set up ratios. We let p = the minimum number of grams of protein a 180-lb person should eat each day.

Translate. If we assume the rates of protein intake are the same, the ratios are the same and we have an equation.

$$\text{Protein} \to \frac{44}{120} = \frac{p}{180} \leftarrow \text{Protein}$$
$$\text{Weight} \to \phantom{\frac{44}{120}} \phantom{\frac{p}{180}} \leftarrow \text{Weight}$$

Solve. We solve the proportion.

$$360 \cdot \frac{44}{120} = 360 \cdot \frac{p}{180} \quad \text{Multiplying by the LCM, 360}$$
$$3 \cdot 44 = 2 \cdot p$$
$$132 = 2p$$
$$66 = p$$

Check. $\frac{44}{120} = \frac{4 \cdot 11}{4 \cdot 30} = \frac{\cancel{4} \cdot 11}{\cancel{4} \cdot 30} = \frac{11}{30}$ and

$\frac{66}{180} = \frac{6 \cdot 11}{6 \cdot 30} = \frac{\cancel{6} \cdot 11}{\cancel{6} \cdot 30} = \frac{11}{30}$. The ratios are the same.

State. A 180-lb person should eat a minimum of 66 g of protein each day.

29. *Familiarize*. 10 cc of human blood contains 1.2 grams of hemoglobin, and we wish to find how many grams of hemoglobin are contained in 16 cc of the same blood. We can set up ratios. Let H = the amount of hemoglobin in 16 cc of the same blood.

Translate. Assuming the two ratios are the same, we can translate to a proportion.

$$\text{Grams} \to \frac{H}{16} = \frac{1.2}{10} \leftarrow \text{Grams}$$
$$\text{cm}^3 \to \phantom{\frac{H}{16}} \phantom{\frac{1.2}{10}} \leftarrow \text{cm}^3$$

Solve. We solve the proportion.

We multiply by 16 to get H alone.

$$16 \cdot \frac{H}{16} = 16 \cdot \frac{1.2}{10}$$
$$H = \frac{19.2}{10}$$
$$H = 1.92$$

Check.
$$\frac{1.92}{16} = 0.12 \qquad \frac{1.2}{10} = 0.12$$
The ratios are the same.

State. 16 cc of the same blood would contain 1.92 grams of hemoglobin.

31. *Familiarize*. Let h = the amount of honey, in pounds, that 35,000 trips to flowers would produce.

Translate. We translate to a proportion.

$$\text{Honey} \to \frac{1}{20,000} = \frac{h}{35,000} \leftarrow \text{Honey}$$
$$\text{Trips} \to \phantom{\frac{1}{20,000}} \phantom{\frac{h}{35,000}} \leftarrow \text{Trips}$$

Solve. We solve the proportion.

$$35,000 \cdot \frac{1}{20,000} = 35,000 \cdot \frac{h}{35,000}$$
$$1.75 = h$$

Check. $\frac{1}{20,000} = 0.00005$ and $\frac{1.75}{35,000} = 0.00005$.
The ratios are the same.

State. 35,000 trips to gather nectar will produce 1.75 lb of honey.

33. *Familiarize*. The ratio of the weight of copper to the weight of zinc in a U.S. penny is $\frac{1}{39}$, and we wish to find how much copper is needed if 50 kg of zinc is being turned into pennies. We can set up a second ratio to go with the one we already have. Let C = the amount of copper needed, in kg, if 50 kg of zinc is being turned into pennies.

Translate. We translate to a proportion.

$$\frac{1}{39} = \frac{C}{50}$$

Solve. We solve the proportion.

$$50 \cdot \frac{1}{39} = 50 \cdot \frac{C}{50}$$
$$\frac{50}{39} = C, \text{ or}$$
$$1\frac{11}{39} = C$$

Check. $\frac{50/39}{50} = \frac{1}{39}$, so the ratios are the same.

State. $1\frac{11}{39}$ kg of copper is needed if 50 kg of zinc is turned into pennies.

35. (a) $\frac{72}{217} \approx 0.332$

Suzuki's batting average was 0.332.

(b) Let h = the number of hits Suzuki would get in the 162-game season. We translate to a proportion and solve it.

$$\frac{72}{48} = \frac{h}{162}$$
$$162 \cdot \frac{72}{48} = 162 \cdot \frac{h}{162}$$
$$243 \approx h$$

Suzuki would get 243 hits in the 162-game season.

(c) Let h = the number of hits Suzuki would get if he batted 700 times. We translate to a proportion and solve it.

$$\frac{72}{217} = \frac{h}{700}$$
$$700 \cdot \frac{72}{217} = 700 \cdot \frac{h}{700}$$
$$232 \approx h$$

Suzuki would get 232 hits if he batted 700 times.

37. Let h = the head circumference, in inches. We translate to a proportion and solve it.

$$\frac{6\frac{3}{4}}{21\frac{1}{5}} = \frac{7}{h}$$

$$6\frac{3}{4} \cdot h = 21\frac{1}{5} \cdot 7$$
$$\frac{27}{4} \cdot h = \frac{106}{5} \cdot 7$$
$$h = \frac{4}{27} \cdot \frac{106}{5} \cdot 7$$
$$h \approx 22$$

The head circumference is 22 in.

Now let c = the head circumference, in centimeters. We translate to a proportion and solve it.

$$\frac{6\frac{3}{4}}{53.8} = \frac{7}{c}$$

$$\frac{6.75}{53.8} = \frac{7}{c} \quad \left(6\frac{3}{4} = 6.75\right)$$

$$6.75 \cdot c = 53.8 \cdot 7$$

$$c = \frac{53.8 \cdot 7}{6.75}$$

$$c \approx 55.8$$

The head circumference is 55.8 cm.

39. Let h = the hat size. We translate to a proportion and solve it.

$$\frac{6\frac{3}{4}}{21\frac{1}{5}} = \frac{h}{22\frac{4}{5}}$$

$$6\frac{3}{4} \cdot 22\frac{4}{5} = 21\frac{1}{5} \cdot h$$

$$\frac{27}{4} \cdot \frac{114}{5} = \frac{106}{5} \cdot h$$

$$\frac{5}{106} \cdot \frac{27}{4} \cdot \frac{114}{5} = h$$

$$7.26 \approx h$$

$$7\frac{1}{4} \approx h$$

The hat size is $7\frac{1}{4}$.

Now let c = the head circumference, in centimeters. We translate to a proportion and solve it. We use the hat size found above in the translation.

$$\frac{6\frac{3}{4}}{53.8} = \frac{7\frac{1}{4}}{c}$$

$$\frac{6.75}{53.8} = \frac{7.25}{c}$$

$$6.75 \cdot c = 53.8 \cdot 7.25$$

$$c = \frac{53.8 \cdot 7.25}{6.75}$$

$$c \approx 57.8$$

The head circumference is 57.8 cm. (Answers may vary slightly depending on when rounding occurs.)

41. Let h = the hat size. We translate to a proportion and solve it.

$$\frac{6\frac{3}{4}}{53.8} = \frac{h}{59.8}$$

$$\frac{6.75}{53.8} = \frac{h}{59.8}$$

$$59.8 \cdot \frac{6.75}{53.8} = h$$

$$7.5 \approx h, \text{ or}$$

$$7\frac{1}{2} \approx h$$

The hat size is $7\frac{1}{2}$.

Now let c = the head circumference, in inches. We translate to a proportion and solve it. We use the hat size found above in the translation.

$$\frac{6\frac{3}{4}}{21\frac{1}{5}} = \frac{7\frac{1}{2}}{c}$$

$$6\frac{3}{4} \cdot c = 21\frac{1}{5} \cdot 7\frac{1}{2}$$

$$\frac{27}{4} \cdot c = \frac{106}{5} \cdot \frac{15}{2}$$

$$c = \frac{4}{27} \cdot \frac{106}{5} \cdot \frac{15}{2}$$

$$c \approx 23.6, \text{ or}$$

$$c \approx 23\frac{3}{5}$$

The head circumference is $23\frac{3}{5}$ in.

43. *Familiarize.* The ratio of trout tagged to the total trout population, P, is $\frac{112}{P}$. Of the 82 trout checked later, 32 were tagged. The ratio of trout tagged to trout checked is $\frac{32}{82}$.

Translate. Assuming the two ratios are the same, we can translate to a proportion.

$$\begin{array}{ccc} \text{Trout tagged} & & \text{Tagged trout} \\ \text{originally} \longrightarrow & \frac{112}{P} = \frac{32}{82} & \longleftarrow \text{caught later} \\ \text{Trout} \longrightarrow & & \longleftarrow \text{Trout caught} \\ \text{population} & & \text{later} \end{array}$$

Solve. We solve the equation.

$$82P \cdot \frac{112}{P} = 82P \cdot \frac{32}{82} \quad \begin{array}{l} \text{Multiplying by the LCM,} \\ 82P \end{array}$$

$$82 \cdot 112 = P \cdot 32$$

$$9184 = 32P$$

$$287 = P$$

Check.

$$\frac{112}{287} \approx 0.390 \text{ and } \frac{32}{82} \approx 0.390.$$

The ratios are the same.

State. The trout population is 287.

45. *Familiarize*. A sample of 144 firecrackers contained 9 duds, and we wish to find how many duds could be expected in a sample of 3200 firecrackers. We can set up ratios, letting $d =$ the number of duds expected in a sample of 3200 firecrackers.

***Translate*.** Assuming the rates of occurrence of duds are the same, we can translate to a proportion.

$$\text{Duds} \rightarrow \frac{9}{144} = \frac{d}{3200} \leftarrow \text{Duds}$$
$$\text{Sample size} \rightarrow \qquad\qquad \leftarrow \text{Sample size}$$

***Solve*.** We solve the equation. We multiply by 3200 to get d alone.

$$3200 \cdot \frac{9}{144} = 3200 \cdot \frac{d}{3200}$$
$$\frac{28,800}{144} = d$$
$$200 = d$$

***Check*.**
$$\frac{9}{144} = 0.0625 \quad \text{and} \quad \frac{200}{3200} = 0.0625$$
The ratios are the same.

***State*.** You would expect 200 duds in a sample of 3200 firecrackers.

47. *Familiarize*. The ratio of the weight of an object on Mars to the weight of an object on earth is 0.4 to 1.

a) We wish to find how much a 12-ton rocket would weigh on Mars.

b) We wish to find how much a 120-lb astronaut would weigh on Mars.

We can set up ratios. We let $r =$ the weight of a 12-ton rocket and $a =$ the weight of a 120-lb astronaut on Mars.

***Translate*.** Assuming the ratios are the same, we can translate to proportions.

a)
$$\text{Weight on Mars} \rightarrow \frac{0.4}{1} = \frac{r}{12} \leftarrow \text{Weight on Mars}$$
$$\text{Weight on earth} \rightarrow \qquad\qquad \leftarrow \text{Weight on earth}$$

b)
$$\text{Weight on Mars} \rightarrow \frac{0.4}{1} = \frac{a}{120} \leftarrow \text{Weight on Mars}$$
$$\text{Weight on earth} \rightarrow \qquad\qquad \leftarrow \text{Weight on earth}$$

***Solve*.** We solve each proportion.

a) $\quad \dfrac{0.4}{1} = \dfrac{r}{12}$ $\qquad$ b) $\quad \dfrac{0.4}{1} = \dfrac{1}{120}$

$\quad 12(0.4) = r \qquad\qquad 120(0.4) = a$

$\qquad 4.8 = r \qquad\qquad\qquad 48 = a$

***Check*.** $\dfrac{0.4}{1} = 0.4$, $\dfrac{4.8}{12} = 0.4$, and $\dfrac{48}{120} = 0.4$. The ratios are the same.

***State*.** a) A 12-ton rocket would weigh 4.8 tons on Mars.

b) A 120-lb astronaut would weigh 48 lb on Mars.

49. We write a proportion and then solve it.

$$\frac{b}{6} = \frac{7}{4}$$
$$b = \frac{7}{4} \cdot 6 \qquad \text{Multiplying by 6}$$
$$b = \frac{42}{4}$$
$$b = \frac{21}{2}, \text{ or } 10.5$$

$\left(\text{Note that the proportions } \dfrac{6}{b} = \dfrac{4}{7}, \dfrac{b}{7} = \dfrac{6}{4}, \text{ or } \dfrac{7}{b} = \dfrac{4}{6} \text{ could also be used.}\right)$

51. We write a proportion and then solve it.

$$\frac{4}{f} = \frac{6}{4}$$
$$4f \cdot \frac{4}{f} = 4f \cdot \frac{6}{4}$$
$$16 = 6f$$
$$\frac{8}{3} = f \qquad \text{Simplifying}$$

$\left(\text{One of the following proportions could also be used: } \dfrac{f}{4} = \dfrac{4}{6}, \dfrac{4}{f} = \dfrac{9}{6}, \dfrac{f}{4} = \dfrac{6}{9}, \dfrac{4}{9} = \dfrac{f}{6}, \dfrac{9}{4} = \dfrac{6}{f}\right)$

53. We write a proportion and then solve it.

$$\frac{h}{7} = \frac{10}{6}$$
$$h = \frac{10}{6} \cdot 7 \quad \text{Multiplying by 7}$$
$$h = \frac{70}{6}$$
$$h = \frac{35}{3} \qquad \text{Simplifying}$$

$\left(\text{Note that the proportions } \dfrac{7}{h} = \dfrac{6}{10}, \dfrac{h}{10} = \dfrac{7}{6}, \text{ or } \dfrac{10}{h} = \dfrac{6}{7} \text{ could also be used.}\right)$

55. We write a proportion and then solve it.

$$\frac{4}{10} = \frac{6}{l}$$
$$10l \cdot \frac{4}{10} = 10l \cdot \frac{6}{l}$$
$$4l = 60$$
$$l = 15 \text{ ft}$$

$\left(\text{One of the following proportions could also be used: } \dfrac{4}{6} = \dfrac{10}{l}, \dfrac{10}{4} = \dfrac{l}{6}, \text{ or } \dfrac{6}{4} = \dfrac{l}{10}\right)$

57. Discussion and Writing Exercise

59. $x^5 \cdot x^6 = x^{5+6} = x^{11}$

61. $x^{-5} \cdot x^{-6} = x^{-5+(-6)} = x^{-11} = \dfrac{1}{x^{11}}$

63. Graph: $y = 2x - 6$.

We select some x-values and compute y-values.

If $x = 1$, then $y = 2 \cdot 1 - 6 = -4$.

If $x = 3$, then $y = 2 \cdot 3 - 6 = 0$.

If $x = 5$, then $y = 2 \cdot 5 - 6 = 4$.

x	y	(x, y)
1	-4	$(1, -4)$
3	0	$(3, 0)$
5	4	$(5, 4)$

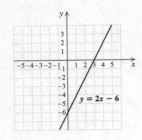

65. Graph: $3x + 2y = 12$.

We can replace either variable with a number and then calculate the other coordinate. We will find the intercepts and one other point.

If $y = 0$, we have:

$$3x + 2 \cdot 0 = 12$$
$$3x = 12$$
$$x = 4$$

The x-intercept is $(4, 0)$.

If $x = 0$, we have:

$$3 \cdot 0 + 2y = 12$$
$$2y = 12$$
$$y = 6$$

The y-intercept is $(0, 6)$.

If $y = -3$, we have:

$$3x + 2(-3) = 12$$
$$3x - 6 = 12$$
$$3x = 18$$
$$x = 6$$

The point $(6, -3)$ is on the graph.

We plot these points and draw a line through them.

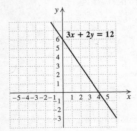

67. Graph: $y = -\dfrac{3}{4}x + 2$.

We select some x-values and compute y-values. We use multiples of 4 to avoid fractions.

If $x = -4$, then $y = -\dfrac{3}{4}(-4) + 2 = 5$.

If $x = 0$, then $y = -\dfrac{3}{4} \cdot 0 + 2 = 2$.

If $x = 4$, then $y = -\dfrac{3}{4} \cdot 4 + 2 = -1$.

x	y	(x, y)
-4	5	$(-4, 5)$
0	2	$(0, 2)$
4	-1	$(4, -1)$

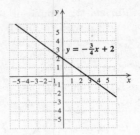

69. *Familiarize*. Let $t =$ the time it would take for Ann to complete the report working alone. Then $t + 6 =$ the time it would take Betty to complete the report working alone. In 1 hour they would complete $\dfrac{1}{t} + \dfrac{1}{t+6}$ of the report and in 4 hours they would complete $4\left(\dfrac{1}{t} + \dfrac{1}{t+6}\right)$, or $\dfrac{4}{t} + \dfrac{4}{t+6}$ of the report.

***Translate*.** In 4 hours one entire job is done, so we have

$$\frac{4}{t} + \frac{4}{t+6} = 1.$$

***Solve*.** We solve the equation.

$$\frac{4}{t} + \frac{4}{t+6} = 1, \text{ LCM} = t(t+6)$$

$$t(t+6)\left(\frac{4}{t} + \frac{4}{t+6}\right) = t(t+6) \cdot 1$$

$$t(t+6) \cdot \frac{4}{t} + t(t+6) \cdot \frac{4}{t+6} = t^2 + 6t$$

$$4(t+6) + 4t = t^2 + 6t$$

$$4t + 24 + 4t = t^2 + 6t$$

$$0 = t^2 - 2t - 24$$

$$0 = (t - 6)(t + 4)$$

$$t - 6 = 0 \ or \ t + 4 = 0$$
$$t = 6 \ or \qquad t = -4$$

***Check*.** The time cannot be negative, so we check only 6. If it takes Ann 6 hr to complete the report, then it would take Betty $6 + 6$, or 12 hr, to complete the report. In 4 hr Ann does $4 \cdot \dfrac{1}{6}$, or $\dfrac{2}{3}$, of the report, Betty does $4 \cdot \dfrac{1}{12}$, or $\dfrac{1}{3}$, of the report, and together they do $\dfrac{2}{3} + \dfrac{1}{3}$, or 1 entire job. The answer checks.

***State*.** It would take Ann 6 hr and Betty 12 hr to complete the report working alone.

71. *Familiarize*. Let $t =$ the number of minutes after 5:00 at which the hands of the clock will first be together. While the minute hand moves through t minutes, the hour hand moves through $t/12$ minutes. At 5:00 the hour hand is on the 25-minute mark. We wish to find when a move of the minute hand through t minutes is equal to $25 + t/12$ minutes.

***Translate*.** We use the last sentence of the familiarization step to write an equation.

$$t = 25 + \frac{t}{12}$$

Solve. We solve the equation.

$$t = 25 + \frac{t}{12}$$

$$12 \cdot t = 12\left(25 + \frac{t}{12}\right)$$

$$12t = 300 + t \qquad \text{Multiplying by 12}$$

$$11t = 300$$

$$t = \frac{300}{11} \text{ or } 27\frac{3}{11}$$

Check. At $27\frac{3}{11}$ minutes after 5:00, the minute hand is at the $27\frac{3}{11}$-minutes mark and the hour hand is at the $25 + \frac{27\frac{3}{11}}{12}$-minute mark. Simplifying $25 + \frac{27\frac{3}{11}}{12}$, we get

$$25 + \frac{\frac{300}{11}}{12} = 25 + \frac{300}{11} \cdot \frac{1}{12} = 25 + \frac{25}{11} = 25 + 2\frac{3}{11} = 27\frac{3}{11}.$$

Thus, the hands are together.

State. The hands are first together $27\frac{3}{11}$ minutes after 5:00.

73.
$$\frac{t}{a} + \frac{t}{b} = 1, \text{ LCM} = ab$$

$$ab\left(\frac{t}{a} + \frac{t}{b}\right) = ab \cdot 1$$

$$ab \cdot \frac{t}{a} + ab \cdot \frac{t}{b} = ab$$

$$bt + at = ab$$

$$t(b + a) = ab$$

$$t = \frac{ab}{b + a}$$

Exercise Set 14.8

1.
$$\frac{1 + \frac{9}{16}}{1 - \frac{3}{4}} \qquad \text{LCM of the denominators is 16.}$$

$$= \frac{1 + \frac{9}{16}}{1 - \frac{3}{4}} \cdot \frac{16}{16} \qquad \text{Multiplying by 1 using } \frac{16}{16}$$

$$= \frac{\left(1 + \frac{9}{16}\right)16}{\left(1 - \frac{3}{4}\right)16} \qquad \text{Multiplying numerator and denominator by 16}$$

$$= \frac{1(16) + \frac{9}{16}(16)}{1(16) - \frac{3}{4}(16)}$$

$$= \frac{16 + 9}{16 - 12}$$

$$= \frac{25}{4}$$

3.
$$\frac{1 - \frac{3}{5}}{1 + \frac{1}{5}}$$

$$= \frac{1 \cdot \frac{5}{5} - \frac{3}{5}}{1 \cdot \frac{5}{5} + \frac{1}{5}} \qquad \text{Getting a common denominator in numerator and in denominator}$$

$$= \frac{\frac{5}{5} - \frac{3}{5}}{\frac{5}{5} + \frac{1}{5}}$$

$$= \frac{\frac{2}{5}}{\frac{6}{5}} \qquad \text{Subtracting in numerator; adding in denominator}$$

$$= \frac{2}{5} \cdot \frac{5}{6} \qquad \text{Multiplying by the reciprocal of the divisor}$$

$$= \frac{2 \cdot 5}{5 \cdot 2 \cdot 3}$$

$$= \frac{\cancel{2} \cdot \cancel{5} \cdot 1}{\cancel{5} \cdot \cancel{2} \cdot 3}$$

$$= \frac{1}{3}$$

5.
$$\frac{\frac{1}{2} + \frac{3}{4}}{\frac{5}{8} - \frac{5}{6}} = \frac{\frac{1}{2} \cdot \frac{2}{2} + \frac{3}{4}}{\frac{5}{8} \cdot \frac{3}{3} - \frac{5}{6} \cdot \frac{4}{4}} \qquad \text{Getting a common denominator in numerator and in denominator}$$

$$= \frac{\frac{2}{4} + \frac{3}{4}}{\frac{15}{24} - \frac{20}{24}}$$

$$= \frac{\frac{5}{4}}{\frac{-5}{24}} \qquad \text{Adding in numerator; subtracting in denominator}$$

$$= \frac{5}{4} \cdot \frac{24}{-5} \qquad \text{Multiplying by the reciprocal of the divisor}$$

$$= \frac{5 \cdot 4 \cdot 6}{4 \cdot (-1) \cdot 5}$$

$$= \frac{\cancel{5} \cdot \cancel{4} \cdot 6}{\cancel{4} \cdot (-1) \cdot \cancel{5}}$$

$$= -6$$

7. $\dfrac{\dfrac{1}{x}+3}{\dfrac{1}{x}-5}$ LCM of the denominators is x.

$=\dfrac{\dfrac{1}{x}+3}{\dfrac{1}{x}-5}\cdot\dfrac{x}{x}$ Multiplying by 1 using $\dfrac{x}{x}$

$=\dfrac{\left(\dfrac{1}{x}+3\right)x}{\left(\dfrac{1}{x}-5\right)x}$

$=\dfrac{\dfrac{1}{x}\cdot x+3\cdot x}{\dfrac{1}{x}\cdot x-5\cdot x}$

$=\dfrac{1+3x}{1-5x}$

9. $\dfrac{4-\dfrac{1}{x^2}}{2-\dfrac{1}{x}}$ LCM of the denominators is x^2.

$=\dfrac{4-\dfrac{1}{x^2}}{2-\dfrac{1}{x}}\cdot\dfrac{x^2}{x^2}$

$=\dfrac{\left(4-\dfrac{1}{x^2}\right)x^2}{\left(2-\dfrac{1}{x}\right)x^2}$

$=\dfrac{4\cdot x^2-\dfrac{1}{x^2}\cdot x^2}{2\cdot x^2-\dfrac{1}{x}\cdot x^2}$

$=\dfrac{4x^2-1}{2x^2-x}$

$=\dfrac{(2x+1)(2x-1)}{x(2x-1)}$ Factoring numerator and denominator

$=\dfrac{(2x+1)(2x-1)}{x(2x-1)}$

$=\dfrac{2x+1}{x}$

11. $\dfrac{8+\dfrac{8}{d}}{1+\dfrac{1}{d}}=\dfrac{8\cdot\dfrac{d}{d}+\dfrac{8}{d}}{1\cdot\dfrac{d}{d}+\dfrac{1}{d}}$

$=\dfrac{\dfrac{8d+8}{d}}{\dfrac{d+1}{d}}$

$=\dfrac{8d+8}{d}\cdot\dfrac{d}{d+1}$

$=\dfrac{8(d+1)(d)}{d(d+1)}$

$=\dfrac{8(d+1)(d)}{d(d+1)(1)}$

$=8$

13. $\dfrac{\dfrac{x}{8}-\dfrac{8}{x}}{\dfrac{1}{8}+\dfrac{1}{x}}$ LCM of the denominators is $8x$.

$=\dfrac{\dfrac{x}{8}-\dfrac{8}{x}}{\dfrac{1}{8}+\dfrac{1}{x}}\cdot\dfrac{8x}{8x}$

$=\dfrac{\left(\dfrac{x}{8}-\dfrac{8}{x}\right)8x}{\left(\dfrac{1}{8}+\dfrac{1}{x}\right)8x}$

$=\dfrac{\dfrac{x}{8}(8x)-\dfrac{8}{x}(8x)}{\dfrac{1}{8}(8x)+\dfrac{1}{x}(8x)}$

$=\dfrac{x^2-64}{x+8}$

$=\dfrac{(x+8)(x-8)}{x+8}$

$=\dfrac{(x+8)(x-8)}{1(x+8)}$

$=x-8$

15. $\dfrac{1+\dfrac{1}{y}}{1-\dfrac{1}{y^2}}=\dfrac{1\cdot\dfrac{y}{y}+\dfrac{1}{y}}{1\cdot\dfrac{y^2}{y^2}-\dfrac{1}{y^2}}$

$=\dfrac{\dfrac{y+1}{y}}{\dfrac{y^2-1}{y^2}}$

$=\dfrac{y+1}{y}\cdot\dfrac{y^2}{y^2-1}$

$=\dfrac{(y+1)y\cdot y}{y(y+1)(y-1)}$

$=\dfrac{(y+1)y\cdot y}{y(y+1)(y-1)}$

$=\dfrac{y}{y-1}$

17. $\dfrac{\dfrac{1}{5} - \dfrac{1}{a}}{\dfrac{5-a}{5}}$ LCM of the denominators is $5a$.

$= \dfrac{\dfrac{1}{5} - \dfrac{1}{a}}{\dfrac{5-a}{5}} \cdot \dfrac{5a}{5a}$

$= \dfrac{\left(\dfrac{1}{5} - \dfrac{1}{a}\right)5a}{\left(\dfrac{5-a}{5}\right)5a}$

$= \dfrac{\dfrac{1}{5}(5a) - \dfrac{1}{a}(5a)}{a(5-a)}$

$= \dfrac{a-5}{5a - a^2}$

$= \dfrac{a-5}{-a(-5+a)}$

$= \dfrac{1(a-5)}{-a(a-5)}$

$= -\dfrac{1}{a}$

19. $\dfrac{\dfrac{1}{a} + \dfrac{1}{b}}{\dfrac{1}{a^2} - \dfrac{1}{b^2}}$ LCM of the denominators is a^2b^2.

$= \dfrac{\dfrac{1}{a} + \dfrac{1}{b}}{\dfrac{1}{a^2} - \dfrac{1}{b^2}} \cdot \dfrac{a^2b^2}{a^2b^2}$

$= \dfrac{\left(\dfrac{1}{a} + \dfrac{1}{b}\right) \cdot a^2b^2}{\left(\dfrac{1}{a^2} - \dfrac{1}{b^2}\right) \cdot a^2b^2}$

$= \dfrac{\dfrac{1}{a} \cdot a^2b^2 + \dfrac{1}{b} \cdot a^2b^2}{\dfrac{1}{a^2} \cdot a^2b^2 - \dfrac{1}{b^2} \cdot a^2b^2}$

$= \dfrac{ab^2 + a^2b}{b^2 - a^2}$

$= \dfrac{ab(b+a)}{(b+a)(b-a)}$

$= \dfrac{ab(b+a)}{(b+a)(b-a)}$

$= \dfrac{ab}{b-a}$

21. $\dfrac{\dfrac{p}{q} + \dfrac{q}{p}}{\dfrac{1}{p} + \dfrac{1}{q}}$ LCM of the denominators is pq.

$= \dfrac{\left(\dfrac{p}{q} + \dfrac{q}{p}\right) \cdot pq}{\left(\dfrac{1}{p} + \dfrac{1}{q}\right) \cdot pq}$

$= \dfrac{\dfrac{p}{q} \cdot pq + \dfrac{q}{p} \cdot pq}{\dfrac{1}{p} \cdot pq + \dfrac{1}{q} \cdot pq}$

$= \dfrac{p^2 + q^2}{q + p}$

23. $\dfrac{\dfrac{2}{a} + \dfrac{4}{a^2}}{\dfrac{5}{a^3} - \dfrac{3}{a}}$ LCD is a^3

$= \dfrac{\dfrac{2}{a} + \dfrac{4}{a^2}}{\dfrac{5}{a^3} - \dfrac{3}{a}} \cdot \dfrac{a^3}{a^3}$

$= \dfrac{\dfrac{2}{a} \cdot a^3 + \dfrac{4}{a^2} \cdot a^3}{\dfrac{5}{a^3} \cdot a^3 - \dfrac{3}{a} \cdot a^3}$

$= \dfrac{2a^2 + 4a}{5 - 3a^2}$

(Although the numerator can be factored, doing so will not enable us to simplify further.)

25. $\dfrac{\dfrac{2}{7a^4} - \dfrac{1}{14a}}{\dfrac{3}{5a^2} + \dfrac{2}{15a}} = \dfrac{\dfrac{2}{7a^4} \cdot \dfrac{2}{2} - \dfrac{1}{14a} \cdot \dfrac{a^3}{a^3}}{\dfrac{3}{5a^2} \cdot \dfrac{3}{3} + \dfrac{2}{15a} \cdot \dfrac{a}{a}}$

$= \dfrac{\dfrac{4 - a^3}{14a^4}}{\dfrac{9 + 2a}{15a^2}}$

$= \dfrac{4 - a^3}{14a^4} \cdot \dfrac{15a^2}{9 + 2a}$

$= \dfrac{15 \cdot a^2(4 - a^3)}{14a^2 \cdot a^2(9 + 2a)}$

$= \dfrac{15(4 - a^3)}{14a^2(9 + 2a)}$, or $\dfrac{60 - 15a^3}{126a^2 + 28a^3}$

27.
$$\frac{\frac{a}{b}+\frac{c}{d}}{\frac{b}{a}+\frac{d}{c}} = \frac{\frac{a}{b}\cdot\frac{d}{d}+\frac{c}{d}\cdot\frac{b}{b}}{\frac{b}{a}\cdot\frac{c}{c}+\frac{d}{c}\cdot\frac{a}{a}}$$

$$= \frac{\frac{ad+bc}{bd}}{\frac{bc+ad}{ac}}$$

$$= \frac{ad+bc}{bd}\cdot\frac{ac}{bc+ad}$$

$$= \frac{ac(ad+bc)}{bd(bc+ad)}$$

$$= \frac{ac}{bd}\cdot\frac{ad+bc}{bc+ad}$$

$$= \frac{ac}{bd}\cdot 1$$

$$= \frac{ac}{bd}$$

29.
$$\frac{\frac{x}{5y^3}+\frac{3}{10y}}{\frac{3}{10y}+\frac{x}{5y^3}}$$

Observe that, by the commutative law of addition, the numerator and denominator are equivalent, so the result is 1. We could also simplify this expression as follows:

$$\frac{\frac{x}{5y^3}+\frac{3}{10y}}{\frac{3}{10y}+\frac{x}{5y^3}} = \frac{\frac{x}{5y^3}+\frac{3}{10y}}{\frac{3}{10y}+\frac{x}{5y^3}}\cdot\frac{10y^3}{10y^3}$$

$$= \frac{\frac{x}{5y^3}\cdot 10y^3+\frac{3}{10y}\cdot 10y^3}{\frac{3}{10y}\cdot 10y^3+\frac{x}{5y^3}\cdot 10y^3}$$

$$= \frac{2x+3y^2}{3y^2+2x}$$

$$= 1$$

31.
$$\frac{\frac{3}{x+1}+\frac{1}{x}}{\frac{2}{x+1}+\frac{3}{x}} = \frac{\frac{3}{x+1}+\frac{1}{x}}{\frac{2}{x+1}+\frac{3}{x}}\cdot\frac{x(x+1)}{x(x+1)}$$

$$= \frac{\frac{3}{x+1}\cdot x(x+1)+\frac{1}{x}\cdot x(x+1)}{\frac{2}{x+1}\cdot x(x+1)+\frac{3}{x}\cdot x(x+1)}$$

$$= \frac{3x+x+1}{2x+3(x+1)}$$

$$= \frac{4x+1}{2x+3x+3}$$

$$= \frac{4x+1}{5x+3}$$

33. Discussion and Writing Exercise

35.
$$(2x^3-4x^2+x-7)+(4x^4+x^3+4x^2+x)$$
$$= 4x^4+3x^3+2x-7$$

37.
$$p^2-10p+25 = p^2-2\cdot p\cdot 5+5^2 \qquad \text{Trinomial square}$$
$$= (p-5)^2$$

39.
$$50p^2-100 = 50(p^2-2) \qquad \text{Factoring out the common factor}$$

Since p^2-2 cannot be factored, we have factored completely.

41. **Familiarize.** Let $w =$ the width of the rectangle. Then $w+3 =$ the length. Recall that the formula for the area of a rectangle is $A = lw$ and the formula for the perimeter of a rectangle is $P = 2l+2w$.

Translate. We substitute in the formula for area.

$$10 = lw$$
$$10 = (w+3)w$$

Solve.

$$10 = (w+3)w$$
$$10 = w^2+3w$$
$$0 = w^2+3w-10$$
$$0 = (w+5)(w-2)$$
$$w+5 = 0 \quad \text{or} \quad w-2 = 0$$
$$w = -5 \quad \text{or} \qquad w = 2$$

Check. Since the width cannot be negative, we only check 2. If $w = 2$, then $w+3 = 2+3$, or 5. Since $2\cdot 5 = 10$, the given area, the answer checks. Now we find the perimeter:

$$P = 2l+2w$$
$$P = 2\cdot 5+2\cdot 2$$
$$P = 10+4$$
$$P = 14$$

We can check this by repeating the calculation.

State. The perimeter is 14 yd.

43.
$$\frac{1}{\frac{2}{x-1}-\frac{1}{3x-2}}$$

$$= \frac{1}{\frac{2}{x-1}-\frac{1}{3x-2}}\cdot\frac{(x-1)(3x-2)}{(x-1)(3x-2)}$$

$$= \frac{(x-1)(3x-2)}{\left(\frac{2}{x-1}-\frac{1}{3x-2}\right)(x-1)(3x-2)}$$

$$= \frac{(x-1)(3x-2)}{\frac{2}{x-1}(x-1)(3x-2)-\frac{1}{3x-2}(x-1)(3x-2)}$$

$$= \frac{(x-1)(3x-2)}{2(3x-2)-(x-1)}$$

$$= \frac{(x-1)(3x-2)}{6x-4-x+1}$$

$$= \frac{(x-1)(3x-2)}{5x-3}$$

45. $1 + \cfrac{1}{1 + \cfrac{1}{1 + \cfrac{1}{1 + \cfrac{1}{x}}}} = 1 + \cfrac{1}{1 + \cfrac{1}{1 + \cfrac{1}{\frac{x+1}{x}}}}$

$$= 1 + \cfrac{1}{1 + \cfrac{1}{1 + \cfrac{x}{x+1}}}$$

$$= 1 + \cfrac{1}{1 + \cfrac{1}{\frac{x+1+x}{x+1}}}$$

$$= 1 + \cfrac{1}{1 + \cfrac{1}{\frac{2x+1}{x+1}}}$$

$$= 1 + \cfrac{1}{1 + \cfrac{x+1}{2x+1}}$$

$$= 1 + \cfrac{1}{\frac{2x+1+x+1}{2x+1}}$$

$$= 1 + \cfrac{1}{\frac{3x+2}{2x+1}}$$

$$= 1 + \frac{2x+1}{3x+2}$$

$$= \frac{3x+2+2x+1}{3x+2}$$

$$= \frac{5x+3}{3x+2}$$

Exercise Set 14.9

1. We substitute to find k.

$y = kx$

$36 = k \cdot 9$ Substituting 36 for y and 9 for x

$\frac{36}{9} = k$

$4 = k$ $\quad k$ is the variation constant.

The equation of the variation is $y = 4x$.

To find the value of y when $x = 20$ we substitute 20 for x in the equation of variation.

$y = 4x$

$y = 4 \cdot 20$

$y = 80$

The value of y is 80 when $x = 20$.

3. We substitute to find k.

$y = kx$

$0.8 = k \cdot 0.5$ Substituting 0.8 for y and 0.5 for x

$\frac{0.8}{0.5} = k$

$\frac{8}{5} = k$ $\quad k$ is the variation constant.

The equation of the variation is $y = \frac{8}{5}x$.

To find the value of y when $x = 20$ we substitute 20 for x in the equation of variation.

$y = \frac{8}{5}x$

$y = \frac{8}{5} \cdot 20$

$y = \frac{160}{5}$

$y = 32$

The value of y is 32 when $x = 20$.

5. We substitute to find k.

$y = kx$

$630 = k \cdot 175$ Substituting 630 for y and 175 for x

$\frac{630}{175} = k$

$3.6 = k$ $\quad k$ is the variation constant.

The equation of the variation is $y = 3.6x$.

To find the value of y when $x = 20$ we substitute 20 for x in the equation of variation.

$y = 3.6x$

$y = 3.6(20)$

$y = 72$

The value of y is 72 when $x = 20$.

7. We substitute to find k.

$y = kx$

$500 = k \cdot 60$ Substituting 500 for y and 60 for x

$\frac{500}{60} = k$

$\frac{25}{3} = k$ $\quad k$ is the variation constant.

The equation of the variation is $y = \frac{25}{3}x$.

To find the value of y when $x = 20$ we substitute 20 for x in the equation of variation.

$y = \frac{25}{3}x$

$y = \frac{25}{3} \cdot 20$

$y = \frac{500}{3}$

The value of y is $\frac{500}{3}$ when $x = 20$.

9. Familiarize and Translate. The problem states that we have direct variation between the variables P and H. Thus, an equation $P = kH$, $k > 0$, applies. As the number of hours increases, the paycheck increases.

Solve.

a) First find an equation of variation.

$$P = kH$$

$$84 = k \cdot 15 \quad \text{Substituting 78.75 for } P \text{ and 15 for } H$$

$$\frac{84}{15} = k$$

$$5.6 = k$$

The equation of variation is $P = 5.6H$.

b) Use the equation to find the pay for 35 hours work.

$$P = 5.6H$$

$$P = 5.6(35) \quad \text{Substituting 35 for } H$$

$$P = 196$$

Check. This check might be done by repeating the computations. We might also do some reasoning about the answer. The paycheck increased from \$84 to \$196. Similarly, the hours increased from 15 to 35.

State. a) The equation of variation is $P = 5.6H$.

b) For 35 hours work, the paycheck is \$196.

11. Familiarize and Translate. The problem states that we have direct variation between the variables C and S. Thus, an equation $C = kS$, $k > 0$, applies. As the depth increases, the cost increases.

Solve.

a) First find an equation of variation.

$$C = kS$$

$$67.5 = k \cdot 6 \quad \text{Substituting 75 for } C \text{ and 6 for } S$$

$$\frac{67.5}{6} = k$$

$$11.25 = k$$

The equation of variation is $C = 11.25S$.

b) Use the equation to find the cost of filling the sandbox to a depth of 9 inches.

$$C = 11.25S$$

$$C = 11.25(9) \quad \text{Substituting 9 for } S$$

$$C = 101.25$$

Check. In addition to repeating the computations, we can also do some reasoning. The depth increased from 6 inches to 9 inches. Similarly, the cost increased from \$67.50 to \$101.25.

State. a) The equation of variation is $C = 11.25S$.

b) The sand will cost \$101.25.

13. Familiarize and Translate. The problem states that we have direct variation between the variables M and E. Thus, an equation $M = kE$, $k > 0$, applies. As the weight on earth increases, the weight on the moon increases.

Solve.

a) First find an equation of variation.

$$M = kE$$

$$32 = k \cdot 192 \quad \text{Substituting 32 for } M \text{ and 192 for } E$$

$$\frac{32}{192} = k$$

$$\frac{1}{6} = k$$

The equation of variation is $M = \frac{1}{6}E$.

b) Use the equation to find how much a 110-lb person would weigh on the moon.

$$M = \frac{1}{6}E$$

$$M = \frac{1}{6} \cdot 110 \quad \text{Substituting 110 for } E$$

$$M = \frac{110}{6}, \text{ or } 18.\overline{3}$$

c) Use the equation to find how much a person who weighs 5 lb on the moon would weigh on Earth.

$$M = \frac{1}{6}E$$

$$5 = \frac{1}{6}E$$

$$30 = E \quad \text{Multiplying by 6}$$

Check. In addition to repeating the computations we can do some reasoning. When the weight on Earth decreased from 192 lb to 110 lb, the weight on the moon decreased from 32 lb to $18.\overline{3}$ lb. Similarly, when the weight on the moon decreased from 32 lb to 5 lb, the weight on Earth decreased from 192 lb to 30 lb.

State. a) The equation of variation is $M = \frac{1}{6}E$.

b) A person who weighs 110 lb on Earth would weigh $18.\overline{3}$ lb on the moon.

c) A person who weighs 5 lb on the moon would weigh 30 lb on Earth.

15. Familiarize and Translate. The problem states that we have direct variation between the variables N and S. Thus, an equation $N = kS$, $k > 0$, applies. As the speed of the internal processor increases, the number of instructions increases.

Solve.

a) First find an equation of variation.

$$N = kS$$

$$2,000,000 = k \cdot 25 \quad \text{Substituting 2,000,000 for } N \text{ and 25 for } S$$

$$\frac{2,000,000}{25} = k$$

$$80,000 = k$$

The equation of variation is $N = 80,000S$.

b) Use the equation to find how many instructions the processor will perform at a speed of 200 megahertz.

$$N = 80,000S$$

$$N = 80,000 \cdot 200 \quad \text{Substituting 200 for } S$$

$$N = 16,000,000$$

Check. In addition to repeating the computations we can do some reasoning. The speed of the processor increased from 25 to 200 megahertz. Similarly, the number of instructions performed per second increased from 2,000,000 to 16,000,000.

State. a) The equation of variation is $N = 80,000S$.

b) The processor will perform 16,000,000 instructions per second running at a speed of 200 megahertz.

17. **Familiarize and Translate**. This problem states that we have direct variation between the variables S and W. Thus, an equation $S = kW$, $k > 0$, applies. As the weight increases, the number of servings increases.

Solve.

a) First find an equation of variation.

$$S = kW$$

$$70 = k \cdot 9 \quad \text{Substituting 70 for } S \text{ and 9 for } W$$

$$\frac{70}{9} = k$$

The equation of variation is $S = \frac{70}{9}W$.

b) Use the equation to find the number of servings from 12 kg of round steak.

$$S = \frac{70}{9}W$$

$$S = \frac{70}{9} \cdot 12 \qquad \text{Substituting 12 for } W$$

$$S = \frac{840}{9}$$

$$S = \frac{280}{3}, \text{ or } 93\frac{1}{3}$$

Check. A check can always be done by repeating the computations. We can also do some reasoning about the answer. When the weight increased from 9 kg to 12 kg, the number of servings increased from 70 to $93\frac{1}{3}$.

State. $93\frac{1}{3}$ servings can be obtained from 12 kg of round steak.

19. We substitute to find k.

$$y = \frac{k}{x}$$

$$3 = \frac{k}{25} \quad \text{Substituting 3 for } y \text{ and 25 for } x$$

$$25 \cdot 3 = k$$

$$75 = k$$

The equation of variation is $y = \frac{75}{x}$.

To find the value of y when $x = 10$ we substitute 10 for x in the equation of variation.

$$y = \frac{75}{x}$$

$$y = \frac{75}{10}$$

$$y = \frac{15}{2}, \text{ or } 7.5$$

The value of y is $\frac{15}{2}$, or 7.5, when $x = 10$.

21. We substitute to find k.

$$y = \frac{k}{x}$$

$$10 = \frac{k}{8} \quad \text{Substituting 10 for } y \text{ and 8 for } x$$

$$8 \cdot 10 = k$$

$$80 = k$$

The equation of variation is $y = \frac{80}{x}$.

To find the value of y when $x = 10$ we substitute 10 for x in the equation of variation.

$$y = \frac{80}{x}$$

$$y = \frac{80}{10}$$

$$y = 8$$

The value of y is 8 when $x = 10$.

23. We substitute to find k.

$$y = \frac{k}{x}$$

$$6.25 = \frac{k}{0.16} \quad \text{Substituting 6.25 for } y \text{ and 0.16 for } x$$

$$0.16(6.25) = k$$

$$1 = k$$

The equation of variation is $y = \frac{1}{x}$.

To find the value of y when $x = 10$ we substitute 10 for x in the equation of variation.

$$y = \frac{1}{x}$$

$$y = \frac{1}{10}$$

The value of y is $\frac{1}{10}$ when $x = 10$.

25. We substitute to find k.

$$y = \frac{k}{x}$$

$$50 = \frac{k}{42} \quad \text{Substituting 50 for } y \text{ and 42 for } x$$

$$42 \cdot 50 = k$$

$$2100 = k$$

The equation of variation is $y = \dfrac{2100}{x}$.

To find the value of y when $x = 10$ we substitute 10 for x in the equation of variation.

$$y = \frac{2100}{x}$$

$$y = \frac{2100}{10}$$

$$y = 210$$

The value of y is 210 when $x = 10$.

27. We substitute to find k.

$$y = \frac{k}{x}$$

$$0.2 = \frac{k}{0.3} \quad \text{Substituting 0.2 for } y \text{ and 0.3 for } x$$

$$0.06 = k$$

The equation of variation is $y = \dfrac{0.06}{x}$.

To find the value of y when $x = 10$ we substitute 10 for x in the equation of variation.

$$y = \frac{0.06}{x}$$

$$y = \frac{0.06}{10}$$

$$y = \frac{6}{1000} \quad \text{Multiplying } \frac{0.06}{10} \text{ by } \frac{100}{100}$$

$$y = \frac{3}{500}, \text{ or } 0.006$$

The value of y is $\dfrac{3}{500}$, or 0.006, when $x = 10$.

29. a) It seems reasonable that, as the number of hours of production increases, the number of compact-disc players produced will increase, so direct variation might apply.

b) **Familiarize.** Let $H =$ the number of hours the production line is working, and let $P =$ the number of compact-disc players produced. An equation $P = kH$, $k > 0$, applies. (See part (a)).

Translate. We write an equation of variation.

Number of players produced varies directly as hours of production. This translates to $P = kH$.

Solve.

a) First we find an equation of variation.

$$P = kH$$

$$15 = k \cdot 8 \quad \text{Substituting 8 for } H \text{ and 15 for } P$$

$$\frac{15}{8} = k$$

The equation of variation is $P = \dfrac{15}{8}H$.

b) Use the equation to find the number of players produced in 37 hr.

$$P = \frac{15}{8}H$$

$$P = \frac{15}{8} \cdot 37 \quad \text{Substituting 37 for } H$$

$$P = \frac{555}{8} = 69\frac{3}{8}$$

Check. In addition to repeating the computations, we can do some reasoning. The number of hours increased from 8 to 37. Similarly, the number of compact disc players produced increased from 15 to $69\frac{3}{8}$.

State. About $69\frac{3}{8}$ compact-disc players can be produced in 37 hr.

31. a) It seems reasonable that, as the number of workers increases, the number of hours required to do the job decreases, so inverse variation might apply.

b) **Familiarize.** Let $T =$ the time required to cook the meal and $N =$ the number of cooks. An equation $T = k/N$, $k > 0$, applies. (See part (a)).

Translate. We write an equation of variation. Time varies inversely as the number of cooks. This translates to $T = \dfrac{k}{N}$.

Solve.

a) First find the equation of variation.

$$T = \frac{k}{N}$$

$$4 = \frac{k}{9} \quad \text{Substituting 4 for } T \text{ and 9 for } N$$

$$36 = k$$

The equation of variation is $T = \dfrac{36}{N}$.

b) Use the equation to find the amount of time it takes 8 cooks to prepare the dinner.

$$T = \frac{36}{N}$$

$$T = \frac{36}{8} \quad \text{Substituting 8 for } N$$

$$T = 4.5$$

Check. The check might be done by repeating the computation. We might also analyze the results. The number of cooks decreased from 9 to 8, and the time increased from 4 hr to 4.5 hr. This is what we would expect with inverse variation.

State. It will take 8 cooks 4.5 hr to prepare the dinner.

33. **Familiarize.** The problem states that we have inverse variation between the variables N and P. Thus, an equation $N = k/P$, $k > 0$, applies. As the miles per gallon rating increases, the number of gallons required to travel the fixed distance decreases.

Translate. We write an equation of variation. Number of gallons varies inversely as miles per gallon rating. This translates to $N = \dfrac{k}{P}$.

Solve.

a) First find an equation of variation.

$$N = \frac{k}{P}$$

$$20 = \frac{k}{14} \quad \text{Substituting 20 for } N \text{ and 14 for } P$$

$$280 = k$$

The equation is $N = \dfrac{280}{P}$.

b) Use the equation to find the number of gallons of gasoline needed for a car that gets 28 mpg.

$$N = \frac{k}{P}$$

$$N = \frac{280}{28} \quad \text{Substituting 28 for } P$$

$$N = 10$$

Check. In addition to repeating the computations, we can analyze the results. The number of miles per gallon increased from 14 to 28, and the number of gallons required decreased from 20 to 10. This is what we would expect with inverse variation.

State. a) The equation of variation is $N = \dfrac{280}{P}$.

b) A car that gets 28 mpg will need 10 gallons of gasoline to travel the fixed distance.

35. *Familiarize*. The problem states that we have inverse variation between the variables I and R. Thus, an equation $I = k/R$, $k > 0$, applies. As the resistance increases, the current decreases.

Translate. We write an equation of variation. Current varies inversely as resistance. This translates to $I = \dfrac{k}{R}$.

Solve.

a) First find an equation of variation.

$$I = \frac{k}{R}$$

$$96 = \frac{k}{20} \quad \text{Substituting 96 for } I \text{ and 20 for } R$$

$$1920 = k$$

The equation of variation is $I = \dfrac{1920}{R}$.

b) Use the equation to find the current when the resistance is 60 ohms.

$$I = \frac{1920}{R}$$

$$I = \frac{1920}{60} \quad \text{Substituting 60 for } R$$

$$I = 32$$

Check. The check might be done by repeating the computations. We might also analyze the results. The resistance increased from 20 ohms to 60 ohms, and the current decreased from 96 amperes to 32 amperes. This is what we would expect with inverse variation.

State. a) The equation of variation is $I = \dfrac{1920}{R}$.

b) The current is 32 amperes when the resistance is 60 ohms.

37. *Familiarize*. The problem states that we have inverse variation between the variables m and n. Thus, an equation $m = k/n$, $k > 0$, applies. As the number of questions increases, the number of minutes allowed for each question decreases.

Translate. We write an equation of variation. Time allowed per question varies inversely as the number of questions.

Solve.

a) First find an equation of variation.

$$m = \frac{k}{n}$$

$$2.5 = \frac{k}{16} \quad \text{Substituting 2.5 for } m \text{ and 16 for } n$$

$$40 = k$$

The equation of variation is $m = \dfrac{40}{n}$.

b) Use the equation to find the number of questions on a quiz when students have 4 min per question.

$$m = \frac{40}{n}$$

$$4 = \frac{40}{n} \quad \text{Substituting 4 for } m$$

$$4n = 40 \quad \text{Multiplying by } n$$

$$n = 10 \quad \text{Dividing by 4}$$

Check. The check might be done by repeating the computations. We might also analyze the results. The time allowed for each question increased from 2.5 min to 4 min, and the number of questions decreased from 16 to 10. This is what we would expect with inverse variation.

State. a) The equation of variation is $m = \dfrac{40}{n}$.

b) There would be 10 questions on a quiz for which students have 4 min per question.

39. *Familiarize*. The problem states that we have inverse variation between the variables A and d. Thus, an equation $A = k/d$, $k > 0$, applies. As the distance increases, the apparent size decreases.

Translate. We write an equation of variation. Apparent size varies inversely as the distance. This translates to $A = \dfrac{k}{d}$.

Solve.

a) First find an equation of variation.

$$A = \frac{k}{d}$$

$$27.5 = \frac{k}{30} \quad \text{Substituting 27.5 for } A \text{ and 30 for } d$$

$$825 = k$$

The equation of variation is $A = \dfrac{825}{d}$.

b) Use the equation to find the apparent size when the distance is 100 ft.

$$A = \frac{825}{d}$$

$$A = \frac{825}{100} \quad \text{Substituting 100 for } d$$

$$A = 8.25$$

Check. The check might be done by repeating the computations. We might also analyze the results. The distance increased from 30 ft to 100 ft, and the apparent size decreased from 27.5 ft to 8.25 ft. This is what we would expect with inverse variation.

State. The flagpole will appear to be 8.25 ft tall when it is 100 ft from the observer.

41. Discussion and Writing Exercise

43. Discussion and Writing Exercise

45.
$$\frac{x+2}{x+5} = \frac{x-4}{x-6}, \text{ LCM is } (x+5)(x-6)$$

$$(x+5)(x-6) \cdot \frac{x+2}{x+5} = (x+5)(x-6) \cdot \frac{x-4}{x-6}$$

$$(x-6)(x+2) = (x+5)(x-4)$$

$$x^2 - 4x - 12 = x^2 + x - 20$$

$$-4x - 12 = x - 20 \quad \text{Subtracting } x^2$$

$$-5x = -8 \quad \begin{array}{l}\text{Subtracting } x \text{ and}\\ \text{adding 12}\end{array}$$

$$x = \frac{8}{5}$$

The number $\dfrac{8}{5}$ checks and is the solution.

47. $x^2 - 25x + 144 = 0$

$$(x-9)(x-16) = 0$$

$$x - 9 = 0 \quad or \quad x - 16 = 0$$

$$x = 9 \quad or \qquad x = 16$$

The solutions are 9 and 16.

49.
$$35x^2 + 8 = 34x$$

$$35x^2 - 34x + 8 = 0$$

$$(7x - 4)(5x - 2) = 0$$

$$7x - 4 = 0 \quad or \quad 5x - 2 = 0$$

$$7x = 4 \quad or \qquad 5x = 2$$

$$x = \frac{4}{7} \quad or \qquad x = \frac{2}{5}$$

The solutions are $\dfrac{4}{7}$ and $\dfrac{2}{5}$.

51. We do the divisions in order from left to right.

$$3^7 \div 3^4 \div 3^3 \div 3 = 3^3 \div 3^3 \div 3$$

$$= 1 \div 3$$

$$= \frac{1}{3}$$

53. $-5^2 + 4 \cdot 6 = -25 + 4 \cdot 6$

$$= -25 + 24$$

$$= -1$$

55.

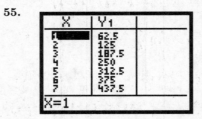

X	Y1	
1	62.5	
2	125	
3	187.5	
4	250	
5	312.5	
6	375	
7	437.5	

$X=1$

The y-values become larger.

57. $P^2 = kt$

59. $P = kV^3$

Chapter 14 Review Exercises

1. $\dfrac{3}{x}$

The denominator is 0 when $x = 0$, so the expression is not defined for the replacement number 0.

2. $\dfrac{4}{x-6}$

To determine the numbers for which the rational expression is not defined, we see the denominator equal to 0 and solve:

$$x - 6 = 0$$

$$x = 6$$

The expression is not defined for the replacement number 6.

3. $\dfrac{x+5}{x^2 - 36}$

To determine the numbers for which the rational expression is not defined, we see the denominator equal to 0 and solve:

$$x^2 - 36 = 0$$

$$(x+6)(x-6) = 0$$

$$x + 6 = 0 \quad or \quad x - 6 = 0$$

$$x = -6 \quad or \qquad x = 6$$

The expression is not defined for the replacement numbers -6 and 6.

4. $\dfrac{x^2 - 3x + 2}{x^2 + x - 30}$

To determine the numbers for which the rational expression is not defined, we see the denominator equal to 0 and solve:

$$x^2 + x - 30 = 0$$

$$(x+6)(x-5) = 0$$

$$x + 6 = 0 \quad or \quad x - 5 = 0$$
$$x = -6 \quad or \quad x = 5$$

The expression is not defined for the replacement numbers -6 and 5.

5. $\dfrac{-4}{(x+2)^2}$

To determine the numbers for which the rational expression is not defined, we see the denominator equal to 0 and solve:

$$(x+2)^2 = 0$$
$$(x+2)(x+2) = 0$$
$$x + 2 = 0 \quad or \quad x + 2 = 0$$
$$x = -2 \quad or \quad x = -2$$

The expression is not defined for the replacement number -2.

6. $\dfrac{x-5}{5}$

Since the denominator is the constant 5, there are no replacement numbers for which the expression is not defined.

7.
$$\frac{4x^2 - 8x}{4x^2 + 4x} = \frac{4x(x-2)}{4x(x+1)}$$
$$= \frac{4x}{4x} \cdot \frac{x-2}{x+1}$$
$$= 1 \cdot \frac{x-2}{x+1}$$
$$= \frac{x-2}{x+1}$$

8.
$$\frac{14x^2 - x - 3}{2x^2 - 7x + 3} = \frac{(2x-1)(7x+3)}{(2x-1)(x-3)}$$
$$= \frac{2x-1}{2x-1} \cdot \frac{7x+3}{x-3}$$
$$= 1 \cdot \frac{7x+3}{x-3}$$
$$= \frac{7x+3}{x-3}$$

9.
$$\frac{(y-5)^2}{y^2 - 25} = \frac{(y-5)(y-5)}{(y+5)(y-5)}$$
$$= \frac{y-5}{y+5} \cdot \frac{y-5}{y-5}$$
$$= \frac{y-5}{y+5} \cdot 1$$
$$= \frac{y-5}{y+5}$$

10.
$$\frac{a^2 - 36}{10a} \cdot \frac{2a}{a+6} = \frac{(a^2-36)(2a)}{10a(a+6)}$$
$$= \frac{(a+6)(a-6) \cdot 2 \cdot a}{2 \cdot 5 \cdot a \cdot (a+6)}$$
$$= \frac{(a+6)(a-6) \cdot \cancel{2} \cdot \cancel{a}}{\cancel{2} \cdot 5 \cdot \cancel{a} \cdot (a+6)}$$
$$= \frac{a-6}{5}$$

11.
$$\frac{6t-6}{2t^2 + t - 1} \cdot \frac{t^2 - 1}{t^2 - 2t + 1}$$
$$= \frac{(6t-6)(t^2-1)}{(2t^2+t-1)(t^2-2t+1)}$$
$$= \frac{6(t-1)(t+1)(t-1)}{(2t-1)(t+1)(t-1)(t-1)}$$
$$= \frac{6(\cancel{t-1})(\cancel{t+1})(\cancel{t-1})}{(2t-1)(\cancel{t+1})(\cancel{t-1})(\cancel{t-1})}$$
$$= \frac{6}{2t-1}$$

12.
$$\frac{10-5t}{3} \div \frac{t-2}{12t} = \frac{10-5t}{3} \cdot \frac{12t}{t-2}$$
$$= \frac{(10-5t)(12t)}{3(t-2)}$$
$$= \frac{5(2-t) \cdot 3 \cdot 4t}{3(t-2)}$$
$$= \frac{5(-1)(t-2) \cdot 3 \cdot 4t}{3(t-2)} \quad 2-t = -1(t-2)$$
$$= \frac{5(-1)(\cancel{t-2}) \cdot \cancel{3} \cdot 4t}{\cancel{3}(\cancel{t-2}) \cdot 1}$$
$$= -20t$$

13.
$$\frac{4x^4}{x^2 - 1} \div \frac{2x^3}{x^2 - 2x + 1} = \frac{4x^4}{x^2-1} \cdot \frac{x^2 - 2x + 1}{2x^3}$$
$$= \frac{4x^4(x^2 - 2x + 1)}{(x^2-1)(2x^3)}$$
$$= \frac{2 \cdot 2 \cdot x \cdot x^3(x-1)(x-1)}{(x+1)(x-1) \cdot 2 \cdot x^3}$$
$$= \frac{\cancel{2} \cdot 2 \cdot x \cdot \cancel{x^3}(\cancel{x-1})(x-1)}{(x+1)(\cancel{x-1}) \cdot \cancel{2} \cdot \cancel{x^3}}$$
$$= \frac{2x(x-1)}{x+1}, \text{ or}$$
$$\frac{2x^2 - 2x}{x+1}$$

14. $3x^2 = 3 \cdot x \cdot x$
$$10xy = 2 \cdot 5 \cdot x \cdot y$$
$$15y^2 = 3 \cdot 5 \cdot y \cdot y$$
$$\text{LCM} = 2 \cdot 3 \cdot 5 \cdot x \cdot x \cdot y \cdot y, \text{ or } 30x^2 y^2$$

15. $a - 2 = a - 2$
$$4a - 8 = 4(a-2)$$
$$\text{LCM} = 4(a-2)$$

16. $y^2 - y - 2 = (y-2)(y+1)$
$$y^2 - 4 = (y+2)(y-2)$$
$$\text{LCM} = (y-2)(y+1)(y+2)$$

17. $\dfrac{x+8}{x+7} + \dfrac{10-4x}{x+7} = \dfrac{x+8+10-4x}{x+7} = \dfrac{-3x+18}{x+7}$

18. $\dfrac{3}{3x-9} + \dfrac{x-2}{3-x} = \dfrac{3}{3(x-3)} + \dfrac{x-2}{3-x}$

$= \dfrac{3}{3(x-3)} + \dfrac{x-2}{3-x} \cdot \dfrac{-1}{-1}$

$= \dfrac{3}{3(x-3)} + \dfrac{-1(x-2)}{-1(3-x)}$

$= \dfrac{3}{3(x-3)} + \dfrac{-x+2}{x-3}$

$= \dfrac{3}{3(x-3)} + \dfrac{-x+2}{x-3} \cdot \dfrac{3}{3}$

$= \dfrac{3}{3(x-3)} + \dfrac{-3x+6}{3(x-3)}$

$= \dfrac{3-3x+6}{3(x-3)}$

$= \dfrac{-3x+9}{3(x-3)}$

$= \dfrac{-3(x-3)}{3(x-3)}$

$= \dfrac{-1 \cdot \cancel{3}(\cancel{x-3})}{1 \cdot \cancel{3}(\cancel{x-3})}$

$= -1$

19. $\dfrac{2a}{a+1} + \dfrac{4a}{a^2-1}$

$= \dfrac{2a}{a+1} + \dfrac{4a}{(a+1)(a-1)}$, LCM is $(a+1)(a-1)$

$= \dfrac{2a}{a+1} \cdot \dfrac{a-1}{a-1} + \dfrac{4a}{(a+1)(a-1)}$

$= \dfrac{2a(a-1)+4a}{(a+1)(a-1)}$

$= \dfrac{2a^2-2a+4a}{(a+1)(a-1)}$

$= \dfrac{2a^2+2a}{(a+1)(a-1)}$

$= \dfrac{2a(a+1)}{(a+1)(a-1)}$

$= \dfrac{2a(\cancel{a+1})}{(\cancel{a+1})(a-1)}$

$= \dfrac{2a}{a-1}$

20. $\dfrac{d^2}{d-c} + \dfrac{c^2}{c-d} = \dfrac{d^2}{d-c} + \dfrac{c^2}{c-d} \cdot \dfrac{-1}{-1}$

$= \dfrac{d^2}{d-c} + \dfrac{-c^2}{d-c}$

$= \dfrac{d^2-c^2}{d-c}$

$= \dfrac{(d+c)(d-c)}{d-c}$

$= \dfrac{(d+c)(\cancel{d-c})}{(\cancel{d-c}) \cdot 1}$

$= d+c$

21. $\dfrac{6x-3}{x^2-x-12} - \dfrac{2x-15}{x^2-x-12} = \dfrac{6x-3-(2x-15)}{x^2-x-12}$

$= \dfrac{6x-3-2x+15}{x^2-x-12}$

$= \dfrac{4x+12}{x^2-x-12}$

$= \dfrac{4(x+3)}{(x-4)(x+3)}$

$= \dfrac{4(\cancel{x+3})}{(x-4)(\cancel{x+3})}$

$= \dfrac{4}{x-4}$

22. $\dfrac{3x-1}{2x} - \dfrac{x-3}{x}$, LCM is $2x$

$= \dfrac{3x-1}{2x} - \dfrac{x-3}{x} \cdot \dfrac{2}{2}$

$= \dfrac{3x-1}{2x} - \dfrac{2(x-3)}{2x}$

$= \dfrac{3x-1-2(x-3)}{2x}$

$= \dfrac{3x-1-2x+6}{2x}$

$= \dfrac{x+5}{2x}$

23. $\dfrac{x+3}{x-2} - \dfrac{x}{2-x} = \dfrac{x+3}{x-2} - \dfrac{x}{2-x} \cdot \dfrac{-1}{-1}$

$= \dfrac{x+3}{x-2} - \dfrac{-x}{x-2}$

$= \dfrac{x+3-(-x)}{x-2}$

$= \dfrac{x+3+x}{x-2}$

$= \dfrac{2x+3}{x-2}$

24. $\dfrac{1}{x^2-25} - \dfrac{x-5}{x^2-4x-5}$

$= \dfrac{1}{(x+5)(x-5)} - \dfrac{x-5}{(x-5)(x+1)}$,

   LCM is $(x+5)(x-5)(x+1)$

$= \dfrac{1}{(x+5)(x-5)} \cdot \dfrac{x+1}{x+1} - \dfrac{x-5}{(x-5)(x+1)} \cdot \dfrac{x+5}{x+5}$

$= \dfrac{x+1}{(x+5)(x-5)(x+1)} - \dfrac{(x-5)(x+5)}{(x+5)(x-5)(x+1)}$

$= \dfrac{x+1-(x^2-25)}{(x+5)(x-5)(x+1)}$

$= \dfrac{x+1-x^2+25}{(x+5)(x-5)(x+1)}$

$= \dfrac{-x^2+x+26}{(x+5)(x-5)(x+1)}$

25.

$$\frac{3x}{x+2} - \frac{x}{x-2} + \frac{8}{x^2-4}$$

$$= \frac{3x}{x+2} - \frac{x}{x-2} + \frac{8}{(x+2)(x-2)}, \text{ LCM is } (x+2)(x-2)$$

$$= \frac{3x}{x+2} \cdot \frac{x-2}{x-2} - \frac{x}{x-2} \cdot \frac{x+2}{x+2} + \frac{8}{(x+2)(x-2)}$$

$$= \frac{3x(x-2)}{(x+2)(x-2)} - \frac{x(x+2)}{(x+2)(x-2)} + \frac{8}{(x+2)(x-2)}$$

$$= \frac{3x(x-2) - x(x+2) + 8}{(x+2)(x-2)}$$

$$= \frac{3x^2 - 6x - x^2 - 2x + 8}{(x+2)(x-2)}$$

$$= \frac{2x^2 - 8x + 8}{(x+2)(x-2)}$$

$$= \frac{2(x^2 - 4x + 4)}{(x+2)(x-2)}$$

$$= \frac{2(x-2)(x-2)}{(x+2)(x-2)}$$

$$= \frac{2(x-2)(x-2)}{(x+2)(x-2)}$$

$$= \frac{2(x-2)}{x+2}$$

26.

$$\frac{\frac{1}{z}+1}{\frac{1}{z^2}-1} \quad \text{LCM of the denominators is } z^2.$$

$$= \frac{\frac{1}{z}+1}{\frac{1}{z^2}-1} \cdot \frac{z^2}{z^2}$$

$$= \frac{\left(\frac{1}{z}+1\right)z^2}{\left(\frac{1}{z^2}-1\right)z^2}$$

$$= \frac{\frac{1}{z} \cdot z^2 + 1 \cdot z^2}{\frac{1}{z^2} \cdot z^2 - 1 \cdot z^2}$$

$$= \frac{z + z^2}{1 - z^2}$$

$$= \frac{z(1+z)}{(1+z)(1-z)}$$

$$= \frac{z(1+z)}{(1+z)(1-z)}$$

$$= \frac{z}{1-z}$$

27.

$$\frac{\frac{c}{d} - \frac{d}{c}}{\frac{1}{c} + \frac{1}{d}}$$

$$= \frac{\frac{c}{d} \cdot \frac{c}{c} - \frac{d}{c} \cdot \frac{d}{d}}{\frac{1}{c} \cdot \frac{d}{d} + \frac{1}{d} \cdot \frac{c}{c}} \quad \text{Getting a common denominator in numerator and in denominator}$$

$$= \frac{\frac{c^2}{cd} - \frac{d^2}{cd}}{\frac{d}{cd} + \frac{c}{cd}}$$

$$= \frac{\frac{c^2 - d^2}{cd}}{\frac{d+c}{cd}}$$

$$= \frac{c^2 - d^2}{cd} \cdot \frac{cd}{d+c}$$

$$= \frac{(c^2 - d^2)cd}{cd(d+c)}$$

$$= \frac{(c+d)(c-d)cd}{cd(d+c)}$$

$$= \frac{(c+d)(c-d)cd}{cd(d+c) \cdot 1}$$

$$= c - d$$

28.

$$\frac{3}{y} - \frac{1}{4} = \frac{1}{y}, \text{ LCM} = 4y$$

$$4y\left(\frac{3}{y} - \frac{1}{4}\right) = 4y \cdot \frac{1}{y}$$

$$4y \cdot \frac{3}{y} - 4y \cdot \frac{1}{4} = 4y \cdot \frac{1}{y}$$

$$12 - y = 4$$

$$-y = -8$$

$$y = 8$$

This checks, so the solution is 8.

29.

$$\frac{15}{x} - \frac{15}{x+2} = 2, \text{ LCM} = x(x+2)$$

$$x(x+2)\left(\frac{15}{x} - \frac{15}{x+2}\right) = x(x+2) \cdot 2$$

$$x(x+2) \cdot \frac{15}{x} - x(x+2) \cdot \frac{15}{x+2} = 2x(x+2)$$

$$15(x+2) - 15x = 2x(x+2)$$

$$15x + 30 - 15x = 2x^2 + 4x$$

$$30 = 2x^2 + 4x$$

$$0 = 2x^2 + 4x - 30$$

$$0 = 2(x^2 + 2x - 15)$$

$$0 = x^2 + 2x - 15 \quad \text{Dividing by 2}$$

$$0 = (x+5)(x-3)$$

$$x + 5 = 0 \quad or \quad x - 3 = 0$$

$$x = -5 \quad or \quad x = 3$$

Both numbers check. The solutions are -5 and 3.

30. *Familiarize*. Let $t =$ the time the job would take if the crews worked together.

Translate. We use the work principle, substituting 9 for a and 12 for b.

$$\frac{t}{a} + \frac{t}{b} = 1$$

$$\frac{t}{9} + \frac{t}{12} = 1 \quad \text{Substituting}$$

Solve.

$$\frac{t}{9} + \frac{t}{12} = 1, \text{ LCM} = 36$$

$$36\left(\frac{t}{9} + \frac{t}{12}\right) = 36 \cdot 1$$

$$36 \cdot \frac{t}{9} + 36 \cdot \frac{t}{12} = 36$$

$$4t + 3t = 36$$

$$7t = 36$$

$$t = \frac{36}{7}, \text{ or } 5\frac{1}{7}$$

Check. In $\frac{36}{7}$ hr, the portion of the job done is
$$\frac{36}{7} \cdot \frac{1}{9} + \frac{36}{7} \cdot \frac{1}{12} = \frac{4}{7} + \frac{3}{7} = 1.$$
The answer checks.

State. The job would take $5\frac{1}{7}$ hr if the crews worked together.

31. *Familiarize*. Let $r =$ the speed of the slower train, in km/h. Then $r + 40 =$ the speed of the faster train. We organize the information in a table.

	Distance	Speed	Time
Slower train	60	r	t
Faster train	70	$r+40$	t

Translate. We use the formula $d = rt$ in each row of the table to obtain two equations.

$$60 = rt,$$
$$70 = (r+40)t$$

Since the times are the same, we solve each equation for t and set the results equal to each other.

$$60 = rt, \text{ so } t = \frac{60}{r}.$$

$$70 = (r+40)t, \text{ so } t = \frac{70}{r+40}.$$

Then we have
$$\frac{60}{r} = \frac{70}{r+40}.$$

Solve.

$$\frac{60}{r} = \frac{70}{r+40}, \text{ LCM} = r(r+40)$$

$$r(r+40) \cdot \frac{60}{r} = r(r+40) \cdot \frac{70}{r+40}$$

$$60(r+40) = 70r$$

$$60r + 2400 = 70r$$

$$2400 = 10r$$

$$240 = r$$

If $r = 240$, then $r + 40 = 240 + 40 = 280$.

Check. If the speeds are 240 km/h and 280 km/h, then the speed of the faster train is 40 km/h faster than the speed of the slower train. At 240 km/h, the slower train travels 60 km in 60/240, or 1/4 hr. At 280 km/h, the faster train travels 70 km in 70/280, or 1/4 hr. Since the times are the same, the answer checks.

State. The speed of the slower train is 240 km/h; the speed of the faster train is 280 km/h.

32. *Familiarize*. Let $r =$ the speed of the slower plane, in mph. Then $r + 80 =$ the speed of the faster plane. We organize the information in a table.

	Distance	Speed	Time
Slower plane	950	r	t
Faster plane	1750	$r+80$	t

Translate. We use the formula $d = rt$ in each row of the table to obtain two equations.

$$950 = rt,$$
$$1750 = (r+80)t$$

Since the times are the same, we solve each equation for t and set the results equal to each other.

$$950 = rt, \text{ so } t = \frac{950}{r}.$$

$$1750 = (r+80)t, \text{ so } t = \frac{1750}{r+80}.$$

Then we have
$$\frac{950}{r} = \frac{1750}{r+80}.$$

Solve.

$$\frac{950}{r} = \frac{1750}{r+80}, \text{ LCM} = r(r+80)$$

$$r(r+80) \cdot \frac{950}{r} = r(r+80) \cdot \frac{1750}{r+80}$$

$$950(r+80) = 1750r$$

$$950r + 76,000 = 1750r$$

$$76,000 = 800r$$

$$95 = r$$

If $95 = r$, then $r + 80 = 95 + 80 = 175$.

Check. If the speeds are 95 mph and 175 mph, then the speed of the faster plane is 80 mph faster than the speed of the slower plane. At 95 mph, the slower plane travels 950 mi in 950/95, or 10 hr. At 175 mph, the faster plane travels 1750 mi in 1750/175, or 10 hr. The times are the same, so the answer checks.

State. The speed of the slower plane is 95 mph; the speed of the faster plane is 175 mph.

33. Familiarize. We can translate to a proportion, letting $d =$ the number of defective calculators that can be expected in a sample of 5000.

Translate.

$$\text{Number defective} \to \frac{8}{250} = \frac{d}{5000} \leftarrow \text{Number defective}$$
$$\text{Sample size} \to \qquad \qquad \leftarrow \text{Sample size}$$

Solve. We solve the proportion.

$$\frac{8}{250} = \frac{d}{5000}$$

$$5000 \cdot \frac{8}{250} = 5000 \cdot \frac{d}{5000}$$

$$160 = d$$

Check.

$$\frac{8}{250} = 0.032 \quad \text{and} \quad \frac{160}{5000} = 0.032.$$

The ratios are the same, so the answer checks.

State. You would expect to find 160 defective calculators in a sample of 5000.

34. a) Let $x =$ the number of cups of onion that would be used. Then we can write and solve a proportion.

$$\frac{6}{13} = \frac{x}{2}$$

$$2 \cdot \frac{6}{13} = 2 \cdot \frac{x}{2}$$

$$\frac{12}{13} = x$$

Thus, $\frac{12}{13}$ cup of onion would be used.

b) Let $c =$ the number of cups of cheese that would be used. We can write and solve a proportion.

$$\frac{5}{7} = \frac{3}{c}$$

$$7c \cdot \frac{5}{7} = 7c \cdot \frac{3}{c}$$

$$5c = 21$$

$$c = \frac{21}{5}, \text{ or } 4\frac{1}{5}$$

Thus, $4\frac{1}{5}$ cups of cheese would be used.

c) Let $c =$ the number of cups of cheese that would be used. We can write and solve a proportion.

$$\frac{9}{14} = \frac{6}{c}$$

$$14c \cdot \frac{9}{14} = 14c \cdot \frac{6}{c}$$

$$9c = 84$$

$$c = \frac{84}{9} = \frac{28}{3}, \text{ or } 9\frac{1}{3}$$

Thus, $9\frac{1}{3}$ cups of cheese would be used.

35. Familiarize. The ratio of blue whales tagged to the total blue whale population, P, is $\frac{500}{P}$. Of the 400 blue whales checked later, 20 were tagged. The ratio of blue whales tagged to blue whales checked is $\frac{20}{400}$.

Translate. Assuming the two ratios are the same, we can translate to a proportion.

$$\begin{array}{c}\text{Whales tagged} \\ \text{originally} \\ \text{Whale} \\ \text{population}\end{array} \longrightarrow \frac{500}{P} = \frac{20}{400} \longleftarrow \begin{array}{c}\text{Tagged whales} \\ \text{caught later} \\ \text{Whales caught} \\ \text{later}\end{array}$$

Solve. We solve the proportion.

$$400P \cdot \frac{500}{P} = 400P \cdot \frac{20}{400} \quad \begin{array}{l}\text{Multiplying by the LCM,} \\ 400P\end{array}$$

$$400 \cdot 500 = P \cdot 20$$

$$200,000 = 20P$$

$$10,000 = P$$

Check.

$$\frac{500}{10,000} = \frac{1}{20} \quad \text{and} \quad \frac{20}{400} = \frac{1}{20}.$$

The ratios are the same.

State. The blue whale population is about 10,000.

36. We write a proportion and solve it.

$$\frac{3.4}{8.5} = \frac{2.4}{x}$$

$$8.5x \cdot \frac{3.4}{8.5} = 8.5x \cdot \frac{2.4}{x}$$

$$3.4x = 20.4$$

$$x = 6$$

(Note that the proportions $\frac{8.5}{3.4} = \frac{x}{2.4}$, $\frac{8.5}{x} = \frac{3.4}{2.4}$, and $\frac{x}{8.5} = \frac{2.4}{3.4}$ could also be used.)

37. We substitute to find k.

$$y = kx$$

$$12 = k \cdot 4$$

$$3 = k$$

The equation of variation is $y = 3x$.

To find the value of y when $x = 20$ we substitute 20 for x in the equation of variation.

$$y = 3x$$

$$y = 3 \cdot 20$$

$$y = 60$$

The value of y is 60 when $x = 20$.

38. We substitute to find k.

$$y = kx$$

$$4 = k \cdot 8$$

$$\frac{1}{2} = k$$

The equation of variation is $y = \frac{1}{2}x$.

To find the value of y when $x = 20$ we substitute 20 for x in the equation of variation.

$$y = \frac{1}{2}x$$
$$y = \frac{1}{2} \cdot 20$$
$$y = 10$$

The value of y is 10 when $x = 20$.

39. We substitute to find k.

$$y = kx$$
$$0.4 = k \cdot 0.5$$
$$\frac{0.4}{0.5} = k$$
$$\frac{4}{5} = k \quad \left(\frac{0.4}{0.5} = \frac{0.4}{0.5} \cdot \frac{10}{10} = \frac{4}{5}\right)$$

The equation of variation is $y = \frac{4}{5}x$.

To find the value of y when $x = 20$ we substitute 20 for x in the equation of variation.

$$y = \frac{4}{5}x$$
$$y = \frac{4}{5} \cdot 20$$
$$y = 16$$

The value of y is 16 when $x = 20$.

40. We substitute to find k.

$$y = \frac{k}{x}$$
$$5 = \frac{k}{6}$$
$$30 = k$$

The equation of variation is $y = \frac{30}{x}$.

To find the value of y when $x = 5$ we substitute 5 for x in the equation of variation.

$$y = \frac{30}{x}$$
$$y = \frac{30}{5}$$
$$y = 6$$

The value of y is 6 when $x = 5$.

41. We substitute to find k.

$$y = \frac{k}{x}$$
$$0.5 = \frac{k}{2}$$
$$1 = k$$

The equation of variation is $y = \frac{1}{x}$.

To find the value of y when $x = 5$ we substitute 5 for x in the equation of variation.

$$y = \frac{1}{x}$$
$$y = \frac{1}{5}$$

The value of y is $\frac{1}{5}$ when $x = 5$.

42. We substitute to find k.

$$y = \frac{k}{x}$$
$$1.3 = \frac{k}{0.5}$$
$$0.65 = k$$

The equation of variation is $y = \frac{0.65}{x}$.

To find the value of y when $x = 5$ we substitute 5 for x in the equation of variation.

$$y = \frac{0.65}{x}$$
$$y = \frac{0.65}{5}$$
$$y = 0.13$$

The value of y is 0.13 when $x = 5$.

43. *Familiarize and Translate*. The problem states that we have direct variation between the variables P and H. Thus, an equation $P = kH$, $k > 0$, applies.

***Solve*.**

a) First we find an equation of variation.

$$P = kH$$
$$165 = k \cdot 20$$
$$8.25 = k$$

The equation of variation is $P = 8.25H$.

b) Use the equation to find the pay for 35 hr of work.

$$P = 8.25H$$
$$P = 8.25(35)$$
$$P = 288.75$$

***Check*.** We can repeat the computations. Also note that when the number of hours worked increased from 20 to 35, the pay increased from \$165.00 to \$288.75 so the answer seems reasonable.

***State*.** The pay for 35 hr of work is \$288.75.

44. *Familiarize and Translate*. Let M = the number of washing machines used and T = the time required to do the laundry. The problem states that we have inverse variation between T and M, so an equation $T = k/M$, $k > 0$, applies.

Solve.

a) First we find an equation of variation.

$$T = \frac{k}{M}$$

$$5 = \frac{k}{2}$$

$$10 = k$$

The equation of variation is $T = \frac{10}{M}$.

b) Use the equation to find the time required to do the laundry if 10 washing machines are used.

$$T = \frac{10}{M}$$

$$T = \frac{10}{10}$$

$$T = 1$$

Check. We can repeat the computations. Also note that as the number of washing machines increased from 5 to 10, the time required to do the laundry decreased from 2 hr to 1 hr, so the answer seems reasonable.

State. It would take 1 hr for 10 washing machines to do the laundry.

45. *Discussion and Writing Exercise.* $\frac{5x + 6}{(x + 2)(x - 2)}$; used to find an equivalent expression for each rational expression with the LCM as the least common denominator

46. *Discussion and Writing Exercise.* $\frac{3x + 10}{(x - 2)(x + 2)}$; used to find an equivalent expression for each rational expression with the LCM as the least common denominator

47. *Discussion and Writing Exercise.* 4; used to clear fractions

48. *Discussion and Writing Exercise.* $\frac{4(x - 2)}{x(x + 4)}$; method 1: used to multiply by 1 using LCM/LCM; method 2: LCM of the denominators in the numerator used to subtract in the numerator and LCM of the denominators in the denominator used to add in the denominator

49.
$$\frac{2a^2 + 5a - 3}{a^2} \cdot \frac{5a^3 + 30a^2}{2a^2 + 7a - 4} \div \frac{a^2 + 6a}{a^2 + 7a + 12}$$

$$= \frac{2a^2 + 5a - 3}{a^2} \cdot \frac{5a^3 + 30a^2}{2a^2 + 7a - 4} \cdot \frac{a^2 + 7a + 12}{a^2 + 6a}$$

$$= \frac{(2a - 1)(a + 3)}{a^2} \cdot \frac{5a^2(a + 6)}{(2a - 1)(a + 4)} \cdot \frac{(a + 3)(a + 4)}{a(a + 6)}$$

$$= \frac{(2a - 1)(a + 3)(5a^2)(a + 6)(a + 3)(a + 4)}{a^2(2a - 1)(a + 4)(a)(a + 6)}$$

$$= \frac{(2a - 1)(a + 3) \cdot 5 \cdot a^2 (a + 6)(a + 3)(a + 4)}{a^2 (2a - 1)(a + 4)(a)(a + 6)}$$

$$= \frac{5(a + 3)^2}{a}$$

50.
$$\frac{12a}{(a - b)(b - c)} - \frac{2a}{(b - a)(c - b)}$$

$$= \frac{12a}{(a-b)(b-c)} - \frac{2a}{-1(a-b)(-1)(b-c)} \quad \begin{array}{l} \text{Factoring } -1 \\ \text{out of } b - a \text{ and} \\ c - b \end{array}$$

$$= \frac{12a}{(a - b)(b - c)} - \frac{2a}{(a - b)(b - c)}$$

$$= \frac{10a}{(a - b)(b - c)}$$

51.
$$\frac{A + B}{B} = \frac{C + D}{D}$$

$$\frac{A}{B} + \frac{B}{B} = \frac{C}{D} + \frac{D}{D}$$

$$\frac{A}{B} + 1 = \frac{C}{D} + 1$$

$$\frac{A}{B} = \frac{C}{D}$$

The two given proportions are equivalent.

Chapter 14 Test

1. $\frac{8}{2x}$

To determine the numbers for which the rational expression is not defined, we set the denominator equal to 0 and solve:

$$2x = 0$$

$$x = 0$$

The expression is not defined for the replacement number 0.

2. $\frac{5}{x + 8}$

To determine the numbers for which the rational expression is not defined, we set the denominator equal to 0 and solve:

$$x + 8 = 0$$

$$x = -8$$

The expression is not defined for the replacement number −8.

3. $\frac{x - 7}{x^2 - 49}$

To determine the numbers for which the rational expression is not defined, we set the denominator equal to 0 and solve:

$$x^2 - 49 = 0$$

$$(x + 7)(x - 7) = 0$$

$$x + 7 = 0 \quad or \quad x - 7 = 0$$

$$x = -7 \quad or \quad x = 7$$

The expression is not defined for the replacement numbers −7 and 7.

4. $\dfrac{x^2 + x - 30}{x^2 - 3x + 2}$

To determine the numbers for which the rational expression is not defined, we set the denominator equal to 0 and solve:

$$x^2 - 3x + 2 = 0$$
$$(x - 1)(x - 2) = 0$$
$$x - 1 = 0 \ \ or \ \ x - 2 = 0$$
$$x = 1 \ \ or \ \ \ \ \ \ x = 2$$

The expression is not defined for the replacement numbers 1 and 2.

5. $\dfrac{11}{(x - 1)^2}$

To determine the numbers for which the rational expression is not defined, we set the denominator equal to 0 and solve:

$$(x - 1)^2 = 0$$
$$(x - 1)(x - 1) = 0$$
$$x - 1 = 0 \ \ or \ \ x - 1 = 0$$
$$x = 1 \ \ or \ \ \ \ \ \ x = 1$$

The expression is not defined for the replacement number 1.

6. $\dfrac{x + 2}{2}$

Since the denominator is the constant 2, there are no replacement numbers for which the expression is not defined.

7. $\dfrac{6x^2 + 17x + 7}{2x^2 + 7x + 3} = \dfrac{(2x + 1)(3x + 7)}{(2x + 1)(x + 3)}$

$$= \dfrac{2x + 1}{2x + 1} \cdot \dfrac{3x + 7}{x + 3}$$

$$= 1 \cdot \dfrac{3x + 7}{x + 3}$$

$$= \dfrac{3x + 7}{x + 3}$$

8. $\dfrac{a^2 - 25}{6a} \cdot \dfrac{3a}{a - 5} = \dfrac{(a^2 - 25)(3a)}{6a(a - 5)}$

$$= \dfrac{(a + 5)(a - 5) \cdot 3 \cdot a}{2 \cdot 3 \cdot a \cdot (a - 5)}$$

$$= \dfrac{(a + 5)(a - 5) \cdot 3 \cdot a}{2 \cdot 3 \cdot a \cdot (a - 5)}$$

$$= \dfrac{a + 5}{2}$$

9. $\dfrac{25x^2 - 1}{9x^2 - 6x} \div \dfrac{5x^2 + 9x - 2}{3x^2 + x - 2}$

$$= \dfrac{25x^2 - 1}{9x^2 - 6x} \cdot \dfrac{3x^2 + x - 2}{5x^2 + 9x - 2}$$

$$= \dfrac{(25x^2 - 1)(3x^2 + x - 2)}{(9x^2 - 6x)(5x^2 + 9x - 2)}$$

$$= \dfrac{(5x + 1)(5x - 1)(3x - 2)(x + 1)}{3x(3x - 2)(5x - 1)(x + 2)}$$

$$= \dfrac{(5x + 1)(5x - 1)(3x - 2)(x + 1)}{3x(3x - 2)(5x - 1)(x + 2)}$$

$$= \dfrac{(5x + 1)(x + 1)}{3x(x + 2)}$$

10. $y^2 - 9 = (y + 3)(y - 3)$

$y^2 + 10y + 21 = (y + 3)(y + 7)$

$y^2 + 4y - 21 = (y - 3)(y + 7)$

$LCM = (y + 3)(y - 3)(y + 7)$

11. $\dfrac{16 + x}{x^3} + \dfrac{7 - 4x}{x^3} = \dfrac{16 + x + 7 - 4x}{x^3} = \dfrac{23 - 3x}{x^3}$

12. $\dfrac{5 - t}{t^2 + 1} - \dfrac{t - 3}{t^2 + 1} = \dfrac{5 - t - (t - 3)}{t^2 + 1}$

$$= \dfrac{5 - t - t + 3}{t^2 + 1}$$

$$= \dfrac{8 - 2t}{t^2 + 1}$$

13. $\dfrac{x - 4}{x - 3} + \dfrac{x - 1}{3 - x} = \dfrac{x - 4}{x - 3} + \dfrac{x - 1}{3 - x} \cdot \dfrac{-1}{-1}$

$$= \dfrac{x - 4}{x - 3} + \dfrac{-x + 1}{x - 3}$$

$$= \dfrac{x - 4 - x + 1}{x - 3}$$

$$= \dfrac{-3}{x - 3}$$

14. $\dfrac{x - 4}{x - 3} - \dfrac{x - 1}{3 - x} = \dfrac{x - 4}{x - 3} - \dfrac{x - 1}{3 - x} \cdot \dfrac{-1}{-1}$

$$= \dfrac{x - 4}{x - 3} - \dfrac{-x + 1}{x - 3}$$

$$= \dfrac{x - 4 - (-x + 1)}{x - 3}$$

$$= \dfrac{x - 4 + x - 1}{x - 3}$$

$$= \dfrac{2x - 5}{x - 3}$$

15. $\dfrac{5}{t-1} + \dfrac{3}{t}$, LCD is $t(t-1)$.

$= \dfrac{5}{t-1} \cdot \dfrac{t}{t} + \dfrac{3}{t} \cdot \dfrac{t-1}{t-1}$

$= \dfrac{5t}{t(t-1)} + \dfrac{3(t-1)}{t(t-1)}$

$= \dfrac{5t}{t(t-1)} + \dfrac{3t-3}{t(t-1)}$

$= \dfrac{5t+3t-3}{t(t-1)}$

$= \dfrac{8t-3}{t(t-1)}$

16. $\dfrac{1}{x^2-16} - \dfrac{x+4}{x^2-3x-4}$

$= \dfrac{1}{(x+4)(x-4)} - \dfrac{x+4}{(x-4)(x+1)}$, LCD is $(x+4)(x-4)(x+1)$.

$= \dfrac{1}{(x+4)(x-4)} \cdot \dfrac{x+1}{x+1} - \dfrac{x+4}{(x-4)(x+1)} \cdot \dfrac{x+4}{x+4}$

$= \dfrac{x+1-(x+4)(x+4)}{(x+4)(x-4)(x+1)}$

$= \dfrac{x+1-(x^2+8x+16)}{(x+4)(x-4)(x+1)}$

$= \dfrac{x+1-x^2-8x-16}{(x+4)(x-4)(x+1)}$

$= \dfrac{-x^2-7x-15}{(x+4)(x-4)(x+1)}$

17. $\dfrac{1}{x-1} + \dfrac{4}{x^2-1} - \dfrac{2}{x^2-2x+1}$

$= \dfrac{1}{x-1} + \dfrac{4}{(x+1)(x-1)} - \dfrac{2}{(x-1)(x-1)}$,
LCD is $(x+1)(x-1)(x-1)$

$= \dfrac{1}{x-1} \cdot \dfrac{(x+1)(x-1)}{(x+1)(x-1)} + \dfrac{4}{(x+1)(x-1)} \cdot \dfrac{x-1}{x-1} -$

$\quad \dfrac{2}{(x-1)(x-1)} \cdot \dfrac{x+1}{x+1}$

$= \dfrac{(x+1)(x-1) + 4(x-1) - 2(x+1)}{(x+1)(x-1)(x-1)}$

$= \dfrac{x^2-1+4x-4-2x-2}{(x+1)(x-1)(x-1)}$

$= \dfrac{x^2+2x-7}{(x+1)(x-1)^2}$

18. We multiply the numerator and the denominator by the LCM of the denominators, y^2.

$\dfrac{9-\dfrac{1}{y^2}}{3-\dfrac{1}{y}} = \dfrac{9-\dfrac{1}{y^2}}{3-\dfrac{1}{y}} \cdot \dfrac{y^2}{y^2}$

$= \dfrac{\left(9-\dfrac{1}{y^2}\right) \cdot y^2}{\left(3-\dfrac{1}{y}\right) \cdot y^2}$

$= \dfrac{9 \cdot y^2 - \dfrac{1}{y^2} \cdot y^2}{3 \cdot y^2 - \dfrac{1}{y} \cdot y^2}$

$= \dfrac{9y^2-1}{3y^2-y}$

$= \dfrac{(3y+1)(3y-1)}{y(3y-1)}$

$= \dfrac{(3y+1)(3\cancel{y-1})}{y(3\cancel{y-1})}$

$= \dfrac{3y+1}{y}$

19. $\dfrac{7}{y} - \dfrac{1}{3} = \dfrac{1}{4}$, LCM is $12y$

$12y\left(\dfrac{7}{y} - \dfrac{1}{3}\right) = 12y \cdot \dfrac{1}{4}$

$12y \cdot \dfrac{7}{y} - 12y \cdot \dfrac{1}{3} = 12y \cdot \dfrac{1}{4}$

$84 - 4y = 3y$

$84 = 7y$

$12 = y$

The number 12 checks, so it is the solution.

20. $\dfrac{15}{x} - \dfrac{15}{x-2} = -2$, LCM is $x(x-2)$

$x(x-2)\left(\dfrac{15}{x} - \dfrac{15}{x-2}\right) = x(x-2)(-2)$

$x(x-2) \cdot \dfrac{15}{x} - x(x-2) \cdot \dfrac{15}{x-2} = -2x(x-2)$

$15(x-2) - 15x = -2x(x-2)$

$15x - 30 - 15x = -2x^2 + 4x$

$-30 = -2x^2 + 4x$

$2x^2 - 4x - 30 = 0$

$2(x^2 - 2x - 15) = 0$

$x^2 - 2x - 15 = 0 \quad$ Dividing by 2

$(x-5)(x+3) = 0$

$x - 5 = 0 \quad or \quad x + 3 = 0$

$x = 5 \quad or \quad x = -3$

Both numbers check. The solutions are 5 and -3.

21. We substitute to find k.

$$y = kx$$
$$6 = k \cdot 3$$
$$2 = k$$

The equation of variation is $y = 2x$.

To find the value of y when $x = 25$ we substitute 25 for x in the equation of variation.

$$y = 2x$$
$$y = 2 \cdot 25$$
$$y = 50$$

The value of y is 50 when $x = 25$.

22. We substitute to find k.

$$y = kx$$
$$1.5 = k \cdot 3$$
$$0.5 = k$$

The equation of variation is $y = 0.5x$.

To find the value of y when $x = 25$ we substitute 25 for x in the equation of variation.

$$y = 0.5x$$
$$y = 0.5(25)$$
$$y = 12.5$$

The value of y is 12.5 when $x = 25$.

23. We substitute to find k.

$$y = \frac{k}{x}$$
$$6 = \frac{k}{3}$$
$$18 = k$$

The equation of variation is $y = \dfrac{18}{x}$.

To find the value of y when $x = 100$ we substitute 100 for x in the equation of variation.

$$y = \frac{18}{x}$$
$$y = \frac{18}{100}$$
$$y = \frac{9}{50}$$

The value of y is $\dfrac{9}{50}$ when $x = 100$.

24. We substitute to find k.

$$y = \frac{k}{x}$$
$$11 = \frac{k}{2}$$
$$22 = k$$

The equation of variation is $y = \dfrac{22}{x}$.

To find the value of y when $x = 100$ we substitute 100 for x in the equation of variation.

$$y = \frac{22}{x}$$
$$y = \frac{22}{100}$$
$$y = \frac{11}{50}$$

The value of y is $\dfrac{11}{50}$ when $x = 100$.

25. *Familiarize and Translate*. The problem states that we have direct variation between the variables d and t. Thus, an equation $d = kt$, $k > 0$, applies.

Solve.

a) First find an equation of variation.

$$d = kt$$
$$60 = k \cdot \frac{1}{2}$$
$$120 = k$$

The equation of variation is $d = 120t$.

b) Use the equation to find the distance the train will travel in 2 hr.

$$d = 120t$$
$$d = 120 \cdot 2$$
$$d = 240$$

Check. We can repeat the computations. Also note that when the time increases from $\dfrac{1}{2}$ hr to 2 hr, the distance traveled increases from 60 km to 240 km so the answer seems reasonable.

State. The train will travel 240 km in 2 hr.

26. *Familiarize and Translate*. Let $T =$ the time required to do the job and let $M =$ the number of concrete mixers used. We have inverse variation between T and M so an equation $T = \dfrac{k}{M}$, $k > 0$, applies.

Solve.

a) First we find an equation of variation.

$$T = \frac{k}{M}$$
$$3 = \frac{k}{2}$$
$$6 = k$$

The equation of variation is $T = \dfrac{6}{M}$.

b) Use the equation to find the time required to do the job when 5 mixers are used.

$$T = \frac{6}{M}$$
$$T = \frac{6}{5}$$
$$T = 1\frac{1}{5}$$

Check. We can repeat the computations. Also note that as the number of mixers increased from 2 to 5, the time required to mix the concrete decreased from 3 hr to $1\dfrac{1}{5}$ hr, so the answer seems reasonable.

State. It would take $1\frac{1}{5}$ hr to do the job if 5 concrete mixers are used.

27. *Familiarize*. We can translate to a proportion, letting d = the number of defective spark plugs that would be expected in a sample of 500.

Translate.

Number defective $\rightarrow$ $\dfrac{4}{125} = \dfrac{d}{500}$ $\leftarrow$ Number defective
Sample size $\rightarrow$ $\qquad\qquad$ $\leftarrow$ Sample size

Solve. We solve the proportion.

$$\frac{4}{125} = \frac{d}{500}$$

$$500 \cdot \frac{4}{125} = 500 \cdot \frac{d}{500}$$

$$16 = d$$

Check.

$$\frac{4}{125} = 0.032 \text{ and } \frac{16}{500} = 0.032.$$

The ratios are the same, so the answer checks.

State. You would expect to find 16 defective spark plugs in a sample of 500.

28. *Familiarize*. The ratio of zebras tagged to the total zebra population, P, is $\dfrac{15}{P}$. Of the 20 zebras checked later, 6 were tagged. The ratio of zebras tagged to zebras checked is $\dfrac{6}{20}$.

Translate. Assuming the two ratios are the same, we can translate to a proportion.

Zebras tagged
originally $\longrightarrow$ $\dfrac{15}{P} = \dfrac{6}{20}$ $\longleftarrow$ Tagged zebras
$\qquad\qquad\qquad\qquad\qquad\qquad$ caught later
Zebra $\longrightarrow$ $\qquad\qquad$ $\longleftarrow$ Zebras caught
population $\qquad\qquad\qquad\qquad\qquad$ later

Solve. We solve the proportion.

$$\frac{15}{P} = \frac{6}{20}$$

$$20P \cdot \frac{15}{P} = 20P \cdot \frac{6}{20}$$

$$300 = 6P$$

$$50 = P$$

Check.

$$\frac{15}{50} = 0.3 \text{ and } \frac{6}{20} = 0.3.$$

The ratios are the same, so the answer checks.

State. The zebra population is 50.

29. *Familiarize*. Let t = the time, in minutes, required to copy the report using both copy machines working together.

Translate. We use the work principle, substituting 20 for a and 30 for b.

$$\frac{t}{a} + \frac{t}{b} = 1$$

$$\frac{t}{20} + \frac{t}{30} = 1$$

Solve.

$$\frac{t}{20} + \frac{t}{30} = 1, \text{ LCM is } 60$$

$$60\left(\frac{t}{20} + \frac{t}{30}\right) = 60 \cdot 1$$

$$60 \cdot \frac{t}{20} + 60 \cdot \frac{t}{30} = 60$$

$$3t + 2t = 60$$

$$5t = 60$$

$$t = 12$$

Check. In 12 min, the portion of the job done is

$$12 \cdot \frac{1}{20} + 12 \cdot \frac{1}{30} = \frac{3}{5} + \frac{2}{5} = 1.$$

The answer checks.

State. It would take 12 min to copy the report using both machines working together.

30. *Familiarize*. Let r = Marilyn's speed, in km/h. Then $r + 20$ = Craig's speed. We organize the information in a table.

	Distance	Speed	Time
Marilyn	225	r	t
Craig	325	$r + 20$	t

Translate. We use the formula $d = rt$ in each row of the table to obtain two equations.

$$225 = rt,$$

$$325 = (r + 20)t$$

Since the times are the same, we solve each equation for t and set the results equal to each other.

$$225 = rt, \text{ so } t = \frac{225}{r}.$$

$$325 = (r + 20)t, \text{ so } t = \frac{325}{r + 20}.$$

Then we have

$$\frac{225}{r} = \frac{325}{r + 20}.$$

Solve.

$$\frac{225}{r} = \frac{325}{r + 20}, \text{ LCM is } r(r + 20)$$

$$r(r + 20) \cdot \frac{225}{r} = r(r + 20) \cdot \frac{325}{r + 20}$$

$$225(r + 20) = 325r$$

$$225r + 4500 = 325r$$

$$4500 = 100r$$

$$45 = r$$

If $r = 45$, then $r + 20 = 45 + 20 = 65$.

Check. If Marilyn's speed is 45 km/h and Craig's speed is 65 km/h, then Craig's speed is 20 km/h faster than Marilyn's. At 45 km/h, Marilyn travels 225 km in 225/45, or 5 hr. At 65 km/h, Craig travels 325 km in 325/65, or 5 hr. Since the times are the same, the answer checks.

State. The speed of Marilyn's car is 45 km/h and the speed of Craig's car is 65 km/h.

31. We write a proportion and solve it.

$$\frac{12}{9} = \frac{20}{x}$$

$$9x \cdot \frac{12}{9} = 9x \cdot \frac{20}{x}$$

$$12x = 180$$

$$x = 15$$

(Note that the proportions $\frac{9}{12} = \frac{x}{20}$, $\frac{12}{20} = \frac{9}{x}$, and $\frac{20}{12} = \frac{x}{9}$ could also be used.)

32. *Familiarize*. Let $r =$ the number of hours it would take Rema to do the job working alone. Then $r + 6 =$ the number of hours it would take Reggie to do the job working alone.

Translate. We use the work principle, substituting $2\frac{6}{7}$, or $\frac{20}{7}$, for t, r for a, and $r + 6$ for b.

$$\frac{t}{a} + \frac{t}{b} = 1$$

$$\frac{\frac{20}{7}}{r} + \frac{\frac{20}{7}}{r + 6} = 1$$

Solve.

$$\frac{\frac{20}{7}}{r} + \frac{\frac{20}{7}}{r + 6} = 1$$

$$\frac{20}{7} \cdot \frac{1}{r} + \frac{20}{7} \cdot \frac{1}{r + 6} = 1$$

$$\frac{20}{7r} + \frac{20}{7(r + 6)} = 1, \text{ LCM is } 7r(r+6)$$

$$7r(r + 6)\left(\frac{20}{7r} + \frac{20}{7(r + 6)}\right) = 7r(r + 6) \cdot 1$$

$$7r(r+6) \cdot \frac{20}{7r} + 7r(r+6) \cdot \frac{20}{7(r+6)} = 7r(r + 6)$$

$$20(r + 6) + 20r = 7r^2 + 42r$$

$$20r + 120 + 20r = 7r^2 + 42r$$

$$40r + 120 = 7r^2 + 42r$$

$$0 = 7r^2 + 2r - 120$$

$$0 = (7r + 30)(r - 4)$$

$$7r + 30 = 0 \quad \text{or} \quad r - 4 = 0$$

$$7r = -30 \quad \text{or} \quad r = 4$$

$$r = -\frac{30}{7} \quad \text{or} \quad r = 4$$

Check. Since the time cannot be negative, $-\frac{30}{7}$ cannot be a solution. If $r = 4$, then $r + 6 = 4 + 6 = 10$. In $2\frac{6}{7}$ hr, or $\frac{20}{7}$ hr, then the portion of the job done is

$$\frac{\frac{20}{7}}{4} + \frac{\frac{20}{7}}{10} = \frac{20}{7} \cdot \frac{1}{4} + \frac{20}{7} \cdot \frac{1}{10} = \frac{5}{7} + \frac{2}{7} = 1.$$

The answer checks.

State. Working alone it would take Rema 4 hr to do the job and it would take Reggie 10 hr.

33.

$$1 + \cfrac{1}{1 + \cfrac{1}{1 + \cfrac{1}{a}}} = 1 + \cfrac{1}{1 + \cfrac{1}{\frac{a}{a} + \frac{1}{a}}}$$

$$= 1 + \cfrac{1}{1 + \cfrac{1}{\frac{a + 1}{a}}}$$

$$= 1 + \cfrac{1}{1 + 1 \cdot \frac{a}{a + 1}}$$

$$= 1 + \cfrac{1}{1 + \frac{a}{a + 1}}$$

$$= 1 + \cfrac{1}{\frac{a + 1}{a + 1} + \frac{a}{a + 1}}$$

$$= 1 + \cfrac{1}{\frac{a + 1 + a}{a + 1}}$$

$$= 1 + \cfrac{1}{\frac{2a + 1}{a + 1}}$$

$$= 1 + 1 \cdot \frac{a + 1}{2a + 1}$$

$$= 1 + \frac{a + 1}{2a + 1}$$

$$= \frac{2a + 1}{2a + 1} + \frac{a + 1}{2a + 1}$$

$$= \frac{2a + 1 + a + 1}{2a + 1}$$

$$= \frac{3a + 2}{2a + 1}$$

Chapter 15

Systems of Equations

Exercise Set 15.1

1. We check by substituting alphabetically 1 for x and 5 for y.

$$
\begin{array}{c|c}
5x - 2y = -5 \\
\hline
5 \cdot 1 - 2 \cdot 5 \ ? \ -5 \\
5 - 10 \\
-5 & \text{TRUE}
\end{array}
\qquad
\begin{array}{c|c}
3x - 7y = -32 \\
\hline
3 \cdot 1 - 7 \cdot 5 \ ? \ -32 \\
3 - 35 \\
-32 & \text{TRUE}
\end{array}
$$

The ordered pair $(1, 5)$ is a solution of both equations, so it is a solution of the system of equations.

3. We check by substituting alphabetically 4 for a and 2 for b.

$$
\begin{array}{c|c}
3b - 2a = -2 \\
\hline
3 \cdot 2 - 2 \cdot 4 \ ? \ -2 \\
6 - 8 \\
-2 & \text{TRUE}
\end{array}
\qquad
\begin{array}{c|c}
b + 2a = 8 \\
\hline
2 + 2 \cdot 4 \ ? \ 8 \\
2 + 8 \\
10 & \text{FALSE}
\end{array}
$$

The ordered pair $(4, 2)$ is not a solution of $b + 2a = 8$, so it is not a solution of the system of equations.

5. We check by substituting alphabetically 15 for x and 20 for y.

$$
\begin{array}{c|c}
3x - 2y = 5 \\
\hline
3 \cdot 15 - 2 \cdot 20 \ ? \ 5 \\
45 - 40 \\
5 & \text{TRUE}
\end{array}
\qquad
\begin{array}{c|c}
6x - 5y = -10 \\
\hline
6 \cdot 15 - 5 \cdot 20 \ ? \ -10 \\
90 - 100 \\
-10 & \text{TRUE}
\end{array}
$$

The ordered pair $(15, 20)$ is a solution of both equations, so it is a solution of the system of equations.

7. We check by substituting alphabetically -1 for x and 1 for y.

$$
\begin{array}{c}
x = -1 \\
\hline
-1 \ ? \ -1 \quad \text{TRUE} \\

\end{array}
\qquad
\begin{array}{c|c}
x - y = -2 \\
\hline
-1 - 1 \ ? \ -2 \\
-2 & \text{TRUE}
\end{array}
$$

The ordered pair $(-1, 1)$ is a solution of both equations, so it is a solution of the system of equations.

9. We check by substituting alphabetically 18 for x and 3 for y.

$$
\begin{array}{c|c}
y = \dfrac{1}{6}x \\
\hline
3 \ ? \ \dfrac{1}{6} \cdot 18 \\
3 & \text{TRUE}
\end{array}
\qquad
\begin{array}{c|c}
2x - y = 33 \\
\hline
2 \cdot 18 \ ? \ 33 \\
36 - 3 \\
33 & \text{TRUE}
\end{array}
$$

The ordered pair $(18, 3)$ is a solution of both equations, so it is a solution of the system of equations.

11. We graph the equations.

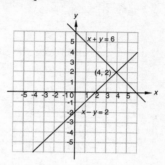

The point of intersection looks as if it has coordinates $(4, 2)$.

Check:

$$
\begin{array}{c|c}
x - y = 2 \\
\hline
4 - 2 \ ? \ 2 \\
2 & \text{TRUE}
\end{array}
\qquad
\begin{array}{c|c}
x + y = 6 \\
\hline
4 + 2 \ ? \ 6 \\
6 & \text{TRUE}
\end{array}
$$

The solution is $(4, 2)$.

13. We graph the equations.

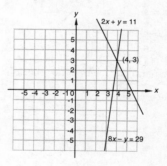

The point of intersection looks as if it has coordinates $(4, 3)$.

Check:

$$
\begin{array}{c|c}
8x - y = 29 \\
\hline
8 \cdot 4 - 3 \ ? \ 29 \\
32 - 3 \\
29 & \text{TRUE}
\end{array}
\qquad
\begin{array}{c|c}
2x + y = 11 \\
\hline
2 \cdot 4 + 3 \ ? \ 11 \\
8 + 3 \\
11 & \text{TRUE}
\end{array}
$$

The solution is $(4, 3)$.

15. We graph the equations.

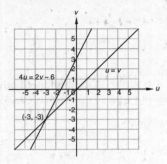

The point of intersection looks as if it has coordinates $(-3, -3)$.

Check:

$$\begin{array}{c|c} u = v & 4u = 2v - 6 \\ \hline -3 \ ? \ -3 \quad \text{TRUE} & 4(-3) \ ? \ 2(-3) - 6 \\ & -12 \ \Big| \ -6 - 6 \\ & \Big| \ -12 \qquad \text{TRUE} \end{array}$$

The solution is $(-3, -3)$.

17. We graph the equations.

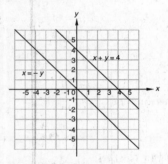

The lines are parallel. There is no solution.

19. We graph the equations.

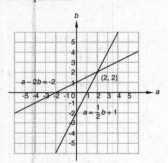

The point of intersection looks as if it has coordinates $(2, 2)$.

Check:

$$\begin{array}{c|c} a = \dfrac{1}{2}b + 1 & a - 2b = -2 \\ \hline 2 \ ? \ \dfrac{1}{2} \cdot 2 + 1 & 2 - 2 \cdot 2 \ ? \ -2 \\ \Big| \ 1 + 1 & 2 - 4 \\ \Big| \ 2 \qquad \text{TRUE} & -2 \ \Big| \quad \text{TRUE} \end{array}$$

The solution is $(2, 2)$.

21. We graph the equations.

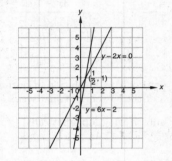

The point of intersection looks as if it has coordinates $\left(\dfrac{1}{2}, 1\right)$.

Check:

$$\begin{array}{c|c} y - 2x = 0 & y = 6x - 2 \\ \hline 1 - 2 \cdot \dfrac{1}{2} \ ? \ 0 & 1 \ ? \ 6 \cdot \dfrac{1}{2} - 2 \\ 1 - 1 \ \Big| & \Big| \ 3 - 2 \\ 0 \ \Big| \quad \text{TRUE} & \Big| \ 1 \qquad \text{TRUE} \end{array}$$

The solution is $\left(\dfrac{1}{2}, 1\right)$.

23. We graph the equations.

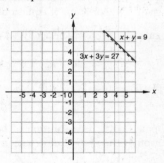

The lines coincide. The system has an infinite number of solutions.

25. We graph the equations.

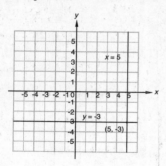

The point of intersection looks as if it has coordinates $(5, -3)$.

Check:

$x = 5$

$5\ ?\ 5$ TRUE

$y = -3$

$-3\ ?\ -3$ TRUE

The solution is $(5, -3)$.

27. Discussion and Writing Exercise

29.

$$\frac{1}{x} - \frac{1}{x^2} + \frac{1}{x+1}, \text{ LCM is } x^2(x+1)$$

$$= \frac{1}{x} \cdot \frac{x(x+1)}{x(x+1)} - \frac{1}{x^2} \cdot \frac{x+1}{x+1} + \frac{1}{x+1} \cdot \frac{x^2}{x^2}$$

$$= \frac{x(x+1) - (x+1) + x^2}{x^2(x+1)}$$

$$= \frac{x^2 + x - x - 1 + x^2}{x^2(x+1)}$$

$$= \frac{2x^2 - 1}{x^2(x+1)}$$

31.

$$\frac{x+2}{x-4} - \frac{x+1}{x+4}, \text{ LCM is } (x-4)(x+4)$$

$$= \frac{x+2}{x-4} \cdot \frac{x+4}{x+4} - \frac{x+1}{x+4} \cdot \frac{x-4}{x-4}$$

$$= \frac{(x+2)(x+4) - (x+1)(x-4)}{(x-4)(x+4)}$$

$$= \frac{x^2 + 6x + 8 - (x^2 - 3x - 4)}{(x-4)(x+4)}$$

$$= \frac{x^2 + 6x + 8 - x^2 + 3x + 4}{(x-4)(x+4)}$$

$$= \frac{9x + 12}{(x-4)(x+4)}$$

33. The polynomial has exactly three terms, so it is a trinomial.

35. The polynomial has exactly one term, so it is a monomial.

37. $(2, -3)$ is a solution of $Ax - 3y = 13$. Substitute 2 for x and -3 for y and solve for A.

$$Ax - 3y = 13$$

$$A \cdot 2 - 3(-3) = 13$$

$$2A + 9 = 13$$

$$2A = 4$$

$$A = 2$$

$(2, -3)$ is a solution of $x - By = 8$. Substitute 2 for x and -3 for y and solve for B.

$$x - By = 8$$

$$2 - B(-3) = 8$$

$$2 + 3B = 8$$

$$3B = 6$$

$$B = 2$$

39. Answers may vary. Any two equations with a solution of $(6, -2)$ will do. One possibility is

$$x + y = 4,$$

$$x - y = 8.$$

41.–47. Left to the student

Exercise Set 15.2

1. $x + y = 10,$ (1)

$y = x + 8$ (2)

We substitute $x + 8$ for y in Equation (1) and solve for x.

$x + y = 10$ Equation (1)

$x + (x + 8) = 10$ Substituting

$2x + 8 = 10$ Collecting like terms

$2x = 2$ Subtracting 8

$x = 1$ Dividing by 2

Next we substitute 1 for x in either equation of the original system and solve for y. We choose Equation (2) since it has y alone on one side.

$y = x + 8$ Equation (2)

$y = 1 + 8$ Substituting

$y = 9$

We check the ordered pair $(1, 9)$.

$x + y = 10$

$1 + 9\ ?\ 10$

10 | TRUE

$y = x + 8$

$9\ ?\ 1 + 8$

9 | TRUE

Since $(1, 9)$ checks in both equations, it is the solution.

3. $y = x - 6,$ (1)

$x + y = -2$ (2)

We substitute $x - 6$ for y in Equation (2) and solve for x.

$x + y = -2$ Equation (2)

$x + (x - 6) = -2$ Substituting

$2x - 6 = -2$ Collecting like terms

$2x = 4$ Adding 6

$x = 2$ Dividing by 2

Next we substitute 2 for x in either equation of the original system and solve for y. We choose Equation (1) since it has y alone on one side.

$$y = x - 6 \quad \text{Equation (1)}$$
$$y = 2 - 6 \quad \text{Substituting}$$
$$y = -4$$

We check the ordered pair $(2, -4)$.

$y = x - 6$	$x + y = -2$
$-4 \ ? \ 2 - 6$	$2 + (-4) \ ? \ -2$
$\ \ \ \ \mid -4 \quad$ TRUE	$-2 \mid$ TRUE

Since $(2, -4)$ checks in both equations, it is the solution.

5. $\ \ y = 2x - 5, \quad (1)$
$\quad \ \ 3y - x = 5 \quad (2)$

We substitute $2x - 5$ for y in Equation (2) and solve for x.

$$3y - x = 5 \quad \text{Equation (2)}$$
$$3(2x - 5) - x = 5 \quad \text{Substituting}$$
$$6x - 15 - x = 5 \quad \text{Removing parentheses}$$
$$5x - 15 = 5 \quad \text{Collecting like terms}$$
$$5x = 20 \quad \text{Adding 15}$$
$$x = 4 \quad \text{Dividing by 5}$$

Next we substitute 4 for x in either equation of the original system and solve for y.

$$y = 2x - 5 \quad \text{Equation (1)}$$
$$y = 2 \cdot 4 - 5 \quad \text{Substituting}$$
$$y = 8 - 5$$
$$y = 3$$

We check the ordered pair $(4, 3)$.

$y = 2x - 5$	$3y - x = 5$
$3 \ ? \ 2 \cdot 4 - 5$	$3 \cdot 3 - 4 \ ? \ 5$
$\ \ \ \mid 8 - 5$	$9 - 4 \mid$
$\ \ \ \mid 3 \quad$ TRUE	$5 \quad$ TRUE

Since $(4, 3)$ checks in both equations, it is the solution.

7. $\ \ x = -2y, \quad (1)$
$\quad \ \ x + 4y = 2 \quad (2)$

We substitute $-2y$ for x in Equation (2) and solve for y.

$$x + 4y = 2 \quad \text{Equation (2)}$$
$$-2y + 4y = 2 \quad \text{Substituting}$$
$$2y = 2 \quad \text{Collecting like terms}$$
$$y = 1 \quad \text{Dividing by 2}$$

Next we substitute 1 for y in either equation of the original system and solve for x.

$$x = -2y \quad \text{Equation (1)}$$
$$x = -2 \cdot 1$$
$$x = -2$$

We check the ordered pair $(-2, 1)$.

$x = -2y$	$3y - x = 5$
$-2 \ ? \ -2 \cdot 1$	$3 \cdot 1 - (-2) \ ? \ 5$
$\ \ \ \ \mid -2 \quad$ TRUE	$3 + 2 \mid$
	$5 \mid$ TRUE

Since $(-2, 1)$ checks in both equations, it is the solution.

9. $\ \ x - y = 6, \quad (1)$
$\quad \ \ x + y = -2 \quad (2)$

We solve Equation (1) for x.

$$x - y = 6 \quad \text{Equation (1)}$$
$$x = y + 6 \quad \text{Adding } y \quad (3)$$

We substitute $y + 6$ for x in Equation (2) and solve for y.

$$x + y = -2 \quad \text{Equation (2)}$$
$$(y + 6) + y = -2 \quad \text{Substituting}$$
$$2y + 6 = -2 \quad \text{Collecting like terms}$$
$$2y = -8 \quad \text{Subtracting 6}$$
$$y = -4 \quad \text{Dividing by 2}$$

Now we substitute -4 for y in Equation (3) and compute x.

$$x = y + 6 = -4 + 6 = 2$$

The ordered pair $(2, -4)$ checks in both equations. It is the solution.

11. $\ \ y - 2x = -6, \quad (1)$
$\quad \ \ 2y - x = 5 \quad (2)$

We solve Equation (1) for y.

$$y - 2x = -6 \quad \text{Equation (1)}$$
$$y = 2x - 6 \quad (3)$$

We substitute $2x - 6$ for y in Equation (2) and solve for x.

$$2y - x = 5 \quad \text{Equation (2)}$$
$$2(2x - 6) - x = 5 \quad \text{Substituting}$$
$$4x - 12 - x = 5 \quad \text{Removing parentheses}$$
$$3x - 12 = 5 \quad \text{Collecting like terms}$$
$$3x = 17 \quad \text{Adding 12}$$
$$x = \frac{17}{3} \quad \text{Dividing by 3}$$

We substitute $\frac{17}{3}$ for x in Equation (3) and compute y.

$$y = 2x - 6 = 2\left(\frac{17}{3}\right) - 6 = \frac{34}{3} - \frac{18}{3} = \frac{16}{3}$$

The ordered pair $\left(\frac{17}{3}, \frac{16}{3}\right)$ checks in both equations. It is the solution.

13. $\ \ 2x + 3y = -2, \quad (1)$
$\quad \ \ 2x - y = 9 \quad (2)$

We solve Equation (2) for y.

$$2x - y = 9 \quad \text{Equation (2)}$$
$$2x = 9 + y \quad \text{Adding } y$$
$$2x - 9 = y \quad \text{Subtracting 9} \quad (3)$$

We substitute $2x - 9$ for y in Equation (1) and solve for x.

$$2x + 3y = -2 \qquad \text{Equation (1)}$$
$$2x + 3(2x - 9) = -2 \qquad \text{Substituting}$$
$$2x + 6x - 27 = -2 \qquad \text{Removing parentheses}$$
$$8x - 27 = -2 \qquad \text{Collecting like terms}$$
$$8x = 25 \qquad \text{Adding 27}$$
$$x = \frac{25}{8} \qquad \text{Dividing by 8}$$

Now we substitute $\dfrac{25}{8}$ for x in Equation (3) and compute y.

$$y = 2x - 9 = 2\left(\frac{25}{8}\right) - 9 = \frac{25}{4} - \frac{36}{4} = -\frac{11}{4}$$

The ordered pair $\left(\dfrac{25}{8}, -\dfrac{11}{4}\right)$ checks in both equations. It is the solution.

15. $x - y = -3$, (1)

$2x + 3y = -6$ (2)

We solve Equation (1) for x.

$$x - y = -3 \qquad \text{Equation (1)}$$
$$x = y - 3 \qquad\qquad (3)$$

We substitute $y - 3$ for x in Equation (2) and solve for y.

$$2x + 3y = -6 \qquad \text{Equation (2)}$$
$$2(y - 3) + 3y = -6 \qquad \text{Substituting}$$
$$2y - 6 + 3y = -6 \qquad \text{Removing parentheses}$$
$$5y - 6 = -6 \qquad \text{Collecting like terms}$$
$$5y = 0 \qquad \text{Adding 6}$$
$$y = 0 \qquad \text{Dividing by 5}$$

Now we substitute 0 for y in Equation (3) and compute x.

$$x = y - 3 = 0 - 3 = -3$$

The ordered pair $(-3, 0)$ checks in both equations. It is the solution.

17. $r - 2s = 0$, (1)

$4r - 3s = 15$ (2)

We solve Equation (1) for r.

$$r - 2s = 0 \qquad \text{Equation (1)}$$
$$r = 2s \qquad\qquad (3)$$

We substitute $2s$ for r in Equation (2) and solve for s.

$$4r - 3s = 15 \qquad \text{Equation (2)}$$
$$4(2s) - 3s = 15 \qquad \text{Substituting}$$
$$8s - 3s = 15 \qquad \text{Removing parentheses}$$
$$5s = 15 \qquad \text{Collecting like terms}$$
$$s = 3 \qquad \text{Dividing by 5}$$

Now we substitute 3 for s in Equation (3) and compute r.

$$r = 2s = 2 \cdot 3 = 6$$

The ordered pair $(6, 3)$ checks in both equations. It is the solution.

19. *Familiarize*. We let $w = $ the width of the court, in ft, and $l = $ the length, in ft. Recall that the perimeter of a rectangle with length l and width w is given by $2l + 2w$.

***Translate*.**

$$\underbrace{\text{The perimeter}}_{2l + 2w} \underbrace{\text{is}}_{=} \underbrace{288 \text{ ft.}}_{288}$$

$$\underbrace{\text{The length}}_{l} \underbrace{\text{is}}_{=} \underbrace{44 \text{ ft}}_{44} \underbrace{\text{more than}}_{+} \underbrace{\text{the width.}}_{w}$$

The resulting system is

$$2l + 2w = 288, \quad (1)$$
$$l = 44 + w. \qquad (2)$$

***Solve*.** We solve the system. Substitute $44 + w$ for l in the first equation and solve for w.

$$2(44 + w) + 2w = 288$$
$$88 + 2w + 2w = 288$$
$$88 + 4w = 288$$
$$4w = 200$$
$$w = 50$$

Now substitute 50 for w in Equation (2).

$$l = 44 + w = 44 + 50 = 94$$

***Check*.** If the length is 94 ft and the width is 50 ft, then the length is 44 ft more than the width and the perimeter is $2 \cdot 94 + 2 \cdot 50$, or $188 + 100$, or 288 ft. The answer checks.

***State*.** The length of the court is 94 ft, and the width is 50 ft.

21. *Familiarize*. We make a drawing. We let $l = $ the length and $w = $ the width, in inches.

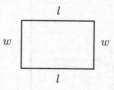

***Translate*.** The perimeter is $2l + 2w$. We translate the first statement.

$$\underbrace{\text{The perimeter}}_{2l + 2w} \underbrace{\text{is}}_{=} \underbrace{10 \text{ in.}}_{10}$$

We translate the second statement.

$$\underbrace{\text{The length}}_{l} \underbrace{\text{is}}_{=} \underbrace{\text{twice the width.}}_{w + 2}$$

The resulting system is

$$2l + 2w = 10, \quad (1)$$
$$l = w + 2. \qquad (2)$$

Solve. We solve the system. We substitute $w + 2$ for l in Equation (1) and solve for w.

$$2l + 2w = 10 \qquad \text{Equation (1)}$$
$$2(w + 2) + 2w = 10 \qquad \text{Substituting}$$
$$2w + 4 + 2w = 10 \qquad \text{Removing parentheses}$$
$$4w + 4 = 10$$
$$4w = 6 \qquad \text{Collecting like terms}$$
$$w = \frac{3}{2}, \text{ or } 1\frac{1}{2}$$

Now we substitute $\frac{3}{2}$ for w in Equation (2) and solve for l.

$$l = w + 2 \qquad \text{Equation (2)}$$
$$l = \frac{3}{2} + 2 \qquad \text{Substituting}$$
$$l = \frac{3}{2} + \frac{4}{2}$$
$$l = \frac{7}{2}, \text{ or } 3\frac{1}{2}$$

Check. A possible solution is a length of $\frac{7}{2}$, or $3\frac{1}{2}$ in. and a width of $\frac{3}{2}$, or $1\frac{1}{2}$ in. The perimeter would be $2 \cdot \frac{7}{2} + 2 \cdot \frac{3}{2}$, or $7 + 3$, or 10 in. Also, the width plus 2 is $\frac{3}{2} + 2$, or $\frac{7}{2}$, which is the length. These numbers check.

State. The length is $3\frac{1}{2}$ in., and the width is $1\frac{1}{2}$ in.

23. Familiarize. Let l = the length, in mi, and w = the width, in mi. Recall that the perimeter of a rectangle with length l and width w is given by $2l + 2w$.

Translate.

The perimeter is 1280 mi.
$$2l + 2w = 1280$$

The width is the length less 90 mi.
$$w = l - 90$$

The resulting system is
$$2l + 2w = 1280, \quad (1)$$
$$w = l - 90. \qquad (2)$$

Solve. We solve the system. Substitute $l - 90$ for w in the first equation and solve for l.

$$2l + 2(l - 90) = 1280$$
$$2l + 2l - 180 = 1280$$
$$4l - 180 = 1280$$
$$4l = 1460$$
$$l = 365$$

Now substitute 365 for l in Equation (2).
$$w = l - 90 = 365 - 90 = 275$$

Check. If the length is 365 mi and the width is 275 mi, then the width is 90 mi less than the length and the perimeter is $2 \cdot 365 + 2 \cdot 275$, or $730 + 550$, or 1280. The answer checks.

State. The length is 365 mi, and the width is 275 mi.

25. Familiarize. Let l = the length in ft, and w = the width, in ft. Recall that the perimeter of a rectangle with length l and width w is given by $2l + 2w$.

Translate.

The perimeter is 120 ft.
$$2l + 2w = 120$$

The length is twice the width.
$$l = 2w$$

The resulting system is
$$2l + 2w = 120, \quad (1)$$
$$l = 2w. \qquad (2)$$

Solve. We solve the system.

Substitute $2w$ for l in Equation (1) and solve for w.

$$2 \cdot 2w + 2w = 120 \quad (1)$$
$$4w + 2w = 120$$
$$6w = 120$$
$$w = 20$$

Now substitute 20 for w in Equation (2).
$$l = 2w \qquad (2)$$
$$l = 2 \cdot 20 \qquad \text{Substituting}$$
$$l = 40$$

Check. If the length is 40 ft and the width is 20 ft, the perimeter would be $2 \cdot 40 + 2 \cdot 20$, or $80 + 40$, or 120 ft. Also, the length is twice the width. These numbers check.

State. The length is 40 ft, and the width is 20 ft.

27. Familiarize. Let l = the length and w = the width, in yards. The perimeter is $l + l + w + w$, or $2l + 2w$.

Translate.

The perimeter is 340 yd.
$$2l + 2w = 340$$

The length is 10 yd less than twice the width.
$$l = 2w - 10$$

The resulting system is
$$2l + 2w = 340, \quad (1)$$
$$l = 2w - 10. \quad (2)$$

Solve. We solve the system. We substitute $2w - 10$ for l in Equation (1) and solve for w.

$$2l + 2w = 340 \quad (1)$$
$$2(w - 10) + 2w = 340$$
$$4w - 20 + 2w = 340$$
$$6w - 20 = 340$$
$$6w = 360$$
$$w = 60$$

Next we substitute 60 for w in Equation (2) and solve for l.

$$l = 2w - 10 = 2 \cdot 60 - 10 = 120 - 10 = 110$$

Check. The perimeter is $2 \cdot 110 + 2 \cdot 60$, or 340 yd. Also 10 yd less than twice the width is $2 \cdot 60 - 10 = 120 - 10 = 110$. The answer checks.

State. The length is 110 yd, and the width is 60 yd.

29. Familiarize. We let $x =$ the larger number and $y =$ the smaller number.

Translate. We translate the first statement.

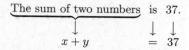

Now we translate the second statement.

One number is 5 more than the other.
$\downarrow$ $\downarrow$ $\downarrow$ $\downarrow$ $\downarrow$
x $=$ 5 $+$ y

The resulting system is

$$x + y = 37, \quad (1)$$
$$x = 5 + y. \quad (2)$$

Solve. We solve the system of equations. We substitute $5 + y$ for x in Equation (1) and solve for y.

$$x + y = 37 \qquad \text{Equation (1)}$$
$$(5 + y) + y = 37 \qquad \text{Substituting}$$
$$5 + 2y = 37 \qquad \text{Collecting like terms}$$
$$2y = 32 \qquad \text{Subtracting 5}$$
$$y = 16 \qquad \text{Dividing by 2}$$

We go back to the original equations and substitute 16 for y. We use Equation (2).

$$x = 5 + y \qquad \text{Equation (2)}$$
$$x = 5 + 16 \qquad \text{Substituting}$$
$$x = 21$$

Check. The sum of 21 and 16 is 37. The number 21 is 5 more than the number 16. These numbers check.

State. The numbers are 21 and 16.

31. Familiarize. Let $x =$ one number and $y =$ the other.

Translate. We reword and translate.

The sum of two numbers is 52.
$\downarrow$ $\downarrow$
$x + y$ $=$ 52

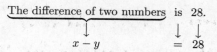

(The second statement could also be translated as $y - x = 28$.)

The resulting system is

$$x + y = 52, \quad (1)$$
$$x - y = 28. \quad (2)$$

Solve. We solve the system. First we solve Equation (2) for x.

$$x - y = 28 \qquad \text{Equation (2)}$$
$$x = y + 28 \qquad \text{Adding } y \qquad (3)$$

We substitute $y + 28$ for x in Equation (1) and solve for y.

$$x + y = 52 \qquad \text{Equation (1)}$$
$$(y + 28) + y = 52 \qquad \text{Substituting}$$
$$2y + 28 = 52 \qquad \text{Collecting like terms}$$
$$2y = 24 \qquad \text{Subtracting 28}$$
$$y = 12 \qquad \text{Dividing by 2}$$

Now we substitute 12 for y in Equation (3) and compute x.

$$x = y + 28 = 12 + 28 = 40$$

Check. The sum of 40 and 12 is 52, and their difference is 28. These numbers check.

State. The numbers are 40 and 12.

33. Familiarize. We let $x =$ the larger number and $y =$ the smaller number.

Translate. We translate the first statement.

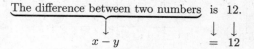

Now we translate the second statement.

Two times the larger number is five times the smaller.
$\downarrow$ $\downarrow$ $\downarrow$
$2x$ $=$ $5y$

The resulting system is

$$x - y = 12, \quad (1)$$
$$2x = 5y. \quad (2)$$

Solve. We solve the system. First we solve Equation (1) for x.

$$x - y = 12 \qquad \text{Equation (1)}$$
$$x = y + 12 \qquad \text{Adding } y \qquad (3)$$

We substitute $y + 12$ for x in Equation (2) and solve for y.

$$2x = 5y \qquad \text{Equation (2)}$$
$$2(y + 12) = 5y \qquad \text{Substituting}$$
$$2y + 24 = 5y \qquad \text{Removing parentheses}$$
$$24 = 3y \qquad \text{Subtracting 2y}$$
$$8 = y \qquad \text{Dividing by 3}$$

Now we substitute 8 for y in Equation (3) and compute x.

$$x = y + 12 = 8 + 12 = 20$$

Check. The difference between 20 and 8 is 12. Two times 20, or 40, is five times 8. These numbers check.

State. The numbers are 20 and 8.

35. Discussion and Writing exercise

37. Graph: $2x - 3y = 6$

To find the x-intercept, let $y = 0$. Then solve for x.

$$2x - 3 \cdot 0 = 6$$
$$2x = 6$$
$$x = 3$$

The x-intercept is $(3, 0)$.

To find the y-intercept, let $x = 0$. Then solve for y.

$$2 \cdot 0 - 3y = 6$$
$$-3y = 6$$
$$y = -2$$

The y-intercept is $(0, -2)$.

We plot these points and draw the line.

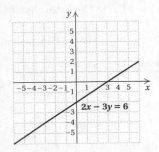

A third point should be used as a check. We let $x = -3$:

$$2(-3) - 3y = 6$$
$$-6 - 3y = 6$$
$$-3y = 12$$
$$y = -4$$

The point $(-3, -4)$ is on the graph, so our graph is probably correct.

39. Graph: $y = 2x - 5$

We select several values for x and compute the corresponding y-values.

When $x = 0$, $y = 2 \cdot 0 - 5 = 0 - 5 = -5$.

When $x = 2$, $y = 2 \cdot 2 - 5 = 4 - 5 = -1$.

When $x = 4$, $y = 2 \cdot 4 - 5 = 8 - 5 = 3$.

x	y	(x, y)
0	-5	$(0, -5)$
2	-1	$(2, -1)$
4	3	$(4, 3)$

We plot these points and draw the line connecting them.

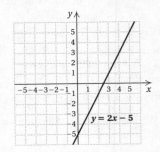

41. $6x^2 - 13x + 6$

The possibilities are $(x+\quad)(6x+\quad)$ and $(2x+\quad)(3x+\quad)$. We look for a pair of factors of the last term, 6, which produces the correct middle term. Since the last term is positive and the middle term is negative, we need only consider negative pairs. The factorization is $(2x - 3)(3x - 2)$.

43. $4x^2 + 3x + 2$

The possibilities are $(x+\quad)(4x+\quad)$ and $(2x+\quad)(2x+\quad)$. We look for a pair of factors of the last term, 2, which produce the correct middle term. Since the last term and the middle term are both positive, we need only consider positive pairs. We find that there is no possibility that works. The trinomial cannot be factored.

45. $\dfrac{x^{-2}}{x^{-5}} = x^{-2-(-5)} = x^3$

47. $x^{-2} \cdot x^{-5} = x^{-2+(-5)} = x^{-7} = \dfrac{1}{x^7}$

49. First put the equations in "$y =$" form by solving for y. We get

$$y_1 = x - 5,$$
$$y_2 = (-1/2)x + 7/2.$$

Then graph these equations in the standard window and use the INTERSECT feature from the CALC menu to find the coordinates of the point of intersection of the graphs. The solution is $(5.\overline{6}, 0.\overline{6})$.

51. First put the equations in "$y =$" form by solving for y. We get

$$y_1 = 2.35x - 5.97,$$
$$y_2 = (1/2.14)x + (4.88/2.14).$$

Then graph these equations in the standard window and use the INTERSECT feature from the CALC menu to find the coordinates of the point of intersection of the graphs. The solution is approximately $(4.38, 4.33)$.

53. *Familiarize.* Let $s =$ the perimeter of a softball diamond, in yards, and $b =$ the perimeter of a baseball diamond, in yards.

Translate.

$$\underbrace{\text{Perimeter of a softball diamond}}_{s} \underset{=}{\text{is}} \underbrace{\frac{2}{3}}_{\frac{2}{3}} \underset{\cdot}{\text{of}} \underbrace{\text{perimeter of a baseball diamond.}}_{b}$$

$$\underbrace{\text{The sum of the perimeters}}_{s+b} \underset{=}{\text{is}} \underbrace{200 \text{ yd.}}_{200}$$

The resulting system is

$$s = \frac{2}{3}b, \quad (1)$$
$$s + b = 200. \quad (2)$$

Solve. We solve the system of equations. We substitute $\frac{2}{3}b$ for s in Equation (2) and solve for b.

$$s + b = 200$$
$$\frac{2}{3}b + b = 200$$
$$\frac{5}{3}b = 200$$
$$\frac{3}{5} \cdot \frac{5}{3}b = \frac{3}{5} \cdot 200$$
$$b = 120$$

Next we substitute 120 for b in Equation (1) and solve for s.

$$s = \frac{2}{3}b = \frac{2}{3} \cdot 120 = 80$$

Each diamond has four sides of equal length, so we divide each perimeter by 4 to find the distance between bases in each sport. For the softball diamond the distance is 80/4, or 20 yd. For the baseball diamond it is 120/4, or 30 yd.

Check. The perimeter of the softball diamond, 80 yd, is $\frac{2}{3}$ of 120 yd, the perimeter of the baseball diamond. The sum of the perimeters is $80 + 120$, or 200 yd. We can also recheck the calculations of the distances between the bases. The answer checks.

State. The distance between bases on a softball diamond is 20 yd and the distance between bases on a baseball diamond is 30 yd.

Exercise Set 15.3

1. $x - y = 7$ (1)
$$\underline{x + y = 5} \quad (2)$$
$$2x \quad\quad = 12 \quad \text{Adding}$$
$$x = 6 \quad \text{Dividing by 2}$$

Substitute 6 for x in either of the original equations and solve for y.

$$x + y = 5 \quad \text{Equation (2)}$$
$$6 + y = 5 \quad \text{Substituting}$$
$$y = -1 \quad \text{Subtracting 6}$$

Check:

$x - y = 7$		$x + y = 5$	
$6 - (-1)\ ?\ 7$		$6 + (-1)\ ?\ 5$	
$6 + 1$		5	TRUE
7	TRUE		

Since $(6, -1)$ checks, it is the solution.

3. $x + y = 8$ (1)
$$\underline{-x + 2y = 7} \quad (2)$$
$$3y = 15 \quad \text{Adding}$$
$$y = 5 \quad \text{Dividing by 3}$$

Substitute 5 for y in either of the original equations and solve for x.

$$x + y = 8 \quad \text{Equation (1)}$$
$$x + 5 = 8 \quad \text{Substituting}$$
$$x = 3$$

Check:

$x + y = 8$		$-x + 2y = 7$	
$3 + 5\ ?\ 8$		$-3 + 2 \cdot 5\ ?\ 7$	
8	TRUE	$-3 + 10$	
		7	TRUE

Since $(3, 5)$ checks, it is the solution.

5. $5x - y = 5$ (1)
$$\underline{3x + y = 11} \quad (2)$$
$$8x \quad\quad = 16 \quad \text{Adding}$$
$$x = 2 \quad \text{Dividing by 8}$$

Substitute 2 for x in either of the original equations and solve for y.

$$3x + y = 11 \quad \text{Equation (2)}$$
$$3 \cdot 2 + y = 11 \quad \text{Substituting}$$
$$6 + y = 11$$
$$y = 5$$

Check:

$5x - y = 5$		$3x + y = 11$	
$5 \cdot 2 - 5\ ?\ 5$		$3 \cdot 2 + 5\ ?\ 11$	
$10 - 5$		$6 + 5$	
5	TRUE	11	TRUE

Since $(2, 5)$ checks, it is the solution.

7. $4a + 3b = 7$ (1)
$$\underline{-4a + b = 5} \quad (2)$$
$$4b = 12 \quad \text{Adding}$$
$$b = 3$$

Substitute 3 for b in either of the original equations and solve for a.

$$4a + 3b = 7 \quad \text{Equation (1)}$$
$$4a + 3 \cdot 3 = 7 \quad \text{Substituting}$$
$$4a + 9 = 7$$
$$4a = -2$$
$$a = -\frac{1}{2}$$

Check:

$4a + 3b = 7$		$-4a + b = 5$	
$4\left(-\frac{1}{2}\right) + 3 \cdot 3 \ ? \ 7$		$-4\left(-\frac{1}{2}\right) + 3 \ ? \ 5$	
$-2 + 9$		$2 + 3$	
7	TRUE	5	TRUE

Since $\left(-\frac{1}{2}, 3\right)$ checks, it is the solution.

9. $8x - 5y = -9 \quad (1)$
  $\underline{3x + 5y = -2} \quad (2)$
  $11x \qquad\quad = -11 \quad \text{Adding}$
  $x = -1$

Substitute -1 for x in either of the original equations and solve for y.

$$3x + 5y = -2 \quad \text{Equation (2)}$$
$$3(-1) + 5y = -2 \quad \text{Substituting}$$
$$-3 + 5y = -2$$
$$5y = 1$$
$$y = \frac{1}{5}$$

Check:

$8x - 5y = -9$		$3x + 5y = -2$	
$8(-1) - 5\left(\frac{1}{5}\right) \ ? \ -9$		$3(-1) + 5\left(\frac{1}{5}\right) \ ? \ -2$	
$-8 - 1$		$-3 + 1$	
-9	TRUE	-2	TRUE

Since $\left(-1, \frac{1}{5}\right)$ checks, it is the solution.

11. $4x - 5y = 7$
  $\underline{-4x + 5y = 7}$
  $\qquad 0 = 14 \quad \text{Adding}$

We obtain a false equation, $0 = 14$, so there is no solution.

13. $x + y = -7, \quad (1)$
  $3x + y = -9 \quad (2)$

We multiply on both sides of Equation (1) by -1 and then add.

  $-x - y = 7 \qquad \text{Multiplying by } -1$
  $\underline{3x + y = -9} \quad \text{Equation (2)}$
  $2x \qquad = -2 \quad \text{Adding}$
  $x = -1$

Substitute -1 for x in one of the original equations and solve for y.

$$x + y = -7 \quad \text{Equation (1)}$$
$$-1 + y = -7 \quad \text{Substituting}$$
$$y = -6$$

Check:

$x + y = -7$		$3x + y = -9$	
$-1 + (-6) \ ? \ -7$		$3(-1) + (-6) \ ? \ -9$	
-7	TRUE	$-3 - 6$	
		-9	TRUE

Since $(-1, -6)$ checks, it is the solution.

15. $3x - y = 8, \quad (1)$
  $x + 2y = 5 \quad (2)$

We multiply on both sides of Equation (1) by 2 and then add.

  $6x - 2y = 16 \quad \text{Multiplying by 2}$
  $\underline{x + 2y = 5} \qquad \text{Equation (2)}$
  $7x \qquad = 21 \quad \text{Adding}$
  $x = 3$

Substitute 3 for x in one of the original equations and solve for y.

$$x + 2y = 5 \quad \text{Equation (2)}$$
$$3 + 2y = 5 \quad \text{Substituting}$$
$$2y = 2$$
$$y = 1$$

Check:

$3x - y = 8$		$x + 2y = 5$	
$3 \cdot 3 - 1 \ ? \ 8$		$3 + 2 \cdot 1 \ ? \ 5$	
$9 - 1$		$3 + 2$	
8	TRUE	5	TRUE

Since $(3, 1)$ checks, it is the solution.

17. $x - y = 5, \quad (1)$
  $4x - 5y = 17 \quad (2)$

We multiply on both sides of Equation (1) by -4 and then add.

  $-4x + 4y = -20 \quad \text{Multiplying by } -4$
  $\underline{4x - 5y = 17} \qquad \text{Equation (2)}$
  $\qquad -y = -3 \quad \text{Adding}$
  $y = 3$

Substitute 3 for y in one of the original equations and solve for x.

$$x - y = 5 \quad \text{Equation (1)}$$
$$x - 3 = 5 \quad \text{Substituting}$$
$$x = 8$$

Check:

$$\frac{x - y = 5}{8 - 3 \; ? \; 5}$$
$$5 \; \bigg| \; \text{TRUE}$$

$$\frac{4x - 5y = 17}{4 \cdot 8 - 5 \cdot 3 \; ? \; 17}$$
$$32 - 15 \; \bigg|$$
$$17 \; \bigg| \; \text{TRUE}$$

Since $(8, 3)$ checks, it is the solution.

19. $2w - 3z = -1$, (1)

$3w + 4z = 24$ (2)

We use the multiplication principle with both equations and then add.

$$\frac{\begin{aligned}8w - 12z &= -4 \quad \text{Multiplying (1) by 4} \\ 9w + 12z &= 72 \quad \text{Multiplying (2) by 3}\end{aligned}}{\begin{aligned}17w \quad\;\; &= 68 \quad \text{Adding} \\ w &= 4\end{aligned}}$$

Substitute 4 for w in one of the original equations and solve for z.

$$3w + 4z = 24 \quad \text{Equation (2)}$$
$$3 \cdot 4 + 4z = 24 \quad \text{Substituting}$$
$$12 + 4z = 24$$
$$4z = 12$$
$$z = 3$$

Check:

$$\frac{2w - 3z = -1}{2 \cdot 4 - 3 \cdot 3 \; ? \; -1}$$
$$8 - 9 \; \bigg|$$
$$-1 \; \bigg| \; \text{TRUE}$$

$$\frac{3w + 4z = 24}{3 \cdot 4 + 4 \cdot 3 \; ? \; 24}$$
$$12 + 12 \; \bigg|$$
$$24 \; \bigg| \; \text{TRUE}$$

Since $(4, 3)$ checks, it is the solution.

21. $2a + 3b = -1$, (1)

$3a + 5b = -2$ (2)

We use the multiplication principle with both equations and then add.

$$\frac{\begin{aligned}-10a - 15b &= 5 \quad \text{Multiplying (1) by } -5 \\ 9a + 15b &= -6 \quad \text{Multiplying (2) by 3}\end{aligned}}{\begin{aligned}-a \quad\quad\;\; &= -1 \quad \text{Adding} \\ a &= 1\end{aligned}}$$

Substitute 1 for a in one of the original equations and solve for b.

$$2a + 3b = -1 \quad \text{Equation (1)}$$
$$2 \cdot 1 + 3b = -1 \quad \text{Substituting}$$
$$2 + 3b = -1$$
$$3b = -3$$
$$b = -1$$

Check:

$$\frac{2a + 3b = -1}{2 \cdot 1 + 3(-1) \; ? \; -1}$$
$$2 - 3 \; \bigg|$$
$$-1 \; \bigg| \; \text{TRUE}$$

$$\frac{3a + 5b = -2}{3 \cdot 1 + 5(-1) \; ? \; -2}$$
$$3 - 5 \; \bigg|$$
$$-2 \; \bigg| \; \text{TRUE}$$

Since $(1, -1)$ checks, it is the solution.

23. $x = 3y$, (1)

$5x + 14 = y$ (2)

We first get each equation in the form $Ax + By = C$.

$$x - 3y = 0, \quad \text{(1a)} \quad \text{Adding } -3y$$
$$5x - y = -14 \quad \text{(2a)} \quad \text{Adding } -y - 14$$

We multiply by -5 on both sides of Equation (1a) and add.

$$\frac{\begin{aligned}-5x + 15y &= 0 \quad \text{Multiplying by } -5 \\ 5x - \quad y &= -14\end{aligned}}{\begin{aligned}14y &= -14 \quad \text{Adding} \\ y &= -1\end{aligned}}$$

Substitute -1 for y in Equation (1) and solve for x.

$$x - 3y = 0$$
$$x - 3(-1) = 0 \quad \text{Substituting}$$
$$x + 3 = 0$$
$$x = -3$$

Check:

$$\frac{x - 3y = 0}{-3 - 3(-1) \; ? \; 0}$$
$$-3 + 3 \; \bigg|$$
$$0 \; \bigg| \; \text{TRUE}$$

$$\frac{5x - y = -14}{5(-3) - (-1) \; ? \; -14}$$
$$-15 + 1 \; \bigg|$$
$$-14 \; \bigg| \; \text{TRUE}$$

Since $(-3, -1)$ checks, it is the solution.

25. $2x + 5y = 16$, (1)

$3x - 2y = 5$ (2)

We use the multiplication principle with both equations and then add.

$$\frac{\begin{aligned}4x + 10y &= 32 \quad \text{Multiplying (1) by 2} \\ 15x - 10y &= 25 \quad \text{Multiplying (2) by 5}\end{aligned}}{\begin{aligned}19x \quad\quad\;\; &= 57 \\ x &= 3\end{aligned}}$$

Substitute 3 for x in one of the original equations and solve for y.

$$2x + 5y = 16 \quad \text{Equation (1)}$$
$$2 \cdot 3 + 5y = 16 \quad \text{Substituting}$$
$$6 + 5y = 16$$
$$5y = 10$$
$$y = 2$$

Check:

$$\frac{2x + 5y = 16}{2 \cdot 3 + 5 \cdot 2 \ ? \ 16}$$
$$6 + 10$$
$$16 \quad \mid \quad \text{TRUE}$$

$$\frac{3x - 2y = 5}{3 \cdot 3 - 2 \cdot 2 \ ? \ 5}$$
$$9 - 4$$
$$5 \quad \mid \quad \text{TRUE}$$

Since $(3, 2)$ checks, it is the solution.

27. $p = 32 + q,$ (1)

$3p = 8q + 6$ (2)

First we write each equation in the form $Ap + Bq = C$.

$p - q = 32,$ (1a)   Subtracting q

$3p - 8q = 6$ (2a)   Subtracting $8q$

Now we multiply both sides of Equation (1a) by -3 and then add.

$$-3p + 3q = -96 \quad \text{Multiplying by } -3$$
$$\underline{3p - 8q = 6} \quad \text{Equation (2a)}$$
$$-5q = -90 \quad \text{Adding}$$
$$q = 18$$

Substitute 18 for q in Equation (1) and solve for p.

$p = 32 + q$

$p = 32 + 18$ Substituting

$p = 50$

Check:

$$\frac{p - q = 32}{50 - 18 \ ? \ 32}$$
$$32 \quad \mid \quad \text{TRUE}$$

$$\frac{3p - 8q = 6}{3 \cdot 50 - 8 \cdot 18 \ ? \ 6}$$
$$150 - 144$$
$$6 \quad \mid \quad \text{TRUE}$$

Since $(50, 18)$ checks, it is the solution.

29. $3x - 2y = 10,$ (1)

$-6x + 4y = -20$ (2)

We multiply by 2 on both sides of Equation (1) and add.

$$6x - 4y = 20$$
$$\underline{-6x + 4y = -20}$$
$$0 = 0$$

We get an obviously true equation, so the system has an infinite number of solutions.

31. $0.06x + 0.05y = 0.07,$

$0.04x - 0.03y = 0.11$

We first multiply each equation by 100 to clear the decimals.

$6x + 5y = 7,$ (1)

$4x - 3y = 11$ (2)

We use the multiplication principle with both equations of the resulting system.

$$18x + 15y = 21 \quad \text{Multiplying (1) by 3}$$
$$\underline{20x - 15y = 55} \quad \text{Multiplying (2) by 5}$$
$$38x \quad = 76 \quad \text{Adding}$$
$$x = 2$$

Substitute 2 for x in Equation (1) and solve for y.

$6x + 5y = 7$

$6 \cdot 2 + 5y = 7$

$12 + 5y = 7$

$5y = -5$

$y = -1$

Check:

$$\frac{0.06x + 0.05y = 0.07}{0.06(2) + 0.05(-1) \ ? \ 0.07}$$
$$0.12 - 0.05$$
$$0.07 \quad \mid \quad \text{TRUE}$$

$$\frac{0.04x - 0.03y = 0.11}{0.04(2) - 0.03(-1) \ ? \ 0.11}$$
$$0.08 + 0.03$$
$$0.11 \quad \mid \quad \text{TRUE}$$

Since $(2, -1)$ checks, it is the solution.

33. $\dfrac{1}{3}x + \dfrac{3}{2}y = \dfrac{5}{4},$

$\dfrac{3}{4}x - \dfrac{5}{6}y = \dfrac{3}{8}$

First we clear the fractions. We multiply on both sides of the first equation by 12 and on both sides of the second equation by 24.

$$12\left(\frac{1}{3}x + \frac{3}{2}y\right) = 12 \cdot \frac{5}{4}$$
$$12 \cdot \frac{1}{3}x + 12 \cdot \frac{3}{2}y = 15$$
$$4x + 18y = 15$$

$$24\left(\frac{3}{4}x - \frac{5}{6}y\right) = 24 \cdot \frac{3}{8}$$
$$24 \cdot \frac{3}{4}x - 24 \cdot \frac{5}{6}y = 9$$
$$18x - 20y = 9$$

The resulting system is

$4x + 18y = 15,$ (1)

$18x - 20y = 9.$ (2)

We use the multiplication principle with both equations.

$$72x + 324y = 270 \quad \text{Multiplying (1) by 18}$$
$$\underline{-72x + 80y = -36} \quad \text{Multiplying (2) by } -4$$
$$404y = 234$$
$$y = \frac{234}{404}, \text{ or } \frac{117}{202}$$

Substitute $\frac{117}{202}$ for y in Equation (1) and solve for x.

$$4x + 18\left(\frac{117}{202}\right) = 15$$

$$4x + \frac{1053}{101} = 15$$

$$4x = \frac{462}{101}$$

$$x = \frac{1}{4} \cdot \frac{462}{101}$$

$$x = \frac{231}{202}$$

The ordered pair $\left(\frac{231}{202}, \frac{117}{202}\right)$ checks in both equations. It is the solution.

35. $-4.5x + 7.5y = 6,$

$\qquad -x + 1.5y = 5$

First we clear the decimals by multiplying by 10 on both sides of each equation.

$$10(-4.5x + 7.5y) = 10 \cdot 6$$

$$-45x + 75y = 60$$

$$10(-x + 1.5y) = 10 \cdot 5$$

$$-10x + 15y = 50$$

The resulting system is

$\qquad -45x + 75y = 60, \quad (1)$

$\qquad -10x + 15y = 50. \quad (2)$

We multiply both sides of Equation (2) by -5 and then add.

$$\begin{array}{ll} -45x + 75y = 60 & \text{Equation (1)} \\ \underline{50x - 75y = -250} & \text{Multiplying by } -5 \\ 5x \qquad\quad = -190 & \text{Adding} \\ x = -38 \end{array}$$

Substitute -38 for x in Equation (2) and solve for y.

$$-10x + 15y = 50$$

$$-10(-38) + 15y = 50$$

$$380 + 15y = 50$$

$$15y = -330$$

$$y = -22$$

The ordered pair $(-38, -22)$ checks in both equations. It is the solution.

37. Discussion and Writing Exercise

39. Parallel lines have the same <u>slope</u> and different <u>y-intercepts</u>.

41. A <u>solution</u> of a system of two equations is an ordered pair that makes both equations true.

43. The graph of $y = b$ is a <u>horizontal</u> line.

45. The equation $y = mx + b$ is called the <u>slope-intercept</u> equation.

47.-55. Left to the student

57.-65. Left to the student

67. $3(x - y) = 9,$

$\qquad x + y = 7$

First we remove parentheses in the first equation.

$$3x - 3y = 9, \quad (1)$$

$$x + y = 7 \quad (2)$$

Then we multiply Equation (2) by 3 and add.

$$\begin{array}{l} 3x - 3y = 9 \\ \underline{3x + 3y = 21} \\ 6x \qquad\quad = 30 \\ x = 5 \end{array}$$

Now we substitute 5 for x in Equation (2) and solve for y.

$$x + y = 7$$

$$5 + y = 7$$

$$y = 2$$

The ordered pair $(5, 2)$ checks and is the solution.

69. $2(5a - 5b) = 10,$

$\qquad -5(6a + 2b) = 10$

First we remove parentheses.

$$10a - 10b = 10, \quad (1)$$

$$-30a - 10b = 10 \quad (2)$$

Then we multiply Equation (2) by -1 and add.

$$\begin{array}{l} 10a - 10b = 10 \\ \underline{30a + 10b = -10} \\ 40a \qquad\quad = 0 \\ a = 0 \end{array}$$

Substitute 0 for a in Equation (1) and solve for b.

$$10 \cdot 0 - 10b = 10$$

$$-10b = 10$$

$$b = -1$$

The ordered pair $(0, -1)$ checks and is the solution.

71. $y = -\dfrac{2}{7}x + 3, \quad (1)$

$\qquad y = \dfrac{4}{5}x + 3 \quad (2)$

Observe that these equations represent lines with different slopes and the same y-intercept. Thus, their point of intersection is the y-intercept, $(0, 3)$ and this is the solution of the system of equations.

We could also solve this system of equations algebraically. First substitute $\dfrac{4}{5}x + 3$ for y in Equation (1) and solve for x.

$$\frac{4}{5}x + 3 = -\frac{2}{7}x + 3$$

$$35\left(\frac{4}{5}x + 3\right) = 35\left(-\frac{2}{7}x + 3\right) \quad \text{Clearing fractions}$$

$$35 \cdot \frac{4}{5}x + 35 \cdot 3 = 35\left(-\frac{2}{7}x\right) + 35 \cdot 3$$

$$28x + 105 = -10x + 105$$

$$28x = -10x$$

$$38x = 0$$

$$x = 0$$

Now substitute 0 for x in one of the original equations and find y. We will use Equation (1).

$$y = -\frac{2}{7}x + 3 = -\frac{2}{7} \cdot 0 + 3 = 0 + 3 = 3$$

The ordered pair $(0, 3)$ checks and is the solution.

73.　$y = ax + b$,　(1)

　　　$y = x + c$　　(2)

Substitute $x + c$ for y in Equation (1) and solve for x.

$$y = ax + b$$

$$x + c = ax + b \quad \text{Substituting}$$

$$x - ax = b - c$$

$$(1 - a)x = b - c$$

$$x = \frac{b - c}{1 - a}$$

Substitute $\frac{b - c}{1 - a}$ for x in Equation (2) and simplify to find y.

$$y = x + c$$

$$y = \frac{b - c}{1 - a} + c$$

$$y = \frac{b - c}{1 - a} + c \cdot \frac{1 - a}{1 - a}$$

$$y = \frac{b - c + c - ac}{1 - a}$$

$$y = \frac{b - ac}{1 - a}$$

The ordered pair $\left(\dfrac{b - c}{1 - a}, \dfrac{b - ac}{1 - a}\right)$ checks and is the solution. This ordered pair could also be expressed as $\left(\dfrac{c - b}{a - 1}, \dfrac{ac - b}{a - 1}\right)$.

Exercise Set 15.4

1. Familiarize. Let x = the number of two-point baskets made and y = the number of three-point baskets made. Then $2x$ points were scored from two-point baskets and $3y$ points were scored from three-point baskets.

Translate. Since a total of 39 baskets were made we have one equation: $x + y = 39$. Since a total of 85 points was scored, we have a second equation: $2x + 3y = 85$.

The resulting system is

$$x + y = 39, \quad (1)$$

$$2x + 3y = 85. \quad (2)$$

Solve. We use the elimination method. First we multiply both sides of Equation (1) by -2 and add.

$$-2x - 2y = -78$$

$$\underline{2x + 3y = 85}$$

$$y = 7$$

Now we substitute 7 for y in Equation (1) and solve for x.

$$x + y = 39$$

$$x + 7 = 39$$

$$x = 32$$

Check. If 32 two-point baskets and 7 three-point baskets are made, then a total of $32 + 7$, or 39, baskets are made. The points scored are $2 \cdot 32 + 3 \cdot 7$, or $64 + 21$, or 85. The answer checks.

State. The Spurs made 32 two-point shots and 7 three-point shots.

3. Familiarize. Let x = the number of 24-exposure rolls that were shot and y = the number of 36-exposure rolls that were processed. It costs $9x$ dollars to process the 24-exposure rolls and $12.60y$ dollars to process the 36-exposure rolls.

Translate. Since 17 rolls of film were processed, we have one equation: $x + y = 17$.

The total cost of processing the film was \$171, so we have a second equation: $9x + 12.6y = 171$. The resulting system is

$$x + y = 17, \quad (1)$$

$$9x + 12.6y = 171. \quad (2)$$

Solve. We use the elimination method. First we multiply both sides of Equation (1) by -9 and add.

$$-9x - 9y = -153$$

$$\underline{9x + 12.6y = 171}$$

$$3.6y = 18$$

$$y = 5$$

Now we substitute 5 for y in Equation (1) and solve for x.

$$x + y = 17$$

$$x + 5 = 17$$

$$x = 12$$

Check. If 12 rolls of 24-exposure film and 5 rolls of 36-exposure film are processed, then $12 + 5$, or 17 rolls are shot. The processing cost is $\$9(12) + \$12.60(5)$, or $\$108 + \$63$, or \$171. The answer checks.

State. 12 rolls of 24-exposure film and 5 rolls of 36-exposure film were processed.

5. Familiarize. We let x = the number of pounds of hay and y = the number of pounds of grain that should be fed to the horse each day. We arrange the information in a table.

Type of feed	Hay	Grain	Mixture
Amount of feed	x	y	15
Percent of protein	6%	12%	8%
Amount of protein in mixture	6%x	12%y	8% × 15, or 1.2 lb

Translate. The first and last rows of the table give us two equations. The total amount of feed is 15 lb, so we have

$$x + y = 15.$$

The amount of protein in the mixture is to be 8% of 15 lb, or 1.2 lb. The amounts of protein from the two feeds are 6%x and 12%y. Thus

$$6\%x + 12\%y = 1.2, \quad \text{or}$$
$$0.06x + 0.12y = 1.2, \quad \text{or}$$
$$6x + 12y = 120 \quad \text{Clearing decimals}$$

The resulting system is

$$x + y = 15, \quad (1)$$
$$6x + 12y = 120. \quad (2)$$

Solve. We use the elimination method. Multiply on both sides of Equation (1) by -6 and then add.

$$-6x - 6y = -90$$
$$\underline{6x + 12y = 120}$$
$$6y = 30$$
$$y = 5$$

We go back to Equation (1) and substitute 5 for y.

$$x + y = 15$$
$$x + 5 = 15$$
$$x = 10$$

Check. The sum of 10 and 5 is 15. Also, 6% of 10 is 0.6 and 12% of 5 is 0.6, and $0.6 + 0.6 = 1.2$. These numbers check.

State. Brianna should feed her horse 10 lb of hay and 5 lb of grain each day.

7. *Familiarize*. Let $x =$ the number of $50 bonds and $y =$ the number of $100 bonds. Then the total value of the $50 bonds is $50x$ and the total value of the $100 bonds is $100y$.

Translate.

$$\underbrace{\text{Total value of bonds}}_{\downarrow} \; \underset{\downarrow}{\text{is}} \; \underset{\downarrow}{\$1250.}$$
$$50x + 100y \quad = \quad 1250$$

$$\underbrace{\text{Number of} \atop \$50 \text{ bonds}}_{\downarrow} \; \underset{\downarrow}{\text{is}} \; \underset{\downarrow}{7} \; \underset{\downarrow}{\text{more} \atop \text{than}} \; \underbrace{\text{number of} \atop \$100 \text{ bonds.}}_{\downarrow}$$
$$x \quad = \quad 7 \quad + \quad y$$

The resulting system is

$$50x + 100y = 1250, \quad (1)$$
$$x = 7 + y. \quad (2)$$

Solve. We use the substitution method, substituting $7 + y$ for x in Equation (1).

$$50x + 100y = 1250 \quad (1)$$
$$50(7 + y) + 100y = 1250$$
$$350 + 50y + 100y = 1250$$
$$350 + 150y = 1250$$
$$150y = 900$$
$$y = 6$$

Now we substitute 6 for y in Equation (2) to find x.

$$x = 7 + y \quad (2)$$
$$x = 7 + 6 = 13$$

Check. If there are 13 $50 bonds and 6 $100 bonds, there are 7 more $50 bonds than $100 bonds. The total value of the bonds is $\$50 \cdot 13 + \$100 \cdot 6$, or $\$650 + \$600$, or $1250. The answer checks.

State. Cassandra has 13 $50 bonds and 6 $100 bonds.

9. *Familiarize*. Let $x =$ the number of cardholders tickets that were sold and $y =$ the number of non-cardholders tickets. We arrange the information in a table.

	Card-holders	Non-card-holders	Total
Price	$2.25	$3	
Number sold	x	y	203
Money taken in	2.25x	3y	$513

Translate. The last two rows of the table give us two equations. The total number of tickets sold was 203, so we have

$$x + y = 203.$$

The total amount of money collected was $513, so we have

$$2.25x + 3y = 513.$$

We can multiply the second equation on both sides by 100 to clear decimals. The resulting system is

$$x + y = 203, \quad (1)$$
$$225x + 300y = 51,300. \quad (2)$$

Solve. We use the elimination method. We multiply on both sides of Equation (1) by -225 and then add.

$$-225x - 225y = -46,675 \quad \text{Multiplying by } -225$$
$$\underline{225x + 300y = 51,300}$$
$$75y = 5625$$
$$y = 75$$

We go back to Equation (1) and substitute 75 for y.

$$x + y = 203$$
$$x + 75 = 203$$
$$x = 128$$

Check. The number of tickets sold was $128 + 75$, or 203. The money collected was $\$2.25(128) + \$3(75)$, or $\$288 + \$225$, or $513. These numbers check.

State. 128 cardholders tickets and 75 non-cardholders tickets were sold.

11. *Familiarize*. Let $a =$ the number of adults and $c =$ the number of children who visited the exhibit. We organize the information in a table.

	Adults	Children	Total
Price	$7	$6	
Number bought	a	c	1630
Receipts	$7a$	$6c$	$11,080

Translate. The last two rows of the table give us two equations. The total number of admissions was 1630, so we have

$$a + c = 1630.$$

The total receipts were $11,080, so we have

$$7a + 6c = 11,080.$$

The resulting system is

$$a + c = 1630, \quad (1)$$
$$7a + 6c = 11,080. \quad (2)$$

Solve. We use the elimination method. We multiply Equation (1) by -7 and add.

$$\begin{aligned} -7a - 7c &= -11,410, \quad \text{Multiplying by } -7 \\ 7a + 6c &= 11,080 \\ \hline -c &= -330 \\ c &= 330 \end{aligned}$$

We go back to Equation (1) and substitute 330 for c.

$$a + c = 1630$$
$$a + 330 = 1630$$
$$a = 1300$$

Check. The total admissions were $1300 + 330$, or 1630. The total receipts were $7 \cdot 1300 + 6 \cdot 330$, or $9100 + $1980, or $11,080. The answer checks.

State. 1300 adults, and 330 children visited the exhibit.

13. *Familiarize*. We complete the table in the text. Note that x represents the number of liters of solution A to be used and y represents the number of liters of solution B.

Type of solution	A	B	Mixture
Amount of solution	x	y	100 L
Percent of acid	50%	80%	68%
Amount of acid in solution	50%x	80%y	68% × 100, or 68 L

Equation from first row:　$x + y = 100$

Equation from second row:　$50\%x + 80\%y = 68$

Translate. The first and third rows of the table give us two equations. Since the total amount of solution is 100 liters, we have

$$x + y = 100.$$

The amount of acid in the mixture is to be 68% of 100, or 68 liters. The amounts of acid from the two solutions are 50%x and 80%y. Thus

$$50\%x + 80\%y = 68,$$
$$\text{or} \quad 0.5x + 0.8y = 68,$$
$$\text{or} \quad 5x + 8y = 680 \quad \text{Clearing decimals}$$

The resulting system is

$$x + y = 100, \quad (1)$$
$$5x + 8y = 680. \quad (2)$$

Solve. We use the elimination method. We multiply on both sides of Equation (1) by -5 and then add.

$$\begin{aligned} -5x - 5y &= -500 \quad \text{Multiplying by } -5 \\ 5x + 8y &= 680 \\ \hline 3y &= 180 \\ y &= 60 \end{aligned}$$

We go back to Equation (1) and substitute 60 for y.

$$x + y = 100$$
$$x + 60 = 100$$
$$x = 40$$

Check. We consider $x = 40$ and $y = 60$. The sum is 100. Now 50% of 40 is 20 and 80% of 60 is 48. These add up to 68. The numbers check.

State. 40 liters of solution A and 60 liters of solution B should be used.

15. *Familiarize*. Let d represent the number of dimes and q the number of quarters. Then, $10d$ represents the value of the dimes in cents, and $25q$ represents the value of the quarters in cents. The total value is $15.25, or 1525¢. The total number of coins is 103.

Translate.

Number of dimes	plus	number of quarters	is	103.
↓	↓	↓	↓	↓
d	+	q	=	103

Value of dimes	plus	value of quarters	is	$15.25.
↓	↓	↓	↓	↓
$10d$	+	$25q$	=	1525

The resulting system is

$$d + q = 103, \quad (1)$$
$$10d + 25q = 1525. \quad (2)$$

Solve. We use the addition method. We multiply Equation (1) by -10 and then add.

$$-10d - 10q = -1030 \quad \text{Multiplying by } -10$$
$$\underline{10d + 25q = 1525}$$
$$15q = 495 \qquad \text{Adding}$$
$$q = 33$$

Now we substitute 33 for q in one of the original equations and solve for d.

$$d + q = 103 \quad (1)$$
$$d + 33 = 103 \quad \text{Substituting}$$
$$d = 70$$

Check. The number of dimes plus the number of quarters is $70+33$, or 103. The total value in cents is $10 \cdot 70 + 25 \cdot 33$, or $700+825$, or 1525. This is equal to \$15.25. This checks.

State. There are 70 dimes and 33 quarters.

17. *Familiarize*. We complete the table in the text. Note that x represents the number of pounds of Brazilian coffee to be used and y represents the number of pounds of Turkish coffee.

Type of coffee	Brazilian	Turkish	Mixture
Cost of coffee	\$19	\$22	\$20
Amount (in pounds)	x	y	300
Mixture	$19x$	$22y$	\$20(300), or \$6000

Equation from second row: $x + y = 300$

Equation from third row: $19x + 22y = 6000$

Translate. The second and third rows of the table give us two equations. Since the total amount of the mixture is 300 lb, we have

$$x + y = 300.$$

The value of the Brazilian coffee is $19x$ (x lb at \$19 per pound), the value of the Turkish coffee is $22y$ (y lb at \$22 per pound), and the value of the mixture is \$20(300) or \$6000. Thus we have

$$19x + 22y = 6000.$$

The resulting system is

$$x + y = 300, \quad (1)$$
$$19x + 22y = 6000. \quad (2)$$

Solve. We use the elimination method. We multiply on both sides of Equation (1) by -19 and then add.

$$-19x - 19y = -5700 \quad \text{Multiplying by } -19$$
$$\underline{19x + 22y = 6000}$$
$$3y = 300$$
$$y = 100$$

We go back to Equation (1) and substitute 100 for y.

$$x + y = 300$$
$$x + 100 = 300$$
$$x = 200$$

Check. The sum of 200 and 100 is 300. The value of the mixture is \$19(200)+\$22(100), or \$3800+\$2200, or \$6000. These values check.

State. 200 lb of Brazilian coffee and 100 lb of Turkish coffee should be used.

19. *Familiarize*. Let x and y represent the number of liters of 28%-fungicide solution and 40%-fungicide solution to be used in the mixture, respectively.

Translate. We organize the given information in a table.

Type of solution	28%	40%	36%
Amount of solution	x	y	300
Percent fungicide	28%	40%	36%
Amount of fungicide in solution	$0.28x$	$0.4y$	0.36(300), or 108

We get a system of equations from the first and third rows of the table.

$$x + y = 300,$$
$$0.28x + 0.4y = 108$$

Clearing decimals we have

$$x + y = 300, \quad (1)$$
$$28x + 40y = 10,800 \quad (2)$$

Solve. We use the elimination method. Multiply Equation (1) by -28 and add.

$$-28x - 28y = -8400$$
$$\underline{28x + 40y = 10,800}$$
$$12y = 2400$$
$$y = 200$$

Now substitute 200 for y in Equation (1) and solve for x.

$$x + y = 300$$
$$x + 200 = 300$$
$$x = 100$$

Check. The sum of 100 and 200 is 300. The amount of fungicide in the mixture is $0.28(100)+0.4(200)$, or $28+80$, or 108 L. These numbers check.

State. 100 L of the 28%-fungicide solution and 200 L of the 40%-fungicide solution should be used in the mixture.

21. *Familiarize*. We let x = the number of pages in large type and y = the number of pages in small type. We arrange the information in a table.

Size of type	Large	Small	Mixture (Book)
Words per page	830	1050	
Number of pages	x	y	12
Number of words	$830x$	$1050y$	11,720

Translate. The last two rows of the table give us two equations. The total number of pages in the document is 12, so we have

$$x + y = 12.$$

The number of words on the pages with large type is $830x$ (x pages with 830 words per page), and the number of words on the pages with small type is $1050y$ (y pages with 1050 words per page). The total number of words is 11,720, so we have

$$830x + 1050y = 11,720.$$

The resulting system is

$$x + y = 12, \qquad (1)$$
$$830x + 1050y = 11,720. \quad (2)$$

Solve. We use the elimination method. We multiply on both sides of Equation (1) by -830 and then add.

$$-830x - 830y = -9,960 \quad \text{Multiplying by } -830$$
$$\underline{830x + 1050y = 11,720}$$
$$220y = 1760$$
$$y = 8$$

We go back to Equation (1) and substitute 8 for y.

$$x + y = 12$$
$$x + 8 = 12$$
$$x = 4$$

Check. The sum of 4 and 8 is 12. The number of words in large type is $830 \cdot 4$, or 3320, and the number of words in small type is $1050 \cdot 8$, or 8400. Then the total number of words is $3320 + 8400$, or 11,720. These numbers check.

State. There were 4 pages in large type and 8 pages in small type.

23. **Familiarize.** Let $x =$ the number of pounds of the 70% mixture and $y =$ the number of pounds of the 45% mixture to be used. We organize the information in the table.

Percent of cashews	70%	45%	
Amount	x	y	60
Mixture	70%x, or 0.7x	45%x, or 0.45x	60(60%) or 60(0.6) or 36

Translate. The last two rows of the table give us two equations. The total weight of the mixture is 60 lb, so we have

$$x + y = 60.$$

The amount of cashews in the mixture is 36 lb, so we have

$$0.7x + 0.45y = 36, \quad \text{or}$$
$$70x + 45y = 3600 \quad \text{Clearing decimals}$$

The resulting system is

$$x + y = 60, \qquad (1)$$
$$70x + 45y = 3600. \quad (2)$$

Solve. We use the elimination method. We multiply Equation (1) by -45 and then add.

$$-45x - 45y = -2700$$
$$\underline{70x + 45y = 3600}$$
$$25x = 900$$
$$x = 36$$

Next we substitute 36 for x in one of the original equations and solve for y.

$$x + y = 60 \quad (1)$$
$$36 + y = 60$$
$$y = 24$$

Check. The total weight of the mixture is 36 lb + 24 lb, or 60 lb. The amount of cashews in the mixture is $0.7(36 \text{ lb}) + 0.45(24 \text{ lb})$, or 36 lb. Since 36 lb is 60% of 60 lb, the answer checks.

State. The new mixture should contain 36 lb of the 70% cashew mixture and 24 lb of the 45% cashew mixture.

25. **Familiarize.** We arrange the information in a table. Let $a =$ the number of type A questions and $b =$ the number of type B questions.

Type of question	A	B	Mixture (Test)
Number	a	b	16
Time	3 min	6 min	
Value	10 points	15 points	
Mixture (Test)	$3a$ min, $10a$ points	$6b$ min, $15b$ points	60 min, 180 points

Translate. The table actually gives us three equations. Since the total number of questions is 16, we have

$$a + b = 16.$$

The total time is 60 min, so we have

$$3a + 6b = 60.$$

The total number of points is 180, so we have

$$10a + 15b = 180.$$

The resulting system is

$$a + b = 16, \qquad (1)$$
$$3a + 6b = 60, \qquad (2)$$
$$10a + 15b = 180. \quad (3)$$

Solve. We will solve the system composed of Equations (1) and (2) and then check to see that this solution also satisfies Equation (3). We multiply equation (1) by -3 and add.

$$-3a - 3b = -48$$
$$\underline{3a + 6b = 60}$$
$$3b = 12$$
$$b = 4$$

Now we substitute 4 for b in Equation (1) and solve for a.

$$a + b = 16$$
$$a + 4 = 16$$
$$a = 12$$

Check. We consider $a = 12$ questions and $b = 4$ questions. The total number of questions is 16. The time required is $3 \cdot 12 + 6 \cdot 4$, or $36 + 24$, or 60 min. The total points are $10 \cdot 12 + 15 \cdot 4$, or $120 + 60$, or 180. These values check.

State. 12 questions of type A and 4 questions of type B were answered. Assuming all the answers were correct, the score was 180 points.

27. Familiarize. Let $k =$ the age of the Kuyatt's house now and $m =$ the age of the Marconi's house now. Eight years ago the houses' ages were $k - 8$ and $m - 8$.

Translate. We reword and translate.

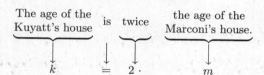

The resulting system is

$$k = 2m, \qquad (1)$$
$$k - 8 = 3(m - 8). \qquad (2)$$

Solve. We use the substitution method. We substitute $2m$ for k in Equation (2) and solve for m.

$$k - 8 = 3(m - 8)$$
$$2m - 8 = 3(m - 8)$$
$$2m - 8 = 3m - 24$$
$$-8 = m - 24$$
$$16 = m$$

We find k by substituting 16 for m in Equation (1).

$$k = 2m$$
$$k = 2 \cdot 16$$
$$k = 32$$

Check. The age of the Kuyatt's house, 32 years, is twice the age of the Marconi's house, 16 years. Eight years ago, when the Kuyatt's house was 24 years old and the Marconi's house was 8 years old, the Kuyatt's house was three times as old as the Marconi's house. These numbers check.

State. The Kuyatt's house is 32 years old, and the Marconi's house is 16 years old.

29. Familiarize. Let $R =$ Randy's age now and $M =$ Mandy's age now. In twelve years their ages will be $R + 12$ and $M + 12$.

Translate. We reword and translate.

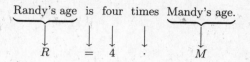

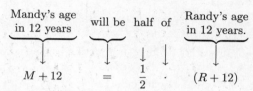

The resulting system is

$$R = 4M, \qquad (1)$$
$$M + 12 = \frac{1}{2}(R + 12). \qquad (2)$$

Solve. We use the substitution method. We substitute $4M$ for R in Equation (2) and solve for M.

$$M + 12 = \frac{1}{2}(R + 12)$$
$$M + 12 = \frac{1}{2}(4M + 12)$$
$$M + 12 = 2M + 6$$
$$12 = M + 6$$
$$6 = M$$

We find R by substituting 6 for M in Equation (1).

$$R = 4M$$
$$R = 4 \cdot 6$$
$$R = 24$$

Check. Randy's age now, 24, is 4 times 6, Mandy's age. In 12 yr, when Randy will be 36 and Mandy 18, Mandy's age will be half of Randy's age. These numbers check.

State. Randy is 24 years old now, and Mandy is 6.

31. Familiarize. Let $x =$ the smaller angle and $y =$ the larger angle.

Translate. We reword the problem.

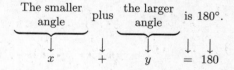

The resulting system is

$$x + y = 180,$$
$$y = 30 + 2x.$$

Solve. We solve the system. We will use the elimination method although we could also easily use the substitution method. First we get the second equation in the form $Ax + By = C$.

$$x + y = 180 \quad (1)$$
$$-2x + y = 30 \quad (2) \text{ Adding } -2x$$

Now we multiply Equation (2) by -1 and add.

$$x + y = 180$$
$$\underline{2x - y = -30}$$
$$3x \quad\quad = 150$$
$$x = 50$$

Then we substitute 50 for x in Equation (1) and solve for y.

$$x + y = 180 \quad \text{Equation (1)}$$
$$50 + y = 180 \quad \text{Substituting}$$
$$y = 130$$

Check. The sum of the angles is $50° + 130°$, or $180°$, so the angles are supplementary. Also, $30°$ more than two times the $50°$ angle is $30° + 2 \cdot 50°$, or $30° + 100°$, or $130°$, the other angle. These numbers check.

State. The angles are $50°$ and $130°$.

33. **Familiarize**. We let $x =$ the larger angle and $y =$ the smaller angle.

Translate. We reword and translate the first statement.

The sum of two angles is $90°$.
$$\underbrace{}$$
$$\downarrow \quad\quad\quad \downarrow \quad \downarrow$$
$$x + y \quad\quad = \quad 90$$

We reword and translate the second statement.

The difference of two angles is $34°$.
$$\underbrace{}$$
$$\downarrow \quad\quad\quad \downarrow \quad \downarrow$$
$$x - y \quad\quad = \quad 34$$

The resulting system is
$$x + y = 90,$$
$$x - y = 34.$$

Solve. We solve the system.
$$x + y = 90, \quad (1)$$
$$\underline{x - y = 34} \quad (2)$$
$$2x \quad\quad = 124 \quad \text{Adding}$$
$$x = 62$$

Now we substitute 62 for x in Equation (1) and solve for y.

$$x + y = 90 \quad \text{Equation (1)}$$
$$62 + y = 90 \quad \text{Substituting}$$
$$y = 28$$

Check. The sum of the angles is $62° + 28°$, or $90°$, so the angles are complementary. The difference of the angles is $62° - 28°$, or $34°$. These numbers check.

State. The angles are $62°$ and $28°$.

35. **Familiarize**. Let $x =$ the number of gallons of 87-octane gas and $y =$ the number of gallons of 93-octane gas that should be used. We arrange the information in a table.

Type of gas	87-octane	93-octane	Mixture
Amount of gas	x	y	18
Octane rating	87	93	89
Mixture	$87x$	$93y$	$18 \cdot 89$, or 1602

Translate. The first and last rows of the table give us a system of equations.

$$x + \quad y = \quad 18, \quad (1)$$
$$87x + 93y = 1602 \quad (2)$$

Solve. We multiply Equation (1) by -87 and then add.

$$-87x - 87y = -1566$$
$$\underline{87x + 93y = 1602}$$
$$6y = 36$$
$$y = 6$$

Then substitute 6 for y in Equation (1) and solve for x.

$$x + y = 18$$
$$x + 6 = 18$$
$$x = 12$$

Check. The total amount of gas is 12 gal $+$ 6 gal, or 18 gal. Also $87(12) + 93(6) = 1044 + 558 = 1602$. The answer checks.

State. 12 gal of 87-octane gas and 6 gal of 93-octane gas should be blended.

37. **Familiarize**. Let $x =$ the number of ounces of Dr. Zeke's cough syrup and $y =$ the number of ounces of Vitabrite cough syrup that should be used. We organize the information in a table.

	Dr. Zeke's	Vitabrite	Mixture
Percent of alcohol	2%	5%	3%
Amount	x	y	80
Mixture	2%x, or 0.02x	5%x, or 0.05y	$3\% \cdot 80$, or $0.03 \cdot 80$, or 2.4

Translate. The last two rows of the table give us a system of equations.

$$x + y = 80,$$
$$0.02x + 0.05y = 2.4.$$

Clearing decimals, we have

$$x + y = 80, \quad (1)$$
$$2x + 5y = 240. \quad (2)$$

Solve. We use the elimination method. First we multiply Equation (1) by -2 and then add.

$$-2x - 2y = -160$$
$$\underline{2x + 5y = 240}$$
$$3y = 80$$
$$y = 26\frac{2}{3}$$

Substitute $26\frac{2}{3}$ for y in one of the original equations and solve for x.

$$x + y = 80$$
$$x + 26\frac{2}{3} = 80$$
$$x = 53\frac{1}{3}$$

Check. The number of ounces in the mixture is $53\frac{1}{3} + 26\frac{2}{3}$, or 80. The amount of alcohol in the mixture is $0.02\left(53\frac{1}{3}\right) + 0.05\left(26\frac{2}{3}\right)$, or 2.4 oz. Since 2.4 oz is 3% of 80 oz, the answer checks.

State. The mixture should contain $53\frac{1}{3}$ oz of Dr. Zeke's cough syrup and $26\frac{2}{3}$ oz of Vitabrite cough syrup.

39. Discussion and Writing Exercise

41. $25x^2 - 81 = (5x)^2 - 9^2$
$$= (5x + 9)(5x - 9)$$

43. $4x^2 + 100 = 4(x^2 + 25)$

45. $y = -2x - 3$

The equation is in the form $y = mx + b$, so the y-intercept is $(0, -3)$.

To find the x-intercept, we let $y = 0$ and solve for x.

$$0 = -2x - 3$$
$$2x = -3$$
$$x = -\frac{3}{2}$$

The x-intercept is $\left(-\frac{3}{2}, 0\right)$.

We plot the intercepts and draw the line.

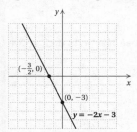

A third point should be used as a check.

For example, let $x = -3$. Then

$$y = -2(-3) - 3 = 6 - 3 = 3.$$

It appears that the point $(-3, 3)$ is on the graph, so the graph is probably correct.

47. $5x - 2y = -10$

To find the y-intercept, let $x = 0$ and solve for y.

$$5 \cdot 0 - 2y = -10$$
$$-2y = -10$$
$$y = 5$$

The y-intercept is $(0, 5)$.

To find the x-intercept, let $y = 0$ and solve for x.

$$5x - 2 \cdot 0 = -10$$
$$5x = -10$$
$$x = -2$$

The x-intercept is $(-2, 0)$.

We plot the intercepts and draw the line.

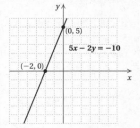

A third point should be used as a check. For example, let $x = -4$. then

$$5(-4) - 2y = -10$$
$$-20 - 2y = -10$$
$$-2y = 10$$
$$y = -5$$

It appears that the point $(-4, -5)$ is on the graph, so the graph is probably correct.

49. $\dfrac{x^2 - 5x + 6}{x^2 - 4} = \dfrac{(x - 3)(x - 2)}{(x + 2)(x - 2)}$

$$= \dfrac{(x - 3)(x - 2)}{(x + 2)(x - 2)}$$

$$= \dfrac{x - 3}{x + 2}$$

51. $\dfrac{x - 2}{x + 3} - \dfrac{2x - 5}{x - 4}$ LCD is $(x + 3)(x - 4)$

$$= \dfrac{x - 2}{x + 3} \cdot \dfrac{x - 4}{x - 4} - \dfrac{2x - 5}{x - 4} \cdot \dfrac{x + 3}{x + 3}$$

$$= \dfrac{(x - 2)(x - 4)}{(x + 3)(x - 4)} - \dfrac{(2x - 5)(x + 3)}{(x - 4)(x + 3)}$$

$$= \dfrac{x^2 - 6x + 8}{(x + 3)(x - 4)} - \dfrac{2x^2 + x - 15}{(x - 4)(x + 3)}$$

$$= \dfrac{x^2 - 6x + 8 - (2x^2 + x - 15)}{(x + 3)(x - 4)}$$

$$= \dfrac{x^2 - 6x + 8 - 2x^2 - x + 15}{(x + 3)(x - 4)}$$

$$= \dfrac{-x^2 - 7x + 23}{(x + 3)(x - 4)}$$

53. Familiarize. We arrange the information in a table. Let x = the number of liters of skim milk and y = the number of liters of 3.2% milk.

Type of milk	4.6%	Skim	3.2% (Mixture)
Amount of milk	100 L	x	y
Percent of butterfat	4.6%	0%	3.2%
Amount of butterfat in milk	4.6% × 100, or 4.6 L	0% · x, or 0 L	3.2%y

Translate. The first and third rows of the table give us two equations.

Amount of milk: $100 + x = y$

Amount of butterfat: $4.6 + 0 = 3.2\%y$, or $4.6 = 0.032y$.

The resulting system is
$$100 + x = y,$$
$$4.6 = 0.032y.$$

Solve. We solve the second equation for y.
$$4.6 = 0.032y$$
$$\frac{4.6}{0.032} = y$$
$$143.75 = y$$

We substitute 143.75 for y in the first equation and solve for x.
$$100 + x = y$$
$$100 + x = 143.75$$
$$x = 43.75$$

Check. We consider $x = 43.75$ L and $y = 143.75$ L. The difference between 143.75 L and 43.75 L is 100 L. There is no butterfat in the skim milk. There are 4.6 liters of butterfat in the 100 liters of the 4.6% milk. Thus there are 4.6 liters of butterfat in the mixture. This checks because 3.2% of 143.75 is 4.6.

State. 43.75 L of skim milk should be used.

55. Familiarize. In a table we arrange the information regarding the solution <u>after</u> some of the 30% solution is drained and replaced with pure antifreeze. We let x represent the amount of the original (30%) solution remaining, and we let y represent the amount of the 30% mixture that is drained and replaced with pure antifreeze.

Type of solution	Original (30%)	Pure antifreeze	Mixture
Amount of solution	x	y	16
Percent of antifreeze	30%	100%	50%
Amount of antifreeze in solution	0.3x	1 · y, or y	0.5(16), or 8

Translate. The table gives us two equations.

Amount of solution: $x + y = 16$

Amount of antifreeze in solution: $0.3x + y = 8$, or $3x + 10y = 80$

The resulting system is
$$x + y = 16, \quad (1)$$
$$3x + 10y = 80. \quad (2)$$

Solve. We multiply Equation (1) by -3 and then add.
$$-3x - 3y = -48$$
$$\underline{3x + 10y = 80}$$
$$7y = 32$$
$$y = \frac{32}{7}, \text{ or } 4\frac{4}{7}$$

Then we substitute $4\frac{4}{7}$ for y in Equation (1) and solve for x.
$$x + y = 16$$
$$x + 4\frac{4}{7} = 16$$
$$x = 11\frac{3}{7}$$

Check. When $x = 11\frac{3}{7}$ L and $y = 4\frac{4}{7}$ L, the total is 16 L. The amount of antifreeze in the mixture is $0.3\left(11\frac{3}{7}\right) + 4\frac{4}{7}$, or $\frac{3}{10} \cdot \frac{80}{7} + \frac{32}{7}$, or $\frac{24}{7} + \frac{32}{7} = \frac{56}{7}$, or 8 L. This is 50% of 16 L, so the numbers check.

State. $4\frac{4}{7}$ of the original mixture should be drained and replaced with pure antifreeze.

57. Familiarize. Let x = the tens digit and y = the units digit. Then the number is $10x + y$.

Translate. The number is six times the sum of its digit, so we have
$$10x + y = 6(x + y)$$
$$10x + y = 6x + 6y$$
$$4x - 5y = 0.$$

The tens digit is 1 more than the units digit so we have
$$x = y + 1.$$

The resulting system is
$$4x - 5y = 0, \quad (1)$$
$$x = y + 1. \quad (2)$$

Solve. First substitute $y + 1$ for x in Equation (1) and solve for y.
$$4x - 5y = 0 \quad (1)$$
$$4(y + 1) - 5y = 0$$
$$4y + 4 - 5y = 0$$
$$-y + 4 = 0$$
$$4 = y$$

Now substitute 4 for y in one of the original equations and solve for x.
$$x = y + 1 \quad (2)$$
$$x = 4 + 1$$
$$x = 5$$

Check. If the number is 54, then the sum of the digits is $5 + 4$, or 9, and $54 = 6 \cdot 9$. Also, the tens digit, 5, is one more than the units digit, 4. The answer checks.

State. The number is 54.

Exercise Set 15.5

1. *Familiarize*. We first make a drawing.

	30 mph	
Slow car	t hours	d miles

	46 mph	
Fast car	t hours	$d + 72$ miles

We let $d =$ the distance the slow car travels. Then $d+72 =$ the distance the fast car travels. We call the time t. We complete the table in the text, filling in the distances as well as the other information.

$$d = r \cdot t$$

	Distance	Speed	Time
Slow car	d	30	t
Fast car	$d + 72$	46	t

Translate. We get an equation $d = rt$ from each row of the table. Thus we have

$$d = 30t, \qquad (1)$$
$$d + 72 = 46t. \quad (2)$$

Solve. We use the substitution method. We substitute $30t$ for d in Equation (2).

$$d + 72 = 46t$$
$$30t + 72 = 46t \quad \text{Substituting}$$
$$72 = 16t \quad \text{Subtracting } 30t$$
$$4.5 = t \quad \text{Dividing by 16}$$

Check. In 4.5 hr the slow car travels $30(4.5)$, or 135 mi, and the fast car travels $46(4.5)$, or 207 mi. Since 207 is 72 more than 135, our result checks.

State. The trains will be 72 mi apart in 4.5 hr.

3. *Familiarize*. First make a drawing.

Station	72 mph	
Slow train	$t + 3$ hours	d miles

Station	120 mph	
Fast train	t hours	d miles

Trains meet here.

From the drawing we see that the distances are the same. Let's call the distance d. Let t represent the time for the faster train and $t+3$ represent the time for the slower train. We complete the table in the text.

$$d = r \cdot t$$

	Distance	Speed	Time
Slow train	d	72	$t + 3$
Fast train	d	120	t

Equation from first row: $d = 72(t + 3)$

Equation from second row: $d = 120t$

Translate. Using $d = rt$ in each row of the table, we get the following system of equations:

$$d = 72(t + 3), \quad (1)$$
$$d = 120t. \qquad (2)$$

Solve. Substitute $120t$ for d in Equation (1) and solve for t.

$$d = 72(t + 3)$$
$$120t = 72(t + 3) \quad \text{Substituting}$$
$$120t = 72t + 216$$
$$48t = 216$$
$$t = \frac{216}{48}$$
$$t = 4.5$$

Check. When $t = 4.5$ hours, the faster train will travel $120(4.5)$, or 540 mi, and the slower train will travel $72(7.5)$, or 540 mi. In both cases we get the distance 540 mi.

State. In 4.5 hours after the second train leaves, the second train will overtake the first train. We can also state the answer as 7.5 hours after the first train leaves.

5. *Familiarize*. We first make a drawing.

With the current		$r + 6$
4 hours		d kilometers

Against the current		$r - 6$
10 hours		d kilometers

From the drawing we see that the distances are the same. Let d represent the distance. Let r represent the speed of the canoe in still water. Then, when the canoe is traveling with the current, its speed is $r + 6$. When it is traveling against the current, its speed is $r - 6$. We complete the table in the text.

$$d = r \cdot t$$

	Distance	Speed	Time
With current	d	$r + 6$	4
Against current	d	$r - 6$	10

Equation from first row: $d = (r + 6)4$

Equation from second row: $d = (r - 6)10$

Translate. Using $d = rt$ in each row of the table, we get the following system of equations:

$$d = (r+6)4, \quad (1)$$
$$d = (r-6)10 \quad (2)$$

Solve. Substitute $(r+6)4$ for d in Equation (2) and solve for r.

$$d = (r-6)10$$
$$(r+6)4 = (r-6)10 \quad \text{Substituting}$$
$$4r + 24 = 10r - 60$$
$$84 = 6r$$
$$14 = r$$

Check. When $r = 14$, $r + 6 = 20$ and $20 \cdot 4 = 80$, the distance. When $r = 14$, $r - 6 = 8$ and $8 \cdot 10 = 80$. In both cases, we get the same distance.

State. The speed of the canoe in still water is 14 km/h.

7. Familiarize. First make a drawing.

Passenger   96 km/h

$t - 2$ hours d kilometers

Freight   64 km/h

t hours d kilometers

Central City   Clear Creek

From the drawing we see that the distances are the same. Let d represent the distance. Let t represent the time for the freight train. Then the time for the passenger train is $t - 2$. We organize the information in a table.

$$d = r \cdot t$$

	Distance	Speed	Time
Passenger	d	96	$t - 2$
Freight	d	64	t

Translate. From each row of the table we get an equation.

$$d = 96(t - 2), \quad (1)$$
$$d = 64t \quad (2)$$

Solve. Substitute $64t$ for d in Equation (1) and solve for t.

$$d = 96(t - 2)$$
$$64t = 96(t - 2) \quad \text{Substituting}$$
$$64t = 96t - 192$$
$$192 = 32t$$
$$6 = t$$

Next we substitute 6 for t in one of the original equations and solve for d.

$$d = 64t \quad \text{Equation (2)}$$
$$d = 64 \cdot 6 \quad \text{Substituting}$$
$$d = 384$$

Check. If the time is 6 hr, then the distance the passenger train travels is $96(6 - 2)$, or 384 km. The freight train travels $64(6)$, or 384 km. The distances are the same.

State. It is 384 km from Central City to Clear Creek.

9. Familiarize. We first make a drawing.

Downstream $r + 6$

3 hours d miles

Upstream $r - 6$

5 hours d miles

We let r represent the speed of the boat in still water and d represent the distance Antoine traveled downstream before he turned back. We organize the information in a table.

$$d = r \cdot t$$

	Distance	Speed	Time
Downstream	d	$r + 6$	3
Upstream	d	$r - 6$	5

Translate. Using $d = rt$ in each row of the table, we get the following system of equations:

$$d = (r+6)3, \quad (1)$$
$$d = (r-6)5 \quad (2)$$

Solve. Substitute $(r+6)3$ for d in Equation (2) and solve for r.

$$d = (r-6)5$$
$$(r+6)3 = (r-6)5 \quad \text{Substituting}$$
$$3r + 18 = 5r - 30$$
$$48 = 2r$$
$$24 = r$$

If $r = 24$, then $d = (r+6)3 = (24+6)3 = 30 \cdot 3 = 90$.

Check. If $r = 24$, then $r + 6 = 24 + 6 = 30$ and $r - 6 = 24 - 6 = 18$. If Antoine travels for 3 hr at 30 mph, then he travels $3 \cdot 30$, or 90 mi, downstream. If he travels for 5 hr at 18 mph, then he also travels $5 \cdot 18$, or 90 mi, upstream. Since the distances are the same, the answer checks.

State. (a) Antoine must travel at a speed of 24 mph.

(b) Antoine traveled 90 mi downstream before he turned back.

11. Familiarize. We first make a drawing.

230 ft/min

Toddler $t + 1$ min d ft

660 ft/min

Mother t min d ft

They meet here.

From the drawing we see that the distances are the same. Let's call the distance d. Let $t = $ the time the mother runs.

Then $t + 1 =$ the time the toddler runs. We arrange the information in a table.

	d	$=r$	$\cdot t$
	Distance	Speed	Time
Toddler	d	230	$t + 1$
Mother	d	660	t

Translate. Using $d = rt$ in each row of the table we get two equations.

$$d = 230(t + 1), \quad (1)$$
$$d = 660t \qquad (2)$$

Solve. Substitute $660t$ for d in Equation (1) and solve for t.

$$d = 230(t + 1)$$
$$660t = 230(t + 1) \quad \text{Substituting}$$
$$660t = 230t + 230$$
$$430t = 230$$
$$t = \frac{230}{430}, \text{ or } \frac{23}{43}$$

Check. When $t = \frac{23}{43}$ the toddler will travel $230\left(1\frac{23}{43}\right)$, or $230 \cdot \frac{66}{43}$, or $\frac{15,180}{43}$ ft and the mother will travel $660 \cdot \frac{23}{43}$, or $\frac{15,180}{43}$ ft. Since the distances are the same, our result checks.

State. The mother will overtake the toddler $\frac{23}{43}$ min after she starts running. We can also state the answer as $1\frac{23}{43}$ min after the toddler starts running.

13. Familiarize. First make a drawing.

Home t hr   45 mph | $(2 - t)$ hr   6 mph   Work

Motorcycle distance | Walking distance

$\longleftarrow$ —————— 25 miles —————— $\longrightarrow$

Let t represent the time the motorcycle was driven. Then $2 - t$ represents the time the rider walked. We organize the information in a table.

	d	$=r$	$\cdot t$
	Distance	Speed	Time
Motorcycling	Motorcycle distance	45	t
Walking	Walking distance	6	$2 - t$
Total	25		

Translate. From the drawing we see that

Motorcycle distance + Walking distance $= 25$

Then using $d = rt$ in each row of the table we get

$$45t + 6(2 - t) = 25$$

Solve. We solve this equation for t.

$$45t + 12 - 6t = 25$$
$$39t + 12 = 25$$
$$39t = 13$$
$$t = \frac{13}{39}$$
$$t = \frac{1}{3}$$

Check. The problem asks us to find how far the motorcycle went before it broke down. If $t = \frac{1}{3}$, then $45t$ (the distance the motorcycle traveled) $= 45 \cdot \frac{1}{3}$, or 15 and $6(2 - t)$ (the distance walked) $= 6\left(2 - \frac{1}{3}\right) = 6 \cdot \frac{5}{3}$, or 10. The total of these distances is 25, so $\frac{1}{3}$ checks.

State. The motorcycle went 15 miles before it broke down.

15. Discussion and Writing Exercise

17. $\dfrac{8x^2}{24x} = \dfrac{8}{24} \cdot \dfrac{x^2}{x} = \dfrac{1}{3} \cdot x^{2-1} = \dfrac{x}{3}$

19. $\dfrac{5a + 15}{10} = \dfrac{5(a + 3)}{5 \cdot 2}$

$\qquad = \dfrac{\cancel{5}(a + 3)}{\cancel{5} \cdot 2}$

$\qquad = \dfrac{a + 3}{2}$

21. $\dfrac{2x^2 - 50}{x^2 - 25} = \dfrac{2(x^2 - 25)}{x^2 - 25} = \dfrac{2}{1} \cdot \dfrac{x^2 - 25}{x^2 - 25} = 2$

23. $\dfrac{x^2 - 3x - 10}{x^2 - 2x - 15} = \dfrac{(x - 5)(x + 2)}{(x - 5)(x + 3)}$

$\qquad = \dfrac{\cancel{(x - 5)}(x + 2)}{\cancel{(x - 5)}(x + 3)}$

$\qquad = \dfrac{x + 2}{x + 3}$

25. $\dfrac{(x^2 + 6x + 9)(x - 2)}{(x^2 - 4)(x + 3)} = \dfrac{(x + 3)(x + 3)(x - 2)}{(x + 2)(x - 2)(x + 3)}$

$\qquad = \dfrac{\cancel{(x + 3)}(x + 3)\cancel{(x - 2)}}{(x + 2)\cancel{(x - 2)}\cancel{(x + 3)}}$

$\qquad = \dfrac{x + 3}{x + 2}$

27. $\dfrac{6x^2 + 18x + 12}{6x^2 - 6} = \dfrac{6(x^2 + 3x + 2)}{6(x^2 - 1)}$

$\qquad = \dfrac{6(x + 1)(x + 2)}{6(x + 1)(x - 1)}$

$\qquad = \dfrac{\cancel{6}\cancel{(x + 1)}(x + 2)}{\cancel{6}\cancel{(x + 1)}(x - 1)}$

$\qquad = \dfrac{x + 2}{x - 1}$

29. *Familiarize*. We arrange the information in a table. Let d = the length of the route and t = Lindbergh's time. Note that 16 hr and 57 min = $16\frac{57}{60}$ hr = 16.95 hr.

$$d = r \cdot t$$

	Distance	Speed	Time
Lindbergh	d	107.4	t
Hughes	d	217.1	$t - 16.95$

Translate. From the rows of the table we get two equations.

$$d = 107.4t, \qquad (1)$$
$$d = 217.1(t - 16.95) \quad (2)$$

Solve. We substitute $107.4t$ for d in Equation (2) and solve for t.

$$d = 217.1(t - 16.95)$$
$$107.4t = 217.1(t - 16.95)$$
$$107.4t = 217.1t - 3679.845$$
$$-109.7t = -3679.845$$
$$t \approx 33.54$$

Now we go back to Equation (1) and substitute 33.54 for t.

$$d = 107.4t$$
$$d = 107.4(33.54)$$
$$d \approx 3602$$

Check. When $t \approx 33.54$, Lindbergh traveled $107.4(33.54) \approx 3602$ mi, and Hughes traveled $217.1(16.59) \approx 3602$ mi. Since the distances are the same, our result checks.

State. The route was 3602 mi long. (Answers may vary slightly due to rounding differences.)

31. *Familiarize*. We arrange the information in a table. Let's call the distance d. When the riverboat is traveling upstream its speed is $12 - 4$, or 8 mph. Its speed traveling downstream is $12 + 4$, or 16 mph.

$$d = r \cdot t$$

	Distance	Speed	Time
Upstream	d	8	Time upstream
Downstream	d	16	Time downstream
Total			1

Translate. From the table we see that (Time upstream) + (Time downstream) = 1. Then using $d = rt$, in the form $\frac{d}{r} = t$, in each row of the table we get

$$\frac{d}{8} + \frac{d}{16} = 1.$$

Solve. We solve the equation. The LCM is 16.

$$\frac{d}{8} + \frac{d}{16} = 1$$
$$16\left(\frac{d}{8} + \frac{d}{16}\right) = 16 \cdot 1$$
$$16 \cdot \frac{d}{8} + 16 \cdot \frac{d}{16} = 16$$
$$2d + d = 16$$
$$3d = 16$$
$$d = \frac{16}{3}, \text{ or } 5\frac{1}{3}$$

Check. When $d = \frac{16}{3}$,

(Time upstream) + (Time downstream)

$$= \frac{\frac{16}{3}}{8} + \frac{\frac{16}{3}}{16}$$
$$= \frac{16}{3} \cdot \frac{1}{8} + \frac{16}{3} \cdot \frac{1}{16}$$
$$= \frac{2}{3} + \frac{1}{3}$$
$$= 1 \text{ hr}$$

Thus the distance of $\frac{16}{3}$ mi, or $5\frac{1}{3}$ mi checks.

State. The pilot should travel $5\frac{1}{3}$ mi upstream before turning around.

Chapter 15 Review Exercises

1. We check by substituting alphabetically 6 for x and -1 for y.

$$\frac{x - y = 3}{6 - (-1) \; ? \; 3}$$
$$\begin{array}{c|c} 6 + 1 & \\ 7 & \text{FALSE} \end{array}$$

Since $(6, -1)$ is not a solution of the first equation, it is not a solution of the system of equations.

2. We check by substituting alphabetically 2 for x and -3 for y.

$$\frac{2x + y = 1}{2 \cdot 2 + (-3) \; ? \; 1} \qquad \frac{x - y = 5}{2 - (-3) \; ? \; 5}$$
$$\begin{array}{c|c} 4 - 3 & \\ 1 & \text{TRUE} \end{array} \qquad \begin{array}{c|c} 2 + 3 & \\ 5 & \text{TRUE} \end{array}$$

The ordered pair $(2, -3)$ is a solution of both equations, so it is a solution of the system of equations.

3. We check by substituting alphabetically -2 for x and 1 for y.

$$\frac{x + 3y = 1}{-2 + 3 \cdot 1 \; ? \; 1} \qquad \frac{2x - y = -5}{2(-2) - 1 \; ? \; -5}$$
$$\begin{array}{c|c} -2 + 3 & \\ 1 & \text{TRUE} \end{array} \qquad \begin{array}{c|c} -4 - 1 & \\ -5 & \text{TRUE} \end{array}$$

The ordered pair $(-2, 1)$ is a solution of both equations, so it is a solution of the system of equations.

4. We check by substituting alphabetically -4 for x and -1 for y.

$$\frac{x - y = 3}{\begin{array}{c|c} -4 - (-1) \ ? \ 3 \\ -4 + 1 \\ -3 \end{array} \ \text{FALSE}}$$

Since $(-4, -1)$ is not a solution of the first equation, it is not a solution of the system of equations.

5. We graph the equations.

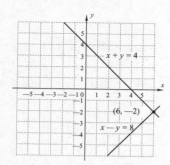

The point of intersection looks as if it has coordinates $(6, -2)$.

Check:

$$\frac{x + y = 4}{\begin{array}{c|c} 6 + (-2) \ ? \ 4 \\ 4 \end{array} \ \text{TRUE}} \qquad \frac{x - y = 8}{\begin{array}{c|c} 6 - (-2) \ ? \ 8 \\ 6 + 2 \\ 8 \end{array} \ \text{TRUE}}$$

The solution is $(6, -2)$.

6. We graph the equations.

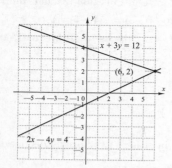

The point of intersection looks as if it has coordinates $(6, 2)$.

Check:

$$\frac{x + 3y = 12}{\begin{array}{c|c} 6 + 3 \cdot 2 \ ? \ 12 \\ 6 + 6 \\ 12 \end{array} \ \text{TRUE}} \qquad \frac{2x - 4y = 4}{\begin{array}{c|c} 2 \cdot 6 - 4 \cdot 2 \ ? \ 8 \\ 12 - 8 \\ 4 \end{array} \ \text{TRUE}}$$

The solution is $(6, 2)$.

7. We graph the equations.

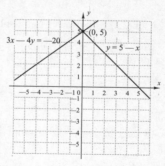

The point of intersection looks as if it has coordinates $(0, 5)$.

Check:

$$\frac{y = 5 - x}{\begin{array}{c|c} 5 \ ? \ 5 - 0 \\ 5 \end{array} \ \text{TRUE}} \qquad \frac{3x - 4y = -20}{\begin{array}{c|c} 3 \cdot 0 - 4 \cdot 5 \ ? \ -20 \\ -20 \end{array} \ \text{TRUE}}$$

The solution is $(0, 5)$.

8. We graph the equations.

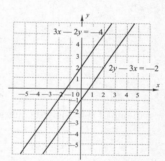

The lines are parallel. There is no solution.

9. $y = 5 - x$,   (1)

     $3x - 4y = -20$ (2)

We substitute $5 - x$ for y in Equation (2) and solve for x.

$$3x - 4y = -20 \quad (2)$$
$$3x - 4(5 - x) = -20$$
$$3x - 20 + 4x = -20$$
$$7x - 20 = -20$$
$$7x = 0$$
$$x = 0$$

Next we substitute 0 for x in one of the original equations and solve for y.

$$y = 5 - x \quad (1)$$
$$y = 5 - 0$$
$$y = 5$$

The ordered pair $(0, 5)$ checks in both equations. It is the solution.

10. $x + y = 6,$ (1)

$y = 3 - 2x$ (2)

We substitute $3 - 2x$ for y in Equation (1) and solve for x.

$$x + y = 6 \quad (1)$$
$$x + (3 - 2x) = 6$$
$$-x + 3 = 6$$
$$-x = 3$$
$$x = -3$$

Now we substitute -3 for x in one of the original equations and solve for y.

$$y = 3 - 2x \quad (2)$$
$$y = 3 - 2(-3)$$
$$y = 3 + 6$$
$$y = 9$$

The ordered pair $(-3, 9)$ checks in both equations. It is the solution.

11. $x - y = 4,$ (1)

$y = 2 - x$ (2)

We substitute $2 - x$ for y in Equation (1) and solve for x.

$$x - y = 4 \quad (1)$$
$$x - (2 - x) = 4$$
$$x - 2 + x = 4$$
$$2x - 2 = 4$$
$$2x = 6$$
$$x = 3$$

Now substitute 3 for x in one of the original equations and solve for y.

$$y = 2 - x \quad (2)$$
$$y = 2 - 3$$
$$y = -1$$

The ordered pair $(3, -1)$ checks in both equations. It is the solution.

12. $s + t = 5,$ (1)

$s = 13 - 3t$ (2)

We substitute $13 - 3t$ for s in Equation (1) and solve for t.

$$s + t = 5 \quad (1)$$
$$(13 - 3t) + t = 5$$
$$13 - 2t = 5$$
$$-2t = -8$$
$$t = 4$$

Now substitute 4 for t in one of the original equations and solve for s.

$$s = 13 - 3t \quad (2)$$
$$s = 13 - 3 \cdot 4$$
$$s = 13 - 12$$
$$s = 1$$

The ordered pair $(1, 4)$ checks in both equations. It is the solution.

13. $x + 2y = 6,$ (1)

$2x + 3y = 8$ (2)

We solve Equation (1) for x.

$$x + 2y = 6 \quad (1)$$
$$x = -2y + 6 \quad (3)$$

We substitute $-2y + 6$ for x in Equation (2) and solve for y.

$$2x + 3y = 8 \quad (2)$$
$$2(-2y + 6) + 3y = 8$$
$$-4y + 12 + 3y = 8$$
$$-y + 12 = 8$$
$$-y = -4$$
$$y = 4$$

Now substitute 4 for y in Equation (3) and compute x.

$$x = -2y + 6 = -2 \cdot 4 + 6 = -8 + 6 = -2$$

The ordered pair $(-2, 4)$ checks in both equations. It is the solution.

14. $3x + y = 1,$ (1)

$x - 2y = 5$ (2)

We solve Equation (2) for x.

$$x - 2y = 5 \quad (1)$$
$$x = 2y + 5 \quad (3)$$

We substitute $2y + 5$ for x in Equation (1) and solve for y.

$$3x + y = 1 \quad (1)$$
$$3(2y + 5) + y = 1$$
$$6y + 15 + y = 1$$
$$7y + 15 = 1$$
$$7y = -14$$
$$y = -2$$

Now substitute -2 for y in Equation (3) and compute x.

$$x = 2y + 5 = 2(-2) + 5 = -4 + 5 = 1$$

The ordered pair $(1, -2)$ checks in both equations. It is the solution.

15. $x + y = 4,$ (1)

$\underline{2x - y = 5} \quad (2)$

$3x = 9$

$x = 3$

Substitute 3 for x in either of the original equations and solve for y.

$$x + y = 4 \quad (1)$$
$$3 + y = 4$$
$$y = 1$$

The ordered pair $(3, 1)$ checks in both equations. It is the solution.

16. $x + 2y = 9,$ (1)

$\underline{3x - 2y = -5} \quad (2)$

$4x = 4$

$x = 1$

Substitute 1 for x in either of the original equations and solve for y.

$$x + 2y = 9 \quad (1)$$
$$1 + 2y = 9 \quad (2)$$
$$2y = 8$$
$$y = 4$$

The ordered pair $(1, 4)$ checks in both equations. It is the solution.

17. $x - y = 8, \quad (1)$
$$\underline{2x + y = 7} \quad (2)$$
$$3x \quad\;\; = 15$$
$$x = 5$$

Substitute 5 for x in either of the original equations and solve for y.

$$2x + y = 7 \quad (2)$$
$$2 \cdot 5 + y = 7$$
$$10 + y = 7$$
$$y = -3$$

The ordered pair $(5, -3)$ checks in both equations. It is the solution.

18. $2x + 3y = 8, \quad (1)$
$$5x + 2y = -2 \quad (2)$$

We use the multiplication principle with both equations and then add.

$$4x + 6y = 16 \quad \text{Multiplying (1) by 2}$$
$$\underline{-15x - 6y = 6} \quad \text{Multiplying (2) by } -3$$
$$-11x \quad\;\; = 22$$
$$x = -2$$

Substitute -2 for x in one of the original equations and solve for y.

$$2x + 3y = 8 \quad (1)$$
$$2(-2) + 3y = 8$$
$$-4 + 3y = 8$$
$$3y = 12$$
$$y = 4$$

The ordered pair $(-2, 4)$ checks in both equations. It is the solution.

19. $5x - 2y = 2, \quad (1)$
$$3x - 7y = 36 \quad (2)$$

We use the multiplication principle with both equations and then add.

$$35x - 14y = 14 \quad \text{Multiplying (1) by 7}$$
$$\underline{-6x + 14y = -72} \quad \text{Multiplying (2) by } -2$$
$$29x \quad\quad\;\; = -58$$
$$x = -2$$

Substitute -2 for x in one of the original equations and solve for y.

$$5x - 2y = 2 \quad (1)$$
$$5(-2) - 2y = 2$$
$$-10 - 2y = 2$$
$$-2y = 12$$
$$y = -6$$

The ordered pair $(-2, -6)$ checks in both equations. It is the solution.

20. $-x - y = -5, \quad (1)$
$$2x - y = 4 \quad (2)$$

We multiply Equation (1) by -1 and then add.

$$x + y = 5$$
$$\underline{2x - y = 4}$$
$$3x \quad\;\; = 9$$
$$x = 3$$

Substitute 3 for x in one of the original equations and solve for y.

$$-x - y = -5 \quad (1)$$
$$-3 - y = -5$$
$$-y = -2$$
$$y = 2$$

The ordered pair $(3, 2)$ checks in both equations. It is the solution.

21. $6x + 2y = 4, \quad (1)$
$$10x + 7y = -8 \quad (2)$$

We use the multiplication principle with both equations and then add.

$$42x + 14y = 28 \quad \text{Multiplying (1) by 7}$$
$$\underline{-20x - 14y = 16} \quad \text{Multiplying (2) by } -2$$
$$22x \quad\quad\;\; = 44$$
$$x = 2$$

Substitute 2 for x in one of the original equations and solve for y.

$$6x + 2y = 4 \quad (1)$$
$$6 \cdot 2 + 2y = 4$$
$$12 + 2y = 4$$
$$2y = -8$$
$$y = -4$$

The ordered pair $(2, -4)$ checks in both equations. It is the solution.

22. $-6x - 2y = 5, \quad (1)$
$$12x + 4y = -10 \quad (2)$$

We multiply Equation (1) by 2 and then add.

$$-12x - 4y = 10$$
$$\underline{12x + 4y = -10}$$
$$0 = 0$$

We get an obviously true equation, so the system has an infinite number of solutions.

23. $\dfrac{2}{3}x + y = -\dfrac{5}{3}$

$\quad\;\; x - \dfrac{1}{3}y = -\dfrac{13}{3}$

First we multiply both sides of each equation to clear the fractions.

$$3\left(\dfrac{2}{3}x + y\right) = 3\left(-\dfrac{5}{3}\right)$$

$$3 \cdot \dfrac{2}{3}x + 3y = -5$$

$$2x + 3y = -5$$

$$3\left(x - \dfrac{1}{3}y\right) = 3\left(-\dfrac{13}{3}\right)$$

$$3x - 3 \cdot \dfrac{1}{3}y = -13$$

$$3x - y = -13$$

The resulting system is

$$2x + 3y = -5, \quad (1)$$
$$3x - y = -13. \quad (2)$$

Now we multiply Equation (2) by 3 and then add.

$$2x + 3y = -5$$
$$\underline{9x - 3y = -39}$$
$$11x \qquad\;\; = -44$$
$$x = -4$$

Substitute -4 for x in Equation (1) and solve for y.

$$2x + 3y = -5$$
$$2(-4) + 3y = -5$$
$$-8 + 3y = -5$$
$$3y = 3$$
$$y = 1$$

The ordered pair $(-4, 1)$ checks in both equations. It is the solution.

24. *Familiarize*. We make a drawing. We let l = the length and w = the width, in cm.

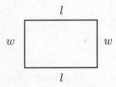

***Translate*.** The perimeter is $2l + 2w$. We translate the first statement.

$$\underbrace{\text{The perimeter}}_{\displaystyle 2l + 2w} \;\;\underbrace{\text{is}}_{\displaystyle =}\;\; \underbrace{\text{96 cm.}}_{\displaystyle 96}$$

We translate the second statement.

$$\underbrace{\text{The length}}_{\displaystyle l} \;\;\underbrace{\text{is}}_{\displaystyle =}\;\; \underbrace{\text{27 cm}}_{\displaystyle 27}\;\; \underbrace{\text{more than}}_{\displaystyle +}\;\; \underbrace{\text{the width.}}_{\displaystyle w}$$

The resulting system is

$$2l + 2w = 96, \quad (1)$$
$$l = 27 + w. \quad (2)$$

***Solve*.** First we substitute $27 + w$ for l in Equation (1) and solve for w.

$$2l + 2w = 96 \quad (1)$$
$$2(27 + w) + 2w = 96$$
$$54 + 2w + 2w = 96$$
$$54 + 4w = 96$$
$$4w = 42$$
$$w = 10.5$$

Now we substitute 10.5 for w in Equation (2) and find l.

$$l = 27 + w = 27 + 10.5 = 37.5$$

***Check*.** If the length is 37.5 cm and the width is 10.5 cm, then the perimeter is $2(37.5) + 2(10.5)$, or $75 + 21$, or 96 cm. Also, the length is 27 cm more than the width. The answer checks.

***State*.** The length of the rectangle is 37.5 cm, and the width is 10.5 cm.

25. *Familiarize*. Let x = the number of orchestra seats sold and y = the number of balcony seats sold. We organize the information in a table.

	Orchestra	Balcony	Total
Price	$25	$18	
Number bought	x	y	508
Receipts	$25x$	$18y$	$11,223

***Translate*.** The last two rows of the table give us a system of equations.

$$x + y = 508, \qquad\qquad (1)$$
$$25x + 18y = 11,223. \quad (2)$$

***Solve*.** First we multiply Equation (1) by -18 and then add.

$$-18x - 18y = -9144$$
$$\underline{25x + 18y = 11,223}$$
$$7x \qquad\;\; = 2079$$
$$x = 297$$

Now we substitute 297 for x in Equation (1) and solve for y.

$$x + y = 508$$
$$297 + y = 508$$
$$y = 211$$

***Check*.** The total number of tickets sold was $297 + 211$, or 508. The total receipts were $\$25 \cdot 297 + \$18 \cdot 211$, or $\$7425 + \$3798$, or $\$11,223$. The answer checks.

***State*.** 297 orchestra seats and 211 balcony seats were sold.

26. *Familiarize*. Let $c =$ the number of liters of Clear Shine and $s =$ the number of liters of Sunstream window cleaner to be used in the mixture. We organize the information in a table.

Type of cleaner	Clear Shine	Sunstream	Mixture
Amount used	c	s	80
Percent of alcohol	30%	60%	45%
Amount of alcohol in solution	$0.3c$	$0.6s$	45% × 80, or 36 L

***Translate*.** The first and third rows of the table give us a system of equations.

$$c + s = 80,$$
$$0.3c + 0.6s = 36$$

After we clear decimals we have

$$c + s = 80, \quad (1)$$
$$3c + 6s = 360. \quad (2)$$

***Solve*.** First we multiply Equation (1) by -3 and then add.

$$-3c - 3s = -240$$
$$\underline{3c + 6s = 360}$$
$$3s = 120$$
$$s = 40$$

Now we substitute 40 for s in Equation (1) and solve for c.

$$c + s = 80$$
$$c + 40 = 80$$
$$c = 40$$

***Check*.** If 40 L of Clear Shine and 40 L of Sunstream are used, then there is 40 L + 40 L, or 80 L, of solution. The amount of alcohol in the solution is $0.3(40) + 0.6(40)$, or $12 + 24$, or 36 L. The answer checks.

***State*.** 40 L of each window cleaner should be used.

27. *Familiarize*. Let $x =$ the weight of the Asian elephant and $y =$ the weight of the African elephant, in kg.

***Translate*.**

African elephant's weight	is	2400 kg	more than	Asian elephant's weight.
↓	↓	↓	↓	↓
y	$=$	2400	$+$	x

Asian elephant's weight	plus	African elephant's weight	is	12,000 kg
↓	↓	↓	↓	↓
x	$+$	y	$=$	12,000

The resulting system is

$$y = 2400 + x, \quad (1)$$
$$x + y = 12,000. \quad (2)$$

***Solve*.** First we substitute $2400 + x$ for y in Equation (2) and solve for x.

$$x + y = 12,000$$
$$x + (2400 + x) = 12,000$$
$$2x + 2400 = 12,000$$
$$2x = 9600$$
$$x = 4800$$

Now substitute 4800 for x in Equation (1) and find y.

$$y = 2400 + x = 2400 + 4800 = 7200$$

***Check*.** If the Asian elephant weighs 7200 kg and the African elephant weighs 4800 kg, then the African elephant weighs 2400 kg more than the Asian elephant and their total weight is $4800 + 7200$, or 12,000 kg. The answer checks.

***State*.** The Asian elephant weighs 4800 kg, and the African elephant weighs 7200 kg.

28. *Familiarize*. Let $x =$ the number of pounds of peanuts and $y =$ the number of pounds of fancy nuts to be used. We organize the information in a table.

Type of nuts	Peanuts	Fancy	Mixture
Cost per pound	$4.50	$7.00	
Amount	x	y	13
Mixture	$4.5x$	$7y$	$71

***Translate*.** The last two rows of the table give us a system of equations.

$$x + y = 13,$$
$$4.5x + 7y = 71$$

After clearing decimals, we have

$$x + y = 13, \quad (1)$$
$$45x + 70y = 710. \quad (2)$$

***Solve*.** First we multiply Equation (1) by -45 and then add.

$$-45x - 45y = -585$$
$$\underline{45x + 70y = 710}$$
$$25y = 125$$
$$y = 5$$

Now substitute 5 for y in one of the original equations and solve for x.

$$x + y = 13 \quad (1)$$
$$x + 5 = 13$$
$$x = 8$$

***Check*.** If 8 lb of peanuts and 5 lb of fancy nuts are used, the mixture weighs 13 lb. The value of the mixture is $\$4.50(8) + \$7.00(5) = \$36 + \$35 = \$71$. The answer checks.

***State*.** 8 lb of peanuts and 5 lb of fancy nuts should be used.

29. Familiarize. Let $x =$ the number of minutes used in the $29.95 plan and $y =$ the number of minutes used in the 7¢ a minute plan. Expressing 7¢ as $0.07, the 7¢ plan costs $0.07y + $3.95 per month.

Translate.

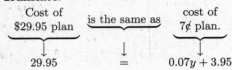

Cost of $29.95 plan	is the same as	cost of 7¢ plan.
29.95	=	$0.07y + 3.95$

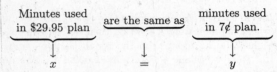

Minutes used in $29.95 plan	are the same as	minutes used in 7¢ plan.
x	=	y

After clearing decimals we have the following system of equations.

$$2995 = 7y + 395, \quad (1)$$
$$x = y \qquad\qquad (2)$$

Solve. First we solve Equation (1) for y.

$$2995 = 7y + 395$$
$$2600 = 7y$$
$$371 \approx y$$

Now substitute 371 for y in Equation (2) and find x.

$$x = y$$
$$x = 371$$

Check. For 371 min the 7¢ plan costs $0.07(371) + $3.95, or $29.92. We rounded the value of y and $29.92 \approx $29.95, so the answer checks.

State. The two plans cost the same for about 371 min.

30. Familiarize. Let $x =$ the number of gallons of 87-octane gas and $y =$ the number of gallons of 95-octane gas to be used. We arrange the information in a table.

Type of gas	87-octane	95-octane	Mixture
Amount	x	y	10
Octane rating	87	95	93
Mixture	$87x$	$95y$	10·93, or 930

Translate. The first and last rows of the table give us a system of equations.

$$x + \ \ y = 10, \quad (1)$$
$$87x + 95y = 930 \quad (2)$$

Solve. We multiply Equation (1) by -87 and then add.

$$-87x - 87y = -870$$
$$\underline{87x + 95y = 930}$$
$$8y = 60$$
$$y = 7.5$$

Now substitute 7.5 for y in one of the original equations and solve for x.

$$x + y = 10 \quad (1)$$
$$x + 7.5 = 10$$
$$x = 2.5$$

Check. If 2.5 gal of 87-octane gas and 7.5 gal of 95-octane gas are used, then there are $2.5 + 7.5$, or 10 gal, of gas in the mixture. Also, $87(2.5) + 95(7.5) = 217.5 + 712.5 = 930$, so the answer checks.

State. 2.5 gal of 87-octane gas and 7.5 gal of 95-octane gas should be used.

31. Familiarize. Let $x =$ Jeff's age now and $y =$ his son's age now. In 13 yr their ages will be $x + 13$ and $y + 13$.

Translate.

Jeff's age	is	three	times	his son's age.
x	=	3	·	y

In 13 yr,

Jeff's age	will be	two	times	his son's age.
$x + 13$	=	2	·	$(y + 13)$

The resulting system is

$$x = 3y, \qquad\qquad (1)$$
$$x + 13 = 2(y + 13). \quad (2)$$

Solve. First we substitute $3y$ for x in Equation (2) and solve for y.

$$x + 13 = 2(y + 13)$$
$$3y + 13 = 2(y + 13)$$
$$3y + 13 = 2y + 26$$
$$y + 13 = 26$$
$$y = 13$$

Now substitute 13 for y in Equation (1) and find x.

$$x = 3y = 3 \cdot 13 = 39$$

Check. If Jeff is 39 years old and his son is 13 years old, then Jeff is three times as old as his son. In 13 yr Jeff's age will be $39 + 13$, or 52, his son's age will be $13 + 13$, or 26, and $52 = 2 \cdot 26$. The answer checks.

State. Jeff is 39 years old now, and his son is 13 years old.

32. Familiarize. Let $x =$ the measure of the larger angle and $y =$ the measure of the smaller angle. Recall that the sum of the measures of complementary angles is 90°.

Translate.

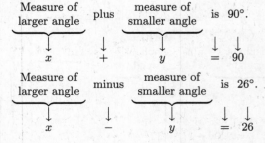

Measure of larger angle	plus	measure of smaller angle	is	90°.
x	+	y	=	90

Measure of larger angle	minus	measure of smaller angle	is	26°.
x	−	y	=	26

The resulting system is

$$x + y = 90, \quad (1)$$
$$x - y = 26. \quad (2)$$

Solve. We add.

$$\begin{array}{r} x + y = 90 \\ x - y = 26 \\ \hline 2x \phantom{{}-y} = 116 \\ x = 58 \end{array}$$

Substitute 58 for x in one of the original equations and solve for y.

$$x + y = 90 \quad (1)$$
$$58 + y = 90$$
$$y = 32$$

Check. $58° + 32° = 90°$ and $58° - 32° = 26°$, so the answer checks.

State. The measures of the angles are 58° and 32°.

33. *Familiarize*. Let $x =$ the measure of the larger angle and $y =$ the measure of the smaller angle. Recall that the sum of the measures of supplementary angles is 180°.

Translate.

Measure of larger angle	plus	measure of smaller angle	is	180°.
↓	↓	↓	↓	↓
x	$+$	y	$=$	180

Measure of larger angle	minus	measure of smaller angle	is	26°.
↓	↓	↓	↓	↓
x	$-$	y	$=$	26

The resulting system is

$$x + y = 180, \quad (1)$$
$$x - y = 26. \quad (2)$$

Solve. We add.

$$\begin{array}{r} x + y = 180 \\ x - y = 26 \\ \hline 2x \phantom{{}-y} = 206 \\ x = 103 \end{array}$$

Now substitute 103 for x in one of the original equations and solve for y.

$$x + y = 180 \quad (1)$$
$$103 + y = 180$$
$$y = 77$$

Check. $103° + 77° = 180°$ and $103° - 77° = 26°$; so the answer checks.

State. The measures of the angles are 103° and 77°.

34. *Familiarize*. Let $r =$ the speed of the airplane in still air, in km/h. Then $r + 15 =$ the speed with a 15 km/h tail wind and $r - 15 =$ the speed against a 15 km/h wind. We fill in the table in the text.

	Distance	Speed	Time
Going	d	$r + 15$	4
Returning	d	$r - 15$	5

Translate. We get an equation $d = rt$ from each row of the table.

$$d = (r + 15)4, \quad (1)$$
$$d = (r - 15)5 \quad (2)$$

Solve. We substitute $(r + 15)4$ for d in Equation (2).

$$d = (r - 15)5$$
$$(r + 15)4 = (r - 15)5$$
$$4r + 60 = 5r - 75$$
$$60 = r - 75$$
$$135 = r$$

Check. The plane's speed with the tail wind is $135 + 15$, or 150 km/h. At that speed, in 4 hr it will travel $150 \cdot 4$, or 600 km. The plane's speed against the wind is $135 - 15$, or 120 km/h. At that speed, in 5 hr it will travel $120 \cdot 5$, or 600 km. Since the distances are the same, the answer checks.

State. The speed of the airplane in still air is 135 km/h.

35. *Familiarize*. Let $t =$ the number of hours the slower car travels before the second car catches up to it. Then $t - 2 =$ the number of hours the faster car travels. We fill in the table in the text.

	Distance	Speed	Time
Slow car	d	55	t
Fast car	d	75	$t - 2$

Translate. We get an equation $d = rt$ from each row of the table.

$$d = 55t, \quad (1)$$
$$d = 75(t - 2) \quad (2)$$

Solve. We substitute $55t$ for d in Equation (2).

$$d = 75(t - 2)$$
$$55t = 75(t - 2)$$
$$55t = 75t - 150$$
$$-20t = -150$$
$$t = 7.5$$

Now substitute 7.5 for t in Equation (1) and find d.

$$d = 55t = 55(7.5) = 412.5$$

Check. From the calculation of d above, we see that the slow car travels 412.5 mi in 7.5 hr. The fast car travels $7.5 - 2$, or 5.5 hr. At a speed of 75 mph, it travels $75(5.5)$, or 412.5 mi. Since the distances are the same, the answer checks.

State. The second car catches up to the first car 412.5 mi from Phoenix.

36. *Discussion and Writing Exercise*. The equations have the same slope but different y-intercepts, so they represent parallel lines. Thus the system of equations has no solution.

37. *Discussion and Writing Exercise.* Answers will vary.

38. *Familiarize.* Let $c =$ the compensation agreed upon for 12 months of work and let $h =$ the value of the horse. After 7 months, Stephanie would be owed $\frac{7}{12}$ of the compensation agreed upon for 12 months of work, or $\frac{7}{12}c$.

Translate.

$$\underbrace{\text{Compensation for 12 months}}_{\downarrow} \;\; \text{is} \;\; \underbrace{\text{a horse}} \;\; \text{plus} \;\; \$2400.$$

$$c \;\;\;\;\;\; = \;\;\; h \;\;\; + \;\;\;\; 2400$$

$$\underbrace{\text{Compensation for 7 months}}_{\downarrow} \;\; \text{is} \;\; \underbrace{\text{a horse}} \;\; \text{plus} \;\; \$1000.$$

$$\frac{7}{12}c \;\;\;\;\;\; = \;\;\; h \;\;\; + \;\;\;\; 1000$$

After clearing the fractions we have the following system of equations.

$$c = h + 2400, \quad (1)$$
$$7c = 12h + 12{,}000 \quad (2)$$

Solve. We substitute $h + 2400$ for c in Equation (2) and solve for h.

$$7c = 12h + 12{,}000$$
$$7(h + 2400) = 12h + 12{,}000$$
$$7h + 16{,}800 = 12h + 12{,}000$$
$$16{,}800 = 5h + 12{,}000$$
$$4800 = 5h$$
$$960 = h$$

Check. If the value of the horse is \$960, then the compensation for 12 months of work is \$960 + \$2400, or \$3360; $\frac{7}{12}$ of \$3360 is \$1960. This is the value of the horse, \$960, plus \$1000, so the answer checks.

State. The value of the horse was \$960.

39. We substitute 6 for x and 2 for y in each equation.

$$2x - Dy = 6 \qquad\qquad Cx + 4y = 14$$
$$2 \cdot 6 - D \cdot 2 = 6 \qquad\quad C \cdot 6 + 4 \cdot 2 = 14$$
$$12 - 2D = 6 \qquad\qquad 6C + 8 = 14$$
$$-2D = -6 \qquad\qquad 6C = 6$$
$$D = 3 \qquad\qquad\qquad C = 1$$

40. $3(x - y) = 4 + x, \quad (1)$
$x = 5y + 2 \qquad\qquad (2)$

Substitute $5y + 2$ for x in Equation (1) and solve for y.

$$3(x - y) = 4 + x$$
$$3(5y + 2 - y) = 4 + 5y + 2$$
$$3(4y + 2) = 5y + 6$$
$$12y + 6 = 5y + 6$$
$$7y + 6 = 6$$
$$7y = 0$$
$$y = 0$$

Now substitute 0 for y in Equation (2) and find x.

$$x = 5y + 2 = 5 \cdot 0 + 2 = 0 + 2 = 2$$

The solution is $(2, 0)$.

41. The line graphed in red contains the points $(0, 0)$ and $(3, 2)$. We find the slope:

$$m = \frac{2 - 0}{3 - 0} = \frac{2}{3}$$

The y-intercept is $(0, 0)$, so the equation of the line is

$$y = \frac{2}{3}x + 0, \text{ or } y = \frac{2}{3}x.$$

The line graphed in blue contains the points $(0, 5)$ and $(3, 2)$. We find the slope:

$$m = \frac{2 - 5}{3 - 0} = \frac{-3}{3} = -1$$

The y-intercept is $(0, 5)$, so the equation of the line is $y = -1 \cdot x + 5$, or $y = -x + 5$.

42. The line graphed in red contains the points $(-3, 0)$ and $(0, -3)$. We find the slope:

$$m = \frac{-3 - 0}{0 - (-3)} = \frac{-3}{3} = -1$$

The y-intercept is $(0, -3)$, so the equation of the line is $y = -1 \cdot x - 3$, or $y = -x - 3$, or $x + y = -3$.

The line graphed in blue contains the points $(0, 4)$ and $(4, 0)$. We find the slope:

$$m = \frac{0 - 4}{4 - 0} = \frac{-4}{4} = -1$$

The y-intercept is $(0, 4)$, so the equation of the line is $y = -1 \cdot x + 4$, or $y = -x + 4$, or $x + y = 4$.

43. *Familiarize.* Let $x =$ the number of rabbits and $y =$ the number of pheasants. Then the rabbits have a total of x heads and $4x$ feet; the pheasants have a total of y heads and $2y$ feet.

Translate.

$$\underbrace{\text{Rabbit heads}} \;\; \text{plus} \;\; \underbrace{\text{pheasant heads}} \;\; \text{is} \;\; \underbrace{\text{35 heads.}}$$
$$x \;\;\;\;\;\; + \;\;\;\;\;\; y \;\;\;\;\;\; = \;\;\;\;\; 35$$

$$\underbrace{\text{Rabbit feet}} \;\; \text{plus} \;\; \underbrace{\text{pheasant feet}} \;\; \text{is} \;\; \underbrace{\text{94 feet.}}$$
$$4x \;\;\;\;\;\; + \;\;\;\;\;\; 2y \;\;\;\;\;\; = \;\;\;\;\; 94$$

The resulting system is

$$x + y = 35, \quad (1)$$
$$4x + 2y = 94. \quad (2)$$

Solve. First we multiply Equation (1) by -2 and then add.

$$-2x - 2y = -70$$
$$\underline{4x + 2y = 94}$$
$$2x = 24$$
$$x = 12$$

Now substitute 12 for x in one of the original equations and solve for y.

$$x + y = 35 \quad (1)$$
$$12 + y = 35$$
$$y = 23$$

Check. If there are 12 rabbits and 23 pheasants, then there are $12 + 23$, or 35, heads and $4 \cdot 12 + 2 \cdot 23$, or $48 + 46$, or 94, feet. The answer checks.

State. There are 12 rabbits and 23 pheasants.

Chapter 15 Test

1. We check by substituting alphabetically -2 for x and -1 for y.

$$\begin{array}{l} x = 4 + 2y \\ \hline -2 \ ? \ 4 + 2(-1) \\ \quad \ \ \big| \ \ 4 - 2 \\ \quad \ \ \big| \ \ 2 \qquad \text{FALSE} \end{array}$$

Since $(-2, -1)$ is not a solution of the first equation, it is not a solution of the system of equations.

2. We graph the equations.

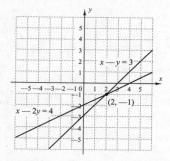

The point of intersection looks as if it has coordinates $(2, -1)$.

Check:

$$\begin{array}{l} x - y = 3 \\ \hline 2 - (-1) \ ? \ 3 \\ \quad 2 + 1 \ \ \big| \\ \qquad \ 3 \ \ \big| \ \text{TRUE} \end{array} \qquad \begin{array}{l} x - 2y = 4 \\ \hline 2 - 2(-1) \ ? \ 4 \\ \quad 2 + 2 \ \ \big| \\ \qquad \ 4 \ \ \big| \ \text{TRUE} \end{array}$$

The solution is $(2, -1)$.

3. $$y = 6 - x, \qquad (1)$$
$$2x - 3y = 22 \quad (2)$$

We substitute $6 - x$ for y in Equation (2) and solve for x.

$$2x - 3y = 22 \quad (2)$$
$$2x - 3(6 - x) = 22$$
$$2x - 18 + 3x = 22$$
$$5x - 18 = 22$$
$$5x = 40$$
$$x = 8$$

Next we substitute 8 for x in one of the original equations and solve for y.

$$y = 6 - x \quad (1)$$
$$y = 6 - 8$$
$$y = -2$$

The ordered pair $(8, -2)$ checks in both equations. It is the solution.

4. $$x + 2y = 5, \quad (1)$$
$$x + y = 2 \qquad (2)$$

We solve Equation (1) for x.

$$x + 2y = 5 \qquad \qquad (1)$$
$$x = -2y + 5 \quad (3)$$

We substitute $-2y + 5$ for x in Equation (2) and solve for y.

$$x + y = 2 \qquad (2)$$
$$-2y + 5 + y = 2$$
$$-y + 5 = 2$$
$$-y = -3$$
$$y = 3$$

Now substitute 3 for y in Equation (3) and compute x.

$$x = -2y + 5 = -2 \cdot 3 + 5 = -6 + 5 = -1.$$

The ordered pair $(-1, 3)$ checks in both equations. It is the solution.

5. $$y = 5x - 2, \quad (1)$$
$$y - 2 = 5x \quad (2)$$

Substitute $5x - 2$ for y in Equation (2) and solve for x.

$$y - 2 = 5x \quad (2)$$
$$5x - 2 - 2 = 5x$$
$$5x - 4 = 5x$$
$$-4 = 0$$

We obtain a false equation, so there is no solution.

6. $$\begin{array}{l} x - y = 6 \qquad (1) \\ \underline{3x + y = -2 \quad (2)} \\ 4x \qquad \ \ = 4 \\ \quad \ x = 1 \end{array}$$

Substitute 1 for x in either of the original equations and solve for y.

$$3x + y = -2 \quad (2)$$
$$3 \cdot 1 + y = -2$$
$$3 + y = -2$$
$$y = -5$$

The ordered pair $(1, -5)$ checks in both equations. It is the solution.

7. $$\frac{1}{2}x - \frac{1}{3}y = 8,$$
$$\frac{2}{3}x + \frac{1}{2}y = 5$$

First we multiply each equation by 6 to clear the fractions.

$$6\left(\frac{1}{2}x - \frac{1}{3} + y\right) = 6 \cdot 8$$

$$6 \cdot \frac{1}{2}x - 6 \cdot \frac{1}{3}y = 48$$

$$3x - 2y = 48$$

$$6\left(\frac{2}{3}x + \frac{1}{2}y\right) = 6 \cdot 5$$

$$6 \cdot \frac{2}{3}x + 6 \cdot \frac{1}{2}y = 30$$

$$4x + 3y = 30$$

The resulting system is

$$3x - 2y = 48, \quad (1)$$

$$4x + 3y = 30. \quad (2)$$

Now we multiply Equation (1) by 3 and Equation (2) by 2 and then add.

$$9x - 6y = 144$$

$$\underline{8x + 6y = 60}$$

$$17x \qquad = 204$$

$$x = 12$$

Next we substitute 12 for x in Equation (2) and solve for y.

$$4x + 3y = 30 \qquad (2)$$

$$4 \cdot 12 + 3y = 30$$

$$48 + 3y = 30$$

$$3y = -18$$

$$y = -6$$

The ordered pair $(12, -6)$ checks in both equations. It is the solution.

8. $\quad -4x - 9y = 4, \quad (1)$

$\quad\quad 6x + 3y = 1 \quad (2)$

Multiply Equation (2) by 3 and then add.

$$-4x - 9y = 4$$

$$\underline{18x + 9y = 3}$$

$$14x \qquad = 7$$

$$x = \frac{1}{2}$$

Now substitute $\frac{1}{2}$ for x in one of the original equations and solve for y.

$$6x + 3y = 1 \qquad (2)$$

$$6 \cdot \frac{1}{2} + 3y = 1$$

$$3 + 3y = 1$$

$$3y = -2$$

$$y = -\frac{2}{3}$$

The ordered pair $\left(\frac{1}{2}, -\frac{2}{3}\right)$ checks in both equations. It is the solution.

9. $\quad 2x + 3y = 13, \quad (1)$

$\quad\quad 3x - 5y = 10 \quad (2)$

Multiply Equation (1) by 5 and Equation (2) by 3 and then add.

$$10x + 15y = 65$$

$$\underline{9x - 15y = 30}$$

$$19x \qquad = 95$$

$$x = 5$$

Now substitute 5 for x in one of the original equations and solve for y.

$$2x + 3y = 13 \qquad (1)$$

$$2 \cdot 5 + 3y = 13$$

$$10 + 3y = 13$$

$$3y = 3$$

$$y = 1$$

The ordered pair $(5, 1)$ checks in both equations. It is the solution.

10. *Familiarize*. Let $l =$ the length and $w =$ the width, in yd. Recall that the perimeter of a rectangle is given by the formula $P = 2l + 2w$.

Translate.

The perimeter is 8266 yd.

$$2l + 2w \quad = \quad 8266$$

The length is 84 yd more than the width.

$$l \quad = \quad 84 \quad + \quad w$$

The resulting system is

$$2l + 2w = 8266, \quad (1)$$

$$l = 84 + w. \qquad (2)$$

Solve. First we substitute $84 + w$ for l in Equation (1) and solve for w.

$$2l + 2w = 8266$$

$$2(84 + w) + 2w = 8266$$

$$168 + 2w + 2w = 8266$$

$$168 + 4w = 8266$$

$$4w = 8098$$

$$w = 2024.5$$

Now substitute 2024.5 for w in Equation (2) and find l.

$$l = 84 + w = 84 + 2024.5 = 2108.5$$

Check. If the length is 2108.5 yd and the width is 2024.5 yd, the length is 84 yd more than the width and the perimeter is $2(2108.5) + 2(2024.5)$, or $4217 + 4049$, or 8266 yd. The answer checks.

State. The length is 2108.5 yd and the width is 2024.5 yd.

11. *Familiarize*. Let a = the amount of solution A and b = the amount of solution B in the mixture, in liters. We organize the information in a table.

Solution	A	B	Mixture
Amount	a	b	60 L
Percent of acid	25%	40%	30%
Amount of acid	0.25a	0.4b	0.3(60) or 18 L

***Translate*.** The first and third rows of the table give us a system of equations.

$$a + b = 60,$$
$$0.25a + 0.4b = 18$$

After clearing decimals, we have

$$a + b = 60, \quad (1)$$
$$25a + 40b = 1800. \quad (2)$$

***Solve*.** First we multiply Equation (1) by -25 and then add.

$$\begin{array}{r} -25a - 25b = -1500 \\ 25a + 40b = 1800 \\ \hline 15b = 300 \\ b = 20 \end{array}$$

Now we substitute 20 for b in one of the original equations and solve for a.

$$a + b = 60 \quad (1)$$
$$a + 20 = 60$$
$$a = 40$$

***Check*.** If 40 L of solution A and 20 L of solution B are used, then there are $40 + 20$, or 60 L, in the mixture. The amount of acid in the mixture is $0.25(40) + 0.4(20)$, or $10 + 8$, or 18 L. The answer checks.

***State*.** 40 L of solution A and 20 L of solution B should be used.

12. *Familiarize*. Let r = the speed of the motorboat in still water, in km/h. Then the speed of the boat with the current is $r + 8$ and the speed against the current is $r - 8$. We organize the information in a table.

	Distance	Speed	Time
With current	d	$r + 8$	2
Against current	d	$r - 8$	3

***Translate*.** We get an equation from each row of the table.

$$d = (r + 8)2 \quad (1)$$
$$d = (r - 8)3 \quad (2)$$

***Solve*.** We substitute $(r + 8)2$ for d in Equation (2) and solve for r.

$$d = (r - 8)3 \quad (2)$$
$$(r + 8)2 = (r - 8)3$$
$$2r + 16 = 3r - 24$$
$$16 = r - 24$$
$$40 = r$$

***Check*.** When $r = 40$, then $r + 8 = 40 + 8 = 48$ and in 2 hr at this speed the boat travels $48 \cdot 2$, or 96 km. Also, $r - 8 = 40 - 8 = 32$, and in 3 hr at this speed the boat travels $32 \cdot 3$, or 96 km. The distances are the same, so the answer checks.

***State*.** The speed of the motorboat in still water is 40 km/h.

13. *Familiarize*. Let c = the receipts from concessions and r = the receipts from the rides.

***Translate*.**

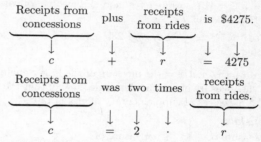

The resulting system is

$$c + r = 4275, \quad (1)$$
$$c = 2r. \quad (2)$$

***Solve*.** First we substitute $2r$ for c in Equation (1) and solve for r.

$$c + r = 4275 \quad (1)$$
$$2r + r = 4275$$
$$3r = 4275$$
$$r = 1425$$

Now we substitute 1425 for r in one of the original equations and find c.

$$c = 2r = 2 \cdot 1425 = 2850$$

***Check*.** If concessions brought in \$2850 and rides brought in \$1425, then the total receipts were $\$2850 + \$1425$, or \$4275. Also, $2 \cdot \$1425 = \$2850$, so concessions brought in twice as much as rides. The answer checks.

***State*.** Concessions brought in \$2850; rides brought in \$1425.

14. *Familiarize*. Let x = the number of acres of hay planted and y = the number of acres of oats planted.

***Translate*.**

Acres of hay plus acres of oats is 650 acres.

$$x \quad + \quad y \quad = \quad 650$$

Acres of hay is 180 more than acres of oats.

$$x \quad = \quad 180 \quad + \quad y$$

The resulting system is

$$x + y = 650, \quad (1)$$
$$x = 180 + y. \quad (2)$$

***Solve*.** First we substitute $180 + y$ for x in Equation (1) and solve for y.

$$x + y = 650 \quad (1)$$
$$(180 + y) + y = 650$$
$$180 + 2y = 650$$
$$2y = 470$$
$$y = 235$$

Now substitute 235 for y in Equation (2) and find x.

$$x = 180 + y = 180 + 235 = 415$$

Check. If 415 acres of hay and 235 acres of oats are planted, a total of $415 + 235$, or 650 acres, is planted. Also, $235 + 180 = 415$, so 180 acres more of hay than of oats are planted. The answer checks.

State. 415 acres of hay and 235 acres of oats should be planted.

15. **Familiarize**. Let $x =$ the measure of the larger angle and $y =$ the measure of the smaller angle. Recall that the sum of the measures of supplementary angles is $180°$.

Translate.

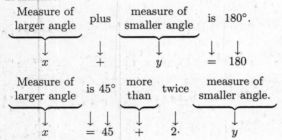

The resulting system is

$$x + y = 180, \quad (1)$$
$$x = 45 + 2y. \quad (2)$$

Solve. First we substitute $45 + 2y$ for x in Equation (1) and solve for y.

$$x + y = 180 \quad (1)$$
$$(45 + 2y) + y = 180$$
$$45 + 3y = 180$$
$$3y = 135$$
$$y = 45$$

Now substitute 45 for y in Equation (2) and find x.

$$x = 45 + 2y = 45 + 2 \cdot 45 = 45 + 90 = 135$$

Check. $135° + 45° = 180°$; and $135° = 45° + 2 \cdot 45°$, so the answer checks.

State. The measures of the angles are $135°$ and $45°$.

16. **Familiarize**. Let $x =$ the number of gallons of 87-octane gas and $y =$ the number of gallons of 93-octane gas that should be used. We organize the information in a table.

Type of gas	87-octane	93-octane	Mixture
Amount	x	y	12
Octane rating	87	93	91
Mixture	$87x$	$93y$	$12 \cdot 91$, or 1092

Translate. The first and last rows of the table give us a system of equations.

$$x + y = 12, \quad (1)$$
$$87x + 93y = 1092 \quad (2)$$

Solve. We multiply Equation (1) by -87 and then add.

$$-87x - 87y = -1044$$
$$\underline{87x + 93y = 1092}$$
$$6y = 48$$
$$y = 8$$

Then substitute 8 for y in one of the original equations and solve for x.

$$x + y = 12$$
$$x + 8 = 12$$
$$x = 4$$

Check. If 4 gal of 87-octane gas and 8 gal of 93-octane gas are used, then there are $4 + 8$, or 12 gal, of gas in the mixture. Also $87 \cdot 4 + 93 \cdot 8 = 348 + 744 = 1092$, so the answer checks.

State. 4 gal of 87-octane gas and 8 gal of 93-octane gas should be used.

17. **Familiarize**. Let $x =$ the number of minutes used in the $2.95 plan and $y =$ the number of minutes used in the $1.95 plan. Expressing 10¢ as $0.10 and 15¢ as $0.15, the $2.95 plan costs $2.95 + $0.10x$ per month and the $1.95 plan costs $1.95 + $0.15y$ per month.

Translate.

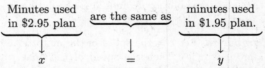

$$2.95 + 0.10x \qquad = \qquad 1.95 + 0.15y$$

Minutes used in $2.95 plan are the same as minutes used in $1.95 plan.

$$x \qquad = \qquad y$$

After clearing decimals we have the following system of equations.

$$295 + 10x = 195 + 15y, \quad (1)$$
$$x = y \qquad\qquad (2)$$

Solve. We substitute x for y in Equation (1) and solve for x.

$$295 + 10x = 195 + 15y \quad (1)$$
$$295 + 10x = 195 + 15x$$
$$295 = 195 + 5x$$
$$100 = 5x$$
$$20 = x$$

Substitute 20 for x in Equation (2) to find y.

$$x = y$$
$$20 = y$$

Check. For 20 min the $2.95 plan costs $2.95 + $0.10(20)$, or $4.95 per month and the $1.95 plan costs

$1.95 + $0.15(20)$, or $4.95 per month. The costs are the same, so the answer checks.

State. The two plans cost the same for 20 min.

18. **Familiarize**. Let $t =$ the number of hours it will take the SUV to catch up with the car. Then $t + 2 =$ the number of hours the car travels. We organize the information in a table.

	Distance	Speed	Time
Car	d	55	$t + 2$
SUV	d	65	t

Translate. We get an equation $d = rt$ from each row of the table.

$$d = 55(t + 2), \quad (1)$$
$$d = 65t \quad\quad\quad (2)$$

Solve. Substitute $65t$ for d in Equation (1) and solve for t.

$$d = 55(t + 2) \quad (1)$$
$$65t = 55(t + 2)$$
$$65t = 55t + 110$$
$$10t = 110$$
$$t = 11$$

Check. At a speed of 55 mph for $11 + 2$, or 13 hr, the car travels $55 \cdot 13$, or 715 mi; at a speed of 65 mph for 11 hr, the SUV travels $65 \cdot 11$, or 715 mi. Since the distances are the same, the answer checks.

State. It will take the SUV 11 hr to catch up to the car.

19. Substitute -2 for x and 3 for y in each equation.

$$Cx - 4y = 7 \qquad\qquad 3x + Dy = 8$$
$$C(-2) - 4 \cdot 3 = 7 \qquad 3(-2) + D \cdot 3 = 8$$
$$-2C - 12 = 7 \qquad\qquad -6 + 3D = 8$$
$$-2C = 19 \qquad\qquad\qquad 3D = 14$$
$$C = -\frac{19}{2} \qquad\qquad\qquad D = \frac{14}{3}$$

20. **Familiarize**. Let $a =$ the number of people ahead of you and $b =$ the number of people behind you. Then in the entire line there are the people ahead of you, the people behind you, and you yourself, or $a + b + 1$.

Translate.

$$\underbrace{\text{Number ahead of you}}_{a} \;\; \underbrace{\text{is}}_{=} \;\; \underbrace{\text{two}}_{2} \;\; \underbrace{\text{more than}}_{+} \;\; \underbrace{\text{number behind you.}}_{b}$$

$$\underbrace{\text{Number in entire line}}_{a + b + 1} \;\; \underbrace{\text{is}}_{=} \;\; \underbrace{\text{three}}_{3} \;\; \underbrace{\text{times}}_{\cdot} \;\; \underbrace{\text{number behind you.}}_{b}$$

The resulting system is

$$a = 2 + b, \qquad (1)$$
$$a + b + 1 = 3b. \qquad (2)$$

Solve. We substitute $2 + b$ for a in Equation (2) and solve for b.

$$a + b + 1 = 3b \quad (2)$$
$$(2 + b) + b + 1 = 3b$$
$$2b + 3 = 3b$$
$$3 = b$$

Now substitute 3 for b in Equation (1) and find a.

$$a = 2 + b = 2 + 3 = 5$$

Check. If there are 5 people ahead of you and 3 people behind you, then the number of people ahead of you is two more than the number behind you. Also, in the entire line there are $5 + 1 + 3$, or 9 people, and $3 \cdot 3 = 9$. The answer checks.

State. There are 5 people ahead of you.

21. The line graphed in red contains the points $(-2, 3)$ and $(3, 0)$.

$$m = \frac{3 - 0}{-2 - 3} = \frac{3}{-5} = -\frac{3}{5}$$

Then the equation of the line is $y = -\frac{3}{5}x + b$. To find b we substitute the coordinates of either point for x and y in the equation. We use $(3, 0)$.

$$y = -\frac{3}{5}x + b$$
$$0 = -\frac{3}{5} \cdot 3 + b$$
$$0 = -\frac{9}{5} + b$$
$$\frac{9}{5} = b$$

The equation of the line is $y = -\frac{3}{5}x + \frac{9}{5}$.

The line graphed in blue contains the points $(-2, 3)$ and $(3, 4)$.

$$m = \frac{3 - 4}{-2 - 3} = \frac{-1}{-5} = \frac{1}{5}$$

Then the equation is $y - \frac{1}{5}x + b$. To find b we substitute the coordinates of either point for x and y in the equation. We use $(3, 4)$.

$$y = \frac{1}{5}x + b$$
$$4 = \frac{1}{5} \cdot 3 + b$$
$$4 = \frac{3}{5} + b$$
$$\frac{17}{5} = b$$

The equation of the line is $y = \frac{1}{5}x + \frac{17}{5}$.

22. The line graphed in red is a vertical line, so its equation is of the form $x = a$. The line contains the point $(3, -2)$, so the equation is $x = 3$.

The line graphed in blue is a horizontal line, so its equation is of the form $y = b$. The line contains the point $(3, -2)$, so the equation is $y = -2$.

Chapter 16

Radical Expressions and Equations

Exercise Set 16.1

1. The square roots of 4 are 2 and -2, because $2^2 = 4$ and $(-2)^2 = 4$.

3. The square roots of 9 are 3 and -3, because $3^2 = 9$ and $(-3)^2 = 9$.

5. The square roots of 100 are 10 and -10, because $10^2 = 100$ and $(-10)^2 = 100$.

7. The square roots of 169 are 13 and -13, because $13^2 = 169$ and $(-13)^2 = 169$.

9. The square roots of 256 are 16 and -16, because $16^2 = 256$ and $(-16)^2 = 256$.

11. $\sqrt{4} = 2$, taking the principal square root.

13. $\sqrt{9} = 3$, so $-\sqrt{9} = -3$.

15. $\sqrt{36} = 6$, so $-\sqrt{36} = -6$.

17. $\sqrt{225} = 15$, so $-\sqrt{225} = -15$.

19. $\sqrt{361} = 19$, taking the principal square root.

21. 2.236

23. 20.785

25. $\sqrt{347.7} \approx 18.647$, so $-\sqrt{347.7} \approx -18.647$.

27. 2.779

29. $\sqrt{8 \cdot 9 \cdot 200} = 120$, so $-\sqrt{8 \cdot 9 \cdot 200} = -120$.

31. a) We substitute 25 into the formula:
$$N = 2.5\sqrt{25} = 2.5(5) = 12.5 \approx 13 \text{ spaces}$$
   b) We substitute 89 into the formula and use a calculator to find an approximation.
$$N = 2.5\sqrt{89} \approx 2.5(9.434) = 23.585 \approx 24 \text{ spaces}$$

33. Substitute 33.5 in the formula.
$$T = 0.144\sqrt{33.5} \approx 0.833 \text{ sec}$$

35. Substitute 40 in the formula.
$$T = 0.144\sqrt{40} \approx 0.911 \text{ sec}$$

37. The radicand is the expression under the radical, 200.

39. The radicand is the expression under the radical, $a - 4$.

41. The radicand is the expression under the radical, $t^2 + 1$.

43. The radicand is the expression under the radical, $\dfrac{3}{x+2}$.

45. No, because the radicand is negative

47. Yes, because the radicand is nonnegative

49. No, because the radicand is negative

51. $\sqrt{c^2} = c$ Since c is assumed to be nonnegative

53. $\sqrt{9x^2} = \sqrt{(3x)^2} = 3x$ Since $3x$ is assumed to be nonnegative

55. $\sqrt{(8p)^2} = 8p$ Since $8p$ is assumed to be nonnegative

57. $\sqrt{(ab)^2} = ab$

59. $\sqrt{(34d)^2} = 34d$

61. $\sqrt{(x+3)^2} = x + 3$

63. $\sqrt{a^2 - 10a + 25} = \sqrt{(a-5)^2} = a - 5$

65. $\sqrt{4a^2 - 20a + 25} = \sqrt{(2a-5)^2} = 2a - 5$

67. Discussion and Writing Exercise

69. **Familiarize.** Let x and y represent the angles. Recall that supplementary angles are angles whose sum is $180°$.

   Translate. We reword the problem.

   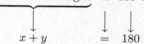

   $$x + y \qquad = \qquad 180$$

   $$x \quad = \quad 2 \quad \cdot \quad y \quad - \quad 3$$

   The resulting system is
   $$x + y = 180, \quad (1)$$
   $$x = 2y - 3. \quad (2)$$

   Solve. We use substitution. We substitute $2y - 3$ for x in Equation (1) and solve for y.
   $$(2y - 3) + y = 180$$
   $$3y - 3 = 180$$
   $$3y = 183$$
   $$y = 61$$

   We substitute 61 for y in Equation (2) to find x.
   $$x = 2(61) - 3 = 122 - 3 = 119$$

   Check. $61° + 119° = 180°$, so the angles are supplementary. Also, $3°$ less than twice $61°$ is $2 \cdot 61° - 3°$, or $122° - 3°$, or $119°$. The numbers check.

   State. The angles are $61°$ and $119°$.

71. *Familiarize*. This problem states that we have direct variation between F and I. Thus, an equation $F = kI$, $k > 0$, applies. As the income increases, the amount spent on food increases.

Translate. We write an equation of variation.

Amount spent on food varies directly as the income.

This translates to $F = kI$.

Solve.

a) First find an equation of variation.

$$F = kI$$

$$10,192 = k \cdot 39,200 \quad \text{Substituting 10,192 for } F \\ \text{and 39,200 for } I$$

$$\frac{10,192}{39,200} = k$$

$$0.26 = k$$

The equation of variation is $F = 0.26I$.

b) We use the equation to find how much a family spends on food when their income is \$41,000.

$$F = 0.26I$$

$$F = 0.26(\$41,000) \quad \text{Substituting \$41,000 for } I$$

$$F = \$10,660$$

Check. Let us do some reasoning about the answer. The income increased from \$39,200 to \$41,000. Similarly, the amount spend on food increased from \$10,192 to \$10,660. This is what we would expect with direct variation.

State. The amount spent on food is \$10,660.

73.

$$\frac{x^2 + 10x - 11}{x^2 - 1} \div \frac{x + 11}{x + 1}$$

$$= \frac{x^2 + 10x - 11}{x^2 - 1} \cdot \frac{x + 1}{x + 11}$$

$$= \frac{(x^2 + 10x - 11)(x + 1)}{(x^2 - 1)(x + 11)}$$

$$= \frac{(x + 11)(x - 1)(x + 1)}{(x + 1)(x - 1)(x + 11)}$$

$$= 1$$

75. To approximate $\sqrt{3}$, locate 3 on the x-axis, move up vertically to the graph, and then move left horizontally to the y-axis to read the approximation.

$$\sqrt{3} \approx 1.7 \quad \text{(Answers may vary.)}$$

To approximate $\sqrt{5}$, locate 5 on the x-axis, move up vertically to the graph, and then move left horizontally to the y-axis to read the approximation.

$$\sqrt{5} \approx 2.2 \quad \text{(Answers may vary.)}$$

To approximate $\sqrt{7}$, locate 7 on the x-axis, move up vertically to the graph, and then move left horizontally to the y-axis to read the approximation.

$$\sqrt{7} \approx 2.6 \quad \text{(Answers may vary.)}$$

77. If $\sqrt{x^2} = 16$, then $x^2 = 256$ since $\sqrt{256} = 16$. Thus $x = 16$ or $x = -16$.

79. If $t^2 = 49$ then the values of t are the square roots of 49, 7 and -7.

Exercise Set 16.2

1. $\sqrt{12} = \sqrt{4 \cdot 3}$ 4 is a perfect square.

$\quad = \sqrt{4}\,\sqrt{3}$ Factoring into a product of radicals

$\quad = 2\sqrt{3}$ Taking the square root

3. $\sqrt{75} = \sqrt{25 \cdot 3}$ 25 is a perfect square.

$\quad = \sqrt{25}\,\sqrt{3}$ Factoring into a product of radicals

$\quad = 5\sqrt{3}$ Taking the square root

5. $\sqrt{20} = \sqrt{4 \cdot 5}$ 4 is a perfect square.

$\quad = \sqrt{4}\,\sqrt{5}$ Factoring into a product of radicals

$\quad = 2\sqrt{5}$ Taking the square root

7. $\sqrt{600} = \sqrt{100 \cdot 6}$ 100 is a perfect square.

$\quad = \sqrt{100} \cdot \sqrt{6}$ Factoring into a product of radicals

$\quad = 10\sqrt{6}$ Taking the square root

9. $\sqrt{486} = \sqrt{81 \cdot 6}$ 81 is a perfect square.

$\quad = \sqrt{81} \cdot \sqrt{6}$ Factoring into a product of radicals

$\quad = 9\sqrt{6}$ Taking the square root

11. $\sqrt{9x} = \sqrt{9 \cdot x} = \sqrt{9}\,\sqrt{x} = 3\sqrt{x}$

13. $\sqrt{48x} = \sqrt{16 \cdot 3 \cdot x} = \sqrt{16}\,\sqrt{3x} = 4\sqrt{3x}$

15. $\sqrt{16a} = \sqrt{16 \cdot a} = \sqrt{16}\,\sqrt{a} = 4\sqrt{a}$

17. $\sqrt{64y^2} = \sqrt{64}\,\sqrt{y^2} = 8y$, or

$\quad \sqrt{64y^2} = \sqrt{(8y)^2} = 8y$

19. $\sqrt{13x^2} = \sqrt{13}\,\sqrt{x^2} = \sqrt{13} \cdot x$, or $x\sqrt{13}$

21. $\sqrt{8t^2} = \sqrt{2 \cdot 4 \cdot t^2} = \sqrt{4}\,\sqrt{t^2}\,\sqrt{2} = 2t\sqrt{2}$

23. $\sqrt{180} = \sqrt{36 \cdot 5} = 6\sqrt{5}$

25. $\sqrt{288y} = \sqrt{144 \cdot 2 \cdot y} = \sqrt{144}\,\sqrt{2y} = 12\sqrt{2y}$

27. $\sqrt{28x^2} = \sqrt{4 \cdot 7 \cdot x^2} = \sqrt{4}\,\sqrt{x^2}\,\sqrt{7} = 2x\sqrt{7}$

29. $\sqrt{x^2 - 6x + 9} = \sqrt{(x - 3)^2} = x - 3$

31. $\sqrt{8x^2 + 8x + 2} = \sqrt{2(4x^2 + 4x + 1)} =$

$\quad \sqrt{2(2x + 1)^2} = \sqrt{2}\,\sqrt{(2x + 1)^2} = \sqrt{2}\,(2x + 1)$

33. $\sqrt{36y + 12y^2 + y^3} = \sqrt{y(36 + 12y + y^2)} =$

$\quad \sqrt{y(6 + y)^2} = \sqrt{y}\,\sqrt{(6 + y)^2} = \sqrt{y}\,(6 + y)$

35. $\sqrt{t^6} = \sqrt{(t^3)^2} = t^3$

37. $\sqrt{x^{12}} = \sqrt{(x^6)^2} = x^6$

39. $\sqrt{x^5} = \sqrt{x^4 \cdot x}$ One factor is a perfect square
$= \sqrt{x^4} \sqrt{x}$
$= \sqrt{(x^2)^2} \sqrt{x}$
$= x^2 \sqrt{x}$

41. $\sqrt{t^{19}} = \sqrt{t^{18} \cdot t} = \sqrt{t^{18}} \sqrt{t} = \sqrt{(t^9)^2} \sqrt{t} = t^9 \sqrt{t}$

43. $\sqrt{(y-2)^8} = \sqrt{[(y-2)^4]^2} = (y-2)^4$

45. $\sqrt{4(x+5)^{10}} = \sqrt{4[(x+5)^5]^2} = \sqrt{4} \sqrt{[(x+5)^5]^2} = 2(x+5)^5$

47. $\sqrt{36m^3} = \sqrt{36 \cdot m^2 \cdot m} = \sqrt{36} \sqrt{m^2} \sqrt{m} = 6m\sqrt{m}$

49. $\sqrt{8a^5} = \sqrt{2 \cdot 4 \cdot a^4 \cdot a} = \sqrt{2 \cdot 4 \cdot (a^2)^2 \cdot a} = \sqrt{4} \sqrt{(a^2)^2} \sqrt{2a} = 2a^2\sqrt{2a}$

51. $\sqrt{104p^{17}} = \sqrt{4 \cdot 26 \cdot p^{16} \cdot p} = \sqrt{4 \cdot 26 \cdot (p^8)^2 \cdot p} = \sqrt{4} \sqrt{(p^8)^2} \sqrt{26p} = 2p^8\sqrt{26p}$

53. $\sqrt{448x^6y^3} = \sqrt{64 \cdot 7 \cdot x^6 \cdot y^2 \cdot y} = \sqrt{64 \cdot 7 \cdot (x^3)^2 \cdot y^2 \cdot y} = \sqrt{64} \sqrt{(x^3)^2} \sqrt{y^2} \sqrt{7y} = 8x^3y \sqrt{7y}$

55. $\sqrt{3} \sqrt{18} = \sqrt{3 \cdot 18}$ Multiplying
$= \sqrt{3 \cdot 3 \cdot 6}$ Looking for perfect-square factors or pairs of factors
$= \sqrt{3 \cdot 3} \sqrt{6}$
$= 3\sqrt{6}$

57. $\sqrt{15} \sqrt{6} = \sqrt{15 \cdot 6}$ Multiplying
$= \sqrt{5 \cdot 3 \cdot 3 \cdot 2}$ Looking for perfect-square factors or pairs of factors
$= \sqrt{3 \cdot 3} \sqrt{5 \cdot 2}$
$= 3\sqrt{10}$

59. $\sqrt{18} \sqrt{14x} = \sqrt{18 \cdot 14x} = \sqrt{3 \cdot 3 \cdot 2 \cdot 2 \cdot 7 \cdot x} = \sqrt{3 \cdot 3} \sqrt{2 \cdot 2} \sqrt{7x} = 3 \cdot 2\sqrt{7x} = 6\sqrt{7x}$

61. $\sqrt{3x} \sqrt{12y} = \sqrt{3x \cdot 12y} = \sqrt{3 \cdot x \cdot 3 \cdot 4 \cdot y} = \sqrt{3 \cdot 3 \cdot 4 \cdot x \cdot y} = \sqrt{3 \cdot 3} \sqrt{4} \sqrt{x \cdot y} = 3 \cdot 2\sqrt{xy} = 6\sqrt{xy}$

63. $\sqrt{13} \sqrt{13} = \sqrt{13 \cdot 13} = 13$

65. $\sqrt{5b} \sqrt{15b} = \sqrt{5b \cdot 15b} = \sqrt{5 \cdot b \cdot 5 \cdot 3 \cdot b} = \sqrt{5 \cdot 5 \cdot b \cdot b \cdot 3} = \sqrt{5 \cdot 5} \sqrt{b \cdot b} \sqrt{3} = 5b\sqrt{3}$

67. $\sqrt{2t} \sqrt{2t} = \sqrt{2t \cdot 2t} = 2t$

69. $\sqrt{ab} \sqrt{ac} = \sqrt{ab \cdot ac} = \sqrt{a \cdot a \cdot b \cdot c} = \sqrt{a \cdot a} \sqrt{b \cdot c} = a\sqrt{bc}$

71. $\sqrt{2x^2y} \sqrt{4xy^2} = \sqrt{2x^2y \cdot 4xy^2} = \sqrt{2 \cdot x^2 \cdot y \cdot 4 \cdot x \cdot y^2} = \sqrt{4} \sqrt{x^2} \sqrt{y^2} \sqrt{2xy} = 2xy\sqrt{2xy}$

73. $\sqrt{18} \sqrt{18} = \sqrt{18 \cdot 18} = 18$

75. $\sqrt{5} \sqrt{2x-1} = \sqrt{5(2x-1)} = \sqrt{10x-5}$

77. $\sqrt{x+2} \sqrt{x+2} = \sqrt{(x+2)^2} = x+2$

79. $\sqrt{18x^2y^3} \sqrt{6xy^4} = \sqrt{18x^2y^3 \cdot 6xy^4} = \sqrt{3 \cdot 6 \cdot x^2 \cdot y^2 \cdot y \cdot 6 \cdot x \cdot y^4} = \sqrt{6 \cdot 6 \cdot x^2 \cdot y^6 \cdot 3 \cdot x \cdot y} = \sqrt{6 \cdot 6} \sqrt{x^2} \sqrt{y^6} \sqrt{3xy} = 6xy^3\sqrt{3xy}$

81. $\sqrt{50x^4y^6} \sqrt{10xy} = \sqrt{50x^4y^6 \cdot 10xy} = \sqrt{5 \cdot 10 \cdot x^4 \cdot y^6 \cdot 10 \cdot x \cdot y} = \sqrt{10 \cdot 10 \cdot x^4 \cdot y^6 \cdot 5 \cdot x \cdot y} = \sqrt{10 \cdot 10} \sqrt{x^4} \sqrt{y^6} \sqrt{5xy} = 10x^2y^3\sqrt{5xy}$

83. $\sqrt{99p^4q^3} \sqrt{22p^5q^2} = \sqrt{99p^4q^3 \cdot 22p^5q^2} = \sqrt{9 \cdot 11 \cdot p^4 \cdot q^2 \cdot q \cdot 2 \cdot 11 \cdot p^4 \cdot p \cdot q^2} = \sqrt{9 \cdot 11 \cdot 11 \cdot p^4 \cdot q^2 \cdot p^4 \cdot q^2 \cdot 2 \cdot q \cdot p} = 3 \cdot 11 \cdot p^2 \cdot q \cdot p^2 \cdot q\sqrt{2pq} = 33p^4q^2\sqrt{2pq}$

85. $\sqrt{24a^2b^3c^4} \sqrt{32a^5b^4c^7} = \sqrt{24a^2b^3c^4 \cdot 32a^5b^4c^7} = \sqrt{4 \cdot 2 \cdot 3 \cdot a^2 \cdot b^2 \cdot b \cdot c^4 \cdot 16 \cdot 2 \cdot a^4 \cdot a \cdot b^4 \cdot c^6 \cdot c} = \sqrt{4 \cdot 2 \cdot 2 \cdot 16 \cdot a^2 \cdot b^2 \cdot c^4 \cdot a^4 \cdot b^4 \cdot c^6 \cdot 3 \cdot b \cdot a \cdot c} = 2 \cdot 2 \cdot 4 \cdot a \cdot b \cdot c^2 \cdot a^2 \cdot b^2 \cdot c^3\sqrt{3abc} = 16a^3b^3c^5\sqrt{3abc}$

87. Discussion and Writing Exercise

89.
$$x - y = -6 \quad (1)$$
$$\underline{x + y = 2 \quad (2)}$$
$$2x = -4 \quad \text{Adding}$$
$$x = -2$$

Now we substitute -2 for x in one of the original equations and solve for y.
$$x + y = 2 \quad \text{Equation (2)}$$
$$-2 + y = 2 \quad \text{Substituting}$$
$$y = 4$$

Since $(-2, 4)$ checks in both equations, it is the solution.

91.
$$3x - 2y = 4, \quad (1)$$
$$2x + 5y = 9 \quad (2)$$

We will us the elimination method. We multiply on both sides of Equation (1) by 5 and on both sides of Equation (2) by 2. Then we add

$$15x - 10y = 20$$
$$\underline{4x + 10y = 18}$$
$$19x = 38$$
$$x = 2$$

Now we substitute 2 for x in one of the original equations and solve for y.
$$2x + 5y = 9 \quad \text{Equation (2)}$$
$$2 \cdot 2 + 5y = 9$$
$$4 + 5y = 9$$
$$5y = 5$$
$$y = 1$$

Since $(2, 1)$ checks, it is the solution.

93. Familiarize. We let $l =$ the length of the rectangle and $w =$ the width. Recall that the perimeter of a rectangle is $2l + 2w$, and the area is lw.

Translate. We translate the first statement.

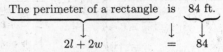

$$2l + 2w = 84$$

Now we translate the second statement.

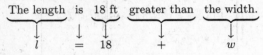

$$l = 18 + w$$

The resulting system is

$$2l + 2w = 84, \quad (1)$$
$$l = 18 + w. \quad (2)$$

Solve. We use the substitution method. We substitute $15 + w$ for l in Equation (1) and solve for w.

$$2l + 2w = 84 \quad \text{Equation (1)}$$
$$2(18 + w) + 2w = 84 \quad \text{Substituting}$$
$$36 + 2w + 2w = 84$$
$$36 + 4w = 84$$
$$4w = 48$$
$$w = 12$$

Substitute 12 for w in Equation (2) and compute l.

$$l = 18 + w = 18 + 12 = 30$$

When $l = 30$ ft and $w = 12$ ft, the area is 30 ft $\cdot$ 12 ft, or 360 ft^2.

Check. When $l = 30$ ft and $w = 12$ ft, the perimeter is $2 \cdot 30$ ft $+ 2 \cdot 12$ ft, or 60 ft $+ 24$ ft, or 84 ft. The length, 30 ft, is 18 ft greater than 12 ft, the width. We recheck the computation of the area.

State. The area of the rectangle is 360 ft^2.

95. Familiarize. Let $x =$ the number of liters of 30% solution and $y =$ the number of liters of 50% solution to be used. We organize the information in a table.

Type of solution	30% insecticide	50% insecticide	Mixture
Amount of solution	x	y	200 L
Percent of insecticide	30%	50%	42%
Amount of insecticide in solution	30%x	50%y	42% × 200, or 84 L

Translate. The first and last rows of the table give us two equations. Since the total amount of solution is 200 L, we have

$$x + y = 200.$$

The amount of insecticide in the mixture is to be 42% of 200, or 84 L. The amounts of insecticide from the two solutions are 30%x and 50%y. Thus

$$30\%x + 50\%y = 84,$$
$$\text{or} \quad 0.3x + 0.5y = 84,$$
$$\text{or} \quad 3x + 5y = 840 \quad \text{Clearing decimals}$$

The resulting system is

$$x + y = 200, \quad (1)$$
$$3x + 5y = 840. \quad (2)$$

Solve. We use the elimination method. Multiply by -3 on both sides of Equation (1) and then add.

$$-3x - 3y = -600$$
$$\underline{3x + 5y = 840}$$
$$2y = 240$$
$$y = 120$$

We go back to Equation (1) and substitute 120 for y.

$$x + y = 200$$
$$x + 120 = 200$$
$$x = 80$$

Check. The sum of 80 L and 120 L is 200 L. Also, 30% of 80 is 24 and 50% of 120 is 60 and $24 + 60 = 84$. The answer checks.

State. 80 L of 30% solution and 120 L of 50% solution should be used.

97. $\sqrt{5x - 5} = \sqrt{5(x - 1)} = \sqrt{5}\sqrt{x - 1}$

99. $\sqrt{x^2 - 36} = \sqrt{(x + 6)(x - 6)} = \sqrt{x + 6}\sqrt{x - 6}$

101. $\sqrt{x^3 - 2x^2} = \sqrt{x^2(x - 2)} = \sqrt{x^2}\sqrt{x - 2} = x\sqrt{x - 2}$

103. $\sqrt{0.25} = \sqrt{(0.5)^2} = 0.5$

105. $\sqrt{2y}\sqrt{3}\sqrt{8y} = \sqrt{2y \cdot 3 \cdot 8y} = \sqrt{2 \cdot y \cdot 3 \cdot 2 \cdot 4 \cdot y} = $
$\sqrt{2 \cdot 2 \cdot 4 \cdot y \cdot y \cdot 3} = \sqrt{2 \cdot 2}\sqrt{4}\sqrt{y \cdot y}\sqrt{3} = $
$2 \cdot 2 \cdot y\sqrt{3} = 4y\sqrt{3}$

107. $\sqrt{27(x + 1)}\sqrt{12y(x + 1)^2}$
$\sqrt{27(x + 1) \cdot 12y(x + 1)^2} = $
$\sqrt{9 \cdot 3 \cdot (x + 1) \cdot 4 \cdot 3 \cdot y(x + 1)^2} = $
$\sqrt{9 \cdot 3 \cdot 3 \cdot 4 \cdot (x + 1)^2 \cdot (x + 1)y} = $
$\sqrt{9}\sqrt{3 \cdot 3}\sqrt{4}\sqrt{(x + 1)^2}\sqrt{(x + 1)y}$
$3 \cdot 3 \cdot 2(x + 1)\sqrt{(x + 1)y} = 18(x + 1)\sqrt{(x + 1)y}$

109. $\sqrt{x}\sqrt{2x}\sqrt{10x^5} = \sqrt{x \cdot 2x \cdot 10x^5} = $
$\sqrt{x \cdot 2 \cdot x \cdot 2 \cdot 5 \cdot x^4 \cdot x} = \sqrt{x \cdot x \cdot 2 \cdot 2 \cdot x^4 \cdot 5 \cdot x} = $
$\sqrt{x \cdot x}\sqrt{2 \cdot 2}\sqrt{x^4}\sqrt{5x} = x \cdot 2 \cdot x^2\sqrt{5x} = 2x^3\sqrt{5x}$

Exercise Set 16.3

1. $\dfrac{\sqrt{18}}{\sqrt{2}} = \sqrt{\dfrac{18}{2}} = \sqrt{9} = 3$

3. $\dfrac{\sqrt{108}}{\sqrt{3}} = \sqrt{\dfrac{108}{3}} = \sqrt{36} = 6$

5. $\dfrac{\sqrt{65}}{\sqrt{13}} = \sqrt{\dfrac{65}{13}} = \sqrt{5}$

7. $\dfrac{\sqrt{3}}{\sqrt{75}} = \sqrt{\dfrac{3}{75}} = \sqrt{\dfrac{1}{25}} = \dfrac{1}{5}$

9. $\dfrac{\sqrt{12}}{\sqrt{75}} = \sqrt{\dfrac{12}{75}} = \sqrt{\dfrac{4}{25}} = \dfrac{2}{5}$

11. $\dfrac{\sqrt{8x}}{\sqrt{2x}} = \sqrt{\dfrac{8x}{2x}} = \sqrt{4} = 2$

13. $\dfrac{\sqrt{63y^3}}{\sqrt{7y}} = \sqrt{\dfrac{63y^3}{7y}} = \sqrt{9y^2} = 3y$

15. $\sqrt{\dfrac{16}{49}} = \dfrac{\sqrt{16}}{\sqrt{49}} = \dfrac{4}{7}$

17. $\sqrt{\dfrac{1}{36}} = \dfrac{\sqrt{1}}{\sqrt{36}} = \dfrac{1}{6}$

19. $-\sqrt{\dfrac{16}{81}} = -\dfrac{\sqrt{16}}{\sqrt{81}} = -\dfrac{4}{9}$

21. $\sqrt{\dfrac{64}{289}} = \dfrac{\sqrt{64}}{\sqrt{289}} = \dfrac{8}{17}$

23. $\sqrt{\dfrac{1690}{1960}} = \sqrt{\dfrac{169 \cdot 10}{196 \cdot 10}} = \sqrt{\dfrac{169}{196} \cdot \dfrac{10}{10}} = \sqrt{\dfrac{169}{196} \cdot 1} =$
$\sqrt{\dfrac{169}{196}} = \dfrac{\sqrt{169}}{\sqrt{196}} = \dfrac{13}{14}$

25. $\sqrt{\dfrac{25}{x^2}} = \dfrac{\sqrt{25}}{\sqrt{x^2}} = \dfrac{5}{x}$

27. $\sqrt{\dfrac{9a^2}{625}} = \dfrac{\sqrt{9a^2}}{\sqrt{625}} = \dfrac{3a}{25}$

29. $\dfrac{\sqrt{50y^{15}}}{\sqrt{2y^{25}}} = \sqrt{\dfrac{50y^{15}}{2y^{25}}} = \sqrt{\dfrac{25}{y^{10}}} = \dfrac{\sqrt{25}}{\sqrt{y^{10}}} = \dfrac{5}{y^5}$

31. $\dfrac{\sqrt{7x^{23}}}{\sqrt{343x^5}} = \sqrt{\dfrac{7x^{23}}{343x^5}} = \sqrt{\dfrac{x^{18}}{49}} = \dfrac{\sqrt{x^{18}}}{\sqrt{49}} = \dfrac{x^9}{7}$

33. $\sqrt{\dfrac{2}{5}} = \sqrt{\dfrac{2}{5} \cdot \dfrac{5}{5}} = \sqrt{\dfrac{10}{25}} = \dfrac{\sqrt{10}}{\sqrt{25}} = \dfrac{\sqrt{10}}{5}$

35. $\sqrt{\dfrac{7}{8}} = \sqrt{\dfrac{7}{8} \cdot \dfrac{2}{2}} = \sqrt{\dfrac{14}{16}} = \dfrac{\sqrt{14}}{\sqrt{16}} = \dfrac{\sqrt{14}}{4}$

37. $\sqrt{\dfrac{1}{12}} = \sqrt{\dfrac{1}{12} \cdot \dfrac{3}{3}} = \sqrt{\dfrac{3}{36}} = \dfrac{\sqrt{3}}{\sqrt{36}} = \dfrac{\sqrt{3}}{6}$

39. $\sqrt{\dfrac{5}{18}} = \sqrt{\dfrac{5}{18} \cdot \dfrac{2}{2}} = \sqrt{\dfrac{10}{36}} = \dfrac{\sqrt{10}}{\sqrt{36}} = \dfrac{\sqrt{10}}{6}$

41. $\dfrac{3}{\sqrt{5}} = \dfrac{3}{\sqrt{5}} \cdot \dfrac{\sqrt{5}}{\sqrt{5}} = \dfrac{3\sqrt{5}}{5}$

43. $\sqrt{\dfrac{8}{3}} = \sqrt{\dfrac{8}{3} \cdot \dfrac{3}{3}} = \sqrt{\dfrac{24}{9}} = \dfrac{\sqrt{4 \cdot 6}}{\sqrt{9}} = \dfrac{\sqrt{4}\sqrt{6}}{\sqrt{9}} = \dfrac{2\sqrt{6}}{3}$

45. $\sqrt{\dfrac{3}{x}} = \sqrt{\dfrac{3}{x} \cdot \dfrac{x}{x}} = \sqrt{\dfrac{3x}{x^2}} = \dfrac{\sqrt{3x}}{\sqrt{x^2}} = \dfrac{\sqrt{3x}}{x}$

47. $\sqrt{\dfrac{x}{y}} = \sqrt{\dfrac{x}{y} \cdot \dfrac{y}{y}} = \sqrt{\dfrac{xy}{y^2}} = \dfrac{\sqrt{xy}}{\sqrt{y^2}} = \dfrac{\sqrt{xy}}{y}$

49. $\sqrt{\dfrac{x^2}{20}} = \sqrt{\dfrac{x^2}{20} \cdot \dfrac{5}{5}} = \sqrt{\dfrac{5x^2}{100}} = \dfrac{\sqrt{x^2 \cdot 5}}{\sqrt{100}} = \dfrac{\sqrt{x^2}\sqrt{5}}{\sqrt{100}} = \dfrac{x\sqrt{5}}{10}$

51. $\dfrac{\sqrt{7}}{\sqrt{2}} = \dfrac{\sqrt{7}}{\sqrt{2}} \cdot \dfrac{\sqrt{2}}{\sqrt{2}} = \dfrac{\sqrt{14}}{2}$

53. $\dfrac{\sqrt{9}}{\sqrt{8}} = \dfrac{\sqrt{9}}{\sqrt{8}} \cdot \dfrac{\sqrt{2}}{\sqrt{2}} = \dfrac{\sqrt{9 \cdot 2}}{\sqrt{16}} = \dfrac{3\sqrt{2}}{4}$

55. $\dfrac{\sqrt{3}}{\sqrt{2}} = \dfrac{\sqrt{3}}{\sqrt{2}} \cdot \dfrac{\sqrt{2}}{\sqrt{2}} = \dfrac{\sqrt{6}}{2}$

57. $\dfrac{2}{\sqrt{2}} = \dfrac{2}{\sqrt{2}} \cdot \dfrac{\sqrt{2}}{\sqrt{2}} = \dfrac{2\sqrt{2}}{2} = \sqrt{2}$

59. $\dfrac{\sqrt{5}}{\sqrt{11}} = \dfrac{\sqrt{5}}{\sqrt{11}} \cdot \dfrac{\sqrt{11}}{\sqrt{11}} = \dfrac{\sqrt{55}}{11}$

61. $\dfrac{\sqrt{7}}{\sqrt{12}} = \dfrac{\sqrt{7}}{\sqrt{12}} \cdot \dfrac{\sqrt{3}}{\sqrt{3}} = \dfrac{\sqrt{21}}{\sqrt{36}} = \dfrac{\sqrt{21}}{6}$

63. $\dfrac{\sqrt{48}}{\sqrt{32}} = \sqrt{\dfrac{48}{32}} = \sqrt{\dfrac{3}{2}} = \sqrt{\dfrac{3}{2} \cdot \dfrac{2}{2}} = \sqrt{\dfrac{6}{4}} = \dfrac{\sqrt{6}}{\sqrt{4}} = \dfrac{\sqrt{6}}{2}$

65. $\dfrac{\sqrt{450}}{\sqrt{18}} = \sqrt{\dfrac{450}{18}} = \sqrt{25} = 5$

67. $\dfrac{\sqrt{3}}{\sqrt{x}} = \dfrac{\sqrt{3}}{\sqrt{x}} \cdot \dfrac{\sqrt{x}}{\sqrt{x}} = \dfrac{\sqrt{3x}}{x}$

69. $\dfrac{4y}{\sqrt{5}} = \dfrac{4y}{\sqrt{5}} \cdot \dfrac{\sqrt{5}}{\sqrt{5}} = \dfrac{4y\sqrt{5}}{5}$

71. $\dfrac{\sqrt{a^3}}{\sqrt{8}} = \dfrac{\sqrt{a^3}}{\sqrt{8}} \cdot \dfrac{\sqrt{2}}{\sqrt{2}} = \dfrac{\sqrt{2a^3}}{\sqrt{16}} = \dfrac{\sqrt{a^2 \cdot 2a}}{\sqrt{16}} = \dfrac{a\sqrt{2a}}{4}$

73. $\dfrac{\sqrt{56}}{\sqrt{12x}} = \sqrt{\dfrac{56}{12x}} = \sqrt{\dfrac{14}{3x}} = \sqrt{\dfrac{14}{3x} \cdot \dfrac{3x}{3x}} = \sqrt{\dfrac{42x}{3x \cdot 3x}} =$
$\dfrac{\sqrt{42x}}{3x}$

75. $\dfrac{\sqrt{27c}}{\sqrt{32c^3}} = \sqrt{\dfrac{27c}{32c^3}} = \sqrt{\dfrac{27}{32c^2}} = \sqrt{\dfrac{27}{32c^2} \cdot \dfrac{2}{2}} = \sqrt{\dfrac{54}{64c^2}} =$
$\sqrt{\dfrac{9 \cdot 6}{64c^2}} = \dfrac{3\sqrt{6}}{8c}$

77. $\dfrac{\sqrt{y^5}}{\sqrt{xy^2}} = \sqrt{\dfrac{y^5}{xy^2}} = \sqrt{\dfrac{y^3}{x}} = \sqrt{\dfrac{y^3}{x} \cdot \dfrac{x}{x}} = \sqrt{\dfrac{xy^3}{x^2}} =$
$\sqrt{\dfrac{y^2 \cdot xy}{x^2}} = \dfrac{y\sqrt{xy}}{x}$

79. $\dfrac{\sqrt{45mn^2}}{\sqrt{32m}} = \sqrt{\dfrac{45mn^2}{32m}} = \sqrt{\dfrac{45n^2}{32}} = \sqrt{\dfrac{45n^2}{32} \cdot \dfrac{2}{2}} =$
$\sqrt{\dfrac{90n^2}{64}} = \dfrac{\sqrt{90n^2}}{\sqrt{64}} = \dfrac{\sqrt{9 \cdot n^2 \cdot 10}}{8} = \dfrac{3n\sqrt{10}}{8}$

81. Discussion and Writing Exercise

83. $x = y + 2$,   (1)

$x + y = 6$ (2)

We substitute $y + 2$ for x in Equation (2) and solve for y.

$$(y + 2) + y = 6$$
$$2y + 2 = 6$$
$$2y = 4$$
$$y = 2$$

Substitute 2 for y in Equation (1) to find x.

$$x = 2 + 2 = 4$$

The ordered pair $(4, 2)$ checks in both equations. It is the solution.

85. $2x - 3y = 7$ (1)

$2x - 3y = 9$ (2)

We multiply Equation (2) by -1 and add.

$$\begin{array}{r} 2x - 3y = 7 \\ -2x + 3y = -9 \\ \hline 0 = -2 \end{array}$$

We get a false equation. The system of equations has no solution.

87. $x + y = -7$ (1)

$$\begin{array}{r} x - y = 2 \quad (2) \\ \hline 2x = -5 \quad \text{Adding} \end{array}$$

$$x = -\frac{5}{2}$$

Substitute $-\dfrac{5}{2}$ for x in Equation (1) to find y.

$$x + y = -7 \quad \text{Equation (1)}$$
$$-\frac{5}{2} + y = -7 \quad \text{Substituting}$$
$$y = -\frac{9}{2}$$

The ordered pair $\left(-\dfrac{5}{2}, -\dfrac{9}{2} \right)$ checks in both equations. It is the solution.

89.
$$\frac{x^2 - 49}{x + 8} \div \frac{x^2 - 14x + 49}{x^2 + 15x + 56}$$
$$= \frac{x^2 - 49}{x + 8} \cdot \frac{x^2 + 15x + 56}{x^2 - 14x + 49}$$
$$= \frac{(x^2 - 49)(x^2 + 15x + 56)}{(x + 8)(x^2 - 14x + 49)}$$
$$= \frac{(x + 7)(x - 7)(x + 7)(x + 8)}{(x + 8)(x - 7)(x - 7)}$$
$$= \frac{(x + 7)(x\!\!\!/\,7)(x + 7)(x\!\!\!/+8)}{(x\!\!\!/+8)(x\!\!\!/\,7)(x - 7)}$$
$$= \frac{(x + 7)(x + 7)}{x - 7}, \text{ or } \frac{(x + 7)^2}{x - 7}$$

91.
$$\frac{a^2 - 25}{6} \div \frac{a + 5}{3} = \frac{a^2 - 25}{6} \cdot \frac{3}{a + 5}$$
$$= \frac{(a^2 - 25) \cdot 3}{6(a + 5)}$$
$$= \frac{(a + 5)(a - 5) \cdot 3}{2 \cdot 3 \cdot (a + 5)}$$
$$= \frac{(a\!\!\!/+5)(a - 5) \cdot \cancel{3}}{2 \cdot \cancel{3} \cdot (a\!\!\!/+5)}$$
$$= \frac{a - 5}{2}$$

93. $(3x - 7)(3x + 7) = (3x)^2 - 7^2 = 9x^2 - 49$

95.
$$9x - 5y + 12x - 4y$$
$$= (9x + 12x) + (-5y - 4y)$$
$$= 21x - 9y$$

97.   2 ft: $T \approx 2(3.14)\sqrt{\dfrac{2}{32}} \approx 6.28\sqrt{\dfrac{1}{16}} \approx 6.28\left(\dfrac{1}{4}\right) \approx$

     1.57 sec

     8 ft: $T \approx 2(3.14)\sqrt{\dfrac{8}{32}} \approx 6.28\sqrt{\dfrac{1}{4}} \approx 6.28\left(\dfrac{1}{2}\right) \approx$

     3.14 sec

     64 ft: $T \approx 2(3.14)\sqrt{\dfrac{64}{32}} \approx 6.28\sqrt{2} \approx$

     $(6.28)(1.414) \approx 8.88$ sec

     100 ft: $T \approx 2(3.14)\sqrt{\dfrac{100}{32}} \approx 6.28\sqrt{\dfrac{50}{16}} \approx \dfrac{6.28\sqrt{50}}{4} \approx$

     $\dfrac{6.28(7.071)}{4} \approx 11.10$ sec

99. $T = 2\pi\sqrt{\dfrac{\dfrac{32}{\pi^2}}{32}} = 2\pi\sqrt{\dfrac{32}{\pi^2} \cdot \dfrac{1}{32}} = 2\pi\sqrt{\dfrac{1}{\pi^2}} = 2\pi\left(\dfrac{1}{\pi}\right) = 2$ sec

The time it takes the pendulum to swing from one side to the other and back is 2 sec, so it takes 1 sec to swing from one side to the other.

101. $\sqrt{\dfrac{5}{1600}} = \dfrac{\sqrt{5}}{\sqrt{1600}} = \dfrac{\sqrt{5}}{40}$

103. $\sqrt{\dfrac{1}{5x^3}} = \sqrt{\dfrac{1}{5x^3} \cdot \dfrac{5x}{5x}} = \sqrt{\dfrac{5x}{25x^4}} = \dfrac{\sqrt{5x}}{\sqrt{25x^4}} = \dfrac{\sqrt{5x}}{5x^2}$

105. $\sqrt{\dfrac{3a}{b}} = \sqrt{\dfrac{3a}{b} \cdot \dfrac{b}{b}} = \sqrt{\dfrac{3ab}{b^2}} = \dfrac{\sqrt{3ab}}{\sqrt{b^2}} = \dfrac{\sqrt{3ab}}{b}$

107. $\sqrt{0.009} = \sqrt{\dfrac{9}{1000}} = \sqrt{\dfrac{9}{1000} \cdot \dfrac{10}{10}} = \sqrt{\dfrac{90}{10,000}} =$

$\dfrac{\sqrt{90}}{\sqrt{10,000}} = \dfrac{\sqrt{9 \cdot 10}}{100} = \dfrac{\sqrt{9}\sqrt{10}}{100} = \dfrac{3\sqrt{10}}{100}$

109. $\sqrt{\dfrac{1}{x^2} - \dfrac{2}{xy} + \dfrac{1}{y^2}}$, LCD is x^2y^2

$= \sqrt{\dfrac{1}{x^2} \cdot \dfrac{y^2}{y^2} - \dfrac{2}{xy} \cdot \dfrac{xy}{xy} + \dfrac{1}{y^2} \cdot \dfrac{x^2}{x^2}}$

$= \sqrt{\dfrac{y^2 - 2xy + x^2}{x^2y^2}}$

$= \sqrt{\dfrac{(y-x)^2}{x^2y^2}}$

$= \dfrac{\sqrt{(y-x)^2}}{\sqrt{x^2y^2}}$

$= \dfrac{y-x}{xy}$

Exercise Set 16.4

1. $7\sqrt{3} + 9\sqrt{3} = (7+9)\sqrt{3}$

$= 16\sqrt{3}$

3. $7\sqrt{5} - 3\sqrt{5} = (7-3)\sqrt{5}$

$= 4\sqrt{5}$

5. $6\sqrt{x} + 7\sqrt{x} = (6+7)\sqrt{x}$

$= 13\sqrt{x}$

7. $4\sqrt{d} - 13\sqrt{d} = (4-13)\sqrt{d}$

$= -9\sqrt{d}$

9. $5\sqrt{8} + 15\sqrt{2} = 5\sqrt{4 \cdot 2} + 15\sqrt{2}$

$= 5 \cdot 2\sqrt{2} + 15\sqrt{2}$

$= 10\sqrt{2} + 15\sqrt{2}$

$= 25\sqrt{2}$

11. $\sqrt{27} - 2\sqrt{3} = \sqrt{9 \cdot 3} - 2\sqrt{3}$

$= 3\sqrt{3} - 2\sqrt{3}$

$= (3-2)\sqrt{3}$

$= 1\sqrt{3}$

$= \sqrt{3}$

13. $\sqrt{45} - \sqrt{20} = \sqrt{9 \cdot 5} - \sqrt{4 \cdot 5}$

$= 3\sqrt{5} - 2\sqrt{5}$

$= (3-2)\sqrt{5}$

$= 1\sqrt{5}$

$= \sqrt{5}$

15. $\sqrt{72} + \sqrt{98} = \sqrt{36 \cdot 2} + \sqrt{49 \cdot 2}$

$= 6\sqrt{2} + 7\sqrt{2}$

$= (6+7)\sqrt{2}$

$= 13\sqrt{2}$

17. $2\sqrt{12} + \sqrt{27} - \sqrt{48} = 2\sqrt{4 \cdot 3} + \sqrt{9 \cdot 3} - \sqrt{16 \cdot 3}$

$= 2 \cdot 2\sqrt{3} + 3\sqrt{3} - 4\sqrt{3}$

$= 4\sqrt{3} + 3\sqrt{3} - 4\sqrt{3}$

$= (4+3-4)\sqrt{3}$

$= 3\sqrt{3}$

19. $\sqrt{18} - 3\sqrt{8} + \sqrt{50} = \sqrt{9 \cdot 2} - 3\sqrt{4 \cdot 2} + \sqrt{25 \cdot 2}$

$= 3\sqrt{2} - 3 \cdot 2\sqrt{2} + 5\sqrt{2}$

$= 3\sqrt{2} - 6\sqrt{2} + 5\sqrt{2}$

$= (3-6+5)\sqrt{2}$

$= 2\sqrt{2}$

21. $2\sqrt{27} - 3\sqrt{48} + 3\sqrt{12} = 2\sqrt{9 \cdot 3} - 3\sqrt{16 \cdot 3} + 3\sqrt{4 \cdot 3}$

$= 2 \cdot 3\sqrt{3} - 3 \cdot 4\sqrt{3} + 3 \cdot 2\sqrt{3}$

$= 6\sqrt{3} - 12\sqrt{3} + 6\sqrt{3}$

$= (6 - 12 + 6)\sqrt{3}$

$= 0\sqrt{3}$

$= 0$

23. $\sqrt{4x} + \sqrt{81x^3} = \sqrt{4 \cdot x} + \sqrt{81 \cdot x^2 \cdot x}$

$= 2\sqrt{x} + 9x\sqrt{x}$

$= (2 + 9x)\sqrt{x}$

25. $\sqrt{27} - \sqrt{12x^2} = \sqrt{9 \cdot 3} - \sqrt{4 \cdot 3 \cdot x^2}$

$= 3\sqrt{3} - 2x\sqrt{3}$

$= (3 - 2x)\sqrt{3}$

27. $\sqrt{8x + 8} + \sqrt{2x + 2} = \sqrt{4(2x + 2)} + \sqrt{2x + 2}$

$= 2\sqrt{2x + 2} + 1\sqrt{2x + 2}$

$= (2 + 1)\sqrt{2x + 2}$

$= 3\sqrt{2x + 2}$

29. $\sqrt{x^5 - x^2} + \sqrt{9x^3 - 9} = \sqrt{x^2(x^3 - 1)} + \sqrt{9(x^3 - 1)}$

$= x\sqrt{x^3 - 1} + 3\sqrt{x^3 - 1}$

$= (x + 3)\sqrt{x^3 - 1}$

31. $4a\sqrt{a^2b} + a\sqrt{a^2b^3} - 5\sqrt{b^3}$

$= 4a\sqrt{a^2 \cdot b} + a\sqrt{a^2 \cdot b^2 \cdot b} - 5\sqrt{b^2 \cdot b}$

$= 4a \cdot a\sqrt{b} + a \cdot a \cdot b\sqrt{b} - 5 \cdot b\sqrt{b}$

$= 4a^2\sqrt{b} + a^2b\sqrt{b} - 5b\sqrt{b}$

$= (4a^2 + a^2b - 5b)\sqrt{b}$

33. $\sqrt{3} - \sqrt{\dfrac{1}{3}} = \sqrt{3} - \sqrt{\dfrac{1}{3} \cdot \dfrac{3}{3}}$

$= \sqrt{3} - \dfrac{\sqrt{3}}{3}$

$= \left(1 - \dfrac{1}{3}\right)\sqrt{3}$

$= \dfrac{2}{3}\sqrt{3}$, or $\dfrac{2\sqrt{3}}{3}$

35. $5\sqrt{2} + 3\sqrt{\dfrac{1}{2}} = 5\sqrt{2} + 3\sqrt{\dfrac{1}{2}\cdot\dfrac{2}{2}}$

$\qquad\qquad = 5\sqrt{2} + \dfrac{3}{2}\sqrt{2}$

$\qquad\qquad = \left(5 + \dfrac{3}{2}\right)\sqrt{2}$

$\qquad\qquad = \dfrac{13}{2}\sqrt{2}, \text{ or } \dfrac{13\sqrt{2}}{2}$

37. $\sqrt{\dfrac{2}{3}} - \sqrt{\dfrac{1}{6}} = \sqrt{\dfrac{2}{3}\cdot\dfrac{3}{3}} - \sqrt{\dfrac{1}{6}\cdot\dfrac{6}{6}}$

$\qquad\qquad = \dfrac{\sqrt{6}}{3} - \dfrac{\sqrt{6}}{6}$

$\qquad\qquad = \left(\dfrac{1}{3} - \dfrac{1}{6}\right)\sqrt{6}$

$\qquad\qquad = \dfrac{1}{6}\sqrt{6}, \text{ or } \dfrac{\sqrt{6}}{6}$

39. $\sqrt{3}(\sqrt{5} - 1) = \sqrt{3}\,\sqrt{5} - \sqrt{3}\cdot 1$

$\qquad\qquad = \sqrt{15} - \sqrt{3}$

41. $(2 + \sqrt{3})(5 - \sqrt{7})$

$\quad = 2\cdot 5 - 2\sqrt{7} + \sqrt{3}\cdot 5 - \sqrt{3}\sqrt{7} \qquad \text{Using FOIL}$

$\quad = 10 - 2\sqrt{7} + 5\sqrt{3} - \sqrt{21}$

43. $(2 - \sqrt{5})^2$

$\quad = 2^2 - 2\cdot 2\cdot\sqrt{5} + (\sqrt{5})^2$

$\qquad\qquad \text{Using } (A - B)^2 = A^2 - 2AB + B^2$

$\quad = 4 - 4\sqrt{5} + 5$

$\quad = 9 - 4\sqrt{5}$

45. $(\sqrt{2} + 8)(\sqrt{2} - 8)$

$\quad = (\sqrt{2})^2 - 8^2 \qquad \text{Using } (A + B)(A - B) = A^2 - B^2$

$\quad = 2 - 64$

$\quad = -62$

47. $(\sqrt{6} - \sqrt{5})(\sqrt{6} + \sqrt{5})$

$\quad = (\sqrt{6})^2 - (\sqrt{5})^2 \qquad \text{Using } (A + B)(A - B) = A^2 - B^2$

$\quad = 6 - 5$

$\quad = 1$

49. $(3\sqrt{5} - 2)(\sqrt{5} + 1)$

$\quad = 3\sqrt{5}\,\sqrt{5} + 3\sqrt{5} - 2\sqrt{5} - 2 \qquad \text{Using FOIL}$

$\quad = 3\cdot 5 + 3\sqrt{5} - 2\sqrt{5} - 2$

$\quad = 15 + \sqrt{5} - 2$

$\quad = 13 + \sqrt{5}$

51. $(\sqrt{x} - \sqrt{y})^2 = (\sqrt{x})^2 - 2\sqrt{x}\,\sqrt{y} + (\sqrt{y})^2$

$\qquad\qquad \text{Using } (A - B)^2 = A^2 - 2AB + B^2$

$\qquad\qquad = x - 2\sqrt{xy} + y$

53. We multiply by 1 using the conjugate of $\sqrt{3} - \sqrt{5}$, which is $\sqrt{3} + \sqrt{5}$, as the numerator and denominator.

$\dfrac{2}{\sqrt{3} - \sqrt{5}} = \dfrac{2}{\sqrt{3} - \sqrt{5}}\cdot\dfrac{\sqrt{3} + \sqrt{5}}{\sqrt{3} + \sqrt{5}} \qquad \text{Multiplying by 1}$

$\qquad = \dfrac{2(\sqrt{3} + \sqrt{5})}{(\sqrt{3} - \sqrt{5})(\sqrt{3} + \sqrt{5})} \qquad \text{Multiplying}$

$\qquad = \dfrac{2\sqrt{3} + 2\sqrt{5}}{(\sqrt{3})^2 - (\sqrt{5})^2} = \dfrac{2\sqrt{3} + 2\sqrt{5}}{3 - 5}$

$\qquad = \dfrac{2\sqrt{3} + 2\sqrt{5}}{-2} = \dfrac{2(\sqrt{3} + \sqrt{5})}{-2}$

$\qquad = -(\sqrt{3} + \sqrt{5}) = -\sqrt{3} - \sqrt{5}$

55. We multiply by 1 using the conjugate of $\sqrt{3} + \sqrt{2}$, which is $\sqrt{3} - \sqrt{2}$, as the numerator and denominator.

$\dfrac{\sqrt{3} - \sqrt{2}}{\sqrt{3} + \sqrt{2}} = \dfrac{\sqrt{3} - \sqrt{2}}{\sqrt{3} + \sqrt{2}}\cdot\dfrac{\sqrt{3} - \sqrt{2}}{\sqrt{3} - \sqrt{2}} \qquad \text{Multiplying by 1}$

$\qquad = \dfrac{(\sqrt{3} - \sqrt{2})^2}{(\sqrt{3} + \sqrt{2})(\sqrt{3} - \sqrt{2})}$

$\qquad = \dfrac{(\sqrt{3})^2 - 2\sqrt{3}\,\sqrt{2} + (\sqrt{2})^2}{(\sqrt{3})^2 - (\sqrt{2})^2}$

$\qquad = \dfrac{3 - 2\sqrt{6} + 2}{3 - 2} = \dfrac{5 - 2\sqrt{6}}{1}$

$\qquad = 5 - 2\sqrt{6}$

57. We multiply by 1 using the conjugate of $\sqrt{10} + 1$, which is $\sqrt{10} - 1$, as the numerator and denominator.

$\dfrac{4}{\sqrt{10} + 1} = \dfrac{4}{\sqrt{10} + 1}\cdot\dfrac{\sqrt{10} - 1}{\sqrt{10} - 1}$

$\qquad = \dfrac{4(\sqrt{10} - 1)}{(\sqrt{10} + 1)(\sqrt{10} - 1)}$

$\qquad = \dfrac{4\sqrt{10} - 4}{(\sqrt{10})^2 - 1^2} = \dfrac{4\sqrt{10} - 4}{10 - 1}$

$\qquad = \dfrac{4\sqrt{10} - 4}{9}$

59. We multiply by 1 using the conjugate of $3 + \sqrt{7}$, which is $3 - \sqrt{7}$, as the numerator and denominator.

$\dfrac{1 - \sqrt{7}}{3 + \sqrt{7}} = \dfrac{1 - \sqrt{7}}{3 + \sqrt{7}}\cdot\dfrac{3 - \sqrt{7}}{3 - \sqrt{7}}$

$\qquad = \dfrac{(1 - \sqrt{7})(3 - \sqrt{7})}{(3 + \sqrt{7})(3 - \sqrt{7})}$

$\qquad = \dfrac{3 - \sqrt{7} - 3\sqrt{7} + \sqrt{7}\,\sqrt{7}}{3^2 - (\sqrt{7})^2}$

$\qquad = \dfrac{3 - \sqrt{7} - 3\sqrt{7} + 7}{9 - 7} = \dfrac{10 - 4\sqrt{7}}{2}$

$\qquad = \dfrac{2(5 - 2\sqrt{7})}{2} = 5 - 2\sqrt{7}$

61. We multiply by 1 using the conjugate of $4 + \sqrt{x}$, which is $4 - \sqrt{x}$, as the numerator and denominator.

$$\frac{3}{4 + \sqrt{x}} = \frac{3}{4 + \sqrt{x}} \cdot \frac{4 - \sqrt{x}}{4 - \sqrt{x}}$$

$$= \frac{3(4 - \sqrt{x})}{(4 + \sqrt{x})(4 - \sqrt{x})}$$

$$= \frac{12 - 3\sqrt{x}}{4^2 - (\sqrt{x})^2}$$

$$= \frac{12 - 3\sqrt{x}}{16 - x}$$

63. We multiply by 1 using the conjugate of $8 - \sqrt{x}$, which is $8 + \sqrt{x}$, as the numerator and denominator.

$$\frac{3 + \sqrt{2}}{8 - \sqrt{x}} = \frac{3 + \sqrt{2}}{8 - \sqrt{x}} \cdot \frac{8 + \sqrt{x}}{8 + \sqrt{x}}$$

$$= \frac{(3 + \sqrt{2})(8 + \sqrt{x})}{(8 - \sqrt{x})(8 + \sqrt{x})}$$

$$= \frac{3 \cdot 8 + 3 \cdot \sqrt{x} + \sqrt{2} \cdot 8 + \sqrt{2} \cdot \sqrt{x}}{8^2 - (\sqrt{x})^2}$$

$$= \frac{24 + 3\sqrt{x} + 8\sqrt{2} + \sqrt{2x}}{64 - x}$$

65. Discussion and Writing Exercise

67.
$$3x + 5 + 2(x - 3) = 4 - 6x$$
$$3x + 5 + 2x - 6 = 4 - 6x$$
$$5x - 1 = 4 - 6x$$
$$11x - 1 = 4$$
$$11x = 5$$
$$x = \frac{5}{11}$$

The solution is $\frac{5}{11}$.

69.
$$x^2 - 5x = 6$$
$$x^2 - 5x - 6 = 0$$
$$(x + 1)(x - 6) = 0$$
$$x + 1 = 0 \quad or \quad x - 6 = 0$$
$$x = -1 \quad or \quad x = 6$$

The solutions are -1 and 6.

71. $\dfrac{7x^9}{27} \cdot \dfrac{9}{7x^3} = \dfrac{63x^9}{189x^3} = \dfrac{63}{189}x^{9-3} = \dfrac{1}{3}x^6$, or $\dfrac{x^6}{3}$

73. *Familiarize*. Let $x =$ the number of liters of Jolly Juice and $y =$ the number of liters of Real Squeeze in the mixture. We organize the given information in a table.

	Jolly Juice	Real Squeeze	Mixture
Amount	x	y	8
Percent real fruit juice	3%	6%	5.4%
Amount of real fruit juice	0.03x	0.06y	0.054(8), or 0.432

Translate. We get two equation from the first and third rows of the table.

$$x + y = 8,$$
$$0.03x + 0.06y = 0.432$$

Clearing decimals gives

$$x + y = 8, \quad (1)$$
$$30x + 60y = 432. \quad (2)$$

Carry out. We use elimination. Multiply Equation (1) by -30 and add.

$$-30x - 30y = -240$$
$$\underline{30x + 60y = 432}$$
$$30y = 192$$
$$y = 6.4$$

Now substitute 6.4 for y in Equation (1) and solve for x.

$$x + y = 8$$
$$x + 6.4 = 8$$
$$x = 1.6$$

Check. The sum of 1.6 and 6.4 is 8. The amount of real fruit juice in this mixture is $0.03(1.6) + 0.06(6.4)$, or $0.048 + 0.384$, or 0.432 L. The answer checks.

State. 1.6 L of Jolly Juice and 6.4 L of Real Squeeze should be used.

75. For $x = -1$, $y = (-1)^3 - 5(-1)^2 + (-1) - 2 = -1 - 5 - 1 - 2 = -9$.
For $x = 0$, $y = 0^3 - 5 \cdot 0^2 + 0 - 2 = 0 - 0 + 0 - 2 = -2$.
For $x = 1$, $y = 1^3 - 5 \cdot 1^2 + 1 - 2 = 1 - 5 + 1 - 2 = -5$.
For $x = 3$, $y = 3^3 - 5 \cdot 3^2 + 3 - 2 = 27 - 45 + 3 - 2 = -17$.
For $x = 4.85$, $y = (4.85)^3 - 5(4.85)^2 + 4.85 - 2 =$
$114.084125 - 117.6125 + 4.85 - 2 = -0.678375$

These values could have been estimated using the graph also.

77. Since $\sqrt{a^2 + b^2} \neq \sqrt{a^2} + \sqrt{b^2}$ for $a = 2$ and $b = 3$, the two expressions are not equivalent.

79. Enter $y_1 = \sqrt{x^2 + 4}$ and $y_2 = \sqrt{x} + 2$ and look at the graphs or a table of values. Since the graphs do not coincide and the values for y_1 and y_2 are different, the given statement is not correct.

81.
$$\frac{1}{3}\sqrt{27} + \sqrt{8} + \sqrt{300} - \sqrt{18} - \sqrt{162}$$
$$= \frac{1}{3}\sqrt{9 \cdot 3} + \sqrt{4 \cdot 2} + \sqrt{100 \cdot 3} - \sqrt{9 \cdot 2} - \sqrt{81 \cdot 2}$$
$$= \frac{1}{3} \cdot 3\sqrt{3} + 2\sqrt{2} + 10\sqrt{3} - 3\sqrt{2} - 9\sqrt{2}$$
$$= \sqrt{3} + 2\sqrt{2} + 10\sqrt{3} - 3\sqrt{2} - 9\sqrt{2}$$
$$= (1 + 10)\sqrt{3} + (2 - 3 - 9)\sqrt{2}$$
$$= 11\sqrt{3} - 10\sqrt{2}$$

83. $(3\sqrt{x + 2})^2 = (3\sqrt{x + 2})(3\sqrt{x + 2})2 = (3 \cdot 3)(\sqrt{x + 2}\sqrt{x + 2}) = 9(x + 2)$

The statement is true.

Exercise Set 16.5

1. $\sqrt{x} = 6$

$(\sqrt{x})^2 = 6^2$ Squaring both sides

$x = 36$ Simplifying

Check: $\dfrac{\sqrt{x} = 6}{\sqrt{36} \;?\; 6}$

         6 | TRUE

The solution is 36.

3. $\sqrt{x} = 4.3$

$(\sqrt{x})^2 = (4.3)^2$ Squaring both sides

$x = 18.49$ Simplifying

Check: $\dfrac{\sqrt{x} = 4.3}{\sqrt{18.49} \;?\; 4.3}$

       4.3 | TRUE

The solution is 18.49.

5. $\sqrt{y + 4} = 13$

$(\sqrt{y + 4})^2 = 13^2$ Squaring both sides

$y + 4 = 169$ Simplifying

$y = 165$ Subtracting 4

Check: $\dfrac{\sqrt{y + 4} = 13}{\sqrt{165 + 4} \;?\; 13}$

       $\sqrt{169}$

        13 | TRUE

The solution is 165.

7. $\sqrt{2x + 4} = 25$

$(\sqrt{2x + 4})^2 = 25^2$ Squaring both sides

$2x + 4 = 625$ Simplifying

$2x = 621$ Subtracting 4

$x = \dfrac{621}{2}$ Dividing by 2

Check: $\sqrt{2x + 4} = 25$

$\sqrt{2 \cdot \dfrac{621}{2} + 4} \;?\; 25$

     $\sqrt{621 + 4}$

       $\sqrt{625}$

         25 | TRUE

The solution is $\dfrac{621}{2}$.

9. $3 + \sqrt{x - 1} = 5$

$\sqrt{x - 1} = 2$ Subtracting 3

$(\sqrt{x - 1})^2 = 2^2$ Squaring both sides

$x - 1 = 4$

$x = 5$

Check: $\dfrac{3 + \sqrt{x - 1} = 5}{3 + \sqrt{5 - 1} \;?\; 5}$

       $3 + \sqrt{4}$

       $3 + 2$

         5 | TRUE

The solution is 5.

11. $6 - 2\sqrt{3n} = 0$

$6 = 2\sqrt{3n}$ Adding $2\sqrt{3n}$

$6^2 = (2\sqrt{3n})^2$ Squaring both sides

$36 = 4 \cdot 3n$

$36 = 12n$

$3 = n$

Check: $\dfrac{6 - 2\sqrt{3n} = 0}{6 - 2\sqrt{3 \cdot 3} \;?\; 0}$

       $6 - 2 \cdot 3$

       $6 - 6$

         0 | TRUE

The solution is 3.

13. $\sqrt{5x - 7} = \sqrt{x + 10}$

$(\sqrt{5x - 7})^2 = (\sqrt{x + 10})^2$ Squaring both sides

$5x - 7 = x + 10$

$4x = 17$

$x = \dfrac{17}{4}$

Check: $\sqrt{5x - 7} = \sqrt{x + 10}$

$\sqrt{5 \cdot \dfrac{17}{4} - 7} \;?\; \sqrt{\dfrac{17}{4} + 10}$

$\sqrt{\dfrac{85}{4} - \dfrac{28}{4}} \quad \sqrt{\dfrac{57}{4}}$

       $\sqrt{\dfrac{57}{4}}$ | TRUE

The solution is $\dfrac{17}{4}$.

15. $\sqrt{x} = -7$

There is no solution. The principal square root of x cannot be negative.

17. $\sqrt{2y + 6} = \sqrt{2y - 5}$

$(\sqrt{2y + 6})^2 = (\sqrt{2y - 5})^2$

$2y + 6 = 2y - 5$

$6 = -5$

The equation $6 = -5$ is false; there is no solution.

19.
$$x - 7 = \sqrt{x - 5}$$
$$(x - 7)^2 = (\sqrt{x - 5})^2$$
$$x^2 - 14x + 49 = x - 5$$
$$x^2 - 15 + 54 = 0$$
$$(x - 9)(x - 6) = 0$$
$$x - 9 = 0 \quad \text{or} \quad x - 6 = 0$$
$$x = 9 \quad \text{or} \quad x = 6$$

Check:

$$\frac{x - 7 = \sqrt{x - 5}}{}$$

$$9 - 7 \;?\; \sqrt{9 - 5}$$
$$2 \;\bigg|\; \sqrt{4}$$
$$\bigg|\; 2 \qquad \text{TRUE}$$

$$\frac{x - 7 = \sqrt{x - 5}}{}$$

$$6 - 7 \;\bigg|\; \sqrt{6 - 5}$$
$$-1 \;\bigg|\; \sqrt{1}$$
$$\bigg|\; 1 \qquad \text{FALSE}$$

The number 9 checks, but 6 does not. The solution is 9.

21.
$$x - 9 = \sqrt{x - 3}$$
$$(x - 9)^2 = (\sqrt{x - 3})^2$$
$$x^2 - 18x + 81 = x - 3$$
$$x^2 - 19x + 84 = 0$$
$$(x - 12)(x - 7) = 0$$
$$x - 12 = 0 \quad \text{or} \quad x - 7 = 0$$
$$x = 12 \quad \text{or} \quad x = 7$$

Check:

$$\frac{x - 9 = \sqrt{x - 3}}{}$$

$$12 - 9 \;?\; \sqrt{12 - 3}$$
$$3 \;\bigg|\; \sqrt{9}$$
$$\bigg|\; 3 \qquad \text{TRUE}$$

$$\frac{x - 9 = \sqrt{x - 3}}{}$$

$$7 - 9 \;?\; \sqrt{7 - 3}$$
$$-2 \;\bigg|\; \sqrt{4}$$
$$\bigg|\; 2 \qquad \text{FALSE}$$

The number 12 checks, but 7 does not. The solution is 12.

23.
$$2\sqrt{x - 1} = x - 1$$
$$(2\sqrt{x - 1})^2 = (x - 1)^2$$
$$4(x - 1) = x^2 - 2x + 1$$
$$4x - 4 = x^2 - 2x + 1$$
$$0 = x^2 - 6x + 5$$
$$0 = (x - 5)(x - 1)$$
$$x - 5 = 0 \quad \text{or} \quad x - 1 = 0$$
$$x = 5 \quad \text{or} \quad x = 1$$

Both numbers check. The solutions are 5 and 1.

25.
$$\sqrt{5x + 21} = x + 3$$
$$(\sqrt{5x + 21})^2 = (x + 3)^2$$
$$5x + 21 = x^2 + 6x + 9$$
$$0 = x^2 + x - 12$$
$$0 = (x + 4)(x - 3)$$
$$x + 4 = 0 \quad \text{or} \quad x - 3 = 0$$
$$x = -4 \quad \text{or} \quad x = 3$$

Check:

$$\frac{\sqrt{5x + 21} = x + 3}{}$$

$$\sqrt{5(-4) + 21} \;?\; -4 + 3$$
$$\sqrt{1} \;\bigg|\; -1$$
$$1 \;\bigg|\; \qquad \text{FALSE}$$

$$\frac{\sqrt{5x + 21} = x + 3}{}$$

$$\sqrt{5 \cdot 3 + 21} \;?\; 3 + 3$$
$$\sqrt{36} \;\bigg|\; 6$$
$$6 \;\bigg|\; \qquad \text{TRUE}$$

The number 3 checks, but -4 does not. The solution is 3.

27.
$$\sqrt{2x - 1} + 2 = x$$
$$\sqrt{2x - 1} = x - 2 \qquad \text{Isolating the radical}$$
$$(\sqrt{2x - 1})^2 = (x - 2)^2$$
$$2x - 1 = x^2 - 4x + 4$$
$$0 = x^2 - 6x + 5$$
$$0 = (x - 5)(x - 1)$$
$$x - 5 = 0 \quad \text{or} \quad x - 1 = 0$$
$$x = 5 \quad \text{or} \quad x = 1$$

Check:

$$\frac{\sqrt{2x - 1} + 2 = x}{}$$

$$\sqrt{2 \cdot 5 - 1} + 2 \;?\; 5$$
$$\sqrt{10 - 1} + 2 \;\bigg|\;$$
$$\sqrt{9} + 2 \;\bigg|\;$$
$$3 + 2 \;\bigg|\;$$
$$5 \;\bigg|\; \qquad \text{TRUE}$$

$$\frac{\sqrt{2x - 1} + 2 = x}{}$$

$$\sqrt{2 \cdot 1 - 1} + 2 \;?\; 1$$
$$\sqrt{2 - 1} + 2 \;\bigg|\;$$
$$\sqrt{1} + 2 \;\bigg|\;$$
$$1 + 2 \;\bigg|\;$$
$$3 \;\bigg|\; \qquad \text{FALSE}$$

The number 5 checks, but 1 does not. The solution is 5.

29. $\sqrt{x^2+6} - x + 3 = 0$

$\qquad \sqrt{x^2+6} = x - 3 \qquad$ Isolating the radical

$\qquad (\sqrt{x^2+6})^2 = (x-3)^2$

$\qquad x^2 + 6 = x^2 - 6x + 9$

$\qquad -3 = -6x \qquad$ Adding $-x^2$ and -9

$\qquad \dfrac{1}{2} = x$

Check: $\qquad \sqrt{x^2+6} - x + 3 = 0$

$\qquad \sqrt{\left(\dfrac{1}{2}\right)^2 + 6} - \dfrac{1}{2} + 3 \; ? \; 0$

$\qquad \sqrt{\dfrac{25}{4}} - \dfrac{1}{2} + 3$

$\qquad \dfrac{5}{2} - \dfrac{1}{2} + 3$

$\qquad\qquad\qquad 5 \quad \bigm| \quad$ FALSE

The number $\dfrac{1}{2}$ does not check. There is no solution.

31. $\sqrt{x^2-4} - x = 6$

$\qquad \sqrt{x^2-4} = x + 6 \qquad$ Isolating the radical

$\qquad (\sqrt{x^2-4})^2 = (x+6)^2$

$\qquad x^2 - 4 = x^2 + 12x + 36$

$\qquad -40 = 12x \qquad$ Adding $-x^2$ and -36

$\qquad -\dfrac{40}{12} = x$

$\qquad -\dfrac{10}{3} = x$

The number $-\dfrac{10}{3}$ checks. It is the solution.

33. $\sqrt{(p+6)(p+1)} - 2 = p + 1$

$\qquad \sqrt{(p+6)(p+1)} = p + 3 \qquad$ Isolating the radical

$\qquad \left(\sqrt{(p+6)(p+1)}\right)^2 = (p+3)^2$

$\qquad (p+6)(p+1) = p^2 + 6p + 9$

$\qquad p^2 + 7p + 6 = p^2 + 6p + 9$

$\qquad p = 3$

The number 3 checks. It is the solution.

35. $\sqrt{4x-10} = \sqrt{2-x}$

$\qquad (\sqrt{4x-10})^2 = (\sqrt{2-x})^2$

$\qquad 4x - 10 = 2 - x$

$\qquad 5x = 12 \qquad$ Adding 10 and x

$\qquad x = \dfrac{12}{5}$

Check: $\qquad \sqrt{4x-10} = \sqrt{2-x}$

$\qquad \sqrt{4 \cdot \dfrac{12}{5} - 10} \; ? \; \sqrt{2 - \dfrac{12}{5}}$

$\qquad \sqrt{\dfrac{48}{5} - 10} \quad \bigm| \quad \sqrt{-\dfrac{2}{5}}$

Since $\sqrt{-\dfrac{2}{5}}$ does not represent a real number, there is no solution that is a real number.

37. $\sqrt{x-5} = 5 - \sqrt{x}$

$\qquad (\sqrt{x-5})^2 = (5-\sqrt{x})^2 \qquad$ Squaring both sides

$\qquad x - 5 = 25 - 10\sqrt{x} + x$

$\qquad -30 = -10\sqrt{x} \qquad$ Isolating the radical

$\qquad 3 = \sqrt{x} \qquad$ Dividing by -10

$\qquad 3^2 = (\sqrt{x})^2 \qquad$ Squaring both sides

$\qquad 9 = x$

The number 9 checks. It is the solution.

39. $\sqrt{y+8} - \sqrt{y} = 2$

$\qquad \sqrt{y+8} = \sqrt{y} + 2 \qquad$ Isolating one radical

$\qquad (\sqrt{y+8})^2 = (\sqrt{y}+2)^2 \qquad$ Squaring both sides

$\qquad y + 8 = y + 4\sqrt{y} + 4$

$\qquad 4 = 4\sqrt{x} \qquad$ Isolating the radical

$\qquad 1 = \sqrt{y} \qquad$ Dividing by 4

$\qquad 1^2 = (\sqrt{y})^2$

$\qquad 1 = y$

The number 1 checks. It is the solution.

41. $\sqrt{x-4} + \sqrt{x+1} = 5$

$\qquad \sqrt{x-4} = 5 - \sqrt{x+1} \qquad$ Isolating one radical

$\qquad (\sqrt{x-4})^2 = (5-\sqrt{x+1})^2$

$\qquad x - 4 = 25 - 10\sqrt{x+1} + x + 1$

$\qquad -30 = -10\sqrt{x+1} \qquad$ Isolating the radical

$\qquad 3 = \sqrt{x+1} \qquad$ Dividing by -10

$\qquad 3^2 = (\sqrt{x+1})^2$

$\qquad 9 = x + 1$

$\qquad 8 = x$

The number 8 checks. It is the solution.

43. $\sqrt{x} - 1 = \sqrt{x-31}$

$\qquad (\sqrt{x}-1)^2 = (\sqrt{x-31})^2$

$\qquad x - 2\sqrt{x} + 1 = x - 31$

$\qquad -2\sqrt{x} = -32$

$\qquad \sqrt{x} = 16$

$\qquad (\sqrt{x})^2 = 16^2$

$\qquad x = 256$

The number 256 checks. It is the solution.

45. Substitute 27,000 for h in the equation.
$$D = \sqrt{2h}$$
$$D = \sqrt{2 \cdot 27,000}$$
$$D = \sqrt{54,000}$$
$$D \approx 232$$

You can see about 232 mi to the horizon.

47. Substitute 180 for D in the equation and solve for h.
$$D = \sqrt{2h}$$
$$180 = \sqrt{2h}$$
$$180^2 = (\sqrt{2h})^2$$
$$32,400 = 2h$$
$$16,200 = h$$

A pilot must fly 16,200 ft above sea level in order to see a horizon that is 180 mi away.

49. For 65 mph, substitute 64 for S in the equation and solve for x.
$$S = 2\sqrt{5x}$$
$$65 = 2\sqrt{5x}$$
$$65^2 = (2\sqrt{5x})^2$$
$$4225 = 4 \cdot 5x$$
$$4225 = 20x$$
$$211.25 = x$$

At 65 mph, a car will skid 211.25 ft.

For 75 mph, substitute 75 for S in the equation and solve for x.
$$S = 2\sqrt{5x}$$
$$75 = 2\sqrt{5x}$$
$$75^2 = (2\sqrt{5x})^2$$
$$5625 = 4 \cdot 5x$$
$$5625 = 20x$$
$$\frac{5625}{20} = x$$
$$281.25 = x$$

At 75 mph, a car will skid 281.25 ft.

51. Discussion and Writing Exercise

53. Parallel lines have the same slope and different y-intercepts.

55. The number c is a principal square root of a if $c^2 = a$ and c is either zero or positive.

57. The quotient rule asserts that when dividing with exponential notation, if the bases are the same, keep the base and subtract the exponent of the denominator from the exponent of the numerator.

59. The quotient rule for radicals asserts that for any nonnegative number A and any positive number B, $\dfrac{\sqrt{A}}{\sqrt{B}} = \sqrt{\dfrac{A}{B}}$.

61.
$$\sqrt{5x^2 + 5} = 5$$
$$(\sqrt{5x^2 + 5})^2 = 5^2$$
$$5x^2 + 5 = 25$$
$$5x^2 - 20 = 0$$
$$5(x^2 - 4) = 0$$
$$5(x + 2)(x - 2) = 0$$
$$x + 2 = 0 \quad or \quad x - 2 = 0$$
$$x = -2 \quad or \quad x = 2$$

Both numbers check, so the solutions are 2 and -2.

63.
$$4 + \sqrt{19 - x} = 6 + \sqrt{4 - x}$$
$$\sqrt{19 - x} = 2 + \sqrt{4 - x} \quad \text{Isolating one radical}$$
$$(\sqrt{19 - x})^2 = (2 + \sqrt{4 - x})^2$$
$$19 - x = 4 + 4\sqrt{4 - x} + (4 - x)$$
$$19 - x = 4\sqrt{4 - x} + 8 - x$$
$$11 = 4\sqrt{4 - x}$$
$$11^2 = (4\sqrt{4 - x})^2$$
$$121 = 16(4 - x)$$
$$121 = 64 - 16x$$
$$57 = -16x$$
$$-\frac{57}{16} = x$$

$-\dfrac{57}{16}$ checks, so it is the solution.

65.
$$\sqrt{x + 3} = \frac{8}{\sqrt{x - 9}}$$
$$(\sqrt{x + 3})^2 = \left(\frac{8}{\sqrt{x - 9}}\right)^2$$
$$x + 3 = \frac{64}{x - 9}$$
$$(x - 9)(x + 3) = 64 \quad \text{Multiplying by } x - 9$$
$$x^2 - 6x - 27 = 64$$
$$x^2 - 6x - 91 = 0$$
$$(x - 13)(x + 7) = 0$$
$$x - 13 = 0 \quad or \quad x + 7 = 0$$
$$x = 13 \quad or \quad x = -7$$

The number 13 checks, but -7 does not. The solution is 13.

67.–69. Left to the student

Exercise Set 16.6

1.
$$a^2 + b^2 = c^2$$
$$8^2 + 15^2 = c^2 \quad \text{Substituting}$$
$$64 + 225 = c^2$$
$$289 = c^2$$
$$\sqrt{289} = c$$
$$17 = c$$

3. $a^2 + b^2 = c^2$

$4^2 + 4^2 = c^2$ Substituting

$16 + 16 = c^2$

$32 = c^2$

$\sqrt{32} = c$ Exact answer

$5.657 \approx c$ Approximation

5. $a^2 + b^2 = c^2$

$5^2 + b^2 = 13^2$

$25 + b^2 = 169$

$b^2 = 144$

$b = 12$

7. $a^2 + b^2 = c^2$

$(4\sqrt{3})^2 + b^2 = 8^2$

$16 \cdot 3 + b^2 = 64$

$48 + b^2 = 64$

$b^2 = 16$

$b = 4$

9. $a^2 + b^2 = c^2$

$10^2 + 24^2 = c^2$

$100 + 576 = c^2$

$676 = c^2$

$26 = c$

11. $a^2 + b^2 = c^2$

$9^2 + b^2 = 15^2$

$81 + b^2 = 225$

$b^2 = 144$

$b = 12$

13. $a^2 + b^2 = c^2$

$a^2 + 1^2 = (\sqrt{5})^2$

$a^2 + 1 = 5$

$a^2 = 4$

$a = 2$

15. $a^2 + b^2 = c^2$

$1^2 + b^2 = (\sqrt{3})^2$

$1 + b^2 = 3$

$b^2 = 2$

$b = \sqrt{2}$ Exact answer

$b \approx 1.414$ Approximation

17. $a^2 + b^2 = c^2$

$a^2 + (5\sqrt{3})^2 = 10^2$

$a^2 + 25 \cdot 3 = 100$

$a^2 + 75 = 100$

$a^2 = 25$

$a = 5$

19. $a^2 + b^2 = c^2$

$(\sqrt{2})^2 + (\sqrt{7})^2 = c^2$

$2 + 7 = c^2$

$9 = c^2$

$3 = c$

21. We use the drawing in the text, labeling the horizontal distance h.

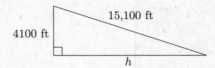

We know that $4100^2 + h^2 = 15,100^2$. We solve this equation.

$16,810,000 + h^2 = 228,010,000$

$h^2 = 211,200,000$

$h = \sqrt{211,200,000}$ ft   Exact answer

$h \approx 14,533$ ft   Approximation

23. We first make a drawing. Let d represent the distance Becky can move away from the building while using the telephone.

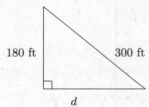

We know that $180^2 + d^2 = 300^2$.

We solve this equation.

$180^2 + d^2 = 300^2$

$32,400 + d^2 = 90,000$

$d^2 = 57,600$

$d = 240$

Becky can use her telephone 240 ft into her backyard.

25. We first make a drawing. We label the diagonal d.

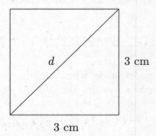

We know that $3^2 + 3^2 = d^2$. We solve this equation.

$$3^2 + 3^2 = d^2$$
$$9 + 9 = d^2$$
$$18 = d^2$$
$$\sqrt{18} \text{ cm} = d \quad \text{Exact answer}$$
$$4.243 \text{ cm} \approx d \quad \text{Approximation}$$

27. We first make a drawing. We label the length of the guy wire w.

We know that $8^2 + 12^2 = w^2$. We solve this equation.
$$8^2 + 12^2 = w^2$$
$$64 + 144 = w^2$$
$$208 = w^2$$
$$\sqrt{208} \text{ ft} = w \quad \text{Exact answer}$$
$$14.422 \text{ ft} \approx w \quad \text{Approximation}$$

29. Discussion and Writing Exercise

31. $5x + 7 = 8y$,
$3x = 8y - 4$

$5x - 8y = -7 \quad (1) \quad$ Rewriting
$3x - 8y = -4 \quad (2) \quad$ the equations

We multiply Equation (2) by -1 and add.

$$\begin{array}{r} 5x - 8y = -7 \\ \underline{-3x + 8y = 4} \\ 2x \quad\quad = -3 \\ x = -\dfrac{3}{2} \end{array}$$

Substitute $-\dfrac{3}{2}$ for x in Equation (1) and solve for y.

$$5x - 8y = -7$$
$$5\left(-\dfrac{3}{2}\right) - 8y = -7$$
$$-\dfrac{15}{2} - 8y = -7$$
$$-8y = \dfrac{1}{2}$$
$$y = -\dfrac{1}{16}$$

The ordered pair $\left(-\dfrac{3}{2}, -\dfrac{1}{16}\right)$ checks. It is the solution.

33. $3x - 4y = -11 \quad (1)$
$5x + 6y = 12 \quad\quad (2)$

We multiply Equation (1) by 3 and Equation (2) by 2, and then we add.

$$\begin{array}{r} 9x - 12y = -33 \\ \underline{10x + 12y = 24} \\ 19x \quad\quad = -9 \\ x = -\dfrac{9}{19} \end{array}$$

Substitute $-\dfrac{9}{19}$ for x in Equation (2) and solve for y.

$$5x + 6y = 12$$
$$5\left(-\dfrac{9}{19}\right) + 6y = 12$$
$$-\dfrac{45}{19} + 6y = 12$$
$$6y = \dfrac{273}{19} \quad \text{Adding } \dfrac{45}{19}$$
$$y = \dfrac{273}{6 \cdot 19} \quad \text{Dividing by 6}$$
$$y = \dfrac{91}{38} \quad \text{Simplifying}$$

The ordered pair $\left(-\dfrac{9}{19}, \dfrac{91}{38}\right)$ checks. It is the solution.

35. Write the equation in the slope-intercept form.

$$4 - x = 3y$$
$$\dfrac{1}{3}(4 - x) = y$$
$$\dfrac{4}{3} - \dfrac{1}{3}x = y, \text{ or}$$
$$y = -\dfrac{1}{3}x + \dfrac{4}{3}$$

The slope is $-\dfrac{1}{3}$.

37.

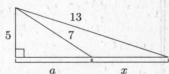

$$a^2 + 5^2 = 7^2$$
$$a^2 + 25 = 49$$
$$a^2 = 24$$
$$a = \sqrt{24}, \text{ or } 2\sqrt{6}$$

$$(a + x)^2 + 5^2 = 13^2$$
$$(2\sqrt{6} + x)^2 + 5^2 = 13^2 \quad \text{Substituting } 2\sqrt{6} \text{ for } a$$
$$(2\sqrt{6} + x)^2 + 25 = 169$$
$$(2\sqrt{6} + x)^2 = 144$$
$$2\sqrt{6} + x = 12 \quad \text{Taking the principal square root}$$
$$x = 12 - 2\sqrt{6}$$
$$x \approx 7.101$$

Chapter 16 Review Exercises

1. The square roots of 64 are 8 and -8, because $8^2 = 64$ and $(-8)^2 = 64$.

2. The square roots of 400 are 20 and -20, because $20^2 = 400$ and $(-20)^2 = 400$.

3. $\sqrt{36} = 6$, taking the principal square root

4. $\sqrt{169} = 13$, so $-\sqrt{169} = -13$.

5. 1.732

6. 9.950

7. $\sqrt{320.12} \approx 17.892$, so $-\sqrt{320.12} \approx -17.892$.

8. 0.742

9. $\sqrt{\dfrac{47.3}{11.2}} \approx 2.055$, so $-\sqrt{\dfrac{47.3}{11.2}} \approx -2.055$.

10. 394.648

11. The radicand is the expression under the radical, $x^2 + 4$.

12. The radicand is the expression under the radical, $5ab^3$.

13. No, because the radicand is negative

14. Yes, because the radicand is nonnegative

15. No, because the radicand is negative

16. No, because the radicand is negative

17. No, because the radicand is negative

18. No, because the radicand is negative

19. $\sqrt{m^2} = m$

20. $\sqrt{(x-4)^2} = x - 4$

21. $\sqrt{3}\sqrt{7} = \sqrt{3 \cdot 7} = \sqrt{21}$

22. $\sqrt{x-3}\sqrt{x+3} = \sqrt{(x-3)(x+3)} = \sqrt{x^2 - 9}$

23. $-\sqrt{48} = -\sqrt{16 \cdot 3} = -\sqrt{16}\sqrt{3} = -4\sqrt{3}$

24. $\sqrt{32t^2} = \sqrt{16 \cdot 2 \cdot t^2} = \sqrt{16}\sqrt{t^2}\sqrt{2} = 4t\sqrt{2}$

25. $\sqrt{t^2 - 49} = \sqrt{(t+7)(t-7)} = \sqrt{t+7}\sqrt{t-7}$

26. $\sqrt{x^2 + 16x + 64} = \sqrt{(x+8)^2} = x + 8$

27. $\sqrt{x^8} = \sqrt{(x^4)^2} = x^4$

28. $\sqrt{m^{15}} = \sqrt{m^{14} \cdot m} = \sqrt{m^{14}}\sqrt{m} = \sqrt{(m^7)^2}\sqrt{m} = m^7\sqrt{m}$

29. $\sqrt{6}\sqrt{10} = \sqrt{6 \cdot 10} = \sqrt{2 \cdot 3 \cdot 2 \cdot 5} = \sqrt{2 \cdot 2}\sqrt{3 \cdot 5} = 2\sqrt{15}$

30. $\sqrt{5x}\sqrt{8x} = \sqrt{5x \cdot 8x} = \sqrt{5 \cdot x \cdot 8 \cdot x} = \sqrt{x \cdot x}\sqrt{5 \cdot 8} = x\sqrt{40}$

31. $\sqrt{5x}\sqrt{10xy^2} = \sqrt{5x \cdot 10xy^2} = \sqrt{5 \cdot x \cdot 2 \cdot 5 \cdot x \cdot y^2} = \sqrt{5 \cdot 5}\sqrt{x \cdot x}\sqrt{y^2}\sqrt{2} = 5xy\sqrt{2}$

32. $\sqrt{20a^3b}\sqrt{5a^2b^2} = \sqrt{20a^3b \cdot 5a^2b^2} = \sqrt{4 \cdot 5 \cdot a^2 \cdot a \cdot b \cdot 5 \cdot a^2 \cdot b^2} = \sqrt{4}\sqrt{5 \cdot 5}\sqrt{a^2}\sqrt{a^2}\sqrt{b^2}\sqrt{ab} = 2 \cdot 5 \cdot a \cdot a \cdot b\sqrt{a \cdot b} = 10a^2b\sqrt{ab}$

33. $\sqrt{\dfrac{25}{64}} = \dfrac{\sqrt{25}}{\sqrt{64}} = \dfrac{5}{8}$

34. $\sqrt{\dfrac{20}{45}} = \sqrt{\dfrac{4}{9}} = \dfrac{\sqrt{4}}{\sqrt{9}} = \dfrac{2}{3}$

35. $\sqrt{\dfrac{49}{t^2}} = \dfrac{\sqrt{49}}{\sqrt{t^2}} = \dfrac{7}{t}$

36. $\sqrt{\dfrac{1}{2}} = \sqrt{\dfrac{1}{2} \cdot \dfrac{2}{2}} = \sqrt{\dfrac{2}{4}} = \dfrac{\sqrt{2}}{\sqrt{4}} = \dfrac{\sqrt{2}}{2}$

37. $\sqrt{\dfrac{1}{8}} = \sqrt{\dfrac{1}{8} \cdot \dfrac{2}{2}} = \sqrt{\dfrac{2}{16}} = \dfrac{\sqrt{2}}{\sqrt{16}} = \dfrac{\sqrt{2}}{4}$

38. $\sqrt{\dfrac{5}{y}} = \sqrt{\dfrac{5}{y} \cdot \dfrac{y}{y}} = \sqrt{\dfrac{5y}{y^2}} = \dfrac{\sqrt{5y}}{\sqrt{y^2}} = \dfrac{\sqrt{5y}}{y}$

39. $\dfrac{2}{\sqrt{3}} = \dfrac{2}{\sqrt{3}} \cdot \dfrac{\sqrt{3}}{\sqrt{3}} = \dfrac{2\sqrt{3}}{3}$

40. $\dfrac{\sqrt{27}}{\sqrt{45}} = \sqrt{\dfrac{27}{45}} = \sqrt{\dfrac{3}{5}} = \sqrt{\dfrac{3}{5} \cdot \dfrac{5}{5}} = \sqrt{\dfrac{15}{25}} = \dfrac{\sqrt{15}}{\sqrt{25}} = \dfrac{\sqrt{15}}{5}$

41. $\dfrac{\sqrt{45x^2y}}{\sqrt{54y}} = \sqrt{\dfrac{45x^2y}{54y}} = \sqrt{\dfrac{5x^2}{6}} = \sqrt{\dfrac{5x^2}{6} \cdot \dfrac{6}{6}} = \sqrt{\dfrac{30x^2}{36}} = \dfrac{\sqrt{30x^2}}{\sqrt{36}} = \dfrac{x\sqrt{30}}{6}$

42. $\dfrac{4}{2 + \sqrt{3}} = \dfrac{4}{2 + \sqrt{3}} \cdot \dfrac{2 - \sqrt{3}}{2 - \sqrt{3}}$

$= \dfrac{4(2 - \sqrt{3})}{(2 + \sqrt{3})(2 - \sqrt{3})}$

$= \dfrac{8 - 4\sqrt{3}}{2^2 - (\sqrt{3})^2} = \dfrac{8 - 4\sqrt{3}}{4 - 3}$

$= \dfrac{8 - 4\sqrt{3}}{1} = 8 - 4\sqrt{3}$

43. $10\sqrt{5} + 3\sqrt{5} = (10 + 3)\sqrt{5} = 13\sqrt{5}$

44. $\sqrt{80} - \sqrt{45} = \sqrt{16 \cdot 5} - \sqrt{9 \cdot 5}$

$= 4\sqrt{5} - 3\sqrt{5}$

$= (4 - 3)\sqrt{5}$

$= \sqrt{5}$

45. $3\sqrt{2} - 5\sqrt{\dfrac{1}{2}} = 3\sqrt{2} - 5\sqrt{\dfrac{1}{2} \cdot \dfrac{2}{2}}$

$\qquad = 3\sqrt{2} - 5 \cdot \dfrac{\sqrt{2}}{2}$

$\qquad = 3\sqrt{2} - \dfrac{5}{2}\sqrt{2}$

$\qquad = \left(3 - \dfrac{5}{2}\right)\sqrt{2}$

$\qquad = \dfrac{1}{2}\sqrt{2}$, or $\dfrac{\sqrt{2}}{2}$

46. $(2 + \sqrt{3})^2 = 2^2 + 2 \cdot 2 \cdot \sqrt{3} + (\sqrt{3})^2 = 4 + 4\sqrt{3} + 3 = 7 + 4\sqrt{3}$

47. $(2 + \sqrt{3})(2 - \sqrt{3}) = 2^2 - (\sqrt{3})^2 = 4 - 3 = 1$

48. $\sqrt{x - 3} = 7$

$\qquad (\sqrt{x - 3})^2 = 7^2$

$\qquad x - 3 = 49$

$\qquad x = 52$

The number 52 checks. It is the solution.

49. $\sqrt{5x + 3} = \sqrt{2x - 1}$

$\qquad (\sqrt{5x + 3})^2 = (\sqrt{2x - 1})^2$

$\qquad 5x + 3 = 2x - 1$

$\qquad 3x + 3 = -1$

$\qquad 3x = -4$

$\qquad x = -\dfrac{4}{3}$

The number $-\dfrac{4}{3}$ does not check. There is no solution.

50. $1 + x = \sqrt{1 + 5x}$

$\qquad (1 + x)^2 = (\sqrt{1 + 5x})^2$

$\qquad 1 + 2x + x^2 = 1 + 5x$

$\qquad x^2 - 3x = 0$

$\qquad x(x - 3) = 0$

$x = 0 \ \ or \ \ x - 3 = 0$

$x = 0 \ \ or \ \ \ \ \ \ \ \ x = 3$

Both numbers check. The solutions are 0 and 3.

51. $\sqrt{x} = \sqrt{x - 5} + 1$

$\qquad (\sqrt{x})^2 = (\sqrt{x - 5} + 1)^2$

$\qquad x = x - 5 + 2\sqrt{x - 5} + 1$

$\qquad x = x - 4 + 2\sqrt{x - 5}$

$\qquad 4 = 2\sqrt{x - 5}$

$\qquad 2 = \sqrt{x - 5}$

$\qquad 2^2 = (\sqrt{x - 5})^2$

$\qquad 4 = x - 5$

$\qquad 9 = x$

The number 9 checks. It is the solution.

52. $a^2 + b^2 = c^2$

$\qquad 15^2 + b^2 = 25^2$

$\qquad 225 + b^2 = 625$

$\qquad b^2 = 400$

$\qquad b = 20$

53. $a^2 + b^2 = c^2$

$\qquad 1^2 + (\sqrt{2})^2 = c^2$

$\qquad 1 + 2 = c^2$

$\qquad 3 = c^2$

$\qquad \sqrt{3} = c \qquad$ Exact answer

$\qquad 1.732 \approx c \qquad$ Approximation

54. First we subtract to find the vertical distance of the descent:

$$30{,}000 \text{ ft} - 20{,}000 \text{ ft} = 10{,}000 \text{ ft}$$

Let $c = $ the distance the plane travels during the descent. This is labeled "?" in the drawing in the text. We know that $10{,}000^2 + 50{,}000^2 = c^2$. We solve this equation.

$\qquad 10{,}000^2 + 50{,}000^2 = c^2$

$\qquad 100{,}000{,}000 + 2{,}500{,}000{,}000 = c^2$

$\qquad 2{,}600{,}000{,}000 = c^2$

$\qquad \sqrt{2{,}600{,}00{,}000} \text{ ft} = c$

$\qquad 50{,}990 \text{ ft} \approx c$

55. Let $d = $ the distance each brace reaches vertically. From the drawing in the text we see that $d^2 + 12^2 = 15^2$. We solve this equation.

$\qquad d^2 + 12^2 = 15^2$

$\qquad d^2 + 144 = 225$

$\qquad d^2 = 81$

$\qquad d = 9 \text{ ft}$

56. a) We substitute 200 for L in the formula.

$$r = 2\sqrt{5 \cdot 200} = 2\sqrt{1000} \approx 63 \text{ mph}$$

b) We substitute 90 for r and solve for L.

$\qquad 90 = 2\sqrt{5L}$

$\qquad 90^2 = (2\sqrt{5L})^2$

$\qquad 8100 = 4 \cdot 5L$

$\qquad 8100 = 20L$

$\qquad 405 \text{ ft} = L$

57. *Discussion and Writing Exercise.* It is incorrect to take the square roots of the terms in the numerator individually; that is, $\sqrt{a + b}$ and $\sqrt{a} + \sqrt{b}$ are not equivalent. The following is correct:

$$\sqrt{\dfrac{9 + 100}{25}} = \dfrac{\sqrt{9 + 100}}{\sqrt{25}} = \dfrac{\sqrt{109}}{5}.$$

58. *Discussion and Writing Exercise.*

a) $\sqrt{5x^2} = \sqrt{5}\sqrt{x^2} = \sqrt{5} \cdot |x| = |x|\sqrt{5}$. The given statement is correct.

b) Let $b = 3$. Then $\sqrt{b^2 - 4} = \sqrt{3^2 - 4} = \sqrt{9 - 4} = \sqrt{5}$, but $b - 2 = 3 - 2 = 1$. The given statement is false.

c) Let $x = 3$. Then $\sqrt{x^2 + 16} = \sqrt{3^2 + 16} = \sqrt{9 + 16} = \sqrt{25} = 5$, but $x + 4 = 3 + 4 = 7$. The given statement is false.

59. After $\frac{1}{2}$ hr, the car traveling east at 50 mph has traveled $\frac{1}{2} \cdot 50$, or 25 mi and the car traveling south at 60 mph has traveled $\frac{1}{2} \cdot 60$, or 30 mph. These distances are the legs of a right triangle. The length of the hypotenuse of the triangle is the distance that separates the cars. Let $d =$ this distance. We know that $25^2 + 30^2 = d^2$. We solve this equation.

$$25^2 + 30^2 = d^2$$
$$625 + 900 = d^2$$
$$1525 = d^2$$
$$\sqrt{1525} \text{ mi} = d$$
$$39.051 \text{ mi} \approx d$$

60. $\sqrt{\sqrt{\sqrt{256}}} = \sqrt{\sqrt{16}} = \sqrt{4} = 2$

61.
$$A = \sqrt{a^2 + b^2}$$
$$A^2 = (\sqrt{a^2 + b^2})^2$$
$$A^2 = a^2 + b^2$$
$$A^2 - a^2 = b^2$$

If $b^2 = A^2 - a^2$, then $b = \sqrt{A^2 - a^2}$ or $b = -\sqrt{A^2 - a^2}$, so we have $b = \pm\sqrt{A^2 - a^2}$.

62. Using the drawing in the text, let $a =$ the hypotenuse of the triangle with legs 4 and x and let $b =$ the hypotenuse of the triangle with legs 9 and x. Then we have

$$4^2 + x^2 = a^2, \text{ or } 16 + x^2 = a^2$$

and

$$9^2 + x^2 = b^2, \text{ or } 81 + x^2 = b^2.$$

Note that a and b are also legs of the large triangle with hypotenuse $4 + 9$, or 13. Then we have

$$a^2 + b^2 = 13^2, \text{ or } a^2 + b^2 = 169.$$

Adding the two equations containing x, we have

$$16 + x^2 = a^2$$
$$\underline{81 + x^2 = b^2}$$
$$97 + 2x^2 = a^2 + b^2.$$

We substitute $97 + 2x^2$ for $a^2 + b^2$ in the equation pertaining to the large triangle.

$$a^2 + b^2 = 169$$
$$97 + 2x^2 = 169$$
$$2x^2 = 72$$
$$x^2 = 36$$
$$x = 6$$

Chapter 16 Test

1. The square roots of 81 are 9 and -9, because $9^2 = 81$ and $(-9)^2 = 81$.

2. $\sqrt{64} = 8$, taking the principal square root

3. $\sqrt{25} = 5$, so $-\sqrt{25} = -5$.

4. $\sqrt{116} \approx 10.770$

5. $\sqrt{87.4} \approx 9.349$, so $-\sqrt{87.4} \approx -9.349$.

6. $\sqrt{\dfrac{96 \cdot 38}{214.2}} \approx 4.127$

7. The radicand is the expression under the radical, $4 - y^3$.

8. Yes, because the radicand is nonnegative

9. No, because the radicand is negative

10. $\sqrt{a^2} = a$

11. $\sqrt{36y^2} = \sqrt{(6y)^2} = 6y$

12. $\sqrt{5}\sqrt{6} = \sqrt{5 \cdot 6} = \sqrt{30}$

13. $\sqrt{x - 8}\sqrt{x + 8} = \sqrt{(x - 8)(x + 8)} = \sqrt{x^2 - 64}$

14. $\sqrt{27} = \sqrt{9 \cdot 3} = \sqrt{9}\sqrt{3} = 3\sqrt{3}$

15. $\sqrt{25x - 25} = \sqrt{25(x - 1)} = \sqrt{25}\sqrt{x - 1} = 5\sqrt{x - 1}$

16. $\sqrt{t^5} = \sqrt{t^4 \cdot t} = \sqrt{t^4}\sqrt{t} = \sqrt{(t^2)^2}\sqrt{t} = t^2\sqrt{t}$

17. $\sqrt{5}\sqrt{10} = \sqrt{5 \cdot 10} = \sqrt{5 \cdot 2 \cdot 5} = \sqrt{5 \cdot 5}\sqrt{2} = 5\sqrt{2}$

18. $\sqrt{3ab}\sqrt{6ab^3} = \sqrt{3ab \cdot 6ab^3} = \sqrt{3 \cdot a \cdot b \cdot 2 \cdot 3 \cdot a \cdot b^2 \cdot b} = \sqrt{3 \cdot 3}\sqrt{a \cdot a}\sqrt{b \cdot b}\sqrt{b^2}\sqrt{2} = 3 \cdot a \cdot b \cdot b\sqrt{2} = 3ab^2\sqrt{2}$

19. $\sqrt{\dfrac{27}{12}} = \sqrt{\dfrac{9}{4}} = \dfrac{\sqrt{9}}{\sqrt{4}} = \dfrac{3}{2}$

20. $\sqrt{\dfrac{144}{a^2}} = \dfrac{\sqrt{144}}{\sqrt{a^2}} = \dfrac{12}{a}$

21. $\sqrt{\dfrac{2}{5}} = \sqrt{\dfrac{2}{5} \cdot \dfrac{5}{5}} = \sqrt{\dfrac{10}{25}} = \dfrac{\sqrt{10}}{\sqrt{25}} = \dfrac{\sqrt{10}}{5}$

22. $\sqrt{\dfrac{2x}{y}} = \sqrt{\dfrac{2x}{y} \cdot \dfrac{y}{y}} = \sqrt{\dfrac{2xy}{y^2}} = \dfrac{\sqrt{2xy}}{y}$

23. $\dfrac{\sqrt{27}}{\sqrt{32}} = \sqrt{\dfrac{27}{32}} = \sqrt{\dfrac{27}{32} \cdot \dfrac{2}{2}} = \sqrt{\dfrac{54}{64}} = \dfrac{\sqrt{54}}{\sqrt{64}} = \dfrac{\sqrt{9 \cdot 6}}{8} = \dfrac{3\sqrt{6}}{8}$

24. $\dfrac{\sqrt{35x}}{\sqrt{80xy^2}} = \sqrt{\dfrac{35x}{80xy^2}} = \sqrt{\dfrac{7}{16y^2}} = \dfrac{\sqrt{7}}{\sqrt{16y^2}} = \dfrac{\sqrt{7}}{4y}$

25. $3\sqrt{18} - 5\sqrt{18} = (3 - 5)\sqrt{18} = -2\sqrt{18} = -2\sqrt{9 \cdot 2} = -2 \cdot 3\sqrt{2} = -6\sqrt{2}$

26. $\sqrt{5} + \sqrt{\dfrac{1}{5}} = \sqrt{5} + \sqrt{\dfrac{1}{5} \cdot \dfrac{5}{5}} = \sqrt{5} + \sqrt{\dfrac{5}{25}} =$

$\sqrt{5} + \dfrac{\sqrt{5}}{5} = \left(1 + \dfrac{1}{5}\right)\sqrt{5} = \dfrac{6}{5}\sqrt{5}, \text{ or } \dfrac{6\sqrt{5}}{5}$

27. $(4 - \sqrt{5})^2 = 4^2 - 2 \cdot 4 \cdot \sqrt{5} + (\sqrt{5})^2 = 16 - 8\sqrt{5} + 5 =$
$21 - 8\sqrt{5}$

28. $(4 - \sqrt{5})(4 + \sqrt{5}) = 4^2 - (\sqrt{5})^2 = 16 - 5 = 11$

29. $\dfrac{10}{4 - \sqrt{5}} = \dfrac{10}{4 - \sqrt{5}} \cdot \dfrac{4 + \sqrt{5}}{4 + \sqrt{5}} = \dfrac{10(4 + \sqrt{5})}{4^2 - (\sqrt{5})^2} = \dfrac{40 + 10\sqrt{5}}{16 - 5} =$

$\dfrac{40 + 10\sqrt{5}}{11}$

30. $a^2 + b^2 = c^2$
$8^2 + 4^2 = c^2$
$64 + 16 = c^2$
$80 = c^2$
$\sqrt{80} = c \quad$ Exact answer
$8.944 \approx c \quad$ Approximation

31. $\sqrt{3x} + 2 = 14$
$\sqrt{3x} = 12$
$(\sqrt{3x})^2 = 12^2$
$3x = 144$
$x = 48$

The number 48 checks. It is the solution.

32. $\sqrt{6x + 13} = x + 3$
$(\sqrt{6x + 13})^2 = (x + 3)^2$
$6x + 13 = x^2 + 6x + 9$
$0 = x^2 - 4$
$0 = (x + 2)(x - 2)$
$x + 2 = 0 \quad or \quad x - 2 = 0$
$x = -2 \quad or \quad x = 2$

Both numbers check. The solutions are -2 and 2.

33. $\sqrt{1 - x} + 1 = \sqrt{6 - x}$
$(\sqrt{1 - x} + 1)^2 = (\sqrt{6 - x})^2$
$1 - x + 2\sqrt{1 - x} + 1 = 6 - x$
$2 - x + 2\sqrt{1 - x} = 6 - x$
$2\sqrt{1 - x} = 4$
$\sqrt{1 - x} = 2$
$(\sqrt{1 - x})^2 = 2^2$
$1 - x = 4$
$-x = 3$
$x = -3$

The number -3 checks. It is the solution.

34. a) Substitute 28,000 for h in the formula.
$D = \sqrt{2 \cdot 28,000} = \sqrt{56,000} \approx 237 \text{ mi}$

b) Substitute 261 for D and solve for h.
$261 = \sqrt{2h}$
$261^2 = (\sqrt{2h})^2$
$68,121 = 2h$
$34,060.5 = h$
The airplane is 34,060.5 ft high.

35. Let $d =$ the length of a diagonal, in yd.
$a^2 + b^2 = c^2$
$60^2 + 110^2 = d^2$
$3600 + 12,100 = d^2$
$15,700 = d^2$
$\sqrt{15,700} \text{ yd} = d$
$125.300 \text{ yd} \approx d$

36. $\sqrt{\sqrt{\sqrt{625}}} = \sqrt{\sqrt{25}} = \sqrt{5}$

37. $\sqrt{y^{16n}} = \sqrt{(y^{8n})^2} = y^{8n}$

Chapter 17

Quadratic Equations

1. $x^2 - 3x + 2 = 0$

This equation is already in standard form.

$a = 1, \ b = -3, \ c = 2$

3. $\qquad 7x^2 = 4x - 3$

$7x^2 - 4x + 3 = 0 \qquad$ Standard form

$a = 7, \ b = -4, \ c = 3$

5. $\qquad 5 = -2x^2 + 3x$

$2x^2 - 3x + 5 = 0 \qquad$ Standard form

$a = 2, \ b = -3, \ c = 5$

7. $x^2 + 5x = 0$

$x(x + 5) = 0$

$x = 0 \ $ or $\ x + 5 = 0$

$x = 0 \ $ or $\qquad x = -5$

The solutions are 0 and -5.

9. $3x^2 + 6x = 0$

$3x(x + 2) = 0$

$3x = 0 \ $ or $\ x + 2 = 0$

$x = 0 \ $ or $\qquad x = -2$

The solutions are 0 and -2.

11. $\qquad 5x^2 = 2x$

$5x^2 - 2x = 0$

$x(5x - 2) = 0$

$x = 0 \ $ or $\ 5x - 2 = 0$

$x = 0 \ $ or $\qquad 5x = 2$

$x = 0 \ $ or $\qquad x = \dfrac{2}{5}$

The solutions are 0 and $\dfrac{2}{5}$.

13. $4x^2 + 4x = 0$

$4x(x + 1) = 0$

$4x = 0 \ $ or $\ x + 1 = 0$

$x = 0 \ $ or $\qquad x = -1$

The solutions are 0 and -1.

15. $0 = 10x^2 - 30x$

$0 = 10x(x - 3)$

$10x = 0 \ $ or $\ x - 3 = 0$

$x = 0 \ $ or $\qquad x = 3$

The solutions are 0 and 3.

17. $\qquad 11x = 55x^2$

$0 = 55x^2 - 11x$

$0 = 11x(5x - 1)$

$11x = 0 \ $ or $\ 5x - 1 = 0$

$x = 0 \ $ or $\qquad 5x = 1$

$x = 0 \ $ or $\qquad x = \dfrac{1}{5}$

The solutions are 0 and $\dfrac{1}{5}$.

19. $\qquad 14t^2 = 3t$

$14t^2 - 3t = 0$

$t(14t - 3) = 0$

$t = 0 \ $ or $\ 14t - 3 = 0$

$t = 0 \ $ or $\qquad 14t = 3$

$t = 0 \ $ or $\qquad t = \dfrac{3}{14}$

The solutions are 0 and $\dfrac{3}{14}$.

21. $\qquad 5y^2 - 3y^2 = 72y + 9y$

$\qquad\qquad 2y^2 = 81y$

$2y^2 - 81y = 0$

$y(2y - 81) = 0$

$y = 0 \ $ or $\ 2y - 81 = 0$

$y = 0 \ $ or $\qquad 2y = 81$

$y = 0 \ $ or $\qquad y = \dfrac{81}{2}$

The solutions are 0 and $\dfrac{81}{2}$.

23. $\quad x^2 + 8x - 48 = 0$

$(x + 12)(x - 4) = 0$

$x + 12 = 0 \qquad$ or $\ x - 4 = 0$

$\qquad x = -12 \ $ or $\qquad x = 4$

The solutions are -12 and 4.

25. $\quad 5 + 6x + x^2 = 0$

$(5 + x)(1 + x) = 0$

$5 + x = 0 \qquad$ or $\ 1 + x = 0$

$\qquad x = -5 \ $ or $\qquad x = -1$

The solutions are -5 and -1.

27. $\quad 18 = 7p + p^2$

$0 = p^2 + 7p - 18$

$0 = (p + 9)(p - 2)$

$p + 9 = 0 \qquad$ or $\ p - 2 = 0$

$\qquad p = -9 \ $ or $\qquad p = 2$

The solutions are -9 and 2.

29. $-15 = -8y + y^2$
$0 = y^2 - 8y + 15$
$0 = (y - 5)(y - 3)$
$y - 5 = 0 \quad or \quad y - 3 = 0$
$y = 5 \quad or \qquad y = 3$
The solutions are 5 and 3.

31. $x^2 + 10x + 25 = 0$
$(x + 5)(x + 5) = 0$
$x + 5 = 0 \quad or \quad x + 5 = 0$
$x = -5 \quad or \qquad x = -5$
The solution is -5.

33. $\qquad r^2 = 8r - 16$
$r^2 - 8r + 16 = 0$
$(r - 4)(r - 4) = 0$
$r - 4 = 0 \quad or \quad r - 4 = 0$
$r = 4 \quad or \qquad r = 4$
The solution is 4.

35. $\qquad 6x^2 + x - 2 = 0$
$(3x + 2)(2x - 1) = 0$
$3x + 2 = 0 \quad or \quad 2x - 1 = 0$
$3x = -2 \quad or \qquad 2x = 1$
$x = -\dfrac{2}{3} \quad or \qquad x = \dfrac{1}{2}$
The solutions are $-\dfrac{2}{3}$ and $\dfrac{1}{2}$.

37. $\qquad 3a^2 = 10a + 8$
$3a^2 - 10a - 8 = 0$
$(3a + 2)(a - 4) = 0$
$3a + 2 = 0 \quad or \quad a - 4 = 0$
$3a = -2 \quad or \qquad a = 4$
$a = -\dfrac{2}{3} \quad or \qquad a = 4$
The solutions are $-\dfrac{2}{3}$ and 4.

39. $\qquad 6x^2 - 4x = 10$
$6x^2 - 4x - 10 = 0$
$2(3x^2 - 2x - 5) = 0$
$2(3x - 5)(x + 1) = 0$
$3x - 5 = 0 \quad or \quad x + 1 = 0$
$3x = 5 \quad or \qquad x = -1$
$x = \dfrac{5}{3} \quad or \qquad x = -1$
The solutions are $\dfrac{5}{3}$ and -1.

41. $\qquad 2t^2 + 12t = -10$
$2t^2 + 12t + 10 = 0$
$2(t^2 + 6t + 5) = 0$
$2(t + 5)(t + 1) = 0$

$t + 5 = 0 \quad or \quad t + 1 = 0$
$t = -5 \quad or \qquad t = -1$
The solutions are -5 and -1.

43. $\qquad t(t - 5) = 14$
$t^2 - 5t = 14$
$t^2 - 5t - 14 = 0$
$(t + 2)(t - 7) = 0$
$t + 2 = 0 \quad or \quad t - 7 = 0$
$t = -2 \quad or \qquad t = 7$
The solutions are -2 and 7.

45. $\qquad t(9 + t) = 4(2t + 5)$
$9t + t^2 = 8t + 20$
$t^2 + t - 20 = 0$
$(t + 5)(t - 4) = 0$
$t + 5 = 0 \quad or \quad t - 4 = 0$
$t = -5 \quad or \qquad t = 4$
The solutions are -5 and 4.

47. $16(p - 1) = p(p + 8)$
$16p - 16 = p^2 + 8p$
$0 = p^2 - 8p + 16$
$0 = (p - 4)(p - 4)$
$p - 4 = 0 \quad or \quad p - 4 = 0$
$p = 4 \quad or \qquad p = 4$
The solution is 4.

49. $(t - 1)(t + 3) = t - 1$
$t^2 + 2t - 3 = t - 1$
$t^2 + t - 2 = 0$
$(t + 2)(t - 1) = 0$
$t + 2 = 0 \quad or \quad t - 1 = 0$
$t = -2 \quad or \qquad t = 1$
The solutions are -2 and 1.

51. $\dfrac{24}{x - 2} + \dfrac{24}{x + 2} = 5$
The LCM is $(x - 2)(x + 2)$.

$(x - 2)(x + 2)\left(\dfrac{24}{x - 2} + \dfrac{24}{x + 2}\right) =$
$(x - 2)(x + 2) \cdot 5$

$(x - 2)(x + 2) \cdot \dfrac{24}{x - 2} + (x - 2)(x + 2) \cdot \dfrac{24}{x + 2} =$
$5(x - 2)(x + 2)$
$24(x + 2) + 24(x - 2) =$
$5(x^2 - 4)$
$24x + 48 + 24x - 48 = 5x^2 - 20$
$48x = 5x^2 - 20$
$0 = 5x^2 - 48x - 20$
$0 = (5x + 2)(x - 10)$

$$5x + 2 = 0 \quad or \quad x - 10 = 0$$
$$5x = -2 \quad or \qquad x = 10$$
$$x = -\frac{2}{5} \quad or \qquad x = 10$$

Both numbers check. The solutions are $-\dfrac{2}{5}$ and 10.

53.
$$\frac{1}{x} + \frac{1}{x+6} = \frac{1}{4}$$

The LCM is $4x(x+6)$.

$$4x(x+6)\left(\frac{1}{x} + \frac{1}{x+6}\right) = 4x(x+6) \cdot \frac{1}{4}$$

$$4x(x+6) \cdot \frac{1}{x} + 4x(x+6) \cdot \frac{1}{x+6} = x(x+6)$$

$$4(x+6) + 4x = x(x+6)$$
$$4x + 24 + 4x = x^2 + 6x$$
$$8x + 24 = x^2 + 6x$$
$$0 = x^2 - 2x - 24$$
$$0 = (x-6)(x+4)$$

$$x - 6 = 0 \quad or \quad x + 4 = 0$$
$$x = 6 \quad or \qquad x = -4$$

Both numbers check. The solutions are 6 and -4.

55.
$$1 + \frac{12}{x^2 - 4} = \frac{3}{x - 2}$$

The LCM is $(x+2)(x-2)$.

$$(x+2)(x-2)\left(1 + \frac{12}{(x+2)(x-2)}\right) =$$
$$(x+2)(x-2) \cdot \frac{3}{x-2}$$

$$(x+2)(x-2) \cdot 1 + (x+2)(x-2) \cdot \frac{12}{(x+2)(x-2)} =$$
$$3(x+2)$$

$$x^2 - 4 + 12 = 3x + 6$$
$$x^2 + 8 = 3x + 6$$
$$x^2 - 3x + 2 = 0$$
$$(x-2)(x-1) = 0$$

$$x - 2 = 0 \quad or \quad x - 1 = 0$$
$$x = 2 \quad or \qquad x = 1$$

The number 1 checks, but 2 does not. (It makes the denominators $x^2 - 4$ and $x - 2$ zero.) The solution is 1.

57.
$$\frac{r}{r-1} + \frac{2}{r^2 - 1} = \frac{8}{r+1}$$

The LCM is $(r-1)(r+1)$.

$$(r-1)(r+1)\left(\frac{r}{r-1} + \frac{2}{(r-1)(r+1)}\right) =$$
$$(r-1)(r+1) \cdot \frac{8}{r+1}$$

$$(r-1)(r+1) \cdot \frac{r}{r-1} + (r-1)(r+1) \cdot \frac{2}{(r-1)(r+1)} =$$
$$8(r-1)$$

$$r(r+1) + 2 = 8(r-1)$$
$$r^2 + r + 2 = 8r - 8$$
$$r^2 - 7r + 10 = 0$$
$$(r-5)(r-2) = 0$$

$$r - 5 = 0 \quad or \quad r - 2 = 0$$
$$r = 5 \quad or \qquad r = 2$$

Both numbers check. The solutions are 5 and 2.

59.
$$\frac{x-1}{1-x} = -\frac{x+8}{x-8}$$

The LCM is $(1-x)(x-8)$.

$$(1-x)(x-8) \cdot \frac{x-1}{1-x} = (1-x)(x-8)\left(-\frac{x+8}{x-8}\right)$$
$$(x-8)(x-1) = -(1-x)(x+8)$$
$$x^2 - 9x + 8 = -(x + 8 - x^2 - 8x)$$
$$x^2 - 9x + 8 = -(-x^2 - 7x + 8)$$
$$x^2 - 9x + 8 = x^2 + 7x - 8$$
$$16 = 16x$$
$$1 = x$$

The number 1 does not check. (It makes the denominator $1 - x$ zero.) There is no solution.

61.
$$\frac{5}{y+4} - \frac{3}{y-2} = 4$$

The LCM is $(y+4)(y-2)$.

$$(y+4)(y-2)\left(\frac{5}{y+4} - \frac{3}{y-2}\right) = (y+4)(y-2) \cdot 4$$
$$5(y-2) - 3(y+4) = 4(y^2 + 2y - 8)$$
$$5y - 10 - 3y - 12 = 4y^2 + 8y - 32$$
$$2y - 22 = 4y^2 + 8y - 32$$
$$0 = 4y^2 + 6y - 10$$
$$0 = 2(2y^2 + 3y - 5)$$
$$0 = 2(2y + 5)(y - 1)$$

$$2y + 5 = 0 \quad or \quad y - 1 = 0$$
$$2y = -5 \quad or \qquad y = 1$$
$$y = -\frac{5}{2} \quad or \qquad y = 1$$

The solutions are $-\dfrac{5}{2}$ and 1.

63. Familiarize. We will use the formula
$$d = \frac{n^2 - 3n}{2},$$
where d is the number of diagonals and n is the number of sides.

Translate. We substitute 10 for n.
$$d = \frac{10^2 - 3 \cdot 10}{2}$$

Solve. We do the computation.
$$d = \frac{10^2 - 3 \cdot 10}{2} = \frac{100 - 30}{2} = \frac{70}{2} = 35$$

Check. We can recheck our computation. We can also substitute 35 for d in the original formula and determine whether this yields $n = 10$. Our result checks.

State. A decagon has 35 diagonals.

65. Familiarize. We will use the formula
$$d = \frac{n^2 - 3n}{2},$$
where d is the number of diagonals and n is the number of sides.

Translate. We substitute 14 for d.

$$14 = \frac{n^2 - 3n}{2}$$

Solve. We solve the equation.

$$\frac{n^2 - 3n}{2} = 14$$
$$n^2 - 3n = 28 \quad \text{Multiplying by 2}$$
$$n^2 - 3n - 28 = 0$$
$$(n-7)(n+4) = 0$$
$$n - 7 = 0 \ \text{ or } \ n + 4 = 0$$
$$n = 7 \ \text{ or } \quad n = -4$$

Check. Since the number of sides cannot be negative, -4 cannot be a solution. To check 7, we substitute 7 for n in the original formula and determine if this yields $d = 14$. Our result checks.

State. The polygon has 7 sides.

67. Discussion and Writing Exercise

69. $\sqrt{64} = 8$, taking the principal square root

71. $\sqrt{8} = \sqrt{4 \cdot 2} = \sqrt{4}\sqrt{2} = 2\sqrt{2}$

73. $\sqrt{20} = \sqrt{4 \cdot 5} = \sqrt{4}\sqrt{5} = 2\sqrt{5}$

75. $\sqrt{405} = \sqrt{81 \cdot 5} = \sqrt{81}\sqrt{5} = 9\sqrt{5}$

77. 2.646

79. 1.528

81.
$$4m^2 - (m+1)^2 = 0$$
$$4m^2 - (m^2 + 2m + 1) = 0$$
$$4m^2 - m^2 - 2m - 1 = 0$$
$$3m^2 - 2m - 1 = 0$$
$$(3m + 1)(m - 1) = 0$$
$$3m + 1 = 0 \ \text{ or } \ m - 1 = 0$$
$$3m = -1 \ \text{ or } \quad m = 1$$
$$m = -\frac{1}{3} \ \text{ or } \quad m = 1$$

The solutions are $-\frac{1}{3}$ and 1.

83.
$$\sqrt{5}x^2 - x = 0$$
$$x(\sqrt{5}x - 1) = 0$$
$$x = 0 \ \text{ or } \ \sqrt{5}x - 1 = 0$$
$$x = 0 \ \text{ or } \quad \sqrt{5}x = 1$$
$$x = 0 \ \text{ or } \quad x = \frac{1}{\sqrt{5}}, \text{ or } \frac{\sqrt{5}}{5}$$

The solutions are 0 and $\frac{\sqrt{5}}{5}$.

85. Graph $y_1 = 3x^2 - 7x$ and $y_2 = 20$. Then use the INTERSECT feature to find the first coordinate(s) of the point(s) of intersection. The solutions are 4 and approximately -1.7.

87. Graph $y_1 = 3x^2 + 8x$ and $y_2 = 12x + 15$. Then use the INTERSECT feature to find the first coordinate(s) of the point(s) of intersection. The solutions are 3 and approximately -1.7.

89. Graph $y_1 = (x-2)^2 + 3(x-2)$ and $y_2 = 4$. Then use the INTERSECT feature to find the first coordinate(s) of the point(s) of intersection. The solutions are -2 and 3.

91. Graph $y_1 = 16(x-1)$ and $y_2 = x(x+8)$. Then use the INTERSECT feature to find the first coordinate(s) of the point(s) of intersection. The solution is 4.

Exercise Set 17.2

1. $x^2 = 121$
$$x = 11 \ \text{ or } \ x = -11 \quad \text{Principle of square roots}$$
The solutions are 11 and -11.

3. $5x^2 = 35$
$$x^2 = 7 \quad \text{Dividing by 5}$$
$$x = \sqrt{7} \ \text{ or } \ x = -\sqrt{7} \quad \text{Principle of square roots}$$
The solutions are $\sqrt{7}$ and $-\sqrt{7}$.

5. $5x^2 = 3$
$$x^2 = \frac{3}{5}$$
$$x = \sqrt{\frac{3}{5}} \quad \text{or} \quad x = -\sqrt{\frac{3}{5}} \quad \text{Principle of square roots}$$
$$x = \sqrt{\frac{3}{5} \cdot \frac{5}{5}} \quad \text{or} \quad x = -\sqrt{\frac{3}{5} \cdot \frac{5}{5}} \quad \text{Rationalizing denominators}$$
$$x = \frac{\sqrt{15}}{5} \quad \text{or} \quad x = -\frac{\sqrt{15}}{5}$$
The solutions are $\frac{\sqrt{15}}{5}$ and $-\frac{\sqrt{15}}{5}$.

7. $4x^2 - 25 = 0$
$$4x^2 = 25$$
$$x^2 = \frac{25}{4}$$
$$x = \frac{5}{2} \ \text{ or } \ x = -\frac{5}{2}$$
The solutions are $\frac{5}{2}$ and $-\frac{5}{2}$.

9. $3x^2 - 49 = 0$
$$3x^2 = 49$$
$$x^2 = \frac{49}{3}$$
$$x = \frac{7}{\sqrt{3}} \quad \text{or} \quad x = -\frac{7}{\sqrt{3}}$$
$$x = \frac{7}{\sqrt{3}} \cdot \frac{\sqrt{3}}{\sqrt{3}} \quad \text{or} \quad x = -\frac{7}{\sqrt{3}} \cdot \frac{\sqrt{3}}{\sqrt{3}}$$
$$x = \frac{7\sqrt{3}}{3} \quad \text{or} \quad x = -\frac{7\sqrt{3}}{3}$$
The solutions are $\frac{7\sqrt{3}}{3}$ and $-\frac{7\sqrt{3}}{3}$.

11. $4y^2 - 3 = 9$

$\quad\;\; 4y^2 = 12$

$\quad\;\;\; y^2 = 3$

$y = \sqrt{3} \;\; or \;\; y = -\sqrt{3}$

The solutions are $\sqrt{3}$ and $-\sqrt{3}$.

13. $49y^2 - 64 = 0$

$\quad\;\; 49y^2 = 64$

$\quad\;\;\; y^2 = \dfrac{64}{49}$

$y = \dfrac{8}{7} \;\; or \;\; y = -\dfrac{8}{7}$

The solutions are $\dfrac{8}{7}$ and $-\dfrac{8}{7}$.

15. $(x + 3)^2 = 16$

$x + 3 = 4 \;\; or \;\; x + 3 = -4$ Principle of square roots

$\quad x = 1 \;\; or \;\;\;\;\;\;\;\; x = -7$

The solutions are 1 and -7.

17. $(x + 3)^2 = 21$

$x + 3 = \sqrt{21} \qquad or \;\; x + 3 = -\sqrt{21}$ Principle of square roots

$\quad\;\; x = -3 + \sqrt{21} \;\; or \qquad x = -3 - \sqrt{21}$

The solutions are $-3 + \sqrt{21}$ and $-3 - \sqrt{21}$, or $-3 \pm \sqrt{21}$.

19. $(x + 13)^2 = 8$

$x + 13 = \sqrt{8} \qquad\;\; or \;\; x + 13 = -\sqrt{8}$

$x + 13 = 2\sqrt{2} \qquad or \;\; x + 13 = -2\sqrt{2}$

$\quad\;\; x = -13 + 2\sqrt{2} \;\; or \qquad x = -13 - 2\sqrt{2}$

The solutions are $-13 + 2\sqrt{2}$ and $-13 - 2\sqrt{2}$, or $-13 \pm 2\sqrt{2}$.

21. $(x - 7)^2 = 12$

$x - 7 = \sqrt{12} \qquad or \;\; x - 7 = -\sqrt{12}$

$x - 7 = 2\sqrt{3} \qquad or \;\; x - 7 = -2\sqrt{3}$

$\quad\;\; x = 7 + 2\sqrt{3} \;\; or \qquad x = 7 - 2\sqrt{3}$

The solutions are $7 + 2\sqrt{3}$ and $7 - 2\sqrt{3}$, or $7 \pm 2\sqrt{3}$.

23. $(x + 9)^2 = 34$

$x + 9 = \sqrt{34} \qquad or \;\; x + 9 = -\sqrt{34}$

$\quad\;\; x = -9 + \sqrt{34} \;\; or \qquad x = -9 - \sqrt{34}$

The solutions are $-9 + \sqrt{34}$ and $-9 - \sqrt{34}$, or $-9 \pm \sqrt{34}$.

25. $\left(x + \dfrac{3}{2}\right)^2 = \dfrac{7}{2}$

$x + \dfrac{3}{2} = \sqrt{\dfrac{7}{2}} \qquad\qquad or \;\; x + \dfrac{3}{2} = -\sqrt{\dfrac{7}{2}}$

$\quad\;\; x = -\dfrac{3}{2} + \sqrt{\dfrac{7}{2}} \qquad or \qquad\; x = -\dfrac{3}{2} - \sqrt{\dfrac{7}{2}}$

$\quad\;\; x = -\dfrac{3}{2} + \sqrt{\dfrac{7}{2} \cdot \dfrac{2}{2}} \;\; or \qquad x = -\dfrac{3}{2} - \sqrt{\dfrac{7}{2} \cdot \dfrac{2}{2}}$

$\quad\;\; x = -\dfrac{3}{2} + \dfrac{\sqrt{14}}{2} \qquad or \qquad x = -\dfrac{3}{2} - \dfrac{\sqrt{14}}{2}$

$\quad\;\; x = \dfrac{-3 + \sqrt{14}}{2} \qquad or \qquad x = \dfrac{-3 - \sqrt{14}}{2}$

The solutions are $\dfrac{-3 \pm \sqrt{14}}{2}$.

27. $x^2 - 6x + 9 = 64$

$\quad\;\; (x - 3)^2 = 64$ Factoring the left side

$x - 3 = 8 \;\; or \;\; x - 3 = -8$ Principle of square roots

$\quad\;\; x = 11 \;\; or \qquad x = -5$

The solutions are 11 and -5.

29. $x^2 + 14x + 49 = 64$

$\quad\;\; (x + 7)^2 = 64$ Factoring the left side

$x + 7 = 8 \;\; or \;\; x + 7 = -8$ Principle of square roots

$\quad\;\; x = 1 \;\; or \qquad x = -15$

The solutions are 1 and -15.

31. $x^2 - 6x - 16 = 0$

$x^2 - 6x \qquad\; = 16$ Adding 16

$x^2 - 6x + \;\; 9 = 16 + 9$ Adding 9: $\left(\dfrac{-6}{2}\right)^2 = (-3)^2 = 9$

$\quad\;\; (x - 3)^2 = 25$

$x - 3 = 5 \;\; or \;\; x - 3 = -5$ Principle of square roots

$\quad\;\; x = 8 \;\; or \qquad x = -2$

The solutions are 8 and -2.

33. $x^2 + 22x + \;\; 21 = 0$

$x^2 + 22x \qquad\;\; = -21$ Subtracting 21

$x^2 + 22x + 121 = -21 + 121$ Adding 121: $\left(\dfrac{22}{2}\right)^2 = 11^2 = 121$

$\quad\;\; (x + 11)^2 = 100$

$x + 11 = 10 \;\; or \;\; x + 11 = -10$ Principle of square roots

$\quad\;\; x = -1 \;\; or \qquad x = -21$

The solutions are -1 and -21.

35. $x^2 - 2x - 5 = 0$

$x^2 - 2x \qquad = 5$

$x^2 - 2x + 1 = 5 + 1$ Adding 1: $\left(\dfrac{-2}{2}\right)^2 = (-1)^2 = 1$

$\quad\;\; (x - 1)^2 = 6$

$x - 1 = \sqrt{6} \;\; or \;\; x - 1 = -\sqrt{6}$

$\quad\;\; x = 1 + \sqrt{6} \;\; or \qquad x = 1 - \sqrt{6}$

The solutions are $1 \pm \sqrt{6}$.

37. $x^2 - 22x + 102 = 0$

$x^2 - 22x \qquad\;\; = -102$

$x^2 - 22x + 121 = -102 + 121$ Adding 121: $\left(\dfrac{-22}{2}\right)^2 = (-11)^2 = 121$

$\quad\;\; (x - 11)^2 = 19$

$x - 11 = \sqrt{19} \qquad or \;\; x - 11 = -\sqrt{19}$

$\quad\;\; x = 11 + \sqrt{19} \;\; or \qquad x = 11 - \sqrt{19}$

The solutions are $11 \pm \sqrt{19}$.

39. $x^2 + 10x - 4 = 0$
$x^2 + 10x \qquad = 4$
$x^2 + 10x + 25 = 4 + 25$ Adding 25: $\left(\dfrac{10}{2}\right)^2 =$
$\qquad\qquad\qquad\qquad 5^2 = 25$

$\qquad (x + 5)^2 = 29$

$x + 5 = \sqrt{29} \quad$ or $\quad x + 5 = -\sqrt{29}$
$\qquad x = -5 + \sqrt{29} \quad$ or $\qquad x = -5 - \sqrt{29}$

The solutions are $-5 \pm \sqrt{29}$.

41. $x^2 - 7x - 2 = 0$
$x^2 - 7x \qquad = 2$
$x^2 - 7x + \dfrac{49}{4} = 2 + \dfrac{49}{4}$ Adding $\dfrac{49}{4}$:
$\qquad\qquad\qquad\qquad \left(\dfrac{-7}{2}\right)^2 = \dfrac{49}{4}$

$\qquad \left(x - \dfrac{7}{2}\right)^2 = \dfrac{8}{4} + \dfrac{49}{4} = \dfrac{57}{4}$

$x - \dfrac{7}{2} = \dfrac{\sqrt{57}}{2} \quad$ or $\quad x - \dfrac{7}{2} = -\dfrac{\sqrt{57}}{2}$

$\qquad x = \dfrac{7}{2} + \dfrac{\sqrt{57}}{2} \quad$ or $\qquad x = \dfrac{7}{2} - \dfrac{\sqrt{57}}{2}$

$\qquad x = \dfrac{7 + \sqrt{57}}{2} \quad$ or $\qquad x = \dfrac{7 - \sqrt{57}}{2}$

The solutions are $\dfrac{7 \pm \sqrt{57}}{2}$.

43. $x^2 + 3x - 28 = 0$
$x^2 + 3x \qquad = 28$
$x^2 + 3x + \dfrac{9}{4} = 28 + \dfrac{9}{4}$ Adding $\dfrac{9}{4}$: $\left(\dfrac{3}{2}\right)^2 = \dfrac{9}{4}$

$\qquad \left(x + \dfrac{3}{2}\right)^2 = \dfrac{121}{4}$

$x + \dfrac{3}{2} = \dfrac{11}{2} \quad$ or $\quad x + \dfrac{3}{2} = -\dfrac{11}{2}$

$\qquad x = \dfrac{8}{2} \quad$ or $\qquad x = -\dfrac{14}{2}$

$\qquad x = 4 \quad$ or $\qquad x = -7$

The solutions are 4 and -7.

45. $x^2 + \dfrac{3}{2}x - \dfrac{1}{2} = 0$

$x^2 + \dfrac{3}{2}x \qquad = \dfrac{1}{2}$

$x^2 + \dfrac{3}{2}x + \dfrac{9}{16} = \dfrac{1}{2} + \dfrac{9}{16}$ Adding $\dfrac{9}{16}$: $\left(\dfrac{3/2}{2}\right)^2 =$
$\qquad\qquad\qquad\qquad\qquad\qquad \left(\dfrac{3}{4}\right)^2 = \dfrac{9}{16}$

$\qquad \left(x + \dfrac{3}{4}\right)^2 = \dfrac{17}{16}$

$x + \dfrac{3}{4} = \dfrac{\sqrt{17}}{4} \qquad$ or $\quad x + \dfrac{3}{4} = -\dfrac{\sqrt{17}}{4}$

$\qquad x = -\dfrac{3}{4} + \dfrac{\sqrt{17}}{4} \quad$ or $\qquad x = -\dfrac{3}{4} - \dfrac{\sqrt{17}}{4}$

$\qquad x = \dfrac{-3 + \sqrt{17}}{4} \quad$ or $\qquad x = \dfrac{-3 - \sqrt{17}}{4}$

The solutions are $\dfrac{-3 \pm \sqrt{17}}{4}$.

47. $2x^2 + 3x - 17 = 0$

$\dfrac{1}{2}(2x^2 + 3x - 17) = \dfrac{1}{2} \cdot 0$ Multiplying by $\dfrac{1}{2}$ to make the x^2-coefficient 1

$x^2 + \dfrac{3}{2}x - \dfrac{17}{2} = 0$

$x^2 + \dfrac{3}{2}x \qquad = \dfrac{17}{2}$

$x^2 + \dfrac{3}{2}x + \dfrac{9}{16} = \dfrac{17}{2} + \dfrac{9}{16}$ Adding $\dfrac{9}{16}$: $\left(\dfrac{3/2}{2}\right)^2 =$
$\qquad\qquad\qquad\qquad\qquad\qquad \left(\dfrac{3}{4}\right)^2 = \dfrac{9}{16}$

$\qquad \left(x + \dfrac{3}{4}\right)^2 = \dfrac{145}{16}$

$x + \dfrac{3}{4} = \dfrac{\sqrt{145}}{4} \qquad$ or $\quad x + \dfrac{3}{4} = -\dfrac{\sqrt{145}}{4}$

$\qquad x = \dfrac{-3 + \sqrt{145}}{4} \quad$ or $\qquad x = \dfrac{-3 - \sqrt{145}}{4}$

The solutions are $\dfrac{-3 \pm \sqrt{145}}{4}$.

49. $3x^2 + 4x - 1 = 0$

$\dfrac{1}{3}(3x^2 + 4x - 1) = \dfrac{1}{3} \cdot 0$

$x^2 + \dfrac{4}{3}x - \dfrac{1}{3} = 0$

$x^2 + \dfrac{4}{3}x \qquad = \dfrac{1}{3}$

$x^2 + \dfrac{4}{3}x + \dfrac{4}{9} = \dfrac{1}{3} + \dfrac{4}{9}$

$\qquad \left(x + \dfrac{2}{3}\right)^2 = \dfrac{7}{9}$

$x + \dfrac{2}{3} = \dfrac{\sqrt{7}}{3} \qquad$ or $\quad x + \dfrac{2}{3} = -\dfrac{\sqrt{7}}{3}$

$\qquad x = \dfrac{-2 + \sqrt{7}}{3} \quad$ or $\qquad x = -\dfrac{-2 - \sqrt{7}}{3}$

The solutions are $\dfrac{-2 \pm \sqrt{7}}{3}$.

51. $\qquad 2x^2 = 9x + 5$
$2x^2 - 9x - 5 = 0 \qquad$ Standard form

$\dfrac{1}{2}(2x^2 - 9x - 5) = \dfrac{1}{2} \cdot 0$

$x^2 - \dfrac{9}{2}x - \dfrac{5}{2} = 0$

$x^2 - \dfrac{9}{2}x \qquad = \dfrac{5}{2}$

$x^2 - \dfrac{9}{2}x + \dfrac{81}{16} = \dfrac{5}{2} + \dfrac{81}{16}$

$\qquad \left(x - \dfrac{9}{4}\right)^2 = \dfrac{121}{16}$

$x - \frac{9}{4} = \frac{11}{4}$ *or* $x - \frac{9}{4} = -\frac{11}{4}$

$x = \frac{20}{4}$ *or* $x = -\frac{2}{4}$

$x = 5$ *or* $x = -\frac{1}{2}$

The solutions are 5 and $-\frac{1}{2}$.

53.
$$6x^2 + 11x = 10$$
$$6x^2 + 11x - 10 = 0 \qquad \text{Standard form}$$
$$\frac{1}{6}(6x^2 + 11x - 10) = \frac{1}{6} \cdot 0$$
$$x^2 + \frac{11}{6}x - \frac{5}{3} = 0$$
$$x^2 + \frac{11}{6}x = \frac{5}{3}$$
$$x^2 + \frac{11}{6}x + \frac{121}{144} = \frac{5}{3} + \frac{121}{144}$$
$$\left(x + \frac{11}{12}\right)^2 = \frac{361}{144}$$

$x + \frac{11}{12} = \frac{19}{12}$ *or* $x + \frac{11}{12} = -\frac{19}{12}$

$x = \frac{8}{12}$ *or* $x = -\frac{30}{12}$

$x = \frac{2}{3}$ *or* $x = -\frac{5}{2}$

The solutions are $\frac{2}{3}$ and $-\frac{5}{2}$.

55. *Familiarize.* We will use the formula $s = 16t^2$.

Translate. We substitute 1483 for s.

$$1483 = 16t^2$$

Solve. We solve the equation.

$$1483 = 16t^2$$

$$\frac{1483}{16} = t^2 \qquad \text{Solving for } t^2$$

$$92.6875 = t^2 \qquad \text{Dividing}$$

$\sqrt{92.6875} = t$ *or* $-\sqrt{92.6875} = t$ Principle of square roots

$9.6 \approx t$ *or* $-9.6 \approx t$ Using a calculator and rounding to the nearest tenth

Check. The number -9.6 cannot be a solution, because time cannot be negative in this situation. We substitute 9.6 in the original equation.

$$s = 16(9.6)^2 = 16(92.16) = 1474.56$$

This is close. Remember that we approximated a solution. Thus we have a check.

State. It takes about 9.6 sec for an object to fall to the ground from the top of the Petronas Towers.

57. *Familiarize.* We will use the formula $s = 16t^2$.

Translate. We substitute 311 for s.

$$311 = 16t^2$$

Solve. We solve the equation.

$$311 = 16t^2$$

$$\frac{311}{16} = t^2 \qquad \text{Solving for } t^2$$

$$19.4375 = t^2 \qquad \text{Dividing}$$

$\sqrt{19.4375} = t$ *or* $-\sqrt{19.4375} = t$ Principle of square roots

$4.4 \approx t$ *or* $-4.4 \approx t$ Using a calculator and rounding to the nearest tenth

Check. The number -4.4 cannot be a solution, because time cannot be negative in this situation. We substitute 4.4 in the original equation.

$$s = 16(4.4)^2 = 16(19.36) = 309.76$$

This is close. Remember that we approximated a solution. Thus we have a check.

State. The fall took approximately 4.4 sec.

59. Discussion and Writing Exercise

61. The product rule asserts when multiplying with exponential notation, if the bases are the same, we keep the base and add the exponents.

63. The number -5 is not the principal square root of 25.

65. The quotient rule asserts that when dividing with exponential notation, if the bases are the same, we keep the base and subtract the exponent of the denominator from the exponent of the numerator.

67. The quotient rule for radicals asserts that for any nonnegative radicand A and positive number B, $\frac{\sqrt{A}}{\sqrt{B}} = \sqrt{\frac{A}{B}}$.

69. $x^2 + bx + 36$

The trinomial is a square if the square of one-half the x-coefficient is equal to 36. Thus we have:

$$\left(\frac{b}{2}\right)^2 = 36$$

$$\frac{b^2}{4} = 36$$

$$b^2 = 144$$

$b = 12$ *or* $b = -12$ Principle of square roots

71. $x^2 + bx + 128$

The trinomial is a square if the square of one-half the x-coefficient is equal to 128. Thus we have:

$$\left(\frac{b}{2}\right)^2 = 128$$

$$\frac{b^2}{4} = 128$$

$$b^2 = 512$$

$b = \sqrt{512}$ *or* $b = -\sqrt{512}$

$b = 16\sqrt{2}$ *or* $b = -16\sqrt{2}$

73. $x^2 + bx + c$

The trinomial is a square if the square of one-half the x-coefficient is equal to c. Thus we have:

$$\left(\frac{b}{2}\right)^2 = c$$
$$\frac{b^2}{4} = c$$
$$b^2 = 4c$$

$b = \sqrt{4c} \quad or \quad b = -\sqrt{4c}$
$b = 2\sqrt{c} \quad or \quad b = -2\sqrt{c}$

75. $4.82x^2 = 12{,}000$

$$x^2 = \frac{12{,}000}{4.82}$$

$x = \sqrt{\dfrac{12{,}000}{4.82}} \quad or \quad x = -\sqrt{\dfrac{12{,}000}{4.82}}$ Principle of square roots

$x \approx 49.896 \quad or \quad x \approx -49.896$ Using a calculator and rounding

The solutions are approximately 49.896 and -49.896.

77. $\dfrac{x}{9} = \dfrac{36}{4x}$, LCM is $36x$

$$36x \cdot \frac{x}{9} = 36x \cdot \frac{36}{4x} \qquad \text{Multiplying by } 36x$$
$$4x^2 = 324$$
$$x^2 = 81$$

$x = 9 \quad or \quad x = -9$

Both numbers check. The solutions are 9 and -9.

Exercise Set 17.3

1. $\quad x^2 - 4x = 21$
$x^2 - 4x - 21 = 0 \qquad$ Standard form

We can factor.
$x^2 - 4x - 21 = 0$
$(x - 7)(x + 3) = 0$

$x - 7 = 0 \quad or \quad x + 3 = 0$
$\quad x = 7 \quad or \qquad x = -3$

The solutions are 7 and -3.

3. $\qquad x^2 = 6x - 9$
$x^2 - 6x + 9 = 0 \qquad$ Standard form

We can factor.
$\quad x^2 - 6x + 9 = 0$
$(x - 3)(x - 3) = 0$

$x - 3 = 0 \quad or \quad x - 3 = 0$
$\quad x = 3 \quad or \qquad x = 3$

The solution is 3.

5. $3y^2 - 2y - 8 = 0$

We can factor.
$3y^2 - 2y - 8 = 0$
$(3y + 4)(y - 2) = 0$

$3y + 4 = 0 \quad or \quad y - 2 = 0$
$\quad 3y = -4 \quad or \qquad y = 2$

$$y = -\frac{4}{3} \quad or \qquad y = 2$$

The solutions are $-\dfrac{4}{3}$ and 2.

7. $\qquad 4x^2 + 4x = 15$
$4x^2 + 4x - 15 = 0 \qquad$ Standard form

We can factor.
$\quad 4x^2 + 4x - 15 = 0$
$(2x - 3)(2x + 5) = 0$

$2x - 3 = 0 \quad or \quad 2x + 5 = 0$
$\quad 2x = 3 \quad or \qquad 2x = -5$
$$x = \frac{3}{2} \quad or \qquad x = -\frac{5}{2}$$

The solutions are $\dfrac{3}{2}$ and $-\dfrac{5}{2}$.

9. $\qquad x^2 - 9 = 0 \qquad$ Difference of squares
$(x + 3)(x - 3) = 0$

$x + 3 = 0 \quad or \quad x - 3 = 0$
$\quad x = -3 \quad or \qquad x = 3$

The solutions are -3 and 3.

11. $x^2 - 2x - 2 = 0$

$a = 1, \; b = -2, \; c = -2$

We use the quadratic formula.
$$x = \frac{-(-2) \pm \sqrt{(-2)^2 - 4 \cdot 1 \cdot (-2)}}{2 \cdot 1}$$
$$x = \frac{2 \pm \sqrt{4 + 8}}{2}$$
$$x = \frac{2 \pm \sqrt{12}}{2} = \frac{2 \pm \sqrt{4 \cdot 3}}{2}$$
$$x = \frac{2 \pm 2\sqrt{3}}{2} = \frac{2(1 \pm \sqrt{3})}{2}$$
$$x = 1 \pm \sqrt{3}$$

The solutions are $1 + \sqrt{3}$ and $1 - \sqrt{3}$, or $1 \pm \sqrt{3}$.

13. $y^2 - 10y + 22 = 0$

$a = 1, \; b = -10, \; c = 22$

We use the quadratic formula.
$$y = \frac{-(-10) \pm \sqrt{(-10)^2 - 4 \cdot 1 \cdot 22}}{2 \cdot 1}$$
$$y = \frac{10 \pm \sqrt{100 - 88}}{2}$$
$$y = \frac{10 \pm \sqrt{12}}{2} = \frac{10 \pm \sqrt{4 \cdot 3}}{2}$$
$$y = \frac{10 \pm 2\sqrt{3}}{2} = \frac{2(5 \pm \sqrt{3})}{2}$$
$$y = 5 \pm \sqrt{3}$$

The solutions are $5 + \sqrt{3}$ and $5 - \sqrt{3}$, or $5 \pm \sqrt{3}$.

15. $x^2 + 4x + 4 = 7$

$x^2 + 4x - 3 = 0$ Adding -7 to get standard form

$a = 1,\ b = 4,\ c = -3$

We use the quadratic formula.

$$x = \frac{-4 \pm \sqrt{4^2 - 4 \cdot 1 \cdot (-3)}}{2 \cdot 1} = \frac{-4 \pm \sqrt{16 + 12}}{2}$$

$$x = \frac{-4 \pm \sqrt{28}}{2} = \frac{-4 \pm \sqrt{4 \cdot 7}}{2}$$

$$x = \frac{-4 \pm 2\sqrt{7}}{2} = \frac{2(-2 \pm \sqrt{7})}{2}$$

$$x = -2 \pm \sqrt{7}$$

The solutions are $-2 + \sqrt{7}$ and $-2 - \sqrt{7}$, or $-2 \pm \sqrt{7}$.

17. $3x^2 + 8x + 2 = 0$

$a = 3,\ b = 8,\ c = 2$

We use the quadratic formula.

$$x = \frac{-8 \pm \sqrt{8^2 - 4 \cdot 3 \cdot 2}}{2 \cdot 3} = \frac{-8 \pm \sqrt{64 - 24}}{6}$$

$$x = \frac{-8 \pm \sqrt{40}}{6} = \frac{-8 \pm \sqrt{4 \cdot 10}}{6}$$

$$x = \frac{-8 \pm 2\sqrt{10}}{6} = \frac{2(-4 \pm \sqrt{10})}{2 \cdot 3}$$

$$x = \frac{-4 \pm \sqrt{10}}{3}$$

The solutions are $\dfrac{-4 + \sqrt{10}}{3}$ and $\dfrac{-4 - \sqrt{10}}{3}$, or $\dfrac{-4 \pm \sqrt{10}}{3}$.

19. $2x^2 - 5x = 1$

$2x^2 - 5x - 1 = 0$ Adding -1 to get standard form

$a = 2,\ b = -5,\ c = -1$

We use the quadratic formula.

$$x = \frac{-(-5) \pm \sqrt{(-5)^2 - 4 \cdot 2 \cdot (-1)}}{2 \cdot 2} = \frac{5 \pm \sqrt{25 + 8}}{4}$$

$$x = \frac{5 \pm \sqrt{33}}{4}$$

The solutions are $\dfrac{5 + \sqrt{33}}{4}$ and $\dfrac{5 - \sqrt{33}}{4}$, or $\dfrac{5 \pm \sqrt{33}}{4}$.

21. $2y^2 - 2y - 1 = 0$

$a = 2,\ b = -2,\ c = -1$

We use the quadratic formula.

$$y = \frac{-(-2) \pm \sqrt{(-2)^2 - 4 \cdot 2 \cdot (-1)}}{2 \cdot 2} = \frac{2 \pm \sqrt{4 + 8}}{4}$$

$$y = \frac{2 \pm \sqrt{12}}{4} = \frac{2 \pm \sqrt{4 \cdot 3}}{4}$$

$$y = \frac{2 \pm 2\sqrt{3}}{4} = \frac{2(1 \pm \sqrt{3})}{2 \cdot 2}$$

$$y = \frac{1 \pm \sqrt{3}}{2}$$

The solutions are $\dfrac{1 + \sqrt{3}}{2}$ and $\dfrac{1 - \sqrt{3}}{2}$, or $\dfrac{1 \pm \sqrt{3}}{2}$.

23. $2t^2 + 6t + 5 = 0$

$a = 2,\ b = 6,\ c = 5$

We use the quadratic formula.

$$t = \frac{-6 \pm \sqrt{6^2 - 4 \cdot 2 \cdot 5}}{2 \cdot 2} = \frac{-6 \pm \sqrt{36 - 40}}{4}$$

$$t = \frac{-6 \pm \sqrt{-4}}{4}$$

Since square roots of negative numbers do not exist as real numbers, there are no real-number solutions.

25.
$$3x^2 = 5x + 4$$
$$3x^2 - 5x - 4 = 0$$

$a = 3,\ b = -5,\ c = -4$

We use the quadratic formula.

$$x = \frac{-(-5) \pm \sqrt{(-5)^2 - 4 \cdot 3 \cdot (-4)}}{2 \cdot 3} = \frac{5 \pm \sqrt{25 + 48}}{6}$$

$$x = \frac{5 \pm \sqrt{73}}{6}$$

The solutions are $\dfrac{5 + \sqrt{73}}{6}$ and $\dfrac{5 - \sqrt{73}}{6}$, or $\dfrac{5 \pm \sqrt{73}}{6}$.

27.
$$2y^2 - 6y = 10$$
$$2y^2 - 6y - 10 = 0$$
$$y^2 - 3y - 5 = 0 \quad \text{Multiplying by } \tfrac{1}{2} \text{ to simplify}$$

$a = 1,\ b = -3,\ c = -5$

We use the quadratic formula.

$$y = \frac{-(-3) \pm \sqrt{(-3)^2 - 4 \cdot 1 \cdot (-5)}}{2 \cdot 1} = \frac{3 \pm \sqrt{9 + 20}}{2}$$

$$y = \frac{3 \pm \sqrt{29}}{2}$$

The solutions are $\dfrac{3 + \sqrt{29}}{2}$ and $\dfrac{3 - \sqrt{29}}{2}$, or $\dfrac{3 \pm \sqrt{29}}{2}$.

29.
$$\frac{x^2}{x + 3} - \frac{5}{x + 3} = 0, \quad \text{LCM is } x + 3$$

$$(x + 3)\left(\frac{x^2}{x + 3} - \frac{5}{x + 3}\right) = (x + 3) \cdot 0$$

$$x^2 - 5 = 0$$
$$x^2 = 5$$

$x = \sqrt{5} \ \text{ or } \ x = -\sqrt{5}$ Principle of square roots

Both numbers check. The solutions are $\sqrt{5}$ and $-\sqrt{5}$, or $\pm\sqrt{5}$.

31.
$$x + 2 = \frac{3}{x + 2}$$

$$(x + 2)(x + 2) = (x + 2) \cdot \frac{3}{x + 2} \quad \text{Clearing the fraction}$$

$$x^2 + 4x + 4 = 3$$
$$x^2 + 4x + 1 = 0$$

$a = 1,\ b = 4,\ c = 1$

We use the quadratic formula.

$$x = \frac{-4 \pm \sqrt{4^2 - 4 \cdot 1 \cdot 1}}{2 \cdot 1} = \frac{-4 \pm \sqrt{16 - 4}}{2}$$

$$x = \frac{-4 \pm \sqrt{12}}{2} = \frac{-4 \pm \sqrt{4 \cdot 3}}{2}$$

$$x = \frac{-4 \pm 2\sqrt{3}}{2} = \frac{2(-2 \pm \sqrt{3})}{2}$$

$$x = -2 \pm \sqrt{3}$$

Both numbers check. The solutions are $-2 + \sqrt{3}$ and $-2 - \sqrt{3}$, or $-2 \pm \sqrt{3}$.

33. $\quad \dfrac{1}{x} + \dfrac{1}{x + 1} = \dfrac{1}{3}, \quad$ LCM is $3x(x + 1)$

$$3x(x+1)\left(\frac{1}{x} + \frac{1}{x+1}\right) = 3x(x+1) \cdot \frac{1}{3}$$

$$3(x + 1) + 3x = x(x + 1)$$
$$3x + 3 + 3x = x^2 + x$$
$$6x + 3 = x^2 + x$$
$$0 = x^2 - 5x - 3$$

$a = 1,\ b = -5,\ c = -3$

We use the quadratic formula.

$$x = \frac{-(-5) \pm \sqrt{(-5)^2 - 4 \cdot 1 \cdot (-3)}}{2 \cdot 1} = \frac{5 \pm \sqrt{25 + 12}}{2}$$

$$x = \frac{5 \pm \sqrt{37}}{2}$$

The solutions are $\dfrac{5 + \sqrt{37}}{2}$ and $\dfrac{5 - \sqrt{37}}{2}$, or $\dfrac{5 \pm \sqrt{37}}{2}$.

35. $x^2 - 4x - 7 = 0$

$a = 1,\ b = -4,\ c = -7$

$$x = \frac{-(-4) \pm \sqrt{(-4)^2 - 4 \cdot 1 \cdot (-7)}}{2 \cdot 1}$$

$$x = \frac{4 \pm \sqrt{16 + 28}}{2} = \frac{4 \pm \sqrt{44}}{2}$$

$$x = \frac{4 \pm \sqrt{4 \cdot 11}}{2} = \frac{4 \pm 2\sqrt{11}}{2}$$

$$x = \frac{2(2 \pm \sqrt{11})}{2} = 2 \pm \sqrt{11}$$

Using a calculator, we have:

$2 + \sqrt{11} \approx 5.31662479 \approx 5.3$, and

$2 - \sqrt{11} \approx -1.31662479 \approx -1.3$.

The approximate solutions, to the nearest tenth, are 5.3 and -1.3.

37. $y^2 - 6y - 1 = 0$

$a = 1,\ b = -6,\ c = -1$

$$y = \frac{-(-6) \pm \sqrt{(-6)^2 - 4 \cdot 1 \cdot (-1)}}{2 \cdot 1}$$

$$y = \frac{6 \pm \sqrt{36 + 4}}{2} = \frac{6 \pm \sqrt{40}}{2}$$

$$y = \frac{6 \pm \sqrt{4 \cdot 10}}{2} = \frac{6 \pm 2\sqrt{10}}{2}$$

$$y = \frac{2(3 \pm \sqrt{10})}{2} = 3 \pm \sqrt{10}$$

Using a calculator, we have:

$3 + \sqrt{10} \approx 6.16227766 \approx 6.2$ and

$3 - \sqrt{10} \approx -0.1622776602 \approx -0.2$.

The approximate solutions, to the nearest tenth, are 6.2 and -0.2.

39. $\qquad 4x^2 + 4x = 1$

$4x^2 + 4x - 1 = 0 \qquad$ Standard form

$a = 4,\ b = 4,\ c = -1$

$$x = \frac{-4 \pm \sqrt{4^2 - 4 \cdot 4 \cdot (-1)}}{2 \cdot 4}$$

$$x = \frac{-4 \pm \sqrt{16 + 16}}{8} = \frac{-4 \pm \sqrt{32}}{8}$$

$$x = \frac{-4 \pm \sqrt{16 \cdot 2}}{8} = \frac{-4 \pm 4\sqrt{2}}{8}$$

$$x = \frac{4(-1 \pm \sqrt{2})}{4 \cdot 2} = \frac{-1 \pm \sqrt{2}}{2}$$

Using a calculator, we have:

$\dfrac{-1 + \sqrt{2}}{2} \approx 0.2071067812 \approx 0.2$ and

$\dfrac{-1 - \sqrt{2}}{2} \approx -1.207106781 \approx -1.2$.

The approximate solutions, to the nearest tenth, are 0.2 and -1.2.

41. $3x^2 - 8x + 2 = 0$

$a = 3,\ b = -8,\ c = 2$

$$x = \frac{-(-8) \pm \sqrt{(-8)^2 - 4 \cdot 3 \cdot 2}}{2 \cdot 3}$$

$$x = \frac{8 \pm \sqrt{64 - 24}}{6} = \frac{8 \pm \sqrt{40}}{6}$$

$$x = \frac{8 \pm \sqrt{4 \cdot 10}}{6} = \frac{8 \pm 2\sqrt{10}}{6}$$

$$x = \frac{2(4 \pm \sqrt{10})}{2 \cdot 3} = \frac{4 \pm \sqrt{10}}{3}$$

Using a calculator, we have:

$\dfrac{4 + \sqrt{10}}{3} \approx 2.387425887 \approx 2.4$ and

$\dfrac{4 - \sqrt{10}}{3} \approx 0.2792407799 \approx 0.3$.

The approximate solutions, to the nearest tenth, are 2.4 and 0.3.

43. Discussion and Writing Exercise

45. $\sqrt{40} - 2\sqrt{10} + \sqrt{90} = \sqrt{4 \cdot 10} - 2\sqrt{10} + \sqrt{9 \cdot 10}$
$\qquad\qquad = \sqrt{4}\sqrt{10} - 2\sqrt{10} + \sqrt{9}\sqrt{10}$
$\qquad\qquad = 2\sqrt{10} - 2\sqrt{10} + 3\sqrt{10}$
$\qquad\qquad = (2 - 2 + 3)\sqrt{10}$
$\qquad\qquad = 3\sqrt{10}$

47. $\sqrt{18} + \sqrt{50} - 3\sqrt{8} = \sqrt{9 \cdot 2} + \sqrt{25 \cdot 2} - 3\sqrt{4 \cdot 2}$

$\qquad\qquad\qquad\quad = \sqrt{9}\sqrt{2} + \sqrt{25}\sqrt{2} - 3\sqrt{4}\sqrt{2}$

$\qquad\qquad\qquad\quad = 3\sqrt{2} + 5\sqrt{2} - 3 \cdot 2\sqrt{2}$

$\qquad\qquad\qquad\quad = 3\sqrt{2} + 5\sqrt{2} - 6\sqrt{2}$

$\qquad\qquad\qquad\quad = (3 + 5 - 6)\sqrt{2}$

$\qquad\qquad\qquad\quad = 2\sqrt{2}$

49. $\sqrt{80} = \sqrt{16 \cdot 5} = \sqrt{16}\sqrt{5} = 4\sqrt{5}$

51. $\sqrt{9000x^{10}} = \sqrt{900 \cdot 10 \cdot x^{10}} = \sqrt{900}\sqrt{x^{10}}\sqrt{10} = 30x^5\sqrt{10}$

53.
$\qquad y = \dfrac{k}{x}$ Inverse variation

$\qquad 235 = \dfrac{k}{0.6}$ Substituting 0.6 for x and 235 for y

$\qquad 141 = k$ Constant of variation

$\qquad y = \dfrac{141}{x}$ Equation of variation

55. $5x + x(x - 7) = 0$

$\qquad 5x + x^2 - 7x = 0$

$\qquad\quad x^2 - 2x = 0$ We can factor.

$\qquad\quad x(x - 2) = 0$

$\qquad x = 0 \;\; or \;\; x - 2 = 0$

$\qquad x = 0 \;\; or \qquad x = 2$

The solutions are 0 and 2.

57. $3 - x(x - 3) = 4$

$\qquad 3 - x^2 + 3x = 4$

$\qquad\quad 0 = x^2 - 3x + 1$ Standard form

$a = 1, \; b = -3, \; c = 1$

We use the quadratic formula.

$x = \dfrac{-(-3) \pm \sqrt{(-3)^2 - 4 \cdot 1 \cdot 1}}{2 \cdot 1} = \dfrac{3 \pm \sqrt{9 - 4}}{2}$

$x = \dfrac{3 \pm \sqrt{5}}{2}$

The solutions are $\dfrac{3 + \sqrt{5}}{2}$ and $\dfrac{3 - \sqrt{5}}{2}$, or $\dfrac{3 \pm \sqrt{5}}{2}$.

59. $(y + 4)(y + 3) = 15$

$\qquad y^2 + 7y + 12 = 15$

$\qquad y^2 + 7y - 3 = 0$ Standard form

$a = 1, \; b = 7, \; c = -3$

We use the quadratic formula.

$y = \dfrac{-7 \pm \sqrt{7^2 - 4 \cdot 1 \cdot (-3)}}{2 \cdot 1} = \dfrac{-7 \pm \sqrt{49 + 12}}{2}$

$y = \dfrac{-7 \pm \sqrt{61}}{2}$

The solutions are $\dfrac{-7 + \sqrt{61}}{2}$ and $\dfrac{-7 - \sqrt{61}}{2}$, or

$\dfrac{-7 \pm \sqrt{61}}{2}$.

61. $x^2 + (x + 2)^2 = 7$

$x^2 + x^2 + 4x + 4 = 7$

$\qquad 2x^2 + 4x + 4 = 7$

$\qquad 2x^2 + 4x - 3 = 0$ Standard form

$a = 2, \; b = 4, \; c = -3$

We use the quadratic formula.

$x = \dfrac{-4 \pm \sqrt{4^2 - 4 \cdot 2 \cdot (-3)}}{2 \cdot 2} = \dfrac{-4 \pm \sqrt{16 + 24}}{4}$

$x = \dfrac{-4 \pm \sqrt{40}}{4} = \dfrac{-4 \pm \sqrt{4 \cdot 10}}{4}$

$x = \dfrac{-4 \pm 2\sqrt{10}}{4} = \dfrac{2(-2 \pm \sqrt{10})}{2 \cdot 2}$

$x = \dfrac{-2 \pm \sqrt{10}}{2}$

The solutions are $\dfrac{-2 + \sqrt{10}}{2}$ and $\dfrac{-2\sqrt{10}}{2}$, or $\dfrac{-2 \pm \sqrt{10}}{2}$.

63.–69. Left to the student

Exercise Set 17.4

1.
$\qquad q = \dfrac{VQ}{I}$

$\qquad I \cdot q = I \cdot \dfrac{VQ}{I}$ Multiplying by I

$\qquad Iq = VQ$ Simplifying

$\qquad I = \dfrac{VQ}{q}$ Dividing by q

3.
$\qquad S = \dfrac{kmM}{d^2}$

$\qquad d^2 \cdot S = d^2 \cdot \dfrac{kmM}{d^2}$ Multiplying by d^2

$\qquad d^2 S = kmM$ Simplifying

$\qquad \dfrac{d^2 S}{kM} = m$ Dividing by kM

5.
$\qquad S = \dfrac{kmM}{d^2}$

$\qquad d^2 \cdot S = d^2 \cdot \dfrac{kmM}{d^2}$ Multiplying by d^2

$\qquad d^2 S = kmM$ Simplifying

$\qquad d^2 = \dfrac{kmM}{S}$ Dividing by S

7.
$\qquad T = \dfrac{10t}{W^2}$

$\qquad W^2 \cdot T = W^2 \cdot \dfrac{10t}{W^2}$ Multiplying by W^2

$\qquad W^2 T = 10t$

$\qquad W^2 = \dfrac{10t}{T}$ Dividing by T

$\qquad W = \sqrt{\dfrac{10t}{T}}$ Principle of square roots. Assume W is nonnegative.

9.
$\qquad A = at + bt$

$\qquad A = t(a + b)$ Factoring

$\qquad \dfrac{A}{a + b} = t$ Dividing by $a + b$

11.
$$y = ax + bx + c$$
$$y - c = ax + bx \qquad \text{Subtracting } c$$
$$y - c = x(a + b) \qquad \text{Factoring}$$
$$\frac{y - c}{a + b} = x \qquad \text{Dividing by } a + b$$

13.
$$\frac{t}{a} + \frac{t}{b} = 1$$
$$ab\left(\frac{t}{a} + \frac{t}{b}\right) = ab \cdot 1 \qquad \text{Multiplying by } ab$$
$$ab \cdot \frac{t}{a} + ab \cdot \frac{t}{b} = ab$$
$$bt + at = ab$$
$$bt = ab - at \qquad \text{Subtracting } at$$
$$bt = a(b - t) \qquad \text{Factoring}$$
$$\frac{bt}{b - t} = a \qquad \text{Dividing by } b - t$$

15.
$$\frac{1}{p} + \frac{1}{q} = \frac{1}{f}$$
$$pqf\left(\frac{1}{p} + \frac{1}{q}\right) = pqf \cdot \frac{1}{f} \qquad \text{Multiplying by } pqf$$
$$pqf \cdot \frac{1}{p} + pqf \cdot \frac{1}{q} = pq$$
$$qf + pf = pq$$
$$qf = pq - pf \qquad \text{Subtracting } pf$$
$$qf = p(q - f) \qquad \text{Factoring}$$
$$\frac{qf}{q - f} = p \qquad \text{Dividing by } q - f$$

17.
$$A = \frac{1}{2}bh$$
$$2 \cdot A = 2 \cdot \frac{1}{2}bh \qquad \text{Multiplying by } 2$$
$$2A = bh$$
$$\frac{2A}{h} = b \qquad \text{Dividing by } h$$

19.
$$S = 2\pi r(r + h)$$
$$S = 2\pi r^2 + 2\pi rh \qquad \text{Removing parentheses}$$
$$S - 2\pi r^2 = 2\pi rh \qquad \text{Subtracting } 2\pi r^2$$
$$\frac{S - 2\pi r^2}{2\pi r} = h, \text{ or} \qquad \text{Dividing by } 2\pi r$$
$$\frac{S}{2\pi r} - r = h$$

21.
$$\frac{1}{R} = \frac{1}{r_1} + \frac{1}{r_2}$$
$$Rr_1r_2 \cdot \frac{1}{R} = Rr_1r_2\left(\frac{1}{r_1} + \frac{1}{r_2}\right) \quad \text{Multiplying by } Rr_1r_2$$
$$r_1r_2 = Rr_1r_2 \cdot \frac{1}{r_1} + Rr_1r_2 \cdot \frac{1}{r_2}$$
$$r_1r_2 = Rr_2 + Rr_1$$
$$r_1r_2 = R(r_2 + r_1) \qquad \text{Factoring}$$
$$\frac{r_1r_2}{r_2 + r_1} = R \qquad \text{Dividing by } r_2 + r_1$$

23.
$$P = 17\sqrt{Q}$$
$$\frac{P}{17} = \sqrt{Q} \qquad \text{Isolating the radical}$$
$$\left(\frac{P}{17}\right)^2 = (\sqrt{Q})^2 \qquad \text{Principle of squaring}$$
$$\frac{P^2}{289} = Q \qquad \text{Simplifying}$$

25.
$$v = \sqrt{\frac{2gE}{m}}$$
$$v^2 = \left(\sqrt{\frac{2gE}{m}}\right)^2 \qquad \text{Principle of squaring}$$
$$v^2 = \frac{2gE}{m}$$
$$mv^2 = 2gE \qquad \text{Multipying by } m$$
$$\frac{mv^2}{2g} = E \qquad \text{Dividing by } 2g$$

27.
$$S = 4\pi r^2$$
$$\frac{S}{4\pi} = r^2 \qquad \text{Dividing by } 4\pi$$
$$\sqrt{\frac{S}{4\pi}} = r \qquad \begin{array}{l}\text{Principle of square roots.}\\ \text{Assume } r \text{ is nonnegative.}\end{array}$$
$$\sqrt{\frac{1}{4} \cdot \frac{S}{\pi}} = r$$
$$\frac{1}{2}\sqrt{\frac{S}{\pi}} = r$$

29. $P = kA^2 + mA$
$$0 = kA^2 + mA - P \qquad \text{Standard form}$$
$$a = k, \ b = m, \ c = -P$$
$$A = \frac{-b \pm \sqrt{b^2 - 4ac}}{2a} \qquad \text{Quadratic formula}$$
$$A = \frac{-m \pm \sqrt{m^2 - 4 \cdot k \cdot (-P)}}{2 \cdot k} \qquad \text{Substituting}$$
$$A = \frac{-m + \sqrt{m^2 + 4kP}}{2k} \qquad \text{Using the positive root}$$

31.
$$c^2 = a^2 + b^2$$
$$c^2 - b^2 = a^2$$
$$\sqrt{c^2 - b^2} = a \qquad \begin{array}{l}\text{Principle of square roots.}\\ \text{Assume } a \text{ is nonnegative.}\end{array}$$

33.
$$s = 16t^2$$
$$\frac{s}{16} = t^2$$
$$\sqrt{\frac{s}{16}} = t \qquad \begin{array}{l}\text{Principle of square roots.}\\ \text{Assume } t \text{ is nonnegative.}\end{array}$$
$$\frac{\sqrt{s}}{4} = t$$

35. $A = \pi r^2 + 2\pi rh$

$0 = \pi r^2 + 2\pi hr - A$

$a = \pi, \ b = 2\pi h, \ c = -A$

$r = \dfrac{-b \pm \sqrt{b^2 - 4ac}}{2a}$

$r = \dfrac{-2\pi h \pm \sqrt{(2\pi h)^2 - 4 \cdot \pi \cdot (-A)}}{2 \cdot \pi}$

$r = \dfrac{-2\pi h + \sqrt{4\pi^2 h^2 + 4\pi A}}{2\pi}$ Using the positive root

$r = \dfrac{-2\pi h + \sqrt{4(\pi^2 h^2 + \pi A)}}{2\pi}$

$r = \dfrac{-2\pi h + 2\sqrt{\pi^2 h^2 + \pi A}}{2\pi}$

$r = \dfrac{2\left(-\pi h + \sqrt{\pi^2 h^2 + \pi A}\right)}{2\pi}$

$r = \dfrac{-\pi h + \sqrt{\pi^2 h^2 + \pi A}}{\pi}$

37.

$F = \dfrac{Av^2}{400}$

$400F = Av^2$ Multiplying by 400

$\dfrac{400F}{A} = v^2$ Dividing by A

$\sqrt{\dfrac{400F}{A}} = v$ Principle of square roots. Assume v is nonnegative.

$\sqrt{400 \cdot \dfrac{F}{A}} = v$

$20\sqrt{\dfrac{F}{a}} = v$

39. $c = \sqrt{a^2 + b^2}$

$c^2 = (\sqrt{a^2 + b^2})^2$ Principle of squaring

$c^2 = a^2 + b^2$

$c^2 - b^2 = a^2$

$\sqrt{c^2 - b^2} = a$ Principle of square roots. Assume a is nonnegative.

41. $h = \dfrac{a}{2}\sqrt{3}$

$2h = a\sqrt{3}$

$\dfrac{2h}{\sqrt{3}} = a$

$\dfrac{2h\sqrt{3}}{3} = a$ Rationalizing the denominator

43. $n = aT^2 - 4T + m$

$0 = aT^2 - 4T + m - n$

$a = a, \ b = -4, \ c = m - n$

$T = \dfrac{-b \pm \sqrt{b^2 - 4ac}}{2a}$

$T = \dfrac{-(-4) \pm \sqrt{(-4)^2 - 4 \cdot a \cdot (m - n)}}{2 \cdot a}$

$T = \dfrac{4 + \sqrt{16 - 4a(m - n)}}{2a}$ Using the positive root

$T = \dfrac{4 + \sqrt{4[4 - a(m - n)]}}{2a}$

$T = \dfrac{4 + 2\sqrt{4 - a(m - n)}}{2a}$

$T = \dfrac{2\left(2 + \sqrt{4 - a(m - n)}\right)}{2 \cdot a}$

$T = \dfrac{2 + \sqrt{4 - a(m - n)}}{a}$

45.

$v = 2\sqrt{\dfrac{2kT}{\pi m}}$

$\dfrac{v}{2} = \sqrt{\dfrac{2kT}{\pi m}}$ Isolating the radical

$\left(\dfrac{v}{2}\right)^2 = \left(\sqrt{\dfrac{2kT}{\pi m}}\right)^2$ Principle of squaring

$\dfrac{v^2}{4} = \dfrac{2kT}{\pi m}$

$\dfrac{v^2}{4} \cdot \dfrac{\pi m}{2k} = \dfrac{2kT}{\pi m} \cdot \dfrac{\pi m}{2k}$ Multiplying by $\dfrac{\pi m}{2k}$

$\dfrac{v^2 \pi m}{8k} = T$

47. $3x^2 = d^2$

$x^2 = \dfrac{d^2}{3}$ Dividing by 3

$x = \dfrac{d}{\sqrt{3}}$ Principle of square roots. Assume x is nonnegative.

$x = \dfrac{d}{\sqrt{3}} \cdot \dfrac{\sqrt{3}}{\sqrt{3}}$ Rationalizing the denominator

$x = \dfrac{d\sqrt{3}}{3}$

49. $N = \dfrac{n^2 - n}{2}$

$2N = n^2 - n$ Multiplying by 2

$0 = n^2 - n - 2N$ Finding standard form

$a = 1, \ b = -1, \ c = -2N$

$n = \dfrac{-b \pm \sqrt{b^2 - 4ac}}{2a}$

$n = \dfrac{-(-1) \pm \sqrt{(-1)^2 - 4 \cdot 1 \cdot (-2N)}}{2 \cdot 1}$ Substituting

$n = \dfrac{1 + \sqrt{1 + 8N}}{2}$ Using the positive root

51.
$$S = \frac{a + b}{3b}$$

$$3b \cdot S = 3b \cdot \frac{a + b}{3b}$$

$$3bS = a + b$$

$$3bS - b = a$$

$$b(3S - 1) = a$$

$$b = \frac{a}{3S - 1}$$

53.
$$\frac{A - B}{AB} = Q$$

$$AB \cdot \frac{A - B}{AB} = AB \cdot Q$$

$$A - B = ABQ$$

$$A = ABQ + B$$

$$A = B(AQ + 1)$$

$$\frac{A}{AQ + 1} = B$$

55.
$$S = 180(n - 2)$$

$$S = 180n - 360$$

$$S + 360 = 180n$$

$$\frac{S + 360}{180} = n, \text{ or}$$

$$\frac{S}{180} + 2 = n$$

57.
$$A = P(1 + rt)$$

$$A = P + Prt$$

$$A - P = Prt$$

$$\frac{A - P}{Pr} = t$$

59.
$$\frac{A}{B} = \frac{C}{D}$$

$$BD \cdot \frac{A}{B} = BD \cdot \frac{C}{D}$$

$$AD = BC$$

$$D = \frac{BC}{A}$$

61.
$$C = \frac{Ka - b}{a}$$

$$a \cdot C = a \cdot \frac{Ka - b}{a}$$

$$aC = Ka - b$$

$$aC - Ka = -b$$

$$a(C - K) = -b$$

$$a = \frac{-b}{C - K}, \text{ or}$$

$$a = \frac{b}{K - C}$$

63. Discussion and Writing Exercise

65.
$$a^2 + b^2 = c^2 \qquad \text{Pythagorean equation}$$
$$4^2 + 7^2 = c^2 \qquad \text{Substituting}$$
$$16 + 49 = c^2$$
$$65 = c^2$$
$$\sqrt{65} = c \qquad \text{Exact answer}$$
$$8.062 \approx c \qquad \text{Approximate answer}$$

67.
$$a^2 + b^2 = c^2 \qquad \text{Pythagorean equation}$$
$$4^2 + 5^2 = c^2 \qquad \text{Substituting}$$
$$16 + 25 = c^2$$
$$41 = c^2$$
$$\sqrt{41} = c \qquad \text{Exact answer}$$
$$6.403 \approx c \qquad \text{Approximate answer}$$

69.
$$a^2 + b^2 = c^2 \qquad \text{Pythagorean equation}$$
$$2^2 + b^2 = (8\sqrt{17})^2 \qquad \text{Substituting}$$
$$4 + b^2 = 64 \cdot 17$$
$$4 + b^2 = 1088$$
$$b^2 = 1084$$
$$b = \sqrt{1084} \qquad \text{Exact answer}$$
$$b \approx 32.924 \qquad \text{Approximate answer}$$

71. We make a drawing. Let l = the length of the guy wire.

Then we use the Pythagorean equation.
$$10^2 + 18^2 = l^2$$
$$100 + 324 = l^2$$
$$424 = l^2$$
$$\sqrt{424} = l \qquad \text{Exact answer}$$
$$20.591 \approx l \qquad \text{Approximation}$$

The length of the guy wire is $\sqrt{424}$ ft ≈ 20.591 ft.

73. $\sqrt{3x} \cdot \sqrt{6x} = \sqrt{18x^2} = \sqrt{9 \cdot x^2 \cdot 2} = \sqrt{9}\sqrt{x^2}\sqrt{2} = 3x\sqrt{2}$

75. $3\sqrt{t} \cdot \sqrt{t} = 3\sqrt{t^2} = 3t$

77. a) $C = 2\pi r$

$$\frac{C}{2\pi} = r$$

b) $A = \pi r^2$

$$A = \pi \cdot \left(\frac{C}{2\pi}\right)^2 \qquad \text{Substituting } \frac{C}{2\pi} \text{ for } r$$

$$A = \pi \cdot \frac{C^2}{4\pi^2}$$

$$A = \frac{C^2}{4\pi}$$

79.
$$3ax^2 - x - 3ax + 1 = 0$$
$$3ax^2 + (-1 - 3a)x + 1 = 0$$
$$a = 3a, \ b = -1 - 3a, \ c = 1$$

$$x = \frac{-b \pm \sqrt{b^2 - 4ac}}{2a}$$

$$x = \frac{-(-1 - 3a) \pm \sqrt{(-1 - 3a)^2 - 4 \cdot 3a \cdot 1}}{2 \cdot 3a}$$

$$x = \frac{1 + 3a \pm \sqrt{1 + 6a + 9a^2 - 12a}}{6a}$$

$$x = \frac{1 + 3a \pm \sqrt{9a^2 - 6a + 1}}{6a}$$

$$x = \frac{1 + 3a \pm \sqrt{(3a - 1)^2}}{6a}$$

$$x = \frac{1 + 3a \pm (3a - 1)}{6a}$$

$$x = \frac{1 + 3a + 3a - 1}{6a} \quad or \quad x = \frac{1 + 3a - 3a + 1}{6a}$$

$$x = \frac{6a}{6a} \quad\quad\quad or \quad x = \frac{2}{6a}$$

$$x = 1 \quad\quad\quad or \quad x = \frac{1}{3a}$$

The solutions are 1 and $\frac{1}{3a}$.

Exercise Set 17.5

1. **Familiarize.** Let h = the height of the screen, in inches. Then $h + 27$ = the width.

   Translate. We use the Pythagorean equation.
   $$h^2 + (h + 27)^2 = 70^2.$$

   Solve. We solve the equation.
   $$h^2 + (h + 27)^2 = 70^2$$
   $$h^2 + h^2 + 54h + 729 = 4900$$
   $$2h^2 + 54h + 729 = 4900$$
   $$2h^2 + 54h - 4171 = 0$$

   We use the quadratic formula with $a = 2$, $b = 54$, and $c = -4171$.
   $$h = \frac{-54 \pm \sqrt{54^2 - 4 \cdot 2 \cdot (-4171)}}{2 \cdot 2}$$
   $$= \frac{-54 \pm \sqrt{2916 + 33,368}}{4} = \frac{-54 \pm \sqrt{36,284}}{4}$$
   $$h = \frac{-54 - \sqrt{36,284}}{4} \quad or \quad h = \frac{-54 \pm \sqrt{36,284}}{4}$$
   $$h \approx -61 \quad\quad or \quad h \approx 34$$

   Check. The height of the screen cannot be negative, so -61 cannot be a solution. If $h \approx 34$, then $h + 27 \approx 61$ and $34^2 + 61^2 = 4877 \approx 4900 = 70^2$. The answer checks.

   State. The width of the screen is about 61 in., and the height is about 34 in.

3. **Familiarize.** Using the labels on the drawing in the text we have w = the width of the rectangle and $w + 3$ = the length.

   Translate. Recall that area is length × width. Then we have

$$(w + 3)(w) = 70.$$

   Solve. We solve the equation.
   $$w^2 + 3w = 70$$
   $$w^2 + 3w - 70 = 0$$
   $$(w + 10)(w - 7) = 0$$

   $$w + 10 = 0 \quad or \quad w - 7 = 0$$
   $$w = -10 \quad or \quad\quad w = 7$$

   Check. We know that -10 is not a solution of the original problem, because the width cannot be negative. When $w = 7$, then $w + 3 = 10$, and the area is $10 \cdot 7$, or 70. This checks.

   State. The width of the rectangle is 7 ft, and the length is 10 ft.

5. **Familiarize.** Using the labels on the drawing in the text we have s = the length of the shorter leg, in inches, and $s + 8$ = the length of the longer leg.

   Translate. We use the Pythagorean equation.
   $$s^2 + (s + 8)^2 = (8\sqrt{13})^2$$

   Solve. We solve the equation.
   $$s^2 + (s + 8)^2 = (8\sqrt{13})^2$$
   $$s^2 + s^2 + 16s + 64 = 64 \cdot 13$$
   $$2s^2 + 16s + 64 = 832$$
   $$2s^2 + 16s - 768 = 0$$
   $$s^2 + 8s - 384 = 0 \quad\quad \text{Dividing by 2}$$
   $$(s + 24)(s - 16) = 0$$
   $$s + 24 = 0 \quad or \quad s - 16 = 0$$
   $$s = -24 \quad or \quad\quad s = 16$$

   Check. The length of a leg cannot be negative, so -24 cannot be a solution. If $s = 16$, then $s + 8 = 16 + 8 = 24$ and $16^2 + 24^2 = 832 = (8\sqrt{13})^2$. The answer checks.

   State. The lengths of the legs are 16 in. and 24 in.

7. **Familiarize.** We first make a drawing. We let x represent the length. Then $x - 4$ represents the width.

   Translate. The area is length × width. Thus, we have two expressions for the area of the rectangle: $x(x - 4)$ and 320. This gives us a translation.

   $$x(x - 4) = 320.$$

   Solve. We solve the equation.
   $$x^2 - 4x = 320$$
   $$x^2 - 4x - 320 = 0$$
   $$(x - 20)(x + 16) = 0$$

   $$x - 20 = 0 \quad or \quad x + 16 = 0$$
   $$x = 20 \quad or \quad\quad x = -16$$

   Check. Since the length of a side cannot be negative, -16 does not check. But 20 does check. If the length is 20, then the width is $20 - 4$, or 16. The area is 20×16, or 320. This checks.

State. The length is 20 cm, and the width is 16 cm.

9. Familiarize. We first make a drawing. We let x represent the length of one leg. Then $x + 2$ represents the length of the other leg.

Translate. We use the Pythagorean equation.

$$x^2 + (x+2)^2 = 8^2.$$

Solve. We solve the equation.

$$x^2 + x^2 + 4x + 4 = 64$$
$$2x^2 + 4x + 4 = 64$$
$$2x^2 + 4x - 60 = 0$$
$$x^2 + 2x - 30 = 0 \qquad \text{Dividing by 2}$$

$a = 1, \ b = 2, \ c = -30$

$$x = \frac{-2 \pm \sqrt{2^2 - 4 \cdot 1 \cdot (-30)}}{2 \cdot 1}$$
$$= \frac{-2 \pm \sqrt{4 + 120}}{2} = \frac{-2 \pm \sqrt{124}}{2}$$
$$= \frac{-2 \pm \sqrt{4 \cdot 31}}{2} = \frac{-2 \pm 2\sqrt{31}}{2}$$
$$= \frac{2(-1 \pm \sqrt{31})}{2} = -1 \pm \sqrt{31}$$

Using a calculator or Table 2 we find that $\sqrt{31} \approx 5.568$:

$$-1 + \sqrt{31} \approx -1 + 5.568 \quad or \quad -1 - \sqrt{31} \approx -1 - 5.568$$
$$\approx 4.6 \qquad\qquad or \qquad\qquad \approx -6.6$$

Check. Since the length of a leg cannot be negative, -6.6 does not check. But 4.6 does check. If the shorter leg is 4.6, then the other leg is $4.6 + 2$, or 6.6. Then $4.6^2 + 6.6^2 = 21.16 + 43.56 = 64.72$ and using a calculator, $\sqrt{64.72} \approx 8.04 \approx 8$. Note that our check is not exact since we are using an approximation.

State. One leg is about 4.6 m, and the other is about 6.6 m long.

11. Familiarize. We first make a drawing. We let x represent the width and $x + 2$ the length.

$$\boxed{20 \text{ in}^2} \quad x$$
$$x + 2$$

Translate. The area is length $\times$ width. We have two expressions for the area of the rectangle: $(x + 2)x$ and 20. This gives us a translation.

$$(x + 2)x = 20.$$

Solve. We solve the equation.

$$x^2 + 2x = 20$$
$$x^2 + 2x - 20 = 0$$

$a = 1, \ b = 2, \ c = -20$

$$x = \frac{-2 \pm \sqrt{2^2 - 4 \cdot 1 \cdot (-20)}}{2 \cdot 1}$$
$$= \frac{-2 \pm \sqrt{4 + 80}}{2} = \frac{-2 \pm \sqrt{84}}{2}$$
$$= \frac{-2 \pm \sqrt{4 \cdot 21}}{2} = \frac{-2 \pm 2\sqrt{21}}{2}$$
$$= \frac{2(-1 \pm \sqrt{21})}{2} = -1 \pm \sqrt{21}$$

Using a calculator or Table 2 we find that $\sqrt{21} \approx 4.583$:

$$-1 + \sqrt{21} \approx -1 + 4.583 \quad or \quad -1 - \sqrt{21} \approx -1 - 4.583$$
$$\approx 3.6 \qquad\qquad or \qquad\qquad \approx -5.6$$

Check. Since the length of a side cannot be negative, -5.6 does not check. But 3.6 does check. If the width is 3.6, then the length is $3.6 + 2$, or 5.6. The area is $5.6(3.6)$, or $20.16 \approx 20$. This checks.

State. The length is about 5.6 in., and the width is about 3.6 in.

13. Familiarize. We make a drawing and label it. We let $w = $ the width of the rectangle and $2w = $ the length.

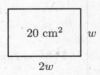

Translate. Recall that area = length $\times$ width. Then we have

$$2w \cdot w = 20.$$

Solve. We solve the equation.

$$2w^2 = 20$$
$$w^2 = 10 \qquad \text{Dividing by 2}$$

$$w = \sqrt{10} \quad or \quad w = -\sqrt{10} \qquad \text{Principle of square roots}$$
$$w \approx 3.2 \quad or \quad w \approx -3.2$$

Check. We know that -3.2 is not a solution of the original problem, because width cannot be negative. When $w \approx 3.2$, then $2w \approx 6.4$ and the area is about $(6.4)(3.2)$, or 20.48. This checks, although the check is not exact since we used an approximation for $\sqrt{10}$.

State. The length is about 6.4 cm, and the width is about 3.2 cm.

15. Familiarize. Using the drawing in the text, we have $x = $ the thickness of the frame, $20 - 2x = $ the width of the picture showing, and $25 - 2x = $ the length of the picture showing.

Translate. Recall that area = length $\times$ width. Then we have

$$(25 - 2x)(20 - 2x) = 266.$$

Solve. We solve the equation.

$$500 - 90x + 4x^2 = 266$$
$$4x^2 - 90x + 234 = 0$$
$$2x^2 - 45x + 117 = 0 \qquad \text{Dividing by 2}$$
$$(2x - 39)(x - 3) = 0$$

$$
\begin{array}{lll}
2x - 39 = 0 & or & x - 3 = 0 \\
2x = 39 & or & x = 3 \\
x = 19.5 & or & x = 3
\end{array}
$$

Check. The number 19.5 cannot be a solution, because when $x = 19.5$ then $20 - 2x = -19$, and the width cannot be negative. When $x = 3$, then $20 - 2x = 20 - 2 \cdot 3$, or 14 and $25 - 2x = 25 - 2 \cdot 3$, or 19 and $19 \cdot 14 = 266$. This checks.

State. The thickness of the frame is 3 cm.

17. Familiarize. Referring to the drawing in the text, we complete the table.

	d	r	t
Upstream	40	$r - 3$	t_1
Downstream	40	$r + 3$	t_2
Total Time			14

Translate. Using $t = d/r$ and the rows of the table, we have
$$t_1 = \frac{40}{r - 3} \text{ and } t_2 = \frac{40}{r + 3}.$$
Since the total time is 14 hr, $t_1 + t_2 = 14$, and we have
$$\frac{40}{r - 3} + \frac{40}{r + 3} = 14.$$

Solve. We solve the equation. We multiply by $(r - 3)(r + 3)$, the LCM of the denominators.

$$(r - 3)(r + 3)\left(\frac{40}{r - 3} + \frac{40}{r + 3}\right) = (r - 3)(r + 3) \cdot 14$$
$$40(r + 3) + 40(r - 3) = 14(r^2 - 9)$$
$$40r + 120 + 40r - 120 = 14r^2 - 126$$
$$80r = 14r^2 - 126$$
$$0 = 14r^2 - 80r - 126$$
$$0 = 7r^2 - 40r - 63$$
$$0 = (7r + 9)(r - 7)$$

$$
\begin{array}{lll}
7r + 9 = 0 & or & r - 7 = 0 \\
7r = -9 & or & r = 7 \\
r = -\dfrac{9}{7} & or & r = 7
\end{array}
$$

Check. Since speed cannot be negative, $-\dfrac{9}{7}$ cannot be a solution. If the speed of the boat is 7 km/h, the speed upstream is $7 - 3$, or 4 km/h, and the speed downstream is $7 + 3$, or 10 km/h. The time upstream is $\dfrac{40}{4}$, or 10 hr. The time downstream is $\dfrac{40}{10}$, or 4 hr. The total time is 14 hr. This checks.

State. The speed of the boat in still water is 7 km/h.

19. Familiarize. Let r represent the speed of the wind. Then the speed of the plane flying with the wind is $300 + r$ and the speed against the wind is $300 - r$. We fill in the table in the text.

	d	r	t
With wind	680	$300 + r$	t_1
Against wind	520	$300 - r$	t_2

Translate. Using $t = d/r$ and the rows of the table, we have
$$t_1 = \frac{680}{300 + r} \text{ and } t_2 = \frac{520}{300 - r}.$$
Since the total time is 4 hr, $t_1 + t_2 = 4$, and we have
$$\frac{680}{300 + r} + \frac{520}{300 - r} = 4.$$

Solve. We solve the equation. We multiply by $(300 + r)(300 - r)$, the LCM of the denominators.

$$(300+r)(300-r)\left(\frac{680}{300+r} + \frac{520}{300-r}\right) = (300+r)(300-r)\cdot 4$$
$$680(300 - r) + 520(300 + r) = 4(90,000 - r^2)$$
$$204,000 - 680r + 156,000 + 520r = 360,000 - 4r^2$$
$$360,000 - 160r = 360,000 - 4r^2$$
$$4r^2 - 160r = 0$$
$$4r(r - 40) = 0$$

$$
\begin{array}{lll}
4r = 0 & or & r - 40 = 0 \\
r = 0 & or & r = 40
\end{array}
$$

Check. If $r = 0$, then the speed of the wind is 0 km/h. That is, there is no wind. In this case the plane travels 680 km in 680/300, or $2\frac{4}{15}$ hr, and it travels 520 km in 520/300, or $1\frac{11}{15}$ hr. The total time is $2\frac{4}{15} + 1\frac{11}{15}$, or 4 hr, so we have one solution. If the speed of the wind is 40 km/h, then the speed of the airplane with the wind is $300 + 40$, or 340 km/h, and the speed against the wind is $300 - 40$, or 260 km/h. The time with the wind is 680/340, or 2 hr, and the time against the wind is 520/260, or 2 hr. The total time is 2 hr + 2 hr, or 4 hr, so we have a second solution.

State. The speed of the wind is 0 km/h (There is no wind.) or 40 km/h.

21. Familiarize. We first make a drawing. We let r represent the speed of the current. Then $10 - r$ is the speed of the boat traveling upstream and $10 + r$ is the speed of the boat traveling downstream.

Upstream
$10 - r$ km/h
⟶
12 km

Downstream
$10 + r$ km/h
⟵
28 km

We summarize the information in a table.

	d	r	t
Upstream	12	$10 - r$	t_1
Downstream	28	$10 + r$	t_2
Total Time			4

Translate. Using $t = d/r$ and the rows of the table, we have

$$t_1 = \frac{12}{10 - r} \text{ and } t_2 = \frac{28}{10 + r}.$$

Since the total time is 4 hr, $t_1 + t_2 = 4$, and we have

$$\frac{12}{10 - r} + \frac{28}{10 + r} = 4.$$

Solve. We solve the equation. We multiply by $(10 - r)(10 + r)$, the LCM of the denominators.

$$(10 - r)(10 + r)\left(\frac{12}{10 - r} + \frac{28}{10 + r}\right) =$$
$$(10 - r)(10 + r) \cdot 4$$
$$12(10 + r) + 28(10 - r) = 4(100 - r^2)$$
$$120 + 12r + 280 - 28r = 400 - 4r^2$$
$$400 - 16r = 400 - 4r^2$$
$$4r^2 - 16r = 0$$
$$4r(r - 4) = 0$$

$$4r = 0 \quad or \quad r - 4 = 0$$
$$r = 0 \quad or \qquad r = 4$$

Check. If $r = 0$, then the speed of the stream is 0 km/h. That is, the stream is still. In this case the boat travels 12 km in 12/10, or 1.2 hr, and it travels 28 km in 28/10, or 2.8 hr. The total time is 1.2 hr + 2.8 hr, or 4 hr, so we have one solution. If the speed of the current is 4 km/h, the speed upstream is $10 - 4$, or 6 km/h, and the speed downstream is $10 + 4$, or 14 km/h. The time upstream is 12/6, or 2 hr. The time downstream is 28/14, or 2 hr. The total time is 2 hr + 2 hr, or 4 hr. This checks also.

State. The speed of the stream is 0 km/h (The stream is still.) or 4 km/h.

23. Familiarize. We first make a drawing. We let r represent the speed of the boat in still water. Then $r - 4$ is the speed of the boat traveling upstream and $r + 4$ is the speed of the boat traveling downstream.

Upstream
$r - 4$ mph

4 mi

Downstream
$r + 4$ mph

12 mi

We summarize the information in a table.

	d	r	t
Upstream	4	$r - 4$	t_1
Downstream	12	$r + 4$	t_2
Total Time			2

Translate. Using $t = d/r$ and the rows of the table, we have

$$t_1 = \frac{4}{r - 4} \text{ and } t_2 = \frac{12}{r + 4}.$$

Since the total time is 2 hr, $t_1 + t_2 = 2$, and we have

$$\frac{4}{r - 4} + \frac{12}{r + 4} = 2.$$

Solve. We solve the equation. We multiply by $(r - 4)(r + 4)$, the LCM of the denominators.

$$(r - 4)(r + 4)\left(\frac{4}{r - 4} + \frac{12}{r + 4}\right) = (r - 4)(r + 4) \cdot 2$$
$$4(r + 4) + 12(r - 4) = 2(r^2 - 16)$$
$$4r + 16 + 12r - 48 = 2r^2 - 32$$
$$16r - 32 = 2r^2 - 32$$
$$0 = 2r^2 - 16r$$
$$0 = 2r(r - 8)$$

$$2r = 0 \quad or \quad r - 8 = 0$$
$$r = 0 \quad or \qquad r = 8$$

Check. If $r = 0$, then the speed upstream, $0 - 4$, would be negative. Since speed cannot be negative, 0 cannot be a solution. If the speed of the boat is 8 mph, the speed upstream is $8 - 4$, or 4 mph, and the speed downstream is $8 + 4$, or 12 mph. The time upstream is $\frac{4}{4}$, or 1 hr. The time downstream is $\frac{12}{12}$, or 1 hr. The total time is 2 hr. This checks.

State. The speed of the boat in still water is 8 mph.

25. Familiarize. We first make a drawing. We let r represent the speed of the stream. Then $9 - r$ represents the speed of the boat traveling upstream and $9 + r$ represents the speed of the boat traveling downstream.

Upstream
$9 - r$ km/h

80 km

Downstream
$9 + r$ km/h

80 km

We summarize the information in a table.

	d	r	t
Upstream	80	$9 - r$	t_1
Downstream	80	$9 + r$	t_2

Translate. Using $t = d/r$ and the rows of the table, we have

$$t_1 = \frac{80}{9 - r} \text{ and } t_2 = \frac{80}{9 + r}.$$

Since the total time is 18 hr, $t_1 + t_2 = 18$, and we have

$$\frac{80}{9 - r} + \frac{80}{9 + r} = 18.$$

Solve. We solve the equation. We multiply by $(9 - r)(9 + r)$, the LCM of the denominators.

$$(9-r)(9+r)\left(\frac{80}{9-r} + \frac{80}{9+r}\right) = (9-r)(9+r)\cdot 18$$
$$80(9+r) + 80(9-r) = 18(81 - r^2)$$
$$720 + 80r + 720 - 80r = 1458 - 18r^2$$
$$1440 = 1458 - 18r^2$$
$$18r^2 = 18$$
$$r^2 = 1$$

$r = 1$ *or* $r = -1$ Principle of square roots

Check. Since speed cannot be negative, -1 cannot be a solution. If the speed of the stream is 1 km/h, the speed upstream is $9 - 1$, or 8 km/h, and the speed downstream is $9 + 1$, or 10 km/h. The time upstream is $\frac{80}{8}$, or 10 hr. The time downstream is $\frac{80}{10}$, or 8 hr. The total time is 18 hr. This checks.

State. The speed of the stream is 1 km/h.

27. Discussion and Writing Exercise

29.
$$5\sqrt{2} + \sqrt{18} = 5\sqrt{2} + \sqrt{9 \cdot 2}$$
$$= 5\sqrt{2} + \sqrt{9}\sqrt{2}$$
$$= 5\sqrt{2} + 3\sqrt{2}$$
$$= (5+3)\sqrt{2}$$
$$= 8\sqrt{2}$$

31.
$$\sqrt{4x^3} - 7\sqrt{x} = \sqrt{4 \cdot x^2 \cdot x} - 7\sqrt{x}$$
$$= \sqrt{4}\sqrt{x^2}\sqrt{x} - 7\sqrt{x}$$
$$= 2x\sqrt{x} - 7\sqrt{x}$$
$$= (2x - 7)\sqrt{x}$$

33.
$$\sqrt{2} + \sqrt{\frac{1}{2}} = \sqrt{2} + \sqrt{\frac{1}{2}\cdot\frac{2}{2}}$$
$$= \sqrt{2} + \sqrt{\frac{2}{4}}$$
$$= \sqrt{2} + \frac{\sqrt{2}}{\sqrt{4}}$$
$$= \sqrt{2} + \frac{\sqrt{2}}{2}$$
$$= (1 + \frac{1}{2})\sqrt{2}$$
$$= \frac{3}{2}\sqrt{2}, \text{ or } \frac{3\sqrt{2}}{2}$$

35.
$$\sqrt{24} + \sqrt{54} - \sqrt{48}$$
$$= \sqrt{4 \cdot 6} + \sqrt{9 \cdot 6} - \sqrt{16 \cdot 3}$$
$$= \sqrt{4} \cdot \sqrt{6} + \sqrt{9} \cdot \sqrt{6} - \sqrt{16} \cdot \sqrt{3}$$
$$= 2\sqrt{6} + 3\sqrt{6} - 4\sqrt{3}$$
$$= 5\sqrt{6} - 4\sqrt{3}$$

37. Familiarize. The radius of a 12-in. pizza is $\frac{12}{2}$, or 6 in. The radius of a d-in. pizza is $\frac{d}{2}$ in. The area of a circle is πr^2.

Translate.

Area of d-in. pizza	is	Area of 12-in. pizza	plus	Area of 12-in. pizza
$\downarrow$	$\downarrow$	$\downarrow$	$\downarrow$	$\downarrow$
$\pi\left(\dfrac{d}{2}\right)^2$	$=$	$\pi \cdot 6^2$	$+$	$\pi \cdot 6^2$

Solve. We solve the equation.
$$\frac{d^2}{4}\pi = 36\pi + 36\pi$$
$$\frac{d^2}{4}\pi = 72\pi$$
$$\frac{d^2}{4} = 72 \qquad \text{Dividing by } \pi$$
$$d^2 = 288$$

$d = \sqrt{288}$ *or* $d = -\sqrt{288}$
$d = 12\sqrt{2}$ *or* $d = -12\sqrt{2}$
$d \approx 16.97$ *or* $d \approx -16.97$ Using a calculator

Check. Since the diameter cannot be negative, -16.97 is not a solution. If $d = 12\sqrt{2}$, or 16.97, then $r = 6\sqrt{2}$ and the area is $\pi(6\sqrt{2})^2$, or 72π. The area of the two 12-in. pizzas is $2 \cdot \pi \cdot 6^2$, or 72π. The value checks.

State. The diameter of the pizza should be $12\sqrt{2}$ in. $\approx$ 16.97 in.

The radius of a 16-in. pizza is $\frac{16}{2}$, or 8 in., so the area is $\pi(8)^2$, or 64π. We found that the area of two 12-in. pizzas is 72π and $72\pi > 64\pi$, so you get more to eat with two 12-in. pizzas than with a 16-in. pizza.

Exercise Set 17.6

1. $y = x^2 + 1$

We first find the vertex. The x-coordinate is
$$-\frac{b}{2a} = -\frac{0}{2 \cdot 1} = 0.$$
We substitute into the equation to find the second coordinate of the vertex.
$$y = x^2 + 1 = 0^2 + 1 = 1$$
The vertex is $(0, 1)$. This is also the y-intercept. The line of symmetry is $x = 0$, the y-axis.

We choose some x-values on both sides of the vertex and graph the parabola.

When $x = 1$, $y = 1^2 + 1 = 1 + 1 = 2$.

When $x = -1$, $y = (-1)^2 + 1 = 1 + 1 = 2$.

When $x = 2$, $y = 2^2 + 1 = 4 + 1 = 5$.

When $x = -2$, $y = (-2)^2 + 1 = 4 + 1 = 5$.

x	y
-2	5
-1	2
0	1 ← Vertex
1	2
2	5
3	10

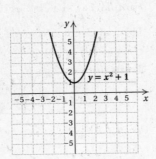

3. $y = -1 \cdot x^2$

Find the vertex. The x-coordinate is
$$-\frac{b}{2a} = -\frac{0}{2(-1)} = 0.$$
The y-coordinate is
$$y = -1 \cdot x^2 = -1 \cdot 0^2 = 0.$$
The vertex is $(0,0)$. This is also the y-intercept. The line of symmetry is $x = 0$, the y-axis.

Choose some x-values on both sides of the vertex and graph the parabola.

When $x = -2$, $y = -1 \cdot (-2)^2 = -1 \cdot 4 = -4$.
When $x = -1$, $y = -1 \cdot (-1)^2 = -1 \cdot 1 = -1$.
When $x = 1$, $y = -1 \cdot 1^2 = -1 \cdot 1 = -1$.
When $x = 2$, $y = -1 \cdot 2^2 = -1 \cdot 4 = -4$.

x	y
0	0 ←Vertex
-2	-4
-1	-1
1	-1
2	-4

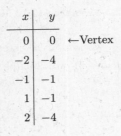

5. $y = -x^2 + 2x$

Find the vertex. The x-coordinate is
$$-\frac{b}{2a} = -\frac{2}{2(-1)} = -(-1) = 1.$$
The y-coordinate is
$$y = -x^2 + 2x = -(1)^2 + 2 \cdot 1 = -1 + 2 = 1.$$
The vertex is $(1,1)$.

We choose some x-values on both sides of the vertex and graph the parabola. We make sure we find y when $x = 0$. This gives us the y-intercept.

x	y
1	1 ←Vertex
0	0 ←y-intercept
-1	-3
2	0
3	-3

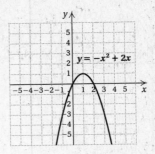

7. $y = 5 - x - x^2$, or $y = -x^2 - x + 5$

Find the vertex. The x-coordinate is
$$-\frac{b}{2a} = -\frac{-1}{2(-1)} = -\frac{1}{2}.$$
The y-coordinate is
$$y = 5 - x - x^2 = 5 - \left(-\frac{1}{2}\right) - \left(-\frac{1}{2}\right)^2 = 5 + \frac{1}{2} - \frac{1}{4} = \frac{21}{4}.$$
The vertex is $\left(-\frac{1}{2}, \frac{21}{4}\right)$.

We choose some x-values on both sides of the vertex and graph the parabola.

x	y
$-\frac{1}{2}$	$\frac{21}{4}$ ←Vertex
0	5 ←y-intercept
-1	5
-2	3
1	3

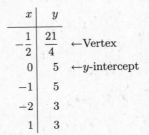

9. $y = x^2 - 2x + 1$

Find the vertex. The x-coordinate is
$$-\frac{b}{2a} = -\frac{-2}{2 \cdot 1} = -(-1) = 1.$$
The y-coordinate is
$$y = x^2 - 2x + 1 = 1^2 - 2 \cdot 1 + 1 = 1 - 2 + 1 = 0.$$
The vertex is $(1, 0)$.

We choose some x-values on both sides of the vertex and graph the parabola.

x	y
1	0 ←Vertex
0	1 ←y-intercept
-1	4
2	1
3	4

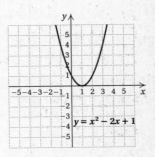

11. $y = -x^2 + 2x + 3$

Find the vertex. The x-coordinate is

$$-\frac{b}{2a} = -\frac{2}{2(-1)} = -(-1) = 1.$$

The y-coordinate is

$$y = -x^2 + 2x + 3 = -(1)^2 + 2 \cdot 1 + 3 = -1 + 2 + 3 = 4.$$

The vertex is $(1, 4)$.

We choose some x-values on both sides of the vertex and graph the parabola.

x	y	
1	4	←Vertex
0	3	←y-intercept
−1	0	
2	3	
3	0	

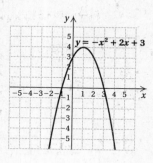

13. $y = -2x^2 - 4x + 1$

Find the vertex. The x-coordinate is

$$-\frac{b}{2a} = -\frac{-4}{2(-2)} = -1.$$

The y-coordinate is

$$y = -2x^2 - 4x + 1 = -2(-1)^2 - 4(-1) + 1 = -2 + 4 + 1 = 3.$$

The vertex is $(-1, 3)$.

We choose some x-values on both sides of the vertex and graph the parabola.

x	y	
−1	3	←Vertex
0	1	←y-intercept
1	−5	
−2	1	
−3	−5	

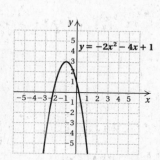

15. $y = 5 - x^2$, or $y = -x^2 + 5$

Find the vertex. The x-coordinate is

$$-\frac{b}{2a} = -\frac{0}{2(-1)} = 0.$$

The y-coordinate is

$$y = 5 - x^2 = 5 - 0^2 = 5.$$

The vertex is $(0, 5)$. This is also the y-intercept.

We choose some x-values on both sides of the vertex and graph the parabola.

x	y	
0	5	←Vertex
−1	4	
−2	1	
1	4	
2	1	

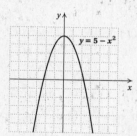

17. $y = \frac{1}{4}x^2$

Find the vertex. The x-coordinate is

$$-\frac{b}{2a} = -\frac{0}{2\left(\frac{1}{4}\right)} = 0.$$

The y-coordinate is

$$y = \frac{1}{4}x^2 = \frac{1}{4} \cdot 0^2 = 0.$$

The vertex is $(0, 0)$. This is also the y-intercept.

We choose some x-values on both sides of the vertex and graph the parabola.

x	y	
0	0	←Vertex
−2	1	
−4	4	
2	1	
4	4	

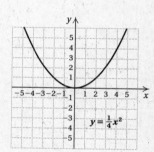

19. $y = -x^2 + x - 1$

Find the vertex. The x-coordinate is

$$-\frac{b}{2a} = -\frac{1}{2(-1)} = -\left(-\frac{1}{2}\right) = \frac{1}{2}.$$

The y-coordinate is

$$y = -x^2 + x - 1 = -\left(\frac{1}{2}\right)^2 + \frac{1}{2} - 1 = -\frac{1}{4} + \frac{1}{2} - 1 = -\frac{3}{4}.$$

The vertex is $\left(\frac{1}{2}, -\frac{3}{4}\right)$.

We choose some x-values on both sides of the vertex and graph the parabola.

x	y	
$\frac{1}{2}$	$-\frac{3}{4}$	←Vertex
0	−1	←y-intercept
−1	−3	
1	−1	
2	−3	

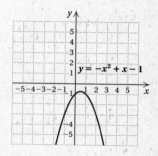

21. $y = -2x^2$

Find the vertex. The x-coordinate is

$$-\frac{b}{2a} = -\frac{0}{2(-2)} = 0.$$

The y-coordinate is

$$y = -2x^2 = -2 \cdot 0^2 = 0.$$

The vertex is $(0,0)$. This is also the y-intercept.

We choose some x-values on both sides of the vertex and graph the parabola.

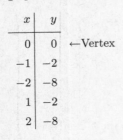

x	y	
0	0	←Vertex
-1	-2	
-2	-8	
1	-2	
2	-8	

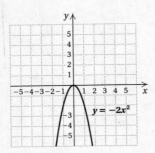

23. $y = x^2 - x - 6$

Find the vertex. The x-coordinate is

$$-\frac{b}{2a} = -\frac{-1}{2 \cdot 1} = -\left(-\frac{1}{2}\right) = \frac{1}{2}.$$

The y-coordinate is

$$y = x^2 - x - 6 = \left(\frac{1}{2}\right)^2 - \frac{1}{2} - 6 = \frac{1}{4} - \frac{1}{2} - 6 = -\frac{25}{4}.$$

The vertex is $\left(\frac{1}{2}, -\frac{25}{4}\right)$.

We choose some x-values on both sides of the vertex and graph the parabola.

x	y	
$\frac{1}{2}$	$-\frac{25}{4}$	←Vertex
0	-6	←y-intercept
-1	-4	
1	-6	
2	-4	

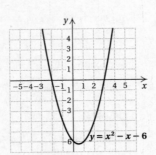

25. $y = x^2 - 2$

To find the x-intercepts we solve the equation $x^2 - 2 = 0$.

$$x^2 - 2 = 0$$
$$x^2 = 2$$

$x = \sqrt{2}$ or $x = -\sqrt{2}$ Principle of square roots

The x-intercepts are $(\sqrt{2}, 0)$ and $(-\sqrt{2}, 0)$.

27. $y = x^2 + 5x$

To find the x-intercepts we solve the equation $x^2 + 5x = 0$.

$$x^2 + 5x = 0$$
$$x(x + 5) = 0$$

$x = 0$ or $x + 5 = 0$
$x = 0$ or $x = -5$

The x-intercepts are $(0,0)$ and $(-5,0)$.

29. $y = 8 - x - x^2$

To find the x-intercepts we solve the equation $8 - x - x^2 = 0$.

$$8 - x - x^2 = 0$$
$$x^2 + x - 8 = 0 \qquad \text{Standard form}$$

$a = 1, \ b = 1, \ c = -8$

$$x = \frac{-1 \pm \sqrt{1^2 - 4 \cdot 1 \cdot (-8)}}{2 \cdot 1}$$

$$x = \frac{-1 \pm \sqrt{33}}{2}$$

The x-intercepts are $\left(\frac{-1 + \sqrt{33}}{2}, 0\right)$ and $\left(\frac{-1 - \sqrt{33}}{2}, 0\right)$.

31. $y = x^2 - 6x + 9$

To find the x-intercepts we solve the equation $x^2 - 6x + 9 = 0$.

$$x^2 - 6x + 9 = 0$$
$$(x - 3)(x - 3) = 0$$

$x - 3 = 0$ or $x - 3 = 0$
$x = 3$ or $x = 3$

The x-intercept is $(3, 0)$.

33. $y = -x^2 - 4x + 1$

To find the x-intercepts we solve the equation $-x^2 - 4x + 1 = 0$.

$$-x^2 - 4x + 1 = 0$$
$$x^2 + 4x - 1 = 0 \qquad \text{Standard form}$$

$a = 1, \ b = 4, \ c = -1$

$$x = \frac{-4 \pm \sqrt{4^2 - 4 \cdot 1 \cdot (-1)}}{2 \cdot 1}$$

$$x = \frac{-4 \pm \sqrt{20}}{2} = \frac{-4 \pm \sqrt{4 \cdot 5}}{2} = \frac{-4 \pm 2\sqrt{5}}{2}$$

$$x = \frac{2(-2 \pm \sqrt{5})}{2} = -2 \pm \sqrt{5}$$

The x-intercepts are $(-2 + \sqrt{5}, 0)$ and $(-2 - \sqrt{5}, 0)$.

35. $y = x^2 + 9$

To find the x-intercepts we solve the equation $x^2 + 9 = 0$.

$$x^2 + 9 = 0$$
$$x^2 = -9$$

The negative number -9 has no real-number square roots. Thus there are no x-intercepts.

37. Discussion and Writing Exercise

39. $\sqrt{8} + \sqrt{50} + \sqrt{98} + \sqrt{128}$
$$= \sqrt{4 \cdot 2} + \sqrt{25 \cdot 2} + \sqrt{49 \cdot 2} + \sqrt{64 \cdot 2}$$
$$= 2\sqrt{2} + 5\sqrt{2} + 7\sqrt{2} + 8\sqrt{2}$$
$$= 22\sqrt{2}$$

41.
$$y = \frac{k}{x}$$
$$12.4 = \frac{k}{2.4} \quad \text{Substituting}$$
$$29.76 = k \quad \text{Variation constant}$$

$$y = \frac{29.76}{x} \quad \text{Equation of variation}$$

43. a) We substitute 128 for H and solve for t:
$$128 = -16t^2 + 96t$$
$$16t^2 - 96t + 128 = 0$$
$$16(t^2 - 6t + 8) = 0$$
$$16(t - 2)(t - 4) = 0$$

$$t - 2 = 0 \quad or \quad t - 4 = 0$$
$$t = 2 \quad or \quad \quad t = 4$$

The projectile is 128 ft from the ground 2 sec after launch and again 4 sec after launch. The graph confirms this.

b) We find the first coordinate of the vertex of the function $H = -16t^2 + 96t$:
$$-\frac{b}{2a} = -\frac{96}{2(-16)} = -\frac{96}{-32} = -(-3) = 3$$

The projectile reaches its maximum height 3 sec after launch. The graph confirms this.

c) We substitute 0 for H and solve for t:
$$0 = -16t^2 + 96t$$
$$0 = -16t(t - 6)$$

$$-16t = 0 \quad or \quad t - 6 = 0$$
$$t = 0 \quad or \quad \quad t = 6$$

At $t = 0$ sec the projectile has not yet been launched. Thus, we use $t = 6$. The projectile returns to the ground 6 sec after launch. The graph confirms this.

45. $y = x^2 + 2x - 3$

$a = 1$, $b = 2$, $c = -3$

$b^2 - 4ac = 2^2 - 4 \cdot 1 \cdot (-3) = 4 + 12 = 16$

Since $b^2 - 4ac = 16 > 0$, the equation $x^2 + 2x - 3 = 0$ has two real-number solutions.

47. $y = -0.02x^2 + 4.7x - 2300$

$a = -0.02$, $b = 4.7$, $c = -2300$

$b^2 - 4ac = (4.7)^2 - 4(-0.02)(-2300) = 22.09 - 184 = -161.91$

Since $b^2 - 4ac = -161.91 < 0$, the equation $-0.02x^2 + 4.7x - 2300 = 0$ has no real solutions.

Exercise Set 17.7

1. Yes; each member of the domain is matched to only one member of the range.

3. Yes; each member of the domain is matched to only one member of the range.

5. No; a member of the domain is matched to more than one member of the range. In fact, each member of the domain is matched to 3 members of the range.

7. Yes; each member of the domain is matched to only one member of the range.

9. This correspondence is a function, because each class member has only one seat number.

11. This correspondence is a function, because each shape has only one number for its area.

13. This correspondence is not a function, because it is reasonable to assume that at least one person has more than one aunt.

The correspondence is a relation, because it is reasonable to assume that each person has at least one aunt.

15. $f(x) = x + 5$

a) $f(4) = 4 + 5 = 9$

b) $f(7) = 7 + 5 = 12$

c) $f(-3) = -3 + 5 = 2$

d) $f(0) = 0 + 5 = 5$

e) $f(2.4) = 2.4 + 5 = 7.4$

f) $f\left(\frac{2}{3}\right) = \frac{2}{3} + 5 = 5\frac{2}{3}$

17. $h(p) = 3p$

a) $h(-7) = 3(-7) = -21$

b) $h(5) = 3 \cdot 5 = 15$

c) $h(14) = 3 \cdot 14 = 42$

d) $h(0) = 3 \cdot 0 = 0$

e) $h\left(\frac{2}{3}\right) = 3 \cdot \frac{2}{3} = \frac{6}{3} = 2$

f) $h(-54.2) = 3(-54.2) = -162.6$

19. $g(s) = 3s + 4$

a) $g(1) = 3 \cdot 1 + 4 = 3 + 4 = 7$

b) $g(-7) = 3(-7) + 4 = -21 + 4 = -17$

c) $g(6.7) = 3(6.7) + 4 = 20.1 + 4 = 24.1$

d) $g(0) = 3 \cdot 0 + 4 = 0 + 4 = 4$

e) $g(-10) = 3(-10) + 4 = -30 + 4 = -26$

f) $g\left(\frac{2}{3}\right) = 3 \cdot \frac{2}{3} + 4 = 2 + 4 = 6$

21. $f(x) = 2x^2 - 3x$

a) $f(0) = 2 \cdot 0^2 - 3 \cdot 0 = 0 - 0 = 0$

b) $f(-1) = 2(-1)^2 - 3(-1) = 2 + 3 = 5$

c) $f(2) = 2 \cdot 2^2 - 3 \cdot 2 = 8 - 6 = 2$

d) $f(10) = 2 \cdot 10^2 - 3 \cdot 10 = 200 - 30 = 170$

e) $f(-5) = 2(-5)^2 - 3(-5) = 50 + 15 = 65$

f) $f(-10) = 2(-10)^2 - 3(-10) = 200 + 30 = 230$

23. $f(x) = |x| + 1$

    a) $f(0) = |0| + 1 = 0 + 1 = 1$

    b) $f(-2) = |-2| + 1 = 2 + 1 = 3$

    c) $f(2) = |2| + 1 = 2 + 1 = 3$

    d) $f(-3) = |-3| + 1 = 3 + 1 = 4$

    e) $f(-10) = |-10| + 1 = 10 + 1 = 11$

    f) $f(22) = |22| + 1 = 22 + 1 = 23$

25. $f(x) = x^3$

    a) $f(0) = 0^3 = 0$

    b) $f(-1) = (-1)^3 = -1$

    c) $f(2) = 2^3 = 8$

    d) $f(10) = 10^3 = 1000$

    e) $f(-5) = (-5)^3 = -125$

    f) $f(-10) = (-10)^3 = -1000$

27. $F(x) = 2.75x + 71.48$

    a) $F(32) = 2.75(32) + 71.48$

                $= 88 + 71.48$

                $= 159.48$ cm

    b) $F(30) = 2.75(30) + 71.48$

                 $= 82.5 + 71.48$

                $= 153.98$ cm

29. $P(d) = 1 + \dfrac{d}{33}$

    $P(20) = 1 + \dfrac{20}{33} = 1\dfrac{20}{33}$ atm

    $P(30) = 1 + \dfrac{30}{33} = 1\dfrac{10}{11}$ atm

    $P(100) = 1 + \dfrac{100}{33} = 1 + 3\dfrac{1}{33} = 4\dfrac{1}{33}$ atm

31. $W(d) = 0.112d$

    $W(16) = 0.112(16) = 1.792$ cm

    $W(25) = 0.112(25) = 2.8$ cm

    $W(100) = 0.112(100) = 11.2$ cm

33. Graph $f(x) = 3x - 1$

Make a list of function values in a table.

When $x = -1$, $f(-1) = 3(-1) - 1 = -3 - 1 = -4$.

When $x = 0$, $f(0) = 3 \cdot 0 - 1 = 0 - 1 = -1$.

When $x = 2$, $f(2) = 3 \cdot 2 - 1 = 6 - 1 = 5$.

x	$f(x)$
-1	-4
0	-1
2	5

Plot these points and connect them.

35. Graph $g(x) = -2x + 3$

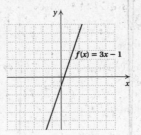

Make a list of function values in a table.

When $x = -1$, $g(-1) = -2(-1) + 3 = 2 + 3 = 5$.

When $x = 0$, $g(0) = -2 \cdot 0 + 3 = 0 + 3 = 3$.

When $x = 3$, $g(3) = -2 \cdot 3 + 3 = -6 + 3 = -3$.

x	$g(x)$
-1	5
0	3
3	-3

Plot these points and connect them.

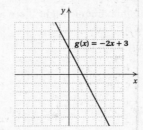

37. Graph $f(x) = \dfrac{1}{2}x + 1$.

Make a list of function values in a table.

When $x = -2$, $f(-2) = \dfrac{1}{2}(-2) + 1 = -1 + 1 = 0$.

When $x = 0$, $f(0) = \dfrac{1}{2} \cdot 0 + 1 = 0 + 1 = 1$.

When $x = 4$, $f(4) = \dfrac{1}{2} \cdot 4 + 1 = 2 + 1 = 3$.

x	$f(x)$
-2	0
0	1
4	3

Plot these points and connect them.

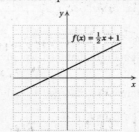

39. Graph $f(x) = 2 - |x|$.

Make a list of function values in a table.

When $x = -4$, $f(-4) = 2 - |-4| = 2 - 4 = -2$.

When $x = 0$, $f(0) = 2 - |0| = 2 - 0 = 2$.

When $x = 3$, $f(3) = 2 - |3| = 2 - 3 = -1$.

x	$f(x)$
-4	-2
0	2
3	-1

Plot these points and connect them.

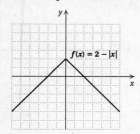

41. Graph $f(x) = x^2$.

Recall from Section 17.6 that the graph is a parabola. Make a list of function values in a table.

When $x = -2$, $f(-2) = (-2)^2 = 4$.

When $x = -1$, $f(-1) = (-1)^2 = 1$.

When $x = 0$, $f(0) = 0^2 = 0$.

When $x = 1$, $f(1) = 1^2 = 1$.

When $x = 2$, $f(2) = 2^2 = 4$.

x	$f(x)$
-2	4
-1	1
0	0
1	1
2	4

Plot these points and connect them.

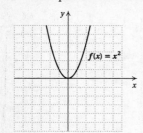

43. Graph $f(x) = x^2 - x - 2$.

Recall from Section 17.6 that the graph is a parabola. Make a list of function values in a table.

When $x = -1$, $f(-1) = (-1)^2 - (-1) - 2 = 1 + 1 - 2 = 0$.

When $x = 0$, $f(0) = 0^2 - 0 - 2 = -2$.

When $x = 1$, $f(1) = 1^2 - 1 - 2 = 1 - 1 - 2 = -2$.

When $x = 2$, $f(2) = 2^2 - 2 - 2 = 4 - 2 - 2 = 0$.

x	$f(x)$
-1	0
0	-2
1	-2
2	0

Plot these points and connect them.

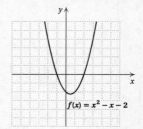

45. We can use the vertical line test:

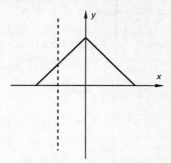

Visualize moving this vertical line across the graph. No vertical line will intersect the graph more than once. Thus, the graph is a graph of a function.

47. We can use the vertical line test:

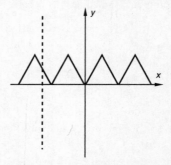

Visualize moving this vertical line across the graph. No vertical line will intersect the graph more than once. Thus, the graph is a graph of a function.

49. We can use the vertical line test.

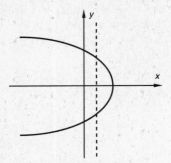

It is possible for a vertical line to intersect the graph more than once. Thus this is not the graph of a function.

51. We can use the vertical line test.

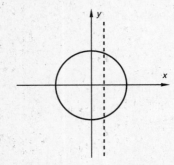

It is possible for a vertical line to intersect the graph more than once. Thus this is not a graph of a function.

53. Locate the point that is directly above 225. Then estimate its second coordinate by moving horizontally from the point to the vertical axis. The rate is about 75 per 10,000 men.

55. Discussion and Writing Exercise

57. The first equation is in slope-intercept form:

$$y = \frac{3}{4}x - 7, \ m = \frac{3}{4}$$

We write the second equation in slope-intercept form.

$$3x + 4y = 7$$
$$4y = -3x + 7$$
$$y = -\frac{3}{4}x + \frac{7}{4}, m = -\frac{3}{4}$$

Since the slopes are different, the equations do not represent parallel lines.

59. $2x - \ y = 6, \quad (1)$
$\quad 4x - 2y = 5 \quad (2)$

We solve Equation (1) for y.

$\quad 2x - y = 6 \quad (1)$
$\quad 2x - 6 = y \quad$ Adding y and -6

Substitute $2x - 6$ for y in Equation (2) and solve for x.

$\quad 4x - 2y = 5 \quad (2)$
$\quad 4x - 2(2x - 6) = 5$
$\quad 4x - 4x + 12 = 5$
$\quad\quad\quad 12 = 5$

We get a false equation, so the system has no solution.

61. Graph $g(x) = x^3$.

Make a list of function values in a table. Then plot the points and connect them.

x	$g(x)$
-2	-8
-1	-1
0	0
1	1
2	8

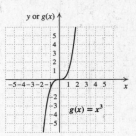

63. Graph $g(x) = |x| + x$.

Make a list of function values in a table. Then plot the points and connect them.

x	$f(x)$
-3	0
-2	0
-1	0
0	0
1	2
2	4
3	6

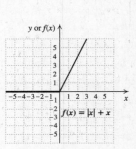

Chapter 17 Review Exercises

1. $8x^2 = 24$
$\quad\ x^2 = 3$
$\quad x = \sqrt{3} \ or \ x = -\sqrt{3}$
The solutions are $\sqrt{3}$ and $-\sqrt{3}$.

2. $40 = 5y^2$
$\quad\ 8 = y^2$
$\quad y = \sqrt{8} \ \ or \ y = -\sqrt{8}$
$\quad y = 2\sqrt{2} \ or \ y = -2\sqrt{2}$
The solutions are $2\sqrt{2}$ and $-2\sqrt{2}$.

3. $\quad 5x^2 - 8x + 3 = 0$
$\quad (5x - 3)(x - 1) = 0$
$\quad 5x - 3 = 0 \ \ or \ \ x - 1 = 0$
$\quad\quad 5x = 3 \ \ or \quad\quad x = 1$
$\quad\quad\ x = \frac{3}{5} \ \ or \quad\quad x = 1$

The solutions are $\frac{3}{5}$ and 1.

4.
$$3y^2 + 5y = 2$$
$$3y^2 + 5y - 2 = 0$$
$$(3y - 1)(y + 2) = 0$$
$$3y - 1 = 0 \quad or \quad y + 2 = 0$$
$$3y = 1 \quad or \qquad y = -2$$
$$y = \frac{1}{3} \quad or \qquad y = -2$$

The solutions are $\frac{1}{3}$ and -2.

5. $(x + 8)^2 = 13$
$$x + 8 = \sqrt{13} \qquad or \quad x + 8 = -\sqrt{13}$$
$$x = -8 + \sqrt{13} \quad or \qquad x = -8 - \sqrt{13}$$

The solutions are $-8 \pm \sqrt{13}$.

6. $9x^2 = 0$
$$x^2 = 0$$
$$x = 0$$

The solution is 0.

7. $5t^2 - 7t = 0$
$$t(5t - 7) = 0$$
$$t = 0 \quad or \quad 5t - 7 = 0$$
$$t = 0 \quad or \qquad 5t = 7$$
$$t = 0 \quad or \qquad t = \frac{7}{5}$$

The solutions are 0 and $\frac{7}{5}$.

8. $x^2 - 2x - 10 = 0$
$$a = 1, \ b = -2, \ c = -10$$
$$x = \frac{-(-2) \pm \sqrt{(-2)^2 - 4 \cdot 1 \cdot (-10)}}{2 \cdot 1}$$
$$x = \frac{2 \pm \sqrt{4 + 40}}{2}$$
$$x = \frac{2 \pm \sqrt{44}}{2} = \frac{2 \pm \sqrt{4 \cdot 11}}{2}$$
$$x = \frac{2 \pm 2\sqrt{11}}{2} = \frac{2(1 \pm \sqrt{11})}{2}$$
$$x = 1 \pm \sqrt{11}$$

The solutions are $1 \pm \sqrt{11}$.

9. $9x^2 - 6x - 9 = 0$
$$a = 9, \ b = -6, \ c = -9$$
$$x = \frac{-(-6) \pm \sqrt{(-6)^2 - 4 \cdot 9 \cdot (-9)}}{2 \cdot 9}$$
$$x = \frac{6 \pm \sqrt{36 + 324}}{18}$$
$$x = \frac{6 \pm \sqrt{360}}{18} = \frac{6 \pm \sqrt{36 \cdot 10}}{18}$$
$$x = \frac{6 \pm 6\sqrt{10}}{18} = \frac{6(1 \pm \sqrt{10})}{3 \cdot 6}$$
$$x = \frac{1 \pm \sqrt{10}}{3}$$

The solutions are $\frac{1 \pm \sqrt{10}}{3}$.

10.
$$x^2 + 6x = 9$$
$$x^2 + 6x - 9 = 0$$
$$a = 1, \ b = 6, \ c = -9$$
$$x = \frac{-6 \pm \sqrt{6^2 - 4 \cdot 1 \cdot (-9)}}{2 \cdot 1}$$
$$x = \frac{-6 \pm \sqrt{36 + 36}}{2}$$
$$x = \frac{-6 \pm \sqrt{72}}{2} = \frac{-6 \pm \sqrt{36 \cdot 2}}{2}$$
$$x = \frac{-6 \pm 6\sqrt{2}}{2} = \frac{2(-3 \pm 3\sqrt{2})}{2}$$
$$x = -3 \pm 3\sqrt{2}$$

The solutions are $-3 \pm 3\sqrt{2}$.

11.
$$1 + 4x^2 = 8x$$
$$4x^2 - 8x + 1 = 0$$
$$a = 4, \ b = -8, \ c = 1$$
$$x = \frac{-(-8) \pm \sqrt{(-8)^2 - 4 \cdot 4 \cdot 1}}{2 \cdot 4}$$
$$x = \frac{8 \pm \sqrt{64 - 16}}{8}$$
$$x = \frac{8 \pm \sqrt{48}}{8} = \frac{8 \pm \sqrt{16 \cdot 3}}{8}$$
$$x = \frac{8 \pm 4\sqrt{3}}{8} = \frac{4(2 \pm \sqrt{3})}{2 \cdot 4}$$
$$x = \frac{2 \pm \sqrt{3}}{2}$$

The solutions are $\frac{2 \pm \sqrt{3}}{2}$.

12. $6 + 3y = y^2$

$$0 = y^2 - 3y - 6$$

$a = 1, b = -3, c = -6$

$$y = \frac{-(-3) \pm \sqrt{(-3)^2 - 4 \cdot 1 \cdot (-6)}}{2 \cdot 1}$$

$$y = \frac{3 \pm \sqrt{9 + 24}}{2} = \frac{3 \pm \sqrt{33}}{2}$$

The solutions are $\frac{3 \pm \sqrt{33}}{2}$.

13. $3m = 4 + 5m^2$

$$0 = 5m^2 - 3m + 4$$

$a = 5, b = -3, c = 4$

$$m = \frac{-(-3) \pm \sqrt{(-3)^2 - 4 \cdot 5 \cdot 4}}{2 \cdot 5}$$

$$m = \frac{3 \pm \sqrt{9 - 80}}{10} = \frac{3 \pm \sqrt{-71}}{10}$$

Since the radicand is negative, there are no real-number solutions.

14. $3x^2 = 4x$

$$3x^2 - 4x = 0$$

$$x(3x - 4) = 0$$

$x = 0 \quad or \quad 3x - 4 = 0$

$x = 0 \quad or \qquad 3x = 4$

$x = 0 \quad or \qquad x = \frac{4}{3}$

The solutions are 0 and $\frac{4}{3}$.

15.
$$\frac{15}{x} - \frac{15}{x+2} = 2, \text{ LCM is } x(x+2)$$

$$x(x+2)\left(\frac{15}{x} - \frac{15}{x+2}\right) = x(x+2)(2)$$

$$x(x+2) \cdot \frac{15}{x} - x(x+2) \cdot \frac{15}{x+2} = 2x(x+2)$$

$$15(x+2) - 15x = 2x^2 + 4x$$

$$15x + 30 - 15x = 2x^2 + 4x$$

$$30 = 2x^2 + 4x$$

$$0 = 2x^2 + 4x - 30$$

$$0 = x^2 + 2x - 15 \qquad \text{Dividing by 2}$$

$$0 = (x+5)(x-3)$$

$x + 5 = 0 \quad or \quad x - 3 = 0$

$\quad x = -5 \quad or \qquad x = 3$

Both numbers check. The solutions are -5 and 3.

16. $x + \frac{1}{x} = 2, \text{ LCM is } x$

$$x\left(x + \frac{1}{x}\right) = x \cdot 2$$

$$x \cdot x + x \cdot \frac{1}{x} = 2x$$

$$x^2 + 1 = 2x$$

$$x^2 - 2x + 1 = 0$$

$$(x-1)^2 = 0$$

$$x - 1 = 0$$

$$x = 1$$

The number 1 checks. It is the solution.

17. $x^2 - 5x + \quad 2 = 0$

$x^2 - 5x \qquad = -2$

$x^2 - 5x + \frac{25}{4} = -2 + \frac{25}{4} \quad \text{Adding } \frac{25}{4}: \left(\frac{-5}{2}\right)^2 = \frac{25}{4}$

$$\left(x - \frac{5}{2}\right)^2 = \frac{17}{4}$$

$x - \frac{5}{2} = \frac{\sqrt{17}}{2} \qquad or \quad x - \frac{5}{2} = -\frac{\sqrt{17}}{2}$

$x = \frac{5}{2} + \frac{\sqrt{17}}{2} \quad or \qquad x = \frac{5}{2} - \frac{\sqrt{17}}{2}$

$x = \frac{5 + \sqrt{17}}{2} \quad or \qquad x = \frac{5 - \sqrt{17}}{2}$

The solutions are $\frac{5 \pm \sqrt{17}}{2}$.

18. $3x^2 - 2x - \quad 5 = 0$

$\frac{1}{3}(3x^2 - 2x - \quad 5) = \frac{1}{3} \cdot 0$

$x^2 - \frac{2}{3}x - \quad \frac{5}{3} = 0$

$x^2 - \frac{2}{3}x \qquad = \frac{5}{3}$

$x^2 - \frac{2}{3}x + \frac{1}{9} = \frac{5}{3} + \frac{1}{9} \quad \text{Adding } \frac{1}{9}:$

$$\left[\frac{1}{2}\left(-\frac{2}{3}\right)\right]^2 = \left(-\frac{1}{3}\right)^2 = \frac{1}{9}$$

$$\left(x - \frac{1}{3}\right)^2 = \frac{16}{9}$$

$x - \frac{1}{3} = \frac{4}{3} \quad or \quad x - \frac{1}{3} = -\frac{4}{3}$

$x = \frac{5}{3} \quad or \qquad x = -1$

The solutions are $\frac{5}{3}$ and -1.

19. From Exercise 17, we know the solutions are $\frac{5 \pm \sqrt{17}}{2}$.

Using a calculator, we have

$$\frac{5 + \sqrt{17}}{2} \approx 4.6 \text{ and } \frac{5 - \sqrt{17}}{2} \approx 0.4.$$

20. $4y^2 + 8y + 1 = 0$

$a = 4, b = 8, c = 1$

$$y = \frac{-8 \pm \sqrt{8^2 - 4 \cdot 4 \cdot 1}}{2 \cdot 4}$$

$$y = \frac{-8 \pm \sqrt{64 - 16}}{8} = \frac{-8 \pm \sqrt{48}}{8}$$

Using a calculator, we have $\dfrac{-8 + \sqrt{48}}{8} \approx -0.1$ and

$\dfrac{-8 - \sqrt{48}}{8} \approx -1.9$.

21.

$$V = \frac{1}{2}\sqrt{1 + \frac{T}{L}}$$

$$V^2 = \left(\frac{1}{2}\sqrt{1 + \frac{T}{L}}\right)^2$$

$$V^2 = \frac{1}{4}\left(1 + \frac{T}{L}\right)$$

$$4 \cdot V^2 = 4 \cdot \frac{1}{4}\left(1 + \frac{T}{L}\right)$$

$$4V^2 = 1 + \frac{T}{L}$$

$$4V^2 - 1 = \frac{T}{L}$$

$$L(4V^2 - 1) = L \cdot \frac{T}{L}$$

$$L(4V^2 - 1) = T$$

22. $y = 2 - x^2$, or $y = -x^2 + 2$

Find the vertex. The x-coordinate is

$$-\frac{b}{2a} = -\frac{0}{2(-1)} = 0.$$

The y-coordinate is

$$y = 2 - x^2 = 2 - 0^2 = 2.$$

The vertex is $(0, 2)$. This is also the y-intercept.

We choose some x-values on both sides of the vertex and graph the parabola.

x	y	
0	2	←Vertex
−1	1	
−2	−2	
1	1	
2	−2	

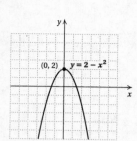

23. $y = x^2 - 4x - 2$

Find the vertex. The x-coordinate is

$$-\frac{b}{2a} = -\frac{-4}{2 \cdot 1} = -(-2) = 2.$$

The y-coordinate is

$$y = 2^2 - 4 \cdot 2 - 2 = 4 - 8 - 2 = -6.$$

The vertex is $(2, -6)$.

We choose some x-values on both sides of the vertex and graph the parabola.

x	y	
2	−6	← Vertex
0	−2	← y-intercept
−1	3	
3	5	
5	3	

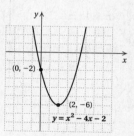

24. $y = 2 - x^2$

To find the x-intercepts we solve the equation $2 - x^2 = 0$.

$$2 - x^2 = 0$$

$$2 = x^2$$

$$x = \sqrt{2} \quad or \quad x = -\sqrt{2}$$

The x-intercepts are $(\sqrt{2}, 0)$ and $(-\sqrt{2}, 0)$.

25. $y = x^2 - 4x - 2$

To find the x-intercepts we solve the equation $x^2 - 4x - 2 = 0$.

$$x = \frac{-(-4) \pm \sqrt{(-4)^2 - 4 \cdot 1(-2)}}{2 \cdot 1}$$

$$x = \frac{4 \pm \sqrt{16 + 8}}{2} = \frac{4 + \sqrt{24}}{2}$$

$$x = \frac{4 \pm \sqrt{4 \cdot 6}}{2} = \frac{4 \pm 2\sqrt{6}}{2}$$

$$x = \frac{2(2 \pm \sqrt{6})}{2} = 2 \pm \sqrt{6}$$

The x-intercepts are $(2 - \sqrt{6}, 0)$ and $(2 + \sqrt{6}, 0)$.

26. *Familiarize*. Using the labels on the drawing in the text, we let a and $a + 3$ represent the lengths of the legs, in cm.

Translate. We use the Pythagorean equation.

$$a^2 + (a + 3)^2 = 5^2$$

Solve.

$$a^2 + (a + 3)^2 = 5^2$$

$$a^2 + a^2 + 6a + 9 = 25$$

$$2a^2 + 6a + 9 = 25$$

$$2a^2 + 6a - 16 = 0$$

$$a^2 + 3a - 8 = 0 \quad \text{Dividing by 2}$$

We use the quadratic formula.

$$a = \frac{-3 \pm \sqrt{3^2 - 4 \cdot 1 \cdot (-8)}}{2 \cdot 1}$$

$$a = \frac{-3 \pm \sqrt{9 + 32}}{2} = \frac{-3 \pm \sqrt{41}}{2}$$

Using a calculator, we have

$$a = \frac{-3 - \sqrt{41}}{2} \approx -4.7 \text{ and } a = \frac{-3 + \sqrt{41}}{2} \approx 1.7.$$

Check. Since the length of a leg cannot be negative, -4.7 cannot be a solution. If $a \approx 1.7$, then $a + 3 \approx 4.7$ and $(1.7)^2 + (4.7)^4 = 24.98 \approx 25 = 5^2$. The answer checks.

State. The lengths of the legs are about 1.7 cm and 4.7 cm.

27. *Familiarize*. Using the labels on the drawing in the text, we let s and $s - 5$ represent the lengths of the legs, in ft.

Translate. We use the Pythagorean equation.
$$s^2 + (s - 5)^2 = 25^2$$

Solve.
$$s^2 + (s - 5)^2 = 25^2$$
$$s^2 + s^2 - 10s + 25 = 625$$
$$2s^2 - 10s + 25 = 625$$
$$2s^2 - 10s - 600 = 0$$
$$s^2 - 5s - 300 = 0 \quad \text{Dividing by 2}$$
$$(s - 20)(s + 15) = 0$$
$$s - 20 = 0 \quad or \quad s + 15 = 0$$
$$s = 20 \quad or \quad s = -15$$

Check. Since the length of a leg of the triangle cannot be negative, -15 cannot be a solution. If $s = 20$, then $s - 5 = 20 - 5 = 15$ and $20^2 + 15^2 = 625 = 25^2$. The answer checks.

State. The height of the ramp is 15 ft.

28. *Familiarize*. We will use the formula $s = 16t^2$.

Translate. We substitute 645 for s.
$$645 = 16t^2$$

Solve.
$$645 = 16t^2$$
$$\frac{645}{16} = t^2$$
$$40.3125 = t^2$$
$$t = \sqrt{40.3125} \quad or \quad t = -\sqrt{40.3125}$$
$$t \approx 6.3 \qquad or \quad t \approx -6.3$$

Check. Time cannot be negative in this application, so -6.3 cannot be a solution. We check 6.3.
$$16(6.3)^2 = 635.04 \approx 645$$

The answer is close, so we have a check. (Remember that we rounded the value of t.)

State. It would take about 6.3 sec for an object to fall to the ground from the top of Lake Point Towers.

29. $f(x) = 2x - 5$
$$f(2) = 2 \cdot 2 - 5 = 4 - 5 = -1$$
$$f(-1) = 2(-1) - 5 = -2 - 5 = -7$$
$$f(3.5) = 2(3.5) - 5 = 7 - 5 = 2$$

30. $g(x) = |x| - 1$
$$g(1) = |1| - 1 = 1 - 1 = 0$$
$$g(-1) = |-1| - 1 = 1 - 1 = 0$$
$$g(-20) = |-20| - 1 = 20 - 1 = 19$$

31. $C(p) = 15p$
$$C(180) = 15 \cdot 180 = 2700 \text{ calories}$$

32. $g(x) = 4 - x$

We find some function values.

When $x = -1$, $g(-1) = 4 - (-1) = 4 + 1 = 5$

When $x = 0$, $g(0) = 4 - 0 = 4$.

When $x = 3$, $g(3) = 4 - 3 = 1$.

x	$g(x)$
-1	5
0	4
3	1

Plot these points and connect them.

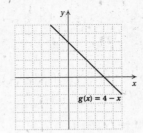

33. $f(x) = x^2 - 3$

Recall from Section 17.6 that the graph is a parabola. We find some function values.

When $x = -3$, $f(-3) = (-3)^2 - 3 = 9 - 3 = 6$.

When $x = -1$, $f(-1) = (-1)^2 - 3 = 1 - 3 = -2$.

When $x = 0$, $f(0) = 0^2 - 3 = 0 - 3 = -3$.

When $x = 1$, $f(1) = 1^2 - 3 = 1 - 3 = -2$.

When $x = 2$, $f(2) = 2^2 - 3 = 4 - 3 = 1$.

x	$f(x)$
-3	6
-1	-2
0	-3
1	-2
2	1

Plot these points and connect them.

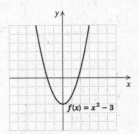

34. $h(x) = |x| - 5$

We find some function values.

When $x = -4$, $h(-4) = |-4| - 5 = 4 - 5 = -1$.

When $x = -2$, $h(-2) = |-2| - 5 = 2 - 5 = -3$.

When $x = 0$, $h(0) = |0| - 5 = 0 - 5 = -5$.

When $x = 1$, $h(1) = |1| - 5 = 1 - 5 = -4$.

When $x = 3$, $h(3) = |3| - 5 = 3 - 5 = -2$.

x	$h(x)$
-4	-1
-2	-3
0	-5
1	-4
3	-2

Plot these points and connect them.

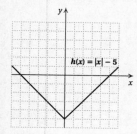

35. $f(x) = x^2 - 2x + 1$

Recall from Section 17.6 that the graph is a parabola.

When $x = -1$, $f(-1) = (-1)^2 - 2(-1) + 1 = 1 + 2 + 1 = 4$.

When $x = 0$, $f(0) = 0^2 - 2 \cdot 0 + 1 = 0 - 0 + 1 = 1$.

When $x = 1$, $f(1) = 1^2 - 2 \cdot 1 + 1 = 1 - 2 + 1 = 0$.

When $x = 2$, $f(2) = 2^2 - 2 \cdot 2 + 1 = 4 - 4 + 1 = 1$.

When $x = 3$, $f(3) = 3^2 - 2 \cdot 3 + 1 = 9 - 6 + 1 = 4$.

x	$f(x)$
-1	4
0	1
1	0
2	1
3	4

Plot these points and connect them.

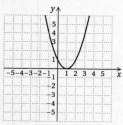

$f(x) = x^2 - 2x + 1$

36. It is possible for a vertical line to intersect the graph more than once, so this is not the graph of a function.

37. No vertical line will intersect the graph more than once, so this is the graph of a function.

38. *Discussion and Writing Exercise.*

Equation	Form	Example
Linear	Equivalent to $x = a$	$3x - 5 = 8$
Quadratic	$ax^2 + bx + c = 0$	$2x^2 - 3x + 1 = 0$
Rational	Contains one or more rational expressions	$\dfrac{x}{3} + \dfrac{4}{x-1} = 1$
Radical	Contains one or more radical expressions	$\sqrt{3x-1} = x - 7$
Systems of equations	$Ax + By = C,$ $Dx + Ey = F$	$4x - 5y = 3,$ $3x + 2y = 1$

39. *Discussion and Writing Exercise.*

a) The third line should be $x = 0$ *or* $x + 20 = 0$; the solution 0 is lost in the given procedure. Also, the last line should be $x = -20$.

b) The addition principle should be used at the outset to get 0 on one side of the equation. Since this was not done in the given procedure, the principle of zero products was not applied correctly.

40. **Familiarize.** Let $x = $ the first integer. Then $x + 1 = $ the second integer.

Translate. If the numbers are positive, then $(x + 1)^2$ is larger than x^2 and we have:

$$\underbrace{\text{Square of larger number}}_{(x+1)^2} \underbrace{\text{minus}}_{-} \underbrace{\text{square of smaller number}}_{x^2} \underbrace{\text{is}}_{=} \underbrace{63.}_{63}$$

If the numbers are negative, then x^2 is larger than $(x + 1)^2$ and we have:

$$\underbrace{\text{Square of smaller number}}_{x^2} \underbrace{\text{minus}}_{-} \underbrace{\text{square of larger number}}_{(x+1)^2} \underbrace{\text{is}}_{=} \underbrace{63.}_{63}$$

Solve. We solve each equation.

$$(x + 1)^2 - x^2 = 63$$
$$x^2 + 2x + 1 - x^2 = 63$$
$$2x + 1 = 63$$
$$2x = 62$$
$$x = 31$$

If $x = 31$, then $x + 1 = 31 + 1 = 32$.

$$x^2 - (x + 1)^2 = 63$$
$$x^2 - (x^2 + 2x + 1) = 63$$
$$x^2 - x^2 - 2x - 1 = 63$$
$$-2x - 1 = 63$$
$$-2x = 64$$
$$x = -32$$

If $x = -32$, then $x + 1 = -32 + 1 = -31$.

Check. 31 and 32 are consecutive integers and $32^2 - 31^2 = 1024 - 961 = 63$. Also, -32 and -31 are consecutive integers and $(-32)^2 - (-31)^2 = 1024 - 961 = 63$. Both pairs of numbers check.

State. The integers are 31 and 32 or -32 and -31.

41. Familiarize. The area of a square with side s is s^2; the area of a circle with radius 5 in. is $\pi \cdot 5^2$, or 25π in^2.

Translate.

$$\underbrace{\text{Area of square}}_{s^2} \quad \underbrace{\text{equals}}_{=} \quad \underbrace{\text{area of circle}}_{25\pi}.$$

Solve.

$$s^2 = 25\pi$$
$$s = 5\sqrt{\pi} \quad or \quad s = -5\sqrt{\pi}$$

Check. Since the length of a side of the square cannot be negative, $-5\sqrt{\pi}$ cannot be a solution. If the length of a side of the square is $5\sqrt{\pi}$, then the area of the square is $(5\sqrt{\pi})^2 = 25\pi$. Since this is also the area of the circle, $5\sqrt{\pi}$ checks.

State. $s = 5\sqrt{\pi}$ in. ≈ 8.9 in.

42. $x - 4\sqrt{x} - 5 = 0$

Let $u = \sqrt{x}$. Then $u^2 = x$. Substitute u for $\sqrt{x}$ and u^2 for x and solve for u.

$$u^2 - 4u - 5 = 0$$
$$(u + 1)(u - 5) = 0$$
$$u + 1 = 0 \quad or \quad u - 5 = 0$$
$$u = -1 \quad or \quad u = 5$$

Now we substitute $\sqrt{x}$ for u and solve for x.

$\sqrt{x} = -1$ has no real-number solutions.

If $u = 5$, then we have:

$$\sqrt{x} = 5$$
$$(\sqrt{x})^2 = 5^2$$
$$x = 25$$

The number 25 checks. It is the solution.

43. The graph of $y = (x+3)^2$ contains the points $(-4, 1)$ and $(-2, 1)$, so the solutions of $(x+3)^2 = 1$ are -4 and -2.

44. The graph of $y = (x+3)^2$ contains the points $(-5, 4)$ and $(-1, 4)$, so the solutions of $(x+3)^2 = 4$ are -5 and -1.

45. The graph of $y = (x+3)^2$ contains the points $(-6, 9)$ and $(0, 9)$, so the solutions of $(x+3)^2 = 9$ are -6 and 0.

46. The graph of $y = (x+3)^2$ contains the point $(-3, 0)$, so the solution of $(x+3)^2 = 0$ is -3.

Chapter 17 Test

1. $7x^2 = 35$
$$x^2 = 5$$
$$x = \sqrt{5} \quad or \quad x = -\sqrt{5}$$
The solutions are $\sqrt{5}$ and $-\sqrt{5}$.

2. $7x^2 + 8x = 0$
$$x(7x + 8) = 0$$

$$x = 0 \quad or \quad 7x + 8 = 0$$
$$x = 0 \quad or \quad 7x = -8$$
$$x = 0 \quad or \quad x = -\frac{8}{7}$$
The solutions are 0 and $-\dfrac{8}{7}$.

3. $48 = t^2 + 2t$
$$0 = t^2 + 2t - 48$$
$$0 = (t + 8)(t - 6)$$
$$t + 8 = 0 \quad or \quad t - 6 = 0$$
$$t = -8 \quad or \quad t = 6$$
The solutions are -8 and 6.

4. $\qquad 3y^2 - 5y = 2$
$$3y^2 - 5y - 2 = 0$$
$$(3y + 1)(y - 2) = 0$$
$$3y + 1 = 0 \quad or \quad y - 2 = 0$$
$$3y = -1 \quad or \quad y = 2$$
$$y = -\frac{1}{3} \quad or \quad y = 2$$
The solutions are $-\dfrac{1}{3}$ and 2.

5. $(x - 8)^2 = 13$
$$x - 8 = \sqrt{13} \quad or \quad x - 8 = -\sqrt{13}$$
$$x = 8 + \sqrt{13} \quad or \quad x = 8 - \sqrt{13}$$
The solutions are $8 \pm \sqrt{13}$.

6. $\qquad x^2 = x + 3$
$$x^2 - x - 3 = 0$$
$$a = 1, b = -1, c = -3$$
$$x = \frac{-(-1) \pm \sqrt{(-1)^2 - 4 \cdot 1 \cdot (-3)}}{2 \cdot 1}$$
$$x = \frac{1 \pm \sqrt{1 + 12}}{2}$$
$$x = \frac{1 \pm \sqrt{13}}{2}$$
The solutions are $\dfrac{1 \pm \sqrt{13}}{2}$.

7. $\qquad m^2 - 3m = 7$
$$m^2 - 3m - 7 = 0$$
$$a = 1, b = -3, c = -7$$
$$m = \frac{-(-3) \pm \sqrt{(-3)^2 - 4 \cdot 1 \cdot (-7)}}{2 \cdot 1}$$
$$m = \frac{3 \pm \sqrt{9 + 28}}{2}$$
$$m = \frac{3 \pm \sqrt{37}}{2}$$
The solutions are $\dfrac{3 \pm \sqrt{37}}{2}$.

8. $10 = 4x + x^2$

$0 = x^2 + 4x - 10$

$a = 1, b = 4, c = -10$

$x = \dfrac{-4 \pm \sqrt{4^2 - 4 \cdot 1 \cdot (-10)}}{2 \cdot 1}$

$x = \dfrac{-4 \pm \sqrt{16 + 40}}{2} = \dfrac{-4 \pm \sqrt{56}}{2}$

$x = \dfrac{-4 \pm \sqrt{4 \cdot 14}}{2} = \dfrac{-4 \pm 2\sqrt{14}}{2}$

$x = \dfrac{2(-2 \pm \sqrt{14})}{2} = -2 \pm \sqrt{14}$

The solutions are $-2 \pm \sqrt{14}$.

9. $3x^2 - 7x + 1 = 0$

$a = 3, b = -7, c = 1$

$x = \dfrac{-(-7) \pm \sqrt{(-7)^2 - 4 \cdot 3 \cdot 1}}{2 \cdot 3}$

$x = \dfrac{7 \pm \sqrt{49 - 12}}{6} = \dfrac{7 \pm \sqrt{37}}{6}$

The solutions are $\dfrac{7 \pm \sqrt{37}}{6}$.

10. $\qquad x - \dfrac{2}{x} = 1$, LCM is x

$x\left(x - \dfrac{2}{x}\right) = x \cdot 1$

$x \cdot x - x \cdot \dfrac{2}{x} = x$

$x^2 - 2 = x$

$x^2 - x - 2 = 0$

$(x - 2)(x + 1) = 0$

$x - 2 = 0 \quad or \quad x + 1 = 0$

$x = 2 \quad or \qquad x = -1$

Both numbers check. The solutions are 2 and -1.

11. $\qquad \dfrac{4}{x} - \dfrac{4}{x + 2} = 1$, LCM is $x(x+2)$

$x(x + 2)\left(\dfrac{4}{x} - \dfrac{4}{x + 2}\right) = x(x + 2) \cdot 1$

$x(x + 2) \cdot \dfrac{4}{x} - x(x + 2) \cdot \dfrac{4}{x + 2} = x(x + 2)$

$4(x + 2) - 4x = x^2 + 2x$

$4x + 8 - 4x = x^2 + 2x$

$8 = x^2 + 2x$

$0 = x^2 + 2x - 8$

$0 = (x + 4)(x - 2)$

$x + 4 = 0 \quad or \quad x - 2 = 0$

$x = -4 \quad or \qquad x = 2$

Both numbers check. The solutions are -4 and 2.

12. $x^2 - 4x - 10 = 0$

$x^2 - 4x \qquad = 10$

$x^2 - 4x + \quad 4 = 10 + 4 \quad$ Adding 4: $\left(\dfrac{-4}{2}\right)^2 = (-2)^2 = 4$

$(x - 2)^2 = 14$

$x - 2 = \sqrt{14} \qquad or \quad x - 2 = -\sqrt{14}$

$x = 2 + \sqrt{14} \quad or \qquad x = 2 - \sqrt{14}$

The solutions are $2 \pm \sqrt{14}$.

13. From Exercise 12 we know that the solutions of the equation are $2 \pm \sqrt{14}$.

Using a calculator, we have

$2 - \sqrt{14} \approx -1.7$ and $2 + \sqrt{14} \approx 5.7$.

14. $d = an^2 + bn$

$0 = an^2 + bn - d$

We will use the quadratic formula with $a = a$, $b = b$, and $c = -d$.

$n = \dfrac{-b \pm \sqrt{b^2 - 4 \cdot a \cdot (-d)}}{2 \cdot a}$

$n = \dfrac{-b + \sqrt{b^2 + 4ad}}{2a} \qquad$ Using the positive square root

15. To find the x-intercepts we solve the following equation.

$-x^2 + x + 5 = 0$

$x^2 - x - 5 = 0 \quad$ Standard form

$a = 1, b = -1, c = -5$

$x = \dfrac{-(-1) \pm \sqrt{(-1)^2 - 4 \cdot 1 \cdot (-5)}}{2 \cdot 1}$

$x = \dfrac{1 \pm \sqrt{1 + 20}}{2} = \dfrac{1 \pm \sqrt{21}}{2}$

The x-intercepts are $\left(\dfrac{1 - \sqrt{21}}{2}, 0\right)$ and $\left(\dfrac{1 + \sqrt{21}}{2}, 0\right)$.

16. $y = 4 - x^2$, or $y = -x^2 + 4$

Find the vertex. The x-coordinate is

$-\dfrac{b}{2a} = -\dfrac{0}{2(-1)} = 0$.

The y-coordinate is

$y = 4 - x^2 = 4 - 0^2 = 4$.

The vertex is $(0, 4)$. This is also the y-intercept.

We choose some x-values on both sides of the vertex and graph the parabola.

x	y	
0	4	←Vertex
-1	3	
-2	0	
1	3	
2	0	

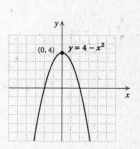

17. $y = -x^2 + x + 5$

Find the vertex. The x-coordinate is

$$-\frac{b}{2a} = -\frac{1}{2(-1)} = -\left(-\frac{1}{2}\right) = \frac{1}{2}.$$

The y-coordinate is

$$y = -\left(\frac{1}{2}\right)^2 + \frac{1}{2} + 5 = -\frac{1}{4} + \frac{1}{2} + 5 = \frac{21}{4}.$$

The vertex is $\left(\frac{1}{2}, \frac{21}{4}\right)$.

We choose some x-values on both sides of the vertex and graph the parabola.

x	y	
$\frac{1}{2}$	$\frac{21}{4}$	← Vertex
0	5	← y-intercept
-2	-1	
-1	3	
2	3	
3	-1	

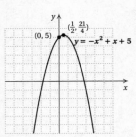

18. $f(x) = \frac{1}{2}x + 1$

$$f(0) = \frac{1}{2} \cdot 0 + 1 = 0 + 1 = 1$$

$$f(1) = \frac{1}{2} \cdot 1 + 1 = \frac{1}{2} + 1 = 1\frac{1}{2}$$

$$f(2) = \frac{1}{2} \cdot 2 + 1 = 1 + 1 = 2$$

19. $g(t) = -2|t| + 3$

$$g(-1) = -2|-1| + 3 = -2 \cdot 1 + 3 = -2 + 3 = 1$$

$$g(0) = -2|0| + 3 = -2 \cdot 0 + 3 = 0 + 3 = 3$$

$$g(3) = -2|3| + 3 = -2 \cdot 3 + 3 = -6 + 3 = -3$$

20. *Familiarize.* Using the labels on the drawing in the text, we let l and $l-4$ represent the length and width of the rug, respectively, in meters. Recall that the area of a rectangle is (length) × (width).

Translate.

$$\underbrace{\text{The area}}_{\downarrow} \underbrace{\text{is}}_{\downarrow} \underbrace{16.25 \text{ m}^2}_{\downarrow}.$$
$$\quad l(l-4) \quad = \quad 16.25$$

Solve.

$$l(l-4) = 16.25$$
$$l^2 - 4l = 16.25$$
$$l^2 - 4l - 16.25 = 0$$

The factorization of $l^2 - 4l - 16.25$ is not readily apparent so we will use the quadratic formula with $a = 1$, $b = -4$, and $c = -16.25$.

$$l = \frac{-(-4) \pm \sqrt{(-4)^2 - 4 \cdot 1 \cdot (-16.25)}}{2 \cdot 1}$$

$$l = \frac{4 \pm \sqrt{16 + 65}}{2} = \frac{4 \pm \sqrt{81}}{2}$$

$$l = \frac{4 \pm 9}{2}$$

$$l = \frac{4 - 9}{2} = \frac{-5}{2} = -\frac{5}{2}, \text{ or } -2.5$$

or

$$l = \frac{4 + 9}{2} = \frac{13}{2} = 6.5$$

Check. Since the length cannot be negative, -2.5 cannot be a solution. If $l = 6.5$, then $l - 4 = 6.5 - 4 = 2.5$ and $6.5(2.5) = 16.25$. Thus, the width is 4 m less than the length and the area is 16.25 m^2. The answer checks.

State. The length of the rug is 6.5 m, and the width is 2.5 m.

21. *Familiarize.* Let $r =$ the speed of the boat in still water. Then $r - 2 =$ the speed upstream and $r + 2 =$ the speed downstream. We organize the information in a table.

	d	r	t
Upstream	44	$r - 2$	t_1
Downstream	52	$r + 2$	t_2

Translate. Using $t = d/r$ and the rows of the table, we have

$$t_1 = \frac{44}{r-2} \text{ and } t_2 = \frac{52}{r+2}.$$

Since the total time is 4 hr, $t_1 + t_2 = 4$, and we have

$$\frac{44}{r-2} + \frac{52}{r+2} = 4.$$

Solve. We solve the equation. We multiply by $(r-2)(r+2)$, the LCM of the denominators.

$$(r-2)(r+2)\left(\frac{44}{r-2} + \frac{52}{r+2}\right) = (r-2)(r+2) \cdot 4$$
$$44(r+2) + 52(r-2) = 4(r^2 - 4)$$
$$44r + 88 + 52r - 104 = 4r^2 - 16$$
$$96r - 16 = 4r^2 - 16$$
$$0 = 4r^2 - 96r$$
$$0 = 4r(r - 24)$$

$$4r = 0 \text{ or } r - 24 = 0$$
$$r = 0 \text{ or } \quad r = 24$$

Check. The boat cannot travel upstream if its speed in still water is 0 km/h. If the speed of the boat in still water is 24 km/h, then it travels at a speed of $24 - 2$, or 22 km/h, upstream and $24 + 2$, or 26 km/h, downstream. At 22 km/h the boat travels 44 km in 44/22, or 2 hr. At 26 km/h it travels 52 km in 52/26, or 2 hr. The total time is $2 + 2$, or 4 hr, so the answer checks.

State. The speed of the boat in still water is 24 km/h.

22. In 2010, $t = 2010 - 1940 = 70$.

$$R(70) = 30.18 - 0.06(70) = 30.18 - 4.2 = 25.98$$

We predict that the record will be 25.98 min in 2010.

23. $h(x) = x - 4$

We find some function values.

When $x = -1$, $h(x) = -1 - 4 = -5$.

When $x = 2$, $h(x) = 2 - 4 = -2$.

When $x = 5$, $h(x) = 5 - 4 = 1$.

x	$h(x)$
-1	-5
2	-2
5	1

Plot these points and connect them.

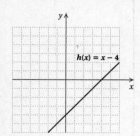

24. $g(x) = x^2 - 4$

Recall from Section 17.6 that the graph is a parabola. We find some function values.

When $x = -3$, $g(x) = (-3)^2 - 4 = 9 - 4 = 5$.

When $x = -1$, $g(x) = (-1)^2 - 4 = 1 - 4 = -3$.

When $x = 0$, $g(x) = 0^2 - 4 = 0 - 4 = -4$.

When $x = 2$, $g(2) = 2^2 - 4 = 4 - 4 = 0$.

When $x = 3$, $g(3) = 3^2 - 4 = 9 - 4 = 5$.

x	$g(x)$
-3	5
-1	-3
0	-4
2	0
3	5

Plot these points and connect them.

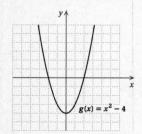

25. No vertical line will intersect the graph more than once, so this is the graph of a function.

26. It is possible for a vertical line to intersect the graph more than once, so this is not the graph of a function.

27. *Familiarize*. We make a drawing. Let $s =$ the length of a side of the square, in feet. Then $s + 5 =$ the length of a diagonal.

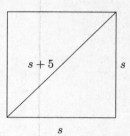

Translate. We use the Pythagorean equation.

$$s^2 + s^2 = (s + 5)^2$$

Solve.

$$s^2 + s^2 = (s + 5)^2$$
$$2s^2 = s^2 + 10s + 25$$
$$s^2 - 10s - 25 = 0$$

We use the quadratic formula with $a = 1$, $b = -10$, and $c = -25$.

$$s = \frac{-(-10) \pm \sqrt{(-10)^2 - 4 \cdot 1 \cdot (-25)}}{2 \cdot 1}$$

$$s = \frac{10 \pm \sqrt{100 + 100}}{2} = \frac{10 \pm \sqrt{200}}{2}$$

$$s = \frac{10 \pm \sqrt{100 \cdot 2}}{2} = \frac{10 \pm 10\sqrt{2}}{2}$$

$$s = \frac{2(5 \pm 5\sqrt{2})}{2} = 5 \pm 5\sqrt{2}$$

Check. Since $5 - 5\sqrt{2}$ is negative, it cannot be the length of a side of the square. If the length of a side is $5 + 5\sqrt{2}$ ft, then $(5 + 5\sqrt{2})^2 + (5 + 5\sqrt{2})^2 = 25 + 50\sqrt{2} + 50 + 25 + 50\sqrt{2} + 50 = 150 + 100\sqrt{2}$. The length of a diagonal is $5 + 5\sqrt{2} + 5$ ft, or $10 + 5\sqrt{2}$, and $(10 + 5\sqrt{2})^2 = 100 + 100\sqrt{2} + 50 = 150 + 100\sqrt{2}$. Since these lengths satisfy the Pythagorean equation, the answer checks.

State. The length of a side of the square is $5 + 5\sqrt{2}$ ft.

28.
$$x - y = 2, \quad (1)$$
$$xy = 4 \quad (2)$$

Solve Equation (1) for y.

$$x - y = 2$$
$$-y = -x + 2$$
$$y = x - 2$$

Substitute $x - 2$ for y in Equation (2) and solve for x.

$$xy = 4$$
$$x(x - 2) = 4$$
$$x^2 - 2x = 4$$
$$x^2 - 2x - 4 = 0$$

We use the quadratic formula with $a = 1$, $b = -2$, and $c = -4$.

$$x = \frac{-(-2) \pm \sqrt{(-2)^2 - 4 \cdot 1 \cdot (-4)}}{2 \cdot 1}$$

$$x = \frac{2 \pm \sqrt{4 + 16}}{2} = \frac{2 \pm \sqrt{20}}{2}$$

$$x = \frac{2 \pm \sqrt{4 \cdot 5}}{2} = \frac{2 \pm 2\sqrt{5}}{2}$$

$$x = \frac{2(1 \pm \sqrt{5})}{2} = 1 \pm \sqrt{5}$$

We are asked to find only x, so we stop here.